Laboratory Models for Foodborne Infections

FOOD MICROBIOLOGY SERIES

Series Editor

Dongyou Liu

Laboratory Models for Foodborne Infections, *edited by Dongyou Liu* (2017)

Food Spoilage Microorganisms: Ecology and Control, *edited by Yanbo Wang* (2017)

Foodborne Viral Pathogens, *edited by Peter A. White, Natalie E. Netzler, and Grant S. Hansman* (2016)

Molecular Biology of Food and Water Borne Mycotoxigenic and Mycotic Fungi, *edited by R. Russell M. Paterson, and Nelson Lima* (2015)

Biology of Foodborne Parasites, *edited by Lihua Xiao, Una Ryan, and Yaoyu Feng* (2015)

Laboratory Models for Foodborne Infections

Edited by
Dongyou Liu
Royal College of Pathologists of Australasia Quality Assurance Programs
New South Wales, Australia

CRC Press
Taylor & Francis Group
Boca Raton London New York

CRC Press is an imprint of the
Taylor & Francis Group, an **informa** business

CRC Press
Taylor & Francis Group
6000 Broken Sound Parkway NW, Suite 300
Boca Raton, FL 33487-2742

Printed on acid-free paper

International Standard Book Number-13: 978-1-4987-2167-7 (Hardback)

Library of Congress Cataloging-in-Publication Data

Names: Names: Liu, Dongyou, editor.
Title: Laboratory models for foodborne infections / [edited by] Dongyou Liu.
Other titles: Food microbiology series.
Description: Boca Raton : CRC Press/Taylor & Francis, 2017. | Series: Food microbiology series | Includes bibliographical references and index.
Identifiers: LCCN 2016040619 | ISBN 9781498721677 (hardback : alk. paper)
Subjects: | MESH: Foodborne Diseases | Models, Animal | Models, Biological | Food Microbiology
Classification: LCC QR201.F62 | NLM WC 268 | DDC 615.9/5293--dc23
LC record available at https://lccn.loc.gov/2016040619

Visit the Taylor & Francis Web site at
http://www.taylorandfrancis.com

and the CRC Press Web site at
http://www.crcpress.com

Contents

Section III Foodborne Infections due to Gram-Negative Bacteria

Section IV Foodborne Infections due to Fungi

Section V Foodborne Infections due to Protozoa

Section VI Foodborne Infections due to Helminths

Preface for Food Microbiology Series

Microorganisms (including viruses, bacteria, molds, yeasts, protozoa, and helminths) represent abundant and diverse forms of life that occupy various ecological niches of earth. Those utilizing food and food products for growth and maintenance are important to human society due not only to their positive and negative impacts on food supply, but also to their potential pathogenicity to human and animal hosts.

On one hand, foodborne microorganisms are known to play a critical role in fermentation and modification of foods, leading to a variety of nutritious food products (e.g., bread, beverage, yogurt, cheese, etc.) that have helped sustain the human civilization from time immemorial. On the other hand, foodborne microorganisms may be responsible for food spoilage, which, albeit a necessary step in keeping up ecological balance, reduces the quality and quantity of foods for human and animal consumption. Furthermore, some foodborne microorganisms are pathogenic to humans and animals, which, besides creating havoc on human health and animal welfare, decrease the availability of meat and other animal-related products.

Food microbiology is a continuously evolving field of biological sciences that addresses issues arising from the interactions between food-/waterborne microorganisms and foods. Topics of relevance to food microbiology include, but are not limited to, adoption of innovative fermentation and other techniques to improve food production; optimization of effective preservation procedures to reduce food spoilage; development of rapid, sensitive, and specific methods to identify and monitor foodborne microbes and toxins, helping alleviate food safety concerns among consumers; use of *-omic* approaches to unravel the pathogenicity of foodborne microbes and toxins; selection of nonpathogenic foodborne microbes as probiotics to inhibit and eliminate pathogenic viruses, bacteria, fungi, and parasites; design and implementation of novel control and prevention strategies against foodborne diseases in human and animal populations.

The *Food Microbiology Series* aims to present a state-of-art coverage on topics central to the understanding of the interactions between food-/waterborne microorganisms and foods. The series consists of individual volumes, each of which focuses on a particular aspect/group of foodborne microbes and toxins, in relation to their biology, ecology, epidemiology, immunology, clinical features, pathogenesis, diagnosis, antibiotic resistance, stress responses, treatment and prevention, etc. The volume editors and the authors are professionals with expertise in their respective fields of food microbiology, and the chapter contributors are scientists directly involved in foodborne microbe and toxin research.

Extending the contents of classical textbooks on food microbiology, this series serves as an indispensable tool for food microbiology researchers, industry food microbiologists, and food regulation authorities wishing to keep abreast with latest developments in food microbiology. In addition, the series offers a reliable reference for undergraduate and graduate students in their pursuit to becoming competent and consummate future food microbiologists. Moreover, the series provides a trustworthy source of information to the general public interested in food safety and other related issues.

Preface

Foodborne infections result from the ingestion of foods and beverages (including drinking water) that are contaminated by pathogenic microorganisms (including viruses, bacteria, fungi, and parasites). While some microbial pathogens stay in the gastrointestinal system and produce toxins (e.g., enterotoxins, exotoxins, and mycotoxins) that are absorbed into the bloodstream, others may directly invade deeper body tissues. Although foodborne infections generally tend to induce mild clinical symptoms (e.g., nausea, vomiting, fever, abdominal cramps, and diarrhea) in immunocompetent individuals, they may have serious consequences in young children and people with suppressed immune functions.

With the increasing consumption of manufactured foods and beverages, foodborne infections are becoming a common and expensive public health problem worldwide. The World Health Organization (WHO) estimates that food-/waterborne diarrheal diseases kill about 2.2 million people (mostly children) annually. Based on FoodNet data collected between 2000 and 2007 by the Centers for Disease Control and Prevention (CDC), 48 million foodborne illness cases (16,000 cases for 100,000 inhabitants) occur in the United States every year, including 128,000 hospitalizations and 3,000 deaths. Interestingly, 31 foodborne pathogens have been implicated in 9.4 of the 48 million foodborne illness cases, with 7 (*Salmonella*, norovirus, *Campylobacter*, *Toxoplasma*, *Escherichia coli* O157:H7, *Listeria*, and *Clostridium perfringens*) accounting for 90% of these illnesses alone. Similarly, an estimated 4.1 million cases of foodborne gastroenteritis were documented in Australia in 2010, with norovirus, pathogenic *E. coli*, *Campylobacter* spp., and nontyphoidal *Salmonella* spp. being the main culprits.

Although proper storage and refrigeration of food play a vital role in the prevention of foodborne infections, other good food safety practices (handwashing, preventing cross-contamination, and maintaining cooking temperatures in the kitchen) are also valuable. In addition, accurate diagnosis and prompt medical intervention are crucial in reducing the mortality due to foodborne infections. However, thorough understanding of host–pathogen interactions and elucidation of molecular mechanisms of pathogenesis are critical for the development of effective vaccines that will lead to ultimate elimination of foodborne infections in human population. Toward this goal, application of laboratory models (including both *in vivo* and *in vitro* models) is essential.

As a part of the *Food Microbiology Series*, this book focuses on the value and utility of various animal and cellular systems (ranging from mice, rats, hamsters, guinea pigs, rabbits, nonhuman primates, birds, zebrafish, frogs, chicken embryo, fruit fly, nematode, and waxworm to established and nonestablished cell lines) in the study of foodborne infections. Written by experts involved in foodborne pathogen research, each chapter presents a state-of-the-art review of laboratory models in the study of a particular foodborne pathogen (of viral, bacterial, fungal, or parasitic origin) in relation to its life cycle, host–pathogen interaction, pathogenesis, immunity, and other related aspects. Besides providing a reliable reference for undergraduates and postgraduates of food microbiology, this book is a valuable guide for scientists using laboratory models in their investigation of foodborne infections.

Given the diversity of foodborne pathogens, a comprehensive book such as this is clearly beyond an individual's capacity. I am fortunate and honored to have a large group of scientists as chapter contributors, whose in-depth knowledge and technical insights on foodborne pathogens have greatly enriched this book. Additionally, the professionalism and dedication of the senior editor, Stephen Zollo, have enhanced its presentation. Finally, the understanding and support from my family—Liling Ma, Brenda, and Cathy—have helped me keep focused during the compilation of this all-inclusive volume.

Dongyou Liu

Editor

Dongyou Liu, PhD, studied veterinary science at Hunan Agricultural University, China, and completed his postgraduate training at the University of Melbourne, Victoria, Australia. Over the past two decades, he has worked at several research and clinical laboratories in Australia and the United States of America, focusing on molecular characterization and virulence determination of microbial pathogens such as ovine footrot bacterium (*Dichelobacter nodosus*), dermatophyte fungi (*Trichophyton*, *Microsporum*, and *Epidermophyton*), and listeriae (*Listeria* spp.), as well as development of nucleic-acid-based quality assurance models for security-sensitive and emerging viral pathogens. He is the author of over 50 original research and review articles in various international journals, a contributor of 165 book chapters, and the editor of *Handbook of Listeria monocytogenes* (2008), *Handbook of Nucleic Acid Purification* (2009), *Molecular Detection of Foodborne Pathogens* (2009), *Molecular Detection of Human Viral Pathogens* (2010), *Molecular Detection of Human Bacterial Pathogens* (2011), *Molecular Detection of Human Fungal Pathogens* (2011), *Molecular Detection of Human Parasitic Pathogens* (2012), *Manual of Security Sensitive Microbes and Toxins* (2014), and *Molecular Detection of Animal Viral Pathogens* (2016), all of which are published by CRC Press. He is also a coeditor of *Molecular Medical Microbiology*, 2nd edition (2014), which was published by Elsevier.

Contributors

Naiyf S. Alharbi
Botany and Microbiology Department
King Saud University
Riyadh, Saudi Arabia

Nalvo F. Almeida
School of Computing
Federal University of Mato Grosso do Sul
Campo Grande, Mato Grosso do Sul, Brazil

María Jesús Andrade
Food Hygiene and Safety
Research Institute of Meat and Meat Products
University of Extremadura
Cáceres, Spain

Yiorgos Apidianakis
Department of Biological Sciences
University of Cyprus
Aglatzia, Cyprus

Emilio Aranda
Food Science and Technology
Agricultural Resources Research Institute
University of Extremadura
Badajoz, Spain

Andrea Cristina Vetö Arnholdt
Laboratório de Biologia do Reconhecer
Centro de Biociências e Biotecnologia
Universidade Estadual do Norte Fluminense Darcy Ribeiro
Campos dos Goytacazes, Rio de Janeiro, Brazil

Miguel A. Asensio
Food Hygiene and Safety
Research Institute of Meat and Meat Products
University of Extremadura
Cáceres, Spain

Mario Julio Avila-Campos
Department of Microbiology
Institute of Biomedical Science
University of São Paulo
São Paulo, Brazil

Krishnaswamy Balamurugan
Department of Biotechnology
Alagappa University
Karaikudi, Tamil Nadu, India

Nikolett Baranyi
Department of Microbiology
University of Szeged
Szeged, Hungary

Soumik Barman
National Institute of Cholera and Enteric Diseases
Beleghata, Kolkata, India

María J. Benito
Food Science and Technology
Agricultural Resources Research Institute
University of Extremadura
Badajoz, Spain

Elena Bermúdez
Food Hygiene and Safety
Research Institute of Meat and Meat Products
University of Extremadura
Cáceres, Spain

Prudhvi Lal Bhukya
Hepatitis Group
National Institute of Virology
Pashan, Pune, India

Thidarut Boonmars
Department of Parasitology
Khon Kaen University
Khon Kaen, Thailand

B. Bricker
Bacterial Diseases of Livestock Research Unit
National Animal Disease Center Agricultural Research Service
United States Department of Agriculture
Ames, Iowa

Jong-Yil Chai
Department of Parasitology and Tropical Medicine
Seoul National University College of Medicine
Seoul, South Korea

Muthusamy Chandrasekaran
Botany and Microbiology Department
King Saud University
Riyadh, Saudi Arabia

Subhashis Chatterjee
Hepatitis Group
National Institute of Virology
Pashan, Pune, India

Sergio Carmona São Clemente
Departamento Tecnologia de Produtos de Origem Animal
Universidade Federal Fluminense
Niterói, Rio de Janeiro, Brazil

Camila H. Coelho
Department of Biology
Georgetown University
Washington, DC

Juan José Córdoba
Food Hygiene and Safety
Research Institute of Meat and Meat Products
University of Extremadura
Cáceres, Spain

María G. Córdoba
Food Science and Technology
Agricultural Resources Research Institute
University of Extremadura
Badajoz, Spain

Renato Augusto DaMatta
Laboratório de Biologia Celular e Tecidual
Centro de Biociências e Biotecnologia
Universidade Estadual do Norte Fluminense Darcy Ribeiro
Campos dos Goytacazes, Rio de Janeiro, Brazil

Flábio R. de Araújo
Embrapa Beef Cattle
Campo Grande, Mato Grosso do Sul, Brazil

Mireya de la Garza
Departamento de Biología Celular
Centro de Investigación y de Estudios Avanzados del Instituto Politécnico Nacional
Mexico City, Mexico

Josué Delgado
Food Hygiene and Safety
Research Institute of Meat and Meat Products
University of Extremadura
Cáceres, Spain

Jonathan J. Dennis
Department of Biological Sciences
University of Alberta
Edmonton, Alberta, Canada

Sarah E.F. D'Orazio
Department of Microbiology, Immunology, & Molecular Genetics
University of Kentucky
Lexington, Kentucky

Doris H. D'Souza
Department of Food Science and Technology
University of Tennessee, Knoxville
Knoxville, Tennessee

Ran Duan
National Institute for Communicable Disease Control and Prevention
Chinese Centre for Disease Control and Prevention
Beijing, People's Republic of China

Sellegounder Durai
Department of Biotechnology
Alagappa University
Karaikudi, Tamil Nadu, India

Maria Augusta Moulin Fantezia
Departamento de Imunobiologia
Universidade Federal Fluminense
Niterói, Rio de Janeiro, Brazil

Israel Figueiredo Jr.
Departamento Materno-Infantil
Universidade Federal Fluminense
Niterói, Rio de Janeiro, Brazil

Alisa Gruden-Movsesijan
Department for Immunology and Immunoparasitology
Institute for the Application of Nuclear Energy
University of Belgrade
Belgrade, Serbia

Wenpeng Gu
National Institute for Communicable Disease Control and Prevention
Chinese Centre for Disease Control and Prevention
Beijing, People's Republic of China

Mónika Homa
Department of Microbiology
University of Szeged
Szeged, Hungary

Sung-Tae Hong
Department of Parasitology and Tropical Medicine
Seoul National University College of Medicine
Seoul, South Korea

Natasa Ilic
Department for Immunology and Immunoparasitology
Institute for the Application of Nuclear Energy
University of Belgrade
Belgrade, Serbia

Victoria Iribarnegaray
Department of Microbiology
Instituto de Investigaciones Biológicas Clemente Estable
Montevideo, Uruguay

Rajagopalaboopathi Jayasudha
Department of Biotechnology
Bharathiar University
Coimbatore, Tamil Nadu, India

Huaiqi Jing
National Institute for Communicable Disease Control and Prevention
Chinese Centre for Disease Control and Prevention
Beijing, People's Republic of China

Somchai Jongwutiwes
Department of Parasitology
Chulalongkorn University
Bangkok, Thailand

Snehal S. Joshi
Department of Food Science and Technology
University of Tennessee, Knoxville
Knoxville, Tennessee

Shine Kadaikunnan
Botany and Microbiology Department
King Saud University
Riyadh, Saudi Arabia

Fatima Kamal
Department of Biological Sciences
University of Alberta
Edmonton, Alberta, Canada

Arumugam Kamaladevi
Department of Biotechnology
Alagappa University
Karaikudi, Tamil Nadu, India

Yuka Kiriyama
Department of Diagnostic Pathology
Fujita Health University
Toyoake, Aichi, Japan

Matthew D. Koci
Prestage Department of Poultry Science
North Carolina State University
Raleigh, North Carolina

Sándor Kocsubé
Department of Microbiology
University of Szeged
Szeged, Hungary

László Kredics
Department of Microbiology
University of Szeged
Szeged, Hungary

Bayissa Chala Legissa
Department of Parasitology and Tropical Medicine
Seoul National University College of Medicine
Seoul, South Korea

Nidia León-Sicairos
Facultad de Medicina
Universidad Autónoma de Sinaloa
Culiacán Sin, Mexico

Junrong Liang
National Institute for Communicable Disease Control and Prevention
Chinese Centre for Disease Control and Prevention
Beijing, People's Republic of China

Baochuan Lin
Center for Bio/Molecular Science & Engineering
US Naval Research Laboratory
Washington, DC

Dongyou Liu
Royal College of Pathologists of Australasia Quality Assurance Programs
St Leonards, New South Wales, Australia

Kavita Lole
Hepatitis Group
National Institute of Virology
Pashan, Pune, India

Aongart Mahittikorn
Department of Protozoology
Mahidol University
Ratchathewi, Bangkok, Thailand

Anthony P. Malanoski
Center for Bio/Molecular Science & Engineering
US Naval Research Laboratory
Washington, DC

Jenny G. Maloney
Department of Biology
Georgetown University
Washington, DC

Palanisamy Manikandan
Aravind Eye Hospital and Postgraduate Institute of Ophthalmology
Coimbatore, Tamil Nadu, India
and
Department of Medical Laboratory Sciences
Majmaah University
Kingdom of Saudi Arabia

Jessica Minnaard
Centro de Investigación y Desarrollo en Criotecnología de Alimentos
Consejo Nacional de Investigaciones Científicas y Técnicas
La Plata, Argentina

Farlen José Bebber Miranda
Laboratório de Imunopatologia
Instituto René Rachou
Belo Horizonte, Minas Gerais, Brazil

Hirotake Mori
Department of Protozoology
Mahidol University
Ratchathewi, Bangkok, Thailand

Antonio Muro
Parasite and Molecular Immunology Laboratory
Universidad de Salamanca
Salamanca, Castilla y León, Spain

Venkatapathy Narendran
Aravind Eye Hospital and Postgraduate Institute of Ophthalmology
Coimbatore, Tamil Nadu, India

Félix Núñez
Food Hygiene and Safety
Research Institute of Meat and Meat Products
University of Extremadura
Cáceres, Spain

S.C. Olsen
Bacterial Diseases of Livestock Research Unit
National Animal Disease Center Agricultural Research Service
United States Department of Agriculture
Ames, Iowa

Takashi Onodera
Research Center for Food Safety
University of Tokyo
Tokyo, Japan

Stavria Panayidou
Department of Biological Sciences
University of Cyprus
Aglatzia, Cyprus

Andrea Patriarca
Departamento de Química Orgánica
Universidad de Buenos Aires
Buenos Aires, Argentina

Pablo F. Pérez
Centro de Investigación y Desarrollo en Criotecnología de Alimentos
Consejo Nacional de Investigaciones Científicas y Técnicas
La Plata, Argentina
and
Cátedra de Microbiología
Universidad Nacional de La Plata
La Plata, Argentina

Danielle L. Peters
Department of Biological Sciences
University of Alberta
Edmonton, Alberta, Canada

Nemani V. Prasadarao
Departments of Pediatrics and Surgery
Children's Hospital Los Angeles
and
Molecular Microbiology and Immunology
University of Southern California
Los Angeles, California

Chaturong Putaporntip
Department of Parasitology
Chulalongkorn University
Bangkok, Thailand

Janaina Ribeiro
Departamento Tecnologia de Produtos de Origem Animal
Universidade Federal Fluminense
Niterói, Rio de Janeiro, Brazil

Alicia Rodríguez
Food Hygiene and Safety
Research Institute of Meat and Meat Products
University of Extremadura
Cáceres, Spain

Mar Rodríguez
Food Hygiene and Safety
Research Institute of Meat and Meat Products
University of Extremadura
Cáceres, Spain

Jose Rojas-Caraballo
Centro de Investigación en Salud para el Trópico
Universidad Cooperativa de Colombia
Santa Marta, Magdalena, Colombia

Ivanna S. Rolny
Cátedra de Microbiología
Universidad Nacional de La Plata
La Plata, Argentina

Akikazu Sakudo
Laboratory of Biometabolic Chemistry
University of the Ryukyus
Nishihara, Okinawa, Japan

Paola Scavone
Department of Microbiology
Instituto de Investigaciones Biológicas Clemente Estable
Montevideo, Uruguay

Stacey L. Schultz-Cherry
Infectious Diseases
St. Jude Children's Research Hospital
Memphis, Tennessee

Jesús Serrano-Luna
Departamento de Biología Celular
Centro de Investigación y de Estudios Avanzados del Instituto Politécnico Nacional
Mexico City, Mexico

Mineko Shibayama
Departamento de Infectómica y Patogénesis Molecular
Centro de Investigación y de Estudios Avanzados del Instituto Politécnico Nacional
Mexico City, Mexico

Coimbatore Subramanian Shobana
Department of Microbiology
PSG College of Arts & Science
Coimbatore, Tamil Nadu, India

Steven M. Singer
Department of Biology
Georgetown University
Washington, DC

Ljiljana Sofronic-Milosavljevic
Department for Immunology and Immunoparasitology
Institute for the Application of Nuclear Energy
University of Belgrade
Belgrade, Serbia

Martin Stahl
Division of Gastroenterology
Child and Family Research Institute and University of British Columbia
Vancouver, British Columbia, Canada

Yoshifumi Takeda
Hideyo Noguchi Memorial Foundation
Inawashiro, Fukushima, Japan

Masae Tatematsu
Department of Diagnostic Pathology
Fujita Health University
Toyoake, Aichi, Japan

Gerlinde Teixeira
Departamento de Imunobiologia
Universidade Federal Fluminense
Niterói, Rio de Janeiro, Brazil

Tetsuya Tsukamoto
Department of Diagnostic Pathology
Fujita Health University
Toyoake, Aichi, Japan

Csaba Vágvölgyi
Department of Microbiology
University of Szeged
Szeged, Hungary
and
Botany and Microbiology Department
King Saud University
Riyadh, Saudi Arabia

Bruce A. Vallance
Division of Gastroenterology
Child and Family Research Institute and University of British Columbia
Vancouver, British Columbia, Canada

János Varga
Department of Microbiology
University of Szeged
Szeged, Hungary

Mauricio Afonso Vericimo
Departamento de Imunobiologia
Universidade Federal Fluminense
Niterói, Rio de Janeiro, Brazil

Xin Wang
National Institute for Communicable Disease Control and Prevention
Chinese Centre for Disease Control and Prevention
Beijing, People's Republic of China

Ke Wen
Department of Biomedical Sciences and Pathobiology
Virginia-Maryland College of Veterinary Medicine
Virginia Polytechnic Institute and State University
Blacksburg, Virginia

Lijuan Yuan
Department of Biomedical Sciences and Pathobiology
Virginia-Maryland College of Veterinary Medicine
Virginia Polytechnic Institute and State University
Blacksburg, Virginia

Pablo Zunino
Department of Microbiology
Instituto de Investigaciones Biológicas Clemente Estable
Montevideo, Uruguay

1

Introductory Remarks

Dongyou Liu

CONTENTS

1.1 Preamble

Foodborne disease (also known as foodborne illness or colloquially as foodborne poisoning) is largely attributable to microbial pathogens and their toxins contained in food and food products that are inappropriately prepared or stored before consumption. Once inside the host, these pathogens establish in their predilection sites and cause damages to the host either through direct physical/mechanical destruction or through secretion of toxins and antigens that provoke host innate and acquired immune responses, leading to a range of clinical symptoms (e.g., diarrhea, abdominal cramps, nausea, fever, joint/back aches, and fatigue).

Although foodborne disease is a current buzzword that appears in various popular media outlets with alarming frequency, it has a long and tortuous history. Our awareness of as well as our struggle against foodborne disease goes hand in hand with our attempts to survive and prosper in a constantly changing, and challenging, world, with significant milestones marked by the use of fire, the development of crop cultivation, the luxury of food storage, the evolution of culinary art, the sophistication of sewage system, the observation of disease-causing microbes, the application of refrigeration, and the discovery of antibiotics [1,2].

From scavengers who searched for the scraps left by other predators for survival, humans have made enormous technological advances that overcome the barrier of seasonality for food supply, that reduce the proliferation of foodborne disease, and that enable rapid identification and tracking of foodborne pathogens implicated in any food-related disease outbreaks. Nonetheless, it is still a long way before we can call foodborne pathogens the genie in the bottle, foodborne disease a memory of the past, and foodborne outbreak an absolute nonevent.

1.2 Foodborne Pathogens and Diseases

1.2.1 Foodborne Pathogens

In a narrow sense, foodborne pathogens refer to microbes that contaminate the foods and related products (e.g., pasteurized carrot juice, peanut butter, broccoli powder on a children's snack food, frozen pot pies, canned chili sauce, hot peppers, white and black pepper, raw cookie dough, hazelnuts, fenugreek sprouts, papayas, pine nuts, raw frozen scraped ground tuna, etc.), the consumption of which by humans leads to infections and diseases. In a broader sense, foodborne pathogens include microbes that occur in animals (in farm/zoo animals and pets), the environment (soil, water, and air), and foods, the ingestion, inhalation, and contact of which by humans result in discomfort and illness. Based on the latter premise, this book covers not only microbial pathogens that come along with foods and food products (foodborne), but also those that may occasionally enter into human host via water (water-borne), air (airborne), or direct contact (skin wound), as well as those that cause diseases not through infection, but through production of toxins and antigens that disturb/upset/confuse the gut, neurological, and immune systems of the human host.

As steadfast survivors, microbes (e.g., viruses, bacteria, fungi, and parasites) utilize animals (including humans), foods, beverages, and water as growth or maintenance media. Some microbes remain in humans (e.g., *Salmonella* Typhi and norovirus) or animal reservoirs and contaminate the food supply via excreta, meat, milk, or eggs. Others persist in the environment and contaminate the ecosystems that are fundamental to food production. Some microbes demonstrate the unusual ability to endure extreme temperature, pH, and osmolarity, to sustain for long periods on dry surfaces, food processing plants, and to exploit any temporary weakness in human innate and acquired immune defense networks (as seen in pregnant women, infants, the elderly, and individuals under immune suppressing therapies).

Although a large number of foodborne microbes are known to infect humans and cause diseases of varying severity, those having the most significant impact on human health in terms of prevalence, morbidity, and mortality include *Escherichia coli* O157:H7, *Campylobacter jejuni*, *Salmonella enteritidis* (e.g., serotypes Typhi and Typhimurium), *Staphylococcus aureus*, *Listeria monocytogenes*, *Bacillus cereus*, *Clostridium botulinum*, *Clostridium perfringens*, *Streptococcus pyogenes*, *Vibrio cholerae*, *Vibrio parahaemolyticus*, *Vibrio vulnificus*, *Shigella*, hepatitis A virus, hepatitis E virus, norovirus, rotavirus, *Cryptosporidium*, *Cyclospora cayetanensis*, and *Toxoplasma gondii* [3]. It is notable that a majority of these high-impact foodborne pathogens are bacteria, and the remainder are viruses and parasites. Interestingly, some of these pathogens have emerged only in the past 30 years presumably due to the increased consumption of processed food products, globalization of food trade, and population ageing. For instance, *E. coli* O157:H7 is a shiga-toxin-producing bacterial strain that was first recognized as a human pathogen in foodborne outbreaks associated with ground beef in 1982, producing symptoms ranging from simple diarrhea and hemorrhagic colitis to hemolytic-uremic syndrome (which is characterized by hemolytic anemia, thrombocytopenia, and renal injury) [4,5]. Subsequently, lettuce grown in close proximity to a dairy farm from which wastewater contaminated with animal feces was used to irrigate the plant was linked to a 2006 outbreak of *E. coli* O157:H7 infection in Iowa and Minnesota (see Chapter 21 in this book). Another recently emerged foodborne pathogen of note is *Aeromonas* (mainly *A. hydrophila*, *A. caviae*, and *A. veronii*), which is responsible for both intestinal and extraintestinal diseases in humans (see Chapter 15 in this book). There is no doubt that new foodborne pathogens will likely emerge or reemerge in the future.

Apart from infections with foodborne viruses, bacteria, fungi, and parasites, another important cause of foodborne diseases is toxins or toxic chemicals produced by foodborne bacteria and fungi as well as those associated with shellfish and plants [6]. Toxins originated from foodborne bacteria can be separated into exotoxins (which remain part of the bacteria, and are secreted, or, similar to endotoxins, released during bacterial lysis) and endotoxins (which form part of the bacterial outer membrane, and are released during bacterial lysis). Some well-known foodborne bacterial exotoxins include superantigens from *S. aureus* and *Streptococcus pyogenes*; pore-forming toxins (PFTs) from *E. coli*, *L. monocytogenes*, and *Streptococcus pneumoniae*; heat-stable enterotoxins (ST, exotoxins targeting the intestine) from pathogenic strains of *E. coli*; and botulinum neurotoxin (BoNT) from *C. botulinum*. A notable foodborne bacterial endotoxin is lipopolysaccharide (LPS, which is made up of O antigen, core oligosaccharide, and lipid A) from Gram-negative bacteria. As water-soluble proteins, PFTs induce host membrane damages as amphiphilic surfactants and phospholipases. On the other hand, endotoxins (e.g., LPS) cause severe inflammation, endotoxemia (septic shock), and autoimmune disease. Being the by-products of foodborne fungi, mycotoxins are responsible for alimentary mycotoxicoses in humans through food consumption. The most common foodborne mycotoxins consist of aflatoxins (from *Aspergillus parasiticus* and *Aspergillus flavus*), altertoxins (from *Alternaria*), fumonisins (from *Fusarium moniliforme*), ochratoxins (from *Aspergillus ochraceus*, *Aspergillus carbonarius*, *Penicillium verrucosum*), patulin (from *Aspergillus*, *Penicillium*), and trichothecenes (from *Fusarium*).

1.2.2 Foodborne Diseases

Foodborne diseases usually arise from consumption of improperly handled, prepared, or stored foods that are contaminated with foodborne pathogens and/or toxins. With incubation period of several hours to 1 week, the initial symptoms of foodborne disease consist of diarrhea, vomiting, abdominal cramps, nausea, fever, joint/back aches, and fatigue, which may last for a week or so. However, some foodborne pathogens (e.g., *Streptococcus pyogenes*) may cause a spectrum of clinical diseases, including (1) localized inflammatory lesions; (2) both local and systemic diseases; and (3) immune dysfunction.

In localized inflammatory lesions, inflammation linked to foodborne pathogens is responsible for lesions in various locations, accompanied by other symptoms. As in the case of *Streptococcus pyogenes* (group A *Streptococcus* or GAS) infection of the pharynx (i.e., pharyngitis, or strep sore throat), inflammation in the pharynx and tonsils leads to sore throat, along with sudden-onset fever, headache, nausea, abdominal pain, vomiting, and patchy exudates. Similarly, GAS infection of the skin (i.e., impetigo) results in the formation of pustules that enlarge and rupture to become thick, honey-colored scabs (see Chapter 14 in this book) [7].

In local and systemic diseases, toxins (e.g., streptococcal pyogenic exotoxins) produced by foodborne pathogens (e.g., GAS) induce a local disease with a deep red, finely papular, erythematous rash (strawberry tongue) and exudates in the pharynx (scarlet fever), or cause soft tissue infection at a surgical site (surgical scarlet fever). Additionally, following minor nonpenetrating trauma, suction lipectomy, hysterectomy, vaginal delivery, bunionectomy, and bone pinning, GAS invades and produces streptococcal toxins that contribute to streptococcal toxic shock syndrome (STSS) (see Chapter 14 in this book) [7].

Regarding immune dysfunction, some foodborne pathogens produce antigens that confuse host immune systems, leading to autoimmune diseases. For example, as a sequela of untreated GAS pharyngeal infection, acute rheumatic fever (ARF) results from the activity of antigens produced by GAS that cause inflammation in the joints (arthritis) and the heart (carditis, also known as rheumatic heart disease or RHD). Another sequela of GAS infection is acute poststreptococcal glomerulonephritis (APSGN), which is a disorder of the kidneys mediated by the immune complex, with symptoms ranging from edema, hypertension, and urinary sediment abnormalities to reduced serum complement components (see Chapter 14 in this book) [7].

While all people are at risk for foodborne illness and most recover without any lasting effects, some may show serious long-term consequences such as kidney failure, chronic arthritis, brain and nerve damage, and even death, especially infants and toddlers, the elderly, pregnant women, transplant recipients, and individuals with chronic illnesses (e.g., cancer, diabetes, or HIV/AIDS) and compromised immune systems.

1.3 Laboratory Models

1.3.1 Rationales for Using Laboratory Models

In spite of our unrelenting past efforts, including the implementation of procedures to reduce pre- and postharvest contamination, the introduction of best-practice in food products processing, package and storage, the education of the general public about the danger of and effective prevention measures against foodborne diseases, the application of antibiotic and antimicrobial therapies, and the redirection of public resources into research on the mechanisms of foodborne infections and diseases, the war against foodborne pathogens and diseases is far from being over [3]. Among many other things, we are still uncertain how foodborne pathogens sabotage host immune defense and manipulate host cell machinery for their own gain.

Since the best way to observe a battle is to get close to the battlefield, use of laboratory models (i.e., *in vivo* animal models and *in vitro* culture models) provides a unique opportunity to determine the infectivity, host specificity, and life cycle of foodborne pathogens; to compare the virulence potential of various microbial strains and serotypes; to generate large quantities of pathogenic microbes for detailed analysis;

to examine host immune responses to foodborne pathogens; to uncover the pathological and histological details resulting from foodborne infections; to validate the accuracy of clinical diagnostic techniques; and to evaluate the efficacy of newly developed antimicrobial and vaccine preparations against foodborne pathogens without harming human hosts. This is made possible by the common ancestry of all living organisms, the similarity of anatomical structures and functions (e.g., breathing, digestion, movement, sight, hearing, reproduction, immunity, etc.), the homology of genetic materials, the sharing of hundreds of illnesses, and the conservation of cell biological and developmental pathways among vertebrates as well as between vertebrates and invertebrates [8].

1.3.2 Milestones in the Use of Laboratory Models

Animals have long been employed as laboratory models for investigation of the anatomy, physiology, epidemiology, and disease mechanisms of vertebrates. The available records suggest that in the 6th century BCE, Alcmaeon of Croton examined dogs to demonstrate the brain as the seat of intelligence and sensory integration; Aristotle (384–322 BCE) studied embryogenesis and ontogeny in chickens; after analysis of the cardiovascular system in live animals, Erasistratus (304–258 BCE) postulated that the heart functions as a pump; in the 2nd century, Galen of Pergamum employed live animals for extensive studies of cardiovascular and neuroanatomy; in the 12th century, Avenzoar polished his tracheotomy skill on animals before applying to humans; in the mid-16th century, Servetus and Lusitano identified pulmonary and systemic circulation as two connected but distinct blood circuits in the body through animal experiments; in the 17th century, through comparison of the anatomic and functional properties of the heart and vasculature in eels, chicks, and pigeons, William Harvey provided accurate and detailed descriptions of the function of the cardiovascular and other systems; in the 18th century, Antoine Lavoisier used guinea pigs in a calorimeter to prove respiration as a form of combustion; in the 19th century, Louis Pasteur demonstrated the germ theory of disease using a sheep model of anthrax; in the late 19th century, Emil von Behring observed the effect of diphtheria toxin in guinea pigs that led to the development of an antitoxin against diphtheria in animals and humans.

Another significant milestone in the use of laboratory animals for microbial research was achieved in 1902, after William Castle and Abbie Lathrop generated the DBA ("dilute, brown, and non-agouti") inbred mouse strain and other inbred mice for genetic studies. Between 1910 and 1927, working with the fruit fly *Drosophila melanogaster*, Thomas Hunt Morgan pinpointed chromosomes as the vector of inheritance for genes. In the 1920s, Frederick Banting utilized the isolates of pancreatic secretion to treat dogs with diabetes; in the 1930s, Little and MacDowell produced the first fully inbred mouse (20 brother × sister matings); in the 1940s, John Cade discovered the anticonvulsant properties of lithium using guinea pigs, which helped replace lobotomy or electroconvulsive therapy for the treatment of bipolar disorder (manic depression); also in the 1940s, Jonas Salk isolated the most virulent forms of the polio virus from the rhesus monkey and created a highly effective polio vaccine; in the 1960s, Albert Sabin passed the polio virus through animal hosts (including monkeys) to improve the effectiveness of the Sabin vaccine for mass application; in 1976, Rudolf Jaenisch and colleagues developed the first transgenic mouse; in 1987, Capecchi, Evans, and Smithies developed the first knockout mouse; in 1997, Wilmut and Campbell obtained the first cloned animal (Dolly the sheep) from an adult somatic cell; in 2009, Aron Geurts and colleagues developed the first knockout rat [9].

1.3.3 Characteristics of Laboratory Models

Laboratory models used for the study of foodborne infections are of two main types: *in vivo* animal models, and *in vitro* culture models. The *in vivo* animal models involve vertebrates [nonhuman primates (e.g., rhesus monkey, cynomolgus monkey, chimpanzee, baboon), rodents (e.g., mice, rats, gerbils, hamsters, chinchillas, guinea pigs), rabbits, cats, dogs, pigs, sheep, cattle, chicken, zebrafish (*Danio rerio*), etc.] and invertebrates [fruit fly (*D. melanogaster*), silkworm (*Bombayx mori*), waxworm (*Galleria mellonella*), roundworm (*Caenorhabditis elegans*), protozoa (*Tetrahymena thermophila* or *Tetrahymena pyriformis*), etc.]; the listing order reflects the evolutionary relationship between these animals and humans, with nonhuman primates being most close and roundworm being least close to humans (Table 1.1) [10–14]. The *in vitro* culture models rely on the use various established and non-established cell lines (derived from epithelia, endothelia, macrophage, etc.), embryonated eggs, and organs and tissues from hosts (Table 1.1).

Among various *in vivo* animal models, nonhuman primates (NHPs, with genomes of 2.85–3.09 Gb dispersed in 21–24 chromosome pairs) are the closest relatives to humans (with a genome of 3.23 Gb dispersed in 23 chromosome pairs), and represent ideal models for investigation of foodborne infections and other human diseases, on the basis of biological, physiological, immunological, and genetic similarities. However, because of limited availability, prohibitive cost, and ethical concerns, NHPs are rarely used nowadays [15].

TABLE 1.1

Characteristics of Laboratory Models for Foodborne Infections

Model[a]	Common Species/ Cell Type	Characteristics	Exemplary Application
In Vivo			
Nonhuman primates (family Hominidae, order Primates)	Chimpanzee (*Pan troglodytes*), rhesus monkey (*Macaca mulatta*), cynomolgus monkey (*Macaca fascicularis*), olive baboon (*Papio anubis*)	Chimpanzee has a genome of 3.02 Gb, rhesus monkey 3.09 Gb, cynomolgus monkey (crab-eating macaque, long-tailed macaque, or Java macaque) 2.85 Gb, olive baboon 2.94 Gb. Ideal models for foodborne infections and other human diseases, but limited by availability, cost, and ethical concerns	*Helicobacter pylori*, *L. monocytogenes*, *Mycobacterium*, hepatitis E virus
Mice (family Muridae, order Rodentia)	House mouse (*Mus musculus*) strains: BALB/c/(inbred, albino), C57BL/6/ (inbred, dark brown), athymic nude mice (outbred)	Mice (house mice) possess a genome of 2.67 Gb, are small, readily available, easy to handle, amenable to genetic manipulation, and reproduce quickly, representing an efficient, cost-effective, and widely applicable animal model for experimentation on foodborne infections and other human diseases	*L. monocytogenes*, *S. aureus*, *Salmonella*
Rats (family Muridae, order Rodentia)	Norway rat (*Rattus norvegicus*) (inbred), Wistar rat (outbred, albino), Lewis rat (inbred)	Norway rat (brown rat) has a genome of 2.61 Gb. Developed in 1906, Wistar rat (outbred albino) is the ancestor of most laboratory rats used today, including the Lewis rat. Wistar rat shows albino coloring, a docile behavior, and low fertility, and tolerates crowding	*Salmonella*, *S. aureus*, *Yersinia*, *Acanthamoeba*
Gerbils (family Muridae, order Rodentia)	Mongolian gerbil (*Meriones unguiculatus*) (outbred)	Mongolian gerbil (Mongolian jird) is easy to keep as it adapts to a new setting well	*H. pylori*, *L. monocytogenes*, *Giardia*
Hamsters (family Cricetidae, order Rodentia)	Syrian hamster (*Mesocricetus auratus*) (outbred), Chinese hamster (*Cricetulus griseus*)	Syrian hamster (golden hamster) possesses a genome of 2.50 Gb, Chinese hamster 2.36 Gb. Hamsters have a short life cycle and breed well in captivity; being relatively free from natural diseases, hamsters are susceptible to experimental pathogens	*Mycobacterium*, *Acanthamoeba*
Chinchillas (family Chinchillidae, order Rodentia)	Long-tailed chinchilla (*Chinchilla lanigera*)	Chinchilla has a genome of 2.39 Gb. Being crepuscular rodents, chinchilla is a robust host for experimental study	*L. monocytogenes*, *Yesinia*
Guinea pigs (family Caviidae, order Rodentia)	Hartley Guinea pig (*Cavia porcellus*) (outbred, albino)	Guinea pig has a genome of 2.72 Gb, and shows similarity to humans in disease symptoms, immune response, and pathogenesis	*L. monocytogenes*, *S. aureus*

(*Continued*)

TABLE 1.1 (*Continued*)

Characteristics of Laboratory Models for Foodborne Infections

Model[a]	Common Species/ Cell Type	Characteristics	Exemplary Application
Rabbits (family Leporidae, order Lagomorpha)	New Zealand white rabbit (*Oryctolagus cuniculus*) (outbred)	New Zealand white rabbit possesses a genome of 2.73 Gb, and represents a nonaggressive host for experimental work	*L. monocytogenes*, *Salmonella*
Cats (family Felidae, order Carnivora)	Domestic cat (*Felis catus*)	Domestic cat has a genome of 2.9 Gb, and is useful for modeling some foodborne infections	*S. aureus*
Dogs (family Canidae, order Carnivora)	Domestic dog (*Canis familiaris*)	Domestic dog possesses a genome of 2.25 Gb, and may be used experimentally for a number of foodborne infections	*H. pylori*
Pigs (family Suidae, order Artiodactyla)	Domestic pig (*Sus scrofa domesticus*)	Domestic pig has a genome of 2.5 Gb. Being truly omnivorous, pigs (piglets) show strikingly similar nutritional requirements to those of humans. Pigs practice coprophagy, and represent a useful model for a number of foodborne infections	*L. monocytogenes*, *S. aureus*, *Taenia solium*
Sheep (family Bovidae, order Artiodactyla)	Sheep (*Ovis aries*)	Sheep harbors a genome of 2.61 Gb, and is useful for modeling some foodborne infections	*L. monocytogenes*, bovine spongiform encephalopathy (BSE)
Cattle (family Bovidae, order Artiodactyla)	Cattle (*Bos taurus*)	Cattle possess a genome of 2.69 Gb, and may be used for a number of foodborne infections	*E. coli*, *Taenia saginata*
Chicken (family Phasianidae, order Galliformes)	Domestic chicken (*Gallus gallus domesticus*)	Domestic chickens have a genome of 1.23 Gb, are noted for their rapid growth rate, distinct anatomy, relatively small size, and low cost	*E. coli*, *Aspergillus fumigatus*
Zebrafish (family Cyprinidae, order Cypriniformes)	Zebrafish (*D. rerio*)	Zebrafish possess a genome of 1.4 Gb. Due to small size, zebrafish are easy to house and care for, easy to introduce genetic changes, and easy to observe the impact of any genetic mutation (with transparent early life stages)	*Mycobacterium*
Fruit fly (family Drosophilidae, order Diptera)	Common fruit fly (*D. melanogaster*)	Common fruit fly has a genome of 139 Mb, and shares 75% of known human disease genes. Due to its small size, simple anatomy, high fecundity, and short life cycle (about 30 days at 29°C), the fruit fly is easy and inexpensive to maintain. However, the fruit fly does not have an adaptive immune system and is not an appropriate model for the study of antibody- and lymphocyte-dependent adaptive immune defenses	*L. monocytogenes*, *S. aureus*
Silkworm (family Bombycidae, order Lepidoptera)	Domestic silkworm (*B. mori*)	Domestic silkworm possesses a genome of 397 Mb, and represents a low-cost model for some foodborne infection. It has a body size large enough for easy handling (e.g., injecting sample solution into the hemolymph)	*Pseudomonas aeruginosa*, *S. aureus*

(*Continued*)

TABLE 1.1 (*Continued*)

Characteristics of Laboratory Models for Foodborne Infections

Model[a]	Common Species/ Cell Type	Characteristics	Exemplary Application
Waxworm (family Pyralidae, order Lepidoptera)	Greater wax moth or honeycomb moth (*G. mellonella*)	Despite lacking an adaptive immune response, greater wax moth (waxworm) shows an innate immune response functionally similar to that of mammals, and provides a rapid, inexpensive, and reliable model for certain foodborne infections	*Streptococcus pyogenes*
Roundworm (family Rhabditidae, order Rhabditida)	Soil nematode (*C. elegans*)	Soil nematode possesses a genome of 100 Mb, and lacks an adaptive immune system. It has a short life cycle, simple anatomy, is easy to handle, and has low cost maintenance. *C. elegans* intestine is composed of cells that share striking similarities to human intestinal epithelial cells	*Enterococcus faecalis*, *E. coli*, *S. aureus*
Protozoa (family Tetrahymenidae, order Hymenostomatida)	Ciliated protozoan (*T. thermophila* or *T. pyriformis*)	Ciliated protozoan *T. thermophila* has a genome of 104 Mb. Being able to switch from commensalistic to pathogenic modes of survival, *Tetrahymena* offers a low-cost and easy to handle alternative for modeling foodborne infections	*Aeromonas*, *E. coli*, *Listeria*, *Vibrio*, *Yesinia*
In Vitro			
Epithelial cell lines	Human colorectal cells Caco-2 and HT29, human colonic cell T84, human cervical cell HeLa, African green monkey kidney cell Vero, Madin-Darby canine kidney cell (MDCK)	Easy and low-cost maintenance, high sensitivity, and broad spectrum	*E. coli* O157:H7, *L. monocytogenes*
Endothelial cell lines	Human umbilical vein endothelial cell (HUVEC), human glomerular microvascular endothelial cell (GMVEC)	Easy and low-cost maintenance, high sensitivity, and broad spectrum	*E. coli* O157:H7, *L. monocytogenes*
Macrophage cell lines	Mouse macrophage cell J774, human macrophage cell U937	Easy and low-cost maintenance, high sensitivity, and broad spectrum	*E. coli* O157:H7, *L. monocytogenes*
Embryonated eggs	Chicken eggs	Cost-effective and easy maintenance, ready availability, sterile, and wide ranging fluids and tissues	*C. perfringens*, *Aspergillus fumigatus*
In vivo grown organ cultures (IVOC)	Various organs or tissues	Close to native live tissue; IVOC usage is limited to several hours (when the tissue dies). Further, it is technically challenging and shows sample variability	*E. coli* O157:H7, *Salmonella*, norovirus
Ussing chamber	Epithelial tissues	Ussing chamber detects and quantifies transport and barrier functions of living tissue	*E. coli* O157:H7, *Salmonella*

[a] Animal models are listed (in descending order) according to their evolutionary closeness to humans, with nonhuman primates being the closest and roundworm being the most distant.

On the other hand, as the next in the order of closeness to humans, rodents (especially mice) are increasingly employed as preferred animal models for foodborne infections and other human diseases. Mice possess a genome of 2.67 Gb dispersed in 20 chromosome pairs with ~25,000 genes, 99% of which have human counterparts. Having a relatively small body size, 18-day gestation, 10–15 pups per litter, 7 weeks to sexual maturity, and a 2–3-year lifespan (1 mouse year equals about 30 human years), mice provide an efficient and cost-effective model for human disease research including foodborne infections. It should be noted that mice practice coprophagy, an aspect that may be considered in experimental design for certain disease types.

Mice are highly amenable to manipulation, and can be inbred to yield genetically identical strains, which allows for more accurate and repeatable experiments. Through practice of cesarian birth, flexible-film isolator cages, and irradiated food, mice (and other animal species) can be maintained in completely germ-free conditions or colonized with one or more defined bacterial species (gnotobiotics). In addition, use of genetic selection and manipulation technologies enables insertion of extra genetic materials into genome, creating a variety of transgenic mice (including knockout, knock-in, and humanized mice as well as mice with conditional gene modifications or chromosomal rearrangement) [8].

For example, athymic nude mice are selected for the nude spontaneous mutation (*Foxn1^{nu}*, formerly *Hfh11^{nu}*) (which results in abnormal hair growth) and in the defective development of the thymic epithelium (which abrogates a cell-mediated immunity, despite the presence of T-cell precursors). Homozygous nude mice show partial defect in B cell development probably due to the absence of functional T cells, and their responses to thymus-dependent antigens are primarily limited to IgM due to a defect in helper T-cell activity.

Knockout mice are created by inserting a specific mutation into the endogenous gene. This leads to inactivation/silencing of the gene of interest, suppressing its expression and function. Knock-in mice are created by inserting a transgene into an exact location for overexpression. Both knockout and knock-in animals rely on the use of embryonic stem (ES) cells containing null or point mutations and complex chromosomal rearrangements (e.g., large deletions, translocations, or inversions), which are injected into the host mouse embryo, and subsequently implanted into a foster mother.

Humanized mice are created by inserting human genes (more recently entire human systems) into mice for subsequent expression. For instance, mice with human "immune systems" were generated by implanting either fetal lymphoid tissue or peripheral blood leukocytes into mice with spontaneous severe combined immunodeficiency. Humanized mice are capable of accepting a variety of human cells (blood, immune, cancer, etc.) without rejection.

Mice with conditional gene modifications are created with two different types of genetic alterations: one contains a conditional vector [through inserting recognition sequences for the bacterial Cre recombinase (*loxP* sites) using homologous recombination in ES cells], which functions as an "on switch" for the mutation, and the other contains specific sites (called *loxP*) inserted on either side of a whole gene, or part of a gene, that encodes a certain component of a protein that will be deleted. Similarly, mice with chromosomal rearrangement are created using the Cre/*loxP* recombination system to induce site-specific mutations that display defects resembling those caused by human chromosomal rearrangements (e.g., chromosome deletions, duplications, inversions, translocations, and nested chromosome deletions) [8].

Depending on the levels of simulation to human disease, animal models may be separated into three types: homologous, isomorphic, and predictive. Homologous animals demonstrate identical causes, symptoms, and treatments relative to human diseases; isomorphic animals have identical symptoms and treatments; predictive models share only a couple of aspects of human disease with humans, but nevertheless provide useful predictions about mechanisms of particular disease features. Similarly, depending on the way in which animal disease is induced, animal models may be divided into four categories: experimental, spontaneous, negative, and orphan. Experimental disease models resemble human disease conditions in phenotype or response to treatment but are induced artificially in the laboratory. Spontaneous disease models are analogous to human disease conditions that occur naturally. Negative disease models are essentially control animals, and are used to validate an experimental result. Orphan disease models have no human analog and occur exclusively in the species studied. Furthermore, to examine a particular disease, various approaches may be used. For example, inflammation may be studied via Carrageenan footpad edema (CFE) model, collagen-induced arthritis (CIA) model, pristane-induced arthritis (PIA) model, adjuvant-induced arthritis (AIA) model, ovalbumin-induced arthritis (OIA) model, air pouch model, and delayed-type hypersensitivity (DTH) model [8].

The *in vitro* culture models provide an alternative to the *in vivo* animals for mechanistic studies, by preserving the physiology of the living cell, without the need to sacrifice an animal. The advantages of the *in vitro* culture models include low cost, easy maintenance, relatively high efficiency, and little ethical concern. For instance, Caco-2 cells (of human colonic origin) can differentiate in culture, form brush border membranes, demonstrate transport properties (similar to intestinal epithelia), and express abundant intestinal microvilli, enzymes, and differentiation markers (typical of human small intestinal enterocytes), offering a valuable model for investigation of vectorial epithelial passage by para- and transcellular routes. Apart from the established cell lines (of epithelial, endothelial, and macrophage origins), other cells, organs, and tissues may be obtained from animal and human hosts for *in vitro* modeling. These include enterocyte suspensions, brush border membranes and vesicles, perfused duodenal segment, everted gut sacs, lymphocytes, etc.

When selecting an animal model for research, considerations should include: (1) appropriateness as an analog, (2) transferability of information, (3) genetic uniformity of organisms, (4) background knowledge of biological properties, (5) cost and availability, (6) generalizability of the results, (7) ease of and adaptability to experimental manipulation, (8) ecological consequences, and (9) ethical implications. If possible, three basic principles should be applied: replacement, reduction, and refinement. Replacement aims to use alternatives [e.g., computer models, tissues and cells, "lower-order" animals (cold-blooded animals, invertebrates, bacteria) instead of "higher-order" animals (primates and mammals) for experimentation]. Reduction employs mathematical calculations of statistical power to minimize the number of animals used. For example, by using an alternative way to LD50 for result interpretation, the number of experimental mouse groups for assessing *L. monocytogenes* virulence may be reduced from four to two, with further advantage of obviating the necessity to perform colony forming unit (CFU) estimation [16]. Refinement aims to minimize the suffering of each animal subject through experimental design that is as painless and efficient as possible [8].

1.4 Future Perspective

Despite our nonstopping efforts in the past, foodborne disease continues to savage human society at random and cause particular misery to vulnerable population groups. Naturally, we can point our fingers to the fact that foodborne pathogens have uncanny ability to constantly evolve and develop phenotypic and genetic traits that enable their evasion of host innate and acquired immune defense mechanisms, and their sabotage of our every intervention attempt. However, this does not hide the reality that some obvious gaps exist in our knowledge about the molecular basis of pathogenicity of foodborne organisms. Use of laboratory models including animal and cell culture models has contributed greatly to our past understanding of foodborne pathogens and diseases, and more will have to be learned via this approach in combination with other emerging technologies. The documentation and summation of the existing findings in this area provide a platform from which new insights will be uncovered and innovative mitigation measures will be launched.

REFERENCES

1. Bell A. Foodborne illness: the history of an invisible enemy. Available at https://antiquitynow.org/2014/06/05/foodborne-illness-the-history-of-an-invisible-enemy/.
2. Satin M. History of foodborne disease—Part 1—ancient history. Encyclopedia of Food Safety (Internet). 2014. Available at http://tinyurl.com/kmzt56u.
3. Behravesh CB, Williams IT, Tauxe RV. Emerging foodborne pathogens and problems: expanding prevention efforts before slaughter and harvest. In: *Institute of Medicine (US). Improving Food Safety through a One Health Approach: Workshop Summary.* Washington, DC: National Academies Press (US); 2012. A14. Available at http://www.ncbi.nlm.nih.gov/books/NBK114501/.

4. Riley LW, et al. Hemorrhagic colitis associated with a rare *Escherichia coli* serotype. *New Engl J Med.* 1983;308(12):681, 684–685.
5. Rangel JM, et al. Epidemiology of *Escherichia coli* O157:H7 outbreaks, United States, 1982–2002. *Emerg Infect Dis.* 2005;11(4):603.
6. Henkel JS, Baldwin MR, Barbieri JT. Toxins from bacteria. *Experienta Suppl.* 2010;100:1–29.
7. Walker MJ, et al. Disease manifestations and pathogenic mechanisms of group A *Streptococcus. Clin Microbiol Rev.* 2014;27(2):264–301.
8. McGonigle P, Ruggeri B. Animal models of human disease: challenges in enabling translation. *Biochem Pharmacol.* 2014;87(1):162–171.
9. Ericsson AC, Crim MJ, Franklin CL. A brief history of animal modeling. *Mo Med.* 2013;110(3):201–205.
10. Wolfgang MJ, Golos TG. Nonhuman primate transgenesis: progress and prospects. *Trends Biotechnol.* 2002;20:479–484.
11. Padilla-Carlin DJ, McMurray DN, Hickey AJ. The guinea pig as a model of infectious diseases. *Comp Med.* 2008;58(4):324–340.
12. Meurens F, Summerfield A, Nauwynck H, Saif L, Gerdts V. The pig: a model for human infectious diseases. *Trends Microbiol.* 2012;20(1):50–57.
13. Bou Ghanem EN, Myers-Morales T, D'Orazio SE. A mouse model of foodborne *Listeria monocytogenes* infection. *Curr Protoc Microbiol.* 2013;31:9B.3.1–9B.3.16.
14. Disson O, Lecuit M. In vitro and in vivo models to study human listeriosis: mind the gap. *Microbes Infect.* 2013;15(14–15):971–980.
15. García Y, Díaz-Castro J. Advantages and disadvantages of the animal models v. in vitro studies in iron metabolism: a review. *Animal.* 2013;7(10):1651–1658.
16. Liu D. Listeria monocytogenes: comparative interpretation of mouse virulence assay. *FEMS Microbiol Lett.* 2004;233(1):159–164.

Section I

Foodborne Infections due to Viruses

2

Adenoviruses

Anthony P. Malanoski and Baochuan Lin

CONTENTS

2.1 Introduction

Adenovirus (AdV) is an important human pathogen and is estimated to account for 8% of clinically relevant viral diseases globally.[1] First identified in 1953 as the cause for acute febrile respiratory disease, AdVs are endemic in the pediatric population worldwide, affecting children younger than 5 years old with mild symptoms and generally self-limiting illnesses.[2–4] Self-limiting infections may also occur in adults, but some serotypes have been associated with severe respiratory illness and potentially fatal outbreaks of pneumonia in residential facilities and military bases. The main etiologic agents for these outbreaks are serotype 4 and occasionally serotypes 3, 7, 14, and 21. It is possible that stress and crowding may contribute to AdV transmission and susceptibility.[5] AdVs have serious complications, impacting morbidity and mortality in immunocompromised individuals of any age.[6–8]

With more AdV serotypes being identified, it becomes clear that AdVs cause an array of clinical diseases, including epidemic keratoconjunctivitis (EKC), acute hemorrhagic cystitis, hepatitis, gastroenteritis, myocarditis, and pneumonia. Being one of the most prevalent enteropathogens causing infantile gastroenteritis, enteric AdVs are implicated in sporadic cases as well as in outbreaks of food-borne illness in kindergartens, schools, and hospitals.[9] Gastroenteritis due to AdVs often occurs in children younger than 5 years of age, accounting for ~12% of all enteropathogenic viruses identified, and is most commonly associated with serotypes 40 and 41; however, other types including 1, 2, 7, 9, 10, 18, 19, 22, 31, 42, 52, 58, 65, and 67 have also been reported as etiologic agents of viral gastroenteritis.[9–20] Serotypes 40 and 41 account for 5%–20% of hospital-admitted diarrhea cases in children under 2 years old. As children age, the incidence of AdV gastroenteritis decreases due to increasing levels of population immunity to AdV infection.

AdVs can be easily propagated in cell culture, and there are several cell lines that can be used as laboratory models for AdVs. The primary human embryonic kidney (HEK) cells are the best host for replicating various serotypes of AdVs. The lung epithelial cell line A549 and other epithelial cell lines, such as HEP-2, HeLa, and KB, are also good hosts for AdVs. For enteric AdVs, such as AdV40 and AdV41, the HEK 293 cell offers a convenient laboratory model.[2] In addition, *Sigmodon hispidus* cotton rats and mice, such as C57BL/6N, C57BL/lOScN, CBA/N, and C3H/N strains, were used as animal models to investigate the molecular pathogenesis of pneumonia caused by AdV infection.[21] AdVs have been used as models of virus–cell interaction. Decades of studies have contributed to the extensive understanding of the molecular biology, including life cycles, the host–pathogen interaction, genetics, epidemiology, and pathogenesis of AdVs, which are discussed in this chapter. AdVs continue to be studied as delivery vehicles for gene therapy, vaccination, and cancer treatment, which underscores the importance of understanding these viruses.

2.2 Classification and Morphology

AdVs constitute the Adenoviridae family, which is divided into the genera *Mastadenovirus* and *Aviadenovirus*. The genus *Mastadenovirus* covers viruses of several different animals, including bat, bovine, canine, equine, human, murine, ovine, porcine, simian, and so on, whereas the genus *Aviadenovirus* is limited to viruses of birds.[2] Currently, there are 68 reported human AdVs according to the National Center for Biotechnology Information Taxonomy Browser, representing seven different species or subgroups (A–G). The classification of AdVs was originally based on their hemagglutination patterns and serologic profiles. Recent advancements in sequencing capability have allowed the discovery and classification of new AdVs (types 52–68), where the differentiation of strains is based on bioinformatics analysis of their genomic sequencing (Table 2.1).[2,4,11–15,22–37] The majority of these newly discovered AdVs are products of homologous recombination, a common evolutionary adaptation of AdVs. Among the seven different species, species B can be further divided into B1 and B2 based on their organ tropisms.[22] There is a correlation between the species and their tissue tropisms, which determines the clinical manifestation of AdV infection. Species A, F, and G show tissue tropisms toward the gastrointestinal tract and induce gastroenteritis. Species B1, C, and E mainly cause respiratory illness; species B1, B2, D, and E produce ocular infection, whereas B2 AdVs cause kidney and urinary tract infection.[8,37]

AdVs are nonenveloped double-stranded DNA (ds DNA) viruses with icosahedral shells and nucleoprotein cores, ranging in size from 65 to 100 nm in diameter. The capsid of the viral particle is composed of seven proteins: hexon, three hexon-associated proteins, penton, a penton-associated protein, and fiber. The proteins form 252 capsomeres, which consist of 240 trimers of the major capsid protein hexon and 12 pentons. The fiber protein, which has a length that varies among the different serotypes, embeds in the penton base and projects out from the capsid. These 12 extensions out of the particle serve crucial

TABLE 2.1

Classification of Human Adenoviruses

Species	Hemagglutination	Serotype	References
A	IV	12, 18, 31, 61	2,12
B1	I	3, 7, 16, 21, 50, 66, 68	22–25
B2	I	11, 14, 34, 35, 55	22,26,27
C	III	1, 2, 5, 6, 57	2,28
D	II	8–10, 13, 15, 17, 19, 20, 22–30, 32, 33, 36–39, 42–49, 51, 53, 54, 56, 58, 59, 60a, 62–65, 67	2,13–15,25,29–36,120
E	III	4	2
F	III	40, 41	2
G		52	11

I, complete agglutination of monkey erythrocytes; II, complete agglutination of rat erythrocytes; III, partial agglutination of rat erythrocytes; IV, little or no agglutination.

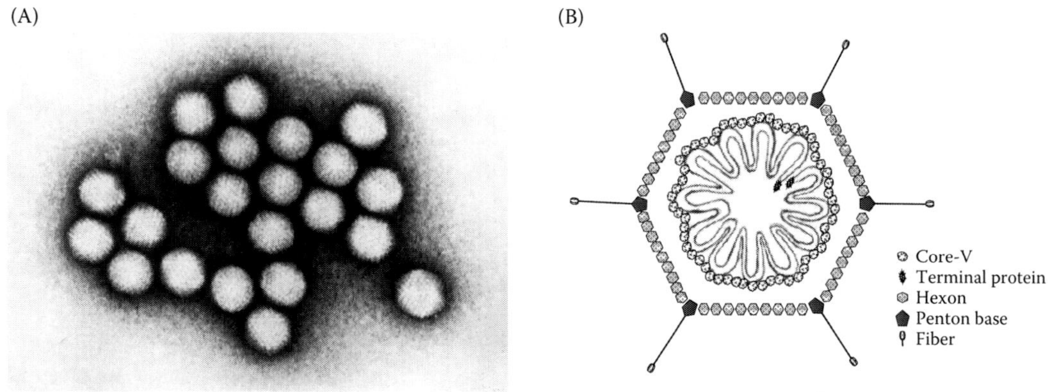

FIGURE 2.1 Image and schematic depiction of the structure of adenovirus. (A) Transmission electron micrograph (TEM) of adenovirus virions. Visible at this high magnification are the capsomeres, which are hexagonally shaped, also called hexons. Hexons comprised the outer shell of the adenovirus known as a capsid. These adenoviruses displayed an icosahedral symmetry, characterized by 12 vertices and 20 facets. Each virion was 70–80 nm and exhibited no spikes. Image by Dr. G. William Gary, Jr. was downloaded from the public health image library (ID #237) of Centers for Disease Control and Prevention, USA. (B) Schematic depiction of the icosahedral capsid of human adenovirus (HAdV) showing the major structural proteins, hexon, penton based, and fiber in the virion. The ds DNA genome is shown as a dark line inside the capsid. (Modified from Russell, W.C., *J. Gen. Virol.*, 90, 1–20, 2009.)

functions in cell attachment and entry (Figure 2.1).[2,6] All the reported AdVs encode a single fiber protein except AdV40, AdV41, and AdV52, which encode two fiber proteins. Fiber proteins mediate viral entry by interacting with receptors on susceptible cells; the expression of two fiber proteins might expand the range of susceptible cells for these AdVs.[2,11] The nucleoprotein core of AdVs contains the viral genome and five structural proteins, V, VII, Mu (μ), IVa2, and a terminal protein (TP).[2,6]

2.3 Genome Organization and Conserved Features

The AdV genome consists of a linear ds DNA of ~35 kb encoding more than 30 structural and nonstructural proteins in most serotypes. All AdV serotypes have the same general organization; that is, the genes encoding specific functions are located at the same position on the viral chromosome. However, DNA sequence homology among the strains within a human AdV species ranges from 48% to 99%, with the highest homology among strains of species C. The homology between species is <20%.[2,4] The 5′ and 3′ ends of the human AdV genome consist of inverted terminal repeats (ITRs) ranging from 40 to 212 nucleotides, which contain the origin of DNA replication and are hallmark motifs of the HAdV genome. All HAdV genomes contain five early (E) transcribed genes (E1A, E1B, E2–E4), two intermediate genes (IX, IVa2), and five late (L) genes (L1–L5) (Figure 2.2). The early genes are mainly involved in the control of DNA replication, transcription, transportation of Ad mRNAs (messenger ribonucleic acids) to the cytoplasm, and immunoregulation; the intermediate genes control DNA packing and also function as transcription factors, whereas the late genes are coded for viral structural proteins.[38]

The E1A gene is the first to be transcribed and expressed after infection, and multiple E1A proteins are generated through alternative splicing, which regulate the transcription of viral and cellular genes. The E1B gene encodes proteins that block apoptosis and potentiate viral replication. The E2 region is composed of two transcripts, E2A and E2B, and produces proteins required for viral replications, which include DNA-binding protein (DBP), terminal protein precursor (pTP), and DNA polymerase (pol).[29,39–41] Proteins encoded by E3 gene are not essential for viral growth, but function as modulators of the host immune response.[38,42] The E4 gene encodes proteins with various functions, including promoting viral gene expression and replication, regulating cellular transcription factor E2F, and controlling protein phosphorylation.[43] The two intermediate genes produce proteins, IX and IVa2, which control DNA

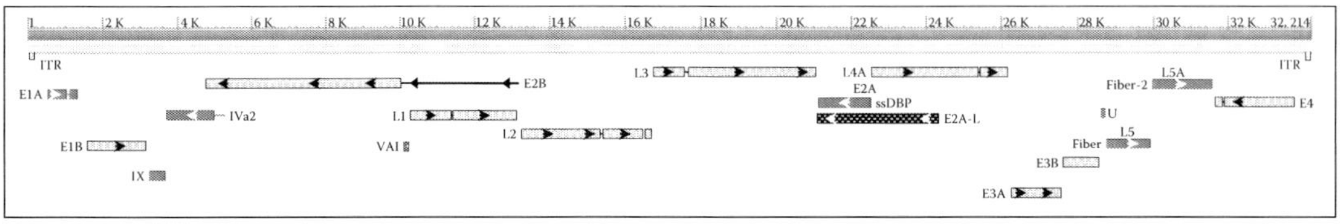

FIGURE 2.2 Genome structure of human AdV-type 40. The genome structure was constructed based on the sequence deposited in GenBank (NC_001454). The half-filled rectangle indicates a gene, while the fully filled rectangle indicates CDS. The arrows indicate the direction of the gene that is transcribed. VAI, virus-associated RNA I; ssDBP, single-stranded DNA-binding protein.

packing and also function as transcription factors. The five late genes encode viral proteins involved in capsid production for mature virions, assembly, and viral maturation, such as 52-kDa protein, IIIa, penton base (III), V, VII, X, minor capsid protein (VI), hexon, 23-kDa protease, 100-kDa protein, 33-kDa protein, VIII, and fiber.[2,39,40]

There are 16 genus-common proteins; that is, there is homology in all genera in AdV genomes, which are probably rooted in a common ancestral AdV. These proteins primarily function in DNA replication, DNA encapsidation, and viral particle structural formation. The genus-common proteins involved in DNA replication are DBP, pTP, and pol, and two proteins are involved in DNA encapsidation, 52-kDa protein and IVa2. Viral proteins, such as IIIa, III, VII, X, VI, hexon, protease, 100-kDa protein, 33-kDa protein, pVIII, and fiber, necessary for virion formation are also genus common.[41] Additionally, U exon is also considered as a genus-common feature, even though it has been lost in certain members of the *Mastadenovirus* genus (PAdV-5 and MAdV-1). U exon locates between the E3 region and the *fiber* gene, encodes the U exon protein (UXP, ~24 kDa), which initiates in U exon, continues in frame in the DBP leader, and extends throughout the DBP coding region but with a different reading frame from DBP that has functions relevant to DNA replication or RNA transcription.[41,44] For HAdVs, most genus-specific genes are located within the E1 and E4 regions.

2.4 Protein Function

There are seven structural proteins that form the capsid of the viral particle: hexon (polypeptide II), hexon-associated proteins (pVI, pVIII, pIX), L1 peripentonal hexon-associated protein (pIIIa), fiber (polypeptide IV), and penton (polypeptide III). Inside the capsid, the nucleoprotein core contains a viral genome and five additional proteins (pV, pVII, μ, IVa2, and TP) as well as a viral protease.[37]

Hexon, the main building block of AdV capsid, forms homotrimers, which in general are referred to as hexon capsomeres and arrange in a pseudohexagonal symmetry. The size of the hexon monomer varies with the serotype ranging from 917 to 978 amino acids, with the largest one belonging to HAdV6.[28] Each viral particle consists of 240 capsomeres, which form the facets of the icosahedral virus shell. The N- and C-terminals of hexon can adopt different conformations and provide interactions between hexons or between hexons and hexon-associated proteins pIIIa and pVIII. In the base of each hexon molecule, an eight-stranded β-barrel arranges in a "jellyroll" topology, which provides the means for interacting with neighboring capsomeres.[37,45,46] Hexon capsomeres are further stabilized by minor coat proteins pIIIa, pVI, pVIII, and pIX, also called hexon-associated proteins. In addition, these minor coat proteins also serve as extensions of major capsid proteins (i.e., hexon, penton, and fiber), which mediate the protein–protein interactions and stabilization of all capsomeres.[2,47]

Minor coat protein pIIIa is a highly helical protein and serves to stabilize the vertex region by tethering peripentonal hexons and penton bases via its N-terminal domain. In addition, pIIIa also interacts with pVIII, which tethers the peripenton hexons to the hexons in the central plate of the facet. Studies also indicated that pIIIa promotes correct genome packaging during viral assembly by interacting with the putative scaffold protein L1 52/55 kDa via its N-terminal domain as well as signaling for vertex and genome release during uncoating.[45,48] Similar to pIIIa, pVI also plays multiple roles during AdV infection in addition to stabilizing hexon capsomeres and bridging the core to the icosahedral shell. In the early phase of viral infection, the N-terminal amphipathic α-helix is responsible for inducing curvature of the endosome membrane, which leads to membrane disruption, thereby allowing the virus to escape into the cell's cytosol. pVI also contains a ubiquitin ligase-interacting motif (PPxY), which not only facilitates microtubule-dependent trafficking toward the nucleus, but also activates E1A promoter and leads to initiation of adenoviral gene expression. In later phases of AdV infection, the nuclear export signals of pVI interact with importin α/β, which mediates the nuclear import of hexon and the subsequent proteolytic removal of C-terminal nuclear export signal transition pVI to support infectious viral particle assembly.[45,49–51] Minor coat protein pVIII is not as well characterized, and its only known function is stabilizing capsomeres. pIX, located on the outer part of the capsid, also has multiple functions in the life cycle of AdV infection. Upon viral entry and escaping from endosome, pIX is responsible for interacting with kinesin-1, which results in capsid dismantling and promotion of transport of the viral genome into nucleus. Subsequently, the C-terminal leucine repeat is important for enhancing gene expression from the E1A promoter in the early stage of infection, and the N-terminal region is required for incorporation into viral capsid at the later stage of infection. pIX also plays a role in virus-induced nuclear reorganization, that is, the formation of specific nuclear structures, which appear as dispersed nuclear globules on immunofluorescence staining and as clear amorphous spherical inclusions on the electron microscopy of cells that are in the later stages of infection. It is suggested that the reorganization of nuclear proteins provides a favorable environment for viral replication. Evidence also suggests that pIX may play a role in modulating the viral tropism and/or interfering with the immune response.[45,52–54]

In addition to hexon capsomeres, the penton base and fiber proteins are also the main building blocks of AdV capsid and form penton capsomeres, which locate at 12 vertices of the icosahedral facets. The penton capsomere consists of a homopentameric penton base and homotrimeric fibers, with the fiber molecules projecting from the penton base at each vertex. Penton and fiber molecules are the key players in AdV attachment and cell entry.[45,55] The C-terminal knob domain of fiber molecules initiates the interaction with cell receptors. Once the cell binding is initiated, the integrins (coreceptor) $\alpha_v\beta_3$ and $\alpha_v\beta_5$ bind to conserved RGD domain at penton base, which in turn starts the receptor-mediated endocytosis with the exception of species F enteric AdVs, which neither have the RGD motif nor use integrin for cell infection.[56] The size of the penton protein varies with the serotype, ranging from 493 to 594 amino acids. The most significant size variation of the major capsid proteins of AdVs is observed in fiber protein, which ranges from 319 to 587 amino acids (aa). Serotypes within species A and C have long fibers (>500 aa), serotypes within B and D species have short fibers (<400 aa), and species E has intermediate length of fibers (425 aa). Interestingly, serotypes belonging to species F and G have two fibers (long and short fibers) and all cause gastroenteritis. For species F, the long fiber has been shown to bind to the AdV receptor, while the function of the short fiber remains unknown, but it likely plays a role in enteric AdV tropism and cell infection.[57]

Capsid core proteins include pV, pVII, Mu, TP, pIVa2, and the protease, and all but the protease are associated with the viral genome. The minor core protein pV, containing a basic arginine-rich domain, interacts with viral DNA. Functional studies indicated that pV is not essential for viral replication and infectious viral particle assembly in cancerous cells, but it mediates the translocation of nucleophosmin 1/NPM1/B23.1 from the nucleoli to the nucleoplasm that is essential for productive infection in normal cells.[37,58,59] The major core protein pVII, with four basic lysine- and arginine-rich domains, serves as a histone-like structure that viral DNA wraps around. pVII is synthesized as a precursor protein (preVII) and is matured by the removal of the N-terminal 24 amino acids via the viral protease. Interestingly, preVII is involved in condensing the viral DNA during particle assembly, but only mature pVII is present in the viral nucleoprotein core. In addition, pVII also mediates virus DNA import into the nucleus through the transportin pathway during infection, potentially through its nuclear and nucleolar localization signals. Furthermore, pVII interacts with pIVa2 and L1 52/55 kDa during viral particle assembly.[37,60,61] pIVa2 performs multiple tasks during the life cycle of an AdV. Encapsidation of viral DNA requires the binding of pIVa2 to the packaging domain as a multimeric complex with the L4 22-kDa protein and the L1 52/55-kDa protein, as well as pVII. The binding of pIVa2 to the packaging domain initiates genome recognition and recruits the L4 22-kDa protein to the packaging domain. In addition, pIVa2 forms a dimer with the L4 33-kDa protein and functions as a transcription activator of the viral major late promoter and regulates the synthesis of the majority of late structural proteins of the virus.[37,62–64] Core protein Mu (μ), generated from a precursor protein (79 aa) by the removal of the N- and C-terminal amino acids via the viral protease, is a function partner of pV and pVII in condensing the viral genome and DNA encapsidation.[65,66] TP is a 55-kDa protein, covalently linked to the 5′ end of ITR, which allows the circularization of viral DNA and enhances *in vitro* DNA replication.[67–69] The TP is synthesized as an 80-kDa pTP, which forms a heterodimer with pol that plays a key role in DNA replication by serving as a primer for initiating viral DNA replication. Furthermore, pTP is a key participant in the AdV replication complex consisting of pTP, pol, and DBP as well as cellular factors NFI, Oct-1, and type I topoisomerase NFII. This replication complex is tightly bound to the nuclear matrix through interaction with the pTP, thereby anchoring the viral genome to the nuclear matrix.[68,70–72] Protease plays a key role in several steps of the viral life cycle, including uncoating the viral particles upon viral entry into cells and enabling the cleavage of core precursor proteins that facilitates the formation of infectious viral particles and the cleavage of cytokeratins leading to host cell lysis. It is synthesized in an inactive form and becomes partially activated by binding to the viral DNA, which then cleaves out an 11-amino-acid viral peptide (pVIc, GVQSLKRRRCF) from the C-terminal of the precursor of pVI. pVIc binds to the protease and fully activates its function.[73–78]

The majority of nonstructural proteins of AdVs are encoded by the early-transcribed genes. E1A is the first gene expressed upon viral infection and plays an essential role in regulating the viral transcription of genes necessary for replication and reprograming the cellular transcription to facilitate viral replication. Two major E1A proteins, 12S and 13S E1A, are synthesized via alternative splicing of mRNA transcripts, which results in a 46 aa conserved region (CR3) that is unique in 13S E1A. Despite their similarity, 13S E1A is primarily responsible for activating the viral gene expression via CR3, while 12S E1A represses transcription, which is achieved via temporal regulation.[79–82] E1A proteins activate transcription through a TATA motif, YY1 recognition sites, by interacting with a variety of cellular transcription factors and relieving transcription repression.[2] In addition to modulating transcription, E1A proteins also regulate signaling pathways and 26S proteasome, interfere with pathways involved in immune regulation, as well as involve the evasion of T-cell immunity to ensure viral survival and infectivity.[82]

E1B gene products are required for efficient viral replication during AdV infection. At least five E1B gene products were identified in Ad5, which were generated through alternative splicing. One of these proteins, E1B-55-kDa protein, in conjunction with E4 open reading frame (ORF) 6 protein (E4ORF6) forms an E3 ubiquitin ligase complex consisting of cellular proteins cullin 5 (Cul5), RING-box 1 (Rbx1), and elongins B and C. This viral ubiquitin ligase regulates the degradation of cellular proteins, such as p53, the MRE11-Rad50-NBS1 (MRN) DNA damage recognition/repair complex, and DNA ligase IV, that have a detrimental effect on viral replication during the early stage of viral infection. In addition, the E1B-55-kDa/E4ORF6 ubiquitin ligase complex also prevents DNA damage repair mechanisms by interfering with the MRN complex. During AdV infection, promyelocytic leukemia (PML) bodies, consisting of several critical DNA response proteins including components of the MRN complex, are disrupted and relocated to nuclear track-like structures. The MRN complex is initially relocalized into PML-containing nuclear tracks by another E4 protein, E4ORF3, where it subsequently binds to the E1B-55-kDa protein. The interaction of E1B-55 kDa and E4ORF3 enables the localization of E1B-55 kDa to the PML-containing nuclear track, and forms complexes containing MRN, E1B-55 kDa, E4ORF3, and E4ORF6 proteins, which are then transported out of the nucleus into cytoplasmic aggresomes to further MRN proteasomal degradation by the E1B-55K/E4orf6 ubiquitin ligase complex. E1B-55K protein can also contribute to the deregulation of the cell cycle during lytic infection. During the later stages of viral infection, the E1B-55-kDa/E4orf6 ubiquitin ligase complex inhibits the export of host cell mRNA while promoting the export of late viral mRNA from the nucleus and its translation.[83,84] Another E1B gene product, E1B-19-kDa protein (small T antigen), is the first viral BCL-2 homolog discovered, and it inhibits apoptosis triggered by AdV infection by blocking the BIK/NBK pathway to ensure successful virus production.[85,86]

E2 encodes three replication proteins, DBP, pol, and pTP. These proteins along with three cellular transcription factors, NFI, Oct-1, and type I topoisomerase, constitute the AdV DNA replication system.[2] DBP, a product of the E2A gene, binds dsDNA, ssDNA, and RNA without apparent sequence specificity. It is involved in initiating DNA replication and is also essential for elongation by enhancing the activity of pol, unwinding dsDNA, removing secondary structures, and protecting ssDNA. DBP is a multifunctional protein involved in DNA replication, transcription regulation, mRNA stability, transformation, virion assembly, and host range determination. During DNA replication, DBP first stimulates the initiation by increasing the binding of NFI and recruiting the pTP–pol complex to the replication origin. Then, DBP unwinds the parental strand and enhances the progress of the polymerase by binding to the displaced strand and protecting it from nuclease digestion, which enhances strand displacement activity during the elongation phase. Finally, DBP enhances the rewinding of complementary displaced strands.[87–89] AdV pol is a DNA-dependent DNA-pol encoded by E2B transcript with a molecular weight of 140 kDa and belongs to a distinct family of protein priming pol, eukaryotic pol α family, that uses a protein primer for the initiation of replication. It also shows 3′→5′ exonuclease activity for proofreading. AdV pol forms a stable heterodimer with pTP, protein primer, which initiates the viral DNA replication by binding to NFI and Oct-1 and results in a pre-initiation complex which then binds the viral core origin.[90–92]

The E3 region is distinct for each species varying in length and number of ORFs representing areas of major sequence divergence.[93] Common to all species, E3 encodes three proteins, gp19K, receptor internalization and degradation (RIDα/β, 10.4 kDa/14.5 kDa), and 14.7-kDa protein, which function in modulating host immune response and AdV death protein (ADP, 11.6 kDa), which is expressed in the later stages of infection.[38,94] The main function of gp19K is to suppress the cell surface expression of class I major histocompatibility complex (MHC) by inhibiting the transport of class I MHC from the endoplasmic reticulum (ER) to the plasma membrane and the processing of peptides by tapasin, thus decreasing the amount of peptide presented by class I MHC in infected cells. The E3 14.7-kDa protein inhibits apoptosis induced by TNF-α via reducing the secretion of arachidonic acid, while RIDα/β inhibits apoptosis induced by Fas-L and TRAIL. RIDα/β not only destroys the receptors for Fas-L and TRAIL to prevent cell death caused by these factors but also inhibits apoptosis induced by TNF-α through the same mechanism as the E3 14.7-kDa protein.[38] In contrast, ADP is proapoptotic and mediates the release of AdV by lysing the infected cells at very late stages of infection.[94]

E4 is predicted to encode seven different polypeptides that are required for lytic growth, and all but one are expressed. Both ORF3 and ORF6 are involved in viral DNA replication by interacting with E1B-55-kDa protein, as described earlier (see section above).[83] Both ORF3 and ORF6 promote viral gene expression by facilitating the accumulation of relevant mRNAs at a posttranscriptional level in the cytoplasm and stabilizing the unprocessed late RNA in the nucleus. Evidence suggests that the E1B-55-kDa–E4ORF6 complex causes the redistribution of cellular factors necessary for RNA biogenesis and that ORF3 directly affects the distribution of essential transcription/replication factors in the nucleus. In addition, ORF6 interacts with p53 to inactivate its function, which is crucial for successful infection by AdV.[43] Besides ORF3 and ORF6, studies suggested that ORF1 is a transforming protein and may stimulate quiescent cells to promote lytic infection and oncogenesis. E4ORF6/7 forms dimers to interact and modulate the activity of the cellular transcription factor E2F and subsequently activate cellular factors that are important for the S phase of cell cycle.[43,95] The function of E4ORF4 is to negatively regulate E1A and E4 transcription via protein phosphorylation by interacting with protein phosphatase (PP)2A, which creates a regulatory loop limiting cytotoxicity in the early stages of infection. The functional information for E4ORF2 and E4ORF3/4 still awaits discovery.[43]

AdVs possess two intermediate (also called delay early) genes *IVa2* and *IX*, which encode pIVa2 and pIX, respectively. pIVa2 interacts with L1 52/55-kDa protein to activate a major late promoter which is required for viral DNA packaging.[96] pIX is a minor coat protein and its function has been described in the previous section. In addition, pIX also stimulates an AdV major late promoter containing TATA box elements and contributes to the efficient induction of the late transcription unit.[97]

One of the nonstructural proteins is encoded by L4 gene. The L4-100-kDa protein inhibits the translation of cellular mRNA by eliminating the cap-dependent translation pathway via binding to eukaryotic initiation factor 4G (eIF4G) and displacing Mnk1 from eIF4G. It also promotes translation of late viral mRNAs through ribosome shunting, which leads to nuclear accumulation of viral products for capsid assembly. The underlying mechanism for these functions is the interaction of L4-100 kDa with the tripartite leader (TL) sequence present in all the late viral transcripts and eIF4G, which is the scaffolding element of the cap-dependent translation initiation complex. In addition, L4-100-kDa protein not only acts as a chaperone facilitating hexon trimerization but also assists in nuclear accumulation of hexon trimers and in the scaffolding process of AdV capsid.[98–101] Another function of L4-100kDa is to prevent apoptosis induced by granzyme B in infected cells. Granzyme B, one of the lymphocyte granule serine proteases, catalyzes the cleavage and activation of several caspases which induce apoptotic events in infected cells. By interacting with granzyme B, the L4-100-kDa protein prevents access of substrates to the proteinase catalytic site through steric hindrance and inhibits its activity.[102] A unique feature of L4-100kDa is its posttranslational modification by arginine methyltransferase 1 (RPMT1). The methylation of L4-100kDa has regulatory effects on modulating its interaction with hexon and TL mRNA as well as on promoting the capsid assembly.[101]

A conserved ORF in *Mastadenovirus* genus, U exon, located between E3 and L5, was first identified in Ad40 genome sequence, and later identified to be a 24-kDa protein, UXP. UXP is a late protein encoded in an overlapping reading frame with DBP and part of the 100-kDa protein and has been found in all species C serotypes so far. The characterization of UXP indicated that it localizes within the nucleoli, nucleus, and the periphery of the replication center, suggesting its involvement in DNA replication or RNA transcription.[44,103]

2.5 Life Cycle

2.5.1 Cell Entry and Replication

As nonenveloped viruses, AdVs require engaging host cell factors, that is, protein or carbohydrate receptors, to gain cell entry via interactions between the viral fiber protein and host cell receptor. Host cell factors for species A, C, E, and F have been identified as coxsackievirus and AdV receptor (CAR), and CD46 or desmoglein-2 (DSG-2) are used by species B. The majority of species D uses CAR except subtypes 8, 19, and 37, which use sialic acid and GD1a glycan for attachment.[104] After initial attachment, the interaction of the penton base (via RGD motif) and $\alpha_v\beta_3$ or $\alpha_v\beta_5$ integrins is the required second step for cell entry, which triggers clathrin-mediated endocytosis of virions. Upon entry into cells, capsids are partially disassembled and escape from the endosome, which are then transported along microtubules to the perinuclear envelope, where further disintegration of the capsid allows import of the viral genome into the nucleus (Figure 2.3).[55] It should be noted that the relative importance of the fiber–CAR interactions and penton base attachment mechanism in the determination of successful infection depends on the length of the fiber protein. For example, fibers in group C species that are consistently on the longer side required less of the penton base attachment mechanism for successful infection.[38]

DNA replication of HAdVs employs a unique mechanism similar to the protein-primed DNA replication of several bacteriophages and requires three viral proteins—DBP, pTP, and pol—as well as three cellular factors—NFI, Oct-1, and type I topoisomerase. Located within the ITRs of HAdV genomes are DNA replication sites, which consist of core and auxiliary origins (the terminal nucleotides form the minimal origin and the remainder of ITR). For initiating DNA replication, the formation of a pre-initiation complex consisting of DBP, pTP, pol, NFI, and Oct-1 is needed. The interactions of NFI with pol and Oct-1 with pTP enable the pTP/pol dimer to be recruited to the replication origin. NFI and Oct-1 also bind to the auxiliary origin, which enhances the initiation and changes the origin structure of the DBP. The covalent addition of dCMP to pTP was followed by the formation of a pTP–trinucleotide intermediate, pTP–CAT, which uses nucleotides 4–6 as a template. This pTP–CAT intermediate then jumps back three bases and becomes paired with template residues 1–3. This jumping-back mechanism explains the presence of a short 3-bp (or sometimes 2- or 4-bp) repeat sequence in the first 10 bp of all HAdV origins. Either during or shortly after jumping back, pol dissociates from pTP and in conjunction with the action of DBP elongates the protein primer, leading to the formation of a new duplex genome and the displacement of the nontemplate strand. Full-length replication of the genome requires the presence of NFII/topoisomerase I. In the final stages of replication, pTP is cleaved by a viral protease to TP, resulting in the progeny DNA, which is subsequently packaged in virions.[71,89,92]

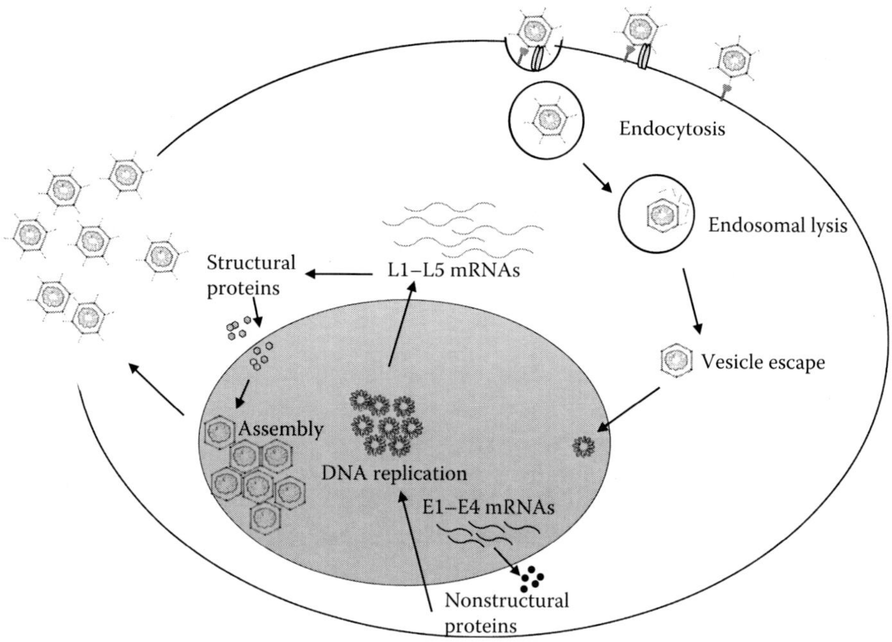

FIGURE 2.3 The life cycle of HAdV. First, the virus attaches to a primary receptor on the cell surface via fiber. This is followed by penton base interaction with $\alpha_v\beta_3$ and $\alpha_v\beta_5$ integrins, which facilitates clathrin-mediated endocytosis. Inside the endosome, the virion begins to dissociate in the low pH environment and releases the vertex proteins including pVI, which has been implicated in the disruption of the endosomal membrane, allowing the virion to escape from the endosome. The partially disassembled virion is transported to the nuclear pore complex. At the nuclear pore, the viral DNA is imported into the nucleus allowing the transcription and replication of the virion. At the final stage, the viral proteinase cleaves its multiple substrates and an icosahedral structure is formed, which contains the full-length DNA. (Modified from Nemerow, G.R. et al., *Virology* 384, 380–388, 2009; Shenk, T.E., Adenoviridae. in *Fields Virology*, Vol. 2. eds. Knipe, D.M., Howley, P.M., Lippincott Williams & Wilkins, Philadelphia, 2265–2300, 2001.)

2.5.2 DNA Packaging and Virion Assembly

The assembly of infectious AdV particles requires *cis*-acting packaging sequence at the left end of the AdV genome, and genetic analysis has identified seven functionally redundant repeated A/T-rich motifs, A repeats 1–7, from left to right that direct the packaging of the DNA. Among these repeats, A1, A2, A5, and A6 are functionally dominant with a bipartite consensus: the $TTTGN_8CG$ sequence. Three proteins, IVa2, L1 52/55 kDa, and L4 22 kDa, have been identified to bind to the packaging domain and trigger the interaction with the structural proteins, hexon, IIIa, penton, fiber, pVI, and VIII, as well as nonstructural proteins (scaffolding proteins), L4 100 and 33 kDa, L1 52/55 kDa, and IVa2, forming a "procapsid" structure with the viral DNA. Encapsidation of the complete viral genome into the procapsid is driven by a putative packaging motor through the portal complex, presumably an ATP-dependent incorporation of the remaining viral DNA. The mechanism is still unclear, but pV and pVII are incorporated into intermediate viral particles containing full-length DNA. At the final stage of virion maturation, viral proteinase cleaves its multiple substrates and forms an icosahedral structure, which contains full-length DNA and has a density of 1.34 g/cc in cesium chloride, but does not contain proteins L4 100 and 33 kDa.[64,105]

2.6 Pathogenesis

While AdVs generally infect the mucosal epithelium, individual serotypes differ in their tissue specificity and cause a number of distinct clinical syndromes, including gastroenteritis, respiratory disease, conjunctivitis, hemorrhagic cystitis, and exanthema. Thus, the pathogenicity of AdVs varies depending on the species, serotype, and organ specificity. In general, AdV-infected cells degenerate, and the infected respiratory epithelial cells have enlarged nuclei containing inclusion bodies.[3] Most of our understanding on AdV pathogenesis comes from studies using *S. hispidus* cotton rats and mice for respiratory illness. For gastroenteritis, it is likely that AdV infection causes destruction of functional mature cells, which in turn disrupts the water readsorption from the gut and consequently diarrhea ensues. In response to the damage, the villi retract and the crypt cells undergo rapid division to repopulate the villi with young and undifferentiated cells which are resistant to virus infection. However, the malabsorption continues until these undifferentiated cells mature and develop the necessary functional capabilities.[106,107]

2.7 Host–Pathogen Interaction

In addition to the interaction of structural proteins and host cell receptors to gain entry, AdVs also employ various strategies interacting with cytokines to modulate the innate and the adaptive host immune defenses.[108] Interferons (IFNs) provide the first line of defense against viral infection by binding to their receptors, which enhances transcription of cellular factors that restrict viral growth and signal to the adaptive immune system the presence of a viral pathogen. For AdVs, E1A is responsible for the anti-IFN effect by blocking the IFN signal transduction pathway mediated by IFN-stimulated gene factor 3 (ISGF3) via binding to p300, one of the transcriptional adaptor proteins that is required for the activation of IFN-regulated genes.[109,110] Similarly, AdE1A also interferes with the expression and signal transduction of interleukin (IL)-6, an inflammatory cytokine, via signal transduction and activation of transcription (STAT) family members.[111]

Tumor necrosis factor (TNF), produced by activated macrophages in response to inflammatory stimuli, can directly inhibit virus replication and lyse virus-infected cells *in vitro*. There are several viral proteins that interact with TNF during its life cycle to provide an optimal viral replication environment. The various functions of E1A proteins are associated with their binding with various cellular factors. Among these, the binding of E1A with cellular factors p300 and p105 Rb (retinoblastoma gene product) resulted in cell cycle dysregulation, that is, the activation of quiescent cells for optimal viral replication; however, these interactions also induce TNF sensitivity and potential apoptosis. The E1A-induced TNF sensitivity and apoptosis are countered by E1B 19-kDa protein, which inhibits the cytolysis activity of TNF by blocking TNF-mediated signal transduction or by counteracting cytotoxicity through activating the transcription of protective gene products, such as Mn superoxide dismutase. Furthermore, E3 10.4, 14.5, and 14.7-kDa proteins provide TNF resistance via a 10.4-kDa/14.5-kDa heterodimer or 14.7-kDa protein. The 10.4-kDa/14.5-kDa heterodimer also downregulates the epidermal growth factor receptor (EGF-R) by stimulating endosome-mediated internalization of the receptor, which induces cell proliferation.[42,108,112]

Cytotoxic CD8+ T cells are critical in controlling primary viral infections via major histocompatibility complex class I proteins (MHC-I). MHC-I binds to proteolytic cleaved viral peptides and is transported to the cell surface, termed as antigen processing and presentation, which will be recognized by receptors on CD8+ T cells, and as a result, the T cells are triggered to kill the virus-infected cells.[108,113] To evade this reaction, species B–E AdVs encode the E3 19-kDa protein, which binds to MHC-I and anchors it to ER, preventing its transport to cell surface, thereby inhibiting the lysis of virus-infected cells by CD8+ T cells. Alternatively, species A and F AdVs do not code for E3 19 kDa, but use 13S E1A instead to downregulate MHC-I. 13S E1A binds to cellular factor p105-NFκB1, the precursor of p50 NFκB1, resulting in reduced amounts of NFκB and KBF1, which are needed to activate MHC-I expression, and consequently, a lesser amount of MHC-I is expressed.[108,114]

2.8 Immunity

Murine pneumonia models (nonpermissive infection with HAdVs) have been used to study the cytokine response induced by AdV infections. During early-phase infiltration, the infected mouse lungs showed high titers of TNF-α, IL-1, and IL-6, but only IL-6 was found in the peripheral blood. IL-6 reached maximum titers 6–24 h after infection, whereas maximum levels of TNF-α and IL-1 were attained 2–3 days after infection.[21] IL-8, a proinflammatory cytokine, secreted by airway macrophages and the epithelial cells within the lung epithelium, was also induced by AdV infections. The higher titer of IL-8 increases the protein synthesis and apical localization of CAR in polarized cells, which in turn tethers the infiltrating neutrophils on the apical surface of polarized epithelia. The adherence of neutrophils on the apical surface provides for amplification of AdV infection.[115] IL-12 is also induced by AdV infections, which may be important in innate immune activation and the subsequent adaptive immune response to AdV infection; however, the exact mechanism remains unclear.[116] AdV infections also induce the production of high levels of type I IFN. The increased levels of type I IFN enhance the transcription of IFN-stimulated genes; however, the E1A gene products specifically suppressed this transcription.[70,109] Recent studies also indicated that antimicrobial peptides, such as defensins and cathelicidins, play a role during AdV infection. Defensins inhibit the infectivity of wild-type AdV via blocking the uncoating events of AdVs, thus preventing genome exposure and nuclear entry by α-defensin HD5 and HNP1, while cathelicidins are upregulated by cytokines induced by AdV, which are chemotactic for neutrophils, monocytes, and T cells, thus causing the inflammatory infiltrate to be seen during AdV infection.[117]

2.9 Conclusions and Future Perspectives

AdV infections mainly cause childhood illness with mild symptoms and are generally self-limiting with the exception of AdV caused acute respiratory distress in residential facilities and military bases. Over six decades of studies have (1) established that AdVs are useful model systems for uncovering fundamental aspects of cell and molecular biology; (2) provided important insights into AdV life cycle as well as its interactions with components of the host immune system; and (3) shown that AdVs are useful vectors in gene transfer for gene therapy, vaccination, and cancer treatment. However, most of the studies are focused on core virus functions and respiratory disease-related aspects of AdV infection. Little is known about the characteristics specific to AdV gastroenteric infections. So much is known about the virus because it is easily grown and serves the useful role of a model system, but less research has been focused toward clinical questions as infections most often produce mild effects.

Without the rising population of immunosuppressed potential hosts, there would be little urgency to further study the remaining unresolved aspects of AdVs; however, it has been recognized that AdV infections are responsible for significant morbidity and mortality among immunocompromised patients. Severe gastroenteritis caused by AdVs has been reported in immunocompromised individuals.[118,119] Without any further understanding on AdV-caused gastroenteritis, only symptomatic treatment is possible. New research to better characterize the molecular features that are specific to the gastroenteric infectious AdVs may allow effective antiviral agents to be developed or may help to determine effective preventive measures for the infections.

REFERENCES

1. Rubin, B.A. Clinical picture and epidemiology of adenovirus infections (a review). *Acta Microbiol Hung* 40, 303–323 (1993).
2. Shenk, T.E. Adenoviridae. in *Fields Virology*, Vol. 2 (eds. Knipe, D.M. & Howley, P.M.) 2265–2300 (Lippincott Williams & Wilkins, Philadelphia, 2001).
3. Horwitz, M.S. Adenoviruses. in *Fields Virology*, Vol. 2 (eds. Knipe, D.M. & Howley, P.M.) 2301–2326 (Lippincott Williams & Wilkins, Philadelphia, 2001).
4. Ghebremedhin, B. Human adenovirus: viral pathogen with increasing importance. *Eur J Microbiol Immunol (Bp)* 4, 26–33 (2014).
5. Kajon, A.E. et al. Molecular epidemiology of adenovirus type 4 infections in US military recruits in the postvaccination era (1997–2003). *J Infect Dis* 196, 67–75 (2007).
6. Nemerow, G.R., Pache, L., Reddy, V. & Stewart, P.L. Insights into adenovirus host cell interactions from structural studies. *Virology* 384, 380–388 (2009).
7. Hoffman, J.A. Adenoviral disease in pediatric solid organ transplant recipients. *Pediatr Transplant* 10, 17–25 (2006).

8. Leen, A.M. & Rooney, C.M. Adenovirus as an emerging pathogen in immunocompromised patients. *Br J Haematol* 128, 135–144 (2005).
9. Dey, S.K. et al. Molecular epidemiology of adenovirus infection among infants and children with acute gastroenteritis in Dhaka City, Bangladesh. *Infect Genet Evol* 9, 518–522 (2009).
10. Ushijima, H. et al. A study of adenovirus gastroenteritis in the Tokyo area. *Eur J Pediatr* 147, 90–92 (1988).
11. Jones, M.S., II et al. New adenovirus species found in a patient presenting with gastroenteritis. *J Virol* 81, 5978–5984 (2007).
12. Matsushima, Y., Shimizu, H., Phan, T.G. & Ushijima, H. Genomic characterization of a novel human adenovirus type 31 recombinant in the hexon gene. *J Gen Virol* 92, 2770–2775 (2011).
13. Matsushima, Y. et al. Novel human adenovirus strain, Bangladesh. *Emerg Infect Dis* 18, 846–848 (2012).
14. Matsushima, Y. et al. Genome sequence of a novel virus of the species human adenovirus D associated with acute gastroenteritis. *Genome Announc* 1, e00068-12 (2013).
15. Liu, E.B. et al. Genetic analysis of a novel human adenovirus with a serologically unique hexon and a recombinant fiber gene. *PLoS One* 6, e24491 (2011).
16. Wood, D.J. Adenovirus gastroenteritis. *Br Med J (Clin Res Ed)* 296, 229–230 (1988).
17. Adrian, T., Wigand, R. & Richter, J. Gastroenteritis in infants, associated with a genome type of adenovirus 31 and with combined rotavirus and adenovirus 31 infection. *Eur J Pediatr* 146, 38–40 (1987).
18. Filho, E.P. et al. Adenoviruses associated with acute gastroenteritis in hospitalized and community children up to 5 years old in Rio de Janeiro and Salvador, Brazil. *J Med Microbiol* 56, 313–319 (2007).
19. Moyo, S.J. et al. Prevalence and molecular characterisation of human adenovirus in diarrhoeic children in Tanzania; a case control study. *BMC Infect Dis* 14, 666 (2014).
20. Magwalivha, M. et al. High prevalence of species D human adenoviruses in fecal specimens from urban Kenyan children with diarrhea. *J Med Virol* 82, 77–84 (2010).
21. Ginsberg, H.S. et al. A mouse model for investigating the molecular pathogenesis of adenovirus pneumonia. *Proc Natl Acad Sci USA* 88, 1651–1655 (1991).
22. Segerman, A., Arnberg, N., Erikson, A., Lindman, K. & Wadell, G. There are two different species B adenovirus receptors: sBAR, common to species B1 and B2 adenoviruses, and sB2AR, exclusively used by species B2 adenoviruses. *J Virol* 77, 1157–1162 (2003).
23. Dehghan, S. et al. Five genome sequences of subspecies B1 human adenoviruses associated with acute respiratory disease. *J Virol* 86, 635–636 (2012).
24. Dehghan, S. et al. Genomics and bioinformatics of human adenovirus type 68 strain Arg 827/04. November 15, 2012 edn (National Center for Biotechnology Information, U.S. National Library of Medicine, Bethesda, MD, 2012).
25. De Jong, J.C. et al. Adenoviruses from human immunodeficiency virus-infected individuals, including two strains that represent new candidate serotypes Ad50 and Ad51 of species B1 and D, respectively. *J Clin Microbiol* 37, 3940–3945 (1999).
26. Yang, Z. et al. Genomic analyses of recombinant adenovirus type 11a in China. *J Clin Microbiol* 47, 3082–3090 (2009).
27. Lu, Q.B. et al. Epidemiology of human adenovirus and molecular characterization of human adenovirus 55 in China, 2009–2012. *Influenza Other Respir Viruses* 8, 302–308 (2014).
28. Walsh, M.P. et al. Computational analysis of two species C human adenoviruses provides evidence of a novel virus. *J Clin Microbiol* 49, 3482–3490 (2011).
29. Robinson, C.M. et al. Computational analysis and identification of an emergent human adenovirus pathogen implicated in a respiratory fatality. *Virology* 409, 141–147 (2011).
30. Singh, G. et al. Homologous recombination in E3 genes of human adenovirus species D. *J Virol* 87, 12481–12488 (2013).
31. Kaneko, H. et al. Recombination analysis of intermediate human adenovirus type 53 in Japan by complete genome sequence. *J Gen Virol* 92, 1251–1259 (2011).
32. Alissa Alkhalaf, M., Al Qurashi, Y.M., Guiver, M. & Cooper, R.J. Genome sequences of three species D adenoviruses isolated from AIDS patients. *Genome Announc* 2(1), e01267-13 (2014).
33. Singh, G. et al. Overreliance on the hexon gene, leading to misclassification of human adenoviruses. *J Virol* 86, 4693–4695 (2012).
34. Robinson, C.M. et al. Genomics and bioinformatics of human adenovirus type 19 Pittsburgh strain 88. February 19, 2012 edn (National Center for Biotechnology Information, U.S. National Library of Medicine, Bethesda, MD, 2012).

35. Robinson, C.M. et al. Predicting the next eye pathogen: analysis of a novel adenovirus. *mBio* 4, e00595–12 (2013).
36. Liu, E.B. et al. Computational and serologic analysis of novel and known viruses in species human adenovirus D in which serology and genomics do not correlate. *PLoS One* 7, e33212 (2012).
37. Russell, W.C. Adenoviruses: update on structure and function. *J Gen Virol* 90, 1–20 (2009).
38. Horwitz, M.S. Function of adenovirus E3 proteins and their interactions with immunoregulatory cell proteins. *J Gene Med* 6 (Suppl 1), S172–S183 (2004).
39. Robinson, C.M., Shariati, F., Gillaspy, A.F., Dyer, D.W. & Chodosh, J. Genomic and bioinformatics analysis of human adenovirus type 37: new insights into corneal tropism. *BMC Genomics* 9, 213 (2008).
40. Lauer, K.P. et al. Natural variation among human adenoviruses: genome sequence and annotation of human adenovirus serotype 1. *J Gen Virol* 85, 2615–2625 (2004).
41. Davison, A.J., Benko, M. & Harrach, B. Genetic content and evolution of adenoviruses. *J Gen Virol* 84, 2895–2908 (2003).
42. Windheim, M., Hilgendorf, A. & Burgert, H.G. Immune evasion by adenovirus E3 proteins: exploitation of intracellular trafficking pathways. *Curr Top Microbiol Immunol* 273, 29–85 (2004).
43. Leppard, K.N. E4 gene function in adenovirus, adenovirus vector and adeno-associated virus infections. *J Gen Virol* 78 (Pt 9), 2131–2138 (1997).
44. Tollefson, A.E., Ying, B., Doronin, K., Sidor, P.D. & Wold, W.S. Identification of a new human adenovirus protein encoded by a novel late l-strand transcription unit. *J Virol* 81, 12918–12926 (2007).
45. San Martin, C. Latest insights on adenovirus structure and assembly. *Viruses* 4, 847–877 (2012).
46. Roberts, M.M., White, J.L., Grutter, M.G. & Burnett, R.M. Three-dimensional structure of the adenovirus major coat protein hexon. *Science* 232, 1148–1151 (1986).
47. Liu, H. et al. Atomic structure of human adenovirus by cryo-EM reveals interactions among protein networks. *Science* 329, 1038–1043 (2010).
48. Ma, H.C. & Hearing, P. Adenovirus structural protein IIIa is involved in the serotype specificity of viral DNA packaging. *J Virol* 85, 7849–7855 (2011).
49. Wodrich, H. et al. Switch from capsid protein import to adenovirus assembly by cleavage of nuclear transport signals. *EMBO J* 22, 6245–6255 (2003).
50. Maier, O., Galan, D.L., Wodrich, H. & Wiethoff, C.M. An N-terminal domain of adenovirus protein VI fragments membranes by inducing positive membrane curvature. *Virology* 402, 11–19 (2010).
51. Schreiner, S. et al. Transcriptional activation of the adenoviral genome is mediated by capsid protein VI. *PLoS Pathog* 8, e1002549 (2012).
52. Parks, R.J. Adenovirus protein IX: a new look at an old protein. *Mol Ther* 11, 19–25 (2005).
53. Rosa-Calatrava, M., Grave, L., Puvion-Dutilleul, F., Chatton, B. & Kedinger, C. Functional analysis of adenovirus protein IX identifies domains involved in capsid stability, transcriptional activity, and nuclear reorganization. *J Virol* 75, 7131–7141 (2001).
54. Strunze, S. et al. Kinesin-1-mediated capsid disassembly and disruption of the nuclear pore complex promote virus infection. *Cell Host Microbe* 10, 210–223 (2011).
55. Veesler, D. et al. Single-particle EM reveals plasticity of interactions between the adenovirus penton base and integrin $\alpha_V\beta_3$. *Proc Natl Acad Sci USA* 111, 8815–8819 (2014).
56. Zubieta, C., Schoehn, G., Chroboczek, J. & Cusack, S. The structure of the human adenovirus 2 penton. *Mol Cell* 17, 121–135 (2005).
57. Seiradake, E. & Cusack, S. Crystal structure of enteric adenovirus serotype 41 short fiber head. *J Virol* 79, 14088–14094 (2005).
58. Ugai, H. et al. Adenoviral protein V promotes a process of viral assembly through nucleophosmin 1. *Virology* 432, 283–295 (2012).
59. Matthews, D.A. Adenovirus protein V induces redistribution of nucleolin and B23 from nucleolus to cytoplasm. *J Virol* 75, 1031–1038 (2001).
60. Hindley, C.E., Lawrence, F.J. & Matthews, D.A. A role for transportin in the nuclear import of adenovirus core proteins and DNA. *Traffic* 8, 1313–1322 (2007).
61. Lee, T.W., Blair, G.E. & Matthews, D.A. Adenovirus core protein VII contains distinct sequences that mediate targeting to the nucleus and nucleolus, and colocalization with human chromosomes. *J Gen Virol* 84, 3423–3428 (2003).
62. Ahi, Y.S., Vemula, S.V. & Mittal, S.K. Adenoviral E2 IVa2 protein interacts with L4 33K protein and E2 DNA-binding protein. *J Gen Virol* 94, 1325–1334 (2013).

63. Yang, T.C. & Maluf, N.K. Characterization of the non-specific DNA binding properties of the adenoviral IVa2 protein. *Biophys Chem* 193–194, 1–8 (2014).
64. Ewing, S.G., Byrd, S.A., Christensen, J.B., Tyler, R.E. & Imperiale, M.J. Ternary complex formation on the adenovirus packaging sequence by the IVa2 and L4 22-kilodalton proteins. *J Virol* 81, 12450–12457 (2007).
65. Lee, T.W. et al. Precursor of human adenovirus core polypeptide Mu targets the nucleolus and modulates the expression of E2 proteins. *J Gen Virol* 85, 185–196 (2004).
66. Ugai, H., Borovjagin, A.V., Le, L.P., Wang, M. & Curiel, D.T. Thermostability/infectivity defect caused by deletion of the core protein V gene in human adenovirus type 5 is rescued by thermo-selectable mutations in the core protein X precursor. *J Mol Biol* 366, 1142–1160 (2007).
67. Ruben, M., Bacchetti, S. & Graham, F. Covalently closed circles of adenovirus 5 DNA. *Nature* 301, 172–174 (1983).
68. Webster, A., Leith, I.R., Nicholson, J., Hounsell, J. & Hay, R.T. Role of preterminal protein processing in adenovirus replication. *J Virol* 71, 6381–6389 (1997).
69. Pronk, R. & van der Vliet, P.C. The adenovirus terminal protein influences binding of replication proteins and changes the origin structure. *Nucleic Acids Res* 21, 2293–2300 (1993).
70. Joung, I., Angeletti, P.C. & Engler, J.A. Functional implications in apoptosis by interferon inducible gene product 1-8D, the binding protein to adenovirus preterminal protein. *J Microbiol* 41, 295–299 (2003).
71. de Jong, R.N., Meijer, L.A. & van der Vliet, P.C. DNA binding properties of the adenovirus DNA replication priming protein pTP. *Nucleic Acids Res* 31, 3274–3286 (2003).
72. Botting, C.H. & Hay, R.T. Role of conserved residues in the activity of adenovirus preterminal protein. *J Gen Virol* 82, 1917–1927 (2001).
73. Grosche, P. et al. Structure-based design and optimization of potent inhibitors of the adenoviral protease. *Bioorg Med Chem Lett* 25, 438–443 (2015).
74. Mangel, W.F., Baniecki, M.L. & McGrath, W.J. Specific interactions of the adenovirus proteinase with the viral DNA, an 11-amino-acid viral peptide, and the cellular protein actin. *Cell Mol Life Sci* 60, 2347–2355 (2003).
75. Baniecki, M.L., McGrath, W.J. & Mangel, W.F. Regulation of a viral proteinase by a peptide and DNA in one-dimensional space: III. Atomic resolution structure of the nascent form of the adenovirus proteinase. *J Biol Chem* 288, 2081–2091 (2013).
76. Blainey, P.C. et al. Regulation of a viral proteinase by a peptide and DNA in one-dimensional space: IV. Viral proteinase slides along DNA to locate and process its substrates. *J Biol Chem* 288, 2092–2102 (2013).
77. Graziano, V. et al. Regulation of a viral proteinase by a peptide and DNA in one-dimensional space: II. Adenovirus proteinase is activated in an unusual one-dimensional biochemical reaction. *J Biol Chem* 288, 2068–2080 (2013).
78. Graziano, V. et al. Regulation of a viral proteinase by a peptide and DNA in one-dimensional space: I. Binding to DNA AND to hexon of the precursor to protein VI, pVI, of human adenovirus. *J Biol Chem* 288, 2059–2067 (2013).
79. Radko, S. et al. Adenovirus E1A targets the DREF nuclear factor to regulate virus gene expression, DNA replication, and growth. *J Virol* 88, 13469–13481 (2014).
80. Padmanabhan, R., Tanimoto, A. & Sasaguri, Y. Transactivation of human *cdc2* promoter by adenovirus E1A. *Curr Top Microbiol Immunol* 272, 365–397 (2003).
81. Berscheminski, J., Groitl, P., Dobner, T., Wimmer, P. & Schreiner, S. The adenoviral oncogene E1A-13S interacts with a specific isoform of the tumor suppressor PML to enhance viral transcription. *J Virol* 87, 965–977 (2013).
82. Gallimore, P.H. & Turnell, A.S. Adenovirus E1A: remodelling the host cell, a life or death experience. *Oncogene* 20, 7824–7835 (2001).
83. Blackford, A.N. & Grand, R.J. Adenovirus E1B 55-kilodalton protein: multiple roles in viral infection and cell transformation. *J Virol* 83, 4000–4012 (2009).
84. Cheng, C.Y. et al. Role of E1B55K in E4orf6/E1B55K E3 ligase complexes formed by different human adenovirus serotypes. *J Virol* 87, 6232–6245 (2013).
85. Tarakanova, V.L. & Wold, W.S. Adenovirus E1A and E1B-19K proteins protect human hepatoma cells from transforming growth factor β1-induced apoptosis. *Virus Res* 147, 67–76 (2010).
86. Chiou, S.K., Tseng, C.C., Rao, L. & White, E. Functional complementation of the adenovirus E1B 19-kilodalton protein with Bcl-2 in the inhibition of apoptosis in infected cells. *J Virol* 68, 6553–6566 (1994).

87. van Breukelen, B., Kanellopoulos, P.N., Tucker, P.A. & van der Vliet, P.C. The formation of a flexible DNA-binding protein chain is required for efficient DNA unwinding and adenovirus DNA chain elongation. *J Biol Chem* 275, 40897–40903 (2000).
88. Dekker, J. et al. Multimerization of the adenovirus DNA-binding protein is the driving force for ATP-independent DNA unwinding during strand displacement synthesis. *EMBO J* 16, 1455–1463 (1997).
89. de Jong, R.N., van der Vliet, P.C. & Brenkman, A.B. Adenovirus DNA replication: protein priming, jumping back and the role of the DNA binding protein DBP. *Curr Top Microbiol Immunol* 272, 187–211 (2003).
90. de Jong, R.N. & van der Vliet, P.C. Mechanism of DNA replication in eukaryotic cells: cellular host factors stimulating adenovirus DNA replication. *Gene* 236, 1–12 (1999).
91. Brenkman, A.B., Breure, E.C. & van der Vliet, P.C. Molecular architecture of adenovirus DNA polymerase and location of the protein primer. *J Virol* 76, 8200–8207 (2002).
92. Hoeben, R.C. & Uil, T.G. Adenovirus DNA replication. *Cold Spring Harb Perspect Biol* 5, a013003 (2013).
93. Signas, C., Akusjarvi, G. & Pettersson, U. Region E3 of human adenoviruses: differences between the oncogenic adenovirus-3 and the non-oncogenic adenovirus-2. *Gene* 50, 173–184 (1986).
94. Tollefson, A.E. et al. The adenovirus death protein (E3–11.6K) is required at very late stages of infection for efficient cell lysis and release of adenovirus from infected cells. *J Virol* 70, 2296–2306 (1996).
95. Johnson, D.G., Ohtani, K. & Nevins, J.R. Autoregulatory control of E2F1 expression in response to positive and negative regulators of cell cycle progression. *Genes Dev* 8, 1514–1525 (1994).
96. Zhang, W., Low, J.A., Christensen, J.B. & Imperiale, M.J. Role for the adenovirus IVa2 protein in packaging of viral DNA. *J Virol* 75, 10446–10454 (2001).
97. Lutz, P., Rosa-Calatrava, M. & Kedinger, C. The product of the adenovirus intermediate gene IX is a transcriptional activator. *J Virol* 71, 5102–5109 (1997).
98. Hayes, B.W., Telling, G.C., Myat, M.M., Williams, J.F. & Flint, S.J. The adenovirus L4 100-kilodalton protein is necessary for efficient translation of viral late mRNA species. *J Virol* 64, 2732–2742 (1990).
99. Riley, D. & Flint, S.J. RNA-binding properties of a translational activator, the adenovirus L4 100-kilodalton protein. *J Virol* 67, 3586–3595 (1993).
100. Hong, S.S. et al. The 100K-chaperone protein from adenovirus serotype 2 (subgroup C) assists in trimerization and nuclear localization of hexons from subgroups C and B adenoviruses. *J Mol Biol* 352, 125–138 (2005).
101. Koyuncu, O.O. & Dobner, T. Arginine methylation of human adenovirus type 5 L4 100-kilodalton protein is required for efficient virus production. *J Virol* 83, 4778–4790 (2009).
102. Andrade, F., Casciola-Rosen, L.A. & Rosen, A. A novel domain in adenovirus L4-100K is required for stable binding and efficient inhibition of human granzyme B: possible interaction with a species-specific exosite. *Mol Cell Biol* 23, 6315–6326 (2003).
103. Ying, B., Tollefson, A.E. & Wold, W.S. Identification of a previously unrecognized promoter that drives expression of the UXP transcription unit in the human adenovirus type 5 genome. *J Virol* 84, 11470–11478 (2010).
104. Cupelli, K. & Stehle, T. Viral attachment strategies: the many faces of adenoviruses. *Curr Opin Virol* 1, 84–91 (2011).
105. Ostapchuk, P. & Hearing, P. Control of adenovirus packaging. *J Cell Biochem* 96, 25–35 (2005).
106. Carter, M.J. Enterically infecting viruses: pathogenicity, transmission and significance for food and waterborne infection. *J Appl Microbiol* 98, 1354–1380 (2005).
107. Mavromichalis, J. et al. Intestinal damage in rotavirus and adenovirus gastroenteritis assessed by d-xylose malabsorption. *Arch Dis Child* 52, 589–591 (1977).
108. Hayder, H. & Mullbacher, A. Molecular basis of immune evasion strategies by adenoviruses. *Immunol Cell Biol* 74, 504–512 (1996).
109. Gutch, M.J. & Reich, N.C. Repression of the interferon signal transduction pathway by the adenovirus E1A oncogene. *Proc Natl Acad Sci USA* 88, 7913–7917 (1991).
110. Arany, Z., Newsome, D., Oldread, E., Livingston, D.M. & Eckner, R. A family of transcriptional adaptor proteins targeted by the E1A oncoprotein. *Nature* 374, 81–84 (1995).
111. Janaswami, P.M., Kalvakolanu, D.V., Zhang, Y. & Sen, G.C. Transcriptional repression of interleukin-6 gene by adenoviral E1A proteins. *J Biol Chem* 267, 24886–24891 (1992).
112. White, E. et al. The 19-kilodalton adenovirus E1B transforming protein inhibits programmed cell death and prevents cytolysis by tumor necrosis factor α. *Mol Cell Biol* 12, 2570–2580 (1992).

113. Lázaro, S., Gamarra, D. & Del Val, M. Proteolytic enzymes involved in MHC class I antigen processing: a guerrilla army that partners with the proteasome. *Mol Immunol* 68, 72–76 (2015).
114. Schouten, G.J., van der Eb, A.J. & Zantema, A. Downregulation of MHC class I expression due to interference with p105-NFκB1 processing by Ad12E1A. *EMBO J* 14, 1498–1507 (1995).
115. Kotha, P.L. et al. Adenovirus entry from the apical surface of polarized epithelia is facilitated by the host innate immune response. *PLoS Pathog* 11, e1004696 (2015).
116. Coutelier, J.P., Van Broeck, J. & Wolf, S.F. Interleukin-12 gene expression after viral infection in the mouse. *J Virol* 69, 1955–1958 (1995).
117. Gregory, S.M., Nazir, S.A. & Metcalf, J.P. Implications of the innate immune response to adenovirus and adenoviral vectors. *Future Virol* 6, 357–374 (2011).
118. Johansson, M.E., Wirgart, B.Z., Grillner, L. & Bjork, O. Severe gastroenteritis in an immunocompromised child caused by adenovirus type 5. *Pediatr Infect Dis J* 9, 449–450 (1990).
119. Schofield, K.P., Morris, D.J., Bailey, A.S., de Jong, J.C. & Corbitt, G. Gastroenteritis due to adenovirus type 41 in an adult with chronic lymphocytic leukemia. *Clin Infect Dis* 19, 311–312 (1994).
120. Ishiko, H. et al. Novel human adenovirus causing nosocomial epidemic keratoconjunctivitis. *J Clin Microbiol* 46, 2002–2008 (2008).

3

Astrovirus

Matthew D. Koci and Stacey L. Schultz-Cherry

CONTENTS

3.1 Introduction

Astroviruses are small, round, nonenveloped, positive-sense single-stranded RNA viruses. They were named for their "star-like" appearance when visualized by negative staining electron microscopy in the feces of infants suffering from gastroenteritis.[1,2] Since their original discovery, astroviruses have been identified in almost all animal species examined.[3] In the vast majority of these species, astrovirus infection is transmitted through the fecal–oral route and causes acute gastroenteritis. In humans, the classical astroviruses are associated with mild to moderate diarrhea and are recognized as one of the leading causes of enteritis in children and the immunocompromised.[4,5] More recently, astroviruses have been associated with encephalitis.[6,7] In other species, especially poultry, diarrhea can be severe and can involve other organ systems including the liver and kidneys.[8,9] Historically, astroviruses have been thought to be species-specific, with only limited evidence of interspecies transmission.[10] However, there is increasing evidence that astroviruses can infect multiple species. Convincing evidence for this was first found for avian astroviruses,[11–17] but recent reports suggest the same may be true for mammalian astroviruses,[18–20] and that there may even be transmission between birds and mammals.[21] More recently, astrovirus genotypes associated with human infections were detected in nonhuman primates, strongly supporting the suggestion that astroviruses can cross species barriers.[21a]

Since the Astroviridae family was established by the ICTV in 1995, there have been considerable changes in our understanding of the diversity, molecular evolution, and host range of the family. In just the last decade, the numbers of host species detected with astroviruses have increased more than fourfold.[22] Additionally our appreciation of the diversity of astrovirus genotypes capable of infecting a given host has led to a reevaluation of their classification system. Currently, Astroviridae is divided into two genera (Table 3.1), *Mamastrovirus* (MAstV) and *Avastrovirus* (AAstV), representing viruses that affect mammals and avian species, respectively. Within each of these two genera, viruses are classified by distinct genotypic differences within their polymerase and/or capsid genomic sequences.[23]

Based on this nomenclature system, there are 19 species of MAstV recognized to infect humans (MAstV 1–19). Of these species, MAstV 1 encompasses the human astroviruses (HAstVs) first recognized and most extensively studied.[22] This group of classical HAstVs contains eight distinct serotypes (HAstV-1 to HAstV-8), which are known to have a worldwide distribution and are recognized as one of

TABLE 3.1
Astroviridae

Genus	Species	Host(s)
Mamastrovirus (MAstV)	MAstV 1	Human
	MAstV 2	Cat, cheetah
	MAstV 3	Pig
	MAstV 4	California sea lion
	MAstV 5	Dog
	MAstV 6	Human
	MAstV 7	Bottlenose dolphin
	MAstV 8	Human
	MAstV 9	Human
	MAstV 10	Mink
	MAstV 11	California sea lion
	MAstV 12	Bat
	MAstV 13	Sheep
	MAstV 14	Bat
	MAstV 15	Bat
	MAstV 16	Bat
	MAstV 17	Bat
	MAstV 18	Bat
	MAstV 19	Bat
	Unassigned	Pig, sheep, mouse, rat, cattle, deer
Avastrovirus (AAstV)	AAvstV 1	Turkey, chicken
	AAvstV 2	Turkey, guinea fowl, duck
	AAvstV 3	Chicken, turkey
	Unassigned	Dove, pigeon, heron, pintail

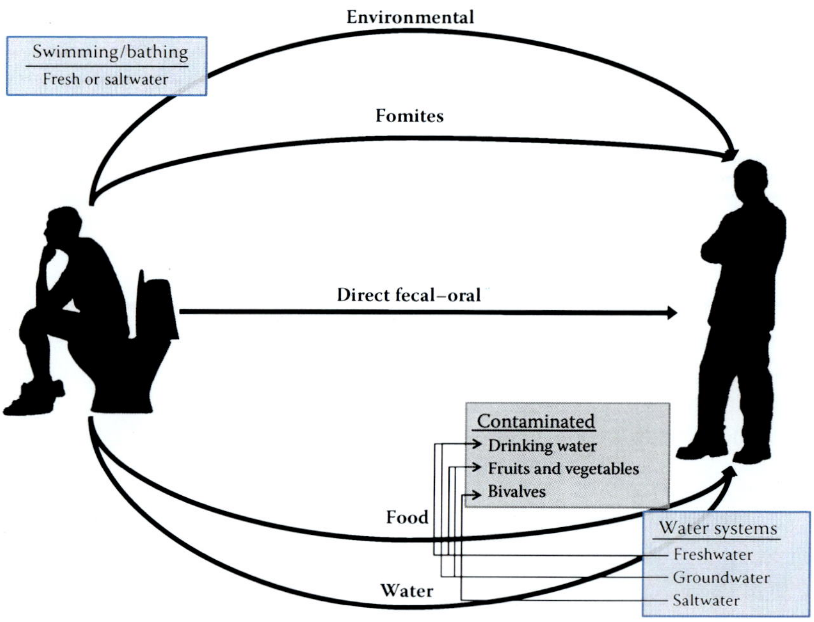

FIGURE 3.1 Direct and indirect routes of astrovirus transmission. As the cause of gastroenteritis, the direct fecal–oral transmission of astroviruses is well established, as is its ability to be transmitted by fomites; however, astroviruses are environmentally very stable, allowing them to be transmitted indirectly to new hosts via water and the environment.

the top three causes of diarrhea in children under 2 years of age.[24–26] The other HAstVs found in MAstV were identified through pathogen discovery surveys of fecal material.[27–32]

Astroviruses are transmitted through the fecal–oral route (Figure 3.1), as demonstrated by several volunteer studies.[33,34] This direct transmission of fecal material from an infected person to a new host, or through fomites is thought to be the major mechanism for the spread of astroviruses through a population; however, astroviruses are highly stable in the environment.[35,36] This stability allows them to spread through wastewater, groundwater, saltwater, and even drinking water.[37–45] These astrovirus-contaminated waters can then infect a new patient following drinking, eating fruits and vegetables treated with these waters,[46] consuming bivalves harvested from these waters,[38,39,47] or even bathing, swimming, or surfing in astrovirus-contaminated waters.[48,49] Given this link between astrovirus stability and water, it is especially important to recognize that most municipal water treatment systems do not filter out, or inactivate astroviruses as part of their process, as studies have demonstrated by isolating astroviruses from household tap water.[4,50,51]

3.2 Life Cycle

After ingestion, the virus passes through the gastrointestinal tract where it is likely activated by host trypsin-like proteases (Figure 3.2).[52–56] The virus then binds to an as yet unidentified receptor(s) on host cells and enters intestinal epithelial cells of the small intestine (host cells for most of the known astroviruses) through clathrin-mediated endocytosis.[57] Following its entry, the virus releases its positive-sense single-stranded RNA genome into the cytoplasm, which is immediately recognized by the host cell as mRNA, leading to the expression of a polyprotein containing the nonstructural proteins (nsPs) and the viral RNA-dependent RNA polymerase (RdRp).[58] Once these proteins have been produced, they quickly

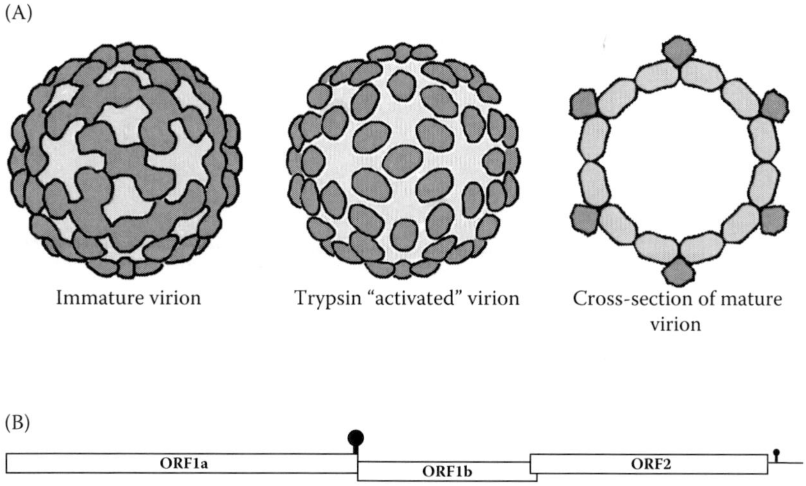

FIGURE 3.2 Model of astrovirus virion (A) and diagram of astrovirus genome (B). The astrovirus virion is a nonenveloped capsid ranging in size from 28 to 41 nm depending on the species and propagation conditions. Studies of HAstV-8 by Dryden et al. suggest that astroviruses are released as immature virions, which become highly infectious mature virions following digestion by trypsin.[112] The mature virion surface is comprised of multiple spike-like projections (dark gray, VP27 and VP25), extending from the shell (light gray, VP34), giving the virus its distinctive five- or six-pointed star-like appearance when visualized under negative staining electron microscopy. Inside the capsid, the astrovirus genome is comprised of single-stranded, positive-sense RNA ranging from 6.4 to 7.7 kb in size. The genome contains three open reading frames: ORF1a encodes the nonstructural proteins; ORF1b encodes the RNA-dependent RNA polymerase; and ORF2 encodes the capsid protein. The retrovirus-like frameshift structure formed between ORF1a and ORF1b allows these two ORFs to be translated as one polyprotein and is one of the molecular hallmarks that distinguishes the astrovirus replication strategy from other small, round RNA viruses.

establish replication complex vesicles.[59] As part of this process, the virus also produces a subgenomic RNA encoding the viral capsid protein (ORF2) (Figure 3.2).[60] The subgenomic message is then used to produce ample copies of the virus structural protein, while the genomic RNA is used to produce more nsPs and RdRp, as well as a template for the progeny genome.[58] Once enough capsid and progeny genomes have been produced, the daughter viruses are assembled. This initial assembly step involves the full 90-kDa capsid protein, which is then digested by host caspases to yield progeny viruses made up of 70-kDa capsid proteins.[55,61–63] These immature progeny viruses are then released from the infected cell, where they will be further digested by host trypsin-like proteases, activating them and allowing them to infect new cells.[52–56]

3.3 Pathogenesis

Classical HAstVs cause gastrointestinal disease in young children, aged adults, and the immunocompromised,[5,64–68] but are only rarely associated with the disease in normal healthy adults.[33,69,70] Molecular epidemiology and serology studies suggest that humans are exposed to HAstVs regularly and antibodies against HAstVs develop early in life.[22] In fact, preexisting anti-HAstV antibody levels seem to explain the biphasic age distribution of HAstV infections, as well as the sporadic development of HAstV-mediated diarrhea during adult volunteer studies.[54,71,72]

In susceptible people, following an incubation period of 4–5 days,[73] HAstV infection leads to mild, watery diarrhea, lasting 48–72 h.[5,22] Clinical symptoms can also include vomiting, fever, and abdominal pain. Overall, astrovirus-induced disease is generally less severe than that caused by rotavirus or calicivirus infections, and rarely requires hospitalization in otherwise healthy people.[5,22]

While HAstVs are regarded as a disease of the intestinal tract, recent evidence suggests that they can induce viremia, potentially causing more severe infections in immunocompromised patients.[74–77] This concept of extraintestinal astrovirus was first reported in studies examining the pathology of avian astroviruses,[12,74,78–80] and was more recently demonstrated in a newly established immunodeficient mouse model of astrovirus infection[81] as well as in immunocompetent mice orally inoculated with murine astrovirus (unpublished observations).

Despite the well-established disease burden caused by HAstVs, the mechanisms involved in astrovirus-mediated disease in humans are still poorly understood. Much of what has been reported regarding the pathophysiological changes induced by astrovirus infection has come from animal models; however, the changes noted and the cell types involved appear to vary considerably between models.

Studies using experimentally infected lambs reported atrophy of the villi of the jejunum and ileum, with astroviruses found in the mature enterocytes.[82] In astrovirus-infected calves, the virus was found in the M cells and enterocytes associated with Peyer's patches in the jejunum and ileum.[83] In ducks and chickens, astroviruses are more commonly associated with other tissues, causing fetal hepatitis in ducks and mild nephritis in chickens.[12,78,84,85] The young turkey model, however, has been the most extensively used to understand astrovirus pathogenesis.[74,79,86–92] These studies demonstrate that astrovirus infection leads to very few, if any, remarkable changes in the intestinal histology. No significant inflammation, flux of immune cells, or changes in the villus height, width, or surface area were observed.[74,79,88,91] While not wholly unique to astroviruses, the absence of histological changes correlating with the onset of diarrhea is uncommon among diarrhea-inducing agents. These observations in turkeys have been supported by histological analysis of small intestine biopsy samples from an immunosuppressed child suffering from persistent diarrhea.[93]

The lack of histological changes consistent with the severity of diarrhea observed has led to a more specific examination of the changes in intestinal function to determine how astroviruses cause diseases. These studies have demonstrated decreased expression of digestive enzymes, specifically maltase and sucrose, in the brush border of the small intestine, suggesting the development of malabsorption of carbohydrates, leading to osmotic diarrhea (Figures 3.3 and 3.4).[91,92] More recent studies suggest that astrovirus infection leads to a rearrangement of actin in intestinal epithelial cells.[88,94] These changes were associated with decreases in the electrical resistance (decreased barrier function) in the jejunum that allowed for increased movement of ions but not larger macromolecules (Figure 3.3).[88] In addition,

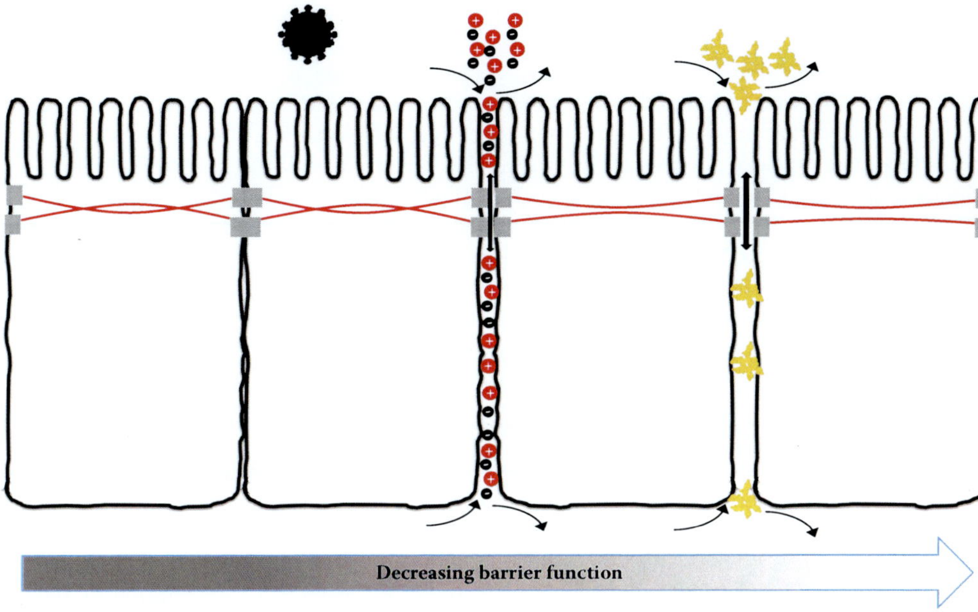

FIGURE 3.3 Astrovirus infection and exposure to astrovirus capsid protein lead to changes in intestinal barrier function. Under normal conditions, the tight junctions (gray boxes) between neighboring cells form water-tight barriers, are held in place and regulated by actin (red lines). Following astrovirus replication, or even just on exposure to astrovirus capsid protein, the actin at the apical ends of epithelial cells is rearranged. The more significant this rearrangement is, the greater the change in barrier function. Smaller changes can allow for the diffusion of solutes and ions across the epithelium, while still larger disruptions allow diffusion of macromolecules.

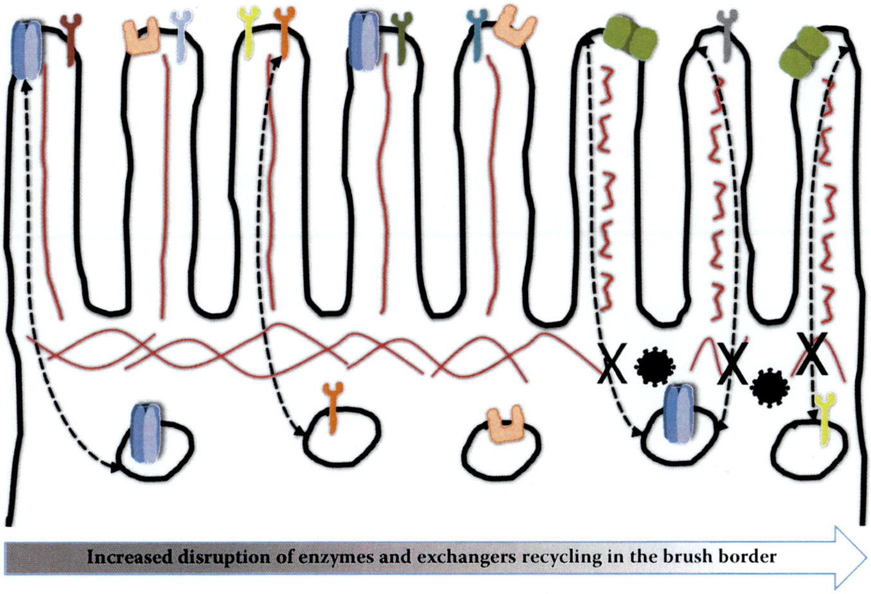

FIGURE 3.4 Astrovirus-induced changes in the apical expression of digestive enzymes and exchangers. Following infection or exposure to astrovirus capsid protein, the apical expression of digestive enzymes such as maltase and the NHE-3 have been reported to be significantly decreased, resulting in malabsorption and osmotic diarrhea. While the specific mechanism(s) involved is not known, it is likely that the astrovirus-induced disruption to actin (red lines) prevents the normal recycling of vesicles and proteins whose function is regulated by a near-constant shuttling in and out of the brush border, and they are more dramatically affected by astrovirus-induced actin rearrangement.

there was a decrease in the absorption of sodium ions that correlated with decreased apical expression of the sodium/hydrogen exchanger-3 (NHE-3) molecule in the jejunum.[88] Collectively, these findings suggested an infection established by sodium malabsorption, leading to osmotic diarrhea. The changes in the turkey jejunum were similar, but not completely consistent with studies of HAstV *in vitro* where viral infection as well as exposure to inactivated HAstV capsid was sufficient to induce actin rearrangement leading to tight-junction disruption, allowing macromolecules to diffuse across the barrier.[88,94] The observation that the capsid protein alone is capable of inducing barrier changes *in vitro* suggests that it possesses enterotoxin-like activity and could be capable of inducing even diarrhea. Current experiments using turkey,[94a] as well as other animal models (unpublished observation), have confirmed this hypothesis, but they also suggest that the potential impact a nonhuman astrovirus may have on a human disease could be greatly underappreciated if replication is not required to induce clinical signs.[95]

3.4 Immunity

Epidemiological evidence suggests that immunity to HAstV is largely mediated through antibodies. Clinical signs of astrovirus infection are most pronounced, if evident at all, in those with no preexisting antibodies or those who are immunocompromised due to age, HIV status, or on immunosuppressive therapy.[67,71,72] The levels of anti-HAstV antibodies observed in normal healthy adults, in some studies, suggest that we may be under frequent exposure to these pathogens.[33,34,96] In addition to antibodies, anti-HAstV-specific T cells have been recovered from intestinal biopsy samples from a normal healthy adult[97,98] suggesting that both the cellular and humoral arms of the adaptive immune system respond to HAstV infections. The relative importance of each in clearing the pathogen and preventing subsequent infection is still unknown.

Given that astrovirus infection is most severe when the adaptive immune system is least active, many researchers have focused on the role of the innate immune response to understand how the body fights the infection under these conditions. Most interestingly, recent studies by Guix et al. have demonstrated that HAstVs can inhibit the production of type I interferon in the early phase of replication allowing for greater viral production. This effect seems to correlate with mutations in one of the mature nsP proteins (nsP1a/4), suggesting that some strains and some genotypes may have different abilities to modulate the interferon response.[99]

This is not the only evidence that astroviruses are able to circumvent the innate immune response. The astrovirus capsid protein can inhibit activation of the lectin and classical complement pathways.[100–103] Amino acids in the N-terminal region of the capsid interact with the C1 and MBL molecules, preventing the formation of C3 convertase and subsequent cleavage of C3 into C3a and C3b. The blockage of these complement activation pathways prevents the formation of a very powerful pro-inflammatory and immune-activating signal,[71] one that host cells use to detect the presence of numerous RNA, DNA, and bacterial infections.[104] The specific role the inhibition of the complement pathway serves in astrovirus pathology is unclear, but it is easy to speculate that it helps to mute the inflammatory signal and is at least partially responsible for the astrovirus's immune evasion *in vivo*.[71]

As with astrovirus pathogenesis, most of our understanding of the immune response to astroviruses *in vivo* comes from the turkey model. In this model, however, the adaptive immune response seems to play little to no role in controlling the infection,[79,86] and in fact some researchers have suggested that it induces an immune dysfunction that can persist long after the virus has cleared.[105–107] Macrophages isolated from astrovirus-infected turkeys demonstrate reduced phagocytosis and the ability to kill engulfed bacteria in *ex vivo* assays, along with the reduced production of pro-inflammatory cytokines.[108–111] Lymphocytes from astrovirus-infected turkeys also demonstrated reduced proliferation in response to mitogenic stimulation. The apparent inhibition of these immune cells could be a response to increased levels of active TGFβ in the serum of infected turkeys.[79] The mechanism leading to the activation of TGFβ in these animals is unclear but given its role as a potent anti-inflammatory cytokine, it is likely key to the limited role the adaptive system plays in response to astrovirus infection.

In spite of numerous immune factors being downregulated following astrovirus infection, the treatment of avian macrophages with astrovirus capsid was seen to stimulate the expression of inducible nitric oxide synthase (iNOS) and its antimicrobial mediator NO.[86] The modulation of this response in an *in ovo* culture system demonstrated that the iNOS response could modulate virus replication; however, there was no evidence that macrophages responded to the sites of infection.[79,86,88] Subsequent studies demonstrated that the intestinal epithelial cells can and do express iNOS following astrovirus infection suggesting that in the absence of professional immune cells the intestinal cells are capable of mounting their own response.[87]

3.5 Future Perspectives

Over the past decade, we have learned about the diversity of astroviruses, the diseases they can cause, and that their ability to move between species is far more complicated than what was originally thought. These revelations underscore the long-held idea that humans are exposed to astroviruses on a frequent basis, and highlight the need for greater attention to fully appreciate their role in gastroenteritis. Given their ubiquity and environmental stability, astroviruses may serve as an ideal indicator virus to assess the sterility and safety of water and food preparation systems. Any treatment system capable of inactivating and removing astroviruses would likely be devoid of other, more episodic, foodborne pathogens. Studies are underway to develop desperately needed small-animal models of infection as well as improved *in vitro* culture systems. This is especially important given that the majority of astrovirus strains are currently unculturable. Given the rapid increase in the number of host species in which astroviruses have been detected, evidence of possible zoonotic/reverse zoonotic infections, and the possibility that the astrovirus capsid protein may be a unique enterotoxin, astrovirus-associated disease is likely to increase in the future.

REFERENCES

1. Appleton, H. & Higgins, P.G. Letter: viruses and gastroenteritis in infants. *Lancet* 1, 1297 (1975).
2. Madeley, C.R. & Cosgrove, B.P. Letter: 28 nm particles in faeces in infantile gastroenteritis. *Lancet* 2, 451–452 (1975).
3. De Benedictis, P., Schultz-Cherry, S., Burnham, A. & Cattoli, G. Astrovirus infections in humans and animals—molecular biology, genetic diversity, and interspecies transmissions. *Infect Genet Evol* 11, 1529–1544 (2011).
4. Bosch, A., Guix, S. & Pintó, R. Epidemiology of human astroviruses. in *Astrovirus Research* (ed. Schultz-Cherry, S.) 1–18 (Springer, New York, 2013).
5. Meliopoulos, V. & Schultz-Cherry, S. Astrovirus pathogenesis. in *Astrovirus Research* (ed. Schultz-Cherry, S.) 65–77 (Springer, New York, 2013).
6. Naccache, S.N. et al. Diagnosis of neuroinvasive astrovirus infection in an immunocompromised adult with encephalitis by unbiased next-generation sequencing. *Clin Infect Dis* 60, 919–923 (2015).
7. Quan, P.L. et al. Astrovirus encephalitis in boy with X-linked agammaglobulinemia. *Emerg Infect Dis* 16, 918–925 (2010).
8. Cattoli, G., Chu, D. & Peiris, M. Astrovirus infections in animal mammalian species. in *Astrovirus Research* (ed. Schultz-Cherry, S.) 135–149 (Springer, New York, 2013).
9. Pantin-Jackwood, M., Todd, D. & Koci, M. Avian astroviruses. in *Astrovirus Research* (ed. Schultz-Cherry, S.) 151–180 (Springer, New York, 2013).
10. Lukashov, V.V. & Goudsmit, J. Evolutionary relationships among Astroviridae. *J Gen Virol* 83, 1397–1405 (2002).
11. Cattoli, G. et al. Co-circulation of distinct genetic lineages of astroviruses in turkeys and guinea fowl. *Arch Virol* 152, 595–602 (2007).
12. Imada, T. et al. Avian nephritis virus (ANV) as a new member of the family Astroviridae and construction of infectious ANV cDNA. *J Virol* 74, 8487–8493 (2000).
13. Todd, D. et al. Capsid protein sequence diversity of avian nephritis virus. *Avian Pathol* 40, 249–259 (2011).
14. Toffan, A. et al. Experimental infection of poults and guinea fowl with genetically distinct avian astroviruses. *Avian Pathol* 41, 429–435 (2012).

15. Zhao, W. et al. Sequence analyses of the representative Chinese-prevalent strain of avian nephritis virus in healthy chicken flocks. *Avian Dis* 55, 65–69 (2011).
16. Zhao, W. et al. Complete sequence and genetic characterization of pigeon avian nephritis virus, a member of the family Astroviridae. *Arch Virol* 156, 1559–1565 (2011).
17. Zhao, W. et al. Detection of astrovirus infection in pigeons (*Columbia livia*) during an outbreak of diarrhoea. *Avian Pathol* 40, 361–365 (2011).
18. Mendenhall, I.H., Smith, G.J. & Vijaykrishna, D. Ecological drivers of virus evolution: astrovirus as a case study. *J Virol* 89, 6978–6981 (2015).
19. Monini, M. et al. Detection and molecular characterization of zoonotic viruses in swine fecal samples in Italian pig herds. *Arch Virol* 160, 2547–2556 (2015).
20. Nagai, M. et al. Full genome analysis of bovine astrovirus from fecal samples of cattle in Japan: identification of possible interspecies transmission of bovine astrovirus. *Arch Virol* 160, 2491–2501 (2015).
21. Pankovics, P., Boros, A., Kiss, T., Delwart, E. & Reuter, G. Detection of a mammalian-like astrovirus in bird, European roller (*Coracias garrulus*). *Infect Genet Evol* 34, 114–121 (2015).
21a. Karlsson, E.A. et al. Non-human primates harbor diverse mammalian and avian astroviruses including those associated with human infections. *PLoS Pathog* 11(11), e1005225 (2015). doi:10.1371/journal.ppat.1005225.
22. Bosch, A., Pinto, R.M. & Guix, S. Human astroviruses. *Clin Microbiol Rev* 27, 1048–1074 (2014).
23. Guix, S., Bosch, A. & Pintó, R. Astrovirus taxonomy. in *Astrovirus Research* (ed. Schultz-Cherry, S.) 97–118 (Springer, New York, 2013).
24. Ham, H. et al. Prevalence of human astrovirus in patients with acute gastroenteritis. *Ann Lab Med* 34, 145–147 (2014).
25. Koopmans, M.P., Bijen, M.H., Monroe, S.S. & Vinje, J. Age-stratified seroprevalence of neutralizing antibodies to astrovirus types 1 to 7 in humans in the Netherlands. *Clin Diagn Lab Immunol* 5, 33–37 (1998).
26. Willcocks, M.M., Kurtz, J.B., Lee, T.W. & Carter, M.J. Prevalence of human astrovirus serotype 4: capsid protein sequence and comparison with other strains. *Epidemiol Infect* 114, 385–391 (1995).
27. Finkbeiner, S.R. et al. Human stool contains a previously unrecognized diversity of novel astroviruses. *Virol J* 6, 161 (2009).
28. Hu, B. et al. Detection of diverse novel astroviruses from small mammals in China. *J Gen Virol* 95, 2442–2449 (2014).
29. Jiang, H. et al. Comparison of novel MLB-clade, VA-clade and classic human astroviruses highlights constrained evolution of the classic human astrovirus nonstructural genes. *Virology* 436, 8–14 (2013).
30. Kapoor, A. et al. Multiple novel astrovirus species in human stool. *J Gen Virol* 90, 2965–2972 (2009).
31. Pativada, M., Bhattacharya, R. & Krishnan, T. Novel human astrovirus strains showing multiple recombinations within highly conserved ORF1b detected from hospitalized acute watery diarrhea cases in Kolkata, India. *Infect Genet Evol* 20, 284–291 (2013).
32. Finkbeiner, S. R. & Holtz, L. R. New human astroviruses. in *Astrovirus Research* (ed. Schultz-Cherry, S.) 119–133 (Springer, New York, 2013).
33. Kurtz, J.B., Lee, T.W., Craig, J.W. & Reed, S.E. Astrovirus infection in volunteers. *J Med Virol* 3, 221–230 (1979).
34. Midthun, K. et al. Characterization and seroepidemiology of a type 5 astrovirus associated with an outbreak of gastroenteritis in Marin County, California. *J Clin Microbiol* 31, 955–962 (1993).
35. Espinosa, A.C. et al. Infectivity and genome persistence of rotavirus and astrovirus in groundwater and surface water. *Water Res* 42, 2618–2628 (2008).
36. Schultz-Cherry, S., King, D.J. & Koci, M.D. Inactivation of an astrovirus associated with poult enteritis mortality syndrome. *Avian Dis* 45, 76–82 (2001).
37. Le Cann, P., Ranarijaona, S., Monpoeho, S., Le Guyader, F. & Ferre, V. Quantification of human astroviruses in sewage using real-time RT-PCR. *Res Microbiol* 155, 11–15 (2004).
38. Le Guyader, F., Haugarreau, L., Miossec, L., Dubois, E. & Pommepuy, M. Three-year study to assess human enteric viruses in shellfish. *Appl Environ Microbiol* 66, 3241–3248 (2000).
39. Le Guyader, F.S. et al. Aichi virus, norovirus, astrovirus, enterovirus, and rotavirus involved in clinical cases from a French oyster-related gastroenteritis outbreak. *J Clin Microbiol* 46, 4011–4017 (2008).

40. Morsy El-Senousy, W., Guix, S., Abid, I., Pinto, R.M. & Bosch, A. Removal of astrovirus from water and sewage treatment plants, evaluated by a competitive reverse transcription-PCR. *Appl Environ Microbiol* 73, 164–167 (2007).
41. Nadan, S., Walter, J.E., Grabow, W.O., Mitchell, D.K. & Taylor, M.B. Molecular characterization of astroviruses by reverse transcriptase PCR and sequence analysis: comparison of clinical and environmental isolates from South Africa. *Appl Environ Microbiol* 69, 747–753 (2003).
42. Pinto, R.M. et al. Astrovirus detection in wastewater samples. *Water Sci Technol* 43, 73–76 (2001).
43. Prevost, B. et al. Deciphering the diversities of astroviruses and noroviruses in wastewater treatment plant effluents by a high-throughput sequencing method. *Appl Environ Microbiol* 81, 7215–7222 (2015).
44. Prevost, B. et al. Large scale survey of enteric viruses in river and waste water underlines the health status of the local population. *Environ Int* 79, 42–50 (2015).
45. Steyer, A., Torkar, K.G., Gutierrez-Aguirre, I. & Poljsak-Prijatelj, M. High prevalence of enteric viruses in untreated individual drinking water sources and surface water in Slovenia. *Int J Hyg Environ Health* 214, 392–398 (2011).
46. Pintó, R.M. & Bosch, A. Rethinking virus detection in food. in *Food-Borne Viruses: Progress and Challenges* (eds. Koopmans, M., Cliver, D.O. & Bosch, A.) (American Society of Microbiology, Washington, DC, 2008).
47. Gallimore, C.I., Cheesbrough, J.S., Lamden, K., Bingham, C. & Gray, J.J. Multiple norovirus genotypes characterised from an oyster-associated outbreak of gastroenteritis. *Int J Food Microbiol* 103, 323–330 (2005).
48. Maunula, L., Kalso, S., Von Bonsdorff, C.H. & Ponka, A. Wading pool water contaminated with both noroviruses and astroviruses as the source of a gastroenteritis outbreak. *Epidemiol Infect* 132, 737–743 (2004).
49. Myint, S., Manley, R. & Cubitt, D. Viruses in bathing waters. *Lancet* 343, 1640–1641 (1994).
50. Abad, F.X., Pinto, R.M., Villena, C., Gajardo, R. & Bosch, A. Astrovirus survival in drinking water. *Appl Environ Microbiol* 63, 3119–3122 (1997).
51. Superti, F. et al. In vitro effect of synthetic flavanoids on astrovirus infection. *Antiviral Res* 13, 201–208 (1990).
52. Bass, D.M. & Qiu, S. Proteolytic processing of the astrovirus capsid. *J Virol* 74, 1810–1814 (2000).
53. Lee, T.W. & Kurtz, J.B. Prevalence of human astrovirus serotypes in the Oxford region 1976–92, with evidence for two new serotypes. *Epidemiol Infect* 112, 187–193 (1994).
54. Mendez, E. & Arias, C.F. Astroviruses. in *Fields Virology*, Vol. 1 (eds. Knipe, D.M. & Howley, P.M.) 981–1000 (Lippincott Williams & Wilkins, Baltimore, 2007).
55. Mendez, E., Fernandez-Luna, T., Lopez, S., Mendez-Toss, M. & Arias, C.F. Proteolytic processing of a serotype 8 human astrovirus ORF2 polyprotein. *J Virol* 76, 7996–8002 (2002).
56. Sanchez-Fauquier, A. et al. Characterization of a human astrovirus serotype 2 structural protein (VP26) that contains an epitope involved in virus neutralization. *Virology* 201, 312–320 (1994).
57. Donelli, G., Superti, F., Tinari, A. & Marziano, M.L. Mechanism of astrovirus entry into Graham 293 cells. *J Med Virol* 38, 271–277 (1992).
58. Méndez, E., Murillo, A., Velázquez, R., Burnham, A. & Arias, C. Replication cycle of astroviruses. in *Astrovirus Research* (ed. Schultz-Cherry, S.) 19–45 (Springer, New York, 2013).
59. Guix, S., Caballero, S., Bosch, A. & Pinto, R.M. C-terminal nsP1a protein of human astrovirus colocalizes with the endoplasmic reticulum and viral RNA. *J Virol* 78, 13627–13636 (2004).
60. Monroe, S.S., Jiang, B., Stine, S.E., Koopmans, M. & Glass, R.I. Subgenomic RNA sequence of human astrovirus supports classification of Astroviridae as a new family of RNA viruses. *J Virol* 67, 3611–3614 (1993).
61. Banos-Lara Mdel, R. & Mendez, E. Role of individual caspases induced by astrovirus on the processing of its structural protein and its release from the cell through a non-lytic mechanism. *Virology* 401, 322–332 (2010).
62. Guix, S., Bosch, A., Ribes, E., Dora Martinez, L. & Pinto, R.M. Apoptosis in astrovirus-infected CaCo-2 cells. *Virology* 319, 249–261 (2004).
63. Mendez, E., Salas-Ocampo, E. & Arias, C.F. Caspases mediate processing of the capsid precursor and cell release of human astroviruses. *J Virol* 78, 8601–8608 (2004).

64. Cubitt, W.D., Mitchell, D.K., Carter, M.J., Willcocks, M.M. & Holzel, H. Application of electronmicroscopy, enzyme immunoassay, and RT-PCR to monitor an outbreak of astrovirus type 1 in a paediatric bone marrow transplant unit. *J Med Virol* 57, 313–321 (1999).
65. Gallimore, C.I. et al. Use of a heminested reverse transcriptase PCR assay for detection of astrovirus in environmental swabs from an outbreak of gastroenteritis in a pediatric primary immunodeficiency unit. *J Clin Microbiol* 43, 3890–3894 (2005).
66. Khamrin, P. et al. Multiple astrovirus MLB1, MLB2, VA2 clades and classic human astrovirus in children with acute gastroenteritis in Japan. *J Med Virol* 88, 356–360 (2015).
67. Liste, M.B. et al. Enteric virus infections and diarrhea in healthy and human immunodeficiency virus-infected children. *J Clin Microbiol* 38, 2873–2877 (2000).
68. Walter, J.E. & Mitchell, D.K. Role of astroviruses in childhood diarrhea. *Curr Opin Pediatr* 12, 275–279 (2000).
69. Belliot, G., Laveran, H. & Monroe, S.S. Capsid protein composition of reference strains and wild isolates of human astroviruses. *Virus Res* 49, 49–57 (1997).
70. Pager, C.T. & Steele, A.D. Astrovirus-associated diarrhea in South African adults. *Clin Infect Dis* 35, 1452–1453 (2002).
71. Krishna, N. K., Koci, M. D. & Guix, S. Immune responses. in *Astrovirus Research* (ed. Schultz-Cherry, S.) 79–95 (Springer, New York, 2013).
72. Kurtz, J. & Lee, T. Astrovirus gastroenteritis age distribution of antibody. *Med Microbiol Immunol (Berl)* 166, 227–230 (1978).
73. Lee, R.M. et al. Incubation periods of viral gastroenteritis: a systematic review. *BMC Infect Dis* 13, 446 (2013).
74. Behling-Kelly, E. et al. Localization of astrovirus in experimentally infected turkeys as determined by in situ hybridization. *Vet Pathol* 39, 595–598 (2002).
75. Brnic, D. et al. Porcine astrovirus viremia and high genetic variability in pigs on large holdings in Croatia. *Infect Genet Evol* 14, 258–264 (2013).
76. Holtz, L.R. et al. Astrovirus MLB2 viremia in febrile child. *Emerg Infect Dis* 17, 2050–2052 (2011).
77. Wunderli, W. et al. Astrovirus infection in hospitalized infants with severe combined immunodeficiency after allogeneic hematopoietic stem cell transplantation. *PLoS One* 6, e27483 (2011).
78. Gough, R.E., Collins, M.S., Borland, E. & Keymer, L.F. Astrovirus-like particles associated with hepatitis in ducklings. *Vet Rec* 114, 279 (1984).
79. Koci, M.D. et al. Astrovirus induces diarrhea in the absence of inflammation and cell death. *J Virol* 77, 11798–11808 (2003).
80. Schultz-Cherry, S. et al. Identifying agent(s) associated with poult enteritis mortality syndrome: importance of the thymus. *Avian Dis* 44, 256–265 (2000).
81. Yokoyama, C.C. et al. Adaptive immunity restricts replication of novel murine astroviruses. *J Virol* 86, 12262–12270 (2012).
82. Gray, E.W., Angus, K.W. & Snodgrass, D.R. Ultrastructure of the small intestine in astrovirus-infected lambs. *J Gen Virol* 49, 71–82 (1980).
83. Woode, G.N., Pohlenz, J.F., Gourley, N.E. & Fagerland, J.A. Astrovirus and Breda virus infections of dome cell epithelium of bovine ileum. *J Clin Microbiol* 19, 623–630 (1984).
84. Fu, Y. et al. The complete sequence of a duck astrovirus associated with fatal hepatitis in ducklings. *J Gen Virol* 90 (Pt 5), 1104–1108 (2009).
85. Todd, D. et al. Identification of chicken enterovirus-like viruses, duck hepatitis virus type 2 and duck hepatitis virus type 3 as astroviruses. *Avian Pathol* 38, 21–30 (2009).
86. Koci, M.D., Kelley, L.A., Larsen, D. & Schultz-Cherry, S. Astrovirus-induced synthesis of nitric oxide contributes to virus control during infection. *J Virol* 78, 1564–1574 (2004).
87. Meyerhoff, R.R., Nighot, P.K., Ali, R.A., Blikslager, A.T. & Koci, M.D. Characterization of turkey inducible nitric oxide synthase and identification of its expression in the intestinal epithelium following astrovirus infection. *Comp Immunol Microbiol Infect Dis* 35, 63–69 (2012).
88. Nighot, P.K., Moeser, A., Ali, R.A., Blikslager, A.T. & Koci, M.D. Astrovirus infection induces sodium malabsorption and redistributes sodium hydrogen exchanger expression. *Virology* 401, 146–154 (2010).
89. Pantin-Jackwood, M.J., Spackman, E. & Day, J.M. Pathogenesis of type 2 turkey astroviruses with variant capsid genes in 2-day-old specific pathogen free poults. *Avian Pathol* 37, 193–201 (2008).

90. Tang, Y., Murgia, M.V., Ward, L. & Saif, Y.M. Pathogenicity of turkey astroviruses in turkey embryos and poults. *Avian Dis* 50, 526–531 (2006).
91. Thouvenelle, M.L., Haynes, J.S. & Reynolds, D.L. Astrovirus infection in hatchling turkeys: histologic, morphometric, and ultrastructural findings. *Avian Dis* 39, 328–336 (1995).
92. Thouvenelle, M.L., Haynes, J.S., Sell, J.L. & Reynolds, D.L. Astrovirus infection in hatchling turkeys: alterations in intestinal maltase activity. *Avian Dis* 39, 343–348 (1995).
93. Sebire, N.J. et al. Pathology of astrovirus associated diarrhoea in a paediatric bone marrow transplant recipient. *J Clin Pathol* 57, 1001–1003 (2004).
94. Moser, L.A., Carter, M. & Schultz-Cherry, S. Astrovirus increases epithelial barrier permeability independently of viral replication. *J Virol* 81, 11937–11945 (2007).
94a. Meliopoulos, V.A. et al. Oral administration of astrovirus capsid protein is sufficient to induce acute diarrhea in vivo. *mBio* 7(6), e01494–16 (2016). doi:10.1128/mBio.01494-16.
95. Meliopoulos, V.A. et al. Detection of antibodies against Turkey astrovirus in humans. *PLoS One* 9, e96934 (2014).
96. Burbelo, P.D. et al. Serological studies confirm the novel astrovirus HMOAstV-C as a highly prevalent human infectious agent. *PLoS One* 6, e22576 (2011).
97. Molberg, O. et al. HLA restriction patterns of gliadin- and astrovirus-specific CD4+ T cells isolated in parallel from the small intestine of celiac disease patients. *Tissue Antigens* 52, 407–415 (1998).
98. Molberg, O. et al. CD4+ T cells with specific reactivity against astrovirus isolated from normal human small intestine. *Gastroenterology* 114, 115–122 (1998).
99. Guix, S. et al. Type I interferon response is delayed in human astrovirus infections. *PLoS One* 10, e0123087 (2015).
100. Bonaparte, R.S. et al. Human astrovirus coat protein inhibits serum complement activation via C1, the first component of the classical pathway. *J Virol* 82, 817–827 (2008).
101. Gronemus, J.Q. et al. Potent inhibition of the classical pathway of complement by a novel C1q-binding peptide derived from the human astrovirus coat protein. *Mol Immunol* 48, 305–313 (2010).
102. Hair, P.S. et al. Human astrovirus coat protein binds C1q and MBL and inhibits the classical and lectin pathways of complement activation. *Mol Immunol* 47, 792–798 (2010).
103. Krishna, N.K. & Cunnion, K.M. Human astrovirus coat protein: a novel C1 inhibitor. *Adv Exp Med Biol* 632, 237–251 (2008).
104. Tam, J.C., Bidgood, S.R., McEwan, W.A. & James, L.C. Intracellular sensing of complement C3 activates cell autonomous immunity. *Science* 345, 1256070 (2014).
105. Koci, M.D. Immunity and resistance to astrovirus infection. *Viral Immunol* 18, 11–16 (2005).
106. Qureshi, M.A., Edens, F.W. & Havenstein, G.B. Immune system dysfunction during exposure to poult enteritis and mortality syndrome agents. *Poult Sci* 76, 564–569 (1997).
107. Qureshi, M.A., Yu, M. & Saif, Y.M. A novel "small round virus" inducing poult enteritis and mortality syndrome and associated immune alterations. *Avian Dis* 44, 275–283 (2000).
108. Heggen, C.L., Qureshi, M.A., Edens, F.W. & Barnes, H.J. Alterations in macrophage-produced cytokines and nitrite associated with poult enteritis and mortality syndrome. *Avian Dis* 44, 59–65 (2000).
109. Heggen, C.L., Qureshi, M.A., Edens, F.W., Barnes, H.J. & Havenstein, G.B. Alterations in the lymphocytic and mononuclear phagocytic systems of turkey poults associated with exposure to poult enteritis and mortality syndrome. *Avian Dis* 42, 711–720 (1998).
110. Qureshi, M.A., Edens, F.W., Ali, R.A. & Saif, Y.M. Alteration in macrophage-mediated cytokines/metabolite production after challenge with a "small round virus" of turkeys. in *Poultry Science Association 89th Annual Meeting* 64 (Montreal, Canada, 2000).
111. Qureshi, M.A., Saif, Y.M., Heggen-Peay, C.L., Edens, F.W. & Havenstein, G.B. Induction of functional defects in macrophages by a poult enteritis and mortality syndrome-associated turkey astrovirus. *Avian Dis* 45, 853–861 (2001).
112. Dryden, K.A. et al. Immature and mature human astrovirus: structure, conformational changes, and similarities to hepatitis E virus. *J Mol Biol* 422, 650–658 (2012).

4

Hepatitis E Virus

Kavita Lole, Prudhvi Lal Bhukya, and Subhashis Chatterjee

CONTENTS

4.1 Introduction

Hepatitis E virus (HEV) is a major causative agent of acute viral hepatitis in developing countries. HEV is primarily transmitted via the fecal–oral route through contaminated water and is endemic in several countries in Asia, Africa, and Central America[1–3] due to poor sanitary conditions. It is also highly endemic in India, Bangladesh, Egypt, Mexico, and China, where hepatitis E can occur either in the form of sporadic cases or occasional large outbreaks. Hepatitis E represents the major cause of waterborne outbreaks in these countries, and contamination of water with sewage has been a common feature preceding such outbreaks.[4–10] HEV is believed to be the most common cause of sporadic acute hepatitis (>50%–60% cases) in adults in regions where the virus is endemic.[11–13] There have also been large outbreaks among displaced persons in Sudan, Chad, and Uganda.[14–17]

Although, previously, sporadic cases of hepatitis E in developed countries were attributed to travel to hyperendemic areas, in recent years, a surprisingly higher number of cases of autochthonous transmission of hepatitis E with high seroprevalence have been documented from the United States, Canada, Europe (including UK, France, the Netherlands, Austria, Denmark, Spain, Italy, Greece, and Germany), and Asia–Pacific countries (Japan, Taiwan, Korea, Hong Kong, Australia, and New Zealand).[18,19] It is estimated that about 2 billion people live in HEV-endemic areas. A global disease burden study estimated that HEV genotypes 1 and 2 account for approximately 20.1 million incidents of HEV infections, 3.4 million cases of symptomatic disease, 70,000 deaths, and 3,000 stillbirths. Clinical HEV infection is most common in adolescents and young adults, but also occurs to a lesser extent in children.[20–22] During epidemics of HEV, the ratio of clinical to subclinical or anicteric infection was shown to be 1:26 in pregnant women, 1:4 in children, and approximately 1:10 in adults.[20,23] In developing countries, although HEV is endemic, seroprevalence usually remains below 40% as against hepatitis A virus, which shows almost 100% seropositivity in adults.[1]

Nucleotide sequence analysis of HEV genomes from different geographical isolates has revealed wide genetic diversity among HEV strains forming four distinct phylogenetic clusters. Of these, genotype 1 and 2 strains are restricted to humans. Genotype 1 strains are associated with large outbreaks of acute hepatitis E in humans in Asia. Genotype 2 includes one Mexican strain and several African strains. Genotype 3 is distributed worldwide. Distribution of genotype 2 virus is restricted to few countries and is not a major problem in comparison with the more widely distributed genotype 1 virus. Genotype 3 is also associated with sporadic, cluster, and chronic cases of hepatitis E in humans, mostly in industrialized countries. Genotype 3 strains have also been isolated from domestic pigs, wild pigs, deer, mongoose, and rabbits.[24–26] Wang et al.[27] identified an additional HEV strain in the sera of Chinese patients with acute hepatitis, classified as genotype 4. Genotype 4 virus can also infect pigs.[28,29] Genotype 3 and 4 strains have been found to infect a broader range of hosts including deer and boars.[26,30–32] Intra- and intergenotype recombination has been recently documented.[33]

HEV is enzootic in domestic pigs even though the virus may not be endemic in human population or belongs to a different genotype than that is circulating in humans in endemic settings. Detection of HEV RNA and antibodies in wild boar populations has been reported from Japan, Germany, Italy, Spain, the Netherlands, and Australia.[34–39] In addition, several divergent strains of HEV have been isolated from animals such as bats,[40] rats,[31] wild boar,[41] ferrets,[42] mongoose,[25] fish, and cutthroat trout.[43]

HEV strains within genotypes 1 and 2 are less divergent as compared with HEV strains from genotypes 3 and 4. These HEV genotypes have been classified further into 24 subgenotypes.[44,45] Currently, four putative genera have been proposed in the family Hepeviridae, one comprising human HEV genotypes and closely related animal viruses and the other three including viruses of rodent (rat), chiropteran (bat), and avian (chicken) origins.[40] HEV isolates from rats share approximately 59.9% and 49.9% sequence identities with human and avian HEVs, respectively, while the ferret HEV sequences share 72.3% identity with the rat HEV and have been proposed to be grouped in the genus *Ortho hepevirus*.[42,46] Bat HEV strains from Africa, Central America, and Europe have a very high sequence divergence (53% at amino acid level) from known HEV isolates and are suggested to be classified in a separate genus *Chiropteranhepevirus*.[47,48] Avian HEV strains share 50%–60% nucleotide sequence identity with swine or human HEV strains, and recently it was proposed that all three known genotypes of avian HEV in chickens (genotype 1 in Australia and Korea, genotype 2 in the United States, and genotype 3 in Europe and China) should be grouped into a new genus *Avihepevirus*.[47,49,50] The cutthroat trout HEV from the United States exhibits only 13%–27% sequence homology leading to a proposal of another genus *Piscihepevirus*.[43,47] Interestingly, the range of different animal species that appear to harbor HEV/HEV-like agents is continuously extending, indicating the ubiquity of HEV/HEV-like agents in different species from many regions. We expect many more additions in this family in the

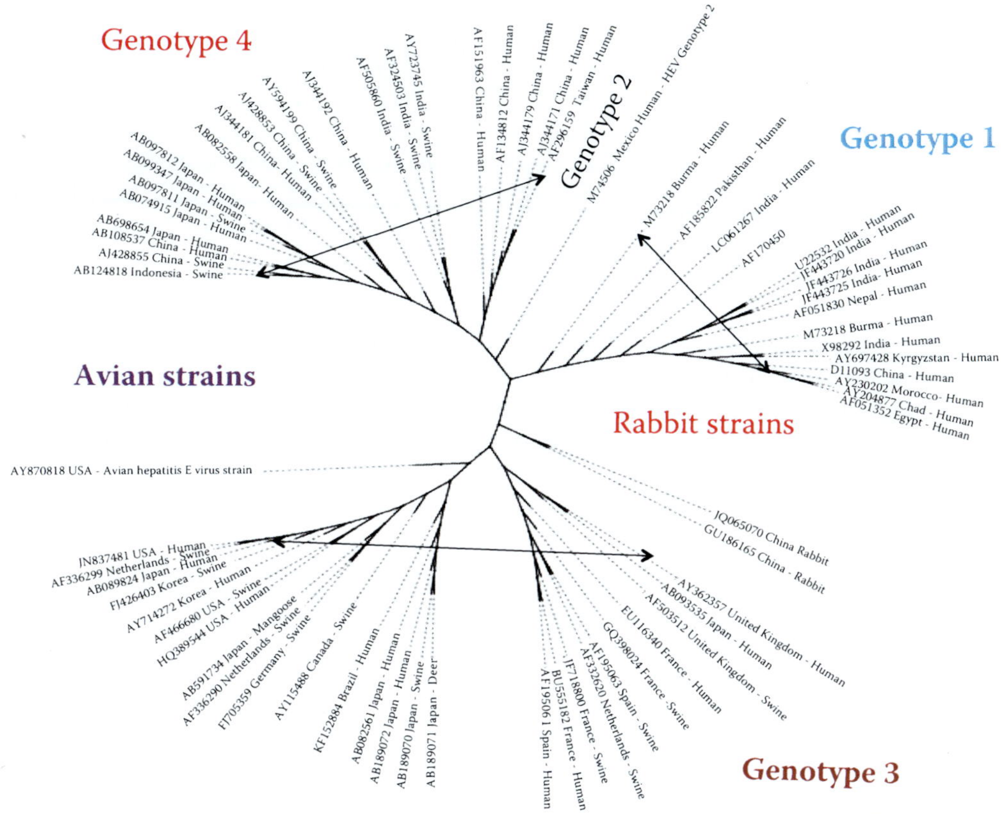

FIGURE 4.1 HEV phylogenetic tree: Full genome/capsid region nucleotide sequence analysis of global hepatitis E virus isolates. Genotypes 1 and 2 (HEV-1 and HEV-2) circulate among humans, primarily in Africa and Asia, while genotypes 3 and 4 (HEV-3 and HEV-4) have animal reservoirs. Recently discovered rabbit strains appear to form a closely related clade. Alignment was done using MEGA 7 and tree by using software available on www.itol.embl.de.

coming years. A phylogenetic tree based on a portion of the nucleotide sequence encoding the capsid protein is shown in Figure 4.1.

4.1.1 HEV Transmission

The virus is relatively stable in the environment[51] and is sensitive to heat, chlorination, and ultraviolet light.[52,53] In endemic regions, HEV remains in circulation due to the excretion of the virus by humans and other animals in the acute phase of infection and environment acts as the major reservoir of the virus.[54] Contamination of drinking and irrigation water mainly occurs due to improper sewage disposal, leading to epidemics in developing countries. Water scarcity in summers or floods during rainy season can increase the possibility of the virus spread. Surface water bodies can easily get contaminated with human and animal wastes. Inadequate treatment and release of sewage in rivers has been documented, with 4.9% HEV RNA positivity in 403 water samples from different locations of a river in one Indian city (Mutha River, Pune).[55] Waterborne route could also be important for genotype 3 and 4 HEVs. Genotype 3 virus has been detected in waste/runoff waters from pig slaughterhouses, swine manure, and swine slurry storage facilities.[56,57] An increased risk of hepatitis E in sewage treatment plant workers has been documented.[58,59]

Person-to-person transmission of HEV is rare in both epidemic[60] and sporadic settings.[61] Occasionally, nosocomial spread has also been reported.[62] Vertical transmission from mother to infant is also known to occur.[63] In endemic as well as nonendemic settings, transfusion-associated hepatitis E infection is possible[64–69]; however, this route of transmission is not so common.[70,71] Plasma products have been documented as the source of HEV infections.[72,73] Easy and rapid HEV infectivity laboratory assays to evaluate virus inactivation processes would be useful for monitoring plasma-derived products.

Close phylogenetic relationships between human and animal viruses prove that HEV is transmitted from animals to humans. Domestic pigs are the major animal reservoirs of HEV. Hepatitis E is now a recognized zoonotic disease, with swine and other likely animals serving as the reservoir for these viruses.[74,75] Naturally acquired anti-HEV antibodies have been shown in many animal species including swine, sheep, cattle, goats, horses, macaques, cats, dogs, rats, and mice, raising the possibility of these animals being the reservoirs and causing zoonotic infections.[10,76,77] Occupational exposure to infected pigs poses the risk of acquiring HEV infection.[56,78–80]

Consumption of undercooked or raw pig liver, pork, and game meat (wild boar, deer, or rabbit) plays a significant role in HEV transmission in industrialized countries. Studies from Japan have documented foodborne HEV infections that occurred due to the consumption of undercooked pork[81,82] and wild boar or deer meat.[83–86] HEV infections due to wild boar meat have also been reported from Germany[86] and South Korea.[32] Consumption of pig liver figatellu sausages has been documented to cause human HEV infections in France, as seen from the identical HEV sequences from the infected individuals and sausages from the local grocery stores.[87,88] Infectious nature of HEV particles in the sausages was demonstrated by Berto et al.[89] HEV presence in pig liver or meat has also been reported from countries such as India,[90] United States,[91] Germany,[92] Canada,[93] Democratic Republic of Congo,[94] Japan,[95] Italy,[96] and the Netherlands.[97]

HEV has been detected in varied shellfish collected from European and Asian countries. Genotype 3 HEV has been detected in mussels and oysters.[98–102] Human HEV infections due to the consumption of shellfish are documented.[103–105] Bivalves are known to concentrate viral particles during their filter feeding process, and as they are mostly consumed raw or cooked for a very short period, they pose a serious risk of HEV infection. Viral RNA sequences have been detected in soft fruits like strawberries and green leafy vegetables such as lettuce, with irrigation water being the suspected contamination source.[106,107] The concentration of the viable virus in an environment or food is an important factor in the outcome of clinical hepatitis E infection. HEV is known to remain stable in frozen conditions even after 10 years. HEV remains viable after heating to 56°C for 1 h,[108] and a cooking temperature of 71°C for 20 min is required for the complete inactivation of the virus.[109] Use of contaminated water for drinking, cooking, cleaning, and irrigation[107] can spread the virus.

4.1.2 HEV Infection in Humans

Hepatitis E is generally a self-limiting disease which lasts for 4–6 weeks. Entry of the virus into the host primarily occurs by the oral route. Infection is presumed to be initiated via cells lining the alimentary tract. The virus then enters the liver, presumably via the portal vein, where it replicates in the cytoplasm of hepatocytes without causing any direct catalytic effect, and is released into the bile and blood. The incubation period for hepatitis E infection ranges from 15 to 60 days with a mean of 40 days as seen from the two studies wherein the clinical course of HEV infection was monitored in two human volunteers who ingested the virus from fecal suspensions.[110,111] HEV was detected in feces by immuno electron microscopy (IEM) on day 28, and the liver enzyme level peaked on day 42.[110] Viremia was detected from day 22 onward up to the alanine aminotransferase (ALT) peak. Anti-HEV antibody was first detected on day 41 and was detectable until 2 years later.[111]

HEV infection can lead to liver injury of variable severity. Clinically, HEV infection manifests in different ways from asymptomatic infection, acute hepatitis to fulminant hepatitis. The disease occurs in two stages: the prodromal (pre-icteric) stage and the icteric stage. The illness is often clinically and biochemically similar to other hepatotropic viruses, such as hepatitis A virus or hepatitis B virus. Typical symptoms of HEV infection include fever, fatigue, loss of appetite, nausea, vomiting, abdominal pain, jaundice, dark urine, clay-colored stool, and enlargement of liver. Few patients with hepatitis E infection develop prolonged cholestasis, characterized by persistence of jaundice, marked itching, and elevated alkaline phosphates.[112] The prodromal symptoms usually subside at the onset of clinical jaundice; however, some patients may not show visible signs of jaundice despite experiencing severe symptoms. The ALT level peaks at the onset of symptoms followed by conjugated hyperbilirubinemia. The level of

ALT, however, does not correlate with the degree of liver cell damage (peak levels vary from 1000 to 2000 U/L at the onset). Peak serum total bilirubin levels range from 5 to 25 mg/dL; both conjugated and unconjugated fractions are increased. ALT progressively diminishes during the recovery phase. Patients with anicteric or subclinical acute hepatitis E infection only exhibit elevated ALT levels, which is useful in the early diagnosis of clinically suspected cases, along with the presence of anti-HEV IgM in the serum. Liver damage results, presumably mediated by host cellular immune responses. The course of hepatitis E in human subjects strongly suggests that HEV, like other hepatitis viruses, is noncytopathic in most circumstances, and liver disease is probably caused by immunological and hormonal factors.[113–116]

The severity of hepatitis E is considered to be dose-dependent, and associated host factors such as chronic liver disease or alcohol overuse may enhance it further.[3] Patients with HEV infection superimposed with other viral or nonviral chronic liver disease (acute on chronic liver disease) are at the higher risk of having a poor outcome.[117,118] Case fatality rates in hepatitis E epidemics range from 0.1% to 4%, but pregnant women, especially during the third trimester, are at a higher risk of severe disease, with case fatality rates of 10%–25%.[119–121] Hepatitis E is also associated with premature delivery, with low birth weight and an increased risk of perinatal mortality.[122] HEV transmission from mother to fetus was reported in 33% of cases; the virus has been detected in human colostrum as well.[117,123]

The outcome of hepatitis E infection during pregnancy is probably influenced by nutritional status, host immune response, and hormonal factors.[113,115,116] It is not yet understood how HEV infections during pregnancy are less severe in Egypt[124] while they are more severe with higher mortality rates in sub-Saharan Africa and South and South-East Asia,[115] although genotype 1 HEV predominates in these regions. Thus, severity during pregnancy is not the unique feature of genotype 1 or 2 viral infections, but there seem to be additional contributing factors that influence the disease severity. Only occasional cases of hepatitis E during pregnancy with less severity have been reported from low-endemic, developed countries.[125–127] Thus, understanding HEV pathogenesis during pregnancy still remains a big challenge.

There are certain observations to suggest that HEV strains may vary in their virulence. Genotype 3 strain (3c) circulating in Europe is known to cause mostly subclinical infections.[128] A comparative study of genotype 3 and 4 HEV-infected individuals in Japan revealed that genotype 4 HEV is associated with significantly higher ALT and total bilirubin levels, higher viral loads, and an aggressive disease course.[129] Similar observations have been reported by Jeblaoui et al.[130] in a study from France that the clinical presentation is more severe in patients with HEV4 infections than in patients with HEV3 infections. Purcell et al.[131] examined the relative virulence of human genotypes 1 and 2 and swine genotype 3 in rhesus monkeys and found that genotype 3 is significantly less virulent as compared with human genotypes 1 and 2. There are studies indicating the association of point mutations in the HEV genome with disease severity; however, the underlying mechanism still remains unknown. Silent substitutions of U at the nucleotide 3148 (in the helicase domain) and C at the nucleotide 5907 (in the capsid gene) in the genomes of HEV strains of genotypes 3 and 4 have been documented to have an association with fulminant hepatitis and disease severity in patients. C5907 mutation was also associated with high viral loads in the infected individuals.[132,133] Takahashi et al.[134] have reported the association of V239A mutation (in the helicase domain) with increased virulence of genotype 3 virus. Similarly, L1110F and V1120I mutations in the helicase domain of genotype 1 sequences were found to be associated with fulminant hepatitis cases.[135] Shukla et al.[136] have recently reported the integration of human S17 rRNA sequence into the genotype 3 HEV genome in a patient with chronic HEV infection. This patient had both neurologic and hepatic symptoms. Nguyen et al.[137] have also reported the integration of S19 in an isolate analyzed from a chronic hepatitis E patient with organ transplant in the United States. These integrations in HEV genomes probably altered tissue specificity and pathogenicity of the virus.

HEV normally does not lead to chronic infection in immunocompetent individuals. However, chronic HEV infections have been observed in immunocompromised organ transplant recipients[138,139] and in patients with other conditions of immunosuppression such as HIV infection[140] and hematological malignancies.[141,142] HEV infection results in persistent viral shedding and may rapidly progress to liver cirrhosis and may also cause extrahepatic manifestations such as neurological disorders and impaired renal functions in these patients.[143] Chronic HEV infections were thought to exclusively involve genotype 3 viral strains, however, recently a report from China documented the chronic course of HEV infection with genotype 4 virus in a boy who was receiving chemotherapy for an acute lymphoblastic leukemia.[144]

A study from India showed no evidence of chronic hepatitis E in a cohort of kidney transplant recipients, suggesting that HEV genotype 1 may not cause chronic hepatitis E.[145] However, there is a need for more data from other endemic regions to confirm these findings.

4.1.3 Clinical Course/Manifestations of HEV Infection

In macaques, the incubation period is generally 3–8 weeks, but shorter and longer incubation periods have been observed.[146–149] In intravenously inoculated macaques, the expression of HEV antigens can be seen in hepatocytes 7 days postinfection with a parallel appearance in bile and feces, suggesting the release of the virus from hepatocytes into bile and then into feces.[150] Peak viremia and peak shedding of HEV into the feces occurs during the incubation period and early acute phase of the disease.[146,151–153] HEV RNA could be detected in serum, bile, and feces several days before the elevation of serum ALT levels.[146–148] Histopathological changes in the liver were observed in parallel with ALT elevation. High mortality (~20% particularly in the third trimester) of hepatitis E during human pregnancy was not reproducible in pregnant rhesus macaques.[152,153] HEV RNA detection in experimentally infected chimpanzees revealed viral genomic sequences in serum and stools from the very beginning of the infection and a sudden drop of the viral titer with the development of an antibody response.[154] All known mammalian HEV genotypes (1–4) and swine HEV genotypes 3 (United States)[155,156] and 4 (India)[157,158] have been transmitted to nonhuman primates. The course of hepatitis E in human subjects and in experimentally infected primates strongly suggests that HEV, like other hepatitis viruses, is noncytopathic in most circumstances and liver disease is probably caused by immunological factors.[113–116]

Swine is both a reservoir and a host of HEV. Both natural and experimental HEV (genotypes 3 and 4) infections in swine are asymptomatic with only mild microscopic lesions in the liver and associated lymph.[35,157–160] There is a transient viremia while fecal shedding continues from 3 to 7 weeks and anti-HEV antibodies are seen in 4–6 weeks.[28,156,159] The acquisition of anti-HEV is age dependent in pigs, the majority acquiring antibodies by 4–5 months of age.[159,161]

Avian HEV also shows mostly subclinical infections with mortality rates up to 0.3%–1.0%.[162–164] Clinical signs may include egg drop in some flocks up to 20%, enlargement of the liver and spleen, and acute death of affected birds. Since avian HEV does not infect humans, it is not a concern for food and environmental safety.

4.1.4 Immune Response

4.1.4.1 Innate Immune Response

After entry into the host, HEV needs to overcome the host innate immune response and establish infection in the host. In the past decade, remarkable progress has been made in understanding strategies of HEV in combating host innate immune responses. HEV interferes with type 1 interferon (IFN) induction and IFN-activated signaling. HEV inhibits IFN-α signaling and manages to replicate in the presence of IFN-α. There is a study demonstrating the reduced ubiquitination and degradation of IkBα in hepatoma cells expressing the open reading frame (ORF) 2 protein, affecting the nuclear translocation of NF-κB. Recently, the macro domain and papain-like cysteine protease (PCP) domain from HEV ORF1 were identified as the putative IFN antagonists by Nan et al.[165] In contrast, HEV ORF3 was shown to enhance RIG-I-mediated signaling leading to enhanced IFN-α synthesis, indicating complex strategies of the virus during its establishment in the host cells.[166]

A microarray-based gene expression analysis of serial liver biopsy samples from HEV and HCV in chimpanzees has shown attenuated adaptive immune response in HEV-infected animals as compared with HCV.[167] However, there was a robust innate immune response correlating well with viremia in both HCV- and HEV-infected chimpanzees. A major component of the host response to HEV infection was type I interferon-induced genes (ISGs). HEV infection was completely resolved without any recurrence in the infected chimpanzees, indicating that HEV is highly susceptible to ISGs. This indicated that HEV infection is mostly taken care of by the robust innate immune response. Altered peripheral frequencies and activation status of NK and NKT cells in HEV-infected individuals suggest that innate immunity plays a role in the pathogenesis of hepatitis E.[168] It was speculated that the reduced NK cell levels during pregnancy could contribute toward the increased severity of hepatitis E during pregnancy. Robust nonspecific IFN-γ-producing T-cell response in acute hepatitis E patients suggesting innate immune mechanisms involving NK/NKT/T reg cells has been documented.[169] Higher frequencies of CD4+ CD25+ Foxp3+ regulatory T cells, elevated levels of IL-10, and suppressive functionality of the regulatory T cells in acute hepatitis E patients have been documented.[170] Overall, innate immune responses seem to play a major role during HEV infection.

4.1.4.2 HEV-Induced Adaptive Immune Response

HEV usually causes mild disease or subclinical infections. Icteric symptoms in hepatitis E coincide with the rise in anti-HEV antibodies. Experimental animal infections and human studies suggest that liver damage mainly occurs due to immune response rather than by the virus itself. The immunohistochemistry of liver biopsies from patients with HEV-induced acute liver failure revealed significant infiltration of CD8+ T cells containing granzyme suggesting their role in liver damage.[171] In comparison to patients with acute hepatitis E and healthy controls, patients with fulminant hepatitis E have a less marked expansion of HEV-specific IFN-γ or TNF-α-secreting CD4+ T cells. Husain et al.[172] provided evidence for the activation of effector T cells during acute hepatitis E. Srivastava et al.[173] observed less-marked cellular immune responses and a heightened antiviral humoral response in severe HEV infection. In contrast, significantly higher levels of both Th1 (IFN-γ, IL-2, and TNF-α) and Th2 (IL-10) cytokines and higher anti-HEV IgM and IgG titers were recorded in fulminant hepatic failure (FHF) patients as compared with acute hepatitis E patients by Saravanabalaji et al.[174] The existence of a Th2 bias in pregnant women with acute hepatitis E infection is reported.[18] Impairment of HEV-specific T-cell responses was shown to be associated with chronic HEV infection in immunocompromised organ transplant patients.[175] Association of TNF-α and IFN-γ cytokine gene polymorphism with the susceptibility and clinical outcome of hepatitis E has been shown.[176] Authors noted an overrepresentation of the T (vs. A) allele at IFN-γ+874, which is linked to higher IFN-γ production, among symptomatic cases. These observations suggest a probable role of higher acquired anti-HEV response in serious complications in FHF with HEV.

4.1.5 Diagnosis

Hepatitis E diagnosis mainly depends on the clinical features and exclusion of other causes of acute hepatitis, especially hepatitis A, in the endemic regions. The presence of anti-HEV IgM is a marker of acute infection, while anti-HEV IgG alone is a marker of past infection. HEV was first identified by immunoelectron microscopy.[110] Serological as well as molecular assays were developed eventually after successful cloning of the HEV genome.[177,178] Anti-HEV IgM appears during the early acute phase of illness and may be detected as early as 4 days after the onset of jaundice and lasts for up to 2–5 months.[179] Anti-HEV IgG has been shown to persist for long periods of time (>14 years) and provides protection against subsequent infections. Anti-HEV IgA has also been detected in the serum of naturally infected individuals.[180] For practical purposes, HEV IgM can be detected from blood to confirm the clinical diagnosis. Detection of viral RNA in HEV IgM-positive patients provides additional confirmation and further information about the virus strain, enabling molecular epidemiological studies. In immunocompromised patients, HEV RNA testing is essential for diagnosis since patients have impaired immune response. There are several HEV RNA detection assays developed during the past decade.[181–185] However, the assay developed by Jothikumar et al.[185] seems to be very commonly used. For optimal diagnosis, a combination of serological and nucleic acid-based assays has been recommended.[186,187]

HEV ORF2 protein expressed in SF9 cells has been shown to have glycosylation and performs well in anti-HEV IgM and IgG diagnostic assays. All the four genotypes of HEV belong to a single serotype,[188] thus suggesting that diagnostic antigens from a single HEV genotype should detect antibodies against HEV strains of different genotypes. Broad cross immunoreactivity among HEV genotypes has been documented.[189–192] Commercially available assays for hepatitis E diagnosis have problems of sensitivity and specificity, and many studies have shown poor concordance of these tests.[193,194] There is a need to have a validated assay that would have reasonable sensitivity and specificity in detecting infections with different HEV genotypes. Such an assay would be extremely useful in generating worldwide seroprevalence data to understand the realistic disease burden of hepatitis E.

4.1.6 HEV Vaccine

Being refractory to cell culture, the conventional approach of developing killed or attenuated virus-based vaccines for HEV was not possible and hence recombinant DNA technology has been employed. Four mammalian HEV genotypes are believed to represent a single serotype. Cross-challenge studies in rhesus monkeys using various strains of genotype 1 HEV have shown that animals are protected from the disease up to 5 years after primary infection, indicating the possible utility of the vaccine to prevent hepatitis E.[1,12,64,195] HEV capsid (ORF2) protein is highly immunogenic and has been used as the target antigen for vaccine development. Several approaches based on recombinant DNA technology have been attempted to develop candidate subunit vaccines, either using a complete or partial ORF2 protein. A 56-kDa (112–607 amino acids) recombinant protein-based candidate vaccine phase III clinical trial, with 95% efficacy, was successfully completed in 2007 by the NIH group.[196] This subunit vaccine protected against genotype 1 (homologous) as well as genotype 2 and 3 HEV (heterologous) challenges in the monkey model.[197] However, since then, there has been no progress in the commercialization of this vaccine. The second candidate vaccine that was found to be safe and immunogenic was based on the bacterially expressed genotype 1 HEV ORF2 protein (p239, 368–606 amino acids). Phase III clinical trial of this vaccine showed >95% efficacy and cross protection against the heterologous genotype 4 virus.[198] The vaccine (Hecolin) is currently licensed and commercially available only in China.

4.1.7 Antiviral Therapy against HEV

There has been no established therapy for treating acute HEV infection and associated disease. Infections with genotype 1 and 2 HEVs are generally self-limiting and therefore do not require specific therapy in the majority of infections. Similarly, the disease course of genotype 3 infections is generally self-limiting illness that lasts for 4–6 weeks in immunocompetent individuals and hence generally does not require specific treatment. Therapy is essential only when patients are at the risk of having severe manifestations of the disease. Recent evidences of the chronic course of HEV genotype 3 infections in immunocompromised patients with rapid progression into cirrhosis showed the need to control the virus replication. This has increased the awareness for this ignored disease, and several studies have been initiated to understand the mechanisms leading to persistent infection and for the testing of different antiviral therapies in such patients. Pegylated IFN-α-2a has been successfully used for treating chronic hepatitis E in transplant recipients.[199] However, IFN-α-2a treatment is known to have side effects and may also increase the risk of graft rejection. Ribavirin is emerging as the treatment of choice for chronic HEV infection as it results in sustained virological response; however, there are treatment failures.[200–202] Ribavirin needs to be administered for at least 3 months in chronically infected individuals. Ribavirin monotherapy has also been used in immunocompetent patients suffering from severe acute hepatitis E.[203,204]

In HEV-endemic regions, genotype 1 HEV infections in patients with chronic liver disease have been known to worsen rapidly to a condition known as acute-on-chronic liver failure (ACLF), leading to very high mortality. Several studies have reported HEV as one of the leading causes of ACLF in Asia and Africa, where HEV is endemic.[205–208] Goyal et al.[209] have recently demonstrated the effectiveness of ribavirin therapy in treating genotype 1 HEV infections. However, the evaluation of ribavirin therapy needs to be carried out in a well-designed study, with a large number of patients in the active viremic phase. Overall, considering the limitations of the existing therapies, alternative antivirals are needed. Table 4.1 describes the overall characteristics of different HEV genotypes.

TABLE 4.1

Characteristics of HEV Genotypes 1–4

Characteristics	Genotype 1	Genotype 2	Genotype 3	Genotype 4
Distribution of virus in humans	Asia, Africa, and the Middle East	Mexico, West Africa	North America, Europe, Latin America, and Japan	China, Taiwan, and South-East Asia
Distribution of virus in animals	Not reported and identified	Not reported and identified	Widespread and reported in all continents	China, Taiwan, and India, with a few recent reports from Europe and North America
Interspecies transmission	Only human to human and no interspecies transmission observed	Only human to human and no interspecies transmission observed	Animal to human (pigs, wild boar, and deer)	Animal to human (pigs, wild boar)
Waterborne transmission	Yes, frequent (source is from human feces)	Yes, frequent (source is from human feces)	Not reported	Not reported
Food-borne transmission	Not reported and recognized	Not reported and recognized	Reported from contaminated animal meat	Reported from contaminated animal meat
Zoonotic transmission	Not reported	Not reported	Yes	Yes
Mortality among pregnant women	High	Not reported	Not reported	Not reported
Infection and attack rate with respect to age	Most common in young adults (15–44 years)	Most common in young adults (15–44 years)	Middle age and older above 55 years, males, and immunocompromised persons	Middle age and older above 55 years
Outbreaks	Common	Smaller scale outbreaks	Uncommon	Uncommon
Symptoms	Fever, fatigue, loss of appetite, nausea, vomiting, abdominal pain, jaundice, dark urine, clay-colored stool, and joint pain	Fever, fatigue, loss of appetite, nausea, vomiting, abdominal pain, jaundice, dark urine, clay-colored stool, and joint pain	Fever, fatigue, loss of appetite, nausea, vomiting, abdominal pain, jaundice, dark urine, clay-colored stool, and joint pain	Fever, fatigue, loss of appetite, nausea, vomiting, abdominal pain, jaundice, dark urine, clay-colored stool, and joint pain
Diagnosis	Anti-HEV test	Anti-HEV test	Anti-HEV test	Anti-HEV test
Prevention	Good sanitation and the availability of clean drinking water	Good sanitation and the availability of clean drinking water	Avoiding raw pork and other contaminated animal meat can reduce the risk	Avoiding raw pork and other contaminated animal meat can reduce the risk

4.1.8 HEV Biology

HEV is a nonenveloped, spherical particle of approximately 30–34 nm in diameter.[110] The HEV genome is a positive-sense single-stranded RNA of approximately 7.2 kb with a 5′-methylguanine cap and 3′-poly (A) stretch. Viral RNA contains short 5′ (27–35 nt) and 3′ (65–74 nt) untranslated regions (UTRs) and three ORFs, ORF1, ORF2, and ORF3.[178] ORF1 encodes the nonstructural polyprotein, ORF2 encodes

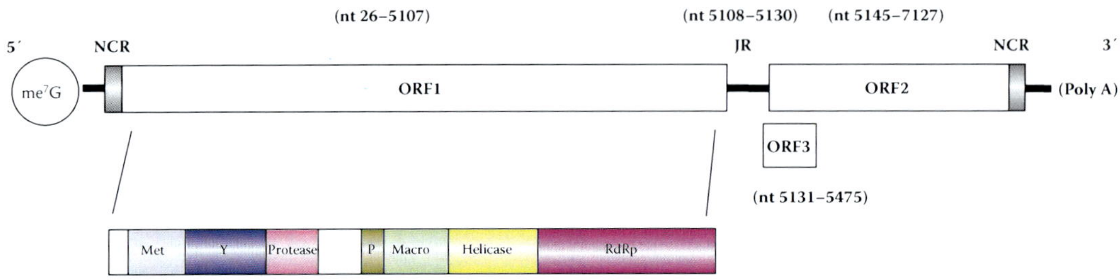

FIGURE 4.2 Schematic diagram of the HEV genome: The three ORFs are labeled and shown as boxes with the putative ORF1 domains indicated outside the box. JR, junction region; me^{7}G, 5′ Cap; NCR, noncoding region; ORF, open reading frame; P, Proline-rich hinge region; RdRp, RNA-dependent RNA-polymerase; Y, Y domain (nucleotide numbering done as per prototype genotype 1 sequence SAR55).

the capsid protein, and ORF3 encodes a multifunctional protein. ORF2 and ORF3 are proposed to be translated from a single bicistronic messenger ribonucleic acid (mRNA) and overlap with each other.[210,211] Figure 4.2 shows a schematic diagram of the HEV genome.

4.1.8.1 Proteins Encoded by the HEV Genome

4.1.8.1.1 Nonstructural Proteins

The ORF1 encodes for a 1693 amino acid polypeptide (180 kDa) (for the genotype 1 SAR-55 strain), predicted to encode different domains required for virus replication such as methyltransferase (MeT), a "Y" domain, PCP, a proline-rich region that contains a hypervariable region (HVR), an "X" domain, RNA helicase (Hel), and RNA-dependent RNA-polymerase (RdRp).[212]

4.1.8.1.2 Methyltransferase

It is the first functional domain of ORF1.[213] Recombinant protein-encompassing amino acids 1–979 from HEV ORF1 were shown to have guanine-7-methyltransferase as well as guanyltransferase activities. The presence of an m7G cap at the 5′ end of the HEV genomic RNA confirmed the functional role of HEV methyltransferase.[214] The source of RNA triphosphatase activity, that is required for 5′ cap formation, was later confirmed to be a HEV helicase protein.[215]

4.1.8.1.3 PCP

The most intriguing question of HEV biology that has not yet been answered is whether the ORF1 protein (pORF1) is processed into functional biochemical subunits or whether it remains as a single protein. The postulated role of PCP is the polyprotein processing of ORF1 polyprotein. The above hypothesis was put to test by many researchers, but none could show the evidence of ORF1 processing and the role of PCP in polyprotein processing.[216–221] A recent study from our group showed that HEV MeT-PCP protein recognizes the LXGG motif and shows *in vitro* deubiquitination (DUB) activity. The protein also showed deISGylation of cellular proteins, suggesting its probable role in combating cellular antiviral innate immunity during the HEV life cycle.[222] Paliwal et al.[223] recently demonstrated the role of PCP in ORF1 polyprotein processing.

4.1.8.1.4 Helicase

The putative RNA helicase of HEV was shown to contain all the seven conserved motifs found in SF-1 helicases and was proposed to have the NTPase as well as RNA-binding domains.[212] Functionality of HEV helicase was recently demonstrated by our group. HEV Hel exhibited NTPase and RNA unwinding activities. Enzyme hydrolyzed all rNTPs efficiently; dATP and dCTP were hydrolyzed more efficiently as compared with dGTP and dTTP. Enzyme showed the unwinding of only RNA duplexes with 5′ overhangs showing a 5′-to-3′ polarity.[224]

4.1.8.1.5 RdRp

HEV RdRp contains eight conserved motifs (motif I–VIII) that are similar to the RdRps from other positive-sense RNA viruses. RdRp containing polypeptide, encoded by 3546–5106 nt in ORF1, was shown to interact with the 3′NCR of HEV genomic RNA and synthesizes the complementary strand *in vitro*.[225] Rehman et al.[226] demonstrated the localization of RdRp to the endoplasmic reticulum (ER), suggesting that HEV replicates probably in the ER in the cytosolic compartment of the cells.

4.1.8.1.6 Macro Domain (X-Domain)

It is proposed that viral macro domains bind viral poly(A) tails and recruit poly(ADP-ribose)-modified cellular factors to the site of replication. Polypeptide encompassing amino acids 775–960 (Burmese strain, genotype 1) of HEV ORF1 was expressed in *Escherichia coli*. The purified protein showed ADPR-1″ phosphatase activity and *in vitro* binding to poly(ADP-ribose) and ADP ribose.[227,228] The affinity for poly(ADP-ribose) was high as compared with ADP ribose. However, the significance of these functions is not yet clear.

4.1.8.1.7 HVR

The hypervariable region overlaps with the proline-rich sequence located between the N-terminus of the X-domain and the C-terminal portion of the putative PCP domain. It varies both in length and sequence among different HEV strains. HEV can tolerate small deletions in the HVR[229]; however, the replication efficiency of HVR deletion mutants was found to be reduced when tested in Huh7 cells.[230] HVR sequences were found to be interchangeable between different HEV genotypes, resulting in differential replication efficiencies. From these observations, it was suggested that the HVR probably could influence the efficiency of HEV replication by interacting with viral and/or host factors.

4.1.8.1.8 The Capsid Protein

ORF2 codes for the structural (capsid) protein (pORF2) of the predicted mass of approximately 72 kDa (660 amino acids) and has three putative glycosylation sites at the asparagine residues, Asn137, Asn310, and Asn562. The pORF2 amino-terminal signal sequence and all three *N*-linked glycosylation sites are universally conserved in all isolates of human HEV, suggesting the functional significance of glycosylation.[231] The N-terminal part of pORF2 contains a putative ER-directing signal peptide and is rich in codons for arginine and highly basic in charge.[178] It is synthesized as a precursor and co-translationally translocated into the ER, where it is processed into the mature protein through signal peptide cleavage.[235] ORF2 was reported to occur in glycosylated and nonglycosylated forms (88 and 72 kDa, respectively) when expressed in animal cells in culture.[232] The proposed encapsidation of the viral genome by pORF2 was experimentally demonstrated by Surjit et al.[233] by the specific binding of N-terminal 111 amino acids. Many constructs corresponding to the short and long regions of ORF2 have been expressed in *E. coli*. These include the p239 (amino acids 268–607), E2 (amino acids 394–606), E2a (amino acids 459–660), and E2s (amino acids 455–602). All of these constructs fold to form higher-order structures. By carrying out truncations at the N-terminus, C-terminus, or both, it was proved that amino acids 126–601 are essential elements required for the initiation of the VLP assembly. X-ray studies on ORF2 crystals obtained recently showed that capsid protein monomers are composed of three linear domains: the shell domain (S) (amino acids 129–319), the middle domain (M) (amino acids 320–455), and the protruding domain (P) (amino acids 456–606), harboring the key viral neutralizing domain.[234,235]

4.1.8.1.9 ORF3 Protein

HEV ORF3 is translated from the bicistronic subgenomic RNA.[211] It overlaps with ORF2 at its 3′ end and is most variable among the different HEV strains. It is a small 114-amino acid phosphoprotein that remains associated with the cytoskeleton.[236] ORF3 contains two large hydrophobic domains—D1 (amino acids 7–23) and D2 (amino acids 28–53) at the N-terminus—and two proline-rich domains—P1 (amino acids 66–77) and P2 (amino acids 95–111) toward the end.[237,238] Of these, the cysteine-rich D1 domain was shown to be required for the association of ORF3 to the cytoskeleton[244] to bind microtubules[239] and mitogen-activated protein kinase (MAPK)-phosphates,[240] while the D2 domain was shown to bind to the hemopexin and proposed to aid in viral infection by affecting cellular iron homeostasis.[241] The C-terminal region of the ORF3 protein is multifunctional and appears to be involved in virion morphogenesis and pathogenesis. Using an *in vitro* replication system, it was shown that the ORF3 protein is not essential for replication in cell lines;[242,243] however, intrahepatic inoculations of these constructs in rhesus macaques revealed that ORF3 is essential for the establishment of infection.[211] ORF3 protein was detected on the surface of HEV particles produced in cell culture, and these ORF3-coated viral particles were seen to be lighter than uncoated particles, indicating a probable interaction with lipids.[244] ORF3 coat was present on the virus particles present in the serum, but not in the feces of HEV-infected individuals, suggesting that the lipid-ORF3 "coat" was shed as the virus passed through the enteric system due to exposure to bile salts. An intact PSAP motif was shown to be required for the formation of membrane-associated HEV particles with the ORF3 protein on their surface.[244] Taken together, these results suggest that the ORF3 protein is a multifunctional protein that appears to play an important role in both HEV replication and pathogenesis.

4.1.8.2 cis-Regulatory Genomic Regions

4.1.8.2.1 5′NCR

The 5′NCR of the HEV genome is only 25–27 nt in length, has a methylated cap, and plays a role in the initiation of HEV replication. Mapping experiments have revealed that ORF2 binds to the 76 nt region at the 5′ end of the HEV genome. This region includes the 51 nt sequence, conserved across alphaviruses. Secondary structure predictions and the location of the ORF2 binding region within the HEV genome indicate that this interaction may play a role in viral encapsidation.[233]

4.1.8.2.2 3′NCR

The interaction of viral RdRp with 3′NCR and the adjacent region containing two stem-loop (SL) structures (SL1 and SL2) and its importance in virus replication were documented by Agrawal et al.[225] Using gel shift assays, they could demonstrate that RdRp bound specifically to 3′NCR, while 3′NCR without a poly A tail failed to bind with RdRp. Similarly, a single nucleotide change in the SL2 of the genotype 1 viral genome significantly affected virus replication, indicating the important role of both poly A tail and SL structures in HEV replication.[244,245]

4.1.8.2.3 Putative Subgenomic Promoter

A junction region (JR) between ORF1 and ORF2 has a highly conserved SL structure found to be important for HEV replication.[210,246] The putative subgenomic promoter required for the synthesis of the subgenomic mRNA is located in this region of the HEV genome. Graff et al.[245] reported that silent mutations in this region abolish the synthesis of ORF2 and ORF3 proteins, thus giving the first evidence of this region being the regulatory region. An in-depth study involving genotype 3 swine HEV and mutants with changes in the junction region identified a double SL structure of approximately 50 nucleotides in this region.[246]

4.1.8.2.4 Subgenomic RNA

Graff et al.[211] detected two positive-sense RNA species, approximately 7.3 and 2.2 kb in size, in HEV replicating cells. The 2.2-kb subgenomic RNA was capped, and its 5′ end matched the nucleotide number 5122 in genotype 1, SAR55 HEV genome. With these results, this group proposed the ORF2 start at 5145 and 5131 as the ORF3 start, and the translation of ORF3 and ORF3 proteins from the same bicistronic subgenomic mRNA. This model was confirmed by using the intrahepatic inoculation of wild-type and mutant genotype 3 swine HEV replicons into pigs by Huang et al.[210]

4.1.9 HEV Life Cycle

The proposed model of life cycle is mainly based on the existing knowledge of well-characterized similar positive-sense RNA viruses. The target cells are hepatocytes; however, several extra hepatic tissues also support replication. Virus capsid protein is believed to be involved in binding to a cell receptor for entry. It is suggested that the C-terminal portion of pORF2 may bind to heat shock cognate protein 70 (HSPC70) on the cell surface to initiate cell entry.[247] Heparan sulfate proteoglycans (HSPGs) appear to be required as attachment factors.[248] The intracellular trafficking following entry is also poorly understood; HSP90 and tubulin appear to be involved in this process.[249] A recent study has documented that HEV enters through clathrin-mediated endocytosis.[250] However, it is still not clear whether these are specific receptors or just attachment factors. Holla et al.[251] have recently reported that following internalization, the HEV-LP initially traffics into Rab5-positive compartments *en route* to acidic lysosomal compartments. HEV entry requires dynamin-2, clathrin, membrane cholesterol, and actin, but is independent of factors associated with macropinocytosis.

Once viral RNA is released in the cytosol, ORF1 translation is initiated by the cap-dependent recruitment of ribosomes. It is not clear whether the ORF1 encoded nonstructural polyprotein functions as a single protein or whether it is processed into individual functional units. However, the regions predicted to encode viral methyltransferase, helicase, and RdRp produce functionally active proteins when expressed in heterologous systems. The genomic RNA is copied into a negative-sense RNA intermediate by viral RdRp. These RNA intermediates then serve as templates for the synthesis of genomic as well as subgenomic positive-sense RNA species. Genomic intermediates have been detected in replicon-transfected cells[220] in the livers of experimentally infected macaques[252] and pigs.[155] It was demonstrated using a replicon system that there is an alternate cycle of positive- and negative-sense RNA synthesis.[253] Single subgenomic RNA is translated to synthesize ORF2 and ORF3 proteins.[210,211,245] The ORF2 protein packages the genomic positive-sense RNA into progeny virions. Immunocapture PCR analysis showed the association of ORF3 on the surface of the cell-culture-generated HEV, which also showed a lower density than the ORF3 deficient virus. These observations suggested that the ORF3 protein is present on the virion surface in association with cellular lipids and probably plays some role during viral egress.[254] Emerson et al.[243] and Nagashima et al.[255] have documented the importance of the PSAP motif within the P2 domain of the ORF3 protein in the virus egress and the probable role of SRC homology 3 signaling pathways in HEV maturation and egress. PSAP motif is required for the formation of membrane-associated HEV particles with ORF3 on their surface, which is mediated by cellular Tsg101 protein.[256,257] These findings suggest that HEV follows the vacuolar protein sorting pathway and uses cellular proteins such as Tsg101 for its release from infected cells.

4.2 Tools That Can Be Used for Studying HEV

Molecular mechanisms of HEV replication, cell surface receptors, tissue/species specificity of different viral strains, and immunopathogenesis of HEV are still not understood. Researchers have used tools such as established cell lines, primary cells lines, and animal models including nonhuman primates, pigs, rabbits, rats, and chickens to understand the different aspects of HEV biology. Although not very efficient, these models have helped in shedding some light on the various aspects of virus pathogenesis, immune response, virus–host interactions, and virus replication.

4.2.1 Use of Infectious cDNA Clones of HEV

HEV has been a difficult virus to grow in cell culture. Reverse genetics approach is a tool that can be very useful in such situations. Direct genetic manipulation using infectious cDNA clones has been an indispensable tool for studying HEV replication and pathogenesis. The development of infectious cDNA clones has allowed the exploration of the structural and functional relationship of HEV genes. The successful development of a genotype 1 full-genome cDNA clone encompassing the complete HEV genome from an Indian isolate (Accession Number AF076239) was performed by Panda et al.[217] Transfection studies were carried out in HepG2 cells using *in vitro* transcribed uncapped RNA transcripts. The presence of negative-sense RNA, indicative of viral replication, was demonstrated in the transfected cells. The culture supernatant from the transfected cells was able to produce HEV infection in rhesus monkeys following intravenous injection, indicating the successful generation of viable HEV particles following the transfection of cells with *in vitro* synthesized HEV genomic RNA.

The second report of an infectious clone from the Sar-55 strain of genotype 1 HEV (Pakistan) was reported by Emerson et al.[244] This recombinant genome could establish infection in rhesus macaques and chimpanzees, and developed symptoms like hepatitis. They used capped transcripts for these studies. It was controversial whether capping of RNA transcripts was essential for the infectivity since a previous study by Panda et al. showed the successful generation of infectious particles using uncapped RNA transcripts. Subsequently, Emerson et al.[258] reported that HEV capped full-genome transcripts are 32–38 times more efficiently translated than their uncapped counterparts. Several primate cell lines such as PLC/PRF/5, Huh-7, Caco-2, HepG2/3CA, FKRHK, Vero, and AGMK were shown to support HEV replication; however, none of the nonprimate cells were able to support replication, indicating the requirement of species-specific factors for HEV replication. Further studies from the same group using an infectious cDNA tool demonstrated that the highly conserved regions of HEV ORF3 harbor *cis*-reactive regulatory elements and that mutations in this region eliminate ORF2 and ORF3 synthesis. A single unique nucleotide change within the stem structure at the 3′ end of the HEV genome similarly reduced the efficiency of replication in both Huh-7 cells and rhesus macaques.[245,257] Quantitative measurement of HEV replication became possible due to the development of fused ORF2-GFP, ORF2-luciferase subgenomic HEV replicons.[246,258,259]

Successful replication of a virus in a host is a complex phenomenon. With the wider range of host species for genotype 3 and 4 HEVs, it was evident that these viruses are comparatively flexible in their host specificity than genotype 1 and 2 viruses which have a very narrow host range. It is still not understood how HEV determines its host specificity at the molecular level. A recent report proposes that the host restriction for the genotype 1 virus could be due to efficiency of the virus to synthesize the ORF2 protein in the given host.[260] Lack of compatibility between cell surface receptor and receptor-binding region (456–605 amino acids) in the capsid protein of the virus was suggested as the deciding factor. This group used genotype 3 virus replicon (P6), developed from the Kernow C-1 virus,[261] for developing chimeric genotype 1/3 viral genomes. P6 is known to have a 171-nucleotide insertion (a sequence from the human S17 ribosomal protein encoding gene) in the HVR region of the viral genome. It was previously reported by this group that the P6 virus, that was isolated from a chronic HEV case, can cross the species barrier.[136] Ability of the S17 sequence to alter species specificity was confirmed by inserting this sequence in SAR55 (genotype 1). This insertion enabled the replication of the chimeric virus in LLC-PK (pig) cells, which otherwise was refractory to SAR55 1 virus infection.

Pigs being the most promising model for HEV studies, full-length infectious cDNA clones were also developed from genotype 3 swine HEV.[262] Capped transcripts from these clones were demonstrated to be competent for forming infectious virions in Huh-7 cells, and these virions were able to establish infection in pigs. Rescue of a genotype 4 human HEV from cloned cDNA in Huh7 cells and the infectivity of this virus in HepG2/C3A cells were reported by Cordoba et al.[263] Capped *in vitro* transcripts from this clone could also establish infection in pigs. Zhu et al.[264] recently reported cloning of the genotype 4 virus and successful infection in rats.

Considering chickens as the practical animal model for HEV pathogenesis and replication studies, cDNA clones from avian viruses were developed.[265] Direct intrahepatic inoculation of RNA transcripts from these infectious cDNA clones could establish infections in chickens and showed fecal virus shedding, viremia, seroconversion, and histopathological lesions characteristic of avian HEV infection.

4.2.2 HEV Cell-Culture Models

HEV has proven to be very refractory to cell culture. Establishment of an efficient cell-culture system that will facilitate HEV propagation is very critical for understanding HEV biology. HEV propagation has been attempted by many researchers in primary hepatocytes from nonhuman primates (chimpanzees, cynomolgus macaques, tamarins, and African green monkeys)[266,267] in cell lines, such as human normal embryonic liver cells (WRL68), human hepatoma cell lines (PLC/PRF/5, HepG2, and Huh-7 cells), human colon carcinoma cells (Caco-2), human embryo lung diploid cells (2BS), human lung embryonic fibroblast cells (MRC-5), human lung cancer cells (A549), human chorio carcinoma cells (HCCM), African green monkey kidney cells (Vero), and Rhesus monkey kidney cells (LLC-MK2).[268–273] However, these cell-culture systems either were nonpermissive or failed to support the generation of an infectious progeny virus or were not very efficient.

4.2.2.1 Genotype 3 Virus Culture Systems

Recently, an efficient cell-culture system for HEV replication has been reported. A fecal sample taken from a sporadic acute hepatitis E case from Japan with a high HEV load (the JE03–1760F strain of genotype 3, 2×10^7 copies/ml per 10% fecal suspension) was used as an inoculum to evaluate the replication in 21 different established cell lines derived from humans, monkeys, cows, dogs, rats, and mice, including three human hepatocellular carcinoma cell lines (Huh7, HepG2, and PLC/PRF/5). The virus efficiently replicated in two cell lines, PLC/PRF/5 and A549 from human lung cancer.[274] PLC/PRF/5 and A549 cells showed the presence of HEV on days 12 and 14 after inoculation. HEV progeny virus released in cell-culture supernatants could be passaged >50 times in both cell lines to achieve cell titers as high as 10^{-9}–10^{10} copies/mL. The successful culturing of the virus was attributed to the high viral load of the fecal sample, which possibly helped selecting viral variant from the pool which efficiently adapted to the cell culture as it had favorable mutations. The JE03–1760F strain harbored the 29 unique point mutations, which were not possessed by any of the reported genotype 3 HEV strains. Yamada et al.[254] constructed an infectious cDNA clone of the virus successfully.

This cell-culture system was utilized to assess the thermal stability of HEV, and it was observed that HEV incubated at 56°C for 30 min remained infectious, while the virus incubated at higher than 70°C was completely inactivated. Further, it was used to show that convalescent serum samples obtained from patients infected with HEV genotypes 1, 3, or 4 can neutralize genotype 3 virus, indicating the broad cross-reactivity of anti-HEV antibodies. Serum samples obtained from patients 8.7 to 24.0 years after the onset of HEV infection also prevented the propagation of HEV in PLC/PRF/5 cells, suggesting the presence of long-lasting anti-HEV antibodies with a neutralizing activity.[274] However, the use of this system for various applications is yet to be established. It is known that HEV does not cause any cytopathic changes in the cell culture irrespective of the cells used.

4.2.2.2 Kernow C1 Virus Strain

A genotype 3 virus strain, Kernow C-1, isolated from a chronically infected patient was shown to grow efficiently in HepG2/C3A cells by Shukla et al.[136] This virus contained a 58-amino acid human S17 ribosomal protein sequence insert in the HVR1 region of the viral genome. An infectious cDNA clone of the virus was developed to analyze the significance of insertion.[261] Their mutagenesis studies showed that the S17 insertion was mainly responsible for the cell-culture adaptation. Introduction of 54 synonymous mutations into the insert had no detectable effect, thus implicating the protein, rather than the RNA, as the important component responsible for altered abilities of the virus replication. Both the sequence length and the amino acid composition of the insert were important in increasing the replication efficiency. Substitution of the S17 sequence by a different ribosomal protein sequence or by GTPase-activating protein sequence resulted in the partial enhancement of virus replication. On insertion of this S17 sequence in genotype 1 cDNA clone, the chimeric virus acquired the ability to cross the species barrier and replicated in hamster cells (BHK-21), though with a low efficiency.

4.2.2.3 3D-Cell-Culture System

Berto et al.[89] used a rotating wall vessel for the culture of PLC/PRF/5 cells in a three-dimensional (3D) configuration and infected them with HEV obtained from the liver of an experimentally infected pig containing 1.38×10^6 genome equivalents/mL of the virus. This study demonstrated that HEV can replicate efficiently in these cells for up to 5 months as seen by the presence of a virus in the supernatants. The virions generated from the culture were able to infect fresh 3D cultures, showing that this method is able to produce infectious hepatitis E virions. Since these researchers were not able to cultivate genotype 3 HEV on the monolayer cultures of HepG2/C3A and PLC/PRF5 cells as described by Tanaka et al.,[274] it suggested the possibility of these cells being permissive for only selective viral strains and not for all. Overall, existing HEV cell-culture systems need to be evaluated with a wide range of HEV strains.

4.2.2.4 Primary Hepatocytes as the Culture System

The propagation and production of HEV *in vitro* have been attempted by many researchers in primary hepatocytes from nonhuman primates as well as in cells from other tissues (chimpanzees, cynomolgus macaques, tamarins, and African green monkeys), and in Rhesus monkey kidney cells (LLC-MK2).[20,159,243,249,266–273] Overall, the results concluded that HEV is refractory to the cell culture. However, recent results from Rogee et al.[275] are encouraging. Their group used a human hepatoma-derived cell line, HepaRG, and a porcine embryonic stem-cell-derived cell line, PICM-19, for HEV replication studies. These cells have morphological and functional properties similar to primary hepatocytes. These researchers used matrigel-embedded cell cultures to retain a hepatocyte-like morphology and liver-specific gene expression patterns of the cells. These cells supported the HEV replication and release of virions. However, the virus release in the supernatants of infected cell cultures was not as efficient as reported by Tanaka et al. (10^3 vs. 10^7). Using these two cell lines, HEV suspensions heated at 56°C for 60 min were shown to be noninfectious. This result is different from previous studies, where HEV genotype 1 or HEV genotype 3 present in the pork liver or stool sample were found to be infectious after the 56°C treatment.[108,273,274] Since this culture system was unable to generate high titers of infectious progeny virions, its usefulness as a model system would be limited.

4.2.2.5 Genotype 4 Virus Culture System

After the successful development of a culture system for genotype 3 virus by Okamoto's laboratory in Japan, their group opted for a similar strategy to test the genotype 4 virus culture in PLC/PRF/5 and A549 cell lines. A HEV RNA-positive fecal sample from a fulminant hepatitis E patient was used as a virus source for culturing experiments (the HE-JF5/15F strain of genotype 4, 1.3×10^7 copies/mL).[276] The genotype 4 HE-JF5/15F virus strain was found to be even better in its replication efficiency as compared with the JE03–1760F strain. This culture system will be very useful in delineating the role of viral factors with respect to the recently observed association of genotype 4 virus with fulminant hepatitis E cases in Japan.

Zhang et al.[277] have recently reported a swine cell-culture system for genotype 4 virus. They used rectal swabs/liver as the virus source and infected swine cells (IBRS-2, derived from swine kidney, ATCCCRL-1835) and human cells, A549. HEV RNA was detectable for up to 12 passages inIBRS-2 cells and 24 passages in A549 cells. Visible cytopathic effects could be seen from passages 8–12 in infected IBRS-2 cells, and passages 22–24 in infected A549 cells. However, the authors did not check for the release of infectious virions in the culture supernatants.

4.2.2.6 Propagation of HEV Strains Circulating in Blood

The cell-culture systems developed for genotype 3 and 4 viruses used primarily fecal samples as the virus source. In order to check whether the virus present in the serum would grow as efficiently as the virus that is shed in feces, Takahashi and his group[278] studied the replication of HEV strains from serum samples in PLC/PRF/5 and A549 cells and found that HEV strains of genotypes 1, 3, or 4 can replicate efficiently in PLC/PRF/5 and A549 cells. HEV strains in all serum samples tested, with or without concurrent HEV antibodies, were successfully propagated in cultured cells when inoculated at an HEV load of $\geq 10^5$ copies/well of a 6-well plate. Progeny viruses in the culture supernatant could be successfully passaged in the same cells, indicating their infectious nature. Viral concentrations of $<10^5$ particles were not efficient, which could be a major hurdle in using this culture system. Moreover, it is not easy to get viremic serum samples due to transient viremia during HEV infection.

Overall, it is observed that HEV does not cause any cytopathic changes in the cell culture irrespective of the cells used. Use of current cell-culture systems for various applications is yet to be established. Further, the reproducibility of these culture systems in different laboratories is apparently not uniform. Robust cell-culture systems for genotype 1 and 2 viruses are still lacking. Virus infectivity assays are still dependent on either PCR-based assays, IFA or FACS, which would mean that there is a requirement for a well-equipped laboratory for studying these viruses. Considering the abundance of these viruses in developing countries and the major need of such tools there, we still need to develop practical HEV study tools. Table 4.2 gives an overview of different cell-culture models used in HEV studies.

TABLE 4.2

Occurrence of HEV in Animal Species

Animal Species	HEV Genotype	Serological Evidence (Anti HEV)	Experimental Susceptibility	Applications as Animal Model for HEV Studies
Cynomolgus monkey	Human 1,2 Rabbit HEV	Positive	Yes	Pathogenesis, molecular biology, and cross-species infection
Rhesus monkeys	Human 1,2,3,4 Swine 3,4 Avian HEV	Positive	Yes	Vaccine, pathogenesis, molecular biology, and cross-species infection
Chimpanzee	Human 1,2,3,4 Swine 3,4	Positive	Yes	Pathogenesis, molecular biology, and cross-species infection
Tamarins	Human 1,2	Positive	Yes	Not reported
Owl monkey	Human 1,2	Positive	Yes	Cross-species infection
Vervets	Human 1,2	Positive		Infection studies
Squirrel monkeys	Human 1,2	Positive	Yes	Not reported
Domestic swine	Genotype 3,4	Positive	Yes	Vaccine, pathogenesis, molecular biology, and cross-species infection
Wild boar	Genotype 3,4	Positive	Yes	Vaccine, pathogenesis, molecular biology, cross-species infection
Rabbit	Genotype 3	Positive	Yes	Vaccine, pathogenesis, and cross-species infection
Chicken	Avian HEV	Positive	Yes	Vaccine, pathogenesis, molecular biology, and cross-species infection
Turkey	Avian HEV	Positive	Yes	No reports
Rodents	Unknown	Positive	Yes	Vaccine, cross-species infection
Rat	Human 1,2,3,4 Swine 3 Avian HEV	Positive	Yes	Molecular biology and pathogenesis studies
Balb/c mice	Swine 4	Positive	Yes	Molecular biology and pathogenesis studies
C57BL/6	Human 1 Swine 3,4	Positive	Yes	
Mongolian gerbil	Genotype 4	Positive	Yes	Pathogenesis, molecular biology, and cross-species infection
Deer	Genotype 3,4	Positive	Not reported	Not reported
Goat	Genetically divergent from the known HEV strains	Positive	Not reported	Not reported

(Continued)

TABLE 4.2 (*Continued*)
Occurrence of HEV in Animal Species

Animal Species	HEV Genotype	Serological Evidence (Anti HEV)	Experimental Susceptibility	Applications as Animal Model for HEV Studies
Dog	Genetically divergent from the known HEV strains	Positive	Not reported	Not reported
Cats	Genetically divergent from the known HEV strains	Positive	Not reported	Not reported
Bats	Bat HEV	Positive	Not reported	Not reported
Ferrets	Genetically divergent from the known HEV strains	Positive	Not reported	Not reported
Fish	Genotype 3	Positive	Not reported	Not reported
Camel	Unknown new HEV genotype	Positive	Not reported	Not reported
Sheep	Genotype 4	Positive	Not reported	Not reported
Foxes	Divergent from known HEV genotypes	Positive	Not reported	Not reported
Horses	Genotype 3	Positive	Not reported	Not reported

4.2.3 HEV Animal Models

4.2.3.1 Nonhuman Primates

Rhesus/cynomolgus macaques and chimpanzees have been used to understand HEV infectivity and pathogenesis. Animal species differ in virus excretion, liver enzyme elevation, and histopathological changes in the liver.[131,148,224] There are diverse outcomes of experimental infections done in different nonhuman primate species. Tamarins were found to be inconsistent in developing HEV infection, while chimpanzees,[20] pig-tailed macaques,[148] vervets,[148] owl monkeys,[147] squirrel monkeys,[148] and patas monkeys were found to be susceptible. All known mammalian HEV genotypes (1–4) and swine HEV genotypes 3 (United States)[155,156] and 4 (India)[157,158] have been transmitted to nonhuman primates. Among primates, chimpanzees, cynomolgus, and rhesus macaques were found to be the best animal models[110,148] since experimental HEV infections result in elevated liver enzymes and histopathologic changes in the liver of these animals similar to hepatitis E patients. Although chimpanzees have been used to study the clinical course, and host gene responses to HEV infection,[167] the use of this model system is not practical.[110,148,279] Although HEV is transmitted naturally via the feco–oral route, this route is not preferred for experimental infections in primates as this requires a much higher virus dose. Most studies in nonhuman primates have utilized the intravenous route for virus inoculation. Direct intrahepatic inoculation of chimpanzees and rhesus monkeys with RNA transcripts from full-length HEV cDNA clones has helped in establishing a direct link between the disease and the genetic material of HEV by inducing virologic, pathologic, and serologic characteristics typical of hepatitis E infection.

The disease course of HEV infection in nonhuman primates was comparable to that in humans, without leading to chronicity. The incubation period to peak serum levels of liver enzymes is generally 3–8 weeks, but shorter and longer incubation periods have been observed.[110,149] Enzyme elevations are usually unimodal; however bimodal curves have also been observed.[146–149] Peak viremia and peak shedding of HEV into the feces occurs during the incubation period and early acute phase of the disease.[110,148,151] Detection of HEV antigens in the liver generally precedes or parallels viremia and fecal shedding.[150,151] The clinical presentation of hepatitis E infection in nonhuman primates was dose dependent. Severity of infection was directly related to the infectivity titer of the challenge virus. For demonstration of consistent hepatitis E infection in nonhuman primates, challenge virus doses at least 1000 times greater than the minimum dose needed for infection were required. Also, the oral route of inoculation was not efficient and required a 10,000-fold higher challenge virus than that was used for the intravenous inoculation of nonhuman primates.[195] The observed high mortality (approximately 20% particularly in the third trimester) of hepatitis E during human pregnancy was not reproducible in pregnant rhesus macaques.[153,280]

Rhesus monkeys infected with one Indian HEV isolate were found to be immune 1.5–2.75 years after the primary infection when challenged with closely related Indian strains.[1] Further, a long-term serological follow-up of HEV immunity in experimentally infected rhesus monkeys indicated that anti-HEV IgG titers declined to undetectable levels, but still protected the animals from homologous and heterologous HEV challenge 5 years after the initial infection.[64] It was previously documented that naturally acquired anti-HEV in nonhuman primates renders them resistant to experimental infection.[153]

In general, rhesus monkeys and cynomolgus macaques were found to be most suitable model systems for hepatitis E vaccine efficacy studies.[195,281–286] Macaques immunized with an HEV candidate vaccine developed using one genotype showed protection against homologous and heterologous HEV challenges. The noteworthy vaccine studies that were taken further for phase 2 and phase 3 trials in populations of hepatitis E endemic countries were from Nepal and China.

Cumbersome procurement procedures, limited animal resources, and ethical concerns have limited the use of nonhuman primate models in HEV research today.

4.2.3.2 Pigs

HEV research and animal model development got a boost after Meng et al.[159] discovered a novel strain of the virus in pigs. As discussed earlier, swine serve as a major reservoir for zoonotic genotype 3 and 4 HEVs. Wild boars are assumed to be a natural reservoir of HEV. Although attempts to infect pigs with genotype 1 or 2 of human HEVs were not successful,[155] they were susceptible to genotype 3 and 4 human HEV strains under experimental conditions.[156,287] The pig model system has been successfully used to study the structural and functional relationship of HEV genes and in understanding the mechanism of HEV replication, pathogenesis, and species specificity.

Pathogenesis study with the specific-pathogen-free (SPF) pig model using a genotype 3 human HEV strain revealed more severe and persistent hepatic lesions (multifocal lymphoplasmacytic hepatitis and focal hepatocellular necrosis) than those infected with the swine HEV strain.[156] Extra hepatic sites of HEV replication observed were the small intestine, lymph nodes, and colon.[288,289] Pigs also shed virus in feces, became viremic, and seroconverted to IgG anti-HEV when infected with human HEV genotype 4 (Taiwanese strain).[287] The HEV pig model was used to determine the infectivity of the first infectious clone of HEV genotype 3.[262] This model also helped in the identification of the actual initiation site of the HEV ORF3. Recently, HEV genotype 3 mutants with deletions in the HVR were shown to be infectious in pigs, indicating dispensability of the HVR during HEV replication and infection. Thus, pigs seem to be a promising model of HEV infection, as they are relatively inexpensive to house, feed, maintain, and breed than nonhuman primates. However, since they do not reproduce hepatic disease with overt clinical symptoms, their usefulness in understanding HEV pathogenesis will be limited.

4.2.3.3 Rodents

The rat strain of HEV was first identified in Hamburg, Germany, in 2009.[41] Huang et al.[290] have reported the successful infection of Balb/c nude mice with a strain of genotype 4 HEV recovered from a pig; however, Li et al.[291] failed to infect C57BL/6 mice with genotype 1, 3, and 4 HEV strains. Wistar rats could be infected with the human strain of HEV,[292] but subsequent studies aimed to confirm that the transmission of genotypes 1, 2, and 4 of human HEV and an avian HEV was unsuccessful. Mongolian gerbils (desert rats) were recently shown to be susceptible to experimental infection with a genotype 4 strain of HEV.[293] Viremia and fecal virus shedding were detected in inoculated gerbils, and HEV antigen was detected in the liver, kidneys, spleen, and small intestine. Naturally infected rats do not show any HEV-related illness. In experimental infections, laboratory rats infected with rat HEV showed seroconversion and fecal shedding of the virus without any apparent clinical signs. Histopathological analysis of tissues revealed mild portal inflammation, foci of parenchymal necrosis, and aggregates of lymphocytes, and Kupffer cells within the lobules, indicating evidence of mild hepatitis consistent with human acute HEV infection. Availability of a small animal model would give a boost to HEV research. Although laboratory mice and rats have been found to be susceptible to HEV, the lack of reproducibility is the major drawback, and more work is needed in optimizing this model.

4.2.3.4 Rabbits

The potential of rabbits as an animal model was tested after the discovery of genotype 3 virus in farmed rabbits from the Gansu province in China.[26] Subsequent to these findings, HEV strains related to genotype 3 were detected in farms as well as in wild rabbits in the United States and France. Rabbits inoculated with rabbit HEV shed virus in the feces, become viremic, and have elevated levels of serum ALT with irregular multifocal lymphohistiocytic infiltrates and local hepatocellular necrosis.[294] Rabbits experimentally inoculated with human HEV genotype 4 showed viremia and fecal virus shedding in 2/9 rabbits but in none inoculated with genotype 1 human HEV.[294] This indicated that the rabbit model of HEV infection may be useful for studying genotype 3 and 4 HEVs; however, its use for studying genotypes 1 and 2 would be limited.

Han et al.[295] utilized the rabbit model to see whether they can establish an animal model for chronic HEV infections. On infecting SPF rabbits with a homologous rabbit HEV isolate and a heterologous genotype 4 swine HEV, they observed the development of chronic hepatitis and an associated liver fibrosis in some rabbits infected with homologous HEV. Persistent fecal shedding of the virus and elevated liver enzymes were evident for more than 6 months after infection in the chronic infections. The detection of both the positive-/negative-sense RNA and HEV antigen in the extra hepatic tissues such as brain, stomach, duodenum, and kidney further confirmed similarity with human chronic HEV infection. However, the underlying mechanism was not clear. In humans, HEV can establish chronic infections only in immunocompromised individuals, this chronic disease pathology was not observed in the rabbits infected with the heterologous genotype 4 swine HEV. Rabbit HEV has 31 amino acid insertions in the X-domain, which probably broadened the tissue specificity of the virus as previously documented by Shukla et al.[261] in the case of chronic HEV infection in human patients.

4.2.3.5 Ferrets

HEV was first detected in ferrets (*Mustela putorius*) in Netherlands[43] followed by detection in the United States. Phylogenetic analysis has shown that ferret HEV strains are closer to rat strains. Nucleotide sequence analyses indicated that the ferret HEV genome shares the highest nucleotide sequence identity (72.3%) with rat HEV. A putative ORF4 was observed in the ferret HEV genome similar to the rat HEV genome. Information about ferret HEV epidemiology, distribution, transmission, and pathogenesis is still not available. Usefulness of ferret HEV as an animal model needs to be evaluated considering its small size and extensive use as an animal model for studying the pathogenesis of respiratory viruses.

4.2.3.6 Chickens

Haqshenas et al.[296] discovered avian HEV and facilitated the use of chickens as a model system for avian HEV infection. Avian HEV has been shown to cross species barriers and infect turkeys.[163] Although chickens are not susceptible to human strains, upon inoculation with avian HEV, infection results in subcapsular hemorrhages, focal lymphocytic periphlebitis, and phlebitis in the liver; extra hepatic sites of HEV replication are also evident.[297] A recent study by Kwon et al.[265] showed that an infectious cDNA clone of the variant avian HEV (isolated from healthy chicken) could still induce histological liver lesions. This study also proved that RNA transcripts may serve as an alternative for live virus in animal pathogenesis studies.

Avian ORF2 capsid protein expressed in *E. coli* having a similar antigenic structure as that of human HEV containing major neutralizing epitopes was assessed for immunogenicity in chickens. All the tested mock-immunized control chickens developed typical avian HEV infection characterized by viremia, fecal virus shedding, and seroconversion to avian HEV antibodies, while none of the tested chickens immunized with avian HEV capsid protein had detectable viremia, fecal virus shedding, or observable gross hepatitis lesions. The results from this study suggested that immunization of chickens with the avian HEV recombinant ORF2 capsid protein with aluminum as the adjuvant can induce protective immunity against avian HEV infection.[298] Overall, avian HEV seem to be useful for HEV studies.

Animal models such as chimpanzees, rhesus/cynomolgus macaques, owl monkeys, rodents, pigs, chickens, and rabbits have been used for studying HEV biology. Although nonhuman primates and pigs seem to be useful animal models, ethical issues, high prices, and difficulties in handling, manipulating, and housing these animals are major hurdles. Rats seem to be a promising laboratory model due to their small size and easy handling. Table 4.2 shows different animal models being used for HEV studies.

4.3 Conclusions

HEV transmission occurs primarily due to contaminated water in developing countries while transmission via uncooked or partially cooked pig meat and meat products is mainly responsible for hepatitis E infection in developed countries. For waterborne transmission, effective prevention and control depends on ensuring a safe drinking water supply, adequate sanitation, and proper personal and environmental hygiene. For the prevention of food-borne infections, a stringent screening of meat products is needed. HEV infections in pigs are mostly silent and difficult to detect with traditional meat-screening procedures. In such situations, a quick on-site diagnostic test is needed. PCR-based detection is comparatively time consuming and needs well-equipped laboratories.

The proper and timely diagnosis of human HEV infections in nonendemic regions is technically challenging. Lack of consistency of serological tests and viral load quantification in terms of sensitivity and specificity are the limiting factors. Future research should focus on developing broadly applicable serological and molecular assays for hepatitis E diagnosis.

The unsolved issues such as the relative importance of HEV transmission pathways, inactivation properties of the virus, and dose–response relationship of HEV infection need to be addressed. It is essential to know the factors that lead to clinical hepatitis E infection in humans. For this, there is a need for a practical animal model that would mimic the natural course of HEV infection in humans and its outcome. With the ever expanding host range of HEV, we hope to find new animal strains that can be used as model systems. Development of a reliable cell-culture system(s) that would support the complete life cycle of HEV will help in knowing the virus better. Most therapeutic approaches and discovery efforts to suppress HEV propagation will not progress unless we have these basic tools.

REFERENCES

1. Arankalle VA, et al., Age-specific prevalence of antibodies to hepatitis A and E viruses in Pune, India, 1982 and 1992, *J Infect Dis*, 171, 447–450, 1995.
2. Chandra V, Taneja S, Kalia M, Jameel S, Molecular biology and pathogenesis of hepatitis E virus, *J Biosci*, 33, 451–464, 2008.
3. Purcell RH, Emerson SU, Animal models of hepatitis A and E, *ILAR J*, 42, 161–177, 2001.
4. Vishwanathan R, Infectious hepatitis in Delhi (1955–56): a critical study: epidemiology, *Indian J Med Res*, 45, 49–58, 1957.
5. Sreenivasan MA, et al., Epidemiological investigations of an outbreak of infectious hepatitis in Ahmedabad city during 1975–76, *Indian J Med Res*, 67, 197–206, 1978.
6. Khuroo MS, Study of an epidemic of non-A, non-B hepatitis possibility of another human hepatitis virus distinct from post-transfusion non-A, non-B type, *Am J Med*, 68, 818–824, 1980.

7. Wong DC, et al., Epidemic and endemic hepatitis in India: evidence for a non-A, non-B hepatitis etiology, *Lancet*, 2, 876–879, 1980.
8. Chadha MS, Arankalle VA, Banerjee K, Follow up of cases of enterically transmitted non-A, non-B hepatitis, *J Assoc Phys India*, 39, 651–652, 1991.
9. Naik SR, Aggarwal R, Salunke PN, Mehrotra NN, A large waterborne viral hepatitis E epidemic in Kanpur, India, *Bull World Health Organ*, 70, 597–604, 1992.
10. Arankalle VA, Goverdhan MK, Banerjee K, Antibodies against hepatitis E virus in Old World monkeys, *J Viral Hepat*, 1, 125–129, 1994.
11. Khuroo MS, Duermeyer W, Zargar SA, Ahanger MA, Shah MA, Acute sporadic non-A, non-B hepatitis in India, *Am J Epidemiol*, 118, 360–364, 1983.
12. Arankalle VA, et al., Aetiology of acute sporadic non-A, non-B viral hepatitis in India, *J Med Virol*, 40, 121–125, 1993.
13. Chadha MS, Walimbe AM, Chobe LP, Arankalle VA, Comparison of etiology of sporadic acute and fulminant viral hepatitis in hospitalized patients in Pune, India during 1978–81 and 1994–97, *Indian Gastroenterol*, 22, 11–15, 2003.
14. Kim JH, Nelson KE, Panzner U, Kasture Y, Labrique AB, Wierzba TF, A systematic review of the epidemiology of hepatitis E virus in Africa, *BMC Infect Dis*, 14, 308, 2014.
15. Teshale EH, et al., A large outbreak of hepatitis E in northern Uganda, *Emerg Infect Dis*, 16, 126–129, 2010.
16. MMWR, Investigation of hepatitis E outbreak among refugees—Upper Nile, South Sudan, 2012–2013, *MMWR*, 62(29), 581–586, 2013.
17. Guthmann JP, et al., A large outbreak of hepatitis E among a displaced population in Darfur, Sudan, 2004: the role of water treatment methods, *Clin Infect Dis*, 42, 1685–1691, 2006.
18. Purcell RH, Emerson SU, Hepatitis E: an emerging awareness of an old disease, *J Hepatol*, 48, 495–503, 2008.
19. Okamoto H, Genetic variability and evolution of hepatitis E virus, *Virus Res*, 127, 216–228, 2007.
20. Arankalle, VA, Chadha, MS, Mehendale, SM, Banerjee, K, Outbreak of enterically transmitted non-A, non-B hepatitis among schoolchildren, *Lancet*, 2, 1199–1200, 1988.
21. Hyams KC, et al., Acute sporadic hepatitis E in Sudanese children: analysis based on a new western blot assay, *J Infect Dis*, 165, 1001–1005, 1992.
22. Arankalle VA, Chadha MS, Dama BM, Tsarev SA, Purcell RH, Banerjee K, Role of immune serum globulins in pregnant women during an epidemic of hepatitis E, *J Viral Hepat*, 5, 199–204, 1998.
23. Clayson ET, et al., Rates of hepatitis E virus infection and disease among adolescents and adults in Kathmandu, Nepal, *J Infect Dis,* 176, 763–766, 1997.
24. Cossaboom C, Cordoba L, Dryman B, Meng X-J, Hepatitis E virus in rabbits, Virginia, USA, *Emerg Infect Dis*, 17, 2047–2049, 2011.
25. Nakamura M, et al., Hepatitis E virus infection in wild mongooses of Okinawa, Japan: demonstration of anti-HEV antibodies and a full genome nucleotide sequence, *Hepatol Res*, 34, 137–140, 2006.
26. Zhao C, et al., A novel genotype of hepatitis E virus, *J Med Virol*, 81, 1371–1379, 2009.
27. Wang Y, et al., A divergent genotype of hepatitis E virus in Chinese patients with acute hepatitis, *J Gen Virol,* 80, 169–177, 1999.
28. Meng XJ, From barnyard to food table: the omnipresence of hepatitis E virus and risk for zoonotic infection and food safety, *Virus Res*, 161(1), 23–30, 2011.
29. Wang YC, et al., Prevalence, isolation, and partial sequence analysis of hepatitis E virus from domestic animals in China, *J Med Virol*, 67, 516–521, 2002.
30. Balayan MS, Epidemiology of hepatitis E virus infection, *J Viral Hepat*, 4, 155–165, 1997.
31. Johne R, Plenge-Bönig A, Hess M, Ulrich RG, Reetz J, Schielke A, Detection of a novel hepatitis E-like virus in faeces of wild rats using a nested broad-spectrum RT-PC, *J Gen Virol*, 91, 750–758, 2010.
32. Sonoda H, et al., Prevalence of hepatitis E virus (HEV) infection in wild boars and deer and genetic identification of a genotype 3 HEV from a boar in Japan, *J Clin Microbiol*, 42, 5371–5374, 2004.
33. Wang H, et al., Recombination analysis reveals a double recombination event in hepatitis E virus, *Virol J*, 7, 129, 2010.
34. Chandler JD, Riddell MA, Li F, Love RJ, Anderson DA, Serological evidence for swine hepatitis E virus infection in Australian pig herds, *Vet Microbiol*, 68, 95–105, 1999.
35. de Deus N, et al., Hepatitis E virus infection dynamics and organic distribution in naturally infected pigs in a farrow-to-finish farm, *Vet Microbiol*, 132, 19–28, 2008.

36. Kaci S, Nockler K, Johne, R, Detection of hepatitis E virus in archived German wild boar serum samples, *Vet Microbiol*, 128, 380–385, 2008.
37. Martelli, F, et al., Detection of hepatitis E virus (HEV) in a demographic managed wild boar (*Sus scrofa scrofa*) population in Italy, *Vet Microbiol*, 126, 74–81, 2008.
38. Michitaka K, et al., Prevalence of hepatitis E virus among wild boar in the Ehime area of western Japan, *Hepatol Res*, 37, 214–220, 2007.
39. Rutjes SA, et al., Seroprevalence and molecular detection of hepatitis E virus in wild boar and red deer in the Netherlands, *J Virol Methods*, 168, 197–206, 2010.
40. Drexler J, et al., Bats worldwide carry hepatitis E virus-related viruses that form a putative novel genus within the family *Hepeviridae*, *J Virol*, 86, 9134–9147, 2012.
41. Takahashi M, et al., Analysis of the full-length genome of a hepatitis E virus isolate obtained from a wild boar in Japan that is classifiable into a novel genotype, *J Gen Virol*, 92, 902–908, 2011.
42. Raj V, et al., Novel hepatitis E virus in ferrets, the Netherlands, *Emerg Infect Dis*, 18, 1369–1370, 2012.
43. Batts W, Yun S, Hedrick R, Winton J, A novel member of the family *Hepeviridae* from cutthroat trout (*Oncorhynchus clarkii*), *Virus Res*, 158, 116–123, 2011.
44. Lu L, Li C, Hagedorn CH, Phylogenetic analysis of global hepatitis E virus sequences: genetic diversity, subtypes and zoonosis, *Rev Med Virol*, 16, 5–36, 2006.
45. Purdy MA, Khudyakov YE, The molecular epidemiology of hepatitis E virus infection, *Virus Res*, 161, 31–39, 2011.
46. Johne R, et al., Novel hepatitis E virus genotype in Norway rats, Germany, *Emerg Infect Dis*, 16, 1452–1455, 2011.
47. Meng X-J, Zoonotic and foodborne transmission of hepatitis E virus, *Semin Liv Dis*, 33, 41–49, 2013.
48. Yugo DM, Cossaboom CM, Meng XJ, Naturally occurring animal models of human hepatitis E virus infection, *ILAR J*, 55, 187–199, 2014.
49. Marek A, Bilic I, Prokofieva I, Hess M, Phylogenetic analysis of avian hepatitis E virus samples from European and Australian chicken flocks supports the existence of a different genus within the *Hepeviridae* comprising at least three different genotypes, *Vet Microbiol*, 145, 54–61, 2010.
50. Kwon H, Sung H, Meng XJ, Serological prevalence, genetic identification, and characterization of the first strains of avian hepatitis E virus from chickens in Korea, *Virus Genes*, 45, 237–245, 2012.
51. Clemente-Casares P, et al., Hepatitis E virus epidemiology in industrialized countries, *Emerg Infect Dis*, 9, 448–454, 2003.
52. Albinana-Gimenez N, et al., Distribution of human polyomaviruses, adenoviruses, and hepatitis E virus in the environment and in a drinking water treatment plant, *Environ Sci Technol*, 40, 7416–7422, 2006.
53. Gironés R, Desire for information in the elderly: interactions with patients, family, and physicians, *J Canc Educ*, doi:10.1007/s13187-014-0760-5, 2014.
54. Teo CG, Much meat, much malady: changing perceptions of the epidemiology of hepatitis E, *Clin Microbiol Infect*, 16, 24–32, 2010.
55. Parashar D, Khalkar P, Arankalle VA, Survival of hepatitis A and E viruses in soil samples, *Clin Microbiol Infect*, 17, e1–e4, 2011.
56. Rutjes SA, et al., Sources of hepatitis E virus genotype 3 in the Netherlands, *Emerg Infect Dis*, 15, 381–387, 2009.
57. McCreary C, Martelli F, Grierson S, Ostanello F, Nevel A, Banks M, Excretion of hepatitis E virus by pigs of different ages and its presence in slurry stores in the United Kingdom, *Vet Rec*, 163, 261–265, 2008.
58. El-Esnawy N, Examination for hepatitis E virus in wastewater treatment plants and workers by nested RT-PCR and ELISA, *J Egypt Public Health Assoc*, 75, 219–231, 2000.
59. Vaidya SR, Tilekar BN, Walimbe AM, Arankalle VA, Increased risk of hepatitis E in sewage workers from India, *J Occup Environ Med*, 45, 1167–1170, 2003.
60. Arankalle VA, Chadha MS, Mehendale SM, Tungatkar SP, Epidemic hepatitis E: serological evidence for lack of intrafamilial spread, *Indian J Gastroenterol*, 19, 24–28, 2000.
61. Aggarwal R, Naik SR, Hepatitis E: intrafamilial transmission versus waterborne spread, *J Hepatol*, 21, 718–723, 1994.
62. Robson SC, Adams S, Brink N, Woodruff B, Bradley D, Hospital outbreak of hepatitis E, *Lancet*, 339, 1424–1425, 1992.
63. Khuroo MS, Kamili S, Aetiology and prognostic factors in acute liver failure in India, *J Viral Hepat*, 10, 224–231, 2003.

64. Arankalle VA, Paranjape S, Emerson SU, Purcell RH, Walimbe AM, Phylogenetic analysis of hepatitis E virus isolates from India (1976–1993), *J Gen Virol*, 80, 1691–1700, 1999.
65. Boxall E, et al., Transfusion-transmitted hepatitis E in a "non-hyperendemic" country, *Transfus Med*, 16, 79–83, 2006.
66. Colson P, Coze C, Gallian P, Henry M, De Micco P, Tamalet C, Transfusion-associated hepatitis E, France, *Emerg Infect Dis*, 13, 648–649, 2007.
67. Matsubayashi K, et al., A case of transfusion-transmitted hepatitis E caused by blood from a donor infected with hepatitis E virus via zoonotic food-borne route, *Transfusion*, 48, 1368–3675, 2008.
68. Mitsui T, et al., Prevalence of hepatitis E virus infection among hemodialysis patients in Japan: evidence for infection with a genotype 3 HEV by blood transfusion, *J Med Virol*, 74, 563–572, 2004.
69. Tamura A, et al., Persistent infection of hepatitis E virus transmitted by blood transfusion in a patient with T-cell lymphoma, *Hepatol Res*, 37, 113–120, 2007.
70. Psichogiou M, et al., Hepatitis E virus infection in individuals at high risk of transmission of non-A, non-B hepatitis and sexually transmitted diseases, *Scand J Infect Dis*, 28, 443–445, 1996.
71. Barzilai A, et al., Hepatitis E virus infection in hemophiliacs, *J Med Virol*, 46, 153–156, 1995.
72. Hauser L, et al., Hepatitis E transmission by transfusion of intercept blood system-treated plasma, *Blood,* 123, 796–797, 2014.
73. Hewitt PE, et al., Hepatitis E virus in blood components: a prevalence and transmission study in southeast England, *Lancet*, doi:10.1016/S0140-6736(14) 61034–61035, 2014.
74. Meng XJ, Hepatitis E virus: animal reservoirs and zoonotic risk, *Vet Microbiol*, 140, 256–265, 2010.
75. Bouquet J, et al., Close similarity between sequences of hepatitis E virus recovered from humans and swine, France, 2008–2009, *Emerg Infect Dis*, 17, 2018–2025, 2011.
76. Balayan MS, Usmanov R, Zamyatina N, Brief report: experimental hepatitis E infection in domestic pigs, *J Med Virol*, 32, 58–59, 1990.
77. Meng X-J, Zoonotic and xenozoonotic risks of the hepatitis E virus, *Infect Dis Rev*, 2, 35–41, 2000.
78. Vulcano A, et al., HEV prevalence in the general population and among workers at zoonotic risk in Latium Region, *Ann Ig*, 19, 181–186, 2007.
79. Bouwknegt M, et al., Estimation of hepatitis E virus transmission among pigs due to contact-exposure, *Vet Res*, 39, 40, 2008.
80. Galiana C, Fernandez-Barredo S, Garcia A, Gomez MT, Perez-Gracia MT, Occupational exposure to hepatitis E virus (HEV) in swine workers, *Am J Trop Med Hyg*, 78, 1012–1015, 2008.
81. Yazaki Y, et al., Sporadic acute or fulminant hepatitis E in Hokkaido, Japan, may be food-borne, as suggested by the presence of hepatitis E virus in pig liver as food, *J Gen Virol*, 84, 2351–2357, 2003.
82. Miyashita K, et al., Three cases of acute or fulminant hepatitis E caused by ingestion of pork meat and entrails in Hokkaido, Japan: zoonotic food-borne transmission of hepatitis E virus and public health concerns, *Hepatol Res,* 42, 870–878, 2012.
83. Li TC, et al., Hepatitis E virus transmission from wild boar meat, *Emerg Infect Dis*, 11, 1958–1960, 2005.
84. Matsuda H, Okada K, Takahashi K, Mishiro S, Severe hepatitis E virus infection after ingestion of uncooked liver from a wild boar, *J Infect Dis*, 188, 944, 2003.
85. Tamada Y, Yano K, Yatsuhashi H, Inoue O, Mawatari F, Ishibashi H, Consumption of wild boar linked to cases of hepatitis E, *J Hepatol*, 40, 869–870, 2004.
86. Hijioka S, Sato Y, Iwashita Y, Indou Y, A case of acute hepatitis E who had a history of frequent ingestion of raw meat and viscera from wild deer and boars, *Nihon Shokakibyo Gakkai Zasshi*, 102, 723–728, 2005.
87. Colson P, et al., Pig liver sausage as a source of hepatitis E virus transmission to humans, *J Infect Dis*, 202, 825–834, 2010.
88. Renou C, et al., Association Nationale des Hépato-Gastroentérologues des Hôpitaux Généraux (ANGH), Prospective study of hepatitis E virus infection among pregnant women in France, *Virol J*, 11, 68, 2014.
89. Berto A, et al., Hepatitis E virus in pork liver sausage, France, *Emerg Infect Dis*, 19, 264–266, 2013.
90. Kulkarni MA, Arankalle VA, The detection and characterization of hepatitis E virus in pig livers from retail markets of India, *J Med Virol*, 80, 1387–1390, 2008.
91. Feagins AR, Opriessnig T, Guenette DK, Halbur PG, Meng XJ, Detection and characterization of infectious hepatitis E virus from commercial pig livers sold in local grocery stores in the USA, *J Gen Virol*, 88, 912–917, 2007.

92. Wenzel JJ, Preiss J, Schemmerer M, Huber B, Plentz A, Jilg W, Detection of hepatitis E virus (HEV) from porcine livers in southeastern Germany and high sequence homology to human HEV isolate, *J Clin Virol*, 52, 50–54, 2011.
93. Wilhelma B, et al., Preliminary molecular epidemiological 2 investigation of hepatitis E virus 3 sequences from Québec, Canada, *Prev Vet Med*, 118, 359–369, 2014.
94. Kaba M, et al., Hepatitis E virus infection in sheltered homeless persons, France, *Emerg Infect Dis*, 16, 1761–1763, 2010.
95. Okano H, et al., Characterization of sporadic acute hepatitis E and comparison of hepatitis E virus genomes in acute hepatitis patients and pig liver sold as food in Mie, Japan, *Hepatol Res*, 44, E63–E76, 2014.
96. Di Bartolo I, Angeloni G, Ponterio E, Ostanello F, Ruggeri FM, Detection of hepatitis E virus in pork liver sausages, *Int J Food Microbiol*, 193, 29–33, 2015.
97. Bouwknegt M, et al., Hepatitis E virus RNA in commercial porcine livers in the Netherlands, *J Food Prot*, 70, 2889–2895, 2007.
98. Donia D, et al., Presence of hepatitis E RNA in mussels used as bio-monitors of viral marine pollution, *J Virol Methods,* 186, 198–202, 2012.
99. Crossan C, Baker PJ, Craft J, Takeuchi Y, Dalton HR, Scobie L, Hepatitis E virus genotype 3 in shellfish, United Kingdom, *Emerg Infect Dis*, 18, 2085–2087, 2012.
100. Song YJ, et al., Analysis of complete genome sequences of swine hepatitis E virus and possible risk factors for transmission of HEV to humans in Korea, *J Med Virol*, 82, 583–591, 2010.
101. Li TC, Miyamura T, Takeda N, Detection of hepatitis E virus RNA from the bivalve Yamato-Shijimi (*Corbicula japonica*) in Japan, *Am J Trop Med Hyg*, 76, 170–172, 2007.
102. Cacopardo B, Russo R, Preiser W, Benanti F, Brancati G, Nunnari A, Acute hepatitis E in Catania (eastern Sicily) 1980–1994, the role of hepatitis E virus, *Infection*, 25, 313–316, 1997.
103. Thorne ET, Williams E, Diseases and endangered species: the blackfooted ferret as a recent example, *Conserv Biol*, 2, 6673, 1988.
104. Koizumi Y, et al., Infection of a Japanese patient by genotype 4 hepatitis E virus while traveling in Vietnam, *J Clin Microbiol*, 42, 3883–3885, 2004.
105. Said B, et al., Hepatitis E Incident Investigation Team, Hepatitis E outbreak on cruise ship, *Emerg Infect Dis*, 15, 1738–1744, 2009.
106. Brassard J, Gagné MJ, Généreux M, Côté C, Detection of human food-borne and zoonotic viruses on irrigated, field-grown strawberries, *Appl Environ Microbiol*, 78, 3763–3766, 2012.
107. Kokkinos P, et al., Harmonised investigation of the occurrence of human enteric viruses in the leafy green vegetable supply chain in three European countries, *Food Environ Virol*, 4, 179–191, 2012.
108. Emerson SU, Arankalle VA, Purcell RH, Thermal stability of hepatitis E virus, *J Infect Dis*, 192, 930–933, 2005.
109. Barnaud E, Rogee S, Garry P, Rose N, Pavio N, Thermal inactivation of infectious hepatitis E virus in experimentally contaminated food, *Appl Environ Microbiol*, 78, 5153–5159, 2012.
110. Balayan MS, et al., Evidence for a virus in non-A, non-B hepatitis transmitted via the fecal-oral route, *Intervirology*, 20, 23–31, 1983.
111. Chauhan A, Jameel S, Dilawari JB, Chawla YK, Kaur U, Ganguly NK, Hepatitis E virus transmission to a volunteer, *Lancet*, 341, 149–150, 1993.
112. Aggarwal R, Jameel S, Hepatitis E, *Hepatology*, 54, 2218–2222, 2011.
113. Pal R, Aggarwal R, Naik SR, Das V, Das S, Naik S, Immunological alterations in pregnant women with acute hepatitis E, *J Gastroenterol Hepatol*, 20, 1094–1101, 2005.
114. Bhatia V, Singhal A, Panda SK, Acharya SK, A 20-year single-center experience with acute liver failure during pregnancy: is the prognosis really worse, *Hepatology*, 48, 1577–1585, 2008.
115. Navaneethan U, Al Mohajer M, Shata MT, Hepatitis E and pregnancy: understanding the pathogenesis, *Liver Int*, 28, 1190–1199, 2008.
116. Kar P, et al., Does hepatitis E viral load and genotypes influence the final outcome of acute liver failure during pregnancy? *Am J Gastroenterol*, 103, 2495–2501, 2008.
117. Kumar A, Beniwal M, Kar P, Sharma JB, Murthy NS, Hepatitis E in pregnancy, *Int J Gynecol Obstet*, 85, 240–241, 2004.
118. Kumar Acharya S, et al., Hepatitis E virus (HEV) infection in patients with cirrhosis is associated with rapid decompensation and death, *J Hepatol*, 46, 387–394, 2007.

119. Khuroo MS, Teli MR, Skidmore S, Sofi MA, Khuroo MI, Incidence and severity of viral hepatitis in pregnancy, *Am J Med*, 70, 252–255, 1981.
120. Myint H, et al., A clinical and epidemiological study of an epidemic of non-A, non-B hepatitis in Rangoon, *Am J Trop Med Hyg*, 34, 1183–1189, 1985.
121. Jilani N, et al., Hepatitis E virus infection and fulminant hepatic failure during pregnancy, *J Gastroenterol Hepatol*, 22, 676–682, 2007.
122. Patra S, Kumar A, Trivedi SS, Puri M, Sarin SK, Maternal and fetal outcomes in pregnant women with acute hepatitis E virus infection, *Ann Intern Med*, 147, 28–33, 2007.
123. Chibber RM, Usmani MA, Al-Sibai MH, Should HEV infected mothers breast feed? *Arch Gynecol Obstet*, 270, 15–20, 2004.
124. Stoszek SK, et al., Prevalence of and risk factors for hepatitis C in rural pregnant Egyptian women, *Trans R Soc Trop Med Hyg*, 100, 102–107, 2006.
125. Andersson G, Carlbring P, Grimlund A, Predicting treatment outcome in Internet versus face to face treatment of panic disorder, *Comput Human Behav*, 24, 1790–1801, 2008.
126. Anty R, et al., First case report of an acute genotype 3 hepatitis E infected pregnant woman living in South-Eastern France, *J Clin Virol*, 54, 76–78, 2012.
127. Mateos Lindemann ML, et al., Fulminant hepatitis E in a woman taking oral contraceptive medication, *Am J Trop Med Hyg*, 82, 12–15, 2010.
128. Renou C, Pariente A, Cadranel JF, Nicand E, Pavio N, Clinically silent forms may partly explain the rarity of acute cases of autochthonous genotype 3c hepatitis E infection in France, *J Clin Virol*, 51(2), 139–141, 2011.
129. Mizuo H, et al., Possible risk factors for the transmission of hepatitis E virus and for the severe form of hepatitis E acquired locally in Hokkaido, Japan, *J Med Virol*, 76, 341–349, 2005.
130. Jeblaoui A, Haim-Boukobza S, Pause A, Mokhtari C, Nicand E, Roque-Afonso A, Emerging hepatitis E genotype 4 infection in France, *J Hepatol*, 58, S405, 2013.
131. Purcell RH, et al., Pathobiology of hepatitis E: lessons learned from primate models, *Emerg Microbes Infect*, 2, e9, doi:10.1038/emi.2013.9, 2013.
132. Inoue J, et al., Analysis of the full-length genome of genotype 4 hepatitis E virus isolates from patients with fulminant or acute self-limited hepatitis E, *J Med Virol*, 78, 476–484, 2006.
133. Inoue J, et al., Nucleotide substitutions of hepatitis E virus genomes associated with fulminant hepatitis and disease severity, *Tohoku J Exp Med*, 218, 279–284, 2009.
134. Takahashi K, et al., Virulent strain of hepatitis E virus genotype 3, Japan, *Emerg Infect Dis*, 15, 704–709, 2009.
135. Mishra N, Walimbe AM, Arankalle VA, Hepatitis E virus from India exhibits significant amino acid mutations in fulminant hepatic failure patients, *Virus Genes*, 46, 47–53, 2013.
136. Shukla P, et al., Cross-species infections of cultured cells by hepatitis E virus and discovery of an infectious virus-host recombinant, *Proc Natl Acad Sci USA*, 108, 2438–2443, 2011.
137. Nguyen HT, et al., A naturally occurring human/hepatitis E recombinant virus predominates in serum but not in feces of a chronic hepatitis E patient and has a growth advantage in cell culture, *J Gen Virol*, 93, 526–530, 2012.
138. Haagsma EB, et al., Chronic hepatitis E virus infection in liver transplant recipients, *Liver Transplant*, 14, 547–553, 2008.
139. Kamar N, et al., Hepatitis E virus and chronic hepatitis in organ-transplant recipients, *N Engl J Med*, 358, 811–817, 2008.
140. Dalton HR, Bendall R, Keane F, Tedder R, Ijaz S, Persistent carriage of hepatitis E virus in patients with HIV infection, *N Engl J Med*, 361, 1025–1027, 2009.
141. le Coutre P, Meisel H, Hofmann J, Reactivation of hepatitis E infection in a patient with acute lymphoblastic leukaemia after allogeneic stem cell transplantation, *Gut*, 58, 699–702, 2009.
142. Ollier L, Tieulie N, Sanderson F, Chronic hepatitis after hepatitis E virus infection in a patient with non-Hodgkin lymphoma taking Rituximab, *Ann Intern Med*, 150, 430–431, 2009.
143. Yugo DM, Meng XJ, Hepatitis E virus: foodborne, waterborne and zoonotic transmission, *Int J Environ Res Public Health*, 10, 4507–4533, 2013.
144. Geng Y, et al., Persistent hepatitis E virus genotype 4 infection in a child with acute lymphoblastic leukemia, *Hepatitis Monthly*, 14, e15618, 2014.

145. Naik A, Gupta N, Goel D, Ippagunta SK, Sharma RK, Aggarwal R, Lack of evidence of hepatitis E virus infection among renal transplant recipients in a disease-endemic area, *J Viral Hepat*, 20, e138–e140, 2013.
146. Bradley DW, et al., Enterically transmitted non-A, non-B hepatitis: serial passage of disease in cynomolgus macaques and tamarins and recovery of disease-associated 27- to 34-nm virus like particles, *Proc Natl Acad Sci USA*, 84, 6277–6281, 1987.
147. Ticehurst J, et al., Infection of owl monkeys (*Aotus trivirgatus*) and cynomolgus monkeys (*Macaca fascicularis*) with hepatitis E virus from Mexico, *J Infect Dis*, 165, 835–845, 1992.
148. Tsarev SA, et al., Variation in course of hepatitis E in experimentally infected cynomolgus monkeys, *J Infect Dis*, 167, 1302–1306, 1993.
149. Uchida T, et al., Serial transmission of a putative causative virus of enterically transmitted non-A, non-B hepatitis to *Macaca fascicularis* and *Macaca mulatto*, *Jpn J Exp Med*, 60, 13–21, 1990.
150. Krawczynski K, Bradley DW, Enterically transmitted non-A, non-B hepatitis: identification of virus-associated antigen in experimentally infected cynomolgus macaques, *J Infect Dis*, 159, 1042–1049, 1989.
151. Longer CF, et al., Experimental hepatitis E: pathogenesis in cynomolgus macaques (*Macaca fascicularis*), *J Infect Dis*, 168, 602–609, 1993.
152. Arankalle VA, et al., Etiological association of a virus-like particle with enterically transmitted non-A, non-B hepatitis, *Lancet*, 1, 550–554, 1988.
153. Tsarev SA, et al., Experimental hepatitis E in pregnant rhesus monkeys: failure to transmit hepatitis E virus (HEV) to offspring and evidence of naturally acquired antibodies to HEV, *J Infect Dis*, 172, 31–37, 1995.
154. Li RC, et al., Seroprevalence of hepatitis E virus infection, rural southern People's Republic of China, *Emerg Infect Dis*, 12(11), 1682–1688, 2006.
155. Meng XJ, et al., Genetic and experimental evidence for cross-species infection by swine hepatitis E virus, *J Virol*, 72, 9714–9721, 1998.
156. Halbur PG, et al., Comparative pathogenesis of infection of pigs with hepatitis E viruses recovered from a pig and a human, *J Clin Microbiol*, 39, 918–923, 2001.
157. Arankalle VA, Chobe LP, Joshi MV, Chadha MS, Kundu B, Walimbe AM, Human and swine hepatitis E viruses from western India belong to different genotypes, *J Hepatol*, 36 417–425, 2002.
158. Arankalle VA, Chobe LP, Chadha MS, Type-IV Indian swine HEV infects rhesus monkeys, *J Viral Hepat*, 13, 742–745, 2006.
159. Meng J, Dubreuil P, Pillot J, A new PCR-based seroneutralization assay in cell culture for diagnosis of hepatitis E, *J Clin Microbiol*, 35, 1373–1377, 1997.
160. Leblanc D, et al., Presence of hepatitis E virus in a naturally infected swine herd from nursery to slaughter, *Int J Food Microbiol*, 117, 160–166, 2007.
161. Arankalle VA, Chobe LP, Walimbe AM, Yergolkar PN, Jacob GP, Swine HEV infection in South India and phylogenetic analysis (1985–1999), *J Med Virol*, 69, 391–396, 2003.
162. Huang FF, et al., Heterogeneity and seroprevalence of a newly identified avian hepatitis E virus from chickens in the United States, *J Clin Microbiol*, 40, 4197–4202, 2002.
163. Sun ZF, et al., Generation and infectivity titration of an infectious stock of avian hepatitis E virus (HEV) in chickens and cross-species infection of turkeys with avian HEV, *J Clin Microbiol*, 42, 2658–2662, 2004.
164. Peralta B, et al., Anti-HEV antibodies in domestic animal species and rodents from Spain using a genotype 3-based ELISA, *Vet Microbiol*, 137, 66–73, 2009.
165. Nan Y, Yu Y, Ma Z, Khattar SK, Fredericksen B, Zhang YJ, Hepatitis E virus inhibits type I interferon induction by ORF1 product, *J Virol*, 88, 11924–11932, 2014.
166. Nan Y, et al., Enhancement of interferon induction by ORF3 product of hepatitis E virus, *J Virol*, 88, 8696–8705, 2014.
167. Yu C, et al., Pathogenesis of hepatitis E virus and hepatitis C virus in chimpanzees: similarities and differences, *J Virol*, 84, 11264–11278, 2010.
168. Srivastava R, Aggarwal R, Bhagat MR, Chowdhury A, Naik S, Alterations in natural killer cells and natural killer T cells during acute viral hepatitis E, *J Viral Hepat*, 15, 910–916, 2008.
169. Tripathy AS, Das R, Rathod SB, Gurav YK, Arankalle VA, Peripheral T regulatory cells and cytokines in hepatitis E infection, *Eur J Clin Microbiol Infect Dis*, 31, 179–184, 2012.

170. Rathod SB, Thanapati S, Arankalle VA, Tripathy AS, Suppressive activity and altered conventional phenotype markers/mediators of regulatory T cells in patients with self-limiting hepatitis E, *J Viral Hepat*, 21, 141–151, 2014.
171. Prabhu SB, et al., Study of cellular immune response against hepatitis E virus (HEV), *J Viral Hepat*, 18, 587–594, 2011.
172. Husain MM, Aggarwal R, Kumar D, Jameel S, Naik S, Effector T cells immune reactivity among patients with acute hepatitis E, *J Viral Hepat*, 18, e603–e608, 2011.
173. Srivastava R, Aggarwal R, Sachdeva S, Alam MI, Jameel S, Naik S, Adaptive immune responses during acute uncomplicated and fulminant hepatitis E, *J Gastroenterol Hepatol*, 26, 306–311, 2011.
174. Saravanabalaji S, Tripathy AS, Dhoot RR, Chadha MS, Kakrani AL, Arankalle VA, Viral load, antibody titers and recombinant open reading frame 2 protein-induced TH1/TH2 cytokines and cellular immune responses in self-limiting and fulminant hepatitis E, *Intervirology*, 52, 78–85, 2009.
175. Suneetha PV, et al., Effect of peptide pools on effector functions of antigen-specific CD8+ T cells, *J Immunol Methods*, 342, 33–48, 2009.
176. Mishra N, Arankalle VA, Association of polymorphisms in the promoter regions of TNF-α (–308) with susceptibility to hepatitis E virus and TNF-α (–1031) and IFN-γ (+874) genes with clinical outcome of hepatitis E infection in India, *J Hepatol*, 55, 1227–1234, 2011.
177. Dawson GJ, Chau KH, Cabal CM, Yarbough PO, Reyes GR, Mushahwar IK, Solid-phase enzyme-linked immunosorbent assay for hepatitis E virus IgG and IgM antibodies utilizing recombinant antigens and synthetic peptides, *J Virol Methods*, 38, 175–186, 1992.
178. Tam AW, et al., Hepatitis E virus (HEV): molecular cloning and sequencing of the full-length viral genome, *Virology*, 185, 120–131, 1991.
179. Favorov MO, et al., Serologic identification of hepatitis E virus infections in epidemic and endemic settings, *J Med Virol*, 36, 246–250, 1992.
180. Chau KH, Dawson GJ, Bile KM, Magnius LO, Sjogren MH, Mushahwar IK, Detection of IgA class antibody to hepatitis E virus in serum samples from patients with hepatitis E virus infection, *J Med Virol*, 40, 334–338, 1993.
181. Orrù G, Masia G, Orrù G, Romanò L, Piras V, Coppola RC, Detection and quantitation of hepatitis E virus in human faeces by real-time quantitative PCR, *J Virol Methods*, 118, 77–82, 2004.
182. Enouf V, et al., Validation of single real-time TaqMan PCR assay for the detection and quantitation of four major genotypes of hepatitis E virus in clinical specimens, *J Med Virol*, 78, 1076–1082, 2006.
183. Gyarmati P, Mohammed N, Norder H, Blomberg J, Belák S, Widén F, Universal detection of hepatitis E virus by two real-time PCR assays: TaqMan and primer-probe energy transfer, *J Virol Methods*, 146, 226–235, 2007.
184. Zhao C, et al., Comparison of real-time fluorescent RT-PCR and conventional RT-PCR for the detection of hepatitis E virus genotypes prevalent in China, *J Med Virol*, 79, 1966–1973, 2007.
185. Jothikumar N, Cromeans TL, Robertson BH, Meng XJ, Hill VR, A broadly reactive one-step real-time RT-PCR assay for rapid and sensitive detection of hepatitis E virus, *J Virol Methods*, 131, 65–71, 2006.
186. Huang S, et al., Profile of acute infectious markers in sporadic hepatitis E, *PLoS One*, 5, e13560, 2010.
187. Baylis SA, Hanschmann KM, Blumel J, Nubling CM, Standardization of hepatitis E virus (HEV) nucleic acid amplification technique-based assays: an initial study to evaluate a panel of HEV strains and investigate laboratory performance, *J Clin Microbiol*, 49, 1234–1239, 2011.
188. Emerson SU, Purcell RH, Hepatitis E virus, *Rev Med Virol*, 13, 145–154, 2003.
189. Arankalle VA, Lole KS, Deshmukh TM, Chobe LP, Gandhe SS, Evaluation of human (genotype 1) and swine (genotype 4)-ORF2-based ELISAs for anti-HEV IgM and IgG detection in an endemic country and search for type 4 human HEV infections, *J Viral Hepat*, 14, 435–445, 2007.
190. Herremans M, et al., Swine-like hepatitis E viruses are a cause of unexplained hepatitis in the Netherlands, *J Viral Hepat*, 14, 140–146, 2007.
191. Yarbrough PO, et al., Hepatitis E virus: identification of type-common epitopes, *J Virol*, 65, 5790–5797, 1991.
192. Zhou YH, Purcell RH, Emerson SU, An ELISA for putative neutralizing antibodies to hepatitis E virus detects antibodies to genotypes 1, 2, 3, and 4, *Vaccine*, 22, 2578–2585, 2004.
193. Drobeniuc J, et al., Serologic assays specific to immunoglobulin M antibodies against hepatitis E virus: pangenotypic evaluation of performances, *Clin Infect Dis*, 1(51), e24–e27, 2010.

194. Abravanel F, et al., Performance of anti-HEV assays for diagnosing acute hepatitis E in immunocompromised patients, *J Clin Virol*, 58, 624–628, 2013.
195. Tsarev SA, et al., Infectivity titration of a prototype strain of hepatitis E virus in cynomolgus monkeys, *J Med Virol*, 43, 135–142, 1994.
196. Shrestha MP, et al., Safety and efficacy of a recombinant hepatitis E vaccine, *N Engl J Med*, 356, 895–903, 2007.
197. Purcell RH, et al., Pre-clinical immunogenicity and efficacy trial of a recombinant hepatitis E vaccine, *Vaccine*, 21, 2607–2615, 2003.
198. Zhu FC, et al., Efficacy and safety of a recombinant hepatitis E vaccine in healthy adults: a large-scale, randomised, double-blind placebo-controlled, phase 3 trial, *Lancet*, 376, 895–902, 2010.
199. Kamar N, et al., Hepatitis E virus-induced neurological symptoms in a kidney-transplant patient with chronic hepatitis, *Am J Transplant*, 10, 1321–1324, 2010.
200. Neukam K, et al., Chronic hepatitis E in HIV patients: rapid progression to cirrhosis and response to oral ribavirin, *Clin Infect Dis*, 57, 465–468, 2013.
201. Mallet V, et al., Brief communication: case reports of ribavirin treatment for chronic hepatitis E, *Ann Intern Med*, 153, 85–89, 2010.
202. Pischke S, Behrendt P, Bock C, Jilg W, Manns MP, Wedemeyer H, Hepatitis E in Germany—an underreported infectious disease, *Deutsches Ärzteblatt*, 111, 577–583, 2014.
203. Dalton HR, Keane FE, Bendall R, Mathew J, Ijaz S, Treatment of chronic hepatitis E in a patient with HIV infection, *Ann Intern Med*, 155, 479–480, 2011.
204. Peron JM, Dalton H, Izopet J, Kamar N, Acute autochthonous hepatitis E in western patients with underlying chronic liver disease: a role for ribavirin? *J Hepatol*, 54, 1323–1324, 2011.
205. Kumar A, Saraswat VA, Hepatitis E and acute-on-chronic liver failure, *J Clin Exp Hepatol*, 3, 225–230, 2013.
206. Mahtab EA, et al., Podoplanin deficient mice show a rhoa-related hypoplasia of the sinus venosus myocardium including the sinoatrial node, *Dev Dyn*, 238, 183–193, 2009.
207. Duseja A, et al., Nonalcoholic fatty liver in a developing country is responsible for significant liver disease, *Hepatology*, 52, 2248–2249, 2010.
208. El Sayed Zaki M, Othman W, Role of hepatitis E infection in acute on chronic liver failure in Egyptian patients, *Liver Int*, 31, 1001–1005, 2011.
209. Goyal R, Kumar A, Panda SK, Paul SB, A Charaya SK, Ribavirin therapy for hepatitis E virus induced acute on chronic liver failure: a preliminary report, *Antivir Ther*, 17, 1091–1096, 2012.
210. Huang YW, Opriessnig T, Halbur PG, Meng XJ, Initiation at the third in-frame AUG codon of open reading frame 3 of the hepatitis E virus is essential for viral infectivity in vivo, *J Virol*, 81, 3018–3026, 2007.
211. Graff J, Torian U, Nguyen H, Emerson SU, A bicistronic subgenomic mRNA encodes both the ORF2 and ORF3 proteins of hepatitis E virus, *J Virol*, 80, 5919–5926, 2006.
212. Koonin EV, Gorbalenya AE, Purdy MA, Rozanov MN, Reyes GR, Bradley DW, Computer-assisted assignment of functional domain in the nonstructural polyprotein of hepatitis E virus: delineation of an additional group of positive-strand RNA plant and animal viruses, *Proc Natl Acad Sci USA*, 89, 8259–8263, 1992.
213. Magden J, et al., Virus-specific mRNA capping enzyme encoded by hepatitis E virus, *J Virol*, 75, 6249–6255, 2001.
214. Kabrane-Lazizi Y, et al., Evidence for widespread infection of wild rats with hepatitis E virus in the United States, *Am J Trop Med Hyg*, 61, 331–335, 1999.
215. Karpe YA, Lole KS, NTPase and 5′ to 3′ RNA duplex-unwinding activities of the hepatitis E virus helicase domain, *J Virol*, 84, 3595–3602, 2010.
216. Ansari IH, et al., Cloning, sequencing, and expression of the hepatitis E virus (HEV) nonstructural open reading frame 1 (ORF1), *J Med Virol*, 60, 275–283, 2000.
217. Panda SK, Ansari IH, Durgapal H, Agrawal S, Jameel S, The in vitro-synthesized RNA from a cDNA clone of hepatitis E virus is infectious, *J Virol*, 74, 2430–2437, 2000.
218. Ropp SL, Tam AW, Beames B, Purdy M, Frey TK, Expression of the hepatitis E virus ORF1, *Arch Virol*, 145, 1321–1337, 2000.
219. Sehgal D, Thomas S, Chakraborty M, Jameel S, Expression and processing of the hepatitis E virus ORF1 nonstructural polyprotein, *Virol J*, 3, 38, 2006.
220. Suppiah S, Zhou Y, Frey TK, Lack of processing of the expressed ORF1 gene product of hepatitis E virus, *Virol J*, 8, 245, 2011.

221. Perttila J, Spuul P, Ahola T, Early secretory pathway localization and lack of processing for hepatitis E virus replication protein pORF1, *J Gen Virol*, 94, 807–816, 2013.
222. Karpe YA, Lole KS, Deubiquitination activity associated with hepatitis E virus putative papain-like cysteine protease, *J Gen Virol*, 92, 2088–2092, 2011.
223. Paliwal D, Panda SK, Kapur N, Varma SP, Durgapal H, Hepatitis E virus (HEV) protease: a chymotrypsin-like enzyme that processes both non-structural (pORF1) and capsid (pORF2) protein, *J Gen Virol*, 95, 1689–1700, 2014.
224. Karpe YA, Lole KS, RNA 5′-triphosphatase activity of the hepatitis E virus helicase domain, *J Virol*, 84, 9637–9641, 2010.
225. Agrawal S, Gupta D, Panda SK, The 3′ end of hepatitis E virus (HEV) genome binds specifically to the viral RNA dependent RNA polymerase (RdRp), *Virology*, 282, 87–101, 2001.
226. Rehman S, Kapur N, Durgapal N, Panda SK, Subcellular localization of hepatitis E virus (HEV) replicase, *Virology*, 370, 77–92, 2008.
227. Egloff M-P, et al., Structural and functional basis for ADP-ribose and poly(ADP-ribose) binding by viral macro domains, *J Virol*, 80, 8493–8502, 2006.
228. Neuvonen M, Ahola T, Differential activities of cellular and viral macro domain proteins in binding of ADP-ribose metabolites, *J Mol Biol*, 385, 212–225, 2009.
229. Pudupakam RS, Huang YW, Opriessnig T, Halbur PG, Pierson FW, Meng XJ, Deletions of the hypervariable region (HVR) in open reading frame 1 of hepatitis E virus do not abolish virus infectivity: evidence for attenuation of HVR deletion mutants in vivo, *J Virol*, 83, 384–395, 2009.
230. Pudupakam RS, et al., Mutational analysis of the hyper variable region of hepatitis E virus reveals its involvement in the efficiency of viral RNA replication, *J Virol*, 85, 10031–10040, 2011.
231. Zafrullah M, Ozdener MH, Kumar R, Panda SK, Jameel S, Mutational analysis of glycosylation, membrane translocation, and cell surface expression of the hepatitis E virus ORF2 protein, *J Virol*, 73, 4074–4082, 1999.
232. Jameel S, Zafrullah M, Ozdener MH, Panda SK, Expression in animal cells and characterization of the hepatitis E virus structural proteins, *J Virol*, 70, 207–216, 1996.
233. Surjit M, Jameel S, Lal SK, The ORF2 protein of hepatitis E virus binds the 5′ region of viral RNA, *J Virol*, 78, 320–328, 2004.
234. Yamashita, T, et al., Biological and immunological characteristics of hepatitis E virus-like particles based on the crystal structure, *Proc Natl Acad Sci USA*, 106, 12986–12991, 2009.
235. Guu TS, et al., Structure of the hepatitis E virus-like particle suggests mechanisms for virus assembly and receptor binding, *Proc Natl Acad Sci USA*, 106, 12992–12997, 2009.
236. Zafrullah M, Ozdener MH, Panda SK, Jameel S, The ORF3 protein of hepatitis E virus is a phosphoprotein that associates with the cytoskeleton, *J Virol*, 71, 9045–9053, 1997.
237. Ahmad I, Holla RP, Jameel S, Molecular virology of hepatitis E virus, *Virus Res*, 161, 47–58, 2011.
238. Holla RP, Ahmad I, Ahmad Z, Jameel S, Molecular virology of hepatitis E virus, *Semin Liver Dis*, 33, 3–14, 2013.
239. Kannan H, Fan S, Patel D, Bossis I, Zhang YJ, The hepatitis E virus open reading frame 3 product interacts with microtubules and interferes with their dynamics, *J Virol*, 83, 6375–6382, 2009.
240. Kar-Roy A, Korkaya H, Oberoi R, Lal S K, Jameel S, The hepatitis E virus open reading frame 3 protein activates ERK through binding and inhibition of the MAPK phosphatase, *J Biol Chem*, 279, 28345–28357, 2004.
241. Ratra R, Kar-Roy A, Lal SK, The ORF3 protein of hepatitis E virus interacts with hemopexin by means of its 26 amino acid N-terminal hydrophobic domain II, *Biochemistry*, 47, 1957–1969, 2008.
242. Emerson SU, Nguyen H, Torian U, Purcell RH, ORF3 protein of hepatitis E virus is not required for replication, virion assembly, or infection of hepatoma cells in vitro, *J Virol*, 80, 10457–10464, 2006.
243. Emerson SU, Nguyen HT, Torian U, Burke D, Engle R, Purcell RH, Release of genotype 1 hepatitis E virus from cultured hepatoma and polarized intestinal cells depends on open reading frame 3 protein and requires an intact PXXP motif, *J Virol*, 84, 9059–9069, 2010.
244. Emerson SU, et al., Recombinant hepatitis E virus genomes infectious for primates: importance of capping and discovery of a *cis*-reactive element, *Proc Natl Acad Sci USA*, 98, 15270–15275, 2001.
245. Graff J, et al., In vitro and in vivo mutational analysis of the 3′-terminal regions of hepatitis E virus genomes and replicons, *J Virol*, 79, 1017–1026, 2005.

246. Cao D, Huang YW, Meng XJ, The nucleotides on the stem–loop RNA structure in the junction region of the hepatitis E virus genome are critical for virus replication, *J Virol*, 84, 13040–13044, 2010.
247. Zhou Y, Emerson SU, P.302 Heat shock cognate protein 70 may mediate the entry of hepatitis E virus into host cells, *J Clin Virol*, 36(Suppl 2), S155, 2006.
248. Kalia M, Chandra V, Rahman SA, Sehgal D, Jameel S, Heparan sulfate proteoglycans are required for cellular binding of the hepatitis E virus ORF2 capsid protein and for viral infection, *J Virol*, 83, 12714–12724, 2009.
249. Zheng ZZ, et al., Role of heat-shock protein 90 in hepatitis E virus capsid trafficking, *J Gen Virol*, 91, 1728–1736, 2010.
250. Kapur N, Thakral D, Durgapal H, Panda SK, Hepatitis E virus enters liver cells through receptor-dependent clathrin-mediated endocytosis, *J Viral Hepat*, 19, 436–448, 2012.
251. Holla P, Ahmad I, Ahmed Z, Jameel S, Hepatitis E virus enters liver cells through a dynamin-2, clathrin and membrane cholesterol-dependent pathway, *Traffic*, 16, 398–416, 2015.
252. Nanda SK, Panda SK, Durgapal H, Jameel S, Detection of the negative strand of hepatitis E virus RNA in the livers of experimentally infected rhesus monkeys: evidence for viral replication, *J Med Virol*, 42, 237–240, 1994.
253. Varma SP, Kumar A, Kapur N, Durgapal H, Acharya SK, Panda SK, Hepatitis E virus replication involves alternating negative- and positive-sense RNA synthesis, *J Gen Virol*, 92, 572–581, 2011.
254. Yamada K, et al., Construction of an infectious cDNA clone of hepatitis E virus strain JE03–1760F that can propagate efficiently in cultured cells, *J Gen Virol*, 90, 457–462, 2009.
255. Nagashima S, et al., A PSAP motif in the ORF3 protein of hepatitis E virus is necessary for virion release from infected cells, *J Gen Virol*, 92, 269–278, 2011.
256. Nagashima S, et al., Tumour susceptibility gene 101 and the vacuolar protein sorting pathway are required for release of hepatitis E virions, *J Gen Virol*, 92, 2838–2848, 2011.
257. Graff J, et al., The open reading frame 3 gene of hepatitis E virus contains a cis-reactive element and encodes a protein required for infection of macaques, *J Virol*, 79, 6680–6689, 2005.
258. Emerson SU, et al. (Eds.), *Virus Taxonomy*, VIIIth Report of the ICTV, Elsevier/Academic Press, London, 851–855, 2004.
259. Thakral D, Nayak B, Rehman S, Durgapal H, Panda SK, Replication of a recombinant hepatitis E virus genome tagged with reporter genes and generation of a short-term cell line producing viral RNA and proteins, *J Gen Virol*, 86, 1189–1200, 2005.
260. Nguyen, HT, Shukla P, Torian U, Faulk K, Emerson SU, Hepatitis E virus genotype 1 infection of swine kidney cells in vitro is inhibited at multiple levels, *J Virol*, 88, 868–877, 2014.
261. Shukla P, et al., Adaptation of a genotype 3 hepatitis E virus to efficient growth in cell culture depends on an inserted human gene segment acquired by recombination, *J Virol*, 86, 5697–5707, 2012.
262. Huang YW, Haqshenas G, Kasorndorkbua C, Halbur PG, Emerson SU, Meng XJ, Capped RNA transcripts of full-length cDNA clones of swine hepatitis E virus are replication competent when transfected into Huh7 cells and infectious when intrahepatically inoculated into pigs, *J Virol*, 79, 1552–1558, 2005.
263. Cordoba L, et al., Rescue of a genotype 4 human hepatitis E virus from cloned cDNA and characterization of intergenotypic chimeric viruses in cultured human liver cells and in pigs, *J Gen Virol*, 93, 2183–2194, 2012.
264. Zhu Y, et al., Infectivity of a genotype 4 hepatitis E virus cDNA clone by intrahepatic inoculation of laboratory rats, *Vet Microbiol*, 166, 405–411, 2013.
265. Kwon HM, et al., Construction of an infectious cDNA clone of avian hepatitis E virus (avian HEV) recovered from a clinically healthy chicken in the United States and characterization of its pathogenicity in specific-pathogen-free chickens, *Vet Microbiol*, 147, 310–319, 2011.
266. Tam AW, et al., In vitro propagation and production of hepatitis E virus from in vivo-infected primary macaque hepatocytes, *Virology*, 215, 1–9, 1996.
267. Tam AW, et al., In vitro infection and replication of hepatitis E virus in primary cynomolgus macaque hepatocytes, *Virology*, 238, 94–102, 1997.
268. Huang R, et al., Cell culture of sporadic hepatitis E virus in China, *Clin Diagn Lab Immunol*, 6, 729–733, 1999.
269. Huang RT, Li DR, Wei J, Huang XR, Yuan XT, Tian X, Isolation and identification of hepatitis E virus in Xinjiang, China, *J Gen Virol*, 73, 1143–1148, 1992.
270. Kazachkov Yu, et al., Hepatitis E virus in cultivated cells, *Arch Virol*, 127, 399–402, 1992.

271. Li DR, Huang RT, Pang JJ, Yuan XT, Li XY, Biological feature and genome analysis of HEV isolated by cell-culture. In: Buisson Y, Coursaget P, Kane M (Eds.), *Enterically-Transmitted Hepatitis Viruses*. La Simarre, Joue-les-Tours, 349–361, 1996.
272. Meng J, Guinet R, Pillot J, Infection of PLC/PRF-5 cells with the hepatitis E virus. In: Buisson Y, Coursaget P, Kane M. (Eds.), *Enterically-Transmitted Hepatitis Viruses*, La Simarre, Joue-les-Tours, 336–345, 1996.
273. Wei S, Walsh P, Huang R, To SS, 93G, a novel sporadic strain of hepatitis E virus in South China isolated by cell culture, *J Med Virol*, 61, 311–318, 2000.
274. Tanaka T, Takahashi M, Kusano E, Okamoto H, Development and evaluation of an efficient cell-culture system for hepatitis E virus, *J Gen Virol*, 88, 903, 2007.
275. Rogee S, et al., New models of hepatitis E virus replication in human and porcine hepatocyte cell lines, *J Gen Virol*, 94, 549–558, 2013.
276. Tanaka T, et al., Development and characterization of a genotype 4 hepatitis E virus cell culture system using a HE-Jf5/15f strain recovered from a fulminant hepatitis patient, *J Clin Microbiol*, 47, 1906–1910, 2009.
277. Zhang HY, Chen DS, Wu YQ, He QG, Chen HC, Liu ZF, Both swine and human cells are capable to support the replication of swine hepatitis E virus type 4 in vitro, *Virus Res*, 158, 289–293, 2011.
278. Takahashi M, et al., Hepatitis E Virus (HEV) strains in serum samples can replicate efficiently in cultured cells despite the coexistence of HEV antibodies: characterization of HEV virions in blood circulation, *J Clin Microbiol*, 48, 1112–1125, 2010.
279. McCaustland KA, et al., Hepatitis E virus infection in chimpanzees: a retrospective analysis, *Arch Virol*, 145, 1909–1918, 2000.
280. Arankalle VA, Favorov MO, Chadha MS, Phule DM, Banerjee K, Rhesus monkeys infected with hepatitis E virus (HEV) from the former USSR are immune to subsequent challenge with an Indian strain of HEV, *Acta Virol*, 37, 515–518, 1993.
281. Tsarev SA, et al., Recombinant vaccine against hepatitis E: dose response and protection against heterologous challenge, *Vaccine*, 15, 1834–1838, 1997.
282. Zhang M, et al., Recombinant vaccine against hepatitis E: duration of protective immunity in rhesus macaques, *Vaccine*, 20, 3285–3291, 2002.
283. Zhou YH, Purcell RH, Emerson SU, Truncated ORF2 protein contains the most immunogenic site on ORF2: antibody responses to non-vaccine sequences following challenge of vaccinated and non-vaccinated macaques with hepatitis E virus, *Vaccine*, 23, 3157–3165, 2005.
284. Li TC, Suzuki Y, Ami Y, Dhole TN, Miyamura T, Takeda N, Protection of cynomolgus monkeys against HEV infection by oral administration of recombinant hepatitis E virus-like particles, *Vaccine*, 22, 370–377, 2004.
285. Im SW, et al., A bacterially expressed peptide prevents experimental infection of primates by the hepatitis E virus, *Vaccine*, 19, 3726–3732, 2001.
286. Ma H, et al., Immunogenicity and efficacy of a bacterially expressed HEV ORF3 peptide, assessed by experimental infection of primates, *Arch Virol*, 154, 1641–1648, 2009.
287. Feagins AR, Opriessnig T, Huang YW, Halbur PG, Meng XJ, Cross-species infection of specific-pathogen-free pigs by a genotype 4 strain of human hepatitis E virus, *J Med Virol*, 80, 1379–1386, 2008.
288. Williams TP, et al., Evidence of extra hepatic sites of replication of the hepatitis E virus in a swine model, *J Clin Microbiol*, 39 3040–3046, 2001.
289. Huang FF, Pierson FW, Toth TE, Meng XJ, Construction and characterization of infectious cDNA clones of a chicken strain of hepatitis E virus (HEV), avian HEV, *J Gen Virol*, 86, 2585–2593, 2005.
290. Huang F, et al., Experimental infection of Balb/c nude mice with hepatitis E virus, *BMC Infect Dis*, 9, 93, doi: 10.1186/1471-2334-9-93, 2009.
291. Li TC, Suzaki Y, Ami Y, Tsunemitsu H, Miyamura T, Takeda N, Mice are not susceptible to hepatitis E virus infection, *J Vet Med Sci*, 70, 1359–1362, 2008.
292. Maneerat Y, Clayson ET, Myint KS, Young GD, Innis BL, Experimental infection of the laboratory rat with the hepatitis E virus, *J Med Virol*, 48, 121–128, 1996.
293. Li WG, et al., Experimental infection of Mongolian gerbils by a genotype 4 strain of swine hepatitis E virus, *J Med Virol*, 81, 1591–1596, 2009.
294. Ma H, et al., Experimental infection of rabbits with rabbit and genotypes 1 and 4 hepatitis E viruses, *PLoS One*, 5, e9160, 2010.

295. Han J, et al., SPF rabbits infected with rabbit hepatitis E virus isolate experimentally showing the chronicity of hepatitis, *PLoS One*, 9, e99861, 2014.
296. Haqshenas G, Shivaprasad HL, Woolcock PR, Read DH, Meng XJ, Genetic identification and characterization of a novel virus related to human hepatitis E virus from chickens with hepatitis-splenomegaly syndrome in the United States, *J Gen Virol*, 82, 2449–2462, 2001.
297. Billam P, Pierson FW, Li W, LeRoith T, Duncan RB, Meng XJ, Development and validation of a negative-strand-specific reverse transcription-PCR assay for detection of a chicken, *J Clin Microbiol*, 46, 2630–2634, 2008.
298. Guo H, Zhou EM, Sun ZF, Meng XJ, Protection of chickens against avian hepatitis E virus (avian HEV) infection by immunization with recombinant avian HEV capsid protein, *Vaccine*, 25, 2892–2899, 2007.

5

Noroviruses: Laboratory Surrogates for Determining Survival and Inactivation

Doris H. D'Souza and Snehal S. Joshi

CONTENTS

5.1 Introduction

Noroviruses (NoVs) belong to the Caliciviridae family that currently comprises of five genera: *Norovirus*, *Sapovirus*, *Lagovirus*, *Vesivirus*, and *Nebovirus*, with potentially additional genera as new sequence information becomes available on unclassified viruses [1,2]. The name "Norovirus" was shortened from the initial "Norwalk-like" virus term that was associated with an epidemic outbreak that occurred in Norwalk, Ohio, in 1968 in an elementary school [3]. Infected school children upon returning home transmitted the secondary infection to family members resulting in 50% of students and teachers developing nonbacterial gastroenteritis.

NoVs are small viruses about 27–32 nm in size and round in structure with an icosahedral symmetry. The human norovirus (HNoV) genome contains a single-stranded positive-sense RNA about 7.6 kb in length that is enclosed in a capsid without an envelope [3]. The capsid is made of 90 capsomers protruding from the shell that has 90 dimers of capsid protein. The genome has three open reading frames (ORFs). ORF1 (nucleotides 146–5359) is about 5 kb in size and encodes a ~200 kDa nonstructural polyprotein. This nonstructural protein is cleaved to produce the N-terminal protein, the enzyme nucleoside triphosphatase, a 3A-like protein, a genome-linked viral protein (VpG), a 3C-like protease, and RNA-dependent RNA-polymerase (RdRp) [4]. ORF2 (nucleotides 5346–6935) is ~1.8 kb in size and encodes the 57 kDa major structural capsid viral protein VP1; ORF3 (nucleotides 6938–7573) is ~0.6 kb in size and encodes a small 22 kDa minor viral structural protein, VP2, reported to package the genome into virions [5,6]. The NoV genus at the time of this submission, is composed of five genogroups based on sequence analysis: genogroup I (GI) (prototype Norwalk virus); GII (prototype Snow Mountain virus); GIII (prototype bovine enteric calicivirus); GIV (prototype Alphatron and Ft. Lauderdale viruses); and GV (prototype Murine NoV) [7,8].

From the five genogroups of NoV, genogroups GI, GII, and GIV specifically infect humans (and are commonly referred to as human noroviruses or HNoV) and as indicated above, GIII includes bovine enteric calicivirus that infects cattle, and GV is associated with the infection of mice. There have been at least 32 genetic clusters identified based on the amino acid-sequence similarity between these genogroups [1,9]. Eight genotypes are recognized in GI and 19 genotypes are recognized in GII, and of these, HNoV genotype GII.4 has been the most prevalent during the past few decades. Indeed, the majority of HNoV outbreaks are reported to be caused by the GII.4 genocluster and its variants, where in 2002, a GII.4 variant, called the Farmington Hill strain, was responsible for 80% of acute HNoV outbreaks in the United States [10]. In addition, the Hunter strain GII.4 variant was found to be circulating in Australia, Europe, and Asia in 2004, and then the Sakai strain (reported in Southeast Asia) and Minerva strain (found in the United States and the Netherlands) were found to be circulating in 2006, replacing the Hunter GII.4 variant [11]. Another HNoV GII.6 strain (previously not reported in California) classified as Seacroft was reported to be responsible for causing illness in 30 individuals at a local university in Los Angeles, California, in October 2008 [12]. It is also noteworthy to report that outbreaks of the HNoV genogroup GII.4 in health-care settings in the United States, Europe, and Oceania were responsible for 70%–80% of all HNoV outbreaks [13]. The HNoV strains continue to remain a worldwide concern and are continually evolving, with these emerging HNoV GII.4 variants also becoming virulent and known to cause death in the elderly and immunocompromised [14].

HNoVs have a very short incubation period of 18–48 h, with typically self-limiting infection that lasts for up to 72 h, characterized by mild gastroenteritis symptoms that include nausea, vomiting, abdominal cramps, fever, and malaise. "Winter vomiting disease" is the common seasonal syndrome associated with HNoV infection, with peak outbreaks in winter, even though HNoV transmission reportedly occurs all year round [15]. It is also important to keep in mind that from a public health aspect, infected individuals can shed viruses for 72 h (and even longer, greater than a week) after signs of the first symptoms appear [16]. Outbreaks have been reported in restaurants, cruise ships, health-care settings, closed environments, and nursing homes, either due to the consumption of at-risk ready-to-eat or undercooked contaminated foods, such as seafood, shellfish, produce, deli items, and bakery products, and also due to environmental transmission and person-to-person transmission [17]. Therefore, adequate control strategies, proper sanitation, food processing, storage, and hygienic practices of workers are important to prevent further transmission especially in food environments and health-care settings.

5.2 HNoVs and the Use of Surrogates

HNoVs are epidemiologically significant as they are considered the leading cause of nonbacterial gastroenterititis to date worldwide. Given the prevalence of HNoV outbreaks worldwide and the spread of emerging viruses that cause severe symptoms even leading to death, and the fact that HNoVs cannot be cultivated in cell culture in the laboratory to date, cultivable viral surrogates are used to determine the infectivity and inactivation of HNoVs. In fact, with regard to the cultivation and propagation of HNoVs, even 3D cell-culture models using Int-407 or Caco-2 cells did not support HNoV replication [18]. However, more recently, it was reported that Histoblood group antigen (HBGA)-expressing enteric bacteria were required for the HNoV infection of B cells, though validation of HNoV replication in these cells is still pending, and the replication of HNoV needs to be demonstrated and validated by researchers from other laboratories [19].

Due to the numerous issues and challenges facing the cultivation of HNoVs, cultivable animal viral surrogates such as feline calicivirus (FCV-F9), followed by murine norovirus (MNV-1) and then Tulane virus (TV) and porcine enteric sapovirus have been considered as alternate HNoV surrogates to enable progress in research toward prophylactic therapies and control measures against HNoVs. Initially, during environmental studies, bacteriophage MS2 (viruses that infect bacteria) was also used as a surrogate to determine survival and persistence.

5.3 Human Feeding Studies and Associated Drawbacks

Indeed, it is apparent that humans are the most relevant models to study HNoV biology and replication. However, for these feeding or challenge studies, volunteers are needed along with privacy protection, requiring appropriate institutional and board approvals, labor, cost, and time. Also, health-care facilities should be easily accessible with a team of skilled technicians, scientists, researchers, and health-care professionals. In addition, as noted by other researchers [20], sampling difficulties and expenses, biocontainment due to the presence of live and infectious viruses, fecal and sera sample collection and storage, invasive biopsies, and monitoring of the health of eligible volunteers need to be also considered as some limitations. It must be noted that feces and sera provide valuable data to understand HNoV survival, persistence, prevalence, disease symptoms, mechanisms of associated diarrhea, immune responses, and diagnostics and also have potential to develop vaccines and control strategies [20]. In fact, human volunteer studies played a key role in resistance marker identification (related to HBGA blood groups) and understanding the immune selection mechanism responsible for the persistence of HNoV GII.4 in human populations [20].

Thus, the only source of live infectious HNoVs remains the human population during outbreaks and/or human volunteer feeding or challenge studies. A randomized, double-blinded clinical trial was carried out to determine the inactivation of Norwalk virus (HNoV genogroup I.1) by high hydrostatic pressure processing (HPP) in virus-seeded oysters that were ingested by adults [21]. Healthy, positive-secretor adults who consumed oysters artificially seeded with Norwalk virus (8FIIb, 1.0×10^4 genomic equivalent copies) that were treated with 600 MPa HPP at 6°C for 5 min (but not 400 MPa at 6°C or 25°C) did not show any HNoV infection, as determined using reverse transcription-PCR of their stool or vomitus samples, indicating the complete inactivation of HNoV in seeded oysters [21]. Thus, these researchers showed the use of HPP as a potential intervention to inactivate infectious HNoV in oysters for the commercial shellfish industry. In another earlier clinical trial carried out in Australia, shellfish testing positive for HNoV were depurated and subsequently fed to volunteers who became ill, which demonstrated that commercial depuration was not effective in eliminating HNoV in contaminated oysters [22]. Human feeding studies do provide reliable results on the effectiveness of inactivation methods, and represent the best model for determination of the inactivation, pathogenesis, and immunity of HNoVs for research in the absence of cell-culture infectivity assays and inability to discriminate infectious from noninfectious HNoVs. Richards has suggested and recommended a movement from surrogate research to human volunteer studies to identify practical and valid processing methods for improving food safety [23]. However, due to research-funding limitations, extensive labor, and the time associated with human feeding studies, other models and surrogates have to be relied on to determine the best approaches for HNoV inactivation.

5.4 Animal Viruses as HNoV Surrogates and Model Systems to Study HNoVs

5.4.1 Feline Calicivirus

FCV-F9 belongs to the genus *Vesivirus* of the Caliciviridae family and is responsible for acute oral and upper respiratory tract disease in cats characterized by oral ulcerations, limping syndrome, and ocular and nasal discharge [24]. It is typically transmitted via the nasal, oral, or conjunctival routes [25]. This virus has been extensively utilized as a model surrogate for HNoVs and was the first virus surrogate to be used to conduct research on HNoV inactivation, since HNoVs are considered unculturable in the laboratory to date as mentioned above [26]. FCV-F9 was considered a suitable model system as it is a nonenveloped single-stranded RNA virus with an approximate diameter of 35–39 nm and icosahedral symmetry, being similar to HNoVs in these aspects. Also, the RNA genome is 7.5 kb in size and has three ORFs [27]. However, there are many issues with this surrogate that question its suitability as a surrogate and model system. FCV-F9 is transmitted through the respiratory route unlike the fecal–oral route of HNoVs. Additionally, given its respiratory route of transmission and sensitivity to the low pH of 2.0–4.0 encountered in the gastrointestinal tract [6], unlike HNoVs that are known to be resistant to low pH conditions, alternate surrogates and model systems are being researched.

FCV-F9 has been utilized as a surrogate in determining improved inactivation strategies against HNoVs in foods and on contact surfaces, such as chemical, nonthermal, and thermal treatments (refer to Tables 5.1 through 5.3). Broad-spectrum contact surface disinfectants have been evaluated using

TABLE 5.1

Effect of Chemicals on the Feline Calicivirus (FCV-F9) Used as a Cultivable HNoV Surrogate to Develop Disinfection and Control Strategies

Virus	Treatment	Reduction (log PFU/mL)	References
Feline calicivirus	Hydrogen peroxide vapors; 25 mL volume:		[70,109]
	70% ethanol, 5 min	2.6	
	Formaldehyde; 0.7%, 60 min at 20°C	4	[110]
	Glutaraldehyde; 0.5%, 1 min at room temperature	5	[26]
	Bacoban WB (quaternary ammonium compound); 2%; 240 min at 20°C	4	[110]
	Benzalkonium chloride; 0.25 mg/mL, 2 h at room temperature	3.1	[28]
	β-Propiolactone; 0.1%, 60 min at 22°C	5.2	[111]
	Free chlorine; 500 ppm, 10 min at room temperature	1.9	[112]

TABLE 5.2

Effects of Surface Inactivation Methods Using Feline Calicivirus (FCV-F9) as a Cultivable HNoV Surrogate

Virus	Medium	Treatment	Reduction	References
Feline calicivirus	Plastic surfaces	Steam (130°C)–ultrasound 30–50 kHz for 3 s	4.8 log	[36]
	Low protein virus stock	Gamma radiation	5.9 log PFU	[113]
		1 μM methylene blue; human plasma, 5 min illumination	3.9 log	[114]
	Virus + cell-culture media	200 MPa—0°C, 4 min	4.4 log	[115]
		250 MPa—0°C, 4 min	4.8 log	
		300 MPa—10°C, 3 min	5 log	
	Glass dishes	ClO_2 gas (1.7 mg/h at 25°C for 5 h)	3 log	[116]

TABLE 5.3

Evaluation of Heat Treatments on Feline Calicivirus (FCV-F9) as a HNoV Surrogate

Virus	Medium	Inactivation Parameters	References
Feline calicivirus	Buffer	56°C; *D*-value 6.36	[73]
		65°C; *D*-value 0.32	
		72°C; *D*-value 0.11	
	Seafood	56°C; *D*-value 3.33	[75]
		65°C; *D*-value 0.33	
		72°C; *D*-value 0.07	
	Spinach	56°C; *D*-value 5.83	[40]
		65°C; *D*-value 0.27	
		72°C; *D*-value 0.15	
	Virus stock	56°C; *D*-value 3.473	[6]
		63°C; *D*-value 0.435	
		72°C; *D*-value 0.166	

FCV-F9 for their antiviral properties. Potassium peroxymonosulfate (KPMS) at concentrations of 5, 10, and 20 mg/mL was reported to reduce FCV-F9 titers to undetectable levels from an initial 5 and 7 log PFU/mL within 2 h at room temperature [28]. Similarly, benzalkonium chloride (BAC) at 0.1, 0.25, and 0.5 mg/mL was reported to reduce FCV-F9 by 2.87, 3.08, and 3.25 log PFU/mL from an initial 7 log PFU/mL after 2 h at room temperature [28]. D'Souza and Su reported that FCV-F9 could be inactivated to non-detectable levels (6.84 log PFU/mL reduction) after a contact time of 30 s with 10% bleach (0.6% sodium hypochlorite, 5000 ppm available chlorine) and 2% trisodium phosphate (TSP) [29]. A neutral solution of electrochemically activated anolyte Ecasol was reported to reduce FCV-F9 titers by more than 5 log within 1 min of treatment at room temperature [30].

FCV-F9 was also used as a model virus in a study to determine the effect of chlorine dioxide (ClO_2) gas against the virus inoculated on glass surfaces in a test room (39 m^3), where a constant low-concentration ClO_2 gas was produced [31]. These researchers found that a low-concentration ClO_2 gas (mean 0.05 ppmv, 0.14 mg/m^3) inactivated FCV-F9 (by >2 logs) in the wet state on glass dishes within 5 h. Other researchers used FCV-F9 as a cultivable HNoV surrogate and model to determine the effect of aqueous ClO_2 at concentrations ranging from 0.2 to 0.8 mg/L at pH 7 and 20°C, and found that at a concentration of 0.8 mg/L ClO_2, a complete elimination of FCV-F9 was obtained in 2 min, while lower concentrations of 0.2 mg/L required 30 min for inactivation [32]. FCV-F9 was also used as a model to determine the inactivation of enteric viruses by ozone treatment, where for a 4-log (99.99%) inactivation of FCV at 5°C and pH 7 about <0.01–0.03 mg/L min of ozone was required [the *Ct* value which is described as the concentration of ozone multiplied by the virus contact time, using application of the efficiency factor Hom (EFH) model] [33]. In another study using FCV-F9 as an HNoV surrogate, ozone treatment at 6.25 ppm for 4 min showed >6 log $TCID_{50}$/mL reduction of FCV-F9 in water and ~2-log $TCID_{50}$/mL on lettuce and green onions [34].

High-pressure processing and ultraviolet (UV) light treatments are commonly used processing methods in the food industry. A study reported a decrease of 3.48, 3.82, and 4.62 log $TCID_{50}$ in FCV-F9 titers on lettuce after exposure to UV light at doses of 16, 40, and 75 mW s/cm^2, respectively [35]. High HPP treatments of 500 and 600 MPa were reported to reduce FCV-F9 titers to undetectable levels after 1, 3, 5, and 7 min, while lower pressures of 300 and 400 MPa caused 1.13 and 0.55 log decreases after 1 min, respectively [32]. A combination of pressurized steam (130°C) and high-power ultrasound (30–50 kHz) was reported to reduce FCV-F9 titers on plastic surfaces by >99.99% corresponding to >4.8-log after 3 s of treatment [36]. FCV-F9 in milk showed reductions of ≥4 log PFU/mL after high-pressure homogenization at 300 MPa, ~1.3 log PFU/mL at 250 MPa, and insignificant reduction at ≤200 MPa [37]. These researchers showed that in orange juice, FCV-F9 was reduced by ≥4 and ~1 log PFU/mL only after high-pressure homogenization at 300 and 250 MPa, respectively [37]. FCV-F9 in PBS at titers of ~4 log PFU/mL after a treatment at 20 kHz of high-intensity ultrasound (HIU) for 5 min could be reduced to undetectable levels [38].

FCV-F9 has also been used as a model/surrogate for small, round-structured viruses on lettuce, where the lettuce was inoculated with FCV-F9 by immersion to simulate contamination from irrigation or wash water. Then the lettuce was subjected to electron beam irradiation at various dose levels [39]. These researchers reported the D_{10}-value of FCV-F9 on lettuce to be 2.95 kGy. When FCV-F9 was used as a model to determine heat inactivation of HNoVs in spinach, the *D*-values calculated from the first-order model (50°C–72°C) ranged from 0.15 to 17.39 min [40].

Significant research has been conducted on natural oils and extracts for their antiviral properties, including 0.25% natural mulberry (*Morus alba*) juice that caused a 50% reduction in FCV-F9 titers [41]. Essential oil thymol at 0.5% and 1% was reported to reduce FCV-F9 titers to undetectable levels from an initial 6 log $TCID_{50}$ after 2 h at 37°C [42]. Black raspberry juice showed antiviral activity when Crandell Reese feline kidney cells were pretreated with the juice or directly or FCV-F9 itself [43]. After treatment with cranberry juice and cranberry proanthocyanidins at 0.30, 0.60, and 1.20 mg/mL for 1 h at room temperature, FCV-F9 from initial 5 log PFU/mL was reported to be reduced to undetectable levels [44]. When FCV-F9 was used as a surrogate to determine the effect of grape-seed extract (GSE) as a wash to disinfect lettuce and jalapeno peppers, FCV-F9 was reported to be reduced by 2.33, 2.58, and 2.71 log PFU on lettuce, and 2.20, 2.74, and 3.05 log PFU on peppers after 1 min using 0.25, 0.50, and 1 mg/mL GSE, respectively [45]. Thus, FCV-F9 has also been extensively used as a model to understand antiviral options.

The survivability/persistence of FCV is surprisingly high on surfaces and was detectable for up to 28 days at room temperature [26]. FCV-F9 was detected on three surfaces (formica, stainless steel, and ceramic) for up to 7 days postinoculation at room temperature [46]. A persistence/survival study of FCV-F9 at storage (7°C) and room temperatures over 70 days showed that the infectivity of FCV-F9 at room temperature decreased after 6 h, 1 day, and 7 days by 1 log, 3 log, and to below the detection limit (~4.5 log), respectively [47]. However, FCV-F9 was shown to survive longer at 7°C, with only a reduction of 2 log after 7 days and a titer decrease of 3 log after 70 days being reported [47]. Similarly, FCV-F9 at 6 log PFU/mL was reported to be reduced to undetectable levels after 14 days in orange as well as pomegranate juice and by ~3 logs after 21 days in milk at 4°C [48]. However, FCV-F9 was reported to be completely reduced to undetectable levels after 1 day in the blend of orange and pomegranate juice at 4°C [48]. In another study, FCV-F9 (at initial titers of ~5 log PFU/mL) was reported to be reduced to undetectable levels after 1 day in blueberry juice at 4°C [49].

Thus, FCV-F9 has been widely used as a surrogate for HNoVs due to its ease of cultivation *in vitro*, its relatively straightforward infectivity assays to develop and understand antiviral agents and for environmental persistence and stability studies, and because it is readily manipulated genetically for use as a model for replication and translation [20]. The development of the FCV reverse genetics system made FCV a desirable model to study the molecular mechanism of calicivirus translation and genome replication [50]. Infection with FCV was reported to lead to the inhibition of cellular protein synthesis associated with the cleavage of host translation initiation factors [51] and also to the first report and identification of a functional protein receptor molecule for a calicivirus [52]. Subsequent studies with FCV showed conformational changes in the viral capsid protein upon receptor binding and FCV was used to identify the first functional host cell factor–viral genome interaction required for FCV replication. Similar interactions are reported to occur for HNoVs [20]. However, when compared to the other surrogates including MNV-1 and the newer cultivable TV, FCV does not appear to be the most robust surrogate model for the determination of HNoV survival and inactivation, specifically because it causes a respiratory or systemic disease in some cases, unlike the gastrointestinal illness caused by HNoVs.

5.4.2 Murine Norovirus

MNV-1 is a member of the Caliciviridae family and since it falls under the *Norovirus* genus, it shares closer biological and molecular properties with HNoVs as compared to FCV-F9 [53]. Besides, MNV-1 can replicate to high titers in tissue culture having a dendritic cell and macrophage tropism and the infectious assays are relatively straightforward as well. MNV-1 is clearly an intestinal pathogen spread via the fecal–oral route, and MNV infection exacerbates the onset of inflammatory bowel disease similar to HNoVs [20]. These properties make MNV-1 a suitable model or surrogate to study HNoVs. MNV-1 was isolated from mice brain tissue with clinical symptoms including diarrhea, fever, nausea, and abdominal pain [53,54]. Similar to HNoVs, with transmission via the fecal–oral route, MNV-1 is shed for up to 8 weeks after infection and/or inoculation [2]. It is the only NoV that currently replicates in cell culture in a reproducible manner, which has made HNoV surrogate studies adaptable and somewhat economical [54]. On an average the size of this nonenveloped virus lies between 28 and 35 nm in diameter, with an icosahedral symmetry. MNV-1 contains a single-stranded positive-sense RNA with three ORFs, where ORF1 encodes a predicted 187.5-kDa polyprotein that comprises of the 2C helicase, 3C protease, and 3D polymerase motifs typically observed in other caliciviruses and picornaviruses, while ORF2 encodes a 58.9-kDa capsid protein which self-assembles into virus-like particles (VLPs) in baculovirus expression systems, and ORF3 encodes a putative 22.1-kDa basic protein [54].

As a laboratory model to understand HNoV survival, MNV-1 has been shown to be more stable and survive longer under various environmental conditions than FCV-F9. Upon exposure to a low pH of 2 at 37°C for 30 min, MNV-1 titers were reported to be reduced by <1 log [6]. MNV-1 titers were found to decrease by a mere 0.09 log PFU/mL/day in environmental waters at 25°C over 3–5 weeks [55]. In a study conducted for its survival in the hospital environment, MNV-1 infectivity was reduced by <2 log over 40 days at both −20°C and 4°C on the surface of gauze and diaper materials; however, at a higher temperature of 30°C, a higher reduction of 5 log was reported after 24 days [56]. On stainless steel, under wet and dry conditions, MNV-1 was shown to survive for 7 days with <2 log PFU/mL reduction at 4°C, with similar reductions at room temperature [6]. MNV-1 is also known to survive freezing temperatures of −21°C, and no change in infectivity on frozen onions and frozen spinach was reported after 6 months [57]. MNV-1 was found to survive in commercial blueberry juice (BJ) even after 21 days at 4°C [37]. In orange juice and milk, MNV-1 titers were not reported to be reduced after 21 days; however, in pomegranate juice a reduction of 1.4 log PFU/mL was reported. MNV-1 was reduced to undetectable levels after 7 days in the orange and pomegranate juice blend at 4°C [48]. Escudero et al. also showed that infectious MNV-1 could be detected until 21 days on inoculated stainless steel, formica, or ceramic surfaces, and did not persist on lettuce stored at room temperature or 4°C [58]. Taken together, these survival studies provide an indication of the ease of transmission of HNoVs in the environment.

MNV-1 has also been studied as a laboratory model to understand the effect of various inactivation methods against HNoV (Tables 5.4 through 5.6). With free chlorine concentrations of 0.193 mg/L, MNV infective titers were found to be reduced to undetectable levels within 2 min of contact time at 5°C and 1 min of contact time at 20°C [59]. Upon exposure to 0.255 mg/L ClO_2, MNV-1 was reduced to undetectable levels (<3.5 log) within 1 min of contact time at 5°C and 30 s of contact time at 20°C [59]. It

TABLE 5.4

Inactivation of Murine Norovirus (MNV-1) as an HNoV Surrogate by Physical and Chemical Approaches

Virus	Treatment	Reduction	References
Murine norovirus	15 ppm active chlorine	1.4 log	[60]
	0.193 mg/L free chlorine; 2 min at 5°C	5 log PFU/mL	[59]
	0.255 mg/L ClO_2; 1 min at 5°C	3.5-log PFU/mL	[59]
	100 ppm peroxyacetic-biocide; 2 min, room temperature	2.3 log RT-PCR units	[60]
	2.5% liquid H_2O_2; 5 min, room temperature	1 log PFU/mL	[61]
	6.25 ppm ozone; 10 min, room temperature	2.5 log PFU/plant	[86]
	2% trisodium phosphate; 30 s, room temperature	3 log PFU/mL	[86]
	0.2 mg/mL benzalkonium chloride; 2 h, room temperature	5 log PFU/mL	[28]
	5 mg/mL potassium peroxymonosulfate; 2 h, room temperature	5 log PFU/mL	[28]
	5% levulinic acid and 2% sodium dodecyl sulfate; 1 min	1.50 log PFU/mL	[65]
	Tween 20 (50 ppm) and chlorine (200 ppm); 1 h	3 log PFU/mL	[66]
	450-MPa; high hydrostatic pressure in cell-culture media	6.85 log PFU	[117]
	400-MPa; high hydrostatic pressure; oysters	4.05 log PFU	[69]
	300 MPa; high-pressure homogenization; blueberry juice	0.71 log PFU/mL	[37]
	UV at 25 mJ/cm^2; 200 s	3.6 log	[56]
	Sonication; 20 kHz	4 log PFU/mL	[38]

TABLE 5.5

Thermal Inactivation of Murine Norovirus (MNV-1) as a Cultivable Surrogate for HNoVs

Murine norovirus	Buffered cell-culture media	56°C; *D*-value 3.74 65°C; *D*-value 0.77 72°C; *D*-value 0.25	[73]
	Seafood	56°C; *D*-value 6.12 65°C; *D*-value 0.14 72°C; *D*-value 0.18	[75]
	Spinach	56°C; *D*-value 3.29 65°C; *D*-value 0.40 72°C; *D*-value 0.16	[40]
	Virus stock in cell-culture media, 72°C for 25 s	1 log PFU/mL	[6]
	Soft shelled clams; 90 s at 90°C	3.33 log cycles	[76]
	60°C for 5 min; buffer	3.03 log $TCID_{50}$/mL	[77]
	60°C for 5 min	0.78 log $TCID_{50}$/mL	

TABLE 5.6

Inactivation of Murine Norovirus (MNV-1) as a Cultivable HNoV Surrogate Using Natural Plant Extracts

Murine norovirus	Grape seed extract (0.35 mg/mL); 37°C for 2 h	0.82 log PFU/mL	[78]
	Cranberry proanthocyanidins (0.6 mg/mL); 2 h, room temperature	2.9 log PFU/mL	[44]
	Pomegranate juice; 1 h, room temperature	1.32 log PFU/mL	[79]
	Pomegranate polyphenols (4 mg/mL); 1 h, room temperature	1.30 log PFU/mL	[79]
	Black raspberry juice (6%); 1 h at 37°C	99% reduction in plaque titers	[43]
	Aqueous hibiscus extracts (40 mg/mL); 6 h, 37°C	1.78 log PFU/mL	[80]
	Oregano oil (4%); 24 h, room temperature	1.07 log $TCID_{50}$	[82]

was reported that 15 ppm of active chlorine could reduce MNV-1 by 1.4 log units (detected by PCR) and the application of 100 ppm peroxyacetic-based biocide on lettuce as a produce wash could decrease MNV-1 titers by 2.3 log RT-PCR units in 2 min [60]. Liquid hydrogen peroxide (H_2O_2) at 2.5% when used as an antiviral produce wash was reported to reduce MNV-1 by ~1 log after 5 min [61]. Ozone is known to decompose in the water phase of foods, though ozone might be able to diffuse and act on viruses as shown in a study where the internalized MNV-1 in green onions was reduced by 2.5 log PFU/plant after an ozone treatment of 6.25 ppm for 10 min [62]. When the green onion plants that were inoculated with MNV-1 were sprayed with calcium hypochlorite (150 ppm, 4°C) or ozone (6.25 ppm for 10 min), a reduction of 2 and 2.5 log PFU/plant, respectively, from an initial titer of 4.92 log PFU/plant was reported [63]. TSP at 2% and 5% after 30 s as a produce wash for lettuce and peppers was reported to reduce MNV-1 by 3 log PFU/mL and to undetectable levels, respectively, from an initial titer of 5 log PFU/mL [64]. BAC, KPMS, tannic acid (TA), and gallic acid (GA) were also evaluated for their antiviral effect using MNV-1 as a laboratory model, and MNV-1 (5 log PFU/mL) was reported to be reduced to undetectable levels with BAC at 0.2, 0.5, and 1 mg/mL after 2 h at room temperature [28]. KPMS at 2.5 and 5 mg/mL was reported to reduce low-titer MNV-1 (5 log PFU/mL) to undetectable levels and high-titer MNV-1 (7 log PFU/mL) by 0.92 and 3.44 log PFU/mL, respectively, after 2 h at room temperature. TA at 0.1 mg/mL and GA at 0.2 and 0.4 mg/mL were reported to have caused no reduction in MNV-1 titers after 2 h at room temperature [28].

Another novel sanitizer consisting of 0.5% levulinic acid in combination with an anionic detergent sodium dodecyl sulfate (SDS) at 0.5% was shown to decrease MNV-1 titers by 3 log PFU/mL after 1 min, while a solution of 5% levulinic acid and 2% SDS was reported to reduce MNV-1 titer by 1.50 log PFU/mL after 1 min and by 3.3 log PFU/mL after 5 min on stainless steel surfaces [65]. Another study explored the effects of combinations of surfactants (SDS) and polysorbates (Tween 20, Tween 65, and Tween 80) along with the traditional chlorine (200 ppm) washes against MNV-1 from fresh produce [66]. The study reported that the combination of Tween 20 (50 ppm) and chlorine (200 ppm) after 1 h at room temperature was most effective, with MNV-1 being reduced by 3 log PFU/mL from an initial titer of ~6.5 log PFU/mL.

MNV-1 was also used as a laboratory model and surrogate for HNoV to determine inactivation by physical and chemical inactivation methods that include high HPP, high-pressure homogenization, UV light, and ozone. The high hydrostatic pressure inactivation of viruses is shown to be due to changes in the capsid proteins or by disruption of capsid proteins and thereby their inability to bind to host cells [67,68]. MNV-1 titers after a high hydrostatic pressure treatment of 450-MPa were shown to decrease by 6.85 log PFU in cell-culture media [69]. A reduction of 4.05 log was reported in MNV-1-contaminated oysters after 5 min at 5°C at 400-MPa high hydrostatic pressure [69]. The high hydrostatic pressure treatment of 300 MPa was shown to inactivate MNV-1 completely from initial titers of ~7–8 log PFU/mL after 1 min [70], while MNV-1 in inoculated clams was reported to be reduced by a mere 1 log after high hydrostatic pressure treatment of 400 MPa for 5 min [71]. When the effect of UV treatment on MNV-1 infectivity was evaluated, 3.6 log reductions after 200 s with UV at 25 mJ/cm^2 were obtained [56]. When green onions were inoculated with MNV-1 and treated with UV (240 mJ s/cm^2), 1.2 log PFU reduction per plant was obtained from an initial titer of 4.92 log PFU/plant [62]. These researchers also showed that high hydrostatic pressures (500 MPa for 5 min at 20°C) could decrease MNV-1 to undetectable levels. When MNV-1 was used as a surrogate to understand the effect of electron beam irradiation against enteric viruses, the electron-beam dose required to reduce MNV-1 titers by 90% (D_{10}-value) in whole oysters was reported to be 4.05 ± 0.63 kGy [72].

When sonication was studied as another physical inactivation method for HNoV, MNV-1 titers were found to be decreased by ~4 log PFU/mL after a 30-min treatment at 20 kHz [44]. When exposed to high-pressure homogenization, MNV-1 in blueberry juice was reported to be reduced by 0.33 log PFU/mL at 250 MPa and by 0.71 log PFU/mL at 300 MPa [37]. Studies carried out by Bozkurt et al. aimed to characterize the thermal inactivation kinetics of MNV-1 as a laboratory surrogate to understand the inactivation of HNoV at 50°C, 56°C, 60°C, 65°C, and 72°C in buffer as well as different food samples including spinach and blue mussels [40,73,74]. The *D*-values for MNV-1 using the capillary tube method at 50°C, 56°C, 60°C, 65°C, and 72°C were reported to be 34.49, 3.65, 0.57, 0.30, and 0.15 min, respectively [73]. In buffered cell-culture media, *D*-values of 36.28, 3.74, 1.09, 0.77, and 0.25 at 50°C, 56°C, 60°C, 65°C, and 72°C, respectively, were reported [74]. In blue mussels, the reported *D*-values were 20.19, 6.12, 2.64, 0.41, and 0.18 at temperatures of 50°C, 56°C, 60°C, 65°C, and 72°C, respectively [75]. In spinach, the *D*-values were comparatively lower than in blue mussels, where values of 14.57, 3.29, 0.98, 0.40, and 0.16 were obtained at 50°C, 56°C, 60°C, 65°C, and 72°C, respectively [40]. The effect of heat treatment on MNV-1-infected soft shell clams was studied using RT-PCR as a method of detection that showed a reduction of 3.33 log cycles after 90 s at 90°C and 5.47 log cycles after 180 s at 90°C [76]. MNV-1 was reported to be reduced by 3.03, 3.69, 4.35, 5.05, and 5.88 log $TCID_{50}$/mL in suspensions at temperature–time combinations of 60°C for 5 min, 60°C for 15 min, 60°C for 30 min, 85°C for 3 min, and 85°C for 6 min, respectively, and in dried mussels, titer reductions of 0.78, 2.00, 3.35, 1.95, and 3.20 log, respectively, were reported [77].

A variety of natural plant and fruit extracts have been examined for their antiviral properties against enteric viruses using MNV-1 as a cultivable laboratory surrogate for HNoVs. GSE at 0.25, 0.5, and 1.0 mg/mL was shown to reduce MNV-1 by 0.82, 1.35, and 1.73 log PFU/mL, respectively, at 37°C for 2 h [78]. Cranberry proanthocyanidins were also reported to decrease MNV-1 titers by 1.6, 2.4, and 2.9 log PFU/mL after 0, 10, and 60 min, respectively [44]. Su et al. showed that commercial pomegranate juice could reduce MNV-1 by 1.32 log PFU/mL, while pomegranate polyphenols at 4 mg/mL reduced MNV-1 by 1.30 log PFU/mL after 1 h at room temperature [79]. Black raspberry juice (6%) was also shown to have antiviral effects on MNV-1 causing reduction in plaque titers by 99% after 1 h [43]. Cotreatment (simultaneous addition of treatment and virus to the host cells) was demonstrated to exert maximal antiviral activity on MNV-1, suggesting that the juice exerted its antiviral effect by inhibiting viral attachment to host cells, though gallic acid and quercetin were not shown to have any significant effect [43]. Aqueous hibiscus extracts were shown to reduce MNV-1 by 1.89 and 1.78 log PFU/mL after 6 h with 100 and 40 mg/mL of the extract, respectively, and to undetectable levels after 24 h with both 100 and 40 mg/mL of the extract at 37°C, respectively [80]. When host cells RAW 264.7 were pretreated with Korean Red Ginseng at 5, 6.7, and 10 μg/mL, a reduction in the infectivity of MNV-1 by 0.38 ± 0.41, 0.73 ± 0.19, and 1.48 ± 0.27 log $TCID_{50}$/mL from an initial titer of 6.7 ± 0.2 log $TCID_{50}$/mL after 24 h at 37°C was reported [81]. Oregano oil at 4% was reported to cause a reduction of 1.07 log $TCID_{50}$ in MNV-1 titers after 24 h, while carvacrol caused a 3.87 log $TCID_{50}$ decrease after 1 h at room temperature [82].

As immune knockout mice are readily available, MNV-1 is often used as a model to study the immune responses to HNoV infection and to examine the host responses required for norovirus pathogenesis, clearance and establishment of immunity, and insights for the future development of vaccines [21]. MNV-1 has also gained popularity as the surrogate or model system of choice to study HNoVs due to the development of two reverse genetics systems, one based on the polymerase II-driven expression of viral RNA and the second based on the T7 RNA polymerase-driven expression of viral RNA with the recovery of fully infectious virions by both systems that are used to determine the fundamentals of norovirus genome translation and replication [21]. The cleavage of the MNV protease–polymerase precursor form of RdRp into separate units and RNA secondary structures was shown to be essential for replication, while the correct sequence of the last nucleotide of the MNV genome, immediately upstream of the poly A tail was found to be essential for the recovery of infectious virus. Terminal sialic acid moieties on gangliosides function as an attachment receptor for MNV on murine macrophages [20,83].

However, the limitations of using MNV-1 as a model and surrogate need to be addressed. MNV-1 causes a persistent disease in its host that is different from HNoV as mice do not routinely develop diarrhea and cannot vomit, while HNoV causes acute rapid infection and subsequent shedding and clearance with sometimes longer shedding periods that can last for more than a week. MNV-1 is less genetically variable than the variable HNoV genome that is associated with repeated infection and short-term immunity [20]. The relevance of the apparent immune-cell-tropism of MNV-1 is currently unknown as HNoV tropism *in vivo* remains unclear to date [20]. Therefore, research on suitable cultivable surrogates and model systems for understanding HNoV infections and inactivation is ongoing.

5.4.3 Tulane Virus

Tulane virus (TV) is a cultivable enteric calicivirus and is also known to recognize the same receptors (human histoblood group antigens) as HNoVs [84]. TV is 36 nm in size with a genome size of about 6.7 kb single-stranded RNA with a poly(A) tail comprising of one of the shortest known genomes in the Caliciviridae family and belongs to the *Recovirus* genus. The viral genome that is enclosed in a nonenveloped capsid is divided into three ORFs, where ORF1 encodes for one nonstructural polyprotein, ORF2 encodes for the capsid protein, and ORF3 encodes for a minor structural protein [85]. TV is closely genetically related to HNoVs and pairwise homology has revealed the highest amino acid identity with HNoVs among other members of the Caliciviridae family [85]. TV was isolated from stool samples of monkeys without typical symptoms of gastrointestinal disease, questioning the ability of monkeys to be robust animal models. TV is known to bind to all B antigens (types 1–4) and type 3 of the A antigen. Since TV can be easily cultured and grown *in vitro*, it remains an important cultivable surrogate for HNoVs and a strong candidate as a laboratory model to study HNoVs.

TV has been used as a cultivable HNoV surrogate to determine inactivation and survival under different environmental stresses and conditions that include different pH, temperature, and chlorine levels [86]. Hirneisen and Kniel reported that TV could be reduced by 1.59, 0.96, and 0.82 log PFU/mL at pH values 2.0, 3.0, and 4.0, after 30 min at room temperature, respectively [62]. At temperatures of 50°C, 55°C, 60°C, and 65°C for 2 min, TV was reported to be reduced by 1.79, 1.83, 2.90, and 3.07 log PFU/mL, respectively [62]. Upon chlorine treatments of 0.2, 2, 20, and 200 ppm, TV was reported to be reduced by 1.33, 2.11, 1.53, and 2.93 log PFU/mL after 5 min. TV was also examined for its survival on spinach plants with and without exposure to sunlight [62]. It was found to persist on whole spinach plants and on both semisavoy and smooth spinach plants maintained in a greenhouse chamber for 7 days [62]. To assess the role of sunlight, these researchers exposed the spinach plants to UVA/UVB using a lamp to mimic sunlight for 10 h/day for 7 days; however, they did not find any significant difference in TV titers from the initial titer values.

The survival and dissemination of TV was also studied by inoculating the roots of romaine lettuce with TV and growing them for 2 weeks in hydroponic feed water [87]. These researchers detected about 3 log PFU/g of TV infectious particles in leaves and shoots after 2 days postinoculation; however, the levels of detected infectious particles in leaves and shoots reached up to 6 log PFU/g after 7 days postinoculation. Thus, TV was found to internalize in the roots of plants of lettuce and can be disseminated to the leaves and root shoots. It was also found to persist on alfalfa seeds for over 50 days at 22°C, where TV was reported to be reduced from an initial titer of 3.87–0.85 log PFU/g [88].

TV was also used to determine the effects of high hydrostatic pressure on its inactivation in cell-culture media, blueberries, and oysters [89]. These researchers showed that in cell-culture media, a treatment of 350 MPa caused a reduction of 3.8 and 2.4 log PFU/mL of TV at pH 7 and 4, respectively, after 2 min at 21°C, while 600 MPa at temperatures of 4°C, 21°C, and 35°C was shown to not cause any reduction of TV on dry blueberries after 2 min. However, it was shown that complete inactivation of TV occurred when blueberries were immersed in phosphate-buffered saline after 300 MPa treatment for 2 min at 4°C and also at 400 MPa for 2 min at 4°C. In oysters, TV reductions of 2.9 and 2 log PFU/mL were reported after a treatment of 250 MPa at 4°C and 21°C, respectively.

TV was also used as a surrogate to understand the effects of ionizing radiation by an electron beam, as a nonthermal processing approach on the viral infectivity of spiked produce, and to determine the mechanism of inactivation [90]. TV was reported to be reduced from 7 log PFU/mL to undetectable levels with 16 kGy or higher doses in strawberries and lettuce, while with lower target doses of 4 kGy, reductions of 1.5 and 1.8 log PFU/mL were reported in PBS and cell-culture media (Opti-MEM), respectively (Table 5.7). Transmission electron microscopy (TEM) did not reveal any significant damage to

TABLE 5.7

Thermal and Nonthermal Approaches for the Inactivation of Tulane Virus (TV) as a Cultivable HNoV Surrogate

Virus	Treatment	Reduction	References
Tulane virus	50°C, 2 min	1.79 log PFU/mL	[86]
	55°C, 2 min	1.83	
	60°C, 2 min	2.90	
	65°C, 2 min	3.07	
	High hydrostatic pressure; 350 MPa; cell-culture media	3.8 log PFU/mL	[89]
	High hydrostatic pressure; 600 MPa; dry blueberries	No reduction	[89]
	High hydrostatic pressure; 600 MPa; oysters	2.9 log PFU/mL	[89]
	16 kGy electron beam; strawberries	7 log PFU/mL	[90]
	70% ethanol; 20 s	5 log $TCID_{50}$	[91]
	UVC; 60 mJ/cm^2; 4 min	4 log $TCID_{50}$	[91]
	300 ppm free chlorine; 10 min	4 log $TCID_{50}$	[91]

the treated viral particles, though the numbers of detectable virions were reported to be reduced in the treated samples. Sodium dodecyl sulfate–polyacrylamide gel electrophoresis (SDS–PAGE) of the treated viral particles showed degraded viral proteins compared to the controls posttreatment [90]. TV was also used as an HNoV surrogate to determine inactivation by 50%–70% ethanol [91], where TV was reduced by 5 log $TCID_{50}$ after 20 s treatment, and heating at 63°C for 5 min and 56°C for 30 min, UVC exposure of 60 mJ/cm^2 for 4 min, and 300 ppm of free chlorine for 10 min were also reported to cause a reduction of 4 log $TCID_{50}$ [91].

TV infectivity was also determined in the presence of bacterial cell-free supernatants (CFS), showing reduction by 0.44, 0.37, 0.38, and 0.81 logs with CFS of *Bacillus coagulans*, *Escherichia coli*, *Enterococcus faecalis*, *Staphylococcus epidermidis* grown in tryptic soy broth (TSB), when host cells were overlaid with CFS at 37°C for 44 h [92]. These results should help toward understanding the factors affecting viral infections and how colonization of bacteria in the gut or the native microbiome can inhibit HNoV or other viral infections through competitive attachment [92].

Replication and biological studies have demonstrated that an authentic 5′ end of TV RNA or the N-terminal protein appears to be essential for infectivity, indicating that while nonstructural proteins appear to be sufficient for TV genomic RNA replication and translation, structural proteins are still needed to generate progeny viruses [20].

However, TV causes an uncharacterized disease in its rhesus macaque host and represents a separate genus than HNoV. Nonhuman primates such as common marmosets, cotton-top tamarins, and rhesus macaques serve as a good alternate model to study HNoV biology, as they seem to be susceptible to the HNoV GI Norwalk virus and shed virus after infection, though without exhibiting any disease symptoms [20]. Thus, based on the information provided above, TV may be another appropriate alternate model to study HNoV pathogenesis, survival, and transmission.

5.4.4 Porcine Sapovirus

Porcine sapovirus belongs to the genus *Sapovirus* within the Caliciviridae family, and the porcine sapovirus Cowden strain has been successfully propagated in the continuous porcine kidney cell line (LLC-PK) [93]. The presence of bile acid or intestinal content fluid filtrate obtained from uninfected gnotobiotic pigs was found to be necessary during preinfection for viral growth in cell-culture media [94]. Human sapoviruses are known to cause gastroenteritis in humans with similar symptoms to HNoVs. However, similar to HNoVs, they currently remain unculturable in the laboratory. From an epidemiological perspective, the HNoVs are considered to be more significant due to their prevalence/number of outbreaks as compared to *Sapoviruses*. The properties of porcine sapovirus that make it a suitable surrogate are that it is an enteric virus, is genetically related to HNoVs, is cultivable in the lab, replicates in intestinal cells of pigs, and causes gastroenteritis in piglets [95–97].

TABLE 5.8
Evaluation of Various Inactivation Methods on Porcine Enteric Virus (PEV) as a Cultivable Surrogate

Virus	Treatment	Reduction	References
Porcine enteric virus	56°C, 20 min	2 log $TCID_{50}$	[70]
	70% and 90% ethanol, 1 min	0.38×10^6 $TCID_{50}$	[70]
	200 ppm chlorine, 5 min	0.4 log $TCID_{50}$	[70]
	1000 ppm Chlorine, 5 min	1.3 log $TCID_{50}$	
	High hydrostatic pressure; 700 MPa	7 log $TCID_{50}$	[70]
	High hydrostatic pressure; 400 MPa	4 log $TCID_{50}$	

Porcine enteric caliciviruses (PEC) are reported to infect pigs of all ages, but cause diarrhea only in young piglets. Gnotobiotic pigs can become infected with HNoVs and therefore have been reported to be used to study and model antibody and cytokine responses [20]. PEC is subclinical in pigs, yet may play an important role in evolutionary modeling if they are a natural reservoir or are involved in zoonotic disease transfer and have been used as a model for tissue culture–derived enteric calicivirus attenuation [20]. To gain insight into PEC replication, a reverse genetics system that was developed for PEC showed that capped RNA transcripts derived from a cDNA clone were fully infectious [20].

Porcine sapovirus appears to be suitable as a cultivable HNoV for determining inactivation by various methods (Table 5.8) as PEC infectious (using $TCID_{50}$ assays) titers were reported to have no significant changes after exposure to pH 4–8 at room temperature for 1 h and showed <l-log reduction at pH 3 [88]. Porcine sapovirus was also reported to attach to lettuce leaves at its capsid isoelectric point (pH 5.0), and even remained infectious on lettuce after 1 week of storage at 4°C [88]. These researchers also showed that porcine sapovirus and HNoV shared similar resistance to heat and chlorine treatment. However, the stability of porcine sapovirus to food-processing technologies is yet to be determined. Hence, further studies using PEC as a model or surrogate for HNoVs and human sapoviruses are ongoing.

5.4.5 Bovine Noroviruses

Bovine NoVs are reported to be less closely related to HNoVs than porcine viruses, and the prototype bovine NoV strain Newbury agent 2 was identified in calves suffering from diarrhea. The overall similarity of bovine NoVs and their potential zoonotic transfer make them an alternative model for *in vivo* studies, although recently NoVs have also been identified in sheep [20].

5.4.6 Rabbit Caliciviruses

Rabbit hemorrhagic disease virus (RHDV), rabbit calicivirus (RCV) and, European brown hare syndrome virus (EBHSV) are caliciviruses of the *Lagovirus* genus, mainly of concern in America and China, that most often cause fatal disease with symptoms such as hepatitis and hemorrhages, and are virulent viruses of rabbits [20]. Their suitability as model systems for HNoVs is unclear as they cause different disease symptoms than HNoVs, while RCV is nonpathogenic and has not been completely characterized.

However, for RHDV, an animal model, a tissue culture system, and a reverse genetics system are reportedly available, and it has been demonstrated that the minor capsid protein VP2 of RHDV does not play a role in virus infectivity and also that the 3′ poly A tail has no effect on virus replication *in vitro* [20]. Thus, RHDV may be considered for use in an alternate animal model system to study HNoV biology.

5.4.7 Norovirus-Like Particles

Norovirus-like particles have also been used as a model and alternate surrogate in the laboratory for inactivation studies on HNoVs in order to investigate the structural and biochemical characteristics of

HNoVs after inactivation treatments. The recombinant expression of the VP1 major capsid protein of HNoV has been shown to result in self-assembly to form empty, noninfectious VLPs that have similar morphology and antigenic properties to the native infectious virus particle [114]. The protruding (P) domain of the major structural capsid protein that forms the outermost surface of the capsid contains the elements required for viral capsid binding to host carbohydrate receptors [89,95,98]. A variety of expression systems have been reported to be used for the expression of VLPs, including baculovirus insect cell and transgenic plant expression systems [99–102]. When the P protein is expressed in *E. coli*, subviral particles, namely P particles, are formed that may be used as a surrogate to study the binding ability of HNoVs, and it is the VLP receptor binding ability that has been proposed as a unique indicator for virus survival and infectivity [95], as well as for the determination of the damage of VLPs by using electron microscopy and SDS–PAGE techniques. These VLPs have been used in research to demonstrate the ability of gamma irradiation to disrupt the VLP structure causing degradation of the VP1 protein [61].

VLPs have also been used as a surrogate for determining the interaction/binding of HNoV with fresh produce, seafood, and other at-risk foods [95]. HNoV P particles, using saliva binding enzyme-linked immunosorbent assays, were used as an HNoV surrogate, and the presence of cranberry and pomegranate juices, showed reduced HNoV binding [103]. However, some fresh produce extracts that included strawberry, blackberry, blueberry, cherry tomato, spinach, romaine lettuce, or raspberry did not demonstrate any effect on the binding ability of these particles [103].

5.5 Bacteriophage MS2

Bacteriophage MS2 is a bacterial virus, commonly found in sewage and wastewaters, that infects the bacteria *E. coli* [in particular, American Type Culture Collection (ATCC) 15597B]. Bacteriophage MS2 is a single-stranded icosahedral RNA virus, between 27 and 34 nm in diameter, that belongs to the Leviviridae family in group 1 of the RNA coliphages and is adapted to the intestinal tract [104].

Bacteriophage MS2 has been used by various researchers as a gastrointestinal virus surrogate to determine disinfection and inactivation efficacy as well as survival and transmission in the environment, including the use of electrochemical oxidants (ECOs) for surface disinfection, due to its ease of propagation and relatively simplistic assays [105]. ECOs were generated using battery power to electrolyze brine (NaCl) solutions, which contain both chlorine [HOCl, OCl(−)] and reactive oxygen species (e.g., ·OH, O_3, H_2O_2, and O_2–), and at free available chlorine concentrations of 2500 ppm for a 30-s contact time, infective MS2 bacteriophage was reported to be inactivated by >7 log compared to the MNV-1 inactivation of ~2 log [105]. When these researchers compared genomic RNA inactivation, MS2 RNA inactivation was reported to be around 5 log compared to the MNV-1 RNA inactivation of around 1.5 log, with comparable inactivation efficacy to household bleach at similar free available chlorine concentrations. MS2 was reported to be reduced by ≥6 log PFU/mL at 5% TSP, and a 4.5 log PFU/mL reduction was reported to be achieved with 1% TSP after a contact time of 30 s [29]. Lee and Ko reported that MS2 was highly resistant to disinfection by UVA, but the addition of TiO_2 enhanced the efficacy of UVA, where a UVA dose of 1379 mJ/cm^2 resulted in a 4 log reduction, while UVB alone inactivated both MS2 by 4 log with a dose of 367 mJ/cm^2 [59]. A study of the virucidal inactivation efficacy of an in-house-designed atmospheric pressure, nonthermal plasma jet operated at varying helium–oxygen feed gas concentrations used MS2 bacteriophage as a surrogate and model for HNoV, showed that the inactivation rate constant increased with increasing oxygen concentrations up to 0.75% [106]. These researchers reported that 3 log (99.9%) reductions in infectious MS2 were obtained after 3 min of exposure in a helium–oxygen (99.25%:0.75%) gas mixture, with increased reduction after 9 min exposure by >7 log. Black et al. found that MS2 was inactivated by <1 log PFU/mL at hydrostatic pressures of 500 MPa for 5 min at 20°C, showing its resistance as compared to other surrogates and making it a suitable surrogate and model virus for hydrostatic pressure studies [107]. Dawson et al. studied the survivability/persistence of MS2 bacteriophage on fresh iceberg lettuce, baton carrot, cabbage, spring onion, curly leaf parsley, capsicum pepper, tomato, cucumber, raspberries, and strawberries and showed that MS2 survived for a very long time and even extended the shelf life of the produce, where <1 log PFU/mL reduction was observed after 50 days at 4°C and 8°C [104]. In a study determining the survival of HNoV in juices, bacteriophage MS2

(~6 log PFU/mL) was used as a surrogate and model that showed significant reduction (1.93 log PFU/mL) after 2 days and was undetectable after 7 days in blueberry juice at 4°C [48].

5.6 Conclusions

HNoVs are a worldwide public health and food safety concern. However, in the absence of reproducible cell-culture-based infectivity assays and due to the lack of available vaccines to date, cultivable surrogates and model systems are used in order to determine survival, transmission, and inactivation methods to prevent outbreaks and minimize the risk of HNoV contamination. Typically, cultivable surrogates that mimic the characteristics of HNoVs are genetically related and are used in laboratory settings. These surrogates include cultivable animal viruses from the Caliciviridae family as well as bacteriophages and noroviral-like particles. The application of these surrogates in environmental studies and to enhance food safety remains ongoing. However, the ideal situation of having reproducible infectivity assays for HNoVs appears to be a challenge and research efforts continue to be made toward attaining this goal.

It is evident that a reproducible and validated cell-culture system for the propagation of HNoVs is needed to understand HNoV biology, replication, and molecular aspects. Toward this end, researchers have still to validate a B-cell replication system for HNoVs [19]. Virus entry and genome uncoating are considered the "vital missing link" for HNoV propagation [20]. Despite the availability of a variety of animal models and surrogates for understanding HNoVs, there does not appear to be one single overall "ideal" model system, whose suitability varies and depends on the topic being researched. As indicated and described in detail by Vashist et al., for water-related and environmental studies, bacteriophages such as MS2 and phiX-174 have been used as indicators and surrogates in the past to determine survival and transmission routes. For studying pathogenesis, animal caliciviruses appear to be most suitable, yet none of them reproduce all the HNoV disease aspects. Most animal caliciviruses appear to be suitable models for genome translation and replication studies, and MNV-1 appears to be most closely related or suitable for HNoVs [20]. With regards to vaccines and antivirals, progress seems to be limited and slow, and small molecule inhibitors, natural plant polyphenols, and antisense RNA technologies show potential for future application (further suggested reading on NoVs in reference [20,108]). Thus, in conclusion, research emphasis should continue to be placed on the propagation of HNoV in cell-culture systems *in vitro* to obtain reproducible HNoV infectivity assays.

Acknowledgments

The authors gratefully acknowledge the support provided by the University of Tennessee-Institute of Agriculture. The authors also wish to thank Carrie Yard for her assistance with the formatting some of the references during the final stage of this chapter submission.

REFERENCES

1. Scipioni, A., et al., Animal noroviruses. *Vet J*, 2008. 178(1): 32–45.
2. Green, K., et al., Taxonomy of the caliciviruses. *J Infect Dis*, 2000. 181(Suppl 2): S322–S330.
3. Kapikian, A.Z., The discovery of the 27-nm Norwalk virus: an historic perspective. *J Infect Dis*, 2000. 181(Suppl 2): S295–S302.
4. Seah, E.L., J.A. Marshall, and P.J. Wright, Open reading frame 1 of the Norwalk-like virus Camberwell: completion of sequence and expression in mammalian cells. *J Virol*, 1999. 73(12): 10531–10535.
5. D'Souza, D.H. and S.S. Joshi, Foodborne viruses of human health concern, in *The Encyclopedia of Food and Health*, vol. 3, Caballero, B., Finglas, P., and Toldrá, F., (Eds). 2016, Academic Press, Oxford, pp. 87–93.
6. Cannon, J.L., et al., Surrogates for the study of norovirus stability and inactivation in the environment: a comparison of murine norovirus and feline calicivirus. *J Food Prot*, 2006. 69(11): 2761–2765.
7. Zheng, D.P., et al., Norovirus classification and proposed strain nomenclature. *Virology*, 2006. 346(2): 312–323.

8. Xerry, J., et al., Genetic characterization of genogroup I norovirus in outbreaks of gastroenteritis. *J Clin Microbiol*, 2010. 48(7): 2560–2562.
9. Zheng, D.P., et al., Molecular epidemiology of genogroup II-genotype 4 noroviruses in the United States between 1994 and 2006. *J Clin Microbiol*, 2010. 48(1): 168–177.
10. Fankhauser, R.L., et al., Epidemiologic and molecular trends of "Norwalk-like viruses" associated with outbreaks of gastroenteritis in the United States. *J Infect Dis*, 2002. 186(1): 1–7.
11. Donaldson, E.F., et al., Norovirus pathogenesis: mechanisms of persistence and immune evasion in human populations. *Immunol Rev*, 2008. 225(1): 190–211.
12. CDC, Norovirus outbreaks on three college campuses—California, Michigan, and Wisconsin, 2008. *MMWR*, 2009. 58(39): 1095–1100.
13. Pang, X.L., et al., Influence of novel norovirus GII.4 variants on gastroenteritis outbreak dynamics in Alberta and the Northern Territories, Canada between 2000 and 2008. *PLoS One*, 2010. 5(7): e11599.
14. Siebenga, J.J., et al., Norovirus illness is a global problem: emergence and spread of norovirus GII.4 variants, 2001–2007. *J Infect Dis*, 2009. 200(5): 802.
15. Lopman, B.A., et al., A summertime peak of "winter vomiting disease": surveillance of noroviruses in England and Wales, 1995 to 2002. *BMC Public Health*, 2003. 3: 1–4.
16. Malek, M., et al., Outbreak of norovirus infection among river rafters associated with packaged delicatessen meat, Grand Canyon, 2005. *Clin Infect Dis*, 2009. 48(1): 31–37.
17. Escudero-Abarca, B., et al., Molecular methods used to estimate thermal inactivation of a prototype human norovirus: more heat resistant than previously believed? *Food Microbiol*, 2014. 41: 91–95.
18. Papafragkou, E., et al., Challenges of culturing human norovirus in three-dimensional organoid intestinal cell culture models. *PLoS One*, 2013. 8(6): e63485.
19. Jones, M.K., et al., Enteric bacteria promote human and mouse norovirus infection of B cells. *Science*, 2014. 346(6210): 755–759.
20. Vashist, S., et al., Model systems for the study of human norovirus biology. *Future Virol*, 2009. 4(4): 353–367.
21. Leon, J., et al., Randomized, double-blinded clinical trial for human norovirus inactivation in oysters by high hydrostatic pressure processing. *Appl Environ Microbiol*, 2011. 77(15): 5476–5482.
22. Grohmann, G., et al., Norwalk virus gastroenteritis in volunteers consuming depurated oysters. *Aust J Exp Biol Med Sci*, 1981. 59(2): 219–228.
23. Richards, G.P., Critical review of norovirus surrogates in food safety research: rationale for considering volunteer studies. *Food Environ Virol*, 2012. 4(1): 6–13.
24. Thiel, H.-J. and M. König, Caliciviruses: an overview. *Vet Microbiol*, 1999. 69(1): 55–62.
25. Radford, A.D., et al., Feline calicivirus. *Vet Res*, 2007. 38(2): 319–335.
26. Doultree, J.C., et al., Inactivation of feline calicivirus, a Norwalk virus surrogate. *J Hosp Infect*, 1999. 41(1): 51–57.
27. Greening, G.E., et al., Molecular epidemiology of norovirus gastroenteritis outbreaks in New Zealand from 2002–2009. *J Med Virol*, 2012. 84(9): 1449–1458.
28. Su, X. and D.H. D'Souza, Inactivation of human norovirus surrogates by benzalkonium chloride, potassium peroxymonosulfate, tannic acid, and gallic acid. *Foodborne Pathog Dis*, 2012. 9(9): 829–834.
29. D'Souza, D.H. and X. Su, Efficacy of chemical treatments against murine norovirus, feline calicivirus, and MS2 bacteriophage. *Foodborne Pathog Dis*, 2010. 7(3): 319–326.
30. Chander, Y., et al., Antiviral activity of Ecasol against feline calicivirus, a surrogate of human norovirus. *J Infect Public Health*, 2012. 5(6): 420–424.
31. Morino, H., et al., Effect of low-concentration chlorine dioxide gas against bacteria and viruses on a glass surface in wet environments. *Lett Appl Microbiol*, 2011. 53(6): 628–634.
32. Tibollo, S., et al., High hydrostatic pressure activity on the disinfection of clams artificially contaminated with feline calicivirus. *Ann Ig*, 2013. 25(3): 201–208.
33. Thurston-Enriquez, J.A., et al., Inactivation of enteric adenovirus and feline calicivirus by ozone. *Water Res*, 2005. 39(15): 3650–3656.
34. Hirneisen, K.A. and K.E. Kniel, Inactivation of internalized and surface contaminated enteric viruses in green onions. *Int J Food Microbiol*, 2013. 166(2): 201–206.
35. Fino, V.R. and K.E. Kniel, UV light inactivation of hepatitis A virus, Aichi virus, and feline calicivirus on strawberries, green onions, and lettuce. *J Food Prot*, 2008. 71(5): 908–913.

36. Schultz, A., et al., Inactivation of norovirus surrogates on surfaces and raspberries by steam-ultrasound treatment. *J Food Prot*, 2012. 75(2): 376–381.
37. Horm, K.M., et al., Survival and inactivation of human norovirus surrogates in blueberry juice by high-pressure homogenization. *Foodborne Pathog Dis*, 2012. 9(11): 974–979.
38. Su, X., S. Zivanovic, and D. D'Souza, Inactivation of human enteric virus surrogates by high-intensity ultrasound. *Foodborne Pathog Dis*, 2010. 7(9): 1055–1061.
39. Zhou, F., et al., Inactivation of feline calicivirus as a surrogate for norovirus on lettuce by electron beam irradiation. *J Food Prot*, 2011. 74(9): 1500–1503.
40. Bozkurt, H., D.H. D'Souza, and P.M. Davidson, Thermal inactivation of human norovirus surrogates in spinach and measurement of its uncertainty. *J Food Prot*, 2014. 77(2): 276–283.
41. Lee, J.H., et al., Antiviral effects of mulberry (*Morus alba*) juice and its fractions on foodborne viral surrogates. *Foodborne Pathog Dis*, 2014. 11(3): 224–229.
42. Sanchez, G. and R. Aznar, Evaluation of natural compounds of plant origin for inactivation of enteric viruses. *Food Environ Virol*, 2015: 183–187.
43. Oh, M., et al., Antiviral effects of black raspberry (*Rubus coreanus*) juice on foodborne viral surrogates. *Foodborne Pathog Dis*, 2012. 9(10): 915–921.
44. Su, X., A.B. Howell, and D.H. D'Souza, Antiviral effects of cranberry juice and cranberry proanthocyanidins on foodborne viral surrogates—a time dependence study in vitro. *Food Microbiol*, 2010. 27(8): 985.
45. Su, X. and D.H. D'Souza, Grape seed extract for foodborne virus reduction on produce. *Food Microbiol*, 2013. 34(1): 1–6.
46. D'Souza, D.H., et al., Persistence of caliciviruses on environmental surfaces and their transfer to food. *Int J Food Microbiol*, 2006. 108(1): 84–91.
47. Mormann, S., et al., Tenacity of human norovirus and the surrogates feline calicivirus and murine norovirus during long-term storage on common nonporous food contact surfaces. *J Food Prot*, 2015. 78(1): 224–229.
48. Horm, K.M. and D.H. D'Souza, Survival of human norovirus surrogates in milk, orange, and pomegranate juice, and juice blends at refrigeration (4°C). *Food Microbiol*, 2011. 28(5): 1054–1061.
49. Horm, K., F. Harte, and D. D'Souza, Human norovirus surrogate reduction in milk and juice blends by high pressure homogenization. *J Food Prot*, 2012. 75(11): 1984–1990.
50. Sosnovtsev, S. and K. Green, RNA transcripts derived from a cloned full-length copy of the feline calicivirus genome do not require VPG for infectivity. *Virology*, 1995. 210(2): 383–390.
51. Willcocks, M.M., M.J. Carter, and L.O. Roberts, Cleavage of eukaryotic initiation factor elF4G and inhibition of host-cell protein synthesis during feline calicivirus infection. *J Gen Virol*, 2004. 85(5): 1125–1130.
52. Makino, A., et al., Junctional adhesion molecule 1 is a functional receptor for feline calicivirus. *J Virol*, 2006. 80(9): 4482–4490.
53. Karst, S.M., et al., STAT1-dependent innate immunity to a Norwalk-like virus. *Science*, 2003. 299(5612): 1575–1578.
54. Wobus, C.E., L.B. Thackray, and H.W. Virgin, 4th, Murine norovirus: a model system to study norovirus biology and pathogenesis. *J Virol*, 2006. 80(11): 5104–5112.
55. Bae, J. and K.J. Schwab, Evaluation of murine norovirus, feline calicivirus, poliovirus, and MS2 as surrogates for human norovirus in a model of viral persistence in surface water and groundwater. *Appl Environ Microbiol*, 2008. 74(2): 477–484.
56. Lee, J., K. Zoh, and G. Ko, Inactivation and UV disinfection of murine norovirus with TiO_2 under various environmental conditions. *Appl Environ Microbiol*, 2008. 74(7): 2111–2117.
57. Baert, L., et al., Survival and transfer of murine norovirus 1, a surrogate for human noroviruses, during the production process of deep-frozen onions and spinach. *J Food Prot*, 2008. 71(8): 1590–1597.
58. Escudero, B., et al., Persistence and transferability of noroviruses on and between common surfaces and foods. *J Food Prot*, 2012. 75(5): 927–935.
59. Lim, M.Y., J.M. Kim, and G. Ko, Disinfection kinetics of murine norovirus using chlorine and chlorine dioxide. *Water Res*, 2010. 44(10): 3243–3251.
60. Fraisse, A., et al., Comparison of chlorine and peroxyacetic-based disinfectant to inactivate feline calicivirus, murine norovirus and hepatitis A virus on lettuce. *Int J Food Microbiol*, 2011. 151(1): 98–104.

61. Herdt, J. and H. Feng, Aqueous antimicrobial treatments to improve fresh and fresh-cut produce safety, in *Microbial Safety of Fresh Produce*, Fan, X., et al., (Eds). 2009, IFT Press, Wiley-Blackwell, Ames, IA, pp. 167–190.
62. Hirneisen, K. and K. Kniel, Norovirus surrogate survival on spinach during preharvest growth. *Phytopathology*, 2013. 103(4): 389–394.
63. Hirneisen, K.A., S.M. Markland, and K.E. Kniel, Ozone inactivation of norovirus surrogates on fresh produce. *J Food Prot*, 2011. 74(5): 836–839.
64. Su, X. and D.H. D'Souza, Trisodium phosphate for foodborne virus reduction on produce. *Foodborne Pathog Dis*, 2011. 8(6): 713–717.
65. Cannon, J.L., et al., Efficacy of a levulinic acid plus sodium dodecyl sulfate-based sanitizer on inactivation of human norovirus surrogates. *J Food Prot*, 2012. 75(8): 1532–1535.
66. Predmore, A. and J. Li, Enhanced removal of a human norovirus surrogate from fresh vegetables and fruits by a combination of surfactants and sanitizers. *Appl Environ Microbiol*, 2011. 77(14): 4829–4838.
67. Tang, Q., et al., Mechanism of inactivation of murine norovirus-1 by high pressure processing. *Int J Food Microbiol*, 2010. 137(2–3): 186–189.
68. Kingsley, D.H., et al., Inactivation of hepatitis A virus and a calicivirus by high hydrostatic pressure. *J Food Prot*, 2002. 65(10): 1605–1609.
69. Kingsley, D.H., et al., Inactivation of a norovirus by high-pressure processing. *Appl Environ Microbiol*, 2007. 73(2): 581–585.
70. Cromeans, T., et al., Comprehensive comparison of cultivable norovirus surrogates in response to different inactivation and disinfection treatments. *Appl Environ Microbiol*, 2014. 80(18): 5743–5751.
71. Arcangeli, G., et al., Effect of high hydrostatic pressure on murine norovirus in Manila clams. *Lett Appl Microbiol*, 2012. 54(4): 325–329.
72. Praveen, C., et al., Susceptibility of murine norovirus and hepatitis A virus to electron beam irradiation in oysters and quantifying the reduction in potential infection risks. *Appl Environ Microbiol*, 2013. 79(12): 3796–3801.
73. Bozkurt, H., D.H. D'Souza, and P.M. Davidson, Determination of the thermal inactivation kinetics of the human norovirus surrogates, murine norovirus and feline calicivirus. *J Food Prot*, 2013. 76(1): 79–84.
74. Bozkurt, H., D.H. D'Souza, and P.M. Davidson, A comparison of the thermal inactivation kinetics of human norovirus surrogates and hepatitis A virus in buffered cell culture medium. *Food Microbiol*, 2014. 42: 212–217.
75. Bozkurt, H., et al., Thermal inactivation kinetic modeling of human norovirus surrogates in blue mussel (*Mytilus edulis*) homogenate. *Int J Food Microbiol*, 2014. 172: 130–136.
76. Sow, H., et al., Heat inactivation of hepatitis A virus and a norovirus surrogate in soft-shell clams (*Mya arenaria*). *Foodborne Pathog Dis*, 2011. 8(3): 387–393.
77. Park, S.Y., et al., Thermal inactivation of murine norovirus-1 in suspension and in dried mussels (*Mytilus edulis*). *J Food Safety*, 2014. 34(3): 193–198.
78. Su, X. and D.H. D'Souza, Grape seed extract for control of human enteric viruses. *Appl Environ Microbiol*, 2011. 77(12): 3982–3987.
79. Su, X., M.Y. Sangster, and D.H. D'Souza, In vitro effects of pomegranate juice and pomegranate polyphenols on foodborne viral surrogates. *Foodborne Pathog Dis*, 2010. 7(12): 1473–1479.
80. Joshi, S.S., L. Dice, and D.H. D'Souza, Aqueous extracts of *Hibiscus sabdariffa* calyces decrease hepatitis A virus and human norovirus surrogate titers. *Food Environ Virol*, 2015 7(4): 366–373.
81. Lee, M.H., et al., Antiviral effect of Korean red ginseng extract and ginsenosides on murine norovirus and feline calicivirus as surrogates for human norovirus. *J Ginseng Res*, 2011. 35(4): 429–435.
82. Gilling, D.H., et al., Antiviral efficacy and mechanisms of action of oregano essential oil and its primary component carvacrol against murine norovirus. *J Appl Microbiol*, 2014. 116(5): 1149–1163.
83. Taube, S., et al., Ganglioside-linked terminal sialic acid moieties on murine macrophages function as attachment receptors for murine noroviruses. *J Virol*, 2009. 83(9): 4092–4101.
84. Zhang, D., et al., Tulane virus recognizes the A type 3 and B histo-blood group antigens. *J Virol*, 2015. 89(2): 1419–1427.
85. Farkas, T., et al., Characterization of a rhesus monkey calicivirus representing a new genus of Caliciviridae. *J Virol*, 2008. 82(11): 5408–5416.

86. Hirneisen, K.A. and K.E. Kniel, Comparing human norovirus surrogates: murine norovirus and Tulane virus. *J Food Prot*, 2013. 76(1): 139–143.
87. Dicaprio, E., et al., Internalization and dissemination of human norovirus and animal caliciviruses in hydroponically grown romaine lettuce. *Appl Environ Microbiol*, 2012. 78(17): 6143–6152.
88. Wang, Q., et al., Survival of murine norovirus, Tulane virus, and hepatitis A virus on alfalfa seeds and sprouts during storage and germination. *Appl Environ Microbiol*, 2013. 79(22): 7021–7027.
89. Li, X., et al., Pressure inactivation of Tulane virus, a candidate surrogate for human norovirus and its potential application in food industry. *Int J Food Microbiol*, 2013. 162(1): 37.
90. Predmore, A., et al., Electron beam inactivation of Tulane virus on fresh produce, and mechanism of inactivation of human norovirus surrogates by electron beam irradiation. *Int J Food Microbiol*, 2015. 198: 28–36.
91. Tian, P., et al., Inactivation of the Tulane virus, a novel surrogate for the human norovirus. *J Food Prot*, 2013. 76(4): 712–718.
92. Shearer, A., D.G. Hoover, and K. Kniel, Effect of bacterial cell-free supernatants on infectivity of norovirus surrogates. *J Food Prot*, 2014. 77(1): 145–149.
93. Saif, L.J., et al., Rotavirus-like, calicivirus-like, and 23-nm virus-like particles associated with diarrhea in young pigs. *J Clin Microbiol*, 1980. 12(1): 105–111.
94. Chang, K.-O., et al., Bile acids are essential for porcine enteric calicivirus replication in association with down-regulation of signal transducer and activator of transcription 1. *Proc Natl Acad Sci USA*, 2004. 101(23): 8733–8738.
95. Li, J., et al., New interventions against human norovirus: progress, opportunities, and challenges. *Annu Rev Food Sci Technol*, 2012. 3: 331–352.
96. Flynn, W.T. and L.J. Saif, Serial propagation of porcine enteric calicivirus-like virus in primary porcine kidney cell cultures. *J Clin Microbiol*, 1988. 26(2): 206–212.
97. Guo, M., et al., Comparative pathogenesis of tissue culture-adapted and wild-type Cowden porcine enteric calicivirus (PEC) in gnotobiotic pigs and induction of diarrhea by intravenous inoculation of wild-type PEC. *J Virol*, 2001. 75(19): 9239–9251.
98. Tan, M., R.S. Hegde, and X. Jiang, The P domain of norovirus capsid protein forms dimer and binds to histo-blood group antigen receptors. *J Virol*, 2004. 78(12): 6233–6242.
99. Mason, H.S., et al., Expression of Norwalk virus capsid protein in transgenic tobacco and potato and its oral immunogenicity in mice. *Proc Natl Acad Sci USA*, 1996. 93(11): 5335–5340.
100. Santi, L., et al., An efficient plant viral expression system generating orally immunogenic Norwalk virus-like particles. *Vaccine*, 2008. 26(15): 1846–1854.
101. Jiang, X., et al., Expression, self-assembly, and antigenicity of the Norwalk virus capsid protein. *J Virol*, 1992. 66(11): 6527–6532.
102. Koho, T., et al., Production and characterization of virus-like particles and the P domain protein of GII. 4 norovirus. *J Virol Methods*, 2012. 179(1): 1–7.
103. Li, D., et al., Effects of a variety of food extracts and juices on the specific binding ability of norovirus GII.4 P particles. *J Food Prot*, 2012. 7(75): 1350–1354.
104. Dawson, D., et al., Survival of viruses on fresh produce, using MS2 as a surrogate for norovirus. *J Appl Microbiol*, 2005. 98(1): 203–209.
105. Julian, T.R., J.M. Trumble, and K.J. Schwab, Evaluating efficacy of field-generated electrochemical oxidants on disinfection of fomites using bacteriophage MS2 and mouse norovirus MNV-1 as pathogenic virus surrogates. *Food Environ Virol*, 2014. 6(2): 145–155.
106. Alshraiedeh, N., et al., Atmospheric pressure, nonthermal plasma inactivation of MS2 bacteriophage: effect of oxygen concentration on virucidal activity. *J Appl Microbiol*, 2013. 115(6): 1420–1426.
107. Black, E., et al., Coliphage as pressure surrogates for enteric viruses in foods. *Innov Food Sci Emerg Technol*, 2010. 11(2): 239–244.
108. Jaykus, L., D.H. D'Souza, and C.L. Moe, Foodborne viral pathogens, in *Food Microbiology, Fundamentals and Frontiers*, Doyle, M.P. and Buchanan, R.L., (Eds). 2014, ASM Press, Washington, DC.
109. Goyal, S.M., et al., Evaluating the virucidal efficacy of hydrogen peroxide vapour. *J Hosp Infect*, 2014. 86(4): 255–259.
110. Steinmann, J., Evaluation of effectiveness of Bacoban WB against feline calicivirus as surrogate for human norovirus. 2005; Available from: http://www.bacoban.de/downloads/zertifikate/Bacoban%20WB/Norovirus_EN14476_EN.pdf.

111. Nims, R. and M. Plavsic, Inactivation of caliciviruses. *Pharmaceuticals (Basel)*, 2013. 6(3): 358–392.
112. Park, G.W. and M.D. Sobsey, Simultaneous comparison of murine norovirus, feline calicivirus, coliphage MS2, and GII.4 norovirus to evaluate the efficacy of sodium hypochlorite against human norovirus on a fecally soiled stainless steel surface. *Foodborne Pathog Dis*, 2011. 8(9): 1005–1010.
113. De Roda Husman, A.M., et al., Calicivirus inactivation by nonionizing (253.7-nanometer-wavelength [UV]) and ionizing (gamma) radiation. *Appl Environ Microbiol*, 2004. 70(9): 5089–5093.
114. Mohr, H., B. Lambrecht, and A. Selz, Photodynamic virus inactivation of blood components. *Immunol Invest*, 1995. 24(1–2): 73–85.
115. Chen, H., D.G. Hoover, and D.H. Kingsley, Temperature and treatment time influence high hydrostatic pressure inactivation of feline calicivirus, a norovirus surrogate. *J Food Prot*, 2005. 68(11): 2389–2394.
116. Morino, H., et al., Inactivation of feline calicivirus by chlorine dioxide gas-generating gel. *Yakugaku Zasshi*, 2013. 133(9): 1017–1022.
117. Kingsley, D.H., H. Chen, and D.G. Hoover, Inactivation of selected picornaviruses by high hydrostatic pressure. *Virus Res*, 2004. 102(2): 221–224.

6

Rotavirus

Lijuan Yuan and Ke Wen

CONTENTS

6.1 Introduction

Rotaviruses are recognized as the leading cause of severe gastroenteritis in infants and young children worldwide since the discovery of human rotavirus (HRV) in the early 1970s.[1–3] Rotaviruses are transmitted via the fecal oral route, are highly contagious and stable in the environment,[4] and post a high risk of foodborne infections. Rotavirus is a nonenveloped, 70-nm icosahedral virus in the family of Reoviridae. It has a genome consisting of 11 double-stranded RNA segments surrounded by a distinctive three-layered protein capsid.[5] These 11 segments encode six structural proteins, VP1–4, VP6, and VP7, and six nonstructural proteins, NSP1–6.[6] Among the 11 viral genes, the properties of the proteins encoded by genes 3, 4, 5, 9, and 10 are known to be related to rotavirus virulence in the host.[7] Gene 3 encodes the capping enzyme that affects the level of viral RNA replication.[8,9] Genes 4 and 9 produce the outer capsid proteins VP4 and VP7 required to initiate infection, and they induce virus neutralizing antibodies independently from each other.[10,11] Gene 5 codes NSP1 that functions as an interferon antagonist.[12–14] Gene 10 codes for the nonstructural protein NSP4, which regulates calcium homeostasis and virus replication and acts as an enterotoxin.[7] NSP2 (encoded by gene segment 8) is also involved in virulence, especially in mice.[15]

Seven rotavirus groups (A to G) are known based on the antigenicity of VP6 and sequences of genomic RNA.[16] Among them, Groups A, B, and C are associated with illness in humans and animals, and Groups D through G with illness in animals only. Group A rotaviruses are the most common cause of illness in infants and young children worldwide. The antigenicity of Group A rotaviruses is highly diverse, and about 19 G (based on VP7) and 28 [P] (based on VP4) sero/genotypes have been identified.[17] Serotypes G1, G3, G4, and G9 are responsible for 90% of all HRV infections in North America and Europe but for less than 70% infections in Africa.[18] Genotypes P[8] and P[4] account for over 90% of circulating P types worldwide, whereas in Africa, the relative frequency of these two P types is lower and P[6] accounts for around a third of all detected P types.[18] There are potentially "2^{11}" different combinations of G and P proteins that can be generated; however, the actual number of G and P combinations is less than the possible number because most combinations do not survive subsequent rounds of replication in the host due to lack of fitness. The different rotaviruses circulating in humans are categorized as (1) common human genotypes (G1P[8], G2P[4], G3P[8], G4P[8]); (2) reassortants among human genotypes (G1P[4], G2P[8], G4P[4]); (3) reassortants between animal and human genotypes (G1P[9], G4P[6], G9P[8], G12P[8]); and (4) likely zoonotic introductions (G9P[6], G9P[11], G10P[11], G12P[6]).[17,19] Among these, G1, G2, G3, and G4 in combination with P[4] and P[8] represented over 88% of strains worldwide based on studies published between 1998 and 2004.[18] The data from the surveillance networks of the WHO also indicate that G1P[8], G9P[8], and G2P[4] account for 75% of samples genotyped in North America, Europe, southeast Asia, and western Pacific regions.

Rotavirus replicates in the small intestine and causes gastroenteritis. The symptoms of rotavirus infection are severe watery diarrhea, fever, and vomiting, leading to fluid and electrolyte disequilibrium. Severe diarrhea without fluid and electrolyte replacement may result in other secondary complications (e.g., renal failure and death).[20] Before the implementation of rotavirus vaccines in many developed and some developing countries starting from 2006, it was estimated that rotavirus infections annually accounted for more than 125 million cases of diarrhea (20%–50% of hospitalizations for gastroenteritis) and 600,000–870,000 deaths (around 20% of all diarrhea-associated deaths) in children under 5 years of age worldwide.[21] In the United States, rotavirus infections caused an estimated 410,000 physician visits and 50,000–70,000 hospitalizations each year before rotavirus vaccines were implemented in the vaccination program.[22] Following implementation of rotavirus vaccination in 2006, all-cause acute gastroenteritis hospitalization rates among US children younger than 5 years declined by 31%–55% in each of the postvaccine years from 2008 to 2012[23] and the laboratory detection rate of rotavirus declined by 57.8%–89.9% in each of the seven postvaccine years from 2007 to 2014.[24]

Because rotaviruses are highly contagious and stable in the environment, exposure in closed environments usually results in rapid, widespread infections among susceptible individuals. Outbreaks of rotavirus infection are common in childcare centers, and rotavirus is one of the most common causes of nosocomial infection, especially in pediatric hospitals.[25,26] In spite of improved sanitation and water supply, rotavirus infections remain an important cause of pediatric hospitalization in many countries where rotavirus vaccines have not been implemented.

Given that rotaviruses are the single most important cause of severe acute gastroenteritis in infants and young children, the worldwide impact of rotaviruses on public health has led to major research efforts in the past 40 years in understanding rotavirus replication, host–pathogen interactions, pathogenesis, and immunity, and in vaccine development to control rotavirus diarrhea. Animal models including mice, rats, rabbits, pigs, lambs, calves, and nonhuman primates have been used extensively in these studies. Among these animal models, gnotobiotic (Gn) piglets were among the first animal models used to study HRV right after the virus was associated with infantile gastroenteritis. Since then, the model has continuously contributed to our understanding of rotavirus pathogenesis and immunity. Various mouse models with different genetic backgrounds and immune competence are used extensively for studies of rotavirus–host interactions, mechanisms of host range restriction, and protective immunity. In this chapter, animal models pertaining to the establishment and applications in studies of rotavirus–host interactions, pathogenesis, immunity, and vaccine development are reviewed with emphasis on mouse and Gn pig models. The advantages, disadvantages, and limitations of the animal models are discussed. Cell lines and tissue culture models that support rotavirus replication and have been used in rotavirus research are summarized.

6.2 Overview of Animal Models for Human Rotavirus (HRV) Research

Mice are resistant to HRV infection due to the species barrier. However, several murine rotaviruses isolated from mouse colonies are used for studies of rotavirus biology and immunity as homologous model systems. The diversity of murine rotaviruses is more limited as compared with HRV. All the known murine rotaviruses belong to one single G type, G3, and two P types, P[16] or P[20]. The epidemic diarrhea of infant mice (EDIM) strain and all other murine rotavirus strains (EW, EB, EC, EL, EMcN, YR-1, YR-2) are G3P[16], except for EHP which is G3P[20].[27]

In comparison to murine rotaviruses, porcine rotaviruses are more closely related to HRV in genetic composition (gene constellation). There is a common origin between Wa-like HRV strains and porcine rotavirus strains.[28] The interspecies reassortment between porcine rotaviruses and HRV and zoonotic transmissions of reassortant porcine rotaviruses to human patients are frequently reported in the literature.[29–33] This close relationship forms the basis for the establishment of the Gn pig model of HRV infection and disease. The mortality of rotavirus infection in pigs is usually low, and the severity of disease is dependent on the rotavirus strain and the age of inoculation, with more severe disease and mortality in pigs inoculated with porcine rotavirus strains than with HRV.[34] However, colostrum-deprived neonatal Gn pigs can be readily infected by a non-cell-culture-adapted virulent HRV Wa strain (G1P[8]) and M strain G3P[8]) and develop gastroenteritis similar to human infants and young children.[35–37] It is important to note that only Gn pigs develop diarrhea for up to 8 weeks of age from HRV infection. No conventional laboratory animals such as mice, rats, and rabbits develop rotavirus diarrhea for nearly as long. This makes the Gn pig model the only animal model of HRV disease that allows for the assessment of active immunity against diarrhea over an extended period.

Newborn Gn calves and lambs are also susceptible to HRV infection.[38,39] After oral inoculation of calves with a "reovirus-like agent" of human infantile gastroenteritis, the same rotavirus inoculum used in Gn pigs,[39] fecal virus shedding was detected for 2–7 days using electron microscopy. Diarrhea (for less than 24 h) was observed in seven calves on the second to fourth serial passage of HRV in calves but in none of the four animals inoculated with the first passage.[40] Rotavirus infection in newborn calves leads to a change in the villus epithelium from columnar to cuboidal, causing villi to become stunted and shortened.[41] The Gn calf model was then used to test the feasibility of heterotypic protective immunity against HRV. Gn calves were infected *in utero* with the bovine rotavirus NCDV strain and challenged with the HRV D strain or NCDV strain shortly after birth.[42] Infection *in utero* with bovine rotavirus induced resistance to diarrhea caused by HRV as well as the homologous bovine virus. This finding led to the attempts of using bovine rotavirus as a candidate vaccine for human infants.[43,44] Gn lambs are limited to studies of immunity to lamb and bovine rotavirus infection and vaccines.[45–47]

Nonhuman primates are animals genetically most closely related to humans. They are naturally infected with simian rotaviruses; these viruses isolated so far include SA11 (G3P5B[2] or G3P6[1]), RRV MMU18006 (G3P5B[3]), YK-1 (G3P[3]), TUCH (G3P23[24]), and PTRV (G8P6[1]).[48–51] However, the genetic and antigenic relationships between HRV and simian rotaviruses are more distant than those between HRV and porcine rotaviruses.[28] Various simian rotaviruses isolated from different nonhuman primates and several HRV isolates have been tested or suggested for use in animal challenge models for studies of rotavirus pathogenesis and immunity, but all with limited success.[48,49,51–57]

Other small laboratory models for rotavirus research are rabbits and rats.[58,59] In rabbits, like in mice, rotavirus disease is age restricted.[60] In rabbits infected with lapine rotavirus between 1 week and 2 months of age, diarrhea was observed only within the first 2 weeks of age, but histopathological changes including villous shortening and fusion, increased vacuolation of epithelial cells, and mononuclear cell infiltration of the lamina propria were observed throughout the small intestine. Five-day-old rats are susceptible to various group A animal rotaviruses and HRV infections, and most of the viruses resulted in diarrhea in rats that lasted from 1 to 10 days.[59] The severity of disease and spread of infection to naive rat littermates differed depending on the rotavirus strain. The rabbit model was used for assessing host range restriction, active immunity, and protection after infection or vaccination with virus or virus-like particles (VLPs),[61–64] and the neonatal rat model of rotavirus infection was used to determine the kinetics of viremia, spread, and pathology of rotavirus in extraintestinal organs.[59,65]

6.2.1 History of Mouse Models for Rotavirus Research

In the landmark study by Eydelloth et al.,[66] CD-1 mice from Charles River Breeding Laboratories (Wilmington, MA) were used for the investigation of rotavirus replication and immune responses. Mice orally inoculated with the murine EDIM virus, a murine rotavirus first reported in 1957 as causing diarrhea in 3- to 11-day-old mice,[67,68] were used to examine the kinetics of rotavirus replication, the effect of age on rotavirus infection and diseases, and the relationship of viral replication to the immune response and to the development and resolution of disease. The study demonstrated the following: (1) Younger mice (1, 7, and 14 days old) were more susceptible to EDIM infection than older mice; when the mice were inoculated at 28 days of age, only minimal viral replication was detected; (2) mice older than 7 days can still be infected with EDIM, but they do not develop diarrhea as seen in the younger mice beyond 14 days of age; (3) the increase in intestinal antibody levels at 7 days postinoculation coincided with a rapid decline in the intestinal EDIM virus antigen in animals inoculated at 7 days of age; (4) mice inoculated at 7 or 14 days of age developed the highest titers of serum antibodies; these titers reached peak levels at days 10 and 7 postinoculation, respectively.[66]

Thus, in mice under 15 days of age, homologous rotavirus infection causes diarrhea and lethargy, but only mild intestinal lesions consisting of vacuolization of villous tip epithelial cells with little or no villous atrophy. An adult mouse model of rotavirus infection was later established with the EDIM virus using infection as the endpoint in the BALB/c strain of mice.[69] This adult mouse model has been extensively used to study the determinants of immunity against rotavirus infection. Overall, the majority of mechanistic studies in history have been conducted with the adult mouse model of murine rotavirus infection.[70]

However, caution needs to be taken when using the murine rotavirus for challenge studies in mice with different genetic backgrounds and using virus shedding as the readout. Rotavirus fecal shedding detected after the EDIM challenge of mouse strains other than BALB/c (H2-d), for example, C57BL/6 (H2-b), and genetically modified mice on a C57BL/6 background, has been found to be less reliable and sometimes undetectable.[71] The reasons for the different infectivities of murine rotaviruses could be the difference in available rotavirus receptors on intestinal epithelial cells or the difference in the innate immunity of the mouse strains.[72] Such factors can affect the extrapolation of mechanisms of adaptive immunity induced by rotavirus vaccines using mouse models.[73] Conclusions drawn with a specific mouse and/or rotavirus vaccine set cannot be generalized to other sets.[72] Another murine rotavirus strain, EMcN, was later reported to shed in much greater quantities in adult BALB/c mice than in EDIM. In addition, in contrast to the EDIM strain, EMcN was shown to consistently shed in large quantities in adult C57BL/6 mice and in genetically modified mice of this background.[71] Therefore, rotavirus studies on adult mice with the C57BL/6 background should use the EMcN strain as the challenge virus.

Neonatal mice at 5–7 days of age can develop diarrhea after inoculation with high doses of a heterologous rotavirus, including human, simian, or bovine rotaviruses,[74] but virus shedding is not consistently detected[75] and virus replication is not required for the development of diarrhea. Neonatal mice inoculated with recombinant NSP4 develop diarrhea comparable to that after challenge with live virus, suggesting the enterotoxigenic properties of rotavirus NSP4.[76,77]

In addition to wild-type mice, various gene knockout adult mice that have defective adaptive immune systems (i.e., Rag-2 mice devoid of both T and B cells, β2m mice that lack cytotoxic T cell responses, JHD mice that lack B cell responses, and IgA-knockout mice that have no detectable IgA in the serum or in any secretions) have been extensively used in studying determinants of adaptive protective immunity. These studies have produced important knowledge regarding the role of humoral and cellular immunity in protection against rotavirus reinfection.[78–85] Other gene knockout mice that have defective innate immune pathways (i.e., TLR3, MAVS, STAT1, IFN-α/βR, IFN-γR, and IFN-λR knockout) have provided critical tools for determining the role of innate immune responses in controlling homologous versus heterologous rotavirus, intestinal versus extraintestinal rotavirus replication, virus systemic dissemination, and age-dependent susceptibility.[86–89] For comprehensive reviews on rotavirus studies using mouse models prior to 2007, see Feng et al.,[90] Franco et al.,[91,92] and Ward et al.[27,93]

6.2.2 History of the Gnotobiotic (Gn) Pig Model for HRV Research

Although mouse models for rotavirus research are relatively easy to set up and do not require specially designed laboratory facilities for husbandry, mice older than 15 days do not develop diarrhea after rotavirus infection and the pathogenesis of rotavirus in mice is different from that in humans.[27,94] Animal models of HRV diseases, not subclinical infection, are more desirable, especially for preclinical studies on the safety and efficacy of rotavirus vaccines. Gn pigs have successfully fulfilled this need. Germfree isolators provide a sterile environment free of bacteria, virus, fungus, and so on. Once an agent, such as rotavirus, is introduced into the isolator, it is no longer truly germfree and is referred to as being gnotobiotic. In the literature, however, the words germfree pig, Gn pig, and isolator pig are used interchangeably.

The first study showing that Gn pigs were productively infected with HRV and developed diarrhea and virus shedding was reported by Dr. Albert Kapikian's group in 1976.[39,95] Nineteen of twenty-one piglets developed diarrhea after being inoculated at 1–4 days of age with the rotavirus isolate acquired from human infants with acute gastroenteritis in 1974, which is the original Wa strain HRV that belongs to G1P1A[8] serotypes with the genome constellation of G1-P[8]-I1-R1-C1-M1-A1-N1-T1-E1-H1.[96] Virus particles were observed in pig intestinal contents and/or fecal samples by electron microscopy. The total duration of rotavirus shedding was 2–6 days after the onset of diarrhea. The onset of diarrhea was 2–7 days after virus inoculation. The same infectivity was reproduced in Gn pigs for up to five serial passages.[39] Two years later, another group from Australia reported that three HRV isolates induced only subclinical infection in newborn Gn pigs and failed to infect Gn calves and lambs.[97] The different outcome is likely due to the different HRV strains used in the studies. Gn pigs can be infected with G1 (Wa strain) and G3 (M strain) HRV and develop diarrhea, but not G2 (DS-1 strain) HRV[36,98]; however, the genotypes of the viruses used in the Australian study were not reported.

The Gn pig model of HRV infection and diarrhea was formally established in 1996 using the same virus isolate, the HRV Wa strain, from Wyatt and Kapikian's group.[37,99,100] The virulence of Wa HRV in Gn pigs was maintained by passaging the virus solely through the intestinal contents of infected neonatal Gn pigs. Over the past two decades, the Gn pig model of Wa HRV infection and diarrhea has been the key element for many studies published in over 50 peer-reviewed journal articles, testifying to the usefulness of this unique animal model in rotavirus research.[37,101–150]

Recently, the human gut microbiota (HGM) transplanted Gn pig model for HRV infection and diarrhea has been established.[105] The gut microbiota of the recipient pigs was similar to that of the human donor.[102] The HGM Gn pigs more closely mimic the intestine environment of human infants and provide a more robust model for virus–host interaction studies and for evaluating intervention approaches to reduce rotavirus diarrhea.[151]

6.2.3 History of Nonhuman Primate Models for HRV Research

An early study showed that diarrhea developed in five colostrum-deprived newborn rhesus monkeys (*Macaca mulatta*) inoculated orally on the first day of their life with a "reovirus-like agent" of infantile gastroenteritis (later named as Wa HRV).[54] As reported, the onset of diarrhea ranged from 2 to 5 days; virus particles were detected in stools by electron microscopy in association with illness, and virus shedding lasted 1–3 days. The virus derived from monkeys that developed illness following inoculation was infectious for other monkeys but did not induce diarrhea.[54]

Similarly, after oral administration of HRV (strain or genotype unknown) to six newborn cynomolgus monkeys (*Macaca fascicularis*) naturally delivered and normally nursed, five developed diarrhea.[55] Virus excretion was observed in the stool of four animals. This virus was transmitted to four out of six other monkeys, but only caused diarrhea in one animal. Another attempt was made to infect nonhuman primates with a HRV G8P[6] isolate.[56] Rotavirus was detected in the stool of all the rotavirus-inoculated vervets and in one out of the two rotavirus-inoculated baboons; however, none showed any clinical disease. Therefore, nonhuman primates have not become a principal animal model for HRV research. The advantages, disadvantages, and limitations of the three most commonly used animal models (mice, Gn pigs, and nonhuman primates) are listed in Table 6.1.

TABLE 6.1

Advantages and Disadvantages of the Major Animal Models Used in Rotavirus Research

Animal Species and Age	Advantages	Disadvantages
Neonatal mice (between 1 and 14 days of age)	• Small size • Inbred; homogeneous response; allow immune cell transfer experiments • Availability of many genetically modified strains (gene knockout mice) • Availability of molecular biology and immunology reagents • Produce clinical signs of disease (diarrhea) for a limited time after infection by murine rotavirus and high-dose heterologous rotavirus strains • Can be used to study passive protective immunity against rotavirus diarrhea by immunizing dams	• Resistant to infection by HRV • Different from humans in gastrointestinal physiology and immune system • Different histopathology from humans after rotavirus infection • Cannot be used to study infection- or vaccine-induced active immunity against rotavirus diarrhea due to age-dependent disease • Maternal antibodies can interfere with rotavirus infectivity and development of immune responses
Adult mice (older than 15 days of age)	• Small size • Inbred; homogeneous response; allow immune cell transfer experiments • Availability of many genetically modified strains (gene knockout mice) • Availability of molecular biology and immunology reagents • Can be used to study rotavirus infection or vaccine-induced protective immunity using virus shedding as the readout	• Resistant to infection by HRV • Different from humans in gastrointestinal physiology and immune system • Mouse strains with different genetic backgrounds display different susceptibility to infection for the same rotavirus strain, and shed different titers of rotavirus • Cannot be used to study infection- or vaccine-induced active immunity against rotavirus diarrhea
Gnotobiotic pigs	• Susceptible to infection by HRV (Wa and M strains) without prior adaptation and develop diarrhea up to 8 week of age • Display clinical signs of disease (diarrhea) and histopathological changes similar to human infants • Similarity to human infants in physiology, immune system, body size, milk diet, and metabolism • Large amount of cells can be isolated from lymphoid tissues for comprehensive immunological analysis • Devoid of maternal antibodies and other maternal and environmental confounding factors • Can be colonized with human gut microbiota to more closely mimic humans	• Outbred; variability in host response • Although increasing, the availability of reagents is still less than that for mice and humans • Expensive • Special facilities and trainings are required
Nonhuman primates	• High similarity with humans in genomics, gastrointestinal physiology, and immune system • Neonates are susceptible to HRV infection and diarrhea • Broad availability of immunology and molecular biology reagents	• Fail to induce clinical signs of disease by second passage HRV (after the HRV isolate is passaged in nonhuman primates) • Outbred; high variability in host response • Varying levels of maternal antibodies • Ethical concerns • Very expensive • Special facilities and trainings are required

6.3 Rotavirus Replication and Pathogenesis

6.3.1 Recent Major Advancements Made Using Mouse Models

Rotavirus primarily infects and replicates in villous mature epithelial cells of the small intestine. An innate immune response triggered by rotavirus infection is the first line of defense. One of the most distinct advantages of using the mouse model for research is the availability of a wide variety of genetically modified (gene knockout) mouse strains that allow the identification of molecular mechanisms in the studies. Using knockout mouse models of murine rotavirus infection, studies have clearly identified the role of innate immune effectors induced by homologous rotavirus that control the virus replication after oral inoculation.[87,152] In the study by Angel et al.,[152] type I IFN (IFN-α/β) receptor or type II IFN (IFN-γ) knockout suckling mice developed diarrhea with similar characteristics and duration and had comparable quantities of viral antigen in the intestines as wild-type mice. Furthermore, type I IFN receptor knockout adult mice infected with murine rotavirus also shed equivalent quantities of viral antigen and with similar kinetics as the wild-type mice. Thus, type I and II IFNs are not major inhibitors of rotavirus diarrhea or replication in mice. Further, in another study,[87] both suckling and adult IFN-λ receptor knockout mice had increased susceptibility to EDIM rotavirus infection, whereas IFN-α/β receptor knockout mice were similar to wild-type mice. In addition, the systemic treatment (injection) of suckling mice with IFN-λ repressed rotavirus replication in the gut, whereas treatment with type I IFN was not effective. These studies clearly demonstrated that type III IFN (IFN-λ), not type I or II IFN, plays a critical role in the intestinal epithelial antiviral host defense in mice and that this function is nonredundant in the mucosal antiviral innate immune system.[87]

Homologous and heterologous rotaviruses have different replication efficiencies in the intestine. In general, heterologous rotaviruses are less virulent, or totally attenuated, compared to homologous rotaviruses. The host species-specific restriction of rotavirus replication and virulence forms the basis for heterologous attenuated oral rotavirus vaccines. These include RotaShield, RotaTaq, and LLR, in which an animal rotavirus that is naturally restricted for replication and virulence in humans is used either as the vaccine (LLR)[153] or as the genetic backbone to produce the reassortant rotaviruses included in the multivalent vaccine (RotaShield and RotaTaq).[154] To identify the genetic basis and mechanism for the host range restriction, intestinal replication of a series of EW × RRV reassortants in suckling mice was used to identify rotavirus genes that influence rotavirus replication in the intestine.[155] It was found that permissive replication of a homologous murine rotavirus is primarily regulated by VP4 and NSP1, and high-efficiency replication in the mouse intestine requires a constellation of murine genes encoding VP3, NSP2, and NSP3 along with NSP1. VP4 acts as a determinant of host range restriction from having modest to very strong effects, depending on the species origin of the heterologous VP4.[155]

The most dominant host factor in rotavirus pathogenesis is age: Only newborn mice, rats, and rabbits less than 2 weeks of age develop diarrhea after rotavirus infection. On the other hand, newborn humans infected with rotavirus rarely have symptomatic disease. This protection is thought to be mediated primarily by the transplacental transfer of maternal antibodies,[156] but recent studies have indicated that the age-dependent pathogenesis of rotavirus may also derive from the age-dependent expression of TLR3[89] and *N*-acetyllactosamine (LacNAc), a precursor of the human histo-blood group antigen (HBGA) on intestinal epithelial cells.[157] A study by Pott et al.[89] investigated the age-dependent mechanisms of the intestinal epithelial innate immune response to rotavirus infection. TLR3 expression is low in the epithelium of wild-type suckling mice but is strongly increased during the postnatal period. TLR3 expression is inversely correlated with rotavirus susceptibility, viral shedding, and histological damage. After oral infection with EDIM rotavirus, TLR3 or the adaptor molecule TRIF knockout adult mice, but not neonatal knockout mice, had significantly higher viral shedding and decreased epithelial expression of proinflammatory and antiviral genes as compared to wild-type mice. This study identified the age-dependent expression of TLR3 as a factor of rotavirus susceptibility.[89]

For most HRV strains, the spike protein VP8* binds to the human HBGAs as a potential receptor or attachment factor similar to that of human noroviruses[158]; strains in the P[4] and P[8] genotypes share the common antigens of Lewis b (Le^b) and H type 1, while strains of the P[6] genotype bind the H type 1 antigen only. Notably, only rotavirus strains in the P[11] genotype have a preference for neonates (human 116E and bovine B223 strains). *In vitro* studies have suggested that the P[11] rotavirus specifically recognizes LacNAc that are likely expressed on the intestinal epithelial cells of only neonates and infants, not in older children or adults, as potential receptors for P[11] rotaviruses.[157] Rotavirus can also infect adults, but severe symptomatic disease is relatively uncommon and can result from immunosuppression,[159] infection with an unusual virus strain, or extremely high doses of rotavirus.[7] Further studies are needed to better understand the age-dependent susceptibility and pathogenesis of rotaviruses, and to fully identify the host genetic factors in susceptibility to rotavirus illness and the mechanisms behind asymptomatic rotavirus shedding in adults.

6.3.2 Major Advancements Made Using the Gn Pig Model

Pathological changes induced by HRV infections in the intestines of humans have been examined in only a limited numbers of studies due to the difficulty of collecting specimens. In pigs, the macroscopic changes induced by porcine rotavirus infection demonstrate the thinning of the intestinal wall, and the microscopic changes include villus atrophy, villus blunting, and conversion to a cuboidal epithelium.[15] Based on duodenal biopsies from children with acute rotavirus infection, the histopathologic changes are similar to those found in piglets.[37,160,161] The general progression of events in rotavirus infections in humans and pigs consists of virus replication in and destruction of the mature differentiated villous epithelial cells of the small intestine, leading to necrosis and desquamation of these cells.[162,163] As a consequence, crypt cells invade the villus surface to cause a decrease in the digestive and absorptive capacities of the intestine and generate malabsorptive and maldigestive diarrhea.[164] In neonatal Gn pigs infected with virulent Wa HRV, diarrhea developed at 13 h postinoculation and correlated with the presence of viral antigens within villous epithelial cells.[37] Villous atrophy was observed at 24–48 h postinoculation, coincident with the peak of virus replication. Diarrhea and virus shedding persisted for 4–7 postinoculation days. Recovery from disease correlated with the return of morphologically normal villi.[37]

Rotavirus pathogenesis is multifactorial. Both viral and host factors affect the outcome of rotavirus infection. Using porcine SB-1A × HRV DS-1 reassortants in neonatal Gn pigs and using diarrhea as the readout (replication was not measured), VP3, VP4, VP7, and NSP4 together were associated with the capacity of the porcine strain to induce diarrhea in neonatal Gn pigs. Reassortants that possessed only one, two, or three of these porcine rotavirus genes on a background of HRV genes failed to induce diarrhea.[98]

Rotavirus primarily infects intestinal villus enterocytes. However, all infected individuals and animals undergo at least a short period of viremia, and rotavirus can be detected in several other tissues in addition to the intestine.[165–167] A study using Gn pigs provided clear evidence that virulent HRV causes transient viremia and upper respiratory tract infection in addition to gastrointestinal infection.[128] In Gn pigs, after oral, intranasal, or gavage inoculation with virulent Wa HRV, all pigs had viremia along with nasal and rectal virus shedding and diarrhea.[128] The HRV in serum was confirmed to be infectious by orally inoculating native Gn pigs. Serum-inoculated pigs developed diarrhea and fecal and nasal virus shedding. It is important to note that pigs inoculated intravenously with serum or intestinal contents from the viremic virulent HRV-inoculated pigs developed diarrhea, virus shedding, and viremia, similar to the orally inoculated pigs, suggesting that intestinal infection might be initiated from the basolateral side of the epithelial cells via viremia.[128]

The Gn pig model can be used to address many unaddressed questions in rotavirus research, such as the influence of interactions between gut microbiota and rotavirus in virus replication and pathogenesis. In addition, genetically modified pigs have become available.[168,169] This opens up a new toolbox for using the Gn pig model in mechanistic studies of rotavirus pathogenesis and immunity.

6.4 Rotavirus Immunity and Vaccine Development

6.4.1 Mouse Models in the Study of Rotavirus Protective Immunity and Vaccines

In the history of rotavirus immunity research, mouse models have contributed to a number of milestone studies that have greatly improved our understanding of the mechanisms of rotavirus protective immunity. In sucking mice orally inoculated with a heterologous simian or bovine rotavirus (RRV or NCDV) strain or a homologous murine rotavirus (wild-type or tissue culture-adapted EHP) strain and challenged after 6 weeks with a virulent murine rotavirus (wild-type strain ECW), a fecal IgA antibody induced by homologous infection was identified as the correlate of protection.[74] A murine model with "backpack tumor" transplantation was used to determine the protective effects of antibodies against VP4 (the outer capsid viral protein) and VP6 (the major inner capsid viral protein). The study indicated that *in vivo* intracellular viral inactivation by anti-VP6 secretory IgA during transcytosis is a mechanism of host defense against rotavirus infection.[170]

Various gene knockout mice have been used to identify the determinants of rotavirus protective immunity,[78–85] including Rag-2 (devoid of both T and B cells), β2m (lack cytotoxic T cell responses), JHD (lack B cell responses), and IgA knockout mice (no detectable IgA in the serum or in any secretions). These studies have examined the roles of various components of humoral and cellular immunity in the resolution of primary infection or protection against rotavirus reinfection. The findings indicate that the IgA antibody, CD4, and CD8 T cells are all important in mediating protective immunity; however, none of the cellular effectors is absolutely indispensable in mediating protective immunity against rotavirus infection, suggesting the redundant nature of the host immune defense system of mice. The B-cell-dependent humoral immunity appears to be the main mechanism of protection from rotavirus infection.[171]

The effectiveness of various novel rotavirus vaccines including DNA plasmids,[172] VLP,[173] and subunit[174] vaccines was first evaluated in mouse models of rotavirus infection for protection against virus shedding before it was further evaluated in the Gn pig model of rotavirus diarrhea.[104,127,136]

6.4.2 Gnotobiotic Pig Model in the Study of Rotavirus Immunity and Vaccines

6.4.2.1 Unique Features of the Gn Pig Model

Hysterectomy-derived colostrum-deprived Gn pigs have been used extensively in rotavirus research to understand protective immunity and for the evaluation of rotavirus vaccine efficacy.[130] Gn pigs are kept in sterile isolator units and are raised on sterilized milk supplements.[175] They are susceptible to HRV disease for at least 8 weeks.[138] The minimal infectious dose of virulent Wa HRV in Gn pigs is one plaque or fluorescent forming unit (PFU or FFU) or less,[37] similar to that of porcine rotaviruses.[176] However, after HRV infection of Gn pigs, clinical disease is usually milder compared to that produced with porcine rotaviruses and mortality is uncommon. Gn pigs from birth to 8 weeks of age develop diarrhea in 1–3 days, and diarrhea usually continues for 3–5 days after oral inoculation with HRV.[37,110] In comparison to other animal models such as mice, rats, and rabbits, Gn pigs present a number of important advantages (Table 6.1). Although the rodents serve as useful models for the evaluation of immune responses to rotaviruses, older mice, rats, and rabbits are refractory to disease, and therefore, the readout is only for protection against viral infection; it is not useful for the evaluation of active immunity to clinical disease. In comparison, Gn pigs are susceptible for up to at least 8 weeks of age to infection and diarrhea when inoculated with several HRV strains.[99] Although the evaluation of relatively short-term protection (3–8 weeks) is a limitation of this model, no other animal model is susceptible to HRV infection and clinical disease, even for this limited period. Another important feature regarding Gn pigs is that exposure to extraneous enteric pathogens, especially the wild-type rotaviruses as a confounding variable, is excluded; therefore, true primary immune responses to a single rotavirus infection or vaccination can be evaluated. Gn pigs are also valuable in the identification of virulence factors that are often masked in studies using conventional animals.[177]

Pigs are monogastric and closely resemble humans physiologically and immunologically.[178,179] The ontogeny of the porcine immune system, especially the development of immunoglobulins, antibody repertoire, and B cells, has been extensively reviewed by Butler and colleagues.[179–181] The development of lymphoid organs from birth to the adult stage in pigs and T- and B-cell immune responses have been documented and show great similarity to humans.[182] Because of the physiological and immunological similarities to humans, the pig is an important large animal model for human biomedical research.[177,178,183] The ability to isolate large numbers of lymphocytes following the disruption of solid organs offers considerable advantages for the study of the intestinal and systemic immune systems.[112,114,134]

Although there are many similarities between porcine and human physiology and immune systems, there are also differences that need to be considered. One significant difference between humans and pigs is that humans acquire maternal antibodies transplacentally, whereas the placenta of pigs acts as a barrier to the transfer of macromolecules, including maternal antibodies and cytokines. Unlike mice and humans, no maternal antibodies are transferred from the mother to the porcine fetus during intrauterine development due to the special structure of the placenta wall in swine; hence, Gn pigs are agammaglobulinemic but are immunocompetent at birth.[184] Piglets acquire immunoglobulins (Ig) solely by intestinal absorption of colostral Ig for about 36 h after birth before gut closure occurs.[185] This feature permits basic studies of the ontogeny of neonatal immune responses and a true primary antibody response to rotavirus to be evaluated. This is also an advantage for using this model to study the effects of maternal antibodies and other maternal immune regulators on rotavirus vaccine efficacy because the levels and titers of those introduced into the circulation or intestines of piglets can be manipulated experimentally.[125,144,145,150]

6.4.2.2 HRV Vaccines and Therapeutics Evaluated in the Gn Pig Model of Wa HRV Infection and Diarrhea

The Gn pig model has been widely accepted to be a valuable model for studying HRV pathogenesis and as an ideal model for preclinical testing of the safety and efficacy of rotavirus candidate vaccines.[99,130] The Gn pig model of Wa HRV infection and diarrhea has been used in evaluating numerous rotavirus candidate vaccines including live attenuated (Wa), reassortants (RotaTaq, LLR), inactivated (Wa, CDC-9), recombinant protein (P2-VP8*), DNA plasmid (VP6), and VLP (2/6 and 2/6/7) vaccines with different adjuvants (LT-R192G, ISCOM, aluminum phosphate), immunostimulating supplements (probiotics and rice bran) and immunization routes (oral, intramuscular, intranasal, intradermal by gene gun), and with or without maternal antibodies.[103,104,118,124,125,127,133,136,138,141,144,186,187] The main criteria used in these studies to establish the efficacy of HRV vaccines include the protection of immunized animals from fecal virus shedding and rotavirus-induced diarrhea (onset, duration, and fecal consistence scores). Immune responses associated with protection against rotavirus diarrhea identified in these studies include intestinal IgA antibody secreting cells[141] and memory B cells,[134] intestinal IFN-γ-producing T cells,[188] and serum IgA, intestinal IgA, and intestinal IgG antibodies.[139] Through these studies, the Gn pig model has established its role as the most reliable animal model for the preclinical evaluation of rotavirus vaccines and therapeutics.

The Gn pig model has been used to evaluate the therapeutic effect of passive antibodies and dietary rice bran supplementation on protection against rotavirus diarrhea.[103,109] Oral administration of VP6-specific llama-derived single domain nanoantibodies was shown to be an effective treatment against virulent Wa HRV-induced diarrhea.[109] Rice bran demonstrated strong effects on the stimulation of nonspecific and HRV-specific immune responses and protection against HRV diarrhea.[103] In addition, rice bran significantly enhanced the growth and colonization of probiotic *Lactobacillus rhamnosus* GG (LGG) and *Escherichia coli* Nissle 1917 (EcN) in the intestine of Gn pigs, protected against damage to intestinal epithelium while maintaining intestinal homeostasis, maintained intestine permeability, enhanced the IFN-γ and IgA protective immune responses during HRV infection, and provided complete protection against HRV diarrhea in LGG- and EcN-colonized pigs.[189]

6.4.2.3 Gnotobiotic Pig Model Colonized with Probiotic Bacteria or Human Gut Microbiota

Gn animal models are an important tool for studying the role of probiotics and commensal microbiota in the development of the mucosal immune system.[190] Despite the differences from conventional animals, Gn pigs offer distinct advantages for studies on the effects and immunomodulating mechanisms of probiotics and gut microbiota on enteric virus infections and vaccines. The most distinctive advantage of using the Gn pig model is that neonatal pigs provide an immunologically and microbiologically naive background that allows the identification of host responses to a single pathogen or a single vaccine in hosts colonized with a clearly defined probiotic strain or gut microbiota.[183]

The mechanisms of probiotic LGG on reducing HRV pathogenesis were examined in Gn pigs fed with high-dose LGG and challenged with virulent Wa HRV.[108,110] LGG treatment before and during HRV gastroenteritis partially prevented virus-induced intestinal tissue damage, reduced autophagy marker expression to normal levels, induced apoptosis,[108] and partially prevented HRV-induced compensatory increases of the adherent junction proteins, α-catenin, and β-catenin; tight junction proteins, occludin, claudin-3, and claudin-4; and leak protein claudin-2 in the ileal epithelium.[110]

The immune-modulating effects of probiotic *Lactobacillus* strains *L. acidophilus* NCFM, *L. reuteri* and LGG on intestinal and systemic B cells, CD4 and CD8 T cells, γ/δ T cell, macrophage/dendritic cell, toll-like receptor, and cytokine responses to virulent HRV infection or the live attenuated Wa HRV vaccine were comprehensively examined in a series of studies on Gn pigs fed with different doses of *L. acidophilus* NCFM or LGG[101,106,113,191] or the mixture of *L. acidophilus* NCFM and *L. reuteri*.[116,117,120–123] *L. acidophilus* NCFM and LGG are shown to have therapeutic as well as immune-stimulating effects, and have dual functions in partial protection against HRV diarrhea and as adjuvants for the HRV vaccine when used in the proper dosage.[101,106] These studies also demonstrated that both the dose (CFU/day) and dosing regimen (total number of days' intake) of probiotics have significant effects on the immune modulatory functions of probiotics. Different LGG dosages differentially modulated immune responses to favor either the mucosal IgA response (five doses) or the T-cell response (nine doses).[101] High-dosage *L. acidophilus* NCFM intake induced regulatory immune responses, similar to the effect of microbiota colonization in infants, which had a negative impact on the immunostimulatory effect of probiotics and abolished its effectiveness as a vaccine adjuvant.[106]

The lack of gut microbiota is a unique feature of the Gn pig model that provides an indispensable tool for the study of the consequences of bacterial colonization, but it can also be a drawback when results from preclinical vaccine studies need to be extrapolated to humans. In normal humans and conventional animals, commensal microbiota helps the host to defend against pathogenic microorganisms.[190] These bacteria provide the colonization resistance against attachment, multiplication, and invasion of pathogenic microorganisms into epithelial cells and their potential circulation in the host. Furthermore, they exert a major influence on the development of the mucosal immune system and regulation of host immune responses. In the very early stage of the postnatal period, the immune responses are biased toward Th2, and the Th1-mediated immune response is lacking.[192,193] Thereafter, intestinal commensal microbiota stimulates the development of both local and systemic immune systems, preferentially Th1 immune responses, but later induces regulatory mechanisms to maintain homeostasis.[194] The intestinal colonization of germfree animals with commensal microbes not only significantly stimulates the development of mucosal and systemic immune systems, but may also promote integrity of the epithelial barrier by regulating tight junctions and protecting it from injury with enhanced proliferation and cytoprotective protein production.[195–199] In addition, commensal bacteria regulate cell migration to reinforce the epithelial barrier.[200] Intestinal colonization also stimulates mucin secretion from goblet cells to limit intestinal infections through binding between mucin and pathogens.[201,202] Intestinal microbiota also drive the development of Th17 cells[203] and Treg cells.[204] Therefore, the lack of gut microbiota in Gn pigs can cause deviations in the pathogenesis of HRV infection and in the patterns of immune responses induced by HRV infection or on vaccination, as compared to those observed in human infants.

In order to identify the influence of microbiota on the Gn pig's response to HRV and to more closely mimic human infants, Gn pigs transplanted with newborn HGM and infected with HRV have been established. This model has been used to test the effects of probiotics on the gut microbiome structure during an HRV infection.[102,105] The development of HRV vaccine-induced immune responses has been compared between the HGM and non-HGM-transplanted Gn pigs. HGM successfully colonized the Gn pig intestine after three oral inoculations. Sequencing of the V4 region of 16S rRNA genes demonstrated that the pigs carried a microbiome similar to that of a C-section delivered human infant.[102] The attenuated Wa HRV vaccine conferred similar overall protection against rotavirus diarrhea and virus shedding in Gn pigs and HGM-transplanted Gn pigs. HGM promoted the development of the neonatal immune system, significantly enhanced IFN-γ-producing T-cell responses, and reduced Treg cell responses in the AttHRV-vaccinated pigs.[105] Given the many advantages of HGM pigs (reviewed by Wang et al.[151]), the HGM pig model will have wide applications for the studies of viral pathogenesis; interactions between host-microbiota, host-pathogens, and microbiota-pathogens; and for the evaluation of vaccines and therapeutics.[143]

6.5 Cell Lines and Organoid/Enteroid Models for Rotavirus Infection

Unlike the notoriously difficult human noroviruses,[205] HRV is very easy to adapt into cell cultures where primary cells can support virus growth directly from fecal specimens.[206] When stool samples from 454 diarrheic episodes were attempted, rotavirus was cultured from 381 of the 423 specimens that do not contain other interfering agents (a 90% success rate).[207] In the study, two initial cell-culture passages were made in primary African green monkey kidney (AGMK) cells, followed by additional passages in MA104 cells. Another study showed that the Wa HRV strain was adapted to cell culture through 14 passages in primary AGMK cells.[100] This passage series was successfully initiated only with a virus that had been serially passaged 11 times in newborn Gn pigs. The virus present in the stool of patient Wa and the virus from the first three passages in Gn pigs could not be propagated in AGMK cells. This cell-culture adaptation process simultaneously started the attenuation process of the virulence of the Wa HRV strain.[37] The virus became fully attenuated for pigs after an additional 27 passages in MA104 cells, and its complete genome sequence was determined.[96] Because it has the same G and P types as the currently licensed Rotarix vaccine, it has been serving as the prototype of the rotavirus vaccine in studies of live oral rotavirus vaccines in Gn pigs.[141,142]

For studies of rotavirus replication, innate immune responses, and mechanisms of immune evasion, various mammalian cell lines have been utilized. Some recent examples are listed in Table 6.2; however, the findings from these studies are beyond the scope of this chapter.

A new, emerging research area in laboratory models for rotavirus is the development of stem-cell-derived human intestinal organoid[208] and human enteroid[209] models for rotavirus infection and as *ex vivo* models of host–pathogen interactions in the gastrointestinal tract. Organoids are generated from induced-pluripotent adult stem cells, whereas enteroids are from intestinal crypts isolated from human surgical specimens or endoscopic biopsies.[209] Human stem cell technology has allowed the *in vitro* re-creation of 3D organoids or enteroids that better mimic human intestinal tissues than cell lines. These models may complement and expand upon the limitations of cell and animal models currently used to study rotavirus infection and diseases.[209]

TABLE 6.2
Cell Lines Used in Rotavirus Research

Cell Line	Rotavirus Strain	References
MA104 cells	Bovine rotavirus RF and UK Simian rotavirus RRV and SA-11 HRV Wa, RV4, M4, Wi, M69 strains, and from clinical stool samples Porcine rotavirus OSU, CRW-8, and YM Murine rotavirus EB	141,210–218
Caco-2	RRV and SA-11 HRV Wa, Wi, and M69	217,219,220
Porcine small intestinal epithelial cell line (IPEC-J2)	Porcine rotavirus OSU HRV Wa	221,222
Vero cells	HRV Wa and CDC-9	223,224
Primary AGMK cells	HRV from clinical stool samples, murine rotavirus EB	100,207,218
Cos-7 cells	Bovine rotavirus RF HRV M4	216,225
HT-29	SA11–4F, Wa, Bovine rotavirus A5–13	226
Stem-cell-derived human intestinal organoids	SA11 and rotavirus from clinical stool samples	208
Human enteroids	Laboratory strains and rotavirus from clinical stool samples	209

6.6 Concluding Remarks

Animal models have been critical in our understanding of rotavirus pathogenesis and immunity and in the preclinical assessment of the safety and efficacy of rotavirus vaccines and therapeutics. The most widely used animal models, namely mice and Gn pigs, present different sets of advantages and limitations, and will continue to be indispensable tools for rotavirus research. Further optimizations of the models, including humanization of the immune system through stem cell transfer, transplantation with HGM from donors representing different health and immune statuses, and genetic modification using CRISPR/Cas9 technology, will further improve the usefulness and reliability of the models for mimicking rotavirus infection in humans.

REFERENCES

1. Davidson, G.P., Bishop, R.F., Townley, R.R. & Holmes, I.H. Importance of a new virus in acute sporadic enteritis in children. *Lancet* 1, 242–6 (1975).
2. Bishop, R.F. et al. An epidemic of diarrhoea in human neonates involving a reovirus-like agent and "enteropathogenic" serotypes of *Escherichia coli*. *J Clin Pathol* 29, 46–9 (1976).
3. Kapikian, A.Z. et al. Human reovirus-like agent as the major pathogen associated with "winter" gastroenteritis in hospitalized infants and young children. *N Engl J Med* 294, 965–72 (1976).
4. Hurst, C.J. & Gerba, C.P. Stability of simian rotavirus in fresh and estuarine water. *Appl Environ Microbiol* 39, 1–5 (1980).
5. McDonald, S.M. & Patton, J.T. Assortment and packaging of the segmented rotavirus genome. *Trends Microbiol* 19, 136–44 (2011).
6. Patton, J.T., Vasquez-Del Carpio, R. & Spencer, E. Replication and transcription of the rotavirus genome. *Curr Pharm Des* 10, 3769–77 (2004).
7. Greenberg, H.B. & Estes, M.K. Rotaviruses: from pathogenesis to vaccination. *Gastroenterology* 136, 1939–51 (2009).
8. Chen, D., Luongo, C.L., Nibert, M.L. & Patton, J.T. Rotavirus open cores catalyze 5′-capping and methylation of exogenous RNA: evidence that VP3 is a methyltransferase. *Virology* 265, 120–30 (1999).
9. Liu, M., Mattion, N.M. & Estes, M.K. Rotavirus VP3 expressed in insect cells possesses guanylyltransferase activity. *Virology* 188, 77–84 (1992).
10. Li, Y.J. et al. Oral vaccination with the porcine rotavirus VP4 outer capsid protein expressed by *Lactococcus lactis* induces specific antibody production. *J Biomed Biotechnol* 2010, 708460 (2010).
11. Ward, R. Mechanisms of protection against rotavirus infection and disease. *Pediatr Infect Dis J* 28, S57–9 (2009).
12. Barro, M. & Patton, J.T. Rotavirus nonstructural protein 1 subverts innate immune response by inducing degradation of IFN regulatory factor 3. *Proc Natl Acad Sci USA* 102, 4114–9 (2005).
13. Pina-Vazquez, C., De Nova-Ocampo, M., Guzman-Leon, S. & Padilla-Noriega, L. Post-translational regulation of rotavirus protein NSP1 expression in mammalian cells. *Arch Virol* 152, 345–68 (2007).
14. Graff, J.W., Mitzel, D.N., Weisend, C.M., Flenniken, M.L. & Hardy, M.E. Interferon regulatory factor 3 is a cellular partner of rotavirus NSP1. *J Virol* 76, 9545–50 (2002).
15. Lundgren, O. & Svensson, L. Pathogenesis of rotavirus diarrhea. *Microbes Infect* 3, 1145–56 (2001).
16. Saif, L.J. & Jiang, B. Nongroup A rotaviruses of humans and animals. *Curr Top Microbiol Immunol* 185, 339–71 (1994).
17. Hyser, J.M. & Estes, M.K. Rotavirus vaccines and pathogenesis: 2008. *Curr Opin Gastroenterol* 25, 36–43 (2009).
18. Santos, N. & Hoshino, Y. Global distribution of rotavirus serotypes/genotypes and its implication for the development and implementation of an effective rotavirus vaccine. *Rev Med Virol* 15, 29–56 (2005).
19. Gray, J. et al. Rotavirus. *J Pediatr Gastroenterol Nutr* 46 Suppl 2, S24–31 (2008).
20. Offit, P.A. & Rubin, D.H. Viral diseases: infections of the gastrointestinal tract. *Compr Ther* 8, 21–6 (1982).
21. Bresee, J.S., Glass, R.I., Ivanoff, B. & Gentsch, J.R. Current status and future priorities for rotavirus vaccine development, evaluation and implementation in developing countries. *Vaccine* 17, 2207–22 (1999).

22. Glass, R.I., Gentsch, J.R. & Ivanoff, B. New lessons for rotavirus vaccines. *Science* 272, 46–8 (1996).
23. Leshem, E. et al. Acute gastroenteritis hospitalizations among US children following implementation of the rotavirus vaccine. *JAMA* 313, 2282–4 (2015).
24. Aliabadi, N., Tate, J.E., Haynes, A.K. & Parashar, U.D. Sustained decrease in laboratory detection of rotavirus after implementation of routine vaccination-United States, 2000–2014. *MMWR Morb Mortal Wkly Rep* 64, 337–42 (2015).
25. O'Ryan, M.L., Matson, D.O., Estes, M.K., Bartlett, A.V. & Pickering, L.K. Molecular epidemiology of rotavirus in children attending day care centers in Houston. *J Infect Dis* 162, 810–6 (1990).
26. Matson, D.O. & Estes, M.K. Impact of rotavirus infection at a large pediatric hospital. *J Infect Dis* 162, 598–604 (1990).
27. Ward, R.L., McNeal, M.M., Farone, M.B. & Farone, A.L. Reoviridae. In *The Mouse in Biomedical Research. Vol. 2, Diseases* (eds. Fox, J.G. et al.), pp. 235–268. Academic Press, New York/London (2007).
28. Matthijnssens, J. et al. Full genome-based classification of rotaviruses reveals a common origin between human Wa-Like and porcine rotavirus strains and human DS-1-like and bovine rotavirus strains. *J Virol* 82, 3204–19 (2008).
29. Papp, H. et al. Zoonotic transmission of reassortant porcine G4P[6] rotaviruses in Hungarian pediatric patients identified sporadically over a 15 year period. *Infect Genet Evol* 19, 71–80 (2013).
30. Zeller, M., Heylen, E., De Coster, S., Van Ranst, M. & Matthijnssens, J. Full genome characterization of a porcine-like human G9P[6] rotavirus strain isolated from an infant in Belgium. *Infect Genet Evol* 12, 1492–500 (2012).
31. My, P.V. et al. Novel porcine-like human G26P[19] rotavirus identified in hospitalized paediatric diarrhoea patients in Ho Chi Minh City, Vietnam. *J Gen Virol* 95, 2727–33 (2014).
32. Steyer, A., Poljsak-Prijatelj, M., Barlic-Maganja, D. & Marin, J. Human, porcine and bovine rotaviruses in Slovenia: evidence of interspecies transmission and genome reassortment. *J Gen Virol* 89, 1690–8 (2008).
33. Zhou, X. et al. Genomic characterization of G3P[6], G4P[6] and G4P[8] human rotaviruses from Wuhan, China: evidence for interspecies transmission and reassortment events. *Infect Genet Evol* 33, 55–71 (2015).
34. Bohl, E.H. et al. Rotavirus as a cause of diarrhea in pigs. *J Am Vet Med Assoc* 172, 458–63 (1978).
35. Cordle, C.T. et al. Passive immune protection from diarrhea caused by rotavirus or *E. coli*: an animal model to demonstrate and quantitate efficacy. *Adv Exp Med Biol* 310, 317–27 (1991).
36. Torres, A. & Ji-Huang, L. Diarrheal response of gnotobiotic pigs after fetal infection and neonatal challenge with homologous and heterologous human rotavirus strains. *J Virol* 60, 1107–12 (1986).
37. Ward, L.A., Rosen, B.I., Yuan, L. & Saif, L.J. Pathogenesis of an attenuated and a virulent strain of group A human rotavirus in neonatal gnotobiotic pigs. *J Gen Virol* 77(Pt 7), 1431–41 (1996).
38. Snodgrass, D.R., Madeley, C.R., Wells, P.W. & Angus, K.W. Human rotavirus in lambs: infection and passive protection. *Infect Immun* 16, 268–70 (1977).
39. Torres-Medina, A., Wyatt, R.G., Mebus, C.A., Underdahl, N.R. & Kapikian, A.Z. Diarrhea caused in gnotobiotic piglets by the reovirus-like agent of human infantile gastroenteritis. *J Infect Dis* 133, 22–7 (1976).
40. Mebus, C.A. et al. Diarrhea in gnotobiotic calves caused by the reovirus-like agent of human infantile gastroenteritis. *Infect Immun* 14, 471–4 (1976).
41. Varshney, K.C. et al. The lesions of rotavirus infection in 1- and 10-day-old gnotobiotic calves. *Vet Pathol* 32, 619–27 (1995).
42. Wyatt, R.G. et al. Rotaviral immunity in gnotobiotic calves: heterologous resistance to human virus induced by bovine virus. *Science* 203, 548–50 (1979).
43. Vesikari, T. et al. Immunogenicity and safety of live oral attenuated bovine rotavirus vaccine strain RIT 4237 in adults and young children. *Lancet* 2, 807–11 (1983).
44. Zissis, G. et al. Protection studies in colostrum-deprived piglets of a bovine rotavirus vaccine candidate using human rotavirus strains for challenge. *J Infect Dis* 148, 1061–8 (1983).
45. Van Pinxteren, L.A., Campbell, I., Clarke, C.J., Snodgrass, D.R. & Bruce, M.G. A single oral dose of inactivated rotavirus and ISCOM matrices induces partial protection in lambs. *Biochem Soc Trans* 25, 340S (1997).

46. van Pinxteren, L.A., Bruce, M.G., Campbell, I., Clarke, C.J. & Snodgrass, D.R. Characterisation of the primary local and systemic immune response in gnotobiotic lambs against rotavirus infection. *Vet Immunol Immunopathol* 64, 349–65 (1998).
47. van Pinxteren, L.A. et al. Effect of oral rotavirus/iscom vaccines on immune responses in gnotobiotic lambs. *Vet Immunol Immunopathol* 71, 53–67 (1999).
48. McNeal, M.M. et al. Development of a rotavirus-shedding model in rhesus macaques, using a homologous wild-type rotavirus of a new P genotype. *J Virol* 79, 944–54 (2005).
49. Westerman, L.E. et al. Isolation and characterization of a new simian rotavirus, YK-1. *Virol J* 3, 40 (2006).
50. Soike, K.F., Gary, G.W. & Gibson, S. Susceptibility of nonhuman primate species to infection by simian rotavirus SA-11. *Am J Vet Res* 41, 1098–1103 (1980).
51. Hoshino, Y. et al. A rotavirus strain isolated from pig-tailed macaque (*Macaca nemestrina*) with diarrhea bears a P6[1]:G8 specificity. *Virology* 345, 1–12 (2006).
52. Westerman, L.E., Xu, J., Jiang, B., McClure, H.M. & Glass, R.I. Experimental infection of pigtailed macaques with a simian rotavirus, YK-1. *J Med Virol* 75, 616–25 (2005).
53. Zhao, W. et al. Evaluation of rotavirus dsRNA load in specimens and body fluids from experimentally infected juvenile macaques by real-time PCR. *Virology* 341, 248–56 (2005).
54. Wyatt, R.G. et al. Induction of diarrhea in colostrum-deprived newborn rhesus monkeys with the human reovirus-like agent of infantile gastroenteritis. *Arch Virol* 50, 17–27 (1976).
55. Majer, M. et al. Diarrhea in newborn cynomologus monkeys infected with human rotavirus. *Infection* 6, 71–2 (1978).
56. Chege, G.K. et al. Experimental infection of non-human primates with a human rotavirus isolate. *Vaccine* 23, 1522–8 (2005).
57. Sestak, K. et al. Defining T-cell-mediated immune responses in rotavirus-infected juvenile rhesus macaques. *J Virol* 78, 10258–64 (2004).
58. Conner, M.E., Estes, M.K. & Graham, D.Y. Rabbit model of rotavirus infection. *J Virol* 62, 1625–33 (1988).
59. Ciarlet, M., Conner, M.E., Finegold, M.J. & Estes, M.K. Group A rotavirus infection and age-dependent diarrheal disease in rats: a new animal model to study the pathophysiology of rotavirus infection. *J Virol* 76, 41–57 (2002).
60. Ciarlet, M. et al. Rotavirus disease, but not infection and development of intestinal histopathological lesions, is age restricted in rabbits. *Virology* 251, 343–60 (1998).
61. Ciarlet, M. et al. Subunit rotavirus vaccine administered parenterally to rabbits induces active protective immunity. *J Virol* 72, 9233–46 (1998).
62. Ciarlet, M., Estes, M.K., Barone, C., Ramig, R.F. & Conner, M.E. Analysis of host range restriction determinants in the rabbit model: comparison of homologous and heterologous rotavirus infections. *J Virol* 72, 2341–51 (1998).
63. Crawford, S.E. et al. Heterotypic protection and induction of a broad heterotypic neutralization response by rotavirus-like particles. *J Virol* 73, 4813–22 (1999).
64. Bertolotti-Ciarlet, A., Ciarlet, M., Crawford, S.E., Conner, M.E. & Estes, M.K. Immunogenicity and protective efficacy of rotavirus 2/6-virus-like particles produced by a dual baculovirus expression vector and administered intramuscularly, intranasally, or orally to mice. *Vaccine* 21, 3885–900 (2003).
65. Crawford, S.E. et al. Rotavirus viremia and extraintestinal viral infection in the neonatal rat model. *J Virol* 80, 4820–32 (2006).
66. Eydelloth, R.S., Vonderfecht, S.L., Sheridan, J.F., Enders, L.D. & Yolken, R.H. Kinetics of viral replication and local and systemic immune responses in experimental rotavirus infection. *J Virol* 50, 947–50 (1984).
67. Kraft, L.M. Studies on the etiology and transmission of epidemic diarrhea of infant mice. *J Exp Med* 106, 743–55 (1957).
68. Kraft, L.M. Observations on the control and natural history of epidemic diarrhea of infant mice (EDIM). *Yale J Biol Med* 31, 121–37 (1958).
69. Ward, R.L., McNeal, M.M. & Sheridan, J.F. Development of an adult mouse model for studies on protection against rotavirus. *J Virol* 64, 5070–5 (1990).
70. Ward, R.L. Mechanisms of protection against rotavirus in humans and mice. *J Infect Dis* 174 Suppl 1, S51–8 (1996).

71. McNeal, M.M., Belli, J., Basu, M., Choi, A.H. & Ward, R.L. Discovery of a new strain of murine rotavirus that is consistently shed in large quantities after oral inoculation of adult mice. *Virology* 320, 1–11 (2004).
72. Choi, A.H., McNeal, M.M., Basu, M. & Ward, R.L. Immunity to homologous rotavirus infection in adult mice: response. *Trends Microbiol* 8, 52 (2000).
73. Franco, M.A. & Greenberg, H.B. Immunity to homologous rotavirus infection in adult mice. *Trends Microbiol* 8, 50–2 (2000).
74. Feng, N., Burns, J.W., Bracy, L. & Greenberg, H.B. Comparison of mucosal and systemic humoral immune responses and subsequent protection in mice orally inoculated with a homologous or a heterologous rotavirus. *J Virol* 68, 7766–73 (1994).
75. Reimerink, J.H. et al. Systemic immune response after rotavirus inoculation of neonatal mice depends on source and level of purification of the virus: implications for the use of heterologous vaccine candidates. *J Gen Virol* 88, 604–12 (2007).
76. Ball, J.M., Tian, P., Zeng, C.Q., Morris, A.P. & Estes, M.K. Age-dependent diarrhea induced by a rotaviral nonstructural glycoprotein. *Science* 272, 101–4 (1996).
77. Horie, Y. et al. Diarrhea induction by rotavirus NSP4 in the homologous mouse model system. *Virology* 262, 398–407 (1999).
78. McNeal, M.M. et al. CD4 T cells are the only lymphocytes needed to protect mice against rotavirus shedding after intranasal immunization with a chimeric VP6 protein and the adjuvant LT(R192G). *J Virol* 76, 560–8 (2002).
79. Franco, M.A., Tin, C. & Greenberg, H.B. CD8+ T cells can mediate almost complete short-term and partial long-term immunity to rotavirus in mice. *J Virol* 71, 4165–70 (1997).
80. Franco, M.A. & Greenberg, H.B. Immunity to rotavirus infection in mice. *J Infect Dis* 179 Suppl 3, S466–9 (1999).
81. Dharakul, T. et al. Immunization with baculovirus-expressed recombinant rotavirus proteins VP1, VP4, VP6, and VP7 induces CD8+ T lymphocytes that mediate clearance of chronic rotavirus infection in SCID mice. *J Virol* 65, 5928–32 (1991).
82. Dharakul, T., Rott, L. & Greenberg, H.B. Recovery from chronic rotavirus infection in mice with severe combined immunodeficiency: virus clearance mediated by adoptive transfer of immune CD8+ T lymphocytes. *J Virol* 64, 4375–82 (1990).
83. Blutt, S.E., Miller, A.D., Salmon, S.L., Metzger, D.W. & Conner, M.E. IgA is important for clearance and critical for protection from rotavirus infection. *Mucosal Immunol* 5, 712–9 (2012).
84. Vancott, J.L., McNeal, M.M., Choi, A.H. & Ward, R.L. The role of interferons in rotavirus infections and protection. *J Interferon Cytokine Res* 23, 163–70 (2003).
85. VanCott, J.L. et al. Role for T cell-independent B cell activity in the resolution of primary rotavirus infection in mice. *Eur J Immunol* 31, 3380–7 (2001).
86. Holloway, G. & Coulson, B.S. Innate cellular responses to rotavirus infection. *J Gen Virol* 94, 1151–60 (2013).
87. Pott, J. et al. IFN-λ determines the intestinal epithelial antiviral host defense. *Proc Natl Acad Sci USA* 108, 7944–9 (2011).
88. Broquet, A.H., Hirata, Y., McAllister, C.S. & Kagnoff, M.F. RIG-I/MDA5/MAVS are required to signal a protective IFN response in rotavirus-infected intestinal epithelium. *J Immunol* 186, 1618–26 (2011).
89. Pott, J. et al. Age-dependent TLR3 expression of the intestinal epithelium contributes to rotavirus susceptibility. *PLoS Pathog* 8, e1002670 (2012).
90. Feng, N., Franco, M.A. & Greenberg, H.B. Murine model of rotavirus infection. *Adv Exp Med Biol* 412, 233–40 (1997).
91. Franco, M.A., Feng, N. & Greenberg, H.B. Rotavirus immunity in the mouse. *Arch Virol Suppl* 12, 141–52 (1996).
92. Franco, M.A., Angel, J. & Greenberg, H.B. Immunity and correlates of protection for rotavirus vaccines. *Vaccine* 24, 2718–31 (2006).
93. Ward, R.L. Possible mechanisms of protection elicited by candidate rotavirus vaccines as determined with the adult mouse model. *Viral Immunol* 16, 17–24 (2003).
94. Boshuizen, J.A. et al. Changes in small intestinal homeostasis, morphology, and gene expression during rotavirus infection of infant mice. *J Virol* 77, 13005–16 (2003).

95. Torres-Medina, A., Wyatt, R.G., Mebus, C.A., Underdahl, N.R. & Kapikian, A.Z. Patterns of shedding of human reovirus-like agent in gnotobiotic newborn piglets with experimentally-induced diarrhea. *Intervirology* 7, 250–55 (1976).
96. Wentzel, J.F., Yuan, L., Rao, S., van Dijk, A.A. & O'Neill, H.G. Consensus sequence determination and elucidation of the evolutionary history of a rotavirus Wa variant reveal a close relationship to various Wa variants derived from the original Wa strain. *Infect Genet Evol* 20, 276–83 (2013).
97. Tzipori, S.R., Makin, T.J. & Smith, M.L. The clinical response of gnotobiotic calves, pigs and lambs to inoculation with human, calf, pig and foal rotavirus isolates. *Aust J Exp Biol Med Sci* 58, 309–18 (1980).
98. Hoshino, Y. et al. Identification of group A rotavirus genes associated with virulence of a porcine rotavirus and host range restriction of a human rotavirus in the gnotobiotic piglet model. *Virology* 209, 274–80 (1995).
99. Saif, L.J., Ward, L.A., Yuan, L., Rosen, B.I. & To, T.L. The gnotobiotic piglet as a model for studies of disease pathogenesis and immunity to human rotaviruses. *Arch Virol Suppl* 12, 153–61 (1996).
100. Wyatt, R.G. et al. Human rotavirus type 2: cultivation in vitro. *Science* 207, 189–91 (1980).
101. Wen, K. et al. *Lactobacillus rhamnosus* GG dosage affects the adjuvanticity and protection against rotavirus diarrhea in gnotobiotic pigs. *J Pediatr Gastroenterol Nutr* 60, 834–843 (2015).
102. Zhang, H. et al. Probiotics and virulent human rotavirus modulate the transplanted human gut microbiota in gnotobiotic pigs. *Gut Pathog* 6, 39 (2014).
103. Yang, X. et al. Dietary rice bran protects against rotavirus diarrhea and promotes Th1-type immune responses to human rotavirus vaccine in gnotobiotic pigs. *Clin Vaccine Immunol* 21, 1396–403 (2014).
104. Wen, X. et al. Inclusion of a universal tetanus toxoid CD4+ T cell epitope P2 significantly enhanced the immunogenicity of recombinant rotavirus ΔVP8* subunit parenteral vaccines. *Vaccine* 32, 4420–7 (2014).
105. Wen, K. et al. Probiotic *Lactobacillus rhamnosus* GG enhanced Th1 cellular immunity but did not affect antibody responses in a human gut microbiota transplanted neonatal gnotobiotic pig model. *PLoS One* 9, e94504 (2014).
106. Liu, F. et al. Dual functions of *Lactobacillus acidophilus* NCFM as protection against rotavirus diarrhea. *J Pediatr Gastroenterol Nutr* 58, 169–76 (2014).
107. Kandasamy, S., Chattha, K.S., Vlasova, A.N., Rajashekara, G. & Saif, L.J. Lactobacilli and bifidobacteria enhance mucosal B cell responses and differentially modulate systemic antibody responses to an oral human rotavirus vaccine in a neonatal gnotobiotic pig disease model. *Gut Microbes* 5, 639–51 (2014).
108. Wu, S. et al. Probiotic *Lactobacillus rhamnosus* GG mono-association suppresses human rotavirus-induced autophagy in the gnotobiotic piglet intestine. *Gut Pathog* 5, 22 (2013).
109. Vega, C.G. et al. Recombinant monovalent llama-derived antibody fragments (VHH) to rotavirus VP6 protect neonatal gnotobiotic piglets against human rotavirus-induced diarrhea. *PLoS Pathog* 9, e1003334 (2013).
110. Liu, F. et al. *Lactobacillus rhamnosus* GG on rotavirus-induced injury of ileal epithelium in gnotobiotic pigs. *J Pediatr Gastroenterol Nutr* 57, 750–8 (2013).
111. Chattha, K.S. et al. Probiotics and colostrum/milk differentially affect neonatal humoral immune responses to oral rotavirus vaccine. *Vaccine* 31, 1916–23 (2013).
112. Wen, K. et al. $CD4^+$ $CD25^-$ $FoxP3^+$ regulatory cells are the predominant responding regulatory T cells after human rotavirus infection or vaccination in gnotobiotic pigs. *Immunology* 137, 160–71 (2012).
113. Wen, K. et al. High dose and low dose *Lactobacillus acidophilus* exerted differential immune modulating effects on T cell immune responses induced by an oral human rotavirus vaccine in gnotobiotic pigs. *Vaccine* 30, 1198–207 (2012).
114. Wen, K. et al. Characterization of immune modulating functions of γδ T cell subsets in a gnotobiotic pig model of human rotavirus infection. *Comp Immunol Microbiol Infect Dis* 35, 289–301 (2012).
115. Vega, C.G. et al. IgY antibodies protect against human rotavirus induced diarrhea in the neonatal gnotobiotic piglet disease model. *PLoS One* 7, e42788 (2012).
116. Azevedo, M.S. et al. *Lactobacillus acidophilus* and *Lactobacillus reuteri* modulate cytokine responses in gnotobiotic pigs infected with human rotavirus. *Benef Microbes* 3, 33–42 (2012).
117. Wen, K. et al. Development of γδ T cell subset responses in gnotobiotic pigs infected with human rotaviruses and colonized with probiotic lactobacilli. *Vet Immunol Immunopathol* 141, 267–75 (2011).

118. Wang, Y. et al. Inactivated rotavirus vaccine induces protective immunity in gnotobiotic piglets. *Vaccine* 28, 5432–6 (2010).
119. Azevedo, M.S. et al. An oral versus intranasal prime/boost regimen using attenuated human rotavirus or VP2 and VP6 virus-like particles with immunostimulating complexes influences protection and antibody-secreting cell responses to rotavirus in a neonatal gnotobiotic pig model. *Clin Vaccine Immunol* 17, 420–8 (2010).
120. Wen, K. et al. Toll-like receptor and innate cytokine responses induced by lactobacilli colonization and human rotavirus infection in gnotobiotic pigs. *Vet Immunol Immunopathol* 127, 304–15 (2009).
121. Zhang, W. et al. Lactic acid bacterial colonization and human rotavirus infection influence distribution and frequencies of monocytes/macrophages and dendritic cells in neonatal gnotobiotic pigs. *Vet Immunol Immunopathol* 121, 222–31 (2008).
122. Zhang, W. et al. Probiotic *Lactobacillus acidophilus* enhances the immunogenicity of an oral rotavirus vaccine in gnotobiotic pigs. *Vaccine* 26, 3655–61 (2008).
123. Zhang, W. et al. Influence of probiotic Lactobacilli colonization on neonatal B cell responses in a gnotobiotic pig model of human rotavirus infection and disease. *Vet Immunol Immunopathol* 122, 175–81 (2008).
124. Yuan, L. et al. Virus-specific intestinal IFN-γ producing T cell responses induced by human rotavirus infection and vaccines are correlated with protection against rotavirus diarrhea in gnotobiotic pigs. *Vaccine* 26, 3322–31 (2008).
125. Nguyen, T.V. et al. High titers of circulating maternal antibodies suppress effector and memory B-cell responses induced by an attenuated rotavirus priming and rotavirus-like particle-immunostimulating complex boosting vaccine regimen. *Clin Vaccine Immunol* 13, 475–85 (2006).
126. Azevedo, M.S. et al. Cytokine responses in gnotobiotic pigs after infection with virulent or attenuated human rotavirus. *J Virol* 80, 372–82 (2006).
127. Yuan, L. et al. Mucosal and systemic antibody responses and protection induced by a prime/boost rotavirus-DNA vaccine in a gnotobiotic pig model. *Vaccine* 23, 3925–36 (2005).
128. Azevedo, M.S. et al. Viremia and nasal and rectal shedding of rotavirus in gnotobiotic pigs inoculated with Wa human rotavirus. *J Virol* 79, 5428–36 (2005).
129. Nguyen, T.V. et al. Protection and antibody responses to oral priming by attenuated human rotavirus followed by oral boosting with 2/6-rotavirus-like particles with immunostimulating complexes in gnotobiotic pigs. *Vaccine* 21, 4059–70 (2003).
130. Yuan, L. & Saif, L.J. Induction of mucosal immune responses and protection against enteric viruses: rotavirus infection of gnotobiotic pigs as a model. *Vet Immunol Immunopathol* 87, 147–60 (2002).
131. Iosef, C. et al. Systemic and intestinal antibody secreting cell responses and protection in gnotobiotic pigs immunized orally with attenuated Wa human rotavirus and Wa 2/6-rotavirus-like-particles associated with immunostimulating complexes. *Vaccine* 20, 1741–53 (2002).
132. Iosef, C., Chang, K.O., Azevedo, M.S. & Saif, L.J. Systemic and intestinal antibody responses to NSP4 enterotoxin of Wa human rotavirus in a gnotobiotic pig model of human rotavirus disease. *J Med Virol* 68, 119–28 (2002).
133. Yuan, L. et al. Protective immunity and antibody-secreting cell responses elicited by combined oral attenuated Wa human rotavirus and intranasal Wa 2/6-VLPs with mutant *Escherichia coli* heat-labile toxin in gnotobiotic pigs. *J Virol* 75, 9229–38 (2001).
134. Yuan, L., Geyer, A. & Saif, L.J. Short-term immunoglobulin A B-cell memory resides in intestinal lymphoid tissues but not in bone marrow of gnotobiotic pigs inoculated with Wa human rotavirus. *Immunology* 103, 188–98 (2001).
135. Chang, K.O., Vandal, O.H., Yuan, L., Hodgins, D.C. & Saif, L.J. Antibody-secreting cell responses to rotavirus proteins in gnotobiotic pigs inoculated with attenuated or virulent human rotavirus. *J Clin Microbiol* 39, 2807–13 (2001).
136. Yuan, L. et al. Intranasal administration of 2/6-rotavirus-like particles with mutant *Escherichia coli* heat-labile toxin (LT-R192G) induces antibody-secreting cell responses but not protective immunity in gnotobiotic pigs. *J Virol* 74, 8843–53 (2000).
137. Parreno, V. et al. Serum and intestinal isotype antibody responses to Wa human rotavirus in gnotobiotic pigs are modulated by maternal antibodies. *J Gen Virol* 80(Pt 6), 1417–28 (1999).
138. Yuan, L., Kang, S.Y., Ward, L.A., To, T.L. & Saif, L.J. Antibody-secreting cell responses and protective immunity assessed in gnotobiotic pigs inoculated orally or intramuscularly with inactivated human rotavirus. *J Virol* 72, 330–8 (1998).

139. To, T.L., Ward, L.A., Yuan, L. & Saif, L.J. Serum and intestinal isotype antibody responses and correlates of protective immunity to human rotavirus in a gnotobiotic pig model of disease. *J Gen Virol* 79, 2661–72 (1998).
140. Saif, L., Yuan, L., Ward, L. & To, T. Comparative studies of the pathogenesis, antibody immune responses, and homologous protection to porcine and human rotaviruses in gnotobiotic piglets. *Adv Exp Med Biol* 412, 397–403 (1997).
141. Yuan, L., Ward, L.A., Rosen, B.I., To, T.L. & Saif, L.J. Systematic and intestinal antibody-secreting cell responses and correlates of protective immunity to human rotavirus in a gnotobiotic pig model of disease. *J Virol* 70, 3075–83 (1996).
142. Ward, L.A., Yuan, L., Rosen, B.I., To, T.L. & Saif, L.J. Development of mucosal and systemic lymphoproliferative responses and protective immunity to human group A rotaviruses in a gnotobiotic pig model. *Clin Diagn Lab Immunol* 3, 342–50 (1996).
143. Yang, X. & Yuan, L. Neonatal gnotobiotic pig models for studying viral pathogenesis, immune responses, and for vaccine evaluation. *Br J Virol* 1, 87–91 (2014).
144. Hodgins, D.C. et al. Effects of maternal antibodies on protection and development of antibody responses to human rotavirus in gnotobiotic pigs. *J Virol* 73, 186–97 (1999).
145. Nguyen, T.V. et al. Low titer maternal antibodies can both enhance and suppress B cell responses to a combined live attenuated human rotavirus and VLP-ISCOM vaccine. *Vaccine* 24, 2302–16 (2006).
146. Azevedo, M.S. et al. Magnitude of serum and intestinal antibody responses induced by sequential replicating and nonreplicating rotavirus vaccines in gnotobiotic pigs and correlation with protection. *Clin Diagn Lab Immunol* 11, 12–20 (2004).
147. Azevedo, M.P., Vlasova, A.N. & Saif, L.J. Human rotavirus virus-like particle vaccines evaluated in a neonatal gnotobiotic pig model of human rotavirus disease. *Expert Rev Vaccines* 12, 169–81 (2013).
148. Gonzalez, A.M. et al. Innate immune responses to human rotavirus in the neonatal gnotobiotic piglet disease model. *Immunology* 131, 242–56 (2010).
149. Gonzalez, A.M. et al. Antibody responses to human rotavirus (HRV) in gnotobiotic pigs following a new prime/boost vaccine strategy using oral attenuated HRV priming and intranasal VP2/6 rotavirus-like particle (VLP) boosting with ISCOM. *Clin Exp Immunol* 135, 361–72 (2004).
150. Nguyen, T.V. et al. Transfer of maternal cytokines to suckling piglets: in vivo and in vitro models with implications for immunomodulation of neonatal immunity. *Vet Immunol Immunopathol* 117, 236–48 (2007).
151. Wang, M. & Donovan, S.M. Human microbiota-associated swine: current progress and future opportunities. *ILAR J* 56, 63–73 (2015).
152. Angel, J., Franco, M.A., Greenberg, H.B. & Bass, D. Lack of a role for type I and type II interferons in the resolution of rotavirus-induced diarrhea and infection in mice. *J Interferon Cytokine Res* 19, 655–9 (1999).
153. Fu, C. et al. Effectiveness of the Lanzhou lamb rotavirus vaccine against gastroenteritis among children. *Vaccine* 31, 154–8 (2012).
154. Ciarlet, M. & Schodel, F. Development of a rotavirus vaccine: clinical safety, immunogenicity, and efficacy of the pentavalent rotavirus vaccine, RotaTeq®. *Vaccine* 27 Suppl 6, G72–81 (2009).
155. Feng, N., Yasukawa, L.L., Sen, A. & Greenberg, H.B. Permissive replication of homologous murine rotavirus in the mouse intestine is primarily regulated by VP4 and NSP1. *J Virol* 87, 8307–16 (2013).
156. Ray, P.G., Kelkar, S.D., Walimbe, A.M., Biniwale, V. & Mehendale, S. Rotavirus immunoglobulin levels among Indian mothers of two socio-economic groups and occurrence of rotavirus infections among their infants up to six months. *J Med Virol* 79, 341–9 (2007).
157. Liu, Y. et al. Poly-LacNAc as an age-specific ligand for rotavirus P[11] in neonates and infants. *PLoS One* 8, e78113 (2013).
158. Huang, P. et al. Spike protein VP8* of human rotavirus recognizes histo-blood group antigens in a type-specific manner. *J Virol* 86, 4833–43 (2012).
159. Liakopoulou, E. et al. Rotavirus as a significant cause of prolonged diarrhoeal illness and morbidity following allogeneic bone marrow transplantation. *Bone Marrow Transplant* 36, 691–4 (2005).
160. Barnes, G.L. & Townley, R.R. Duodenal mucosal damage in 31 infants with gastroenteritis. *Arch Dis Child* 48, 343–9 (1973).
161. Davidson, G.P. & Barnes, G.L. Structural and functional abnormalities of the small intestine in infants and young children with rotavirus enteritis. *Acta Paediatr Scand* 68, 181–6 (1979).

162. Yuan, L., Stevenson, G. & Saif, L. Rotavirus and reovirus. In *Diseases of Swine* (eds. Straw, B.E., Zimmerman, J.J., D'Allaire, S. & Taylor, D.J.), pp. 435–454. Blackwell Publishing, Ames, IA (2006).
163. Conner, M.E. & Ramig, R.F. Viral enteric diseases. In *Viral Pathogenesis* (ed. Nathanson, N.), pp. 713–742. Lippincott-Raven, Philadelphia, PA (1997).
164. Davidson, G.P., Gall, D.G., Petric, M., Butler, D.G. & Hamilton, J.R. Human rotavirus enteritis induced in conventional piglets. Intestinal structure and transport. *J Clin Invest* 60, 1402–9 (1977).
165. Blutt, S.E. et al. Rotavirus antigenemia in children is associated with viremia. *PLoS Med* 4, e121 (2007).
166. Blutt, S.E. & Conner, M.E. Rotavirus: to the gut and beyond! *Curr Opin Gastroenterol* 23, 39–43 (2007).
167. Blutt, S.E., Fenaux, M., Warfield, K.L., Greenberg, H.B. & Conner, M.E. Active viremia in rotavirus-infected mice. *J Virol* 80, 6702–5 (2006).
168. Mendicino, M. et al. Generation of antibody- and B cell-deficient pigs by targeted disruption of the J-region gene segment of the heavy chain locus. *Transgenic Res* 20, 625–41 (2011).
169. Lee, K. et al. Engraftment of human iPS cells and allogeneic porcine cells into pigs with inactivated *RAG2* and accompanying severe combined immunodeficiency. *Proc Natl Acad Sci USA* 111, 7260–5 (2014).
170. Burns, J.W., Siadat-Pajouh, M., Krishnaney, A.A. & Greenberg, H.B. Protective effect of rotavirus VP6-specific IgA monoclonal antibodies that lack neutralizing activity. *Science* 272, 104–7 (1996).
171. Franco, M.A. & Greenberg, H.B. Immunity to rotavirus in T cell deficient mice. *Virology* 238, 169–79 (1997).
172. Chen, S.C. et al. Protective immunity induced by rotavirus DNA vaccines. *Vaccine* 15, 899–902 (1997).
173. O'Neal, C.M., Crawford, S.E., Estes, M.K. & Conner, M.E. Rotavirus virus-like particles administered mucosally induce protective immunity. *J Virol* 71, 8707–17 (1997).
174. Choi, A.H., Basu, M., McNeal, M.M., Clements, J.D. & Ward, R.L. Antibody-independent protection against rotavirus infection of mice stimulated by intranasal immunization with chimeric VP4 or VP6 protein. *J Virol* 73, 7574–81 (1999).
175. Meyer, R.C., Bohl, E.H. & Kohler, E.M. Procurement and maintenance of germ-free swine for microbiological investigations. *Appl Microbiol* 12, 295–300 (1964).
176. Graham, D.Y., Dufour, G.R. & Estes, M.K. Minimal infective dose of rotavirus. *Arch Virol* 92, 261–71 (1987).
177. Butler, J.E. Isolator and other neonatal piglet models in developmental immunology and identification of virulence factors. *Anim Health Res Rev* 10, 35–52 (2009).
178. Butler, J.E. et al. The piglet as a model for B cell and immune system development. *Vet Immunol Immunopathol* 128, 147–70 (2009).
179. Sinkora, M. & Butler, J.E. The ontogeny of the porcine immune system. *Dev Comp Immunol* 33, 273–83 (2009).
180. Butler, J.E., Wertz, N., Deschacht, N. & Kacskovics, I. Porcine IgG: structure, genetics, and evolution. *Immunogenetics* 61, 209–30 (2009).
181. Butler, J.E., Zhao, Y., Sinkora, M., Wertz, N. & Kacskovics, I. Immunoglobulins, antibody repertoire and B cell development. *Dev Comp Immunol* 33, 321–33 (2009).
182. Bianchi, A.T., Zwart, R.J., Jeurissen, S.H. & Moonen-Leusen, H.W. Development of the B- and T-cell compartments in porcine lymphoid organs from birth to adult life: an immunohistological approach. *Vet Immunol Immunopathol* 33, 201–21 (1992).
183. Meurens, F., Summerfield, A., Nauwynck, H., Saif, L. & Gerdts, V. The pig: a model for human infectious diseases. *Trends Microbiol* 20, 50–7 (2012).
184. Butler, J.E. & Sinkora, M. The isolator piglet: a model for studying the development of adaptive immunity. *Immunol Res* 39, 33–51 (2007).
185. Mehrazar, K., Gilman-Sachs, A. & Kim, Y.B. Intestinal absorption of immunologically intact macromolecules in germfree colostrum-deprived piglets maintained on total parenteral nutrition. *JPEN J Parenter Enteral Nutr* 17, 8–15 (1993).
186. Yuan, L. & Saif, L.J. Rotavirus-like particle vaccines evaluated in a pig model of human rotavirus diarrhea and in cattle. In *Viral Gastroenteritis* (eds. Desselberger, U. & Gray, J.), pp. 357–368. Elsevier Sciences, Netherlands (2003).

187. Kandasamy, S., Chattha, K.S., Vlasova, A.N. & Saif, L.J. Prenatal vitamin A deficiency impairs adaptive immune responses to pentavalent rotavirus vaccine (RotaTeq®) in a neonatal gnotobiotic pig model. *Vaccine* 32, 816–24 (2014).
188. Yuan, L. et al. Rotavirus-specific IFN-γ producing and proliferating T cell responses to human rotavirus infection and vaccination in gnotobiotic pigs. *The 4th International Conference on Vaccines for Enteric Diseases,* Apr. 25–27, Lisbon, Portugal (2007).
189. Yang, X. et al. High protective efficacy of rice bran against human rotavirus diarrhea via enhancing probiotic growth, gut barrier function, and innate immunity. *Sci Rep* 5, 15004 (2015).
190. Tlaskalova-Hogenova, H. et al. The role of gut microbiota (commensal bacteria) and the mucosal barrier in the pathogenesis of inflammatory and autoimmune diseases and cancer: contribution of germ-free and gnotobiotic animal models of human diseases. *Cell Mol Immunol* 8, 110–20 (2011).
191. Yuan, L., Wen, K., Liu, F. & G., L. Dose effects of LAB on modulation of rotavirus vaccine induced immune responses. In *Lactic Acid Bacteria—R & D for Food, Health and Livestock Purposes* (ed. Kongo, J.M.). InTech, Rijeka, http://www.intechopen.com/books/lactic-acid-bacteria-r-d-for-food-health-and-livestock-purposes/dose-effects-of-lab-on-modulation-of-rotavirus-vaccine-induced-immune-responses (2013).
192. Smith, P.M. & Garrett, W.S. The gut microbiota and mucosal T cells. *Front Microbiol* 2, 111 (2011).
193. McLoughlin, R.M. & Mills, K.H. Influence of gastrointestinal commensal bacteria on the immune responses that mediate allergy and asthma. *J Allergy Clin Immunol* 127, 1097–107; quiz 1108–9 (2011).
194. Walker, W.A. Initial intestinal colonization in the human infant and immune homeostasis. *Ann Nutr Metab* 63 Suppl 2, 8–15 (2013).
195. Hooper, L.V. et al. Molecular analysis of commensal host-microbial relationships in the intestine. *Science* 291, 881–4 (2001).
196. O'Hara, A.M. & Shanahan, F. The gut flora as a forgotten organ. *EMBO Rep* 7, 688–93 (2006).
197. Ukena, S.N. et al. Probiotic *Escherichia coli* Nissle 1917 inhibits leaky gut by enhancing mucosal integrity. *PLoS One* 2, e1308 (2007).
198. Ewaschuk, J.B. et al. Secreted bioactive factors from *Bifidobacterium infantis* enhance epithelial cell barrier function. *Am J Physiol Gastrointest Liver Physiol* 295, G1025–34 (2008).
199. Rakoff-Nahoum, S., Paglino, J., Eslami-Varzaneh, F., Edberg, S. & Medzhitov, R. Recognition of commensal microflora by toll-like receptors is required for intestinal homeostasis. *Cell* 118, 229–41 (2004).
200. Swanson, P.A. et al. Enteric commensal bacteria potentiate epithelial restitution via reactive oxygen species-mediated inactivation of focal adhesion kinase phosphatases. *Proc Natl Acad Sci USA* 108, 8803–8 (2011).
201. Linden, S.K., Florin, T.H. & McGuckin, M.A. Mucin dynamics in intestinal bacterial infection. *PLoS One* 3, e3952 (2008).
202. Linden, S.K. et al. MUC1 limits *Helicobacter pylori* infection both by steric hindrance and by acting as a releasable decoy. *PLoS Pathog* 5, e1000617 (2009).
203. Atarashi, K., Tanoue, T. & Honda, K. Induction of lamina propria Th17 cells by intestinal commensal bacteria. *Vaccine* 28, 8036–8 (2010).
204. Kunisawa, J. & Kiyono, H. Peaceful mutualism in the gut: revealing key commensal bacteria for the creation and maintenance of immunological homeostasis. *Cell Host Microbe* 9, 83–4 (2011).
205. Vashist, S., Bailey, D., Putics, A. & Goodfellow, I. Model systems for the study of human norovirus biology. *Future Virol* 4, 353–367 (2009).
206. Ward, R.L., Knowlton, D.R. & Pierce, M.J. Efficiency of human rotavirus propagation in cell culture. *J Clin Microbiol* 19, 748–53 (1984).
207. Ward, R.L. et al. Culture adaptation and characterization of group A rotaviruses causing diarrheal illnesses in Bangladesh from 1985 to 1986. *J Clin Microbiol* 29, 1915–23 (1991).
208. Finkbeiner, S.R. et al. Stem cell-derived human intestinal organoids as an infection model for rotaviruses. *mBio* 3, e00159-12 (2012).
209. Foulke-Abel, J. et al. Human enteroids as an ex-vivo model of host–pathogen interactions in the gastrointestinal tract. *Exp Biol Med (Maywood)* 239, 1124–34 (2014).
210. Razavinikoo, H. et al. Activation of calcium/calmodulin-dependent kinase II following bovine rotavirus enterotoxin NSP4 expression. *Iran J Basic Med Sci* 18, 393–7 (2015).

211. Donato, C.M. et al. Characterization of a G1P[8] rotavirus causing an outbreak of gastroenteritis in the Northern Territory, Australia, in the vaccine era. *Emerg Microbes Infect* 3, e47 (2014).
212. Lipson, S.M., Ozen, F.S., Louis, S. & Karthikeyan, L. Comparison of α-glucosyl hesperidin of citrus fruits and epigallocatechin gallate of green tea on the loss of rotavirus infectivity in cell culture. *Front Microbiol* 6, 359 (2015).
213. Torres-Flores, J.M., Silva-Ayala, D., Espinoza, M.A., Lopez, S. & Arias, C.F. The tight junction protein JAM-A functions as coreceptor for rotavirus entry into MA104 cells. *Virology* 475, 172–8 (2015).
214. Teimoori, A., Soleimanjahi, H. & Makvandi, M. Characterization and transferring of human rotavirus double-layered particles in MA104 cells. *Jundishapur J Microbiol* 7, e10375 (2014).
215. Pane, J.A., Dang, V.T., Holloway, G., Webster, N.L. & Coulson, B.S. VP7 of rhesus monkey rotavirus RRV contributes to diabetes acceleration in association with an elevated anti-rotavirus antibody response. *Virology* 468–470, 504–9 (2014).
216. Duponchel, S. et al. Transfection of exogenous rotavirus rearranged RNA segments in cells infected with a WT rotavirus results in subsequent gene rearrangements. *J Gen Virol* 95, 2089–98 (2014).
217. Guerrero, C.A., Murillo, A. & Acosta, O. Inhibition of rotavirus infection in cultured cells by *N*-acetylcysteine, PPARγ agonists and NSAIDs. *Antiviral Res* 96, 1–12 (2012).
218. Tsugawa, T., Tatsumi, M. & Tsutsumi, H. Virulence-associated genome mutations of murine rotavirus identified by alternating serial passages in mice and cell cultures. *J Virol* 88, 5543–58 (2014).
219. Bautista, D., Rodriguez, L.S., Franco, M.A., Angel, J. & Barreto, A. Caco-2 cells infected with rotavirus release extracellular vesicles that express markers of apoptotic bodies and exosomes. *Cell Stress Chaperones* 20, 697–708 (2015).
220. Huang, H., Liao, D., Liang, L., Song, L. & Zhao, W. Genistein inhibits rotavirus replication and upregulates AQP4 expression in rotavirus-infected Caco-2 cells. *Arch Virol* 160, 1421–33 (2015).
221. Zhao, Y. et al. Effect of 25-hydroxyvitamin D_3 on rotavirus replication and gene expressions of RIG-I signalling molecule in porcine rotavirus-infected IPEC-J2 cells. *Arch Anim Nutr* 69, 227–35 (2015).
222. Liu, F. et al. Porcine small intestinal epithelial cell line (IPEC-J2) of rotavirus infection as a new model for the study of innate immune responses to rotaviruses and probiotics. *Viral Immunol* 23, 135–49 (2010).
223. Lee do, K. et al. Probiotic bacteria, *B. longum* and *L. acidophilus* inhibit infection by rotavirus in vitro and decrease the duration of diarrhea in pediatric patients. *Clin Res Hepatol Gastroenterol* 39, 237–44 (2015).
224. Esona, M.D. et al. Molecular characterization of human rotavirus vaccine strain CDC-9 during sequential passages in vero cells. *Hum Vaccines* 6, pii: 10409 (2010).
225. Diaz, Y. et al. Dissecting the Ca^{2+} entry pathways induced by rotavirus infection and NSP4-EGFP expression in Cos-7 cells. *Virus Res* 167, 285–96 (2012).
226. Bagchi, P. et al. Identification of common human host genes involved in pathogenesis of different rotavirus strains: an attempt to recognize probable antiviral targets. *Virus Res* 169, 144–53 (2012).

7

Prions

Akikazu Sakudo and Takashi Onodera

CONTENTS

7.1 Introduction

Transmissible spongiform encephalopathies (TSEs), also known as prion diseases, are fatal neurological disorders that include Creutzfeldt–Jakob disease (CJD) in humans, scrapie in sheep and goats, bovine spongiform encephalopathy (BSE) in cattle, chronic wasting disease (CWD) in cervids, transmissible mink encephalopathy (TME) in mink, feline spongiform encephalopathy (FSE) in cats, and exotic ungulate encephalopathy (EUE) in zoo animals such as kudu, nyala, gemsbok, eland, and oryx [1]. The prion agent associated with each disease is named after the disease itself (e.g., CJD agent in CJD, scrapie agent in scrapie, etc.). A key event in the development of prion diseases is the conversion of the cellular, host-encoded prion protein (PrP^{C}) to its abnormal isoform (PrP^{Sc}) predominantly in the central nervous system (CNS) of the infected host [1]. Numerous compelling observations support the notion that PrP^{Sc} is the main component of prion agents [2,3]. PrP^{Sc}, in contrast to PrP^{C}, is resistant to complete digestion with proteinase K (PK). Consequently, the detection of PK-resistant prion protein (PrP^{res}) is generally used for identifying the presence of a prion agent, because PrP^{res} contains PrP^{Sc}. Therefore, most diagnostic methods for prion diseases are based on the index of PrP^{Sc}.

Several diagnostic methods have been used for prion diseases. These include enzyme-linked immunosorbent assay (ELISA), western blotting, and immunohistochemistry (IHC) [4]. As an index of prion infection, representative changes in the brain are generally assayed. Prion protein (PrP) accumulates in the brain to form deposits. PrP^{res} is biochemically detected after treatment with PK using western blotting (Figure 7.1) and ELISA (Figure 7.2). In the case of ELISA for BSE, a homogenate is usually prepared from the obex region of the brain, which is subsequently treated with PK. The PK-treated sample is then applied to a microtiter plate for absorption and reacted with an anti-PrP antibody. This method is commercially exploited in the Bio-Rad TeSE BSE kit (Bio-Rad, France), Enfer-TSE kit (Abbott Laboratories, USA), and FRELISA BSE Kit (Fujirebio Inc., Japan). Although the extensively employed ELISA is a sensitive and high-throughput method, the large number of false positives remains a significant problem. Therefore, if a result is positive, ELISA must be repeated to validate the finding. If a positive result is obtained again, western blotting and IHC are performed.

Western blotting initially involves separation of the PK-treated proteins by sodium dodecyl sulfate (SDS)–polyacrylamide gel electrophoresis (PAGE) followed by electroblotting onto a suitable membrane support. After blotting, PrP^{res} present as a membrane-bound protein can be detected using an anti-PrP

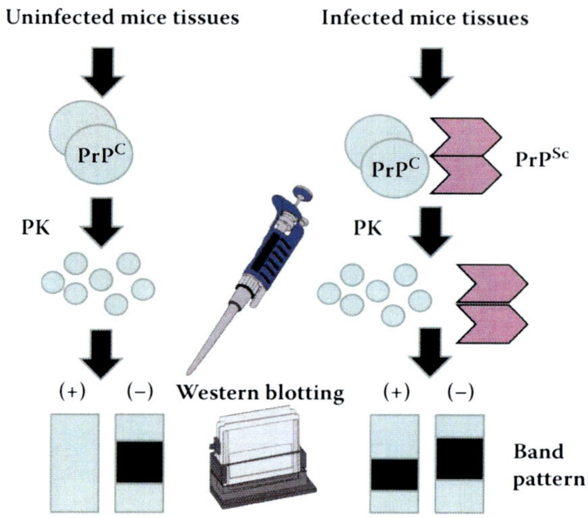

FIGURE 7.1 Western blotting for detection of PrP^{Sc}. Most methods for diagnosing prion diseases are based on the characteristics of the PrP^{Sc}, which is resistant to PK. PK completely degrades the PrP^{C} but only partially digests PrP^{Sc} because the latter forms protease-resistant aggregates. After protease treatment, PrP^{res} is detected by western blotting with an anti-PrP antibody. Western blotting reveals that PK-resistant PrP shifts to a lower molecular weight compared to untreated PrP (−) because the N-terminal region of PrP^{Sc} is susceptible to PK (+). A schematic representation of a Western blot for PrP derived from representative tissues of prion-infected and uninfected mice before (−) and after (+) PK treatment is shown. The tissue of choice for PrP^{Sc} analysis is the brain.

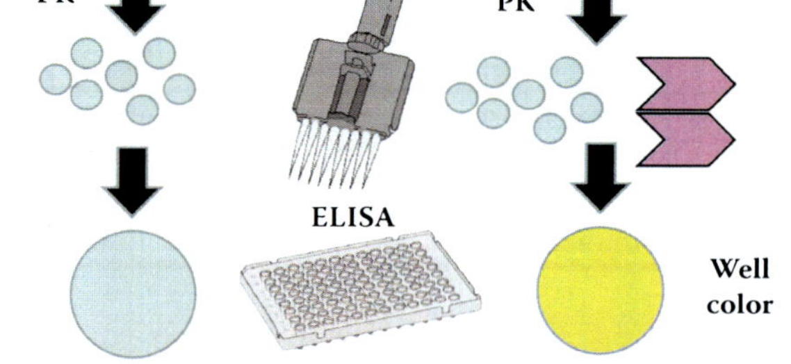

FIGURE 7.2 ELISA for detection of PrP^{Sc}. ELISA diagnosis of prion diseases is based on the characteristics of PrP^{Sc}, which is resistant to PK. PK completely degrades PrP^{C} but only partially digests PrP^{Sc}, because PrP^{Sc} forms protease-resistant aggregates. After PK treatment, ELISA is performed using an anti-PrP antibody to detect PK-resistant PrP (when PrP^{Sc} is present in the sample). Because the N-terminal portion of PrP^{Sc} is digested with PK, an antibody should be chosen that recognizes an alternative region of PrP. In the case of a sandwich ELISA, which is becoming increasingly popular, two discrete anti-PrP antibodies recognizing different epitopes of PrP are used. Specifically, one acts as a capture antibody, which is coated onto the well of the microplate, while the other is a detection antibody conjugated with an enzyme such as horseradish peroxidase (HRP) or alkaline phosphatase (AP). After reaction with the detection antibody, the corresponding substrate for the conjugated enzyme is added. The resulting color change caused by the enzyme reaction is then measured using a microplate reader on the basis of absorbance.

antibody. Importantly, western blotting provides information on both the prion infection and the mobility of peptides, which is influenced by the host genotype and strain of prion [5]. In the case of IHC, the indexes for prion infection are neuronal cell loss, astrocytosis, and vacuolation in addition to PrP accumulation (amyloid plaques) [6]. In IHC-based analyses of brain sections, these changes are examined by light microscopy.

A number of novel, alternative diagnostic methods for prion diseases have also been developed, and recently reviewed [4,7,8]. Readers may refer to these articles for further technical details.

In this chapter, we describe laboratory models used for analyzing prion infections. The laboratory models for prion infections are divided into *in vitro* models, cellular models, and animal models. In all three cases, prion infection is diagnosed by detection of PrP^{Sc} as described earlier.

7.2 *In Vitro* Models

The representative *in vitro* model for prion infection is protein misfolding cyclic amplification (PMCA) (Figure 7.3). PMCA is a PrP^{res} amplification method that is conceptually analogous to DNA amplification by PCR. This *in vitro* model could be used as a surrogate for animal bioassay.

PMCA [9] exploits the fact that PrP^{Sc} can convert PrP^{C} to PrP^{Sc}. Specifically, PMCA enables the *in vitro* amplification of PrP^{res} in an accelerated manner from a small quantity of PrP^{Sc}, which acts as the seed, via sequential cycles of incubation and sonication [10]. In the first phase, small quantities of PrP^{res} are incubated with a large quantity of PrP^{C}, leading to the growth of PrP^{res} aggregates. In the second phase, the PrP^{res} aggregates are broken up by sonication, causing a proliferation in the number of nuclei for further PrP^{res} amplification. By repeated cycles of incubation followed by sonication, the number of seeds increases in an exponential fashion resulting in the amplification of PrP^{res}. Healthy brain homogenate is usually used as the source of PrP^{C}. Optimization of the buffer conditions used for the amplification step may be required. Successful PrP^{res} amplification after PMCA is generally verified by western blotting using an anti-PrP antibody following PK treatment. The levels of PrP^{res} amplified by this method are then correlated with the prion infectivity titer [11]. The results from this analysis indicate that amplified PrP^{res} includes PrP^{Sc}, suggesting that PrP^{Sc} can be amplified by PMCA. Furthermore, PMCA is able to detect

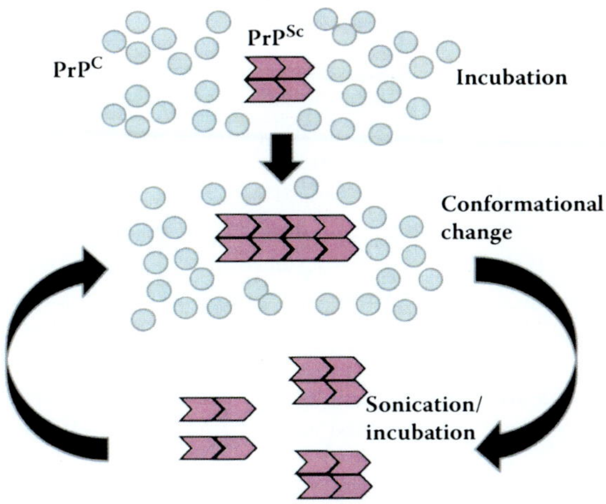

FIGURE 7.3 PMCA for amplification of PrP^{Sc}. Recently, a powerful diagnostic method for prion diseases, known as PMCA, has been developed for the amplification of PrP^{Sc} prions from body fluids such as blood, cerebral spinal fluid, semen, milk, urine, and saliva during the preclinical and clinical phases of prion diseases [61]. In this method, a PrP^{Sc} seed is incubated with the PrP^{C} source. PrP^{Sc} in the presence of excess PrP^{C} induces the formation of large aggregates that are subsequently broken up by sonication. Repeated cycles of this procedure efficiently amplify PrP^{Sc}.

prions in blood [12]. PMCA has been successfully used to diagnose both terminally diseased hamsters and prion-infected presymptomatic hamsters [13] and deer [14]. In summary, this method can be used to amplify PK-resistant PrP derived from various species including cervid, ferret, hamster, mouse, sheep, bovine, and human [12–17].

Although conventional PMCA utilizes normal brain homogenate as a source of PrP^{C}, a novel adaptation of the PMCA procedure is to amplify PrP^{res} and generate highly infectious prions using recombinant PrP derived from *Escherichia coli* [18] or baculovirus [19] in the absence of brain homogenate. This novel procedure requires RNA and lipids, such as 1-palmitoyl-2-oleolyl-sn-glycero-3-phospho(1′-rac-glycerol) (POPG) or phosphatidylethanolamine (PE), to act as conversion cofactors [18,20,21].

Recently, the real-time quaking-induced conversion (RT-QuIC) test [22], which is a modified version of PMCA, has been developed. This method is based on the prion-seeded fibrillization of recombinant PrP. In RT-QuIC reactions, prion-associated seeds induce the amyloid fibril formation of bacterially expressed recombinant PrP in multiwell plates. The resulting amyloid fibrils are then detected by the enhanced fluorescence of an amyloid sensitive dye, thioflavin T, present in the reaction mix. RT-QuIC is known to be highly specific and sensitive for the detection of multiple human and animal prion diseases [23]. Epidemiological surveillance of CJD, which currently relies heavily on autopsy-based diagnosis, could be more efficient, cost-effective, and broadly applicable with RT-QuIC testing for cerebrospinal fluid (CSF) samples that can be obtained without autopsies [24]. The second-generation RT-QuIC assay markedly improves the speed and sensitivity of detecting prion seeds in CSF specimens from CJD patients [24].

In addition, the RT-QuIC assay has been used successfully to detect multiple human, bovine, cervid, ovine hamster, and mouse strains in a variety of biological tissues such as saliva, blood, and nasal fluids [25]. Indeed, numerous CJD diagnostic laboratories around the world are currently implementing and validating RT-QuIC testing for human sCJD CSF using standard conditions [26]. RT-QuIC enables an estimation of the quantity of CWD prion infectivity in tissues, such as the obex, left ventricle, pancreas, jejunum, and spleen, as well as body fluids, such as saliva and urine [27]. Such an assay will contribute to the risk assessment of deer tissues, biological fluids, excreta, or environmental samples obtained from CWD-endemic areas.

7.3 Cell-Culture Models

Several prion-susceptible cell lines have been exploited to generate cell-culture models for prion infection [4,28]. The general cell-culture procedures for prion infection have been described previously [29,30]. Brain homogenates of prion-infected animals are added to the culture medium for 1–2 days before culture passage. Pretreatment by heating at 80°C for 20 min, sonication for 3 min, and filtration through a 0.22 μm filter unit reduce toxicity of the brain homogenate, which displays an efficient rate of infection. As an alternative approach, it is known that immortalized cell groups are observed in prion-infected animals. Thus, primary cultures from a prion-infected brain can act as a source of cells by continuous passage (e.g., ScHB and SMB produced from prion-infected animals) [31]. Consequently, established cell lines derived from prion-infected animals sometimes result in the establishment of persistent prion-infected cells after multiple passages. Indeed, such cell lines of neuronal and nonneuronal origin have been established. After persistent infection, these cells proliferate without any overt cytopathic effect.

Infectivity or PrP^{Sc} production per cell can be assayed to determine the percentage of infection of prions. The infectivity of harvested cells is measured after the brain sample has been injected into experimental animals via intracerebral inoculation. After performing behavioral observations for 1 year, the effective lethal dose (LD_{50}) can be derived from the survival curve of lysate-injected animals. PrP^{Sc} detection in cells and animals is usually performed by western blotting. After confirming persistent infection of prions using these methods, the persistently infected cells are established and used for analysis.

To date, numerous cell lines have been established for prion infection (Table 7.1). The representative neuronal cell, rat pheochromocytoma PC12, can be differentiated by nerve growth factor (NGF) and infected with prions [32], probably due to increased levels of PrP through NGF stimulation. Rabbit kidney–derived RK13 cells overexpressed with ovine PrP are susceptible to sheep scrapie infection [33]. Furthermore, mouse fibroblast cells NIH/3T3 and L929 are also susceptible to mouse-adapted scrapie infection [34]. In other cell types, microglial cells as well as epithelial cells and myoblasts can be successfully infected with prions. Thus, under certain conditions, nonneuronal cells can also be infected with prions. Indeed, this is consistent with the observation that PrP^{Sc} accumulates not only in the CNS but also in other tissues such as the placenta, lymphoreticular system, and muscle tissue [35,36].

The use of the prion-susceptible cell line N2a is attractive because it is easily cultured. The N2a cell line has been extensively used since 1970 to study prion infection [37], although the observed low postinfection titers and rapid attenuation in titers remain problematic [38,39]. The hypothalamic cell line GT1 displays higher PrP^{C} expression levels and is more susceptible to prion infection than other cell lines [40,41]. Although the human neuroblastoma cell line SH-SYS5 was reported to be infected with CJD prions, reproducibility of the result has yet to be confirmed by other laboratories [42]. Therefore, cell-culture systems for CJD prion-infection have not been established. For BSE prion, only two cell lines, the mouse microglial cell line MG20 overexpressing mouse PrP and the rabbit epithelial cell line RK13 overexpressing bank vole PrP, can be infected with mouse-adapted BSE and bank vole-proliferated BSE, respectively.

TABLE 7.1

Representative Cell Lines Used for Prion Infection

Cell Lines	Origins	Prion Strains	References
PC12	Rat pheochromocytoma	139A, ME7	[32,62,63]
C2C12	Mouse skeletal myoblast cell	22L	[64]
CAD	Mouse catecholaminergic cell	RML, 22L, 22F, 79A, 139A, ME7	[44,65–67]
CF10	Mouse brain–derived PrP-deficient cell	22L	[68]
GT1	Mouse hippocampal neuron	Chandler, RML, 139A, 22L, kCJD, FU CJD, M1000	[40,41,69–72]
HaB	Hamster brain–derived cell	Sc237	[73]
HpL3–4	Mouse hippocampal neuronal cell	22L, Chandler	[74,75]
L. fibroblasts	Mouse fibroblast	ME7, Chandler	[76]
L929	Mouse fibroblast	22L, ME7, RML	[34]
MDB	Deer immortalized brain cell	CWD	[77]
MG20	Mouse microglial cell	Chandler, ME7, Obihiro, mouse-adapted BSE	[78]
MNB	Mouse neuroblastoma	Chandler	[39]
MovS	Mouse Schwann cell-like dorsal root ganglia	PG127, SSBP/1, scrapie field isolates	[79,80]
MSC-80	Mouse Schwann cell	Chandler	[81]
N2a	Mouse neuroblastoma	Chandler, RML, 139A, 22L, C506, Fukuoka-1, FU CJD	[38–40,71,82–86]
NIH/3T3	Mouse fibroblast	22L	[34]
RK13	Rabbit kidney epithelial cell	Fukuoka-1, 22L, Chandler, M1000, mo sCJD, voBSE, PG127, LA404, SSBP/1, scrapie field isolates, CWD	[33,69,80,87–91]
ScHB	Hamster brain cell	Scrapie (Chandler)	[31]
SH-SY5Y	Human neuroblastoma	CJD	[42]
SMB	Mouse brain cell	Chandler, 139A, 22F, 79A	[31,92,93]
SN56	Mouse cholinergic septal cell	Chandler, ME7, 22L	[94]

BSE, bovine spongiform encephalopathy; CJD, Creutzfeldt–Jakob disease; kCJD, Kuru CJD; sCJD, Sporadic CJD; CWD, chronic wasting disease; RML, Rocky Mountain Laboratory isolate.

It should be noted that cell lines susceptible to some prion strains demonstrate a remarkable resistance to other strains [43,44]. For example, mouse fibroblast 3T3 cells are only susceptible to prion strain 22L, while mouse fibroblast L929 cells can be infected with prion strain 22L, RML, and ME7 [34,44]. In addition, although most of the cell lines and primary cells have been shown to express PrP^{C} [45,46], only a limited number of these cells are susceptible to prion infection. Moreover, it is intriguing that both neuronal and nonneuronal cell lines are susceptible to prion infection [47]. However, the reason for this difference in susceptibility to prion infection among cell lines and prion strains remains unclear. Such differences may be related to the cellular properties of the receptor expression or possibly cofactors contributing to PrP^{Sc} internalization, such as laminin receptor precursor (LRP/LR) [48], low-density lipoprotein receptor-related protein 1 (LPR1) [49], proteoglycans, and glycosaminoglycans (GAGS) [50]. Indeed, some cells infected with prions secrete exosomes associated with the prion agent, suggesting that exosome release from cells into the cell-culture medium might be related to susceptibility to infection [51,52].

7.4 Animal Models

The most common experimental animal models for studying prion infection are mouse and hamster [53]. For example, infection of a hamster with 263K prion leads to the onset of prion disease within a short incubation period (~100 days). The experimental route of infection is normally intracerebral and intraperitoneal inoculation (Figure 7.4). In the case of intracerebral inoculation, a sample is injected into the cerebral ventricular system of mice using a microsyringe [54]. Intraperitoneal inoculation is the injection of the sample into the peritoneum (body cavity). In some cases, such as during an inactivation or removal assay for prions, stainless steel wires contaminated with prions are often used [55].

To estimate the titer of infectious prion, samples used to inoculate animals should be serially diluted and assayed. The length of the incubation period following inoculation with the serially diluted prions before the onset of the disease is shortened in a dose-dependent manner. Clinical symptoms such as tremors and ataxia are used for the calculation of incubation time. Thus, using this protocol, it is possible to estimate the infectious titer of prions by monitoring clinical TSE signs and comparing mice inoculated with serially diluted prions. In addition to the incubation time, the survival rate after inoculation is also important. By monitoring the survival curve, quantitative bioassays can be performed using biostatistics. Kaplan–Meier estimators have been used to generate survival curves and the results were analyzed by using other tests such as the log-rank test or generalized Wilcoxon test.

TSE infected mice have a prolonged asymptomatic incubation period, lasting between 4 months and the full lifespan of the mice (over 2 years), depending on the particular mouse and prion strains under investigation. However, the incubation time after intracerebral inoculation is uniform for any given mouse and

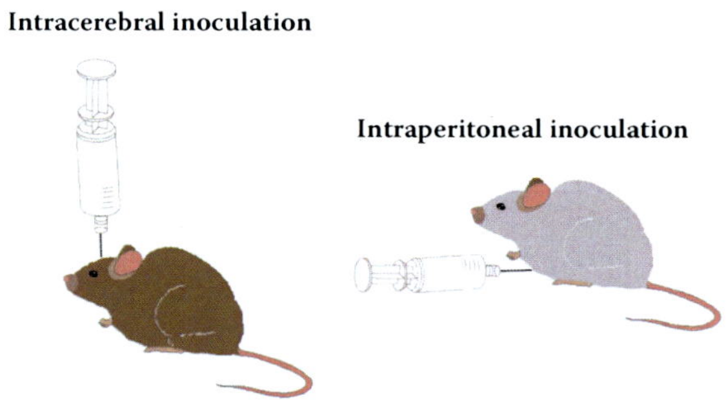

FIGURE 7.4 Animal infection models of prion diseases. Animal infection models of prion diseases generally involve intracerebral or intraperitoneal inoculation. In both cases, the survival time and incubation period are measured as an index of prion infection and infectious titer. Both the survival time and incubation period after intracerebral infection are shorter than after intraperitoneal infection. The survival curve of the mice is then examined by statistical analysis (e.g., the log-rank test).

prion strain combination. Nonetheless, markedly different incubation times are observed when different prion strains are tested against a single inbred mouse strain [56]. Likewise, strikingly different incubation times are observed when a single prion strain is tested against different mouse strains, especially those having different genotypes of the PrP gene (*Prnp*). There are two alleles of mouse *Prnp*, namely *Prnp*^{a} and *Prnp*^{b}, encoding PrP that differ by two amino acid residues at codons 108 and 189 [57]. Mice with different PrP genotypes may have an incubation period that differs by hundreds of days after infection with the same prion strain. Subtle changes in the behavior of infected mice are often seen for a few weeks before the onset of definite neurological signs during the last 2–3 weeks of the incubation period.

Behavioral changes are closely related to neuropathology. Images of histological sections of the brains of infected animals are dramatically different among prion strains [58]. The areas of vacuolation and PrP deposits depend on the prion strain, but also to some extent on PrP and other genetic factors. The so-called "lesion profile" is an objective analysis that involves scoring the area of vacuolation and/or PrP deposits in nine gray matter and three white matter brain areas. Interestingly, each combination of prion strain and mouse strain gives a characteristic lesion profile [56,59,60].

Due to the extended incubation period, animal bioassays take a very long time to complete (e.g., up to 2.5 years for primary transmission in mice, then at least 6 months for each serial mouse passage). In addition, a special biosafety level facility is required, which makes the whole procedure very costly. Therefore, an appropriate sample size (number of animals, type of animals) for each assay should be carefully considered. Given the difficulties and expense of performing such experiments, researchers should consider whether an alternative approach, such as *in vitro* testing or cell-culture assays, can be used.

7.5 Conclusions

Studies of prion biology have until recently used conventional laboratory models for prion infection, cell-culture assays, and bioassays. The development of novel tube-based methods involving PrP^{res} amplification, such as PMCA and RT-QuIC, has revolutionized the analysis of prion diseases [61]. In particular, PMCA and RT-QuIC have dramatically improved the diagnosis of prion diseases by facilitating the sensitive detection of prions. Furthermore, RT-QuIC provides quantitative data on PrP^{res}, while PMCA enables amplification of PrP^{Sc} *in vitro*. Novel future applications of these methods in prion biology are anticipated.

Studies using cell lines suggest that PrP^{C} is not the sole determinant for prion susceptibility. Indeed, other cellular factors may be required for efficient infection. Under certain conditions, loss of host factors for susceptibility to prion infection may occur. Furthermore, most cell lines have restricted prion-strain specificity, that is, matching pairs between the host cell and prion strain. Similarly, constituents of the buffer used for PMCA vary depending on the particular prion strain under investigation. Research into anti-prion drugs using cell lines and PMCA systems must take these factors into consideration, because some of the anti-prion activity might be attributed to a special property of the prion strain or cell line. For example, an anti-prion activity may be caused by an indirect effect on cellular metabolism. Thus, studies using different strains and cell lines are necessary in order to identify an anti-prion compound effective against a broad range of prion strains. Unfortunately, animal bioassays are often time consuming, laborious, and expensive to perform. Nonetheless, animal bioassays are still valuable in determining the biological activity of prions and remain the gold standard laboratory model for prion infections.

Acknowledgments

This work was supported by Grant-in-Aids for Scientific Research from the Ministry of Education, Science, Culture and Technology of Japan (25450447, 24110717, 23780299, 22110514, 20780219, 17780228), Grants-in-Aid from the Research Committee of Prion Disease and Slow Virus Infection, the Ministry of Health, Labour and Welfare of Japan, and Japan Agency for Medical Research and Development (AMED).

REFERENCES

1. Prusiner, S.B. Prions. *Proc Natl Acad Sci USA* 95, 13363–83 (1998).
2. Aguzzi, A. & Polymenidou, M. Mammalian prion biology: one century of evolving concepts. *Cell* 116, 313–27 (2004).
3. Prusiner, S.B. et al. Scrapie agent contains a hydrophobic protein. *Proc Natl Acad Sci USA* 78, 6675–9 (1981).
4. Sakudo, A., Nakamura, I., Ikuta, K. & Onodera, T. Recent developments in prion disease research: diagnostic tools and in vitro cell culture models. *J Vet Med Sci* 69, 329–37 (2007).
5. Thuring, C.M. et al. Discrimination between scrapie and bovine spongiform encephalopathy in sheep by molecular size, immunoreactivity, and glycoprofile of prion protein. *J Clin Microbiol* 42, 972–80 (2004).
6. Imran, M. & Mahmood, S. An overview of animal prion diseases. *Virol J* 8, 493 (2011).
7. Sakudo, A. & Onodera, T. Chapter 98: Bovine spongiform encephalopathy (BSE), In *Molecular Detection of Animal Viral Pathogens* (Liu, D., ed.). Taylor & Francis, CRC Press, Boca Raton, FL, pp. 901–912 (2016).
8. Sakudo, A. & Onodera, T. Chapter 99: Chronic wasting disease (CWD), In *Molecular Detection of Animal Viral Pathogens* (Liu, D., ed.). Taylor & Francis, CRC Press, Boca Raton, FL, pp. 913–921 (2016).
9. Saborio, G.P., Permanne, B. & Soto, C. Sensitive detection of pathological prion protein by cyclic amplification of protein misfolding. *Nature* 411, 810–3 (2001).
10. Castilla, J., Saá, P. & Soto, C. Cyclic amplification of prion protein misfolding, In *Techniques in Prion Research* (Methods and Tools in Biosciences and Medicine) (Lehmann, S. & Grassi, J., eds.). Birkhäuser Verlag, Basel, Switzerland, pp. 198–213, 2005.
11. Castilla, J., Saa, P., Hetz, C. & Soto, C. In vitro generation of infectious scrapie prions. *Cell* 121, 195–206 (2005).
12. Thorne, L. & Terry, L.A. In vitro amplification of PrP^{Sc} derived from the brain and blood of sheep infected with scrapie. *J Gen Virol* 89, 3177–84 (2008).
13. Saa, P., Castilla, J. & Soto, C. Presymptomatic detection of prions in blood. *Science* 313, 92–4 (2006).
14. Rubenstein, R. et al. Prion disease detection, PMCA kinetics, and IgG in urine from sheep naturally/experimentally infected with scrapie and deer with preclinical/clinical chronic wasting disease. *J Virol* 85, 9031–8 (2011).
15. Kurt, T.D., Telling, G.C., Zabel, M.D. & Hoover, E.A. Trans-species amplification of PrP^{CWD} and correlation with rigid loop 170N. *Virology* 387, 235–43 (2009).
16. Murayama, Y. et al. Ultrasensitive detection of PrP^{Sc} in the cerebrospinal fluid and blood of macaques infected with bovine spongiform encephalopathy prion. *J Gen Virol* 95, 2576–88 (2014).
17. Murayama, Y. et al. Efficient in vitro amplification of a mouse-adapted scrapie prion protein. *Neurosci Lett* 413, 270–3 (2007).
18. Wang, F., Wang, X., Yuan, C.G. & Ma, J. Generating a prion with bacterially expressed recombinant prion protein. *Science* 327, 1132–5 (2010).
19. Faburay, B., Tark, D., Kanthasamy, A.G. & Richt, J.A. *In vitro* amplification of scrapie and chronic wasting disease PrP^{res} using baculovirus-expressed recombinant PrP as substrate. *Prion* 8, 393–403 (2014).
20. Deleault, N.R. et al. Isolation of phosphatidylethanolamine as a solitary cofactor for prion formation in the absence of nucleic acids. *Proc Natl Acad Sci USA* 109, 8546–51 (2012).
21. Deleault, N.R. et al. Cofactor molecules maintain infectious conformation and restrict strain properties in purified prions. *Proc Natl Acad Sci USA* 109, E1938–46 (2012).
22. Atarashi, R., Sano, K., Satoh, K. & Nishida, N. Real-time quaking-induced conversion: a highly sensitive assay for prion detection. *Prion* 5, 150–3 (2011).
23. Orru, C.D. et al. Detection and discrimination of classical and atypical L-type bovine spongiform encephalopathy by real-time quaking-induced conversion. *J Clin Microbiol* 53, 1115–20 (2015).
24. Orru, C.D. et al. Rapid and sensitive RT-QuIC detection of human Creutzfeldt–Jakob disease using cerebrospinal fluid. *mBio* 6, e02451-14 (2015).
25. Orru, C.D., Wilham, J.M., Vascellari, S., Hughson, A.G. & Caughey, B. New generation QuIC assays for prion seeding activity. *Prion* 6, 147–52 (2012).
26. McGuire, L.I. et al. Real time quaking-induced conversion analysis of cerebrospinal fluid in sporadic Creutzfeldt–Jakob disease. *Ann Neurol* 72, 278–85 (2012).

27. Henderson, D.M. et al. Quantitative assessment of prion infectivity in tissues and body fluids by real-time quaking-induced conversion. *J Gen Virol* 96, 210–9 (2015).
28. Grassmann, A., Wolf, H., Hofmann, J., Graham, J. & Vorberg, I. Cellular aspects of prion replication in vitro. *Viruses* 5, 374–405 (2013).
29. Beranger, F., Mange, A., Solassol, J. & Lehmann, S. Cell culture models of transmissible spongiform encephalopathies. *Biochem Biophys Res Commun* 289, 311–6 (2001).
30. Lehmann, S., Solassol, J. & Perrier, V. Cell culture models of TSEs, In *Techniques in Prion Research* (Methods and Tools in Biosciences and Medicine) (Lehmann, S. & Grassi, J., eds.). Birkhäuser Verlag, Basel, Switzerland, pp. 72–81, 2005.
31. Clarke, M.C. & Haig, D.A. Evidence for the multiplication of scrapie agent in cell culture. *Nature* 225, 100–1 (1970).
32. Rubenstein, R., Carp, R.I. & Callahan, S.M. In vitro replication of scrapie agent in a neuronal model: infection of PC12 cells. *J Gen Virol* 65 (Pt 12), 2191–8 (1984).
33. Vilette, D. et al. Ex vivo propagation of infectious sheep scrapie agent in heterologous epithelial cells expressing ovine prion protein. *Proc Natl Acad Sci USA* 98, 4055–9 (2001).
34. Vorberg, I., Raines, A., Story, B. & Priola, S.A. Susceptibility of common fibroblast cell lines to transmissible spongiform encephalopathy agents. *J Infect Dis* 189, 431–9 (2004).
35. Thomzig, A., Kratzel, C., Lenz, G., Kruger, D. & Beekes, M. Widespread PrP^{Sc} accumulation in muscles of hamsters orally infected with scrapie. *EMBO Rep* 4, 530–3 (2003).
36. Bosque, P.J. et al. Prions in skeletal muscle. *Proc Natl Acad Sci USA* 99, 3812–7 (2002).
37. Clarke, M.C. & Haig, D.A. Multiplication of scrapie agent in cell culture. *Res Vet Sci* 11, 500–1 (1970).
38. Butler, D.A. et al. Scrapie-infected murine neuroblastoma cells produce protease-resistant prion proteins. *J Virol* 62, 1558–64 (1988).
39. Race, R.E., Fadness, L.H. & Chesebro, B. Characterization of scrapie infection in mouse neuroblastoma cells. *J Gen Virol* 68 (Pt 5), 1391–9 (1987).
40. Nishida, N. et al. Successful transmission of three mouse-adapted scrapie strains to murine neuroblastoma cell lines overexpressing wild-type mouse prion protein. *J Virol* 74, 320–5 (2000).
41. Schatzl, H.M. et al. A hypothalamic neuronal cell line persistently infected with scrapie prions exhibits apoptosis. *J Virol* 71, 8821–31 (1997).
42. Ladogana, A., Liu, Q., Xi, Y.G. & Pocchiari, M. Proteinase-resistant protein in human neuroblastoma cells infected with brain material from Creutzfeldt–Jakob patient. *Lancet* 345, 594–5 (1995).
43. Kleer, C.G. et al. EZH2 is a marker of aggressive breast cancer and promotes neoplastic transformation of breast epithelial cells. *Proc Natl Acad Sci USA* 100, 11606–11 (2003).
44. Mahal, S.P. et al. Prion strain discrimination in cell culture: the cell panel assay. *Proc Natl Acad Sci USA* 104, 20908–13 (2007).
45. Sakudo, A. & Onodera, T. Prion protein (PrP) gene-knockout cell lines: insight into functions of the PrP. *Front Cell Dev Biol* 2, 75 (2015).
46. Sakudo, A. et al. Recent advances in clarifying prion protein functions using knockout mice and derived cell lines. *Mini Rev Med Chem* 6, 589–601 (2006).
47. Solassol, J., Crozet, C. & Lehmann, S. Prion propagation in cultured cells. *Br Med Bull* 66, 87–97 (2003).
48. Gauczynski, S. et al. The 37-kDa/67-kDa laminin receptor acts as the cell-surface receptor for the cellular prion protein. *EMBO J* 20, 5863–75 (2001).
49. Taylor, D.R. & Hooper, N.M. The low-density lipoprotein receptor-related protein 1 (LRP1) mediates the endocytosis of the cellular prion protein. *Biochem J* 402, 17–23 (2007).
50. Holmes, B.B. et al. Heparan sulfate proteoglycans mediate internalization and propagation of specific proteopathic seeds. *Proc Natl Acad Sci USA* 110, E3138–47 (2013).
51. Alais, S. et al. Mouse neuroblastoma cells release prion infectivity associated with exosomal vesicles. *Biol Cell* 100, 603–15 (2008).
52. Fevrier, B. et al. Cells release prions in association with exosomes. *Proc Natl Acad Sci USA* 101, 9683–8 (2004).
53. Bruce, M.E., Boyle, A. & McConnell, I. TSE strain typing in mice, In *Techniques in Prion Research* (Methods and Tools in Biosciences and Medicine) (Lehmann, S. & Grassi, J., eds.). Birkhäuser Verlag, Basel, Switzerland, pp. 132–146, 2005.
54. Inoue, Y. et al. Infection route-independent accumulation of splenic abnormal prion protein. *Jpn J Infect Dis* 58, 78–82 (2005).

55. Fichet, G. et al. Novel methods for disinfection of prion-contaminated medical devices. *Lancet* 364, 521–6 (2004).
56. Bruce, M.E., McConnell, I., Fraser, H. & Dickinson, A.G. The disease characteristics of different strains of scrapie in *Sinc* congenic mouse lines: implications for the nature of the agent and host control of pathogenesis. *J Gen Virol* 72 (Pt 3), 595–603 (1991).
57. Westaway, D. et al. Distinct prion proteins in short and long scrapie incubation period mice. *Cell* 51, 651–62 (1987).
58. Fraser, H. Diversity in the neuropathology of scrapie-like diseases in animals. *Br Med Bull* 49, 792–809 (1993).
59. Fraser, H. & Dickinson, A.G. The sequential development of the brain lesion of scrapie in three strains of mice. *J Comp Pathol* 78, 301–11 (1968).
60. Bruce, M.E., McBride, P.A. & Farquhar, C.F. Precise targeting of the pathology of the sialoglycoprotein, PrP, and vacuolar degeneration in mouse scrapie. *Neurosci Lett* 102, 1–6 (1989).
61. Saa, P. & Cervenakova, L. Protein misfolding cyclic amplification (PMCA): current status and future directions. *Virus Res* 207, 47–61 (2015).
62. Rubenstein, R. et al. Demonstration of scrapie strain diversity in infected PC12 cells. *J Gen Virol* 73 (11), 3027–31 (1992).
63. Rubenstein, R., Deng, H., Scalici, C.L. & Papini, M.C. Alterations in neurotransmitter-related enzyme activity in scrapie-infected PC12 cells. *J Gen Virol* 72 (Pt 6), 1279–85 (1991).
64. Dlakic, W.M., Grigg, E. & Bessen, R.A. Prion infection of muscle cells in vitro. *J Virol* 81, 4615–24 (2007).
65. Dron, M., Dandoy-Dron, F., Farooq Salamat, M.K. & Laude, H. Proteasome inhibitors promote the sequestration of PrP^{Sc} into aggresomes within the cytosol of prion-infected CAD neuronal cells. *J Gen Virol* 90, 2050–60 (2009).
66. Browning, S. et al. Abrogation of complex glycosylation by swainsonine results in strain- and cell-specific inhibition of prion replication. *J Biol Chem* 286, 40962–73 (2011).
67. Julius, C. et al. Transcriptional stability of cultured cells upon prion infection. *J Mol Biol* 375, 1222–33 (2008).
68. McNally, K.L., Ward, A.E. & Priola, S.A. Cells expressing anchorless prion protein are resistant to scrapie infection. *J Virol* 83, 4469–75 (2009).
69. Vella, L.J. et al. Packaging of prions into exosomes is associated with a novel pathway of PrP processing. *J Pathol* 211, 582–90 (2007).
70. Lewis, V. et al. Increased proportions of C1 truncated prion protein protect against cellular M1000 prion infection. *J Neuropathol Exp Neurol* 68, 1125–35 (2009).
71. Arjona, A., Simarro, L., Islinger, F., Nishida, N. & Manuelidis, L. Two Creutzfeldt–Jakob disease agents reproduce prion protein-independent identities in cell cultures. *Proc Natl Acad Sci USA* 101, 8768–73 (2004).
72. Miyazawa, K., Emmerling, K. & Manuelidis, L. High CJD infectivity remains after prion protein is destroyed. *J Cell Biochem* 112, 3630–7 (2011).
73. Taraboulos, A., Serban, D. & Prusiner, S.B. Scrapie prion proteins accumulate in the cytoplasm of persistently infected cultured cells. *J Cell Biol* 110, 2117–32 (1990).
74. Maas, E. et al. Scrapie infection of prion protein-deficient cell line upon ectopic expression of mutant prion proteins. *J Biol Chem* 282, 18702–10 (2007).
75. Sakudo, A., Wu, G., Onodera, T. & Ikuta, K. Octapeptide repeat region of prion protein (PrP) is required at an early stage for production of abnormal prion protein in PrP-deficient neuronal cell line. *Biochem Biophys Res Commun* 365, 164–9 (2008).
76. Clarke, M.C. & Millson, G.C. Infection of a cell line of mouse L fibroblasts with scrapie agent. *Nature* 261, 144–5 (1976).
77. Raymond, G.J. et al. Inhibition of protease-resistant prion protein formation in a transformed deer cell line infected with chronic wasting disease. *J Virol* 80, 596–604 (2006).
78. Iwamaru, Y. et al. Microglial cell line established from prion protein-overexpressing mice is susceptible to various murine prion strains. *J Virol* 81, 1524–7 (2007).
79. Archer, F. et al. Cultured peripheral neuroglial cells are highly permissive to sheep prion infection. *J Virol* 78, 482–90 (2004).

80. Neale, M.H. et al. Infection of cell lines with experimental and natural ovine scrapie agents. *J Virol* 84, 2444–52 (2010).
81. Follet, J. et al. PrP expression and replication by Schwann cells: implications in prion spreading. *J Virol* 76, 2434–9 (2002).
82. Enari, M., Flechsig, E. & Weissmann, C. Scrapie prion protein accumulation by scrapie-infected neuroblastoma cells abrogated by exposure to a prion protein antibody. *Proc Natl Acad Sci USA* 98, 9295–9 (2001).
83. Markovits, P., Dautheville, C., Dormont, D., Dianoux, L. & Latarjet, R. In vitro propagation of the scrapie agent. I. Transformation of mouse glia and neuroblastoma cells after infection with the mouse-adapted scrapie strain c-506. *Acta Neuropathol* 60, 75–80 (1983).
84. Ostlund, P., Lindegren, H., Pettersson, C. & Bedecs, K. Up-regulation of functionally impaired insulin-like growth factor-1 receptor in scrapie-infected neuroblastoma cells. *J Biol Chem* 276, 36110–5 (2001).
85. Race, R. The scrapie agent in vitro. *Curr Top Microbiol Immunol* 172, 181–93 (1991).
86. Scott, M.R., Kohler, R., Foster, D. & Prusiner, S.B. Chimeric prion protein expression in cultured cells and transgenic mice. *Protein Sci* 1, 986–97 (1992).
87. Courageot, M.P. et al. A cell line infectible by prion strains from different species. *J Gen Virol* 89, 341–7 (2008).
88. Lawson, V.A. et al. Mouse-adapted sporadic human Creutzfeldt–Jakob disease prions propagate in cell culture. *Int J Biochem Cell Biol* 40, 2793–801 (2008).
89. Sabuncu, E. et al. PrP polymorphisms tightly control sheep prion replication in cultured cells. *J Virol* 77, 2696–700 (2003).
90. Bian, J. et al. Cell-based quantification of chronic wasting disease prions. *J Virol* 84, 8322–6 (2010).
91. Kim, H.J. et al. Establishment of a cell line persistently infected with chronic wasting disease prions. *J Vet Med Sci* 74, 1377–80 (2012).
92. Birkett, C.R. et al. Scrapie strains maintain biological phenotypes on propagation in a cell line in culture. *EMBO J* 20, 3351–8 (2001).
93. Kanu, N. et al. Transfer of scrapie prion infectivity by cell contact in culture. *Curr Biol* 12, 523–30 (2002).
94. Baron, T.G., Biacabe, A.G., Bencsik, A. & Langeveld, J.P. Transmission of new bovine prion to mice. *Emerg Infect Dis* 12, 1125–8 (2006).

Section II

Foodborne Infections due to Gram-Positive Bacteria

8

Bacillus

Jessica Minnaard, Ivanna S. Rolny, and Pablo F. Pérez

CONTENTS

8.1 Introduction

The ability of prokaryotes to sporulate is one of the evolutionary strategies to cope with stress. Spore-forming microorganisms invest considerable amounts of energy in a refined differentiation process (sporogenesis) that involves the expression of hundreds of genes in order to give a morphologically distinct cell that presents several barriers related to the resistance to physical, chemical, and biological challenges.[1]

Among spore-forming microorganisms, those belonging to the genus *Bacillus* play an important role in several aspects of human activities. Indeed, the ability of fast growth and the diversity of metabolic capabilities along with the capacity to survive in adverse conditions lead to a versatile genus that actively participates in shaping the environment. However, some members of the genus *Bacillus* have evolved pathogenic potential and are thus associated with distinct pathological processes.

The genus *Bacillus* includes spore-forming Gram-positive, low G + C content, rod-shaped bacteria. Members of the genus are ubiquitous and capable of surviving and growing in very different, even extreme, conditions such as high temperatures, high saline concentrations, and low pH.[2]

The early works of Pollender, Brauell, Delafond, and Davaine on anthrax or carbuncle prompted Robert Koch to enunciate the paramount postulates of the germ theory of disease and Louis Pasteur to establish effective immunization for anthrax prevention in the seminal experiment of massive livestock vaccination at Pouilly-le-Fort.[3–5] It is known that some *Bacillus* species constitute a very definite cluster sharing many characteristics. This so-called "cereus group" includes microorganisms that can be identified as *Bacillus cereus* (*B. cereus sensu lato*) on the basis of conventional microbiological procedures, but the use of more refined techniques lead to a more precise identification. The *B. cereus* group encompasses seven species, that is, *Bacillus anthracis*—the etiological agent of anthrax, *B. cereus* (*sensu stricto*), *Bacillus thuringiensis*, *Bacillus mycoides*, *Bacillus pseudomycoides*, *Bacillus weihenstephanensis*, and *Bacillus cytotoxicus*.[6–8] The division between species of the *B. cereus* group is based on phenotypic characteristics related to differences in ecology and pathogenesis. However, adoption of molecular approaches strongly suggests that members of the *B. cereus* group belong to one single species.[9–12] In this chapter, we will include bacterial species stated in the original publications irrespective of the exact taxonomic nomenclature.

It has been demonstrated that the *B. cereus* group diversified along evolution. This evolutionary pathway includes thermotolerant lineages as the older microorganisms evolved into mesophilic members, with a more recent shift to psychrotrophic groups.[13] Interestingly, the food-poisoning risk is not necessarily related to species but to the phylogenetic group.[7,14]

Although the exact cause of many historical epidemics may never be known, there are relevant references compatible with epidemic anthrax. Even though no diagnostic methods were available, descriptions from historians and physicians revealed an illness; their symptoms probably implicated anthrax, that is, the fifth and sixth plagues of Egypt described in the Book of Exodus, the Plague of Athens (430–427 BCE), and the epizootics described in Virgil's Aeneid (70–19 BCE).[15] However, there are other reports that postulated that anthrax is unlikely the basis of the Plague of Athens[16–18] or other biblical plagues.[19]

Beyond controversial historical and epidemiological aspects, it is evident that *B. anthracis* is an important zoonotic pathogen and that several outbreaks are well documented. In the 18th century, there was an outbreak of intestinal anthrax due to consumption of smoked or salted meat from ill animals after an earthquake occurred in Saint Domingue (modern Haiti). In this episode, 15,000 people were killed.[3] Human gastrointestinal anthrax outbreaks are described in different studies, and they are often associated with consumption of foods originating from diseased animals.[20–24] However, unusual sources of contamination such as animal-hide drums have been reported.[24] Intestinal anthrax constitutes the main presentation for livestock that normally ingests spores in foods.[25,26]

The *B. cereus* group represents an example of how natural plasmid transfer within a genus leads to different microorganisms that are allocated to different species.

Indeed, in the same chromosomal genetic background, the presence of PXO1 and PXO2 characterizes *B. anthracis*, a microorganism with high pathogenic potential, and the presence of large transferable plasmids in *B. thuringiensis* enables its production of insecticidal proteins.[27] It is worth noting that *B. cereus* harboring *B. anthracis* toxin genes leads to life-threatening pathology.[28–30] Because of these particularities, members of the group other than *B. anthracis* were not clearly differentiated until the late 1960s. Indeed, aerobic endospore-forming rods were not identified at all or assigned to the *Bacillus subtilis* species.[31] Earlier reports that univocally identified *B. cereus* as the causative agent of foodborne diseases were published between 1950 and 1955.[31]

Foods are suitable culture media for *Bacillus* spp., and thus, growth in high numbers can occur in many steps of food preparation and preservation. In addition, some strains are psychrotrophic.[32] To a lesser extent, other *Bacillus* species have been associated with gastrointestinal illness, for example, *B. subtilis*, *Bacillus pumilus*, *Bacillus licheniformis*, *Bacillus brevis*, and *Bacillus mojavensis*, although these microorganisms are more often involved in food spoilage.[31,33–35]

8.2 Virulence Factors: An Overview

Virulence traits differ between members of the *B. cereus* group. Certainly, *B. anthracis* possesses the highest pathogenic potential because of its ability to produce protein toxins with high biological activity[36] and its potential for phagosomal escape.[37] The virulence of this microorganism is related to the presence of two plasmids, namely, pXO1 and pXO2 encoding for the protective antigen (PA) (pXO1), edema factor (EF) (pXO1), lethal factor (LF) (pXO1), and capsule (pXO2).[25] The expression of toxin genes is regulated by AtxA, a global regulator encoded by *atxA* located in the pathogenicity island within pXO1.[38]

Virulence factors produced by members of the genus *Bacillus* differ between species. There is a well-defined difference between virulence factors of *B. anthracis* and the more diffuse traits of other members of the *B. cereus* group and other *Bacillus* spp. Concerning *B. anthracis*, there are three main routes of infection: inhalational, cutaneous, and digestive. Even though severity differs according to the infection route, the same virulence factors are involved.

The pathogenesis of *B. anthracis* is related to the production of an A_2B toxin that includes a B subunit, that is, PA, and two catalytic subunits, the EF and LF. In addition, it has the ability to synthesize a poly-δ-D-glutamic capsule, an anti phagocytic factor.

On a mechanistic basis, the mode of action of the AB toxin involves binding of PA that, after proteolytic cleavage, binds either EF or LF that are now able to be internalized to trigger the biological effect. EF activates adenylate cyclase, thus leading to the secretion of water and electrolytes. On the other hand, LF cleaves MAPKK, thus interfering with signal transduction.

After secretion and dissemination, PA binds to cell surface receptors, that is, anthrax toxin receptor 1 (TEM8) and anthrax toxin receptor 2 (CMG2). The latter has the higher binding affinity *in vivo*.[36] The binding of LF or EF to PA requires the proteolytic activation of PA. Indeed, PA83 (83 kDa) splits into PA63 and PA20 by furin or furin-like proteases, with PA63 being the receptor for LF and EF. After the proteolytic cleavage of PA83, PA63 oligomerizes on the plasma membrane of the target cell, binds LF or EF, and is further internalized by receptor-mediated endocytosis.

LF inhibits signaling events of the MAPK pathway, that is, those mediated by ERK, p38, and JNK. This inhibition occurs after the proteolytic cleavage of MAPKK.[36] Since this signaling pathway is paramount for cellular homeostasis, LF leads to major adverse effects on the eukaryotic cells. Particularly, macrophages are very susceptible to LF that triggers apoptosis in these cells. However, this effect is species dependent and seems to be inversely correlated with the host's susceptibility to *B. anthracis* infection.[39,40] EF is a calmodulin-dependent adenylate cyclase that converts ATP to AMPc. It has been suggested that the cystic fibrosis transmembrane conductance regulator (CFTR) is involved as it is for *Vibrio cholerae*.[36] The phosphorylation of CFTR leads to the efflux of Cl^- followed by Na^+ and subsequent water movement to the extracellular milieu.

The other main player of *B. anthracis* virulence is a unique poly-δ-D-glutamic capsule that endows the microorganism the ability to disseminate.[41] Even though the mechanisms behind this are not fully understood, the role of the capsule in shielding bacterial surface structures and inhibiting the activity of antimicrobial peptides such as α and β defensins has been emphasized, thus allowing the microorganisms to survive in intracellular compartments.[37]

Concerning other members of the *Bacillus* genus, virulence factors are not well identified and they are still under debate.

The virulence of *B. cereus sensu lato* has been ascribed to several factors, for example, hemolysin BL (HBL),[42,43] nonhemolytic enterotoxin (NHE),[44] enterotoxin FM, cytolysin K, phosphatidylinositol-specific phospholipase C, enterotoxin S, sphingomyelinase, cereolysin O,[45,46] hemolysin II (HLyII),[47] and camelysin.[48] Emetic strains produce cereulide, a thermostable cyclic dodecadepsipeptide.[49,50] The mechanisms behind this plethora of extracellular factors are not completely understood, and the relevance of these factors in the intestinal pathology is still under discussion. Besides these traditional factors, the biological effect of direct bacteria–enterocyte interaction has been demonstrated *in vitro*.[51–53]

Other virulence factors have been described in different species of the *Bacillus* genus. Interestingly, *B. subtilis* and *B. mojavensis* involved in food poisoning have been demonstrated to produce amylosin, a heat-stable ion channel forming toxin, which also reported for *B. amyloliquefaciens*.[54,55] Some strains are able to produce this toxin during growth at refrigeration temperatures.[33] In addition, different factors with biological activity on mammalian cells have been associated to *B. licheniformis* (lichenysin)[56] and *B. pumilus* (pumilacidin).[57]

In the next sections, models that contributed to the understanding of the pathogenesis of *Bacillus* spp. will be discussed in the context of the biological activities they are able to detect.

8.3 *In Vitro* Models

8.3.1 Cultured Eukaryotic Cells

Studies with cultured eukaryotic cells have contributed to the knowledge of the virulence factors of many pathogens. These models mimic specific steps of the interaction between microorganisms and the host, and some cell lines have been demonstrated to generate responses similar to those of the more sophisticated models.

Several eukaryotic cell lines have been employed to assess the virulence factors of *Bacillus* spp. (Table 8.1). Although the modeling of gastrointestinal pathologies can be achieved with cultured enterocytes, other cell lines have successfully been employed (Table 8.1).[46,47,51,58–78]

The experiments with cell-free culture supernatants of *B. cereus* have provided general information about the effects of extracellular factors. One of the first reports showed that filtered-dialyzed spent culture supernatants of strain B-48 grown in a fresh beef infusion were cytotoxic for two cell lines (guinea pig spleen and buccal carcinoma cells) and the primary explants of mouse-embryo tissue.[79] This effect was observed after 1 h of incubation, and this activity was partially conserved after heating filtrates at 60°C for 2 h.[79] In contrast, *B. anthracis* culture filtrates did not detach cells from monolayers but are lethal to rats.[79] These findings demonstrate that the biological activity on cultured cells does not necessarily correlate with the presence of known factors in the spent culture supernatants.

Supernatants from cultures of *B. cereus* induce a decrease in the metabolic activity, rounding-up, cell detachment, and/or death in Caco-2, CHO, Hep-2, HT1080, Raw264.7, McCoy, and human polymorphonuclear (neutrophils) cells.[51,58,61,63,68,69] Heat or trypsin treatment of the supernatants reduced or eliminated the cytotoxic effects in Hep-2 and Caco-2 cells.[63]

Susceptibility to extracellular factors varies between cell lines; that is, the Ped-2E9 cell line was 25- to 58-fold more susceptible than CHO cells when the cytotoxicity of two diarrheal strains was evaluated by assessing the release of alkaline phosphatase and impairment of mitochondrial dehydrogenase activity. Fractions with molecular sizes higher than 30 kDa yielded the highest cytotoxicity values.[59] Only one isolate, identified as *B. cereus*, was active on Ped-2E9 cells as assessed by alkaline phosphatase release. Other microorganisms identified as *B. subtilis* or *Bacillus* spp. were inactive.[59]

Dramatic changes on the surface of Caco-2 cells were observed in the presence of sterile filtrate supernatants from *B. cereus*.[58] These changes include microvilli effacement and correlate with cytoskeleton disorganization (Figure 8.1).

Assays on cultured human enterocyte-like cells (Caco-2) allowed refining studies on the extracellular factors present in spent culture supernatants of *B. cereus*. This cell line has been considered as a reference model for the study of the interaction between intestinal microorganisms and the host since Caco-2 cells spontaneously differentiate in culture and show morphological and functional polarization.[80] In this context, three levels of biological activity were defined on the basis of dose–response curves in detachment vs. supernatant concentration plots. These different behaviors were high, low, and non-detaching strains. Culture filtrate supernatants from high detaching strains are able to completely detach cells from monolayers at least in two serial dilutions. On the other hand, low detaching strains show a dose–response curve in the entire range of concentrations studied, with the highest detachment only at the highest dose. Non-detaching strains do not detach cells at any of the concentrations assayed.[51]

TABLE 8.1

In Vitro, *In Vivo*, and *Ex Vivo* Models Used for the Study of the Biological Activity of *Bacillus* spp.

		Model		
In vitro	Epithelial cells	Caco-2 (human colorectal adenocarcinoma)	HT1080 (human fibrosarcoma cells)	Vero (kidney epithelial cells from an African green monkey)
		CHO (Chinese hamster ovary)	Hep-2 (human laryngeal carcinoma)	Hep-G2 (human hepatocellular carcinoma)
		A549 (human lung carcinoma)	HeLa (cervical cancer from human)	Calu-3 (human lung adenocarcinoma)
		Paju (human neural crest-derived cell line)	Ptk6 *null* colonic epithelial cells	A204 (human rhabdomyosarcoma)
		McCoy (murine fibroblasts)	Huvet (human umbilical vein endothelial cells)	
	Immune cells	Raw264.7 (monocytes from murine origin)	J774 (monocyte/macrophage from murine origin)	Neutrophils (hPMNs)
		Jurkat (human T lymphoblastoid cell line)	Ped-2E9 (lymphocyte from murine origin)	U937 (monocytes from human origin)
Ex vivo			Spermatozoon test	
In vivo		Invertebrates	*Caenorhabditis elegans*	
			Galleria mellonella	
			Bombyx mori	
		Vertebrates	Ligated rabbit or mouse ileal loop	
			Vascular permeability reaction	
			Zebrafish	
			Suncus murinus	
			Mice	
			Rats	
			Monkeys	

This simple experimental approach allowed for the differentiation of the biological activity of *B. cereus* strains, even though this activity is related to several extracellular factors present in spent culture supernatants. In addition, a multivariate analysis demonstrated that high detaching strains were positive for the sequences of the virulence genes *entS*, *nheC*, and *sph*.[51]

Cytotoxicity is strain-dependent[51,58,68] with species-to-species variations.[63] Since nutritional factors influence the expression of virulence factors, the culture medium modifies the values of cytotoxicity, cell adhesion, and invasion.[63]

Fractioning and purification of factors present in the culture supernatants allowed for the identification and further study of several toxins of *B. cereus*. Hemolysin BL (HBL) and nonhemolytic enterotoxin (NHE) are consistently described as key virulence factors. They are tripartite protein toxins expressed in different proportions by members of the *B. cereus* group.[61] HBL components are HBLB (37.5 kDa), HBLL1 (38.2 kDa), and HBLL2 (45 kDa),[81] and NHE components are NHA (41 kDa), NHB (39.8 kDa), and NHC (36.5 kDa).[67,82] HBL is expressed in 42%–73% of food-poisoning-associated strains, whereas NHE is expressed in 97%–99% of strains.[61,83,84] Both toxins are responsible for the cytotoxic effect on several cell lines such as Vero, Hep-2, Hep-G2, Caco-2, Jurkat, RPMI 8226, A204, A549, HUVET, and U937 cells.[62] For NHE, the specificity of the observed biological effects was demonstrated in neutralization experiments with monoclonal antibodies.[85]

The presence of the three components for maximum cytotoxic effect was necessary for the activity of both HBL and NHE.[60,66,81,82] Furthermore, the relative concentrations of the components are relevant

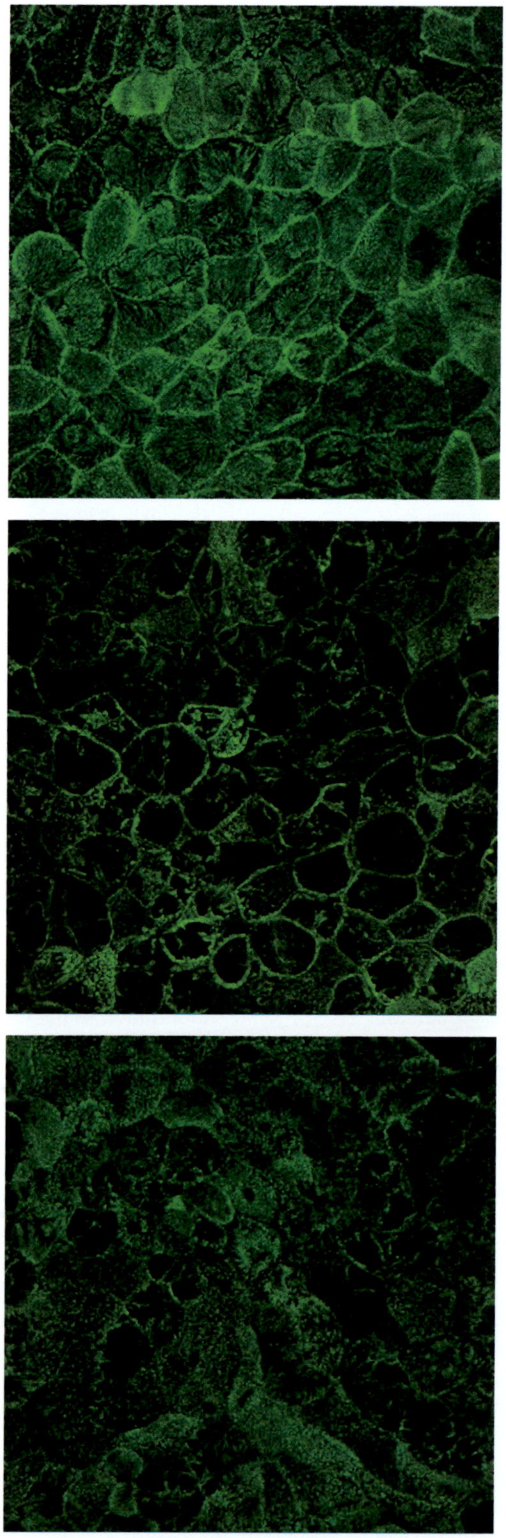

FIGURE 8.1 Disorganization of the F-actin network following incubation of cultured human enterocytes (Caco-2 cells) with spent culture supernatants of *B. cereus*. References: control cells (top), strain 2 (middle), and strain M2 (bottom).

for the observed effects (e.g., inhibition of the protein synthesis of Vero cells by NHE[44,82]). NHEB and HBLL2 are crucial for the cytotoxic effect since when these components were not present significantly, lower cytotoxic effects were observed in some cell lines.[62]

Studies on the structures involved in the binding of toxins to target cells by using CHO and red blood cells demonstrated that for HBL, the binding component is B (Hbl-B) and L1 and L2 bind to Hbl-B.[61,86] On the other hand, the NHE component responsible for the binding to eukaryotic cells is Nhe-C.[61,87] Then, there is a sequential binding of Nhe-B and Nhe-A.[87] It is worth noting that despite its name (nonhemolytic enterotoxin), the NHE toxin also has hemolytic activity.[88]

Even though several extracellular factors have been described in the *B. cereus* group, NHE was proposed as the main virulence factor[84,89] showing the highest cytotoxic effect on Vero (fibroblasts) and HUVEC (endothelial) cells.[62] Albeit to a lesser extent, NHE is also active on U937 cells[62] and is responsible for the impairment of the plasma membrane of Caco-2 cells.[72] Vero cells treated with spent culture supernatants showed blebs, which subsequently burst, and these effects were not observed when supernatants from *nheBC* mutants were used.[72] By using a *nheBC* mutant strain and Ptk6 cells, it has been demonstrated that NHE and sphingomyelinase have a concerted effect.[71]

In relation to the hemolytic action of toxins, *B. cereus* isolates with the ability to produce trimeric toxin HBL lead to a characteristic ring-shaped zone of hemolysis in blood agar plates.[86,90,91] Hemolysis depends on the source of erythrocytes: Sheep or calf erythrocytes are the most susceptible, and human or horse erythrocytes are the least susceptible.[86]

There is a high similarity in the structure and sequence between HBL and NHE toxins,[72] but it is not possible to complement or substitute components to give an active toxin.[61] So far, little is known about the mechanisms involved in the biological effects of these toxins. Studies performed with osmotic protectants such as carbohydrates or polyethylene glycol of increasing size allowed researchers to determine that the mechanism involved in HBL and NHE lysis cells is colloid-osmotic[72,92] leading to pore formation in planar lipid bilayers.[72] Spent culture supernatants lead to the release of intracellular ATP and LDH from Caco-2 and Vero cells; these effects were not detected when supernatants from cultures of *nheBC* mutants were employed.[72] Because of structural similarities between NHE and HBL components with hemolysin E from *E. coli*, it is proposed that the biological effects of both toxins share common mechanisms.[61,93,94]

Cytotoxin K (CytK) is another relevant extracellular factor of *B. cereus* associated with necrotic enteritis. It is a single-component protein toxin of 34 kDa that is hemolytic to rabbit red blood cells (it is a β-barrel pore-forming toxin) and cytotoxic to several cells lines.[8,62,70] However, CytK has lower biological activity as compared with other virulence factors such as NHE and HBL.[62] Caco-2 and Vero cells have been seen to be particularly susceptible to CytK.[62] Toxicity to Caco-2 cells is related to the ability to form nonselective pores in the plasma membrane.[8,70] This effect is similar to that of *C. perfringens* C toxin, although this toxin is cytotoxic and noncytolytic due to the formation of cation selective pores.[70]

CytK, also known as hemolysin IV,[95] has two variants: CytK1 (firstly reported by Lund et al.[8]) and CytK2 which shares 89% identity with CytK1 and has a similar size.[96] CytK2 is hemolytic to bovine erythrocytes and cytotoxic to Caco-2 although to a lesser extent than the original variant, and it did not show toxic effects on Vero-2 cells.[96] Both variants lead to different distributions of pore sizes on lipid bilayers.[96]

HlyII is another single-component protein toxin (42.3 kDa), which is a member of the family of β-barrel pore-forming toxins.[8,97–99] HlyII leads to a decrease of the mitochondrial membrane potential, and it is lytic to Paju and Caco-2 cells.[99] Besides, it has hemolytic activity on erythrocytes of different species, with the highest effects on human and rabbit red blood cells.[99] However, participation of HlyII in diarrhea caused by *B. cereus* has not been demonstrated.[95]

The apoptosis of macrophages (J774 cell line) was induced by purified HlyII, and the involvement of this factor in the observed effect was supported by studies with mutants with deletions in the *hlyII* gene.[47] This effect is related to the ability of HlyII to form pores because macrophages incubated with purified HlyII in the presence of polyethylene glycol abolished apoptosis and preserved the permeability of the cell membrane. There is also a contribution of caspase activation on the observed proapoptotic effect.[47]

Members of the *B. cereus* group are able to produce other pore-forming toxins such as cereolysin O, hemolysin III, and three phospholipases C.[95,100–102] Cereolysin O (or Hemolysin I), a heat-labile protein toxin, is a member of the cholesterol-dependent cytolysin family; its best known members are listeriolysin O and perfringolysin O. They are able to form large pores in the cell membrane.[103] Concerning hemolysin III, it was first reported by Baida and Kuzmin in 1995.[100] This is a heat-labile factor that is not inhibited by cholesterol, but no further reports were found. Sphingomyelinase, phosphatidylcholine–phospholipase C, and phosphatidylinositol–phospholipase C (PI-PLC) have cytolytic activity.[42,46,104–107] Interestingly, PI-PLC from *B. anthracis* inhibits the activation of murine dendritic cells.[108]

The *B. cereus* emetic toxin is cereulide, a heat-stable dodecadepsipeptide preformed in food.[109,110] It is a toxin that causes mitochondria dysfunction due to its activity on K^+-channels.[111,112] Since cereulide is structurally closely related to valinomycin, several studies used this antibiotic as a surrogate molecule.[110,113,114] However, the emetic toxin showed higher affinity for potassium than valinomycin thus leading to a higher biological effect.[113,114]

Cereulide is produced by a non ribosomal peptide synthetase (NRPS) encoded in an operon situated in a virulence plasmid related to the *B. anthracis* toxin plasmid pXO1.[115,116] The gene sequence (*ces*) that encodes the enzymatic machinery required for the synthesis of cereulide also includes genes encoding a putative hydrolase (*cesH*), a phosphopantetheinyl transferase (*cesP*), a type II thioesterase (*cesT*), and a putative ABC transporter (*cesC7D*).[115,116] Because of its small size and low antigenicity, cereulide is difficult to detect using immunological methods. Instead, assays with cultured cells (Hep-2) allow for the determination of its biological activity leading to vacuole formation and impairment of mitochondrial dehydrogenase activity.[109,117–119]

It has been demonstrated that there are different cereulide variants (isocereulides) that differ in toxicity to Hep-2 and lead to different effects on the conductance of lipid bilayers.[75,120] These *in vitro* effects could correlate with the severity of the gastrointestinal disease.[75]

Although it has been established that cereulide doses of around 8 µg/kg body weight can be associated to severe emetic outbreaks,[102] *in vitro* assays in Caco-2 cells showed that there is an accumulative effect with successive subemetic doses of cereulide ranging from 0.125 to 4 ng/mL. Indeed, the effects on viability, proliferation, and release of lactate dehydrogenase were detected at cereulide concentrations of 0.5–5 µg/mL.[74] Interestingly, cereulide treatments also change the expression of proteins such as PCSK9, cathepsin D, apolipoprotein A-I, proapolipoprotein, and apolipoprotein preprotein.[74]

Although Hep-2 cells are most widely used to evaluate the toxicity of cereulide, other cell lines such as HepG2 (inhibition of RNA-synthesis),[76] Hepa-1 (cytotoxicity),[76] fetal porcine Langerhans islets (necrosis),[121] natural killer (NK) cells (apoptosis and swelling of mitochondria)[122] and HeLa, Calu-3, and Paju (dissipation of the mitochondrial membrane potential)[77] have also been used. These assays of cereulide activity correlate with the gold standard assay for this toxin, that is, the boar sperm motility assay (see *ex vivo* assays).

8.3.2 Regulation of Toxin Production

B. cereus exocellular factors HBL, NHE, CytK, proteases, phospholipases, and surface proteins are controlled by PlcR, a pleiotropic regulator activated in the stationary phase by quorum sensing.[103,123–128] This activity depends on the presence of a signal peptide, PAPR, which is exported and reimported by a permease.[129,130]

Other cytotoxic factors such as HLyII are not controlled by PlcR. Instead, this extracellular factor is negatively controlled by the global regulator Ferric Uptake Regulator (Fur) and the transcriptional regulator HlyIIR.[102,131,132] Glucose 6P and iron promote repression by binding to HlyIIR and Fur, respectively.[95] When glucose and iron diminish, the expression of HlyII is activated and induces macrophage apoptosis and red blood cell lysis.[95]

Cereulide synthesis is partially controlled by Spo0A-AbrB regulon involved in the onset of sporulation,[133] and it is negatively regulated by CodY, a regulator involved in response to nutrient limitation in Gram-positive bacteria.[134] Besides, CodY activates enterotoxin synthesis and up-regulates the expression of Plc-R.[134]

Also, environmental factors control the production of extracellular factors in *B. cereus*.[102] Indeed, growth in anaerobic conditions and low oxidation–reduction potential favor HBL and NHE production through the participation of the two-component system, ResDE, and the transcriptional redox regulator, Fnr.[102] *B. cereus* has the ability to adapt to low pH conditions and survive within the gastrointestinal tract,[135,136] but the adaptation to sublethal low pH values inhibits HBL production.[136]

8.3.3 Direct Interactions with Eukaryotic Cells: Other Virulence Factors

Cultured eukaryotic cells were also useful models to gain insight into the effect of the direct interaction of *Bacillus* spp. with the host's cells. Even though the virulence of the members of this genus was traditionally allocated to the production of extracellular factors, it has been demonstrated that adhesion/invasion of the host's cells is also relevant for pathogenesis.[14,51–53,137,138]

An early report from Andersson et al.[137] suggests that the adhesion of *B. cereus* to Caco-2 cells and subsequent spore germination could contribute to virulence. By using microscopy techniques, they found that the adhesion of spores to enterocytes was strain-dependent and that high surface hydrophobicity favored adhesion.[137] Surface properties were also relevant for the interaction of spores with abiotic surfaces.[139] Spore adhesion to Hep-2 cells was also observed.[135] Interestingly, differentiated Caco-2 but not Hep-2 cells induced spore germination.[135]

Since ingested spores can germinate after contact with epithelial cells, a series of studies were conducted in order to gain insight into the relevance of the interaction of vegetative (i.e., harvested at the logarithmic phase of growth) cells of *Bacillus* spp. *B. cereus* vegetative cells have the ability to adhere to (Figure 8.2) and invade Caco-2 and HeLa cells.[51–53,63,138] Calcium depletion experiments demonstrate that invasion is favored by loosening tight junctions but no effects on the adhesion values were observed.[53] Also, the differentiation status of the cells is also relevant since when infection was carried out with undifferentiated Caco-2 cells (4 days in culture), adhesion and invasion values were significantly higher than those obtained with differentiated cells.[53] By conducting experiments in the presence of specific inhibitors of signaling pathways, it was demonstrated that the disassembly of the F-actin cytoskeleton that follows the infection of epithelial cells involves signaling through phosphorylated lipids.[53]

Extracellular factors contribute to prokaryote–eukaryote interactions since infection carried out in the presence of supernatants from 3-h-old cultures increases association (adhesion + invasion) of bacteria with cultured enterocytes.[53]

A multivariate analysis revealed that the different effects observed in *B. cereus*-infected Caco-2 cells can be correlated with the presence of virulence genes, that is, *entS*, *entFM*, *nhe* (A, B, and C), *sph*, *hbl* (A, B, C, and D), *piplC*, and *bceT*.[51]

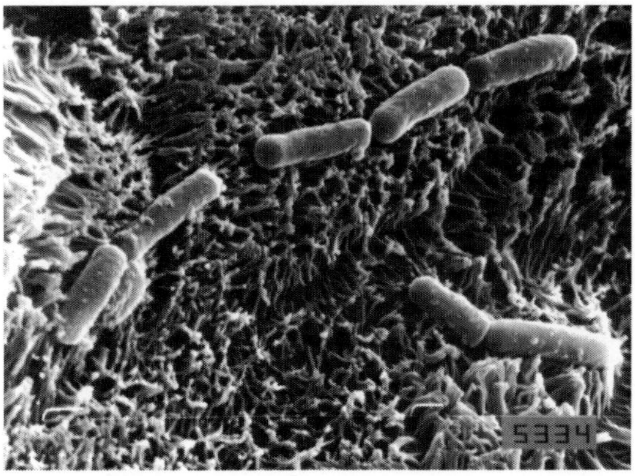

FIGURE 8.2 Adhesion of vegetative cells of *B. cereus* onto cultured human enterocytes (Caco-2 cells).

Adhesion onto the epithelial cells of bacteria of the *B. cereus* group is related to the presence of flagella components (i.e., FLhA) whose expression is not controlled by the PlcR regulator.[138]

Concerning models of professional phagocytic cells, studies with J774 macrophages demonstrate that the metalloprotease InhA1 (main component of the exosporium) is involved in the escape of *Bacillus* from macrophages.[140] These findings are related to the ability of InhA1 to increase membrane permeability.[140]

Spores of *B. anthracis* adhered to a diversity of epithelial cells like A549, CHO, and Caco-2.[141] Interestingly, a model of M-like cells was developed by using cocultures of Caco-2 (epithelial) and Raji B cells (human Burkitt's lymphoma). In this model, there was an efficient translocation of *B. anthracis* as compared with single cultures of Caco-2 cells.[142] These findings encourage further studies on the mechanisms involved in the pathogenesis of intestinal anthrax.

8.3.4 Virulence Factors in Other *Bacillus* spp.

Microorganisms of the genus *Bacillus* often contaminate raw and cooked foods. Even though *B. cereus* is the most studied group, there are other *Bacillus* species having pathogenic potential.[143–145] Models with cultured eukaryotic cells also contributed to the knowledge of this less virulent group of *Bacillus* spp.

Some toxins associated with food borne illness are lichenysin A (*B. licheniformis*), amylopsin (*B. subtilis* and *B. mojavensis*), and pumilacidin (*B. pumilus*).[33,57,145] It is noteworthy that toxins in culture supernatants from species other than those of the *B. cereus* group are less cytotoxic to HT-29 cells[145] or they were not toxic to Vero and Hep-2 cells.[143]

It has been demonstrated that cell-free alcohol-soluble extracts from *B. subtilis* strain depolarized mitochondria of Caco-2 cells. This activity was related to the presence of amylopsin that is also produced by *B. mojavensis*[33] where three surfactin-like compounds were detected.[57] These factors had toxic effects on Vero cells with the inhibition of protein synthesis.[57]

Extracellular factors of *Bacillus megaterium*, a microorganism commonly found in honey and seldom associated with severe pathologies, are cytotoxic to Caco-2 and Hep-2 cells in a strain-dependent manner.[63,146] However, a study performed by Alippi and coworkers[146] demonstrated that most strains are not cytotoxic, since from a total of 53 strains studied, only 7 strains detached Caco-2 cells. Interestingly, some strains of *B. megaterium* have the ability to adhere to and invade both Caco-2 and Hep-2 cells.[63,146]

The heat-stable toxin most well-known from the genus *Bacillus* is cereulide. Nevertheless, *Bacillus firmus*, *Bacillus simplex*, *B. megaterium*, and *B. licheniformis* also produce heat-stable factors.[147] Other members of the *Bacillus* genus (i.e., *B. firmus* and *B. simplex*) lead to vacuolation of Hep-2 cells.[147]

Diagnostic kits were developed on the basis of the knowledge of secreted virulence factors. *Bacillus* Diarrhoeal Enterotoxin Visual Immunoassay (BDE-VIA TM) (3M Tecra, St Paul, MN) detects NheA and NheB components through polyclonal antibodies with detection limits[90] of 2–5 ng mL^{-1}. The Duopath Cereus Enterotoxins assay (EMD Millipore, Darmtadt, Germany) is a lateral flow device that detects the NheB and L2 components of Hbl through monoclonal antibodies with a detection limit of 6 and 20 ng mL^{-1}, respectively.[148] The BCET-RPLA (Oxoid Ltd., Basingstoke Hampshire, UK) detects mainly the L2 component of Hbl with polyclonal antisera by reverse passive latex agglutination.[68,149,150]

Because of the multifactorial character of the virulence of *Bacillus* spp. and some divergences with the results obtained with cultured cells, some authors recommended testing by more than one method.[59,68,69,151–153]

8.4 *Ex Vivo* and *In Vivo* Models

Different *in vivo* and *ex vivo* models of infection have been used to study the pathogenesis of *Bacillus* spp. These models can determine both the importance of the route of infection and the virulence factors involved. Table 8.1 summarizes different models reported in the scientific literature.

8.4.1 *Ex Vivo* Models

Boar sperm motility inhibition bioassay detects the ability of toxins to inhibit sperm motility, which is affected when mitochondria become damaged or the protonmotive force is reduced by dissipation of the electrochemical potential across mitochondrial compartments.[111] The loss of motility of boar spermatozoa after exposure to emetic *B. cereus* toxin (cereulide) or cereulide-containing samples resulted in a sensitive, fast, and inexpensive bioassay for the detection of this toxin that acts as a potassium ionophore.[77,111,154] Indeed, the spermatozoa bioassay has been used to detect the presence of cereulide[77,154] in bacterial cultures and contaminated foods.

Besides cereulide, other toxins showed similar effects on sperm motility. By this method, amylopsin, produced by *B. subtilis* and *B. mojavensis*, that is involved in food-poisoning outbreaks has been detected.[33] Lichenysin, produced by *B. licheniformis*, was also detected through the inhibition of sperm motility. The effect on motility is due to interference in the energy metabolism of cells,[56] and the strong correlation between lichenysin concentration and toxicity to boar sperm cells has been shown.[155] *B. pumilus*, isolated from food-poisoning outbreaks, produced a toxin that induced damage of the cell membrane and collapse of mitochondria, thus provoking the inhibition of motility in boar sperm cells.[156]

Jääskeläinen et al. found that commercial boar sperm was more sensitive to cereulide than commercial bull semen.[77] Additionally, boar sperm cells were as sensitive to cereulide as human cell lines (i.e., HeLa, Caco-2, Calu-3, and Paju), indicating that this bioassay was appropriate for the *in vitro* assessment of the cereulide effect on human cells.[77]

Since it is conducted by microscopic examination, the test of the inhibition of the motility of sperm cells is prone to operator bias. Computer-assisted sperm analysis systems seem to be more objective and offer the advantage of a numerical interpretation of the observed motility changes. Quantification is based on software utilization and permits a real-time observation of the sperm motility.[157]

Quantitative analysis conducted by liquid chromatography connected to ion trap mass spectrometry showed a high correlation with the biological activity of cereulide.[112,114]

8.4.2 *In Vivo* Models

Both invertebrate and vertebrate animal models were used to assess virulence mechanisms of *Bacillus* spp. These experimental approaches add further insight in to the information obtained in the framework of *in vitro* systems. In addition, vertebrate and invertebrate models give complementary data concerning innate and adaptive immune responses.

8.5 Invertebrates

Despite evolutionary distances between mammals and invertebrates, they share common mechanisms concerning detection, signaling, and response to pathogens. These characteristics have encouraged modeling mammals' host–microbe interactions by using different invertebrates.[158]

8.5.1 Nematodes

Caenorhabditis elegans is a bacteriovorous soil nematode that has gained attention as a model for the study of host–microbe interactions.[159] It is a natural host for *Bacillus* spp. and can be involved in the spreading of these microorganisms.[160] In addition, it is considered as a potential vector for human pathogens related to food borne diseases.[161] It is worth noting that the defenses of *C. elegans* against pathogens rely on epithelial cells as key players of its innate immune system.[159] Interestingly, there are similarities between structures of the intestinal epithelia of *C. elegans* and those of mammals.[159]

Since *C. elegans* feeds on a variety of microbes, it is usually maintained on agar plates seeded with *E. coli*. Worms are motile and curvilinear, displaying typical internal structures when in healthy conditions.[162] The effects of pathogens on *C. elegans* are evidenced by assessing the survival ratio, ratio of infected worms, fecundity, development, microbial load, and more refined readouts including transcriptional responses.[159,162]

C. elegans and other nematode species (*Pristionchus pacificus*) have been used to study virulence factors of *Bacillus* spp. isolated from different sources (soil, horse dung, and beetles).[160] There was a strain-dependent effect on the survival ratio of infected nematodes, and there were different responses for the fecundity of the two nematodes studied. Interestingly, virulence patterns depended on both the genetic background of worms and bacteria.[160]

By using this model, Franks et al.[162] determined the correlation between the presence of genes related to tellurite resistance and the virulence of *B. anthracis*. In this study, they demonstrated a correlation between results obtained in this model and resistance to reactive oxygen species and antimicrobial peptides as well as survival in the blood of humans and mice.[162]

8.5.2 Insects

8.5.2.1 Galleria mellonella

Caterpillars of the greater wax moth, *G. mellonella*, have been widely used to mimic gastrointestinal infection caused by bacteria of the genus *Bacillus*.

G. mellonella eggs are incubated at 30°C, and the larvae are reared on beeswax and pollen at 25°C. There are three infection routes: free ingestion, force feeding, or intrahemocoelic injection.[163]

The use of this model allowed for great strides in the knowledge of the factors involved in the virulence of *Bacillus* spp. In this context, studies conducted with *G. mellonella* permitted the identification of virulence factors that both *B. cereus and B. thuringiensis* have in common. These factors include the pleiotropic regulator PlcR, the internaline/siderophore-like protein IlsA, the PlcR-regulated metalloprotease InhA2, the flagellar protein FlhA, and other adaptation and pathogenic genes.[101,124,164,166]

As described previously in this chapter, production of secreted virulence factors by *B. thuringiensis* and *B. cereus* is regulated by the PlcR regulon, which is the major virulence regulator in these species. This regulation has been demonstrated in *G. mellonella* where disruption of the *plcR* gene mitigates the virulence of *B. thuringiensis* and *B. cereus*.[124] These results were also seen in mice, suggesting that extracellular factors that depend on the *plcR* regulon are required for the pathogenesis in both hosts.[124]

In a series of experiments, spores of strains from *B. cereus* of a clinical origin were administered by two methods: free ingestion and force feeding. The survival ratio was evaluated and compared with those of other microorganisms including some human pathogens such as *Listeria monocytogenes*. The lowest survival ratios were obtained with the entomopathogen *Pseudomonas entomophila*. Some *B. cereus* strains lead to low ratios of mortality that were significantly increased in the presence of Cry toxin (entomotoxin from *B. thuringiensis*). In contrast, one attenuated strain of *B. anthracis* and *L. monocytogenes* did not lead to larvae mortality even in the presence of Cry toxin. These results demonstrate that the suitability of this model to assess the biological activity of human pathogens by the oral route is both strain- and species-dependent.[165]

B. weihenstephanensis virulence was assayed for the first time in an *in vivo* model using *G. mellonella*. *B. weihenstephanensis* is a psychrotolerant species of the *B. cereus* group that is potentially pathogenic for humans. In this model, it has been demonstrated that the virulence is temperature-dependent. A comparison between the virulence of *B. cereus* and *B. weihenstephanensis* showed that the virulence of the latter is low at 37°C and maximal at 15°C. Interestingly, these findings correlated with those found *in vitro* with cultured Vero cells.[167]

B. cereus sphingomyelinase (SMase) acquired relevance using this infection model. Doll et al. suggested that SMase is a relevant virulence factor involved in the mortality rate of insect larvae infected with *B. cereus*. In addition, their results have shown that Nhe and SMase complement each other for full virulence.[71]

Furthermore, this infection model, along with assays with cultured eukaryotic cells, has been used to test the virulence of *B. cereus* strains isolated in a severe intestinal infection in neonates.[168] Also, the relevance of InhA1, NprA, and HlyII to differentiate pathogenic from nonpathogenic *B. cereus* strains was assessed in *G. mellonella*.[169] This model was also appropriate to gain insight into the expression of virulent genes *in vivo* by using *in vivo* expression technology (IVET). Through this experimental approach, the relevance of the expression of internalin under iron deprivation was demonstrated.[164]

8.5.2.2 Bombyx mori

The larvae of *Bombyx mori* (silkworm) constitute an alternative model to study host–pathogen interactions. Their relatively big size is appropriate for manipulation although accumulated knowledge on this lepidopteran is not comparable to that of *C. elegans*.[158]

This animal model was used to gain insight into the role of zinc metalloproteases (Inh) in *B. thuringiensis* virulence. These proteases can specifically cleave insect antimicrobial peptides, thus overcoming one of the main defense mechanisms. For the experimental procedures, *B. mori* eggs were incubated at 25°C and the resulting larvae were reared on a commercially available artificial diet.[101] In this study, Fedhila et al.[101] used intrahemocelic inoculation for *B. mori* larvae and force feeding for *G. mellonella*. They also conducted immunization experiments with *E. coli* in order to trigger the production of antimicrobial peptides in larvae. The role of Inh metalloproteases was assessed by using mutants in genes *inhA* and *inh2*. Interestingly, the immunization of larvae did not protect *B. mori* inoculated with *B. thuringiensis* by the intrahemocelic route, and deletions in the genes *inhA* and *inhA2* did not modify the killing activity. In contrast, these deletions significantly decreased the virulence for *G. mellonella* inoculated by the oral route.[101]

8.6 Vertebrates

8.6.1 Rabbit

Early studies used this model to analyze crude cell-free spent culture supernatants of *B. cereus* cultures.[170] Biological activity was detected through two main experimental approaches: accumulation of fluid in ligated rabbit ileal loops (RILs) and vascular permeability reaction (VPR). In the RIL test, samples are injected into ileal loops and fluid accumulation is evaluated after euthanasia. Results obtained with this technique correlated with those obtained in the VPR. This latter assay involves the intradermal injection of samples in the shaved back of rabbits. Next, a Blue Evans solution is administered intravenously (ear vein), and the diameter of blue zones in the inoculation zone is evaluated. In some cases, edema and hemorrhage are also present.[170] It has been suggested that capillaries of blood were affected by *B. cereus* enterotoxin, giving a local and also a systemic response since histological changes were observed in the liver.[171] Comparing with ligated ileal loop assays, the VPR is faster, cheaper, and easier to perform.[172]

Both vegetative cells and cell-free culture filtrates from *B. cereus* cultures induce fluid accumulation. However, ileal loop response did not differentiate strains isolated from food-poisoning outbreaks from those isolated from other sources.[173] Fluid accumulation was also positive for other *Bacillus* species, for example, some *B. thuringiensis* strains. It is important to highlight that several strains that were positive for the ileal loop assays failed to elicit diarrhea neither when injected intraluminally into normal ileum rabbits nor when administered by a stomach catheter.[173] This experimental approach (conducted in rabbits or mice) demonstrated histological changes and, in some instances, lethality in mice.[64,171,173,174]

Ligated RIL has been used to analyze purified toxins or their components. For example, Beecher et al. studied hemolysin BL activity, a three-component toxin associated with diarrheal food poisoning. The authors showed that all three components were necessary for maximal activity.[42] In addition, Agata et al. cloned the *bceT* gene, encoding one of the enterotoxic proteins of *B. cereus*. Supernatants of *E. coli* cultures bearing this gene were positive in the VPR, resulting in fluid accumulation in murine ileal loops (mouse), and had mouse lethal activity.[64]

Several enteropathogenic bacteria that induced fluid accumulation in ligated intestinal loop can also produce a characteristic reaction when injected into the rabbit skin.[172] However, other authors have found no relationship between permeability reaction size and fluid accumulation in ligated RIL.[175]

Rabbit vascular permeability reaction has been used for the study of enterotoxins, either for the analysis of toxins that could be related to outbreaks[176] or for the study of isolated toxins.

Sometimes toxins caused vascular permeability, fluid accumulation, and lethal activity in mice— all requested features for a diarrheal enterotoxin.[64]

8.6.2 Zebrafish

Zebrafish (*Danio rerio*) is a vertebrate organism that shares some level of the functional conservation of genes and signaling pathways with mammals. In addition, developmental particularities of the immune system of zebrafish allow for the study of different levels of complexity of the immune response, that is, innate immune response in embryos and larvae or fully developed adaptive response in adult fishes. It is noteworthy that there are many similarities between components of zebrafish and human immune systems. Other advantages are that each single experiment could include hundreds of embryos for examination of several conditions and that the transparency of some stages of the fish cycle allows for the localization of infecting bacteria. In the framework of intestinal infections, bacteria can be administered by the oral route or static immersion.[177]

There are few reports on the use of this model to study the virulence factors of *Bacillus* spp., and to our knowledge, there is only one study conducted by using the oral route of infection with these microorganisms.[178] In contrast, Bolcome et al. developed a zebrafish embryonic model to assess the effect of *B. anthracis* toxins. They demonstrated an increased endothelial permeability by extravasation of fluorescent microspheres in toxin-injected embryos.[179] Further studies demonstrated the involvement of the MAPK pathway in the biological effects of the lethal toxin. Interestingly, in this study, genetically modified zebrafishes were developed.[180]

8.6.3 Monkeys

The concept of diarrheal and emetic syndromes in foodborne *B. cereus* pathologies started in the 1970s, when most of the *in vivo* studies included the analysis of fluid accumulation in ligated rabbit or mice ileal loop and alteration of vascular permeability.

Nausea and vomiting are pathognomonic signs of gastrointestinal infections and intoxications in humans. However, emetic reflex is restricted to some mammals (e.g., pig, goat, dog, cat, seal, sperm whale, and nonhuman primates) and nonmammals (e.g., some fish, reptiles, amphibians, and birds). Although, overall, some of the above-mentioned animals could be suitable to study emesis following food poisoning, there are intrinsic differences between humans and other animals.[181] Consequently, studies in nonhuman primates fulfil most of the expected clinic outcomes although their implementation is restricted due to ethical considerations.

Early studies in rhesus monkeys were used to analyze the effect (diarrhea and vomiting) of *B. cereus* ingested with foods.[182,183] Along with studies in rabbits (ligated ileal loop), the monkey feeding test permitted the demonstration of factors involved in vomiting, and diarrheal syndromes were distinct entities.[184] Some examples of cats and dogs suffering from diarrhea were described, but these models were not practical for the study of *B. cereus* infection.[173]

8.6.4 Asian House Shrew

As stated before in this chapter, *in vivo* studies on cereulide have been limited because of the paucity of animal models for the emetic syndrome. This prompted researchers to develop an easy-to-handle mammal model of emesis. In this context, the Asian house shrew (*Suncus murinus*) was an attractive alternative due to its susceptibility to various emetic drugs administered orally.[185] This model was used by Agata et al. to induce emesis by the oral and intraperitoneal administration of purified cereulide. Animal response was dose-dependent, and no animal death was observed. As a surrogate molecule, the antibiotic valinomycin was also tested.[109] The authors suggested that the emetic effect of cereulide was caused through the 5-HT receptor and stimulation of the vagus afferent. The same model was used by Isobe et al. to analyze the emetic effect of a synthetic cereulide[186] and by Ueda et al.[187] to validate quantitative analytical methods for cereulide.[187]

8.6.5 Mice

Mice (*Mus musculus*) have long been used for modeling host–pathogen interactions. They constitute an easy-to-handle mammalian animal model that presents a fully developed immune system and shares with humans many mechanisms of detection, signaling, and response to microorganisms. Furthermore, a wide range of tools have been developed for this model and genetic engineering resulted in obtaining strains of mice deficient in key steps of the host response against microorganisms.

Since Koch's early reports on *B. anthracis*, mice were used in studies on this microorganism.[5] Models became more and more refined and mouse strains with specific deficiencies in key steps of the immune response allowed for gaining further insight into the interactions between *B. anthracis* and the host.

As indicated in Section 8.1, the capsule of *B. anthracis* endows bacteria with the ability to evade phagocytosis. It has been shown that A/J mice are susceptible to the noncapsulated Sterne strain of *B. anthracis*. This infection model resembles the situation of conventional mice infected with wild-type strains but, concomitantly, has reduced safety concerns. The susceptibility of A/J mice to noncapsulated *B. anthracis* is due to deficiency in C5, a key component of the complement system. With this strain, Duong et al. developed a murine model of systemic anthrax. Using both spores and vegetative cells, and a subcutaneous infection, all infected animals became moribund or died.[188] Mice showed severe damage at the thoracic nodes, thymus, and spleen. However, there was a remarkable lack of hemorrhage or inflammation in Peyer's patches, mesenteric lymph nodes, and bronchial-associated lymphoid tissue. Interestingly, in this model of infection, pathologies in the gastrointestinal tract were absent in contrast with human systemic anthrax.[188]

A/J mice were also used for the development of a murine model of gastrointestinal anthrax.[189] In this study, infection was done by gavage with suspensions of Sterne strain (vegetative form). Results showed animal death in a dose-dependent manner. It is worth mentioning that bacilli penetrated and grew within the intestinal villi. As a consequence, bacteremia and systemic dissemination were observed with pathological changes similar to those of systemic anthrax. The animals showed a decrease in weight, body temperature, and activity along with evidence of intestinal hemorrhage. However, no immune activation was detected.[189]

A remarkable result obtained in this model was the early compromise of the intestinal epithelium. Disruption of the normal villus structures, villus blunting, and ulceration at the ileum and jejunum showed that gastrointestinal anthrax mainly affects the small intestine instead of the colon. Gastrointestinal tract epithelium was the first site of infection. Xie et al. hypothesized that extracellular factors produced by vegetative bacteria (e.g., LF) could be involved in the early steps of pathogenesis as well as in the immunosuppressive effect of anthrax.[189] An antibiotic treatment (amoxicillin + gentamicin) assayed 24 h after infection resulted in complete recuperation with no deaths of the infected animals. These findings encourage the use of this *in vivo* model of gastrointestinal anthrax for the evaluation of potential therapeutic approaches. Interestingly, A/J mice orally infected with spores of the Sterne strain showed a significant breakdown of the intestinal barrier function. This leads to the systemic dissemination of *B. anthracis* and commensal microorganisms.[190] This model of gastrointestinal anthrax allowed for the demonstration of the role of the MAPK pathway in the impairment of the immune response triggered by *B. anthracis*.

In Balb/cJ mice, a bioluminescent *B. anthracis* strain, unable to produce PA, was used for the real-time analysis of gastrointestinal infection.[26] In this study, the importance of the selection of the inoculation technique was demonstrated. Rigid feeding needles cause abrasions in the laryngopharynx, which in turn leads to systemic dissemination, whereas by using a flexible plastic tube, the infection initiated in Peyer's patches resembles gastrointestinal infection.[26]

The DBA/2 strain is a complement-deficient mouse susceptible to the Sterne strain. Tonry et al. analyzed gastrointestinal infection using this model. To maximize spore survival, neutralization of the stomach acidity was performed before infection. In addition, a thiabendazole paste containing spores was formulated to protect spores from gastric acidity. Intragastric inoculation of animals was performed by a catheter syringe. The infected animals showed signs similar to those of human gastrointestinal anthrax, that is, edema and hemorrhage.[142]

Also, the effects of *B. anthracis* toxins were studied in murine models. The effect of purified EF was assayed by intravenous or intraperitoneal injection in Balb/cJ mice. The toxin caused rapid death of mice, even at low doses, and these findings are related to the expression of toxin receptors in various tissues, which in turn correlates with histological damage. The animals showed a marked intestinal dilatation due to the extensive intestinal fluid accumulation in the gastrointestinal tract. Necrotic lesions were observed in the mucosa. This experimental model demonstrated the toxic effects induced by purified EF, suggesting that the death of animals could be related to multiorgan failure.[191] On the other hand, anthrax LF caused intestinal damage in both C57BL/6J and Balb/c mice after being injected intravenously. Likewise, when A/J mice were orally infected with spores of the Sterne strain,[190] a systemic infection with commensal enteric organisms was seen. An early breakdown of the gastrointestinal barrier, villus blunting, hemorrhage, and ulceration, along with the known immunosuppressive effect of LF, provided a potential mechanism for invasion via the enteric route.[192] This toxin causes a dose-dependent disruption of the intestinal epithelial integrity. Interestingly, the effects were partially attenuated by the co administration of antibiotics.

To investigate the toxic activity of *B. cereus*-secreted proteins, DBA/2J mice were inoculated intraperitoneally with filtered spent culture supernatants.[61] Shortly after inoculation (10 min), a high ratio of cells was observed in the peritoneal cavity. When viable bacteria were inoculated, the recruitment of high numbers of neutrophils masked the effect on other cells. The authors concluded that Hbl and Nhe cannot complement each other in spite of the coexpression of these factors.[61]

A murine model of *B. cereus* gastrointestinal infection was developed by Rolny et al.[193] The vegetative *B. cereus* strain B10502 isolated from a food-borne outbreak was used to infect C57BL/6J mice by gavage. To our knowledge, this was the first published data of an *in vivo* model of *B. cereus* gastrointestinal infection. In contrast to murine gastrointestinal anthrax, which produced intestinal hemorrhage, edema, systemic dissemination, and death, mouse infection with *B. cereus* led to a transient passage of the bacteria through the digestive tract. Although self-limiting, *B. cereus* infection modifies the balance of relevant immune cells in different regions of the intestinal mucosa and immune-associated tissues. Ratios and activation statuses of different cell populations were studied in the spleen, Peyer's patches, and mesenteric lymph nodes. These results correlated with an increase in the size of Peyer's patches in infected mice. Additionally, the infected mice showed a higher ratio of intestinal goblet cells and the presence of mononuclear cell infiltrates in the spleen. Cytokine mRNA expression showed a significant increase in IFN-γ in mesenteric lymph nodes. At the same time, a slight increase in IL-12 mRNA and TNF-α mRNA was observed. This study established relevant immune readouts to assess gastrointestinal infection with *B. cereus*.[193]

8.6.6 Rats

The rat model is seldom used for the study of *Bacillus* spp. infections. Human-flora-associated rats (Germfree Sprague–Dawley rats) were used to analyze the effect of *B. cereus* in the gastrointestinal tract. Evaluation of the composition of the indigenous gut flora and examination of whether spores and vegetative cells were able to persist in the gut were carried out. Experiments included administration of irradiated spores, untreated spores, heat-activated spores, and vegetative cells of a known-to-cause food-poisoning *B. cereus* strain.[194] By DGGE analysis, it was demonstrated that when animals were administered with spores, significant changes in the intestinal microbiota were evident. However, culture-dependent approaches did not reveal changes during the passage of *B. cereus*. Using *in vitro* assays and commercial kits for detection of toxins, no enterotoxins were detected in the intestinal tract. In addition, in those infectious conditions, only spores were able to resist the gastrointestinal passage.[194]

8.7 Concluding Remarks

The interaction of the genus *Bacillus* with different hosts leads to various outcomes: from a beneficial relationship[195] to life-threatening pathologies such as anthrax. The use of *in vitro* and *in vivo* models allowed for establishing key steps and mechanisms involved in infection and foodborne outbreaks. Experimental models with cultured eukaryotic cells such as enterocyte-like Caco-2 cells helped unravel extracellular factors involved in the gastrointestinal pathologies as well as to demonstrate the participation of direct bacteria–enterocyte interactions in the virulence. Further insights were obtained with *in vivo* invertebrate and vertebrate models that elucidated many aspects of the pathogenesis in a more complex cellular context. The recent advances in the field of *in vivo* expression technologies herald further gains in our understanding of the control and treatment of foodborne pathologies associated with *Bacillus* spp.

Acknowledgments

Jessica Minnaard is a Research Scientist at the Consejo Nacional de Investigaciones Científicas y Técnicas (CONICET, Argentina), Ivanna S. Rolny is a Research Scientist at the Biological Sciences Department (Facultad de Ciencias Exactas, Universidad Nacional de La Plata, Argentina), and Pablo F. Pérez is a Research Scientist at the Consejo Nacional de Investigaciones Científicas y Técnicas (CONICET, Argentina) and a Microbiology Professor at the Biological Sciences Department (Facultad de Ciencias Exactas, Universidad Nacional de La Plata, Argentina).

REFERENCES

1. Higgins, D. & Dworkin, J. Recent progress in *Bacillus subtilis* sporulation. *FEMS Microbiol Rev* 36, 131–48 (2012).
2. Maughan, H. & Van der Auwera, G. *Bacillus* taxonomy in the genomic era finds phenotypes to be essential though often misleading. *Infect Genet Evol* 11, 789–97 (2011).
3. Morens, D.M. Characterizing a "new" disease: epizootic and epidemic anthrax, 1769–1780. *Am J Public Health* 93, 886–93 (2003).
4. Brock, T.D. *Milestones in Microbiology: 1546 to 1940*. American Society for Microbiology, Washington, DC (1999).
5. Brock, T.D. *Robert Koch: A Life in Medicine and Bacteriology*, p. 363. American Society for Microbiology, Washington, DC (1999).
6. Guinebretiere, M.H. et al. *Bacillus cytotoxicus* sp. nov. is a novel thermotolerant species of the *Bacillus cereus* group occasionally associated with food poisoning. *Int J Syst Evol Microbiol* 63, 31–40 (2013).
7. Guinebretiere, M.H. et al. Ability of *Bacillus cereus* group strains to cause food poisoning varies according to phylogenetic affiliation (groups I to VII) rather than species affiliation. *J Clin Microbiol* 48, 3388–91 (2010).
8. Lund, T., De Buyser, M.L. & Granum, P.E. A new cytotoxin from *Bacillus cereus* that may cause necrotic enteritis. *Mol Microbiol* 38, 254–61 (2000).
9. Helgason, E. et al. *Bacillus anthracis*, *Bacillus cereus*, and *Bacillus thuringiensis*—one species on the basis of genetic evidence. *Appl Environ Microbiol* 66, 2627–30 (2000).
10. Priest, F.G., Barker, M., Baillie, L.W., Holmes, E.C. & Maiden, M.C. Population structure and evolution of the *Bacillus cereus* group. *J Bacteriol* 186, 7959–70 (2004).
11. Rasko, D.A., Altherr, M.R., Han, C.S. & Ravel, J. Genomics of the *Bacillus cereus* group of organisms. *FEMS Microbiol Rev* 29, 303–29 (2005).
12. Tourasse, N.J., Helgason, E., Okstad, O.A., Hegna, I.K. & Kolsto, A.B. The *Bacillus cereus* group: novel aspects of population structure and genome dynamics. *J Appl Microbiol* 101, 579–93 (2006).
13. Guinebretiere, M.H. et al. Ecological diversification in the *Bacillus cereus* group. *Environ Microbiol* 10, 851–65 (2008).
14. Kamar, R. et al. Pathogenic potential of *Bacillus cereus* strains as revealed by phenotypic analysis. *J Clin Microbiol* 51, 320–3 (2013).
15. Sternbach, G. The history of anthrax. *J Emerg Med* 24, 463–7 (2003).
16. Littman, R.J. The plague of Athens: epidemiology and paleopathology. *Mt Sinai J Med* 76, 456–67 (2009).
17. Dagnino, J. What was the plague of Athens? *Rev Chilena Infectol* 28, 374–80 (2011).
18. Papagrigorakis, M.J., Yapijakis, C., Synodinos, P.N. & Baziotopoulou-Valavani, E. DNA examination of ancient dental pulp incriminates typhoid fever as a probable cause of the plague of Athens. *Int J Infect Dis* 10, 206–14 (2006).
19. Ehrenkranz, N.J. & Sampson, D.A. Origin of the old testament plagues: explications and implications. *Yale J Biol Med* 81, 31–42 (2008).
20. Hatami, H., Ramazankhani, A. & Mansoori, F. Two cases of gastrointestinal anthrax with an unusual presentation from Kermanshah (western Iran). *Arch Iran Med* 13, 156–9 (2010).
21. Meric, M., Willke, A., Muezzinoglu, B., Karadenizli, A. & Hosten, T. A case of pneumonia caused by *Bacillus anthracis* secondary to gastrointestinal anthrax. *Int J Infect Dis* 13, e456–8 (2009).

22. Babamahmoodi, F., Aghabarari, F., Arjmand, A. & Ashrafi, G.H. Three rare cases of anthrax arising from the same source. *J Infect* 53, e175–9 (2006).
23. Bravata, D.M. et al. Inhalational, gastrointestinal, and cutaneous anthrax in children: a systematic review of cases: 1900 to 2005. *Arch Pediatr Adolesc Med* 161, 896–905 (2007).
24. CDC. Gastrointestinal anthrax after an animal-hide drumming event—New Hampshire and Massachusetts, 2009. *Morb Mortal Wkly Rep* 59, 872–7 (2010).
25. Owen, J.L., Yang, T. & Mohamadzadeh, M. New insights into gastrointestinal anthrax infection. *Trends Mol Med* 21(3), 154–63 (2015).
26. Glomski, I.J., Piris-Gimenez, A., Huerre, M., Mock, M. & Goossens, P.L. Primary involvement of pharynx and peyer's patch in inhalational and intestinal anthrax. *PLoS Pathog* 3, e76 (2007).
27. Jensen, G.B., Hansen, B.M., Eilenberg, J. & Mahillon, J. The hidden lifestyles of *Bacillus cereus* and relatives. *Environ Microbiol* 5, 631–40 (2003).
28. Avashia, S.B. et al. Fatal pneumonia among metalworkers due to inhalation exposure to *Bacillus cereus* containing *Bacillus anthracis* toxin genes. *Clin Infect Dis* 44, 414–6 (2007).
29. Hoffmaster, A.R. et al. Characterization of *Bacillus cereus* isolates associated with fatal pneumonias: strains are closely related to *Bacillus anthracis* and harbor *B. anthracis* virulence genes. *J Clin Microbiol* 44, 3352–60 (2006).
30. Johnson, S.L. et al. Finished genome sequence of *Bacillus cereus* strain 03BB87, a clinical isolate with *B. anthracis* virulence genes. *Genome Announc* 3, e01446-14 (2015).
31. Logan, N.A. *Bacillus* and relatives in foodborne illness. *J Appl Microbiol* 112, 417–29 (2012).
32. Jan, S. et al. Biodiversity of psychrotrophic bacteria of the *Bacillus cereus* group collected on farm and in egg product industry. *Food Microbiol* 28, 261–5 (2011).
33. Apetroaie-Constantin, C. et al. *Bacillus subtilis* and *B. mojavensis* strains connected to food poisoning produce the heat stable toxin amylosin. *J Appl Microbiol* 106, 1976–85 (2009).
34. Ostensvik, O., From, C., Heidenreich, B., O'Sullivan, K. & Granum, P.E. Cytotoxic *Bacillus* spp. belonging to the *B. cereus* and *B. subtilis* groups in Norwegian surface waters. *J Appl Microbiol* 96, 987–93 (2004).
35. From, C., Pukall, R., Schumann, P., Hormazabal, V. & Granum, P.E. Toxin-producing ability among *Bacillus* spp. outside the *Bacillus cereus* group. *Appl Environ Microbiol* 71, 1178–83 (2005).
36. Liu, S., Moayeri, M. & Leppla, S.H. Anthrax lethal and edema toxins in anthrax pathogenesis. *Trends Microbiol* 22, 317–25 (2014).
37. Tonello, F. & Zornetta, I. *Bacillus anthracis* factors for phagosomal escape. *Toxins (Basel)* 4, 536–53 (2012).
38. Koehler, T.M. *Bacillus anthracis* physiology and genetics. *Mol Aspects Med* 30, 386–96 (2009).
39. Moayeri, M. et al. Inflammasome sensor Nlrp1b-dependent resistance to anthrax is mediated by caspase-1, IL-1 signaling and neutrophil recruitment. *PLoS Pathog* 6, e1001222 (2010).
40. Terra, J.K. et al. Cutting edge: resistance to *Bacillus anthracis* infection mediated by a lethal toxin sensitive allele of Nalp1b/Nlrp1b. *J Immunol* 184, 17–20 (2010).
41. Drysdale, M. et al. Capsule synthesis by *Bacillus anthracis* is required for dissemination in murine inhalation anthrax. *EMBO J* 24, 221–7 (2005).
42. Beecher, D.J., Schoeni, J.L. & Wong, A.C. Enterotoxic activity of hemolysin BL from *Bacillus cereus. Infect Immun* 63, 4423–8 (1995).
43. Beecher, D.J. & Wong, A.C. Tripartite haemolysin BL: isolation and characterization of two distinct homologous sets of components from a single *Bacillus cereus* isolate. *Microbiology* 146 (Pt 6), 1371–80 (2000).
44. Lund, T. & Granum, P.E. Comparison of biological effect of the two different enterotoxin complexes isolated from three different strains of *Bacillus cereus. Microbiology* 143 (Pt 10), 3329–36 (1997).
45. Beecher, D.J., Olsen, T.W., Somers, E.B. & Wong, A.C. Evidence for contribution of tripartite hemolysin BL, phosphatidylcholine-preferring phospholipase C, and collagenase to virulence of *Bacillus cereus* endophthalmitis. *Infect Immun* 68, 5269–76 (2000).
46. Granum, P.E. & Nissen, H. Sphingomyelinase is part of the "enterotoxin complex" produced by *Bacillus cereus. FEMS Microbiol Lett* 110, 97–100 (1993).
47. Tran, S.L. et al. Haemolysin II is a *Bacillus cereus* virulence factor that induces apoptosis of macrophages. *Cell Microbiol* 13, 92–108 (2011).
48. Fricke, B. et al. The cell envelope-bound metalloprotease (camelysin) from *Bacillus cereus* is a possible pathogenic factor. *Biochim Biophys Acta* 1537, 132–46 (2001).

49. Ehling-Schulz, M., Fricker, M. & Scherer, S. Identification of emetic toxin producing *Bacillus cereus* strains by a novel molecular assay. *FEMS Microbiol Lett* 232, 189–95 (2004).
50. Horwood, P.F., Burgess, G.W. & Oakey, H.J. Evidence for non-ribosomal peptide synthetase production of cereulide (the emetic toxin) in *Bacillus cereus*. *FEMS Microbiol Lett* 236, 319–24 (2004).
51. Minnaard, J. et al. Virulence of *Bacillus cereus*: a multivariate analysis. *Int J Food Microbiol* 116, 197–206 (2007).
52. Minnaard, J., Lievin-Le Moal, V., Coconnier, M.H., Servin, A.L. & Perez, P.F. Disassembly of F-actin cytoskeleton after interaction of *Bacillus cereus* with fully differentiated human intestinal Caco-2 cells. *Infect Immun* 72, 3106–12 (2004).
53. Minnaard, J., Rolny, I.S. & Perez, P.F. Interaction between *Bacillus cereus* and cultured human enterocytes: effect of calcium, cell differentiation, and bacterial extracellular factors. *J Food Prot* 76, 820–6 (2013).
54. Mikkola, R. et al. Amylosin from *Bacillus amyloliquefaciens*, a K^+ and Na^+ channel-forming toxic peptide containing a polyene structure. *Toxicon* 49, 1158–71 (2007).
55. Rasimus-Sahari, S. et al. The peptide toxin amylosin of *Bacillus amyloliquefaciens* from moisture-damaged buildings is immunotoxic, induces potassium efflux from mammalian cells, and has antimicrobial activity. *Appl Environ Microbiol* 81, 2939–49 (2015).
56. Salkinoja-Salonen, M.S. et al. Toxigenic strains of *Bacillus licheniformis* related to food poisoning. *Appl Environ Microbiol* 65, 4637–45 (1999).
57. From, C., Hormazabal, V. & Granum, P.E. Food poisoning associated with pumilacidin-producing *Bacillus pumilus* in rice. *Int J Food Microbiol* 115, 319–24 (2007).
58. Minnaard, J., Humen, M. & Perez, P.F. Effect of *Bacillus cereus* exocellular factors on human intestinal epithelial cells. *J Food Prot* 64, 1535–41 (2001).
59. Gray, K.M., Banada, P.P., O'Neal, E. & Bhunia, A.K. Rapid Ped-2E9 cell-based cytotoxicity analysis and genotyping of *Bacillus* species. *J Clin Microbiol* 43, 5865–72 (2005).
60. Lund, T. & Granum, P.E. Characterisation of a non-haemolytic enterotoxin complex from *Bacillus cereus* isolated after a foodborne outbreak. *FEMS Microbiol Lett* 141, 151–6 (1996).
61. Sastalla, I. et al. The *Bacillus cereus* Hbl and Nhe tripartite enterotoxin components assemble sequentially on the surface of target cells and are not interchangeable. *PLoS One* 8, e76955 (2013).
62. Jeßberger, N., Dietrich, R., Bock, S., Didier, A. & Märtlbauer, E. *Bacillus cereus* enterotoxins act as major virulence factors and exhibit distinct cytotoxicity to different human cell lines. *Toxicon* 77, 49–57 (2014).
63. Rowan, N.J. et al. Putative virulence factor expression by clinical and food isolates of *Bacillus* spp. after growth in reconstituted infant milk formulae. *Appl Environ Microbiol* 67, 3873–81 (2001).
64. Agata, N., Ohta, M., Arakawa, Y. & Mori, M. The *bceT* gene of *Bacillus cereus* encodes an enterotoxic protein. *Microbiology* 141, 983–8 (1995).
65. Phung, D., Granum, P.E., Dietrich, R., Martlbauer, E. & Hardy, S.P. Inhibition of cytotoxicity by the Nhe cytotoxin of *Bacillus cereus* through the interaction of dodecyl maltoside with the NheB component. *FEMS Microbiol Lett* 330, 98–104 (2012).
66. Dietrich, R., Fella, C., Strich, S. & Martlbauer, E. Production and characterization of monoclonal antibodies against the hemolysin BL enterotoxin complex produced by *Bacillus cereus*. *Appl Environ Microbiol* 65, 4470–4 (1999).
67. Granum, P.E. & Lund, T. *Bacillus cereus* and its food poisoning toxins. *FEMS Microbiol Lett* 157, 223–8 (1997).
68. Buchanan, R.L. & Frankie, J.S. Evaluation of the Oxoid BCET-RPLA kit for the detection of *Bacillus cereus* diarrheal enterotoxin as compared to cell culture cytotonicity. *J Food Prot* 55, 440–3 (1992).
69. Fletcher, P. & Logan, N.A. Improved cytotoxicity assay for *Bacillus cereus* diarrhoeal enterotoxin. *Lett Appl Microbiol* 28, 394–400 (1999).
70. Hardy, S.P., Lund, T. & Granum, P.E. CytK toxin of *Bacillus cereus* forms pores in planar lipid bilayers and is cytotoxic to intestinal epithelia. *FEMS Microbiol Lett* 197, 47–51 (2001).
71. Doll, V.M., Ehling-Schulz, M. & Vogelmann, R. Concerted action of sphingomyelinase and non-hemolytic enterotoxin in pathogenic *Bacillus cereus*. *PLoS One* 8, e61404 (2013).
72. Fagerlund, A., Lindback, T., Storset, A.K., Granum, P.E. & Hardy, S.P. *Bacillus cereus* Nhe is a pore-forming toxin with structural and functional properties similar to the ClyA (HlyE, SheA) family of haemolysins, able to induce osmotic lysis in epithelia. *Microbiology* 154, 693–704 (2008).

73. Andreeva, Z.I. et al. Purification and cytotoxic properties of *Bacillus cereus* hemolysin II. *Protein Expr Purif* 47, 186–93 (2006).
74. Rajkovic, A. et al. Sub-emetic toxicity of *Bacillus cereus* toxin cereulide on cultured human enterocyte-like Caco-2 cells. *Toxins (Basel)* 6, 2270–90 (2014).
75. Marxen, S. et al. Chemodiversity of cereulide, the emetic toxin of *Bacillus cereus*. *Anal Bioanal Chem* 407, 2439–53 (2015).
76. Andersson, M.A. et al. Toxicological profile of cereulide, the *Bacillus cereus* emetic toxin, in functional assays with human, animal and bacterial cells. *Toxicon* 49, 351–67 (2007).
77. Jääskeläinen, E.L. et al. In vitro assay for human toxicity of cereulide, the emetic mitochondrial toxin produced by food poisoning *Bacillus cereus*. *Toxicol In Vitro* 17, 737–44 (2003).
78. Mikami, T. et al. An improved method for detecting cytostatic toxin (emetic toxin) of *Bacillus cereus* and its application to food samples. *FEMS Microbiol Lett* 119, 53–7 (1994).
79. Bonventre, P.F. Differential cytotoxicity of *Bacillus anthracis* and *Bacillus cereus* culture filtrates. *J Bacteriol* 90, 284–5 (1965).
80. Lievin-Le Moal, V. & Servin, A.L. Pathogenesis of human enterovirulent bacteria: lessons from cultured, fully differentiated human colon cancer cell lines. *Microbiol Mol Biol Rev* 77, 380–439 (2013).
81. Beecher, D.J. & Macmillan, J.D. Characterization of the components of hemolysin BL from *Bacillus cereus*. *Infect Immun* 59, 1778–84 (1991).
82. Lindback, T., Fagerlund, A., Rodland, M.S. & Granum, P.E. Characterization of the *Bacillus cereus* Nhe enterotoxin. *Microbiology* 150, 3959–67 (2004).
83. Guinebretiere, M.H., Broussolle, V. & Nguyen-The, C. Enterotoxigenic profiles of food-poisoning and food-borne *Bacillus cereus* strains. *J Clin Microbiol* 40, 3053–6 (2002).
84. Moravek, M. et al. Determination of the toxic potential of *Bacillus cereus* isolates by quantitative enterotoxin analyses. *FEMS Microbiol Lett* 257, 293–8 (2006).
85. Didier, A. et al. Monoclonal antibodies neutralize *Bacillus cereus* Nhe enterotoxin by inhibiting ordered binding of its three exoprotein components. *Infect Immun* 80, 832–8 (2012).
86. Beecher, D.J. & MacMillan, J.D. A novel bicomponent hemolysin from *Bacillus cereus*. *Infect Immun* 58, 2220–7 (1990).
87. Lindback, T. et al. Cytotoxicity of the *Bacillus cereus* Nhe enterotoxin requires specific binding order of its three exoprotein components. *Infect Immun* 78, 3813–21 (2010).
88. Haug, T.M. et al. Formation of very large conductance channels by *Bacillus cereus* Nhe in Vero and GH(4) cells identifies NheA + B as the inherent pore-forming structure. *J Membr Biol* 237, 1–11 (2010).
89. Dietrich, R., Moravek, M., Burk, C., Granum, P.E. & Martlbauer, E. Production and characterization of antibodies against each of the three subunits of the *Bacillus cereus* nonhemolytic enterotoxin complex. *Appl Environ Microbiol* 71, 8214–20 (2005).
90. Beecher, D.J. & Wong, A.C. Identification and analysis of the antigens detected by two commercial *Bacillus cereus* diarrheal enterotoxin immunoassay kits. *Appl Environ Microbiol* 60, 4614–6 (1994).
91. Ghelardi, E. et al. Identification and characterization of toxigenic *Bacillus cereus* isolates responsible for two food-poisoning outbreaks. *FEMS Microbiol Lett* 208, 129–34 (2002).
92. Beecher, D.J. & Wong, A.C.L. Tripartite hemolysin BL from *Bacillus cereus*. Hemolytic analysis of component interactions and a model for its characteristic paradoxical zone phenomenon. *J Biol Chem* 272, 233–9 (1997).
93. Madegowda, M., Eswaramoorthy, S., Burley, S.K. & Swaminathan, S. X-ray crystal structure of the B component of Hemolysin BL from *Bacillus cereus*. *Proteins* 71, 534–40 (2008).
94. Phung, D. et al. Crystallization and preliminary crystallographic analysis of the NheA component of the Nhe toxin from *Bacillus cereus*. *Acta Crystallogr Sect F Struct Biol Cryst Commun* 68, 1073–6 (2012).
95. Ramarao, N. & Sanchis, V. The pore-forming haemolysins of *Bacillus cereus*: a review. *Toxins (Basel)* 5, 1119–39 (2013).
96. Fagerlund, A., Ween, O., Lund, T., Hardy, S.P. & Granum, P.E. Genetic and functional analysis of the cytK family of genes in *Bacillus cereus*. *Microbiology* 150, 2689–97 (2004).
97. Baida, G., Budarina, Z.I., Kuzmin, N.P. & Solonin, A.S. Complete nucleotide sequence and molecular characterization of hemolysin II gene from *Bacillus cereus*. *FEMS Microbiol Lett* 180, 7–14 (1999).
98. Menestrina, G., Serra, M.D. & Prevost, G. Mode of action of β-barrel pore-forming toxins of the staphylococcal α-hemolysin family. *Toxicon* 39, 1661–72 (2001).

99. Andreeva, Z.I. et al. The properties of *Bacillus cereus* hemolysin II pores depend on environmental conditions. *Biochim Biophys Acta* 1768, 253–63 (2007).
100. Baida, G.E. & Kuzmin, N.P. Cloning and primary structure of a new hemolysin gene from *Bacillus cereus*. *Biochim Biophys Acta* 1264, 151–4 (1995).
101. Fedhila, S., Nel, P. & Lereclus, D. The InhA2 metalloprotease of *Bacillus thuringiensis* strain 407 is required for pathogenicity in insects infected via the oral route. *J Bacteriol* 184, 3296–304 (2002).
102. Stenfors Arnesen, L.P., Fagerlund, A. & Granum, P.E. From soil to gut: *Bacillus cereus* and its food poisoning toxins. *FEMS Microbiol Rev* 32, 579–606 (2008).
103. Alouf, J.E. Cholesterol-binding cytolytic protein toxins. *Int J Med Microbiol* 290, 351–6 (2000).
104. Gilmore, M.S., Cruz-Rodz, A.L., Leimeister-Wachter, M., Kreft, J. & Goebel, W. A *Bacillus cereus* cytolytic determinant, cereolysin AB, which comprises the phospholipase C and sphingomyelinase genes: nucleotide sequence and genetic linkage. *J Bacteriol* 171, 744–53 (1989).
105. Pomerantsev, A.P., Kalnin, K.V., Osorio, M. & Leppla, S.H. Phosphatidylcholine-specific phospholipase C and sphingomyelinase activities in bacteria of the *Bacillus cereus* group. *Infect Immun* 71, 6591–606 (2003).
106. Beecher, D.J. & Wong, A.C. Cooperative, synergistic and antagonistic haemolytic interactions between haemolysin BL, phosphatidylcholine phospholipase C and sphingomyelinase from *Bacillus cereus*. *Microbiology* 146 (Pt 12), 3033–9 (2000).
107. Bottone, E.J. *Bacillus cereus*, a volatile human pathogen. *Clin Microbiol Rev* 23, 382–98 (2010).
108. Zenewicz, L.A., Wei, Z., Goldfine, H. & Shen, H. Phosphatidylinositol-specific phospholipase C of *Bacillus anthracis* down-modulates the immune response. *J Immunol* 174, 8011–6 (2005).
109. Agata, N., Ohta, M., Mori, M. & Isobe, M. A novel dodecadepsipeptide, cereulide, is an emetic toxin of *Bacillus cereus*. *FEMS Microbiol Lett* 129, 17–9 (1995).
110. Agata, N. et al. A novel dodecadepsipeptide, cereulide, isolated from *Bacillus cereus* causes vacuole formation in HEp-2 cells. *FEMS Microbiol Lett* 121, 31–4 (1994).
111. Mikkola, R., Saris, N.E., Grigoriev, P., Andersson, M.A. & Salkinoja-Salonen, M. Ionophoretic properties and mitochondrial effects of cereulide: the emetic toxin of *B. cereus*. *Eur J Biochem* 263, 112–7 (1999).
112. Haggblom, M.M., Apetroaie, C., Andersson, M.A. & Salkinoja-Salonen, M.S. Quantitative analysis of cereulide, the emetic toxin of *Bacillus cereus*, produced under various conditions. *Appl Environ Microbiol* 68, 2479–83 (2002).
113. Teplova, V.V., Mikkola, R., Tonshin, A.A., Saris, N.E. & Salkinoja-Salonen, M.S. The higher toxicity of cereulide relative to valinomycin is due to its higher affinity for potassium at physiological plasma concentration. *Toxicol Appl Pharmacol* 210, 39–46 (2006).
114. Biesta-Peters, E.G. et al. Quantification of the emetic toxin cereulide in food products by liquid chromatography-mass spectrometry using synthetic cereulide as a standard. *Appl Environ Microbiol* 76, 7466–72 (2010).
115. Ehling-Schulz, M. et al. Toxin gene profiling of enterotoxic and emetic *Bacillus cereus*. *FEMS Microbiol Lett* 260, 232–40 (2006).
116. Ehling-Schulz, M. et al. Identification and partial characterization of the nonribosomal peptide synthetase gene responsible for cereulide production in emetic *Bacillus cereus*. *Appl Environ Microbiol* 71, 105–13 (2005).
117. Agata, N., Ohta, M. & Yokoyama, K. Production of *Bacillus cereus* emetic toxin (cereulide) in various foods. *Int J Food Microbiol* 73, 23–7 (2002).
118. Shinagawa, K., Ueno, S., Matsusaka, N. & Sugii, S. In vitro stability in biological activity and antigenicity of the vascular permeability factor produced by *Bacillus cereus*. *J Vet Med Sci* 53, 317–9 (1991).
119. Finlay, W.J., Logan, N.A. & Sutherland, A.D. Semiautomated metabolic staining assay for *Bacillus cereus* emetic toxin. *Appl Environ Microbiol* 65, 1811–2 (1999).
120. Stark, T. et al. Mass spectrometric profiling of *Bacillus cereus* strains and quantitation of the emetic toxin cereulide by means of stable isotope dilution analysis and HEp-2 bioassay. *Anal Bioanal Chem* 405, 191–201 (2013).
121. Virtanen, S.M. et al. In vitro toxicity of cereulide on porcine pancreatic Langerhans islets. *Toxicon* 51, 1029–37 (2008).
122. Paananen, A. et al. Inhibition of human natural killer cell activity by cereulide, an emetic toxin from *Bacillus cereus*. *Clin Exp Immunol* 129, 420–8 (2002).

123. Okstad, O.A. et al. Sequence analysis of three *Bacillus cereus* loci carrying PlcR-regulated genes encoding degradative enzymes and enterotoxin. *Microbiology* 145 (Pt 11), 3129–38 (1999).
124. Salamitou, S. et al. The *plcR* regulon is involved in the opportunistic properties of *Bacillus thuringiensis* and *Bacillus cereus* in mice and insects. *Microbiology* 146, 2825–32 (2000).
125. Gohar, M. et al. Two-dimensional electrophoresis analysis of the extracellular proteome of *Bacillus cereus* reveals the importance of the PlcR regulon. *Proteomics* 2, 784–91 (2002).
126. Slamti, L. et al. Distinct mutations in PlcR explain why some strains of the *Bacillus cereus* group are nonhemolytic. *J Bacteriol* 186, 3531–8 (2004).
127. Slamti, L. & Lereclus, D. Specificity and polymorphism of the PlcR-PapR quorum-sensing system in the *Bacillus cereus* group. *J Bacteriol* 187, 1182–7 (2005).
128. Brillard, J. & Lereclus, D. Characterization of a small PlcR-regulated gene co-expressed with cereolysin O. *BMC Microbiol* 7, 52 (2007).
129. Slamti, L. & Lereclus, D. A cell–cell signaling peptide activates the PlcR virulence regulon in bacteria of the *Bacillus cereus* group. *EMBO J* 21, 4550–9 (2002).
130. Bouillaut, L. et al. Molecular basis for group-specific activation of the virulence regulator PlcR by PapR heptapeptides. *Nucleic Acids Res* 36, 3791–801 (2008).
131. Budarina, Z.I. et al. A new *Bacillus cereus* DNA-binding protein, HlyIIR, negatively regulates expression of *B. cereus* haemolysin II. *Microbiology* 150, 3691–701 (2004).
132. Sineva, E. et al. Iron regulates expression of *Bacillus cereus* hemolysin II via global regulator Fur. *J Bacteriol* 194, 3327–35 (2012).
133. Lucking, G., Dommel, M.K., Scherer, S., Fouet, A. & Ehling-Schulz, M. Cereulide synthesis in emetic *Bacillus cereus* is controlled by the transition state regulator AbrB, but not by the virulence regulator PlcR. *Microbiology* 155, 922–31 (2009).
134. Frenzel, E. et al. CodY orchestrates the expression of virulence determinants in emetic *Bacillus cereus* by impacting key regulatory circuits. *Mol Microbiol* 85, 67–88 (2012).
135. Wijnands, L.M., Dufrenne, J.B., van Leusden, F.M. & Abee, T. Germination of *Bacillus cereus* spores is induced by germinants from differentiated Caco-2 cells, a human cell line mimicking the epithelial cells of the small intestine. *Appl Environ Microbiol* 73, 5052–4 (2007).
136. Berthold-Pluta, A., Pluta, A. & Garbowska, M. The effect of selected factors on the survival of *Bacillus cereus* in the human gastrointestinal tract. *Microb Pathog* 82, 7–14 (2015).
137. Andersson, A., Granum, P.E. & Rönner, U. The adhesion of *Bacillus cereus* spores to epithelial cells might be an additional virulence mechanism. *Int J Food Microbiol* 39, 93–9 (1998).
138. Ramarao, N. & Lereclus, D. Adhesion and cytotoxicity of *Bacillus cereus* and *Bacillus thuringiensis* to epithelial cells are FlhA and PlcR dependent, respectively. *Microbes Infect* 8, 1483–91 (2006).
139. Ronner, U., Husmark, U. & Henriksson, A. Adhesion of bacillus spores in relation to hydrophobicity. *J Appl Bacteriol* 69, 550–6 (1990).
140. Ramarao, N. & Lereclus, D. The InhA1 metalloprotease allows spores of the *B. cereus* group to escape macrophages. *Cell Microbiol* 7, 1357–64 (2005).
141. Bozue, J. et al. *Bacillus anthracis* spores of the *bclA* mutant exhibit increased adherence to epithelial cells, fibroblasts, and endothelial cells but not to macrophages. *Infect Immun* 75, 4498–505 (2007).
142. Tonry, J.H. et al. In vivo murine and in vitro M-like cell models of gastrointestinal anthrax. *Microbes Infect* 15, 37–44 (2013).
143. Lucking, G., Stoeckel, M., Atamer, Z., Hinrichs, J. & Ehling-Schulz, M. Characterization of aerobic spore-forming bacteria associated with industrial dairy processing environments and product spoilage. *Int J Food Microbiol* 166, 270–9 (2013).
144. Fernandez-No, I.C. et al. Detection and quantification of spoilage and pathogenic *Bacillus cereus, Bacillus subtilis* and *Bacillus licheniformis* by real-time PCR. *Food Microbiol* 28, 605–10 (2011).
145. De Bellis, P. et al. Toxigenic potential and heat survival of spore-forming bacteria isolated from bread and ingredients. *Int J Food Microbiol* 197, 30–9 (2015).
146. Lopez, A.C., Minnaard, J., Perez, P.F. & Alippi, A.M. In vitro interaction between *Bacillus megaterium* strains and Caco-2 cells. *Int Microbiol* 16, 27–33 (2013).
147. Taylor, J.M., Sutherland, A.D., Aidoo, K.E. & Logan, N.A. Heat-stable toxin production by strains of *Bacillus cereus, Bacillus firmus, Bacillus megaterium, Bacillus simplex* and *Bacillus licheniformis*. *FEMS Microbiol Lett* 242, 313–7 (2005).

148. Krause, N. et al. Performance characteristics of the Duopath(R) cereus enterotoxins assay for rapid detection of enterotoxinogenic *Bacillus cereus* strains. *Int J Food Microbiol* 144, 322–6 (2010).
149. Buchanan, R.L. & Schultz, F.J. Comparison of the Tecra VIA kit, Oxoid BCET-RPLA kit and CHO cell culture assay for the detection of *Bacillus cereus* diarrhoeal enterotoxin. *Lett Appl Microbiol* 19, 353–6 (1994).
150. Day, T.L., Tatani, S.R., Notermans, S. & Bennett, R.W. A comparison of ELISA and RPLA for detection of *Bacillus cereus* diarrhoeal enterotoxin. *J Appl Bacteriol* 77, 9–13 (1994).
151. Beattie, S.H. & Williams, A.G. Detection of toxigenic strains of *Bacillus cereus* and other *Bacillus* spp. with an improved cytotoxicity assay. *Lett Appl Microbiol* 28, 221–5 (1999).
152. McKillip, J.L. Prevalence and expression of enterotoxins in *Bacillus cereus* and other *Bacillus* spp., a literature review. *Antonie Van Leeuwenhoek* 77, 393–9 (2000).
153. Tallent, S.M., Hait, J.M. & Bennett, R.W. Analysis of *Bacillus cereus* toxicity using PCR, ELISA and a lateral flow device. *J Appl Microbiol* 118, 1068–75 (2015).
154. Andersson, M.A., Mikkola, R., Helin, J., Andersson, M.C. & Salkinoja-Salonen, M. A novel sensitive bioassay for detection of *Bacillus cereus* emetic toxin and related depsipeptide ionophores. *Appl Environ Microbiol* 64, 1338–43 (1998).
155. Madslien, E.H. et al. Lichenysin is produced by most *Bacillus licheniformis* strains. *J Appl Microbiol* 115, 1068–80 (2013).
156. Suominen, I. et al. Toxic *Bacillus pumilus* from indoor air, recycled paper pulp, Norway spruce, food poisoning outbreaks and clinical samples. *Syst Appl Microbiol* 24, 267–76 (2001).
157. Rajkovic, A. et al. Dynamics of boar semen motility inhibition as a semi-quantitative measurement of *Bacillus cereus* emetic toxin (cereulide). *J Microbiol Methods* 65, 525–34 (2006).
158. Garcia-Lara, J., Needham, A.J. & Foster, S.J. Invertebrates as animal models for *Staphylococcus aureus* pathogenesis: a window into host-pathogen interaction. *FEMS Immunol Med Microbiol* 43, 311–23 (2005).
159. Balla, K.M. & Troemel, E.R. *Caenorhabditis elegans* as a model for intracellular pathogen infection. *Cell Microbiol* 15, 1313–22 (2013).
160. Rae, R., Iatsenko, I., Witte, H. & Sommer, R.J. A subset of naturally isolated *Bacillus* strains show extreme virulence to the free-living nematodes *Caenorhabditis elegans* and *Pristionchus pacificus*. *Environ Microbiol* 12, 3007–21 (2010).
161. Anderson, G.L., Caldwell, K.N., Beuchat, L.R. & Williams, P.L. Interaction of a free-living soil nematode, *Caenorhabditis elegans*, with surrogates of foodborne pathogenic bacteria. *J Food Prot* 66, 1543–9 (2003).
162. Franks, S.E. et al. Novel role for the *yceGH* tellurite resistance genes in the pathogenesis of *Bacillus anthracis*. *Infect Immun* 82, 1132–40 (2014).
163. Ramarao, N., Nielsen-Leroux, C. & Lereclus, D. The insect *Galleria mellonella* as a powerful infection model to investigate bacterial pathogenesis. *J Vis Exp* 70, e4392 (2012).
164. Fedhila, S., Daou, N., Lereclus, D. & Nielsen-LeRoux, C. Identification of *Bacillus cereus* internalin and other candidate virulence genes specifically induced during oral infection in insects. *Mol Microbiol* 62, 339–55 (2006).
165. Fedhila, S. et al. Comparative analysis of the virulence of invertebrate and mammalian pathogenic bacteria in the oral insect infection model *Galleria mellonella*. *J Invertebr Pathol* 103, 24–9 (2010).
166. Bouillaut, L. et al. FlhA influences *Bacillus thuringiensis* PlcR-regulated gene transcription, protein production, and virulence. *Appl Environ Microbiol* 71, 8903–10 (2005).
167. Stenfors Arnesen, L., Granum, P.E., Buisson, C., Bohlin, J. & Nielsen-LeRoux, C. Using an insect model to assess correlation between temperature and virulence in *Bacillus weihenstephanensis* and *Bacillus cereus*. *FEMS Microbiol Lett* 317, 196–202 (2011).
168. Decousser, J.W. et al. *Bacillus cereus* and severe intestinal infections in preterm neonates: putative role of pooled breast milk. *Am J Infect Control* 41, 918–21 (2013).
169. Cadot, C. et al. InhA1, NprA, and HlyII as candidates for markers to differentiate pathogenic from nonpathogenic *Bacillus cereus* strains. *J Clin Microbiol* 48, 1358–65 (2010).
170. Shinagawa, K. Analytical methods for *Bacillus cereus* and other *Bacillus* species. *Int J Food Microbiol* 10, 125–41 (1990).
171. Singh, D.K., Narayan, K.G. & Gupta, M.K. Mechanisms of *Bacillus cereus* enteropathy. *Indian J Exp Biol* 30, 324–6 (1992).

172. Glatz, B.A., Spira, W.M. & Goefert, J.M. Alteration of vascular permeability in rabbits by culture filtrates of *Bacillus cereus* and related species. *Infect Immun* 10(2), 299–303 (1974).
173. Spira, W.M. & Goepfert, J.M. *Bacillus cereus*-induced fluid accumulation in rabbit ileal loops. *Appl Microbiol* 24, 341–8 (1972).
174. Shinagawa, K., Sato, K., Konuma, H., Matsusaka, N. & Sugii, S. Fluid accumulation in mouse ligated intestine inoculated with the vascular permeability factor produced by *Bacillus cereus*. *J Vet Med Sci* 53, 167–71 (1991).
175. Hostacká, A., Kosiarová, A., Majtán, V. & Kohútová, S. Toxic properties of *Bacillus cereus* strains isolated from different foodstuffs. *Zentralbl Bakteriol* 276, 303–12 (1992).
176. DeBuono, B.A., Brondum, J., Kramer, J.M., Gilbert, R.J. & Opal, S.M. Plasmid, serotypic, and enterotoxin analysis of *Bacillus cereus* in an outbreak setting. *J Clin Microbiol* 26, 1571–4 (1988).
177. Meijer, A.H. & Spaink, H.P. Host-pathogen interactions made transparent with the zebrafish model. *Curr Drug Targets* 12, 1000–17 (2011).
178. Grisolia, C.K. et al. Acute toxicity and cytotoxicity of *Bacillus thuringiensis* and *Bacillus sphaericus* strains on fish and mouse bone marrow. *Ecotoxicology* 18, 22–6 (2009).
179. Bolcome, R.E., 3rd et al. Anthrax lethal toxin induces cell death-independent permeability in zebrafish vasculature. *Proc Natl Acad Sci USA* 105, 2439–44 (2008).
180. Bolcome, R.E., 3rd & Chan, J. Constitutive MEK1 activation rescues anthrax lethal toxin-induced vascular effects in vivo. *Infect Immun* 78, 5043–53 (2010).
181. Holmes, A.M., Rudd, J.A., Tattersall, F.D., Aziz, Q. & Andrews, P.L. Opportunities for the replacement of animals in the study of nausea and vomiting. *Br J Pharmacol* 157, 865–80 (2009).
182. Turnbull, P.C.B. Studies on the production of enterotoxins by *Bacillus cereus*. *J Clin Pathol* 29, 941–8 (1976).
183. Melling, J., Capel, B.J., Turnbull, P.C.B. & Gilbert, R.J. Identification of a novel enterotoxigenic activity associated with *Bacillus cereus*. *J Clin Pathol* 29, 938–40 (1976).
184. Turnbull, P.C.B., Kramer, J.M., Jorgensen, K., Gibbert, R.J. & Melling, J. Properties and production characteristics of vomiting, diarrheal, and necrotizing toxins of *Bacillus cereus*. *Am J Clin Nutr* 32, 219–28 (1979).
185. Ueno, S., Matsuki, N. & Saito, H. *Suncus murinus*: a new experimental model in emesis research. *Life Sci* 41, 513–8 (1987).
186. Isobe, M., Ishikawa, T., Suwan, S., Agata, N. & Ohta, M. Synthesis and activity of cereulide, a cyclic dodecadepsipeptide lonophore as emetic toxin from *Bacillus cereus*. *Bioorg Med Chem Lett* 5, 2855–8 (1995).
187. Ueda, S., Nakajima, H., Iwase, M., Shinagawa, K. & Kuwabara, Y. LC-MS analysis of the emetic toxin, cereulide, produced by *Bacillus cereus*. *Biocontrol Sci* 17, 191–5 (2012).
188. Duong, S., Chiaraviglio, L. & Kirby, J.E. Histopathology in a murine model of anthrax. *Int J Exp Pathol* 87, 131–7 (2006).
189. Xie, T. et al. A new murine model for gastrointestinal anthrax infection. *PLoS One* 8, e66943 (2013).
190. Lightfoot, Y.L. et al. Colonic immune suppression, barrier dysfunction, and dysbiosis by gastrointestinal *Bacillus anthracis* infection. *PLoS One* 9, e100532 (2014).
191. Firoved, A.M. et al. *Bacillus anthracis* edema toxin causes extensive tissue lesions and rapid lethality in mice. *Am J Pathol* 167, 1309–20 (2005).
192. Sun, C. et al. Anthrax lethal toxin disrupts intestinal barrier function and causes systemic infections with enteric bacteria. *PLoS One* 7, e33583 (2012).
193. Rolny, I.S., Minnaard, J., Racedo, S.M. & Perez, P.F. Murine model of *Bacillus cereus* gastrointestinal infection. *J Med Microbiol* 63, 1741–9 (2014).
194. Wilcks, A., Hansen, B.M., Hendriksen, N.B. & Licht, T.R. Fate and effect of ingested *Bacillus cereus* spores and vegetative cells in the intestinal tract of human-flora-associated rats. *FEMS Immunol Med Microbiol* 46, 70–7 (2006).
195. Cutting, S.M. *Bacillus* probiotics. *Food Microbiol* 28, 214–20 (2011).

9

Clostridium

Emilio Aranda, María G. Córdoba, María J. Benito, and Juan José Córdoba

CONTENTS

9.1 Introduction: *Clostridium* Neurotoxins and Enterotoxins

Species of the *Clostridium* genus are widely distributed in the environment (e.g., soil, sewage, and marine sediments) and in the gastrointestinal tract of humans and domestic animals. Although most *Clostridium* species are saprophytes, 34 species have been considered pathogenic to man and animals. Among the main pathogenic species are *Clostridium botulinum* and *Clostridium perfringens* that usually are involved in foodborne outbreaks. *C. botulinum* produces the highly potent botulinum neurotoxin (BoNT), which is responsible for botulism, a severe foodborne disease with a high fatality rate.[1]

C. botulinum belonging to groups I and II pose different types of risks in food processing. Group I strains, the spores of which are highly heat-resistant, are frequently related to insufficiently processed home-preserved foods such as canned vegetables and cured meats. The failure to effectively apply the botulinum cook (121°C/3 min) to canned or bottled foods has led to many outbreaks of foodborne botulism associated with *C. botulinum* group I.[2]

C. botulinum produces seven different neurotoxins, designated A to G, and a new type H was recently reported.[3,4] Dover et al.[5] have described it as a B2 subtype toxin, with a minor toxin that could be described as a new serotype (H) or as a chimeric A/F toxin. BoNT A, B, E, and F cause human botulism, while C and D cause animal botulism. *C. botulinum* group I produces A, B, F, and H BoNT, and genes are variable, with strains possessing from one to three neurotoxin genes, and forming one active, strains form active type A neurotoxin, but not active type B neurotoxin or even two active toxins.[2] C. *botulinum* group II produces type B, E, and F BoNT; it was recently discovered that strains of *C. botulinum* group II type F also contain a fragment of a type B and type E neurotoxin genes.[6]

Other members of the genus *Clostridium* (*Clostridium baratii* and *Clostridium butyricum*) have occasionally been reported to be involved in foodborne botulism and thus should also be considered as potentially foodborne pathogens.

Furthermore, *C. perfringens*, which produces enterotoxins and is responsible for foodborne intoxication,[7] and *Clostridium difficile*, which is involved in nosocomial, or institution-acquired, diarrhea,[8] must be considered as potentially pathogenic clostridia associated to foodborne outbreaks.

C. perfringens produces clostridial toxins activated in the gastrointestinal tract. Among 17 toxins produced by *C. perfringens*, alpha (CPA), beta (CPB), epsilon (ETX), and iota (ITX) toxins are the four major toxins present in five different toxinotypes (A, B, C, D, and E) of the bacterium.[9–11] In addition, *C. perfringens* may produce other toxins such as *C. perfringens* enterotoxin (CPE), *C. perfringens* beta2 toxin (CPB2), *C. perfringens* NetB toxin, and TpeL.[9,12–14] CPE causes *C. perfringens* type A food poisoning as well as antibiotic-associated diarrhea (AAD) and sporadic diarrheas in humans. This toxin is also responsible for enteric diseases and enterotoxemias in animals.[15,16]

Accurate methods for isolating and detecting these pathogenic clostridia in raw materials and processed foods are necessary to take corrective measures throughout processing, avoiding clostridial foodborne cases. In addition, laboratory (animal or cell culture) should be available to investigate these types of foodborne cases.

9.2 Characteristics and Incidence of Foodborne Outbreaks due to *Clostridium* Neurotoxins and Enterotoxins

The Centers for Disease Control and Prevention (CDC) categorizes human botulism cases into four transmission categories: foodborne, infant, wound, and other.[17]

Foodborne botulism results from the ingestion of preformed botulinum toxin in foods. The toxin can be found in food that has not been properly cooked, processed, handled, or canned and is often present in canned food such as vegetables, meat, and seafood products.[18] Foodborne botulism typically presents 18–36 h after ingestion of contaminated food; symptoms of botulism are often preceded by abdominal symptoms such as nausea, vomiting, and diarrhea, which are absent in botulism of the wound.[19] Most foodborne botulism in Europe and North America has been associated with the use of canning and bottling processes to extend food product shelf-life. The failure to effectively apply the botulinum cook (121°C/3 min) to canned or bottled foods has led to many outbreaks of foodborne botulism associated with proteolytic *C. botulinum* or *C. botulinum* group I. Strains of nonproteolytic *C. botulinum* or *C. botulinum* group II forming type B neurotoxin are often associated with foodborne botulism and meat products in Europe, and sometimes with fish products in North America. Botulism outbreaks involving *C. botulinum* group II type E have been most frequently associated with fish and home-prepared foods in the north of Canada and Alaska.[2]

It should be noted that the true incidence of foodborne botulism is likely to be much higher, with underreporting an issue. Foodborne botulism is not reportable in all countries and the efficiency of investigating potential outbreaks also varies from country to country.[20]

Infant botulism (IB) is an infectious intestinal toxemia affecting only infants younger than 1 year. During a transient period in infancy, gut flora may not inhibit intestinal outgrowth of ingested clostridial spores. It occurs when infants (persons less than 1 year of age) ingest *C. botulinum* spores that then germinate and produce the botulinum toxin in the intestines.[21] Swallowed spores of *C. botulinum* (or, rarely, toxigenic *C. butyricum* or *C. baratii*) germinate and temporarily colonize the lumen of the large intestine, whereas vegetative cells produce BoNT.[22] Because honey is a common dietary source of *C. botulinum* spores, infants should never be fed honey.[18] However, it is recognized that *C. botulinum* spores are present in soils and dust, and it is presumed that these may also be sources of infection. Environmental factors and differences in the ability of physicians to recognize this disease may explain the wide variations in the report of IB between regions.[23]

Although IB has been reported in almost all continents, its epidemiological distribution varies widely between continents and between countries. The United States (with approximately 70–110 cases of IB each year), Argentina, Australia, Canada, Italy, and Japan have reported most of the cases.[23] IB is a rare disorder in Europe. From 1976, the year in which IB was first recognized,[24] through 2006, 65 cases were identified in 13 European countries, most of them in Italy and Spain, followed by the UK.[25] In Europe, IB is less common than foodborne botulism; however, IB is the most common form of human botulism recognized in the United States.[25]

Foodborne illness caused by CPEs commonly occurs in industrialized countries.[7] This food poisoning is the second most common bacterial illness in the USA, where about 1 million cases/year are reported.[26] Outbreaks typically involve a large number of victims and are associated with temperature-abused meat or poultry dishes. The high incidence of *C. perfringens* type A food poisoning is well documented. From 1998 to 2010, 289 confirmed outbreaks of *C. perfringens* illness were reported in the United States with 15,208 illnesses, 83 hospitalizations, and eight deaths.[27] Sudden infant death syndrome (SIDS) accounts for unexpected deaths in infants under the age of 1 year. Strains of *C. perfringens* type A have been discovered to be commonly present in the intestines of babies dying with SIDS.[28]

C. difficile belongs to the normal microbiota of the mammalian gastrointestinal tract. *C. difficile* proliferation and infection in the human colon often occur after use of broad-spectrum antibiotics. *C. difficile* destroys the intestinal lining and causes AAD, colitis, and pseudomembranous colitis. The pathogenicity depends on the production of enterotoxin A and cytotoxin B. *C. difficile* toxin A is an enterotoxin that induces fluid accumulation in the bowel, and toxin B is a cytopathic toxin that is extremely lethal. Both toxins are highly unstable and tend to degrade at room temperature.[10]

9.3 Investigation of Foodborne Intoxication by *Clostridium* Neurotoxins

C. botulinum produces well-defined and very distinctive symptoms compared to *C. perfringens* and *C. difficile,* and its foodborne investigation and differentiation relies on the use of laboratory procedures.

9.3.1 Investigation of *C. botulinum* and BoNTs in Food Samples

Isolation, detection, and characterization of *C. botulinum* from food samples by *traditional microbiological culture techniques* require media deoxygenated by heating in a boiling water bath or by a continuous flow of an anaerobic gas mixture, and the use of anaerobic jars and anaerobic workstations is necessary for successful diagnostics.[29] Reddy et al.[30] studied the effect of media, additives, and incubation conditions on the recovery of *C. botulinum* spores, and Mato Rodriguez and Alatossava[31] studied the effects of copper on germination, growth, and sporulation of *Clostridium* spp. Christian et al.[32] evaluated the effect of calcium, magnesium, and manganese spore resistance in different media. Calcium cations give resistance to spores, while high amounts of magnesium cations appear to have a negative effect. Manganese cations in low concentrations are important for the development resistance to heat and pressure treatments, but not heat alone.

To enhance the isolation of *C. botulinum* from clinical samples (e.g., serum and feces), various sample pre-preparation steps may be used including: (1) ethanol pretreatment to recover spores and eliminate vegetative bacteria, (2) heating to discard nonspore-forming bacteria (i.e., 80°C for 10 min for group I spores or 60°C for 10–20 min for group II spores), and (3) treatment with lysozyme (5 μg/mL) or other heat-resistant lytic enzymes to facilitate germination of heat-stressed spores.[29,33] In germination of spores, the effect of spore density system or proximity on the time of germination could be of great utility.[33]

Cultivation of *C. botulinum* in liquid media could be carried out in chopped-meat-glucose-starch medium; cooked-meat medium; broths containing various combinations of tryptone, peptone, glucose, yeast extract, and trypsin; reinforced clostridial medium; and fastidious anaerobe broth.[30,34] Pretreatment with some of the above liquid culture media is of great importance to assure viability of spores, as it has been reported by Stringer et al.[35] However, all of these media are nonselective and thus allow the growth of a range of other bacteria.

To identify *C. botulinum*, unselective plating media such as blood agar and egg yolk agar (EYA)[36] are commonly used, since it could enable typical lipase production by this microorganism. However, other clostridia species have lipase and may therefore confuse the identification.[37] EYA medium alone does not contain any inhibitory compounds, but it has been reported that medium supplemented with cycloserine, sulfamethoxazole, and trimethoprim can be used for identifying select group I *C. botulinum*.[38,39] Selection of the correct incubation temperature is essential to differentiate strains from the different physiologies of groups I and II of *Clostridium*. Group I strains grow optimally at 35°C–37°C, whereas temperatures of 25°C–30°C or even lower favor growth of group II strains.[40]

Quantification of *C. botulinum* by use of plate count procedures in samples containing other bacteria is difficult, since prevalence of *C. botulinum* in naturally contaminated samples is generally low (10–1000 spores/kg) and proper selective media are not available.[29]

Polymerase chain reaction (PCR) represents an alternative to the traditional microbiological culture technique to detect *C. botulinum* in foods. Fakruddin et al.[41] described improvements and alternatives to PCR such as loop-mediated isothermal amplification (LAMP), nucleic acid sequence based amplification (NASBA), self-sustained sequence replication (3SR), and rolling circle amplification (RCA).

PCR detection of *C. botulinum* often targets the BoNT genes although other types of toxin genes are also of diagnostic value.[29] Use of multiplex PCR assays enables the simultaneous detection of two or more types of BoNT genes.[42–44] One such multiplex method was able to discriminate among BoNT serotypes A, B, E, and F, corroborating mouse bioassay results.[45] Furthermore, Peck and colleagues[1] developed culture enrichment methods that when coupled with multiplex PCR can identify strains of *C. botulinum* that are nonproteolytic (BoNT serotypes B, E, and F).[1] However, PCR detection of neurotoxin genes does not provide details on the physiological group and epidemiology of *C. botulinum* isolates. For differentiation between group I (proteolytic) and group II (nonproteolytic) strains of *C. botulinum*, several molecular typing methods are useful.[46–48] Furthermore, Janda and Whitehouse[49] have developed a multilocus PCR followed by electrospray ionization mass spectrometry to identify *C. botulinum*, and Fach et al.[50] have described an innovative molecular detection tool based on the GeneDisc cycler for tracking all types A, B, E, and F. Detection of *C. botulinum* in foods by the above mentioned traditional culture techniques are highly time consuming and require good microbiological expertise. PCR offers the possibility to detect any of the toxic strains regardless of the type of *C. botulinum* and BoNT produced. This strategy is possible due to the use of universal primers that recognize all BoNT genes, but the nucleotide diversity among the BoNT genes may present a potential problem.[51] Nonetheless, BoNTs are generated as part of a progenitor toxin complex, and a conserved component among serotypes is the nontoxic nonhemaglutinin (NTNH). East and Collins[52] demonstrated that the gene encoding NTNH shows a high level of similarity and is present in all strains that produce BoNTs and absent from strains that are nontoxic. Aranda et al.[53] developed a PCR protocol to detect all BoNTs-producing strains by using a single set of degenerate primers. This protocol yields a single PCR product of 1.1 kb in agarose gel, providing a more specific detection than hybridization with a BoNT probe.

Real-time PCR, biosensors, and DNA microarray offer the possibility of continuous and real-time monitoring of the environment for the presence of infectious agents. Therefore, several real-time PCR protocols[54] along with biosensor technology[55] and DNA microarrays[56] have been reported to detect *Clostridium*. More recently, Fenicia et al.[57] proposed a real-time PCR method for detecting and typing BoNT-producing *C. botulinum* types A, B, E, and F in clinical, food, and environmental samples and thus support use of it as an international standard method. Microarray technology for toxin identification of contaminated food has not been widely used. This may be due to the challenge in isolating high-quality RNA samples from clostridia in food matrices. A recent oligonucleotide microarray with 62 different sequences based on known strain variable regions in the genome of *C. botulinum* strain ATCC 3502 was constructed and used to differentiate different *C. botulinum* type A strains.[58] Regions corresponding to BoNT genes of various serotypes, and other markers were observed. Further development of microarray-based assay approaches may provide a means to rapidly identify toxin-producing strains.

In most of the cases of foodborne botulism investigation, detection of *C. botulinum* should be complemented by detection of BoNT toxins. For BoNT detection, several procedures such as enzyme-linked immunosorbent assay (ELISA)[59,60] or mass spectrometric-based endopeptidase methods[61] have been reported. This approach has been successful in identifying BoNT serotypes A, B, E, and F in a variety of food and clinical sample matrices with submouse bioassay sensitivities. To advance this technique even further, a single, high-affinity mAb (4E17.1) that can simultaneously identify BoNT serotypes A, B, E, and F has been developed.[62] Recently, surface plasmon resonance (SPR) sensors have been also applied for detecting BoNTs serotypes A, B, and F.[63,64] This method uses a label-free biosensor assay for BoNT B detection in food and human serum based on protein chip assay. Lastly, Ching et al.[65] have reported the use of an immunochromatographic test strip for the detection of BoNT serotypes A and B and Liu et al.[66] have developed an immunobiochemical assay to detect BoNT. In addition, ELISA and Endopep MS analyses using fluorogenic reporters to detect BoNT in drinking water have been reported.[67,68]

Several protocols such as ELISA and immuno-PCR and ELISA based on protein antibody microarray have been recently reported to detect BoNT neurotoxins A, B, C, D, E, and F in complex clinical, food, and environmental samples at higher sensitivity than mouse bioassay.[69,70] A number of rapid affinity immunochromatography column (AICC) assays for the detection of BoNT serotypes A, B, E, and F in food matrices have been developed. Brunt et al.[71] reported a detection limit for BoNT/A of 0.5 ng, two-fold more sensitive than earlier reported lateral flow methods. For serotypes B, E, and F, the minimum detection limit ranged from 5 to 50 ng. Although not as sensitive as ELISA or mouse bioassays, immunochromatographic methods generally are rapid assays, requiring only 15–30 min to complete, and do not require enrichment steps, making them highly amenable to use in the field.

9.3.2 Investigation of *C. perfringens* in Food Samples

Detection of genes encoding *C. perfringens* toxins or the corresponding toxins is the most accepted criterion in establishing a definitive diagnosis of this microorganism.[72,73] Several procedures including conventional and genomic molecular methods have been described.

Isolation, characterization, and detection of *C. perfringens* from food samples by traditional culture techniques require culture media such as blood agar plates. In this medium, the microorganism produces flat, rough, and translucent colonies, with regular or irregular margins, an inner zone of hemolysis, and an outer zone of less complete clearing. Occasionally strains without inner zone hemolysis are seen.[74] On Nagler agar containing 5%–10% egg yolk, *C. perfringens* generates a characteristic white precipitate as a result of its alpha toxin interacting with the lipids in egg yolk.[10,75]

Furthermore, several solid media have been reported to be suitable for detecting and isolating *C. perfringens* from different environmental samples. These media include neomycin blood agar, Shahidi Ferguson *perfringens* (SFP) agar, sulfite polymyxin sulfadiazine (SPS) agar, oleandomycin polymyxin sulfadiazine *perfringens* (OPSP), agar, tryptone sulfite neomycin (TSN) agar, and tryptose sulfite cycloserine (TSC) agar with or without egg yolk.[36,76] Isolation of typical colonies from media is followed by confirmation of *C. perfringens* colonies with different tests to reveal phenotypic characteristics. Observation of nitrate reduction, gelatine liquefaction, lactose fermentation, or lack of motility in suspected *C. perfringens* isolates is usually enough to differentiate *C. perfringens* from other microorganims.[37,74] Identification of *C. perfringens* may be also carried out by using biochemical test kits such as API.[76] Several standard international methods are available to differentiate *C. perfringens*, including methods published by the Association of Official Analytical Chemists (AOAC), the International Standardization Organization (ISO), and the Nordic Committee on Food Analysis (NCFA).

Many PCR protocols focus on individual *C. perfringens* toxin genes,[77] in particular the *cpe* gene[78,79] and *cpa* gene.[80] In addition, multiplex PCR protocols targeting several genes have also been reported.[81–84] Baums et al.[85] developed a reliable species-specific multiplex PCR for the detection of the *cpa, cpb, cpb2, cpe, etx,* and *iap* genes of *C. perfringens* isolates without DNA purification. Furthermore, Joshy et al.[86] reported a multiplex PCR for simultaneous detection of the enterotoxin gene of *C. perfringens* and BoNT genes of *C. botulinum*. In addition to PCR, other approaches such as DNA microarray have also been reported for detecting toxin genes of *C. perfringens*.[72]

Real-time PCR allows for quantifying the copy numbers of plasmid-borne toxin genes[87,88] designed a dual-labeled fluorescence hybridization probe (TaqMan)-based real-time multiplex PCR assay for detection of toxin genes α (*cpa*), β (*cpb*), ι (*iap*), ε (*etx*), $\beta 2$ (*cpb2*), and enterotoxin (*cpe*) of *C. perfringens* directly from cattle feces. Similarity, three real-time fluorogenic (TaqMan) multiplex PCRs have been reported for the detection of six toxic genes of *C. perfringens*.[89,90] Recently, Chon et al.[91] developed an accurate real-time PCR for detection and quantification of *C. perfringens* in foods.

Given that not all *C. perfringens* produce toxins and consequently not all strains are related to diseases or cause food poisonings, the confirmation of toxin productions is the most accepted criterion in conventional diagnosis.[92] Among the techniques for detection of major toxins of *C. perfringens* are the mouse neutralization test (MNT),[92] ELISAs,[93–95] counter-immunoelectrophoresis,[96] and latex agglutination test.[97,98] Lastly, Seyer et al.[99] reported a method to detect *C. perfringens* toxins in complex foods and biological matrixes by immunopurification and ultra performance liquid chromatography-tandem mass spectrometry.

The above methods give information about the presence of pathogenic *Clostridium* species and/or the corresponding toxins in foods. However, no information is supplied about the effect of these microorganisms in live cell that it is required in most of the cases during clostridial foodborne investigation. Bioassays using laboratory animal or cells culture methods report data on effects of neurotoxin or enterotoxins in live organisms. They are based on the ability of suspected food extracts to induce symptoms in laboratory animal models and/or superantigenic action in cell culture models. These methods offer a very valuable alternative for foodborne clostridia investigation and will be explored in a separate section.

9.4 Laboratory Models for Foodborne *Clostridium* Intoxication

9.4.1 Animal Models for Foodborne BoNT

Nowadays, the diagnosis of foodborne diseases caused by the *Clostridium*-produced neurotoxin is carried out by molecular, PCR, or immunological analysis but should be confirmed in all cases by the analysis of stool samples with the mouse neutralization bioassay, using either monovalent, toxin-type-specific botulinum antitoxin or polyvalent antibotulinum antitoxin. Despite many attempts and much research to replace the use of animals, it is still the best assay to model all aspects of BoNT intoxication—binding, translocation, and enzymatic activity[60] (Table 9.1).

The samples used could be the serum of intoxicated people, their feces, or food directly involved. It is desirable to obtain 10–15 mL serum; this allows specific identification of botulinum toxin involved from inoculation serum neutralized all antitoxins, and repetition trial, if necessary. Since it is sometimes impossible to obtain this volume of serum in some patients, especially children, a lower quantity is sufficient to confirm botulism, although in most cases the lower volume of 3 mL cannot provide conclusive results. Approximately 25–50 g of feces sample is required, preferably collected before treatment with antitoxin. However, botulism confirmation has been obtained with lower amounts, and *C. botulinum* has also been isolated from stool after treatment with antitoxin. When performing a rectal lavage due to constipation, it is necessary to use a minimum amount of fluid (preferably sterile water without bacteriostatic) to obtain a sample where the toxin is not diluted. If the patient has received a medication that may interfere with the assay for toxin or cultivation of stool, the laboratory should be notified. For example, it has been shown that anticholinesterase drugs administered orally to patients with myasthenia gravis can interfere with the detection assay used to identify toxin in stool samples. When food samples are directly used, it is important to specify whether it is a commercial or a homemade or handmade product. In all cases, refrigerated samples must be submitted to check whether biosafety standards are met.

There are many studies using mouse bioassays to confirm the presence or type of botulinum toxin[100,101] (Table 9.1), although additional techniques have been used. In this sense, López-Laso et al.,[25] in the diagnosis of some of the cases of infant botulism, used PCR to detect *C. botulinum* neurotoxin genes, and all cases were confirmed by analyses of stool samples with mouse neutralization bioassay. On the other hand, Chun et al.[102] described a steam treatment to reduce or eliminate *C. botulinum* type E viable cells in *Cladophora* mats, thereby breaking the potential transmission route of the toxin up the food chain to animals. They used mouse bioassay and antibody neutralization analysis can help in the identification of *C. botulinum* neurotoxin serotypes.

TABLE 9.1

Animal Models Used for the Detection of *Clostridium* Neurotoxins and Enterotoxins

Animal	Studies	Toxins Tested	References
Mouse	Mouse neutralization bioassay—binding, translocation, and enzymatic activity of neurotoxins	BoNT A, B, C, D, E, and F	25,60
	Inactivation/neutralization effect of botulinum toxin	BoNT A, B, C, D, E, and F	102–109,111,115–117
	Neuroimmunological changes	BoNT A	114
	Changes of ECG pattern of hyperpotassemia, fall of blood pressure, and transient hyperpnea and respiratory depression of *C. perfringens* enterotoxins and death	CPE	121
	Mucosal necrosis of small intestinal loops, without accumulation of fluid in the lumen	CPE	122
	Absorption of CPE from the intestine and formation of CH-1-like complexes in the liver and kidney	CPE	122,123
	Hemorrhagic cecitis *C. difficile* enterotoxins	*C. difficile* Enterotoxin A and B	148–152
Rat	Developing, maintaining, and recovering from neuropathic pain	BoNT A	110
	Neuroimmunological changes	BoNT A	108,112,113
	Changes of ECG pattern of hyperpotassemia, fall of blood pressure, and transient hyperpnea and respiratory depression caused by *C. perfringens* enterotoxins and death	CPE *C. perfringens* Enterotoxin Type A	121
121 Rabbit	Fluid and electrolyte secretion and changes to the mucosa consist mainly of mucosal necrosis and hemorrhage in small intestinal and colonic mucosa	CPE *C. perfringens* Enterotoxin Type A	119,125,128
	Toxin binds to liver and kidney	CPE *C. perfringens* Enterotoxin Type A	122,123
	Hemorrhagic lesions and luminal fluid accumulation in these intestinal loops	*C. perfringens* Enterotoxin CPB and CPE	130
Hamster	Hemorrhagic cecitis *C. difficile* enterotoxins	*C. difficile* Enterotoxin A and B	146,147
Piglets, goats	Endothelial damage of intestinal loops	CPB *in C. perfringens* type C	134,135
Primates	Developed vomiting and diarrhea *C. perfringens* enterotoxins	CPE *C. perfringens* Enterotoxin type A	140

Wohlfarth et al.[103] reported that oral poisoning with BoNT serotype B was capable of activating the immune system. These authors studied two cases of foodborne botulism following a meal consisting of homemade canned beans. Four months later, one person developed a moderate and the other a borderline titer of serum antibodies to BoNT serotype B, detected using an *ex vivo* assay (mouse phrenic-nerve hemidiaphragm assay).

Moreover, bioassays in mice are used not only to confirm the type of *C. botulinum*, but also to investigate of substances that neutralize the effect of Botulinum toxin. Examples in recent years are studies that use animal models to find a small molecular weight (SMW) inhibitor of BoNTs that will neutralize the toxin. There have been numerous publications reporting finding SMW inhibitors to BoNTA,[104–106] but few reports on SMW inhibitors to BoNTB, one report on BoNTE,[107] and none on BoNTF. Recently, Pirazzini et al.[108] reported on several inhibitors that were effective against BoNTA, BoNTC, and BoNTE, but Montgomery et al.[109] found an SMW inhibitor that might be useful against multiple serotypes of BoNT (Table 9.1). They tested two promising SMW inhibitors to BoNTA against BoNTs B, C, E, and F in mice phrenic nerve-hemidiaphragm.

Although most studies are performed in mice, to have a confirmation in different species, additional experiments must be carried out in other animal models, where it is also easier to verify the toxin effects. Thus, Marinelli et al.[110] studied in rats the effects of BoNT/A in developing, maintaining, and recovering from neuropathic pain induced by the ligature of the sciatic nerve. They saw this effect was evident starting 24 h after the administration of BoNT/A, and it was long-lasting—present 81 or 25 days after intraplantar injection of the higher dose in mice (15 pg/paw) and 35 days after injection in rats (75 pg/rat). Moreover, BoNT/A-injected mice showed a quicker recovery of the walking pattern and weight bearing compared to control groups.

Botulinum toxin type A used in the treatment of strabismus and other human diseases characterized by hyperactivity of peripheral cholinergic nerve terminals, it could get some patients are or become resistant to it. This can be overcome by using other botulinum toxins. Eleopra et al.[111] investigated the action of BoNT type D in mouse and human muscles and showed that botulinum toxin type D is poorly effective in inducing human skeletal muscle paralysis but is very potent in mice. Mika et al.[112] studied the effect of BoNT A on sciatic-nerve-injury-induced neuroimmunological changes in rats and found evidence that BoNT/A impedes injury-activated neuronal function in structures distant from the injection site using doses based on an approximate equivalence of 100 U of botox (4.8 ng tox) as reported by Cui et al.,[113] the dose injected in this experiment was 75 pg per paw, which corresponded to about 4–5 U/kg in rats weighing 300–350 g. Also, Kato et al.,[114] using 8 weeks old male animals, showed that BoNT A2 reduces incidence of seizures in mouse models of temporal lobe epilepsy.

Pirazzini et al.[108] studied a Thioredoxin Reductase-thioredoxin Redox System cleaves the interchain disulfide bond of BoNTs on the cytosolic surface of synaptic rat vesicles which blocked the different BoNTs tested within a very similar concentration range, suggesting that the interchain disulfide reduction is closely similar for different BoNTs serotypes. These *in vitro* results were validated *in vivo* using digit abduction score (DAS) assay, a well-established model to compare potency and duration of BoNTs.[115] All tested molecules were very effective in reducing the degree and duration of paralysis induced by the local injection of BoNT/A and BoNT/C. For the first time, it was shown that small molecules effectively prevent the paralytic activity of BoNTs. As a proof of concept, they also tested one of these inhibitors, Ebselen, a compound reported to target both TrxR and Trx,[116] in the lethality assay. Ebselen, preventively administered via intraperitoneal injection, was very effective in protecting animals from a lethal amount of BoNT/A, both by prolongation of the time to death and by reduction of the number of deaths[108] (Table 9.1).

In another report, Pellet et al.[117] analyzed *in vivo* the onset and duration of action of BoNT/A1–5 in female ICR mice (Harlan) that were injected with sublethal amounts of BoNT/A1,/A2,/A3,/A4, or/A5 in 10 μL of GelPhos buffer [30 mM sodium phosphate (pH 6.3) and 0.2% gelatin] into the right gastrocnemius muscle. The amount of toxin injected IM per dilution was confirmed by mouse bioassay. Analyses of several of these studied subtypes revealed distinct characteristics, ranging from differences in cell entry and enzyme kinetics to differences in potency in mice and cell-model-specific potency. In a long-term activity study in cultured primary neurons, it was indicated that BoNT/A1, 2, 4, and 5 have a similar duration of action, whereas BoNT/A3 has a significantly shorter duration of action. This report describes an *in vivo* mouse study, showing that local injection of BoNT/A2 resulted in faster onset of local paralysis than BoNT/A1, 3, 4, and 5, whereas BoNT/A3 resulted in significantly faster recovery of motor-neuron deficiency.

9.4.2 Animal Models for Foodborne *C. perfringens* Enterotoxins

Several animal models have been used to study the role of the different toxins of *C. perfringens* in the pathogenesis of infections produced by this microorganism[9,118–120] (Table 9.1).

The pathodynamics of lethal intoxication in rats and mice by administration of enterotoxin of *C. perfringens* type A was studied using whole animals and isolated organs. Rapid changes of the electrocardiogram (ECG) pattern suggestive of hyperpotassemia, rapid fall of blood pressure, and transient hyperpnea followed by respiratory depression were observed. Analysis of plasma levels of rations revealed hyperpotassemia in both animal species. On the other hand, enterotoxin (up to 100 µg) showed little direct cardiotoxicity on the isolated heart.[121] The *in vivo* effects of CPE have also been studied in rats,[121] although much less extensively than in mice and rabbits. In rats, as in mice, this toxin produces lethality when injected i.v., and death in rats is preceded by respiratory difficulty, ECG alterations, and hyperkalemia, because these effects were accompanied by an increase in liver enzymes (GPT, GOT, and LDH). Sugimoto et al.[121] suggested that CPE-induced hyperkalemia was a consequence of cytotoxic action of CPE on hepatocytes. Mice have also been used to study the intestinal and systemic effects of CPE.[122] In these animals, lethality was associated with a rapid fall in blood pressure, respiratory difficulty, and changes in ECG. In the small intestinal loops of mice, as in rabbits, CPE causes dose- and time-dependent mucosal necrosis. However, toxin administered in this manner does not cause fluid accumulation in mice.[122] Results of experiments inoculating CPE into intestinal loops of mice suggest that death observed in constipated human patients with CPE-positive *C. perfringens* type A infection[123] could have been a consequence of absorption of CPE from the intestine. Mice that were inoculated with CPE in the intestinal loops showed that CPE bound to and formed CH-1-like complexes in the liver and kidney. A mouse intravenous lethality model was used to demonstrate that CPB is the main factor responsible for systemic lethality in type C culture supernatants.[124] In that study, lethality was abolished when culture supernatants were preincubated with a CPB monoclonal antibody, but not when the cultures were incubated with a CPA monoclonal antibody, which confirmed the role of CPB in mouse lethality.[124]

Furthermore, for many years, rabbit intestinal loops have been used, and are still used today, to study the effects of CPE *in vivo*.[125] Most of these rabbit models have been followed to evaluate the effect of CPE in the small intestine, where it causes fluid and electrolyte secretion and produces significant damage to the mucosa of the jejunum and ileum, but less damage in the duodenum.[125–127]

García et al.[119] have also demonstrated that rabbit colon is sensitive to the action of purified CPE, with both fluid secretion and mucosal damage observed. The histological changes caused by CPE in both small intestinal and colonic loops of rabbits consist mainly of mucosal necrosis and hemorrhage. These changes are both time- and dose-dependent.[119,125]

Rabbits have also been used to study the binding of CPE to extraintestinal tissues, which led to the demonstration that this toxin binds to liver and kidney.[122,123] This finding suggests that CPE absorbed from the intestine can be responsible for systemic alterations, which may help explain the lethality observed in some cases of experimental animals and human patients with the disease.[122,123]

A rabbit intestinal loop model was also used to determine the spatial distribution of the effects of CPB along the alimentary canal.[128] In that study, fluid accumulation and necrotizing enteritis were observed only in the small intestine, with the jejunum and ileum being most severely affected. This result is in agreement with natural type C disease in animals and humans, in which the jejunum and ileum are primarily affected.[73,129]

Synergism between CPB and CPE for the virulence of CPE-positive type C strains of *C. perfringens* was demonstrated by using a rabbit-ligated intestinal loop model,[130] inducing significant hemorrhagic lesions and luminal fluid accumulation in these intestinal loops. However, when lysate supernatants of the *cpb* or *cpe* knockout mutants of these strains were inoculated into rabbit-ligated intestinal loops, no significant damage or fluid accumulation were observed. Complementing the *cpe* mutant, or reversing the *cpb* mutation, restored the virulent effects of culture lysates. Purified CPB and CPE, inoculated together at concentrations similar to those found in wild-type CN3758 culture lysates, also produced lesions and fluid accumulation in rabbit intestines. However, when either of these toxins was inoculated independently, only higher doses caused damage to the intestine, suggesting that at low concentration, both toxins act synergistically in the intestine.[130] These experiments provided the first evidence of synergistic toxin activity during intestinal *C. perfringens* infections.

Large animal models have also been used to study the pathogenesis of *C. perfringens* type C disease. Initially, piglets were experimentally used for this purpose.[131–133] However, those experiments were performed using whole cultures or crude culture supernatants, and although the results confirmed that *C. perfringens* type C is a pathogen for piglets, they did not identify the main virulence factor(s) involved in the pathogenesis of those infections. The mechanism of action of CPB in type C infection was recently studied in intestinal loops of piglets.[134] This study indicated that there is a tropism of CPB toward endothelial cells, suggesting that endothelial damage induced by CPB plays a role in the early stages of *C. perfringens* type C enteritis in pigs. Koch's postulates for type C disease were fulfilled in goats,[135] using the same set of *C. perfringens* type C mutants previously used to fulfill those postulates in rabbits and mice.[118,136] The results of the experiments in goats confirmed, this time in a natural host of the disease, the key role of CPB in the pathogenesis of natural *C. perfringens* type C disease.

The effects of i.v. administration of extracts of sporulating cultures of CPE-positive *C. perfringens* type A, and of CPE into ligated loops, also have been studied in calves and lambs, respectively.[137–139] The results of these experiments were variable and included diarrhea in calves and mild mucosal changes in the intestinal loops of lambs.

Nonhuman primates have rarely been used to study the pathogenesis of CPE intoxication and/or enterotoxigenic *C. perfringens* type A-associated disease.[140] In the only published study, cynomolgus monkeys fed purified CPE developed vomiting and diarrhea, while monkeys given CPE-positive *C. perfringens* type A orally developed only diarrhea. These effects were only observed when either CPE or CPE-positive *C. perfringens* type A were given together with sodium bicarbonate to neutralize the low gastric pH. Lethality was not observed in these nonhuman primates.[140]

9.4.3 Animal Models for Foodborne *C. difficile*

Various animal models have been used extensively for *C. difficile* research to study disease pathogenesis. Until recently, the most commonly used *C. difficile* disease model has utilized hamsters. However, mouse and pig models have now been developed that unravel different aspects of *C. difficile* pathology[141] (Table 9.1).

Furthermore, animal models have been developed to study various aspects of *C. difficile* infection, including colonization, disease pathophysiology, intoxication, transmission, recurrence, efficacy testing of potential therapeutics, and the impact of strain variability on all of these factors. Small animals that have been utilized in these studies include mice, hamsters, rats, rabbits, hares, guinea pigs, prairie dogs, and quails.[142,143] More recently, zebrafish embryos have been used to study and compare the effects of *C. difficile* variant toxin B derivatives isolated from a number of *C. difficile* strains.[144,145] Larger animals such as foals, Rhesus monkeys, and gnotobiotic piglets have been used in other studies to study *C. difficile*.[142]

Importantly, many of the animal models used for *C. difficile* research require pretreatment with antibiotics to induce *C. difficile* infection, such as metronidazole, vancomycin, kanamycin, gentamicin, clindamycin, and cefoperazone, by drinking a single antibiotic or cocktail with water or by intraperitoneal injection.[141]

C. difficile pathogenesis understanding has steadily increased since the development of hamster and mouse models of *C. difficile* infection. The hamster model provides the foundation for *C. difficile* research and is a useful and important model for studying *C. difficile*. After pretreatment with clindamycin and challenge with toxigenic strains of *C. difficile*, hamsters develop hemorrhagic cecitis, which presents as diarrhea or "wet tail" as well as other fulminant disease symptoms including ruffled fur, hunching, and lethargy, leading to death.[146,147] Histopathological analysis of cecal tissue isolated from diseased hamsters shows mucosal ulceration associated with polymorphonuclear leucocyte influx and tissue hemorrhage. However, the site of *C. difficile* infection in hamsters differs from that in humans, as infection occurs in the cecum of hamsters but in the colon of humans.[147]

The development of new mouse models, combined with wide access to mouse-specific reagents and tools, is offering new opportunities to study subtle features of disease. Three different mouse *C. difficile* infection models have been described. The first employs gnotobiotic/germ-free mice,[148,149] the second uses a cocktail of antibiotics to disrupt the normal gut microbial communities and predispose the mice to infection,[150] and the third uses a single antibiotic to induce susceptibility to *C. difficile* infection.[151,152] However, due to the notable interspecies differences in susceptibility to *C. difficile* infection and the severity of disease outcomes, caution should be exercised when comparing results obtained using different animal models or when extrapolating those results to humans.

Thus, the choice of appropriated animal model is critical when endeavoring to study *C. difficile* infection and it is likely that in coming years the use of various animal models will be reevaluated, particularly as more become known about the complex relationship between *C. difficile*, host cell receptors, the gut microbiota, and the host immune system.

9.5 Cell Culture Models

Cell-based assays measure BoNT receptor biding, translocation, and enzymatic activity and can be alternatives to animal model and particularly to the mouse bioassay. A number of different neuronal and nonneuronal derived cell lines have been generated for use in BoNT assays (Table 9.2). These include rat spinal cord cells,[153] chick embryo neuronal cells,[154] neuroblastoma cells N2A,[155] and BE(2)-M17 cells.[156] The read out for most of the cell-based assays for detection of BoNT/A is the cleavage of SNAP-25. Antibodies for SNAP-25 allow immunoblot detection of cleavage products, specifically detecting a decrease in size of endogenous SNAP-25 protein.

Some studies in recent years use these types of cell culture, modifications of these, or new cell cultures. Lyman et al.[157] said no cell-based platform recapitulates the sensitivity of primary spinal cord neurons to botulinum toxins (BoNTs), suggesting that neurogenic cell lines do not accurately model the mechanisms of toxin internalization and activity. They developed a biologically relevant cell-culture model for BoNT intoxication, in which glutamatergic neurons (ESNs), derived from murine embryonic stem cells (ESCs), are highly sensitive to multiple BoNT serotypes. This study showed that ESNs were susceptible to BoNT/A, /B, and /E with EC50's of 0.81, 12.3, and 67.4 pM, respectively.

On the other hand, several continuous cell lines (Neuro-2a, SK-N-SH, M17, SHSY5Y, PC12 NT2) have been tested for sensitivity to BoNT/A and are being used as research models[158,159] (Table 9.2). However, most of them lack the sensitivity necessary to compete with the commonly used mouse bioassay.[160] Regina et al.[161] demonstrated that mouse neuroblastoma rat glioma hybrid cell line NG108–15 was the most sensitive cell line for detecting BoNT/A1. Even nowadays, human-derived cell systems have been reported as a useful alternative. In this sense, Fonfria et al.[162] compared results from two rodent-derived [dorsal root ganglia (DRG) and cortical neurons (CTX)] and two human-derived neuronal cellular models (SiMa cells and iCell neurons) with data obtained from rat spinal cord neurons (SCN). The rank order of BoNT/A1 sensitivity was: SCN = CTX = iCells > DRG > SiMa cells. Thus the iCell neurons had high

TABLE 9.2

Cell Culture Models Used for the Studies of *Clostridium* Neurotoxins and Enterotoxins

Cell Culture	Studies	Toxins Tested	References
Rat spinal cord cells			153,163
Chick embryo neuronal cell	Receptor binding, translocation, and enzymatic activity of neurotoxins	BoNT A, B, C, D, E, and F	154
Neuroblastoma cells N2A			155
BE(2)-M17 cells			156
Murine embryonic stem cells		BoNT A, B, and E	157
Rat spinal cord cells			153
Neuro-2a, SK-N-SH, M17, SHSY5Y, PC12 NT2	Sensitivity to BoNT	BoNT A	158,159
Neuroblastoma cells NG108–15			161
Rodent ganglia cell			162
Cortical neurons			
Human neuronal cell			
Human intestinal epithelial Caco-2 cells	Cytotoxicity of *C. perfringens* enterotoxins	*C. perfringens* Enterotoxin type A	165–167
Human colonic epithelial cell lines Caco-2, T84, HT-29	Cytotoxicity of *C. difficile* enterotoxins	*C. difficile* Enterotoxin A and B	169,170
Human monocytic THP-1 cells	Inflammatory cytokine release stimulation of *C. difficile* enterotoxins	*C. difficile* Enterotoxin A and B	168,171–173

BoNT/A1 sensitivity comparable with rat SCN, but another human cell model, the SiMa cells, had the lowest BoNT/A1 sensitivity of all the models tested.

Delaflotte et al.[163] developed a procedure to isolate, enrich, and culture spinal motor neurons from rat embryonic spinal cords and to assess sensitivity to BoNT/A1 (Table 9.2). This model obtained a highly enriched culture derived from neonatal rat ventral spinal cord and provides a physiologically relevant model system to understand the biology and characteristics of BoNTs acting at cholinergic terminals. These cultures' exhibited high sensitivity an EC50 (median effective concentration) below 0.5 pM to BoNT/A1, which is more than five times the sensitivity measured in the standard rat spinal cord neuron cultures and could be an ideal model to assess differences between BoNT serotypes and subtypes.

Regarding *C. perfringens* foodborne investigation, cell culture assays are possible alternatives to replace *in vivo* neutralization tests. Animal experiments still play a central role in this kind of quality control test; however, potency tests such as clostridial toxoids require a very high number of animals for quality control testing.[164] In this sense, Allaart et al.[165] demonstrated that CPE expression is necessary and sufficient for *C. perfringens* strains SM101 and F4969 to cause fluid accumulation and GI damage in a rabbit ileal loop model; however, it remains unclear whether CPE is indispensable for bacterial cytotoxicity *in vitro* (Table 9.2).

The significance of toxins in the induction of *in vitro* cytotoxicity has been investigated using human intestinal epithelial Caco-2 cells infected with toxin-gene-harboring *C. perfringens* strains and their mutants or anti-toxin antibody.[165,166] These authors revealed that β2 toxin is not involved in Caco-2 cell cytotoxicity during infection with a cpb2-harboring *C. perfringens* strain. Yasugi et al.[167] examined the cytotoxicity of cpe-harboring *C. perfringens* isolates cocultured with human intestinal epithelial Caco-2 cells. The food poisoning strains showed severe cytotoxicity during sporulation and CPE production, but not during vegetative cell growth. While Caco-2 cells were intact during coculturing with cpe-null mutant derivative of strain SM101 (a food poisoning strain carrying a chromosomal cpe gene), the wild-type level cytotoxicity was observed with cpe-complemented strain. In contrast, both wild-type and cpe-null mutant derivative of the non-foodborne strain F4969 induced Caco-2 cell death during both vegetative and sporulation growth. The Caco-2 cell cytotoxicity caused by *C. perfringens* strain SM101 is considered to be exclusively dependent on CPE production.

Regarding *C. difficile* investigation, the human colonic epithelial cell lines Caco-2, T84, HT-29, and human monocytic THP-1 cells have been used to study direct effects on intestinal cells[168–170] (Table 9.2).

In human intestinal epithelial cell line, *C. difficile* enterotoxins initiate colonic inflammation in humans by injuring epithelial cells and inducing production of IL8.[169] Detachment of these cells from the basement membrane leads to cell death by apoptosis. IL8 produced by the injured epithelial cells, as well as responses by the exposed lamina propria cells (especially macrophages that lie just below the basement membrane 43), to toxins A and B and other luminal contents would initiate a cascade of events characterized by migration of circulating polymorphonuclear cells into the mucosa to mediate tissue damage and induce diarrhea. Most interestingly, colonic epithelial cells from some adults may be resistant to the effects of even high concentrations of toxin A, possibly explaining why some people are resistant to disease.

In cells of the monocyte lineage, *C. difficile* toxins A and B stimulate inflammatory cytokine release, including TNF-α, IL-1β, IL-6, and IL-8,[171–173] and induce necrosis in human monocytes and in THP-1 human monocytic cells[168] (Table 9.2).

Toxins A and B share 63% amino acid homology[174] and possess similar domains. These toxins were shown to block small GTP-binding proteins. In monocytes, toxins induce IL-8 production and necrosis by unknown mechanisms. When these methods are used in foodborne investigation, confirmation of toxin production is necessary.

9.6 Conclusions

Being the most widely distributed pathogenic species within the *Clostridium* genus, *C. botulinum*, *C. perfringens*, and *C. difficile* are frequently involved in foodborne outbreaks. Accurate methods for isolating and detecting these pathogenic clostridia in raw material and processed foods are necessary to take corrective measures throughout food processing that reduce potential clostridial cases. For isolation, detection, and characterization of *C. botulinum*, *C. perfringens*, and *C. difficile* from food samples, traditional microbiological culture techniques and molecular methods (PCR and qPCR) are available. When these methods are used in foodborne investigation, confirmation of toxin production by rapid methods such as ELISA, immuno-PCR, or ELISA based on protein antibody microarray is important. While these methods reveal the presence of pathogenic *Clostridium* species and/or the corresponding toxins in food, no information is available about the effect of these microorganisms in live cells, which is required in most of the cases during clostridial foodborne investigation. Bioassays using laboratory animal or cells culture methods help determine the effects of neurotoxin or enterotoxins in live organisms. Based on the ability of a suspected food extracts to induce symptoms in laboratory animal models and/or superantigenic action in cell culture models, these methods offer a very valuable alternative for foodborne clostridia investigation. The mouse neutralization bioassay, using either monovalent, toxin-type-specific botulinum antitoxin or polyvalent antibotulinum antitoxin is the most valuable method for foodborne botulism investigation. Rabbit intestinal loops, rat, and mice laboratory models are available for *C. perfringens* foodborne investigation. For *C. difficile* investigation, hamster, mouse, and pigs may be utilized. In addition, cell lines provide an alternative to laboratory animal models for foodborne clostridial cases investigation. For example, rat spinal cord, chick embryo neuronal, and neuroblastoma cell lines are available for accurate investigation of foodborne botulism. For the study of pathogenic *C. perfringens* and *C. difficile*, human intestinal epithelial lines such as Caco-2 cells have been reported as sensitive method.

Acknowledgments

This work has been funded by the Spanish Instituto Nacional de Investigación Agraria y Agroalimentaria (INIA) and the Spanish Comisión Interministerial de Ciencia y Tecnología with the projects RTA-2013-00070-C03-03 and Carnisenusa CSD2007-00016, Consolider Ingenio 2010, and GRU09162 of the Junta de Extremadura and FEDER.

REFERENCES

1. Peck, M.W. et al. Development and application of a new method for specific and sensitive enumeration of spores of nonproteolytic *Clostridium botulinum* types B, E, and F in foods and food materials. *Appl. Environ. Microbiol.*, 76, 6607–14, 2010.
2. Carter, A.T. and Peck, M.W. Genomes, neurotoxins and biology of *Clostridium botulinum* group I and group II. *Res. Microbiol.*, 166, 303–17, 2015.
3. Barash, J.R. and Arnon, S.S. A novel strain of *Clostridium botulinum* that produces type B and type H botulinum toxins. *J. Infect. Dis.*, 209, 183–91, 2014.
4. Johnson, E.A. Validity of botulinum neurotoxin serotype H. *J. Infect. Dis.*, 210, 992–3, 2014.
5. Dover, N., Barash, J.R., Hill, K.K., Xie, G. and Arnon, S.S. Molecular characterization of a novel botulinum neurotoxin type H gene. *J. Infect. Dis.*, 209, 192–202, 2014.
6. Carter, A.T., Stringer, S.C., Webb, M.D. and Peck, M.W. The type F6 neurotoxin gene cluster locus of Group II *Clostridium botulinum* has evolved by successive disruption of two different ancestral precursors. *Gen. Biol. Evol.*, 5, 1032–7, 2013.
7. Lindström, M. et al. Novel insights into the epidemiology of *Clostridium perfringens* type A food poisoning. *Food Microbiol.*, 28, 192, 2011.
8. Gillers, S. et al. Microscale sample preparation for PCR of *C. difficile* infected stool. *J. Microbiol. Methods*, 78, 203, 2009.
9. Li, J. et al. Toxin plasmids of *Clostridium perfringens*. *Microbiol. Mol. Biol. Rev.*, 77, 208–33, 2013.
10. Heikinheimo, A., Lindstrom, M., Liu, D. and Korkeala, H. *Clostridium*. In: *Molecular Detection of Foodborne Pathogens*, Ch. 11, Liu, D. (Ed.), CRC Press, Taylor & Francis Group, LLC, Boca Raton, FL, 2009.
11. Hatheway, C.L. Toxigenic clostridia. *Clin. Microbiol. Rev.*, 3, 66, 1990.
12. McClane, B.A. et al. The enterotoxic clostridia. In: *The Prokaryotes*, Dworkin, M. et al. (Eds.), Springer, New York, pp. 688–752, 2006.

13. Keyburn, A.L. et al. NetB, a pore-forming toxin from necrotic enteritis strains of *Clostridium perfringens*. *Toxins*, 2, 1913, 2008.
14. Smedley, J.G., Fisher, D.J., Sayeed, S., Chakrabarti, G. and McClane, B.A. The enteric toxins of *Clostridium perfringens*. *Rev. Physiol. Biochem. Pharmacol.*, 152, 183–204, 2004.
15. Songer, J.G. and Uzal, F.A. Clostridial enteric infections in pigs. *J. Vet. Diagn. Invest.*, 17, 528, 2005.
16. McClane, B.A. and Rood, J.I. Clostridial toxins involved in human enteric and histotoxic infections. In: *Clostridia: Biotechnology and Medical Applications*, Bahl, H. and Dürre, P. (Eds.), Wiley-VCH Verlag GmbH, Weinheim, 2001.
17. Center for Disease Control and Prevention (CDC). National Botulism Surveillance, last updated May 3, 2016, accessed June 30, 2016, http://www.cdc.gov/botulism/surveillance.html.
18. Food and Drug Administration (FDA). *Bad Bug Book: Foodborne Pathogenic Microorganisms and Natural Toxins*, 2nd ed., US Food and Drug Administration, Washington, DC, p. 108, 2012.
19. Aston, S. and Beeching, N. Botulism. In: *Hunters Tropical Medicine and Emerging Infectious Diseases*, Magill, A., Ryan, E.T., Solomon, T. and Hill, D. (Eds.), 9th ed., vol. 1, Saunders, Philadelphia, PA, pp. 511–3, 2012.
20. Therre, H. Botulism in the European Union. *Euro Surveill.*, 4, 2, 1999.
21. Barash, J.R., Hsia, J.K. and Arnon, S.S. Presence of soil-dwelling clostridia in commercial powdered infant formulas. *J. Pediatr.*, 156, 402–8, 2010.
22. Arnon, S.S., Schechter, R., Maslanka, S.E., Jewell, N.P. and Hatheway, C.L. Human botulism immune globulin for the treatment of infant botulism. *N. Engl. J. Med.*, 354, 462–71, 2006.
23. Koepke, R., Sobel, J. and Arnon, S.S. Global occurrence of infant botulism, 1976–2006. *Pediatrics*, 122, 73–82, 2008.
24. Midura, T.F. and Arnon, S.S. Infant botulism: Identification of *Clostridium botulinum* and its toxins in faeces. *Lancet*, 2, 934–6, 1976.
25. López-Laso, E. et al. Infant botulism in Andalusia (southern Spain). *Eur. J. Paediatr. Neurol.*, 18, 321–6, 2014.
26. Scallan, E. et al. Foodborne illness acquired in the United States—major pathogens. *Emerg. Infect. Dis.*, 17, 7–15, 2011.
27. Grass, J.E., Gould, L.H. and Mahon, B.E. Epidemiology of foodborne disease outbreaks caused by *Clostridium perfringens*, United States, 1998–2010. *Foodborne Pathog. Dis.*, 10, 131–6, 2013, doi: 10.1089/fpd.2012.1316..
28. Mage, D.T., Cohen, M. and Donner, M. Sudden infant death syndrome. *N. Engl. J. Med.*, 24, 2581, 2009.
29. Lindström, M. and Korkeala, H. Laboratory diagnostics of botulism. *Clin. Microbiol. Rev.*, 19, 298, 2006.
30. Reddy, N.R., Tetzloff, R.C. and Skinner, G.E. Effect of media, additives, and incubation conditions on the recovery of high pressure and heat-injured *Clostridium botulinum* spores. *Food Microbiol.*, 27, 613, 2010.
31. Mato Rodriguez, L. and Alatossava, T. Effects of copper on germination, growth and sporulation of *Clostridium tyrobutyricum*. *Food Microbiol.*, 27, 434, 2010.
32. Lenz, C. A. and Vogel, R.F. Effect of sporulation medium and its divalent cation content on the heat and high pressure resistance of *Clostridium botulinum* type E spores. *Food Microbiol.*, 44, 156–67, 2014.
33. Webb, M.D. et al. Does proximity to neighbours affect germination of spores of non-proteolytic *Clostridium botulinum*? *Food Microbiol.*, 32, 104, 2012.
34. Saeed, E.M.A. Studies on iolation and identification of *Clostridium botulinum* investigating field samples specially from equine grass sickness cases. Doctoral thesis, University of Göttingen, Göttingen, 2005.
35. Stringer, S.C., Webb, M.D. and Peck, M.W. Lag time variability in individual spores of *Clostridium botulinum*. *Food Microbiol.*, 28, 228, 2011.
36. Hauschild, A.H.W. and Hilsheimer, R. Enumeration of *Clostridium botulinum* spores in meats by a pour-plate procedure. *Can. J. Microbiol.*, 23, 829, 1977.
37. Cato, E., George, W.L. and Finegold, S. Genus *Clostridium*. In: *Bergey's Manual of Systematic Bacteriology*, Sneath, P.H.A. et al. (Eds.), Williams and Wilkins, Baltimore, MD, 1986.
38. Mills, D.C., Midura, T.F. and Arnon, S.S. Improved selective medium for the isolation of lipase-positive *Clostridium botulinum* from feces of human infants. *J. Clin. Microbiol.*, 2, 947, 1985.
39. Silas, J.C. et al. Selective and differential medium for detecting *Clostridium botulinum*. *Appl. Environ. Microbiol.*, 50, 1110, 1985.
40. Smith, L.D.S. and Sugiyama, H. *Botulism: The Organism, Its Toxins, the Disease*, Charles C. Thomas, Springfield, IL, 1988.

41. Fakruddin, M. et al. Nucleic acid amplification: Alternative methods of polymerase chain reaction. *J. Pharm. Bioallied Sci.*, 5(4), 245–52, 2013.
42. Lindström, M. et al. Multiplex PCR assay for detection and identification of *Clostridium botulinum* types A, B, E, and F in food and fecal material. *Appl. Environ. Microbiol.*, 67, 5694, 2001.
43. Kasai, Y. et al. Quantitative duplex PCR of *Clostridium botulinum* types A and B neurotoxin genes. *J. Food Hyg. Soc. Jpn.*, 48, 19, 2007.
44. Sakuma, T. et al. Rapid and simple detection of *Clostridium botulinum* types A and B by loop-mediated isothermal amplification. *J. Appl. Microbiol.*, 106, 1252, 2009.
45. De Medici, D. et al. Multiplex PCR for detection of botulinum neurotoxin-producing clostridia in clinical, food, and environmental samples. *Appl. Environ. Microbiol.*, 75, 6457–61, 2009.
46. Hyytiä, E. et al. Characterisation of *Clostridium botulinum* groups I and II by randomly amplified polymorphic DNA analysis and repetitive element sequence-based PCR. *Int. J. Food Microbiol.*, 48, 179, 1999.
47. Paul, C. et al. A unique restriction site in the *flaA* gene allows rapid differentiation of group I and group II *Clostridium botulinum* strains by PCR-restriction fragments length polymorphism analysis. *J. Food Prot.*, 70, 2133, 2007.
48. Dahlsten, E. et al. PCR assay for differentiating between group I (proteolytic) and group II (nonproteolytic) strains of *Clostridium botulinum. Int. J. Food Microbiol.*, 124, 108, 2008.
49. Janda, J.M. and Whitehouse, C.A. Usefulness of multilocus polymerase chain reaction followed by electrospray ionization mass spectrometry to identify a diverse panel of bacterial isolates. *Diagn. Microbiol. Infect. Dis.*, 63, 403, 2009.
50. Fach, P. et al. An innovative molecular detection tool for tracking and tracing *Clostridium botulinum* types A, B, E, F and other botulinum neurotoxin producing clostridia based on the GeneDisc cycler. *Int. J. Food Microbiol.*, 145, S145, 2011.
51. Hauser, D. et al. Botulinal neurotoxin C1 complex genes, clostridial neurotoxin homology, and genetic transfer in *Clostridium botulinum. Toxicon*, 33, 515, 1995.
52. East, A.K. and Collins, M.D. Conserved structure of genes encoding components of botulinum neurotoxin complex M and the sequence of the gene coding for the nontoxic component in nonproteolytic *Clostridium botulinum* type F. *Curr. Microbiol.*, 29, 69, 1994.
53. Aranda, E. et al. Detection of *Clostridium botulinum* types A, B, E, and F in foods by PCR and DNA probe. *Lett. Appl. Microbiol.*, 25, 186, 1997.
54. Raphael, B.H. and Andreadis, J.D. Real-time PCR detection of the nontoxic nonhemagglutinin gene as a rapid screening method for bacterial isolates harboring the botulinum neurotoxin (A-G) gene complex. *J. Microbiol. Methods*, 71, 343, 2007.
55. Dover, J.E. et al. Recent advances in peptide probe-based biosensors for detection of infectious agents. *J. Microbiol. Methods*, 78, 10, 2009.
56. Yoo, M. et al. High-throughput identification of clinically important bacterial pathogens using DNA microarray. *Mol. Cell. Probe*, 23, 171, 2009.
57. Fenicia, L. et al. Towards an international standard for detection and typing botulinum neurotoxin-producing Clostridia types A, B, E and F in food, feed and environmental samples: A European ring trial study to evaluate a real-time PCR assay. *Int. J. Food Microbiol.*, 145, S152, 2011.
58. Raphael, B.H., Joseph, L.A., McCroskey, L.M., Luquez, C. and Maslanka, S.E. Detection and differentiation of *Clostridium botulinum* type A strains using a focused DNA microarray. *Mol. Cell Probes*, 24, 146–53, 2010.
59. Ferreira, J.L. et al. Detection of botulinal neurotoxins A, B, E, and F by amplified enzyme-linked immunosorbent assay: Collaborative study. *J. AOAC Int.*, 86, 314, 2003.
60. Grate, J.W., Ozanich, R.M., Jr., Warner, M.G., Marks, J.D. and Bruckner-Lea, C.J. Advances in assays and analytical approaches for botulinum-toxin detection. *Trends Anal. Chem.*, 29, 1137–56, 2010.
61. Kalb, S.R. et al. The use of Endopep-MS for the detection of botulinum toxins A, B, E, and F in serum and stool samples. *Anal. Biochem.*, 351, 84, 2006.
62. Kalb, S.R. et al. Extraction of BoNT/A, /B, /E, and /F with a single, high affinity monoclonal antibody for detection of botulinum neurotoxin by Endopep-MS. *PLoS One*, 5, 122–37, 2010.
63. Ladd, J. et al. Detection of botulinum neurotoxins in buffer and honey using a surface plasmon resonance (SPR) sensor. *Sens. Actuators, B*, 130, 129, 2008.
64. Ferracci, G. et al. A label-free biosensor assay for botulinum neurotoxin B in food and human serum. *Anal. Biochem.*, 410, 281, 2011.

65. Ching, K.H. et al. Rapid and selective detection of botulinum neurotoxin serotype-A and -B with a single immunochromatographic test strip. *J. Immunol. Methods*, 380, 23, 2012.
66. Liu, Y.Y.B. et al. A functional dual-coated (FDC) microtiter plate method to replace the botulinum toxin LD_{50} test. *Anal. Biochem.*, 425, 28, 2012.
67. Ruge, D.R. et al. Detection of six serotypes of botulinum neurotoxin using fluorogenic reporters. *Anal. Biochem.*, 411, 200–9, 2011.
68. Raphael, B.H. et al. Ultrafiltration improves ELISA and Endopep MS analysis of botulinum neurotoxin type A in drinking water. *J. Microbiol. Methods*, 90, 267, 2012.
69. Rajkovic, A. et al. Detection of *Clostridium botulinum* neurotoxins A and B in milk by ELISA and immuno-PCR at higher sensitivity than mouse bio-assay. *Food Anal. Methods*, 5, 319, 2012.
70. Zhang, Y. et al. Simultaneous and sensitive detection of six serotypes of botulinum neurotoxin using enzyme-linked immunosorbent assay-based protein antibody microarrays. *Anal. Biochem.*, 430, 185, 2012.
71. Brunt, J., Webb, M.D. and Peck, M.W. Rapid affinity immunochromatography column-based tests for sensitive detection of *Clostridium botulinum* neurotoxins and *Escherichia coli* O157. *Appl. Environ. Microbiol.*, 76, 4143–50, 2010.
72. Al-Khaldi, S.F. et al. Identification and characterization of *Clostridium perfringens* using single target DNA microarray chip. *Int. J. Food Microbiol.*, 91, 289, 2004.
73. Uzal, F.A. Diagnosis of *Clostridium perfringens* intestinal infections in sheep and goats. *Anaerobe*, 10, 135–43, 2004.
74. Córdoba, J.J. et al. *Clostridium*. In: *Molecular Detection of Human Bacterial Pathogens*, Liu, D. (Ed.), CRC Press, Taylor & Francis Group, Boca Raton, FL, p. 367, 2011.
75. Borriello, S.P. and Aktories, K. *Clostridium perfringens*, *Clostridium difficile*, and other *Clostridium* species. In: *Topley and Wilson's Microbiology and Microbial Infections*, Mahy, B.W.J. et al. (Eds.), John Wiley & Sons, Chichester, 2010.
76. Labbé, R. *Clostridium perfringens*. In: *The Microbiological Safety and Quality of Food*, Aspen Publishers, Gaithersburg, MD, 2000.
77. Wu, J. et al. Detection and toxin typing of *Clostridium perfringens* in formalin-fixed, paraffin-embedded tissue samples by PCR. *J. Clin. Microbiol.*, 47, 807, 2009.
78. Nakamura, M. et al. PCR identification of the plasmid-borne enterotoxin gene (*cpe*) in *Clostridium perfringens* strains isolated from food poisoning outbreaks. *Int. J. Food Microbiol.*, 294, 261, 2004.
79. Tang, Y. et al. Detection, cloning, and sequencing of the enterotoxin gene of *Clostridium perfringens* type C isolated from goat. *Turkish J. Vet. Anim. Sci.*, 36, 153, 2012.
80. Shanmugasamy, M. and Rajeswar, J. Alpha toxin specific PCR for detection of toxigenic strains of *Clostridium perfringens* in poultry. *Vet. World*, 5, 365, 2012.
81. Meer, R. and Songer, G. Multiplex polymerase chain reaction assay for genotyping *Clostridium perfringens*. *Am. J. Vet. Res.*, 58, 702, 1997.
82. Heikinheimo, A. and Korkeala, H. Multiplex PCR assay for toxinotyping *Clostridium perfringens* isolates obtained from Finnish broiler chickens. *Lett. Appl. Microbiol.*, 40, 407, 2005.
83. Uzal, F.A. and Songer, J.G. Diagnosis of *Clostridium perfringens* intestinal infections in sheep and goats. *J. Vet. Diag. Invest.*, 20, 253, 2008.
84. Goncuoglu, E.M. et al. Molecular typing of *Clostridium perfringens* isolated from turkey meat by multiplex PCR. *Lett. Appl. Microbiol.*, 47, 31, 2009.
85. Baums, C.G. et al. Diagnostic multiplex PCR for toxin genotyping of *Clostridium perfringens* isolates. *Vet. Microbiol.*, 100, 11, 2004.
86. Joshy, L., Chaurdhry, R. and Chandel, D.S. Multiplex PCR for the detection of *Clostridium botulinum* and *C. perfringens* toxin genes. *Indian J. Med. Res.*, 128, 206, 2008.
87. Feng, Y. et al. Identification of changes in the composition of ileal bacterial microbiota of broiler chickens infected with *Clostridium perfringens*. *Vet. Microbiol.*, 140, 116, 2010.
88. Gurjar, A.A. et al. Real-time multiplex PCR assay for rapid detection and toxintyping of *Clostridium perfringens* toxin producing strains in feces of dairy cattle. *Mol. Cell. Probe*, 22, 90, 2008.
89. Albini, S. et al. Real-time multiplex PCR assays for reliable detection of *Clostridium perfringens* toxin genes in animal isolates. *Vet. Microbiol.*, 127, 179, 2008.
90. Abildgaard, L. et al. Sequence variation in the α-toxin encoding *plc* gene of *Clostridium perfringens* strains isolated from diseased and healthy chickens. *Vet. Microbiol.*, 136, 293, 2009.

91. Chon, J. et al. Development of real-time PCR for the detection of *Clostridium perfringens* in meats and vegetables. *J. Microbiol. Biotechnol.*, 22, 530, 2012.
92. Sterne, M. and Batty, I. Criteria for diagnosing clostridial infection. In: *Pathogenic Clostridia*, Sterne, M. and Batty, L. (Eds.), Butterworths, London, p. 79, 1975.
93. Piyankarage, R.H. et al. Sandwich enzyme-linked immunosorbent assay by using monoclonal antibody for detection of *Clostridium perfringens* enterotoxin. *J. Vet. Med. Sci.*, 61, 45, 1999.
94. El Idrissi, A.H. and Ward, G.E. Development of double sandwich ELISA for *Clostridium perfringens* beta and epsilon toxins. *Vet. Microbiol.*, 31, 89, 1992.
95. Nagahama, M. et al. Enzyme-linked immunosorbent assay for rapid detection of toxins from *Clostridium perfringens*. *FEMS Microbiol. Lett.*, 84, 41, 1991.
96. Petit, L., Gibert, M. and Popoff, M.R. Detection of enterotoxin of *Clostridium perfringens*. In: *Encyclopedia of Food Microbiology*, Batt, C.A., Patel, P. and Robinson, R.K. (Eds.), Academic Press, London, 1999.
97. Marks, S.L. et al. Evaluation of methods to diagnose *Clostridium perfringens* associated diarrhea in dogs. *J. Am. Vet. Med. Assoc.*, 214, 357, 1999.
98. Berry, P.R. et al. Evaluation of ELISA, RPLA, and Vero cell assays for detecting *Clostridium perfringens* enterotoxin in faecal specimens. *J. Clin. Pathol.*, 41, 458, 1988.
99. Seyer, A. et al. Rapid quantification of clostridial epsilon toxin in complex food and biological matrixes by immunopurification and ultraperformance liquid chromatography-tandem mass spectrometry. *Anal. Chem.*, 84, 5103, 2012.
100. CDC (Center for Disease Control and Prevention). *Botulism in the United States (1899–1996). Handbook for Epidemiologists, Clinicians and Laboratory Workers.* U.S. Department of Health and Human Services, CDC, Atlanta, GA, 1998.
101. CFSAN (Center for Food Safety and Applied Nutrition). *Bacteriological Analytical Manual (BAM).* U.S. Food and Drug Administration, Washington, DC, 2001.
102. Chun, C.L. et al. Prevalence of toxin-producing *Clostridium botulinum* associated with the macroalga *Cladophora* in three Great Lakes: Growth and management. *Sci. Total Environ.*, 511, 523–9, 2015.
103. Wohlfarth, K. et al. Foodborne botulism in 2 adults not treated by antiserum. *Toxicon*, 93, S65, 2015.
104. Cai, S., Lindo, P., Park, J.B., Vasa, K. and Singh, B.R. The identification and biochemical characterization of drug-like compounds that inhibit botulinum neurotoxin serotype A endopeptidase activity. *Toxicon*, 55, 818–26, 2010.
105. Capkova, K., Hixon, M.S., Pellett, S., Barbieri, J.T., Johnson, E.A. and Janda, K.D. Benzylidene cyclopentenediones: First irreversible inhibitors against botulinum neurotoxin A's zinc endopeptidase. *Bioorg. Med. Chem. Lett.*, 20, 206–8, 2010.
106. Eichhorn, T., Dolimbek, B.Z., Deeg, K., Efferth, T. and Atassi, M.Z. Inhibition in vivo of the activity of botulinum neurotoxin A by small molecules selected by virtual screening. *Toxicon*, 60, 1180–90, 2012.
107. Kumar, G., Agarwal, R. and Swaminathan, S. Discovery of a fluorine class of compounds as inhibitors of botulinum neurotoxin serotype E by virtual screening. *Chem. Commun.*, 48, 2412–4, 2012.
108. Pirazzini, M. et al. Thioredoxin and its reductase are present on synaptic vesicles, and their inhibition prevents the paralysis induced by botulinum neurotoxins. *Cell Rep.*, 8, 1870–8, 2015.
109. Montgomery, V.A. et al. Ex vivo inhibition of *Clostridium botulinum* neurotoxin types B, C, E, and F by small molecular weight inhibitors. *Toxicon*, 8, 12–9, 2015.
110. Marinelli, S. et al. Botulinum neurotoxin type A counteracts neuropathic pain and facilitates functional recovery after peripheral nerve injury in animal models. *Neuroscience*, 171, 316–28, 2010.
111. Eleopra, R. et al. Botulinum neurotoxin serotype D is poorly effective in humans: An in vivo electrophysiological study. *Clin. Neurophysiol.*, 124, 999–1004, 2013.
112. Mika, J. et al. The effect of botulinum neurotoxin A on sciatic nerve injury-induced neuroimmunological changes in rat dorsal root ganglia and spinal cord. *Neuroscience*, 175, 358–66, 2011.
113. Cui, M., Khanijou, S., Rubino, J. and Aoki, K.R. Subcutaneous administration of botulinum toxin A reduces formalin-induced pain. *Pain*, 107, 125–33, 2004.
114. Kato, K. et al. Botulinum neurotoxin A2 reduces incidence of seizures in mouse models of temporal lobe epilepsy. *Toxicon*, 74, 109–11, 2013.
115. Broide, R.S. et al. The rat Digit Abduction Score (DAS) assay: A physiological model for assessing botulinum neurotoxin-induced skeletal muscle paralysis. *Toxicon*, 71, 18–24, 2013.

116. Zhao, R. Masayasu, H. and Holmgren, A. Ebselen: A substrate for human thioredoxin reductase strongly stimulating its hydroperoxide reductase activity and a superfast thioredoxin oxidant. *Proc. Natl. Acad. Sci. USA*, 99, 8579–84, 2002.
117. Pellett, S. et al. In vivo onset and duration of action varies for botulinum neurotoxin A subtypes 1–5. *Toxicon*, 107, 37–42, 2015.
118. Sayeed, S. et al. Beta toxin is essential for the intestinal virulence of *Clostridium perfringens* type C disease isolate CN3685 in a rabbit ileal loop model. *Mol. Microbiol.*, 67, 15–30, 2008.
119. Garcia, J.P. et al. Comparative neuropathology of ovine enterotoxemia produced by *Clostridium perfringens* type D wild-type strain CN1020 and its genetically modified derivatives. *Vet. Pathol.*, 52(3), 465–75, 2015.
120. Uzal, F.A. et al. Towards an understanding of the role of *Clostridium perfringens* toxins in human and animal disease. *Future Microbiol.*, 9, 361–77, 2014.
121. Sugimoto, N., Chen, Y.-M., Yue, S., Morihiro, L. and Lee, C.-Y. Pathodynamics of intoxication in rats and mice by enterotoxin of *Clostridium perfringens* type A. *Toxicon*, 29, 751–9, 1991.
122. Caserta, J.A., Robertson, S.L., Saputo, J., Shrestha, A., McClane, B.A. and Uzal, F.A. Development and application of a mouse intestinal loop model to study the in vivo action of *Clostridium perfringens* enterotoxin. *Infect. Immun.*, 79, 3020–7, 2011.
123. Bos, J. et al. Fatal necrotizing colitis following a foodborne outbreak of enterotoxigenic *Clostridium perfringens* type A infection. *Clin. Infect. Dis.*, 40, 78–83, 2005.
124. Fisher, D.J. et al. Dissecting the contributions of *Clostridium perfringens* type C toxins to lethality in the mouse intravenous injection model. *Infect. Immun.*, 74, 5200–10, 2006.
125. Smedley, J.G. et al. Noncytotoxic *Clostridium perfringens* enterotoxin (CPE) variants localize CPE intestinal binding and demonstrate a relationship between CPE-induced cytotoxicity and enterotoxicity. *Infect. Immun.*, 76, 3793–800, 2008.
126. Katahira, J., Inoue, N., Horiguchi, Y., Matsuda, M. and Sugimoto, N. Molecular cloning and functional characterization of the receptor for *Clostridium perfringens* enterotoxin. *J. Cell. Biol.*, 136, 1239–47, 1997.
127. Katahira, J., Sugiyama, H., Inoue, N., Horiguchi, Y., Matsuda, M. and Sugimoto, N. *Clostridium perfringens* enterotoxin utilizes two structurally related membrane proteins as functional receptors in vivo. *J. Biol. Chem.*, 272, 26652–8, 1997.
128. Vidal, J.E., McClane, B.A., Saputo, J., Parker, J. and Uzal, F.A. Effects of *Clostridium perfringens* beta-toxin on the rabbit small intestine and colon. *Infect. Immun.*, 76, 4396–404, 2008.
129. Matsuda, T. et al. Enteritis necroticans "pigbel" in a Japanese diabetic adult. *Pathol. Int.*, 57, 622–6, 2007.
130. Ma, M. et al. Synergistic effects of *Clostridium perfringens* enterotoxin and beta toxin in rabbit small intestinal loops. *Infect. Immun.*, 82, 2958–70, 2014.
131. Johannsen, U., Erwerth, W., Kunz, G. and Kohler, B. *Clostridium perfringens* type C enterotoxemia (necrotizing enteritis) in suckling pigs. 1. Study of the experimental production of disease by *Clostridium perfringens* type C poisoning and infection (experimental set-up, clinical aspects, pathological findings). *Arch. Exp. Vet. Med.*, 40, 811–25, 1986.
132. Johannsen, U., Menger, S., Erwerth, W. and Kohler, B. *Clostridium perfringens* type C enterotoxemia (necrotizing enteritis) in suckling pigs. 2. Light and electron microscopic studies of the pathology and pathogenesis of experimental *Clostridium perfringens* type C toxin poisoning. *Arch. Exp. Vet. Med.*, 40, 881–94, 1986.
133. Johannsen, U., Menger, S., Erwerth, W. and Kohler, B. *Clostridium perfringens* type C enterotoxemia (necrotizing enteritis) of suckling pigs. 3. Light and electron microscopic studies of the pathology and pathogenesis of experimental *Clostridium perfringens* type C infection. *Arch. Exp. Vet. Med.*, 40, 895–909, 1986.
134. Schumacher, V.L., Martel, A., Pasmans, F., Van Immerseel, F. and Posthaus, H. Endothelial binding of beta toxin to small intestinal mucosal endothelial cells in early stages of experimentally induced *Clostridium perfringens* type C enteritis in pigs. *Vet. Pathol.*, 50, 626–9, 2013.
135. Garcia, J.P. et al. The effect of *Clostridium perfringens* type C strain CN3685 and its isogenic beta toxin null mutant in goats. *Vet. Microbiol.*, 157, 412–9, 2012.
136. Uzal, F.A. et al. Development and application of new mouse models to study the pathogenesis of *Clostridium perfringens* type C enterotoxemias. *Infect. Immun.*, 77, 5291–9, 2009.

137. Hauschild, A.H.W., Hilsheimer, R. and Rogers, C.G. Experimental enteritis with food poisoning and classical strains of *Clostridium perfringens* type A in lambs. *J. Infect. Dis.*, 117, 379–86, 1967.
138. Niilo, L. Mechanism of action of the enteropathogenic factor of *Clostridium perfringens* type A. *Infect. Immunol.*, 3, 100–6, 1970.
139. Niilo, L. *Clostridium perfringens* type C enterotoxemia. *Can. Vet. J.*, 29, 658–64, 1988.
140. Uemura, T., Sakaguchi, G., Ito, T., Okazawa, K. and Sakais, S. Experimental diarrhea in cynomolgus monkeys by oral administration with *Clostridium perfringens* type A viable cells or enterotoxin. *Jpn. J. Med. Sci. Biol.*, 28, 165–77, 1975.
141. Hutton, M.L., Mackin, K.E., Chakravorty, A. and Lyras, D. Small animal models for the study of *Clostridium difficile* disease pathogenesis. *FEMS Microbiol. Lett.*, 352, 140–9, 2014.
142. Best, E.L., Freeman, J. and Wilcox, M.H. Models for the study of *Clostridium difficile* infection. *Gut Microbes*, 3, 145–67, 2012.
143. Lawley, T.D. and Young, V.B. Murine models to study *Clostridium difficile* infection and transmission. *Anaerobe*, 24, 94–7, 2013.
144. Hamm, E.E., Voth, D.E. and Ballard, J.D. Identification of *Clostridium difficile* toxin B cardiotoxicity using a zebrafish embryo model of intoxication. *Proc. Natl. Acad. Sci. USA*, 103, 14176–81, 2006.
145. Lanis, J.M., Barua, S. and Ballard, J.D. Variations in TcdB activity and the hypervirulence of emerging strains of *Clostridium difficile. PLoS Pathog.*, 6, 1001–61, 2010.
146. Bartlett, J.G., Onderdonk, A.B., Cisneros, R.L. and Kasper, D.L. Clindamycin-associated colitis due to a toxin-producing species of *Clostridium* in hamsters. *J. Infect. Dis.*, 136, 701–5, 1977.
147. Price, A., Larson, H. and Crow, J. Morphology of experimental antibiotic-associated enterocolitis in the hamster: A model for human pseudomembranous colitis and antibiotic-associated diarrhea. *Gut*, 20, 467–75, 1979.
148. Pawlowski, S.W., Calabrese, G. and Kolling, G.L. Murine model of *Clostridium difficile* infection with aged gnotobiotic C57BL/6 mice and a BI/NAP1 strain. *J. Infect. Dis.*, 202, 1708–12, 2010.
149. Reeves, A.E., Koenigsknecht, M.J., Bergin, I.L. and Young, V.B. Suppression of *Clostridium difficile* in the gastrointestinal tracts of germfree mice inoculated with a murine isolate from the family Lachnospiraceae. *Infect Immun.*, 80, 3786–94, 2012.
150. Chen, X. et al. A mouse model of *Clostridium difficile*-associated disease. *Gastroenterology*, 135, 1984–92, 2008.
151. Theriot, C.M., Koumpouras, C.C., Carlson, P.E., Bergin, I.I., Aronoff, D.M. and Young, V.B. Cefoperazone-treated mice as an experimental platform to assess differential virulence of *Clostridium difficile* strains. *Gut Microbes*, 2, 326–34, 2011.
152. Buffie, C.G., Jarchum, I. and Equinda, M. Profound alterations of intestinal microbiota following a single dose of clindamycin results in sustained susceptibility to *Clostridium difficile*-induced colitis. *Infect. Immun.*, 80, 62–73, 2012.
153. Pellett, S., Tepp, W.H., Clancy, C.M., Borodic, G.E. and Johnson, E.A. A neuronal cell-based botulinum neurotoxin assay for highly sensitive and specific detection of neutralizing serum antibodies. *FEBS Lett.*, 581, 4803–8, 2007.
154. Stahl, A.M., Ruthel, G., Torres-Melendez, E., Kenny, T.A., Panchal, R.G. and Bavari, S. Primary cultures of embryonic chicken neurons for sensitive cell-based assay of botulinum eurotoxin: Implications for therapeutic discovery. *J. Biomol. Screen.*, 12, 370–7, 2007.
155. Eubanks, L.M. et al. An in vitro and in vivo disconnect uncovered through high-throughput identification of botulinum neurotoxin A antagonists. *Proc. Natl. Acad. Sci. USA*, 104, 2602–7, 2007.
156. Hale, M., Oyler, G., Swaminathan, S. and Ahmed, S.A. Basic tetrapeptides as potent intracellular inhibitors of type A botulinum neurotoxin protease activity. *J. Biol. Chem.*, 286, 1802–11, 2011.
157. Lyman, M.E., Tuznik, K.M. and McNutt, P. Methods to produce, culture and evaluate embryonic stem-cell-derived neurons as a research tool for botulinum toxin and the black widow spider venom latrotoxin. *Toxicon*, 68, 91, 2013.
158. Rasetti-Escargueil, C., Machado, C.B., Preneta-Blanc, R., Fleck, R.A. and Sesardic, D. Enhanced sensitivity to botulinum type A neurotoxin of human neuroblastoma SH-SY5Y cells after differentiation into mature neuronal cells. *Botulinum J.*, 2, 30, 2011.
159. Tegenge, M.A., Bohnel, H., Gessler, F. and Bicker, G. Neurotransmitter vesicle release from human model neurons (NT2) is sensitive to botulinum toxin A. *Cell Mol. Neurobiol.*, 32, 1021–9, 2012.

160. Hatheway, C.L. Botulism. In: *Laboratory Diagnosis of Infectious Diseases: Principles and Practice*, Balows, A., Hausler, W.H., Ohashi, M. and Turano, M.A. (Eds.), Springer-Verlag, New York, pp. 111–33, 1988.
161. Whitemarsh, R.C.M. et al. Model for studying *Clostridium botulinum* neurotoxin using differentiated motor neuron-like NG108-15 cells. *Biochem. Biophys. Res. Commun.*, 427, 426–30, 2012.
162. Fonfria, E. et al. Rodent- and human-derived cell-based models for the study of botulinum neurotoxin A1 (BoNT/A1). *Toxicon*, 93, S27, 2015.
163. Delaflotte, S., Carré, D., Krupp, J. and Huchet, M. Development of an enriched and highly BoNT/A1-sensitive motor neuron culture. *Toxicon*, 93, S20, 2015.
164. Borrmann, E., Schulze, F., Cussler, K., Hänel, I. and Diller, R. Development of a cell culture assay for the quantitative determination of vaccination-induced antibodies in rabbit sera against *Clostridium perfringens* epsilon toxin and *Clostridium novyi* alpha toxin. *Vet. Microbiol.*, 114, 41–50, 2006.
165. Allaart, J.G., van Asten, A.J., Vernooij, J.C. and Grone, A. Beta2 toxin is not involved in in vitro cell cytotoxicity caused by human and porcine *cpb2*-harbouring *Clostridium perfringens*. *Vet. Microbiol.*, 171, 132–8, 2014.
166. Vidal, J.E., Saputo, M., Garcia, J., Uzal, F.A. and McClane, B.A. Evidence that the Agr-like quorum sensing system regulates the toxin production, cytotoxicity and pathogenicity of *Clostridium perfringens* type C isolate CN3685. *Mol. Microbiol.*, 83, 179–94, 2012.
167. Yasugi, M. et al. In vitro cytotoxicity induced by *Clostridium perfringens* isolate carrying a chromosomal *cpe* gene is exclusively dependent on sporulation and enterotoxin production. *Microb. Pathog.*, 85, 1–10, 2015.
168. Warny, M. et al. p38 MAP kinase activation by *Clostridium difficile* toxin A mediates monocyte necrosis, IL-8 production, and enteritis. *J. Clin. Invest.*, 105, 1147–56, 2000.
169. Mahida, S., Makh, S., Hyde, T., Gray, S.P. and Borriello, Y.R. Effect of *Clostridium difficile* toxin A on human intestinal epithelial cells: Induction of interleukin 8 production and apoptosis after cell detachment. *Gut*, 38, 337–47, 1996.
170. Hecht, G., Pothoulakis, C., LaMont, J.T. and Madara, J.L. *Clostridium difficile* toxin A perturbs cytoskeletal structure and tight junction permeability of cultured human intestinal epithelial monolayers. *J. Clin. Invest.*, 82, 1516–24, 1988.
171. Miller, P.D., Pothoulakis, C., Baeker, T.R., LaMont, J.T. and Rothstein, T.L. Macrophage-dependent stimulation of T cell depleted spleen cells by *Clostridium difficile* toxin A and calcium ionophore. *Cell Immunol.*, 126, 155–63, 1990.
172. Flegel, W.A. et al. Cytokine response by human monocytes to *Clostridium difficile* toxin A and toxin B. *Infect. Immun.*, 5, 3659–66, 1991.
173. Linevsky, J.K. et al. IL-8 release and neutrophil activation by *Clostridium difficile* toxin-exposed monocytes. *Am. J. Physiol.*, 273, G1333–40, 1997.
174. Eichel-Streiber, C., Laufenberg-Feldmann, R., Sartingen, S., Schulze, J. and Sauerborn, M. Cloning of *Clostridium difficile* toxin B gene and demonstration of high N-terminal homology between toxin A and B. *Med. Microbiol. Immunol.*, 179, 271–9, 1990.

10

Enterococcus

Dongyou Liu

CONTENTS

10.1 Introduction

First observed by Thiercelin in 1899 as saprophytic Gram-positive diplococcus in the gastrointestinal (GI) tract of humans, *Enterococcus* (initially named "Enterocoque" to emphasize its morphology and its intestinal origin) was found to be associated with diarrhea, septicemia, and acute endocarditis. In the early 1900s, similar organisms (then known as *Streptococcus faecalis* and *Streptococcus faecium*) were described. In 1937, a classification scheme that separated streptococci into four groups—pyogenic, viridans, lactic, and enterococcus—was proposed, with the "enterococcal group" covering streptococci that grow between 10°C and 45°C, at pH 9.6, in 6.5% NaCl, and survive at 60°C for 30 min. This classification scheme correlated with a serological scheme developed by Lancefield in the 1930s, in which the "enterococcal group" react with group D antisera (thus the so-called group D streptococci), while nonenterococcal streptococci react with antiserum groups A, B, C, E, F, or G. Following the description of additional members in the enterococcal group, a proposal to create the genus *Enterococcus* was made in 1970 and formally accepted in 1984 upon DNA–DNA and DNA–rRNA hybridization analyses [1].

10.1.1 Classification, Morphology, and Genomics

10.1.1.1 Classification

Taxonomically, the genus *Enterococcus* (from Greek *éntero*, "intestine" and *coccos*, "granule") is classified in the family Enterococcaceae, order Lactobacillales, class Bacilli, phylum Firmicutes, and domain Bacteria. As one of the four genera (i.e., *Enterococcus*, *Melissococcus*, *Tetragenococcus*, and *Vagococcus*) within the family Enterococcaceae, the genus *Enterococcus* encompasses over 50 recognized species (*Enterococcus alcedinis*, *Enterococcus aquimarinus*, *Enterococcus asini*, *Enterococcus avium*, *Enterococcus caccae*, *Enterococcus camelliae*, *Enterococcus canintestini*, *Enterococcus canis*, *Enterococcus casseliflavus*, *Enterococcus cecorum*, *Enterococcus columbae*, *Enterococcus devriesei*, *Enterococcus diestrammenae*, *Enterococcus dispar*, *Enterococcus durans*, *Enterococcus eurekensis*, *Enterococcus faecalis*, *Enterococcus faecium*, *Enterococcus gallinarum*, *Enterococcus gilvus*, *Enterococcus haemoperoxidus*, *Enterococcus hermanniensis*, *Enterococcus hirae*, *Enterococcus italicus*, *Enterococcus lactis*, *Enterococcus lemanii*, *Enterococcus malodoratus*, *Enterococcus moraviensis*, *Enterococcus mundtii*, *Enterococcus olivae*, *Enterococcus pallens*, *Enterococcus phoeniculicola*, *Enterococcus plantarum*, *Enterococcus pseudoavium*, *Enterococcus quebecensis*, *Enterococcus raffinosus*, *Enterococcus ratti*, *Enterococcus rivorum*, *Enterococcus rotai*, *Enterococcus saccharolyticus*, *Enterococcus silesiacus*, *Enterococcus solitarius*, *Enterococcus sulfureus*, *Enterococcus termitis*, *Enterococcus thailandicus*, *Enterococcus ureasiticus*, *Enterococcus ureilyticus*, *Enterococcus viikkiensis*, *Enterococcus villorum*, and *Enterococcus xiangfangensis*). On the bases of their varied tropisms, habitats, metabolic, and phenotypic characteristics, members of the genus *Enterococcus* may be separated into at least five groups, namely, *E. faecalis* group (*E. faecalis*, *E. haemoperoxidus*, *E. moraviensis*, *E. silesiacus*, *E. termitis*, and *E. caccae*), *E. faecium* group (*E. faecium*, *E. durans*, *E. hirae*, *E. mundtii*, *E. villorum*, *E. canis*, *E. ratti*, *E. asini*, *E. phoeniculicola*, *E. canintestini*, and *E. thailandicus*), *E. avium* group (*E. avium*, *E. pseudoavium*, *E. malodoratus*, *E. raffinosus*, *E. gilvus*, *E. pallens*, *E. hermanniensis*, *E. devriesei*, and *E. viikkiensis*), *E. gallinarum* group (*E. gallinarum* and *E. casseliflavus*), and *E. cecorum* group (*E. cecorum* and *E. columbae*) in addition to ungrouped *E. saccharolyticus*, *E. aquimarinus*, *E. sulfurous*, *E. dispar*, *E. italicus*, and *E. camelliae* [2]. Of these, *E. faecalis*, *E. faecium*, *E. durans*, *E. mundtii*, *E. avium*, *E. raffinosus*, *E. gallinarum*, and *E. casseliflavus* have been linked to human diseases.

It is notable that being abundantly present as commensal organisms in the intestines of humans, *E. faecalis* and *E. faecium* are implicated in various human nosocomial infections [e.g., urinary tract infections (UTIs), endocarditis, bacteremia, neonatal infections, central nervous system (CNS) infections, and abdominal and pelvic infections] and have accounted for 90%–95% and 5%–10% of clinical isolations, respectively, until the mid-1990s [3]. With the emergence and spread of antibiotic resistance (especially vancomycin and ampicillin) in recent decades, the proportion of *E. faecium*-related infections has shown a marked increase [1]. Other *Enterococcus* species (e.g., *E. avium*, *E. gallinarum*, *E. casseliflavus*, *E. hirae*, *E. mundtii*, and *E. raffinosus*, along with *E. sanguinicola*, *E. gilvus*, *E. pallens*, and *E. canintestini*) are occasionally implicated in human infections, of which *E. gallinarum* and *E. casseliflavus* are noted for their intrinsic resistance to vancomycin, an antibiotic used from the mid-1980s to treat the aminoglycoside-resistant enterococcal infections. In addition, *E. caccae* has been detected in human feces, while the remaining *Enterococcus* species commonly occur in animals, insects, traditional fermented food and dairy products, plants, soil, and water [2].

10.1.1.2 Morphological and Biochemical Characteristics

Enterococci are spherical or ovoid in shape and appear in pairs (diplococci) or short chains. As Gram-positive, non-spore-forming facultative anaerobes (with the capability of cellular respiration in both oxygen-rich and oxygen-poor environments) and obligately fermentative chemoorganotrophs, enterococci grow optimally at 35°C, with a growth range from 10°C to 45°C. In addition, they grow in broth containing 6.5% NaCl, hydrolyze esculin in the presence of 40% bile salts, and withstand heating at 60°C for 30 min [1]. Although enterococci are considered catalase-negative, some species may appear catalase-positive with weak effervescence (pseudocatalase). Being homofermentative, enterococci produce lactic acid as the end product of glucose fermentation without gas. On sheep's blood agar, enterococci typically exhibit gamma-hemolysis. Apart from *E. cecorum*, *E. columbae*, *E. pallens*, and *E. saccharolyticus*, *E. devriesei* (variable among strains), *E. canintestini*, *E. termitis*, and *E. viikiensis*, most enterococci are capable of hydrolyzing pyrrolidonyl-β-naphthylamide and producing leucine aminopeptidase. While some species (e.g., *E. gallinarum* and *E. casseliflavus*) are motile, others (e.g., *E. asini* and *E. phoeniculicola*) are not. Some species, especially those found among plants (e.g., *E. sulfureus*, *E. casseliflavus*, and *E. mundtii*), produce yellow pigments [1].

10.1.1.3 Genomics

Phylogentic comparison of ~1.4 kb of the 16S rRNA gene demonstrates a closer genetic relatedness of *Enterococcus* to *Vagococcus*, *Tetragenococcus*, and *Carnobacterium* than to *Streptococcus* and *Lactococcus*.

Multilocus sequence typing (MLST) targeting the internal regions of seven housekeeping genes (*gdh*, *gyd*, *pstS*, *gki*, *aroE*, *xpt*, and *yqiL*) permits assignment of *E. faecalis* strains into sequence types (ST). MLST analysis of *E. faecalis* strains isolated worldwide from the early 1900s to 2006 suggested that acquired antibiotic resistance is enriched in the clonal complex 2 (CC2), CC8, and CC9 lineages, whereas examination of more recently isolated strains from Europe (between 2006 and 2009) indicated that multidrug resistance is enriched in the CC2, CC16, and CC87 lineages. Similarly, MLST analysis of internal gene fragments (between 395 and 583 nucleotides [nt] in size) of seven housekeeping genes (*atpA*, *ddl*, *gdh*, *purK*, *gyd*, *pstS*, and *adk*) in *E. faecium* resulted in the identification of a large cluster of clinical *E. faecium* isolates that was first termed lineage C1 and later renamed clonal complex 17 (CC17), after ST17, which are resistant to ampicillin and ciprofloxacin and are enriched for several genes with putative roles in virulence (e.g., the large surface protein *esp*, and carbohydrate metabolism), with the presence of IS*16* being the most prominent marker. In addition, Bayesian analysis of genetic population structure (BAPS) analysis of 491 distinct STs found among 1720 *E. faecium* isolates identified 13 groups of related *E. faecium* strains. Phylogenetic analysis based on concatenated MLST gene sequences of isolates contained in the two largest BAPS groups (BAPS 2–1 and BAPS 3–3) revealed that currently circulating clinical isolates belong to three different lineages (lineage-17, lineage-18, and lineage-78), which stem from at least three different ancestral strains and have independently acquired genes that characterize clinical isolates through convergent evolution [4].

The genomes of enterococci are found to range from 2.7 to 3.6 Mb across species, with the G + C content of DNA between 37 and 45 mol% [5]. Examination of *E. faecalis* V583 genome (3.36 Mb)—a bloodstream infection (BSI)-derived, vancomycin-resistant (vanB phenotype) ST6/CC2 lineage strain—revealed the presence of a large amount of mobile DNA (accounting for >25% of the genomic content), including seven predicted prophages, multiple integrated plasmids, IS elements, and genomic islands (including the pathogenicity-associated island or PAI), a *vanB*-type transposon that confers vancomycin resistance, and three extrachromosomal plasmids [pTEF1 (66.3 kb), pTEF2 (57.7 kb), and pTEF3 (18.0 kb)] [6,7]. This highlights the propensity of enterococci to acquire and disseminate mobile elements, such as those that encode antibiotic resistance genes. Other notable traits identified include a capsule, a novel adhesin termed enterococcal surface protein (Esp), a bile acid hydrolase, and an operon for the enterococcal cytolysin [8].

10.1.2 Biology and Epidemiology

Enterococci are common colonizers of the GI tract of humans and other mammals, birds, reptiles, and insects, and are routinely isolated from soil and sediments, beach sand, aquatic and terrestrial vegetation, and ambient waters (rivers, streams, and creeks) [9,10].

Although first implicated in infective endocarditis (IE) in 1899, *Enterococcus* emerged in the 1970s as a leading cause of multidrug-resistant (MDR), hospital-acquired infections. Currently, *Enterococcus* [dominated by *E. faecalis* (80%–90%) and *E. faecium* (10%–15%)] represents the third most common nosocomial pathogen (12% of all hospital infections), causing IE, catheter-related BSIs, UTIs, wound infections, endophthalmitis, and peritonitis.

Patient's own endogenous flora was once thought as the main source of enterococcal infection; however, detailed studies pointed to the transmission of pathogenic enterococci among patients in hospital settings through the hands of health-care workers (e.g., direct inoculation onto intravenous or urinary catheters). Enterococci can persist on hands for as long as 60 min after inoculation onto hands, and as long as 4 months on inanimate surfaces.

Risk factors for MDR enterococcal infections include patients with neutropenia, undergoing transplantation, chronic renal failure, ICU stay, prior antibiotic use (including cephalosporins), bladder catheterization, expression of cytolysin as an enterococcal virulence determinant, and prolonged hospitalization.

The emergence and dissemination of MDR enterococci (especially *E. faecium*) exhibiting resistance to ampicillin, vancomycin, and aminoglycosides have reduced the number of therapeutic options. There is evidence that resistance is also emerging to newer agents used to treat vancomycin-resistant enterococci (VRE) infections (e.g., linezolid, quinupristin/dalfopristin, and daptomycin) [1,11–13].

10.1.3 Clinical Features and Pathogenesis

As commensals of human GI tract, enterococci do not usually cause disease inside the intestine. However, once outside of the gut, they may become pathogenic, inducing a variety of clinical syndromes, ranging from UTIs, intra-abdominal, pelvic, and soft tissue infections, bacteremia, endocarditis, to other uncommon infections [3].

UTIs are the most common type of enterococcal infections that are likely acquired in hospital or long-term care settings (e.g., ICU). Lower UTIs (e.g., cystitis, prostatitis, and epididymitis) frequently occur in older men, and upper UTIs in older men may lead to bacteremia [3].

Intra-abdominal, pelvic, and soft tissue infections are the next common enterococcal infections. Peritonitis (an infection of the abdominal lining) may occur in conjunction with liver cirrhosis or in patients who receive chronic peritoneal dialysis. In addition, enterococci may be detected in cultures from decubiti and foot ulcers [3].

Bacteremia and endocarditis are the more common and serious manifestations of enterococcal infections, with between 5% and 15% of cases of infectious endocarditis (heart valve infections) attributing to enterococci (particularly *E. faecalis*). Enterococcal endocarditis typically occurs in older persons and usually evolves from bacteremia, which in turn originates from the genitourinary or GI tract infections [3].

Other uncommon infections due to enterococci include meningitis, hematogenous osteomyelitis, septic arthritis, pneumonia, and chronic bacterial prostatitis. Enterococcal meningitis usually occurs as a rare complication of neurosurgery [3].

Enterococci are remarkably thermotolerant (growing at 10°C–45°C), pH-tolerant (pH 4.8–9.6), salt-resistant (up to 28% NaCl), and capable of surviving at 60°C for 30 min and extended desiccation. These properties enable the bacterium to flourish in a multitude of environments, including the human gut, and abiotic surfaces in hospitals (e.g., bedrails and medical station keyboards). The ability of *Enterococcus* to enter into a viable but noncultivable state and form a biofilm also facilitates its environmental persistence and provides protection against the host immune response and antibiotic intervention.

Enterococci are known to encode a number of virulence-associated proteins that are essential for their pathogenicity. These include Clp ATPase (encoded by the *clpB* gene, with critical role in protein folding, assembly, and degradation of proteins), methionine sulfoxide reductase A (involved in antioxidant repair), elatinase (GelE, a secreted zinc metalloproteinase), hyaluronidase, cytotoxin (a secreted two-peptide lytic toxin encoded on a pheromone-responsive plasmid on pathogenicity island), enterococcal surface protein (Esp, a surface adhesin that controls adherence to host tissues), the protein Ace (a collagen- and laminin-binding microbial surface component recognizing adhesive matrix molecules or MSCRAMM), and enterococcal aggregation substance (AS, a pheromone-responsive, surface-bound, plasmid-encoded protein), etc. [14–20]. In particular, proteases and enzymes secreted by enterococci not only degrade host tissues into peptide nutrients for the bacteria, but also cause direct destruction of both cells and the extracellular matrix in the host tissue, activate host proteolytic cascades, and interfere with host inflammatory processes [21,22].

10.1.4 Identification

Enterococci grow slowly in culture media and require further conventional biochemical tests or commercial test systems [e.g., the BD GeneOhm VanR (BD Diagnostics, Spark, MD) and Xpert *vanA/vanB* (Cepheid, Sunnydale, CA)] for identification. As many enterococcal species vary by only one phenotypic trait, the biochemical approach is far from being straightforward. For VRE, chromogenic media (e.g., CHROM-agar, chromID, and Spectra VRE media) reduce turnaround times through early visual identification of colonies, although use of standard broth macrodilution or disk diffusion method for vancomycin-susceptibility testing requires additional time [23].

While 16S rRNA sequence analysis is valuable for identification of many *Enterococcus* species (using 97% identity as the threshold), it does not provide adequate resolution between *E. casseliflavus* and *E. gallinarum* (which share 99.9% identity). Therefore, other molecular methods are required for routine identification of enterococcal isolates. These include amplification and sequencing of the domain V of the 23S rRNA gene, rRNA or tRNA intergenic spacers, the D-ala:D-ala ligase genes (*ddl*), the manganese-dependent superoxide dismutase (*sodA*) genes, the chaperonin 60 (*cpn60*) gene, the *Enterococcus* protein A (*efaA*) gene, genes encoding the RNA polymerase α-subunit (*rpoA*), the phenylalanyl-tRNA synthase (*pheS*), and the elongation factor Tu (*tufA*), in addition to the application of ribotyping, repetitive extragenic palindromic PCR (REP-PCR) or BOX-PCR, pulsed-field gel electrophoresis (PFGE), and multilocus sequence typing (MLST) [3].

PFGE represents the current standard in the clinical identification of *Enterococcus* spp. and strain typing, while ribotyping techniques allow accurate discrimination among species with reduced cost. Interestingly, MLST demonstrates an accuracy equivalent to that of PFGE for the identification of organisms to the subspecies level. de Been et al. [24] developed a core genome MLST (cgMLST) scheme for *E. faecium* that is capable of distinguishing between epidemiologically related and unrelated isolates, even between those of identical ST. Hullahalli et al. [25] showed that clustered, regularly interspaced short palindromic repeats (CRISPR) 2 analysis is an inexpensive alternative to MLST for assessing clonality among *E. faecalis* isolates, and can be used in conjunction with MLST to identify recombination events occurring between STs. Other novel approaches for enterococcal identification include quantitative PCR (qPCR) targeting the *esp* gene of *E. faecium* (esp_{fm}), DNA microarrays, and use of bacteriophages specific to certain *Enterococcus* strains.

10.1.5 Treatment and Prevention

For susceptible *Enterococcus* isolates, β-lactam agents (ampicillin and penicillin/aminopenicillin) are the drugs of choice for treating monomicrobial enterococcal infections (e.g., UTIs and nonendocarditis bacteremia). Treatment of polymicrobial infections (e.g., skin and subcutaneous infections and intraabdominal or pelvic infections) relies on a combination of ampicillin and other broad-spectrum antibiotics or a β-lactamase inhibitor (e.g., ampicillin-clavulanic acid or piperacillin-tazobactam). A glycopeptide (vancomycin or teicoplanin) may be employed as a single agent to treat simple enterococcal infections in patients with a serious allergy to penicillins. A combination of a glycopeptide and an aminoglycoside (gentamicin or streptomycin) may be considered for patients with serious allergies. Additionally, nitrofurantoin may be used to treat lower-tract urinary infections [3].

As *Enterococcus* species are intrinsically resistant to many antimicrobial agents (e.g., cephalosporins, clindamycin, semisynthetic penicillinase-stable penicillins, and aminoglycosides) and have the capacity to acquire resistance genes and mutations, treatment of severe infections with VRE or MDR enterococci is challenging [1,11,12,26,27]. Interestingly, most *E. faecalis* isolates expressing vancomycin resistance remain susceptible to ampicillin, whereas many *E. faecium* isolates are resistant to both. Quinupristin/dalfopristin [a combination agent consisting of streptogramin A (70% dalfopristin) and B (30% quinupristin) effective against *E. faecium*] and linezolid are FDA-approved agents for VRE. With intrinsic activity against enterococci, nitrofurantoin, fosfomycin, and doxycycline are potential oral options for treating simple VRE infections. Other antibiotic alternatives include daptomycin (DAP, a lipopeptide antibiotic), tigecycline, lipoglycopeptides (oritavancin, telavancin, and dalvabancin), and rifampicin [28,29].

The intervention measures for controlling the spread of VRE in health-care settings range from (1) active periodic surveillance (cultures or molecular tests) of highest risk carriage patients, (2) decontamination of the hands of health-care workers before and after all patients contact, (3) adherence to barrier precautions (gloves and gowns) and hand antisepsis after glove removal, to (4) thorough terminal cleaning of rooms occupied by patient with VRE and daily cleaning of high-touch items such as bedside rails, tables, toilets, and handles [3].

10.2 Laboratory Models

10.2.1 Animal Models

Rabbits, rats, and mice have proven to be useful for the investigation in the molecular mechanisms of an antibiotic therapy against enterococcal infections [30].

Utilizing a rabbit subdermal abscess model, Frank et al. [31] identified three *in vivo*-activated genes in the *E. faecalis* OG1RF chromosome that encode glutamate 5-kinase (*proB* [EF0038]), the transcriptional regulator EbrA (*ebrA* [EF1809]), and the membrane metalloprotease Eep (*eep* [EF2380]). They showed that the ΔebrA strain was fully virulent, the ΔproB strain was slightly attenuated, and the Δeep strain was severely attenuated in a rabbit model of endocarditis. In a separate study, Frank et al. [32] employed a rabbit model of endocarditis to test *E. faecalis* strains with transposon insertions or in-frame deletions in biofilm-associated loci: *ahrC*, *argR*, *atlA*, *opuBC*, *pyrC*, *recN*, and *sepF*. They noted that while only the *ahrC* mutant was significantly attenuated in endocarditis, the transcriptional regulator AhrC and the protease Eep were also required for full virulence in murine catheter-associated urinary tract infection (CAUTI), confirming AhrC and Eep as enterococcal biofilm-associated virulence factors. More recently, Frank et al. [33] used Wistar rat model of acute foreign body osteomyelitis to examine the relationship between biofilm formation and development of antimicrobial resistance, and found that surface colonization alone is sufficient for *E. faecalis* cells to acquire the biofilm antimicrobial resistance phenotype.

Eguchi et al. [34] assessed the pharmacodynamics of SMP-601 (also known as PTZ601, PZ-601, or SM-216601, a novel parenteral carbapenem with potent activity against MDR Gram-positive pathogens) against vancomycin-resistant *E. faecium* (VREF) in neutropenic murine thigh infection model, and found that SMP-601 had a sufficient therapeutic effect against VREF infections at relatively low exposure conditions. In a rat model of plastic-catheter-induced left-sided enterococcal endocarditis, DAP monotherapy (20 mg/kg twice daily) demonstrated superior efficacy for the treatment of penicillin-resistant *E. faecalis* in comparison with vancomycin. Similarly, DAP (12 mg/kg every 8 h) was effective against vancomycin-resistant and gentamicin-susceptible *E. faecium* [35,36].

Given the evolutionary conservation of both ancient innate host defenses and bacterial virulence mechanisms, invertebrate hosts (e.g., the greater wax moth *Galleria mellonella*, the free-living bacteriovorus nematode *Caenorhabditis elegans*, and the common fruit fly *Drosophila melanogaster*) offer valuable models to study innate immunity and host–pathogen interactions and help unravel the pathobiologic details in enterococcal infections [37–41].

Infection of the greater wax moth *G. mellonella* larvae with *E. hirae* led to the identification of several novel enterococcal virulence factors in the insect hemolymph (e.g., GelE, Clp ATPase). Because of its capacity to escape from the intestinal lumen to the body cavity during the larval to pupal transition, *E. hirae* activates the host defense responses, yielding important clues in the pathogenic mechanisms of enterococcal infection. As *G. mellonella* tolerates relatively high temperatures (37°C or higher), this simple nonmammalian model is clearly more relevant to the study of human entercoccal infections than *C. elegans* and *D. melanogaster* (maximum 25°C) [39].

The small roundworm *C. elegans* possesses 20 nonrenewing intestinal epithelial cells (IECs) that play a key role in the mediation of intestinal immunity. As *C. elegans* IECs are morphologically similar to mammalian IECs, including a "brush border" of microvilli, this worm represents an excellent model for the study of gut immunity during enterococcal infections. Additionally, the relatively small size and the ability to survive in liquid culture make *C. elegans* a cost-effective model for disease investigation including enterococcal infections [42,43].

The common fruit fly *D. melanogaster* allows natural colonization of *E. faecalis* in the intestine and thus offers a useful model for investigations related to its biology, immunity, and pathogenesis [44]. Indeed, examination of *E. faecalis* quorum regulatory system genes *LrgAB* and *SprE*, and bacteriocin EF1097 in *D. melanogaster* confirmed their role in enhancing *E. faecalis* infectivity and toxicity [45].

10.2.2 *In Vitro* Models

Simulated endocardial vegetations (SEVs) prepared from enterococcal bacteria offer a useful model for assessing the efficacy of antibiotic regimens. For example, DAP at dose equivalent to 6–8 mg/kg/day was shown to exert rapid bactericidal activity at 8 h against a VREF isolate using a SEV model. Similarly, DAP at dose of 10–12 mg/kg was sufficient to kill both of vancomycin-resistant *E. faecalis* and *E. faecium* at 96 h [46].

10.3 Conclusion

The genus *Enterococcus* comprises over 50 species of Gram-positive, non-spore-forming bacteria that mostly occur as commensal organisms in the GI tract of mammals and birds. However, when enterococci escape from the gut and establish in other body sites, they may cause a diverse range of clinical diseases, including UTIs, intra-abdominal, pelvic, and soft tissue infections, bacteremia, endocarditis, and other uncommon infections. Further, with the ability to acquire and develop multidrug resistance traits, enterococci have become a serious problem in health-care settings in recent decades, not only putting millions of people at risk, but also exerting a significant economic burden on the society. In order to devise improved treatment and control strategies against enterococcal infections, it is important to clearly define the molecular basis of disease progression and host–microbe interactions. Toward this end, application of various laboratory models (both vertebrate and invertebrate; *in vivo* and *in vitro*) is invaluable.

REFERENCES

1. Cetinkaya Y, Falk P, Mayhall CG. Vancomycin-resistant enterococci. *Clin Microbiol Rev* 2000;13(4):686–707.
2. Byappanahalli MN, Nevers MB, Korajkic A, Staley ZR, Harwood VJ. Enterococci in the environment. *Microbiol Mol Biol Rev* 2012;76(4):685–706.
3. Vu J, Carvalho J. *Enterococcus*: review of its physiology, pathogenesis, diseases and the challenges it poses for clinical microbiology. *Front Biol* 2011;6:357.
4. Palmer KL, van Schaik W, Willems RJL, Gilmore MS. Enterococcal genomics. In: Gilmore MS, Clewell DB, Ike Y, Shankar N, editors. *Enterococci: From Commensals to Leading Causes of Drug Resistant Infection*. Boston: Massachusetts Eye and Ear Infirmary, 2014.
5. Bourgogne A, et al. Large scale variation in *Enterococcus faecalis* illustrated by the genome analysis of strain OG1RF. *Genome Biol* 2008;9:R110.
6. Shankar N, Baghdayan AS, Gilmore MS. Modulation of virulence within a pathogenicity island in vancomycin-resistant *Enterococcus faecalis*. *Nature* 2002;417:746–50.
7. Paulsen IT, et al. Role of mobile DNA in the evolution of vancomycin-resistant *Enterococcus faecalis*. *Science* 2003;299:2071–4.
8. Vebo HC, Solheim M, Snipen L, Nes IF, Brede DA. Comparative genomic analysis of pathogenic and probiotic *Enterococcus faecalis* isolates, and their transcriptional responses to growth in human urine. *PLoS One* 2010;5:e12489.
9. Fisher K, Phillips C. The ecology, epidemiology and virulence of *Enterococcus*. *Microbiology* 2009;155:1749–57.
10. Hammerum AM, Lester CH, Heuer OE. Antimicrobial-resistant enterococci in animals and meat: a human health hazard? *Foodborne Pathog Dis* 2010;7(10):1137–46.
11. Arias CA, Murray BE. Emergence and management of drug-resistant enterococcal infections. *Expert Rev Anti Infect Ther* 2008;6(5):637–55.
12. Arias CA, Murray BE. The rise of the *Enterococcus:* beyond vancomycin resistance. *Nat Rev Microbiol* 2012;10:266–78.
13. Shenoy ES, Paras ML, Noubary F, Walensky RP Hooper DC. Natural history of colonization with methicillin-resistant *Staphylococcus aureus* (MRSA) and vancomycin-resistant *Enterococcus* (VRE): a systematic review. *BMC Infect Dis* 2014;14:177.

14. Giard JC, et al. The stress proteome of *Enterococcus faecalis. Electrophoresis* 2001;22:2947–54.
15. Cox CR, Coburn PS, Gilmore MS. *Enterococcal cytolysin*: a novel two component peptide system that serves as a bacterial defense against eukaryotic and prokaryotic cells. *Curr Protein Peptide Sci* 2005;6:77–84.
16. Park SY, Kim KM, Lee JH, Seo SJ, Lee IH. Extracellular gelatinase of *Enterococcus faecalis* destroys a defense system in insect hemolymph and human serum. *Infect Immun* 2007;75:1861–9.
17. Heikens E, et al. Contribution of the enterococcal surface protein Esp to pathogenesis of *Enterococcus faecium* endocarditis. *Microbes Infect* 2011;13:1185–90.
18. Michaux C. SlyA is a transcriptional regulator involved in the virulence of *Enterococcus faecalis. Infect Immun* 2011;79:2638–45.
19. Lebreton F, et al. AsrR is an oxidative stress sensing regulator modulating *Enterococcus faecium* opportunistic traits, antimicrobial resistance, and pathogenicity. *PLoS Pathog* 2012;8:e1002834.
20. Van Tyne D, Martin MJ, Gilmore MS. Structure, function, and biology of the *Enterococcus faecalis* cytolysin. *Toxins* 2013;5:895–911.
21. Sava IG, Heikens E, Huebner J. Pathogenesis and immunity in enterococcal infections. *Clin Microb Infect* 2010;16:533–40.
22. Gaca AO, Abranches J, Kajfasz JK, Lemos JA. Global transcriptional analysis of the stringent response in *Enterococcus faecalis. Microbiology* 2012;158:1994–2004.
23. Devriese LA, Pot B, Collins MD. Phenotypic identification of the genus *Enterococcus* and differentiation of phylogenetically distinct enterococcal species and species groups. *J Appl Bacteriol* 1993;75:399–408.
24. de Been M, et al. Core genome multilocus sequence typing scheme for high-resolution typing of *Enterococcus faecium. J Clin Microbiol* 2015;53(12):3788–97.
25. Hullahalli K, Rodrigues M, Schmidt BD, Li X, Bhardwaj P, Palmer KL. Comparative analysis of the orphan CRISPR2 locus in 242 *Enterococcus faecalis* strains. *PLoS One* 2015;10(9):e0138890.
26. Crouzet L, Rigottier-Gois L, Serror P. Potential use of probiotic and commensal bacteria as non-antibiotic strategies against vancomycin-resistant enterococci. *FEMS Microbiol Lett* 2015;362(8):fnv012.
27. Luther MK, Rice LB, LaPlante KL. Ampicillin plus ceftaroline, cefepime, or ceftriaxone demonstrate equivalent activity in a high inoculum *Enterococcus faecalis* infection model. *Antimicrob Agents Chemother* 2016;60(9):3178–82.
28. Ramos MC, Grayson ML, Eliopoulos GM, Bayer AS. Comparison of daptomycin, vancomycin, and ampicillin–gentamicin for treatment of experimental endocarditis caused by penicillin-resistant enterococci. *Antimicrob Agents Chemother* 1992;36:1864–9.
29. Gallagher JC, Pérez ME, Marino EA, LoCastro LG, Abrardo LA, MacDougall C. Daptomycin therapy for vancomycin-resistant enterococcal bacteremia: a retrospective case series of 30 patients. *Pharmacotherapy* 2009;29:792–9.
30. Gutschik E. The *Enterococcus* endocarditis model in experimental animals and its relevance to human infection. *J Antimicrob Chemother* 1993;31(Suppl D):87–95.
31. Frank KL, Barnes AM, Grindle SM, Manias DA, Schlievert PM, Dunny GM. Use of recombinase-based *in vivo* expression technology to characterize *Enterococcus faecalis* gene expression during infection identifies *in vivo*-expressed antisense RNAs and implicates the protease Eep in pathogenesis. *Infect Immun* 2012;80(2):539–49.
32. Frank KL, et al. AhrC and Eep are biofilm infection-associated virulence factors in *Enterococcus faecalis. Infect Immun* 2013;81(5):1696–708.
33. Frank KL, et al. Evaluation of the *Enterococcus faecalis* biofilm-associated virulence factors AhrC and Eep in rat foreign body osteomyelitis and in vitro biofilm-associated antimicrobial resistance. *PLoS One* 2015;10(6):e0130187.
34. Eguchi K, Kanazawa K, Eriguchi Y, Ueda Y. Pharmacodynamics of SMP-601 (PTZ601) against vancomycin-resistant *Enterococcus faecium* and methicillin-resistant *Staphylococcus aureus* in neutropenic murine thigh infection models. *Antimicrob Agents Chemother* 2009;53(8):3391–8.
35. Vouillamoz J, Moreillon P, Giddey M, Entenza JM. Efficacy of daptomycin in the treatment of experimental endocarditis due to susceptible and multidrug-resistant enterococci. *J Antimicrob Chemother* 2006;58:1208–14.
36. Munita JM, Murray BE, Arias CA. Daptomycin for the treatment of bacteraemia due to vancomycin-resistant enterococci. *Int J Antimicrob Agents* 2014;44(5):387–95.

37. Gaspar F, et al. Virulence of *Enterococcus faecalis* dairy strains in an insect model: the role of *fsrB* and *gelE*. *Microbiology* 2009;155:3564–71.
38. Apidianakis Y, Rahme LG. *Drosophila melanogaster* as a model for human intestinal infection and pathology. *Dis Models Mech* 2011;4:21–30.
39. Lebreton F, et al. *Galleria mellonella* as a model for studying *Enterococcus faecium* host persistence. *J Mol Microbiol Biotechnol* 2011;21(3–4):191–6.
40. Panayidou S, Ioannidou E, Apidianakis Y. Human pathogenic bacteria, fungi, and viruses in *Drosophila*: disease modeling, lessons, and shortcomings. *Virulence* 2014;5(2):253–69.
41. Yuen GJ, Ausubel FM. *Enterococcus* infection biology: lessons from invertebrate host models. *J Microbiol* 2014;52(3):200–10.
42. Sifri CD, et al. Virulence effect of *Enterococcus faecalis* protease genes and the quorum-sensing locus *fsr* in *Caenorhabditis elegans* and mice. *Infect Immun* 2002;70:5647–50.
43. Maadani A, Fox KA, Mylonakis E, Garsin DA. *Enterococcus faecalis* mutations affecting virulence in the *Caenorhabditis elegans* model host. *Infect Immun* 2007;75:2634–37.
44. Teixeira N, et al. *Drosophila* host model reveals new *Enterococcus faecalis* quorum-sensing associated virulence factors. *PLoS One* 2013;8:e64740.
45. Cox CR, Gilmore MS. Native microbial colonization of *Drosophila melanogaster* and its use as a model of *Enterococcus faecalis* pathogenesis. *Infect Immun* 2007;75:1565–76.
46. Hall AD, Steed ME, Arias CA, Murray BE, Rybak MJ. Evaluation of standard- and high-dose daptomycin versus linezolid against vancomycin-resistant *Enterococcus* isolates in an in vitro pharmacokinetic/pharmacodynamic model with simulated endocardial vegetations. *Antimicrob Agents Chemother* 2012;56:3174–80.

11

Listeria monocytogenes

Sarah E.F. D'Orazio

CONTENTS

11.1 Listeriosis Is a Significant Public Health Concern

Listeria monocytogenes typically causes infection in humans when ready-to-eat food products are contaminated during processing[1]; the bacteria increase in number during refrigerated storage,[2] and then the food is consumed without adequate heating.[3] Ingestion of *L. monocytogenes* leads to a wide spectrum of clinical outcomes, ranging from self-limiting gastroenteritis to life-threatening systemic infections of the blood, brain, and placenta that have a high mortality rate (25%–30%) even in patients receiving antibiotic treatment.[4] The true incidence of intestinal infection is not known, as most people do not seek medical treatment for mild gastroenteritis, and *L. monocytogenes* is not commonly isolated from stool without the use of a specific enrichment broth.[5] However, it is thought that the infection rate for heavily contaminated foods is likely to be high.[6,7] After a short incubation period that can range from a few days up to several weeks, *L. monocytogenes* disseminates and then crosses the blood–brain barrier or the placenta in pregnant women. A retrospective analysis of recent outbreaks found that gastrointestinal symptoms were noted within 24 h of ingestion and bacteremia occurred within 2 days.[8] However, it took an average of 9 days for central nervous system manifestations to develop, and much longer (17–67 days) for pregnancy-associated cases to be reported.

The high fatality rate associated with listeriosis (20%–30%)[4,9] makes it a significant public health concern for high-risk groups including neonates, pregnant women, and people in other categories that are steadily increasing due to medical advances (the elderly, transplant recipients, and patients with chronic disease). *L. monocytogenes* infections can also occur in young, otherwise healthy individuals; in fact, cases in people over 50 with no known underlying medical issues have increased recently.[10–12] The most deadly outbreak of foodborne disease in the United States in more than a decade occurred in 2011 when cantaloupes from Jensen Farms were contaminated during a washing procedure, and were then widely distributed to at least 28 states.[13] Since then, increased awareness of the risk of *L. monocytogenes* contamination in processed forms of fresh produce has resulted in a significant increase in the number of food recalls by the FDA, with an average of 50–60 due to *L. monocytogenes* per year (www.fda.gov/Safety/Recalls/).

11.2 *L. monocytogenes* Is a Facultative Intracellular Pathogen

L. monocytogenes is a nonspore-forming, Gram-positive bacillus that is readily found in both soil and water samples. A large portion of the *L. monocytogenes* genome is devoted to regulatory proteins, and not surprisingly, the bacterium is able to adapt to changes in environmental temperature, pH, salt concentrations, and nutrient availability.[14] Following transmission to a mammalian host, a large proportion of *L. monocytogenes* continue to replicate extracellularly; however, a vital subset of the bacterium can invade host cells and survive and replicate within this intracellular niche.[15]

L. monocytogenes mediates its own uptake into nonphagocytic cells via a family of surface-exposed proteins called internalins. Internalin A (InlA) binds to E-cadherin and internalin B (InlB) binds to c-Met, and engagement of these receptors results in bacterial internalization via clathrin-mediated endocytosis.[16] *L. monocytogenes* contained within host cell phagocytic or endocytic vacuoles produces listeriolysin O (LLO), a cholesterol-dependent pore-forming toxin that lyses the vacuole, releasing the bacterium into the cytosol.[17–19] *L. monocytogenes* multiplies rapidly in the cytosol and begins to nucleate actin filaments via the surface-exposed protein ActA. *L. monocytogenes* can move through the cytoplasm by actin-based motility towards the cell membrane and can be enveloped in pseudopod-like structures and then phagocytosed by neighboring cells.[20,21] This results in a cell-to-cell spread of *L. monocytogenes* without exposure to the extracellular environment.

11.3 Cell Culture Models

To cause infection, foodborne *L. monocytogenes* must survive passage through the stomach and then colonize some portion of the intestines. If the bacterium does not attach firmly or invade the gut mucosal barrier, it will most likely be expelled in feces within just a few hours. Tissue culture studies have been widely used to investigate these initial steps in the infection process. For example, *in vitro* studies showed that several different proteins expressed on the surface of *L. monocytogenes* can bind mucin.[22,23] This could serve to anchor the bacterium in the mucus layer overlying intestinal epithelial cells. The most common model for studying invasion of enterocytes is Caco-2, a human colorectal adenocarcinoma cell line that differentiates into polarized epithelial-like cells at confluence. Caco-2 cells were employed to demonstrate that the bacterial surface protein InlA can bind to E-cadherin on enterocytes, triggering a signaling cascade that leads to cytoskeletal rearrangements and internalization of the bacteria.[16,24]

The gut epithelium actually consists of several different cell types, and interspersed within the enterocytes are goblet cells, Paneth cells, and M cells. M cells, in particular, are known to be a route of invasion for many different bacteria, including *L. monocytogenes*.[25] Corr et al. differentiated C2Bbel enterocytes into cells that resembled M cells by coculturing the epithelial cells with lymphocytes harvested from Peyer's patches in the small intestine.[26] Although cell culture models such as these have been very useful for studying the dynamics of specific invasion pathways, a significant disadvantage of this approach is that a single cell line cannot replicate the complexity of cell types that are actually present in the gut. An innovative approach that shows great promise in this regard is the use of intestinal organoids. Intestinal stem cells are cultured *in vitro* with the appropriate growth factors and this leads to the formation of "miniguts" that contain fully differentiated cell types.[27,28] Microinjection of bacteria directly into the lumen of intestinal organoids may provide an excellent *in vitro* model system for future studies to better understand how *L. monocytogenes* optimally uses the multiple invasion pathways available in the gut.

11.4 Animal Models

An ideal small animal model of *L. monocytogenes* infection should closely mimic all phases of human listeriosis. Oral challenge is the preferred route of transmission to allow investigation of environmental factors that promote bacterial infectivity and the host innate resistance mechanisms that limit colonization of the gut. Only a small subset of humans exposed to *L. monocytogenes* develops severe systemic infections; thus, dissemination from the gut to peripheral tissues should be a rate-limiting step. Ideally, resistant animals should be able to inhibit bacterial replication rapidly, and there should be a lag period before *L. monocytogenes* cross the blood–brain barrier or the maternal–fetal interface in susceptible animals.

Most of the existing models of *L. monocytogenes* infection have caveats that either present significant technical hurdles or limit the physiologic relevance of the model. One such limitation is the large inoculum size required to establish infection in most laboratory animals. The infectious dose in humans has been estimated at 10^6–10^7 CFU.[29,30] At least 100-fold larger doses (10^8–10^{10} CFU) are typically needed for experimental infections in small animal models. Differences in the stomach pH, bile content, or gastric enzyme composition may account for some of the relative resistance of experimental animals. However, it is also possible that in otherwise healthy individuals, systemic listeriosis only occurs when very large doses of bacteria are ingested. In that case, the range of inocula used in immune competent laboratory animals may, in fact, closely model human transmission of the disease.

Another issue is the species specificity of the interaction between the *L. monocytogenes* surface proteins InlA and InlB and the receptors found on mammalian cells.[31,32] Early studies indicated that InlA was important for invasion of epithelial cells,[24,33] and InlB was needed for uptake in endothelial cells.[34,35] However, more recent work suggests that these two proteins may work in conjunction to allow for efficient internalization of *L. monocytogenes* in nonphagocytic cells.[36,37] InlA has a high affinity for human E-cadherin, and the human Met protein serves as a receptor for InlB, so that both ligand/receptor interactions are functional during human infections.[38] As described below, most small animal models of listeriosis are hampered by a low affinity interaction for either InlA or InlB. Importantly, intestinal infection and fetoplacental transfer do still occur in some of these animals, which suggest that other uptake mechanisms may be able to compensate, but it is not yet clear how accurately these systems mimic human infections.

11.4.1 Rabbits, Sheep, and Goats

L. monocytogenes was first identified in 1926 as the causative agent of a mononuclear leukocytosis in rabbits.[39] Until the 1980s, the bacterium was thought of as mainly a veterinary pathogen, with naturally occurring listeriosis found primarily in sheep and other ruminants.[40] Thus, most early research studies focused on feeding trials to model the disease observed in rabbits and sheep.[41,42] Although some investigators still do use rabbit[43] or goat[44] models to study oral transmission, most effort is now focused on rodent and nonhuman primate models. For a comprehensive review of infection studies in ruminants and other veterinary models, see Hoelzer et al.[45]

11.4.2 Mice and Rats

Rodents are the preferred animal for most infectious disease studies because their small size allows large-scale experiments to be performed affordably in a relatively small space. Mice, in particular, are commonly used because many tools and reagents (antibodies, ELISA kits, transgenic, and knockout strains) exist that facilitate experiments, and most of these are commercially available. Early studies indicated that 10^8–10^{10} CFU were needed to establish gastrointestinal infection in mice, and that intravenous inoculation was more reproducibly lethal in mice[46,47]; however, it is clear that infectious dose is dependent on the mouse strain used. For example, C57BL/6 mice are significantly more resistant than either A/J or BALB/c/By/J mice[48–50] that can be colonized by as few as 10^6 CFU administered orally.

The differential susceptibility of these mouse strains is not likely due to the intestinal microbiota, since fecal transplantation between C57BL/6 and BALB/c/By mice did not change the infection outcomes.[51]

Two different "humanized" mouse strains have been developed in an effort to enhance the efficiency of oral infection in mice. In both cases, modifications were designed to promote interaction between InlA expressed on the bacterial surface and E-cadherin expressed on the intestinal epithelium. In the first mouse strain, human E-cadherin was ectopically expressed under the control of the iFABP promoter, resulting in dual expression of both mouse and human E-cadherin in the small intestine.[52] The second mouse strain is a "knock-in" strain that has a single amino acid substitution (E16P) in murine E-cadherin that allows the protein to serve as a high affinity receptor for InlA.[53] Orally infected E16P mice showed enhanced colonization in the gut compared to wild-type mice; therefore, this mouse strain represents the best option for mimicking the InlA/InlB-mediated invasion events that can occur in the human gastrointestinal tract. However, as with most transgenic mice, the knock-in was generated on a C57BL/6 background, and these mice are innately more resistant to *L. monocytogenes* infection compared to other mouse strains. Although it would be time consuming and expensive, it would be very useful to cross the E16P mutation onto more susceptible strain backgrounds such as BALB/c mice.

Intravenous or intraperitoneal injection of *L. monocytogenes* in rats was frequently used in the 1980s as an infection model to study immune responses against an intracellular pathogen. However, very few studies have examined the oral transmission of listeriosis in rats. Czuprynski and Balish first showed that germfree rats were more readily colonized than conventionally housed rats offered *L. monocytogenes* in their drinking water.[54] Later, Schlech et al. used a feeding tube to deliver the bacteria and found that doses of at least 10^9 CFU were needed to consistently establish intestinal infection in young Sprague–Dawley rats.[55] Rats have the same species barrier that limits InlA-mediated invasion of the gut mucosa as is found in mice,[38] and thus, offer no particular advantage for use as a model organism.

11.4.3 Guinea Pigs

Young guinea pigs have also been used to study *L. monocytogenes* infection, and like mice or rats, doses of 10^8–10^{10} CFU are typically needed to facilitate intestinal colonization.[56,57] Melton-Witt et al. infected guinea pigs with a mixture of 20 signature-tagged strains and showed that the spread from the MLN to the spleen was a rate-limiting step for systemic dissemination, with only one in every 100–1000 bacteria getting beyond this bottleneck.[56] Using this model, the total number of *L. monocytogenes* that reached the spleen or liver was very low, suggesting that the guinea pig is not the ideal model to study the systemic phases of listeriosis.

As with most other rodents, the use of guinea pigs does present a species barrier for internalization of *L. monocytogenes*. Guinea pig E-cadherin does bind efficiently to InlA, but the Met protein in guinea pigs has an amino acid substitution that prevents optimal interaction with InlB.[58] In spite of this limitation, guinea pigs are considered the model of choice to study maternal–fetal transmission of *L. monocytogenes*, largely due to the architecture of their placenta. Both humans and guinea pigs have a hemochorial placenta, with only a single layer of cells separating the fetal and maternal blood supplies. Particularly during the last stages of pregnancy, this means that bloodborne *L. monocytogenes* need to traverse just a single trophoblast to invade the placenta.[59] Williams et al. showed that infection of pregnant guinea pigs with oral doses ranging from 10^4 to 10^8 CFU led to invasion in approximately half of the fetuses.[60,61]

11.4.4 Gerbils

The E-cadherin and Met proteins in gerbils efficiently bind InlA and InlB, respectively.[53] The lack of a species barrier for interaction with these key bacterial surface proteins makes gerbils perhaps the most physiologically relevant rodent model to study listeriosis. Disson et al. orally inoculated gerbils and found significant colonization of both the small and large intestines.[53] Furthermore, they demonstrated 100% lethality in fetuses when pregnant females were infected orally.[53] The inocula used in those studies were 10^9–10^{10} CFU, but lower doses were not tested, so it is possible that the infectious dose in gerbils could be closer to that observed in humans. Blanot et al. also demonstrated that gerbils could develop rhombencephalitis, a localized brainstem infection, that closely mimicked the type of brain infections seen in about 20% of human cases.[62] However, in those studies, the gerbils were infected via the middle ear, so it is not yet clear whether brainstem infections can naturally occur following the oral ingestion of *L. monocytogenes* in gerbils. Unfortunately, a lack of available tools and reagents specific for gerbils has limited the enthusiasm for studying host responses in this model.

11.4.5 Nonhuman Primates

Nonhuman primate models have the distinct disadvantage of being expensive to use, and so only small numbers of animals are usually available for any given study. The clinical course of sporadic listeriosis appears to mimic human disease closely, as evidenced by occasional outbreaks of *L. monocytogenes* infection in nonhuman primate colonies that resulted in either meningoencephalitis or spontaneous abortion in pregnant females.[63,64] The few primate infection studies performed to date have been most useful for approximating the infectious dose of *L. monocytogenes* in humans. For example, Farber et al. found that cynomolgus monkeys fed at least 10^7 CFU of strain Scott A shed *L. monocytogenes* in feces for 3 weeks, but only the animals given 10^9 CFU displayed signs of disease (septicemia and occasional diarrhea).[65] More recently, Smith et al. used a monkey clinical isolate of *L. monocytogenes* to infect pregnant rhesus macaques.[30] In that study, the overall LD_{50} was estimated to be 10^7 CFU, but it was noted that monkeys with a stillborn fetus needed a significantly lower dose to establish an intestinal infection than monkeys that had normal deliveries. This suggests that host susceptibility factors are also likely to play an important role in determining the clinical outcomes following ingestion of *L. monocytogenes*, and the use of an inbred animal model (such as laboratory mice) may be the easiest way to identify these genetic loci.

11.5 *L. monocytogenes* Isolates Vary in Virulence

Based on serological reactions between listerial antigens and specific antibodies, *L. monocytogenes* is separated into at least 12 serovars (1/2a, 1/2b, 1/2c, 3a, 3b, 3c, 4a, 4b, 4c, 4d, 4e, and 7), which tend to show varied virulence potential in human and animal hosts. It is notable that serovars 1/2a, 1/2b, 1/2c, and 4b are responsible for >95% of human clinical listeriosis cases, and are also highly pathogenic to animal hosts such as rodents via intragastric (i.g.) route. Most published studies of oral infection models have utilized one of the two most common strains of *L. monocytogenes*: EGDe and 10403s. An advantage of using one of these strains is that there are many isogenic derivatives (specific gene knockouts and reporter strains) available, and it is easier to compare directly the results of unrelated experiments. However, the choice of which *L. monocytogenes* isolate to use for oral challenge studies in small animal models could greatly alter the course of the ensuing infection. Barbour et al. infected BALB/c mice intragastrically with 66 different isolates of *Listeria* and found that the serovar 4b and 1/2a strains were the most virulent, a pattern that has also been observed in humans.[66] Both *L. monocytogenes* EGDe and 10403s are serovar 1/2a strains, but they have been extensively passaged in the lab. Each is probably a good representative of pathogenic *L. monocytogenes*, but it is important to passage the bacteria through an animal periodically to maintain virulence traits.

Alternatively, the use of strains that have been freshly isolated from particular sources may yield results that more closely mimic specific types of human disease. For example, a recent microarray analysis suggested there may be identifiable differences between strains that primarily cause febrile gastroenteritis and those that lead to more invasive disease.[67] In support of this idea, Jensen et al. tested a strain (La111) that persistently colonized a fish processing plant and demonstrated that the bacteria were not more virulent *in vitro* and they did not persist well in the intestines of orally infected guinea pigs. However, oral infection with strain La111 did result in a significantly greater rate of infected fetuses than similar challenges with other clinical isolates.[68,69] In another example, pregnant mice were significantly more susceptible to oral challenge with *L. monocytogenes* 2203, a strain that was isolated during an outbreak of listeriosis that affected primarily pregnant women.[70,71]

Mouse-adapted strains of *L. monocytogenes* have also been developed to overcome the species barrier to receptor-mediated invasion of epithelial cells in the gut mucosa. The modified InlA expressed by these strains ($InlA^m$) has a similar affinity for mouse E-cadherin as wild-type InlA has for human E-cadherin.[72,73] Foodborne transmission of $InlA^m$-expressing *L. monocytogenes* can be achieved with lower doses (10^6–10^8 CFU) than wild-type strains and results in an infection course that closely mimics all phases of human disease.[49] However, the modified $InlA^m$ protein also acquired the ability to bind N-cadherin in addition to E-cadherin, resulting in an enhanced uptake by villous M cells.[74] Thus, the use of an $InlA^m$-expressing strain may result in altered tropism of bacterial invasion compared with human infections.

11.6 Methods of Oral Transmission

There are multiple ways to inoculate experimental animals orally. One of the first methods employed was to contaminate the drinking water of Swiss mice and then allow the animals to drink from the water bottle for 24 h.[75] This approach was not technically challenging, but had a major disadvantage because it was difficult to control the exact dose of bacteria each mouse swallowed. Due to this limitation, most laboratories began using oral gavage as a method of choice to place a defined dose directly into the stomach of each experimental animal. Depending on the size of the animal, this can be performed by placing a blunted feeding needle or flexible tubing attached to a syringe into the esophagus, or for a more invasive approach, the delivery device can be pushed all the way down to the stomach.

Although i.g. injections allow for more control over the delivery of the inoculum, they may have unintended consequences. First, when *L. monocytogenes* is suspended in a liquid, most of the bacteria transit rapidly through the intestines, and the majority of the inoculum is shed in feces within the first 4 h after infection.[56,76] Rapid passage through the gut may not provide enough exposure to the acidic pH and bile present in the upper gastrointestinal tract to promote induction of transcriptional changes that are known to make *L. monocytogenes* more virulent.[77] I.g. inoculation also has the potential to cause minor physical trauma to the lining of the esophagus, especially when a feeding needle is used, and this may promote direct invasion of the bloodstream in a manner that does not efficiently occur when *Listeria*-contaminated food is ingested. A comparative analysis of multiple previously published studies that utilized i.g. inoculation indicates that some investigators observed rapid dissemination to the spleen and liver in as few as 4 h,[48,52,73,78] while studies found no systemic spread of *L. monocytogenes* until 48 h postinfection.[72,79,80] This suggests that i.g. injections can have variable outcomes that are highly dependent on the technique used by the investigator, and therefore, this method of oral inoculation does not yield reproducible results. In addition, the use of certain types of anesthesia for more invasive routes of inoculation may also alter host susceptibility to infection. In this regard, Czuprynski et al. showed that sodium pentobarbital transiently enhanced the severity of infection in mice that were inoculated by the i.g. route, but isoflurane had no effect on infection rates.[81,82]

Another solution to the problem of controlled dosage is to place the bacterial solution directly into the mouth of the animal. For larger animals such as guinea pigs, it is relatively easy to slowly drip a solution into the oral cavity using a syringe.[56] For smaller animals such as mice, Manohar et al. used a sterile bacterial inoculating loop to place a bacterial solution into the mouth.[83] This approach avoids the trauma of an invasive injection, but still involves a bacterial inoculum suspended in a liquid.

Bou Ghanem et al. recently described a novel model of foodborne listeriosis in which mice are fed a *L. monocytogenes*-contaminated piece of bread.[84,85] This natural feeding model has several advantages. First, the infection method is not invasive, does not cause physical trauma, and does not require specialized skills to perform. This greatly decreases the chance of investigator-dependent variability, making it a much more reproducible model, and one that needs fewer animals to achieve statistical significance. Furthermore, this model allows one to test the effect of different types of food and/or variable food storage conditions on the transmissibility of disease.

It should be noted that there is some confusion in the literature concerning the methods used to achieve oral transmission in experimental animal models. Not all published reports describe the inoculation procedure in detail; some simply state that the animals were "orally infected." Since the choice of inoculation procedure may affect the route used by *L. monocytogenes* to invade the gut barrier and spread systemically, caution should be used when interpreting the results of a study that does not specify the exact technique used.

11.7 Other Variables to Consider

As outlined above, each small animal model of listeriosis has both advantages and disadvantages. Ultimately, the choice of which model to use may depend heavily on the availability of the laboratory animals, or the exact phase of *L. monocytogenes* infection that is being studied. Regardless of which *in vivo* model is chosen, there are other factors that should be taken into account to best mimic human disease.

For example, it has been suggested that patients who take antacids or cimetidine to block acid reflux are more susceptible to developing systemic listeriosis.[86] Thus, neutralization of stomach acid might enhance oral transmission of *L. monocytogenes*. Saklani-Jusforgues et al. showed that buffering an i.g. inoculum with sodium bicarbonate or providing only buffered drinking water to mice significantly increased the number of *L. monocytogenes* found in the stomach 15 min after inoculation.[87] However, longer-term studies using a variety of infection models are less conclusive. Sodium bicarbonate pretreatment did increase the severity of infection in neutropenic mice, but not in mice with normal levels of circulating neutrophils.[88,89] Likewise, other studies found no difference in infectivity in mice pretreated with cimetidine[47] or in monkeys given a calcium carbonate solution 1 h prior to infection.[65] However, Schlech et al. demonstrated that pretreatment with cimetidine did significantly decrease the dose needed to establish infection in rats.[55]

One explanation for the different outcomes in these studies may be the endpoints that each investigator used to define infectivity. It is possible that buffering the stomach pH may increase the number of *L. monocytogenes* that survive and pass through to the lower intestines, and this would be most apparent when examining tissues at very early time points following infection. However, efficient invasion of the gut mucosal barrier in the ileum, cecum, or colon may be dependent on gene expression changes that occur following exposure to an acidic environment.[90,91] Thus, the relatively smaller number of *L. monocytogenes* that might survive passage through the stomach could be "gut adapted" and better able to invade the gut mucosa. If this were the case, then the acid-exposed bacteria would have a growth advantage that would not be apparent until later in the infection.

The vehicle used to administer *L. monocytogenes* may also influence the infectivity rate during oral infection. The processed food products most commonly linked to outbreaks of listeriosis (deli meats, cheeses, and raw milk) all share a relatively high-fat content, and it is possible that gastric secretions induced when ingesting a fatty meal help promote the survival or infectivity of *L. monocytogenes*. Several studies have examined whether a high-fat content can improve the efficiency of establishing infection in animal models. Bou Ghanem et al. found that foodborne infection of mice was most consistent when the bacteria were suspended in melted butter, rather than saline, prior to contamination of food particles.[49] Likewise, Smith et al. demonstrated that stillbirths in pregnant monkeys were more likely to happen when whipping cream, rather than whole or skim milk, was used as the delivery vehicle.[64] These studies would appear to suggest that the consumption of fatty foods could be a risk factor for developing listeriosis. However, the deadly 2011 listeriosis outbreak associated with contaminated cantaloupes indicates that produce can be a vehicle to transmit the disease,[13] and suggests that a high-fat food product is not required to promote infectivity. Further studies examining the growth rate of *L. monocytogenes* in various food products, as well as studies of transmission rates in a small animal model are needed to address this question.

11.8 Future Directions

Orally transmitted models of *L. monocytogenes* infection have not been as widely used as intravenous or intraperitoneal models due to a high degree of innate resistance in many species and significant phenotypic variability among infected animals. It is often difficult to directly compare the results obtained in previously published studies because of significant differences in bacterial strain choice, route of inoculation, and various manipulations investigators have used to enhance the infectivity. Recent improvements, particularly for the mouse model, have greatly improved the efficiency of intestinal infection following oral transmission, and the use of a standardized model would help to greatly advance the field.

Future studies should focus on the natural pathways used by *L. monocytogenes* to disseminate from the gut to peripheral tissues. Of particular relevance will be mechanisms used to cross the blood–brain barrier and the placental barrier in pregnant females. Finally, the vast majority of human *L. monocytogenes* infections take place in the elderly.[92] Studies that directly compare infections in young versus aged animals may shed light on potential therapeutic strategies that could prevent these deadly infections.

Acknowledgments

S.E.F.D. was supported by a grant from the National Institutes of Health (R56 AI091918).

REFERENCES

1. Carpentier, B. & Cerf, O. Review—persistence of *Listeria monocytogenes* in food industry equipment and premises. *Int J Food Microbiol* 145, 1–8 (2011).
2. Tasara, T. & Stephan, R. Cold stress tolerance of *Listeria monocytogenes*: a review of molecular adaptive mechanisms and food safety implications. *J Food Prot* 69, 1473–1484 (2006).
3. Rocourt, J., BenEmbarek, P., Toyofuku, H. & Schlundt, J. Quantitative risk assessment of *Listeria monocytogenes* in ready-to-eat foods: the FAO/WHO approach. *FEMS Immunol Med Microbiol* 35, 263–267 (2003).
4. Wing, E.J. & Gregory, S.H. *Listeria monocytogenes*: clinical and experimental update. *J Infect Dis* 185, S18–S24 (2002).
5. Scallan, E. et al. Foodborne illness acquired in the United States—major pathogens. *Emerg Infect Dis* 17, 7–15 (2011).
6. Dalton, C.B. et al. An outbreak of gastroenteritis and fever due to *Listeria monocytogenes* in milk. *N Engl J Med* 336, 100–105 (1997).
7. Ooi, S.T. & Lorber, B. Gastroenteritis due to *Listeria monocytogenes*. *Clin Infect Dis* 40, 1327–1332 (2005).
8. Goulet, V., King, L.A., Vaillant, V. & de Valk, H. What is the incubation period for listeriosis? *BMC Infect Dis* 13, 11 (2013).
9. Mead, P.S. et al. Food-related illness and death in the United States. *Emerg Infect Dis* 5, 607–625 (1999).
10. Allerberger, F. & Wagner, M. Listeriosis: a resurgent foodborne infection. *Clin Microbiol Infect* 16, 16–23 (2010).
11. Broome, C. Role of foods in sporadic and epidemic listeriosis—a retrospective. *16th International Symposium on Problems of Listeriosis* (Savannah, GA, 2007).
12. Gillespie, I.A. et al. Changing pattern of human listeriosis, England and Wales, 2001–2004. *Emerg Infect Dis* 12, 1361–1366 (2006).
13. McCollum, J.T. et al. Multistate outbreak of listeriosis associated with cantaloupe. *N Engl J Med* 369, 944–953 (2013).
14. Xayarath, B. & Freitag, N.E. Optimizing the balance between host and environmental survival skills: lessons learned from *Listeria monocytogenes*. *Future Microbiol* 7, 839–852 (2012).
15. Jones, G.S. et al. Intracellular *Listeria monocytogenes* comprise a minimal but vital fraction of the intestinal burden following foodborne infection. *Infect Immun* 83, 3146–356 (2015).
16. Pizarro-Cerda, J., Kuhbacher, A. & Cossart, P. Entry of *Listeria monocytogenes* in mammalian epithelial cells: an updated view. *Cold Spring Harb Perspect Med* 2, a010009 (2012).
17. Cossart, P., Pizarro-Cerda, J. & Lecuit, M. Invasion of mammalian cells by *Listeria monocytogenes*: functional mimicry to subvert cellular functions. *Trends Cell Biol* 13, 23–31 (2003).
18. Portnoy, D.A., Auerbach, V. & Glomski, I.J. The cell biology of *Listeria monocytogenes* infection: the intersection of bacterial pathogenesis and cell-mediated immunity. *J Cell Biol* 158, 409–414 (2002).

19. Vazquez-Boland, J.A. et al. *Listeria* pathogenesis and molecular virulence determinants. *Clin Microbiol Rev* 14, 584–640 (2001).
20. Cossart, P. & Bierne, H. The use of host cell machinery in the pathogenesis of *Listeria monocytogenes. Curr Opin Immunol* 13, 96–103 (2001).
21. Tilney, L.G. & Portnoy, D.A. Actin filaments and the growth, movement, and spread of the intracellular bacterial parasite, *Listeria monocytogenes. J Cell Biol* 109, 1597–1608 (1989).
22. Linden, S.K. et al. *Listeria monocytogenes* internalins bind to the human intestinal mucin MUC2. *Arch Microbiol* 190, 101–104 (2008).
23. Mariscotti, J.F., Quereda, J.J., Garcia-Del Portillo, F. & Pucciarelli, M.G. The *Listeria monocytogenes* LPXTG surface protein Lmo1413 is an invasin with capacity to bind mucin. *Int J Med Microbiol* 304, 393–404 (2014).
24. Lecuit, M., Ohayon, H., Braun, L., Mengaud, J. & Cossart, P. Internalin of *Listeria monocytogenes* with an intact leucine-rich repeat region is sufficient to promote internalization. *Infect Immun* 65, 5309–5319 (1997).
25. Jensen, V.B., Harty, J.T. & Jones, B.D. Interactions of the invasive pathogens *Salmonella typhimurium, Listeria monocytogenes*, and *Shigella flexneri* with M cells and murine Peyer's patches. *Infect Immun* 66, 3758–3766 (1998).
26. Corr, S., Hill, C. & Gahan, C.G. An in vitro cell-culture model demonstrates internalin- and hemolysin-independent translocation of *Listeria monocytogenes* across M cells. *Microb Pathog* 41, 241–250 (2006).
27. Sato, T. & Clevers, H. Growing self-organizing mini-guts from a single intestinal stem cell: mechanism and applications. *Science* 340, 1190–1194 (2013).
28. Zhang, Y.G., Wu, S., Xia, Y. & Sun, J. *Salmonella*-infected crypt-derived intestinal organoid culture system for host-bacterial interactions. *Physiol Rep* 2, e12147 (2014).
29. FAO/WHO. Risk assessment of *Listeria monocytogenes* in ready-to-eat foods. In *Microbial Risk Assessment Series No. 4* (Rome, Italy, 2004).
30. Smith, M.A. et al. Dose-response model for *Listeria monocytogenes*-induced stillbirths in nonhuman primates. *Infect Immun* 76, 726–731 (2008).
31. Mengaud, J., Ohayon, H., Gounon, P., Mege, R.M. & Cossart, P. E-cadherin is the receptor for internalin, a surface protein required for entry of *L. monocytogenes* into epithelial cells. *Cell* 84, 923–932 (1996).
32. Shen, Y., Naujokas, M., Park, M. & Ireton, K. InlB-dependent internalization of *Listeria* is mediated by the Met receptor tyrosine kinase. *Cell* 103, 501–510 (2000).
33. Gaillard, J.L., Berche, P., Frehel, C., Gouin, E. & Cossart, P. Entry of *L. monocytogenes* into cells is mediated by internalin, a repeat protein reminiscent of surface antigens from Gram-positive cocci. *Cell* 65, 1127–1141 (1991).
34. Greiffenberg, L. et al. Interaction of *Listeria monocytogenes* with human brain microvascular endothelial cells: InlB-dependent invasion, long-term intracellular growth, and spread from macrophages to endothelial cells. *Infect Immun* 66, 5260–5267 (1998).
35. Parida, S.K. et al. Internalin B is essential for adhesion and mediates the invasion of *Listeria monocytogenes* into human endothelial cells. *Mol Microbiol* 28, 81–93 (1998).
36. Grundler, T. et al. The surface proteins InlA and InlB are interdependently required for polar basolateral invasion by *Listeria monocytogenes* in a human model of the blood-cerebrospinal fluid barrier. *Microbes Infect* 15, 291–301 (2013).
37. Pentecost, M., Kumaran, J., Ghosh, P. & Amieva, M.R. *Listeria monocytogenes* internalin B activates junctional endocytosis to accelerate intestinal invasion. *PLoS Pathog* 6, e1000900 (2010).
38. Lecuit, M. et al. A single amino acid in E-cadherin responsible for host specificity towards the human pathogen *Listeria monocytogenes. EMBO J* 18, 3956–3963 (1999).
39. Murray, E., Webb, R. & Swann, M. A disease of rabbits characterized by large mononuclear leukocytosis, caused by a hitherto undescribed bacillus *Bacterium monocytogenes. J Pathol Bacteriol* 29, 407–439 (1926).
40. Gates, G.A., Blenden, D.C. & Kintner, L.D. Listeric myelitis in sheep. *J Am Vet Med Assoc* 150, 200–204 (1967).
41. Gray, M.L., Singh, C. & Thorp, F., Jr. Abortion, stillbirth, early death of young in rabbits by *Listeria monocytogenes*. II. Oral exposure. *Proc Soc Exp Biol Med* 89, 169–175 (1955).

42. Osebold, J.W. & Inouye, T. Pathogenesis of *Listeria monocytogenes* infections in natural hosts. II. Sheep studies. *J Infect Dis* 95, 67–78 (1954).
43. Belen Lopez, M. et al. Serological response in rabbits to *Listeria monocytogenes* after oral or intragastric inoculation. *FEMS Immunol Med Microbiol* 7, 131–134 (1993).
44. Miettinen, A., Husu, J. & Tuomi, J. Serum antibody response to *Listeria monocytogenes*, listerial excretion, and clinical characteristics in experimentally infected goats. *J Clin Microbiol* 28, 340–343 (1990).
45. Hoelzer, K., Pouillot, R. & Dennis, S. Animal models of listeriosis: a comparative review of the current state of the art and lessons learned. *Vet Res* 43, 18 (2012).
46. Audurier, A., Pardon, P., Marly, J. & Lantier, F. Experimental infection of mice with *Listeria monocytogenes* and *L. innocua*. *Ann Microbiol (Paris)* 131B, 47–57 (1980).
47. Golnazarian, C.A., Donnelly, C.W., Pintauro, S.J. & Howard, D.B. Comparison of infectious dose of *Listeria-Monocytogenes* F5817 as determined for normal versus compromised C57bl/6j mice. *J Food Prot* 52, 696–701 (1989).
48. Czuprynski, C.J., Faith, N.G. & Steinberg, H. A/J mice are susceptible and C57BL/6 mice are resistant to *Listeria monocytogenes* infection by intragastric inoculation. *Infect Immun* 71, 682–689 (2003).
49. Bou Ghanem, E.N. et al. InlA promotes dissemination of *Listeria monocytogenes* to the mesenteric lymph nodes during food borne infection of mice. *PLoS Pathog* 8, e1003015 (2012).
50. Bergmann, S. et al. Influence of internalin A murinisation on host resistance to orally acquired listeriosis in mice. *BMC Microbiol* 13, 90 (2013).
51. Myers-Morales, T., Bussell, K.M. & D'Orazio, S.E.F. Fecal transplantation does not transfer either susceptibility or resistance to food borne listeriosis in C57BL/6 and BALC/c/By mice. *F1000 Res* 2, 177 (2013).
52. Lecuit, M. et al. A transgenic model for listeriosis: role of internalin in crossing the intestinal barrier. *Science* 292, 1722–1725 (2001).
53. Disson, O. et al. Conjugated action of two species-specific invasion proteins for fetoplacental listeriosis. *Nature* 455, 1114–1118 (2008).
54. Czuprynski, C.J. & Balish, E. Pathogenesis of *Listeria monocytogenes* for gnotobiotic rats. *Infect Immun* 32, 323–331 (1981).
55. Schlech, W.F., III, Chase, D.P. & Badley, A. A model of food-borne *Listeria monocytogenes* infection in the Sprague-Dawley rat using gastric inoculation: development and effect of gastric acidity on infective dose. *Int J Food Microbiol* 18, 15–24 (1993).
56. Melton-Witt, J.A., Rafelski, S.M., Portnoy, D.A. & Bakardjiev, A.I. Oral infection with signature-tagged *Listeria monocytogenes* reveals organ-specific growth and dissemination routes in guinea pigs. *Infect Immun* 80, 720–732 (2012).
57. Roldgaard, B.B., Andersen, J.B., Hansen, T.B., Christensen, B.B. & Licht, T.R. Comparison of three *Listeria monocytogenes* strains in a guinea-pig model simulating food-borne exposure. *FEMS Microbiol Lett* 291, 88–94 (2009).
58. Khelef, N., Lecuit, M., Bierne, H. & Cossart, P. Species specificity of the *Listeria monocytogenes* InlB protein. *Cell Microbiol* 8, 457–470 (2006).
59. Leiser, R. & Kaufmann, P. Placental structure: in a comparative aspect. *Exp Clin Endocrinol* 102, 122–134 (1994).
60. Williams, D., Irvin, E.A., Chmielewski, R.A., Frank, J.F. & Smith, M.A. Dose-response of *Listeria monocytogenes* after oral exposure in pregnant guinea pigs. *J Food Prot* 70, 1122–1128 (2007).
61. Williams, D., Dunn, S., Richardson, A., Frank, J.F. & Smith, M.A. Time course of fetal tissue invasion by *Listeria monocytogenes* following an oral inoculation in pregnant guinea pigs. *J Food Prot* 74, 248–253 (2011).
62. Blanot, S. et al. A gerbil model for rhombencephalitis due to *Listeria monocytogenes*. *Microb Pathog* 23, 39–48 (1997).
63. Lemoy, M.J., Lopes, D.A., Reader, J.R., Westworth, D.R. & Tarara, R.P. Meningoencephalitis due to *Listeria monocytogenes* in a pregnant rhesus macaque (*Macaca mulatta*). *Comp Med* 62, 443–447 (2012).
64. Smith, M.A. et al. Nonhuman primate model for *Listeria monocytogenes*-induced stillbirths. *Infect Immun* 71, 1574–1579 (2003).
65. Farber, J.M., Daley, E., Coates, F., Beausoleil, N. & Fournier, J. Feeding trials of *Listeria monocytogenes* with a nonhuman primate model. *J Clin Microbiol* 29, 2606–2608 (1991).

66. Barbour, A.H., Rampling, A. & Hormaeche, C.E. Variation in the infectivity of *Listeria monocytogenes* isolates following intragastric inoculation of mice. *Infect Immun* 69, 4657–4660 (2001).
67. Laksanalamai, P., Jackson, S.A., Mammel, M.K. & Datta, A.R. High density microarray analysis reveals new insights into genetic footprints of *Listeria monocytogenes* strains involved in listeriosis outbreaks. *PLoS One* 7, e32896 (2012).
68. Jensen, A., Williams, D., Irvin, E.A., Gram, L. & Smith, M.A. A processing plant persistent strain of *Listeria monocytogenes* crosses the fetoplacental barrier in a pregnant guinea pig model. *J Food Prot* 71, 1028–1034 (2008).
69. Jensen, A. et al. Processing plant persistent strains of *Listeria monocytogenes* appear to have a lower virulence potential than clinical strains in selected virulence models. *Int J Food Microbiol* 123, 254–261 (2008).
70. MacDonald, P.D. et al. Outbreak of listeriosis among Mexican immigrants as a result of consumption of illicitly produced Mexican-style cheese. *Clin Infect Dis* 40, 677–682 (2005).
71. Poulsen, K.P., Faith, N.G., Steinberg, H. & Czuprynski, C.J. Pregnancy reduces the genetic resistance of C57BL/6 mice to *Listeria monocytogenes* infection by intragastric inoculation. *Microb Pathog* 50, 360–366 (2011).
72. Monk, I.R., Casey, P.G., Hill, C. & Gahan, C.G. Directed evolution and targeted mutagenesis to murinize *Listeria monocytogenes* internalin A for enhanced infectivity in the murine oral infection model. *BMC Microbiol* 10, 318 (2010).
73. Wollert, T. et al. Extending the host range of *Listeria monocytogenes* by rational protein design. *Cell* 129, 891–902 (2007).
74. Tsai, Y.H., Disson, O., Bierne, H. & Lecuit, M. Murinization of internalin extends its receptor repertoire, altering *Listeria monocytogenes* cell tropism and host responses. *PLoS Pathog* 9, e1003381 (2013).
75. Miller, J.K. & Burns, J. Histopathology of *Listeria monocytogenes* after oral feeding to mice. *Appl Microbiol* 19, 772–775 (1970).
76. Hardy, J. et al. Extracellular replication of *Listeria monocytogenes* in the murine gall bladder. *Science* 303, 851–853 (2004).
77. Conte, M.P. et al. Acid tolerance in *Listeria monocytogenes* influences invasiveness of enterocyte-like cells and macrophage-like cells. *Microb Pathog* 29, 137–144 (2000).
78. Gajendran, N. et al. Regional IFNγ expression is insufficient for efficacious control of food-borne bacterial pathogens at the gut epithelial barrier. *Int Immunol* 19, 1075–1081 (2007).
79. Kursar, M. et al. Antigen-specific $CD8^+$ T cell responses in intestinal tissues during murine listeriosis. *Microbes Infect* 6, 8–16 (2004).
80. MacDonald, T.T. & Carter, P.B. Cell-mediated immunity to intestinal infection. *Infect Immun* 28, 516–523 (1980).
81. Czuprynski, C.J., Faith, N.G., Steinberg, H. & Neudeck, B. Sodium pentobarbital anesthesia transiently enhances the severity of infection following intragastric, but not intravenous, inoculation of *Listeria monocytogenes* in mice. *Microb Pathog* 35, 81–86 (2003).
82. Sahaghian, R., Faith, N.G. & Czuprynski, C. Comparison of systemic *Listeria monocytogenes* infection in esophageally inoculated mice anesthetized with isoflurane or pentobarbital. *Lab Anim (NY)* 38, 126–130 (2009).
83. Manohar, M., Baumann, D.O., Bos, N.A. & Cebra, J.J. Gut colonization of mice with actA-negative mutant of *Listeria monocytogenes* can stimulate a humoral mucosal immune response. *Infect Immun* 69, 3542–3549 (2001).
84. Bou Ghanem, E.N., Myers-Morales, T. & D'Orazio, S.E. A mouse model of foodborne *Listeria monocytogenes* infection. *Curr Protoc Microbiol* 31, 9B.3.1–9B.3.16 (2013).
85. Bou Ghanem, E.N., Myers-Morales, T., Jones, G.S. & D'Orazio, S.E.F. Oral transmission of *Listeria monocytogenes* in mice via ingestion of contaminated food. *J Visual Exp* 75, e50381, doi:10.3791/50381 (2013).
86. Ho, J.L., Shands, K.N., Friedland, G., Eckind, P. & Fraser, D.W. An outbreak of type 4b *Listeria monocytogenes* infection involving patients from eight Boston hospitals. *Arch Intern Med* 146, 520–524 (1986).
87. Saklani-Jusforgues, H., Fontan, E. & Goossens, P.L. Effect of acid-adaptation on *Listeria monocytogenes* survival and translocation in a murine intragastric infection model. *FEMS Microbiol Lett* 193, 155–159 (2000).

88. Czuprynski, C.J., Faith, N.G. & Steinberg, H. Ability of the *Listeria monocytogenes* strain Scott A to cause systemic infection in mice infected by the intragastric route. *Appl Environ Microbiol* 68, 2893–2900 (2002).
89. Czuprynski, C.J. & Faith, N.G. Sodium bicarbonate enhances the severity of infection in neutropenic mice orally inoculated with *Listeria monocytogenes* EGD. *Clin Diagn Lab Immunol* 9, 477–481 (2002).
90. O'Driscoll, B., Gahan, C.G. & Hill, C. Adaptive acid tolerance response in *Listeria monocytogenes*: isolation of an acid-tolerant mutant which demonstrates increased virulence. *Appl Environ Microbiol* 62, 1693–1698 (1996).
91. Ramalheira, R. et al. Survival of clinical and food isolates of *Listeria monocytogenes* through simulated gastrointestinal tract conditions. *Foodborne Pathog Dis* 7, 121–128 (2010).
92. Centers for Disease Control and Prevention (CDC). Vital signs: *Listeria* illnesses, deaths, and outbreaks—United States, 2009–2011. *Morb Mortal Wkly Rep* 62, 448–452 (2013).

12

Mycobacterium

Flábio R. de Araújo and Nalvo F. Almeida

CONTENTS

12.1 Introduction

Mycobacterium spp. are intimately linked to human society. Leprosy, caused by *Mycobacterium leprae*, has been described in historical texts since ancient times; archaeological and osteological findings indicate its existence in human populations for thousands of years [1]. Evidences of *Mycobacterium tuberculosis* complex (MTC) infection date from the Neolithic C period, more than 8000 years ago [2]. Evidences of human infection by *Mycobacterium bovis* (a bovine tuberculosis agent) have been described in skeletons from South Siberia, with dates ranging from approximately 1761 to 2199 years before present, placing the remains within the Iron Age period [3].

The zoonotic transmission of *Mycobacterium* spp. has been related, among other means, to the ingestion of contaminated food. Historically, *M. bovis* has been associated with extrapulmonary tuberculosis in infants and children, usually occurring due to the consumption of milk, which had not been pasteurized or boiled, from infected cattle [4].

This chapter presents an overview of foodborne *Mycobacterium* spp., including the background information and recent findings relating to the etiology, life cycle, virulence, immunity, and clinical diseases, as well as genomic aspects such as annotation, genome comparison of closely related organisms, and virulence-related genes.

12.2 Etiology

The genus *Mycobacterium* includes a group of high GC Gram-positive microorganisms, including saprophytic species, and important human and animal pathogens. Pathogenic members are usually characterized by their slow growth in culture, with generation times of 12–24 h, whereas nonpathogenic members grow considerably faster [5]. In this section, we focus on mycobacteria species involved in foodborne diseases.

Mycobacterial species causing tuberculosis in humans and animals belong to the *Mycobacterium tuberculosis* complex (MTBC). The following organisms are considered members of the MTBC: *M. tuberculosis* [6]; *Mycobacterium africanum* [7] and *Mycobacterium canettii* [8], which are mainly human pathogens; *M. bovis* [9] and *Mycobacterium caprae* [10], which are mainly ruminant pathogens; *Mycobacterium microti*, a pathogen of small rodents [11]; *Mycobacterium pinnipedii* [12] from marine mammals; *Mycobacterium mungi* from mongooses [13]; and *Mycobacterium orygis* from oryx [14].

Members of the MTBC have been implicated in foodborne transmission to man. The consumption of contaminated raw dairy products has been recognized as a major cause of transmission of *M. bovis* to humans [15], generally associated with the development of extrapulmonary TB [16]. Another species of the MTBC that infects man is *M. caprae*. Although the transmission to man by raw dairy products has not been proven formally for *M. caprae*, the relatedness of the pathogens and the epidemiological settings suggest that this is probably the case [17].

Interestingly, there are pre-Columbian evidences of zoonotic infection in Peruvian human skeletons, revealing that a member of the *M. tuberculosis* complex caused human disease before contact. The ancient strains are distinct from known human-adapted forms and are most closely related to those adapted to seals and sea lions (*M. pinnipedii*) [18], raising the possibility of foodborne transmission by ingestion of sea lions' meat, hides, and bones, which was common practice in this region [19].

Nontuberculous mycobacteria (NTM) are ubiquitous organisms and were believed to represent environmental contamination. These organisms are a significant cause of infection in both immunocompetent [20] and immunocompromised humans [21].

M. avium subsp. *paratuberculosis* (MAP), which belongs to the *Mycobacterium avium* complex (MAC), is a well-known NTM pathogen causing Johne's disease (or paratuberculosis), a chronic progressive, infectious granulomatous enteritis that principally affects ruminants [21]. There is a possible association between MAP and Crohn's disease, a human inflammatory bowel disease. MAP is also a potential human foodborne pathogen because it survives pasteurization treatments [22].

A series of other NTM have been found to be associated with food: *M. nonchromogenicum*, *M. peregrinum*, *M. smegmatis*, *M. neoaurum*, *M. fortuitum*, *M. chelonae*, *M. flavescens*, *M. kansasii*, *M. scrofulaceum*, *M. genavense*, *M. simiae*, and *M. szulgai*, which were detected in milk [23,24], and *M. chelonae* and *M. kansasii*, which were detected in water [25,26].

12.3 Life Cycle

As reviewed by Scherr and Nguyen [27], the developmental stages of mycobacterial life cycle are better defined in *M. tuberculosis*. This pathogen infects human macrophages within granulomatous structures of the lung, where they may undergo morphological alterations into filamentous cells within the infected host cells. Later, host immune responses suppress the vegetative growth, inducing the formation of *persisters* that are able to survive in a dormant state for years (the latent TB stage). These dormant cells of *M. tuberculosis* may also become shorter, and some transform into a spherical, ovoid morphology characterized by a thickened cell wall with altered chemical characteristics. Probably after a weakness in the host immune defense, *M. tuberculosis* reactivates vegetative growth from the dormant state to cause *active TB*. In this state of the disease, many *M. tuberculosis* cells may escape the host cells to replicate extracellularly in the TB cavity within the lung. These actively growing cells also become transmissible through aerosolized droplets expelled from the lung of TB patients, and continue their life cycle in the newly infected patients who inhale the droplets.

12.4 Virulence

Despite the high overall phylogenetic relationship among species of the MTBC, they show wide variability in their phenotypes, especially in their virulence characteristics. The same occurs with MAP members. This section points out some important, recent works on MTBC and MAP virulence issues present in the literature.

12.4.1 Two *M. bovis* Strains and Their Virulence Genes

Using next-generation whole genome sequencing (WGS), two *M. bovis* strains (04-303 and 534) were recently sequenced. The first one was isolated from a wild boar and the other from cattle, both in Argentina. These strains have been used in the experimental BALB/c model of progressive pulmonary tuberculosis [28]. Mice were infected with 04-303, resulting in a mild inflammatory response, followed by sudden pneumonia with extensive necrosis and high mortality. On the other hand, mice infected with 534 presented limited tissue damage and higher survival. Based on these results, 04-303 was designated to be a virulent strain and 534 an attenuated strain.

In a transcriptome analysis of both strains [29], a total of 49 differentially expressed genes were detected, 35 of which were at a higher level in 04-303 and 14 in strain 534. Their expression levels dramatically changed during *in vitro* culture, with the same behavior inside bovine macrophages. The analyses at the nucleotide level revealed that these genes were conserved in both strains, suggesting that the differential expression could be related to mutations in regulatory regions.

In a genome analysis, a comparison of a set of virulence-related genes present in these genomes has been made by using Blastn and Blastx [30] and having *M. bovis* AF2122/97 as a reference genome (Castelão A.B. et al., unpublished data). A better understanding of the virulence factors of these strains may be essential to help researchers in developing new vaccines and diagnostics. In this analysis, 345 virulence genes, as reviewed by Forrellad et al. [31] and/or listed in the TubercuList database (http://genolist.pasteur.fr/TubercuList), were inspected in both strains 04-303 and 534, by looking for SNPs. The SNPs were then classified as synonymous or nonsynonymous. Strain 04-303 has three synonymous mutations and eight nonsynonymous ones. On the other side, strain 534 presented six and nine mutations, respectively. Mutations found in virulence genes, like *hrcA*, *hpx*, and *esat6* (on the 04-303 side) and *irtA*, *pks10*, *clgR*, and *tetR* (on the 534 side), agree with the attenuated phenotype of strain 534 while only being in partial consonance with the hyper-virulent phenotype of the 04-303 strain. Special attention should be given to the mutation found in *hrcA* from 04-303, a heat shock protein transcriptional repressor that could lead to an overexpression of heat shock proteins that can be related with virulence (Castelão A.B. et al., unpublished data).

12.4.2 PE and PPE Genes

By doing genome comparisons (showed in Section 12.7), we could find the presence of the PE (Pro-Glu) and PPE (Pro-Pro-Glu) protein families, involved in immune stimulation and virulence in all seven compared genomes. Besides cell wall proteins, these polymorphic gene families have been indicated as a major factor for differences among these species [32,33].

The total numbers of PE/PPE proteins found in MtH37Rv and MbAF2122/97 were 97 and 99, respectively. On the other hand, in MAP strains, the numbers were 42, 45, 43, 49, and 45 in MapMAP4, MapK10, Ma104, MapS397, and MapS5, respectively. Because these gene families used to be more frequent in pathogenic members of the *Mycobacterium* genus, this small number of PE/PPE proteins in MAP strains could suggest differences in the immunity response, compared to the MTBC species. A total of 18 families containing at least one gene of each one of the MAP strains and not containing genes of the MTBC strains have been found, showing a distinguishable group in MAP strains of this important protein class.

12.4.3 Mycobactin

Because some *Mycobacterium* species secrete iron-chelating siderophores as virulence factors in the iron-limiting environments of their hosts to compete for ferric iron, they need to produce mycobactin, a siderophore responsible for transporting iron into cells. One important phenotypic of MAP when compared with the MTBC and other MAC species is its inability to produce mycobactin [34,35]. Iron transport into cells is associated with a 10-gene cluster *mbtA-J*. In MAP species, the first gene of this cluster is shorter than the corresponding one in the MTBC and MAC. Because MbtA is responsible for initiating mycobactin production, this truncation reported suggests that the cascade leading to mycobactin production may be attenuated or disrupted in MAP [35].

12.4.4 Mce Genes

Virulence genes that are important for the entry and persistence of bacteria in the host deserve our attention. The mammalian cell entry (Mce) gene has this property and has been identified in the MTBC species. Four copies of the Mce gene are present in *M. tuberculosis*, and there are eight homologs of the Mce gene in *M. avium* subsp. *paratuberculosis* K-10 [35]. Because there are Mce operons in both pathogenic and nonpathogenic mycobacteria, it implies that the mere presence of these genes does not give to the species the ability to be virulent. However, the role of this operon in virulence may be determined by its expression under specific conditions.

12.5 Immunity

In this section, we briefly review the model of immune response against *M. bovis*. Following infection, innate immune responses are important for recruiting immune cells and establishing early lesion formation; however, these responses do little to limit infection [36]. On the other hand, the adaptive cell-mediated immune response by Th1 CD4 T cells and their soluble mediators are essential for controlling the disease [37]. Additional cells and soluble mediators that contribute to immunity include CD8 T cells [38], IL-17 [39], nitric oxide [40], and IL-2 [41].

12.6 Foodborne Illnesses Associated with *Mycobacterium* Species

Foodborne diseases are a public health problem worldwide. With regards to *Mycobacterium* spp., prior to mandatory milk pasteurization in many countries, *M. bovis* accounted for about 25% of tuberculosis cases in children [42]. In Great Britain, human consumption of infected cows' milk led to an estimated 2500 deaths and more than 50,000 new cases of TB per year in the early 1900s [43]. In Latin America, *M. bovis* is involved in 2% pulmonary and 8% extrapulmonary human tuberculosis cases [44].

The significance of the public health threats from zoonotic tuberculosis resulted in the adoption of a resolution by the World Organization for Animal Health (Office International des Epizooties; OIE) in 1983, calling for the eradication of *M. bovis* for public health and economic reasons, adoption of stringent meat inspection and pasteurization or boiling of milk for human consumption, and continued research into bovine tuberculosis, particularly in the improvement of diagnostic tests [45].

Experiments to determine the efficacy of high temperature, short time (HTST) pasteurization of milk in terms of inactivation of pathogenic microorganisms were mainly performed between 1930 and 1960. Among the target organisms were *M. bovis* and *M. tuberculosis*. As a result, the Codex Alimentarius prescribes that the HTST treatment of milk should lead to a significant reduction of pathogenic microorganisms during milk pasteurization [46].

While standard HTST pasteurization treatment results in the efficient inactivation of *M. bovis* and *M. caprae* [46], NTB (including MAP) are more resistant to heat treatment and pasteurized milk may be considered one source for NTM human infection [23,46].

Both *M. bovis* and MAP have already been detected in many types of cheese in different countries. In Brazil, MAP was detected in Coalho cheese by PCR and culture [47]. In Mexico, MTBC DNA was found in fresh cheeses that were obtained from municipal markets in the state of Hidalgo [48]. In Switzerland, MAP DNA was detected in 4.2% of the raw milk cheese samples collected at the retail level, although no viable MAP cells could be cultured [49].

Zoonotic TB has reemerged in some places of the world such as the Mexico–EUA border among immigrants from regions where bovine TB is endemic, associated with the consumption of soft fresh cheeses [50–53]. An epidemiological investigation conducted by the Centers for Disease Control and Prevention (CDC) with 35 human cases (1.12% of all analyzed TB cases) of *M. bovis* infection in New York City (NYC) also suggested that the fresh cheese brought to NYC from Mexico was a likely source of infection [54]. In Brazil, this association has also been found. In a recent study, three of 189 patients (1.6%) diagnosed with TB exhibited a coinfection of *M. bovis*–*M. tuberculosis* associated with the consumption of homemade cheeses, processed from raw milk [55].

Some artisanal cheeses are made with raw milk, followed by a ripening period, to ensure safety, since the process of ripening can contribute significantly to the reduction of pathogens in such products [56]. Nevertheless, MAP could be cultured from Cheddar cheeses prepared from pasteurized milk artificially contaminated with MAP strains after a 27-week ripening period [57]. The survival of *M. bovis* during the ripening of cheeses has also been reported. This bacterium was shown to be viable in cheeses for different periods: in Camembert for 47 [58], 60 [59], and more than 180 [60] days; in Edam for 60 days [59]; in Cheddar for 220 days [60]; in Gruyere for more than 22 days; and in Swiss Tilsitier for more than 305 days [58]. During the ripening of blue cheese made from raw milk with tubercle bacilli (104/mL), a decrease in numbers was observed during the first and second weeks, but bacilli were still present after three to four months [61]. Guinea pigs developed TB when inoculated with a 3-month-old Emmental cheese artificially contaminated with *M. bovis* [59].

12.7 Genomics

This section provides a short description and also important information related to genome annotation and the whole genome comparison of some currently available sequenced strains of *Mycobacterium* spp., with special attention to MAP and MTBC. Table 12.1 shows Genbank information and basic features, such as the chromosome length, %GC, and total number of CDS of the genomes used in this analysis. *M. tuberculosis* H37Rv was included as an outgroup in the comparison, since it is closely related with *M. bovis* and *M. avium* spp. and, on the other hand, branched from the parent group before any other one in this set of genomes.

M. tuberculosis H37Rv and *M. bovis* AF2122/97 are genomes from the MTBC. The other ones are genomes from MAP. Specifically, there are two major groups in MAP isolates, known as Type S (Sheep Type) and Type C (Cattle Type). MapS5, MapS397, and Map104 are of Type S and MpaK10 and MapMAP4 are of Type C.

Figure 12.1 shows a maximum-likelihood phylogeny of all seven genomes. Even using only MtH37Rv as the outgroup, the tree clearly shows a separation of MbAF2122 and MtH37Rv from all other

TABLE 12.1

Genbank Information and Some Basic Features of the Genomes Used in This Comparative Analysis

Taxon Name	Short Name	Taxon ID	Accession Number	Chromosome Length	CDS	% GC	References
M. avium subsp. *paratuberculosis* K-10	MapK10	262316	NC_002944	4,829,781	4350	69	[35]
M. avium subsp. *paratuberculosis* MAP4	MapMAP4	1199187	CP005928	4,829,424	4326	69	[62]
M. avium subsp. *paratuberculosis* S397	MapS397	1010838	GCA_000219085	4,815,461	4619	69	[63]
M. avium subsp. *paratuberculosis* S5	MapS5	1247747	GCA_000330785	4,799,927	4288	69	[64]
M. avium 104	Ma104	243243	NC_008595	5,475,491	5120	69	TIGR
M. bovis AF2122/97	MbAF2122	233413	NC_002945	4,345,492	3918	66	[65]
M. tuberculosis H37Rv	MtH37Rv	83332	NC_000962	4,411,532	3906	66	[66]

M. avium taxa, ratifying the differences among genomes in MTBC and MAP. Besides, strains of Type C and Type S could not be clearly clustered in the tree, showing that it is not easy to type these strains using only sequence comparisons in the protein level. The tree was built using first OrthoMCL [67] to find families with orthologous proteins. From all the 5599 families found by OrthoMCL, 2195 contain exactly one representative member of each genome and at most one member of MtH37Rv (the outgroup). Each one of these families was aligned by using MUSCLE [68] and filtered by Gblocks [69], in order to remove noninformative sites. The alignments were concatenated, and the resulting alignment, with 695,953 columns, was used as input in RAxML [70] software and the PROTCATWAGF evolution model to build the tree. All bootstrap support values were 100% and were obtained with 100 replicates. FigTree [71] was used to draw the tree.

In another whole genome comparison method, suggested in Ref. [72], one can calculate a genomic-distance index, called MUMi, taking into account both criteria of diversity, which are based on DNA maximal unique matches shared by two genomes. The matrix of MUMi indexes of our seven genomes is shown in Table 12.2.

A second tree (data not shown) was obtained by using this matrix as input for a neighbor-joining distance-based phylogeny method [73]. The distance-based topology obtained completely agrees with the tree shown in Figure 12.1.

By using the software PanGP [74], based on the output families from OrthoMCL, core-genome and pan-genome rarefaction curves of all five MAP genomes have been built. They are shown in Figure 12.2.

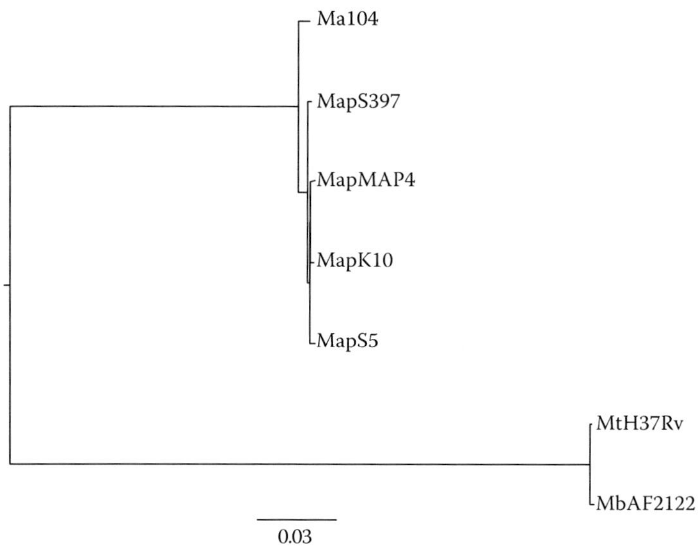

FIGURE 12.1 Phylogenetic whole genome tree, based on protein families of the seven genomes showed in Table 12.1.

TABLE 12.2

MUMi Indexes of the Seven Genomes Used in Our Comparison

	Ma104	MapK10	MapMAP4	MapS397	MapS5	MbAF2122
MapK10	0.1520					
MapMAP4	0.1521	0.0010				
MapS397	0.1448	0.0252	0.0255			
MapS5	0.1558	0.0112	0.0118	0.0296		
MbAF2122	0.9060	0.9011	0.9011	0.9008	0.9015	
MtH37Rv	0.9061	0.9013	0.9013	0.9010	0.9017	0.0208

Note: Pairs of closely related genomes have small indexes.

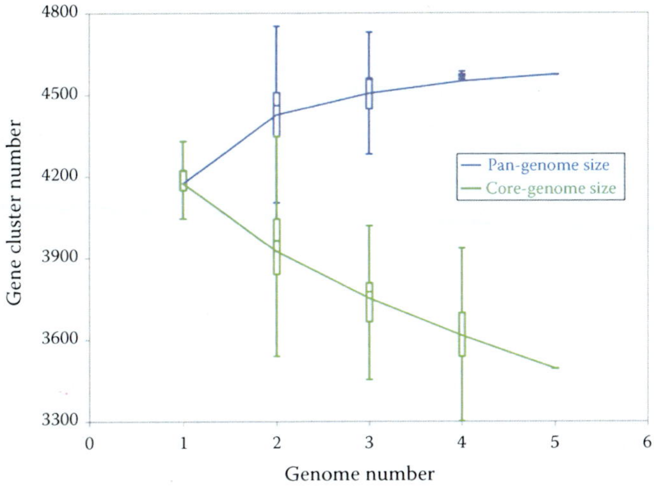

FIGURE 12.2 Core- and pan-genome sizes according to the number of genomes considered in the dataset. For each *k* in the *X*-axis, all possible combinations of *k* genomes (among five) are taken and, for each one of these combinations, pan and core numbers are plotted.

Both curves are far from reaching a plateau, suggesting that currently there is little data on the characteristics of isolates of Type C and Type S. Thus, new genome-wide data would improve the identification of genes involved in functional phenotypes.

12.7.1 Genome Rearrangements

Large genomic rearrangements, like long insertions, deletions, inversions, translocations, and duplications, are also mutational mechanisms that can cause phenotypic and genotypic differences among species and subspecies. It is known that they may lead to distinct virulence and resistance characteristics. Large inversions, for example, may be used by mycobacteria to modify the expression levels of specific genes in order to acquire some advantages during infection [75].

12.7.2 Repeat Sequences

Another important source of variation in mycobacteria is repeat sequences. Variable number of tandem repeat (VNTR) is based on the analysis of DNA segments containing tandem repeated sequences in which the number of copies of the repeated sequence varies among strains. These repeats are 15–100 bp long and dispersed at multiple locations in the genome sequence [76]. Mycobacterial Interspersed Repetitive Units (MIRUs) are a particular case of VNTR. They are intergenic regions and are reported as 12-character designations, each character corresponding to the number of repeats at one of the 12 MIRU loci. Short-sequence repeats (SSRs) are 2–5 bp long tandem repeats and are also present in mycobacteria. All of these repeat sequences have been extensively used for typing [77–79] and also for inferring epidemiologic issues of *Mycobacterium* isolates [80].

12.8 Conclusion

Several reasons make *Mycobacterium* an important bacterial model for foodborne infection. The consumption of contaminated raw dairy products is recognized as a significant cause of transmission of *M. bovis* to humans. *M. caprae* is probably also another species that infects humans by the same means. *M. avium* subsp. *paratuberculosis* is a nontuberculosis mycobacterial human foodborne pathogen, with the ability to survive pasteurization treatments. In addition, several other *Mycobacterium* spp. detected in milk and water have been implicated in foodborne diseases.

Acknowledgments

We thank the Conselho Nacional de Desenvolvimento Científico e Tecnológico (grants 443235/2014-7, 305857/2013-4), Fundect-MS (grants TO 0096/12, TO 0085/15, and TO 007/2015) and Embrapa (03.14.00.054.00.00, 02.13.16.002.00.00) for partial funding.

REFERENCES

1. Inskip, S.A., et al. Osteological, biomolecular and geochemical examination of an early Anglo-Saxon case of lepromatous leprosy. *PLoS One* 10, e0124282 (2015).
2. Hershkovitz, I., et al. Tuberculosis origin: The neolithic scenario. *Tuberculosis* 95, S122–S126 (2015).
3. Taylor, G.M., Murphy, E., Hopkins, R., Rutland, P. & Chistov, Y. First report of *Mycobacterium bovis* DNA in human remains from the Iron Age. *Microbiology* 153, 1243–1249 (2007).
4. Thoen, C., Lobue, P. & de Kantor, I. The importance of *Mycobacterium bovis* as a zoonosis. *Vet. Microbiol.* 112, 339–345 (2006).
5. Chacon, O., Bermudez, L.E. & Barletta, R.G. Johne's disease, inflammatory bowel disease, and *Mycobacterium paratuberculosis*. *Annu. Rev. Microbiol.* 58, 329–363 (2004).
6. Koch, R. Die Aetiologie der Tuberkulose. *Berl. Klin. Wochenschr.* 19, 221–230 (1882).
7. Castets, M. & Sarrat, H. Experimental study of the virulence of *Mycobacterium africanum* (preliminary note). *Bull. Soc. Med. Afr. Noire Lang. Fr.* 14, 693–696 (1969).
8. van Soolingen, D., et al. A novel pathogenic taxon of the *Mycobacterium tuberculosis* complex, *canetti*: Characterization of an exceptional isolate from Africa. *Int. J. Syst. Bacteriol.* 47, 1236–1245 (1997).
9. Karson, A.G. & Lessel, E.F. *Mycobacterium bovis* nom. nov. *Int. J. Syst. Evol. Microbiol.* 20, 273–282 (1970).
10. Aranaz, A., Cousins, D., Mateos, A. & Domínguez, L. Elevation of *Mycobacterium tuberculosis* subsp. *caprae* to species rank as *Mycobacterium caprae* comb. nov., sp. nov. *Int. J. Syst. Evol. Microbiol.* 53, 1785–1789 (2003).
11. Wells, A.Q. & Oxon, D.M. Tuberculosis in wild voles. *Lancet* 1, 1221 (1937).
12. Cousins, D.V., et al. Tuberculosis in seals caused by a novel member of the *Mycobacterium tuberculosis* complex: *Mycobacterium pinnipedii* sp. nov. *Int. J. Syst. Evol. Microbiol.* 53, 1305–1314 (2003).
13. Alexander, K.A., et al. Novel *Mycobacterium tuberculosis* complex pathogen, *M. mungi*. *Emerg. Infect. Dis.* 16, 1296–1299 (2010).
14. van Ingen, J., et al. Characterization of *Mycobacterium orygis* as *M. tuberculosis* complex subspecies. *Emerg. Infect. Dis.* 18, 653–655 (2012).
15. Bezerra, A.V., dos Reis, E.M., Rodrigues, R.O., Cenci, A., Cerva, C. & Mayer, F.Q. Detection of *Mycobacterium tuberculosis* and *Mycobacterium avium* complexes by real-time PCR in bovine milk from Brazilian dairy farms. *J. Food Prot.* 78, 1037–1042 (2015).
16. Grange, J.M. & Yates, M.D. Zoonotic aspects of *Mycobacterium bovis* infection. *Vet. Microbiol.* 40, 137–151 (1994).
17. Prodinger, W.M., Indra, A., Koksalan, O.K., Kilicaslan, Z. & Richter, E. *Mycobacterium caprae* infection in humans. *Expert Rev. Anti Infect. Ther.* 12, 1501–1513 (2014).
18. Bos, K.I., et al. Pre-Columbian mycobacterial genomes reveal seals as a source of New World human tuberculosis. *Nature* 23, 494–497 (2015).
19. Standen, V.G., Santoro, C.M. & Arriaza, B.T. Síntesis y propuestas para el período arcaico en la costa del extremo norte de Chile. *Chungará* 36, 201–212 (2004).
20. Chayakulkeeree, M. & Naksanguan, T. Epidemiology and clinical characteristic of mycobacterial infections in human immunodeficiency virus-infected patients in Siriraj Hospital. *J. Med. Assoc. Thai.* 98, 238–244 (2015).
21. Rindi, L. & Garzelli, C. Genetic diversity and phylogeny of *Mycobacterium avium*. *Infect. Genet. Evol.* 21, 375–383 (2014).
22. Naser, A.S. & Thanigachalam, S. *Mycobacterium avium* subspecies *paratuberculosis* causes Crohn's disease in some inflammatory bowel disease patients. *World J. Gastroenterol.* 20, 7403–7415 (2014).
23. Sgarioni, S.A., et al. Occurrence of *Mycobacterium bovis* and non-tuberculous mycobacteria (NTM) in raw and pasteurized milk in the northwestern region of Paraná, Brazil. *Braz. J. Microbiol.* 45, 707–711 (2014).

24. Aydın, F.E., Ulger, M., Emekdaş, G., Aslan, G. & Günal, S. Isolation and identification of *Mycobacterium bovis* and non-tuberculous mycobacteria in raw milk samples in Mersin province. *Mikrobiyol. Bul.* 46, 283–289 (2012).
25. Cafri, U., Aslan, G., Direkel, S., Tarhan, G., Ceyhan, I. & Emekdaş, G. Identification and isolation of non-tuberculous mycobacteria from environmental samples. *Mikrobiyol. Bul.* 44, 395–403 (2010).
26. Briancesco, R., et al. An Italian investigation on non-tuberculous mycobacteria in an urban water supply. *Ann. Ig.* 26, 264–271 (2014).
27. Scherr, N. & Nguyen, L. *Mycobacterium* versus *Streptomyces*—We are different, we are the same. *Curr. Opin. Microbiol.* 12, 699–707 (2009).
28. Aguilar León, D., et al. *Mycobacterium bovis* with different genotypes and from different hosts induce dissimilar immunopathological lesions in a mouse model of tuberculosis. *Clin. Exp. Immunol.* 157, 139–147 (2009).
29. Blanco, F.C., et al. Differential transcriptome profiles of attenuated and hypervirulent strains of *Mycobacterium bovis*. *Microbes Infect.* 11, 956–963 (2009).
30. Altschul, S.F., Gish, W., Miller, W., Myers, E.W. & Lipman, D.J. Basic local alignment search tool. *J. Mol. Biol.* 215, 403–410 (1990).
31. Forrellad, M.A., et al. Virulence factors of the *Mycobacterium tuberculosis* complex. *Virulence* 4, 3–66 (2013).
32. Marri, P.R., Bannantine, J.P. & Golding, G.B. Comparative genomics of metabolic pathways in *Mycobacterium* species: Gene duplication, gene decay and lateral gene transfer. *FEMS Microbiol. Rev.* 30, 906–925 (2006).
33. Mackenzie, N., Alexander, D.C., Turenne, C.Y., Behr, M.A. & De-Buck, J.M. Genomic comparison of PE and PPE genes in the *Mycobacterium avium* complex. *J. Clin. Microbiol.* 47, 1002–1011 (2009).
34. Quadri, L.E., Sello, J., Keating, T.A., Weinreb, P.H. & Walsh, C.T. Identification of a *Mycobacterium tuberculosis* gene cluster encoding the biosynthetic enzymes for assembly of the virulence-conferring siderophore mycobactin. *Chem. Biol.* 5, 631–645 (1998).
35. Li, L., et al. The complete genome sequence of *Mycobacterium avium* subspecies *paratuberculosis*. *Proc. Natl. Acad. Sci. USA* 102, 12344–12349 (2005).
36. McGill, J.L., Sacco, R.E., Baldwin, C.L., Telfer, J.C., Palmer, M.V. & Waters, W.R. The role of gamma delta T cells in immunity to *Mycobacterium bovis* infection in cattle. *Vet. Immunol. Immunopathol.* 159, 133–143 (2014).
37. Pollock, J.M., Welsh, M.D. & McNair, J. Immune responses in bovine tuberculosis: Towards new strategies for the diagnosis and control of disease. *Vet. Immunol. Immunopathol.* 108, 37–43 (2005).
38. Villarreal-Ramos, B., McAulay, M., Chance, V., Martin, M., Morgan, J. & Howard, C.J. Investigation of the role of $CD8^+$ T cells in bovine tuberculosis in vivo. *Infect. Immun.* 71, 4297–4303 (2003).
39. Aranday-Cortes, E., et al. Upregulation of IL-17A, CXCL9 and CXCL10 in early-stage granulomas induced by *Mycobacterium bovis* in cattle. *Transbound. Emerg. Dis.* 60, 525–537 (2013).
40. Waters, W.R., Palmer, M.V., Whipple, D.L., Carlson, M.P. & Nonnecke, B.J. Diagnostic implications of antigen-induced gamma interferon, nitric oxide, and tumor necrosis factor alpha production by peripheral blood mononuclear cells from *Mycobacterium bovis* infected cattle. *Clin. Diagn. Lab. Immunol.* 10, 960–966 (2003).
41. Buddle, B.M., Pollock, J.M., Skinner, M.A. & Wedlock, D.N. Development of vaccines to control bovine tuberculosis in cattle and relationship to vaccine development for other intracellular pathogens. *Int. J. Parasitol.* 33, 555–566 (2003).
42. Roswurm, J.D. & Ranney, A.F. Sharpening the attack on bovine tuberculosis. *Am. J. Public Health* 63, 884–886 (1973).
43. Bovine TB Eradication Programme for England. Defra publications. http://www.defra.gov.uk/publications/2011/07/19/pb13601-bovine-tb-eradication-programme (2011).
44. Cosivi, O., et al. Zoonotic tuberculosis due to *Mycobacterium bovis* in developing countries. *Emerg. Infect. Dis.* 4, 59–70 (1998).
45. Kleeberg, H.H. Human tuberculosis of bovine origin in relation to public health. *Rev. Sci. Tech. Off. Int. Epiz.* 3, 11–32 (1984).
46. Hammer, P., Kiesner, C. & Walte, H.-G.C. Short communication: Effect of homogenization on heat inactivation of *Mycobacterium avium* subspecies *paratuberculosis* in milk. *J. Dairy Sci.* 97, 2045–2048 (2014).

47. Faria, A.C., et al. Short communication: Viable *Mycobacterium avium* subspecies *paratuberculosis* in retail artisanal Coalho cheese from Northeastern Brazil. *J. Dairy Sci.* 97, 4111–4114 (2014).
48. Pereira-Suárez, A.L., et al. Detection of *Mycobacterium tuberculosis* complex by PCR in fresh cheese from local markets in Hidalgo, Mexico. *J. Food Prot.* 77, 849–852 (2014).
49. Stephan, R., Schumacher, S., Tasara, T. & Grant, I.R. Prevalence of *Mycobacterium avium* subspecies *paratuberculosis* in Swiss raw milk cheeses collected at the retail level. *J. Dairy Sci.* 90, 3590–3595 (2007).
50. Dankner, W.M., Waecker, N.J., Essey, M.A., Moser, K., Thompson, M. & Davis, C.E. *Mycobacterium bovis* infections in San Diego: A clinic epidemiologic study of 73 patients and a historical review of a forgotten pathogen. *Medicine* 72, 11–37 (1993).
51. Lobue, P.A., Betacourt, W., Peter, C. & Moser, K.S. Epidemiology of *Mycobacterium bovis* disease in San Diego County, 1994–2000. *Int. J. Tuberc. Lung Dis.* 7, 180–185 (2003).
52. Harris, N.B., et al. Recovery of *Mycobacterium bovis* from soft fresh cheese originating in Mexico. *Appl. Environ. Microbiol.* 73, 1025–1028 (2007).
53. Rodwell, T.C., et al. Tracing the origins of *Mycobacterium bovis* tuberculosis in humans in the USA to cattle in Mexico using spoligotyping. *Int. J. Infect. Dis.* 14, e129–e135 (2010).
54. Centers for Disease Control and Prevention (CDC). Human tuberculosis caused by *Mycobacterium bovis*—New York City, 2001–2004. *Morb. Mortal. Wkly. Rep.* 54, 605–608 (2005).
55. Silva, M., et al. Tuberculosis in patients co-infected with *Mycobacterium bovis* and *Mycobacterium tuberculosis* in an urban area of Brazil. *Mem. Inst. Oswaldo Cruz.* 108, 321–327 (2013).
56. Martins, J.M., Galinari, É., Pimentel-Filho, N.J., Ribeiro, J.I., Furtado, M.M. & Ferreira, C.L. Determining the minimum ripening time of artisanal Minas cheese, a traditional Brazilian cheese. *Braz. J. Microbiol.* 46, 219–230 (2015).
57. Donaghy, J.A., Totton, N.L. & Rowe, M.T. Persistence of *Mycobacterium paratuberculosis* during manufacture and ripening of Cheddar cheese. *Appl. Environ. Microbiol.* 70, 4899–4905 (2004).
58. Kastli, P. & Binz, M. Die Lebensfähigkeit von *Mycobacterium tuberculosis* in verschiedenen Käsesorten. *Milchwissenschaft* 4, 391–394 (1949).
59. Hahn, H. Is manufacture of Emmental cheese from raw milk safe from the viewpoint of public health? *Tierärztl. Umsch.* 14, 254–256 (1959).
60. Hammer, B.W. & Babel, F.J. *Dairy Bacteriology*, 4th ed. John Wiley & Sons Inc., New York (1957).
61. Lafont, J. & Lafont, P. Some modifications in Koch's bacillus during ripening of blue cheese. *Bull. Acad. Vet. Fr.* 53, 457–461 (1981).
62. Bannantine, J.P., Li, L., Mwangi, M., Cote, R., Garay, J.A. & Kapur, V. Complete genome sequence of *Mycobacterium avium* subsp. *paratuberculosis*, isolated from human breast milk. *Genome Announc.* 2, e01252-13 (2014).
63. Bannantine, J.P., et al. Genome sequencing of ovine isolates of *Mycobacterium avium* subspecies *paratuberculosis* offers insights into host association. *BMC Genomics* 13, 89 (2012).
64. Singh, S.V., et al. Genome sequence of the "Indian bison type" biotype of *Mycobacterium avium* subsp. *paratuberculosis* strain S5. *Genome Announc.* 1, e00005-13 (2013).
65. Garnier, T. & Eiglmeier, K. The complete genome sequence of *Mycobacterium bovis*. *Proc. Natl. Acad. Sci. USA* 100, 7877–7882 (2003).
66. Camus, J.C., Pryor, M.J., Médigue, C. & Cole, S.T. Re-annotation of the genome sequence of *Mycobacterium tuberculosis* H37Rv. *Microbiology* 148, 2967–2973 (2002).
67. Li, L., Stoeckert, C.J. & Roos, D.S. OrthoMCL: Identification of ortholog groups for eukaryotic genomes. *Genome Res.* 13, 2178–2189 (2003).
68. Edgar, R.C. MUSCLE: Multiple sequence alignment with high accuracy and high throughput. *Nucleic Acids Res.* 32, 1792–1797 (2004).
69. Castresana, J. Selection of conserved blocks from multiple alignments for their use in phylogenetic analysis. *Mol. Biol. Evol.* 17, 540–552 (2000).
70. Stamatakis, A. RAxML-VI-HPC: Maximum likelihood-based phylogenetic analyses with thousands of taxa and mixed models. *Bioinformatics* 22, 2688–2690 (2006).
71. Rambaut, A. FigTree v1.4.0. http://tree.bio.ed.ac.uk/software/figtree (2014).
72. Deloger, M., El-Karoui, M. & Petit, M.A. A genomic distance based on MUM indicates discontinuity between most bacterial species and genera. *J. Bacteriol.* 191, 91–99 (2009).
73. Felsenstein, J. PHYLIP—Phylogeny inference package (version 3.2). *Cladistics* 5, 164–166 (1989).

74. Zhao, Y., et al. PanGP: A tool for quickly analyzing bacterial pan-genome profile. *Bioinformatics* 30, 1297–1299 (2014).
75. Stevenson, K. Genetic diversity of *Mycobacterium avium* subspecies *paratuberculosis* and the influence of strain type on infection and pathogenesis: A review. *Vet. Res.* 46, 64 (2015).
76. Centers for Disease Control and Prevention. Guide to the application of genotyping to tuberculosis prevention and control—Division of tuberculosis elimination. http://www.cdc.gov/tb/programs/genotyping (2012).
77. Allix-Béguec, C., Harmsen, D., Weniger, T., Supply, P. & Niemann, S. Evaluation and strategy for use of MIRU-VNTRplus, a multifunctional database for online analysis of genotyping data and phylogenetic identification of *Mycobacterium tuberculosis* complex isolates. *J. Clin. Microbiol.* 48, 2692–2699 (2008).
78. Weniger, T., Krawczyk, J., Supply, P., Niemann, S. & Harmsen, D. MIRU-VNTRplus: A web tool for polyphasic genotyping of *Mycobacterium tuberculosis* complex bacteria. *Nucleic Acids Res.* 38, W326–W331 (2010).
79. Zumárraga, M., et al. Understanding the relationship between *Mycobacterium bovis* spoligotypes from cattle in Latin American countries. *Res. Vet. Sci.* 94, 9–21 (2013).
80. Barandiaran, S., Pérez, A., Gioffré, A., Vivot, M., Cataldi, A. & Zumárraga, M.M. Tuberculosis in swine co-infected with *Mycobacterium avium* subsp. *hominissuis* and *Mycobacterium bovis* in a cluster from Argentina. *Epidemiol. Infect.* 143, 966–974 (2015).

13

Staphylococcus

Mar Rodríguez, Alicia Rodríguez, María Jesús Andrade, Elena Bermúdez, and Juan José Córdoba

CONTENTS

13.1 Introduction

Staphylococci are Gram-positive and catalase-positive ubiquitous bacteria found on the skin and mucous membranes of warm-blooded animals and humans. They can also be isolated from environmental sources such as soil, air, and water and from a wide range of foodstuffs including dry-cured meat products and cheeses.[1] To date, 52 species of staphylococci have been described.[2,3] Staphylococci are grouped into coagulase-positive (CPS) and coagulase-negative (CNS) staphylococci, according to their ability to coagulate rabbit plasma.[4] *Staphylococcus aureus* is the most frequently characterized CPS and a well-known etiological factor of a variety of infections including superficial skin inflammations, systemic infections, and septicemia.[4,5] Furthermore, *S. aureus* is a recognized causative agent of staphylococcal foodborne poisoning and has been considered the only representative of the genus *Staphylococcus* capable of producing enterotoxins (SEs).[4,6,7] However, some CNS strains showing enterotoxigenic capacity have been found in sheep and goat milk and cheeses,[8,9] Spanish dry-cured hams,[10,11] and even in starter cultures.[12] Enterotoxigenicity of these strains could pose considerable risk for food safety as, in some cases, they are introduced as starter cultures at high levels in many animal-derived products due to their positive impact on fermentation processes and sensory characteristics of products.[4,13]

13.2 Characteristics and Incidence of Foodborne Intoxications due to Enterotoxin-Producing Staphylococci

Staphylococcal foodborne poisoning is an intoxication produced by consumption of foods containing enough amounts of preformed enterotoxins.[14] It is among the leading causes of foodborne outbreaks in the European Union[15] and the United States.[16] However, the true incidence of foodborne disease could be a lot higher as sporadic foodborne illness caused by enterotoxin-producing staphylococci is not reportable.[17] Several reasons such as misdiagnosis, improper laboratory analysis, and lack of routine surveillance of clinical stool specimens for *S. aureus* or its enterotoxins have been reported as the cause of the low incidence of *S. aureus* foodborne disease.[17]

Although enterotoxigenic staphylococci are thermally destroyed, the cooked food may contain SEs because such toxins are thermostable and cannot be eliminated by heat processing.[18] Besides *Staphylococcus* spp. are usually very tolerant to NaCl and grow well in NaCl concentrations above 10%.[19] In addition, *Staphylococcus* spp. are tolerant to water activity reduction, being able to survive in ripened products when water activity is higher than 0.87.[11] Thus, foods that have been frequently involved in staphylococcal intoxication include meat and meat products, poultry and egg products, milk and dairy products, salads, bakery products, particularly cream-filled pastries and cakes, and sandwich fillings.[14,20]

Symptoms of staphylococcal intoxication appear in patients between 2 and 8 h after food consumption, and include nausea, vomiting, retching, and abdominal cramping, with or without diarrhea.[21,22] Vomiting is the most frequently observed symptom.[23] The disease is usually self-limiting and typically resolves within 24–48 h after onset.[14] Occasionally, it can be severe enough to warrant hospitalization, particularly when infants, elderly, or debilitated people are concerned.[24] The amount of SEs required to produce food-borne poisoning in humans is difficult to determine. Reliable results from the examination of food implicated in food-poisoning outbreaks are difficult to obtain because normally the enterotoxin is not uniformly distributed in the food and it is impossible to know how much food have been consumed by each person.[23]

TABLE 13.1

Biological Characteristics of Staphylococcal Enterotoxins and Enterotoxin-Like Toxins

Toxin	Genetic Element	Molecular Weight (kDa)	Superantigenic Activity	Emetic Activity	
				Monkey	House Musk Shrew
SEA	Prophage	27.1	+	+	+
SEB	Chromosome, plasmid, pathogenicity island	28.3	+	+	+
SEC1	Pathogenicity island	27.5	+	+	NE
SEC2	Pathogenicity island	27.6	+	NE	+
SEC3	Pathogenicity island	27.6	+	+	NE
SED	Plasmid	26.4	+	NE	+
SEE	Prophage	26.4	+	NE	+
SEG	*egc*, chromosome	27.0	+	+	+
SEH	Transposon	25.2	+	+	+
SEI	*egc*, chromosome	24.9	+	+	+
SElJ	Plasmids	28.6	+	NE	NE
SEK	Pathogenicity island	25.5	+	+	NE
SEL	Pathogenicity island	25.2	+	+	NE
SEM	*egc*, chromosome	24.8	+	+	NE
SEN	*egc*, chromosome	26.1	+	+	NE
SEO	*egc*, chromosome	26.8	+	+	NE
SEP	Prophage	26.6	+	+	+
SEQ	Pathogenicity island	25.1	+	+	NE
SER	Plasmid	27.1	+	+	+
SES	Plasmid	26.3	+	+	+
SET	Plasmid	22.6	+	+	+
SElU	*egc*, chromosome	27.2	+	NE	NE
SElV	*egc*, chromosome	27.6	+	NE	NE
SElX	Chromosome	19.3	+	NE	NE

Source: Adapted from Hennekinne, J.A., et al., *FEMS Microbiol. Rev.*, 36, 815, 2012; Omoe, K., et al., *Infect. Immun.*, 81, 3627, 2013; Hu, D.L. and Nakane, A. *Eur. J. Pharmacol.*, 722, 95, 2014.

NE, not examined.

Enterotoxin-producing staphylococci, mainly *S. aureus* strains, have been reported as producers of more than 20 different SEs[4,25,26] (Table 13.1). SEs were named on the basis of their emetic activities following oral administration in a primate model. Several SEs were designated as SE-like (SEl) since they either lack emetic properties or their emetic activities have not been tested in this model yet.[4,27] To date, 18 new types of SEs and SEl (SEG, SEH, SEI, SElJ, SElK, SElL, SElM, SElN, SElO, SElP, SElQ, SER, SES, SET, SElU, SElU2, SElV, and SElX) and five classical types [SEA, SEB, SEC (with the SC1, SC2, and SC3), SED, and SEE] have been reported.[4,6,28–31] Recently, the emetic activity of SElK, SElL, SElM, SElN, SElO, SElP, and SElQ has been demonstrated in a primate model.[28]

SEs are toxins of 20–30 kDa that have two separate biological activities: they act on the gastrointestinal tract and as a superantigen on the immune system that trigger a strong T-cell activation.[26,32] Among SEs, the SEB is nowadays classified as a potential bioweapon due to the fact that it has all the characteristics of an ideal biological agent.[25] Three mechanisms have been proposed to explain how SEs cause enteric illness: (1) the release of proinflammatory cytokines as a result of SE-induced superantigenic T-cell proliferation, (2) the binding of SEs to intestinal mast cells that leads to degranulation, and (3) a direct effect upon the intestinal epithelium affecting gut transit.[33]

SEs are resistant to environmental conditions (freezing, drying, heat treatment, and low pH), that easily destroy the enterotoxin-producing staphylococci strain. SEs are also resistant to proteolytic enzymes, retaining their activity in the digestive tract after ingestion.[6,19]

Within SEs, the SEA is the most commonly reported enterotoxin in foods, and its predominance (>75% of foodborne poisoning outbreaks) is well documented in different countries, followed by SED, SEC, and SEB.[14,34] Foodborne outbreaks where the remaining SEs are involved have been rarely reported. One possible reason could be that techniques to check for enterotoxigenic staphylococci are usually based on the detection of the SEs using commercial immunological kits, but these methods also have limitations. Besides, they are only suitable for the classical enterotoxins (SEA–SEE), while kits to detect recently reported enterotoxins are not available.[35,36] Current studies collected evidence of two outbreaks, suggesting that newly described SEs can be involved in foodborne poisoning.[36,37] In general, detection of SEs in foods is often difficult because of the small SE amounts present in foods and their proteic nature. Different techniques including animal models have been reported to accurately investigate staphylococcal foodborne disease.

13.3 Diagnosis of Staphylococcal Foodborne Poisoning

The diagnosis of staphylococcal foodborne poisoning is generally confirmed by at least one of the following methods: (1) recovery of 10^5 *S. aureus* cells/g from food remnants, (2) detection of SEs in food remnants, or (3) isolation of *S. aureus* of the same phage type from both patients and food remnants.[38] Conclusive diagnosis of staphylococcal foodborne poisoning is mainly based on the demonstration of SEs in the food.[39]

Staphylococcus strains are usually enumerated by using conventional microbiological techniques employed to detect CPS in food samples, according to EN ISO 6888-1,[40] EN ISO 6888-2,[41] or EN ISO 6888-3.[42] However, they are time consuming since they take up to 5–6 days to be performed. Moreover, it may not be enough for the reliable identification of all the possible enterotoxigenic staphylococci (atypical *S. aureus*, other CPS and CNS). Molecular biology methods have been used as an alternative to conventional microbiological techniques (often PCR and real-time PCR) to detect CPS in food samples. PCR-based methods detect genes encoding enterotoxins in strains of *Staphylococcus* spp. However, these techniques have a major limitation: the results inform the presence or absence of genes encoding SEs, but do not provide any information on their production. These methods therefore cannot be used as sole methods for confirming *Staphylococcus* spp. as the causative agent in an outbreak.[25] In addition, in some cases the confirmation of staphylococcal foodborne poisoning is difficult because *S. aureus* is heat or water activity reduction sensitive, whereas SEs are not. Thus, in heat-treated or in ripened food matrices with less than 0.87 water activity, *S. aureus* may be eliminated without inactivating SEs. In such cases, it is not possible to characterize a foodborne poisoning outbreak by detecting and enumerating *Staphylococcus* spp. in food remnants.[6]

There are three types of methods to detect bacterial toxins in foods—immunological tools, mass spectrometry-based methods, and bioassays.

Immunological tools are the most commonly used methods for detecting SEs in foods. Commercially available kits have been developed according to two different principles: (1) enzyme immunoassay comprising ELISA (TECRA Kit, TRANSIA PLATE, and RIDASCREEN SET) and enzyme-linked fluorescent assay (ELFA) (VIDAS SET and VIDASTM SET2), and (2) latex agglutination (SET-RPLA). It is widely recognized that the use of immunological methods to detect contaminants in food matrices is a difficult task, mainly because of the lack of specificity and sensitivity of the marketed kits. Moreover, only antibodies against SEA to SEE, SEG, SEH, and SElQ were available until recently. The immunological tools do not detect the other SEs.[25] The main drawback of current methods to detect enterotoxins based on specific polyclonal or monoclonal antibodies remains their high cost.[43]

As an alternative to immunological tools, other strategies based on *physicochemical techniques* have been developed. Among these, mass spectrometry method,[44] based on the use of immunoaffinity capture and matrix-assisted laser desorption/ionization time-of-flight mass spectrometry (MALDI-TOF-MS) or electrospray ionization (ESI),[45,46] biomolecular interaction analysis mass spectrometry (BIA-MS),[47] and high-performance liquid chromatography (HPLC)[48] have been proposed. These methods allow an accurate detection of SEs in foods but require investing in expensive HPLC, MALDI-TOF-MS, or BIA-MS equipments.

Bioassays are based on the ability of a suspected food extract to induce symptoms such as vomiting or gastrointestinal signs in laboratory animal models and/or superantigenic action in cell culture models.[25] These methods offer a very valuable alternative for foodborne staphylococcal investigation and will be explorated in a separate section.

13.4 Laboratory Models for Study of Staphylococcal Foodborne Poisoning

13.4.1 Animal Models

Although humans are ideal model to study human pathogens, it is not possible to use humans owing to safety, ethical, and expense-related concerns. However, human volunteers have been used in certain nonfatal diseases. Animal models are frequently utilized as a substitute. Among them, nonhuman primates (monkey, baboon, and chimpanzee) are ideal to mimic many diseases. Again, ethical and expense-related considerations limit their widespread applications.

Rodents are most commonly used as animal models and include mice, rats, rabbits, hamsters, and guinea pigs. Other animals including ferrets, pig or piglets, dogs, and cats are also utilized. Careful considerations while choosing an animal model include pathogens should infect animals by the same route as humans and exhibit a similar colonization pattern, similar tissue distribution patterns, and same degree of virulence as in humans.[49]

As previously described, the emetic activity of SEs is one of the main factors to consider it in the "category of SEs." The International Nomenclature Committee for Staphylococcal Superantigens (INCSS) emphasizes the relevance of staphylococcal foodborne poisoning (emetic activity). To be categorized as SEs, the toxins must demonstrate emetic activity via the oral route in a primate model. Lack of progress in elucidating the mechanism of the emetic activity of SEs can be partially attributed to the lack of convenient and appropriate animal models.[50] To date, the emetic activity of 17 SE types has been tested in monkey and only 12 SE types in house musk shrew models (Table 13.1). This is because the commonly used animal models, with the exception of monkeys, house musk shrews, and ferrets, are relatively insensitive to SEs, unless the toxin is injected intraperitoneally or intravenously. Emesis is the most readily observable reaction to SEs and animals without a vomiting mechanism, such as rodents, were of little value as test subjects.[51]

The most susceptible animal species to develop human-like enterotoxigenic disease are nonhuman primate models, mainly *cynomolgus* and *rhesus monkeys* (Table 13.2). Assays are performed by administering solutions of the enterotoxins in 10 mL of sterile distilled water to monkeys (2–3 kg) and fed to monkeys at a dose of 10 or 100 μg/kg by nasogastric intubation. The animals are observed for 5 h after

TABLE 13.2

Animal Models Used for the Detection of Staphylococcal Enterotoxins

Animal	Symptoms, Pathology	SE Tested	References
Cynomolgus monkey (*Macaca fascicularis*)	Emetic activity	A, B, C, D, E, K, L, M, N, O, P, Q, R, S, T	28,52,55,107,108
	Cytokine and chemokine response	B	109
Rhesus monkey (*Macaca mulatta*)	Emetic activity	A, B, C, D, E, G, H, I	53–55,110,111
	Cytokine response	A, B	112
Pigtail monkey (*Macaca nemestrina*)	Emetic activity	C1	80,81
House musk shrew (*Suncus murinus*)	Emetic activity	A, B, C, C2, D, E, G, H, I, P, R, S, T,	52,59–61
	Intestinal loop assay	A, C	61
Ferret (*Mustela putorius furo*)	Emetic activity	B, C2	66,67
Pig/piglet	Emetic activity	A, B	68,69
	Cytokine response	B	70
	Superantigen activity	B	71
Goat	Clinical symptomatology	B	72
Dog	Acute hemodynamic and gastrointestinal changes	Unknown	73
Cat	Emetic activity	A, B, C2	74,75
	Pyrogenic activity	A, B	76
Rabbit	Intestinal loop assay	A, C	61
	Pyrogenic activity	A, C2, K, Q, L	75,78–81
	Superantigenicity		
	Lethality	K	79
Rat	Gastrointestinal changes	A, B	84,85
	Superantigen activity	B	86,87
Mouse	Cytokine response	A, B, C3, K, Q	89–93,95
	Toxic shock	B	82

the oral administration of the toxin in parallel with real-time recording using a video camera. The number of vomiting events, the time until the first vomiting event (latency period), and behavioral changes are recorded. To minimize the effect of previous intoxication, there was at the minimum a 2-week interval between the intoxication experiments. A response in at least two animals is accepted as a positive reaction.[28,51] Monkeys have been considered to be the primary animal model.

Regarding the sensitivity of this method, the emetic dose for monkeys is somewhat variable for the different SEs. The least amount of SEA that was observed to produce an emetic reaction when given intragastrically was 5 μg/kg of weight, with SEB, SEC, and SEE requiring 10 μg/kg and SED requiring 20 μg/kg. The minimum dose required for the other SEs was not determined.[51] Studies with the new SEs reveal that higher concentrations are needed to induced emetic reactions in monkeys, with values ranging from 30 to 150 μg/kg.[28,52–54]

In addition, it has also been found that some monkeys responded only to some SEs, while other monkeys were sensitive to almost all the SEs, suggesting that the emetic responses of monkeys vary on an individual basis.[28,55]

Other researchers have shown that monkeys have developed tolerance to the emetic effects of SEs if repeated doses are given during a short period of time.[51,56]

Finally, this animal model has been used to demonstrate that the abdominal viscera is the site of action for the induction of the emetic reaction and that the SEs stimulate the vomiting center of the medulla oblongata through the vagus and sympathetic nerves.[57] The use of monkeys in researching SEs is severely restricted by the high cost, availability of these animals, and ethical considerations.[50,58] All these reasons have limited their use for routine testing.

The *house musk shrew* has been described as a suitable small animal model for research on the emetic response to various emetic SEs[52,59–61] (Table 13.2). The emetic assays are performed with house musk shrews weighing from 40 to 70 g. The shrews are housed under controlled conditions of illumination (12/12 h light/dark cycle). Purified SEs are diluted in 0.01 M phosphate-buffered saline (PBS; pH 7.2), and 200 μL volumes of SEs at an appropriate dilution were administered perorally or intraperitoneally to the house musk shrews. The animals are observed for emesis for 3 h after the administration of SEs. The number and times of vomiting, the time to the first vomiting episode, and any behavioral changes were recorded.[62] Vomiting occurred within 14–130 min after administration.

SEA provoked more potent emetic response *in vivo*. The 50% emetic dose of SEA by peroral administration is 10 times higher than by intraperitoneal administration (32 and 3 μg/kg of body weight, respectively).[61,62] Studies with other SEs reveal that higher concentrations are needed to induce emetic reactions in house musk shrews, with values ranging from 10 to 1000 μg/animal.[52,59,60] It is noteworthy that different types of SEs have different emetic activities in house musk shrews.[50]

Interestingly, in contrast to the emetic responses, diarrhea was observed in none of the animals tested, although SEA and SEC induced emetic responses. The results showed that the administration of SEA provoked a potent emetic response *in vivo* and showed high superantigenic activity *in vitro*, but did not induce diarrheal symptom in the animal model.[61]

This model has been useful for vaccine development against SEA, using a recombinantly attenuated SEA devoid of superantigen and emetic activity.[63] Furthermore, sera from these vaccinated animals inhibit SEA-induced proliferation of naïve shrew splenocytes (*in vitro*) as well as emesis.[52]

Another animal used as a model system for SEs detection has been the *ferret* (Table 13.2). Ferrets were used as the emetic model for finding antagonists to treat emesis induced by anticancer therapy[64] and respond to the full spectrum of agents known to induce emesis in humans.[65] An advantage of the ferret for the investigation of foodborne poisoning is that the morphology and physiology of its gastrointestinal tract have many features in common with the human gastrointestinal tract.[66]

For foodborne staphylococcal investigation, the emetic assays are performed with ferrets weighing 700–735 g. Prior to dosing with SEs, animals are deprived of food for 24 h, but are allowed free access to water. SEs are given to groups of five animals at doses of 1–5 mg into the stomach via an oral dosing tube.[66,67] Changes in the body temperature and the activity and the incidence of retching, vomiting, and defecation were monitored over a period of 3 h. This ferret model has been used to check oral SEB and SEC2 intoxications.[66,67] However, in the ferret, the dose of SEs required to elicit emesis is higher than that required to elicit it in primates. Possible explanations for requiring this large dose of SEs include receptor differences and/or more efficient degradation of SEs in the gut of ferrets versus humans or monkeys.[52]

Pig[68] and *piglet*[69] models have also been used to reproduce some of the features associated with staphylococcal intoxication in humans. However, it is not a widely used animal model. It has been only applied for SEA and SEB foodborne investigation (Table 13.2). In this animal model, several breeds (Yorkshire pig, Hampshire pig), crossbreeds, and pigs of different ages (piglets, weanling pigs—0.9 to 9.1-kg weight range) have been used.[68–71] For the analysis, each pig is placed in a separate cage with commercial pig feed and water provided *ad libitum*. Pigs are maintained under controlled lighting (15-h light and 9-h dark cycle) and temperature (20.5°C ± 0.75°C) conditions. The SE can be administered orally, by duodenal catheterization, or intravenously. In these conditions, vomiting occurs 90–180 min after emetic doses of SEs. The 50% emetic dose of SEA was between 20 and 50 μg. The lowest dose inducing emesis in weanling pigs was from 10 to 20 μg.[68–71] Pigs are somewhat more resistant than monkeys to the emetic effects of SEs.[68] This animal model has been used to determine the efficacy of immunization with vaccine against SEB too.[70,71]

Goats are used for studying *in vivo* gastrointestinal effects caused by SEB[72] (Table 13.2). This enterotoxin induces colic and watery diarrhea and a more pronounced increase in blood urea nitrogen and temperature in this animal.[52,72] For analysis, goats weighing between 21 and 45 kg are kept indoors and fed a diet of hay and pelleted concentrate. Water is provided *ad libitum*. Dose levels of SEB are between 0.02

and 0.5 µg/kg and are administered by nasal catheters into the rumen. This animal does not have emetic activity, but SEB-induced inhibition of rumen contractions, which is a consequence of internal vomiting.[72]

Dogs and cats have been used too, but only in some of the staphylococcal toxin investigations (Table 13.2). The *dog* model system has been used to study gastrointestinal and hemodynamic changes caused by intravenous and intraintestinal administration of SEs. They were observed by an autopsy 24 h after injection to make a histological study of injuries.[73]

The *cat* model has been widely used to study emetic[74,75] and pyrogenic[76] activities. For these analyses, cats aged about 8–10 weeks old and, with a mean body weight of 500 g, are caged separately in 12-h light/dark cycle at a temperature of 22°C–26°C with food and water available at all times. Two-milliliter volumes of SEA, SEB, and SEC2 are administered intraperitoneally to the cats. The animals are observed for emesis and diarrhea for up to 6 h after the intraperitoneal administration. The number and times of vomiting and diarrhea and the time to the first response episode are recorded for the foodborne staphylococcal investigation.[75]

The main disadvantage of using those models remain similar to the monkey model—high cost and short supply in the available tools for the study of SEs-associated immunopathology.[77]

Other animals like rabbits, rats, and mice also have low susceptibility to SEs; their response to SEs are not specific or they are not vomit-competent species.[50] These animals are used mainly as a model for the study of superantigenicity, pyrogenicity, capacity to enhance endotoxin shock, and lethality (Table 13.2).

The *rabbit* model was used to compare *in vivo* toxicity induced by SEC2 using intravenous injection to test the pyrogenic activity. Rectal temperatures of rabbits were measured with indwelling rectal thermometers and recorded for 4 h after pyrogen administration.[75] Pyrogen assays have also been used in rabbits that are administered with SEA,[78] SEK,[79] SEL,[80] and SEQ.[81]

Rabbit intestinal loop assay has been used to test the diarrheagenic activity of SEA and SEC.[61] New Zealand white and Dutch belted are the most frequently utilized rabbit strains for testing SEs.[52]

Rodents are frequently used as models because of their inbred homogeneity. In addition, large numbers of animals with the corresponding results are available in a relatively short time. However, mice are poor responders to SEs as the affinity of these toxins to mouse major histocompatibility complex class II (MHC class II) is much lower than that for human MHC class II.[82] Moreover, rat- and mouse-based models are regarded as being substantially less sensitive to SEB intoxication than monkey models.[83] Therefore, the use of potentiating agents such as D-galactosamine, actinomycin D, lipopolysaccharide (LPS), viruses, or even protozoa is required. Thus, lower amounts of these protein toxins elicit a quantifiable form of toxic shock useful for therapeutic and vaccine developments.[52]

Rat has been reported in a model of intestinal inflammation postweaning, based on the systemic administration of SEB[84] and SEA[85] (Table 13.2) and for evaluating the SEB superantigenic activity by intravenous administration.[86,87]

Mice are useful for basic toxin studies and discovery of therapeutics/vaccines for combating staphylococcal superantigen-induced shock. However, mice lack an emetic response and are thus not very appropriate for investigation of SEs foodborne poisoning.[82]

Transgenic mice with human MHC class II were found to be an ideal animal model for examining the biological effects of superantigens, as they responded to much lower doses of toxins due to the higher affinity binding of SEs to human MHC class II molecules.[82,88,89] For this purpose, several mouse strains (BALB/c, NMRI, C57BL/6, C3H/HeJ, C3H/OuJ, HLA-DR3, knockout mouse, etc.) weighing 18–20 g are used.[52]

Major studies using mice as animal model have focused on comparing the responses of T-cells in the gut-associated lymphoid tissue to different SEs administered by oral, intragastric, or intraperitoneal routes. SEA, SEB, SEC3 and SEK, and SEQ have been assessed by using this animal model[89–95] (Table 13.2).

13.4.2 Cell Culture Models

Cells derived from animal tissues are attractive models for studying pathogenesis. There are two types of cells: primary and secondary. Primary cells are mortal, consist of mixed cell types, and are short-lived. Secondary cells are immortal and consist of one type of cell.

There are several advantages of cell culture models over animal models: (1) it is a simple and controlled model to study host–SEs interaction, (2) it is easy to run experiments, (3) cells are able to multiply rapidly—thus the experiment can be conducted quickly, (4) the secondary cells are immortal if nutrients and proper culturing conditions are provided, (5) it is relatively inexpensive compared to animal model, specially secondary cells, and (6) it is ethically more acceptable than assays using animals.[49]

However, there are several limitations: (1) cultured mammalian cells are generally derived from tumor cells, and therefore, genetic aberrations have occurred in these cells, may lose traits of original tissue and lose tissue-specific receptor, (2) cultured cells consist of only one type of cell; therefore, interaction with concerted host cell cannot be studied with cell culture models, and (3) lack of mucus and other secretory components, which normally interact. Thus, animal models are often used to confirm or verify the findings from the *in vitro* cell culture models.[49]

The gastrointestinal injuries associated with SEs foodborne poisoning have been extensively studied for a number of years using various animal models (Section 13.4.1), but in particular, insight into how SEs breach the epithelial barrier is scarce and has been mainly studied in epithelial cell lines.[33] The binding of SEA and SEB toxins onto the surface of the enterocyte microvillus, mediated by binding to digalactosylceramide residues, has been demonstrated. This research further confirmed that the toxin entry into the enterocytes via apical endocytosis within the endosomes and the pathological events following SEs introduction into the intestinal area are due to the combined effect of SEs that disrupt the epithelial barrier by inducing enterocyte-cytopathic toxins produced by *S. aureus*.[96,97]

SEs act as superantigens that target the immune system, inducing massive T-cell activation, cytokine release, and systemic shock.[98] Examination of these SEs for the ability of causing mammalian cell damage provides a means to assay these toxins. Recognition of the superantigen–MHC II complex by T-cell receptor (TCR) results in cell signaling, proliferation, and subsequent release of cytokines/chemokines.[33,52] Immune cell activation by superantigens and subsequent cellular changes are similar to those of conventional antigens and require three important signals: (1) from superantigen interaction with TCR and activation of protein tyrosine kinases; (2) engagement of costimulatory molecules on antigen-presenting cells and T-cells, upon superantigen binding that optimizes T-cell activation; and (3) interleukin (IL)-1, tumor necrosis factor α (TNFα), interferon gamma (IFNγ), IL-2, IL-6, and chemokines, specifically monocyte chemoattractant protein-1 (MCP-1), which are induced directly by superantigens and represent the third signal for T-cell activation.[52] Clearly, SEs-based activation of cells involves a multifactorial event encompassing multiple host molecules.

Several types of cells are available for superantigenic activity bioassays induced by SEs. Thus, human peripheral blood mononuclear cells (PBMCs) and human B lymphoblastoid cells have been used for SEB, SEC2, SEP, SER, SES, SET, and SElX-induced immune response analysis.[31,52,60,67,83,99–101] Superantigenic effect was explored by treating T-lymphocytes isolated from thymus of rats or mice using various doses of SEA, SEG, SEI, SEK, SEM, SEN, SEO, and SEQ.[30,95,102,103] Splenocyte assays have been used for measuring the superantigenic activity of SEA, SEH, and SEL.[32,80,104,105]

Moreover, the interaction between SEs and MHC II molecules was studied using secondary cell models as the human B cell lymphoma Raji and human colorectal cancer cells.[67,106]

13.5 Conclusions

SEs, produced by the *Staphylococcus* genus, mainly *S. aureus*, belong to the family of superantigens that induce potent emesis and are involved in foodborne poisoning. Staphylococcal foodborne poisoning occurs as a consequence of consumption of food containing enough amounts of preformed enterotoxins. Vomiting is the most frequently observed symptom.

Enterotoxin-producing staphylococci have been reported as producers of 23 types of SEs. They are characterized by superantigenic activity and by induction of emesis. Emetic activity is the main factor that has to be met so as to be considered in the "category of SEs." The diagnosis of staphylococcal foodborne poisoning could be confirmed by detecting bacterial toxins in foods by immunological tools or mass spectrometry-based methods. However, additional laboratory models are needed to evaluate the emetic activity (just animal models) as superantigenic activity (both animal and cell models). Lack of progress in elucidating the mechanism of the emetic activity of SEs can be attributed to the lack of convenient and appropriate animal models. Thus, the commonly used animal models are relatively insensitive to the SEs.

The most susceptible animal species are monkeys, followed by house musk shrews and ferrets. Moreover, pigs, piglets, goats, dogs, and cats have been used. The main disadvantage of using these later models remains similar to the monkey models—high cost and short supply in the available tools for the study of SE-associated immunopathology.

Others animals such as rabbits, rats, and mice are less susceptible to SEs or their response to SEs are not specific or they are not vomit-competent species. These animals are mainly used as a model for the study of superantigenicity, pyrogenicity, capacity to enhance endotoxin shock, and lethality. The house musk shrew and the ferret appear to be valuable animal models for studying the emetic activity of SEs. Comparative studies using monkeys, house musk shrews, and ferrets to assess the emetic activity of SEs will lead to an understanding of the molecular basis of the emesis caused by SEs.[59]

Cells derived from animal tissue are attractive models for studying superantigenic activity. There are several advantages of cell culture over animal model: simple host–SEs interaction, easy, rapid, relatively inexpensive compared to animal models, and ethically more acceptable to assays using animals. But this is not an appropriate model for testing the emetic activity of SEs.

Finally, animal models could be useful for additional identification of SEs or the identification of staphylococcal strains that produce an unidentified enterotoxin. For this reason, when other methods become available for all SEs involved in staphylococcal foodborne poisoning, experimental models will not be employed for routine analysis, but only in special cases to confirm outbreak due to SEs.

Acknowledgments

This work has been funded by the Spanish Instituto Nacional de Investigación Agraria y Agroalimentaria (INIA) and the Spanish Comisión Interministerial de Ciencia y Tecnología with the projects RTA-2013-00070-C03-03 and Carnisenusa CSD2007-00016, Consolider Ingenio 2010, and GRU09162 of the Junta de Extremadura and FEDER.

REFERENCES

1. Irlinger, F., Safety assessment of dairy microorganisms: Coagulase-negative staphylococci, *Int. J. Food Microbiol.*, 126, 302, 2008.
2. National Center for Biotechnology Information (NCBI), http://www.ncbi.nlm.nih.gov/Taxonomy/Browser/wwwtax.cgi?mode=Root. Accessed April 30, 2015.
3. Tong, S.Y.C. et al., Novel staphylococcal species that form part of a *Staphylococcus aureus*-related complex: The non-pigmented *Staphylococcus argenteus* sp. nov. and the non-human primate-associated *Staphylococcus schweitzeri* sp. nov., *Int. J. Syst. Evol. Microbiol.*, 65, 15, 2015.
4. Podkowik, M. et al., Enterotoxigenic potential of coagulase-negative staphylococci, *Int. J. Food Microbiol.*, 163, 34, 2013.
5. Wertheim, H.F. et al., The role of nasal carriage in *Staphylococcus aureus* infections, *Lancet Infect. Dis.*, 5, 751, 2005.
6. Hennekinne, J.A., De Buyser, M.L. and Dragacci, S., *Staphylococcus aureus* and its food poisoning toxins: Characterization and outbreak investigation, *FEMS Microbiol. Rev.*, 36, 815, 2012.
7. Le Loir, Y., Baron, F. and Gautier, M., *Staphylococcus aureus* and food poisoning, *Genet. Mol. Res.*, 2, 63, 2003.
8. Bautista, L. et al., A quantitative study of enterotoxin production by sheep milk staphylococci, *Appl. Environ. Microbiol.*, 54, 566, 1988.
9. Vernozy-Rozand, C. et al., Enterotoxin production by coagulase-negative staphylococci isolated from goats' milk and cheese, *Int. J. Food Microbiol.*, 30, 271, 1996.
10. Marín, M.E., de la Rosa, M.C. and Cornejo, I., Enterotoxigenicity of *Staphylococcus* strains isolated from Spanish dry-cured hams, *Appl. Environ. Microbiol.*, 58, 1067, 1992.

11. Rodríguez, M. et al., Gram-positive, catalase-positive cocci from dry cured Iberian ham and their enterotoxigenic potential, *Appl. Environ. Microbiol.*, 62, 1897, 1996.
12. Zell, C. et al., Characterization of toxin production of coagulase-negative staphylococci isolated from food and starter cultures, *Int. J. Food Microbiol.*, 127, 246, 2008.
13. Even, S. et al., Low occurrence of safety hazards in coagulase negative staphylococci isolated from fermented foodstuffs, *Int. J. Food Microbiol.*, 139, 87, 2010.
14. Argudin, M.A., Mendoza, M.C. and Rodicio, M.R., Food poisoning and *Staphylococcus aureus* enterotoxins, *Toxins*, 2, 1751, 2010.
15. EFSA, The European Union summary report on trends and sources of zoonoses, zoonotic agents and food-borne outbreaks in 2013, *EFSA J.*, 13, 3991, 2015.
16. Scallan, E., Foodborne illness acquired in the United States—Major pathogens, *Emerg. Infect. Dis.*, 17, 7, 2011.
17. Kadariya, J., Smith, T.C. and Thapaliya, D., *Staphylococcus aureus* and staphylococcal food-borne disease: An ongoing challenge in public health, *BioMed Res. Int.*, 2014, 9, Article ID 827965, 2014.
18. Schmitt, M., Schuler-Schmid, U. and Scmidt-Lorenz, W., Temperature limits of growth, TNase, and enterotoxin production of *Staphylococcus aureus* strains isolated from foods, *Int. J. Food Microbiol.*, 11, 1, 1990.
19. Bergdoll, M.S., *Staphylococcus aureus*, in *Foodborne Bacterial Pathogens*, p. 463, Doyle, M.P. (Ed.), Marcel Dekker, New York, 1989.
20. Aydin, A., Sudagidan, M. and Muratoglu, K., Prevalence of staphylococcal enterotoxins, toxin genes and genetic-relatedness of foodborne *Staphylococcus aureus* strains isolated in the Marmara region of Turkey, *Int. J. Food Microbiol.*, 148, 99, 2011.
21. Balaban, N. and Rasooly, A., Staphylococcal enterotoxins, *Int. J. Food Microbiol.*, 61, 1, 2000.
22. Tranter, H.S., Foodborne staphylococcal illness, *Lancet*, 336, 1044, 1990.
23. Bergdoll, M.S. and Wong, A.C.L., Staphylococcal intoxications, in *Infections and Intoxications*, p. 523, Riemann H.P. and Cliver, D.O. (Eds.), Academic Press, San Diego, CA, 2006.
24. Murray, R.J., Recognition and management of *Staphylococcus aureus* toxin-mediated disease, *Intern. Med. J.*, 2, S106, 2005.
25. Hennekinne, J.A. and Dragacii, S., Staphylococcal enterotoxins, in *Security Sensitive Microbes and Toxins*, Liu, D. (Ed.), p. 533, CRC Press Taylor & Francis Group, Boca Raton, FL, 2014.
26. Otto, M., *Staphylococcus aureus* toxins, *Curr. Opin. Microbiol.*, 17, 32, 2014.
27. Lina, G. et al., International Nomenclature Committee for Staphylococcal Superantigens, Standard nomenclature for the superantigens expressed by *Staphylococcus, J. Infect. Dis.*, 189, 2334, 2004.
28. Omoe, K. et al., Emetic potentials of newly identified staphylococcal enterotoxin-like toxins, *Infect. Immun.*, 81, 3627, 2013.
29. Ortega, E. et al., Multiple roles of *Staphylococcus aureus* enterotoxins: Pathogenicity, superantigenic activity, and correlation to antibiotic resistance, *Toxins*, 2, 2117, 2010.
30. Thomas, D.Y. et al., Staphylococcal enterotoxin-like toxins U2 and V, two new staphylococcal superantigens arising from recombination within the enterotoxin gene cluster, *Infect. Immun.*, 74, 4724, 2006.
31. Wilson, G.J. et al., A novel core genome-encoded superantigen contributes to lethality of community-associated MRSA necrotizing pneumonia, *PLoS Pathog.*, 7, e1002271, 2011.
32. Rasooly, R. and Hernlem, B.J., CD154 as a potential early molecular biomarker for rapid quantification analysis of active *Staphylococcus* enterotoxin A, *FEMS Immunol. Med. Microbiol.*, 64, 169, 2012.
33. Principato, M.A. and Qian, B.F., Staphylococcal enterotoxins in the etiopathogenesis of mucosal autoimmunity within the gastrointestinal tract, *Toxins*, 6, 1471, 2014.
34. Hennekinne, J.A. et al., Intralaboratory validation according to the EN ISO 16 140 Standard of the Vidas SET2 detection kit for use in official controls of staphylococcal enterotoxins in milk products, *J. Appl. Microbiol.*, 102, 1261, 2007.
35. Nagaraj, S. et al., Development and evaluation of a novel combinatorial selective enrichment and multiplex PCR technique for molecular detection of major virulence-associated genes of enterotoxigenic *Staphylococcus aureus* in food samples, *J. Appl. Microbiol.*, 116, 435, 2013.
36. Johler, S. et al., Further evidence for staphylococcal food poisoning outbreaks caused by *egc*-encoded enterotoxins, *Toxins*, 7, 997, 2015.
37. Baumgartner, A., Niederhauser, I. and Johler, S., Virulence and resistance gene profiles of *Staphylococcus aureus* strains isolated from ready-to-eat foods, *J. Food Prot.*, 77, 1232, 2014.

38. Bryan, F.L., Guzewich, J.J. and Todd, E.C.D., Surveillance of foodborne disease II. Summary and presentation of descriptive data and epidemiologic patterns; their value and limitations, *J. Food Prot.*, 60, 567, 1997.
39. Kérouanton, A. et al., Characterization of *Staphylococcus aureus* strains associated with food poisoning outbreaks in France, *Int. J. Food Microbiol.*, 115, 369, 2007.
40. Anonymous, EN ISO 6888-1, Microbiology of food and animal feeding stuffs—Horizontal method for the enumeration of coagulase-positive staphylococci (*Staphylococcus aureus* and other species)—Part 1: Technique using Baird-Parker agar medium, European Committee for Standardization, Brussels, 1999.
41. Anonymous, EN ISO 6888-2, Microbiology of food and animal feeding stuffs—Horizontal method for the enumeration of coagulase-positive staphylococci (*Staphylococcus aureus* and other species)—Part 2: Technique using rabbit plasma fibrinogen agar medium, European Committee for Standardization, Brussels, 1999.
42. Anonymous, EN ISO 6888-3, Microbiology of food and animal feeding stuffs—Horizontal method for the enumeration of coagulase-positive staphylococci (*Staphylococcus aureus* and other species)—Part 3: Detection and MPN technique for low numbers, European Committee for Standardization, Brussels, 2003.
43. Sospedra, I., Soriano, J.M. and Mañés, J., Enterotoxinomics: The omic sciences in the study of staphylococcal toxins analyzed in food matrices, *Food Res. Int.*, 54, 1052, 2013.
44. Dupuis, A. et al., Protein Standard Absolute Quantification (PSAQ) for improved investigation of staphylococcal food poisoning outbreaks, *Proteomics*, 8, 4633, 2008.
45. Schlosser, G. et al., Coupling immunomagnetic separation on magnetic beads with matrix-assisted laser desorption ionization-time of flight mass spectrometry for detection of staphylococcal enterotoxin, *Appl. Environ. Microbiol.*, 73, 6945, 2007.
46. Sospedra, I. et al., Analysis of staphylococcal enterotoxin A in milk by matrix-assisted laser desorption/ionization-time of flight mass spectrometry, *Anal. Bioanal. Chem.*, 400, 1525, 2011.
47. Nedelkov, D., Rasooly, A., and Nelson, R.W., Multitoxin biosensor mass spectrometry analysis: A new approach for rapid, real-time, sensitive analysis of staphylococcal toxins in food, *Int. J. Food Microbiol.*, 60, 1, 2000.
48. Sospedra, I. et al., Rapid whole protein quantification of staphylococcal enterotoxin B by liquid chromatography, *Food Chem.*, 133, 163, 2012.
49. Bhunia, A.K., Animal and cell culture models to study foodborne pathogen interaction, in *Foodborne Microbial Pathogens, Mechanisms and Pathogenesis*, Heldman, D.R. (Ed.), p. 113, Springer, New York, 2008.
50. Hu, D.L. and Nakane, A., Mechanisms of staphylococcal enterotoxin-induced emesis, *Eur. J. Pharmacol.*, 722, 95, 2014.
51. Bergdoll, M.S. and Wong, A.C.L. Staphylococcal intoxications, in *Foodborne Infection and Intoxication*, p. 523, Riemann, H.P. and Cliver, D.O. (Eds.), Academic Press, London, UK, 2006.
52. Ono, H.K. et al., Identification and characterization of two novel staphylococcal enterotoxins, types S and T, *Infect. Immun.*, 76, 4999, 2008.
53. Munson, S.H. et al., Identification and characterization of staphylococcal enterotoxin types G and I from *Staphylococcus aureus*, *Infect. Immun.*, 66, 3337, 1998.
54. Su, Y.C. and Wong, A.C.L., Identification and purification of a new staphylococcal enterotoxin, H., *Appl. Environ. Microbiol.*, 61, 1438, 1995.
55. Bergdoll, M.S., Monkey feeding test for staphylococcal enterotoxin, *Methods Enzymol.*, 165, 324, 1988.
56. Sugiyama, H., Bergdoll, M.S. and Dack, G.M., Early development of a temporary resistance to the emetic action of staphylococcal enterotoxin, *J. Infect. Dis.*, 111, 233, 1962.
57. Ono, H.K. et al., Submucosal mast cells in the gastrointestinal tract are a target of staphylococcal enterotoxin type A, *FEMS Immunol. Med. Microbiol.*, 64, 392, 2012.
58. Krakauer, T. and Stiles, B.G., The staphylococcal enterotoxin (SE) family SEB and siblings, *Virulence*, 4, 759, 2013.
59. Hu, D.L. et al., Induction of emetic response to staphylococcal enterotoxins in the house musk shrew (*Suncus murinus*), *Infect. Immun.*, 71, 567, 2003.
60. Omoe, K. et al., Characterization of novel staphylococcal enterotoxin-like toxin type P, *Infect. Immun.*, 73, 5540, 2005.
61. Maina, E.K. et al., Staphylococcal enterotoxin A has potent superantigenic and emetic activities but not diarrheagenic activity, *Int. J. Med. Microbiol.*, 302, 88, 2012.

62. Hu, D.L. et al., Emesis in the shrew mouse (*Suncus murinus*) induced by peroral and intraperitoneal administration of staphylococcal enterotoxin A, *J. Food Protect.*, 62, 1350, 1999.
63. Hu, D.L. et al., Immunization with a nontoxic mutant of staphylococcal enterotoxin A, SEAD227A, protects against enterotoxin-induced emesis in house musk shrews, *J. Infect. Dis.*, 199, 302, 2009.
64. Andrews, P.L.R., and Davies C.J., Physiology of emesis induced by anti-cancer therapy, in *Serotonin and the Scientific Basis of Anti-emetic Therapy*, Reynolds, D.J.M., Andrews, P.L.R., and Davis, C.J. (Eds.), p. 25, Oxford Clinical Communications, Oxford, UK, 1995.
65. King, G.L., Animal models in the study of vomiting, *Can. J. Physiol. Pharmacol.*, 68, 260, 1990.
66. Wright, A., Andrews, P.L.R. and Titball, R.W., Induction of emetic, pyrexic, and behavioral effects of *Staphylococcus aureus* enterotoxin B in the ferret, *Infect. Immun.*, 68, 2386, 2000.
67. Hui, J. et al., *Staphylococcus aureus* enterotoxin C2 mutants: Biological activity assay in vitro, *J. Ind. Microbiol. Biotechnol.*, 35, 975, 2008.
68. Taylor, S.L. et al., Emetic action of staphylococcal enterotoxin A on weanling pigs, *Infect. Immun.*, 36, 1263, 1982.
69. Van Gessel, Y.A. et al., Functional piglet model for the clinical syndrome and post-mortem findings induced by staphylococcal enterotoxin B, *Exp. Biol. Med. (Maywood)*, 229, 1061, 2004.
70. Hudson, L.C. et al., Sublethal staphylococcal enterotoxin B challenge model in pigs to evaluate protection following immunization with a soybean-derived vaccine, *Clin. Vaccine Immunol.*, 20, 24, 2013.
71. Inskeep, T.K. et al., Oral vaccine formulations stimulate mucosal and systemic antibody responses against staphylococcal enterotoxin B in a piglet model, *Clin. Vaccine Immunol.*, 17, 1163, 2010.
72. Van Miert, A.S., Van Duin, C.T. and Schotman, A.J., Comparative observations of fever and associated clinical hematological and blood biochemical changes after intravenous administration of staphylococcal enterotoxins B and F (toxic shock syndrome toxin-1) in goats, *Infect. Immun.*, 46, 354, 1984.
73. Kocandrle, V., Houttuin, E. and Prohaska, J.V., Acute hemodynamic and gastrointestinal changes produced by staphylococcal exotoxin and enterotoxin in dogs, *J. Surg. Res.*, 6, 50, 1966.
74. Clark, W.G., Vanderhooft, G.F. and Borison, H.L., Emetic effect of purified staphylococcal enterotoxin in cats, *Proc. Soc. Exp. Biol. Med.*, 111, 205, 1962.
75. Wang, X. et al., Enhancement of superantigen activity and antitumor response of staphylococcal enterotoxin C2 by site-directed mutagenesis, *Cancer Immunol. Immunother.*, 58, 677, 2009.
76. Clark, W.G. and Page, J.S., Pyrogenic responses to staphylococcal enterotoxins A and B in cats, *J. Bacteriol.*, 96, 1940, 1968.
77. Pinchuk, I.V., Beswick, E.J. and Reyes, V.E., Staphylococcal enterotoxins, *Toxins*, 2, 2177, 2010.
78. Huang, W.T., Lin, MT. and Won, S.J., Staphylococcal enterotoxin A-induced fever is associated with increased circulating levels of cytokines in rabbits, *Infect. Immun.*, 65, 2656, 1997.
79. Orwin, P.M. et al., Biochemical and biological properties of staphylococcal enterotoxin K, *Infect. Immun.*, 69, 360, 2001.
80. Orwin, P.M. et al., Characterization of *Staphylococcus aureus* enterotoxin L, *Infect. Immun.*, 71, 2916, 2003.
81. Orwin, P.M. et al., Characterization of a novel staphylococcal enterotoxin-like superantigen, a member of the group V subfamily of pyrogenic toxins, *Biochemistry*, 41, 14033, 2002.
82. Krakauer, T., Buckley, M. and Fisher, D., Murine models of staphylococcal enterotoxin B-induced toxic shock, *Mil. Med.*, 175, 917, 2010.
83. Lindsay, C.D. and Griffiths, G.D., Addressing bioterrorism concerns: Options for investigating the mechanism of action of *Staphylococcus aureus* enterotoxin B, *Hum. Exp. Toxicol.*, 32, 606, 2013.
84. Pérez-Bosque, A. and Moretó, M., A rat model of mild intestinal inflammation induced by *Staphylococcus aureus* enterotoxin B, *Proc. Nutr. Soc.*, 69, 447, 2010.
85. Beery, J.T. et al., Effects of staphylococcal enterotoxin A on the rat gastrointestinal tract, *Infect. Immun.*, 44, 234, 1984.
86. Yang, W.X. et al., Decreased Vβ8.2 T-cells in neonatal rats exposed prenatally to staphylococcal enterotoxin B are further deleted by restimulation in an in vitro cultured thymus, *Mol. Med. Rep.*, 10, 989, 2014.
87. Zhang, T. et al., Staphylococcal enterotoxin B administration during pregnancy imprints the increased CD4:CD8 T-cell ratio in the peripheral blood from neonatal to adult offspring rats, *J. Med. Microbiol.*, 64, 1, 2015.
88. Roy, C.J. et al., Human leukocyte antigen-DQ8 transgenic mice: A model to examine the toxicity of aerosolized staphylococcal enterotoxin B, *Infect. Immun.*, 73, 2452, 2005.

89. Tilahun, M.E. et al., Chimeric anti-staphylococcal enterotoxin B antibodies and lovastatin act synergistically to provide in vivo protection against lethal doses of SEB, *PLoS One*, 6, e-27203, 2011.
90. Spiekermann, G.M. and Nagler-Anderson, C., Oral administration of the bacterial superantigen staphylococcal enterotoxin B induces activation and cytokine production by T cells in murine gut-associated lymphoid tissue, *J. Immunol.*, 161, 5825, 1998.
91. Stiles, B.G. et al., Correlation of temperature and toxicity in murine studies of staphylococcal enterotoxins and toxic shock syndrome toxin 1, *Infect. Immun.*, 67, 1521, 1999.
92. Chen, J.Y. et al., Increased susceptibility to staphylococcal enterotoxin B intoxication in mice primed with actinomycin D, *Infect. Immun.*, 62, 4626, 1994.
93. Miethke, T. et al., T cell-mediated lethal shock triggered in mice by the superantigen staphylococcal enterotoxin B: Critical role of tumor necrosis factor, *J. Exp. Med.*, 175, 91, 1992.
94. Huzella, L.M. et al., Central roles for IL-2 and MCP-1 following intranasal exposure to SEB: A new mouse model, *Res. Vet. Sci.*, 86, 241, 2009.
95. Kang, H. et al., Superantigenicity analysis of staphylococcal enterotoxins SElK and SElQ in a mouse model, *RSC Adv.*, 5, 29684, 2015.
96. Danielsen, E.M., Hansen, G.H. and Karlsdottir, E., *Staphylococcus aureus* enterotoxins A and B: Binding to the enterocyte brush border and uptake by perturbation of the apical endocytic membrane traffic, *Histochem. Cell Biol.*, 139, 513, 2013.
97. Edwards, L.A. et al., Enterotoxin-producing staphylococci cause intestinal inflammation by a combination of direct epithelial cytopathy and superantigen-mediated T cell activation, *Inflamm. Bowel Dis.*, 18, 624, 2012.
98. Fraser, J. et al., Superantigens-powerful modifiers of the immune system, *Mol. Med. Today*, 6, 125, 2000.
99. Krakauer, T., Sulfasalazine attenuates staphylococcal enterotoxin B-induced immune responses, *Toxins*, 7, 553, 2015.
100. Boles, J.W. et al., Generation of protective immunity by inactivated recombinant staphylococcal enterotoxin B vaccine in nonhuman primates and identification of correlates of immunity, *Clin. Immunol.*, 108, 51, 2003.
101. Omoe, K. et al., Biological properties of staphylococcal enterotoxin-like toxin type R, *Infect. Immun.*, 72, 3664, 2004.
102. René-Trouillefou, M. et al., Staphylococcal enterotoxin A: Partial unfolding caused by high pressure or denaturing agents enhances superantigenicity, *Biochim. Biophys. Acta*, 1804, 1322, 2010.
103. Jarraud, S. et al., *egc*, a highly prevalent operon of enterotoxin gene, forms a putative nursery of superantigens in *Staphylococcus aureus*, *J. Immunol.*, 166, 669, 2001.
104. Rasooly, R. and Hernlem, B.J., Quantitative analysis of *Staphylococcus* enterotoxin a by differential expression of IFN-γ in splenocyte and $CD4^+$ T-cells, *Sensors*, 14, 8869, 2014.
105. Liu, Y. et al., Staphylococcal enterotoxin H induced apoptosis of bovine mammary epithelial cells in vitro, *Toxins*, 6, 3552, 2014.
106. Herrmann, T., Accolla, R.S. and MacDonald, H.R., Different staphylococcal enterotoxins bind preferentially to distinct major histocompatibility complex class II isotypes, *Eur. J. Immunol.*, 19, 2171–1989.
107. Reck, B. et al., Protection against the staphylococcal enterotoxin-induced intestinal disorder in the monkey by anti-idiotypic antibodies, *Proc. Natl. Acad. Sci. USA*, 85, 3170, 1988.
108. Reiser, R.F. et al., Identification, purification, and some physicochemical properties of staphylococcal enterotoxin C3, *Infect. Immun.*, 45, 625, 1984.
109. He, C., Narayanan, P.K. and Fort, M.M., Assessment of the performance of three multiplex array panels for the detection of circulating cytokines and chemokines in naive, LPS, and SEB-treated cynomolgus macaques, *Toxicol. Pathol.*, 42, 286, 2014.
110. Stiles, J.W. and Denniston, J.C., Response of the rhesus monkey, *Macaca mulatta*, to continuously infused staphylococcal enterotoxin B, *Lab. Invest.*, 25, 617, 1971.
111. Sheahan, D.G. et al., The effect of staphylococcal enterotoxin on the epithelial mucosubstances of the small intestine of rhesus monkeys, *Am. J. Pathol.*, 60, 1, 1970.
112. Krakauer, T., et al., Superantigen-induced cytokine release from whole-blood cell culture as a functional measure of drug efficacy after oral dosing in nonhuman primates, *Res. Vet. Sci.*, 83, 182, 2007.

14

Streptococcus

Dongyou Liu

CONTENTS

14.1 Introduction

The genus *Streptococcus* comprises a large group of Gram-positive bacteria that have been known to cause human diseases from ancient times. A quick review of historical records indicates that in 1553, a rash (then termed "rossalia," with the whole body covered by numerous spots of large and small, fiery, and red) that differs from measles was first described by Giovanni Filippo Ingrassias of Italy; in 1565, a sore throat epidemic was mentioned by Johann Weyer of the Netherlands; in 1578, scarlet fever (showing general weariness, headache, redness of the eyes, sore throat, and fever) was noted by Jean Cottyar; in 1628, an epidemic showing scarlatinal desquamation, arthritis, and postscarlatinal dropsy and ascites was documented by Daniel Sennert; in 1675, the term "scarlatina" was first used by Sydenham to describe a disease that is distinct from other exanthemas (e.g., measles); in the 1840s, childbed fever was linked by Ignac Semmelweis of Hungary to medical personnel failing to wash their hands and who then transmitted the disease to patients; in 1874, an organism isolated from patients with erysipelas and wound infections was named "streptococcus" (Greek *streptos*, a chain; *coccos*, a berry) by Theodor Billroth of Austria due to its formation of short chains; in 1879, streptococcus was confirmed by Louis Pasteur as the etiological agent of puerperal fever, which caused the highest mortality rates of women and newborns at that time; in 1884, the bacterium isolated from suppurative lesions was defined by Friedrich Julius Rosenbach as *Streptococcus pyogenes* (Greek *pyo*, pus, and *genes*, forming); shortly afterward, the previously proposed species names of pyogenes, eryespaltis, scarlatinae, and puerperalis were united under the single name *Streptococcus pyogenes* by Andrews and Christie; in 1909, serotype-specific immunity against streptococci was reported by Meakins; in the 1920s, hemolytic streptococci that produced a secreted toxin (known as scarlet fever toxin or Dick toxin) were identified by George and Gladys Dick as the causative agent of sore throat that is accompanied by scarlet fever [1].

To date, more than 50 species have been recognized in the genus *Streptococcus*, of which nine (*S. pyogenes*, *S. agalactiae*, *S. equisimilis*, *S. bovis*, *S. anginosus*, *S. sanguinis*, *S. mitis*, *S. mutans*, and *S. pneumoniae*) have been implicated in human infections. In particular, *S. pyogenes* (commonly referred to as group A *Streptococcus* or GAS) is responsible for a majority of human streptococcal diseases, with clinical manifestations ranging from pharyngitis, impetigo, cellulitis, scarlet fever, puerperal sepsis, bacteremia, pneumonia, streptococcal toxic shock syndrome (STSS), necrotizing fasciitis, acute rheumatic fever (ARF), rheumatic heart disease (RHD), to acute poststreptococcal glomerulonephritis (APSGN). All together, over half a million deaths per year worldwide are attributable to *S. pyogenes* [2]. This chapter will focus on *S. pyogenes*, beginning with a brief overview on its classification, morphology, genomics, biology, epidemiology, clinical features, pathogenesis, diagnosis, treatment, and prevention, followed by discussion on laboratory models applied to *S. pyogenes* research.

14.1.1 Classification and Morphology

14.1.1.1 Classification

Classified in the family Streptococcaceae, order Lactobacillales, class Bacilli, phylum Firmicutes, domain Bacteria, the genus *Streptococcus* consists of more than 50 recognized species of facultative anaerobic Gram-positive cocci (GPC), namely, *S. agalactiae*, *S. anginosus*, *S. bovis*, *S. canis*, *S. constellatus*, *S. downei*, *S. dysgalactiae*, *S. equinus*, *S. ferus*, *S. infantarius*, *S. iniae*, *S. intermedius*, *S. milleri*, *S. mitis*, *S. mutans*, *S. oralis*, *S. orisratti*, *S. parasanguinis*, *S. peroris*, *S. pneumoniae*, *S. pseudopneumoniae*, *S. pyogenes*, *S. ratti*, *S. salivarius*, *S. tigurinus*, *S. thermophilus*, *S. sanguinis*, *S. sobrinus*, *S. suis*, *S. uberis*, *S. vestibularis*, *S. viridans*, and *S. zooepidemicus* [3].

Based on the hemolytic properties detected on blood agar plates, members of the genus *Streptococcus* are distinguished into three categories: α-, β-, and γ-hemolysis. α-*Hemolytic* species produce hydrogen peroxide that oxidizes hemoglobin within red blood cells to become green methemoglobin, leading to a greenish and dark zone surrounding and underneath the colonies with short chains (as in the case of *S. pneumoniae* and *S. viridans* or viridans streptococci; viridans, from Latin *vĭrĭdis*, green). β-*Hemolytic* species produce streptolysins (e.g., streptolysin O and streptolysin S) that cause complete lysis of red blood cells in blood agar plate, giving a clear (lightened/yellow and transparent) zone around and underneath the colonies with short chains (as in the case of *S. haemolyticus*). More specifically, streptolysin O (SLO, an oxygen-sensitive cytotoxin secreted by most GAS) interacts with cholesterol in the membrane of eukaryotic cells (red and white blood cells, macrophages, and platelets), causing β-hemolysis under the surface of blood agar. Streptolysin S (SLS, an oxygen-stable cytotoxin also secreted by most GAS strains) induces clearing on the surface of blood agar. It is notable that some weakly β-hemolytic species (e.g., *S. agalactiae*, *Clostridium perfringens*, and *Listeria monocytogenes*) cause intense β-hemolysis when grown together with a strain of *Staphylococcus* (so called the CAMP test). γ-*Hemolytic* species cause no hemolysis (no change) on blood agar (as in the case of enterococci) [3].

Streptococcus isolates may also be subgrouped by serological typing schemes targeting M protein, T-antigens (from pili), and serum opacity factor (SOF), which are LPXTG-linked (or similar) surface proteins with high levels of antigenic heterogeneity and are direct determinants of host tissue site preferences of infection [4].

Developed by Lancefield in 1933, serological typing scheme using group-specific antisera to M protein (a surface protein that is responsible for colony's matte appearance) differentiates *Streptococcus* β-hemolytic species into Lancefield groups A to X (excluding I and J), with strains from human diseases classified as group A, those from bovine and dairy sources as group B, those from other animal sources as group C, and so on. GAS consists of *S. pyogenes*, which causes both noninvasive and invasive infections in humans. The noninvasive infections are more common but less severe, as exemplified by streptococcal pharyngitis (strep throat), impetigo, and scarlet fever. The invasive infections are more severe but less common, as exemplified by STSS, necrotizing fasciitis, pneumonia, and bacteremia. Complications of GAS infections include ARF and acute glomerulonephritis. Affecting the joints, kidneys, and heart valves, rheumatic fever results from damages caused by the antibodies that are generated by the host immune system against the untreated GAS infection. These antibodies cross-react with other proteins in the body, leading to self-inflicted attack and damage. Indeed, the group A *S. pyogenes* strains can be further subdivided into >200 M-types using a combination of serological and molecular typing methods. Group B *Streptococcus* (GBS) consists of *S. agalactiae*, which causes pneumonia, meningitis, and occasional systemic bacteremia in neonates and the elderly. As it also colonizes the intestines and the female reproductive tract, this bacterium may contribute to premature rupture of membranes during pregnancy and subsequent transmission to the infant. Group C Streptococci include *S. equi* (causing strangles in horses), *S. zooepidemicus* (infecting cattle and horses, with *S. equi* being a clonal descendent or biovar of the ancestral *S. zooepidemicus*), and *S. dysgalactiae* (causing pharyngitis and other pyogenic infections similar to GAS). Group D streptococci consist of enterococcal and nonenterococcal strains. The former have been reclassified and placed in the genus *Enterococcus* (including *E. faecalis*, *E. faecium*, *E. durans*, and *E. avium*, with *E. faecalis* being sometimes α-hemolytic and *E. faecium* being sometimes β-hemolytic) (see Chapter 10); the latter include *S. bovis* and *S. equinus*. Group F streptococci (so-called "minute hemolytic streptococci") are represented by *S. anginosus* or the *S. milleri* group. Group G streptococci are usually, but not exclusively, β-hemolytic and represented by *S. canis* that typically occurs in animals, but may cause infection in humans. Group H streptococci are found in canines and rarely cause illness unless humans have direct contact with a canine (mouth-to-mouth or canine licking a human hand) [3].

The serotyping scheme based on T-antigens (trypsin-resistant surface antigens, which are contained within extended surface pili composed of covalently linked polymers of two or three distinct gene products) divides *S. pyogenes* strains into ~20 T-serotypes. However, many *S. pyogenes* strains have multiple T-types (e.g., T3/13/B, T8/25/Imp19) [4].

Another classification scheme proposed by Sherman in 1937 incorporating Lancefield grouping and other criteria divides streptococci into four groups: pyogenic, viridans, lactic, and enterococci. The pyogenic division comprises β-hemolytic strains of Lancefield groups A, B, C, E, F, and G. The viridans division consists of non-β-hemolytic streptococci that are not tolerant to high-pH growth conditions and salt and do not grow at 10°C. The lactic division includes strains of dairy origin, which are nonhuman pathogenic. This group differs from the pyogenic group by being non-β-hemolytic, growing at 10°C but not at 45°C, and failure to grow in broth with 6.5% NaCl. The lactic division was reclassified as the *Lactococcus* genus in the mid-1980s. The enterococci division includes Lancefield group D strains, which are now known as the genus *Enterococcus*. Members of this genus have the capacity to grow in broths at high pH (9.6), high salt concentrations (6.5% NaCl), and a wide temperature range (10°C–45°C), with some enterococci demonstrating β-hemolytic property (see Chapter 10) [3].

Examination of *Streptococcus* 16S rRNA sequences also permits discrimination of streptococci into six groups: *S. anginosus*, *S. bovis*, *S. mitis* (including *S. pneumoniae*), *S. mutans*, *S. pyogenes*, and *S. salivarius*.

Further, sequencing analysis of the *emm* gene that encodes the mature M protein molecule provides additional confirmation on the validity of the traditional M serological typing scheme. Specifically, the *emm* sequence-based typing divides *S. pyogenes* strains into five pattern groups (A–E), with *emm* pattern A–C groups accounting for 47% of pharyngitis isolates, but only 8% of impetigo isolates; *emm* pattern D group accounting for 50% of impetigo isolates, but only <2% of pharyngitis isolates; and *emm* pattern E group accounting for almost equal fractions of throat and skin infections (52% and 42%, respectively). Not surprisingly, *emm* pattern A–C groups are referred to as "throat specialists," *emm* pattern D group as "skin specialists," and *emm* pattern E group as "generalists" [4].

Moreover, genetic analysis of pilus genes (particularly the FCT region, with FCT standing for *F*ibronectin- and *C*ollagen-binding proteins and *T*-antigen) uncovers the molecular basis of the serotyping scheme targeting T-antigens. However, not all *S. pyogenes* strains express pili, and those with a negative regulator may be T-nontypable [4].

Multilocus sequence typing (MLST) targeting seven core housekeeping genes (glucose kinase, *gki*; glutamine transporter protein, *gtr*; glutamate racemase, *murI*; DNA mismatch repair protein, *mutS*; transketolase, *recP*; xanthine phosphoribosyl transferase, *xpt*; and acetyl coenzyme A acetyltransferase, *yqiL*) offers another valuable approach for determination of *S. pyogenes* strains, with clones of *S. pyogenes* being defined by their sequence type (ST) (http://pubmlst.org/spyogenes/). Interestingly, while most *emm* pattern A–C strains correspond to same clone (ST) or clonal complex, *emm* patterns D and E strains correlate with distant ST [4].

14.1.1.2 Morphology

Streptococci are Gram-positive, facultative anaerobic cocci that often grow in chains or in pairs (due to the fact that cell division occurs along a single axis). *Streptococcus* colonies on agar plates are small, smooth, and moist in appearance. Streptococcal cell wall is largely composed of group-specific carbohydrate (the M protein), with Lancefield group A being a polymer of rhamnose and *N*-acetylglucosamine. Apart from the M protein, other proteins present on the cell surface include T-protein, SOF, C5a peptidase, collagen-like protein Scl1, GRAB, and protein F. In GAS, the pili (fimbriae) appear as long, flexible rods protruding up to 3 µm from the cell surface. Structurally, GAS pili are heteropolymers forming a pilus shaft, which is composed of the major pilin protein subunit (Spy0128) and assembled through a series of transpeptidase reactions catalyzed by a class B accessory sortase, SrtC (Spy0129) [3].

14.1.1.3 Genomics

The genomes of *Streptococcus* species are of 1.8–2.3 Mb in size with GC content of about 38.5%, 1700–2300 protein-coding sequences (CDSs), 5–6 rRNA, and 57–67 tRNA encoding genes. Specifically, *S. pyogenes*, *S. agalactiae*, *S. pneumoniae*, and *S. mutans* possess genomes of 1,852,442, 2,211,488, 2,160,837, and 2,030,921 bp, respectively. The genome of GAS contains genes that encode proteins secreted into the extracellular fluid during growth (e.g., erythrogenic toxin, streptolysin O, streptolysin S, proteinase, streptokinase, DNase, RNase, hyaluronidase, CAMP factor, streptococcal inhibitor of complement, immunogenic secreted protein, and superantigens). These proteins are mostly virulence factors with important roles in colonization, invasion, spreading, and pathogenesis. The genome of pathogenic GAS strains also harbors a pathogenicity island located in the FCT region, which encodes fibronectin-binding proteins, collagen-binding proteins, and T-antigens (pilus subunit genes). Among GAS isolates, the FCT region displays considerable genetic diversity, with nine different FCT variants identified [4]. Additionally, the *S. pyogenes* genome consists of some metabolic pathways (e.g., complete glycolytic pathway, fatty acid synthesis, nucleotide synthesis and transport, and carbohydrate transport and metabolism), but lacks a complete tricarboxylic acid cycle pathway. Many *Streptococcus* species contain bacteriophages, with 18 prophages (ranging from 38 to 41 kb in size, encoding from 42 to 66 genes each) being described in *S. pneumoniae*.

14.1.2 Biology and Epidemiology

Apart from nine *Streptococcus* species (*S. pyogenes*, *S. agalactiae*, *S. equisimilis*, *S. bovis*, *S. anginosus*, *S. sanguinis*, *S. mitis*, *S. mutans*, and *S. pneumoniae*) that are implicated in human infections, most other streptococcal species are nonpathogenic and form part of the commensal human microbiota of the mouth, skin, intestine, and upper respiratory tract.

S. pyogenes (GAS) is not considered part of normal flora in humans, despite its presence in 5%–20% of healthy/asymptomatic individuals. As an exclusively human pathogen, GAS enters into human host via oral cavity, skin, and wounds, and typically attaches to the epithelial surfaces of the throat and skin, as well as the vagina and rectum, on which it forms large microcolony aggregates. Additionally, GAS may organize as biofilms and hide inside nonphagocytic cells. These enhance its survival under unfavorable or stressful conditions including antimicrobial therapy, and contribute to its spread to other individuals [5,6].

GAS is commonly transmitted through two routes: foodborne or airborne. Foodborne GAS epidemics often result from consumption of inappropriately prepared/cooked raw milk, cold salads, eggs, mayonnaise, tuna, potatoes, cheese, conch, and other ingredients that are inadvertently contaminated by GAS from infected animals (e.g., cows with streptococcal mastitis) or food handlers (who have sore throat, or have infected skin lesions on hands/arms, or are asymptomatic carriers) [7–9]. Airborne GAS epidemics are due to inhalation of respiratory/saliva droplets from carriers who disseminate the bacterium via sneeze or cough. Another means for GAS to spread is through person-to-person skin contact. External factors favoring GAS disease outbreaks include crowded settings (e.g., military training centers), mass consumption of contaminated foods, and hospital acquisition (e.g., puerperal sepsis). Intrinsic factors include the emergence of dominant clones [e.g., M1T1 clone that acquires three regions of heterologous DNA: a 36-kb chromosomal region encoding the toxins SLO and NAD-glycohydrolase and two bacteriophages encoding the DNase Sda1 and the superantigen SpeA (streptococcal pyrogenic exotoxin A); M3 clone that acquires prophages encoding phospholipase A2 and SpeA and the duplication of four amino acids in the N-terminal region of the M3 protein].

Interestingly, *S. pyogenes* pharyngitis tends to predominate in temperate regions, with seasonal peak in winter, whereas *S. pyogenes* impetigo is mostly present in tropical and subtropical regions, with seasonal peak in summer. In addition, there are notable links of certain *emm* types to particular clinical diseases, including the association of *emm* types 1, 3, 5, 6, 12, 14, 17, 19, and 24 with pharyngitis; *emm* types 33, 41, 42, 52, 53, and 70 with impetigo; *emm* types 1, 3, 5, 6, 11, 12, 14, 17, 18, 19, 24, 27, 29, 30, 32, and 41 with ARF; *emm* types 1, 4, 12, 49, 55, 57, and 60 with APSGN; *emm* type 28 with puerperal sepsis; *emm* types 1, 3, and 28 with necrotizing fasciitis; and *emm* types 1 and 3 with STSS [2].

14.1.3 Clinical Features and Pathogenesis

Human pathogenic species within the genus *Streptococcus* include *S. pyogenes* (GAS, pharyngitis, scarlet fever, impetigo, pyoderma), *S. agalactiae* (GBS, neonatal meningitis, sepsis), *S. equisimilis* (endocarditis, bacteremia, pneumonia, meningitis, respiratory infections), *S. bovis* (biliary or urinary tract infections, endocarditis), *S. anginosus* (subcutaneous/organ abscesses, meningitis, respiratory infections), *S. sanguinis* (endocarditis, dental caries), *S. mitis* (endocarditis), *S. mutans* (dental caries), and *S. pneumoniae* (pneumonia) [10–13].

In individuals with competent immune functions, *S. pyogenes* (GAS) often induces mild and self-healing purulent symptoms in mucosal membranes and skin (e.g., pharyngitis, impetigo) that may disappear in 7–14 days. In patients with predisposed conditions [e.g., immune-suppression, diabetes, or specific MHC class II cell surface receptor (HLA-DR) subtypes], GAS may breach the epithelial barrier, and cause invasive and life-threatening diseases [e.g., cellulitis, puerperal sepsis (childbed fever), necrotizing fasciitis, STSS, RHD, septic arthritis, pneumonia, meningitis, abscess, osteomyelitis, endocarditis, peritonitis], with 8%–23% of patients succumbing within 7 days of infection [2].

14.1.3.1 Pharyngitis

Commonly known as "strep sore throat," GAS pharyngitis presents with sudden-onset fever accompanied by sore throat, and obvious inflammation in the pharynx and tonsils. Other symptoms may include general malaise, headache, nausea, abdominal pain, vomiting, patchy exudates, and cervical lymph node adenopathy. Patients with uncomplicated GAS pharyngitis usually recover within 7–14 days [2].

14.1.3.2 Scarlet Fever

When pharyngitis involves a GAS strain that generates bacteriophage-encoded streptococcal pyogenic exotoxins (notably SpeA), scarlet fever (also known as scarlatina) may emerge, with a deep red, finely papular, erythematous rash (strawberry tongue) and exudative pharyngitis. Scarlet fever tends to occur in children of 4–8 years but rarely in adults. Pharyngitis and soft tissue infection at a surgical site (surgical scarlet fever) are most commonly associated with scarlet fever [2].

14.1.3.3 Impetigo

As a contagious GAS infection of the skin, impetigo (or pyoderma, which is a term for a localized purulent infection of the skin and is used synonymously with streptococcal impetigo and impetigo contagiosa) shows pustules that enlarge and rupture to form thick, honey-colored scabs. Spread through direct skin contact, impetigo mainly affects children with poor hygiene and crowded living conditions in tropical and subtropical climate. When GAS moves to deeper layers of the skin, erysipelas and cellulitis may appear. Further spreading of GAS to the fascia may lead to necrotizing fasciitis (flesh-eating disease), which is severely invasive, potentially fatal disease in the absence of immediate surgical and medical intervention. Another severe invasive disease due to GAS is STSS, which often develops within 24–72 h of minor nonpenetrating trauma resulting in hematoma, deep bruise to the calf, or following muscle strain, in addition to suction lipectomy, hysterectomy, vaginal delivery, bunionectomy, and bone pinning [2].

14.1.3.4 Acute Rheumatic Fever (ARF)

Being a sequela of untreated GAS pharyngeal infection, ARF is a systemic disease that shows inflammation of the joints (arthritis) and heart (carditis, also known as rheumatic heart disease or RHD), along with neurologic manifestation (i.e., Sydenham chorea). Other manifestations may include erythema marginatum and subcutaneous nodules. ARF/RHD largely results from inflammatory changes occurring in cardiac tissue, joints, brain, blood vessels, and skin, with hemodynamic changes in cardiac tissue leading to decompensatory cardiac failure. Once closely linked to childbed fever, nonpasteurized milk, surgical wards, schools, and day care centers, the incidence of ARF has decreased dramatically with the routine use of penicillin in the treatment of GAS nowadays [14].

14.1.3.5 Acute Poststreptococcal Glomerulonephritis (APSGN)

Another sequela of GAS infection is APSGN, which is an immune-complex-mediated disorder of the kidneys, with symptoms ranging from edema, hypertension, urinary sediment abnormalities, and reduced serum complement components. APSGN is particularly prevalent in children in less developed countries, but rarely causes long-term renal damage when proper supportive care is provided [14].

14.1.3.6 Pediatric Autoimmune Neuropsychiatric Disorders Associated with Streptococcal Infections (PANDAS)

PANDAS is a rare child disease tentatively linked to GAS infection, in the form of obsessive–compulsive disorders and Tourette's syndrome, due possibly to the effects of GAS-induced antineuronal antibodies. PANDAS symptoms appear to be similar to neuropsychiatric symptoms of Sydenham's chorea, which is a disease associated with a prior group A streptococcal infection [14].

GAS expresses a number of surface proteins and virulence factors that facilitate its attachment and entry into human host cells, and escape from human immune surveillance. For example, GAS produces M protein, Cpa, Eno, Epf, and firbronectin-binding proteins for extracellular matrix and serum protein binding via microbial surface components recognizing adhesive matrix molecules (MSCRAMMS); secretes EndoS, Mac, C5a peptidase, capsule protein, and Sic for degradation/inhibition of host immunoglobulin and complement factor; utilizes plasminogen/plasmin binding and streptokinase Ska activity for dysregulation of coagulation; generates SLS, SLO, streptococcal pyrogenic exotoxins A and C (SpeA and SpeC, which are superantigens responsible for scarlet fever and STSS), and Nga for cytotoxic and cytolytic activity toward various host cell types, including neutrophils, platelets, and subcellular organelles [15,16].

14.1.4 Diagnosis

Isolation of GAS from throat, food specimens, or swabs from kitchen utensils and surfaces represents a useful initial step for the diagnosis of streptococcal pharyngitis and hemolytic streptococci.

GAS may be grown in a highly nutritious medium consisting of meat extract, peptones, dextrose, and salts (e.g., 5% horse-blood agar), with the formation of colonies showing β-hemolysis. The medium may be modified to enhance the production of some GAS proteins. For instance, addition of a digested RNA fraction increases streptolysin S production; supplement with reducing agents (e.g., glutathione) favors streptolysin O production; a slightly acidic pH strengthens the production of the cysteine proteinase precursor; and addition of hyaluronate improves the production of hyaluronidase [3].

The resulting GAS isolates are then typed by using serotype-specific antisera raised against the M protein (an immunodominant surface antigen and key virulence determinant), T-antigens (trypsin-resistant surface antigens, which are contained within extended surface pili), or SOF (an LPXTG-anchored, multifunctional surface protein that binds fibronectin and enzymatically disrupts the structure of high-density lipoproteins present in blood). Further analysis with pulsed-field gel electrophoresis (PFGE) using *Sma*I is also valuable.

More recently, typing of GAS strains is conducted according to the sequence of the 5′ variable region of the *emm* gene encoding the M protein, leading to the identification of >200 *emm* types. Of these, *emm* pattern A–C strains are considered as "throat specialists" (causing pharyngitis), *emm* pattern D strains as "skin specialists" (causing impetigo), and *emm* pattern E strains as "generalists" (causing both pharyngitis and impetigo) [4].

Use of MLST technique allows definition of *S. pyogenes* clones by their ST (http://pubmlst.org/spyogenes/). Interestingly, most *emm* pattern A–C strains closely correspond to a single clone (ST) or clonal complex [4].

14.1.5 Treatment and Prevention

To avoid the development of secondary autoimmune sequelae (e.g., RHD or glomerulonephritis), it is necessary to implement antibiotic therapy for GAS infections. Despite continued use over the past few decades, GAS remains extremely sensitive to penicillin-based antibiotics. However, GAS resistance to macrolides, clindamycin, and lincosamide has been observed in some parts of the world. The recommended treatment regimen for GAS infections consists of a 10-day course of oral penicillin or intramuscular benzathine penicillin. For penicillin-allergic individuals, erythromycin may be administered.

For the prevention of foodborne streptococcal pharyngitis, individuals with acute pharyngitis (with sneeze or cough) or skin lesions should be excluded from food handling, use of bare hands in food handling should be banned, food (especially prepared in large quantities) must be properly cooked and stored, and unpasteurized milk or milk products should not be used.

Owing to the fact that GAS strains demonstrate considerable serotype diversity and antigenic variability, a safe and effective vaccine for human streptococcal infections is still unavailable. To date, GAS proteins and molecules that have been assessed as vaccine candidates consist of cell-wall-anchored proteins, membrane-associated proteins, secreted proteins, anchorless proteins, and the group A carbohydrate molecule. A number of protein antigens have shown promise in vaccine trials involving various animal models. These include fibronectin-binding protein A (FbaA), protein F1/SfbI, serum opacity factor (SOF/SfbII), fibronectin-binding protein 54 (FBP54), R28 protein, streptococcal protective antigen (Spa), C5a peptidase (ScpA), streptococcal hemoprotein receptor (Shr), streptococcal pyrogenic exotoxin B (SpeB), streptococcal secreted esterase (Sse), streptococcal immunoglobulin protein 35 (Sib35), *Streptococcus pyogenes* cell envelope protease (SpyCEP), arginine deiminase (ADI), and trigger factor (TF).

14.2 Laboratory Models

As a strictly human pathogen, *S. pyogenes* is implicated in a range of clinical diseases, including (1) local, lesional diseases (inflammation) in soft tissue, (2) both local and systemic diseases associated with streptococcal toxins, and (3) immune dysfunction due to streptococcal antigens. To complicate the situation further, *S. pyogenes* encompasses a diversity of strains with varied virulence potential. In spite of our long-standing efforts to uncover the secrets about this ever-present pathogen, our understanding of GAS and its pathogenesis and immune evasion is far from being adequate. Use of laboratory models is vital in helping address this conundrum. Indeed, over the past two decades, a number of *in vivo* and *in vitro* models based on both laboratory animals and cell lines have been exploited, with the ultimate goal of elucidating the molecular basis of *S. pyogenes* pathogenesis and other perplexing issues [17–20].

14.2.1 Animal Models

14.2.1.1 Rodents

Being small, relatively inexpensive to maintain, easy to handle, along with rapid reproduction and short life span, rodents (e.g., mice and rats) represent ideal animal models for disease investigations including GAS infections. It is no surprise that a diverse range of model systems have been established using rodents.

14.2.1.1.1 Murine Subcutaneous Ulcer Model

Murine subcutaneous ulcer model involves injection of 10^6–10^8 CFU of *S. pyogenes* strain into the subcutaneous tissue of a mouse flank (inbred BALB/c C57BL6 or outbred hairless SKH1), leading to a localized inflammatory lesion (characterized by the recruitment of neutrophils and macrophages and the production of cytokines IL-12, interferon-γ, and TNF-α) in soft tissue within 8–12 h, with development of ulceration and eschar by 18–24 h and resolution of lesion by 14 days postinfection. This model offers a useful platform to determine the virulence potential of individual *S. pyogenes* strains according to the variations in lesion size and resolution time, weight change, and numbers of colony forming units (CFU) from recovered bacteria [20].

14.2.1.1.2 Murine Subcutaneous Air Sac Model

Being a derivative of murine subcutaneous ulcer model, murine subcutaneous air sac model involves injection of air under the skin (to create an air pouch) followed by injection of *S. pyogenes* bacteria. This model permits easy recovery of host inflammatory cells in the sac lumen by lavage for subsequent analysis by FACS and other *in vitro* methods, and facilitates analysis of host innate immune response to *S. pyogenes* [20].

14.2.1.1.3 Murine Impetigo Model

This model relies on a humanized mouse (hu-mouse) that is created by engrafting human epidermal tissue from neonatal foreskin onto the flanks of the SCID mouse (which is incapable of rejecting the tissue grafts due to the absence of an adaptive immune response). *S. pyogenes* bacteria are topically applied to the grafts that are superficially damaged by cross-wise cuts with a scalpel blade, and then occluded with a bandage, leading to an impetigo-like lesion [characterized by erosion of the stratum corneum, infiltration of murine polymorphonuclear leukocytes (PMNs), formation of pus surrounded by clumps of proliferating streptococci]. Through a semiquantitative visual assessment of histopathology or enumeration of CFU, the virulence of *S. pyogenes* is ascertained [20].

14.2.1.1.4 Murine Systemic Disease Model

Inoculation of mice with 10^5–10^6 CFU of *S. pyogenes* bacteria via intravenously (IV), intraperitoneally (IP), intranasally (IN), or intratracheally (IT) facilitates examination of different types of *S. pyogenes*-induced systemic disease. Compared to IN infection, IT inoculation bypasses the initial site of upper airway colonization and allows quick development of clinical signs in infected mice. *S. pyogenes* virulence is determined by lethality and the kinetics of clearance in the spleen, liver, and lungs (via CFU enumeration) [5].

14.2.1.1.5 Murine Implanted Chamber Model

Murine implanted chamber model involves surgical implantation of a "cage" made of steel or Teflon into the subcutaneous tissues on mouse flank. The healing of the incision over the course of several weeks results in the formation of fibrous capsule around the implant. Subsequent injection of *S. pyogenes* bacteria through the skin and into the inside of the cage allows the influx of nutrients and the release of streptococcal products without streptococcal cells. This scenario resembles a glomerulonephritis-like disease (characterized by deposition of streptokinase and host complement components in the glomerular basement membrane), facilitating the monitoring of toxin production *in vivo*. A rabbit implanted chamber model has been also described for modeling STSS [20].

14.2.1.1.6 Murine Footpad Model

In this model, *S. pyogenes* bacteria labeled with fluorescent dyes are injected into LysM-eGFP mouse footpad, whose macrophages and neutrophils express eGFP; the velocity and meandering of neutrophils are immediately monitored by two-photon microscopy [20].

14.2.1.1.7 Mouse Oropharyngeal Colonization Model

In this model, mice are infected by IN inoculation into one nostril with 10^7 CFU of mouse-pathogenic, streptomycin-resistant *S. pyogenes* strain in a 2.5 μL normal saline drop; colonization of the upper airway and oropharynx is then monitored by throat swabs after resuspension in saline and plating onto selective media [(Todd Hewitt yeast extract) THY agar with 1000 μg/mL streptomycin] [20].

14.2.1.1.8 Murine Nasopharyngeal-Associated Lymphatic Tissue Colonization Model

This model exploits the ability of murine nasal-associated lymphoid tissue (NALT) (in the form of lymphoid lobes located in the lateral nasopharyngeal wall) to allow colonization of *S. pyogenes* strain 591 following intranasal inoculation and to develop a specific immune response [20].

14.2.1.1.9 Murine Vaginal Colonization Model

In this model, C57BL/6 or BALB/c mice are administered with up to 0.5 mg ethinyl estradiol dissolved in sterile sesame oil 24–48 h (to synchronize mice into the estrous phase of the estrous cycle, which enhances epithelial cell proliferation and minimal inflammatory cell infiltrate and renders mice receptive to bacterial colonization) prior to vaginal inoculation of ~1×10^6 CFU of streptomycin-resistant *S. pyogenes* in 20 μL of PBS. Vaginal lavages with sterile PBS are stained with a modified Wright–Giemsa stain, and epithelial cells and leukocytes are counted under microscope. In addition, dilution and plating of vaginal lavages onto selective media (e.g., THY agar, supplemented with 1000 μg/mL streptomycin) allows determination of streptococcal colony counts. Further, cytokine and other inflammatory cell markers in vaginal lavages are assessed by ELISA [20].

14.2.1.1.10 Rat Oropharyngeal Colonization Model

This model involves treatment of Fischer CDF344 rats with streptomycin (1000 μg/mL) in drinking water (to disrupt their normal flora) before IN (20 μL) or oral (50 μL) inoculation with streptomycin-resistant *S. pyogenes* for pharyngeal colonization; the bacteria are then recovered from throat swabs and CFU enumerated [20].

14.2.1.1.11 Rat Autoimmune Carditis Model

In this model, Lewis rats immunized with *S. pyogenes* M protein (which cross-reacts with rat heart tissue) develop a myocarditis and a valvulitis, with the formation of Anitschkow cell-containing granulomas (called Aschoff bodies) and infiltration of CD4+ and CD8+ T cells into the valvular lesions, that resemble human rheumatic disease [20].

In addition, sera from Lewis rats immunized with a whole cell lysate of *S. pyogenes* (showing quantifiable alterations in behavioral and motor functions) cross-react with brain tissue *in vivo* and with dopamine and serotonin receptors *in vitro*. Direct perfusion of purified IgG from these cross-reacting sera into the striatum region of the brain could reproduce symptomatology of PANDAS [17,18,21].

14.2.1.2 Chinchilla

Chinchilla otitis media model involves inoculation of 10^5 CFU *S. pyogenes* via transbullar injection into the ear of a chinchilla, with tympanic membrane and inner ear inflammation as well as serous middle ear effusion evident by day 2 as examined by otoscopy. The infection can be monitored through enumeration of CFUs recovered from macroscopic structures and also serous effusion, and observation of animal mortality over the 7-day period [20].

14.2.1.3 Nonhuman Primates

Nonhuman primate pharyngitis model offers a more reliable approach for assessing *S. pyogene*-induced pharyngitis than other animals such as rodents. Several nonhuman primate species (e.g., baboon, rhesus monkey, chimpanzee, and cynomolgus macaque) appear to be susceptible to *S. pyogenes* colonization in the oropharynx, with the production of type-specific M protein antibodies in serum [20].

Nonhuman primate sepsis model involves intravenous infusion of baboons (*Papio cynocephalus cynocephalus*) under light anesthesia with 10^{10} CFU *S. pyogenes* over a 2-h period followed by analyses of physiology and blood chemistry over a 10-h period. The infection is evaluated by body temperature, heart rate, mean systemic arterial blood pressure, serum chemistry, and cytokine profiles. It appears that blocking the cytokine TNF-α with a therapeutic monoclonal antibody helps improve both mean arterial blood pressure and survival and reduce hypotension and multiorgan failure. This model is valuable for the study of streptococcal toxic-shock syndrome [20].

14.2.1.4 Insects

Silkworm model involves injection of approximately 9×10^8 CFU *S. pyogenes* into the hemolymph (comparable to the bloodstream) of invertebrate silkworm (*Bombyx mori*) followed by monitoring survival for 5 days at 27°C. Wax worm model differs from silkworm model in that wax worm (*Galleria mellonella*) allows infection to be conducted at 37°C [20].

14.2.1.5 Caenorhabditis elegans *Nematode*

C. elegans is an invertebrate animal that can be used to analyze *S. pyogenes* virulence in plate or liquid format, as >80% of *C. elegans* organisms are killed by *S. pyogenes*-produced hydrogen peroxide within 24h.

14.2.1.6 Zebrafish

Zebrafish myonecrosis model involves inoculation with *S. pyogenes* via IM injection into the dorsal muscle of zebrafish (*Danio rerio*), leading to the formation of a hypopigmented lesion at the site of injection within 24h (indicative of extensive muscle necrosis as revealed by histology) and the death (resulting from toxic shock) of the animal at 36–96h postinjection. Colonization may be evaluated by dissecting the muscle tissue and the spleen for serial dilution plating [22].

14.2.2 *In Vitro* Models

Using several mammalian cell lines (e.g., mouse fibroblast cell line, L cells, as well as HeLa and HEp-2 cells), Miyoshi-Akiyama et al. [23] noticed that high-virulence GAS isolates associated with STSS have lower ability to adhere to mammalian cells than low-virulence isolates.

In a separate study, human keratinocytes are shown to allow *in vitro* formation of GAS biofilms that have superior capacity to colonize the oropharynx *in vivo* compared to broth-grown GAS and provide an effective means to evaluate GAS colonization, invasive disease, and natural transformation [5].

14.3 Conclusion

Constituting a key member of the Gram-positive bacterial genus *Streptococcus*, *S. pyogenes* (commonly called group A *Streptococcus* or GAS) is renowned for its serotype diversity, virulence variation, and ferocious infectivity. Adapted exclusively to human host, pathogenic *S. pyogenes* strains are capable of inducing localized inflammatory lesions (e.g., pharyngitis, impetigo), secreting streptococcal toxins that cause both local and systemic diseases (e.g., pneumonia, sepsis, STSS), and producing streptococcal antigens that confuse host immune systems, leading to autoimmune diseases (e.g., ARF, RHD, APSGN). Given its medical and public health significance, *S. pyogenes* has been extensively studied, and a plethora of laboratory models (both *in vitro* and *in vivo*) have been applied to elucidate the molecular mechanisms of its pathogenicity and to aid in the design of innovative measures for its control and prevention.

REFERENCES

1. Ferretti J, Köhler W. History of streptococcal research. In: Ferretti JJ, Stevens DL, Fischetti VA, editors. *Streptococcus pyogenes: Basic Biology to Clinical Manifestations*. Oklahoma City, OK: University of Oklahoma Health Sciences Center; 2016.
2. Walker MJ, et al. Disease manifestations and pathogenic mechanisms of group A *Streptococcus*. *Clin Microbiol Rev*. 2014;27(2):264–301.
3. Facklam R. What happened to the streptococci: Overview of taxonomic and nomenclature changes. *Clin Microbiol Rev*. 2002;15(4):613–30.
4. Bessen DE. Molecular basis of serotyping and the underlying genetic organization of *Streptococcus pyogenes*. In: Ferretti JJ, Stevens DL, Fischetti VA, editors. *Streptococcus pyogenes: Basic Biology to Clinical Manifestations*. Oklahoma City, OK: University of Oklahoma Health Sciences Center; 2016.
5. Marks LR, Mashburn-Warren L, Federle MJ, Hakansson AP. *Streptococcus pyogenes* biofilm growth in vitro and in vivo and its role in colonization, virulence, and genetic exchange. *J Infect Dis*. 2014;210(1):25–34.
6. Fiedler T, Köller T, Kreikemeyer B. *Streptococcus pyogenes* biofilms-formation, biology, and clinical relevance. *Front Cell Infect Microbiol*. 2015;5:15.
7. Katzenell U, Shemer J, Bar-Dayan Y. Streptococcal contamination of food: An unusual cause of epidemic pharyngitis. *Epidemiol Infect*. 2001;127(2):179–84.
8. Falkenhorst G, et al. Outbreak of group A streptococcal throat infection: Don't forget to ask about food. *Epidemiol Infect*. 2008;136(9):1165–71.
9. Okamoto F, et al. A foodborne outbreak of group A streptococcal infection in Fukuoka Prefecture, Japan. *Jpn J Infect Dis*. 2014;67(4):321–2.
10. van der Mee-Marquet N, et al. Molecular characterization of temporally and geographically matched *Streptococcus agalactiae* strains isolated from food products and bloodstream infections. *Foodborne Pathog Dis*. 2009;6(10):1177–83.
11. Lemos JA, Quivey RG Jr, Koo H, Abranches J. *Streptococcus mutans*: A new Gram-positive paradigm? *Microbiology*. 2013;159(3):436–45.
12. Leung V, Dufour D, Lévesque CM. Death and survival in *Streptococcus mutans*: Differing outcomes of a quorum-sensing signaling peptide. *Front Microbiol*. 2015;6:1176.
13. Watanabe S, Takemoto N, Ogura K, Miyoshi-Akiyama T. Severe invasive streptococcal infection by *Streptococcus pyogenes* and *Streptococcus dysgalactiae* subsp. *equisimilis*. *Microbiol Immunol*. 2016;60(1):1–9.
14. Cunningham MW. Rheumatic fever, autoimmunity, and molecular mimicry: The streptococcal connection. *Int Rev Immunol*. 2014;33(4):314–29.
15. Fieber C, Kovarik P. Responses of innate immune cells to group A *Streptococcus*. *Front Cell Infect Microbiol*. 2014;4:140.

16. LaRock CN, Nizet V. Inflammasome/IL-1β responses to streptococcal pathogens. *Front Immunol.* 2015;6:518.
17. Hornig M, Lipkin WI. Immune-mediated animal models of Tourette syndrome. *Neurosci Biobehav Rev.* 2013;37(6):1120–38.
18. Macrì S, Onori MP, Roessner V, Laviola G. Animal models recapitulating the multifactorial origin of Tourette syndrome. *Int Rev Neurobiol.* 2013;112:211–37.
19. Brosnahan AJ. Animal models used to study superantigen-mediated diseases. *Methods Mol Biol.* 2016;1396:1–17.
20. Watson ME Jr, Neely MN, Caparon MG. Animal models of *Streptococcus pyogenes* infection. In: Ferretti JJ, Stevens DL, Fischetti VA, editors. *Streptococcus pyogenes: Basic Biology to Clinical Manifestations [Internet].* Oklahoma City, OK: University of Oklahoma Health Sciences Center; 2016.
21. Rush CM, Govan BL, Sikder S, Williams NL, Ketheesan N. Animal models to investigate the pathogenesis of rheumatic heart disease. *Front Pediatr.* 2014;2:116.
22. Saralahti A, Rämet M. Zebrafish and streptococcal infections. *Scand J Immunol.* 2015;82(3):174–83.
23. Miyoshi-Akiyama T, Zhao J, Uchiyama T, Yagi J, Kirikae T. Positive correlation between low adhesion of group A *Streptococcus* to mammalian cells and virulence in a mouse model. *FEMS Microbiol Lett.* 2009;293(1):107–14.

Section III

Foodborne Infections due to Gram-Negative Bacteria

15

Aeromonas

Dongyou Liu

CONTENTS

15.1 Introduction

Aeromonas was first described in 1890 by Zimmermann as part of the genus *Bacillus* (*Bacillus punctatus*). In the following year, an *Aeromonas* strain was isolated (then named *Bacillus hydrophilus fuscus*, which is now known as *Aeromonas hydrophila*) by Sanarelli from the blood of infected frog with hemorrhagic exudation in the abdominal and peritoneal cavities (so-called "red leg" disease). In 1936, the genus *Aeromonas* (Greek *aer aeros*, air, gas; Greek *monas*, unit, monad; *Aeromonas*, gas-producing unit) was created by Kluyver and van Neil to cover this group of bacteria. In 1951, association of the genus *Aeromonas* with human infection (fulminant metastatic myositis) was confirmed with the recovery of *Aeromonas* from autopsy samples. In 1968, involvement of the genus *Aeromonas* in other human diseases (e.g., septicemia associated with Laennec's cirrhosis) was documented. In 1986, the genus *Aeromonas* was shown to be phylogenetically distinct from vibrios, and was transferred from the family Vibrionaceae to a newly established family Aeromonadaceae. To date, about 30 species have been identified in the genus *Aeromonas*, and some of them are responsible for a variety of infections in humans and animals.

15.1.1 Classification, Morphology, and Genomics

15.1.1.1 Classification

Encompassing a diverse group of Gram-negative, non-spore-forming, cocco bacillary species, the genus *Aeromonas* is classified taxonomically in the family Aeromonadaceae, order Aeromonadales, class Gammaproteobacteria, phylum Proteobacteria, and domain Bacteria. Being one of the five genera (i.e., *Aeromonas*, *Oceanimonas*, *Oceanisphaera*, *Tolumonas*, and *Zobellella*) within the family Aeromonadaceae, the genus *Aeromonas* shares remarkable biochemical (e.g., cytochrome oxidase), ecological (aquatic), and pathological similarities to members of the families Vibrionaceae (*Vibrio*) and Enterobacteriaceae (*Plesiomonas*), and indeed was once included in the family Vibrionaceae. It was only in 1986 when the genus *Aeromonas* became a member of the newly established Aeromonodaceae family after detailed examination of rRNA and housekeeping gene sequences [1].

Currently, about 30 species are recognized in the genus *Aeromonas*, including *Aeromonas aquatica*, *Aeromonas australiensis*, *Aeromonas bestiarum* (HG2, formerly *Aeromonas hydrophila* genomospecies 2), *Aeromonas bivalvium*, *Aeromonas cavernicola*, *Aeromonas caviae* (HG4, synonym *Aeromonas punctata*), *Aeromonas dhakensis* (synonyms *Aeromonas aquariorum*, *Aeromonas hydrophila* subsp. *dhakensis*), *Aeromonas diversa* (HG13, synonym *Aeromonas* group 501), *Aeromonas encheleia* (HG16), *Aeromonas eucrenophila* (HG6), *Aeromonas finlandensis*, *Aeromonas fluvialis*, *Aeromonas hydrophila* (HG1, synonyms *Bacillus hydrophilus fuscus*, *Bacillus hydrophilus*, *Proteus hydrophilus*, *Bacterium hydrophilum*, *Pseudomonas hydrophila*), *Aeromonas jandaei* (HG9), *Aeromonas lacus*, *Aeromonas media* (HG5A, HG5B), *Aeromonas molluscorum*, *Aeromonas piscicola*, *Aeromonas popoffii* (HG17), *Aeromonas rivuli*, *Aeromonas salmonicida* (HG3), *Aeromonas sanarellii*, *Aeromonas schubertii* (HG12), *Aeromonas simiae*, *Aeromonas sobria* (HG7), *Aeromonas taiwanensis*, *Aeromonas tecta*, *Aeromonas trota* (HG14, synonym *Aeromonas enteropelogenes*), and *Aeromonas veronii* (HG8, HG10, synonyms *Aeromonas ichthiosmia*, *Aeromonas allosaccharophila*, *Aeromonas culicicola*) [1–3]. Interestingly, *Aeromonas sharmana* (which unlike other members of the genus, is negative for nitrate reductase, lysine or ornithine decarboxylase or arginine dihydrolase, and lacks deoxyribonuclease activity) is now considered to be non-*Aeromonas* although it may still fall within the family Aeromonadaceae [1].

Based on their growth and biochemical characteristics, members of the genus *Aeromonas* may be distinguished into two major groupings: mesophilic and psychrophilic. The mesophilic group (e.g., *A. hydrophila*) contains motile strains that grow optimally at 35°C–37°C and cause diseases in humans. The psychrophilic group (e.g., *A. salmonicida*) includes nonmotile strains that grow optimally at 22°C–25°C and cause diseases in fish. Further genetic analysis led to the subdivision of the mesophilic group into three biochemically distinct phenospecies (*A. hydrophila*, *A. sobria*, and *A. caviae*). Additionally, considering their biochemical and genetic resemblances, *A. hydrophila sensu stricto*, *A. bestiarum*, and *A. salmonicida* are collectively referred to as the *A. hydrophila* complex; and *A. caviae sensu stricto*, *A. media*, and *A. eucrenophila* are known as the *A. caviae* complex. Further, using DNA–DNA hybridization techniques, *Aeromonas* species may be separated into 16 hybridization groups (HG, which refers to phenotypically distinguishable species) in direct contrast to phenospecies (which refers to a single heterogeneous species that may be composed of multiple HGs). Moreover, 16S rRNA restriction fragment length polymorphism (RFLP) methods have been developed to classify *Aeromonas* strains into genomospecies [4].

15.1.1.2 Morphology

Aeromonas is a Gram-negative, cocco bacillary or rod-shaped bacterium 0.3–1.0 μm × 1.0–3.5 μm in size, which appears singly, in pairs, and occasionally in short chains. On blood agar, *Aeromonas* forms round, raised, opaque colonies of 1–3 mm in diameter, with color changing from grayish (due to β-hemolysis) to dark green after 3 days (except *A. caviae*). Being a facultative anaerobe, *Aeromonas* is oxidase and catalase positive (best tested on media without a fermentable sugar, such as MacConkey agar).

Other morphological features include the possession of a single polar flagellum (especially motile strains, although some species may form peritrichous or lateral flagella or no flagella in solid media) and pili (fimbriae, which are surface appendages that facilitate attachment to host cells), which are either short rigid (S/R type) or long wavy flexible (L/W type). Most motile aeromonads do not have capsule, although *A. hydrophila* serotypes O:11 and O:34 are known to produce capsule when grown in a glucose-rich medium, which has a potential role in virulence.

15.1.1.3 Genomics

The genome sequences of a number of *Aeromonas* species are available in GenBank. The sizes of *Aeromonas* genomes appear to be in the range of 3.9–5.18 Mb, with GC contents of 57.6%–63.1% and 3609–4794 genes (Table 15.1). Many *Aeromonas* species harbor pseudogenes, while a few also possess plasmids (e.g., *A. salmonicida* subsp. *salmonicida* strain A449). In addition, the genomes of *Aeromonas* spp. encode a variety of virulence factors (including structural components, extracellular factors, secretion systems, iron-acquisition, and quorum-sensing mechanisms) that actively participate in host adherence, colonization, and infection [5–8]. Phylogenomic network analysis of three clinically important *Aeromonas* species (*A. hydrophila*, *A. veronii*, *A. caviae*) highlights the influence of homologous recombination and lateral gene transfer in the evolution of *Aeromonas* spp. [9].

TABLE 15.1

Genomic Features of *Aeromonas* Species

Species	Strain	HG	Genome (Mb)	GC Content (%)	Genes	Plasmid (kb)
A. aquatica			4.58	61.2	4163	
A. australiensis			4.11	58.1	3735	
A. bestiarum		HG2	4.69	60.6	4153	
A. bivalvium			4.3	62.2	3912	
A. caviae	8LM	HG4	4.47	61.8	4076	30
A. dhakensis	AAK1		4.76	61.8	4268	
A. encheleia		HG16	4.47	61.9	4065	
A. eucrenophila		HG6	4.54	61.1	4081	
A. finlandiensis			4.72	58.6	4376	
A. fluvialis			3.9	58.3	3609	
A. hydrophila	ATCC 7966	HG1	4.74	61.5	4284	
A. lacus			4.39	59	4005	
A. media	WS	HG5	4.78	60.7	4340	
A. molluscorum	848		4.24	59.2	4033	
A. piscicola			5.18	59.2	4794	
A. popoffii		HG17	4.76	58.6	4329	
A. rivuli			4.52	60	4230	
A. salmonicida subsp. *salmonicida*	A449	HG3	4.7	58.5	4320	168 and 175
A. sanarellii			4.19	63.1	3823	
A. schubertii		HG12	4.4	61.5	4287	
A. simiae			3.99	61.3	3756	
A. sobria		HG7	4.68	57.6	4212	
A. tecta			4.76	60.1	4329	
A. veronii	B565	HG8/HG10	4.66	58.6	4057	

15.1.2 Biology and Epidemiology

Typically, aeromonads establish infections in humans through oral ingestion of contaminated food or water (intestinal infection) [10]. Direct mucocutaneous contact with or exposure of open wound to environmental mud or water offers another way for aeromonads to enter human hosts (extraintestinal infections). In *Aeromonas* gastroenteritis, after enduring the detrimental effects of gastric acidity, ingested bacteria settle in the small or large intestine, and release enterotoxigenic molecules (leading to enteritis), or invade the gastrointestinal epithelium (causing dysentery or colitis).

With the capacity to tolerate the temperature range of 0°C–45°C, pH range of 4.5–9.0, and sodium chloride concentrations of 0%–4%, *Aeromonas* species are ubiquitously distributed in the environment (freshwater, estuarine, brackish, and saltwaters, and soil) and in foods (beef, pork, lamb, poultry, fish, shellfish, raw milk, dairy products, and fresh produce), and they have been isolated from humans, horses, pigs, sheep, cows, domesticated pets, birds, invertebrate, ticks, and insects [11,12]. It is of interest to note that *A. veronii* is commonly found in raw surface water, *A. hydrophila* in ozonated water, and *A. caviae* (*A. puntacta*) and *A. media* in waste water.

Aeromonad infections often occur in the community settings, as well as in health-care facility settings (e.g., indwelling-device-related infections). While all people are susceptible to *Aeromonas*-associated gastroenteritis, young children (<5 years of age) and older adults (>60 years of age) are particularly vulnerable. Individuals with underlying diseases (e.g., leukemia, carcinoma, diabetes, hepatitis, cirrhosis) or suppressed immune functions are notably prone to *Aeromonas* systemic infections [13]. Additional risk factors for *Aeromonas* infections include traumas and near-drowning events related to recreational activities (e.g., boating, fishing, and diving) and bites from reptiles, snakes, and bears [14,15].

15.1.3 Clinical Features and Pathogenesis

Aeromonads have been implicated in (1) gastrointestinal infections (secretory, dysentery, chronic, and choleraic) and (2) extraintestinal infections (wound and soft tissue infections, blood-borne infections, and miscellaneous infections), although the bacteria may be found in the stools of up to 4% of asymptomatic individuals [16–18].

15.1.3.1 Gastrointestinal Infections

The most common gastrointestinal infection is gastroenteritis, with symptoms ranging from fever, vomiting, and abdominal cramp to diarrhea (secretory, dysentery, chronic, and choleric). The secretory form of *Aeromonas* gastroenteritis is most common and manifests with low-grade fever and abdominal pain, and watery diarrhea, with mild to moderate dehydration. The watery diarrhea is generally a self-limiting illness, lasting a few days to a few weeks. The dysenteric form of *Aeromonas* gastroenteritis is less common and shows cramping abdominal pain and mucus (leukocytes) and blood in stools (similar to shigellosis). The chronic form of *Aeromonas* gastroenteritis lasts for more than 2 months and is frequently associated with *A. caviae* and *A. hydrophila* infections. The choleraic form of *Aeromonas* gastroenteritis is a rare, cholera-like disease. With the spread of gastrointestinal infections, appendicitis, peritonitis, pancreatitis, and acute cholangitis may emerge [19]. Complications linked to *Aeromonas* gastroenteritis include small bowel obstruction, ileal ulceration, intramural intestinal hemorrhage with small bowel obstruction, refractory inflammatory bowel disease, acute renal failure, and hemolytic-uremic syndrome (HUS). Among 14 *Aeromonas* species involved in human illness, six (*A. hydrophila*, *A. caviae*, *A. veronii*, *A. schubertii*, *A. jandaei*, and *A. trota*) are shown to cause human diarrhea [1].

15.1.3.2 Extraintestinal Infections

Extraintestinal infections include wound and soft tissue infections (cellulitis, abscesses), blood-borne infections (bacteremia/septicemia, pneumonia, empyema, septic arthritis, necrotizing fasciitis, myonecrosis, endocarditis, meningitis), and miscellaneous infections (hepatobiliary tract infections, osteomyelitis, endophthalmitis, keratitis, corneal ulcer, infections of bones and joints, the respiratory and urogenital tracts) [20–23]. Application of leeches (which may harbor aeromonads symbiotically) to tissue flaps or replantation areas during plastic or reconstructive surgery to relieve venous congestion may be a potential cause of *Aeromonas* infections (e.g., cellulitis) [24,25]. *Aeromonas* bacteremia/septicemia (attributed mainly to *A. hydrophila*, *A. dhakensis*, *A. veronii*, or *A. caviae*) often displays fever, jaundice, abdominal pain, septic shock, and dyspnea and tends to occur in severely immunocompromised individuals (e.g., those with myeloproliferative disorders, Laennec's cirrhosis, chronic liver disease, neoplasia, biliary disease, acute myeloid leukemia, myelodysplastic syndromes, non-Hodgkin's lymphoma, acute lymphocytic leukemia, diabetes mellitus, renal problems, cardiac anomalies, aplastic anemia, thalassemia, multiple myeloma, and Waldenstrom's macroglobulinemia) [26]. *Aeromonas* biliary tract infections (due to *A. hydrophila* and to a lesser extent, *A. caviae* and *A. veronii*) often occur in patients with biliary tract obstruction or stasis due to hepatobiliary cancer or stones, and most of them are polymicrobial in nature (e.g., *Escherichia coli*, *Klebsiella pneumoniae*, *Enterococcus*, and *Staphylococcus aureus*).

A. hydrophila and *A. veronii* may be also a cause of hemorrhagic septicemia in carp, tilapia, perch, catfish, and salmon; red sore disease in bass and carp; ulcerative infections in catfish, cod, carp, and goby; ulcerative stomatitis in snakes and lizards; "red leg" disease in frogs; septicemia in dogs; and septic arthritis in calves. *A. salmonicida sensu stricto* is linked to fish furunculosis, particularly in salmonids, with presentation ranging from septicemia, hemorrhages at the bases of fins, inappetence, melanosis, lethargy, and slight exophthalmia to hemorrhaging in muscle and internal organs.

Aeromonas spp. produce a number of virulence factors and proteins that enhance their pathogenicity. These include invasins, adhesins, outer membrane proteins, S-layer proteins, lipopolysaccharide (LPS), proteases [thermostable metalloprotease (TSMP) and thermolabile serine protease (TLSP)], lipases, glycerophospholipid cholesterol acyl-transferase (GCAT), superoxide dismutase, hemolysins, cytotoxic and cytotonic enterotoxins [*Aeromonas* cytotoxic enterotoxin (Act), *Aeromonas* heat-labile (56°C) cytotonic enterotoxin (Alt), and *Aeromonas* heat-stable cytotonic enterotoxin (Ast)], as well as type III secretion system (AscV and AscF-G) [27–31].

15.1.4 Diagnosis

Laboratory identification of aeromonads is facilitated by *in vitro* isolation of *Aeromonas* spp. from stool or other gastrointestinal samples. Aeromonads grow readily on routine enteric isolation media (MacConkey, xylose lysine deoxycholate (XLD), Hektoen enteric (HE), *Salmonella-Shigella* (SS), and deoxycholate-citrate (DC) agars). Most (90%) of the *Aeromonas* species from humans produce β-hemolysis on sheep blood agar, with the exception of *A. popoffii* and *A. trota* (0% and 50%, respectively) [32]. Identification at the genus level is achieved by positive oxidase test, fermentation of D-glucose, motility (most *Aeromonas* species are motile apart from *A. salmonicida* and *A. media*), the absence of growth in 6.5% sodium chloride, and resistance to the vibriostatic agent O/129 (150 μg). However, phenotypic identification to species level may require additional work involving the use of selective and differential media [33].

A number of selective media (supplemented with ampicillin and/or inhibitors such as bile salts, brilliant green, and sodium lauryl sulfate) have been used for recovery of *Aeromonas* species. For example, ampicillin dextrin agar (ADA) is useful for isolating most *Aeromonas* spp. apart from *A. trota* and certain strains of *A. caviae*, due to their sensitivity to ampicillin. Comparative analysis indicated that enrichment in alkaline peptone water (APW) and consecutive plating in two selective media [ampicillin–sheep-blood agar supplemented with 30 μg/mL of ampicillin ASBA 30 or bile salts–Irgasan–brilliant green (BIBG) agar] enable qualitative isolation of *Aeromonas* species from meat and fish. In BIBG, bile salts and brilliant green inhibit the growth of Gram-positive bacteria while Irgasan inhibits the growth of Gram-negative bacteria, which possess a type A nitratase. In addition, BIBG also detects the fermentation of xylose in some aeromonads, as production of acid during xylose fermentation retains BIBG's original, purple-red color (instead of green-yellow color). Other selective media include *Aeromonas* agar (AA, containing irgasan and D-xylose, which aeromonads do not ferment), and cefsulodin–irgasan–novobiocin (CIN) agar (in which *Aeromonas* forms a bull's-eye-like colony due to fermentation of D-mannitol) [33].

Aeromonas isolates surviving the selective process can be further distinguished via differentiation media (e.g., detection of amylase and pattern of carbohydrate fermentation in *Aeromonas* species). For example, the modified BIBG medium (mBIBG) has a higher pH (up to 8.7), together with the replacement of xylose by soluble starch as a carbon source, allowing identification of mesophilic aeromonads that are not able to ferment xylose. Further differentiation of aeromonads from vibrios and plesiomonads is their resistance to the vibriostatic agent O/129 (2,4-diamino-6,7-diisopropylpteridine), growth on nutrient agar without salt supplementation, and inability to grow on media containing 6%–6.5% sodium chloride. A combination of selective and differential media enables to classify clinical *Aeromonas* isolates into the *A. hydrophila* group (*A. hydrophila*, *A. bestiarum*, and *A. salmonicida*), the *A. caviae* group (*A. caviae*, *A. media*, and *A. eucrenophila*), and the *A. sobria* group (*A. veronii*, *A. jandei*, *A. schubertii*, and *A. trota*). Although use of the Aerokey II system allows an accurate and reliable phenotypic identification of *A. hydrophila*, *A. caviae*, *A. veronii* biovar sobria, *A. veronii* biovar veronii, *A. schubertii*, *A. jandaei*, and *A. trota*, it is unable to detect many newly described taxa. Similarly, the Vitek 2 does not differentiate between *A. hydrophila* and *A. caviae* [33].

Molecular methods targeting 16S rRNA, hemolysin (*ahh1*), aerolysin (*aerA*), and *asa1* genes allow precise determination of *Aeromonas* species identity and virulence potential [34–39]. Analysis of housekeeping genes (e.g., *dnaK*, *gltA*, *gyrB*, *radA*, *rpoB*, *tsf*, and *zipA*) provides an effective tool for study of taxonomic and phylogenetic relationships among *Aeromonas* species [4,5,40–43]. Indeed, use of a multiplex PCR directed toward the *gyrB* and *rpoB* genes enabled simultaneous identification of four *Aeromonas* species (*A. hydrophila*, *A. media*, *A. veronii*, and *A. caviae*) [44].

15.1.5 Treatment and Prevention

Treatment is s necessary for patients with chronic diarrheal disease or systemic infection. Fortunately, major *Aeromonas* species (e.g., *A. hydrophila*, *A. caviae*, *A. veronii*) appear to be susceptible to fluoroquinolones (e.g., levofloxacin, gatifloxacin, ciprofloxacin, and moxifloxacin), cefotaxime, and trimethoprim–sulfamethoxazole, although some *Aeromonas* isolates show resistance to penicillins, most cephalosporins, erythromycin, nalidixic acid, ciprofloxacin, norfloxacin, chloramphenicol, tetracycline, cotrimoxazole, ampicillin, ceftazidime, and meropenem [45–50].

For specific diseases associated with *Aeromonas* infections, therapeutic approaches may be considered for those presented with: chronic diarrhea (trimethoprim–sulfamethoxazole or ciprofloxacin), biliary tract infections (drainage and fluoroquinolones), spontaneous bacterial peritonitis (cefotaxime, fluoroquinolone, cephalosporin, and tetracycline), soft-tissue infections (cephalosporin with or without doxycycline, aztreonam, or fluoroquinolone as well as surgical debridement for invasive soft-tissue infections), bacteremia (cefotaxime, ceftriaxone, or ceftazidime, with or without minocycline; alternatively, aztreonam, cefepime, or fluoroquinolone such as ciprofloxacin or levofloxacin), pneumonia (cephalosporins, aminoglycosides, fluoroquinolones, and imipenem), *Aeromonas* meningitis (cefotaxime, ceftriaxone, or meropenem), endocarditis (an aminoglycoside plus a β-lactam; alternatively, a cephalosporin such as cefotaxime or ceftriaxone, in conjunction with an aminoglycoside such as gentamicin), eye infections (chloramphenicol, trimethoprim-sulfamethoxazole, and tetracyclines), and severe, invasive eye infections (trimethoprim-sulfamethoxazole) [51,52].

Prevention of *Aeromonas* infections should focus on avoiding physical contact with marine microorganisms or wild water or incidental ingestion of contaminated food or water, and eating uncooked seafood.

15.2 Laboratory Models

15.2.1 Animal Models

15.2.1.1 Rodents

Intramuscular inoculation into BALB/c mice provides a useful way to assess the pathogenicity of *Aeromonas* isolates (e.g., *A. hydrophila*, *A. veronii*, and *A. caviae*) in soft tissue infections. It appears that *A. hydrophila* is capable of causing more severe muscle damage (e.g., fragmentation of muscle fibers, edema of myocytes, and infiltration of inflammatory cells) than *A. caviae*. Additionally, mice undergoing streptomycin pretreatment allow transient *Aeromonas* colonization and enable determination of the colonization rates of different isolates. Indeed, *A. hydrophila*, *A. veronii*, and *A. caviae* exhibit relatively high rates of mouse colon tissue colonization than *A. salmonicida*, *A. encheleia*, and *A. allosaccharophila* [53]. Furthermore, intraperitoneal injection of immunocompromised mice or gastric lavage of neonatal mice offer alternative approach for investigating of *Aeromonas* pathogenicity relating to *Aeromonas*-induced gastroenteritis as well as extraintestinal infections [54,55].

Rats (*Rattus norvegicus*) undergoing clindamycin pretreatment develop a self-limited, loose stool (evidence of enteritis) after oral feeding with *A. hydrophila*, suggesting that antibiotic usage represents a predisposing risk factor to *Aeromonas* infection [56].

15.2.1.2 Zebrafish

Zebrafish represents a useful model for assessing the virulence of and host immune responses against *Aeromonas* strains, and its value for examining the host–pathogen interactions at the molecular level may be limited. Zebrafish succumbing to *Aeromonas* infection often displays clinical signs typical of hemorrhagic septicemia [57].

15.2.1.3 Caenorhabditis elegans *Nematode*

C. elegans is an invertebrate that has simple immune systems, is genetic tractability, is amenable to high-throughput experiments, and can be used in liquid-toxic (LT) assay to assess the virulence of *Aeromonas* species [58]. In *C. elegans* model, wound isolates of *A. dhakensis* demonstrates a higher virulence than wound isolates of *A. hydrophila*. The main advantages of *C. elegans* model include rapid generation time, large progeny, and ease of observation.

15.2.1.4 Tetrahymena *Protozoa*

Tetrahymena spp. (*T. thermophila* and *T. pyriformis*) are freshwater free-living *ciliate protozoa* that grow readily in culture media between 12°C and 41°C without a CO_2-enriched atmosphere, and thus offer an economical and permissive model to evaluate the virulence is aquatic *Aeromonas* isolates. *Aeromonas* isolate is considered virulent when relative survival is >60%, whereas *Aeromonas* isolate is considered avirulent when relative survival is <40%. Similarly, *Aeromonas* isolate is regarded as virulent when relative survival of *T. thermophila* is >40% after coculture with *Aeromonas*, whereas *Aeromonas* isolate is classified as avirulent when relative survival of *T. thermophila* is >50% after coculture with *Aeromonas* [57]. In addition, highly virulent *Aeromonas* strains grow well in *T. thermophila*, causing deformation to and even lysis of *T. thermophila*, whereas avirulent *Aeromonas* strains are largely phagocytozed by *T. thermophila*, causing no obvious damage to *T. thermophila* [59].

Other animal models used to study *Aeromonas* strains include crayfish (*Pacifastacus leniusculus*), mealworm larvae (*Tenebrio molitor*), leech, and blue gourami. Using crayfish and mealworm larvae models, the role of LPS (O-antigen and external core) in the virulence of *A. hydrophila* was confirmed [60].

15.2.2 *In Vitro* Models

Various cell lines (mouse C2C12 fibroblast cell and human intestinal epithelial cell Caco-2) have been used to assess the virulence and cytotoxicity of *Aeromonas* species. Using Caco-2 cells, it was shown that c-jun and c-fos are upregulated upon incubation with virulent *Aeromonas* isolates and thus provide a predictive indicator of *Aeromonas* virulence [61]. Further, it was observed that *A. dhakensis* isolates exhibited more potent cytotoxicity in human normal skin fibroblast cells than *A. hydrophila*, while *A. veronii* isolates showed higher cytotoxicity in C2C12 cells than *A. caviae* [58].

15.3 Conclusion

The genus *Aeromonas* encompasses a large group of Gram-negative, rod-shaped bacteria that are commonly present in aquatic environments (marine water, fresh water, and sediments), vegetation, and various animals. Besides causing diseases in fish and other animals, *Aeromonas* species are also responsible for opportunistic foodborne or wound-related infections in humans, including gastroenteritis and extraintestinal infections (e.g., serious wound infections in healthy individuals, primary and secondary septicemia in immunocompromised individual, as well as peritonitis, meningitis, and infections of the eye, joints, and bones). Pathogenic *Aeromonas* species are known to generate a number of virulence factors that aid their invasion of host cells and sabotage host immune surveillance. Molecular assays targeting the virulence genes, housekeeping genes, and 16S rRNA genes of *Aeromonas* spp. provide a rapid and precise means for their detection, identification, and epidemiological tracking and contribute to the early implementation of antimicrobial therapy. Nonetheless, despite our concerted efforts in the past, an effective vaccine against *Aeromonas* infections is still unavailable. Therefore, further research is urgently required to reveal the intricacy of host–bacterial interactions, and pinpoint the weak links in *Aeromonas* physiobiology and pathogenicity. Undoubtedly, laboratory models will form an indispensable part of this endeavor.

REFERENCES

1. Janda JM, Abbott SL. The genus *Aeromonas*: taxonomy, pathogenicity, and infection. *Clin Microbiol Rev* 2010;23(1):35–73.
2. List of Prokaryotic Names with Standing in Nomenclature. Genus *Aeromonas*. Available at http://www.bacterio.net/aeromonas.html, Last Updated 2015, Accessed May 31, 2016.
3. Beaz-Hidalgo R, et al. *Aeromonas aquatica* sp. nov., *Aeromonas finlandiensis* sp. nov. and *Aeromonas lacus* sp. nov. isolated from Finnish waters associated with cyanobacterial blooms. *Syst Appl Microbiol* 2015;38:161–8.
4. Figueras MJ, Soler L, Chacón MR, Guarro J, Martínez-Murcia AJ. Extended method for discrimination of *Aeromonas* spp. by 16S rDNA RFLP analysis. *Int J Syst Evol Microbiol* 2000;50(6):2069–73.
5. Roger F, Marchandin H, Jumas-Bilak E, Kodjo A, colBVH study group, Lamy B. Multilocus genetics to reconstruct aeromonad evolution. *BMC Microbiol* 2012;12:62.
6. Ruiz-Ruiz JM, Aguilera-Arreola MG, Castro-Escarpulli G. Markers of pathogenicity islands in strains of *Aeromonas* species of clinical and environmental origin. *Indian J Med Microbiol* 2012;30(4):467–9.
7. Beaz-Hidalgo R, Figueras MJ. *Aeromonas* spp. whole genomes and virulence factors implicated in fish disease. *J Fish Dis* 2013;36(4):371–88.
8. Dallaire-Dufresne S, Tanaka KH, Trudel MV, Lafaille A, Charette SJ. Virulence, genomic features, and plasticity of *Aeromonas salmonicida* subsp. *salmonicida*, the causative agent of fish furunculosis. *Vet Microbiol* 2014;169(1–2):1–7.
9. Ghatak S, et al. Pan-genome analysis of *Aeromonas hydrophila*, *Aeromonas veronii* and *Aeromonas caviae* indicates phylogenomic diversity and greater pathogenic potential for *Aeromonas hydrophila*. *Antonie van Leeuwenhoek* 2016;109(7):945–56.
10. Zhang Q, Shi GQ, Tang GP, Zou ZT, Yao GH, Zeng G. A foodborne outbreak of *Aeromonas hydrophila* in a college, Xingyi City, Guizhou, China, 2012. *Western Pac Surveill Response J* 2012;3(4):39–43.
11. Fontes MC, Martins C, Martínez-Murcia AJ, Saavedra MJ. Phylogenetic diversity of *Aeromonas* from "alheira," a traditional Portuguese meat product. *Foodborne Pathog Dis* 2012;9(8):713–8.
12. Jahid IK, Ha SD. Inactivation kinetics of various chemical disinfectants on *Aeromonas hydrophila* planktonic cells and biofilms. *Foodborne Pathog Dis* 2014;11(5):346–53.
13. Ghenghesh KS, Ahmed SF, El-Khalek RA, Al-Gendy A, Klena J. *Aeromonas*-associated infections in developing countries. *J Infect Dev Ctries* 2008;2(2):81–98.
14. Igbinosa IH, Igumbor EU, Aghdasi F, Tom M, Okoh AI. Emerging *Aeromonas* species infections and their significance in public health. *Sci World J* 2012;2012:625023.

15. Diaz JH. Skin and soft tissue infections following marine injuries and exposures in travelers. *J Travel Med* 2014;21(3):207–13.
16. Chao CM, Gau SJ, Lai CC. Empyema caused by *Aeromonas* species in Taiwan. *Am J Trop Med Hyg* 2012;87(5):933–5.
17. Chao CM, Lai CC, Tang HJ, Ko WC, Hsueh PR. Skin and soft-tissue infections caused by *Aeromonas* species. *Eur J Clin Microbiol Infect Dis* 2013;32(4):543–7.
18. Chen PL, et al. *Aeromonas* stool isolates from individuals with or without diarrhea in southern Taiwan: predominance of *Aeromonas veronii*. *J Microbiol Immunol Infect* 2015;48(6):618–24.
19. Huang D, Zhao Y, Jiang Y, Li Z, Yang W, Chen G. Spontaneous bacterial peritonitis caused by *Aeromonas caviae* in a patient with cirrhosis. *Zhong Nan Da Xue Xue Bao Yi Xue Ban* 2015;40(3):341–4.
20. Cui H, Hao S, Arous E. A distinct cause of necrotizing fasciitis: *Aeromonas veronii* biovar *sobria*. *Surg Infect* (*Larchmt*) 2007;8(5):523–8.
21. Danaher PJ, Mueller WP. *Aeromonas hydrophila* septic arthritis. *Mil Med* 2011;176(12):1444–6.
22. Liakopoulos V, et al. *Aeromonas hydrophila* as a causative organism in peritoneal dialysis-related peritonitis: case report and review of the literature. *Clin Nephrol* 2011;75(Suppl 1):65–8.
23. Issa N, Napolitano LM. *Aeromonas* pneumonia in a trauma patient requiring extracorporeal membrane oxygenation for severe acute respiratory distress syndrome: case report and literature review. *Surg Infect (Larchmt)* 2011;12(3):241–5.
24. Tena D, et al. Surgical site infection due to *Aeromonas* species: report of nine cases and literature review. *Scand J Infect Dis* 2009;41(3):164–70.
25. Patel KM, Svestka M, Sinkin J, Ruff P, 4th. Ciprofloxacin-resistant *Aeromonas hydrophila* infection following leech therapy: a case report and review of the literature. *J Plast Reconstr Aesthet Surg* 2013;66(1):e20–2.
26. Turner P, Willemse C, Phakaudom K, Zin TW, Nosten F, McGready R. *Aeromonas* spp. bacteremia in pregnant women, Thailand–Myanmar border, 2011. *Emerg Infect Dis* 2012;18(9):1522–3.
27. Edberg SC, Browne FA, Allen MJ. Issues for microbial regulation: *Aeromonas* as a model. *Crit Rev Microbiol* 2007;33(1):89–100.
28. Chopra AK, Graf J, Horneman AJ, Johnson JA. Virulence factor-activity relationships (VFAR) with specific emphasis on *Aeromonas* species (spp.). *J Water Health* 2009;7(Suppl 1):S29–54.
29. Singh V, Chaudhary DK, Mani I, Jain R, Mishra BN. Development of diagnostic and vaccine markers through cloning, expression, and regulation of putative virulence-protein-encoding genes of *Aeromonas hydrophila*. *J Microbiol* 2013;51(3):275–82.
30. Gashgari RM, Selim SA. Detection and characterization of antimicrobial resistance and putative virulence genes in *Aeromonas veronii* biovar *Sobria* isolated from gilthead sea bream (*Sparus aurata* L.). *Foodborne Pathog Dis* 2015;12(9):806–11.
31. Ji Y, et al. Contribution of nuclease to the pathogenesis of *Aeromonas hydrophila*. *Virulence* 2015;6(5):515–22.
32. Chen PL, Ko WC, Wu CJ. Complexity of β-lactamases among clinical *Aeromonas* isolates and its clinical implications. *J Microbiol Immunol Infect* 2012;45(6):398–403.
33. Abbott SL, Cheung WK, Janda JM. The genus *Aeromonas*: biochemical characteristics, atypical reactions, and phenotypic identification schemes. *J Clin Microbiol* 2003;41(6):2348–57.
34. Balakrishna K, Murali HS, Batra HV. Detection of toxigenic strains of *Aeromonas* species in foods by a multiplex PCR assay. *Indian J Microbiol* 2010;50(2):139–44.
35. Mendes-Marques CL, Nascimento LM, Theophilo GN, Hofer E, Melo Neto OP, Leal NC. Molecular characterization of *Aeromonas* spp. and *Vibrio cholerae* O1 isolated during a diarrhea outbreak. *Rev Inst Med Trop Sao Paulo* 2012;54(6):299–30.
36. Mendes-Marques CL, Hofer E, Leal NC. Development of duplex-PCR for identification of *Aeromonas* species. *Rev Soc Bras Med Trop* 2013;46(3):355–7.
37. Senderovich Y, Ken-Dror S, Vainblat I, Blau D, Izhaki I, Halpern M. A molecular study on the prevalence and virulence potential of *Aeromonas* spp. recovered from patients suffering from diarrhea in Israel. *PLoS One* 2012;7(2):e30070.
38. Sreedharan K, Philip R, Singh IS. Characterization and virulence potential of phenotypically diverse *Aeromonas veronii* isolates recovered from moribund freshwater ornamental fishes of Kerala, India. *Antonie van Leeuwenhoek* 2013;103(1):53–67.

39. Meng S, Wang Y, Wang Y, Liu D, Ye C. Development of cross-priming amplification assays for rapid and sensitive detection of *Aeromonas hydrophila*. *Lett Appl Microbiol* 2015;61(2):171–8.
40. Parker JL, Shaw JG. *Aeromonas* spp. clinical microbiology and disease. *J Infect* 2011;62(2):109–18.
41. Ye Y, et al. Resistance characterization, virulence factors, and ERIC-PCR fingerprinting of *Aeromonas veronii* strains isolated from diseased *Trionyx sinensis*. *Foodborne Pathog Dis* 2012;9(11):1053–5.
42. Rather MA, Willayat MM, Wani SA, Munshi ZH, Hussain SA. A multiplex PCR for detection of enterotoxin genes in *Aeromonas* species isolated from foods of animal origin and human diarrhoeal samples. *J Appl Microbiol* 2014;117(6):1721–9.
43. Fernandes AM, et al. Development of highly sensitive electrochemical genosensor based on multiwalled carbon nanotubes-chitosan-bismuth and lead sulfide nanoparticles for the detection of pathogenic *Aeromonas*. *Biosens Bioelectron* 2015;63:399–406.
44. Persson S, Al-Shuweli S, Yapici S, Jensen JN, Olsen KE. Identification of clinical *Aeromonas* species by *rpoB* and *gyrB* sequencing and development of a multiplex PCR method for detection of *Aeromonas hydrophila*, *A. caviae*, *A. veronii*, and *A. media*. *J Clin Microbiol* 2015;53(2):653–6.
45. Ko WC, Chiang SR, Lee HC, Tang HJ, Wang YY, Chuang YC. In vitro and in vivo activities of fluoroquinolones against *Aeromonas hydrophila*. *Antimicrob Agents Chemother* 2003;47(7):2217–22.
46. Figueira V, Vaz-Moreira I, Silva M, Manaia CM. Diversity and antibiotic resistance of *Aeromonas* spp. in drinking and waste water treatment plants. *Water Res* 2011;45(17):5599–611.
47. Kadlec K, et al. Molecular basis of sulfonamide and trimethoprim resistance in fish-pathogenic *Aeromonas* isolates. *Appl Environ Microbiol* 2011;77(20):7147–50.
48. Carvalho MJ, Martínez-Murcia A, Esteves AC, Correia A, Saavedra MJ. Phylogenetic diversity, antibiotic resistance and virulence traits of *Aeromonas* spp. from untreated waters for human consumption. *Int J Food Microbiol* 2012;159(3):230–9.
49. Shakir Z, et al. Molecular characterization of fluoroquinolone-resistant *Aeromonas* spp. isolated from imported shrimp. *Appl Environ Microbiol* 2012;78(22):8137–41.
50. Aires A, Dias CS, Rosa EA, Saavedra MJ. Antimicrobial susceptibility of *Aeromonas* spp. isolated from pig ileum segments to natural isothiocyanates. *Med Chem* 2013;9(6):861–6.
51. Jones BL, Wilcox MH. *Aeromonas* infections and their treatment. *J Antimicrob Chemother* 1995;35(4):453–61.
52. Chen PL, Wu CJ, Ko WC. *Aeromonas* species. Available at http://www.antimicrobe.org/b74.asp, Accessed May 31, 2016.
53. Lye DJ. A mouse model for characterization of gastrointestinal colonization rates among environmental *Aeromonas* isolates. *Curr Microbiol* 2009;58(5):454–8.
54. Lye DJ, Rodgers MR, Stelma G, Vesper SJ, Hayes SL. Characterization of *Aeromonas* virulence using an immunocompromised mouse model. *Curr Microbiol* 2007;54(3):195–8.
55. Lye DJ. Gastrointestinal colonization rates for human clinical isolates of *Aeromonas veronii* using a mousemodel. *Curr Microbiol* 2011;63(4):332–6.
56. Haberberger RL Jr, Yonushonis WP, Daise RL, Mikhail IA, Ishak EA. Re-examination of *Rattus norvegicus* as an animal model for *Aeromonas*-associated enteritis in man. *Experientia* 1991;47(5):426–9.
57. Pang MD, Lin XQ, Hu M, Li J, Lu CP, Liu YJ. *Tetrahymena*: an alternative model host for evaluating virulence of *Aeromonas* strains. *PLoS One* 2012;7(11):e48922.
58. Chen PL, et al. Virulence diversity among bacteremic *Aeromonas* isolates: ex vivo, animal, and clinical evidences. *PLoS One* 2014;9(11):e111213.
59. Li J, Zhang XL, Liu YJ, Lu CP. Development of an *Aeromonas hydrophila* infection model using the protozoan *Tetrahymena thermophila*. *FEMS Microbiol Lett* 2011;316(2):160–8.
60. Noonin C, Jiravanichpaisal P, Söderhäll I, Merino S, Tomás JM, Söderhäll K. Melanization and pathogenicity in the insect, *Tenebrio molitor*, and the crustacean, *Pacifastacus leniusculus*, by *Aeromonas hydrophila* AH-3. *PLoS One* 2010;5(12):e15728.
61. Hayes SL, Waltmann M, Donohue M, Lye DJ, Vesper SJ. Predicting virulence of *Aeromonas* isolates based on changes in transcription of c-jun and c-fos in human tissue culture cells. *J Appl Microbiol* 2009;107(3):964–9.

16

Bacteroides

Mario Julio Avila-Campos

CONTENTS

16.1 Introduction

Taxonomically, the genus *Bacteroides* belongs to the phylum Bacteroidetes, class Bacteroidia, order Bacteroidales, and family Bacteroidaceae. This genus covers a large number of species pathogenic to humans and animals. *Bacteroides* species are strictly anaerobes and some are aerotolerant, and are considered the most important constituent of the human resident microbiota. Species of this genus are Gram-negative, non-spore-forming rods, living in intestines and giving the host several nutrients produced by the breaking down of food. In disequilibrium in the gastrointestinal ecosystem, they can produce several infections of endogenous nature, such as periodontal diseases, lung and brain infections, urinary or genital infections, and intraabdominal abscesses, and recently they have been linked to colon

TABLE 16.1

Phenotypic Characteristics of the *Bacteroides fragilis* Group and *Parabacteroides*

					Fermentation				
Species	**Growth in 20% Bile**	**Indol**	**Catalase**	**Esculin Hydrolysis**	**Arabinose**	**Cellobiose**	**Rhamnose**	**Sucrose**	**Trehalose**
B. fragilis group	+	V	V	+	$-^{+}$	+	V	+	+
B. fragilis	+	–	+	+	–	$+^{-}$	–	+	–
B. caccae	+	–	+	+	+	$+^{-}$	$+^{-}$	+	+
B. ovatus	+	+	+	+	+	+	+	+	+
B. stercoris	+	+	–	+	$-^{+}$	$-^{+}$	+	+	–
B. thetaiotaomicron	+	+	+	+	+	$+^{-}$	+	+	+
B. uniformis	$+^{w}$	+	V	+	+	+	$-^{w}$	+	$-^{w}$
B. vulgatus	+	–	$-^{+}$	–	+	–	+	$+^{-}$	
P. merdae	+	–	$-^{+}$	+	$-^{+}$	V	+	+	+
P. distasonis	+	–	$-^{+}$	+	$-^{+}$	+	V	+	+

+, positive; –, negative; $-^{+}$, some species positive; $+^{-}$, some species negative; $+^{w}$ or $-^{w}$, some species weak positive or weak negative; V, variable.

cancer. *Bacteroides* spp. show a great variability to the antimicrobial susceptibility, and several species are resistant to various antimicrobial drugs expressing resistance genes or other mechanisms, such as efflux pumps.[1] The key phenotypic characteristics of the species of the *B. fragilis* group are summarized in Table 16.1.

16.2 Taxonomic and Ecological Aspects

The genus *Bacteroides* has undergone major taxonomic revisions in the past few years, and most species suffer changes frequently. Suitable taxonomic changes have become important to clinicians and microbiologists, since these changes can be used as an indicator of virulence or antimicrobial resistance. It is known that some species of the *Bacteroides* genus, such as *B. thetaiotaomicron*, are much more resistant to antimicrobials than *B. fragilis*, for example. In 1990, various species were moved to other genera, including *Porphyromonas* and *Prevotella*, and others were included, such as *Dialister*, *Megamonas*, *Mitsuokella*, *Tannerella*, *Tissierella*, and *Alistipes*. By using culture-independent techniques such as 16S rRNA gene sequencing, a number of new species have been added to the genus *Bacteroides*, achieving more than 20 bacterial species.[2,3] In 2005, new species added to this genus included *Bacteroides goldsteinii*, *Bacteroides nordii*, and *Bacteroides salyersai*, *Bacteroides plebeius* and *Bacteroides coprocola* (isolated from human feces), and *Bacteroides massiliensis* (isolated from the blood of a newborn). Subsequently, *B. goldsteinii*, *Bacteroides distasonis*, and *Bacteroides merdae* were moved to the genus *Parabacteroides*.[4] The occurrence of species of the *B. fragilis* group in the fecal microbiota of children with diarrhea is shown in Table 16.2.

16.3 Biology and Virulence Factors

Most *Bacteroides* spp. have a G-C composition of 40%–48%; for example, *B. fragilis* NCTC 9343 shows 43.19% G-C content, and sequencing studies have demonstrated an influence of environmental factors on biology of these microorganisms.

B. fragilis constitutes 0.5% of the human colonic microbiota, but it is the most commonly isolated anaerobic pathogen. Virulence factors produced by *B. fragilis* can be divided into three categories: (1) adherence to epithelial tissues, (2) protection from the host immune response, and (3) destruction of tissues. The presence of fimbriae and agglutinins causes *B. fragilis* allows its colonization on the host's tissue; the polysaccharide capsule, LPS, and various bacterial enzymes work against the host's immune response, and its histolytic enzymes collaborate in tissue destruction.[5] In addition, *Bacteroides* spp. are able to produce bacteriocin-like substances that inhibit the growth of other species or genera, as observed in Figure 16.1.

TABLE 16.2

Occurrence of the *Bacteroides fragilis* Group in 170 Stool Samples from Children with and without Acute Diarrhea

	Diarrhea		No Diarrhea	
Bacterial Isolates	**No.**	**%**	**No.**	**%**
Non-ETBF	67	72.82	42	53.8
ETBF	2	2.17	0	0
B. uniformis	7	7.6	1	1.3
B. vulgatus	4	4.3	13	16.7
B. thetaiotaomicron	1	1.1	1	1.3
B. ovatus	0	0	4	5.1
Parabacteroides distasonis	11	11.9	17	21.8

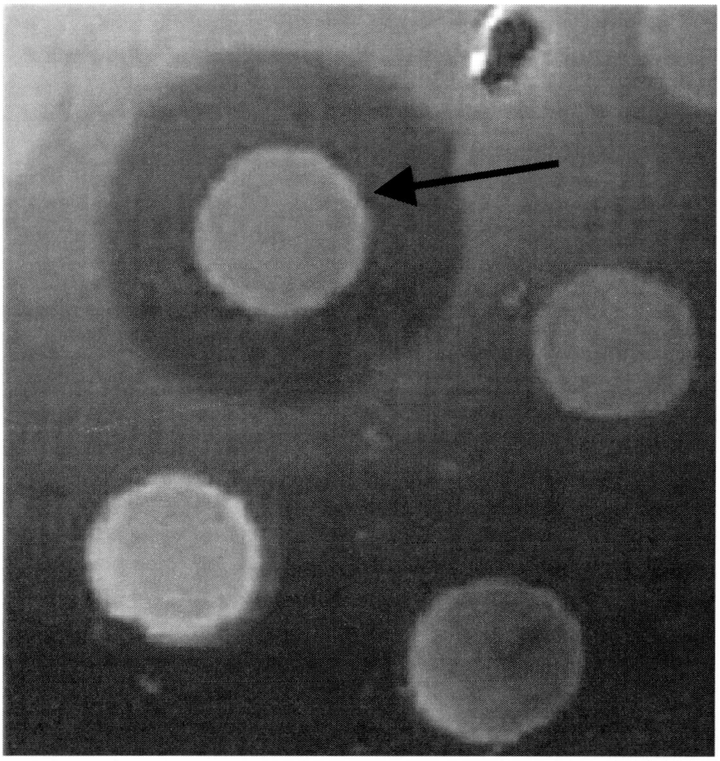

FIGURE 16.1 Bacteriocin-producing *B. fragilis* against *B. vulgatus*. Arrow shows the inhibition halo.

16.3.1 Capsule

The capsule of *B. fragilis* is largely responsible for abscess formation due to its protection afforded to the bacteria against the host's immune response. These abscesses can produce intestinal obstruction, fistula, bacteremia, and disseminated infection. Encapsulated *B. fragilis* has been used to produce experimental abscess in animal model, and it has been reported that responses to other polysaccharide antigens are T-cell independent, but abscess formation induced by *B. fragilis* is dependent on the involvement of T cells. *B. fragilis* produces a polysaccharide capsule of varying molecular weights (PS-A, PS-B, and PS-C). By using electron microscopy, three capsule variants within an individual strain of *B. fragilis* might be observed: large and small capsules, and noncapsulate strain.

16.3.2 Evasion from the Host's Immune Response

The ability to evade the host's immune response contributes to the bacterial virulence. Encapsulated *B. fragilis* is more resistant to the action of peritoneal macrophages, the first host immunologic defense response to the rupture of the intestine or any other compromise of the peritoneal cavity. The production of nitric oxide (NO) is an important biological mediator in the living organism that is synthesized from L-arginine using NADPH and molecular oxygen. However, the overproduction of NO, which is catalyzed by the inducible nitric oxide synthase (iNOS), a soluble enzyme active in its dimeric form, is cytotoxic, and after exposure to cytokines such as gamma interferon it has a microbicidal, cell necrosis, or death role.

16.3.3 Enzymes

Histolytic enzymes, such as hyaluronidase and chondroitin sulfatase present in some *B. fragilis*, can attack the host's extracellular matrix. Proteases of *B. fragilis* have also been implicated in destroying brush border enzymes acting on the microvillus membranes for the final digestion of food and absorption of nutrients. Other enzymes, hemolysins (HlyA and HlyB), and neuraminidase encoded by the *nanH* gene are also important. Neuraminidase is found in many pathogenic bacteria and is generally considered as a virulence factor; this enzyme catalyzes the removal of the sialic acid from epithelial cells and immunoactive proteins such as IgG. Several *B. fragilis* strains have neuraminidase activity, and it has been suggested that this activity plays a role in the bacterial attachment to animal cells and to the hemagglutination.[5]

16.3.4 Enterotoxin

The *B. fragilis* enterotoxin (BFT) is a zinc metalloprotease that destroys the adherence tight junctions in intestinal epithelium by cleaving E-cadherin, resulting in rearrangements of the actin cytoskeleton of the epithelial cells, which causes diarrhea. Enterotoxigenic *B. fragilis* (ETBF) strains encode three isotypes of BFT on distinct *bft* loci, carried on a 6-kb genome segment that is unique and called the *B. fragilis* pathogenicity island, and the presence of the *bft* gene is generally detected by PCR techniques.[6–9] This pathogenicity island is flanked by genes encoding mobilization proteins and may be transmissible to nontoxigenic strains (non-ETBF) (Figure 16.2). Recent studies have shown that the BFT has a possible role as a carcinogen in colorectal cancer.[10]

16.3.5 Endotoxin/LPS

LPS in *B. fragilis* is 10–1000 times less toxic than that in *Escherichia coli*. This endotoxin displays toxicity, and the exposure to antibiotics enhances its production many times higher in *B. fragilis* than in the other species of the *B. fragilis* group. This may partly explain its association with clinical infections and mortality.

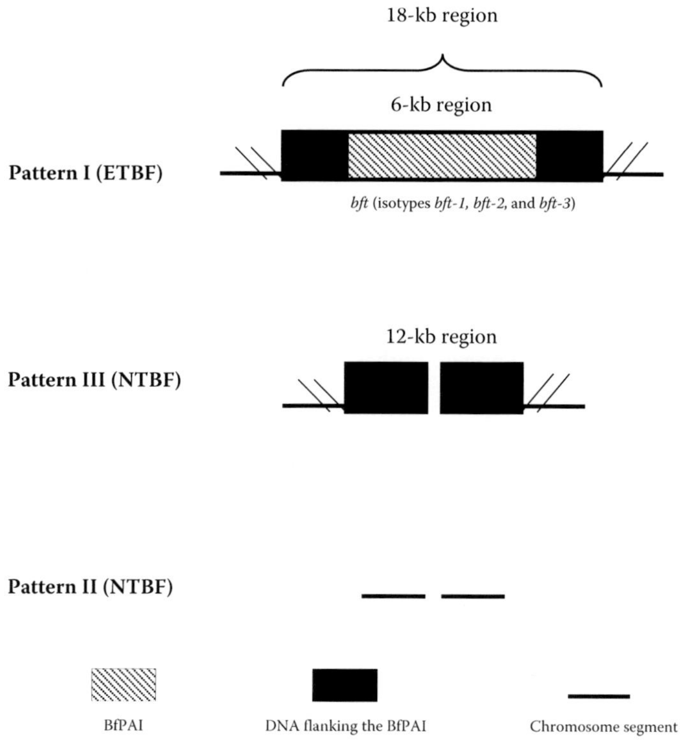

FIGURE 16.2 Pathogenicity island (6-kb region) anchored in chromosomal DNA from enterotoxigenic *B. fragilis*, and their genetic profile. (Adapted from Franco, A.A. et al., *Mol. Microbiol.*, 45, 1067–1077, 2002.)

16.4 Pathogenesis

Capsule present in anaerobes is considered an important virulence factor. Other factors related to the virulence of anaerobes include mucosal damage and resistance against the host immune response. ETBF has been implicated in inflammatory intestinal diseases.[11–13] Studies show that intestinal colonization with ETBF leads to acute or chronic intestinal inflammations explaining its potential pathogenicity. The BFT alters the F-actin structure, resulting in disruption of the epithelial barrier-producing diarrhea.[14,15]

Studies have implicated the presence of *E. coli* and *Bacteroides vulgatus* in the development of Crohn's disease and ulcerative colitis; however, the relationship of this bacterial association with these diseases remains unclear.

Adhesion to the epithelial surface is considered a prerequisite for pathogenicity in most bacteria, and this attachment may be selective for different cell types.[5,16] Establishing a site in the host is critical to the role of *Bacteroides* spp. and *Parabacteroides* spp., both as commensals on the mucosal surfaces of the intestinal epithelium and as pathogens causing abscesses or other infectious processes. The adhesion and invasion processes of intestinal *Bacteroidales* isolated from fecal microbiota of children with diarrhea are shown in Table 16.3.

These organisms contain a variety of cell surface molecules that are either critical or advantageous for colonization, including adhesins, hemagglutinins, a polysaccharide capsule, fimbriae, and proteases. *B. fragilis* is one of the most important species of intestinal *Bacteroidales* and possesses a complex capsular polysaccharide composed of at least eight different polysaccharides, which appear to be antigenically diverse. Also, *B. fragilis* may express three different types of capsules, large or small, and an electron-dense layer. This heterogeneous nature of encapsulating structures within individual strains could explain the controversial observations in the literature with respect to the *B. fragilis* surface structures related to adhesion. Similar surface structures have been seen in other intestinal species, such as *B. thetaiotaomicron*.

16.5 Diagnosis

Species of *Bacteroides* are considered resident members in intestinal microbiota, and they are mostly recovered in intra- and extra-abdominal infections. Most anaerobic infections originate from the host's resident microbiota and are considered infections of an endogenous nature. Conditions such as low blood supply into an affected site can predispose the host to anaerobic infection, as well as trauma, foreign body, malignancy, surgery, edema, shock, colitis, and vascular disease. Mixed infections with aerobic or facultative organisms make the local tissue conditions favorable for the growth of anaerobic bacteria. Anaerobic bacteria are the most common residents of the skin and mucous membrane surfaces and natural cavities' microbiota. Anaerobes belonging to the resident microbiota of the oral cavity can be recovered from various infections adjacent to that area, such as cervical lymphadenitis, subcutaneous

TABLE 16.3

Adhesion and Invasion Assays of 114 Strains of the Intestinal *Bacteroidales* Isolated from Children with Acute Diarrhea

Species	No. of Isolates	Adhesion-Positive (%)	Invasion-Positive (%)
Bacteroides fragilis	39	30 (77)	13 (33.3)
B. vulgatus	8	6 (75)	4 (50)
B. uniformis	6	5 (83.3)	5 (83.3)
B. ovatus	5	4 (80)	0
B. eggerthii	0	0	0
B. thetatiotaomicron	0	0	0
Parabacteroides distasonis	6	4 (66.6)	2 (33.3)

abscesses, and burns in proximity to the oral cavity, human and animal bites, tonsillar and retropharyngeal abscesses, chronic sinusitis, chronic otitis media, and periodontal abscess. Species of anaerobic Gram-negative rods including pigmented *Prevotella*, *Porphyromonas*, *B. fragilis* group, *Fusobacterium*, and Gram-positive anaerobic cocci are associated with several infections, such as peritonitis, liver abscess, intra-abdominal abscesses, neonatal infections, and recently to colorectal cancer.[17,18] In addition, quantitative real-time PCR has provided a convenient, dependable, and rapid method to study the diversity of the presence of the *bft* subtypes and the significance of ETBF in clinical infections.[12]

16.6 Susceptibility to Antimicrobials

Because *Bacteroides* spp. are most commonly found in mixed infections, β-lactams added or not with β-lactamase inhibitors, carbapenems, clindamycin, and metronidazole are often the antibiotics of choice.[19,20] Clindamycin and metronidazole along with or in combination with some fluoroquinolones are also used. These species are highly resistant to a broad range of antibiotics, including heavy metals,[21] and their resistance to antimicrobials can vary in accordance with geographical locations. The resistance to active drugs, such as imipenem, piperacillin-tazobactam, ampicillin-sulbactam, and metronidazole, is observed in some strains.[17,22] Considering that a distinct profile of drug resistance is observed in different countries, it is possible that this resistance is geographically dependent. The resistance profiles of some *Bacteroidales* species are shown in Table 16.4, and the distribution of resistance genes and β-lactamase production in *Bacteroides* spp. and *P. distasonis* are shown in Table 16.5.

16.6.1 Mechanisms of Antimicrobial Resistance

Species of the genus *Bacteroides* display varied resistance mechanisms to antibiotics. *B. fragilis* is intrinsically resistant to several classes of antibiotics, but the mechanisms are poorly understood. The antimicrobial resistance can be due to altered target binding affinity, decreased penetration due to permeability or efflux changes, or by the presence of inactivating enzymes.[1,23] *Bacteroides* species isolated from clinical samples have exhibited increasing resistance to many antibiotics, including cefoxitin, clindamycin, metronidazole, carbapenems, and fluoroquinolones, such as gatifloxacin, levofloxacin, and moxifloxacin, but the newer ones, including sitafloxacin, clinafloxacin, and garenoxacin, have shown a good activity.[1]

16.6.2 β-Lactam Agents

Most *Bacteroides* spp. are resistant to β-lactam agents as they produce β-lactamase commonly mediated by the chromosomal *cepA* gene. However, this enzyme is inhibited by the most commonly used β-lactamase inhibitors (sulbactam, clavulanic acid, and tazobactam).[24]

16.6.3 Carbapenems

Resistance to carbapenems remains rare, and the genes *cfiA* and *ccrA*, both encoding class B metallo-β-lactamases, are able to degrade these agents. Some strains of *Bacteroides* contain "silent" carbapenemase genes, and expression levels are dependent on promoter-containing insertion sequences (IS). Alterations in penicillin-binding proteins (PBP) or porins, and efflux pumps are other bacterial resistance mechanisms for carbapenems.

16.6.4 PBP

Penicillin and other β-lactams are cell-wall-active drugs. The structure of the β-lactam antibiotics facilitates their binding to the active site of PBP in an irreversible inhibition of proteins blocking the transpeptidation of the peptidoglycan layer. In *B. fragilis*, the sequenced genome showed the presence of seven putative PBP genes, and it was difficult to correlate the MIC_{50} since this is not the only mechanism responsible for the resistance.[25]

TABLE 16.4

Resistance Profiles of Intestinal *Bacteroidales* Species to Eight Antibiotics

	% Resistance[b]						
	B. fragilis	***B. vulgatus***	***B. uniformis***	***B. ovatus***	***B. thetaiotaomicron***	***P. distasonis***	***Bacteroides* spp. and *Parabacteroides* sp.**
Antibiotics[a]	**(*n* = 66)**	**(*n* = 14)**	**(*n* = 7)**	**(*n* = 7)**	**(*n* = 2)**	**(*n* = 16)**	**(*n* = 144)**
Amoxicillin	92.4	85.7	100	100	100	93.7	93
Amoxicillin/clavulanic acid	40.9	42.8	0	85.7	100	68.7	47.3
Ampicillin	98.4	92.8	100	100	100	87.7	96.4
Cephalexin	100	92.8	100	100	100	100	99
Cefoxitin	7.5	0	14.2	0	0	12.5	23
Clindamycin	31.8	100	0	71.4	0	43.7	34.2
Penicillin	100	92.8	100	100	100	100	99
Tetracycline	59	50	28.5	71.4	0	43.7	53.5

[a] Breakpoints in accordance with CLSI (2007). Amoxicillin, amoxicillin/clavulanic acid, cephalexin, cefoxitin, clindamycin, penicillin, and tetracycline: 8 μg/ml; ampicillin: 4 μg/ml.
[b] All strains were susceptible to imipenem and metronidazole.

TABLE 16.5

Distribution of Resistance Genes and β-Lactamase Production in Intestinal *Bacteroides* spp. and *Parabacteroides distasonis*

	Genes					β-Lactamase Production
	cfiA	*cepA*	*ermF*	*tetQ*	*nim*	
Species (No.)	**No. (%)**	**No. (%)**	**No. (%)**	**No. (%)**	**No. (%)**	**No. (%)**
B. fragilis (64)	51 (77.2)	53 (80.3)	16 (24.2)	54 (81.8)	5 (7.5)	60 (90.9)
B. vulgatus (14)	5 (35.7)	11 (78.5)	5 (35.7)	7 (50)	1 (7.14)	13 (92.8)
B. uniformis (7)	6 (85.7)	7 (100)	0 (0)	5 (71.4)	2 (28.5)	7 (100)
B. ovatus (7)	1 (14.2)	1 (14.2)	1 (14.2)	6 (85.7)	0 (0)	7 (100)
B. eggerthii (2)	0 (0)	2 (100)	1 (50)	2 (100)	0 (0)	2 (100)
B. thetaiotaomicron (2)	2 (100)	2 (100)	2 (100)	2 (100)	0 (0)	1 (50)
P. distasonis (16)	6 (37.5)	11 (68.7)	6 (37.5)	15 (93.7)	1 (6.25)	15 (93.7)

16.6.5 Outer Membrane Proteins or Porins

Some data have shown a correlation between alterations in porins and antimicrobial resistance in *Bacteroides* spp., but it is not totally elucidated.

16.6.6 Aminoglycosides

Bacteroides species are intrinsically resistant to aminoglycosides since the uptake of this drug requires free oxygen or a nitrate-dependent electron transport chain, and it is lacking in anaerobes.

16.6.7 Macrolides, Lincosamides, and Chloramphenicol

These antimicrobials inhibit the protein synthesis by binding to the 50S subunit of the bacterial ribosome. Genes conferring resistance to clindamycin and erythromycin in *Bacteroides* are similar.

16.6.8 Tetracycline

Tetracyclines inhibit bacterial protein synthesis by blocking the attachment of the tRNA-amino acid to the ribosome. Studies show that 80%–90% of the clinical isolates of *Bacteroides* are resistant to this drug, due to the presence of the gene *tetQ* that is the most commonly involved one in this resistance. Also, the efflux pump for tetracycline has also been described as a resistance mechanism in this microorganism.

16.6.9 Nitroimidazoles

Metronidazole is the most commonly used drug for anaerobic infections, and resistance to this drug is rare. Metronidazole-resistant *B. fragilis* strains have been reported, and their resistance is due to the presence of *nim* genes or associated genes to IS; however, this gene can be "silent" unless activated by IS elements.[26,27] In *Bacteroides* strains, seven *nim* genes (*nimA* to *nimG*) can be found, and each one is associated with both a distinct mobile genetic element and a specific activating IS element.[28]

16.6.10 Quinolones

Quinolones inhibit DNA gyrase and DNA topoisomerase IV, which act in bacterial DNA replication, and gene mutations producing these enzymes are the most common causes of quinolone resistance.[29]

Mutations in *gyrA* causing fluoroquinolone resistance in *B. fragilis* have been identified. Studies have shown that the quinolone resistance in clinical isolates and in mutants is due to the increased activity of the efflux pumps belonging to the RND family.

16.6.11 Plasmids

These genetic elements are very common in *Bacteroides* species and are found in 20%–50% of strains. Genes conferring resistance to different classes of antibiotics have been found on plasmids in *Bacteroides*. Resistance genes *nimA* to *nim-F* and *cfiA* have been observed in clinical isolates and identified on transferable plasmids. Intestinal *Bacteroidales* species often display genetic elements, such as plasmids and/or transposons, because of their predominant number and a close contact with other intestinal bacteria. Clinical isolates of *Bacteroidales* often harbor cryptic plasmids (from 3 to 7 kb of size) with no defined phenotypic or genotypic association.[25]

16.6.12 Multidrug Resistance

Little is known about efflux pumps in anaerobic bacteria. In *Bacteroides* spp., particularly in *B. fragilis*, 16 homologs of RND pumps (*bmeABC1* to *bmeABC 16*) called *B. fragilis* multidrug efflux have been described. A MATE-type efflux system has also been characterized in *B. thetaiotaomicron*.[28] This system (*B. fragilis* multidrug efflux) shows that (1) each operon is formed for three components, (2) the *bmeC10* outer membrane component forms part of a contiguous gene with the *bmeB10* pump gene, (3) there are two functional pump genes (*bmeB11* and *bmeB11*) which are transcribed separately in *bme11*, and (4) at least 15 of the 16 genes are transcribed.[1]

16.7 Laboratory Models

Laboratories studying anaerobic bacteria use some principles for isolating and identifying *Bacteroides* species. Most of the laboratories refer to the *Wadsworth-KTL Anaerobic Bacteriology Manual* and the *Manual of Clinical Microbiology*. The processing for clinical specimens is important to the success of isolation, and they are as follows: (1) aseptic collection of the specimen avoiding contamination with resident microbiota, (2) oxygen-free transport medium system, (3) species of *Bacteroides* grown on selective medium *Bacteroides* bile esculin agar, and (4) *B. fragilis* is resistant to 20% bile, kanamycin, vancomycin, and colistin. Several PCR schemes are also used to identify *Bacteroides* species according to the genus and/or species level, such as using group-specific primers to the β-isopropylmalate dehydrogenase gene *leuB* for rapid diagnosis from infections.[30] In addition, a multiplex PCR assay with group- and species-specific primers to identify rapidly 10 species of the *B. fragilis* group has also been developed.[31]

This bacterium can cause a wide range of human diseases including abscesses, diabetic foot, diarrhea, and sepsis. Studies *in vivo* have shown that *B. fragilis* toxin (BFT) produced by ETBF induces both a fluid secretion and exfoliation of intestinal epithelial cells. In certain human intestinal carcinoma cell lines, particularly HT29/C1, this toxin causes the rounding-up of cells and the rearrangement of the F-actin cytoskeleton.[32,33]

Inoculation of ETBF strains harboring different gene subtypes in germ-free mice produced histopathological alterations mainly in the distal ileum–cecum–colon transition area. Moreover, inoculation in the cecum can produce a superficial erosion and inflammatory infiltration in the lamina propria. It is known that BFT toxin stimulates morphological changes in intestinal epithelial cell lines, such as HT29, HT29/C1, Caco-2, T84, MDCK, and HCT-8.

Studies have suggested that strains carrying different *bft* gene subtypes may have a different pathogenic potential; however, more studies are needed to evaluate the pathogenesis of each subtype in different animal hosts.[34] Certainly, they would provide a better understanding of these organisms in ecological and pathogenic terms.

16.8 Conclusions

Species of *Bacteroides*, including *Bacteroides fragilis*, in an unbalanced gastrointestinal ecosystem, can produce several infections of endogenous nature, and recently have been linked to colon cancer. The *B. fragilis* group is considered as the most commonly isolated anaerobic pathogen. Several virulence factors are produced by *B. fragilis*. The presence of adhesins allows its colonization of the host's tissues, and *Bacteroides* spp. are able to produce bacteriocin-like substances that inhibit the growth of other bacterial species or genera.

ETBF has been implicated in inflammatory intestinal diseases, and its intestinal colonization results in acute or chronic intestinal inflammations. Species of *Bacteroides* are considered resident intestinal microbiota, but they are often recovered from intra- and extra-abdominal infections. *Bacteroides* spp. show great resistance to various antimicrobial drugs expressing resistance genes. *B. fragilis* is intrinsically resistant to several classes of antibiotics, but the mechanisms are poorly understood. Most *Bacteroides* spp. are resistant to β-lactam by producing β-lactamase commonly mediated by the chromosomal *cepA* gene.

Studies *in vivo* have shown that the inoculation of ETBF strains harboring different gene subtypes in germ-free mice produced histopathological alterations, since the BFT toxin stimulates morphological changes in different intestinal epithelial cells. The *in vivo* evaluation of pathogenic potentials of the subtypes of *B. fragilis* could uncover important details about the biology and pathogenesis of this important anaerobic bacterial group.

REFERENCES

1. Wexler, H. M. 2007. *Bacteroides*: the good, the bad, and the nitty-gritty. *Clin. Mirobiol. Rev.* 20: 593–621.
2. Song, Y.; Liu, C.; McTeague, M.; Finegold, S. M. 2004. "*Bacteroides nordii*" sp. nov. and "*Bacteroides salyersae*" sp. nov. isolated from clinical specimens of human intestinal origin. *J. Clin. Microbiol.* 42: 5565–5570.
3. Song, Y.; Liu, C.; Bolanos, M.; Lee, J.; McTeague, M.; Finegold, S. M. 2005. Evaluation of 16S rRNA sequencing and reevaluation of a short biochemical scheme for identification of clinically significant *Bacteroides* species. *J. Clin. Microbiol.* 43: 1531–1537.
4. Sakamoto, M.; Benno, Y. 2006. Reclassification of *Bacteroides distasonis*, *Bacteroides goldsteinii* and *Bacteroides merdae* as *Parabacteroides distasonis* gen. nov., comb. nov., *Parabacteroides goldsteinii* comb. nov. and *Parabacteroides merdae* comb. nov. *Int. J. Syst. Evol. Microbiol.* 56: 1599–1605.
5. Nakano, V.; Piazza, R. M. F.; Avila-Campos, M. J. 2006. A rapid assay of the sialidase activity in species of the *Bacteroides fragilis* group by using peanut lectin hemagglutination. *Anaerobe* 12: 238–241.
6. Almeida, F. S.; Nakano, V.; Avila-Campos, M. J. 2007. Occurrence of enterotoxigenic and nonenterotoxigenic *Bacteroides fragilis* in calves and evaluation of their antimicrobial susceptibility. *FEMS Microbiol. Lett.* 272: 15–21.
7. Avila-Campos, M. J.; Liu, C.; Song, Y.; Rowlinson, M.-C.; Finegold, S. M. 2007. Determination of *bft* gene subtypes in *Bacteroides fragilis* clinical isolates. *J. Clin. Microbiol.* 45: 1336–1338.
8. Bressane, M. A.; Durigon, L. E.; Avila-Campos, M. J. 2001. Prevalence of the *Bacteroides fragilis* group and enterotoxigenic *Bacteroides fragilis* in immunodeficient children. *Anaerobe* 7: 277–281.
9. Claros, M. C.; et al. 2006. Characterization of the *Bacteroides fragilis* pathogenicity island in human blood culture isolates. *Anaerobe* 12: 17–22.
10. Toprak, N. U.; et al. 2006. A possible role of *Bacteroides fragilis* enterotoxin in the aetiology of colorectal cancer. *Clin. Microbiol. Infect.* 12: 782–786.
11. Kryzanowsky, F.; Avila-Campos, M. J. 2003. Detection of non-enterotoxigenic and enterotoxigenic *Bacteroides fragilis* in stool samples from children in São Paulo, Brazil. *Rev. Inst. Med. Trop. São Paulo* 45: 1–4.
12. Merino, V. R. C.; Nakano, V.; Liu, C.; Song, Y.; Finegold, S. M.; Avila-Campos, M. J. 2011. Quantitative detection of enterotoxigenic *Bacteroides fragilis* subtypes isolated from children with and without diarrhea. *J. Clin. Microbiol.* 49: 416–418.

13. Nakano, V.; Gomes, T. A. T.; Vieira, M. A. M.; Ferreira, R. C.; Avila-Campos, M. J. 2007. *bft* gene subtyping in enterotoxigenic *Bacteroides fragilis* isolated from children with acute diarrhea. *Anaerobe* 13: 1–5.
14. Franco, A. A.; Cheng, R. K.; Goodman, A.; Sears, C. L. 2002. Modulation of *bft* expression of the *Bacteroides fragilis* pathogenicity island and its flanking region. *Mol. Microbiol.* 45: 1067–1077.
15. Wexler, H. M.; Read, E. K.; Tomzynski, T. J. 2002. Identification of an OmpA protein from *Bacteroides fragilis*: *ompA* gene sequence, OmpA amino acid sequence and predictions of protein structure. *Anaerobe* 8: 180–191.
16. Nakano, V.; et al. 2008. Adherence and invasion of *Bacteroidales* isolated from the human intestinal tract. *Clin. Microbiol. Infect.* 14: 955–963.
17. Diniz, C. G.; Farias, L. M.; Carvalho, M. A.; Rocha, E. R.; Smith, C. J. 2004. Differential gene expression in a *Bacteroides fragilis* metronidazole resistant mutant. *J. Antimicrob. Chemother.* 54: 100–108.
18. Fukugaiti, M. H.; Ignacio, A.; Fernandes, M. R.; Júnior, U. R.; Nakano, V.; Avila-Campos, M. J. 2015. High occurrence of *Fusobacterium nucleatum* and *Clostridium difficile* in the intestinal microbiota of colorectal carcinoma patients. *Braz. J. Microbiol.* 46(4): 1135–1140.
19. Behra-Miellet, J.; Calvet, L.; Dubreuil, L. A. 2004. A *Bacteroides thetaiotamicron* porin that could take part in resistance to β-lactams. *Int. J. Antimicrob. Agents* 24: 135–143.
20. Brazier, J. S.; Stubbs, S. L.; Duerden, B. I. 1999. Metronidazole resistance among clinical isolates belonging to the *Bacteroides fragilis* group: time to be concerned? *J. Antimicrob. Chemother.* 44: 580–581.
21. Ignacio, A.; Nakano, V.; Avila-Campos, M. J. 2015. Intestinal *Bacteroides vulgatus* showing resistance to metals. *Appl. Med. Res.* 1: 43–47.
22. Hecht, D. W. 2004. Prevalence of antibiotic resistance in anaerobic bacteria: worrisome developments. *Clin. Infect. Dis.* 39: 92–97.
23. Schapiro, J. M.; Gupta, R.; Stefansson, E.; Fang, F. C.; Limaye, A. P. 2004. Isolation of metronidazole-resistant *Bacteroides fragilis* carrying the *nimA* nitroreductase gene from a patient in Washington State. *J. Clin. Microbiol.* 42: 4127–4129.
24. Nakano, V.; Nascimento-Silva, A.; Merino, V. R. C.; Wexler, H. M.; Avila-Campos, M. J. 2011. Antimicrobial resistance and prevalence of resistance genes in intestinal *Bacteroidales* strains. *Clinics* 66: 543–547.
25. Nakano, V.; Padilla, G.; Valle, M. M.; Avila-Campos, M. J. 2004. Plasmid-related β-lactamase production in *Bacteroides fragilis* strains. *Res. Microbiol.* 155: 843–846.
26. Pumbwe, L.; Chang, A.; Smith, R. L.; Wexler, H. M. 2007. BmeRABC5 is a multidrug efflux system that can confer metronidazole resistance in *Bacteroides fragilis*. *Microb. Drug Resist.* 13: 96–101.
27. Pumbwe, L.; Glass, D.; Wexler, H. M. 2006. Efflux pumps over expression in multiple-antibiotic-resistant mutants of *Bacteroides fragilis*. *Antimicrob. Agents Chemother.* 50: 3150–3153.
28. Ueda, O.; Wexler, H. M.; Hirai, K.; Shibata, Y.; Yoshimura, F.; Fujimura, S. 2005. Sixteen homologs of the mex-type multidrug resistance efflux pump in *Bacteroides fragilis*. *Antimicrob. Agents Chemother.* 49: 2807–2815.
29. Miyamae, S.; Ueda, O.; Yoshimura, F.; Hwang, J.; Tanaka, Y.; Nikaido, H. 2001. A MATE family multidrug efflux transporter pumps out fluoroquinolones in *Bacteroides thetaiotaomicron*. *Antimicrob. Agents Chemother.* 45: 3341–3346.
30. Miki, T.; et al. 2005. Simultaneous detection of *Bacteroides fragilis* group species by *leuB*-directed PCR. *J. Med. Invest.* 52: 101–108.
31. Liu, C.; Song, Y.; McTeague, M.; Vu, A. W.; Wexler, H.; Finegold, S. M. 2003. Rapid identification of the species of the *Bacteroides fragilis* group by multiplex PCR assays using group- and species-specific primers. *FEMS Microbiol. Lett.* 222: 9–16.
32. Chambers, F. G.; Koshy, S. S.; Saidi, R. F.; Clark, D. P.; Moore, R. D.; Sears, C. L. 1997. *Bacteroides fragilis* toxin exhibits polar activity on monolayers of human intestinal epithelial cells (T84 cells) in vitro. *Infect. Immun.* 67: 3561–3570.
33. Donelli, G.; Fabbri, A.; Fiorentini, C. 1996. *Bacteroides fragilis* enterotoxin induces cytoskeletal changes and surfaces blebbing in HT-29 cells. *Infect. Immun.* 64: 113–119.
34. Nakano, V.; Gomes, D. A.; Arantes, R. M.; Nicoli, J. R.; Avila-Campos, M. J. 2006. Evaluation of the pathogenicity of the *Bacteroides fragilis* toxin gene subtypes in gnotobiotic mice. *Curr. Microbiol.* 53: 113–117.

17

Brucella

S.C. Olsen and B. Bricker

CONTENTS

Clinical disease caused by infection with bacteria in the genus *Brucella*, brucellosis, remains one of the most important zoonotic diseases worldwide and is considered to be reemerging in some countries. Although the bacteria causing "Malta fever" was reported in human spleens by Bruce in 1887,[1] the link between human disease and infection in animal reservoirs was not correlated until 1918.[2] The highest prevalence of human disease is currently found in areas of Africa, Asia, Latin America, and the Middle East and is associated with significant public health costs. Although transmission to humans can occur through direct contact with abortions or birth materials from infected animals, indirect contact through consumption of nonpasteurized dairy products is a common route for human infection.

Over half a million human cases are reported worldwide annually. Due to the wide-ranging and nonspecific nature of clinical signs, the true number may be much higher.[3] Estimates of disease prevalence range from <1 per 100,000 people in the United Kingdom, United States, and Australia, to >70 per 100,000 people in some countries in the Middle East and Central Asia.[4] Of particular concern is the observation that children appear to represent a high proportion of human brucellosis cases. Multiple studies have demonstrated that addressing brucellosis in animal reservoirs is the most cost-efficient mechanism for controlling human brucellosis.[5–7] Therefore, most control strategies involve control of livestock brucellosis through vaccination and use of test and removal strategies. Livestock vaccination has proved to successfully decrease human brucellosis and can be especially helpful in developing nations. After initiation of a vaccination in Mongolia,[8,9] the incidence rate of human brucellosis declined from 4.8/10,000 to 0.23/100,000.

17.1 The Pathogen

Members of the genus *Brucella* are small (0.4–3 μm), Gram-negative, coccobacillary organisms within alpha *Proteobacteria* that exist as intracellular pathogens in mammalian hosts. Most species of *Brucella* are obligate pathogens in that they do not exist as commensals nor are they found free-living in the environment. However, a recently identified species of *Brucella* in voles, *Brucella microti*, may have genetic differences that support a free-living as compared to a host-associated life style.[10] Although traditional species of *Brucella* may temporarily be recovered from environmental samples associated with infected animals, environmental persistence is generally not believed to be of epidemiologic importance. Because direct or close contact is generally believed to be required for transmission, maintenance of brucellosis in animal populations generally requires continual infection of susceptible hosts. As humans are essentially dead-end hosts, infection in humans requires exposure to brucellosis from animal hosts.

The *Brucella* genus has traditionally been characterized as having six species: *Brucella abortus*, *Brucella melitensis*, *Brucella suis*, *Brucella canis*, *Brucella ovis*, and *Brucella neotomae*. The high degree of homology at the genomic level has led to the proposition that the genus is actually composed of only one species, *B. melitensis*, with the other classical species proposed as strains of *B. melitensis*.[11] However, host preference, epidemiologic features, and diagnostic benefits obtained by assigning *Brucella* strains to separate "nomenspecies" based on their distinctive phenotypic characteristics is more compelling and has supported retention of the current nomenclature. In the past 20 years, new *Brucella* species from sea mammals,[12–14] voles,[15,16] and a prosthetic breast implant[17] have been added to the genus. Recent isolations from Austrian foxes,[18] African bullfrogs (*Pyxicephalus edulis*),[19] and baboons (*Papio* spp.)[20] may eventually further expand the number of species in the *Brucella* genus. Some of the classical species (*B. melitensis*, *B. abortus*, and *B. suis*) are divided into biovars based on biochemical, phenotypic, and antigenic properties. Although division into biovars has been used for epidemiologic purposes, biotyping can be somewhat subjective because it is based on subtle differences such as requirements for higher CO_2 tensions for growth, production of hydrogen sulfide, growth on media containing dyes (thionin or basic fuchsin), and agglutination with monospecific A and M antisera.

One of the important characteristics of *Brucella* is the number of species in the genus that are capable of causing zoonotic infections in humans. Most virulent strains of *Brucella* express the O-polysaccharide (homopolymer of 4,6-dideoxy-4-formamido-α-D-mannopyranosyl units) on their lipopolysaccharide (LPS) and are known as "smooth" strains. With the exception of *B. canis* which does not express the O side-chain (a "rough" strain), most zoonotic strains of *Brucella* are smooth strains. Cumulative data suggests dosages of virulent *Brucella* species required for 50% infection rates across mucosal surfaces of hosts other than guinea pigs or mice are most likely in the 10^3–10^4 CFU range,[21,22] Although virulent strains of *Brucella* have a predilection for lymphoreticular tissues and the reproductive tract, most also frequently localize in mammary tissues and are shed in milk.

17.2 *Brucella* Species Associated with Foodborne Infections

The preferred host of *B. abortus* is cattle but it can also naturally infect numerous other species including: bison (*Bison bison*), elk (*Cervus elaphus*), camels (*Camelus dromedarius* and *Camelus bactrianus*), yaks (*Bos grunniens*), African buffalo (*Syncerus caffer*), goats, swine, and other species.[23–25] Biovars 1 and 2 have widespread worldwide distribution, while biovar 3 is predominantly found in India, Egypt, and Africa. Several countries in Northern and Central Europe, Canada, Australia, Japan, and New Zealand are considered to be free of this pathogen.

Sheep and goats are the preferred hosts of *B. melitensis*, but this species of *Brucella* is also known to infect cattle,[26] camels,[27] wild ibex (*Capra ibex*),[28] chamois (*Rupicapra rupicapra*),[28] Nile catfish (*Clarias gariepinus*),[29] and other species. *B. melitensis* is considered to be endemic in parts of Central and South America, Africa, Asia, the Middle East, and countries in the Mediterranean region. Areas considered to be free of *B. melitensis* include Canada, the United States, South-East Asia, Northern and Central Europe, Australia, and New Zealand.

Domestic and feral swine are the preferred hosts for biovars 1, 2, and 3 of *B. suis*. Biovars 1 and 3 of *B. suis* can also infect cattle and horses.[22,30] Biovar 2 can establish infection in hares (*Lepus capensis*) and has a geographical distribution in a broad range between Scandinavia and the Balkans. Biovar 4 is predominantly associated with infection in caribou (*Rangifer tarandus*) but has also been found in moose, arctic foxes, and wolves in subarctic areas. Biovar 5 has been exclusively isolated from wild rodents in the former USSR. Porcine brucellosis (biovars 1, 2, and 3) has widespread distribution in domestic and feral swine globally, but prevalence is higher in some areas such as South-East Asia and South America. It should be noted that unlike biovars 1, 3, and 4, biovar 2 of *B. suis* is not considered to be important for causing human brucellosis. The increasing isolations of *B. suis* in cattle in the south-eastern U.S. has public health significance due to the shedding of high numbers of bacteria within milk of infected cattle.[31]

Isolates of marine mammal *Brucella* have been recovered from the *Mysticeti* and *Odontoceti* sub-orders of cetaceans. Genomic data have led to taxonomic classification of these isolates into *Brucella ceti* (predominantly porpoises and dolphins) and *Brucella pennipedalis* (predominantly seals) strains, each of which has several subgroups.[32] Bacteriologic studies have indicated these strains are phenotypically smooth; like the classical *Brucella* strains. Seropositive responses to marine brucellae have been observed in 53 cetaceans, with isolation or identification of the bacteria having occurred in samples from 18 marine mammal species.[14] It is suspected that cetacean brucellosis may be distributed worldwide in the oceans. The ST27 genotype in Pacific *B. ceti* and *B. pinnipedalis* strains, characterized by a specific genetic location for the mobile genetic element IS711, has been associated with zoonotic infections in Peru and New Zealand.[33]

17.3 Microbiological Characteristics of *Brucella*

With the exception of *B. suis* biovar 3 with a single chromosome, most *Brucella* strains have two circular chromosomes encoding approximately 3.2 kb with two replicons. Bacteriophages have been isolated from *Brucella* strains,[34] but plasmids have not. The extent to which *Brucella* are able to undergo exchange of genetic material among themselves or with other bacteria is unknown. It has been proposed that intracellular niches inhabited by *Brucella* during residence in the host limit their opportunities for genetic exchange with other bacteria. It should be emphasized that *Brucella* infections are polyclonal rather than clonal infections, and clinical infection is generally associated with multiple bacteria crossing mucosal barriers and colonizing tissues under *in vivo* conditions.[23]

Brucella are nonmotile and do not have spores. The cell wall of *Brucella* is typical for Gram-negative bacteria. The outer membrane, approximately 4–5 nm in thickness, is composed of asymmetric layers of LPS and phospholipids and is supported by an underlying 3–5 nm layer of peptidoglycan. Some proteins, such as OmpA, are covalently bound to the peptidoglycan layer and stabilize the outer membrane. The hydrophobic region of the membrane provides an anchor for proteins and forms a functional and structural barrier between the periplasm and the exterior of the cell. The periplasmic space varies from 3 to 30 nm. Porins in the outer membrane function as channels to the interior of the cell. Other proteins, such as lipoproteins, are also embedded in the outer membrane.

The LPS of *Brucella* is composed of lipid A, core oligosaccharides, and O-polysaccharides. The structure of LPS in *Brucella* differs from Gram-negative enteric bacteria in that the backbone sugar of the lipid A is different and the LPS has a low phosphate content. The LPS protects the bacteria from cationic peptides, oxygen metabolites, and complement-mediated lysis. The O-polysaccharide on the LPS, expressed by smooth strains, is very immunogenic and can express A and/or M antigens dependent upon the species of *Brucella*. The *Brucella* cell envelope, LPS, lipoproteins, and flagellin display reduced pathogen-associated molecular pattern (PAMP) for recognition by innate immunity, most likely due to hydrophobic moieties of the outer membrane including ornithine lipids. The altered PAMP of *Brucella* contributes to its stealthiness *in vivo* and its failure to induce robust innate immune responses.[35–37]

17.4 Invasion and Cellular Pathogenesis

Brucella penetrate the mucosal epithelium and are transported as free bacteria, or within phagocytic cells, to regional lymph nodes. If bacteria are not localized and killed within regional lymph nodes draining the site of infection, they replicate and spread via blood or lymph to other lymphoreticular tissues and organs such as the spleen, reproductive tract, and/or mammary gland. The bacteremia associated with most *Brucella* species is short, and live bacteria are not readily isolated from blood samples.

For cellular entry, smooth *Brucella* that have not been opsonized by antibodies use the cytoskeleton of the host cell and interact with cholesterol-rich microdomains (lipid rafts) within the plasma membrane that facilitate contact with the host cell and mediate internalization. Lipid rafts contain glycosphingolipids, cholesterol, and glycosyl-phosphatidylinositol-anchored proteins[38] and facilitate membrane-associated sorting events such as the formation of multi-subunit membrane complexes and signaling across membranes and membrane fusion. Besides the plasma membrane, lipid rafts are also found in intracellular organelles and vesicles. The *Brucella* LPS O-polysaccharide appears to be a key molecule for interaction with lipid rafts on host cells, but also prevents complement-mediated bacterial lysis and prevents host cell apoptosis.[39] Opsonization of smooth strains of *Brucella* increases entry 10-fold and occurs through IgG (Fc) and complement (C3b and 4b) receptors on the surface of phagocytes, which diverts smooth bacteria from lipid rafts and targets entry to the phagolysosomal compartment. Receptor-mediated phagocytosis leads to greater killing of internalized *Brucella* by monocytes. In contrast to smooth *Brucella*, rough strains of *Brucella* cannot sustain interactions with lipid rafts and are phagocytosed by either toll-like receptor 4 or mannose receptor recognition of the LPS-deficient bacterial surface leading to rapid targeting to the phagolysosomal compartment where they are generally unable to replicate.

Smooth *Brucella* that enter via lipid rafts quickly traffic through the early endosomal compartment and depart the phagosome to form the modified phagosome (termed brucellosome) by acquiring components of endoplasmic reticulum in a manner similar to autophagosome biogenesis. *Brucella* initially localize within acidified phagosomes[40] where they are exposed to free oxygen radicals generated by the respiratory burst. *Brucella* require acidification of the phagosomal compartment to a $pH < 4.5$ before they display wild-type intracellular replication. The requirement for low pH is transient and only extends through the initial stages of intracellular infection. Localization in an acidified environment induces expression of the *VirB* operon (*virB* 1–10), which controls expression of genes associated with a type IV secretion system. The VirB operon interacts with the endoplasmic reticulum to neutralize the pH of the phagosome.[41] *Brucella*-induced modifications of the phagosome leads to inhibition of phagosome maturation and prevention of fusion with lysosomes. The brucellosome environment provides *Brucella* with conditions of nutrient depletion and limited oxygen availability. It should be noted that under *in vitro* conditions, up to 90% of virulent *Brucella*, and 99% of nonvirulent *Brucella*, may be killed following intracellular entry.[42]

Brucella have multiple mechanisms to detoxify free radicals produced by host phagocytes, cationic peptides, and oxygen metabolites. *Brucella* probably use both stationary and exponential stages for intracellular survival, with stationary-phase physiology providing *Brucella* with benefits for adapting to the harsh conditions encountered in the phagosome and exponential stages associated with replication under favorable conditions. The presence of cytochromes (cytochrome bc1 complex or quinol oxidase) with a high oxygen affinity may represent an important adaptation of *Brucella* to their intracellular survival. Several *Brucella* have been shown to utilize heme as an iron source *in vitro*, and have a critical need for heme during residence in the phagosomal compartment[43] and during replication in trophoblasts.[44] *Brucella* scavenge iron through siderophores such as 2,3-dihydroxybenzoic acid or brucebactin.[45]

In accordance with its stealthy nature, *Brucella* has developed mechanisms to minimize stimulation of pattern recognition receptors (PRRs) by the innate immune system of the host. The *Brucella* cell envelope has high hydrophobicity, and its LPS has a noncanonical structure that elicits a reduced and delayed inflammatory response as compared to other Gram-negative bacteria,[40] and has lower stimulatory activity on TLR4 receptors.[46] The O side-chain on the LPS can form complexes with the MHC Class II molecules that interfere with the ability of macrophages to present exogenous proteins. *Brucella* ornithine-containing lipids and lipoproteins in the outer membrane are poor activators of innate immunity. *Brucella* bacteria are also devoid of many classical structures involved in virulence such as pili, fimbria, capsules, and plasmids that stimulate PRRs. The ability of *Brucella* to prevent phagosome maturation and fusion with lysosomes may interfere with other innate and adaptive immune processes. As proteins have been identified in *Brucella* that demonstrate significant homology with toll-like receptor (TLR) adaptor molecules, these peptides may be a mechanism to interfere with, or subvert TLR signaling.[47] Compared to other Gram-negative bacteria, *Brucella* induces a reduced innate immune response, and a lower rate of maturation and activation of dendritic cells, which may impair development of adaptive immune responses.

17.5 Clinical Disease in Humans

Human infection with *B. abortus* can cause chronic, debilitating clinical illness with nonpathognomonic symptoms[48,49] including fever, night sweats, anorexia, polyarthritis, meningitis, and pneumonia. Incubation periods are variable and can range from a week to months (up to 6 months). Although mortality is uncommon, it primarily results from endocarditis associated with *B. melitensis* infection. The most common manifestations of localized disease are osteoarticular (i.e., peripheral arthritis, sacroiliitis, spondylitis).[50] Human infection occurs across mucosal surfaces by aerosolization into respiratory tissues, oral consumption, or penetration through breaks in the epidermis. Inadvertent exposure to live vaccine strains, most commonly via needle sticks, has also been a frequent source of human infections, especially in the veterinarian profession. In the absence of specific treatment, infection may persist for weeks or months.[51] Relapse of infection after treatment is not uncommon, with relapse occurring within 6 months after therapy and not usually associated with emergence of antibiotic-resistant strains.[52] The duration of the human illness and its long convalescence makes brucellosis both a medical and an economic problem due to costs associated with treatment and the time lost from normal activities.[52] The disease can be insidious and may present in many atypical forms.[52] While rare, human-to-human dissemination of brucellosis through breast milk has occurred.[53–55]

Currently in the United States, human brucellosis is predominantly a disease associated with international travel, or as a foodborne disease caused by ingestion of nonpasteurized dairy products originating from other countries.[56,57]

17.6 Immunity

Currently, no vaccines are available to protect humans from brucellosis. Currently available livestock vaccines are composed of attenuated live bacteria, which can cause clinical disease in humans in the event of inadvertent infection. Long-term protection against brucellosis in livestock is associated with stimulation of cellular immunity, while antibodies are considered to play a minor role in protection. Although specific correlates of protective immunity are currently not known, it is believed that protection is mediated by the Th1 subset of CD4+ lymphocytes and is associated with the production of IFN-γ and other cytokines associated with cellular immunity. CD4+ cells play a central role in coordinating and culminating the adaptive immune response by differentiation into functional subsets, such as the Th1 type, which is associated with the production of interferon-γ (IFN-γ), interleukin-2 (IL-2), and tumor necrosis factor-α (TNF-α). CD4+ cells also provide growth factors and signals for the generation and maintenance of CD8+ T cells. CD8+ cells are considered to be of importance for protective immunity because of their ability to lyse or kill infected cells, thereby releasing *Brucella* from intracellular hiding and exposing it to extracellular bactericidal mechanisms. Cytokines released by CD4+ and CD8+ cells may activate macrophages and dendritic cells, thereby increasing their bactericidal activity against *Brucella*.[58] Although many studies have used qualitative measurements of IFN-γ production as an indication that vaccination induces protective Th1 responses, recent data with other pathogens have suggested that a better indicator of the quality of the immune response after immunization may be through the measurement of increases in antigen-specific polyfunctional T cells capable of producing a traid of relevant cytokines.[59] In addition, the type of memory cell produced may influence the capability for protective responses, cytokine production, and cytotoxic activity.[60]

17.7 Foodborne Brucellosis

17.7.1 Inactivation in Food Products

In regard to survival in food materials, *Brucella* is killed by pasteurization[61] or heat treatment (up to 71°C), but environmental studies indicate the bacteria can survive drying, freezing, or long-term storage at cold temperatures.[62] Souring of milk has negligible effects on the survival of *Brucella*[63] and the bacteria can survive in milk for up to 87 days, water for up to 60 days, less than a week in yogurt, and in ripened cheese for up to 50 days despite the reduction in pH associated with ripening.[64–67] Ripening of cheese at 24°C for 20 days was associated with the inability to recover *B. melitensis*, whereas the bacteria were recovered when cheese was ripened at 4°C.[68] In meat products, processes for preparation of cured/smoked meat can reduce, but not completely kill, *Brucella*.[69]

17.7.2 Transmission by Milk or Cheese Products

Foodborne transmission of *Brucella* to humans is almost exclusively associated with *B. abortus*, *B. suis*, and *B. melitensis* infection. As these strains of *Brucella* have a predilection for the mammary gland with intermittent shedding by caprine, bovine, or water buffalo hosts,[2,70–73] the predominant route for human infection occurs through consumption of dairy products made from nonpasteurized milk. Data suggest that shedding of *B. melitensis* bacteria in goat milk is generally greater (up to 2.5×10^5 bacteria/mL)[71] as compared to shedding of *B. abortus* bacteria in milk from cattle.[2] As chronically infected animals frequently do not demonstrate any clinical signs, shedding of *Brucella* in milk can be intermittent and occur in the absence of symptoms of infection. In cattle, excretion of *B. abortus* in milk was more consistent and abundant during later parts of lactation.[73] It should be noted that similar localization in mammary tissue occurs in humans, as shedding of *B. melitensis* in human breast milk has been noted in a number of reports.[53–55]

Although swine are not a source for milk products, *B. suis* frequently infects cattle and is shed in high numbers of milk. Transmission of *B. suis* from feral swine to cattle is a significant and growing issue in the south and southeastern parts of the United States. Consumption of nonpasteurized bovine milk containing *B. suis* has caused clinical disease in numerous people on several occasions.[74,75] This is a concern due to recent efforts in many states to allow the sale of nonpasteurized milk products. It should be noted that The American Academy of Pediatrics has issued a policy statement that raw or unpasteurized milk or milk products pose a health risk for foodborne infection by numerous agents (including *Brucella*), and scientific evidence does not support the argument that pasteurization alters the nutritional content of milk.[76]

In most countries in which brucellosis has been effectively controlled in domestic livestock, human infection is associated with international travel to countries with endemic disease, or importation of nonpasteurized milk products.[77–79] However, in the absence of travel for tourism, ethnic or cultural habits are a very important factor for influencing human infections with *Brucella* spp., with epidemiologic data from some countries indicating higher prevalence of brucellosis in certain ethnic groups.[80–82] Although some infections in these ethnic groups may be related to foodborne infection in the country of origin, many appear to be related to unregulated movement of unpasteurized milk products into a country.[83–86] Unregulated international movement of unpasteurized milk products may to some extent reflect cultural food preferences.[84–86]

17.7.3 Meat or Fish Products

As *Brucella* can be recovered from raw meat,[87] foodborne infection could also occur by handling or consumption of raw or undercooked meat products. In general, bacterial loads in animal muscle are at lower levels as compared to levels in unpasteurized milk products.[88] This route of foodborne transmission is of greatest concern in cultures where consumption of raw or undercooked meat occurs.[89,90]

Ingestion of fish or marine products may also be a source for foodborne infection with *Brucella*. Three cases of human infection with the ST27 genotype of Pacific *B. ceti* and *B. pinnipedalis* strains have occurred in individuals with a history of consumption of raw shellfish or fish.[33,91] The isolation of *B. melitensis* biovar 3 from Nile catfish may also indicate a possible route for foodborne transmission to humans. The presence of *B. melitensis* in this species was hypothesized to have spilled over from endemic infection in ruminants in the area.[29]

17.7.4 Laboratory Models

Due to their high susceptibility to infection with *Brucella*, guinea pigs have historically been used for isolation. For some samples such as cheeses, animal inoculation may be the only means for detecting the presence of brucellae.[92] Guinea pigs are also an accepted model for testing the virulence of *Brucella* strains by assessment of spleen pathology and quantification of infection at 8–12 weeks after inoculation. Lesions of brucellosis are observed in the liver, spleen, lungs, and lymph nodes.[93] Guinea pigs have been used to evaluate effectiveness of antibiotic treatments, and for evaluation of efficacy of vaccines.[94]

Currently, inbred mice are the most common laboratory animal model used to study chronic infection or disease pathogenesis.[95] This is partly because *Brucella* colonize and induce pathologic lesions in liver and spleen tissues of mice that to some extent mimics human infection. Inbred mouse strains differ in susceptibility to infection, with BALB/c and CBA/H being considered relatively more susceptible to infection and C57BL/6 and C57BL/10 more resistant.[95] Intraperitoneal is the most common route of infection for mouse models, but aerosol, oral, and intranasal routes have also been used. It should be noted that murine models generally do not accurately replicate disease pathogenesis in natural hosts where *Brucella* spp. more frequently localize in lymphatic tissues, mammary gland, and reproductive organs, and reproductive losses (i.e., abortion in females and orchitis/epididymitis in males) are common. With the exception of abortion, natural hosts generally do not demonstrate clinical disease. In comparison, humans appear to be aberrant hosts for *Brucella* and often have significant symptoms of clinical disease that are frequently chronic.

Other laboratory models that have been used include rabbits, rats, and nonhuman primates although each has limitations regarding the study of *Brucella* pathogenesis. *Brucella*-induced abortion is rare, but rats can venerally transmit *B. abortus* and offspring can be latently infected.[93] Lesions of necrosis in periplacentomal chorionic epithelium and metritis have been observed in the uterus of pregnant rats infected with *B. abortus*. Rabbits are partially susceptible to *Brucella*, with susceptibility increasing with pregnancy. Nonhuman primate models (*Macaca arctoides* and *Macaca mulatta*) have been infected through oral, subcutaneous, and aerosol routes with virulent strains of *Brucella*. Bacteremia was demonstrated for up to 8 weeks, and lesions of focal granulomatous hepatitis, splenitis, lymphadenitis, and occasional lesions of granulomatous orchitis, epididymitis, and endometritis were reported, similar to lesions in humans infected with *Brucella* reviewed by Silva et al.[95] Current knowledge would suggest that primate models most closely mimic pathogenesis of disease in humans. Although laboratory models continue to have value for basic research, they do not accurately replicate disease pathogenesis in natural hosts of *Brucella*.

17.8 Conclusion

Brucellosis in humans remains a significant zoonosis in many parts of the world and frequently occurs through international travel. In the absence of direct contact with infected animal hosts, transmission of infection to humans almost exclusively occurs through products made from nonpasteurized milk. *Brucella* appear to have a long shelf life in these types of products. In some regions or countries, human brucellosis is associated with certain ethnic groups and may be related to dietary preferences. Although addressing the disease in its natural hosts is the most economic approach, regulatory efforts to prevent the sale or consumption of nonpasteurized milk products would also reduce human brucellosis and benefit public health.

REFERENCES

1. Bruce, D. Note on the discovery of a microorganism in Malta fever. *Practitioner (London)* 39, 161, 1887.
2. Evans, A.C. Further studies on bacterium abortus and related bacteria. II. A comparison of bacterium abortus with bacterium bronchisepticus and with the organism which causes Malta fever. *J. Infect. Dis.* 22, 580, 1918.
3. Godfroid, J., et al. A "One Health" surveillance and control of brucellosis in developing countries: moving away from improvisation. *Comp. Immunol. Microbiol. Infect. Dis.* 36, 241, 2013.
4. Cutler, S.J., Whatmore, A.M., & Commander, N.J. Brucellosis—new aspects of an old disease. *J. Appl. Microbiol.* 98, 1270, 2005.
5. Bernués, A., Manrique, E., & Maza, M.T. Economic evaluation of bovine brucellosis and tuberculosis eradication programmes in a mountain area of Spain. *Prev. Vet. Med.* 30, 137, 1997.
6. Jelastopulu, E., Bikas, C., Petropoulos, C., & Leotsinidis, M. Incidence of human brucellosis in a rural area in Western Greece after the implementation of a vaccination programme against animal brucellosis. *BMC Public Health* 8, 241, 2008.
7. Roth, F., et al. Human health benefits from livestock vaccination for brucellosis: case study. *Bull. World Health Organ.* 81, 867, 2003.
8. Kolar, J. Control of *Brucella melitensis* brucellosis in developing countries. *Ann. Inst. Pasteur Microbiol.* 138, 122, 1987.
9. Roth, F., et al. Human health benefits from livestock vaccination for brucellosis: case study. *Bull. World Health Organ.* 81, 876, 2003.
10. Scholz, H.C., et al. Isolation of *Brucella microti* from soil. *Emerg. Infect. Dis.* 14, 1316, 2008.
11. Vizcaino, N., Cloeckaert, A., Verger, J., Grayon, M., & Fernandez-Lago, L. DNA polymorphism in the genus *Brucella. Microbes Infect.* 2, 1089, 2000.
12. Ewalt, D.R., Payeur, J.B., Martin, B.M., Cummins, D.R., & Miller, W.G. Characteristics of a *Brucella* species from a bottlenose dolphin (*Tursiops truncatus*). *J. Vet. Diagn. Invest.* 6, 448, 1994.
13. Foster, G., Osterman, B.S., Godfroid, J., Jacques, I., & Cloeckaert, A. *Brucella ceti* sp. nov. and *Brucella pinnipedialis* sp. nov. for *Brucella* strains with cetaceans and seals as their preferred hosts. *Int. J. Syst. Evol. Microbiol.* 57, 2688, 2007.
14. Hernandez-Mora, G., Palacios-Alfaro, J.D., & Gonzalez-Barrientos, R. Wildlife reservoirs of brucellosis: *Brucella* in aquatic environments. *Rev. Sci. Tech.* 32, 89, 2013.
15. Hofer, E., et al. Isolation of *Francisella tularensis* and *Brucella suis* from red foxes (*Vulpes vulpes*). *Tierarztl. Umsch.* 65, 229, 2010.
16. Scholz, H.C., et al. Isolation of *Brucella microti* from mandibular lymph nodes of red foxes, *Vulpes vulpes*, in lower Austria. *Vector Borne Zoonotic Dis.* 9, 153, 2009.
17. Scholz, H.C., et al. *Brucella inopinata* sp. nov., isolated form a breast implant infection. *Int. J. Syst. Evol. Microbiol.* 60, 801, 2010.
18. Hofer, E., et al. A potential novel *Brucella* species isolated from mandibular lymph nodes of red foxes in Austria. *Vet. Microbiol.* 155, 93, 2012.
19. Eisenberg, T., et al. Isolation of potentially novel *Brucella* spp. from frogs. *Appl. Environ. Microbiol.* 78, 3753, 2012.
20. Schlabritz-Loutsevitch, N.E., et al. A novel *Brucella* isolate in association with two cases of stillbirth in non-human primates: first report. *J. Med. Primatol.* 38, 70, 2009.
21. Manthei, C.A. Application of research to bovine brucellosis control and eradication programs. *J. Dairy Sci.* 51, 1115, 1968.
22. Teske, S.S., Huang, Y., Tamrakar, S.B., Bartrand, T.A., Weir, M.H., & Haas, C.N. Animal and human dose-response models for *Brucella* species. *Risk Anal.* 31, 1576, 2011.
23. Olsen, S.C., & Palmer, M.V. Advancement of knowledge of *Brucella* over the past 50 years. *Vet. Pathol.* 51, 1076, 2014.
24. Bouvier, G. Distribution geographique de quelques maladies du gibier et des animaux sauvages de la Suisse. *Bull. Off. Int. Epizoot.* 61, 67, 1964.
25. Rollinson, D.H.L. *Brucella* agglutinins in East African game animals. *Vet. Rec.* 74, 904, 1962.
26. Samaha, H., Al-Towaily, M., Khoudair, R.M., & Ashour, H.M. Multicenter study of brucellosis in Egypt. *Emerg. Infect. Dis.* 14, 1916, 2008.
27. Wernery, U. Camelid brucellosis: a review. *Rev. Sci. Tech.* 33, 839, 2014.

28. Gastellu, J., Garin-Bastuji, B., & Oudar, J. Pathologic features of a case of systemic brucellosis due to *Brucella melitensis* biovar 3 in a chamois (*Rupicapra rupicapra*). *Bull. Acad. Natl. Med.* 177, 575, 1993.
29. El-Tras, W.F., Tayell, A.A., Eltholth, M.M., & Guitian, J. *Brucella* infection in fresh water fish: evidence of natural infection of the Nile catfish *Clarias gariepinus*, with *Brucella melitensis*. *Vet. Microbiol.* 141, 321, 2010.
30. Godfroid, J., Garin-Bastuji, B., Saegerman, C., & Blasco, J.M. Brucellosis in terrestrial wildlife. *Rev. Sci. Tech.* 32, 27, 2013.
31. Ewalt, D.R., Payeur, J.B., Rhyan, J.C., & Geer, P.L. *Brucella suis* biovar 1 in naturally infected cattle: a bacteriological, serological and histological study. *J. Vet. Diagn. Invest.* 4, 417, 1997.
32. Foster, G., Osterman, B.S., Godfroid, J., Jacques, I., & Cloeckaert, A. *Brucella ceti* sp. nov. and *Brucella pinnipedialis* sp. nov. for *Brucella* strains with cetaceans and seals as their preferred hosts. *Int. J. Syst. Evol. Microbiol.* 57, 2688, 2007.
33. Cloeckaert, A., Bernardet, N., Koylass, M.S., Whatmore, A.M., & Zygmunt, M.S. Novel IS711 chromosomal location useful for identification of marine mammal *Brucella* genotype ST27, which is associated with zoonotic infection. *J. Clin. Microbiol.* 49, 3954, 2011.
34. Corbel, M.J. *Brucella* phages: advances in the development of a reliable phage typing system for smooth and non-smooth *Brucella* isolates. *Ann. Inst. Pasteur Microbiol.* 138, 70, 1987.
35. Olsen, S.C. Recent developments in livestock and wildlife brucellosis vaccination. *Rev. Sci. Tech.* 32, 207, 2013.
36. Giamartolomei, G.H., & Cassataro, J. Confronting the barriers to develop novel vaccines against brucellosis. *Expert Rev. Vaccines* 10, 1291, 2011.
37. Kaufmann, A., et al. Smooth and rough lipopolysaccharide phenotypes of *Brucella* induce different intracellular trafficking and cytokine/chemokine release in human monocytes. *J. Leukoc. Biol.* 74, 1045, 2003.
38. Dawson, C.E., et al. Isolation and characterization of *Brucella* from the lungworms of a harbor porpoise (*Phocoena phocoena*). *J. Wildl. Dis.* 44, 237, 2008.
39. Jimenez de Bagüés, M.P., Terraza, A., Gross, A., & Domand, J. Different responses of macrophages to smooth and rough *Brucella* spp.: relationship to virulence. *Infect. Immun.* 72, 2429, 2004.
40. Rittig, M.G., Alvarez-Martinez, M.T., Porte, F, Liautard, J.P., & Rouot, B. Intracellular survival of *Brucella* spp. in human monocytes involves conventional uptake but special phagosomes. *Infect. Immun.* 69, 3995, 2001.
41. Celli, J. Surviving inside a macrophage: the many ways of *Brucella*. *Res. Microbiol.* 157, 93, 2006.
42. Porte, F., Liautard, J.P., & Kohler, S. Early acidification of phagosomes containing *Brucella suis* is essential for intracellular survival in murine macrophages. *Infect. Immun.* 67, 4041, 1999.
43. Arellano-Reynoso, B., et al. Cyclic β-1,2-glucan is a *Brucella* virulence factor required for intracellular survival. *Nat. Immunol.* 6, 618, 2005.
44. Cutler, S.J., Whatmore, A.M., & Commander, N.J. Brucellosis—new aspects of an old disease. *J. Appl. Microbiol.* 98, 1270, 2005.
45. Bellaire, B.H., Elzer, P.H., Baldwin, C.L., & Roop, R.M. Production of the siderophore 2,3-dihydroxybenzoic acid is required for wild-type growth of *Brucella abortus* in the presence of erythritol under low-iron conditions *in vitro*. *Infect. Immun.* 71, 2927, 2003.
46. Rittig, M.G., et al. Smooth and rough lipopolysaccharide phenotypes of *Brucella* induce different intracellular trafficking and cytokine/chemokine release in human monocytes. *J. Leukoc. Biol.* 74, 1045, 2003.
47. Oliveira, S.C., Giamartolomei, G.H., & Cassataro, J. Confronting the barriers to develop novel vaccines against brucellosis. *Expert Rev. Vaccines* 10, 1291, 2011.
48. Franco, M.P., Mulder, M., Gilman, R.H., & Smits, H.L. Human brucellosis. *Lancet Infect. Dis.* 7, 775, 2007.
49. Mantur, B.G., Amarnath, S.K., & Shinde, R.S. Review of clinical and laboratory features of human brucellosis. *Indian J. Med. Microbiol.* 25, 188, 2007.
50. Baldi, P.C., & Giambartolemi, G.H. Pathogenesis and pathobiology of zoonotic brucellosis in humans. *Rev. Sci. Tech.* 32, 117, 2013.
51. Ariza, J., et al. Perspectives for the treatment of brucellosis in the 21st century: the Ioannina recommendations. *PLoS Med.* 4, e317, 2007.

52. Corbel, M.J. Brucellosis in humans and animals. World Health Organization. http://www.who.int/csr/resources/publications/Brucellosis.pdf, 2006.
53. Celebi, G., Külah, C., Kilic, S., & Ustündağ, G. Asymptomatic *Brucella* bacteraemia and isolation of *Brucella melitensis* biovar 3 from human breast milk. *Scand. J. Infect. Dis.* 39, 205, 2007.
54. Al-Mafada, S.M., Al-Eissa, Y.A., Saeed, E.S., & Kambal, A.M. Isolation of *Brucella melitensis* from human milk. *J. Infect.* 26, 346, 1993.
55. Arroyo Carrera, I., Lopez Rodriguez, M.J., Sapina, A.M., Lopez Lafuenta, A., & Sacristan, A.R. Probable transmission of brucellosis by breast milk. *J. Trop. Pediatr.* 52, 380, 2006.
56. Doyle, T.J., & Bryan, R.T. Infectious disease morbidity in the US region bordering Mexico, 1990–1998. *J. Infect. Dis.* 182, 1503, 2000.
57. Troy, S.B., Rickman, L.S., & Davis, C.E. Brucellosis in San Diego: epidemiology and species-related differences in acute clinical presentations. *Medicine (Baltimore)* 84, 174, 2005.
58. Olsen, S.C. Recent developments in livestock and wildlife brucellosis vaccination. *Rev. Sci. Tech.* 32, 207, 2013.
59. Orkhon, D., Chimed-Ochir, G., Nansalmaa, M., Kolar, J., & Vounatsou, P. A model of animal-human brucellosis transmission in Mongolia. *Prev. Vet. Med.* 69, 77, 2005.
60. Thakur, A., Pedersen, L.E., & Jungersen, G. Immune markers and correlates of protection for vaccine-induced immune responses. *Vaccine* 30, 4907, 2012.
61. Van den Heever, L.W, Katz, K.W., & Te Brugge, L.A. On the inactivation of *Brucella abortus* in naturally contaminated milk by commercial pasteurisation procedures. *J. South Afr. Vet. Assoc.* 53, 233, 1982.
62. CFSPH-YSU. Porcine and rangiferine brucellosis: *Brucella suis*. Enzootic abortion, contagious abortion, undulant fever. Disease factsheets. *B. suis*. The Centre of Food Security and Public Health, Iowa State University, 2007.
63. Kangumba, J.G., Venter, E.H., & Coetzer, J.A. The effect of activation of the lactoperoxidase system and souring on certain potential human pathogens in cows milk. *J. South Afr. Vet. Assoc.* 68, 130, 1997.
64. Mendez-Gonzalez, K., et al. *Brucella melitensis* survival during manufacture of ripened goat cheese at two temperatures. *Foodborne Pathog. Dis.* 12, 11257, 2011.
65. Plommet, M., Fensterbank, R., Vassal, L., Auclair, J., & Mocquot, G. Survival of *Brucella abortus* in ripened soft cheese made from naturally infected cow's milk. *Le Lait* 68, 115, 1988.
66. Santiago-Rodiquez Mdel, R., et al. Survival of *Brucella abortus aqpX* mutant in fresh and ripened cheeses. *Foodborne Pathog. Dis.* 12, 170, 2015.
67. Falenski, A., Mayer-Scholl, A., Filter, M., Göllner, C., Appel, B., & Nöckler, K. Survival of *Brucella* spp. in mineral water, milk and yogurt. *Int. J. Food Microbiol.* 145, 326, 2011.
68. Mendez-Gonzalez, K.Y., et al. *Brucella melitensis* survival during manufacture of ripened goat cheese at two temperatures. *Foodborne Path. Dis.* 8, 1257, 2011.
69. Hutchings, L.M., McCullough, N.B., Donham, C.B., Eisele, C.W., & Bounnell, D.F. The viability of *Brucella melitensis* in naturally infected cured hams. *Public Health Rep.* 66, 1402, 1951.
70. Smith, J. Excretion of *Br. Abortus* in milk. *J. Compar. Pathol. Ther.* 47, 123, 1934.
71. Fleischner, E.C., Vecki, M., Shaw, E.B., & Meyer, K.F. The pathogenicity of *B. abortus* and *B. melitensis* for monkeys. *J. Infect. Dis.* 29, 663, 1921.
72. Capparelli, R., et al. Heterogeneous shedding of *Brucella abortus* in milk and its effect on the control of animal brucellosis. *J. Appl. Microbiol.* 106, 2041, 2009.
73. Brinley Morgan, W.J., & McDiarmid, A. The excretion of *Brucella abortus* in the milk of experimentally infected cattle. *Res. Vet. Sci.* 1, 53, 1960.
74. Beattie, C.P., & Rice, R.M. Undulant fever due to *Brucella* of the porcine type—*Brucella suis*. *J. Am. Med. Assoc.* 201, 1934, 1970.
75. Borts, I.H., Harris, D.M., Joynt, M.F., Jennings, J.R., & Jordan, C.F. A milk borne epidemic of brucellosis: caused by the porcine type of *Brucella* (*Brucella suis*) in a raw milk supply. *J. Am. Med. Assoc.* 121, 319, 1943.
76. Committee on Infectious Diseases, Committee on Nutrition, American Academy of Pediatrics. Consumption of raw or unpasteurized milk and milk products by pregnant women and children. *Pediatrics* 133, 175, 2014.
77. Arnow, P.M., Smaron, M., & Ormiste, V. Brucellosis in a group of travelers to Spain. *J. Am. Med. Assoc.* 251, 505, 1984.

78. Gautret, P., Benkouiten, S., Gaillard, C., Parola, P., & Brouqui, P. Camel milk-associated infection risk perception and knowledge in French Haji pilgrims. *Vector Borne Zoonotic Dis.* 13, 425, 2013.
79. Di Pierdomenico, A., Borgia, S.M., Richardson, D., & Baqi, M. Brucellosis in a returned traveller. *Can. Med. Assoc. J.* 183, E690, 2011.
80. Chommel, B.B., et al. Changing trends in the epidemiology of human brucellosis in California from 1973 to 1992. A shift toward foodborne transmission. *J. Infect. Dis.* 178, 1216, 1994.
81. Taylor, J.P., & Perdue, J.N. The changing epidemiology of human brucellosis in Texas 1977–1986. *Am. J. Epidemiol.* 130, 160, 1989.
82. Al Dahouk, S., et al. Human brucellosis in a nonendemic country: a report from Germany, 2002 and 2003. *Eur. J. Clin. Microbiol. Infect. Dis.* 24, 450, 2005.
83. Galbraith, N.S., Ross, M.S., DeMowbray, R.R., & Payne, D.J. Outbreak of *Brucella melitensis* type 2 infection in London. *Br. Med. J.* 1, 612, 1969.
84. Thaper, M.K., & Young, E.J. Urban outbreak of goat cheese brucellosis. *Pediatr. Infect. Dis.* 640, 1986.
85. Beutlich, J., et al. Characterization of illegal food items and identification of foodborne pathogens brought into the European Union via two major German airports. *Int. J. Food Microbiol.* 209, 13–19, 2015, doi: 10.1016/j.ijfoodmicro.2014.10.017.
86. Fosgate, G.T., Carpenter, T.E., Chomel, B.B., Case, J.T., DeBess, E.E., & Reilly, K.F. Time-space clustering of human brucellosis, California, 1973–1992. *Emerg. Infect. Dis.* 8, 672, 2002.
87. Hassan Ali, N., Farooqui, A., Khan, A., Khan, A.Y., & Kazmi, S.U. Microbial contamination of raw meat and its environment in retail shops in Karachi, Pakistan. *J. Infect. Dev. Countries* 4, 382, 2010.
88. Pappas, G., Akritidis, N., Bosilkovski, M., & Tsianos, E. Brucellosis. *New Engl. J. Med.* 352, 2325, 2005.
89. Mohd, M.G. Brucellosis in the Gezira area, Central Sudan. *J. Trop. Med. Hyg.* 92, 86, 1989.
90. Chan, J., Baxter, C., & Wenman, W.M. Brucellosis in an inuit child, probably related to caribou meat consumption. *Scand. J. Infect. Dis.* 21, 337, 1989.
91. Whatmore, A.M., et al. Marine mammal *Brucella* genotype associated with zoonotic infection. *Emerg. Infect. Dis.* 14, 517, 2008.
92. Alton, G.G., Jones, L.M., Angus, R.D., & Verger, J.M. *Techniques for the brucellosis laboratory.* Institut National de la Recherche Agronomique, Paris, pp. 33, 154, 1988.
93. Islam, M.A., Khatun, M.M., & Baek, B.K. Male rats transmit *Brucella abortus* biotype 1 through sexual intercourse. *Vet. Microbiol.* 165, 475, 2013.
94. Ivanov, A.V., Salmakov, K.M., Olsen, S.C., & Plumb, G.E. A live vaccine from *Brucella abortus* strain 82 for control of cattle brucellosis in the Russian Federation. *Anim. Health Res. Rev.* 12, 113, 2011.
95. Silva, T.M., Costa, E.A., Paixão, T.A., Tsolis, R.M., & Santos, R.L. Laboratory animal models for brucellosis research. *J. Biomed. Biotechnol.* 2011, 518323, 2011.

18

Burkholderia

Danielle L. Peters, Fatima Kamal, and Jonathan J. Dennis

CONTENTS

18.1 Introduction

The genus *Burkholderia* covers a diverse group of Gram-negative β-proteobacteria. Although currently at least 60 species and proposed species exist in the genus *Burkholderia*, very few have been studied extensively. Much of the research to date has focused on the bacteria of the *Burkholderia cepacia* complex (Bcc), *Burkholderia mallei*, *Burkholderia pseudomallei*, and more recently, *Burkholderia gladioli*. The bacteria of the Bcc are pathogens that typically cause serious infections in plants, animals, and humans.[1–3] However, they can also be beneficial in the environment as they fix nitrogen symbiotically for plants, produce antibiotics and antifungals, and have the capacity to degrade organic and xenobiotic compounds.[4–6] *B. mallei* causes "glanders," a rare condition usually associated with horses, but that can also affect humans, whereas *B. pseudomallei*, endemic in Southeast Asia, causes "melioidosis," a serious disease in humans with a wide variety of symptoms.[7] The species *B. gladioli* has been divided into four pathovars: *gladioli*, *alliicola*, *agaricicola*, and *cocovenenans*.[8] The first three *B. gladioli* pathovars listed are primarily plant pathogens,[9–11] but members of these pathovars can occasionally also infect immunocompromised patients with chronic granulomatous disease (CGD), cystic fibrosis (CF), or acquired immune deficiency syndrome (AIDS).[8,12,13] The taxonomic description of *B. gladioli* pvs. *gladioli*, *alliicola*, and *agaricicola*, published in 2003, suggests that they do not produce toxins that are harmful to humans, although some strains have since been shown to synthesize toxoflavin.[8,14] The fourth *B. gladioli* pathovar, *B. gladioli* pv. *cocovenenans*, is distinct from the other pathovars with regards to its epidemiology and pathogenicity. Although the other *Burkholderia* species can be found as contaminants in food and water supplies (including the Bcc, *B. mallei*, and *B. pseudomallei*), *B. gladioli* pv. *cocovenenans* is the only bacterium of the *Burkholderia* genus that is traditionally characterized as a foodborne pathogen.

Strains of *B. gladioli* pv. *cocovenenans* do not cause disease directly, but instead produce toxins that contaminate foods before ingestion by humans, similar to *Clostridium botulinum*. The laboratory infection models used to test *Burkholderia* pathogenicity and virulence are generally useful across the genera, whereas specific models have been devised to measure *B. gladioli* toxin activity. This chapter will briefly present the literature regarding the presence of *B. mallei*, *B. pseudomallei*, the Bcc, and *B. gladioli* in food and water supplies, and subsequently review the laboratory infection models that have been developed for each group of *Burkholderia* pathogenic bacteria.

18.2 *B. mallei*

18.2.1 Overview

B. mallei causes glanders, a zoonosis disease of horses that can also be transmitted to humans. The symptoms vary depending on the route of transmission, but can include pneumonia, skin lesions, and septic shock.[15] Horses generally become infected following ingestion of *B. mallei* introduced into their food and water by other infected horses.[15] Humans may also become infected via ingestion, typically due to drinking untreated water.[16] Glanders is difficult to diagnose, there is no vaccine available to protect against the bacterium, and any potentially successful therapy requires prolonged antibiotic treatment. However, as a result of stringent measures implemented to control the spread of glanders, the disease has not been seen in the Western hemisphere since 1939 (except following accidental laboratory exposure).[17,18] However, because both *B. mallei* and *B. pseudomallei* have been identified as potential food contaminants, there is concern that these species could be released as bioterrorism agents into a country's food supply.[19,20]

18.2.2 Laboratory Infection Models

Several animals have been investigated as possible infection models of glanders. Because *B. mallei* naturally exists in an equid reservoir, and equine glanders share many features with human glanders, disease progression has been well studied in equine models.[15,21] However, working with equids is cost and space prohibitive. To address this problem, the US Army Medical Research Institute of Infectious Diseases (USAMRIID) conducted a comprehensive study in the late 1940s evaluating the susceptibility of several common laboratory animals to *B. mallei* infection.[22] This study, as well as others, showed that cattle, hogs, birds, rats, and rabbits are relatively resistant to *B. mallei* infection, that mice, sheep, monkeys, and camels are moderately susceptible, that guinea pigs are variably susceptible, and that cats, dogs, ferrets, goats, and hamsters are highly susceptible to infection.[23–25] Although the infectious dose of *B. mallei* in humans has not been established, the low incidence of human glanders throughout history suggests that humans are moderately resistant to *B. mallei* infection. The Syrian golden hamster is one of the most common laboratory animals used to experimentally model acute glanders infection and assess *B. mallei* strain virulence.[23,26–30] Although the pathological characteristics of glanders disease progression in hamsters are comparable to that observed in humans, hamsters are significantly more sensitive to *B. mallei* infection, with an approximate LD_{50} of 10 CFU via aerosol or IP routes of infection. Therefore, hamsters are often chosen to model the acute phase of glanders, whereas BALB/c mice, which like humans are only moderately susceptible to *B. mallei* infection, are commonly used to model chronic *B. mallei* infection and disease progression.[23,31,32]

More recently, experiments were performed to establish a nonhuman primate marmoset model of intranasal infection to study *B. mallei* in order to develop effective medical countermeasures.[33] In marmosets inoculated intranasally with moderate doses (10^4–10^5 CFU) of *B. mallei*, 83% developed acute lethal infection within 3–4 days and exhibited clinical signs such as lethargy, rough coat, and conjunctivitis, and *B. mallei* was cultured from the lungs, spleen, and liver of these animals, with lesion foci similar to that characteristic of human glanders.[33] Similarly, Nelson et al.[34] demonstrated that the common marmoset could also be used to model *B. mallei* subcutaneous infection. Marmosets inoculated in the inner thigh with a lower dose 10^2 CFU *B. mallei* reached endpoints 7–10 days postchallenge,[34] with typical clinical signs and tissue histopathology, whereas intranasal inoculation at this dosage did not produce any obvious symptoms of disease. The differences observed in bacterial dosage and survival reflect the different routes of inoculation used: subcutaneous injection likely circumvents physical barriers used to combat infection, such as the respiratory system's mucosal barrier, and allows immediate access to host tissues. These results suggest that, depending on the site of inoculation and dosage used, marmosets can be used to model effectively different aspects of *B. mallei* infection and pathogenesis, producing results similar to infections observed in humans and equids.

Also recently, researchers have begun to utilize low-cost alternative infection models in order to study various aspects of *B. mallei* infection. For example, in order to evaluate 650 potential gene targets in the *B. mallei* "virulome," Schell et al.[35] utilized *Galleria mellonella* (wax moth) larvae as a surrogate host. *B. mallei* was highly pathogenic in this host, and importantly, four previously identified mutants avirulent in hamster and mouse models also proved to be avirulent in the wax moth larvae, thus validating this approach.[35] Similarly, Madagascar hissing (MH) cockroaches possess a number of qualities that make them desirable for use as *B. mallei* surrogate hosts, including low cost, ease of breeding and handling, a competent innate immune system, and the ability to thrive at 37°C.[36] MH cockroaches are highly susceptible to *B. mallei* infection, the LD_{50} being <10 CFU. Similar to the results obtained for *G. mellonella*, *B. mallei* avirulent Type VI secretion system mutants originally identified in rodents were also attenuated in MH cockroaches.[36] Again, this indicates that alternative infection models can be effective and may be useful as an alternative to mammals to study some aspects of host–pathogen interactions and bacterial virulence determinants.

18.3 *B. pseudomallei*

18.3.1 Overview

B. pseudomallei causes melioidosis, a potentially fatal condition with a variety of symptoms including pneumonia, skin lesions, and septic shock.[7] Most melioidosis cases are seen in Southeast Asia (especially Thailand) and northern Australia, although incidents also occur in nonendemic areas such as Brazil.[7,37] In Southeast Asia, approximately 80% of the population tests seropositive by the age of 4 years,[38] although only a small percentage of these seropositive individuals go on to develop melioidosis. This finding suggests that the dose and frequency of inoculum influences the infectivity and severity of the symptoms. In support of this idea, increased rates of melioidosis follow the monsoon and typhoon seasons,[39,40] when prolonged inhalation of storm-generated aerosols of *B. pseudomallei*-contaminated particles would be highest.[7]

Although the first cases of melioidosis were described in 1913, there remains some debate as to how *B. pseudomallei* is transmitted to humans.[41,42] It has been established that infections can occur following inoculation of broken skin or inhalation of water aerosols containing *B. pseudomallei*.[43] Although there is some evidence of infection occurring following ingestion, it is by no means conclusive. Historically, it was thought that melioidosis was spread through the consumption of contaminated food and water, as it can be transmitted in this manner to guinea pigs, rabbits, and rats.[41,44] However, *B. pseudomallei* has not been definitively shown to spread to humans by this route, although a report published in 1998 described the development of melioidosis in five adults in northwestern Australia.[45] Using pulsed-field gel electrophoresis (PFGE) typing, it was determined that all five patients were infected with the same PFGE type and that this type was identical to that of *B. pseudomallei* isolates found in the area's drinking water, suggesting that these cases developed due to ingestion of a single source of *B. pseudomallei*.[45] *B. pseudomallei* can also be isolated from drinking water in areas where melioidosis is not endemic. Zanetti et al.[46] showed that *B. pseudomallei* could be isolated from 7.1% of drinking water samples taken in Bologna, Italy, and that the levels of *B. pseudomallei* were relatively high, ranging from 1020 to 15,000 CFU/100 mL sample. Treatment of water with chlorine will kill *B. pseudomallei* (as well as *B. mallei*) in an experimental setting, but various factors in the environment (including attachment and nutrient limitation) may affect this susceptibility.[47] Antibiotic treatment for acute and chronic melioidosis includes ceftazidime followed by trimethoprim–sulfamethoxazole (TMP–SMX) with or without doxycycline.[48] Despite this, melioidosis has a high mortality rate and is the third most common cause of death from an infectious disease in endemic areas.[49] Probable causes for this include late diagnosis due to nonspecific symptomatology and a high recurrence rate when antibiotics are discontinued early. Given the severity of the disease, it is understandable that there is significant interest in the development of a *B. pseudomallei* vaccine, and the identification of an accurate laboratory infection model for melioidosis.

18.3.2 Laboratory Infection Models

Historically, the most heavily utilized animal model for melioidosis has been rodents.[50] Animal studies have revealed that the severity of melioidosis and the resulting proinflammatory cytokine involvement stimulated by *B. pseudomallei* infection are mouse strain dependent, and mice are typically classified as either models of acute infection (BALB/c) or chronic infection (C57BL/6) based on their sensitivity to infection and inflammatory responses.[50–52] However, as mentioned above, dose and route of infection may also affect the nature of the melioidosis presentation.[53] In either mouse strain, *B. pseudomallei* presents the same lung, liver, and spleen abscess foci as those observed in humans. However, despite these parallels, early studies in mice have been based on intravenous or intraperitoneal exposure, which does not represent the natural route of human infection.[50,54] In contrast, subcutaneous and aerosol mouse infection models both result in a systemic infection with similar clinical presentation as percutaneous inoculation and inhalational melioidosis in humans.[50,55,56]

Similarly, rats have been successfully used to model pulmonary and chronic melioidosis, and have been instrumental in understanding how underlying diabetes increases the risk of *B. pseudomallei* infection.[57] Syrian golden hamsters have also been evaluated as a small animal model of melioidosis but do not closely mimic human infection due to their extreme sensitivity to *B. pseudomallei* infection.[26] However, they have also been instrumental in evaluating *B. pseudomallei* virulence factors, such as which Type VI secretion systems are essential for virulence and intracellularity.[58]

There are limited published data characterizing *B. pseudomallei* infections in animal models other than mice and rats. Miller et al.[22] evaluated the susceptibility and natural environment of *B. pseudomallei* infection in different animal models, particularly rhesus macaques, following subcutaneous infection with graded doses of *B. pseudomallei.* Only the primate receiving the highest dose (10^6 CFU) had clinical and immunological signs of infection, including abscess formation at the site of infection, fever, and elevated white blood cell counts. However, the infection was cleared after 1–2 weeks, suggesting that rhesus macaques are resistant to *B. pseudomallei.*[22] Similarly, an evaluation of intraperitoneal and subcutaneous *B. pseudomallei* infection in Hamadryas baboons resulted in the same lung, liver, spleen, and lymph node tropisms observed in humans.[59] More recently, the pathology of inhaled *B. pseudomallei* in the marmoset, rhesus macaque, and African green monkeys was evaluated.[60,61] The marmoset is exquisitely sensitive to aerosols of *B. pseudomallei* strain K96243; 10^2 CFU was consistently lethal.[60] Disease onset occurred within 24 h and was characterized by fever, decreased activity, and a rapidly progressive septicemia similar to the acute form of melioidosis observed in humans. The disease progression of melioidosis following inhalation in rhesus macaques and African green monkeys is similar to the marmoset.[61] All of these monkeys exhibit a nonspecific fever 24–40 h postinfection, followed by dyspnea and septicemia. Gross pathology was also similar as both rhesus macaques and African green monkeys presented with lung lesions, lymph node enlargement, splenomegaly, and bone marrow lesions. These reports highlight the comparable pathology of *B. pseudomallei* infection in nonhuman primates and humans.

Because melioidosis is well modeled in rodents and primates, initially few alternative infection models were developed to study *B. pseudomallei.* However, with the influx of funding to support research into "medical countermeasures," several new *B. pseudomallei* alternative infection models have been recently developed. One of these models is a measure of invasion and cytotoxicity toward macrophages. Murine macrophages were used to identify *B. pseudomallei* virulence factors that were expressed using an *in vivo* expression technology (IVET) experiment.[62] Researchers found that the *B. pseudomallei* Cluster 1 Type 6 secretion system was expressed 12-fold over cell culture medium, when cultured in the presence of murine macrophages. Interestingly, the T6SS-1 genes appeared to play no role in the survival or growth of *B. pseudomallei* when cultured inside the macrophages. In contrast, *B. pseudomallei*

T6SS-1 Hcp proteins, integral membrane proteins of the T6SS mechanism that are expressed during functional operation of the system, were found to be important in the intracellular behavior of *B. pseudomallei* in RAW 264.7 macrophages. Mutants unable to express the Hcpl protein were only weakly cytotoxic toward murine macrophages, exhibited a growth defect inside of macrophages, and were unable to form multinucleated macrophage cells like wild-type *B. pseudomallei*.[58] In a different murine macrophage cell line (J774A.1), it was shown that different strains of *B. pseudomallei* exhibit different levels of pathogenicity and that these differing strain-dependent levels of pathogenicity were conserved across mice infection models, murine macrophages, and a *G. mellonella* (wax moth larvae) alternative infection model.[63] It was also shown that in these same models, related species *B. oklahomensis* and *B. thailandensis* were significantly less virulent than *B. pseudomallei* strains.[63] Interestingly, another study demonstrated an inverse relationship between *B. pseudomallei* virulence in a murine macrophage J774 A1 phagocytosis assay as compared to a mouse infection model.[54] However, in this case, the mouse model used was an intraperitoneal infection model, which, as mentioned above, does not accurately mimic the human condition, even though it is reproducible.[54]

Besides *G. mellonella*, which has also been used to identify virulence factors of and antibiotic pharmacokinetics against *B. pseudomallei*,[64,65] another invertebrate alternative infection model that has been successfully used to characterize *B. pseudomallei* infection is *Caenorhabditis elegans*.[66] At least five different bacterial mechanisms of killing *C. elegans* have been described, including toxin-mediated killing, persistent or transient killing, intestinal infection, direct invasion, and biofilm formation.[67] Originally, studies suggested that a *B. pseudomallei* toxin was involved in *C. elegans* killing[68]; however, more recent studies suggest that no toxin-mediated killing is involved, but rather direct prolonged contact with the bacterium.[69] How these putative *B. pseudomallei* virulence mechanisms identified in *C. elegans* relate to the human infection condition remains to be determined. Unlike the *G. mellonella* infection model, both *C. elegans* and the tomato leaf[70] infection models are unable to differentiate between *B. pseudomallei* and *B. thailandensis* virulence levels, which may be an important consideration with respect to accurately modeling human melioidosis.

18.4 *Burkholderia cepacia* Complex (Bcc)

18.4.1 Overview

The Bcc is currently made up of 20 closely related *Burkholderia* species.[71,72] These bacteria cause potentially fatal infections in immunocompromised patients, particularly those with CF and CGD.[73,74] Whereas both melioidosis and glanders can have a wide range of symptoms that affect many different systems of the body, Bcc infections tend to be limited to the lungs. Bcc infections may be asymptomatic, may cause a patient's condition to worsen over time, or may develop into a rapidly fatal pneumonia accompanied by damage to the lung tissue and septicemia.[3]

The bacteria of the *B. cepacia* complex are found more commonly in food and water supplies than either *B. pseudomallei* or *B. mallei*. They can be isolated from the rhizospheres of rice, corn, wheat, soybean, and alfalfa plants.[75,76] Bcc bacteria have also been found associated with *Pleurotus ostreatus*, the oyster mushroom, which is the second-most cultivated mushroom in the United States.[77,78] Bcc organisms may cause spoilage in foods as diverse as onions, Swiss cheese, and Parma ham.[1,79,80] However, these bacteria can also prevent food spoilage: a Bcc strain isolated from Washington navel oranges was found to prevent fungal infection of these fruits during storage.[81] Because of the ability of the Bcc to produce novel antifungals and antibiotic compounds and fix nitrogen for certain crops, there has been a great deal of interest in developing these species into agricultural product growth promoters. However, because of the potential risk to immunocompromised persons, the United States Environmental Protection Agency (EPA) has placed a moratorium upon the use of Bcc bacteria for agricultural purposes since 2003.[82]

Despite their occasional identification in food products, including bakery items, onions, cheese, ham, and oranges, in an extensive examination of different foods, Moore et al.[83] were only able to isolate Bcc from unpasteurized milk. Similarly, Bcc species were isolated from unpasteurized milk by Uraz and Çitak[84,85] and Munsch-Alatossava and Alatossava.[86,87] It is unknown how milk becomes contaminated, but possibly it is through transfer of Bcc organisms from soil to the cow udder and/or to milk storage tanks.[88] However, the Bcc are not found in commercially available dairy products and are effectively killed by pasteurization.[83,88] Berriatua et al.[2] found that the Bcc species *Burkholderia cenocepacia* and *Burkholderia vietnamiensis* were responsible for causing subclinical mastitis (inflammation of the udder) in milking sheep. However, when Moore et al.[88] examined milk samples taken from cows with mastitis to determine whether this condition was responsible for the contamination, they were unable to isolate Bcc organisms.

Like *B. pseudomallei*, Bcc organisms can also be isolated from drinking water. Zanetti et al.[46] found that 3.5% of drinking water samples in Bologna, Italy, contained Bcc bacteria. The counts in these samples were very low compared to *B. pseudomallei*, containing between 1 and 19 CFU/100 mL sample. Pegues et al.[89] examined a Bcc isolate from a water jug at a CF summer camp and found that the ribotype of this isolate was different from the ribotypes taken from patients infected at the camp, indicating that this water was not the source of infection. In contrast to these findings, samples isolated by Moore et al.[83] and Vermis et al.[90] from a variety of drinking water samples, including bottled, tap, well, and spring water, were found to not contain Bcc organisms. However, the major contaminants of the NASA space shuttle water systems for more than 24 missions have been bacteria of the Bcc, with more than 20 CFU/100 mL detected.[91]

To date, only one study has suggested a link between food consumption and Bcc infection. Fisher et al.[92] found that Bcc isolates taken from cucumber, lettuce, and cauliflower purchased from salad bars and food stores near two CF centers were of the same ribotype that was most commonly isolated from the patients in these centers. However, it is not possible to discern from these results whether or not these foods acted as vectors of Bcc infection.

18.4.2 Laboratory Infection Models

Like *B. pseudomallei*, various aspects of *B. cepacia* complex bacteria pathogenicity have been modeled using mammals (rodents), including mice and rats. In particular, the intranasal application of Bcc bacteria to mice[93–95] has been used to introduce and assess Bcc lung infections. However, more consistent results have been achieved using bacterial aerosols,[96] and for chronic infections, by implanting Bcc encapsulated in agar beads into the lungs of mice[97] and rats[95,98–100] by the method of Cash et al.[101] Each of these models has provided insights into different aspects of Bcc virulence, with different virulence factors being identified using each different approach.[102]

The rat agar bead model for studying chronic respiratory infections has been used extensively to study virulence properties of *B. cenocepacia* strains.[99,103–105] An avirulent *B. cenocepacia* strain H111, isolated from a human patient without symptoms for 5 years, was also similarly tested in the rat agar bead model.[102] Interestingly, there was no difference in the ability of H111 versus that of K56-2 to cause persistent infection in the rat. However, there was a marked qualitative difference between the strains in the lung histopathology observed in the infected animals. K56-2-infected lungs had extensive inflammatory infiltrates with predominantly polymorphonuclear cells, whereas H111-infected lungs showed only slight signs of inflammation. At higher doses of H111, infections did result in higher levels of lung histopathology, although inflammation was still limited. Taken together, these results suggest that the virulence of Bcc strains in mammalian infection models is strain dependent and that even mouse and rat infection models do not always accurately model the Bcc strain pathogenicity observed in humans.

Similarly, in immunocompromised intranasal mouse lung infections, it was observed that strains of *B. cenocepacia* caused what was described as an acute infection, which stimulated an intense immune response and relatively early clearance of the infection, whereas the presence of *B. multivorans* strains did not induce such an immune response and thus produced a longer-lasting, persistent type of infection.[94] Therefore, not only is the strain of Bcc important in a particular infection model, so too is the Bcc species. Presumably, each different Bcc strain and species has a subset of virulence factors evolved to defend against different types of hosts, and some Bcc strains or species have better developed weaponry against some hosts than others.

Relating to these findings, a large number of alternative infection models have been developed for the analysis of Bcc virulence, including plants, invertebrates, insects, and fish. It is thought that bacterial virulence factors evolved to combat natural predators in the environment and that different factors might be effective against more than one type of organism. Because the innate immune systems of invertebrates and vertebrates share many common well-conserved features, a bacterium able to infect and/or cause disease in one type of host may use similar mechanisms and strategies to infect and/or cause disease in another.[106,107]

Because the Bcc are natural pathogens of onions, causing soft rot,[1] one of the earliest alternative infection models developed for the Bcc permitted the observation of the maceration of onion tissue.[108–110] Such studies identified a plasmid-encoded pectate hydrolase as the enzyme responsible for maceration in type strain *B. cenocepacia* ATCC 25416.[108] But *B. cepacia* and *B. cenocepacia* strains lacking this enzyme can also produce a tissue water soaking phenotype through expression of a plasmid-encoded Type IV secretion system gene cluster.[111] It was also discovered that quorum sensing appears to be involved in onion tissue maceration, as the quorum sensing signal synthase gene *cepI* and regulator gene *cepR* mutants are attenuated for maceration.[112]

Another simple model used to assess Bcc virulence against plants is the alfalfa seedling infection model,[103] in which the yellowing of leaves (chlorosis), stunted root development, and brown tissue destruction on leaves and roots were used as a measure of bacterial virulence. This model has been used with success to ascertain the role of various virulence factors expressed by the Bcc.[103] More recently, the common duckweed (*Lemna minor*) was used for similar purposes, the advantage of duckweed being that it can be grown axenically. Several different virulent factors discovered in other Bcc hosts were also found to be virulent factors in duckweed,[113] confirming duckweed's validity as an infection model. Interestingly, duckweed was also found to be an effective infection model for enteropathogenic *Escherichia coli*.[113]

A large number of Bcc researchers are also now using insect animal models such as wax moth larvae (*G. mellonella*) or fruit flies (*Drosophila melanogaster*) as substitutes for mammalian infection models.[114,115] Such "lower" animal infection models hold the advantage that they typically possess at least an innate immunity system, but do not involve all of the regulatory and ethical requirements needed for "higher" animal experimentation. Although each alternative infection host may not exactly model the same types of tissue found in humans, it is thought that for each different host, a subset of the bacterium's total armamentarium is expressed to overcome that host's defenses. Therefore, at least some of the cellular weapons used by the Bcc to defeat *G. mellonella* would also be expressed when the Bcc bacterium finds itself in the human lung. Since its development,[114] the *G. mellonella* infection model has been used to identify a novel hemolytic toxin,[116] as well as several other Bcc virulence factors,[117–119] and also show that the larvae can be rescued from lethal bacterial infection by the action of bacteriophages.[120] Similarly, *D. melanogaster* has been successfully employed to demonstrate that different strains of the Bcc can kill by infection. Unlike results found for *Pseudomonas aeruginosa* in *D. melanogaster*, species of the *B. cepacia* complex were only able to kill flies by infection after pricking, rather than through ingestion.[115] This suggests that the Bcc are less toxic to fruit flies than other opportunistic CF bacterial pathogens. Nevertheless, the results obtained with the fruit fly model are comparable with those obtained using mammalian infection models, and *B. cenocepacia* K56–2 mutants less virulent in murine hosts or other alternative infection models than wild type are also less virulent in the fruit fly, even when tested using competitive index analyses.[115]

Caenorhabditis elegans, a roundworm nematode, has also been used successfully as a Bcc model for mammalian infection.[121] These studies have shown that there are two modes of killing, a fast mode due to paralysis by an extracellular Bcc toxin and a slow mode due to ingestion and accumulation of bacteria in the nematode intestine.[121] The Bcc quorum sensing system appears to be required for both types of killing,[121] but only the slow mode was shown to be partially due to certain Bcc virulence factors such as AidA.[122] Several different environmental parameters can influence the performance of this model, including worm age,[123] prior diet,[124] diet,[125] Bcc strain,[102,126,127] and Bcc strain mucoidy status.[128] Nevertheless, a number of Bcc virulence factors effective against *C. elegans* have been discovered or confirmed using this model, including the type III secretion system,[129] the phenylacetic acid catabolic pathway,[130] two different Hfq proteins,[1321,132] the pleiotropic regulator of phenazines Pbr,[133] a small regulatory RNA MtvR,[134] a Type 2 secretion system pseudopilin GspJ,[135] and the third chromosome/virulence plasmid pC3 in many Bcc species.[136,137]

The final Bcc alternative infection model to be discussed is perhaps the most recently developed zebrafish embryos.[138] Bcc bacteria survive and multiply in zebrafish macrophages, later disseminating and causing bacteremia and death within 3 days. *B. cenocepacia* strains that are highly virulent in other animal models are also highly virulent for the embryos, whereas less virulent Bcc species were controlled by the fish innate immune system.[138] This model was used to confirm that both the quorum sensing system and the third chromosome are required for full virulence of *B. cenocepacia* in the zebrafish embryos.[136,138]

18.5 *B. gladioli* pv. *cocovenenans*

18.5.1 Overview

B. gladioli pv. *cocovenenans* cells are motile, aerobic, non-spore-forming, nonencapsulated Gram-negative rods. These bacteria form smooth, round colonies (on potato dextrose agar) that are yellow pigmented due to the production of toxoflavin. Growth can occur between 6°C and 41°C, but is optimal between 30°C and 37°C, and toxin production occurs between 22°C and 30°C.[139,140]

The taxonomical classification of the bacteria named *B. gladioli* pv. *cocovenenans* is historically complicated. van Veen and Mertens[141] published the first studies on this bacterium, which they isolated from contaminated "tempe bongkrek," an Indonesian fermented coconut cake. This bacterium was named *Pseudomonas cocovenenans* (cocos = coconut, veneno = to poison), and this nomenclature was retained in early studies.[142–144] In 1963, a similar bacterium was isolated from contaminated fermented corn flour in China[145] and eventually named in English as *Flavobacterium farinofermentans* sp. nov.[146] After assessing a variety of cellular characteristics, Zhao et al.[147,148] determined that *P. cocovenenans* and *P. farinofermentans* belonged to the same species, suggesting that *P. farinofermentans* should be renamed *P. cocovenenans* biovar *farinofermentans*. In 1992, seven *Pseudomonas* species were reclassified and assigned to the new genus *Burkholderia* (including *B. pseudomallei*, *B. mallei*, *B. cepacia*, and *B. gladioli*).[149] Following analysis of the type strain isolated from tempe bongkrek, in 1995 *P. cocovenenans* was subsequently reassigned as *Burkholderia cocovenenans* comb. nov.[150] In 1997, Vandamme et al.[151] found that SDS–PAGE protein patterns of whole-cell proteins of *B. gladioli* and *B. cocovenenans* strains were similar, and further experimentation confirmed this relationship.[152] This led to the reclassification of *B. cocovenenans* as a member of *B. gladioli*, now called *B. gladioli* pv. *cocovenenans*, to join three existing phytopathogenic pathovars of *B. gladioli* (*gladioli*, *alliicola*, and *agaricicola*).[9–11]

18.5.2 Laboratory Infection Models

For nontoxigenic *B. gladioli*, it is most commonly isolated from the respiratory tract.[153–155] These infections are much less serious than Bcc infections, and the impact on patient health is often minimal.[153,154] Septicemia has also been reported in patients with CF, CGD, AIDS, and diabetes.[12,13,156,157] Laboratory infection models would be similar to those used for bacteria of the Bcc, though few studies have been carried out thus far.

In contrast, there are few laboratory models available for toxigenic *B. gladioli*. Unlike most other foodborne illnesses (except perhaps *C. botulinum*), the pathogenic effects of *B. gladioli* pv. *cocovenenans* are mediated solely by toxins and not by the bacteria themselves. When mice are administered either live or heat-killed bacteria, no adverse effects are observed, as opposed to when mice are administered culture supernatants or partially purified toxins.[8,146] *B. gladioli* pv. *cocovenenans* produces two toxins: toxoflavin (also called xanthothricin) and bongkrekic acid or BA (also called bongkrek acid or flavotoxin A). Production of toxoflavin is not restricted to *B. gladioli* pv. *cocovenenans*, as it is also produced by *Streptomyces*, *Burkholderia glumae*, *Burkholderia plantarii*, and phytopathogenic *B. gladioli*.[14,158–160] Although toxoflavin is highly toxic, it is generally not the cause of *B. gladioli* pv. *cocovenenans* intoxication. Instead, BA is both more toxic and present in higher concentrations in contaminated foods.[161,162] The genes encoding BA have recently been identified, and were found to be part of a Type I polyketide synthase (PKS) operon, involving the action of a cytochrome P450 monooxygenase.[163]

BA toxin is one of the most potent respiratory poisons known. Following human consumption of food contaminated with *B. gladioli* pv. *cocovenenans*, intoxication occurs rapidly, usually within 1–10 h.[164] The classical symptoms are hyperglycemia followed by hypoglycemia, spasms, loss of consciousness, and death.[165] Following autopsy, the liver, kidneys, and brain all show signs of damage, and in some cases, there is also damage to the heart, lungs, gastrointestinal tract, and spleen.[164] Laboratory animal models of infection are generally too sensitive to BA toxin to be of much experimental value. When food contaminated with *B. gladioli* pv. *cocovenenans* is fed to dogs, they develop spasms, enter comas, and die within 2–3 h.[164] Dogs and rhesus monkeys fed with *B. gladioli* pv. *cocovenenans* culture supernatants die within 6–33 h and 15.5–35 h, respectively.[146] When mice were given *B. gladioli* pv. *cocovenenans* culture supernatants (0.4 mL/mouse) partially purified by the method of Hu et al.,[166,167] all mice died within 45 min with stiff tail and feet.[8]

With respect to a mechanism of action, BA inhibits oxidative phosphorylation by binding to the adenine nucleotide translocator, which normally shuttles ADP and ATP across the inner mitochondrial membrane.[168] The activity of this transporter is thought to inhibit cellular apoptosis.[169] Because BA inhibits oxidative phosphorylation, the only way for cells to generate ATP is through anaerobic glycolysis.[170] This metabolic changeover is responsible for the symptoms of intoxication. Glycogen stores are broken down to allow for an increased level of glycolysis, leading to hyperglycemia and increased lactic acid levels, but these stores are soon depleted, resulting in hypoglycemia and an inability to regenerate ATP.[171] It is the ATP depletion that is fatal and not the hypoglycemia, as the injection of glucose is insufficient to prevent death.[171]

18.6 Conclusions

The major pathogenic bacteria of the genus *Burkholderia* include the species *B. mallei*, *B. pseudomallei*, the *B. cepacia* complex, and *B. gladioli*. Surprisingly, the diseases that each of these bacteria causes are diverse in terms of host, mechanism of action, prognosis, and outcome. Coordinately, the laboratory systems used to model these diseases and assess the virulence and toxicity of each of these bacterial species are diverse as well. For *B. mallei*, the causative agent of glanders, the primary hosts are equine, and thus, a variety of equine animal models can be used to study glanders disease progression. For *B. pseudomallei*, the modeling of the disease melioidosis is much more difficult, as the disease progression is more complex, including variable human immune interactions, a latency period, and multifactorial local and systemic virulence effects. Moreover, as humans appear to be a primary host, different aspects of the disease are best studied in different animal models. Similarly, the *B. cepacia* complex is best studied using a variety of animal models, due to its multifactorial virulence factor production. In contrast to *B. pseudomallei*, however, modeling *B. cepacia* complex disease typically requires the animals to be immunocompromised in some way, as healthy animals (and humans) are generally able to resist Bcc infections. Finally, nontoxigenic *B. gladioli* disease modeling can be performed with systems similar to those used for the Bcc, whereas toxigenic *B. gladioli* can be studied at the organismal level, or through the purification of the toxin and with the use of specialized cellular or molecular assays. Historically, murine and equine animal models have been used for the study of *Burkholderia* infections and toxicology. In recent years, however, there has been a great deal of interest in alternative host infection models to analyze all or specific components of the diseases caused by *Burkholderia* species. The development of new and better laboratory models of disease caused by *Burkholderia* will bring about a better understanding of the disease progression and virulence factors utilized by each individual *Burkholderia* species, leading to the identification of drug targets and the development of antibacterial strategies against pathogenic *Burkholderia* species.

REFERENCES

1. Burkholder, W. H. 1950. Sour skin, a bacterial rot of onion bulbs. *Phytopathology*. 40:115–117.
2. Berriatua, E., et al. 2001. Outbreak of subclinical mastitis in a flock of dairy sheep associated with *Burkholderia cepacia* complex infection. *J. Clin. Microbiol.* 39:990–994.

3. Isles, A., I. Maclusky, and M. Corey. 1984. *Pseudomonas cepacia* infection in cystic fibrosis: an emerging problem. *J. Pediatr.* 104:206–210.
4. Estrada-De Los Santos, P., R. Bustillos-Cristales, and J. Caballero-Mellado. 2001. *Burkholderia*, a genus rich in plant-associated nitrogen fixers with wide environmental and geographic distribution. *Appl. Environ. Microbiol.* 67:2790–2798.
5. El-Banna, N., and G. Winkelmann. 1998. Pyrrolnitrin from *Burkholderia cepacia*: antibiotic activity against fungi and novel activities against streptomycetes. *J. Appl. Microbiol.* 85:69–78.
6. Folsom, B. R., P. J. Chapman, and P. H. Pritchard. 1990. Phenol and trichloroethylene degradation by *Pseudomonas cepacia* G4: kinetics and interactions between substrates. *Appl. Environ. Microbiol.* 56:1279–1285.
7. Wiersinga, W. J., T. van der Poll, N. L. White, N. P. Day, and S. J. Peacock. 2006. Melioidosis: insights into the pathogenicity of *Burkholderia pseudomallei*. *Nat. Rev. Microbiol.* 4:272–282.
8. Jiao, Z., et al. 2003. Need to differentiate lethal toxin-producing strains of *Burkholderia gladioli*, which cause severe food poisoning: description of *B. gladioli* pathovar *cocovenenans* and an emended description of *B. gladioli*. *Microbiol. Immunol.* 47:915–925.
9. Ballard, R. W., N. J. Palleroni, M. Doudoroff, R. Y. Stanier, and M. Mandel. 1970. Taxonomy of the aerobic pseudomonads: *Pseudomonas cepacia*, *P. marginata*, *P. alliicola* and *P. caryophylli*. *J. Gen. Microbiol.* 60:199–214.
10. Ko, S. S., W. N. Chang, J. F. Wang, S. J. Cherng, and S. Shanmugasundaram. 2002. Storage variability among short-day onion cultivars under high temperature and high relative humidity, and its relationship with disease incidence and bulb characteristics. *J. Am. Soc. Hortic. Sci.* 127:848–854.
11. Gill, W. M., and A. Tsuneda. 1997. The interaction of the soft rot bacterium *Pseudomonas gladioli* pv. *agaricicola* with Japanese cultivated mushrooms. *Can. J. Microbiol.* 43:639–648.
12. Ross, J. P., S. M. Holland, V. J. Gill, E. S. DeCarlo, and J. I. Gallin. 1995. Severe *Burkholderia* (*Pseudomonas*) *gladioli* infection in chronic granulomatous disease: report of two successfully treated cases. *Clin. Infect. Dis.* 21:1291–1293.
13. Graves, M., T. Robin, A. M. Chipman, J. Wong, S. Khashe, and J. M. Janda. 1997. Four additional cases of *Burkholderia gladioli* infection with microbiological correlates and review. *Clin. Infect. Dis.* 25:838–842.
14. Ura, H., N. Furuya, K. Iiyama, M. Hidaka, K. Tsuchiya, and N. Matsuyama. 2006. *Burkholderia gladioli* associated with symptoms of bacterial grain rot and leaf-sheath browning of rice plants. *J. Gen. Plant Pathol.* 72:98–103.
15. Whitlock, G. C., D. Mark Estes, and A. G. Torres. 2007. Glanders: off to the races with *Burkholderia mallei*. *FEMS Microbiol. Lett.* 277:115–122.
16. Gilad, J., I. Harary, T. Dushnitsky, D. Schwartz, and Y. Amsalem. 2007. *Burkholderia mallei* and *Burkholderia pseudomallei* as bioterrorism agents: national aspects of emergency preparedness. *Israel Med. Assoc. J.* 9:499–503.
17. Wilson, G. S., and A. Miles. 1975. *Topley and Wilson's Principles of Bacteriology, Virology and Immunity*, 6th Edition. Edward Arnold, London.
18. Srinivasan, A., et al. 2001. Glanders in a military research microbiologist. *New Engl. J. Med.* 345:256–258.
19. Rotz, L. D., A. S. Khan, S. R. Lillibridge, S. M. Ostroff, and J. M. Hughes. 2002. Public health assessment of potential biological terrorism agents. *Emerg. Infect. Dis.* 8:225–230.
20. Takhistov, P., and C. M. Bryant. 2006. Protecting the food supply. *Food Technol.* 60:34–43.
21. Schadewaldt, H. 1975. Discovery of glanders bacillus. *Dtsch. Med. Wochenschr.* 100:2292–2295.
22. Miller, W. R., L. Pannell, L. Cravitz, W. A. Tanner, and T. Rosebury. 1948. Studies on certain biological characteristics of *Malleomyces mallei* and *Malleomyces pseudomallei*, II: virulence and infectivity for animals. *J. Bacteriol.* 55:127–135.
23. Bondi, S. K., and J. B. Goldberg. 2008. Strategies towards vaccine against *Burkholderia mallei* and *Burkholderia pseudomallei*. *Expert Rev. Vaccines.* 7:1357–1365.
24. Dembek, Z. F. 2007. *Medical Aspects of Biological Warfare*. Borden Institute, Washington, DC.
25. Wagg, D. M., and D. DeShazer. 2004. Glanders: new insights into an old disease, pp. 209–237. In L. E. Lindler, F. J. Lebeda, and G. W. Korch (eds.), *Biological Weapons Defense: Infectious Diseases and Counterbioterrorism*. Humana Press, Totowa, NJ.
26. Brett, P. J., D. DeShazer, and D. E. Woods. 1997. Characterization of *Burkholderia pseudomallei* and *Burkholderia pseudomallei*-like strains. *Epidemiol. Infect.* 118:137–148.

27. DeShazer, D., D. M. Waag, D. L. Fritz, and D. E. Woods. 2001. Identification of a *Burkholderia mallei* polysaccharide gene cluster by subtractive hybridization and demonstration that the encoded capsule is an essential virulence determinant. *Microb. Pathog.* 30:253–269.
28. Fritz, D. L., P. Vogel, D. R. Brown, and D. M. Waag. 1999. The hamster model of intraperitoneal *Burkholderia mallei* (glanders). *Vet. Pathol.* 36:276–291.
29. Russell, P., et al. 2000. Comparison of efficacy of ciprofloxacin and doxycycline against experimental melioidosis and glanders. *J. Antimicrob. Chemother.* 45:813–818.
30. Ulrich, R. L., and D. DeShazer. 2004. Type III secretion: a virulence factor delivery system essential for the pathogenicity of *Burkholderia mallei*. *Infect. Immun.* 72:1150–1154.
31. Fritz, D. L., P. Vogel, D. R. Brown, D. DeShazer, and D. M. Waag. 2000. Mouse model of sublethal and lethal intraperitoneal glanders (*Burkholderia mallei*). *Vet. Pathol.* 37:626–636.
32. Lever, M. S., et al. 2003. Experimental aerogenic *Burkholderia mallei* (glanders) infection in the BALB/c mouse. *J. Med. Microbiol.* 52:1109–1115.
33. Jelesijevic, T., et al. 2015. Use of the common marmoset to study *Burkholderia mallei* infection. *PLoS One.* 10:e0124181.
34. Nelson, M., et al. 2014. Comparative experimental subcutaneous glanders and melioidosis in the common marmoset (*Callithrix jacchus*). *Int. J. Exp. Pathol.* 95:378–391.
35. Schell, M. A., L. Lipscomb, and D. DeShazer. 2008. Comparative genomics and an insect model rapidly identify novel virulence genes of *Burkholderia mallei*. *J. Bacteriol.* 190:2306–2313.
36. Fisher, N. A., W. J. Ribot, W. Applefeld, and D. DeShazer. 2012. The Madagascar hissing cockroach as a novel surrogate host for *Burkholderia pseudomallei*, *B. mallei* and *B. thailandensis*. *BMC Microbiol.* 12:117.
37. Rolim, D. B., et al. 2005. Melioidosis, northeastern Brazil. *Emerg. Infect. Dis.* 11:1458–1460.
38. Leelarasamee A. 1998. *Burkholderia pseudomallei*: the unbeatable foe? *Southeast Asian J. Trop. Med. Public Health.* 29:410–415.
39. Currie, B. J., and S. P. Jacups. 2003. Intensity of rainfall and severity of melioidosis, Australia. *Emerg. Infect. Dis.* 9:1538–1542.
40. Ko, W. C., et al. 2007. Melioidosis outbreak after typhoon, southern Taiwan. *Emerg. Infect. Dis.* 13:896–898.
41. Whitmore, A. 1913. An account of a glanders-like disease occurring in Rangoon. *J. Hyg.* 13:1–34.
42. Dance, D. A. B. 2000. Ecology of *Burkholderia pseudomallei* and the interactions between environmental *Burkholderia* spp. and human-animal hosts. *Acta Trop.* 74:159–168.
43. Suputtamongkol, Y., et al. 1994. The epidemiology of melioidosis in urban Ratchatani, northeast Thailand. *Int. J. Epidemiol.* 23:1082–1090.
44. Stanton, A. T., and W. Fletcher. 1932. *Studies from the Institute for Medical Research, Federated Malay States, No. 21: Melioidosis.* John Bale, Sons & Danielsson, Ltd., London.
45. Inglis, T. J. J., S. C. Garrow, C. Adams, M. Henderson, and M. Mayo. 1998. Dry-season outbreak of melioidosis in Western Australia. *Lancet.* 352:1600.
46. Zanetti, F., G. De Luca, and S. Stampi. 2000. Recovery of *Burkholderia pseudomallei* and *B. cepacia* from drinking water. *Int. J. Food. Microbiol.* 59:67–72.
47. Rose, L. J., et al. 2005. Chlorine inactivation of bacterial bioterrorism agents. *Appl. Environ. Microbiol.* 71:566–568.
48. White, N. J. 2003. Melioidosis. *Lancet.* 361:1715–1722.
49. Limmathurotsakul, D., et al. 2010. Increasing incidence of human melioidosis in northeast Thailand. *Am. J. Trop. Med. Hyg.* 82:1113–1117.
50. Warawa, J. M. 2010. Evaluation of surrogate animal models of melioidosis. *Front Microbiol.* 1:141.
51. Leakey, A. K., G. C. Ulett, and R. G. Hirst. 1998. BALB/c and C57Bl/6 mice infected with virulent *Burkholderia pseudomallei* provide contrasting animal models for the acute and chronic forms of human melioidosis. *Microb. Pathog.* 24:269–275.
52. Lever, M. S., M. Nelson, A. J. Stagg, R. J. Beedham, and A. J. H. Simpson. 2009. Experimental acute respiratory *Burkholderia pseudomallei* in BALB/c mice. *Int. J. Exp. Pathol.* 90:16–25.
53. Liu, B., G. C. Koo, E. H. Yap, K. L. Chua, and Y. H. Gan. 2002. Model of differential susceptibility to mucosal *Burkholderia pseudomallei* infection. *Infect. Immun.* 70:504–511.
54. Welkos, S. L., et al. 2015. Characterization of *Burkholderia pseudomallei* strains using a murine intraperitoneal infection model and in vitro macrophage assays. *PLoS One.* 10:e0124667.

55. Tan, G. Y., et al. 2008. *Burkholderia pseudomallei* aerosol infection results in differential inflammatory responses in BALB/c and C57Bl/6 mice. *J. Med. Microbiol.* 57:508–515.
56. West, T. E., N. D. Myers, H. D. Liggitt, and S. J. Skerrett. 2012. Murine pulmonary infection and inflammation induced by inhalation of *Burkholderia pseudomallei. Int. J. Exp. Pathol.* 93:421–428.
57. Woods, D. E., A. L. Jones, and P. J. Hill. 1993. Interaction of insulin with *Pseudomonas pseudomallei. Infect. Immun.* 61:4045–4050.
58. Burtnick, M. N., et al. 2011. The cluster 1 type VI secretion system is a major virulence determinant in *Burkholderia pseudomallei. Infect. Immun.* 79:1512–1525.
59. Manzeniuk, I. N., E. A. Galina, V. V. Dorokhin, I. Kalachev IIa, V. N. Borzenkov, and E. A. Svetoch. 1999. *Burkholderia mallei* and *B. pseudomallei*, study of immuno- and pathogenesis of glanders and melioidosis, heterologous vaccines. *Antibiot. Khimioter.* 44:21–26.
60. Nelson, M., et al. 2011. Development of an acute model of inhalational melioidosis in the common marmoset (*Callithrix jacchus*). *Int. J. Exp. Pathol.* 92:428–435.
61. Yeager, J. J., et al. 2012. Natural history of inhalation melioidosis in rhesus macaques (*Macaca mulatta*) and African green monkeys (*Chlorocebus aethiops*). *Infect. Immun.* 80:3332–3340.
62. Shalom, G., J. G. Shaw, and M. S. Thomas. 2007. In vivo expression technology identifies a type VI secretion system locus in *Burkholderia pseudomallei* that is induced upon invasion of macrophages. *Microbiology.* 153:2689–2699.
63. Wand, M. E., C. M. Mueller, R. W. Titball, and S. L. Michell. 2011. Macrophage and *Galleria mellonella* infection models reflect the virulence of natural isolates of *B. pseudomallei, B. thailandensis*, and *B. oklahomensis. BMC Microbiol.* 11:11.
64. Muller, C. M., L. Conjero, N. Spink, M. E. Wand, G. J. Bancroft, and R. W. Titball. 2012. Role of RelA and SpoT in *Burkholderia pseudomallei* virulence and immunity. *Infect. Immun.* 80:3247–3255.
65. Thomas, R. J., et al. 2013. *Galleria mellonella* as a model system to test the pharmacokinetics and efficacy of antibiotics against *Burkholderia pseudomallei. Int. J. Antimicrob. Agents.* 41:330–336.
66. O'Quinn, A. L., E. M. Wiegand, and J. A. Jeddeloh. 2001. *Burkholderia pseudomallei* kills the nematode *Caenorhabditis elegans* using an endotoxin-mediated paralysis. *Cell Microbiol.* 3:381–393.
67. Sifri, C. D., J. Begun, and F. M. Ausubel. 2005. The worm has turned—microbial virulence modeled in *Caenorhabditis elegans. Trends Microbiol.* 13:119–127.
68. Gan, Y. H., et al. 2002. Characterization of *Burkholderia pseudomallei* infection and identification of novel virulence factors using a *Caenorhabditis elegans* host system. *Mol. Microbiol.* 44:1185–1197.
69. Lee, S. H., S. K. Oui, N. M. Mahadi, M. W. Tan, and S. Nathan. 2011. Complete killing of *Caenorhabditis elegans* by *Burkholderia pseudomallei* is dependent on prolonged direct association with the viable pathogen. *PLoS One.* 6:e16707.
70. Lee, Y. H., Y. Chen, X. Ouyang, and Y. H. Gan. 2010. Identification of tomato plant as a novel host for *Burkholderia pseudomallei. BMC Microbiol.* 10:28.
71. Mahenthiralingam, E., T. A. Urban, and J. B. Goldberg. 2005. The multifarious, multireplicon *Burkholderia cepacia* complex. *Nat. Rev. Microbiol.* 3:144–156.
72. De Smet, B., et al. 2015. *Burkholderia stagnalis* sp. nov. and *Burkholderia territorii* sp. nov., two novel *Burkholderia cepacia* complex species from environmental and human sources. *Int. J. Syst. Evol. Microbiol.* 65:2265–2271.
73. Bottone, E. J., S. D. Douglas, A. R. Rausen, and G. T. Keusch. 1975. Association of *Pseudomonas cepacia* with chronic granulomatous disease. *J. Clin. Microbiol.* 1:425–428.
74. Rosenstein, B. J., and D. E. Hall. 1980. Pneumonia and septicemia due to *Pseudomonas cepacia* in a patient with cystic fibrosis. *Johns Hopkins Med. J.* 147:188–189.
75. Ramette, A., and J. M. Tiedje. 2007. Multiscale responses of microbial life to spatial distance and environmental heterogeneity in a patchy ecosystem. *Proc. Natl. Acad. Sci. USA* 104:2761–2766.
76. Zhang, L., and G. Xie. 2007. Diversity and distribution of *Burkholderia cepacia* complex in the rhizosphere of rice and maize. *FEMS Microbiol. Lett.* 266:231–235.
77. Cohen, R., L. Persky, and Y. Hadar. 2002. Biotechnological applications and potential of wood-degrading mushrooms of the genus *Pleurotus. Appl. Microbiol. Biotechnol.* 58:582–594.
78. Yara, R., W. Maccheroni Jr., J. Horii, and J. L. Azevedo. 2006. A bacterium belonging to the *Burkholderia cepacia* complex associated with *Pleurotus ostreatus. J. Microbiol.* 44:263–268.
79. Smith, D., E. Mikolajcik, and J. Lindamood. 1987. Causative organisms and chemical nature of the Swiss cheese rind rot defect. *Cult. Dairy Prod. J.* 22:9–12.

80. Blanco, D., G. Barbieri, P. Mambriani, E. Spotti, and I. Barbuti. 1994. Study of "potato defect" of raw dry-cured ham. *Ind. Conserve.* 69:230–236.
81. Huang, Y., B. J. Deverall, S. C. Morris, and B. L. Wild. 1993. Biocontrol of postharvest orange diseases by a strain of *Pseudomonas cepacia* under semi-commercial conditions. *Postharvest Biol. Technol.* 3:293–304.
82. Environmental Protection Agency. 2003. *Burkholderia cepacia* complex; significant new use rule [online]. Available from http://www.epa.gov/fedrgstr/EPA-TOX/2003/June/Day-13/t15010.htm.
83. Moore, J. E., B. McIlhatton, A. Shaw, P. G. Murphy, and J. S. Elborn. 2001. Occurrence of *Burkholderia cepacia* in foods and waters: clinical implications for patients with cystic fibrosis. *J. Food Protect.* 64:1076–1078.
84. Uraz, G., and S. Çitak. 1998. An investigation about the distribution and isolation of *Pseudomonas* from raw milk samples obtained from different areas. *Turk. J. Agric. Forest.* 22:469–474.
85. Uraz, G., and S. Çitak. 1998. The isolation of *Pseudomonas* and other Gram(–) psychrotrophic bacteria in raw milks. *J. Basic Microbiol.* 38:129–134.
86. Munsch-Alatossava, P., and T. Alatossava. 2006. Phenotypic characterization of raw milk-associated psychrotrophic bacteria. *Microbiol. Res.* 161:334–346.
87. Munsch-Alatossava, P., and T. Alatossava. 2007. Antibiotic resistance of raw-milk-associated psychrotrophic bacteria. *Microbiol. Res.* 162:115–123.
88. Moore, J. E., B. C. Millar, T. Buckley, M. Crowe, and J. S. Elborn. 2002. Effect of high-temperature short-time (HTST) laboratory pasteurization on the survival of *Burkholderia cepacia* complex organisms in whole, low fat and skimmed milks. *J. Dairy Res.* 69:483–490.
89. Pegues, D. A., et al. 1994. Acquisition of *Pseudomonas cepacia* at summer camps for patients with cystic fibrosis. *J. Pediatr.* 124:694–702.
90. Vermis, K., M. Brachkova, P. Vandamme, and H. Nelis. 2003. Isolation of *Burkholderia cepacia* complex genomovars from waters. *Syst. Appl. Microbiol.* 26:595–600.
91. Koenig, D. W., and D. L. Pierson. 1997. Microbiology of the space shuttle water system. *Water Sci. Technol.* 35:59–64.
92. Fisher, M. C., et al. 1993. Source of *Pseudomonas cepacia*: ribotyping of isolates from patients and from the environment. *J. Pediatr.* 123:745–747.
93. Chu, K. K., D. J. Davidson, T. K. Halsey, J. W. Chung, and D. P. Speert. 2002. Differential persistence among genomovars of the *Burkholderia cepacia* complex in a murine model of pulmonary infection. *Infect. Immun.* 70:2715–2720.
94. Chu, K. K., K. L MacDonald, D. J. Davidson, and D. P. Speert. 2004. Persistence of *Burkholderia multivorans* within the pulmonary macrophage in the murine lung. *Infect. Immun.* 72:6142–6147.
95. Sokol, P. A., U. Sajjan, M. B. Visser, S. Gingues, J. Forstner, and C. Kooi. 2003. The CepIR quorum-sensing system contributes to the virulence of *Burkholderia cenocepacia* respiratory infections. *Microbiology.* 149:3649–3658.
96. Semler, D. D., A. D. Goudie, W. H. Finlay, and J. J. Dennis. 2014. Aerosol phage therapy efficacy in *Burkholderia cepacia* complex respiratory infections. *Antimicrob. Agents Chemother.* 58:4005–4013.
97. Cieri, M. V., N. Mayer-Hamblett, A. Griffith, and J. L. Burns. 2002. Correlation between an in vitro invasion assay and a murine model of *Burkholderia cepacia* lung infection. *Infect. Immun.* 70:1081–1086.
98. O'Grady, E. P., and P. A. Sokol. 2011. *Burkholderia cenocepacia* differential gene expression during host-pathogen interactions and adaptation to the host environment. *Front. Cell. Infect. Microbiol.* 1:15.
99. Sokol, P. A., P. Darling, D. E. Woods, E. Mahenthiralingam, and C. Kooi. 1999. Role of ornibactin biosynthesis in the virulence of *Burkholderia cepacia*: characterization of *pvdA*, the gene encoding L-ornithine *N*(5)-oxygenase. *Infect. Immun.* 67:4443–4455.
100. Subramoni, S., D. T. Nguyen, and P. A. Sokol. 2011. *Burkholderia cenocepacia* ShvR-regulated genes that influence colony morphology, biofilm formation, and virulence. *Infect. Immun.* 79:2984–2997.
101. Cash, H. A., D. E. Woods, B. McCullough, W. G. Johanson, Jr., and J. A. Bass. 1979. A rat model of chronic respiratory infection with *Pseudomonas aeruginosa*. *Am. Rev. Respir. Dis.* 119:453–459.
102. Uehlinger, S., et al. 2009. Identification of specific and universal virulence factors in *Burkholderia cenocepacia* strains by using multiple infection hosts. *Infect. Immun.* 77:4102–4110.
103. Bernier, S. P., L. Silo-Suh, D. E. Woods, D. E. Ohman, and P. A. Sokol. 2003. Comparative analysis of plant and animal models for characterization of *Burkholderia cepacia* virulence. *Infect. Immun.* 71:5306–5313.

104. Hunt, T. A., C. Kooi, P. A. Sokol, and M. A. Valvano. 2004. Identification of *Burkholderia cenocepacia* genes required for bacterial survival in vivo. *Infect. Immun.* 72:4010–4022.
105. Visser, M. B., S. Majumdar, E. Hani, and P. A. Sokol. 2004. Importance of the ornibactin and pyochelin siderophore transport systems in *Burkholderia cenocepacia* lung infections. *Infect. Immun.* 72:2850–2857.
106. Hoffmann, J. A., F. C. Kafatos, C. A. Janeway, and R. A. Ezekowitz. 1999. Phylogenetic perspectives in innate immunity. *Science.* 284:1313–1318.
107. Mahajan-Miklos, S., L. G. Rahme, and F. M. Ausubel. 2000. Elucidating the molecular mechanisms of bacterial virulence using non-mammalian hosts. *Mol. Microbiol.* 37:981–988.
108. Gonzalez, C. F., and A. K. Vidaver. 1979. Bacteriocin, plasmid and pectolytic diversity in *Pseudomonas cepacia* of clinical and plant origin. *J. Gen. Microbiol.* 110:161–170.
109. Yohalem, D. S., and J. W. Lorbeer. 1994. Intraspecific metabolic diversity among strains of *Burkholderia cepacia* isolated from decayed onions, soils, and the clinical environment. *Antonie Van Leeuwenhoek.* 65:111–131.
110. Wigley, P., and N. F. Burton. 1999. Genotypic and phenotypic relationships in *Burkholderia cepacia* isolated from cystic fibrosis patients and the environment. *J. Appl. Microbiol.* 86:460–468.
111. Engledow, A. S., E. G. Medrano, E. Mahenthiralingam, J. J. LiPuma, and C. F. Gonzalez. 2004. Involvement of a plasmid-encoded type IV secretion system in the plant tissue watersoaking phenotype of *Burkholderia cenocepacia. J. Bacteriol.* 186:6015–6024.
112. Aguilar, C., I. Bertani, and V. Venturi. 2003. Quorum-sensing system and stationary-phase sigma factor (*rpoS*) of the onion pathogen *Burkholderia cepacia* genomovar I type strain, ATCC 25416. *Appl. Environ. Microbiol.* 69:1739–1747.
113. Thomson, E. L., and J. J. Dennis. 2013. Common duckweed (*Lemna minor*) is a versatile high-throughput infection model for the *Burkholderia cepacia* complex and other pathogenic bacteria. *PLoS One.* 8:e80102.
114. Seed, K. D., and J. J. Dennis. 2008. Development of *Galleria mellonella* as an alternative infection model for the *Burkholderia cepacia* complex. *Infect. Immun.* 76:1267–1275.
115. Castonguay-Vanier, J., L. Vial, J. Tremblay, and E. Déziel. 2010. *Drosophila melanogaster* as a model host for the *Burkholderia cepacia* complex. *PLoS One.* 5:e11467.
116. Thomson, E. L., and J. J. Dennis. 2012. A *Burkholderia cepacia* complex non-ribosomal peptide-synthesized toxin is hemolytic and required for full virulence. *Virulence.* 3:286–298.
117. Ferreira, A. S., et al. 2013. Comparative transcriptomic analysis of the *Burkholderia cepacia* tyrosine kinase *bceF* mutant reveals a role in tolerance to stress, biofilm formation, and virulence. *Appl. Environ. Microbiol.* 79:3009–3020.
118. Silva, I. N., et al. 2011. Mucoid morphotype variation of *Burkholderia multivorans* during chronic cystic fibrosis lung infection is correlated with changes in metabolism, motility, biofilm formation and virulence. *Microbiology.* 157:3124–3137.
119. Silva, I. N., A. C. Tavares, A. S. Ferreira, and L. M. Moreira. 2013. Stress conditions triggering mucoid morphotype variation in *Burkholderia* species and effect on virulence in *Galleria mellonella* and biofilm formation in vitro. *PLoS One.* 8:e82522.
120. Seed, K. D., and J. J. Dennis. 2009. Experimental bacteriophage therapy increases survival of *Galleria mellonella* larvae infected with clinically relevant strains of the *Burkholderia cepacia* complex. *Antimicrob. Agents Chemother.* 53:2205–2208.
121. Köthe, M., et al. 2003. Killing of *Caenorhabditis elegans* by *Burkholderia cepacia* is controlled by the *cep* quorum-sensing system. *Cell Microbiol.* 5:343–351.
122. Huber, B., et al. 2004. Identification of a novel virulence factor in *Burkholderia cenocepacia* H111 required for efficient slow killing of *Caenorhabditis elegans. Infect. Immun.* 72:7220–7230.
123. Laws, T. R., S. V. Harding, M. P. Smith, T. P. Atkins, and R. W. Titball. 2004. Age influences resistance of *Caenorhabditis elegans* to killing by pathogenic bacteria. *FEMS Microbiol. Lett.* 234:281–287.
124. Cooper, V. S., W. A. Carlson, and J. J. Lipuma. 2009. Susceptibility of *Caenorhabditis elegans* to *Burkholderia* infection depends on prior diet and secreted bacterial attractants. *PLoS One.* 4:e7961.
125. Ellis, C. N., and V. S. Cooper. 2010. Experimental adaptation of *Burkholderia cenocepacia* to onion medium reduces host range. *Appl. Environ. Microbiol.* 76:2387–2396.
126. Cardona, S. T., J. Wopperer, L. Eberl, and M. A. Valvano. 2005. Diverse pathogenicity of *Burkholderia cepacia* complex strains in the *Caenorhabditis elegans* host model. *FEMS Microbiol. Lett.* 250:97–104.

127. Springman, A. C., et al. 2009. Genetic diversity and multihost pathogenicity of clinical and environmental strains of *Burkholderia cenocepacia*. *Appl. Environ. Microbiol.* 75:5250–5260.
128. Reddy, K. C., R. C. Hunter, N. Bhatla, D. K. Newman, and D. H. Kim. 2011. *Caenorhabditis elegans* NPR-1-mediated behaviors are suppressed in the presence of mucoid bacteria. *Proc. Natl. Acad. Sci. USA*. 108:12887–12892.
129. Markey, K. M., K. J. Glendinning, J. A. Morgan, C. A. Hart, and C. Winstanley. 2006. *Caenorhabditis elegans* killing assay as an infection model to study the role of type III secretion in *Burkholderia cenocepacia*. *J. Med. Microbiol.* 55:967–969.
130. Law, R. J., J. N. Hamlin, A. Sivro, S. J. McCorrister, G. A. Cardama, and S. T. Cardona. 2008. A functional phenylacetic acid catabolic pathway is required for full pathogenicity of *Burkholderia cenocepacia* in the *Caenorhabditis elegans* host model. *J. Bacteriol.* 190:7209–7218.
131. Sousa, S. A., C. G. Ramos, L. M. Moreira, and J. H. Leitão. 2010. The *hfq* gene is required for stress resistance and full virulence of *Burkholderia cepacia* to the nematode *Caenorhabditis elegans*. *Microbiology*. 156:896–908.
132. Ramos, C. G., S. A. Sousa, A. M. Grilo, J. R. Feliciano, and J. H. Leitão. 2011. The second RNA chaperone, Hfq2, is also required for survival under stress and full virulence of *Burkholderia cenocepacia* J2315. *J. Bacteriol.* 193:1515–1526.
133. Ramos, C. G., S. A. Sousa, A. M. Grilo, L. Eberl, and J. H. Leitão. 2010. The *Burkholderia cenocepacia* K56–2 pleiotropic regulator Pbr, is required for stress resistance and virulence. *Microb. Pathog.* 48:168–177.
134. Ramos, C. G., A. M. Grilo, P. J. da Costa, J. R. Feliciano, and J. H. Leitão. 2013. MtvR is a global small noncoding regulatory RNA in *Burkholderia cenocepacia*. *J. Bacteriol.* 195:3514–3523.
135. Somvanshi, V. S., P. Viswanathan, J. L. Jacobs, M. H. Mulks, G. W. Sundin, and T. A. Ciche. 2010. The type 2 secretion pseudopilin, *gspJ*, is required for multihost pathogenicity of *Burkholderia cenocepacia* AU1054. *Infect. Immun.* 78:4110–4121.
136. Agnoli, K., et al. 2012. Exposing the third chromosome of *Burkholderia cepacia* complex strains as a virulence plasmid. *Mol. Microbiol.* 83:362–378.
137. Agnoli, K., et al. 2014. The third replicon of members of the *Burkholderia cepacia* complex, plasmid pC3, plays a role in stress tolerance. *Appl. Environ. Microbiol.* 80:1340–1348.
138. Vergunst, A. C., A. H. Meijer, S. A. Renshaw, and D. O'Callaghan. 2010. *Burkholderia cenocepacia* creates an intramacrophage replication niche in zebrafish embryos, followed by bacterial dissemination and establishment of systemic infection. *Infect. Immun.* 78:1495–1508.
139. Ko, S. D. 1985. Growth and toxin production of *Pseudomonas cocovenenans*, the so-called "bongkrek bacteria". *ASEAN Food J.* 1:78–84.
140. Zhao, N., C. Qu, E. Wang, and W. Chen. 1995. Phylogenetic evidence for the transfer of *Pseudomonas cocovenenans* (van Damme et al. 1960) to the genus *Burkholderia* as *Burkholderia cocovenenans* (van Damme et al. 1960) comb. nov. *Int. J. Syst. Bacteriol.* 45:600–603.
141. van Veen, A. G., and W. K. Mertens. 1933. On the isolation of a toxic bacterial pigment. *Proc. Acad. Sci. Amsterdam*. 36:666–670.
142. Nugteren, D. H., and W. Berends. 1957. Investigations on bongkrekic acid, the toxine from *Pseudomonas cocovenenans*. *Rec. Trav. Chim. Pays-Bas*. 76:13–27.
143. van Damme, P. A., A. G. Johannes, H. C. Cox, and W. Berends. 1960. On toxoflavin, the yellow poison of *Pseudomonas cocovenenans*. *Rec. Trav. Chim. Pays-Bas*. 79:255–267.
144. Cox, J., E. Kartadarma, and K. Buckle. 2000. *Burkholderia cocovenenans*, pp. 1871–1875. In R. K. Robinson, C. A. Batt, and P. D. Patel (eds.), *Encyclopedia of Food Microbiology*. Academic Press, San Diego.
145. Jin, J. X. 1963. A preliminary study on the pathogenesis of fermented corn meal food poisoning. *Prev. Hyg.* 41:21–26. *Cited in* Zhao, N. X., M. S. Ma, Y. P. Zhang, and D. C. Xu. 1990. Comparative description of *Pseudomonas cocovenenans* (van Damme, Johannes, Cox, and Berends 1960) NCIB 9450T and strains isolated from cases of food poisoning caused by consumption of fermented corn flour in China. *Int. J. Syst. Bacteriol.* 40:452–455.
146. Meng, Z., et al. 1988. Studies on fermented corn flour food poisoning in rural areas of China. II. Isolation and identification of causal microorganisms. *Biomed. Environ. Sci.* 1:105–114.
147. Zhao, N. X., M. S. Ma, N. F. Miao, and Y. P. Zhang. 1988. Comparative study on *Pseudomonas cocovenenans* and *Pseudomonas farinofermentans*. *Chin. J. Microbiol. Immunol.* 8:151–156.

148. Zhao, N. X., M. S. Ma, Y. P. Zhang, and D. C. Xu. 1990. Comparative description of *Pseudomonas cocovenenans* (van Damme, Johannes, Cox, and Berends 1960) NCIB 9450^T and strains isolated from cases of food poisoning caused by consumption of fermented corn flour in China. *Int. J. Syst. Bacteriol.* 40:452–455.
149. Yabuuchi, E., et al. 1992. Proposal of *Burkholderia* gen. nov. and transfer of seven species of the genus *Pseudomonas* homology group II to the new genus, with the type species *Burkholderia cepacia* (Palleroni and Holmes 1981) comb. nov. *Microbiol. Immunol.* 36:1251–1275.
150. Gillis, M., et al. 1995. Polyphasic taxonomy in the genus *Burkholderia* leading to an emended description of the genus and proposition of *Burkholderia vietnamiensis* sp. nov. for N_2-fixing isolates from rice in Vietnam. *Int. J. Syst. Bacteriol.* 45:274–289.
151. Vandamme, P., et al. 1997. Occurrence of multiple genomovars of *Burkholderia cepacia* in cystic fibrosis patients and proposal of *Burkholderia multivorans* sp. nov. *Int. J. Syst. Evol. Microbiol.* 47:1188–1200.
152. Coenye, T., B. Holmes, K. Kersters, J. R. W. Govan, and P. Vandamme. 1999. *Burkholderia cocovenenans* (van Damme et al. 1960) Gillis et al. 1995 and *Burkholderia vandii* Urakami et al. 1994 are junior synonyms of *Burkholderia gladioli* (Severini 1913) Yabuuchi et al. 1993 and *Burkholderia plantarii* (Azegami et al. 1987) Urakami et al. 1994, respectively. *Int. J. Syst. Bacteriol.* 49:37–42.
153. Christenson, J. C., D. F. Welch, G. Mukwaya, M. J. Muszynski, R. E. Weaver, and D. J. Brenner. 1989. Recovery of *Pseudomonas gladioli* from respiratory tract specimens of patients with cystic fibrosis. *J. Clin. Microbiol.* 27:270–273.
154. Kennedy, M. P., et al. 2007. *Burkholderia gladioli*: five year experience in a cystic fibrosis and lung transplantation center. *J. Cyst. Fibros.* 6:267–273.
155. Segonds, C., T. Heulin, N. Marty, and G. Chabanon. 1999. Differentiation of *Burkholderia* species by PCR-restriction fragment length polymorphism analysis of the 16S rRNA gene and application to cystic fibrosis isolates. *J. Clin. Microbiol.* 37:2201–2208.
156. Khan, S. U., S. M. Gordon, P. C. Stillwell, T. J. Kirby, and A. C. Arroliga. 1996. Empyema and bloodstream infection caused by *Burkholderia gladioli* in a patient with cystic fibrosis after lung transplantation, *Pediatr. Infect. Dis. J.* 15:637.
157. Shin, J. H., S. H. Kim, M. G. Shin, S. P. Suh, D. W. Ryang, and M. H. Jeong. 1997. Bacteremia due to *Burkholderia gladioli*: case report, *Clin. Infect. Dis.* 25:1264.
158. Machlowitz, R. A., W. P. Fisher, B. S. McKay, A. A. Tytell, and J. Charney. 1954. Xanthothricin, a new antibiotic. *Antibiot. Chemother.* 4:259–261.
159. Sato, Z., Y. Koiso, S. Iwasaki, I. Matsuda, and A. Shirata. 1989. Toxins produced by *Pseudomonas glumae. Ann. Phytopathol. Soc. Jpn.* 55:353–356. *Cited in* Maeda, Y., et al. 2006. Phylogenetic study and multiplex PCR-based detection of *Burkholderia plantarii*, *Burkholderia glumae* and *Burkholderia gladioli* using gyrB and rpoD sequences. *Int. J. Syst. Evol. Microbiol.* 56:1031–1038.
160. Jeong, Y., J. Kim, S. Kim, Y. Kang, T. Nagamatsu, and I. Hwang. 2003. Toxoflavin produced by *Burkholderia glumae* causing rice grain rot is responsible for inducing bacterial wilt in many field crops. *Plant Dis.* 87:890–895.
161. Lijmbach, G. W. M., H. C. Cox, and W. Berends. 1970. Elucidation of the chemical structure of bongkrekic acid. I. Isolation, purification and properties of bongkrekic acid. *Tetrahedron.* 26:5993–5999.
162. Buckle, K. A., and E. Kartadarma. 1990. Inhibition of bongkrek acid and toxoflavin production in tempe bongkrek containing *Pseudomonas cocovenenans*. *J. Appl. Bacteriol.* 68:571–576.
163. Moebius, N., C. Ross, K. Scherlach, B. Rohm, M. Roth, and C. Hertweck. 2012. Biosynthesis of the respiratory toxin bongkrekic acid in the pathogenic bacterium *Burkholderia gladioli. Chem. Biol.* 19:1164–1174.
164. Meng, Z., et al. 1988. Studies on fermented corn flour poisoning in rural areas of China. I. Epidemiology, clinical manifestations, and pathology. *Biomed. Environ. Sci.* 1:101–104.
165. van Veen, A. G. 1966. The bongkrek toxins, pp. 43–50. In R. I. Mateles, and G. N. Wogan (eds.), *Biochemistry of Some Foodborne Microbial Toxins*. MIT Press, Cambridge, UK.
166. Hu, W. J., G. S. Zhang, and F. S. Chu. 1984. Purification and partial characterization of flavotoxin A. *Appl. Environ. Microbiol.* 48:690–693.

167. Hu, W. J., X. M. Chen, H. D. Meng, and Z. H. Meng. 1989. Fermented corn flour poisoning in rural areas of China. III. Isolation and identification of main toxin produced by causal microorganisms. *Biomed. Environ. Sci.* 2:65–71.
168. Henderson, P. J., and H. A. Lardy. 1970. Bongkrekic acid: an inhibitor of the adenine nucleotide translocase of mitochondria. *J. Biol. Chem.* 245:1319–1326.
169. Zamzami, N., et al. 1996. Inhibitors of permeability transition interfere with the disruption of the mitochondrial transmembrane potential during apoptosis. *FEBS Lett.* 384:53–57.
170. Schwerdt, G., R. Freudinger, C. Schuster, S. Silbernagl, and M. Gekle. 2003. Inhibition of mitochondria prevents cell death in kidney epithelial cells by intra- and extracellular acidification. *Kidney Int.* 63:1725–1735.
171. Welling, W., J. A. Cohen, and W. Berends. 1960. Disturbance of oxidative phosphorylation by an antibioticum produced by *Pseudomonas cocovenenans*. *Biochem. Pharmacol.* 3:122–135.

19

Campylobacter

Martin Stahl and Bruce A. Vallance

CONTENTS

19.1 Introduction

Campylobacter jejuni is a Gram-negative, curved-rod shaped, microaerophilic, enteric bacterial pathogen, and one of the leading causes of gastroenteritis in both the developing and the developed world. This highly infectious bacterium is often acquired through the consumption of contaminated water, raw milk, or meat products (particularly contaminated poultry).[1] Once infected, an individual's symptoms can vary significantly. Milder cases can be largely asymptomatic or involve watery diarrhea. In more severe cases, fever, bloody diarrhea, and abdominal pain can be evident. In most cases, the disease resolves within a few days or in up to 2 weeks; however, under rare and poorly understood circumstances, the infection can trigger the development of autoimmune disorders such as reactive arthritis or Guillain–Barré syndrome,[1] or other complications such as postinfective irritable bowel syndrome.[2] Despite the broad prevalence of this pathogen, the mechanisms by which this organism causes disease in humans remain poorly understood.

Aside from assessing the severity of symptoms, the details of how *C. jejuni* triggers disease symptoms remain poorly defined. We know that a very small infectious dose in humans is capable of transiting the stomach and establishing infection, largely within the colon. Pathogen burdens can be very high, although they drop significantly as the adaptive immune response becomes engaged. The infection usually responds to antibiotic treatment, such as either fluoroquinolones (e.g., ciprofloxacin) or macrolides (e.g., erythromycin), although antibiotic resistance rates in *C. jejuni* have been steadily increasing in recent years.[3] The nature of intestinal colonization by *C. jejuni* in humans is poorly characterized, although numerous *in vitro* studies, as well as *in vivo* animal work, would suggest it is primarily localized to intestinal crypts as well as to the mucus layer that overlies the mucosa. Moreover it is capable of adhering to the intestinal epithelial surface and translocating across the epithelium.[1] Cell invasion by *C. jejuni* has been indicated *in vitro*, *in vivo*, and in clinical samples[1,4]; however, what role cell invasion plays in *C. jejuni* pathogenesis remains unknown. The spread of *C. jejuni* into the blood stream and to systemic sites has been documented in some patients, but it remains relatively atypical.[5] Additionally, long-term complications such as reactive arthritis and Guillain–Barré syndrome are rare and result from antigen mimicry between *C. jejuni* lipooligosaccharides and host gangliosides. To date, no proven animal models exist that mimic these long-term complications.

One of the difficulties in improving our understanding of campylobacteriosis has been the fact that *C. jejuni* behaves quite differently depending on its surroundings. In the wider environment, birds are the main reservoirs for the organism and its preferred hosts[1,6]; however, *C. jejuni* remains a relatively harmless commensal in avian hosts, even as it is capable of growing to high numbers in birds' cecum.[6] Various *Campylobacter* species, including those that are infectious to humans such as *C. jejuni*, are commonly recovered from wild birds,[7,8] although the main reservoirs that serve as a source for human infection are poultry farms.[9] In the densely packed environment of a modern poultry barn, any *Campylobacter* contamination can quickly spread between birds until the entire flock becomes colonized, making *Campylobacter* contamination a significant health concern within the poultry industry.[10] Precisely how *C. jejuni* is able to spread so rapidly and so easily has been somewhat of a mystery, especially considering its rather fastidious growth requirements, including warm, low-oxygen conditions. Evidence suggests that flies[11] or wild birds[6] may serve as sources of contamination, while other sources include runoff from other livestock such as pigs or cattle,[10,12] which themselves can become asymptomatically colonized.[12] Coprophagous behavior of chickens, along with a high density of birds can quickly spread contamination throughout a flock.[13] It has also been suggested that amoeba may serve as a temporary host for *Campylobacter*, allowing for its prolonged survival within contaminated water sources.[14,15] Additional survival mechanisms within the *C. jejuni* cells themselves, such as their observed ability to convert to a nonculturable coccoid form, may also aid in prolonging their survival in the low-nutrient, high-oxygen conditions commonly found in the environment outside of a host,[16] but these mechanisms remain poorly understood.

Besides the commensal-like colonization of poultry, few other animals are susceptible to *C. jejuni* colonization and infection in a fashion similar to human campylobacteriosis.[17] Following exposure to *C. jejuni*, most animals either exhibit only brief, transient colonization, or are colonized asymptomatically. An exception to this involves a connection between spontaneous abortion in sheep and other livestock, and infection by certain *Campylobacter* species, including *Campylobacter fetus* and *C. jejuni*.[18] Humans remain one of the relatively few hosts that exhibit enteric disease in response to exposure to *C. jejuni*, making the use of animal models in the study of *C. jejuni* infection more problematic. Those animal models that are employed usually come with the caveats that they do not entirely replicate the human disease (e.g., poultry and insect models), or that they require some significant modification of the host, such as the construction of mice deficient in key genes designed to make them more susceptible to infection. Some of the more promising infection models currently being used include neonatal piglets, ferrets, and certain knockout mouse strains, whereas models of commensal colonization are typically newly hatched chicks and wild-type mice. Table 19.1 summarizes the current animal models available, along with a brief description of each. In this chapter, we will outline the animal models currently employed for the study of *C. jejuni* in published research, and discuss the advantages and drawbacks of each one, while providing an overview as to what the research community has learned about *C. jejuni* infection from these research models.

19.2 General Features of *C. jejuni* Colonization of Animal Models

Before discussing each animal model in detail, we will start by giving an overview of several common characteristics of *C. jejuni* colonization. Firstly, unless forced into an unusual mode of infection, such as via intraperitoneal (IP) injection for example, the standard route of infection for *C. jejuni* is by oral inoculation. Following inoculation, *C. jejuni* proceeds quickly to the intestine, where it can establish itself. Although there are *Campylobacter* species that can colonize the mouth,[19] and the relatively closely related *Helicobacter pylori* primarily lives within the stomach,[20] *C. jejuni* does not typically remain for long in the upper regions of the GI tract. Even in the lower portions of the small intestine, it does not generally reach high numbers, even if it can establish itself as a stable population. Rather, it proceeds to the large bowel or cecum, which also supports the bulk of the intestinal commensal microbiota. In animals with a cecum (birds, rodents, and pigs), this is the primary site of *C. jejuni* colonization, with

TABLE 19.1

Animal Models for the Study of *Campylobacter jejuni* Infections

Animal Model	Colonization	Pathology	References
Chick	Strong commensal-like colonization of the cecum and colon, with a very small inoculum	Little to no pathology	21,35,49,50–56,60
Piglet	Colonization of the small intestine, cecum, and colon	Significant variation in pathology. Inflammation, distension with gas, and immune cell infiltration	37,42,64,6–71
Ferret	Colonization principally in the colon	Significant inflammation and immune cell infiltration in colonized areas	22,72,75–78
Galleria	No real colonization, although *C. jejuni* may replicate and spread following injection	The mortality and melanization of infected larvae is an indicator of pathology	80,81
Wild-type mouse	Inconsistent colonization	Minimal and inconsistent pathology	84–91
Antibiotic-pretreated or germfree mouse	Consistent colonization of the cecum and colon	Minimal pathology in wild-type mice	26,86,88
IL-10$^{-/-}$ mouse	Consistent colonization of the cecum and colon	Chronic infection, severe inflammation, and immune cell infiltration in the colon and cecum	99,102–107
SIGIRR$^{-/-}$ mouse	Consistent colonization of the cecum and colon	Acute infection, moderate inflammation, and immune cell infiltration	26

the microbes reaching very high numbers (often over 10^9) of colony forming units (CFUs) per gram of luminal content.[21] In animals without a developed cecum (ferrets) and in humans, the primary site of colonization becomes the colon.[22,23]

Focusing more closely, current research suggests that *C. jejuni*'s preferred location within the intestine is the mucus layer,[24,25] which segregates the host epithelium from the bacteria and other contents of the lumen. Histological images of a number of *C. jejuni* animal models, including both chicks[21] and mice,[26] have identified *C. jejuni* as being localized primarily within the mucus layer, particularly within the mucus-filled crypts that line the intestine. Mechanistically, this observation fits with several known aspects of *C. jejuni* colonization and cell biology. Firstly, *C. jejuni* is microaerophilic, meaning it requires low levels of oxygen to grow. This reflects it having only a single class I-type ribonucleotide reductase, which requires oxygen to function.[27] However, *C. jejuni* is also dependent on numerous metabolic enzymes which contain oxygen-sensitive iron–sulfur complexes.[28] Together, these make it unable to survive for extended periods under anaerobic conditions (such as the intestinal lumen), or under higher oxygen conditions. This makes the low oxygen conditions of the intestinal lining its ideal environment as it is distant from the anaerobic lumen, where the commensal anaerobes thrive, yet close enough to be able to access the diffuse oxygen from the highly vascularized epithelium. Furthermore, early research on *C. jejuni* chemotaxis indicated that it was attracted to mucins,[29] which are the primary glycoprotein component of intestinal mucus. Additionally, at least some strains are chemotactic toward L-fucose,[29] a common terminal glycan on mucins. The current hypothesis is that *C. jejuni* recognizes mucin 2 as a chemoattractant to find its way to the mucus layer, using its flagellar motility and corkscrew shape to push its way through intestinal contents to reach its destination.[30,31] In all animal models where the intestine is colonized, mutants in chemotaxis or flagella are either completely or severely impaired for colonization, making motility toward the mucus layer one of the most critical factors for colonization.[32,33]

The second critical factor that governs *C. jejuni* colonization, in almost any host it encounters, is how it deals with the common stresses it almost universally encounters. For example, in a mammalian or avian digestive system, orally inoculated *C. jejuni* would be exposed to the low pH of the stomach, antimicrobial effects of bile salts, oxidative stress, nitrosative stress, competition with the host and microbiota for resources, not to mention the innate immune defenses of the host organism. While many of these stressors may depend on the specific host and situation in which *C. jejuni* finds itself, and thus require more of an explanation than can be addressed here, a few common mechanisms appear to be present in most potential animal hosts that *C. jejuni* can colonize. For example, mutants in the CmeABC multidrug efflux pump, as well as several related pumps and regulators are more sensitive to a number of factors, including bile salts, heavy metals, antimicrobials, and antibiotics, and are significantly impaired for survival inside a living host.[34] Oxidative stress is also a major factor at several stages of colonization, and deletion mutants in key oxidative-stress-related genes including *katA*, *cj1386*, *sodB*, *ahpC*, as well as strains lacking other defense and regulatory genes are severely impaired in gut colonization in a number of *C. jejuni* animal models.[35,36]

Finally, the acquisition of nutrients is a common feature of *C. jejuni*'s survival in almost any environment. A curious feature of *C. jejuni* is its relatively limited repertoire of potential carbon sources. With the exception of certain strains that have acquired an L-fucose metabolic pathway,[37] *C. jejuni* is limited to only a handful of amino acids, 4C-dicarboxylates, and a few molecules that can be easily plugged into the citric acid cycle.[38,39] This makes mutants in almost any of these key metabolic pathways exceptionally poor colonizers, due to a lack of alternate metabolic pathways to draw upon.[40,41] In particular, mutants impaired in the acquisition or metabolism of two of its preferred amino acids, glutamate and serine, are exceptionally poor colonizers.[40,41] Furthermore, the acquisition of a number of metals and other key micronutrients, especially iron, is critical for *C. jejuni*, particularly in a competitive environment such as an animal host. In the case of iron, which is tightly regulated by the host and required by the microbiota, the loss of any of *C. jejuni*'s major iron acquisition pathways or their regulators can result in a substantial decrease in its colonization potential.[42,43]

19.2.1 Cellular Models of Infection

Simple cellular infection models strip away most of the complexity of the whole organism, and allow for the precise study of cellular–bacterial interactions. Due to their simplicity and availability, cellular infection models have been one of the preferred methods for the study of *C. jejuni*. They remain useful, particularly for the study of *C. jejuni* interactions with host cells, but the degree to which they model host colonization and infection is debatable. The most common assays for the study of *C. jejuni* in cellular systems are adherence or invasion assays of colonic epithelial cell lines. The preferred cell lines are usually the polarized Caco-2 or HT29 cells[44]; however, nonpolarized INT 407 cells have been frequently used as well,[45] along with other varieties of established cell lines, depending on the experimental design. The standard assay involves either measuring bacterial adherence to the host cell surface, or using a gentamicin protection assay to quantify the number of internalized bacteria. Increases in either adherence or invasion are usually taken as a sign of increased virulence; however, precise links between these *in vitro* results and infection *in vivo* can often be difficult to establish.[46]

Upon contact with most intestinal epithelial cells, including chicken, mouse, and human cell lines,[47] *C. jejuni* will readily adhere, invade, and persist; however, no intracellular replication has been previously observed.[48] These experimental models are also useful for establishing the mechanisms of cell invasion, as well as for measuring cytokine expression in response to *C. jejuni*, although precisely how it extrapolates to infection of human or animal intestines remains a matter of some debate.

19.2.2 Three-Day-Old Chick Model

One of the most prominently used animal models for *C. jejuni* colonization has been newly hatched chicks.[49] Although early studies indicated some signs of diarrhea in chicks if infections were established early enough (<1 day) after hatching,[50,51] it is generally accepted that the colonization of chicks by *C. jejuni* is commensal in nature, with little detriment to the host.[21,52,53] An infectious dose of approximately 10^7–10^8 CFUs is considered more than sufficient to guarantee infection,[21] with even significantly lower numbers (10^4–10^6 CFUs) often being sufficient.[54] The common age of chicks used for this model is between 1- and 3-days after hatching, although older chicks have been used as well.[54] After inoculation, most studies have found that chicks remain colonized for a few days to a few weeks.[35,54] The number of

CFUs per gram of luminal contents remains relatively constant over this period in the absence of any external manipulation.[21] Any limitations on the duration of an experiment are dependent on the facility and the growth of the birds. While infected chicks are quite small at the beginning of the experiment, they will grow and gain weight rapidly, making housing and handling more problematic over time.

While it has been demonstrated that *C. jejuni* will quickly establish itself within the GI tract of an infected bird, our knowledge of exactly how *C. jejuni* colonizes chicks is relatively limited. The main site of colonization is the cecum and to a lesser extent the cloaca,[54] with relatively low numbers found throughout the rest of the avian GI tract, although there has been little research performed to determine whether other locations of the GI tract can be colonized, or if *C. jejuni* recovered from these regions is just a transient population. Interestingly, several early studies were conducted to better characterize how *C. jejuni* established itself within the cecum. Beery et al. compared histological sections of infected and uninfected 8-day-old chicks.[21] Their primary observation was the presence of large numbers of *C. jejuni* at the base of the chick cecal crypts. A closer examination using transmission electron microscopy clearly enabled the visualization of the helical *C. jejuni* cells in the mucus layer filling the crypts, but they were not associated closely with the microvilli or glycocalyx, suggesting that they were not adhering to the intestinal epithelial cells themselves. In contrast, other studies have identified cell adhesion not only taking place in the chick cecum, but also being an important step in proper colonization.[55] Whether or not cell invasion takes place in the chicken cecum *in vivo* remains uncertain owing to conflicting studies and evidence[53,56]; however, it has been observed to occur in chicken cells *in vitro*.[57] This has led to the belief that *C. jejuni* has the capacity to invade chick cecal epithelial cells; however, additional factors may intervene to prevent this from occurring routinely in an *in vivo* situation.[58] Notwithstanding the lack of conclusive evidence, the cecal mucus layer has been implicated as a modulator of *C. jejuni* pathogenicity. In cell culture experiments, the addition of chicken cecal mucus has been found to reduce cell invasion of both chicken and human epithelial cells, although a mechanism to explain this observation remains unknown.[57,58] It has been suggested that cell invasion may still take place *in vivo*, but infrequently, as an explanation for observations regarding the stimulation of the avian immune response; however, this remains only a hypothesis.[56]

Conflicting information also exists as to whether *C. jejuni* can be found outside of the chick intestinal tract. Some studies have not recovered any *C. jejuni* from other sites throughout the body following inoculation,[21] however, other studies have been able to recover *C. jejuni* from other organs, including the spleen and liver up to 7 days postinoculation.[47] Despite the presence of *C. jejuni* at systemic sites, the chick immune system remains largely unresponsive to *C. jejuni*, with virtually no signs of disease development.[53] The degree to which the chick immune system reacts to *C. jejuni* remains largely unknown, despite the number of studies that have attempted to answer this question. It is of particular interest for researchers attempting to develop a vaccine against *C. jejuni* chicken colonization, which would rely heavily on the ability to stimulate the chicken's own immune system to recognize and clear the colonizing *C. jejuni*, something the immune system does not seem inclined to do under normal circumstances.[59]

Despite the seeming lack of an immune response to *C. jejuni*, evidence would suggest that the chick immune system is not entirely oblivious to the presence of *C. jejuni*. A transcriptomic study identified 270 genes differentially regulated in chicks inoculated by *C. jejuni*, as compared to uninfected controls, with some of these genes involved in inflammatory pathways.[60] Furthermore, studies have identified chTLR4 and chTLR21 expressions as being particularly sensitive to *C. jejuni* colonization.[61,62] Evidence suggests that prior to exposure to *C. jejuni*, the avian innate immune system remains at least somewhat sensitive to *C. jejuni*. Upon initial exposure to *C. jejuni*, there is a measurable spike in production of the chicken equivalent of the human chemokine IL-8 (CXCLi1 and CXCLi2), accompanied by an increase in the numbers of monocytes/macrophages within the cecal tissues.[63] Curiously, this does not lead to an increase in heterophil numbers (roughly the equivalent of a human neutrophil), as one would expect with IL-8. This increase in a key proinflammatory chemokine appears to be temporary, and immune tolerance to *C. jejuni* appears to develop,[53] allowing for a high pathogen burden, but virtually no immune/inflammatory response.

In spite of the lack of immunological responses that resemble human disease, the chick has been widely used as a model for *C. jejuni* colonization. This remains quite relevant due to similarities in basic colonization factors, even in the absence of an immune response, as well as the primary role of poultry as a common source of human infection, making the colonization and spread of *C. jejuni* in chickens relevant in its own right.

19.2.3 Colostrum-Deprived Piglets

The use of colostrum-deprived piglets as a model for *C. jejuni* infection was first defined by Babakhani et al.[64] Although conventionally reared piglets may be susceptible to *Campylobacter* infection early in life, maternal antibodies are usually sufficient to prevent infection. Adult pigs are also known to be frequent carriers of both *C. jejuni* and *Campylobacter coli*[65]; however, the development of symptoms of gastroenteritis in response to these pathogens is rare. Colostrum-deprived piglets lack the protection of maternal antibodies that usually helps prevent infection in piglets. Additionally, at very young ages, they lack a fully mature immune response or a protective, fully developed intestinal microbiota, making them particularly susceptible to infections by enteric pathogens. This vulnerability has been previously exploited for the study of *Escherichia coli*[66] and *Salmonella*[67] infection, among others, and in a number of studies, it has been successfully used as a model for *C. jejuni* infection.[37,42,68]

In the initial characterization of the model by Babakhani et al., the M129 strain of *C. jejuni* was orally inoculated into newborn, colostrum-deprived piglets.[64] Symptomology was somewhat varied, but all the piglets became colonized, and exhibited symptoms including diarrhea, with blood and mucus in their stool. These symptoms were often evident within a day of inoculation, and pathogen burden and disease severity peaked within 3 days postinoculation. This first study is also the only study to undertake a histological analysis of infected piglets. In their analyses, they found signs of inflammation, goblet cell depletion, and immune cell infiltration primarily in the cecum and colon, with minimal signs of damage in the small intestine. Electron microscopy also successfully identified *C. jejuni* cells adherent to the epithelial surface as well as bacteria internalized within epithelial cells.

Continuing studies using the colostrum-deprived model have focused primarily on pathogen burdens, with less attention paid to histology or immune responses within the host piglets. The most common application has been to compare changes in pathogen burden between wild-type *C. jejuni* and isogenic mutants lacking key pathogenicity genes. These have included genes related to nutrient uptake,[37,42] stress-response regulators,[68,69] capsule-modifying enzymes,[70] as well as other genes of interest.[71] In most cases, the strain of choice has been the commonly used NCTC11168 lab strain, which induces limited pathology and mortality in infected piglets.[42] Impaired pathogenesis in a mutant strain is usually concluded based on a decrease in pathogen burdens either by comparison to control piglets infected with the wild-type strain, or through a competitive index from a piglet infected with both the mutant and wild-type strains. In contrast to the NCTC11168 strain, the 81-176 strain is much more pathogenic in the piglet model,[70] and is less frequently used. On the occasions it has been studied, piglets often suffer high mortality within 1–2 days postinoculation.

19.2.4 Ferret Model

Although they are a relatively uncommon and difficult animal species to use as a laboratory animal infection model, domestic ferrets have been used in several studies as a model for human *C. jejuni* infection.[22,72] The first use of these animals as a model for *C. jejuni* infection was published by Fox et al.[22] Prior to this, *Campylobacter*-like organisms (likely *Helicobacter*), had been recovered from ferret gastric lesions,[73] leading to their common use as an animal model in the early study of *Helicobacter* infection.[74] *Campylobacter* infection was further described using domesticated ferrets, and has subsequently been used in several studies to assess the pathogenicity of isogenic *C. jejuni* mutant strains. Recently, a more detailed evaluation of the immune responses and histopathology of *C. jejuni* infection in ferrets has been published.[75]

Despite the difficulty in using ferrets as an animal model of infection, including cost, availability, and a relative lack of experimental tools and expertise relating to ferrets, the ferret model is attractive due to the ability of *C. jejuni* to trigger an acute, self-limiting diarrheal disease without substantial manipulations of the host to render them more susceptible to infection.[22] The exception to this is the need to anaesthetize the animals prior to inoculation, and in some studies treatment with sodium bicarbonate was used prior to inoculation to reduce stomach acidity, and thereby increase the survival of *C. jejuni* through the stomach.[72,75–78] Additionally, some studies used a postinoculation treatment with opium to reduce intestinal peristalsis to give more time for the infection to take hold.

As an infection model, the ferret remains somewhat understudied due to the relatively few researchers who have actually employed it. The initial characterization of the model described an acute diarrheal disease, with mild to moderate diarrhea observed within the first few days postinoculation.[72] Symptoms resolved within a few days, and the infection was cleared within 1–2 weeks. A more detailed description of the ferret immune response to *C. jejuni* was published in 2009.[75] Researchers found a relatively high pathogen burden (10^7–10^{10} CFUs/g) and more pronounced pathology in the colon relative to the small intestine. The pathology observed included substantial increases in immune cell infiltration into the mucosa, and the detection of lactoferrin and blood in the stool. Elevated fecal and systemic IgG levels were detected and the authors linked their presence to the successful clearance of the infection. Furthermore, immunohistochemistry and electron microscopy detected *C. jejuni* adherent to the surface of epithelial cells, with possible evidence of epithelial cell invasion.

19.2.5 *Galleria mellonella*

One of the less intuitive animal models for *C. jejuni* involves the use of the larva of the greater wax moth, *Galleria mellonella*. The use of this animal model has so far been uncommon, and the mechanisms behind *Galleria* infection by *C. jejuni* are poorly characterized; however, the plentiful and low-cost larvae, which are also a commonly used fish bait, make for an attractive infection model.[79,80]

These insects have long been employed for the study of a wide variety of bacterial infections. The bacterial pathogen of choice is injected into the foreleg, where it can infect the insect hemocoel. The relative pathogenicity of the inoculated bacteria is then measured by counting the numbers of resulting dead *Galleria*, accompanied by the development of a distinctive black melanization of the dead and dying larvae.[80,81] In this scenario, the pathogenicity of the bacteria is assessed based on their ability to proliferate and infect the insect's hemocytes, which function in a similar fashion to mammalian neutrophils by phagocytosing and killing engulfed bacteria through a superoxide burst. This makes the *Galleria* insect model a useful means to test a bacterium's ability to defend against a host inflammatory response comprised of neutrophils and/or superoxide production. However, given the fundamental differences between most other aspects of insect biology and a mammalian host, other insights into pathogenesis will likely be limited.

Precisely what can be learned about *C. jejuni* infection using this model is probably limited in its scope. The first uses of this model were published by Champion et al.[81] and by Senior et al.[80] After testing a variety of *C. jejuni* strains and deletion mutants, the mutants showing the largest attenuation in the insect model were those in O-methyl phosphoramidate capsule modification, with mutants defective in the MeOPN capsule modification gene *cj1416* being largely nonvirulent.[82] This observation was supported by further studies by van Alphen et al.[70] An additional study by Gundogdu et al. assessed how a mutant in an oxidative stress regulatory protein (*cj1556*) fared in the insect host, and found them to be much less virulent.[83] Together, these previous studies suggest that assessing bacterial capsule function as well as susceptibility to stresses may be the most effective use of the *C. jejuni* insect model.

19.2.6 Mouse Models

For researchers using animals to model human infections, mice have long been the species of choice. Mice have the advantage of being relatively small, thereby allowing one to house large numbers in a relatively small space, and thus keeping maintenance costs manageable. Mice breed and mature relatively quickly, allowing for the establishment of whole colonies within a matter of months, and for mice to be generated for new experiments relatively quickly. Finally, in many respects, mice are physiologically similar to humans, making them fairly relevant for studying human biology and diseases.

With these advantages in their favor, a variety of research tools have sprung up surrounding mouse research, which are unmatched in any other animal model. Commercial companies have been established to breed mice, providing them on order, thereby eliminating the need for all researchers to maintain their own mouse colonies. Genetic tools have allowed for the creation of mouse strains lacking almost any gene that can be feasibly knocked out. Finally, commercial reagents and assays have been developed specifically for use with mouse models. Many antibodies and diagnostic tools are tailored to mouse biology, allowing for accurate tests and experiments. Although many of the other animal models have some of these tools available for them, none comes close to matching what is available for mice.

With mice already firmly established as research tools, very early experiments on *Campylobacter* sought to apply them to existing mouse infection models. As early as the 1970s, when *Campylobacter* was just starting to be identified as a distinct genus of its own, some initial characterizations of *Campylobacter* spp. colonization in mice were performed.[84] Results from these experiments were often limited and conflicting due largely to a lack of standard tools and techniques at the time. Field et al. reported susceptibility in neonatal mice to intragastric *Campylobacter* infection;[85] however, adult mice were found to be resistant to colonization and infection, unless pretreated with antibiotics.[86] Conversely, using adult HA-ICR mice and a high inoculating dose, Blaser et al. successfully colonized the mice and observed some degree of colonic inflammation and anti-*Campylobacter* IgG production.[87] As further studies were performed, results remained mixed using mice as models for *C. jejuni* infection. Germfree mice, mice with limited flora, or antibiotic pretreated mice were found to be highly susceptible to colonization.[26,86,88] Untreated, conventional mice carrying a full microbiota were found to be colonized in some studies, but in others, colonization either failed or was sporadic, making reproducibility problematic.[89] Symptomology was equally inconsistent between studies. Some studies reported the development of mild to severe diarrhea, and even mortality, in infected mouse strains[90]; however, in other studies, colonization was largely asymptomatic with no significant signs of diarrhea or overt pathology.[86,88] These significant inconsistencies between studies dampened the enthusiasm for the use of mice as a model for *C. jejuni* infection, and by the end of the 1980s, conventional, wild-type mice were used only infrequently as an infection model for *C. jejuni*. Some researchers with better success at colonization continued to use mice as a simple colonization model,[91] but, the success of chicks as an animal model for colonization made them more attractive to those able to procure them. Mouse models, however, remained tempting, once again due to the plethora of tools available, so researchers began trying more creative means of infection, or began testing genetically modified knockout mouse strains.

With regards to the route of infection, IP injection and intranasal challenge have both been employed in mice. IP injection is a common technique, often applied for the study of *Salmonella* Typhimurium, or other pathogens that can cause systemic infections in mice. This technique with *Campylobacter* has been applied on several occasions in mice and it serves the purpose of exposing *C. jejuni* directly to the mouse's immune system.[92,93] Intravenous injection of *C. jejuni* in mice has also been tried, with a similar effect.[92,94] While the relevance of these infection routes has been questioned, they are not entirely artificial, as *C. jejuni* has been found on occasion to cause systemic infection in humans.[95] However, *Campylobacter*-associated bacteremia is rare, and not typical for human infection. Despite this, these models can prove useful in understanding interactions between *C. jejuni* and the mouse immune system.

The intranasal challenge model is perhaps more unusual in concept since *C. jejuni* does not normally affect or even come into contact with the respiratory system. The inspiration for this model came from the effective use of an intranasal challenge using *Shigella*, which proved effective in evaluating potential vaccine candidates.[96] In the case of *C. jejuni*, an intranasal challenge not only results in exposure of the respiratory system, but also quickly leads to *C. jejuni* colonization at systemic sites around the body within a few days, including persistent colonization of the intestine, mesenteric lymph nodes, liver, and spleen.[96] In the years since the first study utilized this technique, it has been repeated on several occasions, including being used to assess the colonization potential of the cell-adhesion-deficient Peb1A mutant and it was used to study cytokine production at colonized systemic sites.[91,97]

Some of the most promising results using mice to study *Campylobacter* infection have come from the use of knockout mice, particularly those deficient in key aspects of their immune systems. An early example of this was a study that assessed *C. jejuni* infection in both SCID- and RAG2-deficient mice, both of which are deficient in cellular and humoral immunity.[88,98] In these cases, pathogen burdens were significantly enhanced compared to immunocompetent BALB/c mice; however, in neither case did the infection trigger diarrhea or significant pathology in a reproducible fashion.

A more successful and widely used approach utilized IL-10 deficient (IL-10$^{-/-}$) mice to mimic the intestinal inflammation caused by a *C. jejuni* infection, in order to assess *C. jejuni* virulence, rather than the simple colonization potential.[99] IL-10 is one of the key cytokines that regulates inflammation.[100] Typically, once inflammation has been induced by infection or injury, the upregulation of IL-10 plays a critical role in its subsequent resolution. High levels of IL-10 can also prevent inflammation from beginning in the first place, making it an important cytokine for the control of unwanted inflammation and the collateral damage it can cause. As expected, mice deficient in this cytokine are very susceptible to developing intestinal inflammation, either from injury, a chemical trigger-like dextran sodium sulfate (DSS),[101] or in response to pathogenic bacteria such as *C. jejuni*. Several studies have used this infection model to assess the ability of different *C. jejuni* isolates from human, animal, and environmental sources to induce inflammation and pathology.[99,102–104] A number of *C. jejuni* mutants, including those lacking Campylobacter invasion antigen (Cia) proteins,[105] showed a reduced potential to trigger inflammation in these mice. Additionally, mice doubly deficient in IL-10 as well as one of the several innate immune receptors such as NOD2, TLR2, and TLR4 were recently used to link signaling by these receptors to the development of inflammation in this model.[106,107]

The increased susceptibility of the IL-10$^{-/-}$ mice has proven very useful as a means of overcoming the seemingly high immune tolerance of mice to *C. jejuni*; however, it comes with a number of complications. With IL-10 being one of the linchpins for controlling and resolving inflammation, IL-10$^{-/-}$ mice are often unable to recover from infection, making *C. jejuni* infection a chronic and ultimately terminal infection, rather than the acute, self-limiting infection typically observed in humans. Furthermore, the sensitivity of these mice to almost any inflammatory trigger often results in the spontaneous development of inflammation, likely in response to their own intestinal microbiota. To limit the development of spontaneous inflammation, IL-10$^{-/-}$ mice are typically kept under germfree conditions,[108] adding to the cost and difficulty of maintaining colonies.

Modifications to the innate immune system have also been investigated as a means of generating an improved mouse infection model for *C. jejuni*, as well as for studying the interactions between *C. jejuni* and innate immune receptors. When Watson et al. infected single and double knockouts of Nrampl and MyD88 with *C. jejuni*,[109] they found that all of them were highly susceptible to *C. jejuni* colonization, with these mouse strains carrying a much higher pathogen burden than wild-type mice, including higher numbers recovered from systemic sites following oral inoculation. However, once again, neither mouse strain developed any particular signs of inflammation in response to this higher pathogen burden.

With a MyD88 knockout being susceptible to asymptomatic colonization, we took the opposite approach, and infected mice lacking single IgG IL-1 related receptor (SIGIRR), a protein highly expressed by the intestinal epithelium, and known to act as a repressor of MyD88-dependent signaling.[26] The resulting mouse exhibits enhanced signaling via MyD88-dependent receptors such as TLRs and IL-1R. Previous work infecting these mice with *Salmonella* Typhimurium and *Citrobacter rodentium* found that their increased innate signaling impaired microbiota-dependent colonization resistance, resulting in increased susceptibility to infection along with increased pathogen burdens and the development of more severe intestinal inflammation in response to infection.[110] Unlike IL-10$^{-/-}$ mice, SIGIRR$^{-/-}$ mice do not develop spontaneous colitis, although they do maintain a slightly higher intestinal inflammatory "tone," reflecting increased basal levels of several key cytokines.[110] Despite being prone to developing more significant pathology in response to infection or DSS colitis, they are still capable of resolving inflammation after a bacterial infection has been cleared. This makes them more relevant for studying the type of acute infection that is normally associated with *C. jejuni*.

When orally inoculated with *C. jejuni*, colonized SIGIRR$^{-/-}$ mice developed noticeable signs of inflammation, primarily within the cecum and proximal colon.[26] This corresponded with the primary site of colonization for *C. jejuni*, with relatively few microbes being recovered from the small bowel or at systemic sites. Inflammation was typically characterized by inflammatory and immune cell infiltration into both the mucosa and submucosa of infected tissues, along with crypt hyperplasia, increased sloughing of epithelial cells, submucosal edema, and in more severe cases the development of ulcers and the loss of crypt structure as the epithelium was damaged/destroyed. Mortality or severe morbidity was not observed, likely due to *C. jejuni's* inability to reach systemic sites in significant numbers. The development of diarrhea was not typical, but the appearance of mucoid or soft stools was common, which differs from typical human infection, in which diarrhea is one of the most prominent sequelae.

Further investigation into *C. jejuni* infection of SIGIRR$^{-/-}$ mice linked most of the inflammatory signaling being triggered by *C. jejuni* to TLR4.[26] TLR4$^{-/-}$ and TLR4/SIGIRR double knockouts were found to display only minimal inflammation and pathology in response to the presence of *C. jejuni*. This was also fairly consistent with other results published with mice lacking both IL-10 and TLR4, where substantially decreased pathology was observed in the absence of TLR4.[106] Conversely, although TLR2 had been previously linked to *C. jejuni*-induced colitis, in SIGIRR$^{-/-}$ mice, TLR2 appeared to have a protective role, with TLR2$^{-/-}$ mice and TLR2/SIGIRR double knockouts both developing more severe colitis following *C. jejuni* infection.[26]

A common refrain among those who use mouse models of *C. jejuni* infection has been the difficulty of establishing consistent infections and colonization numbers, even in knockout mice with more susceptible immune systems. This is believed to reflect the colonization resistance provided by the mouse intestinal microbiota. While most birds, ferrets, and humans can become readily infected by *C. jejuni* without any manipulation to their intestinal microbiota, the piglet model requires a neonatal piglet, lacking a fully developed microbiota. Most other animal species, including mice, can harbor *Campylobacter* within their intestines transiently, but persistent colonization is largely limited by competition from the intestinal microbiota. In order to allow for reliable and consistent mouse colonization, some form of disruption to their intestinal microbiota is necessary. Germfree mice, or mice with a limited microflora are well-known to be very susceptible to *C. jejuni* colonization, even if they typically do not develop significant pathology.[88] The downside of germfree mice, aside from the difficulty of maintaining germfree conditions, is the key role microbiota play in the proper development of the immune system, often leaving germfree mice with an underdeveloped immune system. Fortunately, a temporary disruption of the microbiota through a single antibiotic treatment was sufficient to allow for consistent colonization. We treated mice with a single gavage dose of vancomycin 4h prior to inoculation with *C. jejuni* and achieved reliable colonization and consistent pathogen burdens following infection with a relatively modest 10^7 CFU inoculating dose.[26] Other studies have used a variety of antibiotic cocktails to achieve similar results.[86] In one study, Bereswill et al. "humanized" mice with a representative human microbiota, by clearing out the existing mouse microbiota with a cocktail of antibiotics, followed by inoculation with human fecal samples.[111] The newly "humanized" mice became easily susceptible to *C. jejuni* colonization by several different strains. The authors characterized a number of differences between the mouse and human microbial composition; however, precisely how shifts between mouse and human microbiota opened intestinal niches for *C. jejuni* colonization remains unknown. Stahl et al. also carried out a basic characterization of microbiota shifts following vancomycin pretreatment.[26] Aside from a dramatic drop in overall bacterial numbers, there were dramatic shifts in the relative proportions of Firmicutes and Bacteroidetes, however, how those shifts may influence *C. jejuni* colonization is presently unknown.

With better knowledge regarding the mechanisms underlying the colonization resistance of the murine microbiota and new means for increasing the sensitivity of mice to *C. jejuni* infection, mice may finally become a relevant infection model for *C. jejuni*. What is needed now is new research describing the interactions between *C. jejuni* and the mouse microbiota and immune system to better describe how infection occurs in mice. Once this is accomplished, we can start drawing more parallels between *C. jejuni* infection in mice and humans and employ mouse models to their full potential in understanding the human disease better.

19.3 Conclusions

Campylobacter jejuni is a common foodborne pathogen that can trigger serious diarrhea in infected patients. Proper laboratory models for infection have been difficult to establish due to *C. jejuni* being a commensal or transient colonizer in most potential animal models of infection. Cellular infection models, such as the Caco-2 and HT-29 cell lines have proven to be an effective model for studying cellular interactions. Chicks and germfree mice are useful models for commensal colonization, whereas ferrets, piglets, and certain knockout mouse strains can be used as effective models for human disease. Herein, we have discussed how each of these models has been used, the advantages and disadvantages, and some of what has been learned about *Campylobacter* from each model.

Many of the mechanisms of pathogenesis for *C. jejuni* remain either completely unknown or, at best, poorly characterized. When the first *C. jejuni* genome sequences failed to reveal identifiable pathogenicity factors akin to those already identified in pathogenic *E. coli* and *Salmonella*, a renewed effort was placed on using known animal models to try and elucidate how *C. jejuni* colonizes its hosts and how it in fact causes disease. The lack of animal models that fully replicate the human disease has hindered this process; however, as a better understanding of the interactions between *C. jejuni* and the immune system has started to unfold, better and more relevant animal models are being explored.

REFERENCES

1. Young, K.T., Davis, L.M. & Dirita, V.J. *Campylobacter jejuni*: molecular biology and pathogenesis. *Nat Rev Microbiol* 5, 665–679 (2007).
2. Thabane, M. & Marshall, J.K. Post-infectious irritable bowel syndrome. *World J Gastroenterol* 15, 3591–3596 (2009).
3. Wieczorek, K. & Osek, J. Antimicrobial resistance mechanisms among *Campylobacter. Biomed Res Int* 2013, 340605 (2013).
4. van Spreeuwel, J.P. et al. *Campylobacter* colitis: histological immunohistochemical and ultrastructural findings. *Gut* 26, 945–951 (1985).
5. Allos, B.M. *Campylobacter jejuni* infections: update on emerging issues and trends. *Clin Infect Dis* 32, 1201–1206 (2001).
6. Newell, D.G. & Fearnley, C. Sources of *Campylobacter* colonization in broiler chickens. *Appl Environ Microbiol* 69, 4343–4351 (2003).
7. Van Dyke, M.I., Morton, V.K., McLellan, N.L. & Huck, P.M. The occurrence of *Campylobacter* in river water and waterfowl within a watershed in southern Ontario, Canada. *J Appl Microbiol* 109, 1053–1066 (2010).
8. Waldenstrom, J. et al. *Campylobacter jejuni* colonization in wild birds: results from an infection experiment. *PLoS One* 5, e9082 (2010).
9. Stern, N.J. et al. *Campylobacter* spp. in Icelandic poultry operations and human disease. *Epidemiol Infect* 130, 23–32 (2003).
10. Agunos, A., Waddell, L., Leger, D. & Taboada, E. A systematic review characterizing on-farm sources of *Campylobacter* spp. for broiler chickens. *PLoS One* 9, e104905 (2014).
11. Nichols, G.L. Fly transmission of *Campylobacter. Emerg Infect Dis* 11, 361–364 (2005).
12. Stanley, K. & Jones, K. Cattle and sheep farms as reservoirs of *Campylobacter. J Appl Microbiol* 94(Suppl), 104S–113S (2003).
13. Line, J., Hiett, K. & Conlan, A. Comparison of challenge models for determining the colonization dose of *Campylobacter jejuni* in broiler chicks. *Poult Sci* 87, 1700–1706 (2008).
14. Bui, X.T. et al. Survival of *Campylobacter jejuni* in co-culture with *Acanthamoeba castellanii*: role of amoeba-mediated depletion of dissolved oxygen. *Environ Microbiol* 14, 2034–2047 (2012).
15. Olofsson, J., Axelsson-Olsson, D., Brudin, L., Olsen, B. & Ellstrom, P. *Campylobacter jejuni* actively invades the amoeba *Acanthamoeba polyphaga* and survives within non digestive vacuoles. *PLoS One* 8, e78873 (2013).
16. Ikeda, N. & Karlyshev, A.V. Putative mechanisms and biological role of coccoid form formation in *Campylobacter jejuni. Eur J Microbiol Immunol (Bp)* 2, 41–49 (2012).
17. Newell, D.G. Animal models of *Campylobacter jejuni* colonization and disease and the lessons to be learned from similar *Helicobacter pylori* models. *Symp Ser Soc Appl Microbiol*, 57S–67S (2001).
18. Hedstrom, O.R. et al. Pathology of *Campylobacter jejuni* abortion in sheep. *Vet Pathol* 24, 419–426 (1987).
19. de Vries, J.J., Arents, N.L. & Manson, W.L. *Campylobacter* species isolated from extra-oro-intestinal abscesses: a report of four cases and literature review. *Eur J Clin Microbiol Infect Dis* 27, 1119–1123 (2008).
20. Blaser, M.J. & Atherton, J.C. *Helicobacter pylori* persistence: biology and disease. *J Clin Invest* 113, 321–333 (2004).

21. Beery, J.T., Hugdahl, M.B. & Doyle, M.P. Colonization of gastrointestinal tracts of chicks by *Campylobacter jejuni*. *Appl Environ Microbiol* 54, 2365–2370 (1988).
22. Fox, J.G., Ackerman, J.I., Taylor, N., Claps, M. & Murphy, J.C. *Campylobacter jejuni* infection in the ferret: an animal model of human campylobacteriosis. *Am J Vet Res* 48, 85–90 (1987).
23. Black, R.E., Levine, M.M., Clements, M.L., Hughes, T.P. & Blaser, M.J. Experimental *Campylobacter jejuni* infection in humans. *J Infect Dis* 157, 472–479 (1988).
24. Lee, A., O'Rourke, J.L., Barrington, P.J. & Trust, T.J. Mucus colonization as a determinant of pathogenicity in intestinal infection by *Campylobacter jejuni*: a mouse cecal model. *Infect Immun* 51, 536–546 (1986).
25. Alemka, A., Corcionivoschi, N. & Bourke, B. Defense and adaptation: the complex inter-relationship between *Campylobacter jejuni* and mucus. *Front Cell Infect Microbiol* 2, 15 (2012).
26. Stahl, M. et al. A novel mouse model of *Campylobacter jejuni* gastroenteritis reveals key proinflammatory and tissue protective roles for Toll-like receptor signaling during infection. *PLoS Pathog* 10, e1004264 (2014).
27. Sellars, M.J., Hall, S.J. & Kelly, D.J. Growth of *Campylobacter jejuni* supported by respiration of fumarate, nitrate, nitrite, trimethylamine-*N*-oxide, or dimethyl sulfoxide requires oxygen. *J Bacteriol* 184, 4187–4196 (2002).
28. Kendall, J.J., Barrero-Tobon, A.M., Hendrixson, D.R. & Kelly, D.J. Hemerythrins in the microaerophilic bacterium *Campylobacter jejuni* help protect key iron-sulphur cluster enzymes from oxidative damage. *Environ Microbiol* 16, 1105–1021 (2014).
29. Hugdahl, M.B., Beery, J.T. & Doyle, M.P. Chemotactic behavior of *Campylobacter jejuni*. *Infect Immun* 56, 1560–1156 (1988).
30. Lertsethtakarn, P., Ottemann, K.M. & Hendrixson, D.R. Motility and chemotaxis in *Campylobacter* and *Helicobacter*. *Annu Rev Microbiol* 65, 389–410 (2011).
31. Szymanski, C.M., King, M., Haardt, M. & Armstrong, G.D. *Campylobacter jejuni* motility and invasion of Caco-2 cells. *Infect Immun* 63, 4295–4300 (1995).
32. Hendrixson, D.R. & DiRita, V.J. Identification of *Campylobacter jejuni* genes involved in commensal colonization of the chick gastrointestinal tract. *Mol Microbiol* 52, 471–484 (2004).
33. Kanungpean, D., Kakuda, T. & Takai, S. Participation of CheR and CheB in the chemosensory response of *Campylobacter jejuni*. *Microbiology* 157, 1279–1289 (2011).
34. Lin, J., Sahin, O., Michel, L.O. & Zhang, Q. Critical role of multidrug efflux pump CmeABC in bile resistance and in vivo colonization of *Campylobacter jejuni*. *Infect Immun* 71, 4250–4259 (2003).
35. Flint, A. et al. Phenotypic screening of a targeted mutant library reveals *Campylobacter jejuni* defenses against oxidative stress. *Infect Immun* 82, 2266–2275 (2014).
36. Palyada, K. et al. Characterization of the oxidative stress stimulon and PerR regulon of *Campylobacter jejuni*. *BMC Genomics* 10, 481 (2009).
37. Stahl, M. et al. L-Fucose utilization provides *Campylobacter jejuni* with a competitive advantage. *Proc Natl Acad Sci USA* 108, 7194–7199 (2011).
38. Stahl, M., Butcher, J. & Stintzi, A. Nutrient acquisition and metabolism by *Campylobacter jejuni*. *Front Cell Infect Microbiol* 2, 5 (2012).
39. Hofreuter, D. Defining the metabolic requirements for the growth and colonization capacity of *Campylobacter jejuni*. *Front Cell Infect Microbiol* 4, 137 (2014).
40. Hofreuter, D. et al. Contribution of amino acid catabolism to the tissue specific persistence of *Campylobacter jejuni* in a murine colonization model. *PLoS One* 7, e50699 (2012).
41. Guccione, E. et al. Amino acid-dependent growth of *Campylobacter jejuni*: key roles for aspartase (AspA) under microaerobic and oxygen-limited conditions and identification of AspB (Cj0762), essential for growth on glutamate. *Mol Microbiol* 69, 77–93 (2008).
42. Naikare, H., Palyada, K., Panciera, R., Marlow, D. & Stintzi, A. Major role for FeoB in *Campylobacter jejuni* ferrous iron acquisition, gut colonization, and intracellular survival. *Infect Immun* 74, 5433–5444 (2006).
43. Palyada, K., Threadgill, D. & Stintzi, A. Iron acquisition and regulation in *Campylobacter jejuni*. *J Bacteriol* 186, 4714–4729 (2004).
44. Everest, P.H. et al. Differentiated Caco-2 cells as a model for enteric invasion by *Campylobacter jejuni* and *C. coli*. *J Med Microbiol* 37, 319–325 (1992).

45. Monteville, M.R., Yoon, J.E. & Konkel, M.E. Maximal adherence and invasion of INT 407 cells by *Campylobacter jejuni* requires the CadF outer-membrane protein and microfilament reorganization. *Microbiology* 149, 153–165 (2003).
46. MacCallum, A.J., Harris, D., Haddock, G. & Everest, P.H. *Campylobacter jejuni*-infected human epithelial cell lines vary in their ability to secrete interleukin-8 compared to in vitro-infected primary human intestinal tissue. *Microbiology* 152, 3661–3665 (2006).
47. Van Deun, K. et al. Colonization strategy of *Campylobacter jejuni* results in persistent infection of the chicken gut. *Vet Microbiol* 130, 285–297 (2008).
48. Watson, R.O. & Galan, J.E. *Campylobacter jejuni* survives within epithelial cells by avoiding delivery to lysosomes. *PLoS Pathog* 4, e14 (2008).
49. Davis, L. & DiRita, V. Experimental chick colonization by *Campylobacter jejuni*. *Curr Protoc Microbiol* Chapter 8, Unit 8A.3 (2008).
50. Ruiz-Palacios, G.M., Escamilla, E. & Torres, N. Experimental *Campylobacter* diarrhea in chickens. *Infect Immun* 34, 250–255 (1981).
51. Welkos, S.L. Experimental gastroenteritis in newly-hatched chicks infected with *Campylobacter jejuni*. *J Med Microbiol* 18, 233–248 (1984).
52. Manninen, K.I., Prescott, J.F. & Dohoo, I.R. Pathogenicity of *Campylobacter jejuni* isolates from animals and humans. *Infect Immun* 38, 46–52 (1982).
53. Hermans, D. et al. A tolerogenic mucosal immune response leads to persistent *Campylobacter jejuni* colonization in the chicken gut. *Crit Rev Microbiol* 38, 17–29 (2012).
54. Shanker, S., Lee, A. & Sorrell, T.C. Experimental colonization of broiler chicks with *Campylobacter jejuni*. *Epidemiol Infect* 100, 27–34 (1988).
55. Flanagan, R.C., Neal-McKinney, J.M., Dhillon, A.S., Miller, W.G. & Konkel, M.E. Examination of *Campylobacter jejuni* putative adhesins leads to the identification of a new protein, designated FlpA, required for chicken colonization. *Infect Immun* 77, 2399–2407 (2009).
56. Hermans, D. et al. Colonization factors of *Campylobacter jejuni* in the chicken gut. *Vet Res* 42, 82 (2011).
57. Byrne, C.M., Clyne, M. & Bourke, B. *Campylobacter jejuni* adhere to and invade chicken intestinal epithelial cells in vitro. *Microbiology* 153, 561–569 (2007).
58. Alemka, A. et al. Purified chicken intestinal mucin attenuates *Campylobacter jejuni* pathogenicity in vitro. *J Med Microbiol* 59, 898–903 (2010).
59. de Zoete, M.R., van Putten, J.P. & Wagenaar, J.A. Vaccination of chickens against *Campylobacter*. *Vaccine* 25, 5548–5557 (2007).
60. Shaughnessy, R.G., Meade, K.G., McGivney, B.A., Allan, B. & O'Farrelly, C. Global gene expression analysis of chicken caecal response to *Campylobacter jejuni*. *Vet Immunol Immunopathol* 142, 64–71 (2011).
61. Meade, K.G. et al. Comparative in vivo infection models yield insights on early host immune response to *Campylobacter* in chickens. *Immunogenetics* 61, 101–110 (2009).
62. Shaughnessy, R.G. et al. Innate immune gene expression differentiates the early avian intestinal response between *Salmonella* and *Campylobacter*. *Vet Immunol Immunopathol* 132, 191–198 (2009).
63. Smith, C.K. et al. *Campylobacter* colonization of the chicken induces a proinflammatory response in mucosal tissues. *FEMS Immunol Med Microbiol* 54, 114–121 (2008).
64. Babakhani, F.K., Bradley, G.A. & Joens, L.A. Newborn piglet model for campylobacteriosis. *Infect Immun* 61, 3466–3475 (1993).
65. Munroe, D.L., Prescott, J.F. & Penner, J.L. *Campylobacter jejuni* and *Campylobacter coli* serotypes isolated from chickens, cattle, and pigs. *J Clin Microbiol* 18, 877–881 (1983).
66. Tzipori, S. et al. Studies with enteroaggregative *Escherichia coli* in the gnotobiotic piglet gastroenteritis model. *Infect Immun* 60, 5302–5306 (1992).
67. Boyen, F. et al. Porcine in vitro and in vivo models to assess the virulence of *Salmonella enterica* serovar Typhimurium for pigs. *Lab Anim* 43, 46–52 (2009).
68. Reid, A.N., Pandey, R., Palyada, K., Naikare, H. & Stintzi, A. Identification of *Campylobacter jejuni* genes involved in the response to acidic pH and stomach transit. *Appl Environ Microbiol* 74, 1583–1597 (2008).
69. Dufour, V. et al. Inactivation of the LysR regulator Cj1000 of *Campylobacter jejuni* affects host colonization and respiration. *Microbiology* 159, 1165–1178 (2013).

70. van Alphen, L.B. et al. Biological roles of the O-methyl phosphoramidate capsule modification in *Campylobacter jejuni*. *PLoS One* 9, e87051 (2014).
71. Flint, A., Sun, Y.Q. & Stintzi, A. Cj1386 is an ankyrin-containing protein involved in heme trafficking to catalase in *Campylobacter jejuni*. *J Bacteriol* 194, 334–345 (2012).
72. Bell, J.A. & Manning, D.D. A domestic ferret model of immunity to *Campylobacter jejuni*-induced enteric disease. *Infect Immun* 58, 1848–1852 (1990).
73. Fox, J.G. et al. *Campylobacter*-like organisms isolated from gastric mucosa of ferrets. *Am J Vet Res* 47, 236–239 (1986).
74. Fox, J.G., Otto, G., Murphy, J.C., Taylor, N.S. & Lee, A. Gastric colonization of the ferret with *Helicobacter* species: natural and experimental infections. *Rev Infect Dis* 13(Suppl 8), S671–S680 (1991).
75. Nemelka, K.W. et al. Immune response to and histopathology of *Campylobacter jejuni* infection in ferrets (*Mustela putorius furo*). *Comp Med* 59, 363–371 (2009).
76. Bacon, D.J. et al. A phase-variable capsule is involved in virulence of *Campylobacter jejuni* 81-176. *Mol Microbiol* 40, 769–777 (2001).
77. Goon, S. et al. A σ^{28}-regulated nonflagella gene contributes to virulence of *Campylobacter jejuni* 81-176. *Infect Immun* 74, 769–772 (2006).
78. Yao, R., Burr, D.H. & Guerry, P. CheY-mediated modulation of *Campylobacter jejuni* virulence. *Mol Microbiol* 23, 1021–1031 (1997).
79. Glavis-Bloom, J., Muhammed, M. & Mylonakis, E. Of model hosts and man: using *Caenorhabditis elegans*, *Drosophila melanogaster* and *Galleria mellonella* as model hosts for infectious disease research. *Adv Exp Med Biol* 710, 11–17 (2012).
80. Senior, N.J. et al. *Galleria mellonella* as an infection model for *Campylobacter jejuni* virulence. *J Med Microbiol* 60, 661–669 (2011).
81. Champion, O.L. et al. *Galleria mellonella* as an alternative infection model for *Yersinia pseudotuberculosis*. *Microbiology* 155, 1516–1522 (2009).
82. Champion, O.L. et al. Insect infection model for *Campylobacter jejuni* reveals that *O*-methyl phosphoramidate has insecticidal activity. *J Infect Dis* 201, 776–782 (2010).
83. Gundogdu, O. et al. The *Campylobacter jejuni* transcriptional regulator Cj1556 plays a role in the oxidative and aerobic stress response and is important for bacterial survival in vivo. *J Bacteriol* 193, 4238–4249 (2011).
84. Fernie, D.S. & Park, R.W. The isolation and nature of campylobacters (microaerophilic vibrios) from laboratory and wild rodents. *J Med Microbiol* 10, 325–329 (1977).
85. Field, L.H., Underwood, J.L., Pope, L.M. & Berry, L.J. Intestinal colonization of neonatal animals by *Campylobacter fetus* subsp. *jejuni*. *Infect Immun* 33, 884–892 (1981).
86. Field, L.H., Underwood, J.L. & Berry, L.J. The role of gut flora and animal passage in the colonisation of adult mice with *Campylobacter jejuni*. *J Med Microbiol* 17, 59–66 (1984).
87. Blaser, M.J., Duncan, D.J., Warren, G.H. & Wang, W.L. Experimental *Campylobacter jejuni* infection of adult mice. *Infect Immun* 39, 908–916 (1983).
88. Chang, C. & Miller, J.F. *Campylobacter jejuni* colonization of mice with limited enteric flora. *Infect Immun* 74, 5261–5271 (2006).
89. Jesudason, M.V., Hentges, D.J. & Pongpech, P. Colonization of mice by *Campylobacter jejuni*. *Infect Immun* 57, 2279–2282 (1989).
90. Stanfield, J.T., McCardell, B.A. & Madden, J.M. *Campylobacter* diarrhea in an adult mouse model. *Microb Pathog* 3, 155–165 (1987).
91. Pei, Z. et al. Mutation in the peb1A locus of *Campylobacter jejuni* reduces interactions with epithelial cells and intestinal colonization of mice. *Infect Immun* 66, 938–943 (1998).
92. Pei, Z. & Blaser, M.J. Pathogenesis of *Campylobacter fetus* infections. Role of surface array proteins in virulence in a mouse model. *J Clin Invest* 85, 1036–1043 (1990).
93. Fernandez, H., Vivanco, T. & Eller, G. Expression of invasiveness of *Campylobacter jejuni* ssp. *jejuni* after serial intraperitoneal passages in mice. *J Vet Med B Infect Dis Vet Public Health* 47, 635–639 (2000).
94. Bar, W., Becker, K. & Hewel, C. Systemic spread of *Campylobacter jejuni* after intravenous infections. *FEMS Microbiol Immunol* 1, 263–270 (1989).

95. Blaser, M.J. et al. Extraintestinal *Campylobacter jejuni* and *Campylobacter coli* infections: host factors and strain characteristics. *J Infect Dis* 153, 552–559 (1986).
96. Baqar, S. et al. Murine intranasal challenge model for the study of *Campylobacter* pathogenesis and immunity. *Infect Immun* 64, 4933–4939 (1996).
97. Al-Banna, N.A., Raghupathy, R. & Albert, M.J. Induction of cytokines in different organs after intranasal inoculation of *Campylobacter jejuni* in mice. *Gut Pathog* 4, 23 (2012).
98. Hodgson, A.E., McBride, B.W., Hudson, M.J., Hall, G. & Leach, S.A. Experimental *Campylobacter* infection and diarrhoea in immunodeficient mice. *J Med Microbiol* 47, 799–809 (1998).
99. Mansfield, L.S. et al. C57BL/6 and congenic interleukin-10-deficient mice can serve as models of *Campylobacter jejuni* colonization and enteritis. *Infect Immun* 75, 1099–1115 (2007).
100. Kole, A. & Maloy, K.J. Control of intestinal inflammation by interleukin-10. *Curr Top Microbiol Immunol* 380, 19–38 (2014).
101. Hansen, J.J., Holt, L. & Sartor, R.B. Gene expression patterns in experimental colitis in IL-10-deficient mice. *Inflamm Bowel Dis* 15, 890–899 (2009).
102. Lippert, E. et al. Gnotobiotic IL-10; NF-κB mice develop rapid and severe colitis following *Campylobacter jejuni* infection. *PLoS One* 4, e7413 (2009).
103. Wilson, D.L. et al. Genetic diversity in *Campylobacter jejuni* is associated with differential colonization of broiler chickens and C57BL/6J IL10-deficient mice. *Microbiology* 156, 2046–2057 (2010).
104. Sun, X., Threadgill, D. & Jobin, C. *Campylobacter jejuni* induces colitis through activation of mammalian target of rapamycin signaling. *Gastroenterology* 142, 86–95.e5 (2012).
105. Samuelson, D.R. et al. The *Campylobacter jejuni* CiaD effector protein activates MAP kinase signaling pathways and is required for the development of disease. *Cell Commun Signal* 11, 79 (2013).
106. Haag, L.M. et al. *Campylobacter jejuni* induces acute enterocolitis in gnotobiotic IL-$10^{-/-}$ mice via Toll-like-receptor-2 and -4 signaling. *PLoS One* 7, e40761 (2012).
107. Sun, X. & Jobin, C. Nucleotide-binding oligomerization domain-containing protein 2 controls host response to *Campylobacter jejuni* in $Il10^{-/-}$ mice. *J Infect Dis* 210, 1145–1154 (2014).
108. Karrasch, T., Kim, J.S., Muhlbauer, M., Magness, S.T. & Jobin, C. Gnotobiotic IL-$10^{-/-}$; NF-κB(EGFP) mice reveal the critical role of TLR/NF-κB signaling in commensal bacteria-induced colitis. *J Immunol* 178, 6522–6532 (2007).
109. Watson, R.O., Novik, V., Hofreuter, D., Lara-Tejero, M. & Galan, J.E. A MyD88-deficient mouse model reveals a role for Nramp1 in *Campylobacter jejuni* infection. *Infect Immun* 75, 1994–2003 (2007).
110. Sham, H.P. et al. SIGIRR, a negative regulator of TLR/IL-1R signalling promotes microbiota dependent resistance to colonization by enteric bacterial pathogens. *PLoS Pathog* 9, e1003539 (2013).
111. Bereswill, S. et al. Novel murine infection models provide deep insights into the “menage a trois” of *Campylobacter jejuni*, microbiota and host innate immunity. *PLoS One* 6, e20953 (2011).

20

Cronobacter: *Virulence and Pathogenesis*

Nemani V. Prasadarao

CONTENTS

20.1 Introduction

Originally referred to as yellow-pigmented *Enterobacter cloacae*, it was later classified as a new species *Enterobacter sakazakii*. Subsequent characterization enabled the reclassification of these bacteria into a new genus called *Cronobacter* [1]. *Cronobacter* is composed of a diverse group of Gram-negative bacilli, which includes *Cronobacter sakazakii*, *Cronobacter muytjensii*, *Cronobacter malonaticus*, *Cronobacter dublinensis*, *Cronobacter turicensis*, *Cronobacter universalis*, and *Cronobacter condimenti* [2]. Except *Cronobacter condimenti*, all other *Cronobacter* spp. are associated with human infections. They cause life-threatening infections in neonates due to the consumption of powdered infant formula contaminated with *Cronobacter* [3,4]. Outbreaks of *Cronobacter* infections in neonatal intensive care units have resulted in several CDC warnings, and so efforts are in place to improve health care. Considered as an opportunistic pathogen, *Cronobacter* causes severe illness in neonates, such as necrotizing enterocolitis, bacteremia, and meningitis, often in low-birth-weight preterm infants [5]. Infections due to *Cronobacter* in normal and immunocompromised adults have also been noted, but these are less severe. Based on partial 16S rRNA and *hsp60* sequencing, four cluster groups of *C. sakazakii* have been identified among this diverse group of pathogens [6]. The bacterium can be found in a variety of foods, including dairy-based foods (cheese), dried meats, and rice. Furthermore, it was also detected in environmental sources such as soil, livestock facilities, and food preparation units. Compared to other Enterobacteriaceae family members, *Cronobacter* is highly resistant to heat, dryness, and acidic conditions [7]. It also forms biofilms that function as a protective barrier to withstand environmental stress and obviate immune surveillance of the host.

Several selective media for detecting *Cronobacter* have been developed [8,9]. However, these media are insufficient to support the growth of all strains of *Cronobacter* [10]. To detect *Cronobacter* spp. in powdered infant formula, a one-step enrichment protocol using a chromogenic medium has been designed [11]. More useful molecular-based detection techniques for understanding the epidemiology of *Cronobacter* have also been developed by targeting a number of genes such as 16S rRNA, 16S-23S rRNA intergenic regions, *ompA*, zinc-containing metalloprotease, *dnaG* and *gluA* genes by real-time polymerase chain reaction (PCR) [12–15]. Genes that are unique to different species of *Cronobacter* are particularly useful as candidate markers in molecular detection protocols [16–18]. Due to limitations of individual gene-based techniques, whole-genome sequencing may help identify *Cronobacter* species and assist in comparing the genotypic and phenotypic features of the pathogen being studied.

Although powdered infant formula has been found to be the main source of *Cronobacter* for infecting newborns, transmission from mother to child during the delivery cannot be excluded. It was also reported that plant material may be a natural source of *Cronobacter* spp. [19]. In a study performed to characterize genotypic and phenotypic features of *C. sakazakii* strains obtained in an outbreak in intensive care units in France in 1994 [20], a total of 31 stains were collected, and on the basis of 16S rRNA analysis, 30 strains were confirmed as *C. sakazakii*. Pulsed-field gel electrophoresis (PFGE) typing recognized four different pulsotypes and clearly demonstrated that *C. sakazakii* strains exhibit different genotypic and phenotypic characteristics. This reinforces the necessity for rigorous and careful testing to identify the strains. Genomic sequencing of *C. sakazakii* strain ATCC BAA-894 isolated from contaminated infant formula revealed a single 4.4 Mb chromosome and two plasmids, pESA2 (31 kb) and pESA3 (131 kb) [21]. ATCC BAA-894 strain contains 4392 genes in the core genome, and 21 genes are found to be unique in 5 other *C. sakazakii* strains. Similar to ATCC BAA-894 strain, *C. turicensis* contains a similar genome and 2 plasmids, which also encodes open reading frames (ORFs) for 223 virulence-associated genes [22]. Comparison of pESA3 and pCTU1 (*C. turicensis* plasmid) by *in silico* analysis revealed the presence of two iron-acquisition systems, which may be essential for pathogenesis.

For the pathogenesis of *Cronobacter* spp., the bacteria must attach to intestinal epithelial cells of infants fed with contaminated formula, followed by invasion and hematogenous spread. The adherence capacities of 50 *C. sakazakii* strains have been studied using HEp2 and Caco-2 cell lines and brain microvascular endothelial cells, and two distinctive patterns of binding were observed, diffuse adhesion and localized cluster formation [23]. The adherence of these *Cronobacter* strains appears to be non-fimbrial mediated. Of note, some fimbrial clusters have been identified in the genomes of *Cronobacter*, although *C. sakazakii* is the only bacterium that expresses β-fimbriae, but its role in the pathogenesis is still to be evaluated. The putative virulence factors produced, such as enterotoxin-like compounds, by *Cronobacter* were first assessed by the suckling mouse assay [24]. Analysis of various *Cronobacter* strains from different brands of infant formula revealed that they contain high levels (500-fold) of heat-stable endotoxin (lipopolysaccharide or LPS) [20]. Since these formulae contain a variety of bacterial species, it is not clear what the source of LPS is. However, its presence enhances the translocation of *C. sakazakii* across the gut and blood–brain barrier. Surveillance of microbial burden in food products, growth conditions, and epidemiology would provide clues to developing methods that reduce health risks in vulnerable individuals. Furthermore, understanding the pathophysiology of the diseases caused by *Cronobacter* spp. is also desirable to identify long-term risks in infected neonates and infants. To gain insights into the pathophysiology of *Cronobacter*-induced diseases, a careful selection and usage of animal models is clearly required. Therefore, the purpose of this chapter is to present the animal and tissue culture models used thus far to identify bacterial and host factors that contribute to *Cronobacter* pathogenesis.

20.2 Necrotizing Enterocolitis

20.2.1 A Rat Model of *C. muytjensii*-Induced Necrotizing Enterocolitis

Premature infants are prone to a variety of diseases including necrotizing enterocolitis (NEC), which is a serious gastrointestinal disease and occurs in ~1 in 1000 live births [25]. The etiology of NEC is multifactorial, and the risk factors include prematurity, formula feeding, and abnormal bacterial colonization of intestinal tract. To replicate histopathological manifestations of NEC that mimic human NEC, investigators have used several animal models to study the disease. The pathological features of NEC show the presence of mucosal edema, pneumatosis intestinalis (i.e., the presence of gas within the wall of the intestine), epithelial sloughing/villous atrophy, enterocyte apoptosis, vascular thrombosis, and discontinuous necrotic segments of intestine. A variety of bacterial species are thought to cause NEC such as *Enterobacter*, *Clostridium*, and *Staphylococcus* spp. Since milk-based infant formulas contaminated with *C. sakazakii* have been linked to many NEC outbreaks, a newborn rat model of *Cronobacter*-induced NEC using the combination of hypoxia and formula feeding has been developed [26]. The strain used in these studies, obtained from ATCC (strain 51329), was later classified as *C. muytjensii*. Newborn rats fed with infant formula mixed with 10^5CFU of *C. muytjensii* produced clinical symptoms similar to Grade 3 NEC by day 4 postinfection (Figure 20.1). The pathological analysis of intestinal sections obtained from infected rats showed macroscopic and microscopic features analogous to that seen in human NEC [27]. The formula-fed and hypoxic group of rats only showed lower pathological symptoms and had a mortality rate of 40%, whereas the mortality rate increased to 70% with *Cronobacter* infection. Scanning electron microscopy of intestines of newborn rats revealed that specimens from formula-fed and hypoxic rats exhibited intact villi. *C. muytjensii* infection increased the binding of the bacteria to villi tips and showed enterocyte blebbing and gap formation in the epithelium.

20.2.2 *C. muytjensii* Interaction with Epithelial Cells in Culture

To understand the molecular events induced by *C. muytjensii* in intestinal epithelial cells, an *in vitro* model of infection using IEC-6 cells has been developed. Studies from these experiments revealed that the bacteria bind to IEC-6 cells efficiently in a dose-dependent manner; however, the entry of the bacteria is negligible [28]. Furthermore, the binding of *C. muytjensii* to IEC-6 cells induced apoptosis at 4 h postinfection as assessed by ApopTag staining. Corroborating with the *in vitro* results, newborn rats whose intestines were infected also showed a greater number of apoptotic epithelial cells, whereas the intestines of those rats that were formula fed showed very few apoptotic cells. Follow-up studies revealed that *C. muytjensii* interaction of IEC-6 cells also increased the production of nitric oxide (NO) by 4 h postinfection. This NO production is due to the activation of inducible nitric oxide synthase (iNOS), not endothelial nitric oxide synthase (eNOS), as transfection of IEC-6 cells with siRNA to iNOS significantly inhibited the NO generation. The increased iNOS activity was transcriptionally regulated, which in turn is responsible for the enhanced production of iNOS at the protein level. Furthermore, IEC-6 cells transfected with siRNA to iNOS, not eNOS, and infected with *C. muytjensii* showed no or very few apoptotic cells compared to control or eNOS-transfected cells. Examination of the binding capacities of 24 *C. sakazakii* strains obtained from human and environmental sources exhibited varied binding patterns to IEC-6 cells [29]. Interestingly, *C. sakazakii* strains isolated from humans had higher binding to IEC-6 cells at a frequency of 4%–5% compared to other environmental strains. High-binding

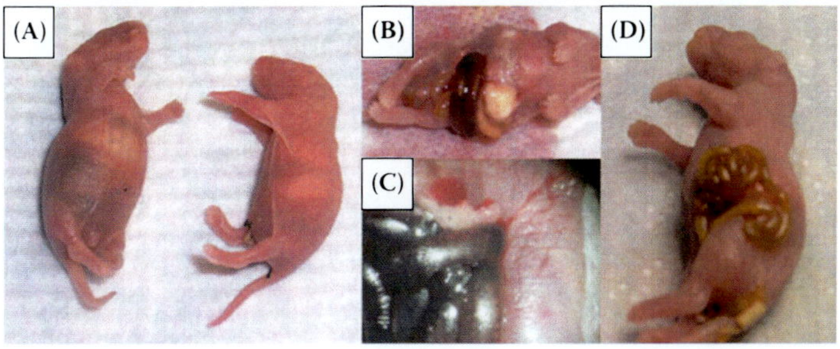

FIGURE 20.1 Neonatal rat model of *Cronobacter muytjensii*-induced necrotizing enterocolitis: Newborn rat pups were fed with Esbilac formula two times and three times daily and were placed under hypoxic conditions with 5% carbon dioxide (FF+H) with (A, left) or without (A, right) *C. muytjensii* infection. FF+H treated rats showed a normal abdominal wall with a milk spot (B). Abdominal discoloration and evidence of clinical peritonitis were observed in FF+H+*C. muytjensii* rats (C). In contrast, control rat pups showed normal intestine (D). (Reprinted with permission from Hunter, C.J., et al., *J. Infect. Dis.*, 198, 586–593, 2008.)

C. sakazakii strains induced more apoptosis in Caco-2 cells than the low-binding strains. Approximately 25% of infected Caco-2 cells were apoptotic. Abnormal colonization of bacteria usually results in the loss of gut barrier integrity, which is an important pathophysiological feature in the development of NEC. Similar to that in human epithelial cells, high-binding *C. sakazakii* infection of Caco-2 cells caused an increase in permeability of the monolayers as assessed by horseradish peroxidase leakage. *C. sakazakii*-induced leakage of the tight junctions in Caco-2 cells were due to the disruption of ZO1 at the periphery of the cells. Moreover, these permeability changes of Caco-2 cells require the activation of PKC, which in turn activates NO production upon *C. sakazakii* infection.

Outer membrane protein A (OmpA) is a 35-kDa surface protein of Gram-negative bacteria whose structure is highly conserved throughout the evolution [30]. OmpA contains eight transmembrane domains and four extracellular loops. However, some minor changes are present in the extracellular loops that help differentiate whether the bacterium is pathogenic or nonpathogenic [31]. Comparison of OmpA sequences among several Gram-negative strains revealed that the extracellular loops of *C. sakazakii* are significantly different from other sequences [30]. OmpA of *C. sakazakii* has been shown to play a critical role in the invasion of INT407 cells with an invasion frequency of 0.08 + 0.002 at a multiplicity of infection of 100 [32,33]. This invasion frequency in epithelial cells is considerably negligible compared to the invasion frequencies of other gut pathogens such as *Salmonella* or *Shigella* [34,35]. The invasion into epithelial cells depends on both microfilament and microtubules [36]. Fibronectin appears to be critical for the binding of *C. sakazakii* to epithelial cells [37]. However, deletion of OmpA in this strain reduced the invasion by 80% but did not reduce the attachment to the cells. Subsequently, a mouse model of infection has been developed in which oral feeding of 10^3 CFU *C. muytjensii* to 3-day-old mice induces NEC-like pathology such as intestinal dilation and bowel discoloration by 48 h postinfection [38]. In contrast, $OmpA^-$ *C. muytjensii* does not cause such injury and the mice survived. In this model, wild-type *C. muytjensii* binds to the intestine efficiently, whereas $OmpA^-$ *C. sakazakii* could not associate with epithelial cells. Of note, prebiotic galacto-oligosaccharides and polydextrose inhibit the binding of *C. sakazakii* 4603 to HEp-2 cells and Caco-2 cells [39].

20.2.3 *C. muytjensii* Interaction with Dendritic Cells

Systemic inflammatory response and local cytokine production contribute to the pathogenesis of NEC. IEC-6 cells infected with *C. muytjensii* led to IL-6 production in a time-dependent manner, which is in agreement with the cytokines observed in the serum of infected rat pups [26]. Initiation and regulation of required immune response to microbial pathogens is a critical function of specialized antigen-presenting cells such as dendritic cells (DCs). The immature DCs in the peripheral tissues recognize pathogenic organisms by capturing antigens or whole bacteria upon contact and undergo maturation by expressing major histocompatibility complex class II, CD40, and CD86 molecules [40]. Supporting the role of DCs, *C. muytjensii* infection of 3-day-old mice resulted in recruitment of a greater number of DCs to the lamina propria, whereas the number of polymorphonuclear cells (PMNs) and macrophages were slightly increased [41]. $OmpA^-$ *C. muytjensii* infection also caused similar DC migration to the lamina propria. The recruited DCs are positive for CD11c/CD103 and negative for F4/80/Gr-1. DCs recruited by wild-type *C. muytjensii* are not mature as they do not express CD40, MHC class II, and CD86 markers, while $OmpA^-$ *C. muytjensii*-infected mice showed mature DCs [38]. How does *C. muytjensii* manipulate DCs to suppress the immune response? The entry of wild-type *C. muytjensii* is critical for suppression of maturation markers and is mediated by DC-specific ICAM grabbing integrin (DC-SIGN) [42]. Concurrent with the requirement of DC-SIGN, DCs pretreated with mannan, which binds mannose-related receptor, and *C. muytjensii* pretreated with His-Mermaid, a C-type lectin, which competes with DC-SIGN, prevented bacterial entry into DCs. Overexpression of DC-SIGN in HeLa cells enabled the cells to take up 50-fold more bacteria, both wild type, and $OmpA^-$ *C. muytjensii*, compared to plasmid-alone transfected cells. The DC-SIGN-mediated entry of *C. muytjensii* did not require the expression of OmpA. However, only wild-type bacteria survived in DCs, indicating that OmpA is necessary for the survival of the bacteria inside DCs. Analysis of signaling events modulated in DCs after infection with *C. muytjensii* revealed that the bacterium suppresses MAP kinases for preventing maturation.

Furthermore, supernatants of myeloid-derived DCs infected with wild-type *C. muytjensii* when incubated with Caco-2 cells grown in Transwell inserts increased the permeability of the monolayers [38]. In contrast, OmpA⁻ *C. muytjensii*-infected DC supernatants showed only minor permeability changes. The supernatant of the wild-type strain also caused disruption of tight junctions, which was further increased by infection. It appears that tight junction disruption is the first step in the loss of epithelial cells in the pathogenesis of NEC. In support of this concept, incubation of Caco-2 cells with the supernatants obtained from DCs infected with *C. muytjensii* enhanced apoptosis of the cells in the presence of the bacteria, which is prevented by the presence of apoptosis inhibitor, ZVAD. The responsible soluble factor in *C. muytjensii*-infected DC supernatants is an anti-inflammatory cytokine, TGFβ. Wild-type strains produced robust quantities of TGFβ from bone-marrow-derived DCs, while OmpA⁻ *C. muytjensii* generated basal levels of the cytokine. Mucosal scrapings obtained from the intestines of infected mice also showed increased *tgfβ* transcript levels. Similarly, intestinal homogenates from *C. muytjensii*-infected mice also revealed higher concentrations of TGFβ. Anti-TGFβ antibodies, but not anit-IL-10 antibodies, when added to the supernatants of *C. muytjensii*-infected DCs prevented the bacteria-induced monolayer permeability of Caco-2 cells and apoptosis. This inhibitory effect was due to the inhibition of iNOS expression in the cells. Upon infection, DCs are recruited to the intestinal lamina propria, which extend their protrusions between enterocytes to sample the pathogen or macromolecules. Using a double-layer model by culturing monocyte-derived macrophages on the top of a Transwell filter insert and bone-marrow-derived DCs at the bottom, it was demonstrated that the presence of DCs increased the production of TGFβ, thereby increasing the permeability of the Caco-2 monolayers. Corroborating the *in vitro* results, depletion of DCs in newborn mice by injecting anti-CD11c antibody at day 1 after birth resulted in resistance to *C. muytjensii*-induced NEC and survival as normal, uninfected mice. DC-depleted mice showed no signs of pathology in intestines. Adoptive transfer of DCs into these mice rendered the animals susceptible to infection again.

20.2.4 The Role of Neutrophils and Macrophages in *Cronobacter* Infection

Neutrophils and macrophages are critical for clearing bacterial pathogens in tissues; however, bacteria may be equipped with strategies to manipulate these two immune cells. In newborn mice infected with *C. muytjensii*, both neutrophils and macrophages are recruited to the intestine in higher numbers compared to the number of cells in uninfected mice; however, these numbers are lower than the number of dendritic cells. Depletion of either neutrophils or macrophages from 2-day-old mice followed by infection with *C. muytjensii* rendered the animals susceptible to infection sooner than normally infected littermates [41]. The intestinal bacterial load was greater and the pathology severe. In the absence of neutrophils or macrophages, the recruitment of DCs increased by 50%, which appears to be responsible for increased severity of the disease. Consistent with the bacterial load and intestinal pathology, the levels of proinflammatory cytokines TNF-α, IL-1β, IL-2, and IL-6 also increased. Furthermore, flagellin, the major protein constituent of flagella, recognizes Toll-like receptor 5 (TLR5) and triggers the production of various cytokines [43]. Flagellins from *C. sakazakii* ST1 and ST4, *C. malonaticus*, *C. muytjensii*, *C. turicensis*, and *C. dublinensis* have been tested for the production of cytokines in monocyte-derived macrophages, and the results have shown that similar levels of IL-8, TNF-α, and IL-10 are produced. Both anti-flagellin and anti-TLR5 antibodies prevented the production of proinflammatory cytokines in these cells, thus confirming the importance of their interaction in the pathogenesis by these bacterial strains.

20.2.5 Prevention of *Cronobacter*-Induced NEC by *Lactobacillus*

Several randomized human clinical trials revealed that oral administration of certain probiotic species can reduce the incidence of NEC although the exact mechanism of protection is unknown [44]. Yogurt is used widely by humans, and its health-promoting effects are ascertained to be due to the presence of probiotics. Pre- and coincubation of IEC-6 cells with *Lactobacillus bulgaricus* and *C. muytjensii* demonstrated that the preincubation of the probiotic inhibited *C. muytjensii*-induced NO production by suppressing the iNOS expression [28]. This effect was due to prevention of the binding of *C. muytjensii* to IEC-6 cell by *L. bulgaricus*. Moreover, the probiotic also inhibited *C. muytjensii*-induced apoptosis of IEC-6 cells. The observations made in IEC-6 were also reflected in infant rats that were orally fed with *L. bulgaricus* and then infected with *C. muytjensii* as they maintained the structural integrity of enterocytes by preventing the attachment of *Cronobacter* under hypoxic conditions. Recent studies have also shown that the conditioned medium of the probiotic *Bifidobacterium infantis* protects neonatal mice from intestinal inflammation when infected with *C. sakazakii* [45]. The probiotic conditioned medium reduced enterocyte apoptosis and mucin production and maintained the integrity of the ileal structure when compared with intestines of mice treated with wild-type bacteria.

20.3 Septicemia and Meningitis

20.3.1 Serum Tolerance of *Cronobacter* spp.

As a foodborne pathogen, *Cronobacter* spp. cause systemic infections by crossing the gastrointestinal barrier, which is followed by invasion of the blood–brain barrier. Therefore, these bacteria should have virulence factors that help to avoid serum killing. Transposon mutagenesis was used to generate a library of *C. sakazakii* ES5, which was then screened for serum tolerance [46]. Several genes encoding surface and membrane proteins, chaperones, and regulatory proteins have been identified. Of these genes, an undescribed Wzt_C superfamily domain consisting a coding region for serum tolerance was identified and confirmed by experiments. Other genes that enhance serum tolerance were also identified, among them YbaJ. Deletion of *ybaJ* element affected the expression of the *fimA* gene, which encodes a structural component of fimbriae.

20.3.2 A Neonatal Rat Model of Meningitis

C. sakazakii strains also cause meningitis in rare situations in newborn infants due to consumption of contaminated infant formula [47,48]. The bacterium must cross the gut and the blood–brain barrier and evade host defenses to cause meningitis [49]. Infection of newborn rats with distinct *C. sakazakii* (*E. sakazakii*) strains via intracranial injection leads to sepsis and chronic inflammation between 6 and 9 days postinfection [50]. Although two different cluster groups of *C. sakazakii* strains have been used, the occurrence of meningitis and the inflammation pattern varies from 30% to 80% except for one strain NTU2, which induced 80% incidence of meningitis and inflammation. Newborn rat brains infected with NTU2 strains showed severe bilateral ventriculitis, astrocytosis, and microhemorrhage. Furthermore, these animals also showed vascular permeability, leading to hydrocephalus. Electron microscopy studies further revealed that *C. sakazakii* NTU57 strain was found in infiltrated neutrophils and macrophages in the brain.

Since the natural route of infection of this pathogen is by gut colonization, the information obtained from the intracranial infection model may have a limitation as regards the extrapolation of these data to human disease. Using 2-day-old rats that were fed *C. muytjensii* orally, which mimics the natural route of infection with the wild-type strain and its OmpA mutant, it was demonstrated that wild-type *C. muytjensii* induced meningitis [50]. Histopathological analysis of brain sections from the infected mice showed that wild-type strain induced migration of glial cells into the cortex and also neutrophil infiltration (Figure 20.2). Focal hemorrhage and neuronal apoptosis were also observed in several areas of the brain. On the other hand, although OmpA$^-$ *C. muytjensii* entered the circulation by 6 h postinfection, the bacterial load was significantly lower compared to wild-type bacteria and could not cause meningitis. The role of OmpA in *C. muytjensii*-induced meningitis was confirmed by introducing a plasmid containing *ompA* gene into OmpA$^-$ *C. muytjensii*, which then regained the capacity to cause disease. OmpA expressing wild-type *C. muytjensii* crossed both the gut and blood–brain barrier very efficiently, whereas OmpA$^-$ *C. muytjensii* is defective in these translocations. Lack of efficient translocation of OmpA$^-$ *C. muytjensii* across the gut barrier is due to inefficient binding of the bacteria to intestinal cells. Intestinal pathology after being fed with wild-type strains showed dilated lumen and focal bacterial

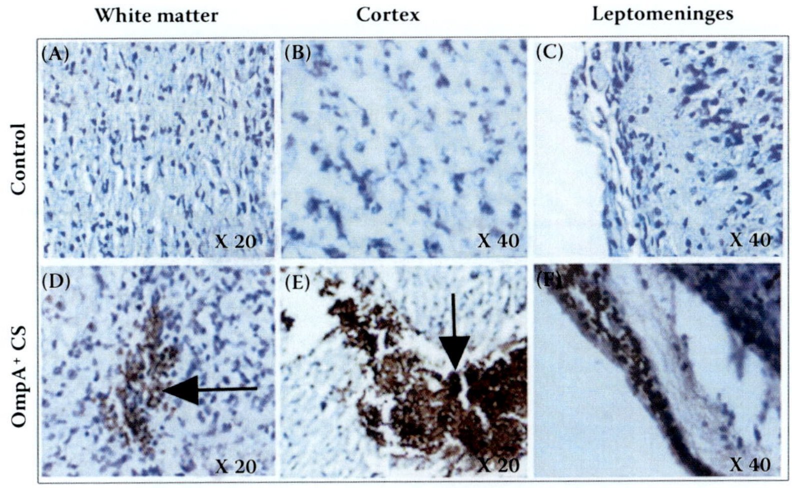

FIGURE 20.2 The presence of OmpA$^+$ *C. muytjensii* in infected newborn rats: *C. muytjensii*-fed newborn rat brains were harvested 48 h postinfection, fixed, sectioned, and stained with the anti-OmpA antibody. Mayer's hematoxylin was used to counterstain the sections. Control pups received saline. Arrows indicate the presence of clusters of bacteria in white matter and the cortex. (Reprinted with permission from Mittal, R., et al., *Lab. Invest.*, 89, 263–277, 2009.)

growth along with microvilli flattening. Furthermore, epithelial degeneration was observed in mucosa along with focal ulceration, reactive stromal changes, and suppurative inflammation. The binding of *C. muytjensii* also caused apoptosis of epithelial cells. Infection of mouse pups with wild-type strain produced TNF-α, IL-1β, IL-6, and IL-10, and chemokine MIP-2 at highest levels by 48 h postinfection. Of note, OmpA$^-$ *C. muytjensii*-infected animals showed very low levels of these cytokines and chemokines except for IL-10, the levels of which increased between 96 and 120 h postinfection. Despite a small number of OmpA$^-$ *C. muytjensii* entering the circulation, further multiplication of the bacteria was not observed. Follow-up serum bactericidal assays showed that wild-type *C. muytjensii* was resistant to killing compared to OmpA$^-$ *C. muytjensii*, which are killed within 1 h postincubation with neonatal rat serum. Other investigators demonstrated that a plasminogen activator (Cpa) mutant of *C. sakazakii* was serum sensitive in comparison with its wild-type strain [51]. Cpa activates plasminogen by inactivating α2-AP and also proteolytically cleaves complement components C3, C3a, and C4b. Lack of serum resistance in OmpA$^-$ *C. muytjensii* could be due to alteration of Cpa expression because of downstream effects of *ompA* deletion. Taken together, it appears that newborn rats infected with *C. muytjensii* without hypoxia also results in NEC-like symptoms followed by penetration of the organism into the blood to cause meningitis.

20.3.3 *Cronobacter* Interaction with Brain Endothelial Cells

The onset of meningitis occurs by hematogenous spread to the central nervous system following the invasion of the blood–brain barrier. Several investigators have used brain microvascular endothelial cells isolated from different sources as an *in vitro* model of the blood–brain barrier. Using rat brain capillary endothelial cells (rBCEC4) in gentamicin protection assays, it was demonstrated that *C. sakazakii* strains NTU2, NTU658, and NTU57 invade the cells with a frequency of 0.43%–1% of input. This invasion frequency is similar to that of the invasion levels exhibited by *E. coli* K1, a meningitic bacteria in these cells. Similar to these studies, *C. muytjensii* also invades human brain microvascular endothelial cells (HBMECs) at a frequency of 0.5%, which increases in a time-dependent manner [52]. In contrast, both epithelial cells and human umbilical vein endothelial cells (HUVECs) were invaded less by *C. muytjensii*. Confocal imaging of HBMECs infected with GFP-expressing *C. muytjensii* (Figure 20.3) and transmission electron microscopy revealed that the bacterium enters and survives inside the cells with a single intact bacterium enclosed in vacuoles. The entry of *C. muytjensii* into HBMECs also requires

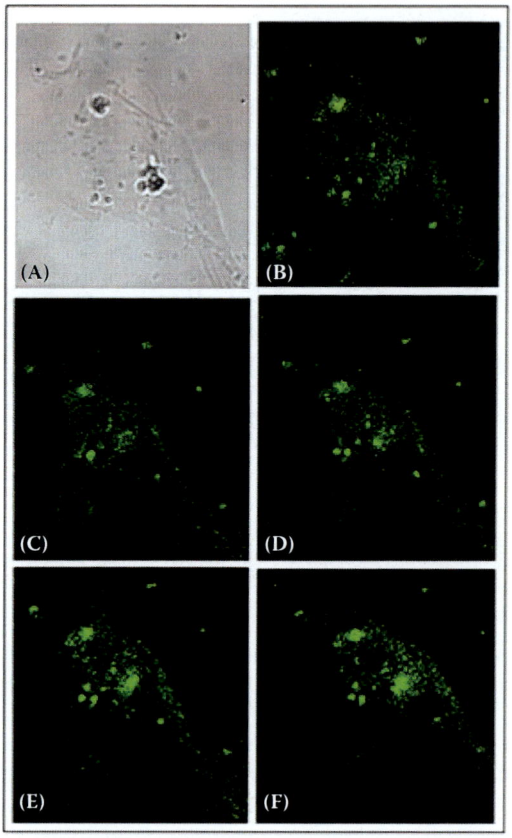

FIGURE 20.3 Invasion of *C. muytjensii* into human brain microvascular endothelial cells (HBMEC): Green-fluorescent-protein-expressing *C. muytjensii* were incubated with HBMEC monolayers for 4 h, washed, and fixed with 2% paraformaldehyde. The monolayers were imaged, and Z-stacks of confocal images were acquired using a Leica laser scanning microscope. The images were taken at 63× magnification. (Reprinted with permission from Singamsetty, V.K., et al., *Microb. Pathog.*, 45, 181–191, 2008.)

the expression of OmpA and microtubule reorganization in the cells. In this entry process, microtubule aggregation is observed beneath the bacterial binding sites but not actin accumulation. Although many bacterial strains utilize microfilament reorganization in eukaryotic cells for internalization, *C. muytjensii* entry appears to be a specialized mechanism. PKC-α plays a central role in cytoskeletal reorganization in the entry of many bacterial pathogens. *C. muytjensii* invasion of HBMECs also depends on PKC-α as overexpression of a dominant negative form of PKC-α, PKC-CAT/KR, in HBMECs prevented the entry of the bacteria. Studies have demonstrated that both PKC-α and PI3-Kinase stabilize microtubules, and in agreement, PI3-kinase activation is also necessary for *C. muytjensii* entry into HBMECs. In contrast, the invasion of *C. sakazakii* strain (ATCC 29544) into HBMECs was prevented by cytochalasin D, an inhibitor of actin microfilaments [53]. However, the invasion also depended on PI3-kinase as *C. sakazakii* ATCC 29544 induced activation of Akt, a downstream substrate for PI3-Kinase.

20.4 *C. sakazakii*: Motility and Biofilm Formation

Cronobacter spp. form biofilms on a variety of surfaces including enteral feeding tubes, which could be a source of infections in low-birth-weight infants [54]. The formation of *Cronobacter* biofilms requires multiple nutritional factors and environments, which increases its antibiotic resistance capacity [55]. A mutant strain of *C. sakazakii*, LWW02, was constructed by deleting a gene encoding heptosyltransferase I [56]. This mutant exhibited slower growth, higher permeability of the membrane, and surface hydrophobicity. The biofilm formation by this mutant was stronger compared to wild-type strains. Furthermore, studies have shown that flagella of *Cronobacter* is responsible for the formation of biofilms and for binding to Caco-2 cells [57]. To study the virulence genes of *C. sakazakii* ATCC 29544, a transposon-mediated random mutant library was generated and screened, leading to the identification of a mutant deficient in invasion in Caco-2 cells [58]. A novel plasmid pCSA2 that contains six ORFs and 4938 bp has been identified. Of the six ORFs, one was predicted to encode methyl-accepting chemotaxis protein (MCP), which contains one MCP domain and two sensor PAS (Per-Arnt-Sim sensory) domains that show similarity with biofilm dispersion protein BdlA of *Pseudomonas aeruginosa*. Deletion of *mcp* gene significantly reduced the invasion in Caco-2 cells, and complementation with a plasmid-containing *mcp* gene enabled the mutant to invade the cells. Lack of invasion into Caco-2 cells is due to reduced adherence to the cells. Corroborating with the role in adherence and invasion for MCP, 3-day-old rats fed with the *mcp*-deleted mutant showed 100-fold reduced efficiency to translocate across the gut and deep into liver and spleen. Interestingly, the deletion of the *mcp* gene in *C. sakazakii* resulted in organisms becoming hypermotile (~31.5 mm in 8 h on 0.3% TSA agar) compared to the wild-type strain (~17.3 mm). As flagella are important for the motility of *Cronobacter*, analysis of genes involved in flagellar assembly revealed that *mcp* deletion enhanced the mRNA levels of *fliA* and *fliC* genes. In addition to the role of the *mcp* gene in motility, it is also required for biofilm formation. Thus, MCP plays a critical role in the virulence of *C. sakazakii* by regulating multiple functions.

20.5 *C. sakazakii* Interaction with *Caenorhabditis elegans*

Besides rat and murine models, several other models have also been used to study bacterial pathogenesis, such as zebrafish and *C. elegans* [59,60]. *C. elegans* is a nonparasitic free-living nematode that is very useful for studying bacterial interaction due to its well-developed genetic and molecular tools to manipulate the organism. Studies by Shivamurthy et al. revealed that *C. elegans* fed with *C. sakazakii* died in liquid conditions with LT_{50} of 134 + 2.8 h, whereas feeding on *E. coli* OP50 (food source of *C. elegans*) had no effect [61]. *C. sakazakii* binding to intestines of *C. elegans* is responsible for killing the organism, which increased with time of infection. Only live *Cronobacter*, not heat-killed ones, were able to colonize and multiply in the intestines, thereby killing the organism. The death of *C. elegans* by *Cronobacter* was due to the apoptosis of the cells of the intestinal lumen, a mechanism that appears to be similar to that of human NEC pathogenesis. Subsequent studies by this group revealed that exposure to *C. sakazakii* LPS led to paralysis, and eventually death, of *C. elegans* [61]. Interestingly, the bacterium modifies the structure of LPS upon interaction with the host to avoid immune response. Map Kinase pathway plays an important role in the immunity of *C. elegans* against *C. sakazakii* LPS.

20.6 Conclusions

Application of suitable animal models to study bacterial infections in such a way that replicates the pathogenesis in humans is critical to developing therapeutic strategies. To date, investigators have used a variety of animal species to understand the pathogenic mechanisms. Commonly present in various food sources and soil, *Cronobacter* spp. have been linked to the onset of NEC. Both newborn rat models under hypoxic conditions and mouse models used to explore the pathogenesis of NEC revealed that *C. sakazakii* and *C. muytjensii* can induce NEC-like symptoms. Further studies are clearly required to fully elucidate the molecular mechanisms involved in the onset of NEC. *C. sakazakii*-induced neonatal meningitis is infrequent but often devastating. The neonatal rat model, which mimics human disease, is extremely useful to identify the bacterial virulence factors and host response to infection, thereby helping to create appropriate treatment strategies.

REFERENCES

1. Iversen C, et al. Identification of "*Cronobacter*" spp. (*Enterobacter sakazakii*). *J Clin Microbiol.* 2007; 45:3814–3816.
2. Healy B, et al. *Cronobacter* (*Enterobacter sakazakii*): an opportunistic foodborne pathogen. *Foodborne Pathog Dis.* 2010; 7:339–350.
3. Drudy D, Mullane NR, Quinn T, Wall PG, Fanning S. *Enterobacter sakazakii*: an emerging pathogen in powdered infant formula. *Clin Infect Dis.* 2006; 42:996–1002.
4. Jaradat ZW, Al Mousa W, Elbetieha A, Al Nabulsi A, Tall BD. *Cronobacter* spp.—opportunistic food-borne pathogens. A review of their virulence and environmental-adaptive traits. *J Med Microbiol.* 2014; 63:1023–1037.
5. Hunter CJ, Bean JF. *Cronobacter*: an emerging opportunistic pathogen associated with neonatal meningitis, sepsis and necrotizing enterocolitis. *J Perinatol.* 2013; 33:581–585.
6. Caubilla-Barron J, et al. Genotypic and phenotypic analysis of *Enterobacter sakazakii* strains from an outbreak resulting in fatalities in a neonatal intensive care unit in France. *J Clin Microbiol.* 2007; 45:3979–3985.
7. Feeney A, Kropp KA, O'Connor R, Sleator RD. *Cronobacter sakazakii*: stress survival and virulence potential in an opportunistic foodborne pathogen. *Gut Microbes.* 2014; 5:711–718.
8. Iversen C, Forsythe SJ. Comparison of media for the isolation of *Enterobacter sakazakii*. *Appl Environ Microbiol.* 2007; 73:48–52.
9. Gičová A, Oriešková M, Oslanecová L, Drahovská H, Kaclíková E. Identification and characterization of *Cronobacter* strains isolated from powdered infant foods. *Lett Appl Microbiol.* 2014; 58:242–247.
10. Yan QQ, Condell O, Power K, Butler F, Tall BD, Fanning S. *Cronobacter* species (formerly known as *Enterobacter sakazakii*) in powdered infant formula: a review of our current understanding of the biology of this bacterium. *J Appl Microbiol.* 2012; 113:1–15
11. O'Brien S, Healy B, Negredo C, Fanning S, Iversen C. Evaluation of a new one-step enrichment in conjunction with a chromogenic medium for the detection of *Cronobacter* spp. (*Enterobacter sakazakii*) in powdered infant formula. *J Food Prot.* 2009; 72:1472–1475.
12. Hassan AA, Akineden O, Kress C, Estuningsih S, Schneider E, Usleber E. Characterization of the gene encoding the 16S rRNA of *Enterobacter sakazakii* and development of a species-specific PCR method. *Int J Food Microbiol.* 2007; 116:214–220.
13. Lehner A, Tasara T, Stephan R. 16S rRNA gene based analysis of *Enterobacter sakazakii* strains from different sources and development of a PCR assay for identification. *BMC Microbiol.* 2004; 4:43.
14. Stoop B, Lehner A, Iversen C, Fanning S, Stephan R. Development and evaluation of *rpoB* based PCR systems to differentiate the six proposed species within the genus *Cronobacter*. *Int J Food Microbiol.* 2009; 136:165–168.
15. Zimmermann J, Schmidt H, Loessner MJ, Weiss A. Development of a rapid detection system for opportunistic pathogenic *Cronobacter* spp. in powdered milk products. *Food Microbiol.* 2014; 42:19–25.
16. Mohan Nair MK, Venkitanarayanan KS. Cloning and sequencing of the *ompA* gene of *Enterobacter sakazakii* and development of an *ompA*-targeted PCR for rapid detection of *Enterobacter sakazakii* in infant formula. *Appl Environ Microbiol.* 2006; 72:2539–2546.
17. Mullane N, et al. Molecular analysis of the *Enterobacter sakazakii* O-antigen gene locus. *Appl Environ Microbiol.* 2008; 74:3783–3894.
18. Healy B, et al. Microarray-based comparative genomic indexing of the *Cronobacter* genus (*Enterobacter sakazakii*). *Int J Food Microbiol.* 2009; 136:159–164.
19. Schmid M, et al. Evidence for a plant-associated natural habitat for *Cronobacter* spp. *Res Microbiol.* 2009; 160:608–614.
20. Townsend S, Hurrell E, Forsythe S. Virulence studies of *Enterobacter sakazakii* isolates associated with a neonatal intensive care unit outbreak. *BMC Microbiol.* 2008; 18;8:64.
21. Kucerova E, et al. Genome sequence of *Cronobacter sakazakii* BAA-894 and comparative genomic hybridization analysis with other *Cronobacter* species. *PLoS One.* 2010; 5:e9556.
22. Stephan R, Lehner A, Tischler P, Rattei T. Complete genome sequence of *Cronobacter turicensis* LMG 23827, a food-borne pathogen causing deaths in neonates. *J Bacteriol.* 2011; 193:309–310.
23. Mange JP, et al. Adhesive properties of *Enterobacter sakazakii* to human epithelial and brain microvascular endothelial cells. *BMC Microbiol.* 2006; 6:58.

24. Pagotto FJ, Nazarowec-White M, Bidawid S, Farber JM. *Enterobacter sakazakii*: infectivity and enterotoxin production in vitro and in vivo. *J Food Prot.* 2003; 66:370–375.
25. Hunter CJ, Petrosyan M, Ford HR, Prasadarao NV. *Enterobacter sakazakii*: an emerging pathogen in infants and neonates. *Surg Infect (Larchmt).* 2008; 9:533–539.
26. Hunter CJ, et al. *Enterobacter sakazakii* enhances epithelial cell injury by inducing apoptosis in a rat model of necrotizing enterocolitis. *J Infect Dis.* 2008; 198:586–593.
27. Hunter CJ, Chokshi N, Ford HR. Evidence vs experience in the surgical management of necrotizing enterocolitis and focal intestinal perforation. *J Perinatol.* 2008; 28(Suppl 1):S14–S17.
28. Hunter CJ, et al. *Lactobacillus bulgaricus* prevents intestinal epithelial cell injury caused by *Enterobacter sakazakii*-induced nitric oxide both in vitro and in the newborn rat model of necrotizing enterocolitis. *Infect Immun.* 2009; 77:1031–1043.
29. Liu Q, Mittal R, Emami CN, Iversen C, Ford HR, Prasadarao NV. Human isolates of *Cronobacter sakazakii* bind efficiently to intestinal epithelial cells in vitro to induce monolayer permeability and apoptosis. *J Surg Res.* 2012; 176:437–447.
30. Krishnan S, Prasadarao NV. Outer membrane protein A and OprF: versatile roles in Gram-negative bacterial infections. *FEBS J.* 2012; 279:919–931.
31. Smith SG, Mahon V, Lambert MA, Fagan RP. A molecular Swiss army knife: OmpA structure, function and expression. *FEMS Microbiol Lett.* 2007; 273:1–11.
32. Kim K, et al. Outer membrane proteins A (OmpA) and X (OmpX) are essential for basolateral invasion of *Cronobacter sakazakii. Appl Environ Microbiol.* 2010; 76:5188–5198.
33. Mohan Nair MK, Venkitanarayanan K. Role of bacterial OmpA and host cytoskeleton in the invasion of human intestinal epithelial cells by *Enterobacter sakazakii. Pediatr Res.* 2007; 62:664–669.
34. García-del Portillo F, Finlay BB. Invasion and intracellular proliferation of *Salmonella* within non-phagocytic cells. *Microbiologia.* 1994; 10:229–238.
35. Sansonetti PJ. Genetic and molecular basis of epithelial cell invasion by *Shigella* species. *Rev Infect Dis.* 1991; 13(Suppl 4):S285–S292.
36. Kim KP, Loessner MJ. *Enterobacter sakazakii* invasion in human intestinal Caco-2 cells requires the host cell cytoskeleton and is enhanced by disruption of tight junction. *Infect Immun.* 2008; 76:562–570.
37. Nair MK, Venkitanarayanan K, Silbart LK, Kim KS. Outer membrane protein A (OmpA) of *Cronobacter sakazakii* binds fibronectin and contributes to invasion of human brain microvascular endothelial cells. *Foodborne Pathog Dis.* 2009; 6:495–501.
38. Emami CN, Mittal R, Wang L, Ford HR, Prasadarao NV. Recruitment of dendritic cells is responsible for intestinal epithelial damage in the pathogenesis of necrotizing enterocolitis by *Cronobacter sakazakii. J Immunol.* 2011; 186:7067–7079.
39. Quintero M, et al. Adherence inhibition of *Cronobacter sakazakii* to intestinal epithelial cells by prebiotic oligosaccharides. *Curr Microbiol.* 2011; 62:1448–1454.
40. Rossi M, Young JW. Human dendritic cells: potent antigen-presenting cells at the crossroads of innate and adaptive immunity. *J Immunol.* 2005; 175:1373–1381.
41. Emami CN, Mittal R, Wang L, Ford HR, Prasadarao NV. Role of neutrophils and macrophages in the pathogenesis of necrotizing enterocolitis caused by *Cronobacter sakazakii. J Surg Res.* 2012; 172:18–28.
42. Mittal R, Bulgheresi S, Emami C, Prasadarao NV. *Enterobacter sakazakii* targets DC-SIGN to induce immunosuppressive responses in dendritic cells by modulating MAPKs. *J Immunol.* 2009; 183:6588–6599.
43. Cruz-Córdova A, et al. Flagella from five *Cronobacter* species induce pro-inflammatory cytokines in macrophage derivatives from human monocytes. *PLoS One.* 2012; 7:e52091.
44. Bernardo WM, Aires FT, Carneiro RM, Sá FP, Rullo VE, Burns DA. Effectiveness of probiotics in the prophylaxis of necrotizing enterocolitis in preterm neonates: a systematic review and meta-analysis. *J Pediatr (Rio J).* 2013; 89:18–24.
45. Weng M, Ganguli K, Zhu W, Shi HN, Walker WA. Conditioned medium from *Bifidobacteria infantis* protects against *Cronobacter sakazakii*-induced intestinal inflammation in newborn mice. *Am J Physiol Gastrointest Liver Physiol.* 2014; 306:G779–G787.
46. Schwizer S, Tasara T, Zurfluh K, Stephan R, Lehner A. Identification of genes involved in serum tolerance in the clinical strain *Cronobacter sakazakii* ES5. *BMC Microbiol.* 2013; 13:38.
47. Joseph S, Forsythe SJ. Predominance of *Cronobacter sakazakii* sequence type 4 in neonatal infections. *Emerg Infect Dis.* 2011; 17:1713–1715.

48. Hariri S, Joseph S, Forsythe SJ. *Cronobacter sakazakii* ST4 strains and neonatal meningitis, United States. *Emerg Infect Dis*. 2013; 19:175–177.
49. Stoll BJ, et al. *Enterobacter sakazakii* invades brain capillary endothelial cells, persists in human macrophage influencing cytokine secretion and induces severe brain pathology in the neonatal rat. *Microbiology*. 2007; 153:3538–3547.
50. Mittal R, Wang Y, Hunter CJ, Gonzalez-Gomez I, Prasadarao NV. Brain damage in newborn rat model of meningitis by *Enterobacter sakazakii*: a role for outer membrane protein A. *Lab Invest*. 2009; 89:263–277.
51. Franco AA, et al. Cpa, the outer membrane protease of *Cronobacter sakazakii*, activates plasminogen and mediates resistance to serum bactericidal activity. *Infect Immun*. 2011; 79:1578–1587.
52. Singamsetty VK, Wang Y, Shimada H, Prasadarao NV. Outer membrane protein A expression in *Enterobacter sakazakii* is required to induce microtubule condensation in human brain microvascular endothelial cells for invasion. *Microb Pathog*. 2008; 45:181–191.
53. Li Q, et al. PI3K-dependent host cell actin rearrangements are required for *Cronobacter sakazakii* invasion of human brain microvascular endothelial cells. *Med Microbiol Immunol*. 2010; 199:333–340.
54. Hurrell E, Kucerova E, Loughlin M, Caubilla-Barron J, Forsythe SJ. Biofilm formation on enteral feeding tubes by *Cronobacter sakazakii*, *Salmonella* serovars and other *Enterobacteriaceae*. *Int J Food Microbiol*. 2009; 136:227–231.
55. Jung JH, Choi NY, Lee SY. Biofilm formation and exopolysaccharide (EPS) production by *Cronobacter sakazakii* depending on environmental conditions. *Food Microbiol*. 2013; 34:70–80.
56. Wang L, Hu X, Tao G, Wang X. Outer membrane defect and stronger biofilm formation caused by inactivation of a gene encoding for heptosyltransferase I in *Cronobacter sakazakii* ATCC BAA-894. *J Appl Microbiol*. 2012; 112:985–997.
57. Hartmann I, Carranza P, Lehner A, Stephan R, Eberl L, Riedel K. Genes involved in *Cronobacter sakazakii* biofilm formation. *Appl Environ Microbiol*. 2010; 76:2251–2261.
58. Choi Y, Kim S, Hwang H, Kim KP, Kang DH, Ryu S. Plasmid-encoded MCP is involved in virulence, motility, and biofilm formation of *Cronobacter sakazakii* ATCC 29544. *Infect Immun*. 2015; 83:197–204.
59. Fehr A, Eshwar AK, Neuhauss SC, Ruetten M, Lehner A, Vaughan L. Evaluation of zebrafish as a model to study the pathogenesis of the opportunistic pathogen *Cronobacter turicensis*. *Emerg Microbes Infect*. 2015; 4:e29.
60. Sivamaruthi BS, Ganguli A, Kumar M, Bhaviya S, Pandian SK, Balamurugan K. *Caenorhabditis elegans* as a model for studying *Cronobacter sakazakii* ATCC BAA-894 pathogenesis. *J Basic Microbiol*. 2011; 51: 540–549.
61. Sivamaruthi BS, Prasanth MI, Balamurugan K. Alterations in *Caenorhabditis elegans* and *Cronobacter sakazakii* lipopolysaccharide during interaction. *Arch Microbiol*. 2015; 197:327–337.

21

Escherichia

Dongyou Liu

CONTENTS

21.1 Introduction

The genus *Escherichia* comprises a small group of Gram-negative, rod-shaped bacterial species that form a part of gut microbiota in mammalian hosts including humans. As part of the well-known and yet still incompletely understood *Escherichia* species, *Escherichia coli* (represented by O157:H4 strain) is an important cause of intestinal (enteritis, diarrhea, and dysentery), or extraintestinal diseases (urinary tract infection, intra-abdominal infection, pneumonia, neonatal meningitis, and sepsis). By first presenting a brief overview on *E. coli* classification, biology, epidemiology, clinical features, diagnosis, treatment, and prevention, this chapter then discusses laboratory models that have been applied to the study of *E. coli*, with the goal of unlocking the secrets of its pathogenic mechanisms and aiding in the development of improved anti-infection strategies.

21.1.1 Classification, Morphology, and Genomics

21.1.1.1 Classification

Covering a group of Gram-negative, rod-shaped, motile, non-spore-forming bacilli (informally known as "coliforms"), the genus *Escherichia* (named after its discoverer Theodor Escherich) is classified taxonomically in the family Enterobacteriaceae, order Enterobacteriales, class Gammaproteobacteria, phylum Proteobacteria. To date, six species are recognized within the genus *Escherichia*, i.e., *Escherichia albertii*, *Escherichia coli* (the type-species, obsolete synonyms: *Bacterium coli* and *Bacillus coli*), *Escherichia fergusonii*, *Escherichia hermannii*, *Escherichia marmotae*, and *Escherichia vulneris*, with two former *Escherichia* species, *Escherichia adecarboxylata* and *Escherichia blattae*, being redesignated as *Leclercia adecarboxylata* and *Shimwellia blattae*, respectively [1,2].

As a key member of the genus, *E. coli* represents a predominant facultative anaerobe in the gastrointestinal tract of warm-blooded animals including humans. Although *E. coli* is largely considered as a harmless commensal species and has been extensively used in molecular biology as a cloning host, some of its strains are known to be involved in various human diseases, such as (1) intestinal disease (gastroenteritis), (2) urinary tract infection (UTI, which may sometimes evolve to hemolytic uremic syndrome or HUS), and (3) blood and central nervous system infections (sepsis/meningitis, especially neonatal meningitis) [3,4]. Nonetheless, other members of the genus *Escherichia* may be occasionally involved in human infections.

On the basis of their unique interactions with eukaryotic cells, adhesion/colonization mechanisms, toxin/virulence factor production, and clinical disease profiles, pathogenic *E. coli* strains may be divided into two categories: intestinal and extraintestinal [1]. The category causing intestinal infections consists of 10 pathotypes: enterotoxigenic *E. coli* (ETEC), enteropathogenic *E. coli* (EPEC), enterohemorrhagic *E. coli* (EHEC), enteroaggregative *E. coli* (EAEC), Shiga-toxin-producing enteroaggregative *E. coli* (STEAEC), enteroinvasive *E. coli* (EIEC), diffusely adhering *E. coli* (DAEC), cell detaching *E. coli* (CDEC), necrotoxic *E. coli* (NTEC), and adherent invasive *E. coli* (AIEC). Of these, ETEC, EPEC, EHEC, EAEC, STEAEC, EIEC, and DAEC are major diarrheagenic *E. coli* pathotypes (Table 21.1); CDEC and NTEC are implicated in a very small number of diarrhea-related cases, and AIEC is an enteric pathogen causing Crohn disease in which diarrhea may present as one of the symptoms [5,6]. The category responsible for extraintestinal infections includes three pathotypes: septicemia-causing *E. coli* (SCEC), neonatal-meningitis-causing *E. coli* (NMEC), and uropathogenic *E. coli* (UPEC), leading to septicemia, pediatric meningitis, and UTIs, respectively [1].

Besides pathotype classification, serological analysis of *E. coli* surface antigens (i.e., somatic O and flagellar H antigens, as well as capsular K and fimbrial F antigens) allows distinction of >200 serogroups (serotypes), which are designated by specific combinations of O and H antigens (e.g., O157:H7). Occasionally, the O-antigen [forming part of lipopolysaccharide (LPS) layer] is "masked" by a heat-labile, acidic polysaccharide capsule (K-antigen). While specific serogroups are often associated with certain clinical syndromes, some may belong to more than one category of *E. coli* pathotypes.

More recently, application of genotyping techniques has helped clearly differentiate *E. coli* strains into six phylogenetic groups (A, B1, B2, D, E, and Shigella). Nonetheless, due to historic considerations, the phylogenetic group Shigella has continually been treated as a separate genus *Shigella* (which consists of the species *Shigella dysenteriae*, *Shigella flexneri*, *Shigella boydii*, *Shigella sonnei*) despite its close phylogenetic relationship with other members of *E. coli* and possibly representing some of the clones of *E. coli*.

21.1.1.2 Morphology

E. coli is a rod-shaped bacterium of about 0.6 μm in diameter and 2 μm in length. The bacterium forms nonspreading black colonies with a characteristic greenish-black metallic sheen on eosin methylene blue (EMB) agar, and deep red colonies on MacConkey agar. Other morphological features of note include:

TABLE 21.1

Characteristics of Major Diarrheagenic *E. coli* Pathotypes

Pathotype	Serogroup[a]	Human Disease	Other Features
Enterotoxigenic (ETEC)	O6, O8, O15, O20, O25, O27, O63, O78, O80, O85, O115, O128ac, O139, O148, O153, O159, O167	Watery diarrhea (without fever) in children <5 years old, and travelers' diarrhea	Colonizing the small intestine, ETEC employs fimbrial adhesins to bind enterocytes and produces two proteinaceous enterotoxins: heat-labile LT enterotoxin (with similarity to cholera toxin in structure and function) and heat-stable ST enterotoxin (inducing cGMP accumulation in the target cells and a subsequent secretion of fluid and electrolytes into the intestinal lumen). ETEC strains are noninvasive and do not leave the small intestinal lumen.
Enteropathogenic (EPEC)	O26, O55, O86, O111, O114, O119, O125, O126, O127, O128, O142, O158	Profuse watery diarrhea in infants	Colonizing the small intestine, EPEC uses an adhesin known as intimin to bind host intestinal cells, leading to rearrangement of actin, A/E lesion, and subsequent diarrhea. Being moderately invasive (with the ability to enter host cells), EPEC produces a number of virulence factors that are similar to those found in Shigella and that can elicit an inflammatory response.
Enterohemorrhagic (EHEC)	O26, O91, O111, O157 among others	Watery diarrhea, hemorrhagic colitis/ bloody diarrhea, HUS	Colonizing the distal ileum and colon, EHEC induces A/E lesion and employs fimbriae for attachment (*E. coli* common pilus, ECP). EHEC produces a phage-encoded Shiga toxin that elicits an intense inflammatory response. Being moderately invasive, EHEC (particularly O157:H7 strain) causes more serious diarrhea (bloody diarrhea without fever) than EPEC, in addition to HUS and sudden kidney failure.
Enteroaggregative (EAEC)	O3, O15, O44, O77, O104, O126	Travelers' diarrhea, HUS (stx^{+}), and persistent diarrhea (infants)	Colonizing the small intestine and/or colon, EAEC (whose fimbriae aggregate tissue culture cells) produces a hemolysin and an ST enterotoxin similar to that of ETEC. EAEC is noninvasive and binds to the intestinal mucosa, leading to watery diarrhea without fever.
Shiga-toxin-producing enteroaggregative (STEAEC)	O104	Food poisoning	STEAEC is an EAEC strain (O104:H4) showing typical EHEC phenotypes (Stx production and strong cell adherence) that causes HUS.
Enteroinvasive (EIEC)	O28ac, O29, O112ac, O124, O136, O143, O144, O152, O164, O167	Shigellosis/bacillary dysentery	Colonizing the colon, EIEC causes a syndrome with profuse diarrhea and high fever, which is identical to shigellosis.
Diffusely adherent (DAEC)	O86, O127, O142, O158	Persistent watery diarrhea (<5 years old)	Colonizing the intestine, DAEC shows a distinct diffuse adherence to epithelial cells (HEp-2 or HeLa). Isolated from children, DAEC harbors Afa/Dr genes.

[a] Some serogroups are associated with more than one diarrheagenic *E. coli* pathotype. For example, serogroup O86 is mainly considered as EPEC, but it has also been implicated in ETEC, EAEC, and DAEC infections.

1. *Fimbriae (pili)*, which are of two kinds, common or conjugative. Common fimbriae (about 100–1000 per cell) comprise mainly an acidic hydrophobic protein called fimbrin and may be divided into seven groups according to the amino acid sequence of their major fimbrin. Conjugative fimbriae (also known as sex pili, usually number one or a few copies per cell) facilitate contact between the donor and recipient bacteria, permitting transfer of DNA during conjugation. Fimbriae are highly antigenic and contain many F antigens.
2. *Flagella*, which are responsible for bacterial mobility. *E. coli* flagella (about 5–10 copies per cell; 5–10 μm in length) are composed of a long filament, a hook, and a basal body and are arranged randomly around the cell surface (a pattern known as peritrichous flagellation). The principal component of *E. coli* flagella is an *N*-methyl-lysine-rich protein (55 kDa in size) known as flagellin, and around 20,000 subunits of flagellin are needed to make the flagellar filament. Flagella are highly antigenic and include a large number of H antigens.
3. *Capsule and outer membrane*. The outer membrane of *E. coli* is composed of a lipid bilayer, which in turn consists of a phospholipid inner leaflet and an LPS outer leaflet, together with several kinds of membrane proteins in the periplasm (the space between the inner and outer leaflets/membranes). In the outer membrane, a glycolipid (called the enterobacterial common antigen or ECA) is found in *E. coli*. In some *E. coli* strains, the outer membrane is covered by a polysaccharide capsule containing K antigens. Under conditions of high osmolarity, low temperature, and low humidity, other polysaccharides such as M antigens (colanic acids, which are polymers of glucose, galactose, fucose, and galacturonic acid) are synthesized.
4. *Periplasm and cell wall*. The periplasm of *E. coli* is osmotically active and contains over 60 known proteins, including binding proteins for amino acids, sugars, vitamins, and ions; degradative enzymes (phosphatases, proteases, and endonucleases); and antibiotic detoxifying enzymes (β-lactamases, alkyl sulfodehydrases, and aminoglycoside phosphorylating enzymes). The cell wall of *E. coli* is responsible for cell shape and rigidity and is composed of a peptidoglycan layer (of one or a few molecules thick) that is anchored to the outer membrane via covalent links to the major membrane lipoprotein and noncovalent links to porins.
5. *Cytoplasmic membrane*. The cytoplasmic membrane of *E. coli* is made up of about 200 distinct proteins (including those involved in peptidoglycan biosynthesis, cell wall elongation, and cell division, as well as penicillin-binding proteins that covalently bind β-lactam antibiotics) and four kinds of phospholipids.
6. *Cytoplasm*. The cytoplasm of *E. coli* contains ribosomes and other organelles in which biological/biochemical activities necessary for growth are carried out. These activities include metabolic fueling (production of energy, reducing power, and precursor metabolites), biosynthesis of building blocks, polymerization into macromolecules, and assembly of cell structures [1].

21.1.1.3 Genomics

The genome of *E. coli* strain MG1655 (a laboratory strain K-12 derivative) consists of a circular DNA molecule 4.64 Mb in length, with 4288 protein-coding genes (organized into 2584 operons), 7 rRNA operons, and 86 tRNA genes, together with a number of transposable genetic elements, repeat elements, cryptic prophages, and bacteriophage remnants. Some pathogenic *E. coli* strains possess larger genomes than that of the commensal K-12 strain, such as enterohemorrhagic *E. coli* strain O157:H7 Sakai (5.50 Mb); enteroaggregative *E. coli* strain O42 (5.36 Mb); and UPEC isolates CFT073 (5.23 Mb), 536 (4.94 Mb), and UTI89 (5.07 Mb).

While many strains (e.g., strain MG1655) do not have plasmid, others possess one to five plasmids, which along with chromosomal pathogenicity islands contribute to the plasticity of the *E. coli* genome. It is notable that all major diarrheagenic *E. coli* pathotypes carry at least one virulence-related property on a plasmid. The plasmid is typically large (>60 megadalton or MDa), of low copy number, of either conjugative or transmissible incompatibility group, and encodes multiple virulence factors. In addition to plasmid-encoded clusters of virulence traits and chromosomal pathogenicity islands, transposon-encoded individual traits (e.g., Shiga toxin) may be also present [7,8].

21.1.2 Biology and Epidemiology

E. coli is a chemoheterotroph with the ability to utilize a variety of sugars or amino acids, and grows rapidly in nutrient rich broths (containing amino acids, nucleosides, sugars, and vitamin precursors, etc.). While some strains found in nature have a single auxotrophic requirement such as thiamin, the growth of many strains may be inhibited by the presence of single amino acids such as serine, valine, or cysteine. *E. coli* tolerates temperature between 8°C and 48°C and grows optimally at 39°C. The bacterium survives between pH 6.0 and 8.0, but is unable to grow in media containing >0.65 M NaCl. In response to changes in the osmotic pressure of the medium, *E. coli* increases its concentration of ions, especially K+ and glutamate.

E. coli is well adapted to the mammalian intestine, including the large intestine, the ileum, and the distal segment of the small intestine. Interestingly, some diarrheagenic *E. coli* strains infect the sparsely colonized small bowel (ETEC, EPEC), while others inhabit in the colonic mucosa (EIEC, EHEC, and probably EAEC), with close proximity to the enterocytes. In addition, some pathotypes (EHEC, EPEC, and EIEC) utilize type 3 secretion system (T3SS) for translocating bacterial proteins (or effectors) directly into the eukaryotic host cell to subvert host cell processes, while others (ETEC, EAEC, STEAEC, DAEC, and AIEC) are non-T3SS dependent and require effective colonization factors and secreted toxins for virulence [9].

Discharged from the intestine into soil and water, *E. coli* does not survive long (a few days). Mammals (typically cattle, chickens, deer, sheep, and pigs) usually become colonized with *E. coli* shortly after birth, possibly during passage through the birth canal, or via the fecal–oral route from the mother or other attendants [10]. In humans, *E. coli* gains entry after exposure to contaminated food, beverages, water, animals, or other persons. Upon ingestion, *E. coli* multiplies rapidly in the large intestine and binds tightly to cells in the intestinal lining, facilitating absorption of the toxin into the small capillaries within the bowel wall and subsequently attaching to globotriaosylceramide (Gb3) receptors.

It is noteworthy that of >200 *E. coli* serotypes identified to date, those producing Shiga toxin (Stx), which is indistinguishable from that produced by *S. dysenteriae* type 1, are frequently found in contaminated foods and beverages (e.g., ground beef, venison, sausages, dried salami, unpasteurized milk and cheese, unpasteurized apple juice and cider, orange juice, alfalfa, parsley, radish sprouts, lettuce, cabbage, spinach, fruit nuts, berry, cookie dough, and water). The most notorious Stx-producing *E. coli* strains are O157:H7 and, to a lesser extent, O121:H19. Strain O157:H7 is responsible for the majority of *E. coli*-related human illnesses such as bloody diarrhea and HUS, the latter being defined by the trilogy of hemolytic anemia (destruction of red blood cells), thrombocytopenia (low platelet count), and acute kidney failure. First recognized as a human pathogen in foodborne outbreaks associated with hamburgers from a fast food chain restaurant in 1982, *E. coli* O157:H7 was later shown to be responsible for a 2006 outbreak in Iowa and Minnesota following consumption of lettuce grown in close proximity to a dairy farm from which wastewater contaminated with animal feces was used for irrigation. Stx-producing *E. coli* strains are hardy organisms that can survive several weeks on surfaces (e.g., countertops) and up to a year in compost. In addition, Stx-producing *E. coli* strains have a very low infectious dose (<50 organisms). Together, these attributes make Stx-producing *E. coli* strains highly dangerous foodborne pathogens.

21.1.3 Clinical Features and Pathogenesis

Pathogenic *E. coli* is implicated in a large number of clinical diseases in humans, including intestinal infection (watery diarrhea or dysentery), UTI, blood infection (septicemia), and meningitis, in addition to pneumonia and intra-abdominal infection (cholecystitis, cholangitis, and peritonitis) [11].

21.1.3.1 ETEC

ETEC is usually acquired from food or water contaminated with human or animal feces. This pathotype causes watery diarrhea of varied severity (ranging from mild, self-limiting disease to severe cholera-like, life-threatening illness). With a sudden onset of watery stool (without blood or inflammatory cells) and vomiting, ETEC infection may lead to dry mouth, rapid pulse, lethargy, decreased skin turgor, decreased blood pressure, muscle cramps, and shock due to progressive loss of fluids (dehydration) and electrolytes (sodium, potassium, chloride, and bicarbonate). The diarrhea is self-limited and generally lasts for only 3–4 days. If hydration is maintained, the patient may recover without any sequelae. ETEC is a major cause of diarrhea in infants younger than 2 years of age in the developing world, and it is also associated with travelers' diarrhea [12].

21.1.3.2 EPEC

EPEC is an attaching and effacing (A/E) pathogen that induces a characteristic A/E lesion (or pedestal-like structure) on the lumenal surfaces of host small intestine, without the production of Shiga toxins. In addition to watery diarrhea containing mucus but not blood, other symptoms associated with EPEC infection include vomiting, fever, malaise, and dehydration. While these symptoms often persist for several days, chronic EPEC infection is sometimes observed. Occasionally, EPEC may cause bloody diarrhea after the colonization of the mid-distal small intestine (ileum). EPEC is separated into two groups according to the presence/absence of EPEC adherence factor plasmid (or pEAF): typical EPEC (tEPEC) and atypical EPEC (aEPEC). tEPEC possesses pEAF, forms localized adherence pattern, and includes classical O26, O55, O86, O111, O119, O125, O126, O127, O128ac, O142, O158, whereas aEPEC lacks pEAF, but harbors other virulence factors and may form diffuse or aggregative adherence patterns. EPEC is a common cause of watery diarrhea among infants in the developing world and is also implicated in sporadic diarrheal outbreaks among infants (e.g., day-care facilities) in the developed world [13].

21.1.3.3 EHEC

EHEC is characterized by the production of Shiga toxins (Stx1 and/or Stx2, leading to its alternative name of STEC) and the formation of A/E lesion, notably in the cecum and ascending colon [14–16]. EHEC typically causes an afebrile bloody colitis (bloody stools with ulcerations of the bowel) known as hemorrhagic colitis (HC), which is characterized by the sudden onset of abdominal pain, severe cramps, and diarrhea within 24 h. In about 10% of patients (e.g., children and the elderly), infection with EHEC O157 may result in HUS, which is defined by acute renal failure, hemolytic anemia, and thrombocytopenia. Colonoscopy examination reveals the presence of edema, erythema (redness), hemorrhage, erosion, and, occasionally, a long ulcer-like lesion, with a marked narrowing of the luminal space. Histologic examination indicates destruction of the surface epithelium, neutrophil infiltration of the lamina propria, and the formation of crypt abscesses. In patients (especially children of <5 years of age and the elderly) with diarrhea and HUS, renal pathology consists of endothelial swelling and glomerular thrombosis with congested rather than ischemic glomeruli. The most prevalent EHEC serotype causing outbreaks in North America and other parts of the world is O157:H7 [17,18]. Besides Shiga toxin, which may account for the severe complications including HUS, O157:H7 expresses several other virulence factors including intimin, translocated intimin receptor (Tir), a T3SS, and enterohemolysin [19]. The genes encoding many of these factors are located on a 44 kb pathogenicity island (also known as the locus of enterocyte effacement or the LEE locus) [20–22]. Cattle act as a primary reservoir for EHEC, although vegetables (lettuce, spinach, and sprouts) and fruits may also serve as vehicles for EHEC outbreaks [23].

21.1.3.4 EAEC

EAEC is noted for forming an AA pattern (which is characterized by prominent, "stacked brick" autoagglutination of the bacterial cells to each other) during its adhesion to HEp-2 cells [24,25]. EAEC infection manifests clinically as watery diarrhea with or without blood and mucus, abdominal pain, nausea, vomiting, and low-grade fever. Since 1980s, EAEC has been recognized as a causative agent of persistent diarrhea in malnourished children in the developing world, and it also causes both outbreaks and sporadic diarrhea among travelers and immunocompromised individuals (specifically HIV-infected patients) in the developed world. Individuals with single-nucleotide polymorphisms (SNPs) in the IL-8 gene promoter and lactoferrin gene may show higher susceptibility to EAEC infection [26]. EAEC is primarily transmitted through contaminated food and water. Food handlers (especially those working in tourist hotels) are important carriers of EAEC. Interestingly, EAEC-induced travelers' diarrhea occurs constantly throughout winter and summer, whereas other diarrheagenic *E. coli* strains such as ETEC, EPEC, and EIEC show lower rates of infection in winter [27].

21.1.3.5 STEAEC

STEAEC is an EAEC strain (O104:H4) that has acquired typical EHEC phenotypes such as Stx production and strong cell adherence. STEAEC O104:H4 is responsible for HUS in a high percentage of patients, with a mortality rate of 1%.

21.1.3.6 EIEC

EIEC is pathogenetically, biochemically, and genetically related to *Shigella* spp. and principally resides in the large intestine. EIEC causes an invasive inflammatory colitis with a watery diarrhea syndrome that is indistinguishable from those caused by other *E. coli* pathotypes and *Shigella* spp. In severe cases, scanty dysenteric stools may contain blood and mucus, and this may occur together with fever and severe cramps.

21.1.3.7 DAEC

DAEC is differentiated from other diarrheic *E. coli* by its diffuse adherence to epithelial cells in the classical laboratory assay of adherence to HEp-2 or HeLa cells. This diffuse adherence pattern is essentially linked to the production of adhesins encoded by a family of *afa/dra/daa*-related operons. DAEC causes a watery diarrhea in adults and children (with increased risk in those aged 18 months to 5 years) [28,29].

21.1.3.8 AIEC

AIEC adheres to intestinal epithelial cells through type 1 pili (especially FimH adhesin variant) that interacts with host glycoprotein CEACAM6 in a mannose-associated manner, gains access to the lamina propria, and replicates in macrophages (which represent an environment with acidic pH, oxidative stress, active proteolytic enzymes, and antimicrobial compounds) [30]. Transient AIEC colonization elicits intestinal inflammation and alters microbiota composition leading to Crohn's disease (CD), with symptoms ranging from fever, fatigue, abdominal pain, cramping, nausea, vomiting, bloody stool, mouth sores, reduced appetite, weight loss, and perianal disease to diarrhea. Host deficiencies in CD patients linked with the increased ability of AIEC LF82 to cause infection include the overexpression of the CEACAM6 and Gp96 receptors in the apical membrane of intestinal epithelial cells (which facilitates AIEC adhesion and invasion), defects in autophagy related to NOD2, ATG16L1, and IRGM function and expression (which impair the ability of host cells to resolve infections), altered bile salts metabolism (which enhances the expression of long polar fimbriae in AIEC, permitting better translocation *via* M cells), and decreased levels of protease meprin (which degrades type 1 pili) [31].

21.1.3.9 SCEC

SCEC is associated with sudden high fever with chills, nausea, vomiting, diarrhea, abdominal pain, hypotension, confusion, anxiety, tachypnea (short of breath), tachycardia (rapid heart rate), and uremia.

21.1.3.10 NMEC

NMEC is responsible for fever, failure to thrive, neurologic signs, jaundice, decreased feeding, periods of apnea, and listlessness in neonates. Infants of <1 month of age may also show irritability, lethargy, vomiting, lack of appetite, and seizures, while those >4 months of age display neck rigidity, tense fontanels, and fever. Older children and adults tend to present with headache, vomiting, confusion, lethargy, seizures, and fever.

21.1.3.11 UPEC

UPEC utilizes type 1 fimbriae (particularly FimH) to interact with urinary tract host epithelia and is among the most prevalent extraintestinal bacteria accounting for 90% of all UTIs, including both cystitis and pyelonephritis. This pathotype has evolved a multitude of virulence factors and strategies to facilitate its growth and persistence within the adverse settings of the host urinary tract [32,33].

21.1.4 Diagnosis

E. coli grown on EMB agar produces black colonies with a diagnostic greenish-black metallic sheen. In addition, being lactose positive, *E. coli* generates deep red colonies on MacConkey agar, as fermentation of lactose decreases the medium's pH and darkens the medium. Other biochemical features of *E. coli* include the ability to reduce nitrates to nitrites and to generate succinate, ethanol, acetate, and carbon dioxide. While most *E. coli* strains are positive for catalase, they are negative for oxidase, citrate, urease, and hydrogen sulfide. Further, *E. coli* is positive for indole production and the methyl red test. Given that about 98% of *E. coli* strains are positive in the indole test, it offers a useful approach to differentiate *E. coli* from other members of the family Enterobacteriaceae.

Use of epithelial cell lines (e.g., HEp-2 and HeLa) provides another means to characterize *E. coli* pathotypes. For instance, HEp-2 cell-adherence assay enables differentiation among EPEC, EAEC, and DAEC isolates, and HeLa cell monolayers facilitate identification of plaque-forming EIEC isolates. Further, Y1 adrenal and Chinese hamster ovarian cells may display morphological changes in the presence of heat-labile enterotoxin (LT)-producing *E. coli.* Infant mouse physiological assay is applicable for heat-stable enterotoxin (ST) identification. Guinea pig may be used for detecting keratoconjunctivitis caused by EIEC isolates. Moreover, Stx-producing *E. coli* may be verified by mammalian cell cultures, which are rapidly killed by the Stx.

By targeting the O (lipopolysaccharide), H (flagellar), K (capsular), and F (fimbrial) antigens, serological procedures (e.g., mannose-resistant agglutination of erythrocytes, passive latex agglutination, enzyme-linked immunosorbent assay, immunoprecipitation in agar, and Biken test) employing polyclonal or monoclonal antibodies offer a more specific way to define *E. coli* strains and pathotypes.

Recent development of nucleic-acid-based techniques makes rapid, sensitive, and specific identification and typing of *E. coli* strains and pathotypes possible. These new-generation methods detect E. coli genes that encode toxins, pili, and other virulence factors. Indeed, polymerase chain reaction (PCR) procedures targeting the *pap*, *afa*, and *sfa* genes have proven valuable for UPEC identification. Application of multiplex and real-time PCR allows for simultaneous detection of several diarrheagenic *E. coli* pathotypes [34–38].

21.1.5 Treatment and Prevention

As most patients with *E. coli* infection recover in a week or two without any long-term sequelae, good supportive care (fluid and electrolyte balance, nutrition) is all that is needed. However, patients who develop HUS (usually 2 weeks after *E. coli* infection) often require hospitalization (3 days to 3 months for children, and even longer for adults). Apart from supportive care, antibiotics may be administered. These range from amoxicillin, cephalosporins, carbapenems, aztreonam, trimethoprim–sulfamethoxazole, ciprofloxacin, and nitrofurantoin to aminoglycosides. It should be noted that some *E. coli* strains may be resistant to β-lactam antibiotics such as penicillins and cephalosporins as well as carbapenem [39].

Considering the widespread dissemination of Shiga-toxin-producing *E. coli* strains nowadays, and their possible transmission via contaminated foods, as well as direct person-to-person/animal-to-person contact, prevention of *E. coli* infection at the individual level should focus on (1) practicing meticulous personal hygiene; (2) proper cleaning, processing, and cooking of fruits or vegetables; (3) avoiding cross-contamination during preparation and cooking of food; (4) avoiding touching/petting farm animals; and (5) avoid drinking any nonchlorinated water.

Application of enterobacteria phage T4 in a mist, spray, or wash on live animals helps reduce potential *E. coli* O157:H7 contamination of meat products. Furthermore, development of effective vaccines will contribute to the control and prevention of *E. coli* infection. In this regard, an *E. coli* O157:H7 O-specific polysaccharide conjugated to recombinant exotoxin A of *Pseudomonas aeruginosa* (O157-rEPA) has been shown to elicit a mitigating immune response in both adults and children 2–5 years of age [40].

21.2 Laboratory Models

In order to uncover the virulence mechanisms of pathogenic *E. coli*, understand bacteria–host cell interactions, determine colonization and attachment factors, investigate host immune response, and evaluate candidate vaccines, use of laboratory models (both *in vivo* and *in vitro*) is critical [41–43].

21.2.1 Animal Models

A variety of animals have been utilized as models for investigating *E. coli* infection and disease. These range from mice, rats, rabbits, chickens, pigs, cows, and dogs to baboons and macaques. For example, gnotobiotic piglets, infant rabbits, calves, chickens, and macaques have proven valuable for assessing characteristic A/E lesions associated with *E. coli* O157:H7. Greyhounds and rabbits are useful for examining naturally occurring HUS-like diseases [i.e., idiopathic cutaneous and renal glomerular vasculopathy of greyhounds (CRVGs) or Alabama rot] [21,44].

21.2.1.1 Rodents

Mice do not show signs of intestinal disease following oral infection with EHEC, although they do develop kidney damage (acute tubular necrosis instead of prothrombotic condition seen in humans) and subsequently die [45]. Pretreatment with antibiotics (e.g., streptomycin), use of germfree animals, or dietary-induced changes (e.g., protein–calorie malnutrition) appears to reduce the natural resistance of mice to EHEC colonization [46,47]. Another useful mouse model is the intact commensal flora (ICF) model for which mice harboring normal flora are utilized. ICF mice infected with O157:H7 enable monitoring of both colonization and disease (e.g., ruffled fur, lethargy, weight loss, renal tubular damage, and mortality) [21,48]. A natural mouse pathogen *Citrobacter rodentium* (which carries a homolog of the LEE locus of EPEC and EHEC and has the capacity to evoke A/E lesions) may be exploited as a surrogate for *E. coli* O157:H7 for evaluation of the virulence mechanisms of EHEC. A mouse intoxication model involving parenteral injection of Stx facilitates *in vivo* can be used for assessment of the effects of Shiga toxin (especially Stx2), which are responsible for damage to renal cortical tubule epithelial cells, paralysis, and mortality (i.e., symptoms associated with HC and HUS) [21].

A neonatal rat model provides a useful method to evaluate systemic infection due to neuropathogenic *E. coli*, as colonization of the gastrointestinal tract leads to dissemination of the pathogen along the gut-lymph-blood-brain course of infection. Similar to human host, neonatal rat shows local inflammation upon *E. coli* penetration of the central nervous system [49]. Additionally, use of a rat model of unilateral *E. coli* epididymitis reveals that despite the eradication of the pathogen by antimicrobial agents accompanied by the reduction of epididymal damage, inflammation in the contralateral epididymis is nevertheless present, which may contribute to impaired fertility [50]. On the whole, rodent models are relatively inexpensive and easy to care for and handle.

21.2.1.2 Rabbits

Suckling New Zealand White (NZW) rabbits are susceptible to oral infection with *E. coli* O157:H7, leading to watery diarrhea in young (5–10 days old) but not older (20 days old) rabbits. Histological abnormalities (e.g., edema, hemorrhage, the presence of an inflammatory infiltrate, and mucosal epithelial apoptosis) are observed in the colon. However, suckling rabbits do not develop signs of renal disease [51]. These data suggest that suckling rabbits may be a useful model for studying the intestinal manifestations of EHEC infection. Nevertheless, older rabbits appear to be recalcitrant to colonization by EHEC [52].

21.2.1.3 *Pigs*

The intestine of gnotobiotic piglets [i.e., cesarean-derived colostrum-deprived (CDCD) piglets] is permissive for EHEC replication, as oral infection of CDCD piglets with 10^{10} CFU of *E. coli* O157:H7 leads to watery diarrhea without blood by 2–4 days postinoculation. Histologic abnormalities include destruction of the mucosal brush border and inflammation (so called A/E lesions) in the cecum and colon. Apart from intestinal disease, gnotobiotic piglets also exhibit signs of central nervous system disease (e.g., vascular damage in the cerebellum, ataxia) and renal lesion upon oral or intraperitoneal administration of O157:H7. The kidneys of infected gnotobiotic piglets show signs of diffuse glomerular endothelial swelling and congestion as well as a narrowing of the capillary vessels. In addition, morphological signs of thrombotic microangiopathy, which is characteristic of HUS in humans, are also noticeable. Naturally born piglets are found to exhibit more severe neurological disease and more quickly succumb to EHEC infection than traditional CDCD animals, although no signs of kidney damage are found. Nonetheless, maintenance of gnotobiotic piglets requires considerable veterinary skill, space, and expense [21].

21.2.1.4 Caenorhabditis elegans *Nematode*

C. elegans is susceptible to EPEC infection and succumbs to the disease via a "slow killing" mechanism that results from bacterial accumulation in the nematode intestine and formation of microcolonies similar to those formed during EPEC infection of cultured epithelial cells.

21.2.2 *In Vitro* Models

In vitro systems [e.g., cell monolayers, transwells, organoids, *in vitro* organ culture (IVOC), and *ex vivo* cultures of biopsies] provide excellent models for elucidation of *E. coli* pathogenesis, including aspects of cell adherence and Stx toxicity. Due to their convenience, rapid growth, uniformity, availability of antibody reagents, and variation in genetic backgrounds, established cell lines represent particularly valuable tools for the study of *E. coli*.

Epithelial cells (e.g., MDCK, Caco-2, T84, and HT29) facilitate evaluation of *E. coli* adherence mechanism and disruption of epithelial barrier function during infection, as they have the ability to mimic polarization by forming tight junctions and brush border. Endothelial cells [human umbilical vein endothelial cells (HUVECs) or human glomerular microvascular endothelial cells (GMVECs)] are valuable for assessing the effect of Stx on vascular cell integrity and cytokine response. Macrophage-like cell lines (e.g., J774 and U937) enable assessment of *E. coli* uptake by professional phagocytes and identification of EPEC EspF, EspJ, and EspH effector functions. In addition, differing from polarized cells of the intestinal epithelium, nonpolarized cell lines (e.g., HeLa cells) are extensively used to study basic aspects of EPEC–host interaction and T3SS function [42].

Transwell system allows for the development of polarized cell monolayers with features characteristic of differentiated cells (e.g., the formation of tight junctions and the expression of unique cellular factors), which are useful for exploring the mechanisms of both *E. coli* O157:H7 binding to and Stx transit across the epithelium.

Organoid system involves cells grown on a scaffold under microgravity conditions that form pieces of tissue-like material. This permits detailed investigation of *E. coli* O157:H7 adherence, formation of A/E lesion, and other forms of host cell damage.

IVOCs involve infection of freshly obtained human intestinal biopsies kept in tissue culture media under high oxygen pressure to delay ischemia and cell death. As the infected tissue is as close as one can get *in vitro* to native live tissue, IVOCs helps uncover insight on *E. coli* O157:H7 adherence to gut mucosal cells and subsequent damage to the intestinal cells. Furthermore, the development of the polarized IVOC model allows better mimicking of the *in vivo* situation during specific apical EPEC infection. Nonetheless, IVOC is limited to several hours until the tissue dies, is technically challenging, and shows sample variability.

21.3 Conclusion

Constituting a key member within the genus *Escherichia*, *E. coli* is a remarkable bacterium that encompasses a large number of strains of varied pathogenic potential. Although many *E. coli* strains are common symbiotic organisms in the gut, some demonstrate propensity to cause intestinal and extraintestinal diseases in humans. To date, 10 pathotypes have been identified among *E. coli* strains involved in intestinal diseases, including ETEC, EPEC, EHEC, EAEC, STEAEC, EIEC, DAEC, CDEC, NTEC, and AIEC [53], whereas three pathotypes have been recognized among *E. coli* strains causing extraintestinal diseases, i.e., SCEC, NMEC, and UPEC. Extensive past research has uncovered valuable insights in the biology, epidemiology, and pathogenesis of *E. coli* infections and contributed to the development of rapid and precise diagnostic techniques [54]. However, much remains unknown or unclear about the molecular and immunological basis of *E. coli* pathogenicity. The application of various laboratory models is unquestionably necessary in generating new clues to virulence strategies employed by *E. coli* organisms [55].

REFERENCES

1. Kaper JB, Nataro JP, Mobley HL. Pathogenic *Escherichia coli*. *Nat Rev Microbiol.* 2004;2:123–140.
2. Liu S, et al. *Escherichia marmotae* sp. nov., isolated from faeces of *Marmota himalayana*. *Int J Syst Evol Microbiol.* 2015;65:2130–2134.
3. Kim KS. Current concepts on the pathogenesis of *Escherichia coli* meningitis: implications for therapy and prevention. *Curr Opin Infect Dis.* 2012;25:273–278.
4. Werber D, et al. Outbreaks of virulent diarrheagenic *Escherichia coli*—are we in control? *BMC Med.* 2012;10:11.
5. Makobe CK, Sang WK, Kikuvi G, Kariuki S. Molecular characterization of virulence factors in diarrhoeagenic *Escherichia coli* isolates from children in Nairobi, Kenya. *J Infect Dev Ctries.* 2012;6(8):598–604.
6. Lozer DM, et al. Genotypic and phenotypic analysis of diarrheagenic *Escherichia coli* strains isolated from Brazilian children living in low socioeconomic level communities. *BMC Infect Dis.* 2013;13:418.
7. Croxen MA, Finlay BB. Molecular mechanisms of *Escherichia coli* pathogenicity. *Nat Rev Microbiol.* 2010;8:26–38.
8. Clements A, Young JC, Constantinou N, Frankel G. Infection strategies of enteric pathogenic *Escherichia coli*. *Gut Microbes.* 2012;3(2):71–87.
9. Keseler IM, et al. EcoCyc: fusing model organism databases with systems biology. *Nucleic Acids Res.* 2013;41:D605–D612.
10. Jafari A, Aslani MM, Bouzari S. *Escherichia coli*: a brief review of diarrheagenic pathotypes and their role in diarrheal diseases in Iran. *Iran J Microbiol.* 2012;4(3):102–117.
11. Koutsoumpas AL, Smyk DS, Bogdanos DP. *E. coli* induced experimental model of primary biliary cirrhosis: at last. *Int J Hepatol.* 2014;2014. doi: 10.1155/2014/848373.
12. Qadri F, Svennerholm A-M, Faruque ASG, Sack RB. Enterotoxigenic *Escherichia coli* in developing countries: epidemiology, microbiology, clinical features, treatment, and prevention. *Clin Microbiol Rev.* 2005;18(3):465–483.
13. Totsika M, et al. Uropathogenic *Escherichia coli* mediated urinary tract infection. *Curr Drug Targets.* 2012;13:1386–1399.
14. Mora A, et al. Characteristics of the Shiga-toxin-producing enteroaggregative *Escherichia coli* O104:H4 German outbreak strain and of STEC strains isolated in Spain. *Int Microbiol.* 2011;14(3):121–141.
15. Farfan MJ, Torres AG. Molecular mechanisms that mediate colonization of Shiga toxin-producing *Escherichia coli* strains. *Infect Immun.* 2012;80:903–913.
16. Melton-Celsa A, Mohawk K, Teel L, O'Brien A. Pathogenesis of Shiga-toxin producing *Escherichia coli*. *Curr Top Microbiol Immunol.* 2012;357:67–103.
17. Preussel K, Hohle M, Stark K, Werber D. Shiga toxin-producing *Escherichia coli* O157 is more likely to lead to hospitalization and death than non-O157 serogroups—except O104. *PLoS One.* 2013;8:e78180.

18. Keithlin J, Sargeant J, Thomas MK, Fazil A. Chronic sequelae of *E. coli* O157: systematic review and meta-analysis of the proportion of *E. coli* O157 cases that develop chronic sequelae. *Foodborne Pathog Dis.* 2014;11:79–95.
19. Mayer CL, Leibowitz CS, Kurosawa S, Stearns-Kurosawa DJ. Shiga toxins and the pathophysiology of hemolytic uremic syndrome in humans and animals. *Toxins.* 2012;4:1261–1287.
20. Lim JY, Yoon JW, Hovde CJ. A brief overview of *Escherichia coli* O157:H7 and its plasmid O157. *J Microbiol Biotechnol.* 2010;20(1):5–14.
21. Mohawk KL, O'Brien AD. Mouse models of *Escherichia coli* O157:H7 infection and Shiga toxin injection. *J Biomed Biotechnol.* 2011;2011. doi: 10.1155/2011/258185.
22. Barnett Foster D. Modulation of the enterohemorrhagic *E. coli* virulence program through the human gastrointestinal tract. *Virulence.* 2013;4(4):315–323.
23. Ferens WA, Hovde CJ. *Escherichia coli* O157:H7: animal reservoir and sources of human infection. *Foodborne Pathog Dis.* 2011;8:465–87.
24. Kaur P, Chakraborti A, Asea A. Enteroaggregative *Escherichia coli*: an emerging enteric foodborne pathogen. *Interdiscip Perspect Infect Dis.* 2010;2010. doi: 10.1155/2010/254159.
25. Muniesa M, Hammerl JA, Hertwig S, Appel B, Brüssow H. Shiga toxin-producing *Escherichia coli* O104:H4: a new challenge for microbiology. *Appl Environ Microbiol.* 2012;78:4065–4073.
26. Beutin L, Martin A. Outbreak of Shiga toxin-producing *Escherichia coli* (STEC) O104:H4 infection in Germany causes a paradigm shift with regard to human pathogenicity of STEC strains. *J Food Prot.* 2012;75(2):408–418.
27. Estrada-Garcia T, Navarro-Garcia F. Enteroaggregative *Escherichia coli* pathotype: a genetically heterogeneous emerging foodborne enteropathogen. *FEMS Immunol Med Microbiol.* 2012;66(3):281–298.
28. Servin AL. Pathogenesis of Afa/Dr diffusely adhering *Escherichia coli. Clin Microbiol Rev.* 2005;18(2):264–292.
29. Mansan-Almeida R, Pereira AL, Giugliano LG. Diffusely adherent *Escherichia coli* strains isolated from children and adults constitute two different populations. *BMC Microbiol.* 2013;13:22.
30. Smith EJ, Thompson AP, O'Driscoll A, Clarke DJ. Pathogenesis of adherent-invasive *Escherichia coli. Future Microbiol.* 2013;8:1289–1300.
31. Martinez-Medina M, Garcia-Gil LJ. *Escherichia coli* in chronic inflammatory bowel diseases: an update on adherent invasive *Escherichia coli* pathogenicity. *World J Gastrointest Pathophysiol.* 2014;5(3):213–227.
32. Agarwal J, Srivastava S, Singh M. Pathogenomics of uropathogenic *Escherichia coli. Indian J Med Microbiol.* 2012;30:141–149.
33. Botkin DJ, Galli L, Sankarapani V, Soler M, Rivas M, Torres AG. Development of a multiplex PCR assay for detection of Shiga toxin-producing *Escherichia coli*, enterohemorrhagic *E. coli*, and enteropathogenic *E. coli* strains. *Front Cell Infect Microbiol.* 2012;2:8.
34. Chandra M, Cheng P, Rondeau G, Porwollik S, McClelland M. A single step multiplex PCR for identification of six diarrheagenic *E. coli* pathotypes and Salmonella. *Int J Med Microbiol.* 2013;303(4):210–216.
35. Wiles TJ, Mulvey MA. The RTX pore-forming toxin α-hemolysin of uropathogenic *Escherichia coli*: progress and perspectives. *Future Microbiol.* 2013;8:73–84.
36. Fialho OB, et al. Detection of diarrheagenic *Escherichia coli* using a two-system multiplex-PCR protocol. *J Clin Lab Anal.* 2013;27(2):155–161.
37. Fujioka M, Otomo Y, Ahsan CR. A novel single-step multiplex polymerase chain reaction assay for the detection of diarrheagenic *Escherichia coli. J Microbiol Methods.* 2013;92(3):289–292.
38. Souza TB, Lozer DM, Kitagawa SM, Spano LC, Silva NP, Scaletsky IC. Real-time multiplex PCR assay and melting curve analysis for identifying diarrheagenic *Escherichia coli. J Clin Microbiol.* 2013;51(3):1031–1033.
39. Bitzan M. Treatment options for HUS secondary to *Escherichia coli* O157:H7. *Kidney Int Suppl.* 2009;112:S62–S66.
40. Toledo CC, Arvidsson I, Karpman D. Cross-reactive protection against enterohemorrhagic *Escherichia coli* infection by enteropathogenic *E. coli* in a mouse model. *Infect Immun.* 2011;79:2224–2233.
41. Garcia-Angulo VA, Kalita A, Torres AG. Advances in the development of enterohemorrhagic *Escherichia coli* vaccines using murine models of infection. *Vaccine.* 2013;31(32):3229–3235.
42. Law RJ, Gur-Arie L, Rosenshine I, Finlay BB. In vitro and in vivo model systems for studying enteropathogenic *Escherichia coli* infections. *Cold Spring Harb Perspect Med.* 2013;3(3):a009977.

43. Kalita A, Hu J, Torres AG. Recent advances in adherence and invasion of pathogenic *Escherichia coli. Curr Opin Infect Dis.* 2014;27(5):459–464.
44. Philipson CW, Bassaganya-Riera J, Hontecillas R. Animal models of enteroaggregative *Escherichia coli* infection. *Gut Microbes.* 2013;4(4):281–291.
45. Mallick EM, et al. A novel murine infection model for Shiga toxin-producing *Escherichia coli. J Clin Invest.* 2012;122:4012–4024.
46. Ritchie JM. Animal models of enterohemorrhagic *Escherichia coli* infection. *Microbiol Spectr.* 2014;2(4). doi: 10.1128/microbiolspec.EHEC-0022-2013.
47. Zotta E, et al. Development of an experimental hemolytic uremic syndrome in rats. *Pediatr Nephrol.* 2008;23:559–567.
48. Mohawk KL, et al. Pathogenesis of *Escherichia coli* O157:H7 strain 86-24 following oral infection of BALB/c mice with an intact commensal flora. *Microb Pathog.* 2010;48:131–142.
49. Dalgakiran F, Witcomb LA, McCarthy AJ, Birchenough GM, Taylor PW. Non-invasive model of neuropathogenic *Escherichia coli* infection in the neonatal rat. *J Vis Exp.* 2014;(92):e52018.
50. Ludwig M, Johannes S, Bergmann M, Failing K, Schiefer HG, Weidner W. Experimental *Escherichia coli* epididymitis in rats: a model to assess the outcome of antibiotic treatment. *BJU Int.* 2002;90(9):933–938.
51. García A, et al. Renal injury is a consistent finding in Dutch belted rabbits experimentally infected with enterohemorrhagic *Escherichia coli. J Infect Dis.* 2006;193:1125–1134.
52. Panda A, et al. *Escherichia coli* O157:H7 infection in Dutch belted and New Zealand white rabbits. *Comp Med.* 2010;60:31–37.
53. Croxen MA, et al. Recent advances in understanding enteric pathogenic *Escherichia coli. Clin Microbiol Rev.* 2013;26(4):822–880.
54. Parsons BD, Zelyas N, Berenger BM, Chui L. Detection, characterization, and typing of Shiga toxin-producing *Escherichia coli. Front Microbiol.* 2016;7:478.
55. Hoogland IC, Houbolt C, van Westerloo DJ, van Gool WA, van de Beek D. Systemic inflammation and microglial activation: systematic review of animal experiments. *J Neuroinflammation.* 2015;12:114.

22

Helicobacter

Tetsuya Tsukamoto, Yuka Kiriyama, and Masae Tatematsu

CONTENTS

22.1 Introduction

Since *Helicobacter pylori* was discovered to play a major causative role in chronic gastritis and gastric carcinogenesis, several attempts have been made to treat and prevent the lesions. In parallel, many experimental studies have confirmed the importance of *H. pylori* in gastric disease processes. Among animal models, mice and Mongolian gerbils offer powerful tools for the analysis of *H. pylori*-associated gastric lesions.[1,2] The *Helicobacter* genus nowadays contains more than 30 species; this chapter focuses on *H. pylori* within the genus and reviews its characteristics, interaction with host environment, and pathogenesis in gastric disease. As gastric cancer remains a significant contributor to cancer-related death, prevention of infection and treatment of gastric cancer represent important aspects of cancer control.

22.2 Discovery of Gastric Microorganisms

When Warren and Marshall successfully isolated a hitherto unidentified Gram-negative bacillus from patients with chronic active gastritis, the role of bacteria as a causative factor in gastric disorders was established.[3,4] Initially named *Campylobacter pyloridis*, this flagellated microorganism thrives under microaerophilic condition. Following the description of several features (including the presence of flagellar sheath and bulb, glycocalyx, urease and catalase, and other biochemical characteristics) that discriminate this bacillus from the *Campylobacter* or *Wolinella* genuses, Goodwin et al.[5] renamed it "*Helicobacter pylori*." According to the recent literature,[6,7] the *Helicobacter* genus currently consists of over 30 gastric and enterohepatic species. Readers may refer to the taxonomy database provided by the National Center for Biotechnology Information for most updated and detailed information about this

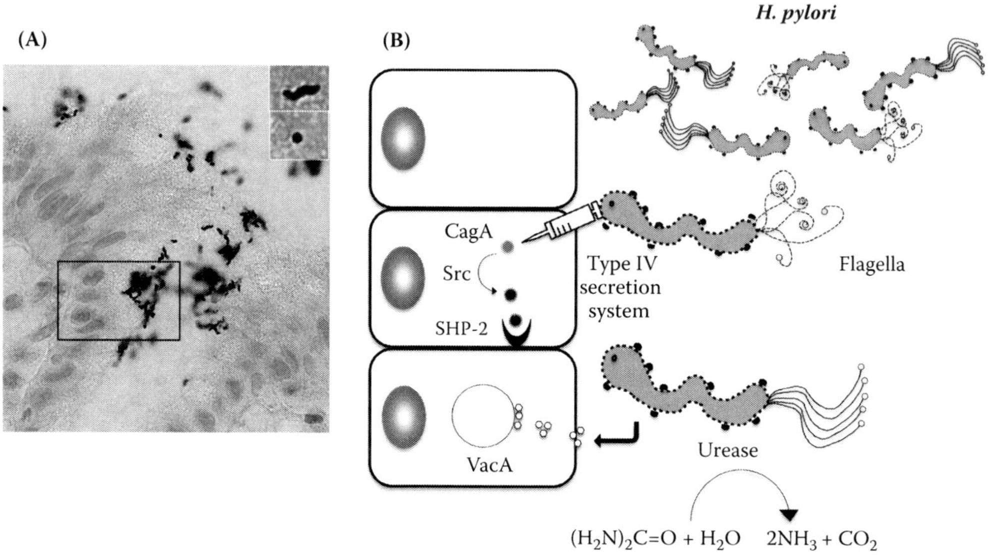

FIGURE 22.1 *H. pylori* and its schematic view. (A) Immunohistochemical view of *H. pylori* in the human stomach colonized in the human gastric surface mucin. Top inset: active bacillary form. Bottom inset: coccoid form in the unsatisfactory condition. The center square is enlarged in (B) as a schematic view. Immunohistochemistry against *H. pylori*. Original magnification, 630×. (B) *H. pylori* utilize flagella to move into gastric surface mucin. CagA is injected through type IV secretion system and tyrosine phosphorylated with Src family kinase. Phosphorylated CagA binds and activates SHP-2 tyrosine phosphatase. VacA is a vacuolating toxin. Bacterial urease catalyzes urea to ammonia to neutralize hydrogen chloride in the acidic environment of the stomach.

bacterial genus.[8] Here, we concentrate mainly on *H. pylori*, since most of the laboratory models were established with this species (Figure 22.1A).

22.3 Gastric Microenvironment of *H. pylori*

In the stomach, gastric glands are made up of surface mucous cells (foveolar epithelium) in the top part and pyloric glands at the bottom region in the antral mucosa. These two kinds of cells produce multilayered histochemically different kinds of mucins, surface mucous cell mucin (SMCM) and pyloric gland mucous cell mucin (GMCM), respectively, in humans[9,10] and Mongolian gerbils.[11] *H. pylori* prefers the SMCM since GMCM contains terminal α1,4-GlcNAc residues attached to core O-glycans, which inhibit biosynthesis of cholesteryl-α-D-glucopyranoside, a major cell wall component of the bacteria.[12] *H. pylori* shows morphologic conversion from bacillary form to coccoid form to adapt to certain worse conditions, as in human stomach possessing adenocarcinomas,[13] by lowering metabolic activity with 1000-fold less adenosine triphosphate (ATP) level[14] (Figure 22.1A).

22.4 Characteristics and Pathogenesis of *H. pylori*

Sequencing analysis revealed that *H. pylori* strain 26695 has a circular genome of 1,667,867 bp. Of these, 91.0% are coding regions with 1590 predicted coding sequences.[15] Global transcriptomic analysis revealed transcriptional organization of the same strain; acid stimulus induced major *H. pylori* virulence loci such as urease (ure) operon or the *cag* pathogenicity island (cagPAI).[16]

CagA is injected into gastric surface epithelial cells through the bacterial type IV secretion system and is tyrosine phosphorylated with Src and Abl[17] at variable EPIYA (Glu-Pro-Ile-Tyr-Ala) motif repeat regions, which show structural diversity among East-Asian and Western countries.[18] *H. pylori* found in the former possess EPIYA-A, B, and D motifs and may contribute to the geographical differences observed in gastric carcinogenesis. Phosphorylated CagA then activates SHP-2 phosphatase, which then dephosphorylates FAK kinase, resulting in impairment of epithelial cell adhesion and morphology.[19] Studies in wild-type CagA transgenic mice have confirmed the oncogenicity of this bacterial protein by demonstrating the development of gastrointestinal epithelial hyperplasia and adenocarcinomas as well as hematopoietic malignancies, whereas phosphorylation-resistant CagA mice did not develop disease.[20]

Broth culture supernatant of *H. pylori* was known to induce vacuolization of the mammalian cells, which was characterized as VacA.[21,22] VacA binds to receptor protein tyrosine phosphatases (RPTPs) α and β on the surface of target cells.[23] Mice deficient in RPTP type Z (RPTPζ or RPTPβ) revealed tolerance to VacA for mucosal damage.[24] Oligomerized VacA enters into the cytoplasm through endocytosis and shows multifunctional toxicity including interference of T cell activation and proliferation, reduction of mitochondria membrane potential, and formation of vacuolation in endosomal/lysosomal compartments[25] (Figure 22.1B).

22.5 Infection Route and Childhood Acquisition of *H. pylori*

It has been demonstrated that transmission of the bacteria in the early childhood and interaction with an infected sibling is an important risk factor. The presence of *H. pylori* in the oral cavity substantiated the oral–oral route of spread. It has also supported the oral–fecal route of spread through contaminated drinking water, especially in areas where living condition and sanitation are poor.[26]

Childhood infection with *H. pylori* is a major concern in the pediatrics field. In the *H. pylori* inoculated and carcinogen administered model, early acquisition of *H. pylori* significantly increased gastric inflammation and carcinogenesis compared to the late counterpart (Figure 22.2).[27] *H. pylori* infection in childhood (mean age 12.1 years) induced chronic gastritis, but this was mostly limited to moderate atrophy without intestinal metaplasia in contrast to that seen in the elderly patients.[28] Thus, earlier eradication of the bacteria should be considered, i.e., before development of atrophy and intestinal metaplasia, for the prevention of gastric cancer later in adult life.[29,30]

22.6 Role of *H. pylori* Infection in Gastric Carcinogenesis

Epidemiological evidence has been accumulated indicating a significant relationship between *H. pylori* infection and chronic atrophic gastritis, intestinal metaplasia, and gastric malignancies including cancer and lymphoma.[31–41] The World Health Organization/International Agency for Research on Cancer (WHO/IARC) defined *H. pylori* as a "definite biological carcinogen" in 1994 in spite of lack of evidence in experimental animals.[42]

Various experimental animal models have been attempted to help clarify the role of *H. pylori*. Lee et al.[43] isolated *Helicobacter felis* (*H. felis*) from cat stomach and inoculated it in germ-free mice, resulting in colonization of the glandular stomach and induction of acute and chronic inflammation. A type of *H. pylori*, designated Sydney strain (SS1), established by screening of fresh clinical isolates, showed high colonizing ability and is currently widely used for mice experiments.[44] The Mongolian gerbil (*Meriones unguiculatus*) model was successfully established to mimic human *H. pylori* infection, resulting in chronic active gastritis, peptic ulcers, and intestinal metaplasia resembling human lesions.[45]

For the gastric carcinogenesis models, chemical carcinogens such as *N*-methyl-*N′*-nitro-*N*-nitrosoguanidine (MNNG) or *N*-methyl-*N*-nitrosourea (MNU) were found to induce adenocarcinomas in rat,[46] mouse,[47] and gerbils.[48] *H. pylori* infection was found to exacerbate the incidence of MNU-[49] and MNNG-[50,51] induced adenocarcinomas in Mongolian gerbils, in which all histological types including differentiated and undifferentiated adenocarcinomas and signet-ring cell carcinoma were found as in human histopathology. Thus, *H. pylori* was confirmed as a strong promoter of gastric carcinogenesis (Figure 22.2).

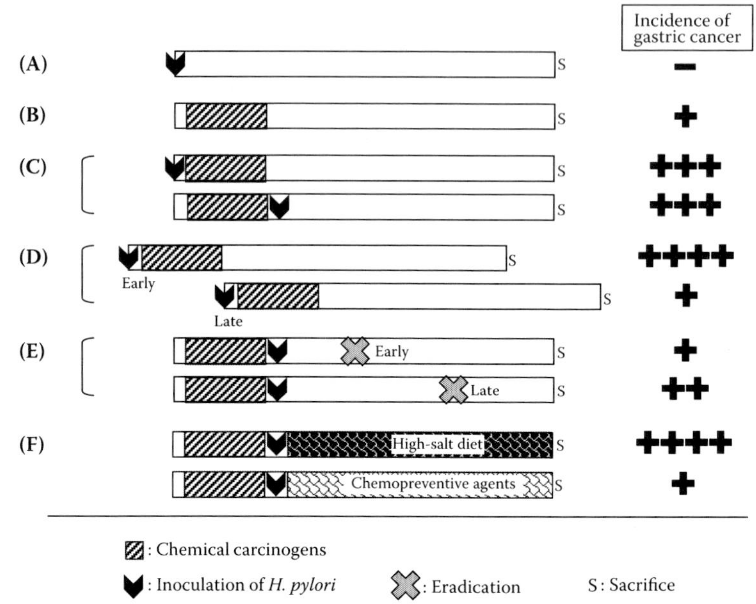

FIGURE 22.2 Modifying factors for *H. pylori*-associated stomach carcinogenesis in the Mongolian gerbil models. (A) *H. pylori* infection alone is rarely carcinogenic. (B) Drinking water containing chemical carcinogens including MNU or MNNG induces stomach cancers. (C) When combined, *H. pylori* become a strong promoter. (D) Earlier infection increases the risk compared with a later event. (E) Earlier eradication of *H. pylori* reduces risk of stomach cancers. (F) A high-salt diet exacerbates inflammation and increases the incidence of *H. pylori*-associated cancer. On the other hand, various natural products and pure chemicals appear to have chemopreventive potential.

Besides the gerbil models, mouse models feature advantages especially for genetic approaches. Transgenic mice (K19-C2mE transgenic mouse) simultaneously expressing cyclooxygenase-2 (COX-2) and microsomal prostaglandin E synthase (mPGES)-1 as well as Wnt1 revealed the importance of these pathways in gastric tumorigenesis with *Helicobacter* infection.[52,53] K19-C2mE mice developed adenocarcinomas not only in pyloric mucosa but also in fundic glands with MNU treatment as well as *H. pylori* infection, and so are considered as a model for increasing proximal malignancies.[54] Interleukin 1β (−/−) knockout mice showed that IL1β induced by *H. pylori* infection affected both inflammatory and epithelial cells and enhanced gastric carcinogenesis.[55]

Long-term infection of *H. pylori* frequently induced neuroendocrine tumors (NETs) (endocrine cell hyperplasia/dysplasia and carcinoid tumors) in a Mongolian gerbil model in association with increased serum gastrin level. Eradication of *H. pylori* prevents development of NETs in the glandular stomach, being associated with reduction in serum gastrin levels.[56] Proton pump inhibitors (PPIs), routinely used for treatment of upper gastrointestinal disorders, have raised some concerns regarding the long-term safety and development of NETs in the stomach. When PPI was administered to *H. pylori*-infected Mongolian gerbils, high dose increased NET development with higher serum gastrin in contrast to no influence at low dose.[57]

22.7 *H. pylori* Infection Induces No Neoplastic but Hyperplastic Lesions in Mongolian Gerbils

Several studies reported that *H. pylori* infection alone induced gastric adenocarcinomas in gerbils with a very high incidence without carcinogen exposure.[58–60] However, several other studies rarely showed development of carcinomas in animals inoculated only with *H. pylori*.[48–51,61] Mongolian gerbils are characteristic for the development of expanded cystic glands in mucosa and submucosa, sometime as deep as subserosa (several months after inoculation of *H. pylori*), and these are designated as heterotopic proliferative glands (HPGs).[62] HPGs consisted of gastric and/or intestinal metaplastic epithelial cells, the latter progressed depending on the duration of *H. pylori*-induced inflammation. HPGs, in turn, dramatically diminished with the eradication of the bacteria, indicating that HPG is not neoplastic but rather hyperplastic. HPGs often resemble differentiated or mucinous adenocarcinomas showing structural abnormality, but lack obvious cellular atypia. Characteristics of HPG include (1) organized polarity of their component cells; (2) differentiation from gastric phenotypic HPG into intestinal phenotypic HPG sometimes associated with mature Paneth cells; (3) formation of large cystic dilatations containing mucin, often with calcification; (4) shedding of epithelial cells and necrosis at the tips of lesions; (5) high-grade inflammation with infiltration of inflammatory cells (neutrophils in acute phase and mononuclear cells in chronic phase); and (6) organized polarity of proliferating zones. These features are quite different from those of well-differentiated adenocarcinomas, which are characterized by obvious cellular atypia. HPGs need to be precisely distinguished from adenocarcinomas (Figure 22.3 and Table 22.1).[1,2]

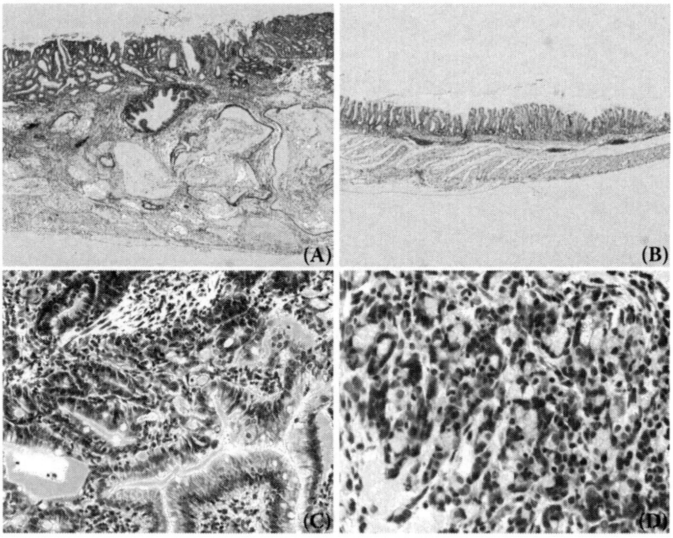

FIGURE 22.3 Gastric lesions in *H. pylori*-infected Mongolian gerbils. (A) Submucosal hyperplastic proliferating glands called HPG after 44-week long inflammation. (B) HPG has drastically disappeared 20 weeks after eradication of the bacteria. Well- (C) and poorly (D) differentiated adenocarcinomas developed with carcinogen treatment. (A–D) Hematoxylin and eosin staining. Original magnification, 25× (A and B); 200× (C); and 400× (D).

TABLE 22.1

Comparison of Submucosal Lesions in *H. pylori*-Infected Mongolian Gerbils

	HPG	Tubular Adenocarcinoma
Gland distribution	Dispersed	Compactly proliferated
Gland shape	Large and cystic	Relatively small
Intracystic material	Eosinophilic, sometimes with calcification	Usually transparent
Lining epithelium of cysts	Shedding of epithelial cells with necrosis at the tip of the cyst	Fully lined with atypical cells.
Intestinal metaplasia	Frequent, sometimes with Paneth cells	Relatively infrequent
Stroma	Severe inflammatory cell infiltrates. Neutrophils in acute phase. Lymphocytes and plasma cells in chronic phase with lymphoid follicle formation.	Desmoplastic reaction
Atypia	No atypia. Organized polarity. Partly degenerated.	Enlarged nuclei with increased chromatin. Loss of polarity.

22.8 Prevention of Gastric Carcinogenesis by Eradication of *H. pylori*

Several studies have been performed to assess whether gastric cancer could be prevented by eradication of *H. pylori* in patients with chronic atrophic gastritis. Some investigated the prophylactic effect of *H. pylori* eradication on the development of gastric carcinomas after endoscopic resection for initial lesions and revealed significant reduction up to 3 years.[63,64] In contrast, longer follow-up studies (more than 7.5 years) in China[65] and Japan[66] showed similar overall incidences in both *H. pylori* eradicated and non-eradicated groups. However, eradication significantly decreased the development of gastric cancer in the subgroup of *H. pylori* carriers without precancerous lesions in the former Chinese trial.[65] Furthermore, in the latter trial, cancer incidence was significantly reduced after eradication only in pepsinogen-test-negative subjects with mild gastritis.[66]

In *H. pylori*-infected, carcinogen-treated Mongolian gerbil models, incidences of gastric cancers after curative treatment for *H. pylori* were significantly lower than in those without eradication.[61] The incidences of gastric cancer were significantly reduced to 6.7%, 27.3%, and 38.2% with eradication at early, middle, and late time points, respectively, compared to 56.3% incidence in the noneradicated group[67] (Figure 22.2).

22.9 Intestinal Metaplasia and Intestinalization of Gastric Cancer

H. pylori infection plays a causative role in the development of chronic atrophic gastritis and intestinal metaplasia. Using gastric and intestinal epithelial cell markers, intestinal metaplasia was divided into two major types: a gastric-and-intestinal mixed type and a solely intestinal type.[68,69] In the former, gastric and intestinal phenotypic markers are localized within a glandular level as well as in a cellular level.[70] In the Mongolian gerbil model, gastric-and-intestinal mixed-type intestinal metaplasia was found to appear first, followed by the solely intestinal type with appearance of Paneth cells during the overall course of *H. pylori* infection in the HPG.[62] It indicated that intestinal metaplasia could be caused by the gradual intestinalization from proper gastric gland cells toward completely intestinal type cells through the intermittent gastric-and-intestinal mixed phenotype.

Stomach cancers could also be classified with gastric and intestinal markers: gastric, gastric-and-intestinal mixed, intestinal, and null types. Although intestinal metaplasia has been considered as a preneoplastic lesion of intestinal type cancer, detailed mechanisms have not yet been clarified.[71] In *H. pylori*-infected gerbils, adenocarcinomas were divided phenotypically into 21 gastric, 24 gastric-and-intestinal mixed, 4 intestinal, and 1 null types. In contrast, noninfected gerbils harbored only gastric-type cancers. Thus, *H. pylori* infection was considered to trigger intestinalization of stomach cancers in parallel with progression of intestinal metaplasia.[2]

22.10 Exacerbating Factors for Gastric Carcinogenesis: Synergistic Effects of *H. pylori* and High-Salt Diet

A large number of case–control and ecological studies have shown salt and salted foods, among various other food ingredients, as probable risk factors for gastric cancer.[72–75] In an experimental study on rats, sodium chloride (NaCl) had been thought to enhance the carcinogenic effects by hindering the gastric mucous barrier.[76] Later, it was revealed that high-salt diet exacerbated *H. pylori*-induced inflammation and gastric carcinogenesis in a Mongolian gerbil model.[77] Food containing various concentrations of salt (2.5%, 5%, and 10% NaCl) increased the incidence of gastric cancer in a dose-dependent manner (33%, 36%, and 63%, respectively) compared to the normal diet group (15%). In contrast, intermittent intragastric injection of saturated NaCl solution induced gastric erosion but did not promote gastric carcinogenesis.[11] A high-salt diet was associated with a more severe inflammatory response including elevation of anti-*H. pylori* antibody titers, serum gastrin levels, and inflammatory cell infiltration in a dose-dependent fashion (Figure 22.2). A high-salt diet upregulated the amount of SMCM, suitable for *H. pylori* colonization, despite no increment in the corresponding mucin core protein, MUC5AC mRNA. Although GMCM with its mucin core protein MUC6 was induced against *H. pylori* infection, high-salt diet decreased the amount of GMCM, which acts against *H. pylori* infection by inhibiting the bacterial cell wall component.[12] In an *H. pylori*-infected MNU-treated mouse model, microarray analysis revealed overexpression of CD177 and Reg3g in a high-salt diet group, the former being associated with better prognosis in human cases.[78] Reduction of salt intake could thus be one of the most important chemopreventive methods for human gastric carcinogenesis.

22.11 Chemoprevention of Gastric Carcinogenesis

Eradication of *H. pylori* is most effective in the earlier stages of infection, as evidenced by studies on the prevention of gastric carcinogenesis carried out in both humans and animals. However, eradication may not be efficient for ones with severe chronic atrophic gastritis. A selective cyclooxygenase 2 (COX-2) inhibitor, etodolac, was attempted to prevent *H. pylori*-infected and MNU-induced gastric carcinogenesis in a Mongolian gerbil model. Etodolac inhibited the development of epithelial cell proliferation, intestinal metaplasia, and gastric cancer in a dose-dependent manner although inflammatory cell infiltration or oxidative DNA damage was not alleviated.[79] In humans, Yanaoka et al.[80] investigated the preventive effects of etodolac on metachronous cancer development after endoscopic resection of early gastric cancer. Etodolac significantly reduced emergence of gastric cancer even with extensive metaplastic gastritis, which indicated that controlling COX-2 function could be an alternative method (Figure 22.2).

Oxidative stress is also a major concern for gastric carcinogenesis because of its ability to damage DNA. A potent antioxidative compound derived from canola oil, 4-vinyl-2,6-dimethoxyphenol (canolol), was effective in reduction of inflammatory cytokines and serum 8-hydroxy-2′-deoxyguanosine (8-OHdG) as well as reduction of the incidence of gastric adenocarcinomas in *H. pylori*-infected and MNU-treated Mongolian gerbils. Canolol was also effective in inhibition of gastric tumorigenesis in K19-C2mE transgenic mice.[81] Thus, the suppression of oxidative stress rather than disappearance of the bacteria would be effective in prevention of gastric cancer (Figure 22.2).

22.12 Conclusion

Discovery of *H. pylori* from the human stomach opened a new avenue to clarify the mechanism of gastric lesions including chronic atrophic gastritis, intestinal metaplasia, and gastric carcinogenesis. A large number of epidemiological studies revealed the positive correlation of *H. pylori* with gastric cancer development. In Mongolian gerbil models, *H. pylori* inoculation caused severe hyperplastic lesions called HPG and showed strong promoting effects in carcinogen-treated animals; *H. pylori* infection alone rarely induced adenocarcinomas. Mouse models are also effective for genetic approach. Eradication of the bacteria was effective in prevention of gastric carcinogenesis, especially in cases of milder gastritis without severe atrophy or intestinal metaplasia, both in human and rodent models. Besides eradication of the bacteria, chemopreventive approaches have been attempted in prevention of gastric neoplasia. It is necessary to establish effective approaches for application in human trials.

REFERENCES

1. Tsukamoto, T., Mizoshita, T. & Tatematsu, M. Animal models of stomach carcinogenesis. *Toxicol Pathol* 35, 636–648 (2007).
2. Tsukamoto, T., Toyoda, T., Mizoshita, T. & Tatematsu, M. *Helicobacter pylori* infection and gastric carcinogenesis in rodent models. *Semin Immunopathol* 35, 177–190 (2013).
3. Warren, J.R. & Marshall, B. Unidentified curved bacilli on gastric epithelium in active chronic gastritis. *Lancet* 321, 1273–1275 (1983).

4. Marshall, B.J. & Warren, J.R. Unidentified curved bacilli in the stomach of patients with gastritis and peptic ulceration. *Lancet* 323, 1311–1315 (1984).
5. Goodwin, C.S. et al. Transfer of *Campylobacter pylori* and *Campylobacter mustelae* to *Helicobacter* gen. nov. as *Helicobacter pylori* comb. nov. and *Helicobacter mustelae* comb. nov., respectively. *Int J Syst Bacteriol* 39, 397–405 (1989).
6. Hannula, M. & Hanninen, M.L. Phylogenetic analysis of *Helicobacter* species based on partial *gyrB* gene sequences. *Int J Syst Evol Microbiol* 57, 444–449 (2007).
7. Mikkonen, T.P., Karenlampi, R.I. & Hanninen, M.L. Phylogenetic analysis of gastric and enterohepatic *Helicobacter* species based on partial HSP60 gene sequences. *Int J Syst Evol Microbiol* 54, 753–758 (2004).
8. National Center for Biotechnology Information. The NCBI taxonomy database: *Helicobacter*. http://www.ncbi.nlm.nih.gov/taxonomy/?term=helicobacter. U.S. National Library of Medicine, Bethesda, MD.
9. Ota, H. & Katsuyama, T. Alternating laminated array of two types of mucin in the human gastric surface mucous layer. *Histochem J* 24, 86–92 (1992).
10. Ota, H. et al. A dual staining method for identifying mucins of different gastric epithelial mucous cells. *Histochem J* 23, 22–28 (1991).
11. Kato, S. et al. High salt diets dose-dependently promote gastric chemical carcinogenesis in *Helicobacter pylori*-infected Mongolian gerbils associated with a shift in mucin production from glandular to surface mucous cells. *Int J Cancer* 119, 1558–1566 (2006).
12. Kawakubo, M. et al. Natural antibiotic function of a human gastric mucin against *Helicobacter pylori* infection. *Science* 305, 1003–1006 (2004).
13. Chan, W.Y. et al. Coccoid forms of *Helicobacter pylori* in the human stomach. *Am J Clin Pathol* 102, 503–507 (1994).
14. Sorberg, M., Nilsson, M., Hanberger, H. & Nilsson, L.E. Morphologic conversion of *Helicobacter pylori* from bacillary to coccoid form. *Eur J Clin Microbiol Infect Dis* 15, 216–219 (1996).
15. Tomb, J.F. et al. The complete genome sequence of the gastric pathogen *Helicobacter pylori*. *Nature* 388, 539–547 (1997).
16. Sharma, C.M. et al. The primary transcriptome of the major human pathogen *Helicobacter pylori*. *Nature* 464, 250–255 (2010).
17. Backert, S. & Selbach, M. Tyrosine-phosphorylated bacterial effector proteins: the enemies within. *Trends Microbiol* 13, 476–484 (2005).
18. Hatakeyama, M. Anthropological and clinical implications for the structural diversity of the *Helicobacter pylori* CagA oncoprotein. *Cancer Sci* 102, 36–43 (2011).
19. Tsutsumi, R., Takahashi, A., Azuma, T., Higashi, H. & Hatakeyama, M. Focal adhesion kinase is a substrate and downstream effector of SHP-2 complexed with *Helicobacter pylori* CagA. *Mol Cell Biol* 26, 261–276 (2006).
20. Ohnishi, N. et al. Transgenic expression of *Helicobacter pylori* CagA induces gastrointestinal and hematopoietic neoplasms in mouse. *Proc Natl Acad Sci USA* 105, 1003–1008 (2008).
21. Cover, T.L. & Blaser, M.J. Purification and characterization of the vacuolating toxin from *Helicobacter pylori*. *J Biol Chem* 267, 10570–10575 (1992).
22. Phadnis, S.H., Ilver, D., Janzon, L., Normark, S. & Westblom, T.U. Pathological significance and molecular characterization of the vacuolating toxin gene of *Helicobacter pylori*. *Infect Immun* 62, 1557–1565 (1994).
23. Wada, A., Yamasaki, E. & Hirayama, T. *Helicobacter pylori* vacuolating cytotoxin, VacA, is responsible for gastric ulceration. *J Biochem* 136, 741–746 (2004).
24. Fujikawa, A. et al. Mice deficient in protein tyrosine phosphatase receptor type Z are resistant to gastric ulcer induction by VacA of *Helicobacter pylori*. *Nat Genet* 33, 375–381 (2003).
25. Cover, T.L. & Blanke, S.R. *Helicobacter pylori* VacA, a paradigm for toxin multifunctionality. *Nat Rev Microbiol* 3, 320–332 (2005).
26. Goh, K.L., Chan, W.K., Shiota, S. & Yamaoka, Y. Epidemiology of *Helicobacter pylori* infection and public health implications. *Helicobacter* 16(Suppl 1), 1–9 (2011).
27. Cao, X. et al. Earlier *Helicobacter pylori* infection increases the risk for the *N*-methyl-*N*-nitrosourea-induced stomach carcinogenesis in Mongolian gerbils. *Jpn J Cancer Res* 93, 1293–1298 (2002).
28. Kato, S., Kikuchi, S. & Nakajima, S. When does gastric atrophy develop in Japanese children? *Helicobacter* 13, 278–281 (2008).

29. Correa, P. et al. Chemoprevention of gastric dysplasia: randomized trial of antioxidant supplements and anti-*Helicobacter pylori* therapy. *J Natl Cancer Inst* 92, 1881–1888 (2000).
30. Malaty, H.M. et al. Age at acquisition of *Helicobacter pylori* infection: a follow-up study from infancy to adulthood. *Lancet* 359, 931–935 (2002).
31. Hu, P.J. et al. *Helicobacter pylori* associated with a high prevalence of duodenal ulcer disease and a low prevalence of gastric cancer in a developing nation. *Gut* 36, 198–202 (1995).
32. Craanen, M.E., Dekker, W., Blok, P., Ferwerda, J. & Tytgat, G.N. Intestinal metaplasia and *Helicobacter pylori*: an endoscopic bioptic study of the gastric antrum. *Gut* 33, 16–20 (1992).
33. Parsonnet, J. et al. *Helicobacter pylori* infection and the risk of gastric carcinoma. *N Engl J Med* 325, 1127–1131 (1991).
34. Nomura, A. et al. *Helicobacter pylori* infection and gastric carcinoma among Japanese Americans in Hawaii. *N Engl J Med* 325, 1132–1136 (1991).
35. Forman, D. et al. Association between infection with *Helicobacter pylori* and risk of gastric cancer: evidence from a prospective investigation. *BMJ* 302, 1302–1305 (1991).
36. Graham, D.Y. et al. Effect of treatment of *Helicobacter pylori* infection on the long-term recurrence of gastric or duodenal ulcer. A randomized, controlled study. *Ann Intern Med* 116, 705–708 (1992).
37. Kuipers, E.J. et al. Long-term sequelae of *Helicobacter pylori* gastritis. *Lancet* 345, 1525–1528 (1995).
38. Asaka, M. et al. Atrophic changes of gastric mucosa are caused by *Helicobacter pylori* infection rather than aging: studies in asymptomatic Japanese adults. *Helicobacter* 1, 52–56 (1996).
39. Huang, J.Q., Sridhar, S., Chen, Y. & Hunt, R.H. Meta-analysis of the relationship between *Helicobacter pylori* seropositivity and gastric cancer. *Gastroenterology* 114, 1169–1179 (1998).
40. Parsonnet, J. et al. *Helicobacter pylori* infection and gastric lymphoma. *N Engl J Med* 330, 1267–1271 (1994).
41. The EUROGAST Study Group. An international association between *Helicobacter pylori* infection and gastric cancer. *Lancet* 341, 1359–1362 (1993).
42. IARC Working Group on the Evaluation of Carcinogenic Risks to Humans. Infection with *Helicobacter pylori*. in *Schistosomes, Liver Flukes and Helicobacter pylori*, Vol. 61, 177–241, World Health Organization/International Agency for Research on Cancer, Lyon (1994).
43. Lee, A., Fox, J.G., Otto, G. & Murphy, J. A small animal model of human *Helicobacter pylori* active chronic gastritis. *Gastroenterology* 99, 1315–1323 (1990).
44. Lee, A. et al. A standardized mouse model of *Helicobacter pylori* infection: introducing the Sydney strain. *Gastroenterology* 112, 1386–1397 (1997).
45. Hirayama, F., Takagi, S., Yokoyama, Y., Iwao, E. & Ikeda, Y. Establishment of gastric *Helicobacter pylori* infection in Mongolian gerbils. *J Gastroenterol* 31(Suppl 9), 24–28 (1996).
46. Sugimura, T. & Fujimura, S. Tumour production in glandular stomach of rat by *N*-methyl-*N′*-nitro-*N*-nitrosoguanidine. *Nature* 216, 943–944 (1967).
47. Tatematsu, M. et al. Induction of adenocarcinomas in the glandular stomach of BALB/c mice treated with N-methyl-N-nitrosourea. *Jpn J Cancer Res* 83, 915–918 (1992).
48. Tatematsu, M. et al. Induction of glandular stomach cancers in *Helicobacter pylori*-sensitive Mongolian gerbils treated with *N*-methyl-*N*-nitrosourea and *N*-methyl-*N′*-nitro-*N*-nitrosoguanidine in drinking water. *Jpn J Cancer Res* 89, 97–104 (1998).
49. Sugiyama, A. et al. *Helicobacter pylori* infection enhances *N*-methyl-*N*-nitrosourea-induced stomach carcinogenesis in the Mongolian gerbil. *Cancer Res* 58, 2067–2069 (1998).
50. Shimizu, N. et al. *Helicobacter pylori* infection enhances glandular stomach carcinogenesis in Mongolian gerbils treated with chemical carcinogens. *Carcinogenesis* 20, 669–676 (1999).
51. Shimizu, N. et al. New animal model of glandular stomach carcinogenesis in Mongolian gerbils infected with *Helicobacter pylori* and treated with a chemical carcinogen. *J Gastroenterol* 34(Suppl 11), 61–66 (1999).
52. Oshima, H., Oshima, M., Inaba, K. & Taketo, M.M. Hyperplastic gastric tumors induced by activated macrophages in COX-2/mPGES-1 transgenic mice. *EMBO J* 23, 1669–1678 (2004).
53. Oshima, H. et al. Carcinogenesis in mouse stomach by simultaneous activation of the Wnt signaling and prostaglandin E_2 pathway. *Gastroenterology* 131, 1086–1095 (2006).
54. Takasu, S. et al. Roles of cyclooxygenase-2 and microsomal prostaglandin E synthase-1 expression and β-catenin activation in gastric carcinogenesis in *N*-methyl-*N*-nitrosourea-treated K19-C2mE transgenic mice. *Cancer Sci* 99, 2356–2364 (2008).

55. Shigematsu, Y. et al. Interleukin-1β induced by *Helicobacter pylori* infection enhances mouse gastric carcinogenesis. *Cancer Lett* 340, 141–147 (2013).
56. Cao, L. et al. Development of carcinoid tumors of the glandular stomach and effects of eradication in *Helicobacter pylori*-infected Mongolian gerbils. *Asian Pac J Cancer Prev* 9, 25–30 (2008).
57. Tsukamoto, H. et al. Long-term high-dose proton pump inhibitor administration to *Helicobacter pylori*-infected Mongolian gerbils enhances neuroendocrine tumor development in the glandular stomach. *Asian Pac J Cancer Prev* 12, 1049–1054 (2011).
58. Honda, S. et al. Development of *Helicobacter pylori*-induced gastric carcinoma in Mongolian gerbils. *Cancer Res* 58, 4255–4259 (1998).
59. Watanabe, T., Tada, M., Nagai, H., Sasaki, S. & Nakao, M. *Helicobacter pylori* infection induces gastric cancer in Mongolian gerbils. *Gastroenterology* 115, 642–648 (1998).
60. Gaddy, J.A. et al. High dietary salt intake exacerbates *Helicobacter pylori*-induced gastric carcinogenesis. *Infect Immun* 81, 2258–2267 (2013).
61. Shimizu, N. et al. Eradication diminishes enhancing effects of *Helicobacter pylori* infection on glandular stomach carcinogenesis in Mongolian gerbils. *Cancer Res* 60, 1512–1514 (2000).
62. Nozaki, K. et al. Reversibility of heterotopic proliferative glands in glandular stomach of *Helicobacter pylori*-infected Mongolian gerbils on eradication. *Jpn J Cancer Res* 93, 374–381 (2002).
63. Uemura, N. et al. Effect of *Helicobacter pylori* eradication on subsequent development of cancer after endoscopic resection of early gastric cancer. *Cancer Epidemiol Biomarkers Prev* 6, 639–642 (1997).
64. Fukase, K. et al. Effect of eradication of *Helicobacter pylori* on incidence of metachronous gastric carcinoma after endoscopic resection of early gastric cancer: an open-label, randomised controlled trial. *Lancet* 372, 392–397 (2008).
65. Wong, B.C. et al. *Helicobacter pylori* eradication to prevent gastric cancer in a high-risk region of China: a randomized controlled trial. *JAMA* 291, 187–194 (2004).
66. Yanaoka, K. et al. Eradication of *Helicobacter pylori* prevents cancer development in subjects with mild gastric atrophy identified by serum pepsinogen levels. *Int J Cancer* 125, 2697–2703 (2009).
67. Nozaki, K. et al. Effect of early eradication on *Helicobacter pylori*-related gastric carcinogenesis in Mongolian gerbils. *Cancer Sci* 94, 235–239 (2003).
68. Inada, K. et al. Gastric and intestinal mixed and solely intestinal types of intestinal metaplasia in the human stomach. *Pathol Int* 47, 831–841 (1997).
69. Tsukamoto, T. et al. Down regulation of a gastric transcription factor, Sox2, and ectopic expression of intestinal homeobox genes, Cdx1 and Cdx2: inverse correlation during progression from gastric/intestinal-mixed to complete intestinal metaplasia. *J Cancer Res Clin Oncol* 130, 135–145 (2004).
70. Niwa, T. et al. Mixed gastric- and intestinal-type metaplasia is formed by cells with dual intestinal and gastric differentiation. *J Histochem Cytochem* 53, 75–85 (2005).
71. Tatematsu, M., Tsukamoto, T. & Inada, K. Stem cells and gastric cancer—role of gastric and intestinal mixed intestinal metaplasia. *Cancer Sci* 94, 135–141 (2003).
72. Tajima, K. & Tominaga, S. Dietary habits and gastro-intestinal cancers: a comparative case-control study of stomach and large intestinal cancers in Nagoya, Japan. *Jpn J Cancer Res* 76, 705–716 (1985).
73. Tsugane, S., Tsuda, M., Gey, F. & Watanabe, S. Cross-sectional study with multiple measurements of biological markers for assessing stomach cancer risks at the population level. *Environ Health Perspect* 98, 207–210 (1992).
74. Joossens, J.V. et al. European Cancer Prevention (ECP) and the INTERSALT Cooperative Research Group. Dietary salt, nitrate and stomach cancer mortality in 24 countries. *Int J Epidemiol* 25, 494–504 (1996).
75. Kono, S. & Hirohata, T. Nutrition and stomach cancer. *Cancer Causes Control* 7, 41–55 (1996).
76. Tatematsu, M., Takahashi, M., Fukushima, S., Hananouchi, M. & Shirai, T. Effects in rats of sodium chloride on experimental gastric cancers induced by *N*-methyl-*N*-nitro-*N*-nitrosoguanidine or 4-nitroquinoline-1-oxide. *J Natl Cancer Inst* 55, 101–106. (1975).
77. Nozaki, K. et al. Synergistic promoting effects of *Helicobacter pylori* infection and high-salt diet on gastric carcinogenesis in Mongolian gerbils. *Jpn J Cancer Res* 93, 1083–1089 (2002).
78. Toyoda, T. et al. Gene expression analysis of a *Helicobacter pylori*-infected and high-salt diet-treated mouse gastric tumor model: identification of CD177 as a novel prognostic factor in patients with gastric cancer. *BMC Gastroenterol* 13, 122 (2013).

79. Magari, H. et al. Inhibitory effect of etodolac, a selective cyclooxygenase-2 inhibitor, on stomach carcinogenesis in *Helicobacter pylori*-infected Mongolian gerbils. *Biochem Biophys Res Commun* 334, 606–612 (2005).
80. Yanaoka, K. et al. Preventive effects of etodolac, a selective cyclooxygenase-2 inhibitor, on cancer development in extensive metaplastic gastritis, a *Helicobacter pylori*-negative precancerous lesion. *Int J Cancer* 126, 1467–1473 (2010).
81. Cao, D. et al. Canolol inhibits gastric tumors initiation and progression through COX-2/PGE2 pathway in K19-C2mE transgenic mice. *PLoS One* 10, e0120938 (2015).

23

Klebsiella: *Caenorhabditis elegans* as a Laboratory Model for *Klebsiella pneumoniae* Infection

Arumugam Kamaladevi and Krishnaswamy Balamurugan

CONTENTS

Klebsiella pneumoniae has emerged as one of the most frequent and potent pathogens causing foodborne infection in humans since 1980. Despite its clinical significance, the absence of appropriate animal models has hampered efforts in identifying the virulence and host factors that determine the pathogenesis and susceptibility of host to gastrointestinal infection. For the past few decades, *Caenorhabditis elegans* has been recognized as a better laboratory animal model based on its potential to closely recapitulate human disease. Additionally, owing to its conserved immune pathways with higher animals and availability of developed transgenic, knockout, and knockdown strains, *C. elegans* has attracted the attention of researchers in various fields as a powerful and versatile model system. *C. elegans* has played a crucial role in elucidating the pathophysiology and immunology of microbial infections and has thus provided valuable insights into the pathogenesis and response of host system, from chronic to acute infection. The information derived from animal model aids researchers to characterize the host response and helps design chemotherapeutic treatment for the public risk associated with infection caused by *K. pneumoniae*-contaminated food commodities. This chapter focuses on *C. elegans* as a model system that has been instrumental in uncovering the clues on host–pathogen interactions during *K. pneumoniae* infection.

23.1 Foodborne Infection: A Global Burden

Foodborne diseases are currently recognized as a global burden by the World Health Organization. The alarming increase in the annual morbidity and mortality has highlighted the importance of proper hygiene in public places. In 2005, about 1.8 million people succumbed to diarrheal diseases following consumption of contaminated food and water [1]. This is not only reported in underdeveloped countries—the same fate of massive death due to foodborne illness has also been reported in developed countries. Annually, there are 76 million cases of foodborne diseases, resulting in 325,000 hospitalizations, and 5000 deaths have been estimated to occur in USA [2,3].

Generally, foodborne infections induce a wide spectrum of symptoms. Although the actual incidence of gastrointestinal infection is largely unknown, the risk of getting infection through heavily contaminated food is probably high [4]. To eradicate the foodborne illness, public health awareness is critical. Since the study based on volunteers feeding does not represent a feasible option, the understanding of *K. pneumoniae* infection is chiefly based on epidemiological data, infection case reports, and use of animal models. An ideal animal model should mimic all phases of the human disease condition, as the host–pathogen interaction in a natural biological condition reveals the host's resistance mechanism and virulence properties of pathogen that promote colonization and subsequent dissemination [4].

23.2 *K. pneumoniae*: A Potent Agent for Foodborne Diseases

Foodborne illness represents an important public health hazard and has a significant economical impact on developing countries worldwide. Approximately 1.5 million neonatal deaths are caused by foodborne diarrheal disease alone. Viruses, bacteria, fungi, and parasites are the causative agents for foodborne infections [5]. Among them, bacteria are implicated in sporadic illness among different populations. Prior to 1990s, the species of *Salmonella* and *Campylobacter* and *Escherichia coli* accounted for most of the outbreaks. Subsequently, *K. pneumoniae* was shown to cause foodborne diseases in humans through contaminated vegetables [6] and hamburgers [7]. Given the increased number and severity of infection, *K. pneumoniae* has been recognized as one of the most important foodborne bacterial pathogens.

K. pneumoniae is one of the most versatile and clinically significant pathogens in causing both nosocomial- and community-acquired infections worldwide. It is a Gram-negative, rod-shaped bacterium belonging to the family Enterobacteriaceae [8]. Although it is found as a common component of normal microflora in the nasopharyngeal and gastrointestinal regions of humans, *K. pneumoniae* can cause a wide variety of diseases, ranging from mild infections to life-threatening diseases, in immunocompromised patients. The diseases caused by *K. pneumoniae* are bowel infections, pneumonia, suppurative infections, bacteremia, septicemia, and inflammation of the urinary tract. In severe infection cases, *K. pneumoniae* causes liver abscess along with meningitis and endophthalmitis, which accounts for a mortality rate of 20%–55% [8]. Due to its severity in causing infection in immunocompromised individuals, it is classified as an opportunistic pathogen. As far as hospital-acquired infection by *K. pneumoniae* is concerned, the prime significance is its ability to evade the innate immune response and cause colonization in the intestinal tract of patients and cause symptoms. Infection with *K. pneumoniae* begins with its initial colonization in the gastrointestinal tract, irrespective of its site of infection [9]. The degree of colonization mainly depends on the potential of the bacterial strain to form biofilms [10] and emerge resistant to antibiotics. The resistance of *K. pneumoniae* toward third-generation cephalosporins and carbapenem were identified in 1983 [11–13]. Since, *K. pneumoniae* is undergoing multidrug resistance, it is necessary to study its pathogenesis in a multicellular host system to approach the infection using an alternate therapy. Thus, researchers paid a great deal of attention to choose a suitable animal model to decipher the host perspectives during bacterial infections.

23.3 Importance of Model Animals in Research

During the 18th century, researchers were fascinated with the availability of huge diversity between model organisms with a variety of purposes. Based on their objectives and aim, they chose the potential model organism. This scientific exploration around the globe paved the way for a variety of theories and classifications (e.g., Linnaean classification and cell theory). Later in the 19th century, researchers understood the various complex processes like metabolism, embryogenesis, neuronal behavior, and disorder by experimental validation using model systems. In addition, much of our knowledge on heredity, development, metabolism, physiology, and several molecular mechanisms have been gained from the studies of model organisms [14]. A model organism is selected on the basis of its amenability for experimental procedure and its potential for extracting a wealth of information on common mechanisms among various life forms. Furthermore, most of the model organisms have been sequenced and provide deep knowledge in respective fields, which enable the researcher to address the serious query raised [14]. Using model organisms for studying pathogenicity of human pathogens did not gain serious interest till Frederick Ausubel used a surrogate model to study host–pathogen interactions [15,16]. The use of model organisms for understanding pathogenesis is not surprising since several of the underlying processes are conserved from yeast to mammals. Additionally, the genes that contribute to important biological pathways have also been found to be conserved across phylogeny. Since most of the pathogens share a common mode of pathogenesis mechanisms in a wide variety of hosts, it is necessary to employ a model host to identify the common and novel virulence factors of pathogen and also to understand the cross talk between a host and pathogen.

23.4 Discovery and Exploration of *C. elegans*

C. elegans was first discovered and described by Emile Maupas, a French librarian zoologist and botanist, in 1900 from soil in Algeria [17]. Initially, Maupas classified this nematode under the *Rhabditis* genus rather than *Caenorhabditis*. Later, in 1952, the new subgenus *Caenorhabditis* was coined by Osche [18], which was later given generic status by Dougherty in 1955 [19]. Surprisingly, the sequence of *C. elegans* has high homology with human genes and many of the genes can code for the function similar to mammalian genes. *C. elegans* is a eukaryotic, free-living soil nematode. It has a complex developmental process including embryogenesis, morphogenesis, and growth mechanism. Besides showing about 35% homology with the human at the genome level, *C. elegans* shares 83% similarity in protein sequence [20]. Its genome has been completely sequenced, and the genome size is approximately 100 Mb. Due to this significant property and smaller genome size, *C. elegans* was introduced to the scientific community by Sydney Brenner as an experimental model in 1960s to explore developmental biology and report his findings on the behavior of the nematode [21]; this attracted huge attention from the researchers. Later, it was used for the study of host–pathogen interactions against *Mycobacterium nematophilum* [22–25], *Drechmeria coniospora* [26], *Pseudomonas aeruginosa* [27–29], *Serratia marcescens* [25,30], *Escherichia coli* [31,32], *Enterococcus faecalis* [33,34], *Staphylococcus aureus* [29,35], *Salmonella* spp. [36–38], *Shigella* spp. [39–41], and *Vibrio* spp. [42–45]. Beside *C. elegans*, other animals have also been used as a model organism to study *K. pneumoniae* pathogenesis and host defense mechanisms.

23.5 Other Model Animals to Study *K. pneumoniae* Infection

Most of the studies on *K. pneumoniae* pathogenesis have been focused on the murine model [46–48]. As *K. pneumoniae* is a prime infectious agent of the respiratory tract, murine models having well-developed respiratory systems have been suggested to be the best option to decipher the response of the host immune system. Injection of *K. pneumoniae* into the murine model alters the level of interleukin-17 (IL-17) [49], tumor necrosis factor alpha, keratinocyte-derived chemokines and bronchoalveolar fluid, and IL-10 [50] during the course of infection. Furthermore, interaction of *K. pneumoniae* with the lung epithelial cells [51] markedly increased the recruitment of neutrophils to the lungs as well as proinflammatory cytokines and chemokines. A recent study has used the wax moth *Galleria mellonella* as a surrogate host model to understand the pathogenesis of *K. pneumoniae* at the molecular level for testing new therapies [52]. In that study, it was reported that *G. mellonella* could distinguish pathogenic and nonpathogenic *Klebsiella* strains. Moreover, infection of *K. pneumoniae* in *G. mellonella* attenuates the host innate immune response by inhibiting the antimicrobial peptides. Furthermore, the study by Wu et al. in HepG2 cells of humans suggested that infection with *K. pneumoniae* attenuates the p38 mitogen-activated protein kinase (p38 MAPK) pathway of host system and that it required lipopolysaccharide for complete pathogenesis during host–pathogen interaction [53]. A recent study by us has used *C. elegans* as a model and reinforced the essential role of Toll-dependent p38 MAPK pathway during *K. pneumoniae* infection [54].

23.6 Characteristic Features of *C. elegans*

Easy maintenance: *C. elegans* is propagated and maintained on agar plate with lawns of nonpathogenic *E. coli* OP50 as food source. Maintenance is less expensive.

Short life cycle: The life cycle of *C. elegans* is temperature dependent and is approximately 3.5–7 days; within this period, each animal can produce around 300 genetically identical progenies.

Easy handling: Stocks can be frozen at −80°C and stored for many years and can be revived whenever required for research purpose and propagated continuously.

Sequenced genome: *C. elegans* is the first complex model organism to have its entire genome sequenced and available on the Wormbase.

Availability of genetic tools: RNAi mechanism is well established in *C. elegans,* and more than 90% of the genome is available as a dsRNA expression library. Generation of transgenic *C. elegans* is made easy by injecting the dsRNA by microinjection method.

Transparency: The transparent body of the *C. elegans* facilitates the characterization of gene expression and enables easy localization of pathogens tagged with fluorescent labels.

Simple anatomy: *C. elegans* has a simple anatomy. It possesses 302 neurons and 959 somatic cells corresponding to the development of lineage, and all these cells have been thoroughly tracked back to the zygote. Hence, it is considered as a simple organism with complex structures such as nervous system, digestive system, and extensive germ line.

23.7 *C. elegans* as a Model for Host–Pathogen Interaction Studies

C. elegans has the ability to survive even after repeated infections as it evokes several immune pathways that activate the host defense against several bacterial infections. Studies in *C. elegans* have shed more light on many pathways of innate immune system that are evolutionary conserved from worms to mammals [55]. Thus, *C. elegans* is a very good model to study both the host immune response and the pathogen's virulence mechanism that facilitate the host–pathogen interface [56]. The orchestrated regulation of immune defense is an important criterion to combat microbial infection and prevent deleterious effects on host cells. Hence, it is particularly important for the mammalian intestine where pathogenic bacteria must be harbored to enable differentiation from bacterial species that could be a part of common flora of humans. One approach to understand the mechanisms of immune defense at the epithelial surface is to use invertebrate hosts like *C. elegans* to decipher the evolutionarily conserved aspects of innate immune defense and pathogen virulence [57].

23.8 Key Differences between Innate Immune System of *C. elegans* and Mammals

Generally, some of the mammalian's innate immune response is not encoded in the *C. elegans* genome, such as homologues of the transcription factor, NF-κB, and adaptor protein MYD88. Additionally, *C. elegans* does not produce any molecules that have homology with the known vertebrate cytokines. Even though the nematode lacks NF-κB, MYD88, and other secretary molecules of the TLR signaling pathway, it mounts an innate immune defense that employ several evolutionarily conserved signaling pathways, including p38 MAPK, β-catenin, and FOXO transcription factors, which interestingly appear to function in synergy to activate effector genes.

23.9 p38 MAPK Pathway in *C. elegans*

The p38 MAPK is a multistep mechanism that includes immune regulators like NSY-1/SEK-1/PMK-1 MAP kinase pathway, which was identified through forward genetics using mutants that had enhanced susceptibility to infection with Gram-negative bacteria *P. aeruginosa* [58]. This pathway is orthologous to the ASK1 (MAP kinase kinase kinase)/MKK3/6 (MAP kinase kinase)/p38 (MAP kinase) pathway in mammals, and its identification in *C. elegans* provided a deeper knowledge on host perspective. The p38 MAP kinase pathway acts cell-autonomously in the intestinal epithelium [59] to coordinate immune defense against a wide variety of ingested pathogens. *C. elegans* having a loss-of-function mutations in *pmk-1* are hypersusceptible to infection by Gram-negative pathogens such as *P. aeruginosa* [58,60], *S. entrica* [61], *Y. pestis* [62], and *S. marcescens* [63]; Gram-positive pathogens *E. faecalis* [63] and *Staphylococcus aureus* [33]; and the fungus *C. albicans* [64,65]. Nematodes' global transcriptional profiling analyses revealed that PMK-1 regulates the expression putative antimicrobial effectors, including C-type lectins, Shk toxins, and CUB-like domain [58], in the absence of microbial challenge, and this is termed as "basal regulation."

23.10 Live *K. pneumoniae* Was More Pathogenic to *C. elegans*

The pathogenicity of *K. pneumoniae* is determined by its ability to affect normal physiological activities like survival, reproduction, and feeding in nematodes. In this study, the physiological activities of the *C. elegans* infected with *K. pneumoniae* were examined and compared with the control nematodes fed with *E. coli* OP50. The control *C. elegans* fed on *E. coli* OP50 showed an intact pharynx [Figure 23.1A(i)] and normal egg formation [Figure 23.1A(ii)], whereas the nematodes fed on *K. pneumoniae* were severely infected, which was evidenced by distended pharynx [Figure 23.1A(iii, iv)], abnormal egg formation and accumulation [Figure 23.1A(v)], and internal hatching with accumulation of bacterial load inside the gastrointestinal region of the exposed nematodes [Figure 23.1A(vi)]. In addition, the severity of *K. pneumoniae* pathogenicity was determined by assessing the rate of killing in nematodes. As shown in Figure 23.1B, *K. pneumoniae* displayed a significant ($p < 0.05$) reduction in the lifespan of nematodes when compared to that of OP50 control. Nematodes fed on *K. pneumoniae* displayed a significantly ($p < 0.05$) decreased mean lifespan of 48 ± 5 h when compared to that of nematodes fed with OP50, which had a mean lifespan of 21 ± 0.5 days (Figure 23.1B). Infection with *K. pneumoniae* also significantly ($p < 0.05$) inhibited egg laying (Figure 23.1C) and feeding (Figure 23.1D) in *C. elegans* [54].

23.11 *K. pneumoniae* Proliferates inside the Host Intestine and Causes Persistent Infection

In confocal laser scanning microscopy (CLSM) images, the intense profile of the pathogen-exposed nematode showed increased colonization in the intestine of the *C. elegans* with longer residence time

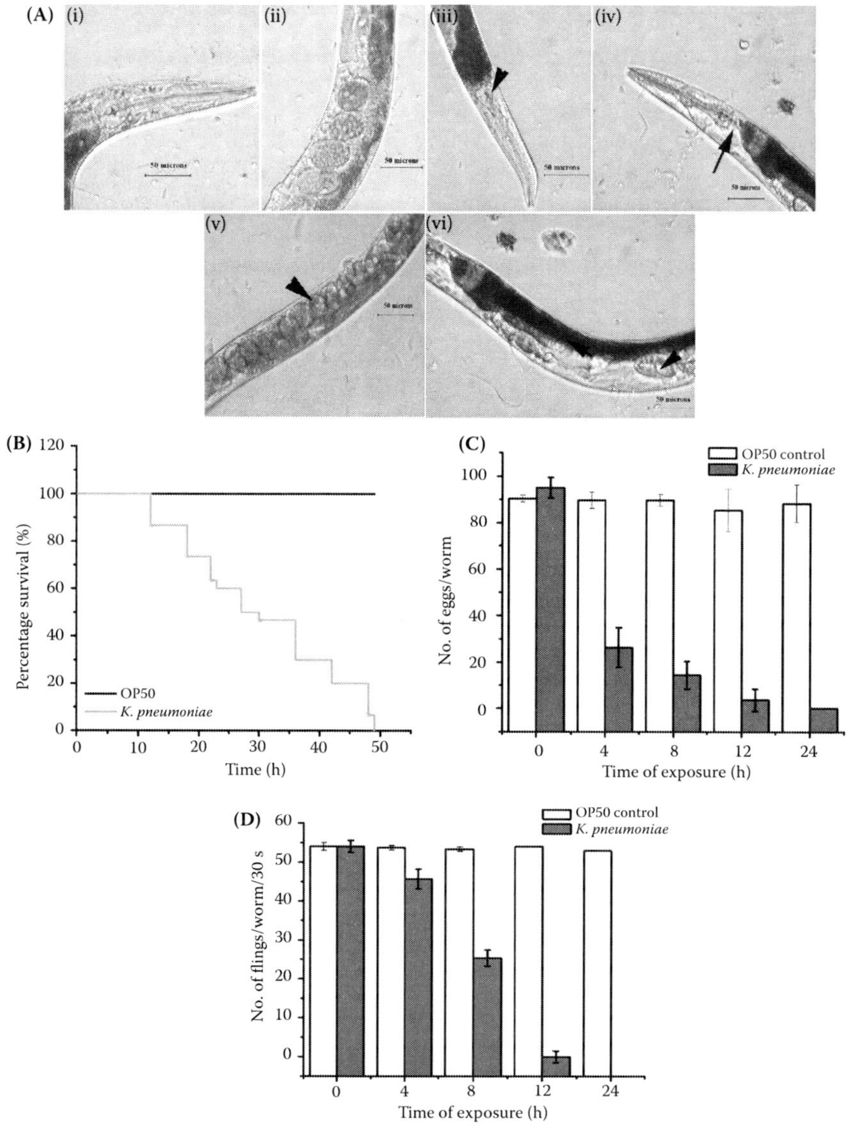

FIGURE 23.1 (A) Microscopic images of *C. elegans* exposed to *K. pneumoniae* for 12 h at 20°C are shown. Wild-type *C. elegans* fed with OP50 showed intact pharynx (i) and (ii) normal eggs. In contrast, nematodes exposed to *K. pneumoniae* showed severe physiological defects such as (iii, iv) distended pharynx (indicated by arrowhead), (v) accumulation of eggs inside the intestine (indicated by arrowhead), (vi) internal hatching, and intestinal colonization (indicated by arrowhead). Physiological assays showing the impact of *K. pneumoniae* infection in *C. elegans*. (B) Survival of *C. elegans* exposed to live *K. pneumoniae*. (C) Cessation of egg laying in nematodes exposed to *K. pneumoniae*. (D) Inhibition of feeding during *K. pneumoniae* infection. Data are presented as mean ± standard deviation of three biological replicates. Statistical analysis was performed by one-way ANOVA followed by Duncan's post hoc analysis.

(Figure 23.2). Measuring the bacterial load inside the intestine of nematodes by colony forming unit (CFU) assay revealed that the number of bacterial cells in *C. elegans* exposed for 4 h was 2.7×10^3, and it was found to be increased to 3.6×10^4, 4.7×10^5, and 5.3×10^6 at 6, 12, and 24 h, respectively (Figure 23.2A). The results of CFU assay suggested that *K. pneumoniae* colonized and proliferated inside the host intestine with an increase in residence time.

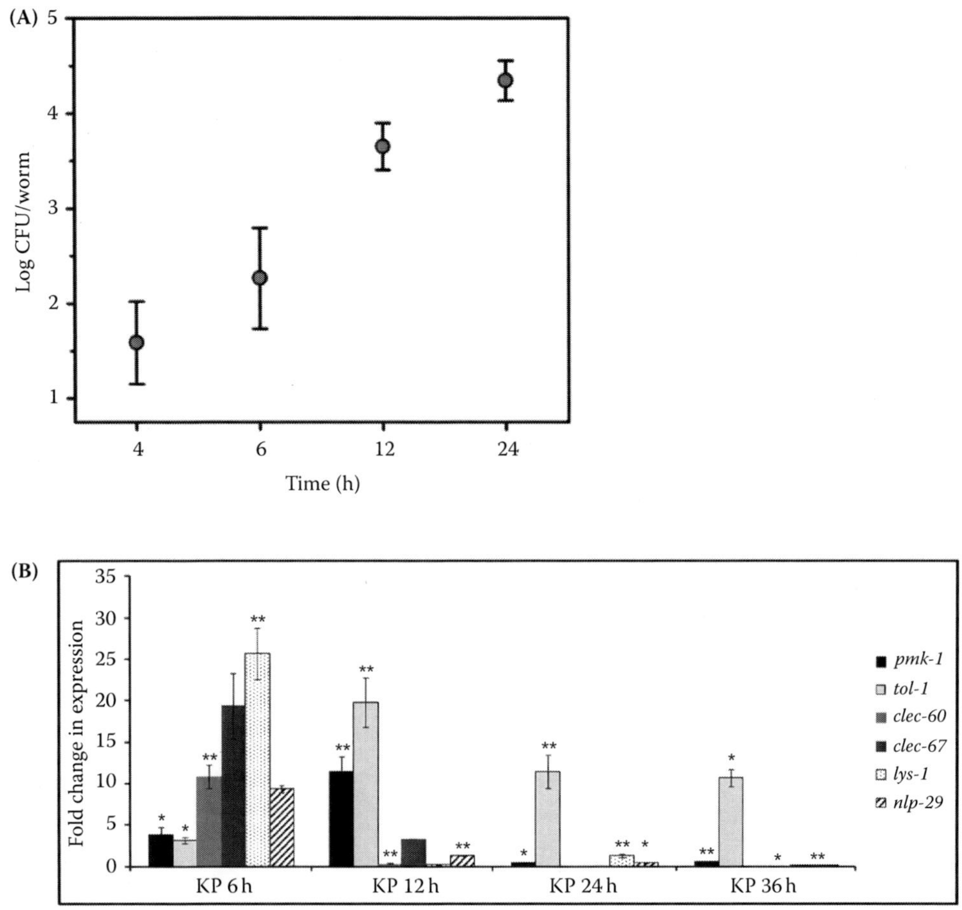

FIGURE 23.2 The histogram showing the CFU of *K. pneumoniae* inside the intestine of *C. elegans* in response to time of infection (A). Changes in the expression profile of N2 nematodes during infection with *K. pneumoniae* (B). (From Kamaladevi, A. and Balamurugan, K., *Pathog. Dis.*, 73(5), ftv021, 2015.)

23.12 Role of p38 MAPK in Immune Defense of Nematode against *K. pneumoniae* Infection

Initial defense of host immune response is the recognition of the pathogen by toll-like receptor (Toll). Toll is encoded by the gene *tol-1* in *C. elegans*. In nematodes exposed to *K. pneumoniae*, the *tol-1* was observed to be significantly ($p < 0.005$) increased. Additionally, the genes *pmk-1*, *clec-60*, *clec-67*, *clec-85*, *lys-1*, and *nlp-29* were upregulated in the initial hours of infection and downregulated gradually during the later hours of infection (Figure 23.2B). The cluster of *clec-60/67* increased the host resistance to infection and increased the survival of nematodes [29,30]. However, the downregulation of *clec-60* and *clec-67* during exposure to *K. pneumoniae* indicated the surrendering of host immune defense to *K. pneumoniae* infection. In particular, *lys-1* and *nlp-29* were reported to be regulated by p38 MAPK pathway during bacterial and fungal infection [62,63]. The increased expression of *pmk-1* along with the increased level of *lys-1* and *nlp-29* in the initial hours of infection and gradual downregulation in the later hours revealed that p38 MAPK pathway may have a role against *K. pneumoniae* (Figure 23.2B).

23.13 Role of Toll-Dependent p38 MAPK Pathway in *C. elegans* against *K. pneumoniae* Infection

To establish the role of p38 MAPK in *C. elegans* against *K. pneumoniae* infection, mutants of this pathway such as AU37 (*sek-1*), KU25 (*pmk-1*), and IG10 (*tol-1*) were examined for their survival during *K. pneumoniae* infection. The mutants of *sek-1, pmk-1,* and *tol-1* exhibited a significantly shorter mean lifespan of 24 ± 5, 33 ± 3, and 30 ± 7 h, respectively, than the N2 nematodes. The hypersusceptibility of mutants to *K. pneumoniae* infection suggested the role of p38 MAPK pathway and toll-like receptor in nematode defense (Figure 23.3A).

To explore the involvement of p38 MAPK pathway and Toll receptor during infection with *K. pneumoniae*, an expression analysis was done in p38 MAPK pathway mutants (*sek-1* and *pmk-1*) and *tol-1* mutant. The expression of *tol-1* was significantly ($p < 0.005$) upregulated in *sek-1* and *pmk-1* mutants (Figure 23.3), suggesting the requirement of toll-like receptor during *K. pneumoniae* exposure. Exposing AU37 to *K. pneumoniae* significantly ($p < 0.005$) downregulated the expression of *pmk-1* during the initial hour of infection itself. However, in *tol-1* mutant IG10, *pmk-1* was not observed to be activated (Figure 23.3B). This indicated that the lack of toll receptor in IG10 during *K. pneumoniae* infection failed to activate the expression of *pmk-1*. These data suggested that *C. elegans* require toll-dependent p38 MAPK pathway to act against *K. pneumoniae* infection. Furthermore, the nonactivation of antimicrobial

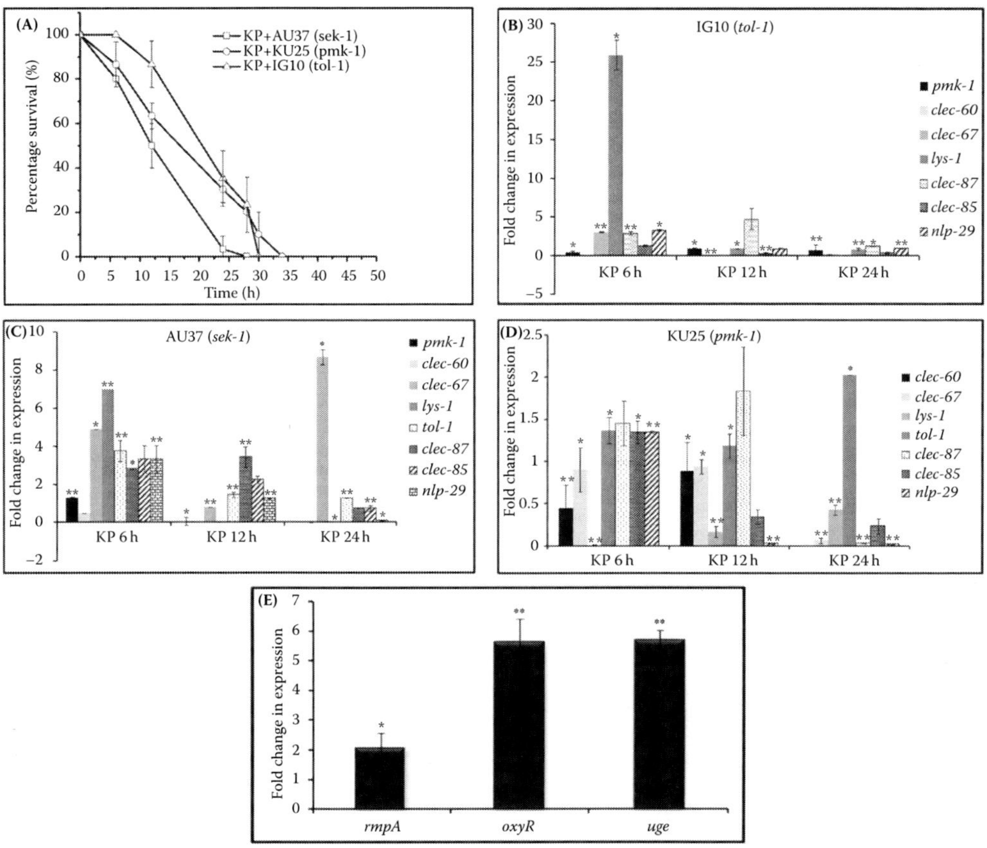

FIGURE 23.3 The increased susceptibility of p38 MAPK pathway mutants against *K. pneumoniae* infection (A). Changes in the expression profile of the *tol-1* mutant IG10 (B), *sek-1* mutant AU37 (C), *pmk-1* mutant KU25 (D), and regulation of selective virulence factors in *K. pneumoniae* during interaction with host *C. elegans* (E). (Figure 23.3A–D were adapted from Kamaladevi, A. and Balamurugan, K., *Pathog. Dis.*, 73(5), ftv021, 2015; Figure 23.3E unpublished data.)

genes, such as *nlp-29* and *lys-1* (Figure 23.3B), that are primarily regulated by p38 MAPK also strongly suggests the role of toll-dependent p38 MAPK pathway against *K. pneumoniae* infection. The antimicrobial genes *clec-60, clec-67, clec-85,* and *clec-87* were significantly ($p < 0.005$) upregulated in the early hours and downregulated at the later hours in the AU37 (*sek-1*) and KU25 (*pmk-1*) mutants during infection, suggesting that *K. pneumoniae* thwarts the host defense and causes a persistent lethal infection (Figure 23.3C and D).

23.14 Upregulation of Virulence Gene in *K. pneumoniae* during Infection

On the other hand, to check the hypothesis that *K. pneumoniae* utilizes the same virulence strategies to infect nematode and mammalian hosts, real-time polymerase chain reaction (PCR) analysis was performed for selective candidate virulent genes. The virulent genes including *rmp*A, *uge,* and *oxy*R have been well established as being essential for colonization [46,47] and to withstand the oxidative stress [48]. Here, the transcriptional analysis showed that the interaction of *K. pneumoniae* with *C. elegans* significantly increased the expression of virulence genes such as *rmpA* ($p < 0.005$), *oxyR* ($p < 0.005$), and *uge* ($p < 0.05$) compare with their respective controls (Figure 23.3E). The overexpression of these reported genes in *K. pneumoniae* during host–pathogen interaction suggested that the *K. pneumoniae* shares a common mode of pathogenesis between nematodes and mammals (unpublished data).

23.15 Conclusion

C. elegans has been considered as an excellent animal model to study the host immune factors and pathogenesis of *K. pneumoniae* under *in vivo* conditions. Concerning the pathogenicity of *K. pneumoniae* to *C. elegans*, the killing of *C. elegans* by *K. pneumoniae* is mediated by the colonization and proliferation inside the host intestine. While *C. elegans* utilizes the Toll-dependent p38 MAPK pathways against *K. pneumoniae* infection, *K. pneumoniae* increases the expression of *rmpA, oxyR,* and *uge* genes during infection. Overall, these interactions underscore the host–pathogen relationship between *C. elegans* and *K. pneumoniae* and will form the basis for identification of potential therapeutic targets against *K. pneumoniae* and for formulation of effective immune therapy in hosts.

Acknowledgments

We thank Caenorhabditis Genetics Center, which is funded by the National Institutes of Health, National Center for Research Resources, for providing *C. elegans* N2 WT, mutant strains, and *E. coli* OP50. Dr. K. Balamurugan thankfully acknowledges the DBT (BT/PR14932/MED/29/233/2010), ICMR (Sanction No: 5/3/3/13/2010-ECD-I), DST-SERC Fast Track Young Scientist Scheme [No. SR/FT/LS-83/2009 (G)], DST-SERB (No. SR/SO/AS-80/2010), UGC Major Research Project [No. 42-222/2013 (SR)], CSIR [No. 37(1460)/11/EMR-II], DST-PURSE [Grant No. SR/S9Z-23/2010/42(G)], FIST (Grant No. SR-FST/LSI-087/2008), and UGC-SAP-DRS-I [Grant No. F.3-28/2011 (SAP-II)], New Delhi, India, for financial assistances. Financial support to AK by DBT in the form of SRF is thankfully acknowledged. The authors gratefully acknowledge the computational and bioinformatics help provided by the Alagappa University Bioinformatics Infrastructure Facility (funded by DBT, GOI; Grant No. BT/BI/25/001/2006).

REFERENCES

1. Hanson, L. A. et al., Estimating global mortality from potentially foodborne diseases: an analysis using vital registration data, *Popul. Health Metrics*, 10, 7, 2012.
2. Scallan, E. et al., Foodborne illness acquired in the United States—major pathogens, *Emerg. Infect. Dis.*, 17, 7, 2011.
3. Mead, P. S. et al., Food-related illness and death in the United States, *Emerg. Infect. Dis.*, 5, 607, 1999.

4. D'Orazio, S. E., Animal models for oral transmission of *Listeria monocytogenes*, *Front. Cell Infect. Microbiol.*, 4, 15, 2014.
5. Atreya, C. D., Major foodborne illness causing viruses and current status of vaccines against the diseases, *Foodborne Pathog. Dis.*, 1, 89, 2004.
6. Puspanadan, S. et al., Detection of *Klebsiella pneumoniae* in raw vegetables using most probable number-polymerase chain reaction (MPN-PCR), *Int. Food Res. J.*, 19, 1757, 2012.
7. Sabota, J. M. et al., A new variant of food poisoning: enteroinvasive *Klebsiella pneumoniae* and *Escherichia coli* sepsis from a contaminated hamburger, *Am. J. Gastroenterol.*, 93, 118, 1998.
8. Podschun, R. and Ullmann, U., *Klebsiella* spp. as nosocomial pathogens: epidemiology, taxonomy, typing methods, and pathogenicity factors, *Clin. Microbiol. Rev.*, 11, 589, 1998.
9. Won, S. Y. et al., Emergence and rapid regional spread of *Klebsiella pneumoniae* carbapenemase-producing Enterobacteriaceae, *Clin. Infect. Dis.*, 53, 532, 2011.
10. Begun, J. et al., Staphylococcal biofilm exopolysaccharide protects against *Caenorhabditis elegans* immune defenses, *PLoS Pathog.*, 3, e57, 2007.
11. White, D. G. et al., Antimicrobial resistance among Gram-negative foodborne bacterial pathogens associated with foods of animal origin, *Foodborne Pathog. Dis.*, 1, 137, 2004.
12. Cao, X. et al., Molecular characterization of clinical multidrug-resistant *Klebsiella pneumoniae* isolates, *Ann. Clin. Microbiol. Antimicrob.*, 13, 16, 2014.
13. Yigit, H. et al., Novel carbapenem-hydrolyzing β-lactamase, KPC-1, from a carbapenem-resistant strain of *Klebsiella pneumoniae*, *Antimicrob. Agents Chemother.*, 45, 1151, 2001.
14. Müller, B. and Grossniklaus, U., Model organisms—a historical perspective, *J. Proteomics*, 73, 2054, 2010.
15. Rahme, L. G. et al., Common virulence factors for bacterial pathogenicity in plants and animals, *Science*, 268, 1899, 1995.
16. Rahme, L. G. et al., Plants and animals share functionally common bacterial virulence factors, *Proc. Natl. Acad. Sci. USA*, 97, 8815, 2000.
17. Maupas, E., Modes et formes de reproduction des nematodes, *Arch. Zool. Exp. Gén.*, 8, 463, 1901.
18. Osche, G., Systematik und phylogenie der gattung Rhabditis (Nematoda), *Zool. Jahrb. Abt. Allg. Zool. Physiol. Tiere*, 81, 190, 1952.
19. Fodor, A. et al., Comparison of a new wild-type *Caenorhabditis briggsae* with laboratory strains of *C. briggsae* and *C. elegans*, *Nematologica*, 29, 203, 1983.
20. Lai, C. H. et al., Identification of novel human genes evolutionarily conserved in *Caenorhabditis elegans* by comparative proteomics, *Genome Res.*, 10, 703, 2000.
21. Brenner, S., The genetics of behaviour, *Br. Med. Bull.*, 29, 269, 1973.
22. Hodgkin, J. et al., A novel bacterial pathogen, *Microbacterium nematophilum*, induces morphological change in the nematode *C. elegans*, *Curr. Biol.*, 10, 1615, 2000.
23. Nicholas, H. R. and Hodgkin, J., Innate immunity: the worm fights back, *Curr. Biol.*, 12, R731, 2002.
24. Gravato-Nobre, M. J. and Hodgkin, J., *Caenorhabditis elegans* as a model for innate immunity to pathogens, *Cell. Microbiol.*, 7, 741, 2005.
25. O'Rourke, D. et al., Genomic clusters, putative pathogen recognition molecules, and antimicrobial genes are induced by infection of *C. elegans* with *M. nematophilum*, *Genome Res.*, 16, 1005, 2006.
26. Jansson, H. B., Adhesion of conidia of *Drechmeria coniospora* to *Caenorhabditis elegans* wild type and mutants, *J. Nematol.*, 26, 430, 1994.
27. Mahajan-Miklos, S. et al., Molecular mechanisms of bacterial virulence elucidated using a *Pseudomonas aeruginosa–Caenorhabditis elegans* pathogenesis model, *Cell*, 96, 47, 1999.
28. Tan, M. W. et al., Killing of *Caenorhabditis elegans* by *Pseudomonas aeruginosa* used to model mammalian bacterial pathogenesis, *Proc. Natl. Acad. Sci. USA*, 96, 715, 1999.
29. Irazoqui, J. E. et al., Distinct pathogenesis and host responses during infection of *C. elegans* by *P. aeruginosa* and *S. aureus*, *PLoS Pathog.*, 6, e1000982, 2010.
30. Mallo, G. et al., Inducible antibacterial defense system in *C. elegans*, *Curr. Biol.*, 12, 1209, 2002.
31. Mellies, J. et al., The global regulator Ler is necessary for enteropathogenic *Escherichia coli* colonization of *Caenorhabditis elegans*, *Infect. Immun.*, 74, 64, 2006.
32. Crossman, L. C. et al., A commensal gone bad: complete genome sequence of the prototypical enterotoxigenic *Escherichia coli* strain H10407, *J. Bacteriol.*, 192, 5822, 2010.
33. Sifri, C. D. et al., *Caenorhabditis elegans* as a model host for *Staphylococcus aureus* pathogenesis, *Infect. Immun.*, 71, 2208, 2003.

34. Van Der Hoeven, R. et al., Ce-Duox1/BLI-3 generated reactive oxygen species trigger protective SKN-1 activity via p38 MAPK signaling during infection in *C. elegans*, *PLoS Pathog.*, 7, e1002453, 2011.
35. JebaMercy, G. et al., Changes in *Caenorhabditis elegans* life span and selective innate immune genes during *Staphylococcus aureus* infection, *Folia Microbiol.*, 56, 373, 2011.
36. Aballay, A. et al., *Salmonella typhimurium* proliferates and establishes a persistent infection in the intestine of *Caenorhabditis elegans*, *Curr. Biol.*, 10, 1539, 2000.
37. Aballay, A. and Ausubel, F. M., Programmed cell death mediated by *ced-3* and *ced-4* protects *Caenorhabditis elegans* from *Salmonella typhimurium*-mediated killing, *Proc. Natl. Acad. Sci. USA*, 98, 2735, 2001.
38. Alegado, R. A. and Tan, M. W., Resistance to antimicrobial peptides contributes to persistence of *Salmonella typhimurium* in the *C. elegans* intestine, *Cell. Microbiol.*, 10, 1259, 2008.
39. Burton, E. A. et al., The *Caenorhabditis elegans* ABL-1 tyrosine kinase is required for *Shigella flexneri* pathogenesis, *Appl. Environ. Microbiol.*, 72, 5043, 2006.
40. Kesika, P. et al., Analysis of *Shigella flexneri*-mediated infections in model organism *Caenorhabditis elegans*, *Scand. J. Infect. Dis.*, 43, 286, 2011.
41. Kesika, P. and Balamurugan, K., Studies on *Shigella boydii* infection in *Caenorhabditis elegans* and bioinformatics analysis of immune regulatory protein interactions, *Biochim. Biophys. Acta*, 1824, 1449, 2012.
42. Zhu, J. and Mekalanos, J. J., Quorum sensing-dependent biofilms enhance colonization in *Vibrio cholerae*, *Dev. Cell*, 5, 647, 2003.
43. Vaitkevicius, K. et al., A *Vibrio cholerae* protease needed for killing of *Caenorhabditis elegans* has a role in protection from natural predator grazing, *Proc. Natl. Acad. Sci. USA*, 103, 9280, 2006.
44. Durai, S. et al., Establishment of a *Caenorhabditis elegans* infection model for *Vibrio alginolyticus*, *J. Basic Microbiol.*, 51, 243, 2011.
45. Durai, S. and Balamurugan, K., Changes in *Caenorhabditis elegans* exposed to *Vibrio parahaemolyticus*, *J. Microbiol. Biotechnol.*, 21, 1026, 2011.
46. Yu, W. L. et al., Association between *rmpA* and *magA* genes and clinical syndromes caused by *Klebsiella pneumoniae* in Taiwan, *Clin. Infect. Dis.*, 42, 1351, 2006.
47. Regué, M. et al., A gene, *uge*, is essential for *Klebsiella pneumoniae* virulence, *Infect. Immun.*, 72, 54, 2004.
48. Srinivasan, V. B. et al., Role of *oxyR*^{KP}, a novel LysR-family transcriptional regulator, in antimicrobial resistance and virulence in *Klebsiella pneumoniae*, *Microbiology*, 159, 1301, 2013.
49. Ye, P. et al., Interleukin-17 and lung host defense against *Klebsiella pneumoniae* infection, *Am. J. Respir. Cell Mol. Biol.*, 25, 335, 2001.
50. Soares, A. C. et al., Impaired host defense to *Klebsiella pneumoniae* infection in mice treated with the PDE4 inhibitor rolipram, *Br. J. Pharmacol.*, 140, 855, 2003.
51. Cortés, G. et al., Role of lung epithelial cells in defense against *Klebsiella pneumoniae* pneumonia, *Infect. Immun.*, 70, 1075, 2002.
52. Insua, J. L. et al., Modeling *Klebsiella pneumoniae* pathogenesis by infection of the wax moth *Galleria mellonella*, *Infect. Immun.*, 81, 3552, 2013.
53. Wu, J. H. et al., Mitogen-activated protein kinase (MAPK) signalling pathways in HepG2 cells infected with a virulent strain of *Klebsiella pneumoniae*, *Cell. Microbiol.*, 8, 1467, 2006.
54. Kamaladevi, A. and Balamurugan, K., Role of PMK-1/p38 MAPK defense in *Caenorhabditis elegans* against *Klebsiella pneumoniae* infection during host–pathogen interaction, *Pathog. Dis.*, 73(5), ftv021, 2015.
55. Kurz, C. L. and Ewbank, J. J., *Caenorhabditis elegans* for the study of host–pathogen interactions, *Trends Microbiol.*, 8, 142, 2000.
56. Marsh, E. K. and May, R. C., *Caenorhabditis elegans*, a model organism for investigating immunity, *Appl. Environ. Microbiol.*, 78, 2075, 2012.
57. Shivers, R. P., Youngman, M. J. and Kim, D. H., Transcriptional responses to pathogens in *Caenorhabditis elegans*, *Curr. Opin. Microbiol.*, 11, 251, 2008.
58. Kim, D. H. et al., A conserved p38 MAP kinase pathway in *Caenorhabditis elegans* innate immunity, *Science*, 297, 623, 2002.
59. Shivers, R. P. et al., Tissue-specific activities of an immune signaling module regulate physiological responses to pathogenic and nutritional bacteria in *C. elegans*, *Cell Host Microbe*, 6, 321, 2009.

60. Troemel, E. R. et al., p38 MAPK regulates expression of immune response genes and contributes to longevity in *C. elegans*, *PLoS Genet.*, 2, e183, 2006.
61. Aballay, A. et al., *Caenorhabditis elegans* innate immune response triggered by *Salmonella enterica* requires intact LPS and is mediated by a MAPK signaling pathway, *Curr. Biol.*, 13, 47, 2003.
62. Bolz, D. D. et al., A conserved PMK-1/p38 MAPK is required in *Caenorhabditis elegans* tissue-specific immune response to *Yersinia pestis* infection, *J. Biol. Chem.*, 285, 10832, 2010.
63. Shivers, R. P. et al., Phosphorylation of the conserved transcription factor ATF-7 by PMK-1 p38 MAPK regulates innate immunity in *Caenorhabditis elegans*, *PLoS Genet.*, 6, e1000892, 2010.
64. Pukkila-Worley, R. et al., *Candida albicans* hyphal formation and virulence assessed using a *Caenorhabditis elegans* infection model, *Eukaryot. Cell*, 8, 1750, 2009.
65. Pukkila-Worley, R. et al., *Candida albicans* infection of *Caenorhabditis elegans* induces antifungal immune defenses, *PLoS Pathog.*, 7, e1002074, 2011.

24

Proteus

Paola Scavone, Victoria Iribarnegaray, and Pablo Zunino

CONTENTS

24.1 Introduction

The genus *Proteus* was named in 1885 by Hauser, inspired by the character of Proteus in Homer's *Odyssey* as he had the power to change his shape to escape different threats. Members of the genus *Proteus* are ubiquitously distributed in the environment, and many of them are human pathogens. In particular, *P. mirabilis* is an important etiological agent of opportunistic and nosocomial infections. It causes urinary tract infections (UTIs), which can be severe in patients under catheterization or with abnormalities in the urinary tract. Pyelonephritis, bacteremia, renal and bladder stones, and catheter obstruction are among the most common complications in *P. mirabilis* infection. Occasionally, it can cause foodborne illness as a result of mishandling during food processing.

In order to uncover its pathogenicity, several virulence factors were expressed *in vitro* and different strategies have been developed to assess the contributions of these factors. For example, application of laboratory models, from pioneer work with cell culture to animal model, has revealed valuable insights into various aspects of *P. mirabilis* infection. Indeed, several conserved antigens have been evaluated as vaccine components using a mouse model of infection. There is no doubt that a thorough understanding of the fundamental aspects of pathogenicity and host–pathogen interaction of *P. mirabilis* infection will be crucial for the formulation of new treatments and preventive strategies.

24.2 *Proteus mirabilis*

Proteus mirabilis is a Gram-negative bacterium, a dimorphic and motile member of the Enterobacteriaceae family. It has the ability to differentiate from short rods into elongated, multinucleate swarm cells that overexpress flagella [1]. It is a commensal bacterium found in the intestinal microbiota and in the environment, including in soil and contaminated water. Generally associated with complicated UTIs, *Proteus* may also cause a wide variety of nosocomial infections, including respiratory infections, affecting different organs (eyes, ears, nose, and skin, among others) [2,3].

Contamination of food by enteric pathogens can occur from the farm if human or animal sewage is used to fertilize soil or if sewage water is used to irrigate crops. Such risks are further increased if the food is mishandled during processing and preparation, a stage where pathogens could multiply exponentially under favorable conditions [4].

Members of the family Enterobacteriaceae have been considered a potent cause of foodborne outbreaks [5]. *P. mirabilis* has been frequently implicated in food deterioration and spoilage. Besides opportunistic infections, it can also cause food poisoning when consumed in contaminated food such as meat, vegetables, and seafood [6]. Indeed, *P. mirabilis* is regularly detected in food. For example, the bacterium was found in 7% of the beef, 10% of the chicken, and 17% of the rice from a cafeteria in Alice Town, South Africa [7]. In China, *P. mirabilis* isolated from chicken products showed a high level of antibiotic resistance, posing a potential threat to public health [8]. Moreover, the presence of bla_{VEB} was confirmed in a nonhuman *P. mirabilis* strain isolated from poultry meat in Bern, Switzerland, in addition to the detection of $bla_{CMY\text{-}2}$ in a sample imported from Austria [9]. CMY-2-positive *P. mirabilis* isolates are commonly found in humans, while VEB-6 isolates are less common. Several reports showed that VEB-6 *P. mirabilis* is responsible for infection in other countries [9]. It is worth noting that *P. mirabilis* strains harboring $bla_{CTXM\text{-}2}$ were recently identified in chicken meat sold in markets in southeast Brazil [10].

24.3 Virulence Factors and Biofilm

24.3.1 Virulence Factors

Proteus mirabilis expresses several virulence factors that are related to pathogenicity and to biofilm formation. Among them are fimbriae that mediate adhesion to different surfaces, flagella-based motility, outer membrane proteins, urease production, hemolysin, proteus toxic agglutinin, and iron acquisition systems [11].

Fimbriae are proteinaceous appendages that protrude from the bacterial cell wall and mediate adhesion to both biotic and abiotic surfaces. *P. mirabilis* genome encodes 17 different fimbrial operons, so far the highest number seen in any bacteria [12]. The most studied fimbriae are MR/P, UCA, PMF, and ATF, while the roles of the other 13 potential fimbriae in *P. mirabilis* pathogenesis are scarcely known. MR/P fimbria is important in UTIs [13], exhibiting reported roles in adhesion [14], in biofilm formation [15], and in eukaryotic cell cytotoxicity and genotoxicity [16]. The MR/P operon includes 10 genes that show several interesting features including an invertible element that modulates the ON/OFF phase variation (MrpI) and a transcriptional regulator (MrpJ) that represses motility, thus allowing the bacteria to control motility and adherence [17]. UCA fimbria facilitates adherence to uroepithelial cells, contributing to infection and also playing a role in biofilm development [18], whereas PMF fimbriae contribute to kidney colonization [19]. While all these fimbriae are involved in infection, ATF, whose expression is optimal at ambient temperature, does not have a role in colonization of the mouse urinary tract. However, it could play a part in catheter colonization, in survival in the environment, and probably in food contamination [20,21]. It has been recently reported that there is a tight regulation of fimbrial expression, due in part to MrpJ, a global regulator of *P. mirabilis* virulence [22]. These authors reported that MrpJ directly represses flhDC and other fimbriae (PMF, Fim8), T6SS, and protease expression. Although it is not well understood how this regulation works, it is clear that a controlled expression of fimbriae is needed to maintain bacterial fitness.

Flagella-mediated motility is related to *P. mirabilis* swarming motility. This behavior is based on sequential rounds of swarm cell differentiation, swarming colony migration, and consolidation with dedifferentiation to a swimmer-cell morphology [23]. Swarmer cells are multinucleate, are elongated (more than 10 µm long), and express several flagella, while swimmer cells are rod shaped, are 1–2 µm long, and have peritrichous flagella [23]. Several studies have revealed a complex regulatory network during swarm cell differentiation. There is a reciprocal regulation between fimbrial and flagella expression. There are 14 additional mrpJ paralogs associated with other fimbrial operons [23,24], and most of them repress motility, inhibiting swarm cell differentiation [24]. Bacteria adopt a filamentous morphology as a result of the sensor activities of flagella in contact, for example, with a urinary catheter. Contact to solid surfaces creates a torsional change in the outer membrane, and this is sensed by an upregulator of the flagellar master operon (Umo) protein, which induces the expression of flagella to produce the highly flagellated cells that are required for swarming [11]. This motility is related with migration across different abiotic surfaces including catheters.

Iron acquisition systems are an important virulence determinant that enhances bacterial colonization of the host cells and survival in the environment [25]. Iron is an essential micronutrient for all living organisms, and its acquisition is vital for bacteria considering that only a minor fraction is available (10^{-18} M). In order to capture iron, bacteria have evolved high-affinity iron-scavenging and uptake systems [25]. The main strategies used by bacteria are the production and uptake of siderophores and the direct utilization of host iron compounds such as transferrin, lactoferrin, or heme-containing molecules [25]. Gram-negative Fe(III) acquisition systems usually consist of an outer membrane receptor, with transport across the outer membrane by TonB/ExbB/ExbD complex, a periplasmic binding protein, and an inner membrane ABC transporter. In *P. mirabilis,* there are at least two gene clusters related to siderophore biosynthesis and ABC transport, three outer membrane proteins induced by iron starvation involved in heme uptake, and a heme receptor [26,27]. One of the clusters related to siderophore biosynthesis is a novel nonribosomal peptide syntheses (NRPS)-independent siderophore (NIS) named proteobactin, as this was first described in a bacterium [28]. The other one contains the *nrp* operon, which has been previously described to be upregulated during iron limitation [29]. This operon is encoded within the high pathogenicity island (HPI) in *P. mirabilis* HI4320 that shows homology compared to the HPI of *Yersinia* spp. [30]. Infection challenges with mutant strains in different genes involved in yersiniabactin-related siderophore showed that it contributes to *P. mirabilis* fitness *in vivo* [31].

Quorum sensing describes the communication between bacterial cells, whereby a coordinated population response is controlled by diffusible molecules produced by individuals, often known as autoinducers. The influence of quorum-sensing molecules in *Proteus* strains is much less known compared to *Pseudomonas* or *Escherichia*. *N*-acyl homoserine lactone (AHL) signaling molecules are utilized by several Gram-negative species to sense population density and coordinate gene expression [32]. *P. mirabilis* lacks a clear AHL synthase (LuxI) homologue and does not seem to produce this type of signaling molecule [12,33]. However, *P. mirabilis* encodes a LuxR family transcriptional regulator and seems to produce compounds with AHL-like activity. Previous studies found that the addition of exogenous AHL to a *P. mirabilis* population had a strain-specific impact on virulence factor expression, swarming, and biofilm formation [34,35]. The quorum-sensing molecule autoinducer 2 (AI-2), encoded by *luxS*, can mediate both intra- and interspecies interactions. *P. mirabilis* possesses a *luxS* homologue and produces AI-2 [36]. However, mutation of *luxS* in *P. mirabilis* str. BB2000 does not significantly affect swarming, virulence factor production, or survival in a mouse model, suggesting that AI-2 does not contribute to pathogenicity [36]. This lack of phenotype might indicate that *P. mirabilis* uses LuxS strictly as part of the activated methyl cycle, particularly as *P. mirabilis* str. HI4320 contains no clear homologue of the Lsr system for sensing and responding to AI-2. However, AI-2 produced by *P. mirabilis* might influence gene expression in other species that use this signaling molecule. Other studies have reported that the presence of a quorum-sensing molecule (*N*-butanoyl homoserine lactone, BHL) increased *P. mirabilis* biofilm thickness and ureolytic activity. Laser interferometric determination of diffusion showed that urea easily diffuses through *P. mirabilis* biofilm, while acetohydroxamic acid (AHA) is blocked. This may suggest that the use of urease inhibitors in catheter-associated UTIs (CAUTIs) may be less effective than in other urease-associated infections [37].

Urease is produced by *P. mirabilis* and is responsible for the hydrolysis of urea to carbon dioxide and ammonia [38]. Urease production is induced by urea and is expressed during growth in urine where it is highly active [39]. In urine, this activity elevates environmental pH and induces crystal formation (calcium crystals and magnesium ammonium phosphate) [39]. These crystals become trapped within the polysaccharides produced by attached bacteria, forming the characteristic crystalline biofilms on catheters [40]. Formation of these crystalline biofilms by *P. mirabilis* could be considered as a protective structure as the host immune system and antibiotics cannot penetrate the biofilm [40]. The activity of this enzyme also causes indirect tissue damage as ammonia becomes toxic for the uroepithelial cells [41].

P. mirabilis produces two toxins that are implicated in tissue damage and dissemination to the kidneys. Hemolysin (HpmA) is a Ca^{+}-dependent pore-forming cytolytic agent that destabilizes the host cell by inserting itself into the cell membrane and causing a Na^{+} efflux [42]. The other is a surface-associated cytotoxic protease proteus toxic agglutinin (Pta) that is functional only in an alkaline pH; it punctures the host cell membrane, causing leakage of the cytosol, osmotic stress, and depolymerization of actin filaments, affecting the integrity of the cell and inducing bladder and kidney damage in the urinary tract [23,43].

One important virulence factor that protects bacteria against immune response is the metalloproteinase ZapA, which cleaves serum and secretory immunoglobulins A1 (IgA1), IgA2, and IgG [44]. ZapA might also cleave complement components C1q and C3; cell matrix components such as collagen, fibronectin, and laminin; and cytoskeletal proteins such as actin and tubulin [45].

24.3.2 Techniques for the Study of Virulence Factors

Generation of mutants with interrupted or deleted genes is a powerful tool for studying the role of virulence factors. One of the most common mutagenesis strategies is to interrupt the gene of interest with an antibiotic cassette by allelic replacement. This has been used in *P. mirabilis* with high success [13,16,19–21,46], and some of the most relevant mutagenesis techniques and genes identified are shown in Table 24.1. The comparison of the mutants with the wild-type strain allows for the study of the effect of the absence of the gene in UTI.

Signature-tagged mutagenesis (STM), described by Hensel et al. in 1995, was designed to detect new virulence genes of the target organism *Salmonella typhimurium* in a murine model of typhoid fever. This technique is a negative selection method in which unique identification tags allow analysis of pools of mutants in mixed populations. In STM, each mutant is tagged with a unique DNA sequence, these allow for coamplification of the DNA tags of a mixed population of mutants by a single polymerase chain reaction (PCR) [56,57]. The tag consists of a short DNA sequence of 40 bp, inserted in a transposon system. DNA tags can be included during allelic replacement (signature-tagged allele replacement) on a systematic genome-wide scale [58]. Detection of the signature tags could be done by dot blot, PCR, or polymorphic tag-length transposon mutagenesis.

In *P. mirabilis,* several articles have described the use of this technique. Burall et al. 2004 [50] used STM in uropathogenic *P. mirabilis* in conjunction with a murine model of ascending UTI to identify

TABLE 24.1

Identification of *P. mirabilis* Microbial Factors Using Different Molecular Tools

Gene	Genetic Tool	References
pmfA	Allelic replacement	[20]
Flagellin genes	Allelic replacement	[46]
Protease	STM	[47]
Mutants with increased sensitivity to antimicrobial peptides	Mini-Tn5 mutagenesis	[48]
Swarming-deficient mutants	Random transposon mutagenesis	[49]
Genes that contribute to pathogenesis of UTI	STM	[50]
64-kDa iron-regulated OMP; heme receptor	Transposon mutagenesis	[27]
Fimbrial mutants	Allelic replacement	[51]
mrpJ	Site-directed mutagenesis	[28]
Virulence genes	STM	[52]
qnrC and *qnrA1*	Site-directed mutagenesis	[53]
fliL	Null mutations	[54]
mrpI; *glnE*; fimbrial subunit; putative MuA-like DNA-binding protein; *rsbA*; *bcr*; putative transmembrane protein; *gltS*; *lrp*; *nirB*; putative lipoprotein	Random transposon mutagenesis	[55]

virulence determinants required for colonization of the urinary tract (Table 24.1). They could identify genes that affected motility, iron acquisition, transcriptional regulation, phosphate transport, urease activity, cell surface structure, and key metabolic pathways [50].

A traditional tool that allows for the identification of gene functions is random transposon mutagenesis. The basis of this technique is in the generation of mutant libraries that harbor a transposon insert, which abolishes the function of the affected gene. Transposons are mobile genetic elements that can move within the genome and can affect the function of gene expression. One of the limitations in identification and separation of nonvirulent mutants from a pool of mutants is that it is time consuming. In *P. mirabilis,* there are many studies that report the use of this technique. Recently, Holling et al. [55] used random transposon mutagenesis to identify genes involved in biofilm formation. They used a mini-Tn5Km2 transposon that was introduced into the wild-type *P. mirabilis* strain B4 by conjugal transfer from the donor organism, *Escherichia coli* S17.1λpir (Table 24.1). The mutants were screened by the crystal violet microtiter plate assay to identify bacterial impairment in catheter blockage. The end adjacent to the transposon was sequenced in order to identify the mutated gene.

The use of null mutations is also an important strategy as it consists of the insertion of a reprogrammed intron carried into a plasmid that finally results in a lack of function of a particular gene. Lee et al. [54] used a *fliL* knockout mutant to gain further insight into the function of FliL (Table 24.1). The mutant was generated by a reprogrammed intron, the intron was inserted between nucleotides (nt) 30 and 31 in *fliL*. The loss of FliL results in cells that cannot swarm on agar surfaces but can swim in broth cultures.

24.3.3 Biofilm Formation and Cell-to-Cell Communication

During the past few decades, biofilms have been widely studied because they cause 65% or more of all infections, being particularly prevalent in device-related infections, infections on body surfaces, and chronic infections [59–61]. Bacterial biofilms are problematic for many food industries including dairy processing, poultry and red meat processing, and brewing among others [62]. This may become a risk of food contamination and transmission of foodborne pathogens [63]. Several other problems are related to biofilm formation in food industry such as slime production, decrease of heat transfer in heat exchanger or condensers, corrosion problems, and hygienic concerns about the sanitation efficacy [64]. Bacterial biofilms are difficult to eliminate from food processing environments and make the design of control strategies a big challenge in the industry [62].

Bacteria are usually regarded as free-living unicellular organisms, but we now know that they predominantly exist as adherent multicellular biofilms in diverse environmental niches [59–61]. The transition from the planktonic state to biofilm growth occurs as a consequence of environmental changes that triggers the disruption of multiple regulatory networks [59–61]. Thus, upon sensing a signal, free planktonic cells will initiate attachment to a surface that will lead to biofilm formation. During this process, dramatic changes occur in gene expression when compared with its planktonic counterparts. Several steps in biofilm formation are common to most bacterial species, but this process can vary from one organism to another. The first stage comprises the attachment to a surface, either biotic or abiotic. Here, flagella play a key role as motility and biofilm development are mutually exclusive events and the transition to sessile occurs in this first stage [65]. Flagella are not only required for propulsion but also have a critical mechanosensory role in surface sensing and thus on the initial stages of surface adhesion that will lead to the formation of biofilm [66]. Lipopolysaccharide (LPS) and O-antigen play a role in *P. mirabilis* surface sensing [67,68].

Five stages in *P. mirabilis* biofilm formation have been described using confocal laser scanning microscopy (CLSM) and different morphologic and topologic descriptors. *P. mirabilis* biofilms on coverslips and Luria-Bertani (LB) broth showed (1) reversible bacterial adhesion to the surface characterized by a slow growth, presence of elongated bacteria, and absence of extracellular matrix (day 1); (2) irreversible bacterial adhesion concomitant to decreasing elongation, and the beginning of extracellular polymer production (days 2–3); (3) accelerated bacterial growth concomitant to continuously decreasing elongation and halting of extracellular polymer production (days 3–4); (4) maturation of biofilm defined by maximum bacterial density and volume, minimum elongation, maximum extracellular material, and highest compaction (day 5); and (5) decreased bacterial density and extracellular material through detachment and dispersion (day 7) [69]. Our group has observed that different *P. mirabilis* mutant strains express flagella in different stages of biofilm formation generally associated with the dispersion step (Figure 24.1).

In other works, the structure of biofilms in artificial urine (AU) and LB broth, a standard laboratory medium, has been studied. Different authors have shown that *P. mirabilis* biofilm formation occurs rapidly in both LB broth and AU. When grown in LB broth, *P. mirabilis* produces large and structurally complex biofilms. In AU, biofilms were less well structurally organized, contained crystalline precipitates that are associated with the urease activity definitive of *P. mirabilis*, and contained significantly higher numbers of the swarmer cell form of *P. mirabilis* [70].

Biofilm development and quorum sensing are closely interconnected processes. Signal molecules (quorum sensing) are important factors in biofilm formation and development [71]. Different studies have demonstrated that quorum-sensing molecules, normally associated with the regulation of virulence factors, could also regulate the development of complex mushroom structures in a *P. aeruginosa* PAO1 biofilm [72]. Disruption of QS systems has been reported to affect the dynamics of biofilm formation in some cases, but the specific mechanisms downstream of the QS regulators are, for the most part, poorly understood [73]. Also, QS antagonists, p-nitrophenyl glycerol and tannic acid, have been shown to inhibit the quorum-sensing system and subsequently inhibit *P. mirabilis* biofilm formation in artificial urine [74].

On the other hand, Stankowska et al. [75] identified the *luxS* gene coding *S*-ribosylhomocysteine lyase responsible for AI-2 synthesis, proving that *P. mirabilis* uses a quorum-sensing communication system.

As mentioned before, *P. mirabilis* can live as single cells (planktonic) or as members of organized microbial communities called biofilms, which are composed of microorganisms and the extracellular matrix-forming polymers they produce [73]. *Proteus* has been shown to produce biofilms in diverse environments, including industrial and clinical settings [76]. Urinary catheters are the medical devices that are most frequently colonized by *Proteus* biofilms. Due to the deposition of crystals within these biofilms, blockage of the urinary catheter can occur. These episodes of bacteriuria can lead to septicemia and shock [77]. In addition, the encrustation of urinary catheters can cause trauma to the bladder mucosa and urethra upon removal of the catheter [77]. Biofilm-based *P. mirabilis* lifestyle provides a protected environment against stresses, such as desiccation, attack by the immune system, protozoa ingestion, and antimicrobials. In the host, biofilm bacteria are protected from the host immune system and treatment by antibiotics [77]. Other works have demonstrated the biofilm ability to induce host hyperinflammation, as shown by elevated levels of proinflammatory cytokines [78] and matrix metalloproteases [79], and excessive numbers of neutrophils [80]. Taking into account that several infections are associated with biofilm formation, we can consider biofilm as another virulence factor.

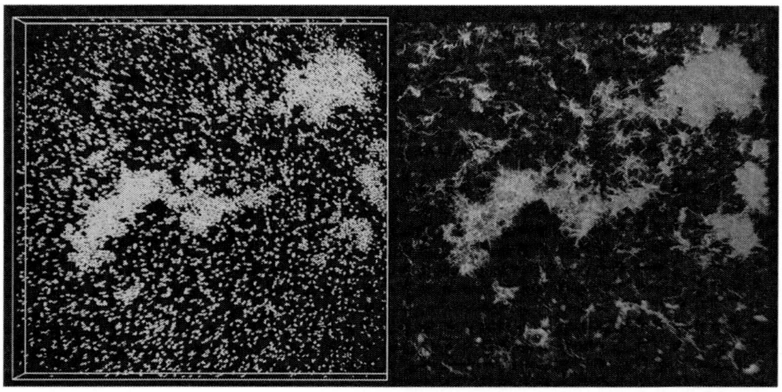

FIGURE 24.1 3D reconstruction of *P. mirabilis* mutant strain that overexpress flagella during biofilm development. Left panel represents 3D model of bacteria, and right panel shows flagella and extracellular matrix. Images were taken in a FV300 Olympus Confocal Microscope.

24.3.4 Techniques for Biofilm Evaluation

A semiquantitative biofilm formation assay was described by O'Toole and Kolter [81]. This assay is based on the ability of cells to adhere to 96-well microtiter dishes made of poly(vinylchloride) (PVC). The assay consists in the inoculation of bacteria in adequate culture medium (100 μL/well) and incubation for 24–48 h at the specific temperature. After that, planktonic bacteria are removed and the bacteria adhered to the well are stained with 1% crystal violet. The plate is subjected to subsequent washes in order to eliminate all the unspecific staining and only the crystal violet associated with bacteria will be solubilized in 95% ethanol, and the absorbance is then determined at 600 nm. The use of polystyrene microtiter plates as a substrate for the formation of biofilms, and subsequent staining with crystal violet, has allowed the isolation of a large number of mutants unable to form biofilms [15,82–85].

Microscopy has evolved as a powerful technique to study microorganisms, and in particular in biofilms, the use of confocal and scanning electron microscopy has become important for understanding such dynamic process.

Confocal laser scanning microscopy (CLSM) has been employed since 1990s and represents one of the most significant advances in optical microscopy ever developed. It is a very useful tool for imaging microbiological samples. It has several advantages that allow many options for sample mounting, high resolution, optical sectioning, analytical precision by virtually eliminating interference from out-of-focus objects, and multichannel imaging [86,87]. Also, the main advantage of CLSM is the reconstruction of 3-dimensional models.

CLSM allows the study of architecture and identification of stages of *P. mirabilis* biofilm formation in a nondestructive manner. Jones et al. [70] compared the structure of *P. mirabilis* biofilms in artificial urine and LB broth. Biofilms in LB formed mushroom structures at 24 h and contained nutrient channels, while AU biofilms were a flat layer without nutrient channels. The authors also observed the presence of swarmer cells protruding out of the biofilm. In another study, the 3D biofilm architecture development in *P. mirabilis* batch culture has been described [69]. Confocal images were used in order to perform 3D reconstruction and calculation of different morphotopological descriptors. Authors stated that *P. mirabilis* biofilm formation followed a five-stage process as mentioned above.

Scanning electron microscopy (SEM) relies on a beam of electrons with high energy generated by a suitable source, typically a tungsten filament. The electrons pass through a system of apertures and electromagnetic lenses to produce a thin beam of electrons. Electrons are emitted from the specimen by the action of the scanning beam and are collected by a detector. This tool was used to study bacterial surface and biofilms and was used by Cox and Hukins [88] to assess the morphology of mineral deposits on encrusted urinary catheters. SEM allows for observation of biofilm formation over catheters of different materials (Figure 24.2). Silicon and latex are common materials for the manufacturing of urinary catheters. SEM also allows the observation of differences among the surface of each material. While silicon is smooth with a repeated pattern, latex seems to be more wrinkled, and this could influence bacterial colonization of the catheter surfaces. However, and taking into account these differences, *P. mirabilis* is able to colonize both surfaces (Figure 24.2). Differences in biofilms generated over both surfaces could be observed, including a brick-shaped biofilm formed over the latex surface.

The difference between SEM and environmental scanning electron microscopy (ESEM) is that ESEM allows the examination of practically any specimen under any gaseous conditions and allows for obtaining images without prior specimen preparation. Holling et al. [55] used this tool to study the surface of catheters colonized by *P. mirabilis* biofilms. This tool permits the imaging of unprocessed, fully hydrated samples, which may provide much insight into the development of *P. mirabilis* biofilms.

24.4 Host–Pathogen Interaction

In vivo approaches are still very useful for the study of bacterial pathogens. Although in the last years efforts have been successful in reducing the use of animals and in refining *in vivo*-based experimental methods, the host environment cannot be easily replaced by *in vitro* techniques.

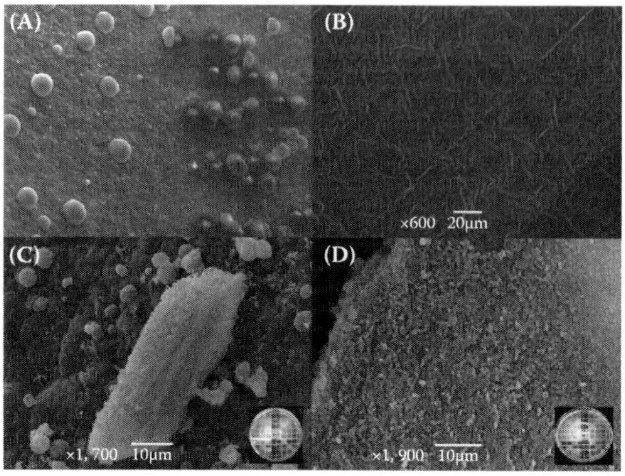

FIGURE 24.2 Scanning electron microscopy of *P. mirabilis* biofilm over catheter material. (Panel A) Latex catheter without bacteria. The round balls were over all the latex surface and correspond to zinc depositions due to manufacturing. (Panel B) Silicon catheter without bacteria. (Panel C and D) *P. mirabilis* growth over latex (C) and silicon (D). In order to allow bacteria to colonize the surface, the catheter bridge technique was employed. Briefly, small sections of the catheter were cut and used as bridges in an agar plate (small pictures in Panel C and D). Images were presented in the 115th General Meeting of the American Society of Microbiology, New Orleans, LA.

Although *P. mirabilis* is a very ubiquitous bacterium, its capacity to induce uropathogenesis is remarkable. Animal models have proven valuable in this regard. A classic ascending UTI described initially by Hagberg et al. [89] for the study of *E. coli* is still being used by several researchers. In this case, bacteria are slowly introduced into the mouse bladder through a transurethral soft catheter so that different aspects of urinary infections can be studied.

This model has been traditionally used to assess the ability of different strains to colonize bladders and kidneys after ascending infection through the urinary tract, typically measured by bacterial counting [21]. Infection of different groups of mice with the different strains or even coinfection of the same animals with the wild type and isogenic mutants is frequently performed to assess bacterial colonization competence.

Different studies have been developed to assess the potential of different *P. mirabilis* UTI treatment- or prevention-based strategies, including vaccination, using this model. A wide range of vaccination strategies to prevent UTI, based on different antigens or immunization routes, have been proposed using this model [90,91].

This model has been recently refined for modern approaches like STM [52], morphological analysis of kidneys colonization by laser confocal microscopy [92], or transcriptome analysis to compare *in vitro* and *in vivo P. mirabilis* gene expression [93].

Also, different modifications have been performed to increase this animal model input. For example, an infection model of permanently catheterized mice has been recently developed, which was initially proposed for the study of *Enterococcus faecalis* biofilms associated with urinary catheters [94]. This new approach will help to elucidate the molecular basis and prevention alternatives of *in vivo* biofilm formation.

The adhesion of pathogenic bacteria to host cells is crucial for the establishment of the infection. Subsequent events include tissue colonization and, in some cases, cellular invasion followed by intracellular multiplication and/or persistence [95]. Colonization of the epithelium is the first step in the pathogenesis of *P. mirabilis* and has been demonstrated by *in vitro* and *in vivo* approaches [3]. However, the mechanisms by which *P. mirabilis* adhere to epithelial cells are not fully elucidated [3,96].

Pioneer work by Chippendale et al. showed that *P. mirabilis* could survive in the epithelial cells, but no significant replication was noticed [97]. Oelschlaeger and Tall found that *P. mirabilis* strains isolated from different sources had varied internalization efficiencies and that microfilaments were important for the process [98]. Bacteria were detected in membrane-bound vacuoles after 3 h of infection using human renal proximal tubular epithelial cells [97].

Torzewska et al. demonstrated that not only could *P. mirabilis* produce urease with the concomitant formation of calcium crystals and magnesium ammonium phosphate precipitates but it also could cause crystallization of urine components inside the cells [99]. The production of crystals inside the cells could induce cell damage or death, but this was not evaluated by the authors. However, in another work, it was demonstrated for the first time that both MR/P fimbriae and flagella mediate genotoxic and cytotoxic effects on eukaryotic cells (Figure 24.3) [16]. It has also been reported that flagella have a role in *P. mirabilis* adhesion [16] (Figure 24.3), as reported in other pathogens [100].

Considering all the information available on eukaryotic cell–*Proteus* interaction, we can conclude that *P. mirabilis* produces several virulence factors that induce damage in cell. The lack of knowledge on the intracellular life of *P. mirabilis* could be explained in part because of the wide range of virulence factors that the bacteria produce. This should be taken into account when cell culture experiments are going to be set up. These factors include toxins such as Pta and HpmA (bacterial toxin hemolysin) that cause cell damage, the expression of MR/P fimbriae that mediate adhesion and cytotoxic/genotoxic damage, and flagella that mediate adhesion and even cell death, among others.

One of the main techniques that allows for visualization of host–pathogen interaction is immunofluorescence. It involves the staining of different cellular structures with antibodies coupled to fluorescent molecules followed by observation with a fluorescent or confocal microscope. The development of this type of technique and the use of time-lapse imaging will contribute to the understanding of infection and to the increase of knowledge of the mechanisms involved in cell invasion.

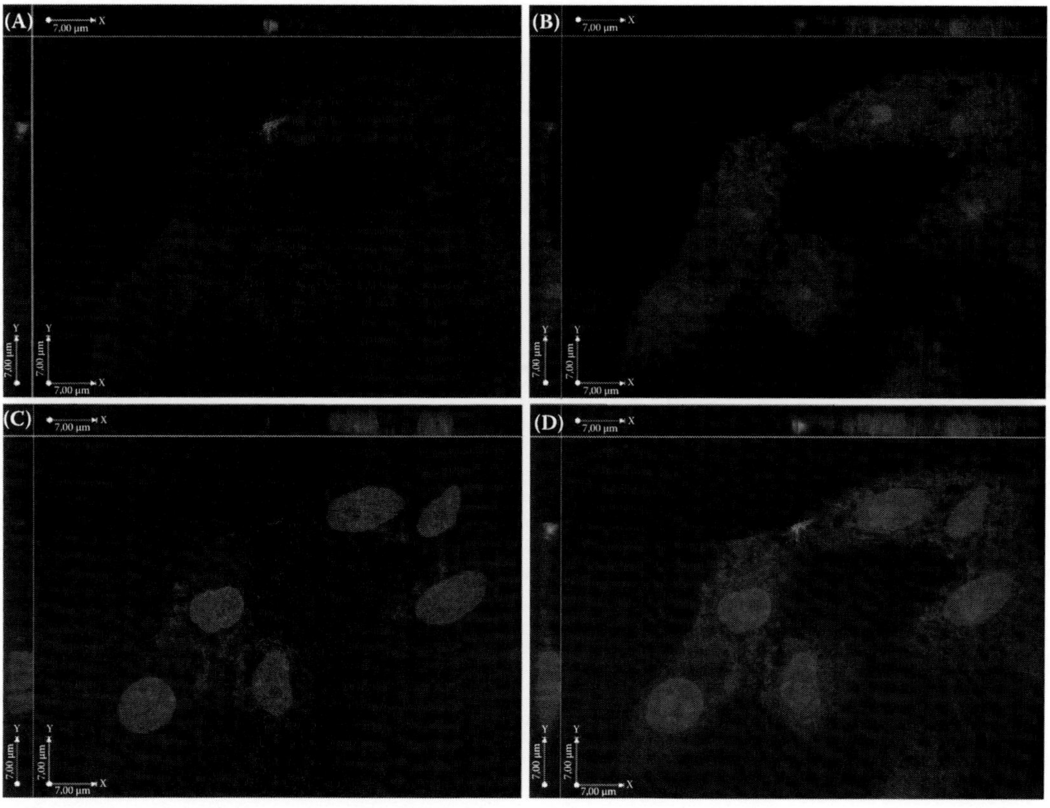

FIGURE 24.3 Confocal microscopy image of *P. mirabilis* wild-type adhesion to Vero cell line (kidney); bacteria is stained with a polyclonal anti-proteus antibody (A), actin is stained with phalloidin-rhodamine (B), and DNA with DAPI (C). Panel D shows a superimposition of the three images (A–C).

24.5 Immunity and Vaccines

Several efforts have been made to generate vaccines against *P. mirabilis* infections. UTIs represent an important problem in health and result in high economical losses; so the development of vaccines is not worth without mentioning that the quality of life of afflicted individuals is also affected. Only in USA, UTIs cause an estimated 11 million doctor visits with an annual cost of 6 billion dollars [11].

UTIs are commonly treated with antibiotics that are effective in noncomplicated infections. The most recommended antibiotics for UTI are trimethoprim–sulfamethoxazole, ciprofloxacin, and ampicillin [101]. In the case of resistant strains, recurrent UTIs, catheterized patients, or those with abnormalities in the urinary tract or with stones, the success of the antibiotic therapy is less effective [102]. The main problem of the antibiotic-based therapy is the fast increase in the resistance to antimicrobial molecules.

This worldwide situation and the magnitude of UTIs clearly justify the efforts to develop effective vaccines.

Early studies showed that initial infection with *P. mirabilis* does not prevent reinfection. Moreover, recovery from a UTI does not generate a protective immune response, and it was also demonstrated that the same strain can be isolated from successive infection episodes [103]. Unlike other mucosal infections such as cholera, where a single exposure to the etiological agent results in protective immunity for life, the resolution of UTIs did not confer protection in case of another episode. In particular, *P. mirabilis* has a high number of conserved surface antigens that do not differ significantly between strains of different origin, which could contribute to the successful development of such strategies [104,105]. There are some vaccines for the prevention of UTIs available in Europe, and these are recommended in patients with uncomplicated recurrent UTI. They are formulated from complex protein extracts from different bacterial strains, producing significant side effects, and are not able to produce strong mucosal responses against antigens related to virulence [106]. Moreover, so far, there are no licenses for UTI vaccines available in the USA. Extensive clinical studies are needed to determine the effectiveness of these vaccines. Various strategies have been developed in recent years to prevent UTIs caused by *P. mirabilis*, directed mainly against bacterial defined subunits. In early studies developed by Pazin and Braude, purified flagella were used to immunize rats by injection. These authors demonstrated that the serum against flagella immobilized the bacteria and prevent the spread of the disease to the other kidney [107]. Moayeri and coworkers showed that immunization of BALB/c mice with outer membrane protein preparations protected the animals against an intravesical homologous challenge [108]. Legnani-Fajardo et al. showed that parenteral immunization with fimbrial preparations protected mice after a hematogenous challenge, both homologous and heterologous [109] strains. In more recent studies carried out in our department, Pellegrino et al. vaccinated mice with purified recombinant antigens of *P. mirabilis* fimbriae (MrpA, PmfA, and UcaA) and observed that MrpA-protected mice when immunized subcutaneously faced hematogenous and ascending challenges [110]. Furthermore, intranasal immunization with the adhesin MrpH or the N-terminal receptor binding domain protects mice against experimental UTI [111]. These authors evaluated four different immunization routes and concluded that the intranasal route produces a response to broader antibodies and specific antibody titers in serum, urine, vaginal douching, and in bile [111,112]. A more recent work showed that a MrpH-FimH recombinant protein induces a significant increase in IgG and IgA in serum, nasal wash, vaginal wash, and urine samples. However, vaccinated mice showed a reduction in bladder and kidney bacterial load, but the differences were not statistically significant [91].

Mucosal vaccines should induce the production of systemic and mucosal antibodies and cellular immune responses. They have the advantage of being safe and inexpensive, and have fewer side effects than systemic vaccines [113]. In our department, we evaluated two mucosal immunization routes (nasal and vesical), the induction of specific antibodies MrpA, UcaA, and PmfA in serum and urine, and protection against an upward experimental UTI in mice. From these results, it was determined that the nasally immunized animals developed a strong antibody response in serum IgG and IgA when immunized with MrpA. Also, a significant decrease in the number of bacteria recovered in kidneys and bladders after experimental infection was reported compared with nonvaccinated animals [114]. However, a correlation between the induction of specific antibodies and reduced bacterial colonization could not be observed. On the other hand, the immune response that develops from entering uropathogens in the urinary tract is particularly complex. The mechanisms of innate immunity would have an important role in the UTI. The role of different mechanisms in the adaptive immune response in resolving urinary infection is currently a matter of debate [11]. During the entrance of uropathogens, an early response in the host that includes recruitment and activation of effector cells at the site of infection results, leading to an inflammatory response whose magnitude and location explains many of the clinical manifestations of UTIs [115]. This response depends on cell receptors present on the uroepithelium that recognize pathogen-associated molecular patterns (PAMPs), which are highly conserved in different organisms, [116] and these can be recognized by the innate immune system that then limits the action of microbial evasion mechanisms and allows for detection of infectious agents from a limited number of cellular receptors. It has been determined that the detection of PAMPs by toll-like receptors (TLRs) is crucial for activation of the innate immune response and control of the adaptive immune response that occurs later [117]. Several studies have shown that bladder epithelial cells, like macrophages, express TLR4, which mediates a rapid cytokine response against low concentrations of LPS [116,118]. In macrophages, the recognition of LPS requires a CD14 coreceptor [119], which is not present in epithelial cells. It has also been demonstrated that TLR4 recognizes fimbriae regardless of the presence of the CD14 [120] coreceptor. The control of UTIs in the bladder depends on TLR4 expression on urothelial and stromal cells and is needed to drive inflammatory responses [121]. Studies of TLR4 expression in humans with UTIs demonstrated that lower TLR4 expression attenuates host responses and promotes asymptomatic carriage of UPEC [122]. TLR5 is essential to mediate flagellin-induced inflammatory responses; it plays a role in control of infection in both the bladder and kidneys [123]. In humans, a polymorphism in *TLR5* gene abrogates the flagellin signaling cascade and has been associated with an increased susceptibility to recurrent UTIs [124].

The last TLR identified and characterized is the TLR11, which is specifically expressed in murine kidney cells [125]. The PAMP that the TLR recognizes was not identified but was observed in an *in vitro* assay as a motif present in *E. coli* uropathogenic strains [126]. Furthermore, in TLR11 knockout mice, a massive kidney infection has been observed by challenging the animals with a uropathogenic strain of *E. coli*. These authors suggest that TLR11 has a role in the innate immune response specifically against uropathogenic bacteria [125]. Human *TLR11* gene is present but not expressed due to the existence of a stop codon in the reading frame. This finding opens new directions such as the absence of TLR being one reason for the particular susceptibility of humans to the ITU.

24.6 Conclusion

P. mirabilis has become an interesting model microorganism since it expresses an impressive variety of virulence factors. Its relation with infection, particularly with UTIs, has increased the necessity for vaccines and new therapeutic agents. Antimicrobial resistance and biofilm development over different surfaces including food make *P. mirabilis* an excellent model for the study of bacterial behavior in different conditions. The study of the basic mechanisms used by this bacterium for survival and multiplication in different environments could contribute to the prevention of infection and even food contamination.

REFERENCES

1. Hoeniger, J., Influence of *p*H on *Proteus* flagella, *J Bacteriol.*, 90, 275–277, 1965.
2. Penner J., The genera *Proteus*, *Providencia*, and *Morganella*, In I. Balows, G. Truper, W. Harder, and K. Schleifer, eds., *The Prokaryotes*, vol. III, 2849–2853, Springer-Verlag KG, Berlin, 1992.
3. Rózalski, A., Sidorczyk, Z., and Kotełko, K., Potential virulence factors of *Proteus* bacilli, *Microbiol Mol Biol Rev.*, 61, 65–89, 1997.
4. Ghosh, M., Wahi, S., Kumar, M., and Ganguli, A., Prevalence of enterotoxigenic *Staphylococcus aureus* and *Shigella* spp. in some raw street vended Indian foods, *Int J Environ Health Res.*, 17, 151–156, 2007.
5. Centinkaya, F., Cibik, G., Soyuteniz, E., Ozkin, C., Kayali, R., and Levent, B., *Shigella* and *Salmonella* contamination in various foodstuffs in Turkey, *J Food Control*, 19, 1059–1063, 2008.

6. Braide, W., Oranusi, S., Udegbunam, L., Oguoma, O., Akobondu, C., and Nwaoguikpe, R., Microbiological quality of an edible caterpillar of an emperor moth, *Bunaea alcinoe*, *J Food Prot.*, 3, 176–180, 2011.
7. Nyenje, M., Odjadjare, C., Tanih, N., Green, E., and Ndip, R., Foodborne pathogens recovered from ready-to-eat foods from roadside cafeterias and retail outlets in Alice, Eastern Cape Province, South Africa: public health implications, *Int J Environ Res Public Health*, 9, 2608–2619, 2012.
8. Wong, M., Wan, H., and Chen, S., Characterization of multidrug-resistant *Proteus mirabilis* isolated from chicken carcasses, *Foodborne Pathog Dis.*, 10, 177–181, 2013.
9. Seiffert, S., Tinguely, R., Lupo, A., Neuwirth, C., Perreten, V., and Endimiani, A., High prevalence of extended-spectrum-cephalosporin-resistant Enterobacteriaceae in poultry meat in Switzerland: emergence of CMY-2- and VEB-6-possessing *Proteus mirabilis*, *Antimicrob Agents Chemother.*, 57, 6406–6408, 2013.
10. Casella, T., et al., Detection of $bla_{CTX\text{-}M}$-type genes in complex class 1 integrons carried by Enterobacteriaceae isolated from retail chicken meat in Brazil, *Int J Food Microbiol.*, 197, 88–91, 2015.
11. Nielubowicz, G. R., and Mobley, H. L., Host–pathogen interactions in urinary tract infection, *Nat Rev Urol.*, 7, 430–441, 2010.
12. Pearson, M. M., et al., Complete genome sequence of uropathogenic *Proteus mirabilis*, a master of both adherence and motility, *J Bacteriol.*, 190, 4027–4037, 2008.
13. Zunino, P., Geymonat, L., Allen, A. G., Preston, A., Sosa, V., and Maskell, D. J., New aspects of the role of MR/P fimbriae in *Proteus mirabilis* urinary tract infection, *FEMS Immunol Med Microbiol.*, 31, 113–120, 2001.
14. Rocha, S. P., et al., Aggregative adherence of uropathogenic *Proteus mirabilis* to cultured epithelial cells, *FEMS Immunol Med Microbiol.*, 51, 319–326, 2007.
15. Jansen, A. M., Lockatell, V., Johnson, D. E., and Mobley, H. L., Mannose-resistant *Proteus*-like fimbriae are produced by most *Proteus mirabilis* strains infecting the urinary tract, dictate the in vivo localization of bacteria, and contribute to biofilm formation, *Infect Immun.*, 72, 7294–7305, 2004.
16. Scavone, P., Villar, S., Umpiérrez, A., and Zunino, P., Role of *Proteus mirabilis* MR/P fimbriae and flagella in adhesion, cytotoxicity and genotoxicity induction in T24 and Vero cells, *Pathog Dis.*, 73, 2015.
17. Li, X., Rasko, D. A., Lockatell, C. V., Johnson, D. E., and Mobley, H. L., Repression of bacterial motility by a novel fimbrial gene product, *EMBO J.*, 20, 4854–4862, 2011.
18. Pellegrino, R., Scavone, P., Umpiérrez, A., Maskell, D. J., and Zunino, P., *Proteus mirabilis* uroepithelial cell adhesin (UCA) fimbria plays a role in the colonization of the urinary tract, *Pathog Dis.*, 67, 104–107, 2013.
19. Zunino, P., Sosa, V., Allen, A. G., Preston, A., Schlapp, G., and Maskell, D. J., *Proteus mirabilis* fimbriae (PMF) are important for both bladder and kidney colonization in mice, *Microbiology*, 149, 3231–3237, 2003.
20. Massad, G., and Mobley, H. L., Genetic organization and complete sequence of the *Proteus mirabilis pmf* fimbrial operon, *Gene*, 150, 101–104, 1994.
21. Zunino, P., Geymonat, L., Allen, A. G., Legnani-Fajardo, C., and Maskell, D. J., Virulence of a *Proteus mirabilis* ATF isogenic mutant is not impaired in a mouse model of ascending urinary tract infection, *FEMS Immunol Med Microbiol.*, 29, 137–143, 2000.
22. Bode, N. J., Debnath, I., Kuan, L., Schulfer, A., Ty, M., and Pearson, M. M., Transcriptional analysis of the MrpJ network: modulation of diverse virulence-associated genes and direct regulation of *mrp* fimbrial and *flhDC* flagellar operons in *Proteus mirabilis*, *Infect Immun.*, 83, 2542–2556, 2015.
23. Armbruster, C. E., and Mobley, H. L. T., Merging mythology and morphology: the multifaceted lifestyle of *Proteus mirabilis*, *Nat Rev Microbiol.*, 10, 743–754, 2012.
24. Hartstein, A. I., Garber, S. B., Ward, T. T., Jones, S. R., and Morthland, V. H., Nosocomial urinary tract infection: a prospective evaluation of 108 catheterized patients, *Infect Control*, 2, 380–386, 1981.
25. Wooldridge, K. G., and Williams, P. H., Iron uptake mechanisms of pathogenic bacteria, *FEMS Microbiol Rev.*, 12, 325–348, 1993.
26. Piccini, C. D., Barbe, F. M., and Legnani-Fajardo, C. L., Identification of iron-regulated outer membrane proteins in uropathogenic *Proteus mirabilis* and its relationship with heme uptake, *FEMS Microbiol Lett.*, 166, 243–248, 1998.

27. Lima, A., Zunino, P., D'Alessandro, B., and Piccini, C., An iron-regulated outer-membrane protein of *Proteus mirabilis* is a haem receptor that plays an important role in urinary tract infection and in in vivo growth, *J Med Microbiol.*, 56, 1600–1607, 2007.
28. Pearson, M. M., and Mobley, H. L. T., Repression of motility during fimbrial expression: identification of fourteen *mrpJ* gene paralogs in *Proteus mirabilis*, *Mol Microbiol.*, 69, 548–558, 2008.
29. Gaisser, S., and Hughes, C., A locus coding for putative non-ribosomal peptide/polyketide synthase functions is mutated in a swarming-defective *Proteus mirabilis* strain, *Mol Gen Genet.*, 253, 415–427, 1997.
30. Flannery, E. L., Mody, L., and Mobley, H. L. T., Identification of a modular pathogenicity island that is widespread among urease-producing uropathogens and shares features with a diverse group of mobile elements, *Infect Immun.*, 77, 4887–4894, 2009.
31. Himpsl, S. D., Pearson, M. M., Arewång, C. J., Nusca, T. D., Sherman, D. H., and Mobley, H. L. T., Proteobactin and a yersiniabactin-related siderophore mediate iron acquisition in *Proteus mirabilis*, *Mol Microbiol.*, 78, 138–157, 2010.
32. Boyer, M., and Wisniewski-Dyé, F., Cell–cell signalling in bacteria: not simply a matter of quorum, *FEMS Microbiol Ecol.*, 70, 1–19, 2009.
33. Belas, R., Schneider, R., and Melch, M., Characterization of *Proteus mirabilis* precocious swarming mutants: identification of *rsbA*, encoding a regulator of swarming behavior, *J Bacteriol.*, 180, 6126–6139, 1998.
34. Bainton, N. J., et al., A general role for the *lux* autoinducer in bacterial cell signalling: control of antibiotic biosynthesis in *Erwinia*, *Gene*, 116, 87–91, 1992.
35. Stankowska, D., Kwinkowski, M., and Kaca, W., Quantification of *Proteus mirabilis* virulence factors and modulation by acylated homoserine lactones, *J Microbiol Immunol Infect.*, 41, 243–253, 2008.
36. Schneider, R., Lockatell, C. V., Johnson, D., and Belas, R., Detection and mutation of a *luxS*-encoded autoinducer in *Proteus mirabilis*, *Microbiology*, 148, 773–782, 2002.
37. Czerwonka, G., Arabski, M., Wąsik, S., Jabłońska-Wawrzycka, A., Rogala, P., and Kaca, W., Morphological changes in *Proteus mirabilis* O18 biofilm under the influence of a urease inhibitor and a homoserine lactone derivative, *Arch Microbiol.*, 196, 169–177, 2014.
38. Griffith, D. P., Musher, D. Á., and Itin, C., Urease: the primary cause of infection-induced urinary stones, *Invest Urol.*, 13, 346–350, 1976.
39. Jones, B. D., and Mobley, H. L., *Proteus mirabilis* urease: nucleotide sequence determination and comparison with jack bean urease, *J Bacteriol.*, 171, 6414–6422, 1989.
40. Flores-Mireles, A. L., Walker, J. N., Caparon, M., and Hultgren, S. J., Urinary tract infections: epidemiology, mechanisms of infection and treatment options, *Nat Rev Microbiol.*, 13, 269–284, 2015.
41. Coker, C., Poore, C. A., Li, X., and Mobley, H. L., Pathogenesis of *Proteus mirabilis* urinary tract infection, *Microbes Infect.*, 2, 1497–1505, 2000.
42. Jacobsen, S. M., Stickler, D. J., Mobley, H. L. T., and Shirtliff, M. E., Complicated catheter-associated urinary tract infections due to *Escherichia coli* and *Proteus mirabilis*, *Clin Microbiol Rev.*, 21, 26–59, 2008.
43. Alamuri, P., and Mobley, H. L., A novel autotransporter of uropathogenic *Proteus mirabilis* is both a cytotoxin and an agglutinin, *Mol Microbiol.*, 68, 997–1017, 2008.
44. Walker, K. E., Moghaddame-Jafari, S., Lockatell, C., Johnson, D., and Belas, R., ZapA, the IgA degrading metalloprotease of *Proteus mirabilis*, is a virulence factor expressed specifically in swarmer cells, *Mol Microbiol.*, 32, 825–836, 1999.
45. Belas, R., Manos, J., and Suvanasuthi, R., *Proteus mirabilis* ZapA metalloprotease degrades a broad spectrum of substrates, including antimicrobial peptides, *Infect Immun.*, 72, 5159–5167, 2004.
46. Legnani-Fajardo, C., Zunino, P., Piccini, C., Allen, A., and Maskell, D., Defined mutants of *Proteus mirabilis* lacking flagella cause ascending urinary tract infection in mice. *Microb Pathog.*, 21, 395–405, 1996.
47. Zhao, H., Li, X., Johnson, D. E., and Mobley, H. L., Identification of protease and rpoN-associated genes of uropathogenic *Proteus mirabilis* by negative selection in a mouse model of ascending urinary tract infection, *Microbiology*, 145, 185–195, 1999.
48. McCoy, A. J., Liu, H., Falla, T. J., and Gunn, J. S., Identification of *Proteus mirabilis* mutants with increased sensitivity to antimicrobial peptides, *Antimicrob Agents Chemother.*, 45, 2030–2037, 2001.

49. Jones, B. V., Young, R., Mahenthiralingam, E., and Stickler, D. J., Ultrastructure of *Proteus mirabilis* swarmer cell rafts and role of swarming in catheter-associated urinary tract infection, *Infect Immun.*, 72, 3941–3950, 2004.
50. Burall, L. S., et al., *Proteus mirabilis* genes that contribute to pathogenesis of urinary tract infection: identification of 25 signature-tagged mutants attenuated at least 100-fold, *Infect Immun.*, 72, 2922–2938, 2004.
51. Zunino, P., Sosa, V., Schlapp, G., Allen, A. G., Preston, A., and Maskell, D. J., Mannose-resistant *Proteus* like and *P. mirabilis* fimbriae have specific and additive roles in *P. mirabilis* urinary tract infections, *FEMS Immunol Med Microbiol.*, 51, 125–133, 2007.
52. Himpsl, S. D., Lockatell, C. V., Hebel, J. R., Johnson, D. E., and Mobley, H. L., Identification of virulence determinants in uropathogenic *Proteus mirabilis* using signature-tagged mutagenesis, *J Med Microbiol.*, 57, 1068–1078, 2008.
53. Guo, Q., et al., A mutational analysis and molecular dynamics simulation of quinolone resistance proteins QnrA1 and QnrC from *Proteus mirabilis*, *BMC Struct Biol.*, 10, 33, 2010.
54. Lee, Y.-Y., Patellis, J., and Belas, R., Activity of *Proteus mirabilis* FliL is viscosity dependent and requires extragenic DNA, *J Bacteriol.*, 195, 823–832, 2013.
55. Holling, N., et al., Elucidating the genetic basis of crystalline biofilm formation in *Proteus mirabilis*, *Infect Immun.*, 82, 1616–1626, 2014.
56. Hensel, M., Shea, J. E., Gleeson, C., Jones, M. D., Dalton, E., and Holden, D. W., Simultaneous identification of bacterial virulence genes by negative selection, *Science*, 269, 400–403, 1995.
57. Mazurkiewicz, P., Tang, C. M., Boone, C., and Holden, D. W., Signature-tagged mutagenesis: barcoding mutants for genome-wide screens, *Nat Rev Genet.*, 7, 929–939, 2006.
58. Winzeler, E. A., et al., Functional characterization of the *S. cerevisiae* genome by gene deletion and parallel analysis, *Science*, 285, 901–906, 1999.
59. Hall-Stoodley, L., Costerton, J. W., and Stoodley, P., Bacterial biofilms: from the natural environment to infectious diseases, *Nat Rev Microbiol.*, 2, 95–108, 2004.
60. Costerton, J. W., Stewart, P. S., and Greenberg, E. P., Bacterial biofilms: a common cause of persistent infections, *Science*, 284, 1318–1322, 1999.
61. O'Toole, G., Kaplan, H. B., and Kolter, R., Biofilm formation as microbial development, *Annu Rev Microbiol.*, 54, 49–79, 2000.
62. Simões, L. C., Simões, M., and Vieira, M. J., Adhesion and biofilm formation on polystyrene by drinking water-isolated bacteria, *Antonie Van Leeuwenhoek*, 98, 317–329, 2010.
63. Lindsay, D., and Von Holy, A., Bacterial biofilms within the clinical setting: what healthcare professionals should know, *J Hosp Infect.*, 64, 313–325, 2006.
64. Kiskó, G., and Szabó-Szabó, O., Biofilm removal of *Pseudomonas* strains using hot water sanitation, *Acta Univ Sapientiae Alimentaria*, 4, 69–79, 2011.
65. Belas, R., Biofilms, flagella, and mechanosensing of surfaces by bacteria, *Trends Microbiol.*, 22, 517–527, 2014.
66. Belas, R., When the swimming gets tough, the tough form a biofilm, *Mol Microbiol.*, 90, 1–5, 2013.
67. Belas, R., Goldman, M., and Ashliman, K., Genetic analysis of *Proteus mirabilis* mutants defective in swarmer cell elongation, *J Bacteriol.*, 177, 823–828, 1995.
68. Morgenstein, R. M., Clemmer, K. M., and Rather, P. N., Loss of the WaaL O-antigen ligase prevents surface activation of the flagellar gene cascade in *Proteus mirabilis*, *J Bacteriol.*, 192, 3213–3221, 2010.
69. Schlapp, G., Scavone, P., Zunino, P., and Härtel, S., Development of 3D architecture of uropathogenic *Proteus mirabilis* batch culture biofilms—a quantitative confocal microscopy approach, *J Microbiol Methods*, 87, 234–240, 2011.
70. Jones, S. M., Yerly, J., Hu, Y., Ceri, H., and Martinuzzi, R., Structure of *Proteus mirabilis* biofilms grown in artificial urine and standard laboratory media, *FEMS Microbiol Lett.*, 268, 16–21, 2007.
71. Viana E. S, Campos M. E, Ponce A. R, Mantovani H. C, and Vanetti M. C., Biofilm formation and acyl homoserine lactone production in *Hafnia alvei* isolated from raw milk, *Biol Res.*, 42, 427–436, 2009.
72. Davies, D. G., Parsek, M. R., Pearson, J. P., Iglewski, B. H., Costerton, J. W., and Greenberg, E. P., The involvement of cell-to-cell signals in the development of a bacterial biofilm, *Science*, 280, 295–298, 1998.
73. Monds, R. D., and O'Toole, G. A., The developmental model of microbial biofilms: ten years of a paradigm up for review, *Trends Microbiol.*, 17, 73–87, 2009.

74. Jones, S. M., Dang, T. T., and Martinuzzi, R., Use of quorum sensing antagonists to deter the formation of crystalline *Proteus mirabilis* biofilms, *Int J Antimicrob Agents*, 34, 360–364, 2009.
75. Stankowska, D., Czerwonka, G., Rozalska, S., Grosicka, M., Dziadek, J., and Kaca, W., Influence of quorum sensing signal molecules on biofilm formation in *Proteus mirabilis* O18, *Folia Microbiol.*, 57, 53–60, 2012.
76. Costerton, J. W., et al., Bacterial biofilms in nature and disease, *Annu Rev Microbiol.*, 41, 435–464, 1987.
77. Stickler, D. J., and Zimakoff, J., Complications of urinary tract infections associated with devices used for long-term bladder management, *J Hosp Infect.*, 28, 177–194, 1994.
78. Trengove, N. J., Langton, S. R., and Stacey, M. C., Biochemical analysis of wound fluid from nonhealing and healing chronic leg ulcers, *Wound Repair Regen.*, 4, 234–239, 1996.
79. Trengove, N. J., et al., Analysis of the acute and chronic wound environments: the role of proteases and their inhibitors, *Wound Repair Regen.*, 7, 442–452, 1999.
80. Diegelmann, R. F., Excessive neutrophils characterize chronic pressure ulcers, *Wound Repair Regen.*, 11, 490–495, 2003.
81. O'Toole, G. A., and Kolter, R., Initiation of biofilm formation in *Pseudomonas fluorescens* WCS365 proceeds via multiple, convergent signalling pathways: a genetic analysis, *Mol Microbiol.*, 28, 449–461, 1998.
82. Kanamaru, S., et al., Increased biofilm formation in *Escherichia coli* isolated from acute prostatitis, *Int J Antimicrob Agents*, 28, 21–25, 2006.
83. Klausen, M., et al., Biofilm formation by *Pseudomonas aeruginosa* wild type, flagella and type IV pili mutants, *Mol Microbiol.*, 48, 1511–1524, 2003.
84. Pratt, L. A., and Kolter, R., Genetic analyses of bacterial biofilm formation. *Curr Opin Microbiol.*, 2, 598–603, 1999.
85. Watnick, P. I., and Kolter, R., Steps in the development of a *Vibrio cholerae* El Tor biofilm, *Mol Microbiol.*, 34, 586–595, 1999.
86. Neu, T. R., and Lawrence, J. R., Investigation of microbial biofilm structure by laser scanning microscopy, In K. Muffler, and R. Ulber, eds., *Productive Biofilms*, 1–51, Springer International Publishing, Berlin, Heidelberg, 2014.
87. Ustione, A., and Piston, D. W., A simple introduction to multiphoton microscopy, *J Microsc.*, 243, 221–226, 2011.
88. Cox, A. J., and Hukins, D. W., Morphology of mineral deposits on encrusted urinary catheters investigated by scanning electron microscopy, *J Urol.*, 142, 1347–1350, 1989.
89. Hagberg, L., Engberg, I., Freter, R., Lam, J., Olling, S., and Eden, C. S., Ascending, unobstructed urinary tract infection in mice caused by pyelonephritogenic *Escherichia coli* of human origin, *Infect Immun.*, 40, 273–283, 1983.
90. Scavone, P., Umpiérrez, A., Maskell, D. J., and Zunino, P., Nasal immunization with attenuated *Salmonella* Typhimurium expressing an MrpA–TetC fusion protein significantly reduces *Proteus mirabilis* colonization in the mouse urinary tract, *J Med Microbiol.*, 60, 899–904, 2011.
91. Habibi, M., Karam, M. R. A., Shokrgozar, M. A., Oloomi, M., Jafari, A., and Bouzari, S., Intranasal immunization with fusion protein MrpH·FimH and MPL adjuvant confers protection against urinary tract infections caused by uropathogenic *Escherichia coli* and *Proteus mirabilis*, *Mol Immunol.*, 64, 285–294, 2015.
92. Jansen, A. M., Lockatell, C. V., Johnson, D. E., and Mobley, H. L., Visualization of *Proteus mirabilis* morphotypes in the urinary tract: the elongated swarmer cell is rarely observed in ascending urinary tract infection, *Infect Immun.*, 71, 3607–3613, 2003.
93. Pearson, M. M., Yep, A., Smith, S. N., and Mobley, H. L., Transcriptome of *Proteus mirabilis* in the murine urinary tract: virulence and nitrogen assimilation gene expression, *Infect Immun.*, 79, 2619–2631, 2011.
94. Guiton, P. S., Hung, C. S., Hancock, L. E., Caparon, M. G., and Hultgren, S. J., Enterococcal biofilm formation and virulence in an optimized murine model of foreign body-associated urinary tract infections, *Infect Immun.*, 78, 4166–4175, 2010.
95. Alamuri, P., Löwer, M., Hiss, J. A., Himpsl, S. D., Schneider, G., and Mobley, H. L. T., Adhesion, invasion, and agglutination mediated by two trimeric autotransporters in the human uropathogen *Proteus mirabilis*, *Infect Immun.*, 78, 4882–4894, 2010.

96. Bahrani, F. K., and Mobley, H. L., *Proteus mirabilis* MR/P fimbrial operon: genetic organization, nucleotide sequence, and conditions for expression, *J Bacteriol.*, 176, 3412–3419, 1994.
97. Chippendale, G. R., Warren, J. W., Trifillis, A. L., and Mobley, H. L., Internalization of *Proteus mirabilis* by human renal epithelial cells, *Infect Immun.*, 62, 3115–3121, 1994.
98. Oelschlaeger, T. A., and Tall, B. D., Uptake pathways of clinical isolates of *Proteus mirabilis* into human epithelial cell lines, *Microb Pathog.*, 21, 1–16, 1996.
99. Torzewska, A., Budzyńska, A., Białczak-Kokot, M., and Różalski, A., In vitro studies of epithelium-associated crystallization caused by uropathogens during urinary calculi development, *Microb Pathog.*, 71, 25–31, 2014.
100. Duan Q, Zhou M, and Zhu L., Flagella and bacterial pathogenicity, *J Basic Microbiol.*, 52, 1–8, 2012.
101. Foxman, B., The epidemiology of urinary tract infection, *Nat Rev Urol.*, 7, 653–660, 2010.
102. Johnson, J. R., and Brown, J. J., Defining inoculation conditions for the mouse model of ascending urinary tract infection that avoid immediate vesicoureteral reflux yet produce renal and bladder infection, *J Infect Dis.*, 173, 746–749, 1996.
103. Russo, T. A., Stapleton, A., Wenderoth, S., Hooton, T. M., and Stamm, W. E., Chromosomal restriction fragment length polymorphism analysis of *Escherichia coli* strains causing recurrent urinary tract infections in young women, *J Infect Dis.*, 172, 440–445, 1995.
104. Mobley, H. L., and Chippendale, G. R., Hemagglutinin, urease, and hemolysin production by *Proteus mirabilis* from clinical sources, *J Infect Dis.*, 161, 525–530, 1990.
105. Mobley, H. L., Vaccines against *Escherichia coli* and *Proteus* urinary infections, In M. M. Levine, G. C. Woodrow, J. B. Kaper, and G. S. Cobon, eds., *New Generation Vaccines*, 2nd edition, 931–944, Marcel Dekker, Inc., New York, 1990.
106. O'Hanley, P., Prospects for urinary tract infection vaccines. In H. L. Mobley, and J. W. Warren, eds., *Urinary Tract Infections: Molecular Pathogenesis and Clinical Management*, 405–425, ASM Press, Washington, DC, 1996.
107. Pazin, G. J., and Braude, A. I., Immobilizing antibodies in pyelonephritis, *J Immunol.*, 102, 1454–1465, 1969.
108. Moayeri, N., Collins, C. M., and O'Hanley, P., Efficacy of a *Proteus mirabilis* outer membrane protein vaccine in preventing experimental *Proteus* pyelonephritis in a BALB/c mouse model, *Infect Immun.*, 59, 3778–3786, 1991.
109. Legnani-Fajardo, C., Zunino, P., Algorta, G., and Laborde, H. F., Antigenic and immunogenic activity of flagella and fimbriae preparations from uropathogenic *Proteus mirabilis*, *Can J Microbiol.*, 37, 325–328, 1991.
110. Pellegrino, R., Galvalisi, U., Scavone, P., Sosa, V., and Zunino, P., Evaluation of *Proteus mirabilis* structural fimbrial proteins as antigens against urinary tract infections, *FEMS Immunol Med Microbiol.*, 36, 103–110, 2003.
111. Li, X., Lockatell, C. V., Johnson, D. E., Lane, M. C., Warren, J. W., and Mobley, H. L., Development of an intranasal vaccine to prevent urinary tract infection by *Proteus mirabilis*, *Infect Immun.*, 72, 66–75, 2004.
112. Li, X., et al., Use of translational fusion of the MrpH fimbrial adhesin-binding domain with the cholera toxin A2 domain, coexpressed with the cholera toxin B subunit, as an intranasal vaccine to prevent experimental urinary tract infection by *Proteus mirabilis*, *Infect Immun.*, 72, 7306–7310, 2004.
113. Otczyk, D. C., and Cripps, A. W., Mucosal immunization: a realistic alternative, *Hum Vaccines*, 6, 978–1006, 2010.
114. Scavone, P., Sosa, V., Pellegrino, R., Galvalisi, U., and Zunino, P., Mucosal vaccination of mice with recombinant *Proteus mirabilis* structural fimbrial proteins, *Microbes Infect.*, 6, 853–860, 2004.
115. Kunin, C. M., *Detection, Prevention, and Management of Urinary Tract Infections*, Lea & Febiger, Philadelphia, 1987.
116. Schilling, J. D., Mulvey, M. A., and Hultgren, S. J., Dynamic interactions between host and pathogen during acute urinary tract infections, *Urology*, 57, 56–61, 2001.
117. Bowie, A. G., and Haga, I. R., The role of Toll-like receptors in the host response to viruses, *Mol Immunol.*, 42, 859–867, 2005.
118. Bäckhed, F., Söderhäll, M., Ekman, P., Normark, S., and Richter-Dahlfors, A., Induction of innate immune responses by *Escherichia coli* and purified lipopolysaccharide correlate with organ-and cell-specific expression of Toll-like receptors within the human urinary tract, *Cell Microbiol.*, 3, 153–158, 2001.

119. Wright, S. D., Ramos, R. A., Tobias, P. S., Ulevitch, R. J., and Mathison, J. C., CD14, a receptor for complexes of lipopolysaccharide (LPS) and LPS binding protein, *Science*, 249, 1431–1433, 1990.
120. Frendéus, B., et al., *Escherichia coli* P fimbriae utilize the Toll-like receptor 4 pathway for cell activation, *Mol Microbiol.*, 40, 37–51, 2001.
121. Schilling, J. D., Martin, S. M., Hung, C. S., Lorenz, R. G., and Hultgren, S. J., Toll-like receptor 4 on stromal and hematopoietic cells mediates innate resistance to uropathogenic *Escherichia coli*, *Proc Natl Acad Sci USA*, 100, 4203–4208, 2003.
122. Hernández, J. G., Sunden, F., Connolly, J., Svanborg, C., and Wullt, B., Genetic control of the variable innate immune response to asymptomatic bacteriuria, *PLoS One*, 6, e28289, 2011.
123. Andersen-Nissen, E., Smith, K. D., Bonneau, R., Strong, R. K., and Aderem, A., A conserved surface on Toll-like receptor 5 recognizes bacterial flagellin, *J Exp Med.*, 204, 393–403, 2007.
124. Hawn, T. R., et al., Toll-like receptor polymorphisms and susceptibility to urinary tract infections in adult women, *PLoS One*, 4, e5990, 2009.
125. Zhang, D., et al., A Toll-like receptor that prevents infection by uropathogenic bacteria, *Science*, 303, 1522–1526, 2004.
126. Lauw, F. N., Caffrey, D. R., and Golenbock, D. T., Of mice and man: TLR11 (finally) finds profiling, *Trends Immunol.*, 26, 509–511, 2005.

25

Pseudomonas aeruginosa

Stavria Panayidou and Yiorgos Apidianakis

CONTENTS

25.1 Introduction

Classified in the family Pseudomonadaceae, order Pseudomonadales, class Gammaproteobacteria, the genus *Pseudomonas* comprises some of the most ubiquitous and diverse Gram-negative bacterial species in nature that are capable of utilizing a wide range of organic compounds and colonizing a variety of ecological niches. Among the members of this genus, *Pseudomonas aeruginosa* is remarkable for its capacity to inhabit diverse environments, including soil and water, and infect multiple organisms, such as insects, plants, and animals.[1–6] *P. aeruginosa* is an important opportunistic human pathogen inflicting predominantly burn, cystic fibrosis (CF), and otherwise immunocompromised patients. It is a frequent cause of nosocomial infections, being the most common pathogen isolated from patients hospitalized for longer than 1 week. One reason for its high prevalence is that it is foodborne—found, for example, in hospital water, food, and feeding tubes—and an efficient intestinal colonizer, especially upon antibiotic treatment and surgical stress.[7] Another reason is its high virulence repertoire, which includes biofilm formation and quorum-sensing controlled factors.[8,9] A third reason is its resistance to antibiotics.

The multifaceted pathogenicity of *P. aeruginosa* in humans necessitates the use of various models of infection and alternative model organisms. Due to ethical considerations and high cost of experimenting with vertebrate animals, invertebrates are widely used as alternative model hosts. In the following sections of this chapter, we describe the mammalian models that recapitulate pivotal aspects

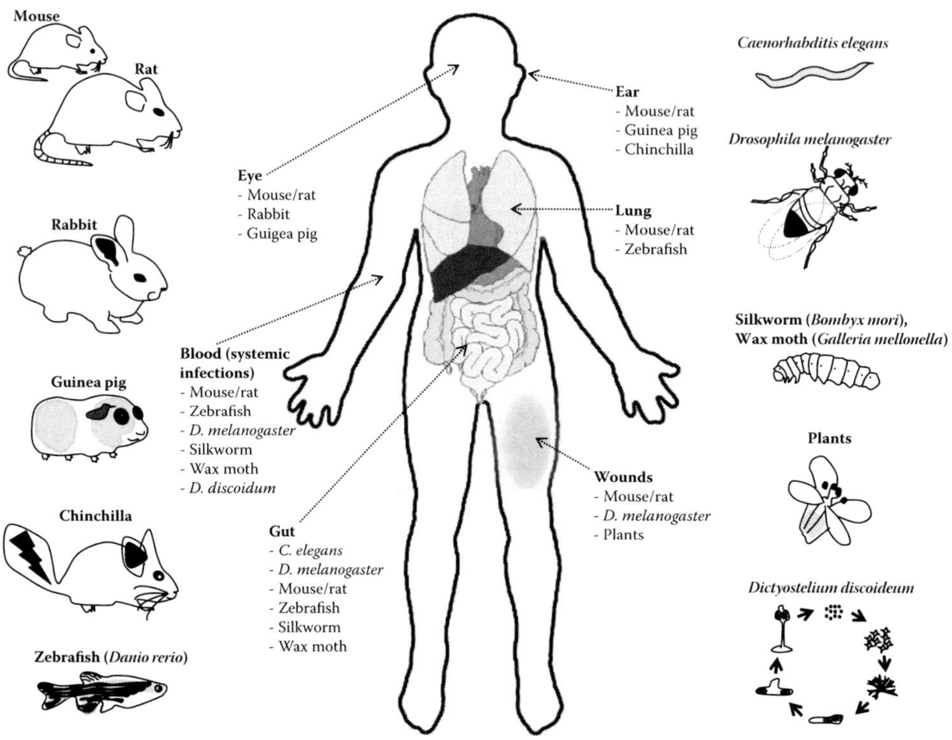

FIGURE 25.1 Model organisms recapitulating aspects of human blood, wound, lung, gut, eye, or ear infection with *P. aeruginosa*.

of severe *P. aeruginosa* pathogenicity, namely, burn and open wound, acute and persistent lung infection, and bacteremia, as well as less severe but potentially dangerous infections of the ear, eye, and intestinal tract. Zebrafish and invertebrate models also recapitulate aspects of wound, systemic, or intestinal/epithelial barrier infection. All established models, including those based on plants, are useful for assessing virulence, the efficacy of various treatments, and the role of host defense to infection (Figure 25.1 and Table 25.1).

25.2 Vertebrate Models

25.2.1 Cystic Fibrosis Mouse Models of Chronic Lung Infection

CF is an inherited disease of the secretory glands that is caused by mutations in the cystic fibrosis transmembrane conductance regulator (*CFTR*) gene.[10] *CFTR* gene mutations prevent the ion channels of the lung and other tissues from moving salt and water into and out of cells. As a result, mucus accumulates in the lungs, trapping bacteria that cause chronic infections.[10,11] *P. aeruginosa* is a common cause of chronic CF lung infection that may persist for decades[12] and leads to mortality in the majority of the cases, due to progressive lung damage. In 1992, only 3 years after the identification of the *CFTR* gene, Snouwaert et al. generated the first CF mouse model.[10,13] Since then, several mouse models of CF and the *P. aeruginosa* lung infection have been developed.[14] Although none of them are ideal, mouse models may provide significant information about the CF pathogenesis and are essential for the preclinical assessment of new therapeutics.[14] A method that mimics very well the human chronic lung infection of CF disease is based on the introduction of *P. aeruginosa*-laden agar/agarose or alginate beads in the mouse lung by transtracheal injection. This model was first described by Cash et al. in 1979 using rats.[15] In this model, agarose beads act as artificial biofilms and protect bacteria from a direct neutrophil attack, facilitating

TABLE 25.1

P. aeruginosa Models of Infection and Indicative Findings Using Each Model

***P. aeruginosa* Infection Models**	**Indicative Findings Using the Model**
Vertebrate Models	
Cystic fibrosis mouse models of chronic lung infection	• Suitable model for testing anti-inflammatory compounds e.g., BIIL 284. • *P. aeruginosa* and *B. cenocepacia* interaction in the mouse lung induces bacterial virulence and concomitant inflammatory response.
Acute lung infection mouse models	• Increased expression of IL-6 is associated with edema formation and decreased lung function. • IL-17 facilitates neutrophil recruitment in the infected lung areas of *P. aeruginosa*-infected mice. • Immunosenescence leads to impaired neutrophil response in the lungs. • IL-27 is implicated in sepsis-induced immunosuppression. • A leukopenic mouse model was developed for testing novel drugs.
P. aeruginosa gut-derived sepsis mouse models	• The probiotic bacterium *B. longum* strain BB536 suppresses the intestinal colonization of *P. aeruginosa.* • IL-1 is critical during gut-derived sepsis. • The bacteriophage strain KPP10 decreases *P. aeruginosa* burden and inflammatory response in the infected mice. • Surgical stress induces *P. aeruginosa* PA-I lectin in the mouse intestine, causing lethal sepsis. • PA-I lectin and cytotoxic exoproducts compromise the intestinal barrier. • PA-I lectin is highly expressed during intestinal ischemia/reperfusion injury in mice and contributes to lethal sepsis. • Surgical hepatectomy leads to low phosphate levels in the mouse intestine, which is sensed by *P. aeruginosa*, which then, enhances its virulence causing gut-derived sepsis.
Burn-wound and open-wound sepsis mouse models	• *P. aeruginosa* strain PAO1 quorum-sensing mutants exhibit reduced virulence in this infection model. • Suitable model for testing new antibacterial agents and treatments. • Targeting of fragellin protein can be a promising approach for the treatment of *P. aeruginosa*-infected burns. • Negative-pressure wound therapy can prevent sepsis. • Low GstA4 expression in the muscle causes susceptibility to infection.
P. aeruginosa keratitis models (mice, rabbits, guinea pigs)	Studies in rabbits: • Numerous antimicrobial treatments against *P. aeruginosa* keratitis have been assessed. • The *P. aeruginosa* proteases, elastase B, and PASP contribute significantly to the pathogenesis of keratitis, whereas alkaline protease has a lesser contribution. Studies in mice: • LiCl and β-catenin promote host resistance against *P. aeruginosa* keratitis. • Mouse TREM-2 suppresses corneal inflammation and promotes resistance to *P. aeruginosa* infection. • MRP8/MRP14 signaling amplify the inflammatory responses and increase corneal susceptibility. • Extracellular matrix protein Lumican, surfactant protein SP-D, and chemokine CXCL10 have protective roles against *P. aeruginosa* keratitis. • *P. aeruginosa* proteases, PASP, and MucD contribute significantly to keratitis pathogenesis. Studies in guinea pigs: • Evaluation of *P. aeruginosa* virulence factors, antimicrobial drugs, treatments, and host defense in the course of infection.
Otitis media models (mice, rats, guinea pigs, and chinchilla)	• Evaluation of *P. aeruginosa* virulence factors, inflammatory responses, and treatments.
Zebrafish (*Danio rerio*) (injection or feeding)	• *P. aeruginosa* strains PA14 and PAO1 are pathogenic to zebrafish embryos. • Infection is influenced by the developmental stage of the host. • CFTR mediates resistance against *P. aeruginosa* infection.

(*Continued*)

TABLE 25.1 (*Continued*)

P. aeruginosa Models of Infection and Indicative Findings Using Each Model

***P. aeruginosa* Infection Models**	**Indicative Findings Using the Model**
Invertebrate Models	
Drosophila melanogaster (feeding, wounding, or injection)	• Humoral and cellular innate immunity is important against infection. • PA14 escapes from the host defenses by suppressing the expression of antimicrobial peptides and muscle genes at the wound site. • Selection of virulence-attenuated mutants, e.g., KerV, which is a conserved virulence factor in Proteobacteria. • Expression of the human lactonase PON1 protects flies from *P. aeruginosa* infection. • CHD1 is important for fly intestinal resistance against *P. aeruginosa*. • JNK signaling pathway synergizes with Ras1 oncogene to induce stem-cell-mediated tumorigenesis and invasion/dissemination in the fly midgut and hindgut, respectively. • The RhlIR and LasIR quorum-sensing systems are important for the full virulence in orally infected flies.
Dictyostelium discoideum (feeding)	• The *rhl* quorum-sensing system is required for the full virulence of *P. aeruginosa* strain PAO1. • *P. aeruginosa* strain PA14 is more virulent than PAO1. • *trpD, pchH, and pchI* mutants are attenuated in virulence not only in *D. discoideum* but also in flies and mice.
Caenorhabditis elegans (feeding)	• *P. aeruginosa* pathogenesis (slow or fast killing) against *C. elegans* depends on the bacterial culture media. • The transcription factor DAF-19, the bZIP transcription factor *zip-2*, and the small organic molecule RPW-24 are important for *C. elegans* immune response against *P. aeruginosa* strain PA14. • HTS assays are developed for identifying novel antimicrobials.
Galleria mellonella (Wax Moth) (feeding or injection)	• Different *P. aeruginosa* strains exhibit different strategies of evading the immune system of *G. mellonella* larvae. • *P. aeruginosa* metalloproteinase elastase B stimulates the humoral immune responses in *G. mellonella*. • *G. mellonella* oral infection model can be useful in investigating *P. aeruginosa* virulence mechanisms.
Silkworm (*Bombyx mori*) (feeding or injection)	• Screening for *P. aeruginosa* virulence factors in the silkworm larvae. • The NO detoxification enzyme NO reductase is important for full virulence of *P. aeruginosa* in the silkworm. • PvdE and ExoS are important virulence factors for *P. aeruginosa* to cross epithelial barriers.
Plants	
• *Arabidopsis thaliana* • Sweet basil (injection)	• *P. aeruginosa* strains PA14 and PAO1 are pathogenic to *A. thaliana* and sweet basil. • The antimicrobial compound rosmarinic acid protects the sweet basil root only from *P. aeruginosa* quorum-sensing mutants unable to form biofilms. • The *P. aeruginosa* virulence factor pyocyanin inhibits the development of *A. thaliana* roots though ethylene-dependent signaling. • PA14 pathogenicity islands, PAPI-1, and PAPI-2 carry many genes, 11 of which are necessary for full virulence in both *A. thaliana* and mice. • Salicylic acid attenuates the virulence of PA14 against *A. thaliana* and *C. elegans*. • *P. aeruginosa*-secreted proteases activate a novel *A. thaliana*-immune signaling pathway. • *P. aeruginosa* alkaline protease AprA cleaves the bacterial fragellin monomers in order to prevent immune recognition in both plants and mammals

the modeling of delayed bacterial clearance. As a consequence, there is a prolonged neutrophil influx into the lung as well as accumulation of cytokines that resembles the *Pseudomonas* lung infection seen in CF patients.[16] Neutrophils are the protagonists of the excessive inflammatory response observed due to bacterial infections in CF patients,[17] but their role in pathogenicity is not clear. Accordingly, many studies are exploring the effect of specific anti-inflammatory compounds, such as BIIL 284, an antagonist of the leukotriene B4 (LTB_4)-receptor.[18] LTB_4 is a product of activated neutrophils and macrophages, and once it makes a complex with its receptor, it triggers NF-κB-dependent inflammatory responses.[18] BIIL 284 had previously caused adverse pulmonary reactions when given to CF patients. Therefore, Döring et al. examined the effect of BIIL 284 treatment in mice infected intratracheally by injection with the *P. aeruginosa* strain PAO1 embedded in agar beads.[18] Interestingly, BIIL 284 treatment led to decreased numbers of neutrophils, and consequently increased bacterial numbers, in the mouse lungs.[18] These observations were accompanied with strong presence of bacteria in the blood of the treated mice, compared with the untreated animals, indicating the important role of neutrophils in controlling bacterial lung infection and preventing sepsis.[18] The data also show the importance of mouse infection models for testing anti-inflammatory drugs before further assessment in clinical trials. Although most studies are focused on a single pathogen each time, CF disease is more complicated and often includes a number of different pathogens. Accordingly, the agar beads model of chronic lung infection has also been used to investigate the interaction between strains of two different opportunistic pathogens, *P. aeruginosa* and *Burkholderia cenocepacia*, in the mouse lung.[8] This coinfection increases the inflammatory response, as compared to the single infections, without any increase in the bacterial load, indicating that *P. aeruginosa* interacts with other bacterial species to increase bacterial virulence.[8]

25.2.2 Acute Lung Infection Mouse Models

Acute microbial lung infection, which mimics the human acute bacterial pneumonia, can occur in mice upon exposure to infectious aerosols or directly by intranasal or intratracheal instillation.[19] Intranasal infection allows the spreading of the bacteria from the upper airways to the intestine and the lower airways.[20] This protocol has been widely used for *P. aeruginosa* virulence factor assessment. Nevertheless, intratracheal instillation delivers much more bacteria into the distal bronchi.[20] Both *P. aeruginosa* acute lung infection and intratracheal instillation murine models can be used to gain insights about the immune responses and the lung function of the infected animals.[21] In mice infected intratracheally with *P. aeruginosa*, there is a correlation among increased interleukin-6 (IL-6) expression, edema formation, and decreased lung function.[21] In addition, the proinflammatory cytokine IL-17 facilitates the recruitment of neutrophils in the infected lung areas of infected mice.[22] On the other hand, immunosenescence leads to impaired neutrophil response, as observed in aged versus young mice subjected to intratracheal infection.[23] Septic mice infected intratracheally lead to the induction of IL-27, which in turn induces immunosuppression.[24] Toward standardization of new therapeutic approaches against human *P. aeruginosa* lung infections, Lawrenz et al. proposed recently a leukopenic (cyclophosphamide-treated) mouse model of lung intratracheal instillation for therapeutic testing of novel drugs against multidrug resistant strains.[25]

25.2.3 *P. aeruginosa* Induced Gut-Derived Sepsis Mouse Models

P. aeruginosa-induced gut-derived sepsis models mimic the pathophysiology of humans, because they involve intestinal colonization, proliferation, and invasion of other host tissues.[26] In order to produce murine gut-derived sepsis, mice receive bacteria in their drinking water as well as antibiotics e.g., ampicillin and streptomycin for a few days. Antibiotics disrupt the intestinal flora of the mice enabling gut colonization with *P. aeruginosa*.[27] To facilitate translocation of *P. aeruginosa* away from the gut, the immunosuppressant cyclophosphamide is administered during infection.[27] Interestingly, intestinal colonization with *P. aeruginosa* is reduced by a major member of probiotic bacterial species, *Bifidobacterium longum* strain BB536, which inhibits *P. aeruginosa* adherence to the intestinal epithelial cells in a murine model of gut-derived sepsis, encouraging its further assessment as a probiotic for immunocompromised patients.[28] Moreover, in IL-1-deficient mice, *P. aeruginosa* load and the inflammatory response are significantly higher in the liver during gut-derived sepsis.[29] This effect is reversed when mice are treated with the bacteriophage strain KPP10, as compared with the phage-untreated mice.[27]

Another way to induce lethal gut-derived sepsis is by surgical stress (30% hepatectomy).[30] Following surgical hepatectomy in mice, *P. aeruginosa* expresses PA-I lectin/adhesin in the intestine of the animals, indicating that pathogens may sense host stress and respond by expressing specific virulence effectors that promote lethal sepsis.[30] The PA-I lectin contributes to damaging the intestinal epithelium barrier by compromising enterocyte tight junctions.[31] Moreover, *P. aeruginosa* senses low phosphate (Pi) levels in the mouse intestine following surgical hepatectomy, promoting lethal gut-derived sepsis.[32] Similarly, mice subjected to intestinal ischemia/reperfusion injury exhibit *P. aeruginosa* PA-I lectin-dependent translocation from the cecum to other organs including liver, lung, and kidney causing lethal sepsis.[33] These, and other studies, suggest that *P. aeruginosa* exhibits enhanced virulence upon stress, surgery, and trauma, all of which may promote intestinal pathologies and systemic bacterial spreading.[7,30–33]

25.2.4 Burn- and Open-Wound Infection Mouse Models

The *P. aeruginosa*-infected burn-wound sepsis model is used to mimic the human burn wound sepsis.[9,34] In 1975, Stieritz and Holder developed a nonlethal thermal injury to examine the pathogenesis of *P. aeruginosa* infection by injecting viable bacteria into the burn skin area.[34] Injection in the burn area caused rapid sepsis,[34] systemic inflammatory response syndrome, and multiple organ dysfunction syndrome.[35] This model allows the investigation of the pathogenicity of various *P. aeruginosa* strains and the identification of virulence factors. For example, Rumbaugh et al. demonstrated that the single quorum-sensing mutants, *lasI*, *lasR*, *rhlI*, exhibited reduced virulence compared to the wild-type PAO1 strain, while the double-mutant *lasI rhlI* was even more attenuated in virulence, suggesting the important role of quorum sensing in virulence in this infection model.[9] Additionally, the burn-wound sepsis model is useful for testing antibacterial agents and evaluating treatments for *P. aeruginosa*-infected burn patients.[36]

P. aeruginosa fragellin is a structural component of flagella and a potent immunostimulant. Barnea et al. examined the effect of anti-fragellin subtype A monoclonal antibody (anti-fla-a) in a *P. aeruginosa*-infected burn-wound mouse sepsis model.[37] Anti-fla-a reduced the mortality and morbidity of the infected mice, showing that targeting fragellin protein can be a promising approach for the treatment of *P. aeruginosa*-infected burns.[37] Moreover, negative-pressure wound therapy may prevent sepsis and decrease mortality by inhibiting the invasion and proliferation of *P. aeruginosa* in the injured tissue of burn-wound septic mice.[38]

In addition to burn-wound, open wound infection models can be used to evaluate and treat infections in the absence of severe systemic stress caused by burns. It involves the removal of ~1 cm^2 of skin from the mouse back and the application of luminescent or GFP-expressing *P. aeruginosa* cells that can be followed longitudinally along with the assessment of mouse survival.[39,40] In a pivotal study, glutathione *S*-transferase A4 (GstA4), a detoxification enzyme against lipid peroxidation byproducts, was found downregulated in human and mouse muscles following burns, while low muscle expression postburn in humans predicted their susceptibility to infection.[41] Moreover open wound infection with *P. aeruginosa* of wild-type and GstA4 mutant mice shows that the mutant mice are more susceptible to infection, indicating the usefulness of the open-wound model to pinpoint genes relevant to both burn- and open-wound infections.[41]

25.2.5 *P. aeruginosa* Keratitis Models (Mice, Rabbits, and Guinea Pigs)

Keratitis is a disease of the cornea that can be due to the infection with various microbes, including bacteria.[42] The characteristics of bacterial keratitis include inflammation with concomitant pain and redness.[42] *P. aeruginosa* is a common cause of bacterial keratitis in humans,[42] but it can also infect the cornea of other mammals including mice, rabbits, and guinea pigs. For example, the guinea pig is a model for evaluating not only antimicrobial drugs, identifying treatments,[43–48] and studying the pathogenesis of *P. aeruginosa* keratitis but also host defense.[49–51] Rabbits, on the other hand, have large eyes similar in size to those of humans and can be used to evaluate several parameters of the disease.[42] The strain most commonly used to model bacterial keratitis and the efficacy of multiple treatments against *P. aeruginosa* is the New Zealand white rabbit.[42,52–59] For example, Chen et al. demonstrated in two separate studies that lithium chloride (LiCl) and β-catenin promote host resistance against *P. aeruginosa* keratitis by reducing the inflammatory responses of the host and by decreasing the bacterial burden.[60,61]

Moreover, *P. aeruginosa* elastase B, protease PASP, MucD, and, to a lesser degree, alkaline protease contribute significantly to the pathogenesis of keratitis.[62–64] Furthermore, induction of the triggering receptor expressed on myeloid cells-2 (TREM-2) upon infection in cornea scrapes triggers PI3K/Akt signaling to confer resistance against *P. aeruginosa* infection.[65] In contrast, induction of myeloid-related protein-8 (MPR8) and MRP14 upon infection in cornea scrapes, despite promoting bacterial clearance, induces inflammation and concomitant susceptibility to infection.[66] Three additional proteins, the extracellular matrix protein Lumican, the Surfactant Protein D (SP-D), and the C-X-C motif chemokine 10 (CXCL10) protect against *P. aeruginosa* keratitis in mice.[67–69]

25.2.6 Otitis Media Models (Mice, Rats, Guinea Pigs, and Chinchilla)

Otitis media (OM) includes a group of inflammatory diseases of the middle ear that can be caused by various conditions including infections by pathogens. Several models have been developed in various animals, including mice, rats, guinea pigs, and chinchilla, to evaluate *P. aeruginosa* biofilm formation, virulence factors, and the role of inflammatory responses and ciprofloxacin-hydrocortisone treatments against OM.[70–75]

25.2.7 Zebrafish (*Danio rerio*)

The zebrafish (*Danio rerio*) is an attractive vertebrate animal model for studying host–pathogen interactions.[5,76] One of its advantages is that, unlike invertebrates, it has an adaptive immune system similar to that of mammals, although most infection models are based on injecting embryos that only have innate immune system.[5,76] Moreover, zebrafish embryos are transparent, which allows the visualization of bacterial infections in real time by using microbes that express fluorescent proteins.[5,76] Rawls et al. took advantage of this transparency to monitor the motility defects of *P. aeruginosa* flagellar mutants within the intestine *in vivo* and in real time and to assess the impact on host immune responses.[77] Another important advantage of zebrafish model is the availability of a wide range of genetic tools that permit generation of ~200 progeny following a single mating.[5,76] Therefore, zebrafish models could help clarify important aspects of the host innate immunity upon bacterial infection. Live bacteria of the wild-type *P. aeruginosa* strains PA14 and PAO1 can kill injected embryos, while quorum sensing and type three secretion system mutants are attenuated in virulence.[76] However, strains significantly attenuated in virulence in late developmental stages are highly lethal in early embryos that lack pivotal innate immunity mechanisms.[76] Furthermore, embryos with reduced *CFTR* gene expression (Cftr morphants) produce less reactive oxygen species (ROS) in their phagocytes and sustain more bacteria during infection.[78] ROS production by phagocytes is known as respiratory burst response and is an important host defense mechanism. Thus, Cftr morphants indicate a connection between the CFTR function and the innate immune response.[78]

25.3 Invertebrate Models

25.3.1 *Drosophila melanogaster*

Drosophila melanogaster (the fruit fly), despite its small size (~2 mm in length), is a great invertebrate model organism that adequately reflects some aspects of the mammalian pathogenesis of infection.[79,80] Its short life cycle and easy rearing allows the production of up to ~50 adult progeny per female fly within

2 weeks, which facilitates the large-scale *in vivo* screening of bacterial mutants. Many human bacterial, fungal, and viral infections can be studied in *Drosophila*.[81] Notwithstanding the lack of adaptive immunity as we know it in mammals, *Drosophila* has similar innate immunity, disease-related signaling pathways, and cellular types to those of mammals. Thus, it is a good model for studying the pathogenicity of microbial infections, including those caused by *P. aeruginosa*.[79,80,82]

There are three most common methods to infect *Drosophila* with *P. aeruginosa*[2,82,83]: (1) the feeding method involves mixing of bacteria with the fly food, which causes intestinal colonization and fly lethality within a few days; (2) the thoracic or abdominal needle pricking infection, that is, an injury being caused using a tungsten needle dipped into a bacterial suspension. Accordingly, bacteria are introduced locally at the wound site and later on spread systemically, killing the flies within 2–4 days; and (3) the injector pumping, which appears similar to the pricking method, but is actually a method of systemic infection and involves the injection of a controlled dose of bacteria directly into the fly hemocoel with a thin glass capillary tip.[83]

The latter two methods have been used to screen *D. melanogaster* for virulence-related mutants of the *P. aeruginosa* strain PA14, for example the virulence-attenuating factor *hudA*[84] and the hypothetical methyltransferase KerV, which is conserved among Proteobacteria.[85] Moreover, NF-κB and JNK signaling pathways are important for flies to resist *P. aeruginosa* infection,[86] although highly virulent *P. aeruginosa* escapes host defenses by suppressing or evading the induction by these pathways that would normally activate antimicrobial peptides systemically and muscle genes at the wound site.[40,87] In addition, transgenic expression of the human lactonase paraoxonase-1 (PON1) in flies protects them from *P. aeruginosa* wound infection by interfering with the bacterial quorum sensing.[88] Thus, human innate immunity factors such as PON1 can be introduced and studied in *Drosophila*.[89]

Using the oral infection model, which recapitulates intestinal colonization and systemic dissemination of *P. aeruginosa*, new aspects of bacterial quorum sensing and intestinal pathology have been revealed. For example, the *Drosophila* chromatin remodeling factor chromo helicase domain protein 1 (CHD1) contributes to fly intestinal resistance to *P. aeruginosa* infection[90] and the quorum-sensing factor *rhlR* contributes to circumvent the fly cellular immune response when bacteria escape the intestine and spread systemically.[91] Actually, both RhlIR and LasIR quorum-sensing systems are required for full virulence in orally infected flies.[92] In addition, intestinal *P. aeruginosa* senses Gram-positive bacterial peptidoglycan to enhance its quorum-sensing-mediated virulence.[93] Strikingly, intestinal *P. aeruginosa* and the quorum-sensing-produced virulence factor pyocyanin induce intestinal stem-cell-mediated regeneration, which facilitates tumorigenesis in the presence of oncogenes or in the absence of tumor suppressor genes.[94] Moreover, the activation of the JNK innate immune signaling pathway in the adult *Drosophila* hindgut cells during *P. aeruginosa* infection synergizes with $Ras1^{V12}$ oncogene expression to induce enterocyte invasion and dissemination to distant sites.[95,96]

25.3.2 *Caenorhabditis elegans*

Caenorhabditis elegans is a small (~1 mm in length) transparent nematode living in the soil that feeds on bacteria. Its life cycle starts with the embryonic stage, followed by four larval stages (L1–L4) and adulthood. Its cellular simplicity and its small generation time facilitates screens related to human pathogens,[97] including *P. aeruginosa*, for the identification of virulence-related genes,[98] host defense factors,[99–101] and antimicrobials.[102,103]

P. aeruginosa can cause different pathologies in *C. elegans* depending on the culture media it grows in.[104–106] *C. elegans* dies slower when exposed to *P. aeruginosa* strain PA14 grown on nematode growth (NG) media due to the accumulation of the bacteria in the gut of the worms.[104] This is known as "slow killing assay" because *C. elegans* succumbs after a few days.[104] In contrast, the nematode dies within a few hours when PA14 is cultured in media of high osmotic strength. This is referred to as "fast killing assay," according to which worms die as a result of diffusible bacterial toxins in their food rather than bacterial growth within them.[105] *P. aeruginosa* strain PAO1 quickly paralyzes and then kills *C. elegans* by using hydrogen cyanide, a poison that could also inflict tissue damage in cystic fibrosis patients.[107] Additionally, the digestive tubes of nematodes fed on PAO1 grown in low-phosphate media become red before they die.[106] This phenomenon named "red death" occurs due to the activation of three systems: the phosphate signaling (PhoB), the MvfR-PQS quorum-sensing system, and the pyoverdine iron acquisition system.[106] Recently, Kirienko et al. established a liquid-based killing assay to show that pyoverdine causes hypoxia-related toxicity to *C. elegans* and that pyoverdine production by the PA14 strain is necessary for killing the worms.[108]

Regarding host defense, many factors important for *C. elegans* innate immunity against *P. aeruginosa* were identified within the last decade, including DAF-19, the ortholog of the human RFX, *zip-2*, a bZIP transcription factor and a small organic molecule, 2N(3chloro-4methylphenyl)-quinazoline-2,4diamine (or RPW-24).[99–101] Moreover, Conery et al. established a high-throughput screening (HTS) protocol in *C. elegans* for the identification of novel anti-infectives against *P. aeruginosa*.[102] Similarly, Zhou et al. developed an HTS assay for secondary metabolites of endophytic fungi using extracts of medicinal plants associated with these fungi to identify bioactive molecules that prolong the survival of *C. elegans* after *P. aeruginosa* infection.[103] With the caveat that HTS hits may not be validated in other systems and that extracts do not provide information on specific chemicals, such studies might serve as a starting point for the discovery of novel therapeutics.

25.3.3 *Dictyostelium discoideum*

Dictyostelium discoideum is a slime mould[109] with two remarkable multistage life cycles.[110,111] One is the asexual cycle, known as social cycle, characterized by the formation of fruiting bodies that release spores.[110,112] Spores give rise to haploid amoebae, which need to feed on bacteria to undergo mitosis.[110] Starving amoebas aggregate, forming new fruiting bodies. If starvation is combined with darkness and humidity, the sexual cycle starts with the fusion of two haploid amoebas of the opposite mate types, which attract and cannibalize surrounding cells forming a macrocyst that releases recombined new amoebas.[110,113] *Dictyostelium* cells naturally live in forest soil, and, by obligingly feeding on bacteria, they can be a natural host of pathogenic bacteria. Thus, they can serve as a great model organism for studying the mechanisms of bacteria–phagocyte interaction. Indeed, *D. discoideum* has been used to investigate the virulence of many human bacterial pathogens, including *P. aeruginosa*.[114] For example, the wild-type *P. aeruginosa* strain PAO1 inhibits *D. discoideum* growth, while *rhl* quorum-sensing system is required for full virulence.[115] Interestingly, the *P. aeruginosa* strain PA14 is more virulent than PAO1 against *D. discoideum*.[116] This is probably because 169 genes are differentially expressed between the 2 strains.[116] Of note, a random mutagenesis screen of the *P. aeruginosa* strain 22D10 identified anthranilate phosphoribosyltransferase gene *trpD* to be important for quorum-sensing function and the siderophore pyochelin genes *pchH* and *pchI* for the induction of the type III secretion system. Importantly, *trpD*, *pchH*, and *pchI* mutants are also attenuated in virulence in the *Drosophila* pricking and feeding assays and the mouse lung acute infection assay.[117]

25.3.4 *Galleria mellonella* (Wax Moth)

The greater wax moth, the lepidopteran *Galleria mellonella*, is a widely used model host for investigating microbial pathogenesis.[118,119] Similar to other insects, its life cycle progresses from egg to larva, pupa, and, finally, adult (moth).[119] The larvae of *G. mellonella* are relatively large in size (1–3 cm), facilitating the injection of bacterial pathogens and antimicrobial compounds.[118,119] An additional asset compared to other invertebrate model hosts is that it survives at the physiologic mammalian temperature (37°C), which favors the growth of many human pathogens.[3]

While *G. mellonella* can be infected orally by feeding, most infection studies use the injection method.[118–120] For example, Andrejko et al. have shown that different strains of *P. aeruginosa*, including two clinical isolates induce innate immunity genes to different extents when injected into hemolymph.[121] Additional data show that injected *P. aeruginosa* escapes the cellular immune responses of *G. mellonella* larvae by causing destruction of their hemocytes.[122] The virulence factors responsible for this strategy are still unknown. However, other studies implicate the *P. aeruginosa* metalloproteinase elastase B in virulence and counteraction of *G. mellonella* immune responses.[123,124]

25.3.5 *Bombyx mori* (Silkworm)

The Silkworm, *Bombyx mori*, is the best-known species of the lepidopteran superfamily Bombycoidea. The life cycle of this insect mainly consists of four developmental stages: egg, larva, pupa, and adult.[125] Numerous immunological, microbiological, and pharmacological studies of many pathogens have been carried out using this simple model showing, for example, that the overactivation of innate immunity in silkworms induces tissue damage followed by host death, resembling sepsis-induced multiorgan failure in humans.[126] Moreover, *P. aeruginosa* mutants and toxins, namely, pyocyanin, gacA, superoxide dismutase (sodB and sodM), nitric oxide (NO)-detoxification enzymes (NO reductase and flavohemoglobin), exotoxin A, and pyoverdine pvdE, have been assessed for virulence against silkworm larvae.[127–133] Among those, pvdE induces ExoS production, which is a bifunctional protein with GAP and ADP-ribosyltransferase activity that facilitates the translocation of the bacteria from the lumen to the hemolymph.[132] Interestingly, the ADP-ribosyltransferase activity of ExoS acts on mouse lung pneumocytes to disrupt the pulmonary–vascular barrier during *P. aeruginosa* acute pneumonia, leading to bacterial dissemination.[134]

25.4 Plants

Arabidopsis thaliana is a popular model plant due to its small size, its short life cycle of only 6 weeks—required for each seed to germinate into a mature plant and produce ~5000 new seeds—and the large number of mutant lines and genetic tools available.[135] It is also a model host for studying various aspects of infection, including host immune responses and microbial virulence strategies.[4] *P. aeruginosa* pathogenicity against *A. thaliana* involves various steps.[136] Infection starts with the syringe-mediated application of the bacteria on the leaf surface. The bacteria then attach and congregate and enter the plant tissues via stomatal openings.[136] Next, bacteria proliferate locally in the substomatal cavity and the intercellular space and destroy the local plant cells before dispersing systemically.[136] *P. aeruginosa* strains PA14 and PAO1 can infect the roots of both *A. thaliana* and sweet basil (*Ocimum basilium*), killing these plants within a week.[137] At the initial stage of infection, the two strains form biofilm structures on the root surfaces of the plants, which is reminiscent of the condition of the lungs of cystic fibrosis patients.[137] Biofilms confer antibiotic resistance and persistent pathogenicity to the organism. Accordingly, the antimicrobial compound rosmarinic acid, a multifunctional caffeic acid ester secreted from the sweet basil root, shows some *in vitro* activity only against *P. aeruginosa* strains unable to form biofilms.[137] Also, *A. thaliana* root development is inflicted by the multihost *P. aeruginosa* virulence factor pyocyanin, which induces the production of ROS and subsequent *A. thaliana* ethylene-dependent signaling.[138] The disaccharide trehalose of *P. aeruginosa* strain PA14 is a virulence factor that promotes pathogenesis only in *A. thaliana* and not in other hosts including mice.[139] However, He et al., demonstrated that PA14 carries two pathogenicity islands, PAPI-1 and PAPI-2, that harbor virulence genes, 11 of which are necessary for full virulence in both *A. thaliana* and in mice.[140] The majority of those genes are present in *P. aeruginosa* clinical isolates,[140] indicating that PA14 could survive in evolutionarily diverse hosts by using conserved functions. Accordingly, Starkey and Rahme have published infectivity protocols of *A. thaliana* and lettuce for screening *P. aeruginosa* bacterial strains to identify virulence factors potentially conserved for pathogenicity against other hosts.[141] A high through put (HTP) *P. aeruginosa*-infection system of *Arabidopsis* seedlings was also proposed by Gopalan and Ausubel for the discovery of potent anti-infective agents.[142] Furthermore, the *A. thaliana* pathogenicity model was used for the identification of host defense mechanisms and factors that repress *P. aeruginosa* virulence. For example, plant-derived salicylic acid, which is a phenolic metabolite, attenuates *P. aeruginosa* virulence against *A. thaliana* and *C. elegans* by downregulating the production of several virulence factors, including pyocyanin, protease, and elastase and by reducing biofilm formation.[143] Remarkably, Cheng et al., discovered a novel *A. thaliana* immune signaling pathway, namely, a mitogen-activated protein kinase (MAPK) cascade that is activated by proteases secreted by *P. aeruginosa*.[144] Nevertheless, *P. aeruginosa* has developed mechanisms to escape from the immune system of both plants and mammals. For example, *P. aeruginosa* alkaline protease AprA cleaves the bacterial fragellin monomers, which are ligands of pattern-recognition receptors in both plants and mammals, preventing the recognition and clearance of the bacteria from the host.[145]

25.5 Conclusions

While all described models are useful, none is ideal. The main asset of mammalian models lies in the close recapitulation of the human disease, especially regarding host response. Nevertheless, zebrafish and invertebrate models also mimic, to a lesser degree, the pathophysiology of systemic, wound, or intestinal/barrier epithelial infections. All models, including plants, are able to elucidate aspects of host defense, including adaptive and innate immunity in the case of mammals and innate immunity in the case of all the rest. Importantly, all models are suitable for screening *P. aeruginosa* virulence factors. As different models may reflect different aspects of infections, involve different virulence factors and require different treatments may be relevant in each case, it is not clear if some models are always superior to the others. Therefore, it is valuable to use as many different models of infection as possible in order to generate the strongest evidence for the relevance of a *P. aeruginosa* gene, and to provide the soundest rationale for the implementation of an anti-*P. aeruginosa* measure.

REFERENCES

1. Filion, G. & Charette, S. J. Assessing *Pseudomonas aeruginosa* virulence using a nonmammalian host: *Dictyostelium discoideum*. *Methods Mol. Biol.* 1149, 671–680 (2014).
2. Haller, S., Limmer, S. & Ferrandon, D. Assessing *Pseudomonas* virulence with a nonmammalian host: *Drosophila melanogaster*. *Methods Mol. Biol.* 1149, 723–740 (2014).
3. Koch, G., Nadal-Jimenez, P., Cool, R. H. & Quax, W. J. Assessing *Pseudomonas* virulence with nonmammalian host: *Galleria mellonella*. *Methods Mol. Biol.* 1149, 681–688 (2014).
4. Baldini, R. L., Starkey, M. & Rahme, L. G. Assessing *Pseudomonas* virulence with the nonmammalian host model: *Arabidopsis thaliana*. *Methods Mol. Biol.* 1149, 689–697 (2014).
5. Llamas, M. A. & van der Sar, A. M. Assessing *Pseudomonas* virulence with nonmammalian host: zebrafish. *Methods Mol. Biol.* 1149, 709–721 (2014).
6. Munder, A. & Tummler, B. Assessing *Pseudomonas* virulence using mammalian models: acute infection model. *Methods Mol. Biol.* 1149, 773–791 (2014).
7. Markou, P. & Apidianakis, Y. Pathogenesis of intestinal *Pseudomonas aeruginosa* infection in patients with cancer. *Front. Cell. Infect. Microbiol.* 3, 115 (2014).
8. Bragonzi, A. et al. Modelling co-infection of the cystic fibrosis lung by *Pseudomonas aeruginosa* and *Burkholderia cenocepacia* reveals influences on biofilm formation and host response. *PLoS One* 7, e52330 (2012).
9. Rumbaugh, K. P., Griswold, J. A., Iglewski, B. H. & Hamood, A. N. Contribution of quorum sensing to the virulence of *Pseudomonas aeruginosa* in burn wound infections. *Infect. Immun.* 67, 5854–5862 (1999).
10. Guilbault, C., Saeed, Z., Downey, G. P. & Radzioch, D. Cystic fibrosis mouse models. *Am. J. Respir. Cell Mol. Biol.* 36, 1–7 (2007).
11. Schwiebert, E. M., Benos, D. J., Egan, M. E., Stutts, M. J. & Guggino, W. B. CFTR is a conductance regulator as well as a chloride channel. *Physiol. Rev.* 79, S145–S166 (1999).
12. Turner, K. H., Wessel, A. K., Palmer, G. C., Murray, J. L. & Whiteley, M. Essential genome of *Pseudomonas aeruginosa* in cystic fibrosis sputum. *Proc. Natl. Acad. Sci. USA* 112, 4110–4115 (2015).
13. Snouwaert, J. N. et al. An animal model for cystic fibrosis made by gene targeting. *Science* 257, 1083–1088 (1992).
14. Bragonzi, A. Murine models of acute and chronic lung infection with cystic fibrosis pathogens. *Int. J. Med. Microbiol.* 300, 584–593 (2010).
15. Cash, H. A., Woods, D. E., McCullough, B., Johanson, W. G., Jr. & Bass, J. A. A rat model of chronic respiratory infection with *Pseudomonas aeruginosa*. *Am. Rev. Respir. Dis.* 119, 453–459 (1979).
16. Kukavica-Ibrulj, I., Facchini, M., Cigana, C., Levesque, R. C. & Bragonzi, A. Assessing *Pseudomonas aeruginosa* virulence and the host response using murine models of acute and chronic lung infection. *Methods Mol. Biol.* 1149, 757–771 (2014).
17. Heeckeren, A. et al. Excessive inflammatory response of cystic fibrosis mice to bronchopulmonary infection with *Pseudomonas aeruginosa*. *J. Clin. Invest.* 100, 2810–2815 (1997).
18. Döring, G. et al. BIIL 284 reduces neutrophil numbers but increases *P. aeruginosa* bacteremia and inflammation in mouse lungs. *J. Cyst. Fibros.* 13, 156–163 (2014).

19. Mizgerd, J. P. & Skerrett, S. J. Animal models of human pneumonia. *Am. J. Physiol. Lung Cell. Mol. Physiol.* 294, L387–L398 (2008).
20. Munder, A. et al. Acute intratracheal *Pseudomonas aeruginosa* infection in cystic fibrosis mice is age-independent. *Respir. Res.* 12, 148. doi: 10.1186/1465-9921-12-148 (2011).
21. Wolbeling, F. et al. Lung function and inflammation during murine *Pseudomonas aeruginosa* airway infection. *Immunobiology* 216, 901–908 (2011).
22. Xu, X. et al. Role of Interleukin-17 in defense against *Pseudomonas aeruginosa* infection in lungs. *Int. J. Clin. Exp. Med.* 7, 809–816 (2014).
23. Chen, M. M., Palmer, J. L., Plackett, T. P., Deburghgraeve, C. R. & Kovacs, E. J. Age-related differences in the neutrophil response to pulmonary *Pseudomonas* infection. *Exp. Gerontol.* 54, 42–46 (2014).
24. Cao, J. et al. IL-27 controls sepsis-induced impairment of lung antibacterial host defence. *Thorax* 69, 926–937 (2014).
25. Lawrenz, M. B. et al. Development and evaluation of murine lung-specific disease models for *Pseudomonas aeruginosa* applicable to therapeutic testing. *Pathog. Dis.* 73(5). pii: ftv025. doi: 10.1093/femspd/ftv025 (2015).
26. Hirakata, Y. et al. Efficacy of erythromycin lactobionate for treating *Pseudomonas aeruginosa* bacteremia in mice. *Antimicrob. Agents Chemother.* 36, 1198–1203 (1992).
27. Watanabe, R. et al. Efficacy of bacteriophage therapy against gut-derived sepsis caused by *Pseudomonas aeruginosa* in mice. *Antimicrob. Agents Chemother.* 51, 446–452 (2007).
28. Matsumoto, T. et al. Oral administration of *Bifidobacterium longum* prevents gut-derived *Pseudomonas aeruginosa* sepsis in mice. *J. Appl. Microbiol.* 104, 672–680 (2008).
29. Horino, T. et al. Interleukin-1-deficient mice exhibit high sensitivity to gut-derived sepsis caused by *Pseudomonas aeruginosa. Cytokine* 30, 339–346 (2005).
30. Alverdy, J. et al. Gut-derived sepsis occurs when the right pathogen with the right virulence genes meets the right host: evidence for in vivo virulence expression in *Pseudomonas aeruginosa. Ann. Surg.* 232, 480–489 (2000).
31. Laughlin, R. S. et al. The key role of *Pseudomonas aeruginosa* PA-I lectin on experimental gut-derived sepsis. *Ann. Surg.* 232, 133–142 (2000).
32. Long, J., Zaborina, O., Holbrook, C., Zaborin, A. & Alverdy, J. Depletion of intestinal phosphate after operative injury activates the virulence of *P. aeruginosa* causing lethal gut-derived sepsis. *Surgery* 144, 189–197 (2008).
33. Fink, D. et al. *Pseudomonas aeruginosa* potentiates the lethal effect of intestinal ischemia-reperfusion injury: the role of in vivo virulence activation. *J. Trauma* 71, 1575–1582 (2011).
34. Stieritz, D. D. & Holder, I. A. Experimental studies of the pathogenesis of infections due to *Pseudomonas aeruginosa*: description of a burned mouse model. *J. Infect. Dis.* 131, 688–691 (1975).
35. Li, N. et al. Systemic inflammatory responses and multiple organ dysfunction syndrome following skin burn wound and *Pseudomonas aeruginosa* infection in mice. *Shock* 40, 152–159 (2013).
36. Barnea, Y. et al. The establishment of a *Pseudomonas aeruginosa*-infected burn-wound sepsis model and the effect of imipenem treatment. *Ann. Plast. Surg.* 56, 674–679 (2006).
37. Barnea, Y. et al. Therapy with anti-flagellin A monoclonal antibody limits *Pseudomonas aeruginosa* invasiveness in a mouse burn wound sepsis model. *Burns* 35, 390–396 (2009).
38. Liu, Y. et al. Negative pressure wound therapy decreases mortality in a murine model of burn-wound sepsis involving *Pseudomonas aeruginosa* infection. *PLoS One* 9, e90494 (2014).
39. Hamblin, M. R., Zahra, T., Contag, C. H., McManus, A. T. & Hasan, T. Optical monitoring and treatment of potentially lethal wound infections in vivo. *J. Infect. Dis.* 187, 1717–1725 (2003).
40. Apidianakis, Y. et al. Involvement of skeletal muscle gene regulatory network in susceptibility to wound infection following trauma. *PLoS One* 2, e1356 (2007).
41. Apidianakis, Y. et al. Down-regulation of *glutathione S-transferase α4* (*hGSTA4*) in the muscle of thermally injured patients is indicative of susceptibility to bacterial infection. *FASEB J.* 26, 730–737 (2012).
42. Marquart, M. E. Animal models of bacterial keratitis. *J. Biomed. Biotechnol.* 2011, 680642 (2011).
43. Davis, S. D. & Chandler, J. W. Experimental keratitis due to *Pseudomonas aeruginosa*: model for evaluation of antimicrobial drugs. *Antimicrob. Agents Chemother.* 8, 350–355 (1975).
44. Davis, S. D., Sarff, L. D. & Hyndiuk, R. A. Antibiotic therapy of experimental *Pseudomonas* keratitis in guinea pigs. *Arch. Ophthalmol.* 95, 1638–1643 (1977).

45. Davis, S. D., Sarff, L. D. & Hyndiuk, R. A. Therapeutic effect of topical antibiotic on untreated eye in experimental keratitis. *Can. J. Ophthalmol.* 13, 273–276 (1978).
46. Davis, S. D., Sarff, L. D. & Hyndiuk, R. A. Bacteriologic cure of experimental *Pseudomonas* keratitis. *Invest. Ophthalmol. Vis. Sci.* 17, 916–918 (1978).
47. Malet, F., Colin, J., Jauch, A. & Abalain, M. L. Bacterial keratitis therapy in guinea pigs with lomefloxacin by initially high-followed by low-dosage regimen. *Ophthalmic Res.* 27, 322–329 (1995).
48. Nanbu, P. N. et al. Heat treatment enhances healing process of experimental *Pseudomonas* corneal ulcer. *Ophthalmic Res.* 36, 218–225 (2004).
49. Van Horn, D. L., Davis, S. D., Hyndiuk, R. A. & Alpren, T. V. Pathogenesis of experimental *Pseudomonas* keratitis in the guinea pig: bacteriologic, clinical, and microscopic observations. *Invest. Ophthalmol. Vis. Sci.* 17, 1076–1086 (1978).
50. Ijiri, Y. et al. The role of *Pseudomonas aeruginosa* elastase in corneal ring abscess formation in pseudomonal keratitis. *Graefes Arch. Clin. Exp. Ophthalmol.* 231, 521–528 (1993).
51. Chusid, M. J. & Davis, S. D. Experimental bacterial keratitis in neutropenic guinea pigs: polymorphonuclear leukocytes in corneal host defense. *Infect. Immun.* 24, 948–952 (1979).
52. Bu, P., Riske, P. S., Zaya, N. E., Carey, R. & Bouchard, C. S. A comparison of topical chlorhexidine, ciprofloxacin, and fortified tobramycin/cefazolin in rabbit models of *Staphylococcus* and *Pseudomonas* keratitis. *J. Ocul. Pharmacol. Ther.* 23, 213–220 (2007).
53. Mah, F. S., Romanowski, E. G., Kowalski, R. P., Yates, K. A. & Gordon, Y. J. Zymar (Gatifloxacin 0.3%) shows excellent Gram-negative activity against *Serratia marcescens* and *Pseudomonas aeruginosa* in a New Zealand white rabbit keratitis model. *Cornea* 26, 585–588 (2007).
54. McCormick, C. et al. Effectiveness of a new tobramycin (0.3%) and dexamethasone (0.05%) formulation in the treatment of experimental *Pseudomonas* keratitis. *Curr. Med. Res. Opin.* 24, 1569–1575 (2008).
55. Sensoy, D. et al. Bioadhesive sulfacetamide sodium microspheres: evaluation of their effectiveness in the treatment of bacterial keratitis caused by *Staphylococcus aureus* and *Pseudomonas aeruginosa* in a rabbit model. *Eur. J. Pharm. Biopharm.* 72, 487–495 (2009).
56. Kowalski, R. P., Romanowski, E. G., Mah, F. S., Shanks, R. M. & Gordon, Y. J. Topical levofloxacin 1.5% overcomes in vitro resistance in rabbit keratitis models. *Acta Ophthalmol.* 88, e120–e125 (2010).
57. Sanders, M. E. et al. Comparison of besifloxacin, gatifloxacin, and moxifloxacin against strains of *Pseudomonas aeruginosa* with different quinolone susceptibility patterns in a rabbit model of keratitis. *Cornea* 30, 83–90 (2011).
58. Li, S. A. et al. Therapeutic potential of the antimicrobial peptide OH-CATH30 for antibiotic-resistant *Pseudomonas aeruginosa* keratitis. *Antimicrob. Agents Chemother.* 58, 3144–3150 (2014).
59. Tajima, K. et al. In vivo challenging of polymyxins and levofloxacin eye drop against multidrug-resistant *Pseudomonas aeruginosa* keratitis. *J. Infect. Chemother.* 20, 343–349 (2014).
60. Chen, K. et al. Lithium chloride promotes host resistance against *Pseudomonas aeruginosa* keratitis. *Mol. Vis.* 19, 1502–1514 (2013).
61. Chen, K. et al. β-Catenin promotes host resistance against *Pseudomonas aeruginosa* keratitis. *J. Infect.* 67, 584–594 (2013).
62. Thibodeaux, B. A., Caballero, A. R., Marquart, M. E., Tommassen, J. & O'Callaghan, R. J. Corneal virulence of *Pseudomonas aeruginosa* elastase B and alkaline protease produced by *Pseudomonas putida*. *Curr. Eye Res.* 32, 373–386 (2007).
63. Tang, A., Caballero, A. R., Marquart, M. E. & O'Callaghan, R. J. *Pseudomonas aeruginosa* small protease (PASP), a keratitis virulence factor. *Invest. Ophthalmol. Vis. Sci.* 54, 2821–2828 (2013).
64. Mochizuki, Y. et al. *Pseudomonas aeruginosa* MucD protease mediates keratitis by inhibiting neutrophil recruitment and promoting bacterial survival. *Invest. Ophthalmol. Vis. Sci.* 55, 240–246 (2014).
65. Sun, M. et al. TREM-2 promotes host resistance against *Pseudomonas aeruginosa* infection by suppressing corneal inflammation via a PI3K/Akt signaling pathway. *Invest. Ophthalmol. Vis. Sci.* 54, 3451–3462 (2013).
66. Deng, Q. et al. MRP8/14 enhances corneal susceptibility to *Pseudomonas aeruginosa* infection by amplifying inflammatory responses. *Invest. Ophthalmol. Vis. Sci.* 54, 1227–1234 (2013).
67. Shao, H., Scott, S. G., Nakata, C., Hamad, A. R. & Chakravarti, S. Extracellular matrix protein lumican promotes clearance and resolution of *Pseudomonas aeruginosa* keratitis in a mouse model. *PLoS One* 8, e54765 (2013).

68. Heimer, S. R., Evans, D. J., Mun, J. J., Stern, M. E. & Fleiszig, S. M. Surfactant protein D contributes to ocular defense against *Pseudomonas aeruginosa* in a murine model of dry eye disease. *PLoS One* 8, e65797 (2013).
69. Yoon, G. S. et al. Interferon regulatory factor-1 in flagellin-induced reprogramming: potential protective role of CXCL10 in cornea innate defense against *Pseudomonas aeruginosa* infection. *Invest. Ophthalmol. Vis. Sci.* 54, 7510–7521 (2013).
70. Trinidad, A. et al. Tissular changes induced by *Pseudomonas aeruginosa* in an otitis media rat model with tubal obstruction. *Acta Otolaryngol.* 127, 132–137 (2007).
71. Trune, D. R. & Zheng, Q. Y. Mouse models for human otitis media. *Brain Res.* 1277, 90–103 (2009).
72. Byrd, M. S. et al. Direct evaluation of *Pseudomonas aeruginosa* biofilm mediators in a chronic infection model. *Infect. Immun.* 79, 3087–3095 (2011).
73. Hou, W., Li, X. & Xiao, H. Formation characteristics and significance of bacteria biofilm in middle ear mucosa of rats with chronic suppurative otitis. *Clin. Ear Nose Esophagus Head Neck Surg. J.* (*Chinese*) 26, 30–33 (2012).
74. Saussereau, E. & Debarbieux, L. Bacteriophages in the experimental treatment of *Pseudomonas aeruginosa* infections in mice. *Adv. Virus Res.* 83, 123–141 (2012).
75. Jang, C. H., Cho, Y. B., Choi, C. H., Lee, J. S. & Kang, S. I. Effect of anti-adhesion barrier solution containing ciprofloxacin-hydrocortisone on abraded mucosa with otitis media. *Int. J. Pediatr. Otorhinolaryngol.* 77, 19–24 (2013).
76. Clatworthy, A. E. et al. *Pseudomonas aeruginosa* infection of zebrafish involves both host and pathogen determinants. *Infect. Immun.* 77, 1293–1303 (2009).
77. Rawls, J. F., Mahowald, M. A., Goodman, A. L., Trent, C. M. & Gordon, J. I. In vivo imaging and genetic analysis link bacterial motility and symbiosis in the zebrafish gut. *Proc. Natl. Acad. Sci. USA* 104, 7622–7627 (2007).
78. Phennicie, R. T., Sullivan, M. J., Singer, J. T., Yoder, J. A. & Kim, C. H. Specific resistance to *Pseudomonas aeruginosa* infection in zebrafish is mediated by the cystic fibrosis transmembrane conductance regulator. *Infect. Immun.* 78, 4542–4550 (2010).
79. Papaioannou, E., Utari, P. D. & Quax, W. J. Choosing an appropriate infection model to study quorum sensing inhibition in *Pseudomonas* infections. *Int. J. Mol. Sci.* 14, 19309–19340 (2013).
80. Apidianakis, Y. & Rahme, L. G. *Drosophila melanogaster* as a model for human intestinal infection and pathology. *Dis. Model. Mech.* 4, 21–30 (2011).
81. Panayidou, S., Ioannidou, E. & Apidianakis, Y. Human pathogenic bacteria, fungi, and viruses in *Drosophila*: disease modeling, lessons, and shortcomings. *Virulence* 5, 253–269 (2014).
82. Fauvarque, M. O. Small flies to tackle big questions: assaying complex bacterial virulence mechanisms using *Drosophila melanogaster*. *Cell. Microbiol.* 16, 824–833 (2014).
83. Apidianakis, Y. & Rahme, L. G. *Drosophila melanogaster* as a model host for studying *Pseudomonas aeruginosa* infection. *Nat. Protoc.* 4, 1285–1294 (2009).
84. Kim, S. H., Park, S. Y., Heo, Y. J. & Cho, Y. H. *Drosophila melanogaster*-based screening for multihost virulence factors of *Pseudomonas aeruginosa* PA14 and identification of a virulence-attenuating factor, HudA. *Infect. Immun.* 76, 4152–4162 (2008).
85. An, D. et al. The pathogenic properties of a novel and conserved gene product, KerV, in proteobacteria. *PLoS One* 4, e7167 (2009).
86. Lau, G. W. et al. The *Drosophila melanogaster* Toll pathway participates in resistance to infection by the Gram-negative human pathogen *Pseudomonas aeruginosa*. *Infect. Immun.* 71, 4059–4066 (2003).
87. Apidianakis, Y. et al. Profiling early infection responses: *Pseudomonas aeruginosa* eludes host defenses by suppressing antimicrobial peptide gene expression. *Proc. Natl. Acad. Sci. USA* 102, 2573–2578 (2005).
88. Stoltz, D. A. et al. *Drosophila* are protected from *Pseudomonas aeruginosa* lethality by transgenic expression of paraoxonase-1. *J. Clin. Invest.* 118, 3123–3131 (2008).
89. Estin, M. L., Stoltz, D. A. & Zabner, J. Paraoxonase 1, quorum sensing, and *P. aeruginosa* infection: a novel model. *Adv. Exp. Med. Biol.* 660, 183–193 (2010).
90. Sebald, J., Morettini, S., Podhraski, V., Lass-Florl, C. & Lusser, A. CHD1 contributes to intestinal resistance against infection by *P. aeruginosa* in *Drosophila melanogaster*. *PLoS One* 7, e43144 (2012).
91. Limmer, S. et al. *Pseudomonas aeruginosa* RhlR is required to neutralize the cellular immune response in a *Drosophila melanogaster* oral infection model. *Proc. Natl. Acad. Sci. USA* 108, 17378–17383 (2011).

92. Lutter, E. I., Purighalla, S., Duong, J. & Storey, D. G. Lethality and cooperation of *Pseudomonas aeruginosa* quorum-sensing mutants in *Drosophila melanogaster* infection models. *Microbiology* 158, 2125–2132 (2012).
93. Korgaonkar, A., Trivedi, U., Rumbaugh, K. P. & Whiteley, M. Community surveillance enhances *Pseudomonas aeruginosa* virulence during polymicrobial infection. *Proc. Natl. Acad. Sci. USA* 110, 1059–1064 (2013).
94. Apidianakis, Y., Pitsouli, C., Perrimon, N. & Rahme, L. Synergy between bacterial infection and genetic predisposition in intestinal dysplasia. *Proc. Natl. Acad. Sci. USA* 106, 20883–20888 (2009).
95. Bangi, E., Pitsouli, C., Rahme, L. G., Cagan, R. & Apidianakis, Y. Immune response to bacteria induces dissemination of Ras-activated *Drosophila* hindgut cells. *EMBO Rep.* 13, 569–576 (2012).
96. Christofi, T. & Apidianakis, Y. Ras-oncogenic *Drosophila* hindgut but not midgut cells use an inflammation-like program to disseminate to distant sites. *Gut Microbes* 4, 54–59 (2013).
97. Marsh, E. K. & May, R. C. *Caenorhabditis elegans*, a model organism for investigating immunity. *Appl. Environ. Microbiol.* 78, 2075–2081 (2012).
98. Feinbaum, R. L. et al. Genome-wide identification of *Pseudomonas aeruginosa* virulence-related genes using a *Caenorhabditis elegans* infection model. *PLoS Pathog.* 8, e1002813 (2012).
99. Estes, K. A., Dunbar, T. L., Powell, J. R., Ausubel, F. M. & Troemel, E. R. bZIP transcription factor *zip*-2 mediates an early response to *Pseudomonas aeruginosa* infection in *Caenorhabditis elegans*. *Proc. Natl. Acad. Sci. USA* 107, 2153–2158 (2010).
100. Pukkila-Worley, R. et al. Stimulation of host immune defenses by a small molecule protects *C. elegans* from bacterial infection. *PLoS Genet.* 8, e1002733 (2012).
101. Xie, Y., Moussaif, M., Choi, S., Xu, L. & Sze, J. Y. RFX transcription factor DAF-19 regulates 5-HT and innate immune responses to pathogenic bacteria in *Caenorhabditis elegans*. *PLoS Genet.* 9, e1003324 (2013).
102. Conery, A. L., Larkins-Ford, J., Ausubel, F. M. & Kirienko, N. V. High-throughput screening for novel anti-infectives using a *C. elegans* pathogenesis model. *Curr. Protoc. Chem. Biol.* 6, 25–37 (2014).
103. Zhou, Y. M. et al. An efficient and novel screening model for assessing the bioactivity of extracts against multidrug-resistant *Pseudomonas aeruginosa* using *Caenorhabditis elegans*. *Biosci. Biotechnol. Biochem.* 75, 1746–1751 (2011).
104. Tan, M. W., Mahajan-Miklos, S. & Ausubel, F. M. Killing of *Caenorhabditis elegans* by *Pseudomonas aeruginosa* used to model mammalian bacterial pathogenesis. *Proc. Natl. Acad. Sci. USA* 96, 715–720 (1999).
105. Mahajan-Miklos, S., Tan, M. W., Rahme, L. G. & Ausubel, F. M. Molecular mechanisms of bacterial virulence elucidated using a *Pseudomonas aeruginosa–Caenorhabditis elegans* pathogenesis model. *Cell* 96, 47–56 (1999).
106. Zaborin, A. et al. Red death in *Caenorhabditis elegans* caused by *Pseudomonas aeruginosa* PAO1. *Proc. Natl. Acad. Sci. USA* 106, 6327–6332 (2009).
107. Gallagher, L. A. & Manoil, C. *Pseudomonas aeruginosa* PAO1 kills *Caenorhabditis elegans* by cyanide poisoning. *J. Bacteriol.* 183, 6207–6214 (2001).
108. Kirienko, N. V. et al. *Pseudomonas aeruginosa* disrupts *Caenorhabditis elegans* iron homeostasis, causing a hypoxic response and death. *Cell Host Microbe* 13, 406–416 (2013).
109. Annesley, S. J. & Fisher, P. R. *Dictyostelium discoideum*—a model for many reasons. *Mol. Cell. Biochem.* 329, 73–91 (2009).
110. Li, S. I. & Purugganan, M. D. The cooperative amoeba: *Dictyostelium* as a model for social evolution. *Trends Genet.* 27, 48–54 (2011).
111. Bozzaro, S., Buracco, S. & Peracino, B. Iron metabolism and resistance to infection by invasive bacteria in the social amoeba *Dictyostelium discoideum*. *Front. Cell. Infect. Microbiol.* 3, 50 (2013).
112. Filosa, M. F. & Dengler, R. E. Ultrastructure of macrocyst formation in the cellular slime mold, *Dictyostelium mucoroides*: extensive phagocytosis of amoebae by a specialized cell. *Dev. Biol.* 29, 1–16 (1972).
113. Gaudet, P., Williams, J. G., Fey, P. & Chisholm, R. L. An anatomy ontology to represent biological knowledge in *Dictyostelium discoideum*. *BMC Genomics* 9, 130. doi:10.1186/1471-2164-9-130 (2008).
114. Bozzaro, S. & Eichinger, L. The professional phagocyte *Dictyostelium discoideum* as a model host for bacterial pathogens. *Curr. Drug Targets* 12, 942–954 (2011).

115. Cosson, P. et al. *Pseudomonas aeruginosa* virulence analyzed in a *Dictyostelium discoideum* host system. *J. Bacteriol.* 184, 3027–3033 (2002).
116. Carilla-Latorre, S. et al. *Dictyostelium* transcriptional responses to *Pseudomonas aeruginosa*: common and specific effects from PAO1 and PA14 strains. *BMC Microbiol.* 8, 109. doi:10.1186/1471-2180-8-109 (2008).
117. Alibaud, L. et al. *Pseudomonas aeruginosa* virulence genes identified in a *Dictyostelium* host model. *Cell. Microbiol.* 10, 729–740 (2008).
118. Ramarao, N., Nielsen-Leroux, C. & Lereclus, D. The insect *Galleria mellonella* as a powerful infection model to investigate bacterial pathogenesis. *J. Vis. Exp.* (70), e4392. doi:10.3791/4392 (2012).
119. Trevijano-Contador, N. & Zaragoza, O. Expanding the use of alternative models to investigate novel aspects of immunity to microbial pathogens. *Virulence* 5, 454–456 (2014).
120. Fedhila, S. et al. Comparative analysis of the virulence of invertebrate and mammalian pathogenic bacteria in the oral insect infection model *Galleria mellonella*. *J. Invertebr. Pathol.* 103, 24–29 (2010).
121. Andrejko, M., Zdybicka-Barabas, A. & Cytrynska, M. Diverse effects of *Galleria mellonella* infection with entomopathogenic and clinical strains of *Pseudomonas aeruginosa*. *J. Invertebr. Pathol.* 115, 14–25 (2014).
122. Mizerska-Dudka, M. & Andrejko, M. *Galleria mellonella* hemocytes destruction after infection with *Pseudomonas aeruginosa*. *J. Basic Microbiol.* 54, 232–246 (2014).
123. Andrejko, M. & Mizerska-Dudka, M. Elastase B of *Pseudomonas aeruginosa* stimulates the humoral immune response in the greater wax moth, *Galleria mellonella*. *J. Invertebr. Pathol.* 107, 16–26 (2011).
124. Andrejko, M. & Mizerska-Dudka, M. Effect of *Pseudomonas aeruginosa* elastase B on level and activity of immune proteins/peptides of *Galleria mellonella* hemolymph. *J. Insect Sci.* 12, 88 (2012).
125. Zhou, L. et al. Developmental changes for the hemolymph metabolome of silkworm (*Bombyx mori* L.). *J. Proteome Res.* 14(5), 2331–2347 (2015).
126. Ishii, K., Hamamoto, H. & Sekimizu, K. Studies of host–pathogen interactions and immune-related drug development using the silkworm: interdisciplinary immunology, microbiology, and pharmacology studies. *Drug Discov. Ther.* 9(4), 238–246 (2015).
127. Chieda, Y. et al. Pathogenicity of *gacA* mutant of *Pseudomonas aeruginosa* PA01 in the silkworm, *Bombyx mori*. *FEMS Microbiol. Lett.* 244, 181–186 (2005).
128. Hossain, M. S. et al. Use of silkworm larvae to study pathogenic bacterial toxins. *J. Biochem.* 140, 439–444 (2006).
129. Iiyama, K. et al. Effect of superoxide dismutase gene inactivation on virulence of *Pseudomonas aeruginosa* PAO1 toward the silkworm, *Bombyx mori*. *Appl. Environ. Microbiol.* 73, 1569–1575 (2007).
130. Chieda, Y. et al. Inactivation of pyocyanin synthesis genes has no effect on the virulence of *Pseudomonas aeruginosa* PAO1 toward the silkworm, *Bombyx mori*. *FEMS Microbiol. Lett.* 278, 101–107 (2008).
131. Chieda, Y. et al. Virulence of an exotoxin A-deficient strain of *Pseudomonas aeruginosa* toward the silkworm, *Bombyx mori*. *Microb. Pathog.* 51, 407–414 (2011).
132. Okuda, J. et al. Complementation of the *exoS* gene in the *pvdE* pyoverdine synthesis gene-deficient mutant of *Pseudomonas aeruginosa* results in recovery of the *pvdE* gene-mediated penetration through the intestinal epithelial cell barrier but not the *pvdE*-mediated virulence in silkworms. *J. Infect. Chemother.* 18, 332–340 (2012).
133. Arai, H. & Iiyama, K. Role of nitric oxide-detoxifying enzymes in the virulence of *Pseudomonas aeruginosa* against the silkworm, *Bombyx mori*. *Biosci. Biotechnol. Biochem.* 77, 198–200 (2013).
134. Rangel, S. M., Diaz, M. H., Knoten, C. A., Zhang, A. & Hauser, A. R. The role of ExoS in dissemination of *Pseudomonas aeruginosa* during pneumonia. *PLoS Pathog.* 11, e1004945 (2015).
135. Hays, J. B. *Arabidopsis thaliana*, a versatile model system for study of eukaryotic genome-maintenance functions. *DNA Repair* (*Amst*) 1, 579–600 (2002).
136. Plotnikova, J. M., Rahme, L. G. & Ausubel, F. M. Pathogenesis of the human opportunistic pathogen *Pseudomonas aeruginosa* PA14 in *Arabidopsis*. *Plant Physiol.* 124, 1766–1774 (2000).
137. Walker, T. S. et al. *Pseudomonas aeruginosa*–plant root interactions. Pathogenicity, biofilm formation, and root exudation. *Plant Physiol.* 134, 320–331 (2004).
138. Ortiz-Castro, R. et al. Pyocyanin, a virulence factor produced by *Pseudomonas aeruginosa*, alters root development through reactive oxygen species and ethylene signaling in *Arabidopsis*. *Mol. Plant Microbe Interact.* 27, 364–378 (2014).

139. Djonovic, S. et al. Trehalose biosynthesis promotes *Pseudomonas aeruginosa* pathogenicity in plants. *PLoS Pathog.* 9, e1003217 (2013).
140. He, J. et al. The broad host range pathogen *Pseudomonas aeruginosa* strain PA14 carries two pathogenicity islands harboring plant and animal virulence genes. *Proc. Natl. Acad. Sci. USA* 101, 2530–2535 (2004).
141. Starkey, M. & Rahme, L. G. Modeling *Pseudomonas aeruginosa* pathogenesis in plant hosts. *Nat. Protoc.* 4, 117–124 (2009).
142. Gopalan, S. & Ausubel, F. M. A high throughput amenable *Arabidopsis-P. aeruginosa* system reveals a rewired regulatory module and the utility to identify potent anti-infectives. *PLoS One* 6, e16381 (2011).
143. Prithiviraj, B. et al. Down regulation of virulence factors of *Pseudomonas aeruginosa* by salicylic acid attenuates its virulence on *Arabidopsis thaliana* and *Caenorhabditis elegans. Infect. Immun.* 73, 5319–5328 (2005).
144. Cheng, Z. et al. Pathogen-secreted proteases activate a novel plant immune pathway. *Nature* 521(7551), 213–216 (2015).
145. Bardoel, B. W. et al. *Pseudomonas* evades immune recognition of flagellin in both mammals and plants. *PLoS Pathog.* 7, e1002206 (2011).

26

Salmonella

Dongyou Liu

CONTENTS

26.1 Introduction

Salmonella is a Gram-negative bacterium that was first observed in the Peyer's patches and spleens of typhoid patients by Karl Eberth in 1880, and the first culture was obtained by Georg Theodor Gaffky in 1884. Following the isolation of an organism (then called "hog-cholera bacillus") from the intestine of a pig by Theobald Smith and Daniel Salmon in 1885, the name *Salmonella enterica* (Choleraesuis) was proposed by Joseph Leon Lignières in 1900 to honor the contribution by the Daniel Salmon's group. In 1892, a pathogen (termed *Bacillus typhimurium*, now known as nontyphoidal *Salmonella typhimurium*) associated with typhoid fever-like disease in mice was isolated by Friedrich Loeffler. Since then, a large number of *Salmonella* serovars have been identified and their roles in the causation of human restricted, invasive, and systemic disease (typhoid fever) and noninvasive gastroenteritis (nontyphoidal *Salmonella* infection) have been clearly determined.

26.1.1 Classification, Morphology, and Genomics

26.1.1.1 Classification

The genus *Salmonella* encompasses a large group of Gram-negative, rod-shaped, facultative anaerobes in the family Enterobacteriaceae, order Enterobacteriales, class Gammaproteobacteria, phylum Proteobacteria, and domain Bacteria. Besides *Salmonella*, the family Enterobacteriaceae includes several medically important pathogens (e.g., *Klebsiella*, *Shigella*, and *Yersinia*).

Based on their differences in genome sequences, two species (*Salmonella bongori* and *Salmonella enterica*) are recognized in the genus *Salmonella*. While only one subspecies (serotype V) exists in the *S. bongori* species (which is found mainly in cold-blooded animals such as reptiles, but may be occasionally involved in human infection), six subspecies [i.e., enterica (serotype I), salamae (serotype II), arizonae (IIIa), diarizonae (IIIb), houtenae (IV), and indica (VI)] are identified in the *S. enterica* species, with the subspecies enterica affecting warm-blooded animals (e.g., mammals) and other subspecies being specific for cold-blooded animals (e.g., reptiles).

Upon examination of the somatic (O) and flagellar (H) antigens on the bacterial surface using serological techniques, 2600 serovars are identified in the *S. enterica* species. Additionally, according to their host specificity and disease manifestations in humans, *Salmonella* serovars are separated into two categories: typhoidal and nontyphoidal. Whereas typhoidal *Salmonella* serovars (TS) demonstrate host specificity and cause invasive bacteremia only in humans, nontyphoidal *Salmonella* serovars (NTS) are present ubiquitously in the environment and cause noninvasive, self-limiting, cross-species gastrointestinal disease.

It is noteworthy that *S. enterica* subsp. *enterica* comprises >2000 genetically similar serovars and accounts for almost all *Salmonella* infections in humans and animals. Belonging to the TS category, *S. enterica* subsp. *enterica* serovars Dublin (bovine), Choleraesuis (swine), and Typhi (abbreviated to *S.* Dublin, *S.* Choleraesuis, and *S.* Typhi, respectively) are host-specific and responsible for systemic (invasive) disease in immunocompetent hosts. Indeed, *S.* Typhi induces typhoid fever in humans (and higher primates), but not in any other mammals. On the other hand, those of the NTS category, *S. enterica* subsp. *enterica* serovars Typhimurium and Enteritidis (abbreviated to *S.* Typhimurium and *S.* Enteritidis, respectively), produce a localized (noninvasive) infection in the intestine (gastroenteritis) of humans and various animal species (e.g., rodents, cattle, and mammals) and are occasionally linked to severe systemic disease in immune-deficient individuals [1].

26.1.1.2 Morphology

Salmonella is rod-shaped, non-spore-forming, predominantly motile enterobacterium of approximately 0.7–1.5 μm in diameter and 2–5 μm in length, with peritrichous flagella around the entire bacterial surface. Typically, *Salmonella* forms transparent and colorless colonies of 1–3 mm, sometimes with dark center, on MacConkey agar (in mixed culture, *Salmonella* colonies tend to clear areas of precipitated bile produced by other organisms); red colonies with or without black centers on xylose lysine desoxycholate (XLD) agar; blue-green to blue colonies with or without large, glossy black centers on Hektoen enteric (HE) agar; and brown, gray, or black colonies on bismuth sulfite (BS) agar, with surrounding medium being brown at first, and turning black with longer incubation, producing the so-called halo effect. Sometimes, *Salmonella* may produce atypical yellow colonies with or without black centers on HE and XLD agars or green colonies with little or no darkening of the surrounding medium on BS agar. Interestingly, colonies formed by *Proteus* may appear similar to those of *Salmonella*; however, *Proteus* colonies may swarm on blood agar (trypticase soy agar + 5% sheep blood), while *Salmonella* colonies do not. Apart from Gallinarum and Pallorum serovars, most *Salmonella* isolates produce motile colonies. It is notable that nonmotile *Salmonella* primary cultures may switch to the motile phase using a Craigie tube or ditch plate. As a chemoorganotroph, *Salmonella* generates energy from organic sources via oxidation and reduction reaction, and has the capacity to survive with or without oxygen (facultative anaerobe). *Salmonella* is positive for catalase, nitrate reduction, methyl red, acid production from glucose, mannitol, and D-mannose [1].

26.1.1.3 Genomics

The genome of *S. bongori* is 4.46 Mb in size, with GC content of 51.3%, and 4010 protein-coding genes. This species possesses only a basic set of ancestral *Salmonella* virulence functions and lacks several *S. enterica*-defining metabolic pathways. However, the *S. bongori* genome contains genes encoding novel effector proteins (*sboA-L*) that may have originated from enteropathogenic *Escherichia coli* (EPEC). Thus, by inheriting the ancestral *Salmonella* virulence gene set, and incorporating virulence determinants that resemble those in EPEC, *S. bongori* may represent an evolutionary link between *E. coli* and *S. enterica* [2].

The genome of typhoid-fever-causing *S. Typhi* strain CT18 consists of a 4.81-Mb chromosome (with GC content of 52%) and two plasmids [including a 218-kb multiple-drug-resistance (MDR) incH1 plasmid (pHCM1), and a 106-kb cryptic plasmid (pHCM2, possibly derived from a virulence plasmid of *Yersinia pestis*)]. Within the chromosome, hundreds of insertions and deletions (ranging in size from single genes to large islands), 4455 coding genes, and 233 pseudogenes (with some corresponding to those related to virulence in *S. Typhimurium*) are found [3].

The genome of gastroenteritis and food-poisoning-inducing *S. Typhimurium* strain LT2 is composed of a 4.86-Mb chromosome (with GC content of 52.2%) and a 94-kb virulence plasmid (pSLT). Within the chromosome, 4631 coding genes and 54 pseudogenes are identified. Compared to those in the *S. typhi* genome, 11% of the coding genes in the *S. Typhimurium* LT2 genome are unique [4].

Salmonella serovars harbor large discrete genomic islands that contain prophage elements and specialized loci termed *Salmonella* pathogenicity islands (SPIs). Many *Salmonella*-specific functions [e.g., complete type III secretion systems encoded by SPI-1 (T3SS-1) and SPI-2 (T3SS-2)] may be traced to the genes located in these large discrete genomic islands, and some of these functions may have been acquired by *S. enterica* after splitting from *S. bongori*. For example, *S. bongori* lacks SPI-2 (essential for optimal replication within macrophages), SPI-6 (encoding a type VI secretion system), SPI-13 (required for survival in chicken macrophages), SPI-14 (encoding an electron transport system), and SPI-16 (bacteriophage remnant carrying genes associated with LPS modification), all of which are present in *S. enterica* [2].

26.1.2 Biology and Epidemiology

As a facultative intracellular pathogen, *Salmonella* usually enters the host via ingestion of contaminated food. After enduring gastric acidity in the stomach, the surviving *Salmonella* migrates to the small intestine and infects epithelial and M cells of Peyer's patches, in which *Salmonella* is phagocytosed by macrophages and dendritic cells and then disseminated to the organs of the reticuloendothelial system (RES) (spleen, liver, bone marrow, etc.) via the lymphatic system and blood. *Salmonella* replication takes place in macrophages within the spleen and liver. Additionally, in the event of intravascular hemolysis, *Salmonella* can replicate in neutrophils and reach high concentrations in the blood. Inside macrophages, *Salmonella* often forms an intracellular vacuole (termed as *Salmonella*-containing vacuole or SCV), which gets juxtaposed to the nucleus by utilizing the microtubule meshwork of the host cell and derives nutrition from the Golgi apparatus. The endotoxins released from the dead salmonellae induce enteritis, gastrointestinal disorder, systemic febrile illness, and ultimately death [1].

Salmonella is commonly found in various water sources (which act as reservoirs), and also in the digestive tracts of humans and animals (especially reptiles) [5]. Feces from infected animals or people represent an important source of water or food contamination. *Salmonella* on the skin of reptiles or amphibians may be passed to animal handlers. *Salmonella* may persist in a contaminated bathroom for weeks, and tolerates freezing, although it is sensitive to UV light and heating at 55°C for 90 min, or 60°C for 12 min.

Both typhoid and nontyphoid *Salmonella* serovars are distributed throughout the world. The etiologic agents of typhoid and paratyphoid fever, *S. Typhi* and *S. Paratyphi* A and B (and occasionally C), collectively known as the enteric fever serovars, thrive in places with a limited sanitation infrastructure that permits transmission between infected humans without involvement of animal hosts. With the ability to spread to the deeper tissues of humans (liver, spleen, and bone marrow), *S. Typhi* [serologically O (lipopolysaccharide) type 09, 012; H (flagellin) type d; and Vi (extracellular capsule) positive] is responsible for a serious invasive bacterial disease of humans, with an annual global burden of approximately 22×10^7 infections and >200,000 deaths [1,6].

Although infection with NTS usually occurs after ingestion of foods that contain a high concentration of bacteria, infection through inhalation of bacteria-laden dust is possible in infants. Important nontyphoidal *Salmonella* serovars include *S.* Typhimurium and *S.* Enteritidis, which are associated with rapid-onset gastroenteritis as well as serious systemic/invasive diseases (e.g., bacteremia, meningitis) in immune-deficient individuals, young children, HIV-positive individuals, or patients who happen to be coinfected with malaria worldwide. In industrialized countries, NTS infection in young infants or the elderly may sometimes lead to severe invasive diseases accompanied by high case fatality rates. In the sub-Saharan Africa, NTS diseases are associated with severe invasive disease in the absence of gastroenteritis [7,8].

26.1.3 Clinical Features and Pathogenesis

26.1.3.1 Clinical Features

Salmonella infection (salmonellosis) in humans is associated with gastrointestinal tract infection, enteric fever, bacteremia, local infection, and the chronic reservoir state. The initial nonspecific symptoms include fever, general weakness, and myalgia. Once the bacteria spread to other parts of the body (bacteremia), localized infections (e.g., arthritis, urinary tract, central nervous system, and bone and soft-tissue infections) or abscesses may appear. Along with deterioration of the function of reticular endothelial cells (after acute salmonellosis or bone infection), back pain or spondylosis may emerge [1].

With an incubation period of about 2 weeks, typhoid fever caused by *Salmonella Typhi, Paratyphi A, Paratyphi B and Paratyphi C* is an invasive disease that tends to occur in immunocompetent individuals. Its main symptoms include fever, headache, a slowed heart rate (bradycardia), and sepsis, but diarrhea is largely absent [9]. During the infection, *Salmonella* goes through the lymphatic system of the intestine into the blood (typhoid form) and migrates to various organs (liver, spleen, kidneys) to form secondary foci (septic form). Histologically, inflammatory infiltrates in the intestine are predominantly mononuclear cells, with few neutrophils. The bacterium is found in histiocytic granulomas (or typhoid nodules) located in the bone marrow, the liver, and the spleen. In a small proportion (approximately 4%) of patients, the bacterium remains in the gallbladder for a period of 1 year after disease resolution. These chronic carriers (or "typhoid Marys") can transmit the disease for the rest of their lives [1,10,11].

Often linked to food poisoning, noninvasive, NTS infection due to *S. Typhimurium* or *S. Enteritidis* is characterized by acute enterocolitis (enteric fever) and gastroenteritis (inflammatory diarrhea), mainly involving immunocompetent individuals in developed countries. With a short incubation period (<1 day), NTS gastroenteritis is localized in the terminal ileum, colon, and mesenteric lymph node, leading to diarrhea, fever, and intestinal inflammatory infiltrates (composed of neutrophils). However, there is evidence that NTS infection may be invasive, presenting with fever, bacteremia, hepatosplenomegaly, and respiratory signs, often without gastroenteritis, involving immunocompromised individuals (as well as those suffering from malnutrition, anemia, and severe malaria), as noted in sub-Saharan Africa. Asymptomatic carriage of NTS may occur infrequently (0.15% in healthy adults, 3.9% in children); therefore, the role of asymptomatic carriers in the maintenance and spread of the disease cannot be ignored [1].

26.1.3.2 Pathogenesis

Typhoidal *Salmonella* serovars breach the intestinal epithelial barrier via phagocytosis and elicit intestinal inflammatory infiltrates consisting primarily of mononuclear cells. After entry into macrophages via macropinocytosis, typhoidal *Salmonella* serovars disseminate via the mononuclear phagocyte system (a network of connective tissue containing immune cells) to other parts of the body, including the liver, spleen, and bone marrow. During the systemic phase of typhoid fever, *Salmonella* may transfer from the liver into the gallbladder, resulting in chronic carrier state [1,12].

On the other hand, NTS serovars enter into M cells through bacterial-mediated endocytosis, which is accompanied by intestinal inflammation (with a massive influx of neutrophils into the terminal ileum and colon). Gaining a growth advantage in the inflamed gut due to their ability to respire on tetrathionate and to grow anaerobically on ethanolamine, NTS serovars disrupt tight junctions between the cells of the intestinal wall, leading to the increased traffic of ions, water, and immune cells into and out of the intestine. A combination of inflammation and the disruption of tight junctions ultimately contributes to vomiting and diarrhea [13].

Salmonella genomes contain many virulence and antimicrobial resistance genes that play crucial parts in host cell invasion (*bapA*, *siiE*, *sopB*), motility (*fliC*), intracellular survival (*sseF*, *sseG*), plasmids (*spvB*, *spvC*), and ion acquisition (*corA*, *mgtA*, *mgtB*). The T3SS (encoded by genes in pathogenicity islands SPI-1 and SPI-2) aids the invasion of nonphagocytic cells, colonization of the intestine, induction of intestinal inflammatory responses and diarrhea, survival and proliferation in macrophages, and establishment of systemic disease [14–16].

26.1.4 Diagnosis

Conventional laboratory identification of *Salmonella* organisms is based on phenotyping methods that assess their morphological, biochemical, and serological characteristics. In particular, the serotyping technique (targeting somatic and flagellar antigens on the bacterial surface) separates *Salmonella* into >2600 serovars in the White–Kauffmann–Le Minor scheme. This provides a useful tool to diagnose and track *Salmonella* implicated in outbreaks, despite its obvious limitations (e.g., slow turnaround, skill demand, and result variability, especially in dealing with rough, monophasic, and nonmotile strains). In addition, on the basis of their ability to ferment certain substrates, *Salmonella* strains and serovars (e.g., Typhi, Paratyphi A, Paratyphi B, Typhimurium) can be biotyped via the phenotypic lead acetate test [17].

With the ongoing development and refinement in molecular detection technologies, a variety of highly efficient and specific procedures have become available for identification, subtyping, and tracking of *Salmonella* strains. These include serotyping and biotyping, virulence and resistance gene profiling, plasmid profiling, and epidemiological tracking of clonality and genetic relatedness [1,18].

26.1.4.1 Serotyping and Biotyping

Polymerase chain reaction (PCR) detection of genes encoding surface antigens (e.g., flagellar antigens) offers a rapid and accurate alternative to conventional serotyping for determination of *Salmonella* strains [19]. Similarly, PCR-based methods allow rapid and reliable identification of *S. paratyphi B* and other biotypes.

26.1.4.2 Virulence and Resistance Gene Profiling

PCR amplification of virulence-associated genes (e.g., *invA* and *spvC* from pathogenicity islands that are involved in adhesion, invasion, intracellular survival, colonization, and systemic infection) facilitates detection, characterization, and monitoring of *Salmonella* serovars, subtypes, and strains [20]. Further, PCR detection of genes associated with antimicrobial resistance (e.g., *aac*, *aad*, *aph*, *strA/B*, bla_{TEM}, bla_{CMY}, *sulI*, *tet(A,B,C,D)*, *dfrA*, etc.) enables accurate profiling of resistance to tetracycline, sulfonamide, streptomycin, nalidixic acid, trimethoprim–sulfamethoxazole, ampicillin, chloramphenicol, cephalothin, kanamycin, ciprofloxacin, gentamicin, cefoxitin, amoxicillin–clavulanate, and amikacin [18].

26.1.4.3 Plasmid Profiling

On the basis of the specific banding patterns of plasmid DNA molecules formed in electrophoretic gel, plasmid profiling reveals details on the mechanisms and modes of transmission of the virulence or antimicrobial resistance determinants among *Salmonella* strains.

26.1.4.4 Epidemiological Tracking

Various gel- and sequence-based genotyping methods may be employed for epidemiological tracking of *Salmonella* strain clonality and genetic relatedness. These include pulsed-field gel electrophoresis (PFGE), PCR-based DNA fingerprinting, multilocus variable-number tandem repeat analysis (MLVA), multilocus sequence typing (MLST), next-generation sequencing (NGS), etc. [18].

PFGE involves digestion of bacterial chromosomal DNA with a rare cutting enzyme (*Xba*I or *Avr*II) and electrophoretic separation of digested DNA fragments using an alternating electric current (pulsed field), resulting in a distinct pattern for each *Salmonella* strain. With sufficiently high discriminative power, PFGE remains the reference method for subtyping *Salmonella* organisms.

PCR-based DNA fingerprinting methods examine the genomic contents of *Salmonella* organisms, with a goal to differentiate among strains and identify the source of the outbreaks. The commonly used methods include random amplified polymorphic DNA (RAPD), enterobacterial repetitive intergenic consensus (ERIC)-PCR, repetitive extragenic palindromic sequences (REP)-PCR, and PCR-based single-nucleotide polymorphism (SNP) typing.

MLVA assesses the variability (in terms of nucleotide sequence and unit size) of the genetic entity called variable number tandem repeat (VNTR), which is present at multiple loci in the *Salmonella* genome. MLVA subtyping of *Salmonella* strains demonstrates serovar specificity, in comparison with PFGE, which is applicable to all serotypes.

MLST exploits the polymorphisms in selected housekeeping genes (*aroC*, *dnaN*, *hemD*, *hisD*, *purE*, *sucA*, and *thrA*) among *Salmonella* strains to determine the sequence types of the organism. MLST subtyping represents a robust technique for epidemiological and evolutionary analyses of *Salmonella* strains.

PCR-based SNP typing is capable of differentiating *Salmonella enteritidis* strains that are indistinguishable by PFGE and phage typing.

NGS is a high-throughput genome sequencing technique that generates whole-genome sequences of *Salmonella* strains in a timely and cost-effective fashion. NGS has proven superior to PFGE for subtyping *Salmonella*.

26.1.5 Treatment and Prevention

26.1.5.1 Treatment

Since many patients infected with *Salmonella* (especially uncomplicated nontyphoidal *Salmonella* gastroenteritis) recover in 3–7 days without treatment, only persons with severe diarrhea may require fluid and electrolyte replacement. However, antibiotic therapy (10–14 day duration) is recommended for persons at increased risk of invasive disease, including infants of <3 months of age and individuals with suppressed immune function. Under this circumstance, patients with typhoid fever may be given antibiotics like chloramphenicol, ampicillin, amoxicillin, and trimethoprim–sulfamethoxazole (if susceptible); or fluoroquinolones (not approved for persons <18 years of age), cefotaxime, and ceftriaxone as alternatives. Surgical intervention may be needed for typhoid fever patients with complication of intestinal perforation or hemorrhage, cholecystitis, endocarditis, arteritis, osteomyelitis, or soft-tissue or spleen abscess. Patients suffering NTS gastroenteritis may be administered antibiotics ampicillin, amoxicillin, and trimethoprim–sulfamethoxazole (if susceptible); or ciprofloxacin, chloramphenicol, and ceftriaxone as alternatives [1].

26.1.5.2 Prevention

Proper preparation and cooking of food offers an effective means to prevent the spreading of *Salmonella* infection. These include washing hands before preparing food; heating food (chicken and beef) to an internal temperature of 75°C for at least 10 min; avoiding drinking raw (unpasteurized) milk; avoiding cross-contamination; washing hands with soap after handling reptiles, amphibians, or birds, or after contact with pet feces.

A number of killed *Salmonella* whole-cell vaccines (through thermal inactivation, or chemical inactivation with acetone, deoxycholate, or formalin) have been tested in mice. In addition, candidate subunit vaccines [targeting protein (purified porins, bulk outer membrane proteins, flagellin protein), lipopolysaccharide, or O-polysaccharide (OPS)] are being developed. Conjugate vaccines that link *Salmonella* OPS to a protein carrier have been shown to enhance the immunogenicity of polysaccharide hapten and increase functional immunological memory [21–25].

26.2 Laboratory Models

Laboratory models offer valuable approaches for elucidating the pathogenesis of human *Salmonella* disease, uncovering insights into immunity to infection, and assessing the efficacy of vaccine preparations.

26.2.1 Animal Models

Due to the fact many *Salmonella* serovars demonstrate broad host range and that the sources of human *Salmonella* infections often come from meat, eggs, and related products, warm-blooded animals of any kind would be suitable models for foodborne *Salmonella* enteritis and systemic typhoid [26].

26.2.1.1 Mice

Mouse models of *Salmonella* infections have a number of obvious advantages, such as relatively low cost, ready availability, and easy access to host genetics and immunological reagents [27].

Mouse colitis model involves streptomycin pretreatment (which reduces the intestinal microbiota in mice) followed by oral administration of NTS *Salmonella* infection, leading to inflammatory colitis and subsequent systemic infection. Interestingly, different mouse strains demonstrate varied susceptibility to *S. Typhimurium* infection, with mouse strain 129sv being resistant and mouse strains C57BL6/J and BALB/c being susceptible (which lack allele of the *Slc11a1* or *Nramp1* gene). Mouse colitis model allows for mechanistic study of intestinal inflammation and *Salmonella* growth in the intestinal lumen. The use of this model has helped clarify the role of flagella, T3SS-1, and T3SS-2 in intestinal inflammation during *S. Typhimurium* infection; highlighted the contribution of intestinal inflammation to luminal outgrowth of *S. Typhimurium*; and revealed the importance of luminal outgrowth of *S. Typhimurium* to transmission of the pathogen [28,29].

The mouse typhoid model involves oral administration with NTS serovar *S.* Typhimurium without streptomycin pretreatment, since oral challenge of mice with the human host-restricted serovars *S.* Typhi and *S.* Paratyphi does not lead to a productive invasive infection [30]. This oral administration leads to an invasive, systemic infection (characterized by a predominantly mononuclear leukocyte infiltrate, with follicular hyperplasia, capillary thrombosis, hemorrhage, and ulcerations at areas of Peyer's patches in moribund animals) of the gut-associated lymphoid tissue and RES that resembles typhoid fever in humans (in terms of entry route, tissue tropism, cellular location of bacteria, immune responses, and profound hepatosplenomegaly) [31]. Nevertheless, without a capsule, *S. Typhimurium* does not show any "stealth" incubation period before the clinical signs and does not express the typhoid toxin and the Vi capsule polysaccharide (ViCPS) that can reduce toll receptor (TLR)-dependent inflammatory responses and prevent complement deposition during *Salmonella* infection. This model has helped verify the role of SPI-1 effector protein GtgE (present in *S. Typhimurium*, but absent in *S. Typhi*) in the proteolytical inactivation of the Rab29 GTPase, permitting intracellular replication. The absence of this effector results in the rapid killing of *S. typhi* inside nonhuman cells [25,28].

Creation of humanized mice provides a sensitive and susceptible model to assess *S. Typhi* infection that resembles the pathogenesis of typhoid in humans. The most commonly used humanized mice include immunodeficient mice (either Rag2$^{-/-}$ $\gamma c^{-/-}$ or nonobese diabetic *scid* IL2rγ^{null}) and TLR11-deficient mice that are engrafted with human hematopoietic stem and progenitor cells (from fetal liver or cord blood) [26].

26.2.1.2 *Guinea Pigs*

Guinea pigs represent a useful model for studying intestinal inflammation that results from *Salmonella* infection, although preconditioning through starvation and opium treatment is necessary prior to experimentation.

26.2.1.3 *Rabbits*

Based on New Zealand white rabbits, this model is useful for evaluating immunogenicity and reactogenicity after oral administration of *Salmonella* serovars. It can also be applied for assessing *S. Typhi* and *S. Paratyphi A* vaccine strains.

26.2.1.4 *Calves*

Calves infected orally or subjected to surgical ligation of ileal loops are useful for examining the intestinal inflammation (a severe diffuse infiltrate composed predominantly of neutrophils, necrosis of the upper mucosa, and formation of a pseudomembrane) and fluid secretion that are associated with *S. Typhimurium*-induced gastroenteritis. Application of this model confirmed the ability of *S. Typhimurium* T3SS-1 in inducing intestinal inflammation and diarrhea, and the contribution of flagella to neutrophil recruitment in the intestinal mucosa during *S. Typhimurium* infection [28,32].

26.2.1.5 *Chicken*

The chick model has been employed to demonstrate the role of *tviA* in enhancing *S. Typhimurium* dissemination to internal organs.

26.2.1.6 *Nonhuman Primates*

Nonhuman primates such as rhesus macaque are susceptible to *S. Typhimurium* infection, which leads to gastroenteritis resembling that in humans. Chimpanzees may develop typhoid fever after oral infection with *S. Typhi.*

26.2.1.7 *Zebrafish*

Being strongly susceptible to *S. Typhimurium*, zebrafish embryos undergo pathological changes (bacteremia and proinflammatory response or cytokine storm) within hours after intravenous injection, resembling those seen in humans [33].

26.2.1.8 C. elegans *Nematode*

This animal model may be utilized for investigation in *Salmonella* pathogenicity and host-pathogen interactions.

26.2.2 *In Vitro* Models

A number of *in vitro* cell models based on established cell lines as well as primary cells have been utilized to unravel the mechanisms of *Salmonella* entry to the host cells and disease outcome. For example, epithelial cells (HeLa, Caco-2, HT-29, Intestine 407, and primary cells), which engulf *Salmonella* by macropinocytosis, provide an excellent platform for determining the essential processes during *Salmonella* invasion of, proliferation in, and apoptosis of epithelial cells. Resembling M cells (which lack glycocalyx) in Peyer's patches, a combination of Caco-2 cells and Raji B cells helps uncover critical details to caveolae-mediated endocytosis that is employed by *Salmonella* in its adhesion and entry into host reticuloendothelial system before being taken up by macrophages. Macrophages that are found in the gut-associated lymphoid tissue may be simulated by murine macrophage like cell line RAW 264.7 and murine macrophages J774-A.1, yielding clues on *Salmonella* intracellular survival. Primary dendritic cells (isolated from bone marrow of animal models or healthy humans) may be employed to investigate the uptake (phagocytosis) and passive dissemination of *Salmonella* to systemic sites. In addition, use of monocytes and granulocytes (such as human monocyte cell line THP-1 and primary cells isolated from animal models) allows study of survival of *Salmonella* in unconventional targets [25].

26.3 Conclusion

The genus *Salmonella* contains some important foodborne pathogens that have the capability to cause both localized infection (gastroenteritis) and systemic infection (typhoid fever) in humans and other animals. With the application of biochemical, serological, and molecular techniques together with the use of various laboratory models (including *in vivo* animal models and *in vitro* epithelial, phagocytic, and other cell models) in the past few decades, the molecular mechanisms of *Salmonella* invasion and intracellular survival in host cells have been largely elucidated, and many rapid and accurate diagnostic techniques for identification and subtyping of *Salmonella* strains have been developed. Nevertheless, much still has to be learnt about the most efficient way to vaccinate humans against *Salmonella* infections. For this reason, laboratory models will continue to play a vital part in helping decipher the host–bacterium interactions, and design improved control strategies against *Salmonella* infections.

REFERENCES

1. Crump JA, Sjölund-Karlsson M, Gordon MA, Parry CM. Epidemiology, clinical presentation, laboratory diagnosis, antimicrobial resistance, and antimicrobial management of invasive *Salmonella* infections. *Clin Microbiol Rev.* 2015;28(4):901–37.
2. Fookes M, et al. *Salmonella bongori* provides insights into the evolution of the Salmonellae. *PLoS Pathog.* 2011;7(8):e1002191.
3. Parkhill J, et al. Complete genome sequence of a multiple drug resistant *Salmonella enterica* serovar Typhi CT18. *Nature.* 2001;413(6858):848–52.
4. McClelland M, et al. Complete genome sequence of *Salmonella enterica* serovar Typhimurium LT2. *Nature.* 2001;413(6858):852–6.
5. Calenge F, Kaiser P, Vignal A, Beaumont C. Genetic control of resistance to salmonellosis and to *Salmonella* carrier-state in fowl: a review. *Genet Sel Evol.* 2010;42:11.
6. de Jong HK, Parry CM, van der Poll T, Wiersinga WJ. Host–pathogen interaction in invasive salmonellosis. *PLoS Pathog.* 2012;8(10):e1002933.
7. Majowicz SE, et al. The global burden of nontyphoidal *Salmonella* gastroenteritis. *Clin Infect Dis.* 2010;50:882–9.
8. Garai P, Gnanadhas DP, Chakravortty D. *Salmonella enterica* serovars Typhimurium and Typhi as model organisms: revealing paradigm of host–pathogen interactions. *Virulence.* 2012;3(4):377–88.
9. Bhan MK, Bahl R, Bhatnagar S. Typhoid and paratyphoid fever. *Lancet.* 2005;366:749–62.
10. Menendez A, et al. *Salmonella* infection of gallbladder epithelial cells drives local inflammation and injury in a model of acute typhoid fever. *J Infect Dis.* 2009;200:1703–13.
11. Gonzalez-Escobedo G, Marshall JM, Gunn JS. Chronic and acute infection of the gall bladder by *Salmonella* Typhi: understanding the carrier state. *Nat Rev Microbiol.* 2011;9(1):9–14.
12. Gunn JS, et al. *Salmonella* chronic carriage: epidemiology, diagnosis, and gallbladder persistence. *Trends Microbiol.* 2014;22(11):648–55.
13. Pilar AV, Reid-Yu SA, Cooper CA, Mulder DT, Coombes BK. Active modification of host inflammation by *Salmonella. Gut Microbes.* 2013;4(2):140–5.

14. Coburn B, Li Y, Owen D, Vallance BA, Finlay BB. *Salmonella enterica* serovar Typhimurium pathogenicity island 2 is necessary for complete virulence in a mouse model of infectious enterocolitis. *Infect Immun*. 2005;73:3219–27.
15. Andrews-Polymenis HL, Bäumler AJ, McCormick BA, Fang FC. Taming the elephant: *Salmonella* biology, pathogenesis, and prevention. *Infect Immun*. 2010;78:2356–69.
16. Broz P, Ohlson MB, Monack DM. Innate immune response to *Salmonella typhimurium*, a model enteric pathogen. *Gut Microbes*. 2012;3(2):62–70.
17. Andrews JR, Ryan ET. Diagnostics for invasive *Salmonella* infections: current challenges and future directions. *Vaccine*. 2015;33(Suppl 3):C8–15.
18. Ngoi ST, Teh CS, Chai LC, Thong KL. Overview of molecular typing tools for the characterization of *Salmonella enterica* in Malaysia. *Biomed Environ Sci*. 2015;28(10):751–64.
19. Shi C, Singh P, Ranieri ML, Wiedmann M, Moreno Switt AI. Molecular methods for serovar determination of *Salmonella*. *Crit Rev Microbiol*. 2015;41(3):309–25.
20. Weening EH, et al. The *Salmonella enterica* serotype Typhimurium *lpf*, *bcf*, *stb*, *stc*, *std*, and *sth* fimbrial operons are required for intestinal persistence in mice. *Infect Immun*. 2005;73:3358–66.
21. Griffin AJ, McSorley SJ. Development of protective immunity to *Salmonella*, a mucosal pathogen with a systemic agenda. *Mucosal Immunol*. 2011;4(4):371–82.
22. Simon R, Tennant SM, Galen JE, Levine MM. Mouse models to assess the efficacy of non-typhoidal *Salmonella* vaccines: revisiting the role of host innate susceptibility and routes of challenge. *Vaccine*. 2011;29(32):5094–106.
23. Simon R, Levine MM. Glycoconjugate vaccine strategies for protection against invasive *Salmonella* infections. *Hum Vaccines Immunother*. 2012;8(4):494–8.
24. McSorley SJ. Immunity to intestinal pathogens: lessons learned from *Salmonella*. *Immunol Rev*. 2014;260(1):168–82.
25. Santos RL. Pathobiology of *Salmonella*, intestinal microbiota, and the host innate immune response. *Front Immunol*. 2014;5:252.
26. Tsolis RM, Xavier MN, Santos RL, Bäumler AJ. How to become a top model: impact of animal experimentation on human *Salmonella* disease research. *Infect Immun*. 2011;79(5):1806–14.
27. Lawley TD, et al. Genome-wide screen for *Salmonella* genes required for long-term systemic infection of the mouse. *PLoS Pathog*. 2006;2:e11.
28. Santos RL, et al. Animal models of *Salmonella* infections: enteritis versus typhoid fever. *Microbes Infect*. 2001;3:1335–44.
29. Barthel M, et al. Pretreatment of mice with streptomycin provides a *Salmonella enterica* serovar Typhimurium colitis model that allows analysis of both pathogen and host. *Infect Immun*. 2003; 71:2839–58.
30. Song J, et al. A mouse model for the human pathogen *Salmonella typhi*. *Cell Host Microbe*. 2010;8:369–76.
31. Brown DE, McCoy MW, Pilonieta MC, Nix RN, Detweiler CS. Chronic murine typhoid fever is a natural model of secondary hemophagocytic lymphohistiocytosis. *PLoS One*. 2010;5:e9441.
32. Tsolis RM, et al. Of mice, calves, and men. Comparison of the mouse typhoid model with other *Salmonella* infections. *Adv Exp Med Biol*. 1999;473:261–74.
33. Torraca V, Masud S, Spaink HP, Meijer AH. Macrophage–pathogen interactions in infectious diseases: new therapeutic insights from the zebrafish host model. *Dis Model Mech*. 2014;7(7):785–97.

27

Shigella

Soumik Barman and Yoshifumi Takeda

CONTENTS

27.1 Shigellosis

27.1.1 Incidence

At the end of the 19th century, young microbiologist Kiyoshi Shiga discovered a unique bacillus causing dysentery during the dysentery epidemic in Japan, and this bacterium, now known as *Shigella dysenteriae* 1, was the first acknowledged member of the genus *Shigella*.[1]

Shigella species are crucial causes of traveler's diarrhea in individuals from industrialized countries visiting developing areas.[2,3] Through accidental hygiene or sanitation breaches, endemic outbreaks of shigellosis may also occur in developed countries.[4] Very recently, the global burden of shigellosis has been reevaluated.[5] Shigellosis generally affects poor populations, especially young children, in the developing world.[5–8]

By invading the large intestinal mucosa, *Shigella* spp. induce major inflammatory destruction and generate dysenteric symptoms in humans and in nonhuman primates.[9] In a few cases, only watery diarrhea is present.[10,11] In humans, remarkably low bacterial loads are able to trigger severe shigellosis.[12]

Suitable antimicrobial therapy of shigellosis reduces the duration of fever and diarrhea. Azithromycin and fluoroquinolones are the first-line choice for treatment of shigellosis.[13] Currently, the emergence of multidrug-resistant *Shigella* variants is the greatest concern among global health problems.[14,15] During the last five decades, extensive research has yielded a large number of vaccine candidates, but no effective vaccines.[16–20] The goal of an effective vaccine against *Shigella* remains elusive.

27.1.2 Bacteriology

The *Shigella* genus includes four species: *Shigella dysenteriae*, *Shigella flexneri*, *Shigella boydii*, and *Shigella sonnei*. Based on the antigenicity of the O-specific polysaccharide chain of their lipopolysaccharide (LPS) molecules, *Shigella* species are further divided into serotypes: 15 for *S. dysenteriae*, 15 for *S. flexneri* (serotypes and subserotypes comprising the recently entitled 7a and 7b subtypes), 20 for *S. boydii*, and 1 for *S. sonnei*.[21] In most tropical countries, shigellosis is endemic and is dominated by a few *S. flexneri* serotypes, as well as *S. dysenteriae* 1.[19] The dominant isolates in Europe and the United States are *S. sonnei* and *S. flexneri*.[19]

Shigella is nonmotile and nonflagellated in nature.[21] Most shigellae are capable of fermenting glucose, with a few exceptions (*S. flexneri* 6, *S. boydii* 13 and 14, and *S. dysenteriae* 3).[22] *S. dysenteriae* is incapable of fermenting mannitol, and this property is used as a key feature for its identification.[21]

For daily laboratory screening of shigellae, two different selective media are generally used, namely Hektoen enteric (HE) agar and xylose lysine desoxycholate (XLD) agar. *Shigella* colonies on XLD agar appear translucent, pink, or red and are smooth in nature. *S. dysenteriae* 1 colonies on XLD agar are commonly very small, unlike those of other *Shigella* species.[23] Colonies of shigellae on HE agar appear green. Serological identification by slide agglutination with polyvalent somatic antigen grouping sera further confirms the serotypes and subserotypes.

27.1.3 Transmission

The most common route of *Shigella* transmission is via fecal–oral route. The bacteria may also be spread by direct contact with an infected person, as shigellae have the ability to survive on human skin for up to 1 h.[21,24] *Shigella* infection requires only a very small inoculum,[12] making person-to-person transmission favorable. *Shigella* can spread rapidly by foodborne transmission and cause large outbreaks.[25–28] Flies may play a role in spreading shigellae from human excrement to foods.[29,30] Shigellae are able to survive for at least 6 months in water at room temperature.[21]

27.1.4 Pathogenesis

The cellular pathogenesis of shigellosis is the result of the complicated actions of a large number of bacterial virulence factors. The essential machinery for bacterial penetration and intracellular endurance is usually encoded in a large "virulence plasmid."[31,32] Loss of the virulence plasmid gives rise to avirulent phenotypes. A type III secretion system (T3SS) is another key factor involved in *Shigella* invasion.[33,34]

Shigella bacilli invade the distal region of the colon and rectum,[35] where they become imprisoned by specialized M-cells. The M-cells deliver the bacterial antigens, LPS, and invasive plasmid antigen (Ipa) proteins to antigen-presenting macrophages and dendritic cells.[36] *Shigella* bacilli are phagocytized by macrophages, but subsequently escape through apoptosis.[37] Before death, the macrophages release pro-inflammatory cytokines interleukin-1b and interleukin-18.[38]

These cytokines trigger a strong inflammatory response and stimulate the migration of neutrophils,[35] which infiltrate the infected site and destabilize the epithelium.[39] The death of infected macrophages and subsequent destabilization allow penetration of more shigellae into the tissue and invasion of epithelial cells through the basal membrane. Survival and replication in macrophages represent the fundamental key to extensive colonization of the intestinal epithelium.[40]

27.2 Animal Models

27.2.1 Guinea Pigs

The oldest animal model for assaying *Shigella* invasion is the guinea pig keratoconjunctivitis model, known as the Sereny test.[41] In this invasion assay, a suspension of live shigellae is generally inoculated into the keratoconjunctival sac of guinea pigs. Invasive shigellae disrupt the conjunctival wall and cause severe conjunctivitis, which is characterized by redness of the eye and corneal opaqueness. Highly pathogenic clinical isolates usually cause acute purulent inflammation, which leads to closing of the eyelids. Noninvasive *Shigella* isolates cannot invade the ocular mucosa. This assay can determine the invasive phenotypes of *Shigella*.

The Sereny test was used to assess immunization efficacy, i.e., determination of protection against ocular infection, by various research groups.[42–48] Although the keratoconjunctivitis assay is regarded as the gold standard for measuring *Shigella* invasion, it does not allow accurate quantification of the inflammatory response.

A lethal enteric infection in the colon was established in 4-day-starved guinea pigs by the combination of calcium carbonate and opium.[49] In a series of experiments, guinea pigs were inoculated orally with live *S. flexneri* 2a ranging from 8×10^5 to 1.4×10^7 CFU, which induced a fatal enteric infection with ulcerative lesions in the colon. However, this model might not be ideal for the purpose of measuring protective efficacy, because the fatal effects appear at a relatively early stage of infection.[20]

The guinea pig colitis model usually induces typical bacillary dysentery after rectal inoculation of *Shigella* species (*S. flexneri* 2a and 5a) without any preparatory treatment like starvation or antibiotic treatment.[50] In this model, guinea pigs were inoculated with four different doses (10^8, 10^9, 10^{10}, and 10^{11} CFU) of *Shigella* through the intrarectal route. Within 24 h of challenge, all doses conferred significant signs of bacillary dysentery. This colitis model has been proven successful in many protective efficacy studies.[50–52] However, recent reports suggested that it has several limitations[53] and methodological complications.[54]

The recently developed guinea pig luminal model was found to be sensitive as it mimics human bacillary dysentery.[54] In this model, the lumen of the colon was separately infected with virulent *S. dysenteriae* 1 and *S. flexneri* 2a (10^9 CFU) after ligation of the distal cecum. The ligation was placed at 4-cm distance from the ileocecal junction. The placement of the surgical tie may facilitate the colonization of shigellae in the distal colon by controlling the flow of the stool from the cecum. Bacillary dysenteric symptoms were recorded within 24 h of infection. The luminal model can be used to demonstrate protective efficacy,[54] but it has become untenable because of a surfeit of surgical complications.

To understand the adhesion and invasion of *Shigella*, *in vitro* studies were performed on colonic explants from guinea pigs, wherein bacterial suspensions were incubated with luminary cuts of the cecal mucosa.[55–57]

27.2.2 Mice

Mice have been the most frequently used small animal species in preclinical studies of shigellosis. The major advantages of mice are the ease of handling, availability of a broad spectrum of reagents, and feasibility of genetic manipulation. Mice are also cost-effective for protective efficacy studies.

During oral inoculation studies, *Shigella* was unintentionally inoculated into the respiratory passage, leading to pulmonary *Shigella* infection.[58,59] Over the last five decades, the murine pulmonary model was used to demonstrate protective efficacy against *Shigella* infection.[59–65] Typically, after immunization through the external nares, mice were generally inculcated with a challenge strain (10^6–10^7 CFU) through the intranasal pathway. Protection was usually demonstrated by the degree of *Shigella* invasion into the lungs and the induction of pulmonary pneumonia. The major lacuna of the pulmonary model is the lack of clinical relevance with respect to the infection site of the pathogen.[20,53]

Another murine infection model was demonstrated via oral inculcation of invasive shigellae in newborn mice.[66] This murine model is extremely sensitive because the neonatal gut is only susceptible to *Shigella* infection by oral inculcation for 3–4 days after birth.[66] Usually, 4-day-old newborn mice were orally inculcated with *S. flexneri* 5a (5×10^9 CFU) for experimental shigellosis. Effective *Shigella* invasion was achieved in the intestinal mucosa of 4-day-old newborn mice, while 5-day-old mice were found to be resistant.

It has been assumed that murine models are inadequate for determination of protective efficacy.[20] However, recent studies have helped overcome this shortcoming.[67,68] The offspring of immunized dams were orally inoculated with invasive shigellae and the level of protection measured by the degree of colonization in the bowel. The murine model developed in these studies can be used as a tool for determination of passive immunization.

To study the interactions of *Shigella* with the human intestine *in vivo*, one group of researchers placed human intestinal xenografts into the subscapular region of 6–8-week-old severe combined immunodeficient (SCID) mice.[69] Inoculation of *Shigella* into these human intestinal xenografts led to the development of severe inflammation and mucosal destruction at 4 h after infection. This xenograft model is useful for understanding intracellular communications, but is somewhat limited and strenuous for the purpose of measuring protective efficacy.[20]

A recent study showed that intraperitoneal inoculation of *S. flexneri* 2a (5×10^8 CFU) into mice provoked acute shigellosis and mimicked human dysentery.[70] The inoculated mice had diarrhea within 2 h of infection. This intraperitoneal model was also shown to be useful for evaluating *Shigella* vaccine candidates.[72]

27.2.3 Rabbits

Rabbits are not naturally susceptible to shigellosis, but may acquire the infection under preparatory conditions. New Zealand white rabbits are usually preferred in the *Shigella* research field because they are commonly available and inexpensive.

For investigations of *Shigella* pathogenesis at the cellular level, the ileal-ligated loops assay has proven beneficial.[71–74] For this, ligated segments of the rabbit small intestine are generally inoculated with various serotypes of *S. dysenteriae*, *S. flexneri*, *S. boydii*, and *S. sonnei*.[71] Within 12–18 h of inoculation (10^9–10^{10} CFU), bacterial invasion and ulceration can be achieved.[71,72] Furthermore, immune responses can be measured in a quantitative manner after infection of the ligated loops.[74] Surgical complications are one of the drawbacks of the ligated loops assay, and it is not useful for protective efficacy studies.

Several studies indicated that acute necrosis occurred in the rabbit ileum after oral inoculation of *Shigella*, under preparatory treatments like starvation, antibiotic treatment, and stomach acid neutralization.[75,76] However, there has been no prominent follow-up of such orogastric "conditioned" rabbit models to date.

Adult rabbits exposed to direct colonic inoculation by cecal bypass also progress to shigellosis.[77] Cecal bypass was achieved by blockade of the cecal lumen above the ileocecal junction with a silk thread ligature, while the ileo-ceco-colic communication was properly maintained. After cecal bypass, bacteria (10^7 CFU) were inoculated into the proximal colon. Onset of dysentery occurred within 24 h of inoculation. The colonic infection model was shown to be worthwhile in various *Shigella* vaccine candidate screenings.[62,78,79] A very recent study clearly demonstrated the usefulness of this model in the assessment of anti-shigella vaccine candidates.[80] Despite encouraging results, the surgical obstacle appears to be the major lacuna of this model.

27.2.4 Pigs and Piglets

Domestic pigs are not susceptible to shigellosis.[81] However, 4-week-old antibiotic-treated pigs were subjected to oral inoculation with invasive *S. dysenteriae* 1 and *S. flexneri* 2a (1×10^{11} CFU). Most of the pigs had no dysenteric symptoms throughout the duration of infection, although signs of bacteria were detected in rectal cultures within 24 h of infection. Histopathological analysis detected tonsillar lesions, which are not clinically relevant for assaying shigellosis.

Jeong et al.[82] developed a unique model for assaying shigellosis using gnotobiotic piglets. Piglets were delivered by cesarean section and kept aseptically in a plastic sterile isolator to maintain their gut sterility. An oral inoculum of *S. dysenteriae* 1 (5×10^9 CFU) induced acute shigellosis within 24 h of infection. The sterile gut of these piglets at 1–3 days of age allowed rapid colonization by *Shigella* organisms. Despite promising results, the piglet model has some limitations. The model became less efficient as maintaining piglets in sterile conditions was not cost-effective, and the model was also incapable of quantifying protective efficacy.

27.2.5 Chicken

Very recently, a research group in China reported invasion of clinically isolated *S. flexneri* 2a into specific pathogen-free (SPF) chicken.[83] SPF chickens at 3 days of age were inoculated by intraperitoneal injection or crop injection. Inoculation of 3×10^9 CFU resulted in death in 100% of the chickens, while crop injection did not cause any intestinal clinical signs of dysentery. The ligated intestinal loop of 1-day-old SPF chicken also showed severe congestion and edema. This study has clinical significance regarding human–poultry cross-infection, although it has a shortcoming regarding vaccine candidate screening.

27.2.6 *Caenorhabditis elegans*

In recent years, the free-living nematode *Caenorhabditis elegans* has been acknowledged as a valuable *in vivo* model for studying host–pathogen interactions. Two independent research groups suggested that *C. elegans* could be used as an *in vivo* model for evaluating the pathogenesis of shigellae.[84,85] These studies established that *S. flexneri* 2a and *S. flexneri* 2b killed *C. elegans* on solid media and in liquid culture, although the exact mechanism remains unknown.

An *S. boydii*-mediated infection assay in *C. elegans* followed by bioinformatics analysis revealed various immune regulators that are important for *Shigella* species-mediated immune responses.[86]

A very recent study by George et al.[87] demonstrated that invasive *S. flexneri* 3a accumulated in the nematode intestinal lumen, produced outer membrane vesicles, and invaded nematode intestinal cells. Further studies revealed that *S. flexneri* 3a interrupted iron homeostasis in nematodes and potentially stimulated a hypoxic response, which could cause death.

27.2.7 Monkeys

In 1965, Formal et al.[88] successfully demonstrated immunogenicity and protective efficacy of attenuated *Shigella* strains against *S. flexneri* 1b, 2a, and 6 infections in rhesus monkeys (*Macaca mulatta*). Both immunization and challenge (5×10^{10} CFU) were performed by the oral route. Later studies by the same group revealed that typical bacillary dysentery could be induced in rhesus monkeys by oral infection with *S. flexneri* without starvation and/or pretreatment with antibiotics.[89,90] Subsequent protective efficacy studies for *Shigella* vaccine candidates also became effective in rhesus monkeys.[91–95] The rhesus monkey model mimics human shigellosis and the human immune response, making it ideal for assembling preclinical data before human trials of vaccine candidates.

Cynomolgus monkeys (*Macaca fascicularis*)[96,97] and *Aotus nancymaae* monkeys[98] were also established as tools for screening *Shigella* vaccine candidates. Despite auspicious results, the cost of using these animals can be prohibitive, especially in developing countries where shigellosis is endemic.

27.2.8 Zebrafish

The zebrafish model opens a new avenue to investigate the significance of innate immunity. It provides a framework where no adaptive responses have yet been developed, because early lymphocytes make their first appearance in 4-day-old larvae and full adaptive immunity requires several weeks to be displayed.[99]

An *in vivo* study of *S. flexneri* interactions with phagocytes and bacterial autophagy was reported very recently.[100] An inoculum of 4×10^3 *S. flexneri* 5a resulted in the death of most zebrafish larvae within 48 h postinfection. Intravenously administered bacteria could survive and replicate in both macrophages and nonimmune cells. The zebrafish larva represents a valuable new host for analysis of *Shigella* infection, although this model cannot define the protective efficacy.

27.2.9 Human Cell Lines

Human cell lines are broadly used as a naive model for understanding the interactions of *Shigella* with its host. The advantages of cell lines are their ease of handling and relatively low costs.

The primary step of shigellosis is invasion of the human colonic mucosa. Host–pathogen interactions can easily be examined *in vitro* using cell lines before animal trials are carried out. Several investigations were performed to disclose host–*Shigella* interactions at the cellular level using HeLa,[101–105] T84,[106–108] THP-1,[109,110] macrophage J774,[111–113] HT-29,[114,115] monoblastic U937,[113,116,117] human embryonic kidney (HEK) 293,[118,119] human colonic Caco-2 epithelial,[120–122] and human CD4$^+$ T (Jurkat)[123] cells. The results identified the relative contributions of antigens, which need to be taken into account for vaccine development.

27.3 Conclusion

The use of various animal models has already improved our understanding of the *in vivo* consequences of intracellular defense mechanisms as well as protective measures against *Shigella* spp. We expect that further application of these powerful models will continue to unravel the intricacy of the virulence strategies employed by shigellae to mitigate host's innate and acquired immune defenses.

REFERENCES

1. Shiga, K. The trend of prevention, therapy and epidemiology of dysentery since the discovery of its causative organism. *N Eng J Med.* 215, 1205–1211 (1936).
2. Niyogi, S.K. Shigellosis. *J Microbiol.* 43, 133–143 (2005).
3. Marchou, B. Traveller's diarrhea: epidemiology, clinical practice guideline for the prevention and treatment. *Presse Med.* 42, 76–81 (2013).
4. Bancroft, J.E., Keifer, S.B. & Keene, W.E. Shigellosis from an interactive fountain: implications for regulation. *J Environ Health.* 73, 16–20 (2010).
5. GBD 2013 Mortality and Causes of Death Collaborators. Global, regional, and national age–sex specific all-cause and cause-specific mortality for 240 causes of death, 1990–2013: a systematic analysis for the Global Burden of Disease Study 2013. *Lancet.* 385, 117–171 (2015).
6. Osorio, M., Bray, M.D. & Walker, R.I. Vaccine potential for inactivated shigellae. *Vaccine.* 25, 1581–1592 (2007).
7. O'Ryan, G.M., Ashkenazi-Hoffnung, L., O'Ryan-Soriano, M.A. & Ashkenazi, S. Management of acute infectious diarrhea for children living in resource-limited settings. *Expert Rev Anti Infect Ther.* 12, 621–632 (2014).
8. Lamberti, L.M., Bourgeois, A.L., Fischer Walker, C.L., Black, R.E. & Sack, D. Estimating diarrheal illness and deaths attributable to *Shigellae* and enterotoxigenic *Escherichia coli* among older children, adolescents, and adults in South Asia and Africa. *PLoS Negl Trop Dis.* 8, e2705 (2014).
9. Sansonetti, P.J. War and peace at mucosal surfaces. *Nat Rev Immunol.* 4, 953–964 (2004).
10. Tobias, J. et al. Involvement of main diarrheagenic *Escherichia coli*, with emphasis on enteroaggregative *E. coli*, in severe non-epidemic pediatric diarrhea in a high-income country. *BMC Infect Dis.* 15, 79 (2015).
11. Zaidi, M.B. & Estrada-García, T. *Shigella*: a highly virulent and elusive pathogen. *Curr Trop Med Rep.* 1, 81–87 (2014).
12. Ferreccio, C. et al. Epidemiologic patterns of acute diarrhea and endemic *Shigella* infections in children in a poor periurban setting in Santiago, Chile. *Am J Epidemiol.* 134, 614–627 (1991).
13. Legros, D. ed. *Guidelines for the Control of Shigellosis, Including Epidemics due to Shigella dysenteriae Type 1.* World Health Organization, Geneva (2005).
14. Omulo, S., Thumbi, S.M., Njenga, M.K. & Call, D.R. A review of 40 years of enteric antimicrobial resistance research in Eastern Africa: what can be done better? *Antimicrob Resist Infect Control.* 4, 1 (2015).
15. Magiorakos, A.P. et al. Multidrug-resistant, extensively drug-resistant and pandrug-resistant bacteria: an international expert proposal for interim standard definitions for acquired resistance. *Clin Microbiol Infect.* 18, 268–281 (2012).

16. Kraehenbuhl, J.P. & Neutra, M.R. Mucosal vaccines: where do we stand? *Curr Top Med Chem.* 13, 2609–2628 (2013).
17. Böhles, N. et al. Vaccines against human diarrheal pathogens: current status and perspectives. *Hum Vaccines Immunother.* 10, 1522–1535 (2014).
18. Barry, E.M. et al. Progress and pitfalls in *Shigella* vaccine research. *Nat Rev Gastroenterol Hepatol.* 10, 245–255 (2013).
19. Levine, M.M., Kotloff, K.L., Barry, E.M., Pasetti, M.F. & Sztein, M.B. Clinical trials of *Shigella* vaccines: two steps forward and one step back on a long, hard road. *Nat Rev Microbiol.* 5, 540–553 (2007).
20. Kweon, M.N. (2008) Shigellosis: the current status of vaccine development. *Curr Opin Infect Dis.* 21, 313–318 (2008).
21. Germani, Y. & Sansonetti, P.J. The genus *Shigella*. In: Dworkin, M. et al., eds. *The Prokaryotes*, Vol. 6, 3rd ed., pp. 99–122. Springer, New York (2006).
22. Coimbra, R.S. et al. Molecular and phenotypic characterization of potentially new *Shigella dysenteriae* serotype. *J Clin Microbiol.* 39, 618–621 (2001).
23. McIntyre, L. ed. *Laboratory Methods for the Diagnosis of Epidemic Dysentery and Cholera.* WHO/CDS/CSR/EDC/99.8, pp. 17–36. Centers for Disease Control and Prevention, Atlanta, GA (1999).
24. Islam, M.S. et al. Detection of non-cultivable *Shigella dysenteriae* type 1 from artificially contaminated volunteer's fingers using fluorescent antibody and PCR techniques. *J Diarrheal Dis Res.* 15, 65–70 (1997).
25. Boyce, J.M. et al. Patterns of *Shigella* infection in families in rural Bangladesh. *Am J Trop Med Hyg.* 31, 1015–1020 (1982).
26. Rennels, M.B. & Levine, M.M. Classical bacterial diarrhea: perspective and update: *Salmonella*, *Shigella*, *Escherichia coli*, *Aeromonas* and *Flesiomonas*. *Pediatr Infect Dis J.* 5, 91–100 (1986).
27. Scallan, E. et al. Foodborne illness acquired in the United States—major pathogens. *Emerg Infect Dis.* 17, 7–15 (2011).
28. Vally, H. et al. Proportion of illness acquired by foodborne transmission for nine enteric pathogens in Australia: an expert elicitation. *Foodborne Pathog Dis.* 11, 727–733 (2014).
29. Hardy, A. & Watts, J. Studies of the acute diarrheal diseases. XVIII: epidemiology. *Public Health Rep.* 63, 363 (1948).
30. Levine, O.S. & Levine, M.M. Houseflies (*Musca domestica*) as mechanical vectors of shigellosis. *Rev Infect Dis.* 13, 688–696 (1991).
31. Sansonetti, P.J., Kopecko, D.J. & Formal, S.B. Involvement of a plasmid in the invasive ability of *Shigella flexneri*. *Infect Immun.* 35, 852–860 (1982).
32. Sasakawa, C., Makino, S., Kamato, K. & Yoshikawa, M. Isolation, characterization, and mapping of Tn5 insertions into the 140 megadalton invasion plasmid defective in the mouse Sereny test in *Shigella flexneri* 2a. *Infect Immun.* 54, 32–36 (1986).
33. Blocker, A. et al. The tripartite type III secreton of *Shigella flexneri* inserts IpaB and IpaC into host membranes. *J Cell Biol.* 147, 683–693 (1999).
34. Blocker, A. et al. Structure and composition of the *Shigella flexneri* "needle complex," a part of its type III secreton. *Mol Microbiol.* 39, 652–663 (2001).
35. Anand, B.S. et al. Rectal histology in acute bacillary dysentery. *Gastroenterology.* 90, 654–660 (1986).
36. Phalipon, A. & Sansonetti, P.J. Shigellosis: innate mechanisms of inflammatory destruction of the intestinal epithelium, adaptive immune response, and vaccine development. *Crit Rev Immunol.* 23, 371–401 (2003).
37. Phalipon, A. & Sansonetti, P.J. *Shigella*'s ways of manipulating the host intestinal innate and adaptive immune system: a tool box for survival? *Immunol Cell Biol.* 85, 119–129 (2007).
38. Chen, Y., Smith, M.R., Thirumalai, K. & Zychlinsky, A. A bacterial invasion induces macrophage apoptosis by binding directly to ICE. *EMBO J.* 15, 3853–3860 (1996).
39. Perdomo, J.J., Gounon, P. & Sansonetti, P.J. Polymorphonuclear leukocyte transmigration promotes invasion of colonic epithelial monolayer by *Shigella flexneri*. *J Clin Invest.* 93, 633–643 (1994).
40. Ashida H. et al. *Shigella* deploy multiple countermeasures against host innate immune responses. *Curr Opin Microbiol.* 14, 16–23 (2011).
41. Sereny, B. Experimental *Shigella* keratoconjunctivitis: a preliminary report. *Acta Microbiol Acad Sci Hung.* 2, 293–296 (1955).

42. Kerekes, L. Immunity of the guinea pig's eye after vaccination with *Shigella sonnei* phase I antigen. *Acta Microbiol Acad Sci Hung.* 23, 353–358 (1976).
43. Linde, K. et al. Live *Shigella flexneri* 2a and *Shigella sonnei* I vaccine candidate strains with two attenuating markers. I. Construction of vaccine candidate strains with retained invasiveness but reduced intracellular multiplication. *Vaccine.* 8, 25–29 (1990).
44. Verma, N.K. & Lindberg, A.A. Construction of aromatic dependent *Shigella flexneri* 2a live vaccine candidate strains: deletion mutations in the *aroA* and the *aroD* genes. *Vaccine.* 9, 6–9 (1991).
45. Hartman, A.B., Powell, C.J., Schultz, C.L., Oaks, E.V. & Eckels, K.H. Small-animal model to measure efficacy and immunogenicity of *Shigella* vaccine strains. *Infect Immun.* 59, 4075–4083 (1991).
46. Hartman, A.B. et al. Local immune response and protection in the guinea pig keratoconjunctivitis model following immunization with *Shigella* vaccines. *Infect Immun.* 62, 412–420 (1994).
47. Mukhopadhaya, A., Mahalanabis, D., Khanam, J. & Chakrabarti, M.K. Protective efficacy of oral immunization with heat-killed *Shigella flexneri* 2a in animal model: study of cross protection, immune response and antigenic recognition. *Vaccine.* 21, 3043–3050 (2003).
48. Li, W. et al. Development of prophylactic recombinant HPV58-attenuated *Shigella* live vector vaccine and evaluation of its protective efficacy and immunogenicity in the guinea pig keratoconjunctivitis model. *Acta Biochim Biophys Sin (Shanghai).* 41, 137–145 (2009).
49. Formal, S.B., Dammin, G.J., Labrec, E.H. & Schneider, H. Experimental *Shigella* infections: characteristics of a fatal infection produced in guinea pigs. *J Bacteriol.* 75, 604–610 (1958).
50. Shim, D.H. et al. New animal model of shigellosis in the guinea pig: its usefulness for protective efficacy studies. *J Immunol.* 178, 2476– 2482 (2007).
51. Barman, S. et al. Live non-invasive *Shigella dysenteriae* 1 strain induces homologous protective immunity in a guinea pig colitis model. *Microbiol Immunol.* 55, 683–693 (2011).
52. Barman, S. et al. Protective immunity by oral immunization with heat-killed *Shigella* strains in a guinea pig colitis model. *Microbiol Immunol.* 57, 762–771 (2013).
53. Kim, Y.J., Yeo, S.G., Park, J.H. & Ko, H.J. *Shigella* vaccine development: prospective animal models and current status. *Curr Pharm Biotechnol.* 14, 903–912 (2013).
54. Barman, S., Saha, D.R., Ramamurthy, T. & Koley, H. Development of a new guinea pig model of shigellosis. *FEMS Immunol Med Microbiol.* 62, 304–314 (2011).
55. Izhar, M., Nuchamowitz, Y. & Mirelman, D. Adherence of *Shigella flexneri* to guinea pig intestinal cells is mediated by a mucosal adhesion. *Infect Immun.* 35, 1110–1118 (1982).
56. Raygoza-Anaya, M., González-Robles, A. & Mora-Galindo, J. In vitro model for the analysis of the interaction between *Shigella flexneri* and the intestinal epithelium. *Arch Invest Med (Mex).* 21, 305–309 (1990).
57. Guhathakurta, B., Sasmal, D. & Datta, A. Adhesion of *Shigella dysenteriae* type 1 and *Shigella flexneri* to guinea-pig colonic epithelial cells in vitro. *J Med Microbiol.* 36, 403–405 (1992).
58. Voino-Yasenetsky, M.V. & Voino-Yasenetskaya, M.K. Experimental pneumonia caused by bacteria of the *Shigella* group. *Acta Morphol Acad Sci Hung.* 11, 439–454 (1962).
59. Mallett, C.P., Van de Verg, L., Collins, H.H. & Hale, T.L. Evaluation of *Shigella* vaccine safety and efficacy in an intranasally challenged mouse model. *Vaccine.* 11, 190–196 (1993).
60. Mallett, C.P. et al. Intransal or intragastric immunization with proteosome-*Shigella* lipopolysaccharide vaccines protects against lethal pneumonia in a murine model of *Shigella* infection. *Infect Immun.* 63, 2382–2386 (1995).
61. Van de Verg, LL et al. Antibody and cytokine responses in a mouse pulmonary model of *Shigella flexneri* serotype 2a infection. *Infect Immun.* 63, 1947–1954 (1995).
62. Koley, H., Barman, S., Roy, N., Saha, D.R. & Kumar, H. The efficacy and immunogenicity of a live transconjugant hybrid strain of *Shigella dysenteriae* type 1 in two animal models. *World J Microbiol Biotechnol.* 25, 679–686 (2009).
63. Shim, D.H. et al. Immunogenicity and protective efficacy offered by a ribosomal-based vaccine from *Shigella flexneri* 2a. *Vaccine.* 25, 4828–4836 (2007).
64. Camacho, A.I. et al. Mucosal immunization with *Shigella flexneri* outer membrane vesicles induced protection in mice. *Vaccine.* 29, 8222–8229 (2011).
65. Pore, D., Mahata, N., Pal, A. & Chakrabarti, M.K. Outer membrane protein A (OmpA) of *Shigella flexneri* 2a, induces protective immune response in a mouse model. *PLoS One.* 6, e22663 (2011).
66. Fernandez, M.I. et al. A newborn mouse model for the study of intestinal pathogenesis of shigellosis. *Cell Microbiol.* 5, 481–491 (2003).

67. Mitra, S., Chakrabarti, M.K. & Koley, H. Multi-serotype outer membrane vesicles of *Shigellae* confer passive protection to the neonatal mice against shigellosis. *Vaccine*. 31, 3163–3173 (2013).
68. Barman, S. et al. Passive immunity with multi-serotype heat-killed *Shigellae* in neonatal mice. *Microbiol Immunol*. 58, 463–466 (2014).
69. Zhang, Z., Jin, L., Champion, G., Seydel, K.B. & Stanley, S.L. Jr. *Shigella* infection in a SCID mouse-human intestinal xenograft model: role for neutrophils in containing bacterial dissemination in human intestine. *Infect Immun*. 69, 3240–3247 (2001)
70. Yang, J.Y. et al. A mouse model of shigellosis by intraperitoneal infection. *J Infect Dis*. 209, 203–215 (2014).
71. Arm, H.G., Floyd, T.M., Faber, J.E. & Hayes, J.R. Use of ligated segments of rabbit small intestine in experimental shigellosis. *J Bacteriol*. 89, 803–809 (1965).
72. Wassef, J.S., Keren, D.F. & Mailloux, J.L. Role of M cells in initial antigen uptake and in ulcer formation in the rabbit intestinal loop model of shigellosis. *Infect Immun*. 57, 858–863 (1989).
73. Sansonetti, P.J., Arondel, J., Cantey, J.R., Prévost, M.C. & Huerre, M. Infection of rabbit Peyer's patches by *Shigella flexneri*: effect of adhesive or invasive bacterial phenotypes on follicle-associated epithelium. *Infect Immun*. 64, 2752–2764 (1996).
74. Schnupf, P. & Sansonetti, P.J. Quantitative RT-PCR profiling of the rabbit immune response: assessment of acute *Shigella flexneri* infection. *PLoS One*. 7, e36446 (2012).
75. Ahmed, Z.U., Sarker, M.R. & Sack, D.A. Protection of adult rabbits and monkeys from lethal shigellosis by oral immunization with a thymine-requiring and temperature-sensitive mutant of *Shigella flexneri* Y. *Vaccine*. 8, 153–158 (1990).
76. Etheridge, M.E., Hoque, A.T. & Sack, D.A. Pathologic study of a rabbit model for shigellosis. *Lab Anim Sci*. 46, 61–66 (1996).
77. Rabbani, G.H. et al. Development of an improved animal model of shigellosis in the adult rabbit by colonic infection with *Shigella flexneri* 2a. *Infect Immun*. 63, 4350–4357 (1995).
78. Chakrabarti, M.K. et al. Killed oral *Shigella* vaccine made from *Shigella flexneri* 2a protects against challenge in the rabbit model of shigellosis. *Acta Paediatr*. 88, 161–165 (1999).
79. Mukhopadhaya, A., Mahalanabis, D., Khanam, J. & Chakrabarti, M.K. Protective efficacy of oral immunization with heat-killed *Shigella flexneri* 2a in animal model: study of cross protection, immune response and antigenic recognition. *Vaccine*. 21, 3043–3050 (2003).
80. Nag, D. et al. Heat killed multi-serotype *Shigella* immunogens induced humoral immunity and protection against heterologous challenge in rabbit model. *Immunobiology*. 220(11), 1275–1283. doi:10.1016/j.imbio.2015.07.002 (2015).
81. Maurelli, A.T. et al. *Shigella* infection as observed in the experimentally inoculated domestic pig, *Sus scrofa domestica*. *Microb Pathog*. 25, 189–196 (1998).
82. Jeong, K.I., Zhang, Q., Nunnari, J. & Tzipori, S. A piglet model of acute gastroenteritis induced by *Shigella dysenteriae* Type 1. *J Infect Dis*. 201, 903–911 (2010).
83. Shi, R. et al. Pathogenicity of *Shigella* in chickens. *PLoS One*. 9, e100264 (2014).
84. Burton, E.A., Pendergast, A.M. & Aballay, A. The *Caenorhabditis elegans* ABL-1 tyrosine kinase is required for *Shigella flexneri* pathogenesis. *Appl Environ Microbiol*. 72, 5043–5051 (2006).
85. Kesika, P., Karutha Pandian, S. & Balamurugan, K. Analysis of *Shigella flexneri*-mediated infections in model organism *Caenorhabditis elegans*. *Scand J Infect Dis*. 43, 286–295 (2011).
86. Kesika, P. & Balamurugan, K. Studies on *Shigella boydii* infection in *Caenorhabditis elegans* and bioinformatics analysis of immune regulatory protein interactions. *Biochim Biophys Acta*. 1824, 1449–1456 (2012).
87. George, D.T. et al. *Shigella flexneri* infection in *Caenorhabditis elegans*: cytopathological examination and identification of host responses. *PLoS One*. 9, e106085 (2014).
88. Formal, S.B., Labrec, E.H., Palmer, A. & Falkow, S. Protection of monkeys against experimental shigellosis with attenuated vaccines. *J Bacteriol*. 90, 63–68 (1965).
89. Formal, S.B., Kent, T.H., Austin, S. & Labrec, E.H. Fluorescent-antibody and histological study of vaccinated and control monkeys challenged with *Shigella flexneri*. *J Bacteriol*. 91, 2368–2376 (1966).
90. Takeuchi, A., Formal, S.B. & Sprinz, H. Experimental acute colitis in the rhesus monkey following peroral infection with *Shigella flexneri*. An electron microscope study. *Am J Pathol*. 52, 503–529 (1968).
91. Rout, W.R., Formal, S.B., Giannella, R.A. & Dammin, G.J. Pathophysiology of *Shigella* diarrhea in the rhesus monkey: intestinal transport, morphological, and bacteriological studies. *Gastroenterology*. 68, 270–278 (1975).

92. Formal, S.B. et al. Oral vaccination of monkeys with an invasive *Escherichia coli* K-12 hybrid expressing *Shigella flexneri* 2a somatic antigen. *Infect Immun.* 46, 465–469 (1984).
93. Kotloff, K.L. et al. Safety, immunogenicity, and efficacy in monkeys and humans of invasive *Escherichia coli* K-12 hybrid vaccine candidates expressing *Shigella flexneri* 2a somatic antigen. *Infect Immun.* 60, 2218–2224 (1992).
94. Collins, T.A. et al. Safety and colonization of two novel *virG(icsA)*-based live *Shigella sonnei* vaccine strains in rhesus macaques (*Macaca mulatta*). *Comp Med.* 58, 88–94 (2008).
95. Islam, D. et al. Evaluation of an intragastric challenge model for *Shigella dysenteriae* 1 in rhesus monkeys (*Macaca mulatta*) for the pre-clinical assessment of *Shigella* vaccine formulations. *APMIS.* 122, 463–475 (2014).
96. Takasaka, M. et al. Isolation and pathogenicity of provisional serovar 1621–54 of "*Shigella*" from imported cynomolgus monkeys. *Jpn J Med Sci Biol.* 36, 27–37 (1983).
97. Shipley, S.T. et al. A challenge model for *Shigella dysenteriae* 1 in cynomolgus monkeys (*Macaca fascicularis*). *Comp Med.* 60, 54–61 (2010).
98. Gregory, M. et al. Development of an *Aotus nancymaae* model for *Shigella* vaccine immunogenicity and efficacy studies. *Infect Immun.* 82, 2027–2036 (2014).
99. Torraca, V., Masud, S., Spaink, H.P. & Meijer, A.H. Macrophage–pathogen interactions in infectious diseases: new therapeutic insights from the zebrafish host model. *Dis Model Mech.* 7, 785–797 (2014).
100. Mostowy, S. et al. The zebrafish as a new model for the in vivo study of *Shigella flexneri* interaction with phagocytes and bacterial autophagy. *PLoS Pathog.* 9, e1003588 (2013).
101. Sansonetti, P.J., Ryter, A., Clerc, P., Maurelli, A.T. & Mounier, J. Multiplication of *Shigella flexneri* within HeLa cells: lysis of the phagocytic vacuole and plasmid-mediated contact hemolysis. *Infect Immun.* 51, 461–469 (1986).
102. Clerc, P., Baudry, B. & Sansonetti, P.J. Plasmid-mediated contact haemolytic activity in *Shigella* species: correlation with penetration into HeLa cells. *Ann Inst Pasteur Microbiol.* 137A, 267–278 (1986).
103. Clerc, P. & Sansonetti, P.J. Entry of *Shigella flexneri* into HeLa cells: evidence for directed phagocytosis involving actin polymerization and myosin accumulation. *Infect Immun.* 55, 2681–2688 (1987).
104. Faherty, C.S. et al. Microarray analysis of *Shigella flexneri*-infected epithelial cells identifies host factors important for apoptosis inhibition. *BMC Genomics.* 11, 272 (2010).
105. Lum, M. & Morona, R. Myosin IIA is essential for *Shigella flexneri* cell-to-cell spread. *Pathog Dis.* 72, 174–187 (2014).
106. Zurawski, D.V., Mitsuhata, C., Mumy, K.L., McCormick, B.A. & Maurelli, A.T. OspF and OspC1 are *Shigella flexneri* type III secretion system effectors that are required for postinvasion aspects of virulence. *Infect Immun.* 74, 5964–5976 (2006).
107. Zurawski, D.V., Mumy, K.L., Faherty, C.S., McCormick, B.A. & Maurelli, A.T. *Shigella flexneri* type III secretion system effectors OspB and OspF target the nucleus to downregulate the host inflammatory response via interactions with retinoblastoma protein. *Mol Microbiol.* 71, 350–368 (2009).
108. Nandakumar, N.S., Pugazhendhi, S. & Ramakrishna, B.S. Effects of enteropathogenic bacteria & lactobacilli on chemokine secretion & Toll like receptor gene expression in two human colonic epithelial cell lines. *Indian J Med Res.* 130, 170–178 (2009).
109. Harrison, L.M., van Haaften, W.C. & Tesh, V.L. Regulation of proinflammatory cytokine expression by Shiga toxin 1 and/or lipopolysaccharides in the human monocytic cell line THP-1. *Infect Immun.* 72, 2618–2627 (2004).
110. Harrison, L.M., van den Hoogen, C., van Haaften, W.C. & Tesh, V.L. Chemokine expression in the monocytic cell line THP-1 in response to purified Shiga toxin 1 and/or lipopolysaccharides. *Infect Immun.* 73, 403–412 (2005).
111. Clerc, P.L., Ryter, A., Mounier, J. & Sansonetti, P.J. Plasmid-mediated early killing of eukaryotic cells by *Shigella flexneri* as studied by infection of J774 macrophages. *Infect Immun.* 55, 521–527 (1987).
112. Fernandez-Prada, C.M. et al. *Shigella flexneri* IpaH(7.8) facilitates escape of virulent bacteria from the endocytic vacuoles of mouse and human macrophages. *Infect Immun.* 68, 3608–3619.
113. Nonaka, T., Kuwabara, T., Mimuro, H., Kuwae, A. & Imajoh-Ohmi, S. *Shigella*-induced necrosis and apoptosis of U937 cells and J774 macrophages. *Microbiology.* 149, 2513–2527 (2003).
114. Dragoi, A.M. & Agaisse, H. The serine/threonine kinase STK11 promotes *Shigella flexneri* dissemination through establishment of cell–cell contacts competent for tyrosine kinase signaling. *Infect Immun.* 82, 4447–4457 (2014).

115. Kuehl, C.J., Dragoi, A.M. & Agaisse, H. The *Shigella flexneri* type 3 secretion system is required for tyrosine kinase-dependent protrusion resolution, and vacuole escape during bacterial dissemination. *PLoS One*. 9, e112738 (2014).
116. Nonaka, T., Kuwae, A., Sasakawa, C. & Imajoh-Ohmi, S. *Shigella flexneri* YSH6000 induces two types of cell death, apoptosis and oncosis, in the differentiated human monoblastic cell line U937. *FEMS Microbiol Lett*. 174, 89–95 (1999).
117. Yu, J., Oragui, E.E., Stephens, A., Kroll, J.S. & Venkatesan, M.M. Inactivation of DsbA alters the behaviour of *Shigella flexneri* towards murine and human-derived macrophage-like cells. *FEMS Microbiol Lett*. 204, 81–88 (2001).
118. Paciello, I. et al. Intracellular *Shigella* remodels its LPS to dampen the innate immune recognition and evade inflammasome activation. *Proc Natl Acad Sci USA*. 110, E4345–E4354 (2013).
119. Rossi, O. et al. Modulation of endotoxicity of *Shigella* generalized modules for membrane antigens (GMMA) by genetic lipid A modifications: relative activation of TLR4 and TLR2 pathways in different mutants. *J Biol Chem*. 289, 24922–24935 (2014).
120. Mounier, J., Vasselon, T., Hellio, R., Lesourd, M. & Sansonetti, P.J. *Shigella flexneri* enters human colonic Caco-2 epithelial cells through the basolateral pole. *Infect Immun*. 60, 237–248 (1992).
121. Pédron, T., Thibault, C. & Sansonetti, P.J. The invasive phenotype of *Shigella flexneri* directs a distinct gene expression pattern in the human intestinal epithelial cell line Caco-2. *J Biol Chem*. 278, 33878–33886 (2003).
122. Lima, C.B., Santos, S.A. & Andrade Júnior, D.R. Hypoxic stress, hepatocytes and Caco-2 viability and susceptibility to *Shigella flexneri* invasion. *Rev Inst Med Trop Sao Paulo*. 55, 341–346 (2013).
123. Konradt, C. et al. The *Shigella flexneri* type three secretion system effector IpgD inhibits T cell migration by manipulating host phosphoinositide metabolism. *Cell Host Microbe*. 9, 263–272 (2011).

28

Vibrio: Caenorhabditis elegans *as a Laboratory Model for* Vibrio *Infections*

Sellegounder Durai and Krishnaswamy Balamurugan

CONTENTS

Vibriosis is a common illness among seafood consumers that is caused by bacteria of the family Vibrionaceae. Among many species within the Vibrionaceae family, *Vibrio cholerae*, *Vibrio parahaemolyticus*, *Vibrio vulnificus*, and *Vibrio alginolyticus* are important foodborne pathogens in humans. Transmitted through uncooked or partially cooked seafood, vibriosis produces a range of symptoms, including cholera, dysentery, diarrhea, gastroenteritis, sepsis, and fasciitis. The virulence, mode of infection, symptoms, and treatment are different for each of these species. Showing variations in virulent factors as well as differences between the wild-type and clinical isolates, the pathogenicity of *Vibrio* species is complex. The nematode *Caenorhabditis elegans* has been the preferred model system for many investigators to study pathogenesis of this disease. The current chapter summarizes the available models for studying *Vibrio* spp. infections, with an emphasis on the utility of *C. elegans* model.

28.1 *V. cholerae*

Vibrio cholerae is facultative anaerobic, Gram-negative, non-spore-forming, curved rod-shaped bacterium measuring 1.4–2.6 μm with the ability for respiratory and fermentative metabolism, which can be defined on the basis of biochemical test and DNA sequence homology analysis. The bacterium is oxidase-positive, reduces nitrate, and moves using a single sheathed polar flagellum. The growth of the bacterium is regulated by altering sodium chloride (NaCl) concentration.[1] *V. cholerae* moves by means of its single polar flagellum, which is driven by the sodium-motive force.[2] The ability of the bacterium to grow in nutrient broth without NaCl makes it unique among *Vibrio* spp. *V. cholera* is a causative agent for the deadly disease cholera, which is considered as a life-threatening intestinal infection. Millions of people are affected with cholera every year, and the disease accounts for thousands of causalities worldwide, most commonly in developing countries, and the number of cholera cases increases day by day.[3,4] Infections spread through contaminated water and seafood. In the lumen of the small intestine, the bacterium begins to multiply persistently, leading to colonization. Recent studies reported the ability of this pathogen to successfully metabolize glucose, thereby facilitating its proliferation in a glucose-rich environment like the human intestine.[5]

V. cholerae initiates infection by adhering to the epithelium while swimming, which is the first step of colonization. Quorum sensing plays an important role in the colonization of the host system by *V. cholerae*.[6] Evolution, growth, and virulence of *V. cholerae* is rigidly controlled by the location of ribosomal protein gene locus and its distance from the origin of replication.[7] During antibiotic treatment, the pathogen modulates its shape from a rod to a sphere, thereby protecting itself from the antibiotics.[8] The availability of model organisms and information about the bacterial genetics has enabled the understanding of the few aspects of *V. cholerae* pathogenicity in humans; however, complete information about the natural habitat of the bacterium and its interaction with host system remains unknown.

28.2 Models to Study *V. cholerae*

Several model organisms have been used to identify the various virulent factors of *V. cholerae* and understand the host immune response against *V. cholerae* infection and effect of secreted cholera toxin. Studied models include rabbit,[9] infant mice,[10,11] adult mouse,[11] *Danio rerio*[12] (Zebrafish), and *Caenorhabditis elegans*,[1] each having its own advantages and disadvantages. The bacterium was administered by parenteral injection as well as oral administration. None of the animal models was found to be successful and reproducible in mimicking the clinical status of cholera infection until the development of the rabbit model.[9,13,14] The model was first developed in India by ligating the intestinal loop of adult and infant rabbit, thus providing insights into the colonization and pathological effect of the cholera toxin.[13] Although this method was extensively used, it is different from the original infection status. Infant animals are easily prone to infection and do not develop immunity; in other words, a model to better study the enteric infection should have a patent intestine and its immune system would be functioning and noncompromised. Two models were developed to overcome the above hurdle using dogs and chinchillas. But, both these models were not extensively used due to high cost in dogs[15] and scarce availability of chinchillas.[16] The other models in which lethal watery diarrhea can be reproduced include the removable intestinal tie adult rabbit diarrhea (RITARD) in dogs and rabbits. These two systems were extensively used for several years and can provide information on the pathogenesis as well as the host immunity status against infection. The direct inoculation of cholera bacteria into the small intestine of rabbits was done in the mid-1950s by Dutta and Habbu to provide reproducible results. This model was very rarely used due to the reluctance of investigators to carry out surgery. In recent years, a nonsurgical method was developed using infant rabbit models. The diarrheal fluid developed in this model was found to be chemically similar to that in human infections but was absent in rabbits infected with mutant strains lacking genes responsible for establishing infection in humans, suggesting that this system is the more reliable for the study of human cholerae infection.[17] Histological, confocal, and scanning electron microscope results in the model revealed the possibility for analyzing the cholera-toxin-mediated mucin secretion from goblet cells in the small intestine. The ability of *V. cholerae ctxAB* mutant to cause mild diarrhea in this model further emphasizes the use of such models to study reactogenicity of live attenuated cholera vaccines.[18]

28.3 *V. alginolyticus*

Vibrio alginolyticus is a common inhabitant of the marine environment in both temperate and tropical waters.[19,20] *V. alginolyticus* is also present in estuarine environments and is frequently isolated from bathing areas as free-living bacteria and is also associated with biotic and abiotic surfaces.[21] *V. alginolyticus* is one of the most common and important pathogen that causes vibriosis in human and marine species, and infection in humans was first recognized in 1973. The complete genome sequence of *V. alginolyticus* ATCC 17749 was revealed recently.[22] Infection of fish by *V. alginolyticus* results in biofilm formation in their intestine, which leads to mortality and potentially significant economic losses. *V. alginolyticus* has been described to cause human infections including wound infection (71%), gastroenteritis (12%), ear infection, and septicemia.[23] Other clinical syndromes reported in association with *V. alginolyticus* infection include chronic diarrhea in a patient with acquired immunodeficiency syndrome (AIDS), conjunctivitis, and posttraumatic intracranial infection.

V. alginolyticus produces many extracellular proteases responsible for interaction between the bacterium and cell hosts (human and fish) and plays an important role in infections.[24,25] The mechanism of pathogenicity induced by *V. alginolyticus* infections is complex and related to several factors, including cytotoxins, enterotoxins, and lytic enzymes.[26,27] The adhesive properties of *Vibrio* spp. are considered as the key factor of bacterial pathogenicity and have been implicated as a risk factor for infection in humans and stressed aquatic animals.[21,28,29] A set of genes regulated during environmental stress are known for their key role in regulating the adhesion process of *V. alginolyticus.*[30] Larger numbers of *Vibrio* spp. strains were frequently isolated from zooplankton, especially from copepods.[21] The ability to form biofilms on biotic and abiotic surfaces and to activate a survival state called "Viable but not cultivable" allow *V. alginolyticus* strains to persist in seawater under environmental stress conditions.[31] In *V. alginolyticus*, genetic analysis reveals that there is no correlation between the virulence of strains and the presence of general virulent genes such as *tlh*, *trh*, *toxR*, *toxRS*, *ctxA*, and VPI, which are present and contribute to the virulence in other *Vibrio* spp. pathogens.[32,33]

Production of multiple virulent extracellular products (ECP), mainly protease, is used to characterize the virulent strains of *V. alginolyticus.*[34,35] Extracellular alkaline serine proteases (Asp) have been demonstrated to be important exotoxins of *V. alginolyticus* strains pathogenic to prawn[35] and fish,[36] and their expressions are subjected to the regulation of a LuxO–LuxR quorum-sensing regulatory circuit.[37] At present, little is known about the relationship between the secretion systems and ECP production as well as their contribution to the pathogenesis in *V. alginolyticus.* A recent study identified and characterized the Tat translocation system in *V. alginolyticus.* The Tat system was strongly involved in the swarming ability and biofilm formation in this bacterium. In addition, the Tat system in *V. alginolyticus* was also required for virulence against the fish model and epithelial cells. These findings suggest that Tat secretion is related to the extracellular protease activity as well as virulence in *V. alginolyticus.*[38]

28.4 Models to Study *V. alginolyticus*

Many studies on *V. alginolyticus* pathogenicity have been focused on marine animals, such as *Epinephelus awoara*, shrimp,[39] *Pagrus majori*,[40] *Pseudosciaena crocea*,[41] *Epinephelus bruneus*,[42] *Haliotis diversicolor* supertexta,[43] *Lutjanus sanguineus*,[44] and *Lutjanus erythropterus*,[45] while few studies also reported pathogenicity of this microbe in humans. Lack of an appropriate model to study *V. alginolyticus* hindered the understanding of the mechanism for many years. A recent study by Liu et al. has established mice as a model for studying *V. alginolyticus*. Results of the study revealed pathogenicity of two different strains of *V. alginolyticus* ATCC 17749 and E0666. Intraperitoneal injection of *V. alginolyticus* caused severe lung and liver damage in mice with high bacterial counts found in heart, lung, and blood of the dead mice. The data suggest that high dose of *V. alginolyticus* injected intraperitoneally absorbs bacteria directly into the blood. IL-1β and IL-6 levels were significantly higher in serum of the *V. alginolyticus*-infected group compared to those in the control. Results suggested that IL-6 and IL-1β are important inflammatory factors that may play a role in the inflammation induced by *V. alginolyticus* infection.[46] A very recent study used high-throughput sequencing to identify differentially expressed genes between normal grouper larvae and larvae with vibriosis.[47]

28.5 *V. parahaemolyticus*

Vibrio parahaemolyticus is a Gram-negative halophilic bacterium that causes acute gastroenteritis in humans.[48] Infection occurs through the consumption of raw or inadequately cooked seafood. Since its discovery about 50 years ago, *V. parahaemolyticus* has been implicated as a major cause of foodborne illness around the world.[49] Clinical manifestations of *V. parahaemolyticus* infections include gastroenteritis, wound infection, and septicemia. Almost all clinical *V. parahaemolyticus* show β-hemolysis in a Wagatsuma agar, and this hemolytic activity is termed the Kanagawa phenomenon (KP).[50] The KP is considered as a good marker to differentiate human pathogenic *V. parahaemolyticus* strains from the nonpathogenic strains. Almost all clinical isolates carry *tdh*, *trh*, or both genes, whereas 1%–2% of environmental isolates may likely possess these genes.[50] Isolates lacking *tdh* and *trh* genes are also isolated from patients with *V. parahaemolyticus* infection symptoms. Both virulent and avirulent strains of *V. parahaemolyticus* are known to occur in a host, leading to *in vivo* serotype diversity.[51] The complete genome sequence of clinical and environmental isolates were reported recently[52,53] followed by draft genome sequence of 14 Canadian *V. parahaemolyticus* clinical isolates that were serologically identified as K group II using polyvalent antisera but were not specifically K serogrouped using monovalent antisera.[53] The use of whole-genome sequencing coupled with phylogeny and multiplex polymerase chain reaction (PCR) assay is also under development for identifying sequence of several *V. parahaemolyticus* isolates.[54] The complete genome sequence followed by pulsed-field electrophoresis showed that some *V. parahaemolyticus* environmental isolates exhibit significant genetic similarity to clinical isolates, which enables them to make the transition from normal aquatic state to a pathogen during infection.[55]

Most of the studies exploring the pathogenesis of *V. parahaemolyticus* on the host focus on TDH and TRH. TDH is responsible for various cytotoxicity effects that include erythrocyte lysis, disruption of the microtubule cytoskeleton, and ion flux disturbance in cultured cells.[56–58] Very less is known about the targets of TRH; a study on the purified TRH toxin indicates their effect on cultured erythrocytes and also result in the fluid accumulation in rabbit ileal loop model.[59] Infection by *V. parahaemolyticus* bacteria results in cell rounding and disruption of epithelial barrier function, which is mediated through the effect of TDH on the ion flow; this loss of barrier results in the excessive loss of body fluid and the condition is termed diarrhea.[50]

The genetic approach for identifying genes regulating the virulent pathway in *V. parahaemolyticus* showed a close association between the urease gene and the *TRH* gene.[60] Hence, it is hypothesized that possession of the gene for *TRH* coincides with the presence of the urease gene among many clinical *V. parahaemolyticus* strains, making urease gene a marker for identifying clinically pathogenic *trh* positive strains.[61] Urease is an enzyme involved in the hydrolysis of urea into ammonia and carbon dioxide and is present across kingdoms in plants, fungi, algae, and bacteria.[62] Many bacterial ureases have been characterized, and the organization of this gene is found to be similar among bacteria with regard to the structural and accessory genes involved in processing nickel ions.[63] Comparing with other *Vibrio* spp., only a small portion of clinical *V. parahaemolyticus* isolates possess the urease gene.[61] But the presence of urease gene is always associated with presence of *trh* gene. Specialized PCR techniques such as the long and accurate PCR have revealed that the distance between these two genes is <8.5 kb.[64]

Type III secretion systems (T3SSs) are considered as one among the major virulent pathways in many Gram-negative pathogens; bacteria use this system to translocate the produced virulent factors to the cytosol of the host cells.[65] The T3SS apparatus is found to be highly conserved across Gram-negative bacteria, and the specific property of the effectors brings about symptomatic effects on the host organism that vary widely.[65] Genome sequence of TDH-positive *V. parahaemolyticus* strains displayed the presence of gene sequences that are similar to T3SS present in chromosome 1 and 2, which are named T3SS1 and T3SS2, respectively.[66] T3SS1 gene sequence was found to be present in most or all of the *V. parahaemolyticus* strains, whereas the T3SS2 was found only in KP-positive strains. Intensive study on the T3SS of *V. parahaemolyticus* revealed that these two systems confer specific phenotypes to the bacteria. Understanding the effector proteins necessary for the activation of T3SS and environmental influence on the regulation of T3SS will provide deeper insight into *V. parahaemolyticus* infection in humans.[64]

28.6 Models to Study *V. parahaemolyticus*

The rabbit or rat ileal loop model has been considered as the best method to detect fluid accumulation due to the enterotoxic activity of *V. parahaemolyticus* TDH *in vivo*.[67,68] The role of the Type III secretion system 1 and 2 was tested using *in vitro* cell lines. Results of the *in vitro* studies have shown that T3SS1 induces cytotoxicity in several cell lines, while T3SS2 can induce cytotoxicity in Caco-2 cells and enterotoxicity *in vivo*.[69,70] The importance of TDH and T3SSs in *V. parahaemolyticus* virulence was recently assessed using a mouse model. These results clarified that T3SS2 rather than TDH was the major contributor to enterotoxicity; however, both T3SS1 and TDH contributed to lethality in a mouse model, suggesting that TDH and/or other proteins(s) that are part of T3SS1 are also likely to contribute to virulence.[71] Two different animal models were used to understand the contributions of T3SS1 and T3SS2 to *V. parahaemolyticus* virulence.[72] Orogastric inoculation of *V. parahaemolyticus* in neonatal pigs demonstrated that T3SS2, but not T3SS1, was required for gastrointestinal disease. The study also showed that T3SS1 was necessary to cause mortality during intrapulmonary inoculation, which raises questions on role of unknown factors related to T3SS1 responsible for lethality.[72] Host intestine damage during *V. parahaemolyticus* infection was studied using infant rabbit model.[73] Enteroinvasiveness of the bacterium has been reported for a rabbit model, in which the organism invaded, colonized, and produced inflammation in the small intestine. In another study, an adult zebrafish model was investigated to assess the overall virulence of *V. parahaemolyticus* strains.[74] The model detected differences in the virulence potential of strains when animals were challenged intraperitoneally, and this was based on survival time. Differences in survival were noted irrespective of the source of isolation of the strain (environmental or clinical) and regardless of the presence or absence of the known virulence factors TDH and TRH, suggesting the influence of additional virulence factors. The model was also effective in comparing differences in virulence between the wild-type *V. parahaemolyticus* strain RIMD2210633 and isogenic pilin mutants.

28.7 *C. elegans* as Model for Studying Bacterial Infection

Caenorhabditis elegans was first discovered and described by Emile Maupas, a French zoologist and botanist, in 1900 from soil in Algeria.[75] Earlier, Maupas used Rhabditis as genus name rather than *Caenorhabditis*. In 1952, the new subgenus *Caenorhabditis* was coined,[76] which was later raised into generic status by Dougherty in 1955. The definition of genera is somewhat arbitrary. Even now, *Caenorhabditis* is a subgenus name, one among many in the *Rhabditis* genus, and *C. elegans* are formally called *Rhabditis (Caenorhabditis) elegans*. *C. elegans* is a eukaryotic and free-living soil nematode. It has a complex developmental process and includes stages of embryogenesis, morphogenesis, and growth to an adult. *C. elegans* demonstrates about 35% of homology to human at genome level and 83% similarity in protein sequence.[77] *C. elegans* was introduced to the scientific community by Sydney Bernner as an experimental model in 1960s in order to explore animal developmental biology when he published his first report on the behavior of the nematode.[78] According to the *C. elegans* Sequencing Consortium, its genome has been completely sequenced, and the genome size is around 100.2 Mb. The availability of the entire genome sequence is useful in genetic studies, allowing researchers to pick a gene of interest to study. Surprisingly, it has been shown frequently that human genes replace their *C. elegans* homologs when introduced into *C. elegans*, and many genes in *C. elegans* can function similar to mammalian genes.

In contrast to other invertebrates, immunity in *C. elegans* solely depends on the innate immune system as it lacks the sophisticated adaptive immune signaling.[79] With this primary innate immune system, the nematode's immune response involves the production of numerous antimicrobial proteins that are conserved in higher vertebrates and humans. Exploring the innate immune signaling pathways in *C. elegans* revealed that many pathways important for development in early stages were reused in the adult stage for immune signaling. The nematode's defense involves three major mechanisms against any microbial infection:[80] (1) avoidance behavior, (2) physical barrier, and (3) inducible signaling pathways.

The neuronal system in *C. elegans* is well equipped in differentiating between the food source and pathogenic bacterium.[81] The repelling behavior against *Serratia marcescens* shed light on the role of several neurons, G-protein-coupled receptors, and TOL-1.[82,83] The cuticle of *C. elegans* is made of collagen and chitin that act as a strong physical barrier that cannot be pierced easily, thereby protecting from the entry of pathogens via the dermal layer.[84,85] The immune system includes pathogen recognition by PRR and other signaling cascades like MAPK, DAF-2, UPR, TGF-β, apoptosis, and necrosis pathways.[86–88]

Forward and reverse genetic studies have identified genes that belong to six signaling cascades, which are activated upon infection with Gram-positive and Gram-negative bacteria, fungi, and bacterial toxins.

The six pathways include ERK MAP kinase,[89] p38 MAP kinase,[90] TGF-β,[91] programmed cell death,[92] DAF-2/DAF-16 insulin-like receptor signaling,[93] and JNK-like MAP kinase.[94]

28.8 *C. elegans* for Host–Pathogen Interaction Studies

Studies in *C. elegans* have shed light on many pathways of the innate immune system that are phylogenetically conserved from worms to humans.[95] *C. elegans* is a very good model to study both the host and the microbe factors that underscore the host–pathogen interaction.[96]

M. nematophilum is one among many pathogens that infect *C. elegans* in its natural environment. It infects *C. elegans* and leads to an unusual and visible tail swelling or deformed anal region (Dar), formerly believed to be a spontaneous heritable and morphological mutation that arises during a regular genetic cross. Later, Hodgkin et al. revealed that *M. nematophilum* establishes a specific rectal infection due to its strong extracellular adherence to the cuticle, resulting in the swelling response and the Dar phenotype.[97,98] The tail-swelling phenotype observed during *M. nematophilum* infection is related to the enhancement of immune pathways ERK and MAPK in *C. elegans* as a response of the innate immune system against the bacterial infection.[89]

Studies with mutant strains that lack the bacterially unswollen (Bus) phenotype revealed that the role of several genes responsible for the Dar phenotype have implications on both the host (capability to elicit a swelling response) and the bacterium (ability to adhere and colonize).[99] The role of genes responsible for normal cuticle development and prevention of bacterial adherence have been reported in previous studies.[100] Quantitative molecular studies identified nearly 68 host genes that are induced upon *M. nematophilum* infection; most of these genes are pathogen receptor molecules including C-type lectin domains and lysozymes.[101]

Drechmeria coniospora is an agriculturally important nematode parasite that infects *C. elegans* through the mouth and penetrates the body by means of the proteinaceous hyphae.[102] *C. elegans* triggers an immune response against the infection through the induction of neuropeptide-like proteins (NLPs).[103] *Nematocida parisii* is a microsporidian parasite of *C. elegans*, and its infection did not induce basic immune-responsible genes known to be vital for other pathogenic infections but alters the components of the immune-signaling pathways (p38 MAPK and insulin signaling/DAF-2 pathways), thereby affecting animal survival.[104] Recent studies have reported the interaction between *C. elegans* and Orsay virus, a small, positive-sense RNA virus belonging to the family Nodaviridae.[105]

28.9 *C. elegans* as a Model for *Vibrio* spp. Infection

Studies with *C. elegans* and *Vibrio* spp. started with the most dreadful pathogen *V. cholerae*.[106] In the marine environment, the bacterium secretes well-characterized toxins including cholera toxin (CT) and toxin coregulated pili (Tcp). Bacterial cell-to-cell communication has a major role in regulating the virulence of the bacterium.[107] Studies with *C. elegans* shed light on the *V. cholerae* metalloprotease prtV, which is involved in establishing infection in higher vertebrates, and this also proved by studying its effect on a monolayer of mammalian intestinal cell lines. Since this protein (PrtV) is found to be essential for infection in humans and *C. elegans*, the results support the use of *C. elegans* as model to study *V. cholera* infection.[1] Another study explored *V. cholerae* cytolysin (VCC), which is one among the accessory *V. cholerae* virulence factors that may contribute to disease pathogenesis in humans. VCC, encoded by hlyA gene, belongs to the most common class of bacterial toxins known as pore-forming toxins (PFTs). *V. cholerae* infects and kills *C. elegans* in a cholerae-toxin-independent manner. VCC is required for lethality, growth retardation, and intestinal cell vacuolation during infection. A microarray analysis was performed in *C. elegans* exposed to *V. cholerae* strains with intact and deleted hlyA genes. Many of the VCC-regulated genes identified, including C-type lectins, Prion-like (glutamine [Q]/asparagine [N]-rich)-domain containing genes, genes regulated by insulin/IGF-1-mediated signaling (IIS) pathway, were previously reported as mediators of innate immune response against other bacteria in *C. elegans*. The protective function of the subset of the genes upregulated by VCC was confirmed using RNAi. By means of a machine learning algorithm called FastMEDUSA, the study identified several putative VCC-induced immune regulatory transcriptional factors and transcription factor binding motifs. Results from the experiments suggest that VCC induces a wide variety of immune-response-related genes during *V. cholerae* infection in *C. elegans.*[108]

A recent study by Durai et al. established *C. elegans* as a model for studying phenotypic changes and regulation of immune-responsible genes in the host against bacterial infection. Preference of *C. elegans* between nonpathogenic *E. coli* OP50 and pathogenic *V. alginolyticus* was well documented in the study. The infection was localized using green fluorescent protein (GFP)-tagged pathogen, and intestinal colonization was reported as the major reason for mortality in *C. elegans.*[109] Since *C. elegans* displayed lawn avoidance behavior to *V. alginolyticus*, a liquid killing assay was developed that comprised both M9 buffer and pathogen liquid culture (in a 3:1 ratio). Infection with *V. alginolyticus* in liquid assay reduced the lifespan of the nematode significantly. However, earlier reports revealed a reduction in lifespan of *C. elegans* exposed to *V. cholerae*, with complete killing of nematode in approximately 5 days,[1] while with *V. vulnificus* infection, the TD_{50} (toxic dose) of the nematode was found to be 9 days.[110] Microscopic image analysis of the infected worm showed colonization of GFP-tagged *V. alginolyticus* in the pharyngeal region and in the intestine of the pathogen-exposed animals. *V. alginolyticus* was able to colonize the gut of *C. elegans* within a short period of time, i.e., after 8 h of infection, and the bacterial load from the colonized animals were found to be ~$3.7 \pm 0.3 \times 10^3$ CFU. It was also observed that a bacterial load of at least 6.5×10^4 CFU of *V. alginolyticus* was required to cause and maintain infection in *C. elegans.* It was also found that *V. alginolyticus* induced a strong inflammatory response in the intestine, which appeared to be more severe than the infection caused by the *V. cholerae*. The persistence of *V. alginolyticus* leads to the damage of the pharyngeal bulb and distention of intestinal lumen, which was confirmed by confocal laser scanning microcopy (CLSM) image analyses.

C. elegans are attracted toward the smell of bacteria for a short period of time irrespective of the bacterial nature. On sensing the pathogens, in this case *V. alginolyticus*, the worms learned to avoid the smell of the bacterium and defended themselves from infection. The lawn-avoidance behavior of *C. elegans* observed in the study could be attributed to its chemosensory response to the pathogen (Figure 28.1).

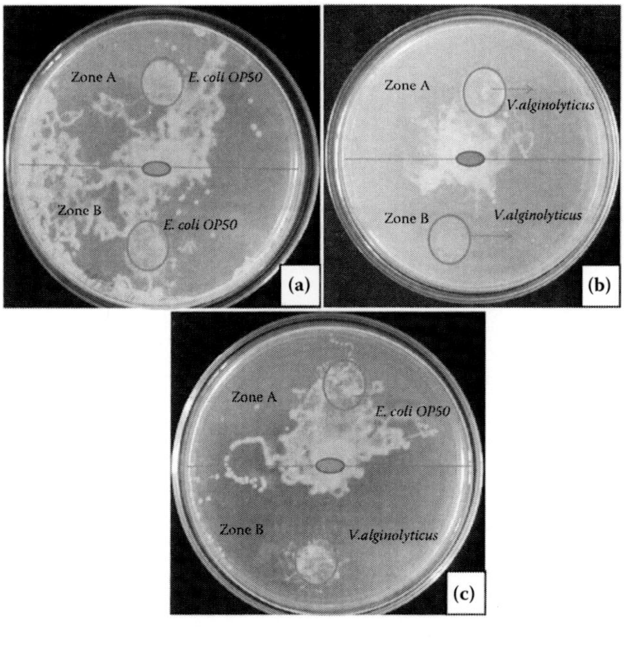

FIGURE 28.1 Chemotaxis behavior: in all cases, animals were tracked on an agar plate. (a) Animals in the presence of food source, *E. coli* OP50 marked at zone A and zone B. (b) Animals in presence of *V. alginolyticus* marked at zone A and zone B. (c) Animals in presence of food source, *E. coli* OP50 marked as zone A and pathogen source, *V. alginolyticus* marked as zone B. (Adapted from Durai, S. et al., *J. Basic Microbiol.*, 51, 243–52, 2011. With permission.)

The immune response of *C. elegans* against *V. alginolyticus* infection was analyzed using quantitative PCR (qPCR). Analogous to the previous studies in *M. nematophilum*, three candidate genes, viz., *lys-7*, *clec-60*, and *clec-87*, were analyzed for their expression during infection by having animals treated with a food source as control.[101] The level of expression of the immune effector gene(s) was significantly higher when compared to the control animals. During the initial stage of infection, the C-type lectin genes *clec-60* and *clec-87* seem to be responsible for the enhancement in adhesion and antibacterial activity.[101] The role of lysozyme gene *lys-7* as a presumptive antimicrobial gene and in induction of immune response as well as its encoded protein in the upregulation to infection with Gram-negative bacterium *S. marcescens* has previously been reported.[111]

Proteome regulation in *C. elegans* against *V. alginolyticus* was also explored by Durai et al.[112] Proteins were separated using two-dimensional differential gel electrophoresis (2D-DIGE), and the differentially regulated proteins were identified using positive matrix factorization (PMF) and matrix-assisted laser desorption/ionization–time of flight (MALDI–TOF)/TOF analysis. The results thus obtained were validated using western blotting for candidate proteins. The corresponding transcriptional regulation was quantified subsequently using real-time PCR. Interaction network for candidate proteins was predicted using the search tool for the retrieval of interacting genes/proteins (STRING), and functional validation was performed using respective mutant strains. Out of the 25 proteins identified, 21 proteins appeared to be upregulated, while 4 were downregulated (Table 28.1). Upregulated proteins included those involved in stress response (PDI-2, HSP-6), immune response (KIN-18, GST-8), and energy production (ATP-2), while proteins involved in structural maintenance (IFB-2) and lipid metabolism (SODH-1) were downregulated. The roles of these players in the host system during *Vibrio* infection were analyzed *in vivo* using wild-type and mutant *C. elegans*. Survival assays using mutants lacking *pdi-2*, *ire-1*, and *xbp-1* displayed enhanced susceptibility to *V. alginolyticus*. Cellular stress generated by *V. alginolyticus* was determined using reactive oxygen species (ROS) assay. This is the first report of proteome changes in

TABLE 28.1

Differentially Expressed Early Responsive Proteins of *C. elegans* against *V. alginolyticus* Infection

S.No.	Protein	Gene
A. Upregulated C. elegans *Protein against* V. alginolyticus *Infection*		
1	Alpha integrin	*Vab-20*
2	Serine/threonine protein kinase	*kin-18*
3	Mitochondrial ribosomal protein, small	*mrp-25*
4	Heat shock protein	*hsp-6*
5	Protein BCCIP homolog	ZK1127.4
6	COP9/signalosome and eIF3 complex shared subunit	*K08F11.3*
7	H1 histone	*hil-1*
8	Glucose-6-phosphate 1 dehydrogenase	*gspd-1*
9	Protein disulfide isomerase	*pdi-2*
10	DNA-directed RNA polymerase subunit II	*phi-14*
11	ATP synthase subunit-2	*atp-2*
12	Glutathione *S*-transferase 8	*gst-8*
13	cGMP-dependent protein kinase	*pkg-1*
14	Eukaryotic translation initiation factor 3 subunit	*eif-3.G*
15	Putative aminopeptidase	*lap-2*
16	Nuclear hormone receptor family	*nhr-44*
17	Probable signal recognition protein	F55C5.8
B. Downregulated C. elegans *Protein against* V. alginolyticus *Infection*		
1	Intermediate filament protein	*ifb-2*
2	Sorbitol dehydrogenase	*sodh-1*
3	Putative selenium-binding protein	R11G10.2

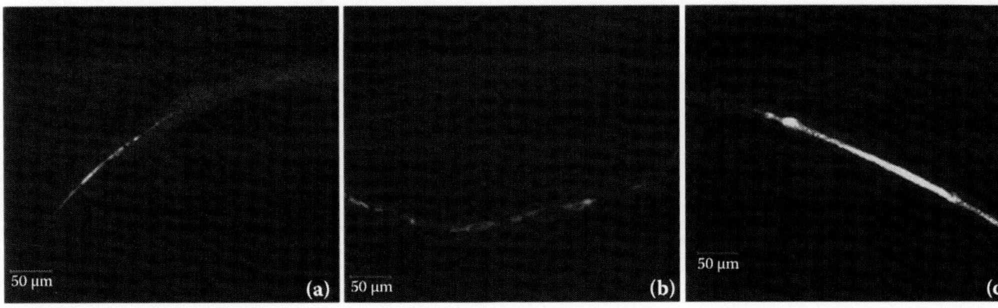

FIGURE 28.2 CLSM images of *C. elegans* exposed to GFP-tagged (a) *E. coli* OP50, (b) *V. parahaemolyticus* CM2, and (c) *V. parahaemolyticus* ATCC.

C. elegans against *V. alginolyticus* challenge and highlights the significance of unfolded protein response (UPR) pathway during bacterial infection.

Changes in *C. elegans* against *V. parahaemolyticus* infection were reported by Durai et al.[113] The study compared the virulence of two *V. parahaemolyticus* strains and proved that intestinal colonization of the bacterium was associated with pharyngeal damage, leading to the death of the nematode in 48 h. The colonization of *V. parahaemolyticus* in *C. elegans* intestine was localized using CLSM, and the intensity of the colonization was measured and compared with respective controls. The 2.5-dimensional topography of the colonized animal clearly showed the colonization of *V. parahaemolyticus* CM2 throughout the intestinal lumen of *C. elegans* (Figure 28.2). The study also showed that transferring the nematodes to a benign food source cleared the intestinal colonization of *V. parahaemolyticus*, which was similar to the earlier observations during *Staphylococcus aureus* infections.[114] The virulence of *V. parahaemolyticus* is associated with the production of two major toxins, TDH and TRH.[115–118] In addition, few strains lacking these virulence genes/factors were also found to be pathogenic, indicating that infection was independent of toxin production.[119] Results from experiments showed that strain CM2 is negative to hemolysin test performed in the Wagatsuma agar using human erythrocytes. Molecular studies for the *tdh* and *trh* supported the hemolysin test. However, the presence of *tlh* favored the notion that CM2 could be a pathogenic strain of *V. parahaemolyticus*. Previous reports show that *tlh* has been used for the molecular-level identifications of the *Vibrio* species.[119–121] *C. elegans* shows a selective response to food source and pathogenic bacteria with the phenomenon of olfactory learning.

28.10 Conclusion

With a mode of transmission through seafood contaminants, vibriosis in humans represents a challenge to researchers. Given that many reoccurring *Vibrio* species containing various virulent factors (including antibiotic resistance genes) are being identified every year, these bacteria pose a significant public risk to human populations. Precautions and proper identification of *Vibrio* contamination in seafood using advanced tools provide an effective way to control and reduce risk of infection. Further elucidation of the molecular mechanisms of *Vibrio* infections and virulent determinants using model organism will help identify potent anti-infective compounds that can be used for combating vibriosis.

Acknowledgements

This work was supported in part by the Grants from UGC, DBT, and Department of Science and Technology to Dr. K. Balamurugan. Financial support to Durai by Lady Tata Memorial Trust's Senior Scholarship is thankfully acknowledged.

REFERENCES

1. Vaitkevicius, K. et al. A *Vibrio cholerae* protease needed for killing of *Caenorhabditis elegans* has a role in protection from natural predator grazing. *Proc Natl Acad Sci USA* 103, 9280–5 (2006).
2. Halang, P., Vorburger, T. & Steuber, J. Serine 26 in the PomB subunit of the flagellar motor is essential for hypermotility of *Vibrio cholerae. PLoS One* 10, e0123518 (2015).
3. Cinar, H.N. et al. *Vibrio cholerae* hemolysin is required for lethality, developmental delay, and intestinal vacuolation in *Caenorhabditis elegans. PLoS One* 5, e11558 (2010).
4. Vanden Broeck, D., Horvath, C. & De Wolf, M.J. *Vibrio cholerae*: Cholera toxin. *Int J Biochem Cell Biol* 39, 1771–5 (2007).
5. Oh, Y.T., Lee, K.M., Bari, W., Raskin, D.M. & Yoon, S.S. (p)ppGpp, a small nucleotide regulator, directs the metabolic fate of glucose in *Vibrio cholerae. J Biol Chem* 290(21), 13178–90 (2015).
6. Jung, S.A., Chapman, C.A. & Ng, W.L. Quadruple quorum-sensing inputs control *Vibrio cholerae* virulence and maintain system robustness. *PLoS Pathog* 11, e1004837 (2015).
7. Soler-Bistue, A. et al. Genomic location of the major ribosomal protein gene locus determines *Vibrio cholerae* global growth and infectivity. *PLoS Genet* 11, e1005156 (2015).
8. Dorr, T., Davis, B.M. & Waldor, M.K. Endopeptidase-mediated beta lactam tolerance. *PLoS Pathog* 11, e1004850 (2015).
9. Spira, W.M., Sack, R.B. & Froehlich, J.L. Simple adult rabbit model for *Vibrio cholerae* and enterotoxigenic *Escherichia coli* diarrhea. *Infect Immun* 32, 739–47 (1981).
10. Klose, K.E. The suckling mouse model of cholera. *Trends Microbiol* 8, 189–91 (2000).
11. Sawasvirojwong, S., Srimanote, P., Chatsudthipong, V. & Muanprasat, C. An adult mouse model of induced diarrhea for studying pathogenesis and potential therapy of cholera. *PLoS Negl Trop Dis* 7, e2293 (2013).
12. Runft, D.L. et al. Zebrafish as a natural host model for *Vibrio cholerae* colonization and transmission. *Appl Environ Microbiol* 80, 1710–7 (2014).
13. De, S.N. & Chatterje, D.N. An experimental study of the mechanism of action of *Vibrio cholerae* on the intestinal mucous membrane. *J Pathol Bacteriol* 66, 559–62 (1953).
14. Dutta, N.K. & Habbu, M.K. Experimental cholera in infant rabbits: A method for chemotherapeutic investigation. *Br J Pharmacol Chemother* 10, 153–9 (1955).
15. Pierce, N.F., Greenough, W.B., III & Carpenter, C.C., Jr. *Vibrio cholerae* enterotoxin and its mode of action. *Bacteriol Rev* 35, 1–13 (1971).
16. Blachman, U., Graboff, S.R., Haag, G.E., Gottfeld, E. & Pickett, M.J. Experimental cholera in chinchillas: The immune response in serum and intestinal secretions to *Vibrio cholerae* and cholera toxin. *Infect Immun* 10, 1098–104 (1974).
17. Finkelstein, R.A. Cholera, *Vibrio cholerae O1 and O139, and Other Pathogenic Vibrios*. University of Texas Medical, Galveston (1996).
18. Millet, Y.A. et al. Insights into *Vibrio cholerae* intestinal colonization from monitoring fluorescently labeled bacteria. *PLoS Pathog* 10, e1004405 (2014).
19. Chakraborty, S., Nair, G.B. & Shinoda, S. Pathogenic vibrios in the natural aquatic environment. *Rev Environ Health* 12, 63–80 (1997).
20. Zanetti, S. et al. Differentiation of *Vibrio alginolyticus* strains isolated from Sardinian waters by ribotyping and a new rapid PCR fingerprinting method. *Appl Environ Microbiol* 65, 1871–5 (1999).
21. Baffone, W. et al. Detection of free-living and plankton-bound vibrios in coastal waters of the Adriatic Sea (Italy) and study of their pathogenicity-associated properties. *Environ Microbiol* 8, 1299–305 (2006).
22. Liu, X.F., Cao, Y., Zhang, H.L., Chen, Y.J. & Hu, C.J. Complete genome sequence of *Vibrio alginolyticus* ATCC 17749^T. *Genome Announc* 3(1), e01500-14 (2015).
23. Hlady, W.G. & Klontz, K.C. The epidemiology of *Vibrio* infections in Florida, 1981–1993. *J Infect Dis* 173, 1176–83 (1996).
24. Hare, P., Scott-Burden, T. & Woods, D.R. Characterization of extracellular alkaline proteases and collagenase induction in *Vibrio alginolyticus. J Gen Microbiol* 129, 1141–7 (1983).
25. Lee, K.K. Pathogenesis studies on *Vibrio alginolyticus* in the grouper, *Epinephelus malabaricus*, Bloch et Schneider. *Microb Pathog* 19, 39–48 (1995).
26. Iyer, L., Vadivelu, J. & Puthucheary, S.D. Detection of virulence associated genes, haemolysin and protease amongst *Vibrio cholerae* isolated in Malaysia. *Epidemiol Infect* 125, 27–34 (2000).

27. Ottaviani, D. et al. Antimicrobial susceptibility of potentially pathogenic halophilic vibrios isolated from seafood. *Int J Antimicrob Agents* 18, 135–40 (2001).
28. Balebona, M.C., Morinigo, M.A. & Borrego, J.J. Role of extracellular products in the pathogenicity of *Vibrio* strains on cultured gilt-head seabream (*Sparus aurata*). *Microbiologia* 11, 439–46 (1995).
29. Zorrilla, I., Morinigo, M.A., Castro, D., Balebona, M.C. & Borrego, J.J. Intraspecific characterization of *Vibrio alginolyticus* isolates recovered from cultured fish in Spain. *J Appl Microbiol* 95, 1106–16 (2003).
30. Kong, W. et al. Investigation of possible molecular mechanisms underlying the regulation of adhesion in *Vibrio alginolyticus* with comparative transcriptome analysis. *Antonie Van Leeuwenhoek* 107, 1197–206 (2015).
31. Ben Kahla-Nakbi, A., Besbes, A., Chaieb, K., Rouabhia, M. & Bakhrouf, A. Survival of *Vibrio alginolyticus* in seawater and retention of virulence of its starved cells. *Mar Environ Res* 64, 469–78 (2007).
32. Deriu, A., Sechi, L.A., Molicotti, P., Spanu, M.L. & Zanetti, S. Virulence genes in halophilic *Vibrio* spp. isolated in common mussels. *New Microbiol* 25, 93–6 (2002).
33. Xie, Z.Y., Hu, C.Q., Chen, C., Zhang, L.P. & Ren, C.H. Investigation of seven *Vibrio* virulence genes among *Vibrio alginolyticus* and *Vibrio parahaemolyticus* strains from the coastal mariculture systems in Guangdong, China. *Lett Appl Microbiol* 41, 202–7 (2005).
34. Lee, K.K., Chen, F.R., Yu, S.R., Yang, T.I. & Liu, P.C. Effects of extracellular products of *Vibrio alginolyticus* on penaeid prawn plasma components. *Lett Appl Microbiol* 25, 98–100 (1997).
35. Lee, K.K., Yu, S.R. & Liu, P.C. Alkaline serine protease is an exotoxin of *Vibrio alginolyticus* in kuruma prawn, *Penaeus japonicus*. *Curr Microbiol* 34, 110–7 (1997).
36. Cai, S.H., Wu, Z.H., Jian, J.C. & Lu, Y.S. Cloning and expression of the gene encoding an extracellular alkaline serine protease from *Vibrio alginolyticus* strain HY9901, the causative agent of vibriosis in *Lutjanus erythopterus* (Bloch). *J Fish Dis* 30, 493–500 (2007).
37. Rui, H., Liu, Q., Ma, Y., Wang, Q. & Zhang, Y. Roles of LuxR in regulating extracellular alkaline serine protease A, extracellular polysaccharide and mobility of *Vibrio alginolyticus*. *FEMS Microbiol Lett* 285, 155–62 (2008).
38. He, H., Wang, Q., Sheng, L., Liu, Q. & Zhang, Y. Functional characterization of *Vibrio alginolyticus* twin-arginine translocation system: Its roles in biofilm formation, extracellular protease activity, and virulence towards fish. *Curr Microbiol* 62, 1193–9 (2012).
39. George, M.R., John, K.R., Iyappan, T. & Jeyaseelan, M.J. Genetic heterogeneity among *Vibrio alginolyticus* isolated from shrimp farms by PCR fingerprinting. *Lett Appl Microbiol* 40, 369–72 (2005).
40. Ye, J. et al. Regulation of *Vibrio alginolyticus* virulence by the LuxS quorum-sensing system. *J Fish Dis* 31, 161–9 (2008).
41. Chen, Q., Yan, Q., Wang, K., Zhuang, Z. & Wang, X. Portal of entry for pathogenic *Vibrio alginolyticus* into large yellow croaker *Pseudosciaena crocea*, and characteristics of bacterial adhesion to mucus. *Dis Aquat Organ* 80, 181–8 (2008).
42. Harikrishnan, R., Balasundaram, C. & Heo, M.S. Poly D,L-lactide-co-glycolic acid-liposome encapsulated ODN on innate immunity in *Epinephelus bruneus* against *Vibrio alginolyticus*. *Vet Immunol Immunopathol* 147, 77–85 (2012).
43. Wu, C.J., Wang, H., Chan, Y.L. & Li, T.L. Passive immune-protection of small abalone against *Vibrio alginolyticus* infection by anti-*Vibrio* IgY-encapsulated feed. *Fish Shellfish Immunol* 30, 1042–8 (2011).
44. Liang, H., Xia, L., Wu, Z., Jian, J. & Lu, Y. Expression, characterization and immunogenicity of flagellin FlaC from *Vibrio alginolyticus* strain HY9901. *Fish Shellfish Immunol* 29, 343–8 (2010).
45. Cai, S.H. et al. Loop-mediated isothermal amplification method for rapid detection of *Vibrio alginolyticus*, the causative agent of vibriosis in mariculture fish. *Lett Appl Microbiol* 50, 480–5 (2010).
46. Liu, X.F. et al. Pathogenic analysis of *Vibrio alginolyticus* infection in a mouse model. *Folia Microbiol (Praha)* 59, 167–71 (2014).
47. Wang, Y.D. et al. Transcriptome analysis of the effect of *Vibrio alginolyticus* infection on the innate immunity-related complement pathway in *Epinephelus coioides*. *BMC Genomics* 15, 1102 (2014).
48. Nakasone, N., Insisengmay, S. & Iwanaga, M. Characterization of the pili isolated from *Vibrio parahaemolyticus* O3:K6. *Southeast Asian J Trop Med Public Health* 31, 360–5 (2000).
49. Barker W.H., Jr. & Gangarosa, E.J. Food poisoning due to *Vibrio parahaemolyticus*. *Ann Rev Med* 25, 75–81 (1974).
50. Park, K.S. et al. Functional characterization of two type III secretion systems of *Vibrio parahaemolyticus*. *Infect Immun* 72, 6659–65 (2004).

51. Bhoopong, P. et al. Variability of properties of *Vibrio parahaemolyticus* strains isolated from individual patients. *J Clin Microbiol* 45, 1544–50 (2007).
52. Ludeke, C.H., Kong, N., Weimer, B.C., Fischer, M. & Jones, J.L. Complete genome sequences of a clinical isolate and an environmental isolate of *Vibrio parahaemolyticus*. *Genome Announc* 3(2), e00216-15 (2015).
53. Ronholm, J., Petronella, N., Kenwell, R. & Banerjee, S. Draft whole-genome sequences of 14 *Vibrio parahaemolyticus* clinical isolates with an ambiguous K serogroup. *Genome Announc* 3(2), e00217-15 (2015).
54. Whistler, C.A. et al. Use of whole genome phylogeny and comparisons in the development of a multiplex-PCR assay to identify sequence type 36 *Vibrio parahaemolyticus*. *J Clin Microbiol* 53(6), 1864–72 (2015).
55. Hazen, T.H. et al. Insights into the environmental reservoir of pathogenic *Vibrio parahaemolyticus* using comparative genomics. *Front Microbiol* 6, 204 (2015).
56. Honda, T. & Iida, T. The pathogenicity of *Vibrio parahaemolyticus* and the role of the thermostable direct haemolysin and related haemolysins. *Rev Med Microbiol* 4, 106–13 (1993).
57. Nishibuchi, M., Fasano, A., Russell, R.G. & Kaper, J.B. Enterotoxigenicity of *Vibrio parahaemolyticus* with and without genes encoding thermostable direct hemolysin. *Infect Immun* 60, 3539–45 (1992).
58. Shirai, H. et al. Molecular epidemiologic evidence for association of thermostable direct hemolysin (TDH) and TDH-related hemolysin of *Vibrio parahaemolyticus* with gastroenteritis. *Infect Immun* 58, 3568–73 (1990).
59. Kishishita, M. et al. Sequence variation in the thermostable direct hemolysin-related hemolysin (trh) gene of *Vibrio parahaemolyticus*. *Appl Environ Microbiol* 58, 2449–57 (1992).
60. Park, K.-S. et al. Genetic characterization of DNA region containing the *trh* and *ure* genes of *Vibrio parahaemolyticus*. *Infect Immun* 68, 5742–48 (2000).
61. Suthienkul, O. et al. Urease production correlates with possession of the *trh* gene in *Vibrio parahaemolyticus* strains isolated in Thailand. *J Infect Dis* 172, 1405–8 (1995).
62. Mobley, H.L. & Hausinger, R.P. Microbial ureases: Significance, regulation, and molecular characterization. *Microbiol Rev* 53, 85–108 (1989).
63. Mobley, H.L., Island, M.D. & Hausinger, R.P. Molecular biology of microbial ureases. *Microbiol Rev* 59, 451–80 (1995).
64. Park, K.-S. et al. Close proximity of the *tdh*, *trh* and *ure* genes on the chromosome of *Vibrio parahaemolyticus*. *Microbiology* 144, 2517–23 (1998).
65. Hueck, C.J. Type III protein secretion systems in bacterial pathogens of animals and plants. *Microbiol Mol Biol Rev* 62, 379–433 (1998).
66. Makino, K. et al. Genome sequence of *Vibrio parahaemolyticus*: A pathogenic mechanism distinct from that of *V. cholerae*. *Lancet* 361, 743–9 (2003).
67. Miyamoto, Y. et al. Simplified purification and biophysicochemical characteristics of Kanagawa phenomenon-associated hemolysin of *Vibrio parahaemolyticus*. *Infect Immun* 28, 567–76 (1980).
68. Takeda, Y. Thermostable direct hemolysin of *Vibrio parahaemolyticus*. *Pharmacol Ther* 19, 123–46 (1982).
69. Kodama, T. et al. Identification of two translocon proteins of *Vibrio parahaemolyticus* type III secretion system 2. *Infect Immun* 76, 4282–9 (2008).
70. Park, K.S. et al. Cytotoxicity and enterotoxicity of the thermostable direct hemolysin-deletion mutants of *Vibrio parahaemolyticus*. *Microbiol Immunol* 48, 313–8 (2004).
71. Hiyoshi, H., Kodama, T., Iida, T. & Honda, T. Contribution of *Vibrio parahaemolyticus* virulence factors to cytotoxicity, enterotoxicity, and lethality in mice. *Infect Immun* 78, 1772–80 (2010).
72. Pineyro, P. et al. Development of two animal models to study the function of *Vibrio parahaemolyticus* type III secretion systems. *Infect Immun* 78, 4551–9 (2010).
73. Ritchie, J.M. et al. Inflammation and disintegration of intestinal villi in an experimental model for *Vibrio parahaemolyticus*-induced diarrhea. *PLoS Pathog* 8, e1002593 (2012).
74. Paranjpye, R.N., Myers, M.S., Yount, E.C. & Thompson, J.L. Zebrafish as a model for *Vibrio parahaemolyticus* virulence. *Microbiology* 159, 2605–15 (2013).
75. Maupas, E. Modes et formes de reproduction des nematodes. *Arch Zool Exp Gen* 8, 463–624 (1901).
76. Osche, G. Systematik und phylogenie der gattung Rhabditis (Nematoda). *Zool Jahrb Abt Allg Zool Physiol Tiere* 81, 190–280 (1952).

77. Lai, C.H., Chou, C.Y., Ch'ang, L.Y., Liu, C.S. & Lin, W. Identification of novel human genes evolutionarily conserved in *Caenorhabditis elegans* by comparative proteomics. *Genome Res* 10, 703–13 (2000).
78. Brenner, S. The genetics of behaviour. *Br Med Bull* 29, 269–71 (1973).
79. Fares, H. & Greenwald, I. Genetic analysis of endocytosis in *Caenorhabditis elegans*: Coelomocyte uptake defective mutants. *Genetics* 159, 133–45 (2001).
80. Barriare, A. & Felix, M.-A. Isolation of *C. elegans* and related nematodes. *WormBook*, 1–19 (2014).
81. Schulenburg, H. & Ewbank, J.J. The genetics of pathogen avoidance in *Caenorhabditis elegans*. *Mol Microbiol* 66, 563–70 (2007).
82. Pradel, E. et al. Detection and avoidance of a natural product from the pathogenic bacterium *Serratia marcescens* by *Caenorhabditis elegans*. *Proc Natl Acad Sci USA* 104, 2295–300 (2007).
83. Pujol, N. et al. A reverse genetic analysis of components of the Toll signaling pathway. *Curr Biol* 16, 1477 (2006).
84. Kim, D.H. et al. A conserved p38 MAP kinase pathway in *Caenorhabditis elegans* innate immunity. *Sci Signal* 297, 623 (2002).
85. Labrousse, A., Chauvet, S., Couillault, C., Kurz, C.L. & Ewbank, J.J. *Caenorhabditis elegans* is a model host for *Salmonella typhimurium*. *Curr Biol* 10, 1543–5 (2000).
86. Aballay, A., Drenkard, E., Hilbun, L.R. & Ausubel, F.M. *Caenorhabditis elegans* innate immune response triggered by *Salmonella enterica* requires intact LPS and is mediated by a MAPK signaling pathway. *Curr Biol* 13, 47–52 (2003).
87. Akira, S. Pathogen recognition by innate immunity and its signaling. *Proc Jpn Acad Ser B Phys Biol Sci* 85, 143–56 (2009).
88. Couillault, C. & Ewbank, J.J. Diverse bacteria are pathogens of *Caenorhabditis elegans*. *Infect Immun* 70, 4705–7 (2002).
89. Nicholas, H.R. & Hodgkin, J. The ERK MAP kinase cascade mediates tail swelling and a protective response to rectal infection in *C. elegans*. *Curr Biol* 14, 1256–61 (2004).
90. Kim, D.H. et al. A conserved p38 MAP kinase pathway in *Caenorhabditis elegans* innate immunity. *Science* 297, 623–6 (2002).
91. Mallo, G.V. et al. Inducible antibacterial defense system in *C. elegans*. *Curr Biol* 12, 1209–14 (2002).
92. Aballay, A. & Ausubel, F.M. Programmed cell death mediated by *ced-3* and *ced-4* protects *Caenorhabditis elegans* from *Salmonella typhimurium*-mediated killing. *Proc Natl Acad Sci USA* 98, 2735–9 (2001).
93. Garsin, D.A. et al. Long-lived *C. elegans daf-2* mutants are resistant to bacterial pathogens. *Science* 300, 1921 (2003).
94. Huffman, D.L. et al. Mitogen-activated protein kinase pathways defend against bacterial pore-forming toxins. *Proc Natl Acad Sci USA* 101, 10995–11000 (2004).
95. Kim, D. Studying host–pathogen interactions and innate immunity in *Caenorhabditis elegans*. *Dis Model Mech* 1, 205–8 (2008).
96. Marsh, E.K. & May, R.C. *Caenorhabditis elegans*, a model organism for investigating immunity. *Appl Environ Microbiol* 78, 2075–81 (2012).
97. Hodgkin, J., Kuwabara, P.E. & Corneliussen, B. A novel bacterial pathogen, *Microbacterium nematophilum*, induces morphological change in the nematode *C. elegans*. *Curr Biol* 10, 1615–8 (2000).
98. Nicholas, H.R. & Hodgkin, J. Innate immunity: The worm fights back. *Curr Biol* 12, R731–2 (2002).
99. Gravato-Nobre, M.J. & Hodgkin, J. *Caenorhabditis elegans* as a model for innate immunity to pathogens. *Cell Microbiol* 7, 741–51 (2005).
100. Palaima, E. et al. The *Caenorhabditis elegans bus-2* mutant reveals a new class of O-glycans affecting bacterial resistance. *J Biol Chem* 285, 17662–72 (2010).
101. O'Rourke, D., Baban, D., Demidova, M., Mott, R. & Hodgkin, J. Genomic clusters, putative pathogen recognition molecules, and antimicrobial genes are induced by infection of *C. elegans* with *M. nematophilum*. *Genome Res* 16, 1005–16 (2006).
102. Jansson, H.B. Adhesion of conidia of *Drechmeria coniospora* to *Caenorhabditis elegans* wild type and mutants. *J Nematol* 26, 430 (1994).
103. Couillault, C. et al. TLR-independent control of innate immunity in *Caenorhabditis elegans* by the TIR domain adaptor protein TIR-1, an ortholog of human SARM. *Nat Immunol* 5, 488–94 (2004).
104. Troemel, E.R. et al. p38 MAPK regulates expression of immune response genes and contributes to longevity in *C. elegans*. *PLoS Genet* 2, e183 (2006).

105. Félix, M.A. et al. Natural and experimental infection of *Caenorhabditis* nematodes by novel viruses related to nodaviruses. *PLoS Biol* 9, e1000586 (2011).
106. Félix, M.A. & Braendle, C. The natural history of *Caenorhabditis elegans*. *Curr Biol* 20, R965–9 (2010).
107. Joelsson, A., Liu, Z. & Zhu, J. Genetic and phenotypic diversity of quorum-sensing systems in clinical and environmental isolates of *Vibrio cholerae*. *Infect Immun* 74, 1141–7 (2006).
108. Sahu, S.N. et al. Genomic analysis of immune response against *Vibrio cholerae* hemolysin in *Caenorhabditis elegans*. *PLoS One* 7, e38200 (2012).
109. Durai, S., Pandian, S.K. & Balamurugan, K. Establishment of a *Caenorhabditis elegans* infection model for *Vibrio alginolyticus*. *J Basic Microbiol* 51, 243–52 (2011).
110. Dhakal, B.K. et al. *Caenorhabditis elegans* as a simple model host for *Vibrio vulnificus* infection. *Biochem Biophys Res Commun* 346, 751–7 (2006).
111. Kurz, C.L. et al. Virulence factors of the human opportunistic pathogen *Serratia marcescens* identified by in vivo screening. *EMBO J* 22, 1451–60 (2003).
112. Durai, S., Singh, N., Kundu, S. & Balamurugan, K. Proteomic investigation of *Vibrio alginolyticus* challenged *Caenorhabditis elegans* revealed regulation of cellular homeostasis proteins and their role in supporting innate immune system. *Proteomics* 14, 1820–32 (2014).
113. Durai, S., Karutha Pandian, S. & Balamurugan, K. Changes in *Caenorhabditis elegans* exposed to *Vibrio parahaemolyticus*. *J Microbiol Biotechnol* 21, 1026–35 (2011).
114. Sifri, C.D., Begun, J., Ausubel, F.M. & Calderwood, S.B. *Caenorhabditis elegans* as a model host for *Staphylococcus aureus* pathogenesis. *Infect Immun* 71, 2208–17 (2003).
115. Ansaruzzaman, M. et al. Pandemic serovars (O3:K6 and O4:K68) of *Vibrio parahaemolyticus* associated with diarrhea in Mozambique: Spread of the pandemic into the African continent. *J Clin Microbiol* 43, 2559–62 (2005).
116. Matsumoto, C. et al. Pandemic spread of an O3:K6 clone of *Vibrio parahaemolyticus* and emergence of related strains evidenced by arbitrarily primed PCR and *toxRS* sequence analyses. *J Clin Microbiol* 38, 578–85 (2000).
117. Nair, G.B. & Hormazabal, J.C. The *Vibrio parahaemolyticus* pandemic. *Rev Chilena Infectol* 22, 125–30 (2005).
118. Okuda, J. et al. Emergence of a unique O3:K6 clone of *Vibrio parahaemolyticus* in Calcutta, India, and isolation of strains from the same clonal group from Southeast Asian travelers arriving in Japan. *J Clin Microbiol* 35, 3150–5 (1997).
119. Lynch, T. et al. *Vibrio parahaemolyticus* disruption of epithelial cell tight junctions occurs independently of toxin production. *Infect Immun* 73, 1275–83 (2005).
120. Schulenburg, H., Hoeppner, M.P., Weiner, J., III & Bornberg-Bauer, E. Specificity of the innate immune system and diversity of C-type lectin domain (CTLD) proteins in the nematode *Caenorhabditis elegans*. *Immunobiology* 213, 237–50 (2008).
121. Wong, D., Bazopoulou, D., Pujol, N., Tavernarakis, N. & Ewbank, J.J. Genome-wide investigation reveals pathogen-specific and shared signatures in the response of *Caenorhabditis elegans* to infection. *Genome Biol* 8, R194 (2007).

29

Yersinia

Xin Wang, Ran Duan, Junrong Liang, Wenpeng Gu, and Huaiqi Jing

CONTENTS

Yersinia is a genus of Gram-negative, rod-shaped, facultative anaerobes in the family Enterobacteriaceae. Of the 17 species identified within the genus, three are pathogenic to humans, namely, the causative agent of plague (or black death)—*Yersinia pestis*—and two enteropathogens—*Yersinia enterocolitica* and *Yersinia pseudotuberculosis*.[1] While *Y. pestis* is generally transmitted by a flea bite, the other two are foodborne pathogens, with *Y. enterocolitica* associated with gastrointestinal syndromes (ranging from acute enteritis to mesenteric lymphadenitis) and *Y. pseudotuberculosis* implicated in mesenteric adenitis and septicemia. Despite their differences in transmission routes and disease types or severity, these three pathogens demonstrate a common tropism for lymphoid tissues and the capacity to resist the nonspecific immune response of the host, in particular phagocytosis and killing by macrophages and polymorphonuclear leukocytes (PMNs).[2–4] *Y. enterocolitica* is recovered from diverse animal sources, from farm animals and domestic pets to free-living and captive wild animals.[1,5] Pigs and dogs are considered to be the primary reservoir of human pathogenic *Y. enterocolitica* strains in the world, and the isolation rate from tonsil samples is the highest.[6–9]

The virulence of pathogenic *Y. enterocolitica* is attributed to a 70-kb pYV (plasmid for Yersinia virulence) plasmid and some chromosome-encoded factors. A similarly sized plasmid, termed pCD, is found in *Y. pestis*,[10] and pYV is also present in *Y. pseudotuberculosis*.[11] Despite this common feature, the diseases caused by the two enteropathogenic yersiniae are chronic in nature and obviously different from plague. The virulent plasmid pYV encodes Yersinia adhesin A (YadA) and the Yersinia outer protein (Yop) virulon, a system consisting of secreted proteins called Yops and their dedicated type III secretion system (T3SS) apparatus called Ysc.[12] The Ysc apparatus forms a channel composed of several proteins.[13] Yop proteins fall into two categories. Some are intracellular effectors, while the others are "translocators" needed to deliver the effectors across the eukaryotic plasma membrane into eukaryotic cells. The translocators (YopB, YopD, LcrV) form a pore in the eukaryotic cell's plasma membrane. The effector Yops are YopE, YopH, YpkA/YopO, YopP/YopJ, YopM, and YopT.[12] The chromosome-encoded virulence factors include *ail* (attachment and invasion locus), *inv* (invasin), and *yst* (*Yersinia* stable toxin). The high pathogenicity island (HPI) bearing an iron acquisition system only exists in the highly pathogenic biotype 1B strains.[14] Many studies have reported the role of biotype 1A strains in foodborne outbreaks, in which the pYV plasmid is absent.[15]

29.1 Mouse Model

29.1.1 Lifestyle and Interaction with Host

After orogastric inoculation of mice, *Y. enterocolitica* selectively invades the Peyer's patches (PP) via M cells.[16–18] This invasion leads to recruitment of PMNs, formation of microabscesses comprising extracellular *Yersinia*, and, finally, complete destruction of the cytoarchitecture of the PP. Subsequently yersiniae disseminate to lymph nodes, spleen, and liver where they form microcolonies/microabscesses, suggesting that *Y. enterocolitica* disseminates via the lymphatic vessels.[16] Similar to *in vivo* observations, *Yersinia* manifests *in vitro* resistance to phagocytosis by macrophages[19,20] and by PMNs.[21–24] Once phagocytosed, *Y. pseudotuberculosis* and *Y. enterocolitica* generally do not survive long. These findings suggest that *Y. enterocolitica* is an extracellular pathogen and that its survival strategy is based on avoiding the nonspecific immune response.

Recent studies showed that in the oral mouse infection model using RFP- (red fluorescent protein) and GFP- (green fluorescent protein) expressing yersiniae, the oral *Y. enterocolitica* infection of mice leads to monoclonal microabscess formation in PP, spleen, and liver. Furthermore, experiments with red and green fluorescing yersiniae revealed that only very few yersiniae were able to invade PP from the gut lumen and that both *Yersinia* and the host contribute to this phenomenon. In contrast to other enteric pathogens such as *Salmonella* and *Shigella*, *Yersinia* is predominantly an extracellular pathogen.[18,25,26] Initially, innate host defenses such as PMNs, macrophages, and natural killer (NK) cells are involved in controlling *Yersinia* infection. But subsequently, a robust adaptive immune response is required to overcome *Yersinia* infection. Specific Abs[27,28] as well as IFN-γ-producing CD4 and CD8 T cells play an essential role in clearing *Yersinia* infection and have been shown to mediate protection in adoptive transfer experiments.[29]

Wang et al. investigated the lethality, cytokine alterations, and histopathological changes of pathogenic *Y. enterocolitica* bioserotypes 1B/O:8 and 2/O:9 in susceptible BALB/c and resistant C57BL/6 mice. They found that the 50% lethal dose (LD_{50}) for the pathogenic *Y. enterocolitica* bioserotype 1B/O:8 was 10^3 CFU in both BALB/c and C57BL/6 mice; while the LD_{50} for the bioserotype 2/O:9 was 10^8 CFU in BALB/c mice and 10^9 CFU in C57BL/6 mice. In BALB/c mice, the bioserotype 2/O:9 induced a higher level of GM-CSF, IL-6, and TNF-α than the bioserotype 1B/O:8, but the status was reversed subsequently. Levels of IFN-γ, IL-1β, IL-2, IL-4, IL-5, IL-10, and IL-12 following infection with the bioserotype 1B/O:8 were always higher than those with the bioserotype 2/O:9 (Figure 29.1). The histopathological changes in the liver and spleen in BALB/c mice infected with the two strains were similar at different times and doses. These observations indicate the different immunological effects and changes in the mouse model due to pathogenic *Y. enterocolitica* 1B/O:8 and 2/O:9 infections.[30]

29.1.2 Virulence of Pathogen

In an oral mouse model, a phospholipase-negative strain was constructed to ascertain whether phospholipase A plays a role in pathogenesis. At a dose of 10^8 CFU/mouse, fewer mutant strains were recovered and less flamed tissue was found from the mesenteric lymph nodes and PP. When given extremely high dose (10^9 CFU/mouse), the numbers of foci and the extent of inflammation were noticeably less as well, confirming the role of phospholipase A in pathogenesis.[31]

The T3SS is a complex system used to deliver bacterial proteins into the cytosol of host cells.[32] The proteins secreted by T3SSs are known as effectors working together to establish an extracellular infection that disrupts the actin cytoskeleton and deregulates signal pathways.[33]

YopM was originally described as Yop48 in *Y. enterocolitica.*[34] According to the LD_{50} test, the yopM mutant has a strongly reduced virulence in mice. Bacterial counts in liver and spleen showed the yopM mutant had a reduced ability to multiply in the host.[35] When three challenge routes were done for YopE mutant, it was avirulent after oral or intraperitoneal infection but virulent after intravenous injection.[36]

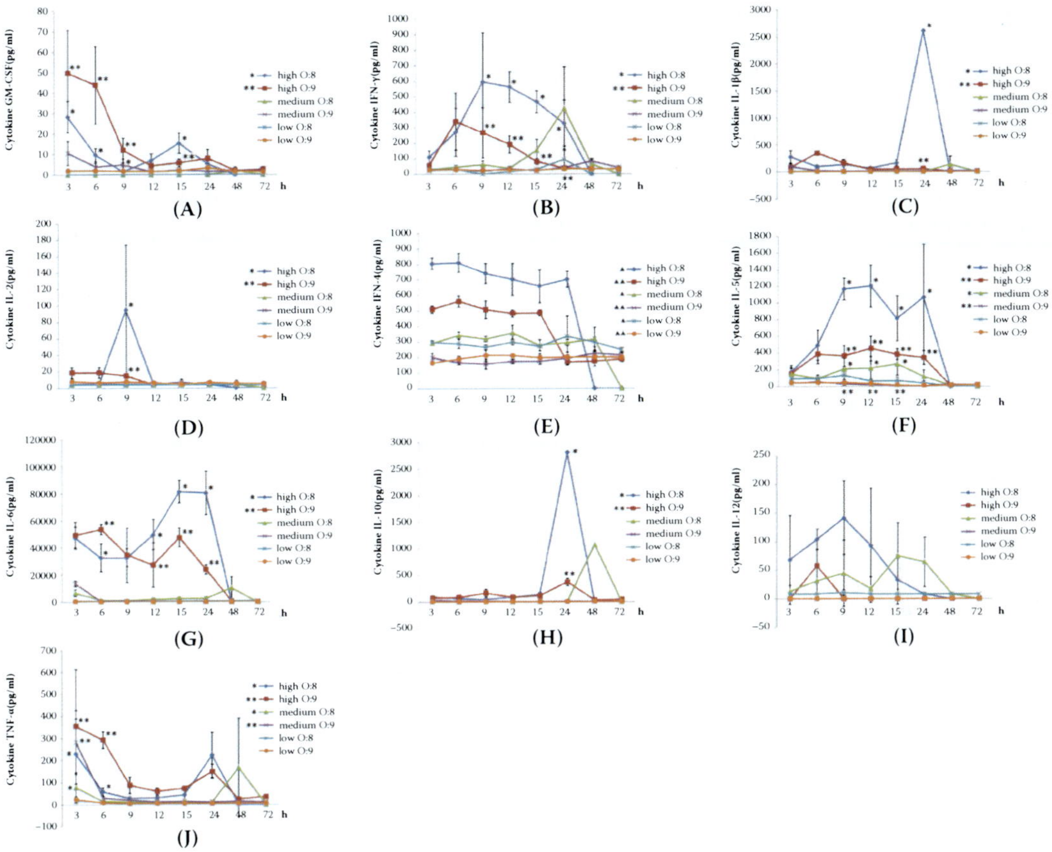

FIGURE 29.1 Comparison of cytokines between two strains in BALB/C mice. (A) GM-CSF, (B) IFN-γ, (C) IL-1β, (D) IL-2, (E) IL-4, (F) IL-5, (G) IL-6, (H) IL-10, (I) IL-12, and (J) TNF-α. *, The cytokine values were statistically significant comparing the two strains in different groups ($p < 0.05$). ▲, All of the cytokine values were statistically significant comparing the two strains in every group ($p < 0.05$). This figure demonstrates the comparison for each cytokine changes between two strains. (From Wang, X. et al., *Mol. Immunol.*, 55, 365–371, 2013. With permission from Elsevier.)

Compared to extensive study of YopP effects *in vitro*, the role of YopP *in vivo* is inadequately investigated. Monack et al. have shown that *Y. pseudotuberculosis* induces apoptosis of Mac-1 cells in mice, which is probably responsible for the attenuated phenotype of a yopJ mutant.[37] In 2015, a study using *Listeria* model shows for the first time that YopP inhibits the development of an effective CD8 T cell response in mice.[38] *In vitro*, this is due to the rapid induction of dendritic cell (DC) apoptosis and maturation,[39] whereas *in vivo* a possible mechanism for this could be the inhibition of Ag presentation by DC.[38]

By using mouse model infected with *Y. pseudotuberculosis*, Rosqvist et al. showed that a strain unable to express YopH has a reduced ability to resist phagocytosis,[20] while the ability to resist phagocytosis could be complemented by introduction of a plasmid carrying only the yopH gene, demonstrating that YopH is indeed involved in the antiphagocytic effect. However, mutation of yopH did not completely abolish the resistance to phagocytosis, suggesting another virulence factor was involved in this phenomenon.

A systematic study of Yops pathogenicity of the highly virulent mouse O:8 strain administered by oral and intravenous routes in mice showed the *yopH*, *yopO*, *yopP*, *yopE*, *yopM*, and *yopQ* mutants had only modest defects in colonization of the small intestine (SI) and PP,[40] consistent with that of *Y. pseudotuberculosis*,[41] indicating that no single effector is absolutely necessary for the colonizing. YopH, YopQ, and YopM are important for inducing systemic disease, whereas YopO, YopP, and YopE are dispensable to reach the spleen and liver. On the other hand, YopT seems to slightly decrease the virulence in both oral and intravenous model.[40]

With the help of the mouse model, the yersinia T3SS has been studied in detail using molecular, cellular, genetic, and biochemical techniques, yielding insight into the efficiency and sophistication of the complex.

Progression of *Y. enterocolitica* infection in mouse model closely mirrors that in the human. In natural infection, *Y. enterocolitica* enters by the oral route.[42] Ruiz-Bravo et al. conducted a comparative experiment with three challenge routes (oral, intraperitoneal, and intravenous) using moderate–virulent O:9 strains with or without pYV and noted considerable differences between oral and intraperitoneal infections by the two strains.[43] This suggests the virulence potential of *Y. enterocolitica* may be influenced by the route of infection. Therefore, at the beginning of an experiment, the route of infection should be chosen prudently.

29.2 Pig Model

In swine, fecal shedding may stop soon after the ingestion of *Y. enterocolitica*, while tonsils may carry the bacterium from hours to months,[44] indicating that tonsils represent a more reliable tissue for the indication of an infection/colonization by *Y. enterocolitica*. As is known, clinical response to *Y. enterocolitica* infection in piglets is related to inoculum size.[45] A dose of 4×10^{10} CFU could cause death[46]; in contrast, a dose of 3×10^8 CFU would provoke a subclinical response as near as possible to a natural infection.[44] The early infection showed a transient bacteremia within the first 3 h postinfection.[45] So if pigs are infected within hours prior to slaughter, carcass of the animal might not be completely cleared of the microorganism and thus result in widespread contamination.[44]

The high prevalence of O:3 strains in fattening pigs indicates serotype- and host-specific colonization strategies. Comparison of O:3 patient isolate and O:8 8081v strain in minipigs and mice models showed that small genetic variations contribute to profound changes of the virulence gene expression pattern.[47] In the O:3 strain, IS1667 insertion (an additional promoter) as well as the more stable variant of RovA (*invA* transcriptional activator protein) toward Lon protease resulted in high expression of invasion at 37°C, indicating a fine adjustment of pathogenicity to pig, with a higher body temperature (Figure 29.2).[47] They

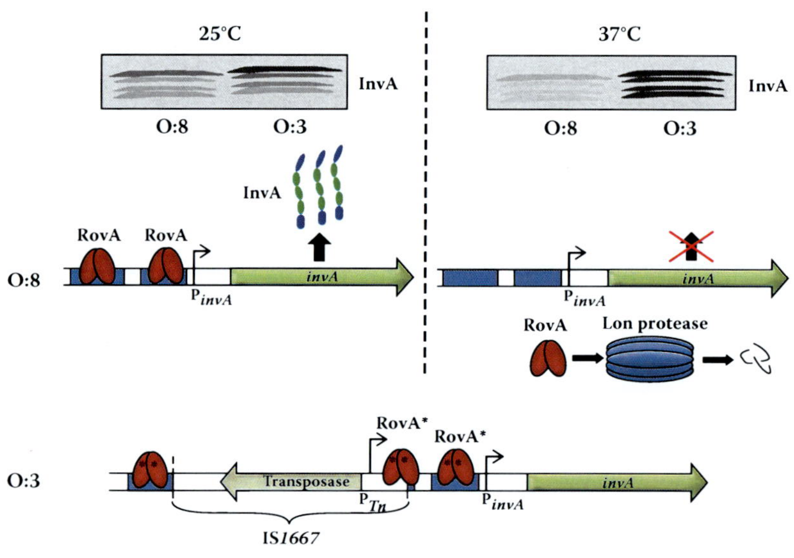

FIGURE 29.2 Different expression of the invasin gene *invA* in *Y. enterocolitica* O:3 and O:8. Whole-cell extracts prepared from overnight cultures of *Y. enterocolitica* O:3 and O:8 grown at 25°C and 37°C were separated on SDS-polyacrylamide gels and analyzed by western blotting using polyclonal antibodies directed against InvA. Overview of the *invA* promoter region of *Y. enterocolitica* O:8 (upper panel) and O:3 (lower panel). The transcriptional start site of the *invA* and the IS1667-specific promoter are indicated by broken arrows. The blue boxes represent the RovA binding sites. The light blue box illustrates the RovA binding site in the IS1667 sequence. The RovA dimer is given in red; the stars illustrate the amino acid substitution that renders the RovA protein less susceptible to the Lon protease. (From Valentin-Weigand, P., et al., *Int. J. Med. Microbiol*, 304, 824–834, 2014. With permission from Elsevier.)

further uncovered why the cell adhesion and invasion properties were restricted to pig-associated O:3 strains. Coinfection studies with O:3 wild and mutant strains demonstrated that small variations within the O:3 genome improved colonization/survival in swine but had only a minor effect on the colonization in mice. In addition, a deletion of the *invA* abolished long-term colonization in the pigs.[48]

As the most important reservoir and asymptomatic carriers of *Y. enterocolitica* prevalent serotype,[49,50] the pig (model) is useful for studies ranging from colonization tropisms to research on the virulence properties of O:3 strains.

29.3 *Caenorhabditis elegans* Model

Caenorhabditis elegans is a validated model to study bacterial pathogenicity. It has been suggested that bacterial–invertebrate interaction has shaped the evolution of microbial pathogenicity.[51] Spanier et al. demonstrated that *Y. enterocolitica* strains W22703 (2/O:9) and WA314 (1B/O:8) colonize and kill *C. elegans* following oral infection and that the insecticidal toxin TcaA, which is expressed only at ambient temperature, is required for full nematocidal activity but alone is not sufficient.[52] In contrast, with *Y. pestis* and *Y. pseudotuberculosis*, both of which use biofilm-dependent mechanisms, biofilm formation is not responsible for the nematocidal activity in *Y. enterocolitica*.[53]

Temperature is a key environmental clue for the expression of yersinia genes.[54,55] Several genetic determinants of *Y. enterocolitica* are repressed at 37°C but induced at a low temperature, suggesting a role in invertebrates.[56,57] However, Yop effectors and YadA are produced only at 37°C, while *C. elegans* cannot be cultivated above 25°C. This is perhaps why Inv, YopE, and YadA did not remarkably contribute to *C. elegans* killing.[52]

Together, these data do not support the feasibility of the *C. elegans* infection model for *Y. enterocolitica* pathogenicity toward mammals.

29.4 Rabbit Model

Rabbits have been a suitable and reproducible animal model to examine the course of intestinal *Yersinia* infection.[58–62] In mouse model, diarrhea is not a major symptom.[63,64] But the infection of the young rabbit resembles that in children: the oral inoculation induces both a clear diarrhea and a systemic invasion.[59,60,65–67] Thus, Delor and Cornells constructed one for establishment of the virulent role of Yst toxin.[68]

The pathogenic and immunogenic potential of parental and mutant *Y. enterocolitica* strains have been successfully evaluated in oral rabbit model, showing O-antigen is required for full virulence. The strain that lacks Wzz protein (the O-antigen chain length determinant) is more attenuated than one that lacks Wzy protein (expressing only one repeat unit of lipopolyasccharide [LPS]).[69]

On the basis of sodA finding in mouse model,[70] Najdenski et al.[71] established a rabbit model to show that the sodA mutant strain not only attenuated in dissemination into various organs but also showed an improved antibody response. In contrast to previous LPS mutant strains, this mutant was noticeably more sensitive to leukocytes. Besides, they first observed yersiniae dissemination to brain in experimental infection. More surprising were the pathomorphological changes not established in tonsils with the virulent strain but with sodA mutant strain. Together, these showed the potential of sodA mutant to be a candidate for immunization.

Overall, the susceptibility of rabbits to *Y. enterocolitica* appeared to be more variable in this study focused on diarrhea manifestation and immunomorphology.

29.5 Cell Models

In cell models, innate host defenses such as PMNs, macrophages, and NK cells are involved in controlling *Yersinia* infection initially, but a robust adaptive immune response is required to overcome *Yersinia* infection subsequently. Specific Abs[27,28] as well as IFN-γ-producing CD4 and CD8 T cells play an essential role in clearing *Yersinia* infection and have been shown to mediate protection in adoptive transfer experiments.[29] Some of the Yops (including YopE, YopH, YopO, and YopM) are delivered into the host cell cytosol where they damage the cytoskeleton and disrupt the signaling network. Infected macrophages displayed general features of apoptosis such as membrane blebbing (apoptotic body formation) and nuclear and cellular shrinkage.

29.5.1 Adhesion and Invasion of Mammalian Cells

The ability to penetrate host cells can be considered crucial to infectious processes caused by enteric pathogens. Specifically, two chromosomal genes, *inv* and *ail*, and one plasmidial gene, *yadA*, of *Y. enterocolitica* are associated with ability to adhere and invade mammalian cells.[72] Invasin (Inv) is the first adhesin required for *Yersinia* to cause infection. Inv is optimally expressed at 26°C or under acidic conditions at 37°C. This protein is thought to be expressed in the cell wall of the bacteria when they reach the small intestine and is required for the initial steps of host colonization and penetration into the intestinal epithelium cells.[73] Another virulence-associated adhesin of enteropathogenic yersiniae is Ail (for attachment and invasion locus). Ail is expressed at 37°C under aerobic conditions. Like Inv, Ail is chromosomally encoded and also mediates the binding and invasion of epithelial cells.[74,75] The major adhesin from enteropathogenic yersiniae is YadA. Different from Inv and Ail, YadA is encoded by the virulence plasmid pYV. YadA can bind to collagen, fibronectin, and laminin and mediate adhesion to epithelial cells and macrophages. Beyond adhesion, YadA can act as an autoagglutinin, has antiphagocytic properties, and serum resistance.[76–78] The heat-stable enterotoxin Yst, encoded by the chromosomal *ystA* gene, is another factor related to *Y. enterocolitica* virulence. Reports suggest that it may cause diarrhea.[79] Yst affects binding and activation of cell-associated guanylate cyclase and elevation of intracellular concentrations of cyclic GMP. Besides *inv*, *ail*, *ystA*, and *virF*, other chromosomal and plasmidial encoded genes are responsible for the expression of virulence determinants in *Y. enterocolitica*, like iron uptake, fimbriae, flagella, and urease, among others.

29.5.2 Modulation of Host Immune Defense Mechanisms

To overcome host defense mechanisms, pathogenic yersiniae translocate a mixture of effector proteins called Yops into the cytosol of eukaryotic cells by their T3SSs to inhibit the innate and adaptive immune system of the host. These major antihost determinants are located on a 70-kb virulence plasmid pYV, which encodes a protein microinjection apparatus called the T3SS and at least six effector proteins (Yops; YopH, YopO/YpkA, YopP/YopJ, YopE, YopM, and YopT).[12,80,81] The main function of these Yops is to inhibit the immune response of the host. At least four Yops (YopH, YopE, YopT, and YopO/YpkA) are involved in inhibiting phagocytosis of yersiniae by disrupting the cytoskeleton of PMNs and macrophages.[82]

A series of studies suggest that *Yersinia* blocks macrophage TNF-α production by downregulation of mitogen-activated protein kinase (MAPK) activities.[83] A recent report from Orth et al. revealed that YopP/YopJ binds and inactivates members of the MAPK kinase superfamily, which function as upstream MAPK activators.[84] The NF-κB system is central to innate immunity. It controls the synthesis of cytokines, acute-phase proteins, and adhesion molecules and mediates cellular survival by prevention of apoptosis. Certain extracellular stimuli, such as TNF-α, genotoxic agents, and ionizing radiation, simultaneously activate proapoptotic and antiapoptotic pathways in eukaryotic cells. NF-κB functions to upregulate synthesis of proteins that counteract the proapoptotic pathways, such as members of the inhibitor of apoptosis protein (IAP), TNFR-associated factor, and Bcl-2 families.[85]

Studies using HeLa cells have shown that YopH dephosphorylates p130cas, paxillin, and the focal adhesion kinase (FAK), leading to disruption of the focal adhesion and a reduced invasin-mediated engulfment.[86,87] Andersson et al. showed that exposure of J774A.1 macrophages to yopH mutant of *Y. pseudotuberculosis* infection resulted in a transient increase in tyrosine phosphorylation of a number of proteins, including paxillin, which is known to be tyrosine phosphorylated upon Fc-receptor-mediated signaling associated with phagocytosis in macrophages.[88]

Studies involving *Y. enterocolitica* and *Y. pseudotuberculosis* demonstrated that *Yersinia* triggers apoptosis of cultured macrophages, and infected macrophages displayed general features of apoptosis, such as membrane blebbing, cellular shrinkage, and DNA fragmentation.[25,89] Monack et al. found that YopJ, the *Y. pseudotuberculosis* homologue of YopP, is required for the induction of the cell death process.[25] YopP/YopJ plays an anti-inflammatory role by preventing the activation of the transcription factor NF-κB.[90,91] It also induces rapid apoptosis of macrophages. The apoptosis process is accompanied by cleavage of the cytosolic protein BID, the release of cytochrome c, and the cleavage of caspase-3 and 7. The release of cytochrome c and the cleavage of BID can both be inhibited by caspase inhibitors, suggesting that YopP/YopJ interferes with a signaling pathway upstream of the mitochondria.[89]

29.6 Conclusion

Y. enterocolitica and *Y. pseudotuberculosis* are foodborne bacteria associated with gastrointestinal syndromes (from acute enteritis to mesenteric lymphadenitis), or mesenteric adenitis, and septicemia. These pathogens generate a number of proteins that help evade host innate immune response and replicate inside the host. Use of mouse models revealed the interaction between the host and the extracellular lifestyle of the pathogen. Application of molecular, cellular, genetic, and biochemical techniques in combination with *in vitro* models yielded valuable insights into efficiency and sophistication of the *Yersinia* T3SS. Comparative studies employing minipig and mouse models uncovered that O:3 strains utilize serotype- and host-specific colonization strategies in fattening pigs. While *C. elegans* infection model is not feasible for pathogenicity study due to inappropriate temperature, rabbit model is susceptible to diarrhea manifestation. Cell models are useful for investigating different stages of yersinia infection: adhesion and invasion, and evasion of host innate and adaptive immune mechanisms.

REFERENCES

1. Bottone, E.J. *Yersinia enterocolitica*: The charisma continues. *Clin Microbiol Rev* 10, 257–76 (1997).
2. Bleves, S. & Cornelis, G.R. How to survive in the host: The *Yersinia* lesson. *Microbes Infect* 2, 1451–60 (2000).
3. Spinner, J.L. et al. Neutrophils are resistant to *Yersinia* YopJ/P-induced apoptosis and are protected from ROS-mediated cell death by the type III secretion system. *PLoS One* 5, e9279 (2010).
4. Cover, T.L. & Aber, R.C. *Yersinia enterocolitica. N Engl J Med* 321, 16–24 (1989).
5. Fukushima, H. & Gomyoda, M. Intestinal carriage of *Yersinia pseudotuberculosis* by wild birds and mammals in Japan. *Appl Environ Microbiol* 57, 1152–5 (1991).
6. Liang, J. et al. Ecology and geographic distribution of *Yersinia enterocolitica* among livestock and wildlife in China. *Vet Microbiol* 178, 125–31 (2015).
7. Wang, X. et al. *Canis lupus familiaris* involved in the transmission of pathogenic *Yersinia* spp. in China. *Vet Microbiol* 172, 339–44 (2014).
8. Liang, J. et al. Prevalence of *Yersinia enterocolitica* in pigs slaughtered in Chinese abattoirs. *Appl Environ Microbiol* 78, 2949–56 (2012).
9. Wang, X. et al. Distribution of pathogenic *Yersinia enterocolitica* in China. *Eur J Clin Microbiol Infect Dis* 28, 1237–44 (2009).
10. Perry, R.D. & Fetherston, J.D. *Yersinia pestis*—Etiologic agent of plague. *Clin Microbiol Rev* 10, 35–66 (1997).
11. Cornelis, G.R. et al. The virulence plasmid of *Yersinia*, an antihost genome. *Microbiol Mol Biol Rev* 62, 1315–52 (1998).
12. Cornelis, G.R. *Yersinia* type III secretion: Send in the effectors. *J Cell Biol* 158, 401–8 (2002).
13. Cornelis, G.R. Molecular and cell biology aspects of plague. *Proc Natl Acad Sci USA* 97, 8778–83 (2000).
14. Dhar, M.S. & Virdi, J.S. Strategies used by *Yersinia enterocolitica* to evade killing by the host: Thinking beyond Yops. *Microbes Infect* 16, 87–95 (2014).
15. Bhagat, N. & Virdi, J.S. The enigma of *Yersinia enterocolitica* biovar 1A. *Crit Rev Microbiol* 37, 25–39 (2011).

16. Autenrieth, I.B. & Firsching, R. Penetration of M cells and destruction of Peyer's patches by *Yersinia enterocolitica*: An ultrastructural and histological study. *J Med Microbiol* 44, 285–94 (1996).
17. Grutzkau, A., Hanski, C., Hahn, H. & Riecken, E.O. Involvement of M cells in the bacterial invasion of Peyer's patches: A common mechanism shared by *Yersinia enterocolitica* and other enteroinvasive bacteria. *Gut* 31, 1011–5 (1990).
18. Hanski, C. et al. Immunohistochemical and electron microscopic study of interaction of *Yersinia enterocolitica* serotype O8 with intestinal mucosa during experimental enteritis. *Infect Immun* 57, 673–8 (1989).
19. Fallman, M. et al. *Yersinia pseudotuberculosis* inhibits Fc receptor-mediated phagocytosis in J774 cells. *Infect Immun* 63, 3117–24 (1995).
20. Rosqvist, R., Bolin, I. & Wolf-Watz, H. Inhibition of phagocytosis in *Yersinia pseudotuberculosis*: A virulence plasmid-encoded ability involving the Yop2b protein. *Infect Immun* 56, 2139–43 (1988).
21. Bacon, G.A. & Burrows, T.W. The basis of virulence in *Pasteurella pestis*: An antigen determining virulence. *Br J Exp Pathol* 37, 481–93 (1956).
22. China, B., N'Guyen, B.T., de Bruyere, M. & Cornelis, G.R. Role of YadA in resistance of *Yersinia enterocolitica* to phagocytosis by human polymorphonuclear leukocytes. *Infect Immun* 62, 1275–81 (1994).
23. Ruckdeschel, K., Roggenkamp, A., Schubert, S. & Heesemann, J. Differential contribution of *Yersinia enterocolitica* virulence factors to evasion of microbicidal action of neutrophils. *Infect Immun* 64, 724–33 (1996).
24. Visser, L.G., Annema, A. & van Furth, R. Role of Yops in inhibition of phagocytosis and killing of opsonized *Yersinia enterocolitica* by human granulocytes. *Infect Immun* 63, 2570–5 (1995).
25. Monack, D.M., Mecsas, J., Ghori, N. & Falkow, S. *Yersinia* signals macrophages to undergo apoptosis and YopJ is necessary for this cell death. *Proc Natl Acad Sci USA* 94, 10385–90 (1997).
26. Lian, C.J., Hwang, W.S. & Pai, C.H. Plasmid-mediated resistance to phagocytosis in *Yersinia enterocolitica*. *Infect Immun* 55, 1176–83 (1987).
27. Vogel, U., Autenrieth, I.B., Berner, R. & Heesemann, J. Role of plasmid-encoded antigens of *Yersinia enterocolitica* in humoral immunity against secondary *Y. enterocolitica* infection in mice. *Microb Pathog* 15, 23–36 (1993).
28. Igwe, E.I. et al. Rational live oral carrier vaccine design by mutating virulence-associated genes of *Yersinia enterocolitica*. *Infect Immun* 67, 5500–7 (1999).
29. Trulzsch, K. et al. *Yersinia* outer protein P inhibits CD8 T cell priming in the mouse infection model. *J Immunol* 174, 4244–51 (2005).
30. Wang, X. et al. Comparison of the cytokine immune response to pathogenic *Yersinia enterocolitica* bioserotype 1B/O:8 and 2/O:9 in susceptible BALB/C and resistant C57BL/6 mice. *Mol Immunol* 55(3–4), 365–71 (2013).
31. Schmiel, D., Wagar, E., Karamanou, L., Weeks, D. & Miller, V. Phospholipase A of *Yersinia enterocolitica* contributes to pathogenesis in a mouse model. *Infect Immun* 66, 3941–51 (1998).
32. Ghosh, P. Process of protein transport by the type III secretion system. *Microbiol Mol Biol Rev* 68, 771–95 (2004).
33. Navarro, L., Alto, N.M. & Dixon, J.E. Functions of the *Yersinia* effector proteins in inhibiting host immune responses. *Curr Opin Microbiol* 8, 21–7 (2005).
34. Cornelis, G., Vanootegem, J.-C. & Sluiters, C. Transcription of the *yop* regulon from *Y. enterocolitica* requires trans acting pYV and chromosomal genes. *Microb Pathog* 2, 367–79 (1987).
35. Mulder, B., Michiels, T., Simonet, M., Sory, M.-P. & Cornelis, G. Identification of additional virulence determinants on the pYV plasmid of *Yersinia enterocolitica* W227. *Infect Immun* 57, 2534–41 (1989).
36. Rosqvist, R., Forsberg, Å., Rimpiläinen, M., Bergman, T. & Wolf-Watz, H. The cytotoxic protein YopE of *Yersinia* obstructs the primary host defence. *Mol Microbiol* 4, 657–67 (1990).
37. Monack, D.M., Mecsas, J., Bouley, D. & Falkow, S. *Yersinia*-induced apoptosis in vivo aids in the establishment of a systemic infection of mice. *J Exp Med* 188, 2127–37 (1998).
38. Trülzsch, K. et al. *Yersinia* outer protein P inhibits CD8 T cell priming in the mouse infection model. *J Immunol* 174, 4244–51 (2005).
39. Ruckdeschel, K., Richter, K., Mannel, O. & Heesemann, J. Arginine-143 of *Yersinia enterocolitica* YopP crucially determines isotype-related NF-κB suppression and apoptosis induction in macrophages. *Infect Immun* 69, 7652–62 (2001).

40. Trülzsch, K., Sporleder, T., Igwe, E.I., Rüssmann, H. & Heesemann, J. Contribution of the major secreted yops of *Yersinia enterocolitica* O:8 to pathogenicity in the mouse infection model. *Infect Immun* 72, 5227–34 (2004).
41. Logsdon, L.K. & Mecsas, J. Requirement of the *Yersinia pseudotuberculosis* effectors YopH and YopE in colonization and persistence in intestinal and lymph tissues. *Infect Immun* 71, 4595–607 (2003).
42. Ruiz-Bravo, A., Jimenez-Valera, M., Sampedro, A. & Moreno, E. Chromosome-mediated resistance of *Yersinia enterocolitica* serotype O9 to intracellular killing by mouse peritoneal macrophages. *FEMS Immunol Med Microbiol* 9, 317–24 (1994).
43. Ruiz-Bravo, A., Moreno, E., Sampedro, A. & Jimenez-Valera, M. Experimental infection of mice with *Yersinia enterocolitica* serotype O9 by oral and parenteral routes: spreading and enterotropism of virulent yersiniae. *Curr Microbiol* 38, 257–63 (1999).
44. Thibodeau, V., Frost, E.H., Chenier, S. & Quessy, S. Presence of *Yersinia enterocolitica* in tissues of orally-inoculated pigs and the tonsils and feces of pigs at slaughter. *Can J Vet Res* 63, 96–100 (1999).
45. Shu, D. et al. Experimental infection of newborn piglets with *Yersinia enterocolitica*: An animal model of enteritis. *N Z Vet J* 43, 50–6 (1995).
46. Robins-Browne, R. et al. The pathogenesis of *Yersinia enterocolitica* infection in gnotobiotic piglets. *J Med Microbiol* 19, 297–308 (1985).
47. Valentin-Weigand, P., Heesemann, J. & Dersch, P. Unique virulence properties of *Yersinia enterocolitica* O:3—An emerging zoonotic pathogen using pigs as preferred reservoir host. *Int J Med Microbiol* 304, 824–34 (2014).
48. Schaake, J. et al. Essential role of invasin for colonization and persistence of *Yersinia enterocolitica* in its natural reservoir host, the pig. *Infect Immun* 82, 960–9 (2014).
49. Bottone, E.J. *Yersinia enterocolitica*: Overview and epidemiologic correlates. *Microbes Infect* 1, 323–33 (1999).
50. Fredriksson-Ahomaa, M. Isolation of enteropathogenic *Yersinia* from non-human sources. in *Advances in Yersinia Research* 97–105 (Springer, New York, 2012).
51. Waterfield, N.R. & Wren, B.W. Invertebrates as a source of emerging human pathogens. *Nat Rev Microbiol* 2, 833–41 (2004).
52. Spanier, B., Starke, M., Higel, F., Scherer, S. & Fuchs, T.M. *Yersinia enterocolitica* infection and *tcaA*-dependent killing of *Caenorhabditis elegans*. *Appl Environ Microbiol* 76, 6277–85 (2010).
53. Darby, C., Hsu, J.W., Ghori, N. & Falkow, S. *Caenorhabditis elegans*: Plague bacteria biofilm blocks food intake. *Nature* 417, 243–4 (2002).
54. Marceau, M. Transcriptional regulation in *Yersinia*: An update. *Curr Issues Mol Biol* 7, 151–78 (2005).
55. Straley, S.C. & Perry, R.D. Environmental modulation of gene expression and pathogenesis in *Yersinia*. *Trends Microbiol* 3, 310–7 (1995).
56. Bresolin, G., Neuhaus, K., Scherer, S. & Fuchs, T.M. Transcriptional analysis of long-term adaptation of *Yersinia enterocolitica* to low-temperature growth. *J Bacteriol* 188, 2945–58 (2006).
57. Heermann, R. & Fuchs, T.M. Comparative analysis of the *Photorhabdus luminescens* and the *Yersinia enterocolitica* genomes: Uncovering candidate genes involved in insect pathogenicity. *BMC Genomics* 9, 40 (2008).
58. Une, T. Studies on the pathogenicity of *Yersinia enterocolitica*. I. Experimental infection in rabbits. *Microbiol Immunol* 21, 341–63 (1977).
59. Pai, C.H., Mors, V. & Seemayer, T.A. Experimental *Yersinia enterocolitica* enteritis in rabbits. *Infect Immun* 28, 238–44 (1980).
60. Heesemann, J., Schröder, J. & Ulrich, M. Analysis of the class-specific immune response to *Yersinia enterocolitica* virulence-associated antigens in oro-gastrically infected rabbits. *Microb Pathog* 5, 437–47 (1988).
61. Heesemann, J., Gaede, K. & Autenrieth, I. Experimental *Yersinia enterocolitica* infection in rodents: A model for human yersiniosis. *APMIS* 101, 417–29 (1993).
62. Vesselinova, A., Najdenski, H., Nikolova, S. & Wesselinova, D. Arthritis after experimental infection with *Yersinia enterocolitica* O:3 in rabbits. *J Vet Med Ser B* 48, 43–53 (2001).
63. Laird, W. & Cavanaugh, D. Correlation of autoagglutination and virulence of yersiniae. *J Clin Microbiol* 11, 430 (1980).
64. Schiemann, D. An enterotoxin-negative strain of *Yersinia enterocolitica* serotype O:3 is capable of producing diarrhea in mice. *Infect Immun* 32, 571–4 (1981).

65. Lian, C.-J., Hwang, W., Kelly, J. & Pai, C. Invasiveness of *Yersinia enterocolitica* lacking the virulence plasmid: An in-vivo study. *J Med Microbiol* 24, 219–26 (1987).
66. O'loughlin, E. et al. Clinical, morphological, and biochemical alterations in acute intestinal yersiniosis. *Pediatr Res* 20, 602–8 (1986).
67. O'Loughlin, E., Pai, C. & Gall, D. Effect of acute *Yersinia enterocolitica* infection on in vivo and in vitro small intestinal solute and fluid absorption in the rabbit. *Gastroenterology* 94, 664–72 (1988).
68. Delor, I. & Cornelis, G.R. Role of *Yersinia enterocolitica* Yst toxin in experimental infection of young rabbits. *Infect Immun* 60, 4269–77 (1992).
69. Najdenski, H., Golkocheva, E., Vesselinova, A., Bengoechea, J. & Skurnik, M. Proper expression of the O-antigen of lipopolysaccharide is essential for the virulence of *Yersinia enterocolitica* O:8 in experimental oral infection of rabbits. *FEMS Immunol Med Microbiol* 38, 97–106 (2003).
70. Roggenkamp, A., Bittner, T., Leitritz, L., Sing, A. & Heesemann, J. Contribution of the Mn-cofactored superoxide dismutase (SodA) to the virulence of *Yersinia enterocolitica* serotype O8. *Infect Immun* 65, 4705–10 (1997).
71. Najdenski, H.M., Golkocheva, E.N., Vesselinova, A.M. & Rüssmann, H. Comparison of the course of infection of virulent *Yersinia enterocolitica* serotype O:8 with an isogenic *sodA* mutant in the peroral rabbit model. *Int J Med Microbiol* 294, 383–93 (2004).
72. Miller, V.L. & Falkow, S. Evidence for two genetic loci in *Yersinia enterocolitica* that can promote invasion of epithelial cells. *Infect Immun* 56, 1242–8 (1988).
73. Pepe, J.C. & Miller, V.L. The *Yersinia enterocolitica inv* gene product is an outer membrane protein that shares epitopes with *Yersinia pseudotuberculosis* invasin. *J Bacteriol* 172, 3780–9 (1990).
74. Pierson, D.E. & Falkow, S. The *ail* gene of *Yersinia enterocolitica* has a role in the ability of the organism to survive serum killing. *Infect Immun* 61, 1846–52 (1993).
75. Goverde, R.L., Jansen, W.H., Brunings, H.A., Huis in 't Veld, J.H. & Mooi, F.R. Digoxigenin-labelled *inv*- and *ail*-probes for the detection and identification of pathogenic *Yersinia enterocolitica* in clinical specimens and naturally contaminated pig samples. *J Appl Bacteriol* 74, 301–13 (1993).
76. Leo, J.C. & Skurnik, M. Adhesins of human pathogens from the genus *Yersinia*. *Adv Exp Med Biol* 715, 1–15 (2011).
77. Paczosa, M.K., Fisher, M.L., Maldonado-Arocho, F.J. & Mecsas, J. *Yersinia pseudotuberculosis* uses Ail and YadA to circumvent neutrophils by directing Yop translocation during lung infection. *Cell Microbiol* 16, 247–68 (2014).
78. Liang, J. et al. Two novel *ail*-positive biotype 1A strains of *Yersinia enterocolitica* isolated in China with unequal adhesion and invasion properties. *Infect Genet Evol* 27, 83–8 (2014).
79. Sulakvelidze, A. Yersiniae other than *Y. enterocolitica*, *Y. pseudotuberculosis*, and *Y. pestis*: The ignored species. *Microbes Infect* 2, 497–513 (2000).
80. Peters, K.N. & Anderson, D.M. Modulation of host cell death pathways by *Yersinia* species and the type III effector YopK. *Adv Exp Med Biol* 954, 229–36 (2012).
81. Heesemann, J., Sing, A. & Trulzsch, K. *Yersinia's* stratagem: Targeting innate and adaptive immune defense. *Curr Opin Microbiol* 9, 55–61 (2006).
82. Gebus, C., Faudry, E., Bohn, Y.S., Elsen, S. & Attree, I. Oligomerization of PcrV and LcrV, protective antigens of *Pseudomonas aeruginosa* and *Yersinia pestis*. *J Biol Chem* 283, 23940–9 (2008).
83. Sodhi, A., Sharma, R.K., Batra, H.V. & Tuteja, U. Recombinant fraction 1 protein of *Yersinia pestis* activates murine peritoneal macrophages in vitro. *Cell Immunol* 229, 52–61 (2004).
84. Orth, K. et al. Inhibition of the mitogen-activated protein kinase kinase superfamily by a *Yersinia* effector. *Science* 285, 1920–3 (1999).
85. Ruckdeschel, K., Richter, K., Mannel, O. & Heesemann, J. Arginine-143 of *Yersinia enterocolitica* YopP crucially determines isotype-related NF-κB suppression and apoptosis induction in macrophages. *Infect Immun* 69, 7652–62 (2001).
86. Black, D.S. & Bliska, J.B. Identification of p130^{Cas} as a substrate of *Yersinia* YopH (Yop51), a bacterial protein tyrosine phosphatase that translocates into mammalian cells and targets focal adhesions. *EMBO J* 16, 2730–44 (1997).
87. Persson, C., Carballeira, N., Wolf-Watz, H. & Fallman, M. The PTPase YopH inhibits uptake of *Yersinia*, tyrosine phosphorylation of p130^{Cas} and FAK, and the associated accumulation of these proteins in peripheral focal adhesions. *EMBO J* 16, 2307–18 (1997).

88. Andersson, K. et al. YopH of *Yersinia pseudotuberculosis* interrupts early phosphotyrosine signalling associated with phagocytosis. *Mol Microbiol* 20, 1057–69 (1996).
89. Mills, S.D. et al. *Yersinia enterocolitica* induces apoptosis in macrophages by a process requiring functional type III secretion and translocation mechanisms and involving YopP, presumably acting as an effector protein. *Proc Natl Acad Sci USA* 94, 12638–43 (1997).
90. Schesser, K. et al. The yopJ locus is required for *Yersinia*-mediated inhibition of NF-κB activation and cytokine expression: YopJ contains a eukaryotic SH2-like domain that is essential for its repressive activity. *Mol Microbiol* 28, 1067–79 (1998).
91. Ruckdeschel, K. et al. *Yersinia enterocolitica* impairs activation of transcription factor NF-κB: Involvement in the induction of programmed cell death and in the suppression of the macrophage tumor necrosis factor α production. *J Exp Med* 187, 1069–79 (1998).

Section IV

Foodborne Infections due to Fungi

30

Alternaria

Alicia Rodríguez, Andrea Patriarca, Mar Rodríguez, María Jesús Andrade, and Juan José Córdoba

CONTENTS

30.1 Introduction

Alternaria is a ubiquitous fungal genus covering 275 species,[1] and it has been typified as saprobic, endophytic, and pathogenic. Currently, 24 *Alternaria* sections, 16 of which are newly described, and 6 monotypic lineages are recognized.[2] *Alternaria* is associated with a variety of foods. Species of *Alternaria* are serious plant pathogens that damage crops in the field and cause postharvest decays. In addition, with the ability to grow at low temperatures and low water activity (a_w),[3,4] *Alternaria* spp. may contaminate and infect fruit and vegetables kept refrigerated.[5,6] The most commonly reported species include *Alternaria alternata*, *Alternaria tenuissima*, *Alternaria arborescens*, *Alternaria radicina*, *Alternaria brassicae*, *Alternaria brassicicola*, and *Alternaria infectoria*. Depending on geographical areas, a specific *Alternaria* species can predominate over others. For instance, *A. tenuissima* is a more frequently isolated species in Argentinean wheat[7] than *A. alternata* and *A. infectoria*, which have been reported as the predominant species in cereals in several studies worldwide.[8–12] *Alternaria* species colonize a range of plants including cereals,[13] tomatoes,[14] oilseeds,[15] apples,[16] berries, grapes, citrus fruits,[17] carrots,[18] and others. Some of these moulds may have a negative impact on foods due to spoilage, causing economic losses and infections.[19] It is well known that *Alternaria* colonization of ripening ears can result in black pointed grain and impact directly on flour color of bread-making wheats,[20] its spores can be allergenic and cause asthma,[21] and some species may synthesize mycotoxins whose high toxicity is detrimental to both human health and animal welfare.

30.2 Foodborne Diseases due to *Alternaria* Mycotoxins

The main cause of *Alternaria* foodborne diseases is due to the production and accumulation of mycotoxins on foods that could lead to acute and, more commonly, chronic effects. Although growth of moulds in foods is not necessarily associated with the formation of mycotoxins,[22] in many cases their presence could lead to the accumulation of these metabolites in foods. Only about 30 of the 120 known secondary metabolites of *Alternaria* are considered toxic to humans and animals, many of them acting as phytotoxins.[23] Only a small proportion of such phytotoxins has been chemically characterized and reported to act as mycotoxins in humans and animals. The most important *Alternaria* mycotoxins are alternariol (AOH), alternariol monomethyl ether (AME), altenuene (ALT), tenuazonic acid (TeA), tentoxin (TEN), and altertoxins I, II, and III (ATX-I, -II, and -III),[24] which belong to three structural classes[25]: (1) dibenzopyrone derivatives (AOH, AME, ALT), (2) perylene derivatives (ATX-I, -II, -III), and (3) tetramic acid derivatives (TeA). The production of important *Alternaria* mycotoxins by the most common *Alternaria* species is shown in Table 30.1.

The environmental conditions in which specific *Alternaria* mycotoxins may be produced have been identified. In general, the conditions for mycotoxin production are slightly narrower in terms of temperature and a_w conditions relative to those for growth. Studies suggest that mycotoxin production by *Alternaria* strains is more variable than their growth.[26–28] There are numerous studies on *A. alternata* growth and evaluation of mycotoxin production in different matrices such as wheat, soya beans, sorghum, and tomatoes.[4,29–32] The optimal conditions for AOH, AME, and TeA production varied with strains and contaminated food. Thus, e.g., for AOH, in soya beans the optimum conditions are at 25°C and 0.98 a_w,[30,31] while in tomatoes the optimum conditions are at 21°C and 0.954 a_w after 28 days.[32] For AME, the optimal production in irradiated soya beans is around 25°C and 0.98 a_w, and no production is found at temperatures higher than 30°C.[30,31] However, the maximum concentration of AME in tomatoes is obtained at 0.954 a_w and 35°C, although high amounts are also produced at 21°C.[32] For TeA, the optimal conditions for production are 0.98 a_w over a range of 25°C–30°C on soya-bean-based media,[27] 0.99–0.93 a_w over 28 days growth periods on sorghum,[31] and 0.982 a_w at 21°C after 28 days of incubation on tomato-based media.[32] In the latter, this mycotoxin is even detected after 28 days of incubation at 0.982 a_w and 6°C. Recently, a study has been performed about the impact of $a_w \times$ temperature conditions on ATX-II by using strains of *A. tenuissima* isolated from Argentinean cereals.[28] In this study, the

TABLE 30.1

Toxic Secondary Metabolites Produced by the Most Important *Alternaria* Species and Foods Involved

Species	Toxic Secondary Metabolites	Foods	References
A. alternata	AOH, AME, ALT, TeA, ATX-I, ATX-II, ATX-III	Tomato, red pepper, fresh fruit, vegetables, peanut, wheat	37,87,88
A. tenuissima	AOH, AME, ATX-I, -III, TeA	Strawberries	28,88
A. arborescens	AOH, AME, TEN	Fruits, tomato, barley grain	87,89
A. radicina	ATX-I, -II, -III, TeA	Umbelliferous plants	88
A. brassicae	AOH, AME	Mustard, opium	90,91
A. brassicicola	Small phytotoxic metabolites (causal agent of black spot disease)	Rape	92–94
A. infectoria	4Z-infectopyrone, novae-zelandin A, novae-zelandin B, phomapyrone A, B, E, G, F, D	Wheat	87

Source: Ostry, V., *World Mycotoxin J.*, 1, 175, 2008.

AOH, alternariol; AME, alternariol methyl ether; ALT, altenuene; TeA, tenuazonic acid; ATX-I, -II, -III, altertoxins-I, -II, and -III.

optimal conditions for ALTX-II production were 0.98 a_w and 30°C, although *A. tenuissima* strains also accumulate significant amounts of this toxin at 34°C.

Taking into account the ecological conditions for mycotoxin production by *Alternaria* species, it is not surprising that these toxins can contaminate a wide range of food and foodstuffs including, e.g., citrus, cereals, olives, tomatoes, apples, seeds, berries, and juices (reviewed by Scott,[33] see Table 30.1). The maximum levels of *Alternaria* mycotoxins reported in marketed food products are in the range of 1–10^3 μg/kg.[24]

Alternaria toxins have recently received increasing attention. The European Commission asked the European Food Safety Authority (EFSA) to review the safety of *Alternaria* toxins in food and feed.[34] EFSA concluded that *Alternaria* toxins are of high concern for public health. In fact, exposure to *Alternaria* toxins has been linked to esophageal cancer in Linxian Province, China, as well as in areas of South Africa where high levels of *A. alternata* contamination have been found in grains.[23,35,36] However, due to the possible co-occurrence of mycotoxins in contaminated grains, this correlation may not have been exclusively ascribed to *Alternaria* toxins. For this reason, it is important to have adequate methods first for analyzing and identifying *Alternaria* species and their toxins and second to examine the toxicity for foodborne mycotoxin-producing *Alternaria.*

30.3 Investigation and Analysis of Toxigenic *Alternaria* Strains

The isolation and identification of *Alternaria* species from food products is based on traditional methods including culturing under standardized conditions and observation of morphological characteristics. Besides being laborious, time consuming, and sometimes restricted to experts in this field, this identification scheme is difficult owing to variations within the same species as the morphological characteristics are very sensitive to the cultural conditions. The most recent taxonomic keys, as reviewed by Simmons[1] in his identification manual, provide descriptions and illustrations of at least 296 taxa, based on the examination of stable isolates in axenic culture. The isolates are cultured in potato carrot agar (PCA) and V8 agar and incubated for 7 days at 23°C under an alternating light/dark cycle consisting of 8 h of cool-white fluorescent daylight and 16 h darkness so as to observe morphological characteristics (Figure 30.1).

An obvious drawback of such traditional methods for detecting potential mycotoxin-producing *Alternaria* is the need of an evaluation of the production of mycotoxins by foodborne strains. The production of secondary metabolites has been used as a means of classification and identification, taking advantage of the enormous potential of this genus to biosynthesize secondary metabolites. Combinations of conidial morphology, genetics, and pathogenicity have been shown to be valuable in *Alternaria* taxonomy.[8]

Many published reports have demonstrated the benefits of using *in vitro* techniques to investigate mycotoxin production by foodborne *Alternaria.* Having been used to assess the mycotoxin production for many years, rice represents an excellent matrix for comparison of toxigenic potential of mould strains isolated from different foods and geographical regions. Mycotoxin production could be evaluated in culture media made with sterilized rice or directly in sterilized rice. However, it should be noted that autoclaved rice has defective mechanical barriers and suppressed defenses because of their thermolability. This provides conditions for *Alternaria* strains to express their whole toxigenic potential, which may or may not be possible under natural or real-food conditions.[37]

Mycotoxin production is strain dependent and additionally conditioned by the food in which the mould strain is growing and by the environmental factors. Thus, it is necessary to evaluate the potential risk of toxin accumulation in the specific food matrix. Culture media based on the food are frequently used as it simplifies the extraction of the toxins, such as tomato medium, wheat agar medium, etc.[28,32] When the real food matrix is used, e.g., cereal grains or legumes, it could be autoclaved[4] or gamma irradiated[31] to eliminate the natural mycobiota. However, as these mild treatments do not destroy mycotoxins, the natural occurrence of the toxins of interest should be verified before their utilization for analyzing the *Alternaria* toxigenic potential.

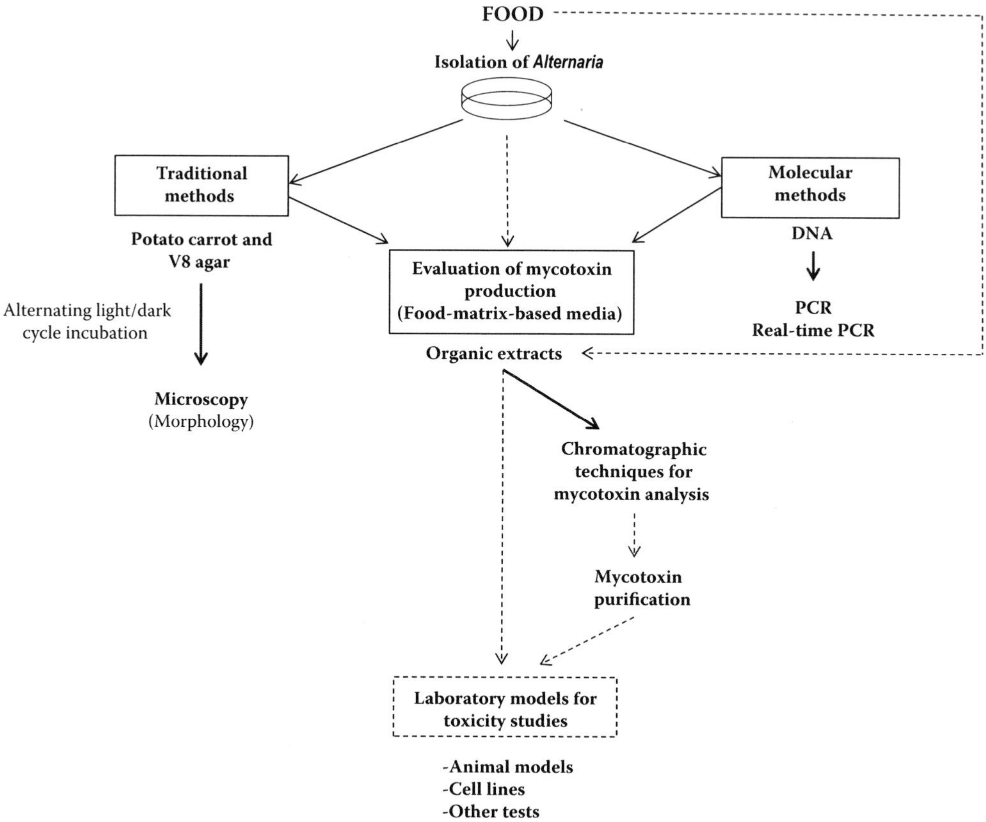

FIGURE 30.1 Procedures for analyzing foodborne *Alternaria*.

For the evaluation of mycotoxin production by *Alternaria*, analytical methods based on chromatography are often used for their extraction and detection from the food matrices (Figure 30.1). Usually, a solvent extraction from solid foods with organic solvents, such as dichloromethane, methanol, acetonitrile, or ethyl acetate, is required, although an acidic extraction or a further acidification step is preferable to increase the recovery for TeA.[33] Cleanup procedures are necessary when a complex matrix is involved. Successive steps of solvent partitioning, solid phase extraction (SPE) columns, or solid phase microextraction are common cleanup techniques used in most food matrices.[24]

Mycotoxin determination has been traditionally carried out by thin-layer chromatography (TLC), gas chromatography (GC), and mainly high-performance liquid chromatography (HPLC), usually with ultraviolet (UV) detection, although fluorescence and electrochemical detection have also been used for certain toxins. Atmospheric pressure chemical ionization (APCI), LC-mass spectroscopy (MS), and LC-MS/MS have been applied to the determination and confirmation of AOH and AME identity at sub ng/mL levels.[33,38] Several multimycotoxin methods have been recently developed, many of which include the detection of *Alternaria* toxins, most of them based on LC-MS/MS systems. A multimethod for detection of 33 mycotoxins (including AOH and AME) in various food matrices has been developed by Spanjer et al.,[39] based on LC-MS/MS using an electrospray ionization interface (ESI) for detection and MS/MS with multiple reaction monitoring (MRM). Rasmussen et al.[40] developed a method for simultaneous detection of 27 mycotoxins (including AOH, AME, TeA, and altersetin) in maize silage. A simple pH-buffered sample extraction has been developed based on the QuEChERS method, and the detection is made by LC-MS/MS without further cleanup. Wang et al.[41] developed an LC-MS/MS method for detecting 17 mycotoxins (including AOH and AME) with application in traditional Chinese medicine products, using a MultiPurification Column for cleanup.

Even though several techniques are effective for *Alternaria* mycotoxin quantification in different food matrices, there are a number of difficulties related to the performance of these methods, such as the efficiency of sample cleanup, low or unequal recoveries for some of the toxins, availability of standards, lack of reference materials for food and feedstuffs, etc. In addition, these methods are time consuming.

As an alternative to traditional methods, nucleic-acid-based techniques are being increasingly applied to examine the taxonomic relationships among *Alternaria* species. Most of them have been focused on small-spored catenulate *Alternaria*, which show little resolution in their molecular phylogeny. However, cladistics analyses of "housekeeping genes" commonly used for other genera, such as the mitochondrial large subunit (mtLSU) ribosomal DNA, internal transcribed spacer (ITS), β-tubulin, translation elongation factor α, calmodulin, actin, chitin synthetase, etc., failed to discriminate among the small-spored species, except for the *A. infectoria* species group.[42–45] There are also genomic techniques to detect, identify, and quantify toxigenic moulds in foodstuffs. So far, there are no molecular methods based on genes involved in the *Alternaria* mycotoxin biosynthesis pathways; however, some methods, which have used unique conserved genes that distinguish toxigenic and nontoxigenic *Alternaria* spp., have been developed successfully. Several molecular methods have been developed to detect the presence of *Alternaria* spores and biomass in foods, such as the one reported by Zur et al.,[46] a polymerase chain reaction (PCR)-based method with primers specific to the ITS1 and ITS2 of the 5.8S rRNA gene of *Alternaria* to detect its presence in commercial tomato products. The main inconvenience of these methods is that viable and nonviable cells cannot be distinguished, thus resulting in an overestimation of the amount of spores that can actually produce mycotoxin in a food product. More recently, Crespo-Sempere et al.[47] developed a method including a pretreatment of samples with nucleic-acid-intercalating dyes (propidium monoazide, PMA) prior to quantitative PCR. PMA selectively penetrates cells with a damaged membrane, inhibiting DNA amplification during PCR. The method, based on a primer pair (*Alt4* and *Alt5*) specific to *Alternaria* spp., allowed quantifying a detection limit of 10^2 spores/g on tomatoes. Even though the tomato matrix had a protective effect on the cells against PMA toxicity, reducing the efficiency to distinguish between viable and nonviable cells, the method is still a suitable tool for quantifying viable *Alternaria* cells, which could be useful for estimating potential risks of mycotoxin contamination. The main drawback of these nucleic-acid-based methods as well as of the traditional identification of *Alternaria* followed by evaluation of mycotoxin production is that they only give information about the negative potential effect derived of the *Alternaria* presence in foods. Although the latter techniques allow taking corrective actions to avoid presence of mycotoxin-producing *Alternaria* on foods, for a more appropriate investigation of foodborne *Alternaria* it is necessary to use laboratory animals or cell system models in which the effect of *Alternaria* extracts or *Alternaria* mycotoxin contaminated foods can be evaluated. This could allow estimating the real risk of the presence of *Alternaria* in foods.

30.4 Laboratory Models for Foodborne *Alternaria* Analysis

The design of extensive biological tests for foodborne *Alternaria* analysis based on mycotoxin effects is limited because of the high number of animals needed and ethical and scientific reasons.[48–50] In addition, such methods have an important limitation related to the toxin amount required for performing the experiments. For this, the total synthesis of AOH, AME, ALT, and iso-ALT has been already established.[51,52] However, due mainly to the mycotoxin production and purification costs, the use of high amounts of toxins for *in vivo* trials is still difficult. Most of the toxicity tests have thus been performed using *Alternaria* culture extracts[53–56] or *in vitro* models focused on cellular systems to obtain data about the toxicity from the different purified mycotoxins synthesized by *Alternaria* species. On the other hand, the extent of *in vivo* mycotoxin absorption could be predicted using *in vitro* permeability and solubility measurements. For this, animal and human cell lines, undifferentiated epithelial cells, have been extensively used for understanding the mechanisms by which some mycotoxins induce different disorders. In Tables 30.2 and 30.3, the most relevant laboratory models described for analyzing foodborne *Alternaria* have been summarized.

TABLE 30.2

Animal Models for Evaluating the Toxic Effects of *Alternaria* Extracts and Produced Mycotoxins

Animals	*Alternaria* Extracts/Mycotoxin	References
Chickens	*Alternaria* extract	55
	TeA	10
Rats	*Alternaria* extract	55
Chicken embryos	*Alternaria* extract	53
Mice	*Alternaria* extract	53,55
	AOH	50
	TeA	48
Dogs	TeA	10

TeA, tenuazonic acid; AOH, alternariol.

TABLE 30.3

Cell Line Systems and Other *In Vitro* Tests Used for Studying the Toxic Effects of *Alternaria* Mycotoxins

Cell Lines/*In Vitro* Assay	Mycotoxins/*Alternaria* Extracts	Studied Toxic Effects	References
Caco-2	AOH/AME	Absorption, metabolism, oxidative stress	59–61
HT-29	AOH/AME	DNA-damaging properties	62–65
	TeA, ATX-I, -II, -III	DNA-damaging properties	66
V79	AOH/AME	Cell proliferation, clastogenic potential, mutagenicity	63,67,69
	ATX-I, -II	Mutagenicity	68,70
H4IIE	ATX-I	Mutagenicity	70
MLC	AOH	Mutagenicity	69
CHL/3T3/L-O2	TeA	Cytotoxicity	71
RAW 264.7	AOH	DNA-damaging properties, autophagy, senescence	72–74
HCT116	AOH	Cell death mode	75
	AOH/AME	Cytotoxicity	76
Ishikawa	AOH	Estrogenic potential	67
H295R	AOH	Alteration in steroid hormone production	77
RGA	AOH	Alterations in androgen, progestogen, glucocorticoid, estrogen nuclear receptor	77
Porcine endometrial	AOH	Oestrogenic effect	78
Hepa-1c1c7/Hepa-1c1c4/Hepa-1c1c12	AOH/AME	Cytotoxicity	79,80
HepG2	AOH	Cytotoxicity	81
A431	AOH/AME/ALT	Viability and cytotoxicity potential	62
MCF-7	AOH	DNA-damaging properties	62
Artemia salina test	AOH, AME, ALT, ATX-I, TeA and *Alternaria* extracts	Toxicity	82,83
Ames *Salmonella* test	*Alternaria* extracts and ATX-I, -II, and -III	Mutagenicity	56,84
	ALT, AOH, AME, ATX-I, TEN, TeA	Mutagenicity	70,85
Human digestive tract simulation	*Alternaria* extracts and AOH and ALT	Bioaccessibility from food	48

AOH, alternariol; AME, alternariol methyl ether; TeA, tenuazonic acid; ATX-I, -II, -III, altertoxins-I, -II and -III; ALT, altenuene.

30.4.1 Animal Laboratory Models

Several animals could be used as model for foodborne *Alternaria* analysis using extracts from *Alternaria* growing on culture media. Culture extracts of these moulds show toxic effects toward chickens and rats,[55] and chicken embryos.[53] In addition, *Alternaria* cultures have also been found to be teratogenic and fetotoxic in mice[53,55] (Table 30.2). The main disadvantage of using these kinds of extracts is their unknown composition, which restricts obtaining of conclusions from the experiments. However, these tests in animal models give general information on the toxicity of *Alternaria* extracts, which could be useful in foodborne *Alternaria* investigation.

Regarding assays of purified *Alternaria* toxins in laboratory animal models, few scientific reports focusing on the combined toxicokinetic and *in vivo* genotoxicity potential have been published. Schuchardt et al.[50] tested the toxicokinetic behavior and *in vivo* genotoxicity of AOH in NMRI mice. The study revealed low systemic absorption, and the target organ toxicity would most likely be restricted to the gastrointestinal tract. In addition, the above study demonstrated no toxic or genotoxic effect of AOH in bone marrow, and the comet assay with liver tissue did not indicate systemic genotoxicity too. TeA has been shown to be subacutely toxic in mice [LD_{50} (lethal dose 50) i.v. 115 mg/kg]. In fact, mice consuming feed contaminated with TeA for 10 months developed alterations in the oesophageal mucosa, suggesting the possibility that progression to esophageal cancer might occur after prolonged exposure.[48] ALT and ATX-I are acutely toxic in mice, with LD_{50} of 50 and 200 mg/kg, respectively[22] (Table 30.2).

TeA is acutely toxic for chickens and dogs. In particular, in chicken feed with increasing TeA levels from sublethal to lethal, the feed efficiency was progressively reduced, the weight gain was suppressed, and the internal hemorrhaging increased.[10] In dogs, a daily ingestion of 10 mg/kg body weight provokes hemorrhaging in several organs[10] (Table 30.2).

Although whichever former animal model could be used for further foodborne *Alternaria* investigation, *in vitro* laboratory models (including cell lines) offer a good alternative to extract adequate conclusions in this kind of investigation. These experiments will be discussed in the next section.

30.4.2 *In Vitro* Laboratory Models

30.4.2.1 Cell Line Systems

In vitro cell line models have to satisfy the following two basic requirements: (1) must be available and easy to handle for high-throughput testing and (2) yield data must support interpretation of results for *in vivo* situation.[57] Only a few cell lines are utilized to examine the toxic effect of *Alternaria* mycotoxins. The main characteristics of these cellular systems and the effect produced by the *Alternaria* mycotoxins and *Alternaria* culture extracts on them will be detailed in this section and summarized in Table 30.3.

In most of the *in vitro* studies of the gut, the human colon tumorigenic cell lines Caco-2 and HT-29 have been used for studies related to intestinal cell function and differentiation.[57,58] Several experiments have been performed to examine the absorption, metabolism, and oxidative stress of AOH and AME in Caco-2 cells.[59–61] There are similarities between the structures of the conjugated metabolites of AOH and AME since they are readily conjugated with glucuronic acid and sulfate.[61] In this study, it was also shown that, based on apparent permeability coefficients, AOH is expected to be extensively and rapidly absorbed from the intestinal lumen *in vivo* and reaches the portal blood both as aglycone and as glucuronide and sulfate.[61] Oxidative stress due to AOH is confirmed by alteration of glutathione (GSH) and the enzymes involved in the redox system after 15, 30, and 60 μM of AOH exposure during 24 h.[60] AOH also causes DNA damage after 24 h in Caco-2 cells.[60] Cytotoxicity of AOH (from 3.125 to 100 μM) was determined during 24, 48, and 72 h of exposure by using different end points.[59] AOH decreases the cell viability and induces a strong oxidative stress in Caco-2 cells.[59]

HT-29 cell line has been mainly used to evaluate the DNA-damaging properties of AOH and AME. Fehr et al.[62] characterized AOH as a poison of topoisomerase I and II, with a certain selectivity for the IIα isoform playing a role in the DNA strand breakage. On the contrary, several studies[63–65] demonstrated that AOH and AME do not have an apparent negative effect on DNA integrity, or at least they do not play a main role in the genotoxic properties. In a toxicity-guided fractionation assay, aimed to identify DNA strand breaking impact compounds in extracts obtained from rice heavily infested with *A. alternata*, the HT-29 cell line was used.[66] TeA, AOH, and AME do not cause significant DNA-damaging effects, while ATX-II contributes to the genotoxic effects of the extracts, showing potent DNA-damaging properties in HT-29 cells. ATX-II does not provoke oxidative stress, while it does influence the cell cycle distribution of HT-29 cells. The inhibition of proliferation of HT-29 cells (seen in a sulforhodamine B assay) matched this interference with the cell cycle, thus arguing for effects on inhibition of cell proliferation rather than cytotoxicity.[66]

Chinese hamster cell line V79 has been developed from lung tissue of a young male Chinese hamster. This cell line is used to investigate DNA damage, the effect on cell proliferation, and the clastogenic potential of AOH. Pfeiffer et al.[63] showed that AOH and AME induce DNA strand breaks in a concentration-dependent manner in the cell line V79. Lehmann et al.[67] demonstrated that AOH inhibits cell proliferation by interference with the cell cycle and induces kinetochore-negative micronucleus (MN) in cultured V79 cells. ATX II is a potent mutagen in the cell line V79, inducing a concentration-dependent increase of mutations at the hypoxanthine guanine phosphoribosyltransferase gene locus at concentrations similar to that of the established mutagen 4-quinoline-N-oxide.[68] However, the mutagenic potency of AOH is at least 50 times lower than that of ATX II.[69] In contrast to AOH and AME, ATX II does not affect the cell cycle of V79 cells. ATX II also causes DNA strand breaks in V79 cells, being more potent than AOH and AME. Schrader et al.[70] demonstrated that nitrosylated ATX-I is mutagenic to V79 cells too. They also used rat hepatoma H4IIE cells to evaluate the mutagenicity of ATX-I, since they retain more metabolic activities and have been used as a model for assessing chemical effects on the induction of cytochrome P450, aryl hydrocarbon hydroxylase, and epoxide hydrolase activities.[70] This study concludes that if nitrosylated ATX-I is similarly toxic to other cell types, in particular esophageal cells, carcinogenesis would be promoted through cell death and the proliferation of neighboring cells to generate replacements. The sensitivity of H4IIE cells also suggested that exposure to nitrosylated ATX-I could lead to liver damage/carcinogenesis.

The mutagenicity of AOH at the thymidine kinase (TK) gene locus in mouse lymphoma L5178 $tk^{+/-}$ (MLC) cells has also been investigated.[69] Concentrations higher than or equal to 10 μM AOH give rise to a significant and concentration-dependent induction of TK mutations in MLC cells. Discrimination between small and large colonies in the TK assay reveals the predominant induction of small colonies, which are indicative for extensive chromosomal deletions and correlate with the induction of micronuclei in MLC cells.

Chinese hamster lung (CHL) cells are also used to evaluate TeA toxicity to mammalian cells. Cell proliferation inhibition increases with extension of toxin exposure time.[71] Total protein content in culture of cells was measured and showed that TeA decreases the total protein content after 72 h toxin exposure with half maximal effective concentration (EC_{50}) value of 56.28 μg/mL. In the same study, 3T3 mouse fibroblasts (3T3 cells) and human hepatocytes (L-O2 cells) showed a similar response. It was concluded that TeA is the most cytotoxic to 3T3 cells, followed by CHL cells and L-O2 cells. At lower concentrations, TeA had lower cytotoxicity to the human hepatocytes, which suggests that such cells may tolerate TeA at low concentrations.

The murine macrophage cell line RAW 264.7 was used to test AOH toxicity as a response to DNA damage and to test if AOH induces autophagy, senescence, abnormal morphology, and cell cycle arrest.[72–74] High AOH concentrations block cell proliferation and increase the level of reactive oxygen species (ROS). However they are not directly linked to each other. It seems that AOH-induced DNA damage and resulting transcriptional changes in the p21, MDM2, and Cyclin B genes contribute to the reduced cell proliferation, while the expression of Sestrin 2 gene would contribute to the oxidant defense.[74] Besides, it seems that the AOH-induced cell cycle arrest, most probably due to DNA damage and incomplete decatenation, is followed by very specific morphological changes.[72] Furthermore, Solhaug et al.[73] found that autophagy and senescence using the RAW264.7 macrophage model are related to stress responses caused by the DNA-damaging AOH.

Human colon carcinoma cells HCT116 are used to evaluate the cell death mode and pathways triggered by AOH 75 and the cytotoxic potential of AOH and AME mixtures.[76] Cells treated with AOH show a loss of cell viability by inducing apoptosis.[75] Bensassi et al.[76] demonstrated that the exposure of HTC116 cells to low cytotoxic AOH doses causes a moderate cytotoxicity. However, when AOH and AME are combined, they exert a significant increase in their toxic potential.

The cell line Ishikawa allows examining the estrogenic potential of AOH.[67] In this study, the estrogenicity of AOH was about 10,000-fold weaker than of the endogenous hormone E2. To identify if AOH may alter steroidogenesis, an *in vitro* screening assay based on measuring alterations in steroid hormone production and the expression of several important genes that encode the various enzymes involved in steroidogenesis using the human adrenocortical carcinoma cell line H295R has been performed by Frizzel et al.[77] They demonstrated that AOH has the ability to interfere with steroidogenesis pathway since it modifies the expression of important genes in steroidogenesis in H295R cells.

Human breast adenocarcinoma cell line RGA is of great relevance to evaluate the effect of AOH on the androgen, progestogen, glucocorticoid, and estrogen nuclear receptors present in those cells.[77] These authors concluded that AOH has a weak estrogenic activity when tested in the estrogen-responsive RGA cells. The sensitivity of AOH to different cell types may be an important factor when considering this mycotoxin oestrogenic effect. In another study performed by Willemsen et al.,[78] the estrogenic influence of the ER α in porcine endometrial cells was not detected.

The evaluation of AOH and AME toxicity in the mouse hepatoma Hepa-1c1c7 cell line with intact aryl hydrocarbon receptor (AhR) signaling and the Hepa-1c1c4 and Hepa-1c1c12 cell lines, which are deficient for the AhR nuclear translocator (ARNT) or the AhR, respectively,[79] has been performed. It was demonstrated that AOH and AME are novel inducers of the AhR/ARNT pathway, which mediates induction of CYP1A1 and apoptosis and might thus contribute to the toxicity of these mycotoxins. Burkhardt et al.[80] used the same three mouse hepatoma (Hepa-1) cell lines with intact and compromised AhR signaling to compare their activities for hydroxylation, methylation, and glucuronidation.

The liver hepatocellular carcinoma HepG2 cell line is widely used to evaluate the cytotoxic effects of AOH.[81] In this study, HepG2 cells were treated at different concentrations over 24, 48, and 72 h. The half maximal inhibitory concentration (IC_{50}) values were from 65 to 96 μM for AOH. Pfeiffer et al.[63] also demonstrated that AOH and AME cause a concentration-dependent induction of DNA strand breaks in HepG2 cells.

The human vulva carcinoma A431 cell line allows testing the viability and cytotoxicity potential of AOH, AME, and ALT as well as to perform a single-cell gel electrophoresis (comet assay) and inmunoband depletion assay.[62] They demonstrated that AOH and AME significantly increase the rate of DNA strand breaks in A431 at micromolar concentrations, whereas ALT does not affect DNA integrity up to 100 μM. The inmunoband depletion assay allowed for the observation that AOH affects topoisomerase, preferentially the IIα isoform. This latter conclusion was also obtained by the same authors[62] when they utilized the human breast adenocarcinoma MCF-7 cell line to evaluate the effects of AOH on topoisomerase I and II by means of relataxion and cleavage assays and decatenation and cleavage assays, respectively.

30.4.2.2 Other In Vitro *Models*

In addition to cell model systems, other *in vitro* models can be used to evaluate the toxicity of foodborne *Alternaria*. Thus, *Artemia salina* is widely used to check the toxicity of purified AOH, AME, ALT, ATX-I, and TeA and different *Alternaria* extracts.[82] In this model, it has been demonstrated that the toxicity of culture extracts is not caused by the presence of the purified *Alternaria* mycotoxins studied. Panigrahi and Dallin[83] evaluated the toxicity of several *Alternaria* mycotoxins (TeA, AOH, AME, ALT, and ATX-I) to *A. salina* by means of a disc assay and scoring larvae mortality by examining under a microscope in a multiwell plate. TeA, AOH, ATX-I, and ALT produced dose-related mortality in the larvae, with the LC_{50} well concentration for the toxins being 75, 100, 200, and 375 μg/mL, respectively. No effect was observed for AME on the larvae, although the authors reported a low solubility of this toxin in the solvents used in this study.

The Ames *Salmonella* test is widely used to evaluate toxicity and can be utilized to test mutagenicity of *Alternaria*. In this test, the crude *Alternaria* extract shows mutagenic activity.[56,84] When this extract was partitioned, the mutagenic fractions corresponded to ATX-I, ATX-II, and ATX-III. The results indicated that all altertoxins were mutagenic in the decreasing order of ATX-III>ATXII>ATX-I, with the potency of ATX-III approximately 10-fold lower than that of aflatoxin B1. Schrader et al.[70]

also examined the major *Alternaria* toxins (ALT, AOH, AME, ATX-I, TEN, and TeA) by the Ames *Salmonella* test, using strains TA98 and TA100 to detect histidine revertants that result from frameshift and base pair mutations, respectively. In the absence of nitrosylation, they found that ATX-1 was positive for mutagenicity in TA98, while the all other toxins were negative. In general, nitrosylation increases mutagenic potencies, although no effect was observed on TEN and TeA mutagenicities. These findings suggest that the presence of mould contamination that contains ATX-I in foodstuff could pose a potent carcinogenic health hazard when combined with a diet high in nitrites and nitrates. In a later work, Schrader et al.[85] reexamined mutagenicity of these toxins with and without nitrosylation, using Ames *Salmonella* strain TA97, sensitive to frameshift mutations at a run of C's, as well as strains TA102 and TA104, reverted by base pair mutations at AT sites and more sensitive to oxidative damage. The results suggested that ATX-I, AME, and AOH induce mutations at AT sites, possibly through oxidative damage. In addition, nitrosylation enhances ATX-I mutagenicity at runs of C's.

Several methods have been developed for simulating the human digestive tract. Dall'Asta et al.[48] used a method proposed by Versantvoort et al.[86] for assessing the bioaccessibility of mycotoxins from food, and another one that involved the use of fecal water obtained as a mixture from healthy volunteers following an omnivore diet. Samples from the simulated digestion assay were taken at the end of the duodenal phase and directly analyzed by LC–MS/MS. Samples from the fecal fermentation assay were taken after 30 min and 24 h of incubation and analyzed by LC–MS/MS after high-speed centrifugation. Their results showed that both AOH and ALT were totally stable along the simulated human gastrointestinal tract. Neither digestive conditions nor human gut microbiome seemed to be able to degrade AOH and ALT, as these were fully recovered in the fecal water after 24 h of incubation.

30.5 Conclusions

Alternaria species are found in a wide range of foods including cereals, fruits, and vegetables due to their ability to grow at low temperatures and low a_w. Although some *Alternaria* spp. may be allergenic and cause asthma, the synthesis of mycotoxins when the environmental conditions are adequate is the most worrying undesirable effect. In spite of the fact that *Alternaria* species produce a huge number of secondary metabolites, only a small proportion is considered toxic to humans and animals. The most important *Alternaria* mycotoxins are AOH, AME, ALT, TeA, TEN, ATX-I, ATX-II, and ATX-III.

The diagnosis of mycotoxin-producing *Alternaria* and *Alternaria* toxins in food could be performed by morphological, biochemical/analytical, and genetic molecular techniques. However, laboratory models based on animals or *in vitro* (cell lines or biological tests) systems are needed to examine the toxic effects of *Alternaria* and their toxins on humans and animals, thus completing the investigation in foodborne *Alternaria*.

Several animal models based on chickens, rats, dogs, and mice could be used to evaluate toxic effects of *Alternaria* culture extracts and some of the main purified *Alternaria* toxins. They may be adapted for foodborne analysis.

In addition, cell lines and other models (Ames *Salmonella* test, *A. salina*, etc.) are also available for this purpose. Ethical and scientific reasons due to the use of high number of animals and the toxin amount required for these experiments are the main limitations on the performance of laboratory toxicity for animal model assays. Cell system models could be a good alternative for foodborne *Alternaria* analysis, with human cell lines such as Caco-2 and HT-29, the Chinese hamster V79 cells, rat H4IIE hepatoma cells, or murine macrophage cell line RAW 264.7 being the most appropriate. Overall, these assays demonstrated that all *Alternaria* toxins are cytotoxic and that AOH and AME are also genotoxic and mutagenic; however, AOH is a weak estrogenic. TeA has been reported to exert cytotoxic properties as well as being acutely toxic. The altertoxins ATX-I, ATX-II, and ATX-III have been shown to be more mutagenic and acutely toxic than AOH and AME.

The toxicity of *Alternaria* mycotoxins still needs a further research. Not enough information is available on the toxicokinetics, including metabolism, of the most toxicologically relevant foodborne *Alternaria* toxins.

Acknowledgments

This work has been funded by the Spanish Comisión Interministerial de Ciencia y Tecnología with the projects AGL2013-45729P and Carnisenusa CSD2007-00016, Consolider Ingenio 2010 and GRU09162 of the Junta de Extremadura and FEDER. Dr. Alicia Rodríguez was supported by a postdoctoral research fellowship (PO12016) from the Gobierno de Extremadura, Consejería de Empleo, Empresa e Innovación, and the Fondo Social Europeo (FSE).

REFERENCES

1. Simmons, E., *Alternaria: An Identification Manual*, CBS Biodiversity Series 6, CBS Fungal Biodiversity Centre, Utrecht, NL, 2007.
2. Woudenberg, J.H.C. et al., *Alternaria* redefined, *Stud. Mycol.*, 75, 171, 2013.
3. Hasan, H.A.H., *Alternaria* mycotoxins in black rot lesion of tomato fruit-conditions and regulation of their production, *Mycopathologia*, 130, 171, 1995.
4. Magan, N., Cayley, G.R. and Lacey, J., The effect of water activity and temperature on mycotoxin production by *Alternaria alternata* in culture and on wheat grain, *Appl. Environ. Microbiol.*, 47, 1113, 1984.
5. Barkai-Golan, R., *Postharvest Diseases of Fruits and Vegetables: Development and Control*, Elsevier, Amsterdam, NL, 2001.
6. Müller, M., *Alternaria* infestation of corn silage and hay, *Zbl. Mikrobiol.*, 146, 481, 1991.
7. Patriarca, A. et al., Mycotoxin production by *Alternaria* strains isolated from Argentinean wheat, *Int. J. Food Microbiol.*, 119, 219, 2007.
8. Andersen, B. et al., Associated field mycobiota on malt barley, *Can. J. Bot.*, 74, 854, 1996.
9. Kosiak, B. et al., *Alternaria* and *Fusarium* in Norwegian grains of reduced quality a matched pair sample study, *Int. J. Food Microbiol.*, 93, 51, 2004.
10. Logrieco, A., Moretti, A. and Solfrizzo, M., *Alternaria* toxins and plant disease: An overview of origin, occurrence and risks, *World Mycotoxin. J.*, 2, 129, 2009.
11. Medina, A. et al., Survey of the mycobiota of Spanish malting barley and evaluation of the mycotoxin producing potential of species of *Alternaria*, *Aspergillus* and *Fusarium*, *Int. J. Food Microbiol.*, 108, 196, 2006.
12. Webley, D.J. et al., *Alternaria* toxins in weather-damaged wheat and sorghum in the 1995–1996 Australian harvest, *Aust. J. Agric. Res.*, 48, 1249, 1997.
13. Li, F.Q. and Yoshizawa, T., *Alternaria* mycotoxins in weathered wheat from China, *J. Agric. Food Chem.*, 48, 2920, 2000.
14. Andersen, B. and Frisvad, J.C., Natural occurrence of fungi and fungal metabolites in moldy tomatoes, *J. Agric. Food Chem.*, 52, 7507, 2004.
15. Vinas, I. et al., Natural occurrence of aflatoxin and *Alternaria* mycotoxins in oilseed rape from Catalonia (Spain)—Incidence of toxigenic strains, *Mycopathologia*, 128, 175, 1994.
16. Robiglio, A.L. and López, S.E., Mycotoxin production by *Alternaria alternata* strains isolated from red-delicious apples in Argentina, *Int. J. Food Microbiol.*, 24, 413, 1995.
17. Tournas, V.H. and Katsoudas, E., Mould and yeast flora in fresh berries, grapes and citrus fruits, *Int. J. Food Microbiol.*, 105, 11, 2005.
18. Solfrizzo, M. et al., Liquid chromatographic determination of *Alternaria* toxins in carrots, *J. AOAC Int.*, 87, 101, 2004.
19. Dagnas, S. and Membré, J.M., Predicting and preventing mold spoilage of food products, *J. Food Protect.*, 76, 538, 2012.
20. Culshaw, F. et al., *Blackpoint of Wheat*, p. 46. Home Grown Cereals Authority, London, 1988.
21. Kustrzeba-Wójcicka, I., et al., *Alternaria alternata* and its allergens: A comprehensive review, *Clin. Rev. Allergy Immunol.*, 47, 354, 2014.
22. Fernández-Cruz, M.L., Mansilla, M.L. and Tadeo, J.L., Mycotoxins in fruits and their processed products: Analysis, occurrence and health implications, *J. Adv. Res.*, 1, 113, 2010.
23. Panigrahi, S., *Alternaria* toxins, in *Handbook of Plant and Fungal Toxicants*, p. 319, D'Mello, J.P.F. (Ed.), CRC Press, Boca Raton, FL, 1997.

24. Ostry, V., *Alternaria* mycotoxins: An overview of chemical characterization, producers, toxicity, analysis and occurrence in foodstuffs, *World Mycotoxin J.*, 1, 175, 2008.
25. Bottalico, A. and Logrieco, A., Toxigenic *Alternaria* species of economic importance, in *Mycotoxins in Agriculture and Food Safety*, p. 65, Sinha, K.K. and Bhatnagar, D. (Eds.), Marcel Dekker, New York, 1998.
26. Magan, N. and Baxter, E.S., Relationship between environmental factors and tenuazonic acid production by *Alternaria* isolates from sorghum, in *Occurrence and Significance of Mycotoxins*, p. 309, Skudamore, K.A. (Ed.), Slough Central Science Laboratories, Slough, UK, 1993.
27. Oviedo, M.S. et al., Effect of environmental factors on tenuazonic acid production by *Alternaria alternata* on soybean-based media, *J. Appl. Microbiol.*, 107, 1186, 2009.
28. Patriarca, A. et al., Temperature and water stress impacts on growth and production of altertoxin-II by strains of *Alternaria tenuissima* from Argentinean wheat, *World Mycotoxin J.*, 7, 329, 2014.
29. Magan, N. and Lacey, J., The effect of water activity and temperature on mycotoxin production by *Alternaria alternata* in culture and on wheat grain, in *Trichothecenes and Other Mycotoxins*, p. 243, Lacey, J. (Ed.), John Wiley & Sons, Chichester, UK, 1985.
30. Oviedo, M.S. et al., Impact of water activity and temperature on growth and alternariol and alternariol monomethyl ether production of *Alternaria alternata* isolated from soybeans, *J. Food Protect.*, 73, 336, 2010.
31. Oviedo, M.S. et al., Influence of water activity and temperature on growth and mycotoxin production by *Alternaria alternata* on irradiated soya beans, *Int. J. Food Microbiol.*, 149, 127, 2011.
32. Pose, G. et al., Water activity and temperature effects on mycotoxin production by *Alternaria alternata* on a synthetic tomato medium, *Int. J. Food Microbiol.*, 142, 348, 2010.
33. Scott, P.M., Analysis of agricultural commodities and foods for *Alternaria* mycotoxins, *J. AOAC Int.*, 84, 1809, 2001.
34. EFSA Panel on Contaminants in the Food Chain, Scientific opinion on the risks for animal and public health related to the presence of *Alternaria* toxins in feed and food, *EFSA J.*, 9, 2407, 2011.
35. Liu, G.T. et al., Etiological role of *Alternaria alternata* in human esophageal cancer, *Chin. Med. J.*, 105, 394, 1992.
36. Liu, G.T. et al., Relationships between *Alternaria alternata* and oesophageal cancer, in *Relevance to Human Cancer of N-Nitroso Compounds, Tobacco Smoke and Mycotoxins*, p. 258, O'Neill, I.K., Chen, J. and Bartsch, H. (Eds.), International Agency for Research on Cancer, Lyon, FR, 1991.
37. Greco, M. et al., Toxigenic *Alternaria* species from Argentinean blueberries, *Int. J. Food Microbiol.*, 154, 187, 2012.
38. Scussel, V.M. et al., Multi-toxins in cork stoppers and fermented beverages by liquid chromatography-mass spectrometry including *Alternaria* toxins, in *XIIth International IUPAC Symposium on Mycotoxins and Phycotoxins*, poster 1356, Istanbul, TR, 2007.
39. Spanjer, M.C., Rensen, P.M. and Scholten, J.M., LC-MS/MS multi-method for mycotoxins after single extraction, with validation data for peanut, pistachio, wheat, maize, cornflakes, raisins and figs, *Food Addit. Contam.*, 25, 472, 2008.
40. Rasmussen, R.R. et al., Multi-mycotoxin analysis of maize silage by LC-MS/MS, *Anal. Bioanal. Chem.*, 397, 765, 2010.
41. Wang, S. et al., Simultaneous determination of seventeen mycotoxins residues in *Puerariae lobatae radix* by liquid chromatography-tandem mass spectrometry, *J. Pharm. Biomed. Anal.*, 98, 201, 2014.
42. Andrew, M., Peever, T.L. and Pryor, B.M., An expanded multilocus phylogeny does not resolve morphological species within the small-spored *Alternaria* species complex, *Mycologia*, 101, 95, 2009.
43. Peever, T.L. et al., Molecular systematics of citrus-associated *Alternaria* species, *Mycologia*, 96, 119, 2004.
44. Pryor, B.M. and Gilbertson, R.L., Molecular phylogenetic relationships amongst *Alternaria* species and related fungi based upon analysis of nuclear ITS and mt SSU rDNA sequences, *Mycol. Res.*, 104, 1312, 2000.
45. Roberts, R.G., Reymond, S.T. and Andersen, B., RAPD fragment pattern analysis and morphological segregation of small-spored *Alternaria* species and species groups, *Mycol. Res.*, 104, 151, 2000.
46. Zur, G. et al., Development of a PCR-based assay for the detection of *Alternaria* fungal contamination in food products, *J. Food Protect.*, 62, 1191, 1999.
47. Crespo-Sempere, A. et al., Propidium monoazide combined with real-time quantitative PCR to quantifiy viable *Alternaria* spp. contamination in tomato products, *Int. J. Food Microbiol.*, 165, 214, 2013.

48. Dall'Asta, C., Cirlini, M. and Falavigna, C., Mycotoxins from *Alternaria*: Toxicological implications, *Adv. Mol. Toxicol.*, 8, 107, 2014.
49. Juan-García, A. et al., Efectos tóxicos de alternariol por ensayos in vitro: Revisión, *Rev. Toxicol.*, 31, 196, 2014.
50. Schuchardt, S., Ziemann, C. and Hansen, T., Combined toxicokinetic and in vivo genotoxicity study on *Alternaria* toxins, EFSA supporting publication, EN-679, 2014.
51. Altemöller, M., Podlech, J. and Fenske, D., Total synthesis of altenuene and isoaltenuene, *Eur. J. Org. Chem.*, 7, 1678, 2006.
52. Koch, K. et al., Total synthesis of alternariol, *J. Org. Chem.*, 70, 3275, 2005.
53. Griffin, G.F. and Chu, F.S., Toxicity of the *Alternaria* metabolites alternariol, alternariol monomethyl ether, altenuene and tenuazonic acid in the chicken embryo assay, *Appl. Environ. Microbiol.*, 46, 1420, 1983.
54. Pero, R.W. et al., Toxicity of metabolites produced by the "*Alternaria*", *Environ. Health Perspect.*, 4, 87, 1973.
55. Sauer, D.B. et al., Toxicity of *Alternaria* metabolites found in weathered sorghum grain at harvest, *J. Agric. Food Chem.*, 26, 1380, 1978.
56. Scott, P.M. and Stoltz, D.R., Mutagens produced by *Alternaria alternata*, *Mutat. Res.*, 78, 33, 1980.
57. Cenčič, A. and Langerholc, T., Functional cell models of the gust and their applications in food microbiology—A review, *Int. J. Food Microbiol.*, 141, S4, 2010.
58. Rousset, M., The human colon carcinoma cell lines HT-29 and Caco-2: Two in vitro models for the study of intestinal differentiation, *Biochimie*, 68, 1035, 1986.
59. Fernández-Blanco, C., Font, G. and Ruiz, M., Oxidative stress of alternariol in Caco-2 cells, *Toxicol. Lett.*, 229, 458, 2014.
60. Fernández-Blanco, C., Font, G. and Ruiz, M., Oxidative DNA damage and disturbance of antioxidant capacity by alternariol in Caco-2 cells, *Toxicol. Lett.*, 235, 61, 2015.
61. Burkhardt, B., Pfeiffer, E. and Metzler, M., Absorption and metabolism of the mycotoxins alternariol and alternariol-9-methyl ether in Caco-2 cells in vitro, *Mycotoxin Res.*, 25, 149, 2009.
62. Fehr, M. et al., Alternariol acts as a topoisomerase poison, preferentially affecting the IIα isoform, *Mol. Nutr. Food Res.*, 53, 441, 2009.
63. Pfeiffer, E., Eschbach, S. and Metzler, M., *Alternaria* toxins: DNA strand-breaking activity in mammalian cells in vitro, *Mycotoxin Res.*, 23, 152, 2007.
64. Tiessen, C. et al., Modulation of the cellular redox status by the *Alternaria* toxins alternariol and alternariol monomethyl ether, *Toxicol. Lett.*, 216, 23, 2013.
65. Schwarz, C., Kreutzer, M. and Marko, D., Minor contribution of alternariol, alternariol monomethyl ether and tenuazonic acid to the genotoxic properties of extracts from *Alternaria alternate* infested rice, *Toxicol. Lett.*, 214, 46, 2012.
66. Schwarz, C. et al., Characterization of a genotoxic impact compound in *Alternaria alternata* infested rice as altertoxin II, *Arch. Toxicol.*, 86, 1911, 2012.
67. Lehmann, L., Wagner, J. and Metzler, M., Estrogenic and clastogenic potential of the mycotoxin alternariol in cultured mammalian cells, *Food Chem. Toxicol.*, 44, 398, 2006.
68. Fleck, S.C. et al., *Alternaria* toxins: Altertoxin II is a much stronger mutagen and DNA strand breaking mycotoxin than alternariol and its methyl ether in cultured mammalian cells, *Toxicol. Lett.*, 214, 27, 2012.
69. Brugger, E.M. et al., Mutagenicity of the mycotoxin alternariol in cultured mammalian cells, *Toxicol. Lett.*, 164, 221, 2006.
70. Schrader, T.J. et al., Examination of *Alternaria alternata* mutagenicity and effects of nitrosylation using the Ames *Salmonella* test, *Teratog. Carcinog. Mutagen.*, 21, 261, 2001.
71. Zhou, B. and Qiang, S., Environmental, genetic and cellular toxicity of tenuazonic acid isolated from *Alternaria alternata*, *Afr. J. Biotechnol.*, 7, 1151, 2008.
72. Solhaug, A. et al., Alternariol induces abnormal nuclear morphology and cell cycle arrest in murine RAW 264.7 macrophages, *Toxicol. Lett.*, 219, 8, 2013.
73. Solhaug, A. et al., Autophagy and senescence, stress responses induces by the DNA-damaging mycotoxin alternariol, *Toxicology*, 326, 119, 2014.
74. Solhaug, A. et al., Mechanisms involved in alternariol-induced cell cycle arrest, *Mutat. Res.*, 738 1, 2012.

75. Bensassi, F. et al., Cell death induced by *Alternaria* mycotoxin alternariol, *Toxicol. In Vitro*, 26, 915, 2012.
76. Bensassi, F. et al., Combined effects of alternariols mixture on human colon carcinoma cells, *Toxicol. Mech. Methods*, 25, 56, 2015.
77. Frizzel, C. et al., An in vitro investigation of endocrine disrupting effects of the mycotoxin alternariol, *Toxicol. Appl. Pharm.*, 271, 64, 2013.
78. Willemsen, P. et al., Use of reporter cell lines for detection of endocrine-disrupter activity, *Anal. Bioanal. Chem.*, 378, 655, 2004.
79. Schreck, I. et al., The *Alternaria* mycotoxins alternariol and alternariol methyl ether induce cytochrome P450 1A1 and apoptosis in murine hepatoma cells dependent on the aryl hydrocarbon receptor, *Arch. Toxicol.*, 86, 625, 2012.
80. Burkhardt, B. et al., Mouse hepatoma cell lines differing in aryl hydrocarbon receptor-mediated signaling have different activities for glucuronidation, *Arch. Toxicol.*, 86, 643, 2012.
81. Juan-García, A. et al., Cytotoxic effects and degradation products of three mycotoxins: Alternariol, 3-acetyl-deoxynivalenol and 15-acetyl-deoxynivalenol in liver hepatocellular carninoma cells, *Toxicol. Lett.*, 235, 8, 2015.
82. Zajwoski, P., Grabarkiewicz-Szcesna, J. and Schmidt, R., Toxicity of mycotoxins produced by four *Alternaria* species to *Artemia salina* larvae, *Mycotoxin Res.*, 7, 11, 1991.
83. Panigrahi, S. and Dallin, S., Toxicity of the *Alternaria* spp. metabolites, tenuazonic acid, alternariol, altertoxin-I, and alternariol monomethyl ether to brine shrimp—(*Artemia salina* L) larvae, *J. Sci. Food Agric.*, 66, 493, 1994.
84. Stack, M.E. and Prival, M.J., Mutagenicity of the *Alternaria* metabolites altertoxins I, II, and III, *Appl. Environ. Microbiol.*, 52, 718, 1986.
85. Schrader, T.J. et al., Further examination of the effects of nitrosylation on *Alternaria alternata* mycotoxin mutagenicity in vitro, *Mutat. Res.*, 606, 61, 2006.
86. Versantvoort, C.H. et al., Applicability of an in vitro digestion model in assessing the bioaccessibility of mycotoxins from food, *Food Chem. Toxicol.*, 43, 31, 2005.
87. Andersen, B. et al., Characterization of *Alternaria* strains from Argentinean blueberry, tomato, walnut and wheat, *Int. J. Food Microbiol.*, 196, 1, 2015.
88. Lee, H.B., Patriarca, A. and Magan, N., *Alternaria* in food: Ecophysiology, mycotoxin production and toxicology, *Mycobiology*, 43, 93, 2015.
89. Ntasiou, P. et al., Identification, characterization and mycotoxigenic ability of *Alternaria* spp. causing core rot of apple fruit in Greece, *Int. J. Food Microbiol.*, 197, 22, 2015.
90. Sharma, S. and Sohal, B.S., Resistance to *Alternaria* blight in *Brassica juncea* (var. RLM619) induced by combination of elicitors and inoculation with *Alternaria brassicae* under controlled environment, *Indian J. Agric. Sci.*, 27, 180, 2014.
91. Pastirčák, M. and Fejér, J., A preliminary survey of fungi on opium poppy in Slovakia, *Acta Hort.*, 1036, 157, 2014.
92. MacDonald, M.V. and Ingram, D.S., Towards the selection in vitro for resistance to *Alternaria brassicicola* (Schw.) Wilts., in *Brassica napus* ssp. oleifera (Metzg.) Sinsk., winter oilseed rape. *New Phytol.*, 104, 621, 1986.
93. MacKinnon, S.L., Keifer, P. and Ayer, W.A., Components from the phytotoxic extract of *Alternaria brassicicola*, a black spot pathogen of canola, *Phytochemistry*, 51, 215, 1999.
94. Otani, H. et al., Production of a host-specific toxin by germinating spores of *Alternaria brassicicola*, *Physiol. Mol. Plant Pathol.*, 52, 285, 1998.

31

Aspergillus

László Kredics, János Varga[†], Rajagopalaboopathi Jayasudha, Sándor Kocsubé, Nikolett Baranyi, Coimbatore Subramanian Shobana, Muthusamy Chandrasekaran, Shine Kadaikunnan, Venkatapathy Narendran, Csaba Vágvölgyi, and Palanisamy Manikandan

CONTENTS

31.1 Introduction

Members of the genus *Aspergillus* are well-known for their detrimental effects on human and animal health. Invasive and superficial infections caused by *Aspergillus* species are more and more frequently reported, and the genus is on the top of the list of filamentous fungal genera of clinical importance. However, clinical infections due to Aspergilli are rarely foodborne. On the other hand, the ability of certain members of the genus to produce various mycotoxins frequently results in serious health problems, both in animals and humans, that are commonly known as *Aspergillus* mycotoxicoses.

This chapter aims to give an overview about the laboratory models used to study *Aspergillus* mycotoxicoses, with a special focus on the most recent publications applying rodent models (rats and mice) and mammalian cell lines, the application of which proved especially fruitful to gain more insight into the toxic and carcinogenic effects of AFB_1, OTA, and other *Aspergillus* mycotoxins.

31.2 The Genus *Aspergillus*

The name *Aspergillus* was first used in 1729 for filamentous fungi with conidial heads in Nova Plantarum Genera by the Italian mycologist and priest Pietro Antonio Micheli as the spore-bearing structures

[†] In memoriam Prof. János Varga

characteristic of the genus reminded him an aspergillum (holy water sprinkler) used in Catholic liturgy. These fungi were later recognized as active decomposers and also as the causal agents of diseases in animals and humans. Today, *Aspergillus* is considered one of the most economically important and most widely distributed filamentous fungal genera on our planet [1]. The genus *Aspergillus* is taxonomically divided into 8 subgenera and more than 20 sections consisting of about 300–350 species [2,3].

Aspergillus species can be both beneficial and harmful for mankind. The greatest economic benefit of the genus is the ability to produce industrial enzymes (e.g., amylases, glycosidases, pectinases, proteases), and organic acids (e.g., citric acid, gluconic acid, itaconic acid) [1]. *Aspergillus oryzae*, *A. sojae*, and *A. tamarii* are known as the "koji moulds" and have been used for centuries in oriental food fermentation processes for the production of soy sauce and sake [1]. Aspergilli can also produce a series of secondary metabolites with useful pharmaceutical and biotechnological properties (e.g., the cholesterol-lowering drug lovastatin produced by *A. terreus* and some other species). However, taxa of this genus may also have serious negative impacts on animal and human health as they cause different diseases called aspergilloses. The most common human pathogen from the genus is *A. fumigatus*, which is responsible for more than 90% of both invasive and noninvasive human aspergilloses; however, other species of the genus are also capable of causing infections [4]. Various fungal diseases like keratitis or onychomycosis, allergic bronchopulmonary aspergillosis (ABPA), and asthma have been reported to be caused by the growth or spores of *Aspergillus* species (among others). Additionally, Aspergilli produce a range of mycotoxins, which can be harmful to animals and humans. *Aspergillus* species can contaminate foods and feeds at different pre- and postharvest, as well as processing or handling stages. The most economically important *Aspergillus* mycotoxins identified as contaminants in foods and feeds are the aflatoxins, ochratoxins, patulin, and fumonisins.

31.3 Mycotoxins of *Aspergillus* Species and Mycotoxicoses

31.3.1 Aflatoxins

The most thoroughly studied mycotoxins produced by *Aspergillus* species are aflatoxins (AFs). Regarding the discovery of AFs, toxicity of animal feeds containing contaminated peanut meal in the early 1960s resulted in the death of more than 100,000 turkeys from acute liver necrosis [5,6]. The toxin-producing fungus was identified as *Aspergillus flavus*, and the toxic agent as AF, named after the first isolated producer. AFs form a group of structurally related difuranocoumarin compounds named as AFs B_1, B_2, G_1, and G_2 based on their fluorescence under ultraviolet light (blue or green) and relative mobility during thin-layer chromatography. Aflatoxin B_1 (AFB_1) is the most efficient known natural carcinogen [7] and is generally the major AF produced by toxigenic strains. The hydroxylated metabolite AFM_1 is primarily found in animal fluids (milk and urine) and tissues as a metabolic product of AFB_1 [8]. Recent data indicate that AFs are produced by more than 20 species belonging to sections Flavi, Nidulantes, and Ochraceorosei of the genus *Aspergillus* [8–11]. Although the ability of AF production was claimed for a series of other fungal species and genera (actually even for bacteria), these observations could not be confirmed [8].

Sterigmatocystin (STC) is a penultimate precursor of AFs, which is itself a toxic and carcinogenic substance produced by many species belonging mainly to sections Nidulantes and Versicolores. STC production also occurs in the phylogenetically unrelated genera *Bipolaris*, *Chaetomium*, and *Humicola* [11].

AFs are primarily produced by *A. flavus* and *A. parasiticus* on agricultural crops like cereals (e.g., corn, rice, wheat), cotton, peanut, pepper, spices, and tree nuts [8]. *A. nomius* was found to be an important producer of AFs in Brazil nuts [12]. The other AF-producing species (e.g., *A. ochraceoroseus*, *A. venezuelensis*, *A. astellata*) are not regarded as potential health hazards, as they produce only small amounts of AFs or are encountered in food products rarely or not at all. STC producers are also frequently identified in various substrates including indoor air, food, and feed [13,14].

AFs are hepatotoxic and hepatocarcinogenic and usually considered as belonging to the most potent naturally occurring carcinogens [15,16]. The diseases caused by AF consumption are known as aflatoxicoses. Acute aflatoxicosis is the result of moderate-to-high level AF consumption. Several deaths were attributed to aflatoxicosis in tropical and subtropical regions of the world, primarily due to the consumption of contaminated corn [17]. Chronic aflatoxicosis results from prolonged ingestion of low-to-moderate levels of AFs and can be characterized by usually subclinical effects that are difficult to recognize. Comprehensive studies indicated that AF is a risk factor for hepatocellular carcinoma in humans, especially in Asia and the sub-Saharan part of Africa [18,19]. The International Agency for Research on Cancer (IARC) classified AFs as group I category carcinogens [20]. Because of its toxicity, the content of AF is restricted in food and feed products in more than 100 countries [21].

31.3.2 Ochratoxins

Ochratoxins (OTs) were discovered in 1965 during a systematic examination of the metabolites of moulds, including *Aspergillus ochraceus* [22]. Ochratoxin A (OTA) is a cyclic pentaketide dihydroisocoumarin derivative linked to an L-β-phenylalanine by an amide bond. The most toxic member of the group is OTA which is a chlorinated pentaketide. Several OTA-derived metabolites have also been identified: OTB is a dechlorinated analogue of OTA, while OTC is a chlorinated ethyl ester derivative. OTα is the isocoumarin core of OTA, while OTβ is a dechlorinated analogue of OTα. Other derivatives include 4-hydroxy-OTA, 10-hydroxy-OTA, OTA methyl ester, OTB ethyl and methyl esters, and several amino acid analogues [23,24].

Several nephropathies affecting animals and humans can be attributed to OTA, including the Danish porcine nephropathy as well as renal disorders detected in other animals [25]. In the case of humans, OTA is often mentioned as the possible causative agent of the Balkan Endemic Nephropathy [26], although recently aristolochic acid was suggested to play a major role in the etiology of this disease [27]. OTA also proved to exhibit immunosuppressive, teratogenic, hepatotoxic, and carcinogenic properties [28]. The IARC classified OTA as a possible human carcinogen in the Group 2B category [20].

Under cooler, temperate climates, OTs are mainly produced by *Penicillium* species (*P. verrucosum* and *P. nordicum*), while in warmer and tropical regions they are produced by a series of *Aspergillus* spp. *Aspergillus* isolates usually produce both OTA and its dechlorinated analogue OTB, while *Penicillium* spp. produce OTA only. In *Aspergillus* section Circumdati, the recently described new species *A. affinis*, *A. fresenii*, *A. occultus*, *A. pulvericola*, *A. sesamicola* [29,30], as well as *A. cretensis*, some isolates of *A. ochraceopetaliformis* (formerly described as *A. flocculosus*), *A. muricatus*, *A. ochraceus*, *A. pseudoelegans*, *A. roseoglobulosus*, *A. steynii*, and *A. westerdijkiae* are able to produce consistently high or medium amounts of OTA. Seven further species of the genus produce OTA inconsistently or in trace amounts only, including *A. ostianus*, *A. melleus*, *A. persii*, *A. salwaensis*, *A. sclerotiorum*, *A. subramanianii*, and *A. westlandensis*. The most relevant species for potential OTA production in rice, coffee, beverages, and other foodstuffs are *A. ochraceus*, *A. steynii*, and *A. westerdijkiae*. In *Aspergillus* section Flavi, *A. albertensis* and *A. alliaceus* are capable of OTA production in synthetic media [31]. Among the representatives of *Aspergillus* section Nigri, *A. carbonarius*, *A. niger*, *A. sclerotioniger*, and *A. welwitschiae* proved to be potential OTA-producing fungi [32,33]. The majority of *A. carbonarius* isolates can produce large amounts of OTA; however, different surveys reported about only 5%–15% of *A. niger* and *A. welwitschiae* isolates producing OTA in smaller quantities [33].

OTs can be detected in several agricultural products including cereals, cocoa, coffee beans, corn, figs, grapes, peanut, red pepper, rice, soybeans, and spices [34]. *A. westerdijkiae* proved to be important in the OTA contamination of coffee beans in Thailand [35], while *A. alliaceus* was found responsible for the OTA contamination of figs in California [36]. Members of *Aspergillus* section Nigri, primarily *A. carbonarius*, are responsible for the OTA contamination of grapes and grape-derived products [37].

31.3.3 Fumonisins

Several species of the genera *Aspergillus* and *Fusarium* (e.g., *Fusarium verticillioides*) are able to produce fumonisins (for details, see Chapter 34 on *Fusarium*). Sequencing the genome of *Aspergillus niger* resulted in the discovery of a region homologous with the fumonisin gene cluster of Fusaria [38]. The production of fumonisins by *A. niger* was confirmed later [39], while recent studies revealed that another black *Aspergillus* species, *A. welwitschiae* can also produce fumonisins [33]. Black Aspergilli produce mainly fumonisins B_2 and B_4, but other analogues have also been identified in smaller quantities [40].

Black Aspergilli and fusaria are both among the saprotrophic fungi that are widely distributed. Among the fumonisin-producing *Aspergillus* species, *A. niger* is an important opportunistic pathogen of grapes, causing raisin mould and bunch or berry rot [41]. It is also able to infect apples, corn, mangoes, onions, peanuts, and dried meat products [42]. In contrast to fusaria, which produces large amounts of fumonisins in plant-extract- (barley, carrots, malt, oat, potatoes, rice)-containing media, Aspergilli produce fumonisins in high quantities on substrates with low water activities [39].

31.3.4 Patulin

The water-soluble lactone patulin (PAT; 4-hydroxy-4*H*-furo[3,2-*c*]pyran-2(6*H*)-one) is produced by many fungi via the polyketide pathway. It was first isolated in 1943 from *Penicillium* species during a screening effort aimed at finding novel antibiotics from fungi [43]. As it was discovered independently in multiple laboratories, various names (e.g., clavacin, expansine, claviformin, clavatin, gigantic acid, myocin C) were given [44]. PAT was the first compound which was examined under the brand name Tercinin in a controlled clinical trial to treat common cold [45]; however, it proved to be unsuccessful, furthermore, even toxic to animals and humans [46].

Although PAT has been reported from up to 30 fungal genera [47], recent studies revealed that it is produced only by a limited number of species from the genera *Aspergillus*, *Byssochlamys*, *Paecilomyces*, and *Penicillium*. According to our recent knowledge, within the genus *Aspergillus* only three representatives of section Clavati (*A. clavatus*, *A. giganteus*, and *A. longivesica*) are capable of producing PAT [48]. The other PAT-producing Aspergilli are rare species and cannot be considered as health hazards in foods and feeds.

PAT provokes congestion and edema of hepatic, pulmonary, and gastrointestinal blood vessels and tissues. Regarding its acute toxicity, PAT mainly induces gastrointestinal disorders with distension, ulceration, and bleeding [49]. It was suggested to be partly responsible for a series of animal intoxications in France, Germany, Hungary, and Japan [50]. Regarding the carcinogenicity of PAT, its subcutaneous injection produced local sarcomas in rats [49].

31.3.5 Other Mycotoxins Produced by Aspergilli

Aspergillus species produce a wide range of other mycotoxins that may contaminate our foods and can be harmful to humans. Among them, the indole-tetramic acid mycotoxin known as cyclopiazonic acid was discovered in 1968 as a metabolite of *Penicillium cyclopium* in groundnuts [51]. A number of further *Penicillium* and *Aspergillus* species are known to produce this mycotoxin, including *P. camemberti*, *P. commune*, *P. dipodomyicola*, *P. griseofulvum*, *P. palitans* [52], *A. flavus*, *A. minisclerotigenes*, *A. oryzae*, *A. pseudotamarii* [9], and *A. lentulus* [53]. These mycotoxin-producing fungi are able to contaminate different agricultural products including various grains and seeds as well as cheese and meat products. Cyclopiazonic acid specifically inhibits calcium-dependent ATPase in the sarcoplasmic reticulum, which results in increased muscle contraction.

The pentaketide-derived mycotoxin citrinin is structurally related to OTs. There are a number of citrinin-producing Aspergilli including members of section Terrei (*A. alabamensis*, *A. allahabadii*, *A. carneus*, *A. flavipes*, *A. floccosus*, *A. niveus*, *and A. terreus*) [54,55]. Certain *Penicillium* species including *P. citrinum* and other members of section Citrina [56], *P. expansum*, *P. radicicola*, and *P. verrucosum*, as well as *Monascus* species are also known producers of this nephrotoxic compound [52]. Citrinin is a common contaminant of grains, food, and feedstuffs (e.g., corn, oats, wheat, wheat bran, fruit juices) [57]. If ingested by animals and humans, citrinin can cause chronic diseases.

Several further *Aspergillus* mycotoxins could also be detected in foods and feeds [16]. The extensive genome sequencing efforts in Aspergilli resulted in the recent clarification of the molecular background behind the production of a series of *Aspergillus* mycotoxins (e.g., gliotoxin [58], kojic acid [59], naphthopyrones, asperfuranone, and orsellinic acid [60]). However, further studies are needed to clarify the possible health effects of these mycotoxins on animals and humans.

31.4 Laboratory Models for Foodborne *Aspergillus* Mycotoxicoses

31.4.1 Invertebrate Models

There is a series of reports available in the literature about the application of invertebrate models for studying the effects of *Aspergillus* mycotoxins. Among the insects, AFB_1 proved to be harmful to *Aedes aegypti*, *Corcyra cephalonica*, *Drosophila melanogaster*, *Heliothis virescens*, *Heterotermes indicolas*, *Locusta migratoria*, *Musca domestica*, *Schistocerca gregaria*, and *Tenebrio molitor* [61]. The use of *Drosophila melanogaster* as a model organism to study the effects of *Aspergillus* mycotoxins has a long history. Studies were published even in the 1970s about the effects of AFB_1 on the development of *D. melanogaster* [62], the induction of recessive lethals by AFB_1 [63], the insecticidal and larvicidal activities of AFB_1 and PAT [61], the application of AFB_1-resistant and -sensitive strains for crossing experiments [64], the variation in sensitivity to AFB_1 among several strains of *D. melanogaster* [65], the effect of AFB_1 on viability, growth, fertility, and crossing over [66], the applicability of AFB_1Cl_2 as a model of AFB_1 in mutagenesis and carcinogenesis [67], the differences displayed by larvae of different ages in their sensitivity to the toxic effects of AFB_1 [68], the genetic background of resistance to AFB_1 toxicity [69], and the alterations in gene expression caused by AFB_1 [70]. Chinnici et al. [71] reported that pretreatments with relatively less toxic mycotoxins, like AFB_2 and STC, enhance the effect of AFB_1-induced toxicity in *D. melanogaster*. Foerster and Würgler [72] carried out *in vitro* studies on AFB_1 metabolism in testes of genetically different strains of *D. melanogaster* and identified AFB_{2a}, AFM_1, and AFR0 among the observed metabolites. AFB_1 was shown to induce a high level of somatic mutagenesis in imaginal disk of *D. melanogaster* larvae, while patulin also elevated the level of somatic mutations [73]. Melone and Chinnici [74] performed selection in *D. melanogaster* for increased resistance to AFB_1 toxicity. Shibahara et al. [75] studied the DNA-damaging potency and genotoxicity of AFM_1 in somatic cells of *D. melanogaster in vivo* and recorded a genotoxic effect for AFM_1 comparable to that of AFB_1. Karekar et al. [76] studied the antimutagenic profile of antioxidants in a *Drosophila* model where AFB_1 was chosen as a positive mutagen. Sidorov et al. [77] applied chemical carcinogens, including AFB_1, for the successful induction of tumors in *D. melanogaster*. Sişman [78] reported that hydrated sodium calcium aluminosilicate could effectively inhibit AFB_1-induced abnormalities in the developmental stages of *D. melanogaster*. Mutlu [79] reported about an increase in mitochondrial DNA copy number in response to OTA-induced mitochondrial DNA damage in *Drosophila*.

Other insect models applied include the navel orangeworm *Amyelois transitella* and the codling moth *Cydia pomonella*, which were used to study the *in vitro* metabolism of AFB_1 [80]. A field strain of *A. transitella* was found to produce the AFB_1 biotransformation products aflatoxicol (AFL), AFB_{2a}, and AFM_1, while a strain of *C. pomonella* produced trace amounts of AFL only, but neither of them produced AFBO, the principal carcinogenic metabolite of AFB_1.

Leung et al. [81] applied *Caenorhabditis elegans* as a model organism to study the role of xenobiotic-metabolizing cytochrome P450 (CYP) enzymes in the generation of genotoxic metabolites from AFB_1. The exposure of *C. elegans* to AFB_1 resulted in significant DNA damage and more efficient inhibition of the growth of xpa-1 nematodes than N2 nematodes. AFB_1-mediated growth inhibition was found to be resulting from CYP-mediated metabolism, as a *C. elegans* strain deficient in emb-8 (the

CYP-nicotinamide adenine dinucleotide phosphate reductase gene) proved to be more resistant to the growth-inhibitory effect of AFB_1 exposure than N2. The results indicated that *C. elegans* can metabolize AFB_1 into DNA-binding metabolites in a CYP-dependent manner. González-Hunt et al. [82] examined the effects of several known toxins including AFB_1 on *C. elegans* and found more damage caused to the mitochondrial than to the nuclear DNA, and AFB_1 also caused dopaminergic neurodegeneration, which supports the role for mitochondrial DNA damage in this process.

31.4.2 Rodent Models

As most popular laboratory animal models for the examination of *Aspergillus* mycotoxins are rodents (principally rats and mice), a broad search strategy was used to collect full text articles and—in cases when full text access was not available—abstracts from the PubMed database. The search was performed in article titles and abstracts using the following keyword combinations in order to find specific literature: aflatoxin OR ochratoxin OR sterigmatocystin OR patulin AND rat OR mouse OR mice. (Although certain *Aspergillus* species are known to produce fumonisins as well, FB_1 and related compounds are discussed in Chapter 34: *Fusarium* of this book.) The search was restricted to papers published in and after 2010 in order to focus on recent literature. Articles and abstracts were screened to include studies with relevant information (exact definition of the rodent model and the mycotoxin studied, as well as details about the dosage and the way of administration).

Table 31.1 summarizes the studies published since 2010 that have used rat models to study *Aspergillus* mycotoxins [83–119]. Among the retrieved records, relevant data were found in 37 publications. The most popular rat models in these studies were Fisher rats (12 studies), Sprague-Dawley rats (12 studies), and Wistar rats (11 studies), but other rat lines (Charles Foster rats, Dark Agouti rats, and hybrids of Sprague-Dawley female rats and a Fisher male rat) were also applied in single studies. Male rats were used in the majority of the studies, with a few exceptions applying both male and female, or just female animals. The age of the model animals ranged from 28 days to 15 months. The most commonly studied *Aspergillus* mycotoxin in rat models was AFB_1 (22 studies), followed by OTA (14 studies), AFG_1, and STC (1–1 study). Depending on the aim of the particular study, the *Aspergillus* mycotoxins were administered through diet, oral or gastric gavage, intraperitoneally, intramuscularly, intratracheally, intravenously, or subcutaneously. The majority of these studies examined AFB_1-induced hepatotoxicity [96,99,102,109,114,115] and hepatocellular carcinoma development [90,101,104,118], as well as the protective effects of different compounds against them [83,88,89,92,105,107,108]. In the case of OTA, many studies focused on its renal accumulation, toxicity, and induction of kidney failure [84,86,93,95,96,103,112]. AFG_1 was examined for lung toxicity in male Sprague-Dawley rats [110], while the potential link between STC and esophageal cancer was studied in Wistar rats [87]. Supriya and Reddy [91] used female Wistar rats to examine the effects of *in utero* exposure to graded doses of AFB_1 on development, behavior, and reproduction and found that AFB_1 severely compromised postnatal development of neonatal rats, caused a delay in the descent of testes, and a reduction in steroidogenesis and spermatogenesis that were accomplished by suppressed reproduction at adulthood. Other studies examined compounds like ginseng extract [108,116] and lycopene [106] for their protective effects against the genotoxicity of AFB_1 and OTA, respectively.

Table 31.2 summarizes the studies published since 2010 that have used mouse models to study *Aspergillus* mycotoxins [120–174]. Among the retrieved records, relevant data were found in 55 publications. The most popular mouse model for studying *Aspergillus* mycotoxins is the albino, laboratory-bred BALB/c strain (16 studies), followed by the black inbred strain C57BL/6 and its mutant and hybrid (e.g., B6C3F1) derivatives (8 studies), the largely noninbred Swiss albino mice (7 studies), and ICR outbred mice (6 studies). Other mouse models used include Kunming, Swiss Webster (CFW), MF1, CF-1, and A/J mice. Male animals were used in more studies; however, the use of females was more frequent than in the case of rat models. The age of the treated mice ranged from newborns to 12-month-old animals. Similar to the studies using rat models, the most commonly studied *Aspergillus* mycotoxin in mouse models was also AFB_1, followed by OTA, PAT, AFM_1, AFG_1, and STC. The mycotoxins were administered to the mice through diet, oral gavage, intraperitoneally, intratracheally, or

TABLE 31.1

Studies Using Rat Models for the Examination of *Aspergillus* Mycotoxins Published Since 2010 Based on PubMed

Rat Model	MT	Dose/Administration	Aim of the Study	Major Findings	References
Male Wistar albino (200–250 g)	AFB_1	1.5 mg/kg intraperitoneally	To study the protective effect of herbal ethanolic extracts of *Ixora coccinea*, *Rhinacanthus nasuta*, and *Spilanthes ciliata* against AFB_1-induced hepatotoxicity	The plant extracts showed significant protection against AFB_1-induced liver damage by lowering the activity of serum enzymes and restoring the decrease in GSH levels	[83]
Weaning male Wistar albino (35–40 g)	OTA	289 μg/kg/d given by gavages through a stomach tube for 28 d	To demonstrate that OTA cytotoxicity can cause kidney failure	Exposure to OTA induced significant lesions to the renal corpuscles, which can reduce the functional capacity of kidney and lead to kidney failure	[84]
Male Wistar Hannover	AFB_1	0.24 mg/kg by gastric gavage for up to 14 d	To investigate the power of conventional primary hepatocyte cultures along with their epigenetically stabilized counterparts in mimicking *in vivo* genotoxic insults	Conventional hepatocyte cultures mimicked the *in vivo* induced changes, which may be suitable for the identification of genotoxic carcinogens that do not require extrahepatic activation	[85]
Male Wistar	OTA	3 mg/kg/d gavage feed	To analyze the effect of OTA on three human and two rat renal proximal tubular models	The carcinogenic effects in three human cell models were significantly reflected in a rat *in vivo* model and partially in two rat *in vitro* models, which clearly evidenced the lack of species specificity between humans and rats in the biological perturbations caused by OTA	[86]
4-w-old male Wistar (40–50 g)	STC	30 μg/kg intraperitoneally injected	To investigate the effects of sterigmatocystin exposure in rats with reflux esophagitis	STC exposure downregulated TAP1 and LMP2, which directly affected the tumor immunity by allowing transformed cells to escape the host immune surveillance, thereby promoting esophageal cancer	[87]
10–12 w old male Wistar albino (200 ± 20 g)	AFB_1	3 mg/kg b.w. intraperitoneally	To evaluate whether antioxidant CAPE relieves oxidative stress in AFB_1-induced liver injury	CAPE induced protective effects on the AFB_1-induced hepatotoxicity by modulating free radical production, biochemical values, and histopathological alterations	[88]
2-m-old male Wistar albino (150 ± 2.54 g)	AFB_1	20 μg/d through gavage for 3 w	To investigate the effect of *Dialium guineense* pulp phenolic extract on AFB_1-induced oxidative imbalance in rat liver	Free and bound phenolic extract of *D. guineense* fruit enhanced the detoxification of AFB_1	[89]
Wistar	AFB_1	100 and 200 μg/kg	To explore the effect of EGb on expressions of Cox-2 and GST-Pi in the pathogenesis of HCC	EGb can regulate the expression of GST-Pi, but it does not seem to have an effect on Cox-2 expression in the liver of rats with the risk of HCC	[90]

(*Continued*)

TABLE 31.1 (*Continued*)

Studies Using Rat Models for the Examination of *Aspergillus* Mycotoxins Published Since 2010 Based on PubMed

Rat Model	MT	Dose/Administration	Aim of the Study	Major Findings	References
Female Wistar (210–220 g)	AFB_1	10, 20, or 50 μg/kg b.w. daily from gestation d 12–19 intramuscularly	To examine the embryonic exposure to graded doses of AFB_1 on development, behavior, and reproduction	*In utero* exposure to AFB_1 severely compromised postnatal development of neonatal rats and caused a delay in testes descent and reduction in steroidogenesis and spermatogenesis that were accomplished by suppressed reproduction at adulthood	[91]
3-w-old male Wistar (approx. 70 g)	AFB_1	200 μg/kg b.w. intraperitoneally	To investigate genotoxic and antigenotoxic effects of PRBE in rats using liver micronucleus assay	PRBE significantly reduced micronucleus formation and mitotic index in the liver of AFB_1-initiated rats	[92]
16-w-old male and female Wistar	OTA	Fed diets with 3 and 5 μg/g b.w. for females and males, respectively	To measure OTA concentration in whole kidneys of rats exposed chronically to dietary toxin	Apparent accumulation of OTA in kidney is due to binding to plasma proteins and long half-life in plasma	[93]
Male Fischer 344	OTA	Daily dietary intake of 50 mg/kg b.w. for up to 2 y	To compare the carcinogenic response between the NTP carcinogenic gavage dose regimen and administration of the same dosage via artificially contaminated diet	Carcinoma incidence (4/34 rats) through dietary exposure is considerably less than in the NTP study (16/51)	[94]
15-m-old male Fischer (445, 486 and 517 g)	OTA	6.25 mg by oral gavage	To evaluate the urinary metabolomic profile through ^{1}H-NMR spectra	^{1}H-NMR metabolomics revealed a progressive cyclic shift in principal components data cluster in predose status and insult phase, returning almost completely to normality after 6 months	[95]
10-w-old male and female F344	AFB_1 and OTA	Single dose of a mixture of 0.5 mg/kg b.w. of AFB_1 and 0.1 mg/kg b.w. of OTA by oral gavage	To validate UHPLC-FLD method with simultaneous extraction and analytical quantification of AFB_1 and OTA in rat plasma, kidney, and liver	UHPLC-FLD has successfully been used in three biological matrices, and important advantages of this method are even 100 μL of plasma or 25 mg of tissue are sufficient for obtaining results	[96]
Male F344 (90–100 g)	AFB_1	Gavaged with 25 μg/rat for 5 d per w for 2 successive w for a cumulative dose of 250 μg/rat	To evaluate 1,4-dioxan-2-one both for its ability to induce early putative preneoplastic lesions and to enhance or promote the growth of foci that have been induced by AFB_1	1,4-dioxan-2-one did not statistically increased focal number or focal volume percent and is not a common carcinogen	[97]

(*Continued*)

TABLE 31.1 (*Continued*)

Studies Using Rat Models for the Examination of *Aspergillus* Mycotoxins Published Since 2010 Based on PubMed

Rat Model	MT	Dose/Administration	Aim of the Study	Major Findings	References
Male F344/N	AFB_1	NTP 2000 feed 1 ppm	To check whether a subset of genes from a previously derived AFB_1 gene signature would be observed in archival RNA from fresh frozen liver, be replicated in FFPE liver RNA, and also be present in other FFPE tissues like the kidney and lung as secondary or alternate targets for carcinoma	Evaluation of gene signatures in archival tissues can be an important toxicological tool for evaluating critical molecular events associated with chemical exposures	[98]
Male F344 (80–120 g)	AFB_1	0.25, 0.75, and 1.5 mg/kg b.w. by oral gavage	To identify mechanisms and potential biomarkers for predicting the development and progression of AFB_1-induced acute hepatotoxicity through integrated analysis of general toxicity studies, transcriptomics, and metabolomics profiles	p53 signaling pathway induced by oxidative damage was the crucial step in AFB_1-induced acute hepatotoxicity, whereas gluconeogenesis and lipid metabolism disorder were found to be the major metabolic effects after acute AFB_1 exposure	[99]
6–7-w-old male F344	OTA	0, 70 or 210 μg/kg b.w. gavage	To explore the relationship between OTA-induced oxidative damage and carcinogenicity in the liver and kidney	OTA caused apparent kidney damage within 13 w but exerted limited effect on oxidative stress parameters	[100]
Male F344	AFB_1	1.5 mg/kg b.w. by gavages	To elucidate the functional complexity of microRNAs in AFB_1-induced hepatocellular tumorigenesis	Abnormally expressed cancer-related miRNAs were identified	[101]
3-w-old (newly weaned) male inbred F344 or outbred Sprague-Dawley	AFB_1	Diets containing 0, 1, 5, 10, or 20 ppm	To develop an experimental animal model of dietary aflatoxin exposure to investigate the relationship between AFB_1 toxin exposure and growth disturbance and to explore candidate mechanisms responsible for this association	AFB_1-exposed animals exhibited dose-dependent wasting and stunting, liver pathology, and suppression of hepatic targets of growth hormone signaling	[102]
6–7 w-old male F344/N	OTA	0, 21, 70, or 210 μg/kg b.w. by oral gavage	To evaluate a range of novel biomarkers of renal toxicity to test their ability in detecting acute and chronic kidney injury	The histopathological alterations induced by OTA were best reflected by changes in urinary Kim-1 marker	[103]

(*Continued*)

TABLE 31.1 (*Continued*)

Studies Using Rat Models for the Examination of *Aspergillus* Mycotoxins Published Since 2010 Based on PubMed

Rat Model	MT	Dose/Administration	Aim of the Study	Major Findings	References
Male F344/N	AFB_1	1 ppm in feed for 90 d	To define gene expression changes that might relate to carcinogenesis produced by AFB_1 exposure prior to onset of malignancy and to begin a high-resolution map of the rat liver transcriptome	49 novel, differentially-expressed transcripts including two unique, unannotated, hepatic AFB_1-responsive transcripts overexpressed by 10- to 25-fold were identified	[104]
Male 344/NHsd	AFB_1	200 µg/kg b.w. via oral gavage daily for 4 w	To assess and characterize the chemoprotective efficacy of synthetic oleanane triterpenoid CDDO-Im against AFB_1-induced HCC	CDDO-Im completely protected against AFB_1-induced liver cancer compared to a 96% incidence in AFB_1 group. With CDDO-Im treatment, integrated level of urinary AFB_1-N^7-guanine was significantly reduced and aflatoxin-NAC, a detoxification product, was consistently elevated after the first AFB_1 dose. Also, the hepatic burden of GST-P-positive foci and the toxicogenomic RNA expression signature were largely absent with CDDO-Im intervention	[105]
Male Sprague-Dawley (230–250 g)	OTA	2.2 mg/kg b.w. orally by gastric gavage	To demonstrate the effects of free radical scavengers MEL and CoQ_{10} on cells damaged by OTA	OTA administration caused oxidative damage, and MEL or CoQ_{10} treatment ameliorated the OTA-induced tissue injuries	[106]
4-w-old male Sprague-Dawley (85–100 g)	AFB_1	150 µg/kg/d for 3 d intraperitoneally	To examine the protective effects of KRG against hepatotoxicity induced by AFB_1 using liver-specific serum marker analysis, histopathology, and terminal deoxynucleotidyl transferase-mediated dUTP nick-end labeling assay	KRG showed protective effects against hepatotoxicity induced by AFB_1 through its antioxidant effects by increasing SOD, CAT, and GPX activity and reducing lipid peroxidation	[107]
3-m-old male Sprague-Dawley (100–120 g)	AFB_1	Fed diets with 2.5 mg/kg b.w.	To evaluate whether AFB_1-induced damage in rats can be counteracted by feeding with WPC and KGE	The genotoxicity of aflatoxins was partly prevented by dietary supplementation with WPC, KGE, or their combination	[108]
Male and female Sprague-Dawley and Fischer CDF	AFB_1	25 µg intraperitoneally	To compare induction of hepatic GSTs by SF and the resultant impact on the formation of AFB_1-DNA adducts in livers of both sexes of two rat strains	A dose-dependent relationship was found for both induction of GST and reduction of AFB_1-N^7-guanine in both rats, no gender-specific responses were observed	[109]

(*Continued*)

TABLE 31.1 (*Continued*)

Studies Using Rat Models for the Examination of *Aspergillus* Mycotoxins Published Since 2010 Based on PubMed

Rat Model	MT	Dose/Administration	Aim of the Study	Major Findings	References
Male Sprague-Dawley (110–130 g, 120.2 ± 5.9 g)	AFG_1	30 μg/kg b.w. intratracheally instilled	To identify the acute toxicity of AFG_1 in AT-II cells	AFG_1 induces structural and functional impairment in AT-II cells involved in the acute toxicity of AFG_1 in lung tissues	[110]
Male Sprague-Dawley (200 ± 20 g)	OTA	0.2 mg/kg b.w. single dose by oral gavage	To develop combinatorial platform of LC–MS/MS and LC–TOF-MS to investigate *in vivo* kinetics and biotransformation of OTA in rats	LC–MS/MS method determined OTA with a run time of 7.0 min. Simultaneously, an LC–TOF-MS method unambiguously identified the metabolites of OTA in a total run time of 14 min	[111]
Male Sprague-Dawley (<200 g)	OTA	0.5 mg/kg b.w. for 14 d by intragastric lavage	To investigate the possible protective effect of lycopene against the renal toxic effects of OTA	Lycopene might be partially protective against OTA-induced nephrotoxicity and oxidative stress as it increases GPx1 activity and GSH levels and decreases apoptotic cell death	[112]
12-w-old male Sprague-Dawley (250–260 g)	OTA	0.5 mg/kg/d by intragastric lavage for 7 and 14 d	To investigate the protective effects of lycopene against the genotoxicity of OTA in rat tissues using the alkaline comet assay	Lycopene provided a protective effect against OTA-induced DNA damage, which was evidenced by decreased tail moment and intensity in both rat kidney and liver cells	[113]
2-m-old Sprague-Dawley	AFB_1	Fed diets with 4, 10, 25, and 50 μg/kg b.w.	To evaluate the toxicological impacts on rats fed with AFB_1-contaminated feedstuffs	Various negative impacts on liver and kidney functions were observed in AFB_1-contaminated groups, which was positively correlated with AFB_1 concentrations	[114]
6–8-w-old male Sprague-Dawley	AFB_1	0.3 mg/kg/b.w. oral or intraperitoneal, intravenous or subcutaneous injection	Transcriptome profiling of rat liver samples treated with 27 chemicals to evaluate the utility of RNA-Seq in safety assessment and toxicity mechanism elucidation	The resulting large toxicogenomics dataset can serve as a resource to characterize various aspects of transcriptomic changes (e.g., alternative splicing) that are byproducts of chemical perturbation	[115]
3-m-old female Sprague-Dawley (100–120 g)	AFB_1 and FB_1	80 μg/kg b.w. AFB_1 or/and 100 μg/kg b.w. FB_1 orally	To evaluate the protective role of PGE against the synergistic effect of subchronic administration of AFB_1 and FB_1 on DNA and gene expression in rat	PGE induced potential protective effects against the oxidative stress and genotoxicity of these mycotoxins	[116]

(*Continued*)

TABLE 31.1 (*Continued*)

Studies Using Rat Models for the Examination of *Aspergillus* Mycotoxins Published Since 2010 Based on PubMed

Rat Model	MT	Dose/Administration	Aim of the Study	Major Findings	References
Hybrids of Sprague-Dawley female and a F344 male	OTA	Dietary intake of 5 ppm OTA	To explore whether the tumor promoter sodium barbiturate could shorten the otherwise long latency between exposure to OTA and tumorigenesis	Female mammary tumorigenesis was advanced by 6 ms in all rats given the OTA-barbiturate	[117]
18- to 20-w-old male Charles Foster albino (200–220 g)	AFB_1	1 mg/kg b.w. of to the 18-h fasted rats on 2 consecutive d intraperitoneally	To describe the histopathological progression of AFB_1-induced HCC in the liver of rats	AFB_1 toxicity activates the oxidative stress proinflammatory pathway, which induces hepatocarcinogenesis	[118]
7-w-old dark Agouti	OTA	Fed with 2.5, 7, 40, and 100 μg/kg wheat	To examine OTA metabolism in the liver and kidney of male and female rats using reduced GSH and NAC conjugates along with nontoxic OTα as biomarkers	The distribution of OTA in male and female rat kidney is not significantly different, but OTA metabolism is greater in male than female rats	[119]

d, day; w, week; m, month; AFB_1, aflatoxin B_1; AT-II, alveolar type II; CAPE, caffeic acid phenethyl ester; CDDO-Im, 1-[2-cyano-3-,12-dioxooleana-1,9(11)-dien-28-oyl]imidazole; CAT, catalase; CoQ10, coenzyme Q10; Cox-2, cyclooxygenase-2; DNA, desoxyribonucleic acid; EgB, *Ginkgo biloba* extract; FB_1, fumonisin B_1; FFPE, formalin-fixed, paraffin-embedded; GPx1, glutathione peroxidase 1; GSH, glutathione; GST, glutathione S-transferase; HCC, hepatocellular carcinoma; KGE, Korean ginseng extract; KRG, Korean red ginseng; LC–MS/MS, liquid chromatography–mass spectrometry/mass spectrometry; LC–TOF-MS, liquid chromatography–time of flight-mass spectrometry; LMP2, low molecular weight protein 2; MEL, melatonin; MT, mycotoxin; NAC, N-acetylcysteine; NMR, nuclear magnetic resonance; NTP, National Toxicology Program; OTA, ochratoxin A; PAT, patulin; PGE, *Panas ginseng* extract; PRBE, purple rice bran extract; RNA, ribonucleic acid; SF, sulforaphane; SOD, superoxide dismutase; STC, sterigmatocystin; TAP1, transporter associated with antigen processing 1; UHPLC-FLD, ultrahigh-performance liquid chromatography with fluorescence detection; WPC, whey-protein concentrates.

TABLE 31.2

Studies Using Mouse Models for the Examination of *Aspergillus* Mycotoxins Published Since 2010 Based on PubMed

Mouse Model	MT	Dose/Administration	Aim of the Study	Major Findings	References
7-w-old male BALB/c (20 g ± 2 g)	OTA	Single dose of 3.5, 7, 35, 70, 289, 578, 1056 μg/kg/b.w. by oral gavage, feed containing increasing amount of OTA (0.5, 1.4, 8, 20 μg/kg/b.w.) every day during 4 w, a single intraperitoneal injection of OTA (2.5 mg/kg) at gestation day 17	To test whether the acute exposure to OTA via food and via exposure *in utero* causes adducts in testicular DNA and to test whether these lesions are identical to those that can be produced in the kidney and testis by the consumption of OTA	*In utero* exposure to OTA caused adducts in the testicular DNA of male offspring that were similar to DNA adducts observed in the kidney and testis of gavage-fed adults, supporting a possible role for OTA in testicular cancer	[120]
6- to 8-w-old female BALB/c ByJ	PAT	5–20 μg by gastric intubation on days 0, 1, 13, 14, 17, and 18, orally on days 18, 21, and 22, 100 ng intranasally on day 22	To investigate the adjuvant activity of mycotoxins on allergic airway inflammation	Patulin increased the allergic immune response in mice by modulating the Th1/Th2 balance via direct effects on IL-12 secretion in DCs and by inducing oxidative stress	[121]
4- to 6-w-old BALB/c athymic, nu/nu, male nude (18.0–22.0 g)	AFB_1	Single dose at 240 μg/kg b.w. once a week intraperitoneally and fed in the drinking water (400 μg/kg d), continuously for 16 w	To evaluate the role of HOC in the development of hepatocellular carcinoma	Oval cells have produced tumors in liver following transfection with hepatitis B virus x gene (HBx) and treatment with AFB_1	[122]
Male BALB/c (20–25 g)	AFB_1	250 μg/kg b.w. intraperitoneally	To evaluate the hepatoprotective effect of CCE on AFB_1-induced liver damage in mice by measuring MDA level, protein carbonyls generation, and expression of Hsp 70 and Hsp 27 in liver	The treatment of CCE showed a total reduction of AFB_1-induced oxidative damage markers and genotoxicity markers and decreased the expressions of proapoptotic proteins p53 and bax	[123]
6- to 8-w-old male BALB/c	AF	625 μg/kg b.w./d given directly into the stomach through a gastric tube	To investigate the antioxidant effect of pumpkin seed oil against the oxidative-stress-inducing potential of aflatoxin	AF exhibited adverse effects on most of the oxidative stress markers, but the administration of pumpkin seed oil diminished aflatoxin-induced adverse effects	[124]

(*Continued*)

TABLE 31.2 (*Continued*)

Studies Using Mouse Models for the Examination of *Aspergillus* Mycotoxins Published Since 2010 Based on PubMed

Mouse Model	MT	Dose/Administration	Aim of the Study	Major Findings	References
3-w-old specific pathogen-free male BALB/c	ST	3, 30, 300, and 3000 μg/kg intraperitoneally	To explore the short-term immunotoxic effects of ST, specifically on FoxP3+ Tregs and pDCs, by observing changes in number/expression of FoxP3+ Tregs, pDCs, and CD4+, CD8+ T cells	The proportion of CD8+ T cells was decreased in the thymus in ST (3 μg/kg) group, while that of CD4+ and CD8+ T cells was increased in the spleen in two treatment groups (3 and 30 μg/kg); the proportion of FoxP3+ Tregs and FoxP3 expressions were all significantly increased in mPBMCs, the thymus, and the spleen. Importantly, the population of pDCs significantly decreased in the thymus but increased in the spleen, which is due to a temporary immune response triggered by the ST inhibition	[125]
BALB/c	ST	3 mg/kg intraperitoneally	To evaluate the putative effects of ST on the expression of TNF-α, IL-6, and IL-12 at mRNA levels in mPBMCs and peritoneal macrophage cells and on the serum TNF-α and IL-6 levels	Downregulation of TNF-α, IL-6, and IL-12 mRNA expression in mPBMCs and peritoneal macrophage cells was observed, and serum TNF-α and IL-6 levels were also decreased	[126]
4- to 6-w-old inbred female BALB/c	AFB_1	0.1 mL at a daily dose of 30, 15, and 10 μL/kg intraperitoneally	To focus on the immunosuppressor activity of AFB_1 and immunostimulatory activity of AGE through the evaluation of Treg counts and the pattern of cytokine production	AGE has increased the level of INF-γ and IL-4 cytokines produced by splenocytes stimulated by specific tumor antigen and decreased the number of Treg cells in the spleen; AFB_1 increased the number of Treg cells in the spleen and decreased cytokine production	[127]
6-w-old male BALB/c (25 ± 0.3 g)	AFM_1	100 μg/kg b.w. orally by gavage	To isolate food-grade probiotic bacteria able to degrade/bind AFM_1 *in vitro* from dairy products and to evaluate whether the same organism(s) could impart a protective role against AFM_1-induced immunotoxicity in exposed mice	*Lactobacillus plantarum* MON03 and *L. rhamnosus* GAF01 were isolated, and both of them could remove AFM_1 *in vitro;* since, *L. rhamnosus* GAF01 showed the highest binding capacity, it was selected for the *in vivo* study. *L. rhamnosus* GAF01 prevented AFM_1-induced effects on total white and red blood cells and lymphocyte subtypes after 15 d of treatment	[128]

(*Continued*)

TABLE 31.2 (*Continued*)

Studies Using Mouse Models for the Examination of *Aspergillus* Mycotoxins Published Since 2010 Based on PubMed

Mouse Model	MT	Dose/Administration	Aim of the Study	Major Findings	References
Nude/BALB/c	AFB_1	1×10^7 cells injected subcutaneously	To investigate the underlying mechanisms of AFB_1-induced effects on ATR function and neoplastic transformation of the cells when B-2A13 cells were exposed to low levels of AFB_1 (0.1–10 nM)	ATR mediated the DNA damage repair response and subsequent neoplastic transformation in P40 B-2A13 cells	[129]
10-w-old female BALB/c (21 ± 2 g)	AFM_1	100 mg/kg b.w orally (by gavage)	To evaluate a new AFM_1-binding/degrading microorganism for biologic detoxification, to examine its ability to degrade AFM_1 in liquid medium and to evaluate its potential for *in vivo* preventive effects against AFM_1-induced immunotoxicity and genotoxicity	The isolated *Lactobacillus plantarum* MON03 (LP) was found to display significant binding ability to AFM_1 in PBS (93%) and was taken for *in vivo* study, where administration of LP with AFM_1 strongly reduced the cytotoxic/genotoxic adverse effects of AFM_1	[130]
4-w-old female BALB/c	AFB_1, ZEA, DON	2.5, 5.0, and 5.0 mg/kg b.w. of AFB_1, ZEA, and DON, respectively, administered orally via gavage needle	To assess the individual and combined toxic effects of AFB_1, ZEA, and DON within the liver	AFB_1, ZEA, and DON induced liver injury and were associated with induced oxidative stress and an upregulation of the apoptotic genes Caspase-3 and Bax, along with a downregulation of the antiapoptotic gene Bcl-2; AFB_1 + DON displayed synergistic hepatotoxic effects, while AFB_1 + ZEA displayed antagonistic hepatotoxic effects	[131]
10-w-old female BALB/c	AFB_1 and AFM_1	0.25 mg/kg b.w. AFB_1 and 0.27 mg/kg AFM_1 administered by oral gavage	To evaluate a new AFB_1- and AFM_1-binding/degrading microorganism for biological detoxification, examine its ability to degrade AFB_1 and AFM_1 in liquid medium, and evaluate its potential for *in vivo* preventive effects against AFB_1- and AFM_1-induced immunomodulation	*Lactobacillus plantarum* MON03 (LP) was isolated and displayed significant binding ability to AFB_1 and AFM_1 in PBS (i.e., 82% and 89%, respectively); *in vivo* study was conducted where LP strongly reduced the adverse effects of each mycotoxin	[132]
BALB/c	AFG_1	100 µg/kg body weight orally	To understand the phenotypic alterations of AT-II cells, which may provide new insight into the immune function of AT-II cells involved in AFG_1-induced pulmonary tumorigenesis	Increased MHC-II expression in alveolar epithelium was observed and associated with enhanced Treg infiltration in mouse lung tissues and also with AFG1-induced inflammation	[133]

(*Continued*)

TABLE 31.2 (*Continued*)

Studies Using Mouse Models for the Examination of *Aspergillus* Mycotoxins Published Since 2010 Based on PubMed

Mouse Model	MT	Dose/Administration	Aim of the Study	Major Findings	References
Female BALB/c (18–20 g)	AFG_1	100 μg/kg b.w. orally using the gavage technique	To examine whether oral gavage of AFG_1 induces chronic lung inflammation and how it contributes to carcinogenesis	Inflammatory responses were heightened in the lung alveolar septum 3 and 6 months after oral administration of AFG_1; 12 months later, AFG_1 induced alveolar epithelial hyperplasia and adenocarcinoma; upregulation of NF-κB, p-STAT3, and Cox-2 was also induced in lung adenocarcinoma	[134]
8- to 12-w-old male BALBb/c (20–25 g)	OTA	0.85 mg/kg b.w. inoculated intraperitoneally	To investigate the protective effects of luteolin, ChlA, and CafA against cyto-genotoxic effects caused by OTA	The tested polyphenols reduced the toxicity caused by OTA on different target cells with good protective effect, ChlA being the compound that showed the best effects	[135]
5- to 6-w-old female Swiss albino	PAT	Topical dose of 400 nmol	To investigate the skin carcinogenic potential of topically applied PAT	Topical application of PAT resulted in cell proliferation, which is mediated by ROS-induced MAPKs signaling pathway, leading to transcriptional activation of downstream target proteins c-fos, c-jun, and transcription factor NFκB, showing that PAT has dermal tumorigenic potential	[136]
Male Swiss (37–40 g)	AF (B_1, B_2, G_1, and G_2)	Orally administered with AF (B_1, B_2, G_1, and G_2 in the ratio of 8:3:2:1) using a feeding needle with a dosage of 750 and 1500 μg/kg b.w in 0.2 mL olive oil/animal/d	To investigate the effect of black tea infusion on AF-induced hepatotoxicity	Significant restoration of AF-induced damages in body weight, organ weight, serum chemistry, and histopathological features was observed in a dose-dependent manner in mice administered AF plus black tea infusion	[137]
Young adult inbred male Swiss albino (32–35 g)	AF	Orally with 25 and 50 μg/0.2 mL olive oil/animal/day (750 and 1500 μg/kg b.w.), respectively, for 30 d	To evaluate AF-induced biochemical changes in the liver and its possible amelioration by black tea extract	Administration of AF caused significant, dose-dependent reduction in DNA, RNA, protein, and glycogen contents; however, cholesterol content and phosphorylase activity were significantly increased; these changes were significantly ameliorated on cotreatment with black tea extract	[138]

(Continued)

TABLE 31.2 (*Continued*)

Studies Using Mouse Models for the Examination of *Aspergillus* Mycotoxins Published Since 2010 Based on PubMed

Mouse Model	MT	Dose/Administration	Aim of the Study	Major Findings	References
6- to 7-w-old female Swiss albino (20 ± 3 g)	OTA	Topical application of 25, 50, 100, and 200 nmol/0.2 mL acetone	To evaluate the effect of a single topical application of OTA at lower doses on skin epidermis and to evaluate the skin-tumor-promoting potential of OTA	OTA has cell proliferative and tumor-promoting potential in mouse skin, which involves EGFR-mediated MAPKs and Akt pathways along with NF-κB and AP-1 transcription factors and that cyclin-D1 and Cox-2 are the target genes responsible for tumor-promoting activity of OTA	[139]
Young male inbred Swiss albino (30–33 g)	OTA	50 µg and 100 µg in 0.2 mL olive oil/animal/d for 45 d orally	To evaluate the ameliorative effects of an aqueous extract of *Emblica officinalis* on OTA-induced lipid peroxidation in the kidney and liver	Administration of *Emblica officinalis* aqueous extract (2 mg/animal/d) and OTA caused a significant amelioration in the OTA-induced lipid peroxidation in liver and kidney	[140]
60-d-old male Swiss albino (30–40 g)	AFB_1	66.6 µg/kg b.w./d orally by gavage	To investigate the protective effect of esculin against pro-oxidant AFB_1-induced nephrotoxicity	Nephrotoxicity induced by AFB_1 was ameliorated by esculin, which is due to its antioxidant, anticarcinogenic, and ROS-scavenging properties	[141]
Male Swiss albino (30 ± 5 g)	AFB_1	2 µg/30 g b.w orally	To investigate the ability of the *Tinospora cordifolia* to scavenge free radicals generated during aflatoxicosis	*T. cordifolia* showed protection against AF-induced nephrotoxicity due to the presence of alkaloids such as a choline, tinosporin, isocolumbin, palmatine, tetrahydropalmatine, and magnoflorine	[142]
6-w-old male ICR	AFB_1	0.75 mg/kg b.w. orally	To examine the biochemical mechanisms associated with the effects of quercetin on AFB_1-mediated liver damage	Quercetin does not directly protect against AFB_1-mediated liver damage *in vivo* but exerts a partial role in promoting antioxidative defense systems and inhibiting lipid peroxidation	[143]
6-w-old male ICR	AFB_1	0.75 mg/kg b.w. orally	To investigate the biochemical mechanisms of the RCMF effects on AFB_1-mediated liver damage	RCMF attenuates AFB_1-mediated damage to the liver, which is at least partially related to the restoration of antioxidant defense systems and an increase in AFB_1–GSH conjugate formation	[144]
4-w-old female ICR	AF, ZEN, DON	Fed diets with 597 µg/kg AF, 729 µg/kg ZEN, and 3.1 mg/kg DON	To investigate the regulation of multiple mycotoxins on the oxidative stress	Naturally contained mycotoxins are toxic *in vivo* and able to induce oxidative stress	[145]

(Continued)

TABLE 31.2 (*Continued*)

Studies Using Mouse Models for the Examination of *Aspergillus* Mycotoxins Published Since 2010 Based on PubMed

Mouse Model	MT	Dose/Administration	Aim of the Study	Major Findings	References
ICR	OTA	1, 5, or 10 μM	To explore the cytotoxic effects exerted by OTA on the blastocyst stage of mouse embryos, on subsequent embryonic attachment, on outgrowth *in vitro*, and following *in vivo* implantation via embryo transfer	*In vitro* exposure to OTA triggers apoptosis and retards early postimplantation development after transfer of embryos to host mice; OTA induces apoptosis-mediated injury of mouse blastocysts via ROS generation and promotes mitochondrion-dependent apoptotic signaling processes that impair subsequent embryonic development	[146]
4-w-old female ICR	DON, ZEN, and AF	Fed diets with 3.875 mg/kg DON, 1897 μg/kg ZEN, and 806 μg/kg AF	To investigate the effects of a mycotoxin-contaminated diet on oocyte quality	Mycotoxin-contaminated diet adversely affected the developmental competence of ovaries and decreased the oocyte quality	[147]
7- to 9-w-old male CD1	OTA	For acute tests, 1 mg/kg b.w. and 10 mg/kg b.w. for 72 h; for chronic test, 0.5 mg/kg b.w. for 21 d via oral gavage	To evaluate OTA-degrading and detoxifying potential of *Cupriavidus basilensis* OR16 strain	OTA was degraded efficiently by *Cupriavidus basilensis* OR1; OTα did not display nephrotoxic effects *in vivo*	[148]
Male Kunming (22 ± 2 g)	PAT	1 mg/kg intraperitoneally	To investigate PAT-induced hepatotoxicity and genotoxicity and the antioxidant and antigenotoxicity efficiency of GTP against PAT-induced toxicity	GTP exerted antioxidative activity in reducing hepatic ROS and TBARS level and increasing GSH content, it also inhibited PAT-induced bone marrow damage, including the formation of micronucleus and chromosomal aberration	[149]
4-w-old male Kunming (20–23 g)	PAT	1 mg/kg intraperitoneally	To investigate the protective effects of selenium supplementation on PAT-induced neurotoxicity	Selenium supplementation increased the activity and expression of GSH-related enzymes and offered significant protection against brain damage induced by PAT	[150]
21- to 28-d-old pathogen-free Carworth Farms white Swiss Webster male (23 ± 1.2 g)	STC	4×10^{-5} mol toxin/kg lung weight by intratracheal instillation	To evaluate lungs intratracheally exposed to a single dose of toxin for histological and histochemical responses at 4 and 12 h postexposure	Time- and toxin-dependent transcription and expression of inflammation-associated genes and inflammatory responses were observed	[151]

(*Continued*)

TABLE 31.2 (*Continued*)

Studies Using Mouse Models for the Examination of *Aspergillus* Mycotoxins Published Since 2010 Based on PubMed

Mouse Model	MT	Dose/Administration	Aim of the Study	Major Findings	References
Swiss Webster	AFB_1	—	To express the putative proteins tGSTAs in an *Escherichia coli* heterologous system and to characterize the functional properties of products of the subunits with respect to their enzymatic activities toward prototype GST substrates and toward enzymatically generated AFBO.	Recombinant tGSTAs detoxified AFBO, whereas their hepatic forms did not, which implies that the hepatic forms of these enzymes are silenced by one or more regulatory mechanisms	[152]
9-w-old male CF-1 (28.5 ± 1.27 g)	PAT	A single dose of 1.0, 2.5, or 3.75 mg/kg b.w. intraperitoneally	To investigate the genotoxic effects of patulin in multiple organs (brain, kidney, liver, and urinary bladder) using an *in vivo* comet assay	PAT induced DNA damage in the brain, liver, and kidneys and demonstrated a high correlation between decreased GSH content and increased lipid peroxidation and DNA damage, suggesting interrelationship between the pro-oxidant and genotoxic effects of PAT; pretreatment administration of *N*-acetyl-cysteine reduced PAT-induced DNA damage	[153]
Female A/J (16–19 g)	AFB_1	50 mg/kg	To examine the effect of *in vivo* treatment with a single tumorigenic dose of AFB_1 on the formation and repair of oxidative DNA damage in lung and liver	2 h posttreatment, AFB_1 increased 8-OHdG levels in mouse lung DNA, but did not alter 8-OHdG levels in liver or 5-OHdC levels in lung or liver; AFB_1 treatment also increased BER activity in mouse lung but did not affect hepatic BER activity; levels of OGG1 immunoreactive protein were increased in both lung and liver; the results are consistent with oxidative DNA damage contributing to the carcinogenicity of AFB_1	[154]
Male MF1 (25–30 g)	PAT	Orally administered with 1 mL of apple juice contaminated with 152.5 ppb of PAT for 6 w	To assess the biochemical and histopathological effects of PAT in apple juice samples collected from different outlets retailing in Jeddah, Kingdom of Saudi Arabia	PAT-contaminated apple juice resulted in liver, kidney, and neurotoxicological effects	[155]

(*Continued*)

TABLE 31.2 (*Continued*)

Studies Using Mouse Models for the Examination of *Aspergillus* Mycotoxins Published Since 2010 Based on PubMed

Mouse Model	MT	Dose/Administration	Aim of the Study	Major Findings	References
Sexually mature female C57BL/6	OTA	Single dose at 3.0 mg/kg b.w. intraperitoneally	To evaluate the possible regulatory effect induced by OTA administration during a critical moment of gestation on expression of some homeotic related genes in order to understand the possible mechanism of OTA in the induction of differentiation defects	Dlx5 is a target for OTA and the inhibition of its function, directly or indirectly, could be the cause of the observed differentiation defects	[156]
16 timed-pregnant *Helicobacter*-free C3H/HeN female, C57BL/6 FL-N/35 crossed with C3H/HeN	AFB_1	7 μg/g body b.w. intraperitoneally	To test whether specific intestinal bacteria promote liver cancer in chemical and viral transgenic mouse models	Intestinal colonization by *Helicobacter hepaticus* was sufficient to promote AF- and Hepatitis C virus transgene-induced HCC; neither bacterial translocation to the liver nor induction of hepatitis was necessary	[157]
$Gnmt^{-/-}$ and wild-type 129/B6 (129sv X C57BL/6)	AFB_1	10 mg/kg of b.w. at 7 d of age and second treatment of 40 μg AFB_1 per mouse at 9 w of age intraperitoneally	To test whether the $Gnmt^{-/-}$ mice are susceptible to AFB_1 carcinogenesis and to test whether Gnmt deficiency may accelerate AFB_1-induced liver tumorigenesis	Liver tumor formation in AFB_1-treated $Gnmt^{-/-}$ mice occurred earlier than in solvent - treated $Gnmt^{-/-}$ mice and AFB_1-treated wild type mice; Gnmt deficiency increased the susceptibility to AFB_1 related HCC; 5 detoxification pathway-related genes: Cyp1a2, Cyp3a44, Cyp2d22, Gsta4, and Abca8a were downregulated in AFB_1-treated $Gnmt^{-/-}$ mice	[158]
Pregnant *gpt* delta C57BL/6J	AFB_1	6 mg/kg for mutation analysis and 5 mg/kg for adduct analysis via intraperitoneal injection or oral gavage on gestation day 14	To study DNA damage by AFB_1 during prenatal development and the risk of genetic disease later in life	Early-life exposure of mycotoxins, especially during the embryonic period, is strikingly more mutagenic than exposure later in life	[159]

(*Continued*)

TABLE 31.2 (*Continued*)

Studies Using Mouse Models for the Examination of *Aspergillus* Mycotoxins Published Since 2010 Based on PubMed

Mouse Model	MT	Dose/Administration	Aim of the Study	Major Findings	References
AlbCre:Keap1$^{flox/flox}$ and GSTA3 knockout mice were crossed to produce compound GSTA3:Alb:Cre:Keap1$^{flox/flox}$ DKO	AFB_1	0.8 mg/kg b.w. orally	To examine whether genetic or pharmacologic activation of Nrf2 signaling could rescue the hypersensitive GSTA3 knockout mouse from the genotoxicity of AFB_1	The inability to rescue GSTA3 knockout mice from AFB_1 genotoxicity through the Nrf2 transcriptional program indicated that Gsta3 is unilaterally responsible for the detoxification of AF in mice	[160]
Male C57BL/6JOlaHsd	AFB_1	1 ppm, ~0.15 mg/kg b.w./d via feed for 7 d	To characterize the potential genotoxic properties of quercetin in the small intestine as well as in the liver by transcriptome analysis in order to provide new additional information to evaluate the safety of quercetin	Supplementation with quercetin at ~350 mg/kg bw/d for 12 w in mice showed no upregulation of genotoxicity-related pathways in liver and small intestine	[161]
HBsAg transgenic mice, C57BL/6J-Tg(Alb1HBV)44Bri/J were crossed to C57BL/6J WT to obtain HBsAg and WT	AFB_1	Single dose of 6 μg/g b.w. intraperitoneally	To explore the option of utilizing mouse models to understand in a systematic and longitudinal manner the molecular pathways that are progressively deregulated by various etiological factors in contributing to HCC formation and report the initial findings in characterizing their validity	The utility of these mouse models provided a rich resource for the longitudinal analysis of molecular changes and biomarkers associated with HCC	[162]
4-d-old B6C3F1	AFB_1	Single dose of 6 mg/kg AFB_1 on d 4 after birth via intraperitoneal injection	To assess the capability of AFB_1 to generate DNA sequence changes including those driven by nonhomologous recombination *in vivo* by analyzing the Spi$^-$ (Sensitive to P2 Inhibition) phenotype resulting from the disruption of *red* and *gam* genes in the λEG10 transgene of *gpt* delta B6C3F1 mice	After 3 w, 10-fold increase in the Spi$^-$ mutant fraction (MF) was observed and after 10 w, a further increase was observed; the MF in the *gpt* gene was also increased at 10 w compared to 3 w; no gender-specific differences were found in the Spi$^-$ or *gpt* MFs; the Spi$^-$ spectrum was dominated by GC to TA transversions, with one exceptionally strong hotspot at position 314, which is the dominant mutation seen in people exposed to AF	[163]

(*Continued*)

TABLE 31.2 (*Continued*)

Studies Using Mouse Models for the Examination of *Aspergillus* Mycotoxins Published Since 2010 Based on PubMed

Mouse Model	MT	Dose/Administration	Aim of the Study	Major Findings	References
4-w-old male ICR mice (N 32)	AFB_1	50 μg/kg b.w. via intraperitoneal injection daily for 45 d	To determine changes in testicular gene expression due to exposure to AFB_1 and to investigate which cell types were affected by treatment with AFB_1	Differential expression of genes was noted during cell differentiation, and extracellular space was observed in the testis; renin was commonly represented amongst many clusters and chosen for further analyses; renin was upregulated in response to AFB_1, especially in males that maintained fertility despite AFB_1 treatment	[164]
7- to 9-w-old male heterozygous p53 knockout mice (p53 (+/–), B6.129-$Trp53^{tm1Brd}$N5) and wild type controls (p53 (+/+))	AFB_1	Fed diets with 0, 0.2 or 1.0 ppm AFB_1	To determine if chronic exposure of mice to low levels of AFB_1 alters NER in lung and liver and to determine if such alterations are affected in p53-deficient mice	Chronic exposure to AFB_1 increased NER activity in wild-type mice, and this response was diminished in heterozygous p53-knockout mice, indicating that loss of one allele of p53 limits the ability of NER to be upregulated in response to DNA damage	[165]
5- to 7-w-old male p53+/– (P53N5-T) and p53 +/+ (P53N5-W)	OTA	Fed diets containing 1, 15, or 40 mg OTA/kg (w 1 and 2) and 0.5, 2, or 10 mg OTA/kg (w 3–26)	To investigate OTA's mode of action (MOA) in p53 heterozygous (p53 +/–) and p53 homozygous (p53 +/+) mouse	OTA induced renal damage, but no tumors were observed in either strain, indicating that p53 heterozygosity conferred little additional sensitivity to OTA; the lowest observed effect level for renal changes in p53 +/– and p53 +/+ mice was 200 mg OTA/kg b.w./day; based on the lack of tumors and the severity of renal and body weight changes at a maximum tolerated dose, the results were interpreted as suggestive of a primarily nongenotoxic (epigenetic) mode of action for OTA carcinogenesis	[166]

(*Continued*)

TABLE 31.2 (*Continued*)

Studies Using Mouse Models for the Examination of *Aspergillus* Mycotoxins Published Since 2010 Based on PubMed

Mouse Model	MT	Dose/Administration	Aim of the Study	Major Findings	References
10-w-old male p53-proficient and p53-deficient mice and p53-proficient and p53-deficient *gpt* delta mice (acquired by intercrossing between p53-deficient mice, *gpt* delta transgenic mice and C57BL/6 mice)	OTA	5 mg/5 mL/kg orally	To investigate the role of p53 in the progression from OTA-induced DNA damage to gene mutations	OTA induced DSBs at the carcinogenic target site in the kidneys; p53 prevented formation of DSBs and following gene mutation, due to p53/p21-mediated cell cycle control	[167]
7- to 9-w-old male heterozygous p53 knockouts and their wild-type controls (B6.129-$Trp53^{tm1Brd}$ N5)	AFB_1	Fed diets with 0, 0.2, or 1.0 ppm	To determine if chronic exposure of mice to AFB_1 alters BER in lung and liver and to determine if p53 heterozygosity affects BER activity	Chronic exposure to AFB_1 did not affect BER in lungs or livers of heterozygous p53-knockout mice. BER activity was lower in livers from p53 wild-type mice exposed to 1.0 ppm AFB_1 versus those exposed to 0.2 ppm AFB_1, an effect that was not attributable to liver cell death or altered levels of OGG1	[168]
Male and female FVB	AFB_1	20 μg/kg b.w. diet	To develope transgenic mice specifically expressing ADTZ in parotid glands by transgenic technology toward AFB_1 detoxification	Transgenic mice specifically expressing ADTZ gene in the parotid gland were successfully developed, and the ADTZ produced in the transgenic mice was shown to reduce the incidence of AFB_1 sepsis	[169]
7- to 10-w-old 129/C57Bl/6 female wild-type, heterozygous (+/−) or homozygous (−/−) ogg1 gene mice	AFB_1	50 mg/kg b.w. intraperitoneally	To determine the effect of ogg1 deficiency on AFB_1-induced oxidatively damaged DNA and tumorigenesis	ogg1 status did not have a significant effect on AFB_1-induced oxidatively damaged DNA or tumorigenesis, but deletion of one or both alleles of ogg1 did increase susceptibility to other aspects of AFB_1 toxicity	[170]
Adult male albino mice (19.75 + 2.0 g)	AFB_1	9 mg/kg b.w. orally	To investigate the effects of AFB_1 and ethanol coexposure on biomarkers of hepatic damage	AFB_1 and ethanol coexposure induced severe oxidative damage to the liver; thus, humans consuming excessive amount of ethanol and diets contaminated with AFB_1 simultaneously may be at greater risk of the hepatotoxic effects of these compounds	[171]

(*Continued*)

TABLE 31.2 (*Continued*)

Studies Using Mouse Models for the Examination of *Aspergillus* Mycotoxins Published Since 2010 Based on PubMed

Mouse Model	MT	Dose/Administration	Aim of the Study	Major Findings	References
ICR female mice (18–22 g)	AF, ZEA DON	Fed with low-dose mycotoxin-containing diet of DON (581 μg/kg), ZEA (285 μg/kg), and AF (121 μg/kg); and fed with high-dose mycotoxin-containing diet of DON (1,163 μg/kg), ZEA (569 μg/kg), and AF (242 μg/kg)	To identify any epigenetic effects of a mycotoxin-containing diet, including altered DNA methylation, H3K9 methylation, H3K27 methylation, and H4K20 methylation on reduced oocyte developmental competence	Diets containing mycotoxins increased the levels of DNA methylation, H3K9me3, and H4K20me3 and decreased the levels of H3K27me3 and H4K20me2; these altered epigenetic modifications may be one of the reasons for the reduced oocyte developmental competence	[172]
Theiler outbred (TO) mice	AFB_1	20 mg/kg b.w. on GD 13 both intraperitoneally and orally	To study the toxicokinetics of AF in pregnant mice	Large amounts of AFB_1 are easily and rapidly absorbed both from the gastrointestinal tract and through the peritoneum	[173]
Pdn/Pdn (Polydactyly Nagoya) mouse (Gli3Pdn)	OTA	2 mg/kg b.w. intraperitoneally on day 7.5 of gestation	To investigate gender-dependent differences in the incidence of NTD induced by OTA in the Pdn/Pdn mouse	A high incidence of NTD in male Pdn/Pdn was induced by OTA treatment, which might be due to complicated altered gene expressions among *Gli3*, *Wnt7b*, *Wnt8b*, *Fez1*, *Barx1*, *Lim1*, *Dmrt1*, *Igf1*, *Fog2*, *Dax1*, and *Sox9*, and in particular, upregulation and gender-dependent difference in *Barx1* and gender-dependent difference in *Sox9* gene expressions was observed	[174]

AF, aflatoxin; AFB_1, aflatoxin B_1; AFG_1, aflatoxin G_1; AFM_1, aflatoxin M_1; AGE, aged garlic extract; AT-II, alveolar type II cells; ADTZ, aflatoxin-detoxifizyme; BER, base excision repair; b.w., body weight; CafA, caffeic acid; CCE, cactus cladode extract; ChlA, chlorogenic acid; Cox-2, cyclooxygenase-2; DC, dendritic cell; DKO, double knockout; DNA, desoxyribonucleic acid; DON, deoxynivalenol; DSB, DNA double strand break; EGFR, epidermal growth factor receptor; FB_1, fumonisin B_1; GSH, glutathione; GSTA3, glutathione S-transferase subunit; GTP, green tea polyphenols; HR, homologous recombination; Hsp, heat shock protein; HCC, hepatocellular carcinoma; HOC, hepatic oval cells; IL, interleukin; MDA, malondialdehyde; mPBMCs, murine peripheral blood mononuclear cells; MT, mycotoxin; NER, nucleotide excision repair; NTD, neural tube defect; OGG1, 8-oxoguanine-glycosylase; 5-OHdC, 5-hydroxy-2′-deoxycytidine; OTA, ochratoxin A; PAT, patulin; pDCs, plasmacytoid dendritic cells; RCMF, *Rhus verniciflua* Stokes chloroform–methanol fraction; ROS, reactive oxygen species; STC, sterigmatocystin; TBARS, thiobarbituric acid-reactive substances; TNF-α, tumor necrosis factor-α; Treg, CD4+ CD25+ FoxP+ regulator cell; ZEA, zearalenone.

subcutaneously. Besides a series of articles dealing with organotoxicity [131,149,155], genotoxicity [149,153,154,159,160,163,167], or hepatocellular carcinoma development [157,158,162], a large number of studies focused on the protective, antioxidant, and immunostimulatory effects of plant-derived metabolites, extracts, and oils against *Aspergillus* mycotoxins [123,124,127,135,137,138,140–144,161], as well as on the protective role of bacteria and their mycotoxin-degrading abilities [128,130,132,148]. In the case of OTA, its cytotoxic effects on embryonic development [146] as well as its possible role in testicular cancer [120] and skin carcinogenesis [139] were also studied in mouse models. The latter has also been examined for PAT [139], and the adjuvant activity on allergic airway inflammation was also studied in the case of this mycotoxin [121]. Zhang et al. [126] examined the negative immunomodulatory effects of STC, while Miller et al. [151] focused on the histological and histochemical responses to STC in lungs. In the case of AFG_1, the link between induced chronic inflammation and lung tumorigenesis was investigated [133,134].

Studies are available also about the application of guinea pigs as laboratory models to study *Aspergillus* mycotoxins; however, the lack of records in the PubMed database since 2007 reflects that the rat and mouse models are more preferred. Earlier studies with guinea pigs examined different aspects of AFB_1 toxicity, including its metabolism [175], the effects of pretreatment with AF on a second AF treatment [176], the complement, bacteriostatic, and enzymatic activities in sera from guinea pigs treated with AF [177], the protective effect of ascorbic acid from acute AFB_1 toxicity [178], the induction of liver fibrosis by AFB_1 in combination with copper [179], the changes in the gastrointestinal tract, cardiovascular function, and drug metabolizing processes induced by AFB_1 intoxication [180], and the DNA damage caused by AFB_1 [181].

Hamster models were much less frequently used in the study of *Aspergillus* mycotoxins. In an early study, Ungar et al. [182] treated 8–10-week-old male Syrian hamsters with AFB_1 in *N-N*-dimethylformamide and studied the pathological effects of both the mycotoxin and the solvent. AFB_1 could be associated with periportal and midzonal necrosis of the liver. More recently, Rajmon et al. [183] studied the influence of repeated low intragastrically administered doses of AFB_1 and T2 toxin on Chinese hamster and found the main liver condition indicators to be influenced by the AFB_1-T2 toxin combination more than by the two mycotoxins alone.

31.4.3 Human and Animal Cell Lines

In order to gain an overview of the recently published articles reporting the application of human and animal cell lines to study *Aspergillus* mycotoxins, we collected full-text papers and abstracts published after 2010 from the PubMed database with the keyword combinations aflatoxin OR ochratoxin OR sterigmatocystin OR patulin AND cell line OR cell culture. Similar to the case of rat and mouse models, the retrieved records were screened for relevant information (exact definition of the mycotoxin and the cell line used).

Table 31.3 summarizes the studies published since 2010 that have used human or animal cell lines as laboratory models for the investigation of *Aspergillus* mycotoxins [184–213]. Among the retrieved records, relevant data were found in 31 publications. Eleven studies applied different liver cell lines (human hepatocellular carcinoma cell lines Huh7, HepG2, Hep3B, or SMMC-7721, HepaRG terminally differentiated hepatic cells derived from a human hepatic progenitor cell line that retains many characteristics of primary human hepatocytes, the HeLa derivative Chang liver cell line, as well as H4IIE rat hepatoma cells) for AFB_1 research. The major outputs of these studies included the development of a stable Huh7 cell line with constitutive expression of human CYP4501A2 [184], the demonstration that AFB_1 is able to activate the pregnane X receptor (PXR), a known regulator of liver xenobiotic metabolism in human hepatocytes [185], the demonstration that AFB_1 stimulates hepatoma cell migration [186], the recognition that AFB_1 exposure decreases the replication of the hepatitis B virus [190], and the detection of an interaction between AFB_1 and FB_1 on activation and expression of CYP1A and its transcription factor Ahr [194]. Another group of studies applied various kidney cell lines (RPTEC/TERT1 immortalized human renal proximal tubular epithelial cells, IHKE immortalized human kidney cells, Vero African green monkey kidney cells, or the porcine kidney cell lines LLC-PK1 and PK-15) to study *Aspergillus* mycotoxins, primarily OTA. Among others, these studies

TABLE 31.3

Studies Using Human and Animal Cell Lines for the Examination of *Aspergillus* Mycotoxins Published Since 2010 Based on PubMed

Applied Cell Line	MT	Applied Concentration	Aim of the Study	Major Findings	References
Huh7 human hepatocellular carcinoma cells	AFB_1	2.5 and 1–10 μM	To develope a stable cell line with constitutive expression of human CYP1A2 in Huh7 cells	A cell line, Huh7–1A2-I-E, with high expression level of CYP1A2 was developed and evaluated for identification of CYP1A2 inhibitors and for studies of cytotoxicity resulting from CYP-mediated drug metabolism; cellular toxicity of AFB_1 in Huh7–1A2-I-E cells could be prevented by furafylline, a CYP1A2 inhibitor	[184]
HepG2 human hepatocellular carcinoma cells	AFB_1 AFM_1 AFG_1	10 μM	To perform a screen of mycotoxins that pose a known environmental threat to human and animal health for the ability to activate PXR function in a human hepatocyte cell line	AFB_1 was able to activate PXR, a known regulator of liver xenobiotic metabolism, in human hepatocytes, and it could upregulate the expression of PXR-dependent genes responsible for AFB_1 biotransformation, including CYP3A4	[185]
HepG2 and SMMC-7721 hepatoma cells and immortalized human Chang liver cells	AFB_1	2.5 μM	To investigate the effects of AFB_1 on key elements in IGF-IR signaling pathway and the effects of AFB_1 on hepatoma cell migration	AFB_1 has stimulated hepatoma cell migration through IGF-IR/IRS2 axis	[186]
HepG2 hepatocellular carcinoma cells	AFB_1	2.5–40 μM	To examine the effects of AFB_1 on UGT mRNA expression in HepG2 cells	AFB_1 has induced UGT2B isoforms rather than UGT1A isoforms, which may closely contribute to the toxicity of AFB_1	[187]
HepG2 hepatocellular carcinoma cells	AFB_1	1, 2.6, 6.5, 16, 41, 102, 256, 640 μM	To develop genetically manipulated, upgraded HepG2 cells (Adv-HepG2) transiently expressing functional levels of three major enzymes (CYP1A2, CYP2C9, and CYP3A4) responsible for oxidative metabolism of drugs to confer drug-metabolic competence	Upgraded cells (Adv-HepG2) were highly able to metabolize the toxin studied in contrast to the reduced metabolic capacity of HepG2 cells; hence, Adv-HepG2 cells are suitable to test hepatotoxicity of bioactivable drugs	[188]
HepG2 hepatocellular carcinoma cells	AFB_1	Concentrations of 35, 75, 150, 300, 600, 1250, 2500, and 5000 nM	To evaluate and establish the role of *Eclipta alba as* MDR reversal agent using multidrug-resistant hepatocellular carcinoma cell line DR-HepG2	EAE significantly inhibited the cell proliferation, and findings revealed that *Eclipta alba* components are effective inhibitors of MDR1/P-gp	[189]

(*Continued*)

TABLE 31.3 (*Continued*)

Studies Using Human and Animal Cell Lines for the Examination of *Aspergillus* Mycotoxins Published Since 2010 Based on PubMed

Applied Cell Line	MT	Applied Concentration	Aim of the Study	Major Findings	References
HepaRG cells	AFB_1	1 and 5 mM	To analyze the effects of HBV infection and replication on DNA damage by AFB_1 and vice-versa in HepaRG, a cell line with hepatocyte-like morphology that metabolizes AFB_1 and supports HBV infection	AFB_1 exposure decreased the HBV replication, whereas DNA damage by AFB_1 and subsequent p53 induction was not affected by the presence of the virus; thus, in HepaRG cell line, AFB_1 and HBV do not cooperate to increase DNA damage by AFB_1	[190]
HepaRG cells	AFB_1	0.05, 0.1, 0.5, and 1.0 μM	To describe an improved methodology for the *in vitro* micronucleus test using an enriched proportion of HepaRG hepatocyte-like cells and based on a complete *in situ* assay, thereby avoiding cell detachment after chemical exposure and allowing cell division after a mitogenic stimulation	HepaRG hepatocytes exposed to various genotoxic compounds requiring or not requiring bioactivation compared favorably with those reported in various other cell types; they supported the view that metabolically competent HepaRG cells have unique potential benefits for testing genotoxic compounds using the *in vitro* micronucleus assay	[191]
Human hepatocytes and 3T3-L1 mouse fibroblast cells	AFB_1	0, 1, 2, and 3 μM	To study the application of IdMOC in the definition of the role of hepatic metabolism on the cytotoxicity of three model toxicants: CPA, AFB, and TMX via the coculturing of the metabolically competent human hepatocytes with the metabolically incompetent mouse 3T3 fibroblasts	The three model toxicants showed three distinct cytotoxic profiles: TMX was cytotoxic to 3T3 cells in the absence of hepatocytes, with slightly lower cytotoxicity towards both 3T3 cells and hepatocytes; AFB was selectively toxic toward the human hepatocytes and relatively noncytotoxic toward 3T3 cells; CPA cytotoxicity to the 3T3 cells was found to be significantly enhanced by the presence of the hepatocytes	[192]
Hep3B human hepatocellular carcinoma cells	STC OTA CTN	10^{-12}–10^{-6} M	To investigate the potential cytogenetic effects of STC, OTA, and CTN alone or in combination, at pM to μM concentrations, on the human hepatocellular cancer cell line Hep3B	STC alone or in combination with OTA and/or CTN had a cytotoxic and cytogenetic potential even at picomolar concentrations on human hepatoma cells *in vitro*	[193]

(*Continued*)

TABLE 31.3 (*Continued*)

Studies Using Human and Animal Cell Lines for the Examination of *Aspergillus* Mycotoxins Published Since 2010 Based on PubMed

Applied Cell Line	MT	Applied Concentration	Aim of the Study	Major Findings	References
H4IIE rat hepatoma cells and spleen cells from male Wistar inbred rats	AFB_1 FB_1	0.31, 0.63, 1.25, 2.50, 5, 10, 20, 30, 40, and 60 μM AFB_1 and 1.56, 3.12, 6.25, 12.50, 25, 50, 100, 150, 200, and 300 μM FB_1	To evaluate the effects of mycotoxins alone or combined on activation and expression of Cyp1A and its transcription factor Ahr in hepatoma cell line H4IIE and rat spleen mononuclear cells	AFB_1 and FB_1, either alone or in combination, increased cyp1A transcription and Cyp1A activity, as well as upregulated the Ahr activity, but the response potency was significantly higher for the mixture, indicating the existence of an interaction between both toxins	[194]
RPTEC/TERT1 human renal proximal tubular epithelial cells	OTA	300 nM	To characterize the metabolome of the human RPTEC/TERT1 cell line and define the response over a 72-hour period to low-level exposure of several compounds that affect kidney using NMR-based metabolic profiling approach	Metabolic profiling of RPTEC/TERT1 cells could report the effect of chemical exposure on multiple cellular pathways at low-level exposure and produced different response profiles for the different compounds, with a greater number of major metabolic effects observed in the toxin-treated cells	[195]
RPTEC/TERT1 human renal proximal tubular epithelial cells	OTA	300 nM	To investigate the effects of several renal chemical carcinogens on primary cilium in proximal tubular epithelial cells	After exposure to two renal carcinogens (OTA and $KBrO_3$), RPTEC/TERT1 cells lost primary cilium; gene expression analysis demonstrated that loss of the cilium was correlated with significant dysregulation of cilia-related genes and further underlined the divergent mechanisms of OTA- and $KBrO_3$-induced deciliation	[196]
LLC-PK1 porcine kidney cells	OTA	2.5–100 μM	To compare AAI and OTA impact on VEGF expression as well as transcription factors regulating its expression in cultured kidney tubulus cells	Both toxins exerted complex effects on various transcription factors, resulting in differential regulation of VEGF expression	[197]
PK-15 porcine kidney cells	AFB_1 ZEA DON FB_1	AFB_1: 5 mM, ZEA: 10 mM, DON: 500 μM, FB_1: 2.5 mM	To evaluate the nephrotoxicity of individual mycotoxins and combinations of AFB_1, ZEA, DON, and FB_1 to livestock using PK-15 cells as a disease model tested via biochemical approaches	Exposure to the mycotoxins AFB_1 and DON could induce ROS generation and cell apoptosis; DON was the most cytotoxic mycotoxin to PK-15 cells followed by AFB_1 and ZEA; ZEA ameliorated ROS production induced by low-dose AFB_1; ZEA or DON displayed synergetic effects with high-dose AFB_1 on ROS production; ZEA also ameliorated AFB_1-induced apoptosis	[198]

(*Continued*)

TABLE 31.3 (*Continued*)

Studies Using Human and Animal Cell Lines for the Examination of *Aspergillus* Mycotoxins Published Since 2010 Based on PubMed

Applied Cell Line	MT	Applied Concentration	Aim of the Study	Major Findings	References
IHKE immortalized human kidney cells	OTA OTα	0.01 nM–50 μM	To confirm the proposed structure of OTα amide and investigate if this compound can also be formed during thermal processing, as it is commonly used in food technology	OTα amide was less cytotoxic compared to OTA; thus, the thermal degradation of OTA to OTα amide might be a detoxification process; complete structure elucidation of OTα amide was performed via NMR/MS	[199]
Vero African green monkey kidney cells	OTA	10, 20, 30, 40, and 50 μM	To investigate the cytoprotective effect of quercetin on OTA-induced toxicity, with specific reference to oxidative stress, intracellular calcium flux, and levels of protective antioxidant enzymes	Quercetin pretreatment prevented the OTA-induced oxidative stress and apoptosis in Vero cell line	[200]
J774A.1 murine macrophage cells	AFB_1 AFB_2 AFG_1	0.01 and 0.1 ng/mL	To study the effect of AFB_1, AFB_2, and AFG_1 exposure, alone and in combination, on the secretion of key pro- and anti-inflammatory cytokines from J774A.1 cells	Exposure of macrophages to low doses of AF resulted in a statistically significant change in the secretion of a number of cytokines; AF exposure affected expression levels of key cell surface markers involved in the inflammatory response	[201]
Jurkat human lymphoblastic T-cells	OTA	10 μM	To gain insights into the molecular mechanisms underlying direct immunotoxicity	Diverse modes of action are involved in direct immunotoxicity, and a set of pathways or genes, rather than one single gene, could be used to screen compounds for direct immunotoxicity	[202]
Jurkat human lymphoblastic T-cells	AFB_1 AFM_1	3, 15, 30, and 50 μM	To investigate effects of AFB_1 and AFM_1 on immune function	Both AFB_1 and AFM_1 produced significant decreases in cell proliferation, whereas only minor effects were noted on IL-2 and IFN-γ cytokines mRNA expression; AFB_1, but not AFM_1, at the highest concentration induced a marked increase in IL-8 mRNA expression	[203]

(*Continued*)

TABLE 31.3 (*Continued*)

Studies Using Human and Animal Cell Lines for the Examination of *Aspergillus* Mycotoxins Published Since 2010 Based on PubMed

Applied Cell Line	MT	Applied Concentration	Aim of the Study	Major Findings	References
Caco-2 human colorectal adenocarcinoma cells H4IIE rat hepatoma cells, HepG2-Luc human hepatoma cells, and MMV-Luc human breast cancer cells	OTA	1 ng/mL, 1.6–100 µg/mL, 4, 40, 400, 4,000, and 40,000 ng/mL	To improve the assessment of the toxicological impact of dietary contaminants (OTA , DON, IMA, and BEN) at realistic human exposure levels, with a special focus on the intestinal compartment	Induction of CYP1A1 activity and inhibition of CYP3A4 activity occurred in Caco-2 cells at a realistic intestinal concentration of IMA, while OTA, DON, and BEN had no effect; some bacterial stress genes were induced in a range of realistic concentrations, following exposure to DON and IMA; BEN clearly provoked an estrogenic receptor agonistic activity in a human estrogen-sensitive reporter cell line	[204]
HCT116, SW620, and Caco-2 human colorectal cells	PAT	1.25, 2.5, 5, 10, and 20 µM	To study the effect of PAT on cancer cells and study its intracellular mechanism	A new anticancer mechanism of PAT, suggesting that EGR-1 phosphorylation by oxidative stress following PAT treatment might trigger EGR-1 binding to ATF3 promoter, which leads to the expression of apoptotic proteins and cell growth arrest of HCT-116 cells	[205]
Caco-2 human colorectal adenocarcinoma cells, MDBK bovine kidney cells, RAW 264.7 murine macrophages	AFB_1, OTA FB_1	AFB_1: 0.0038 ng/mL–3.8 µg/mL, OTA: 0.005 ng/mL–10 µg/mL, FB_1: 0.005 ng/mL–10 µg/mL)	To evaluate the effects of low levels of each mycotoxin in combination at their EU regulatory limits	Individual toxicity showed that OTA was the most cytotoxic mycotoxin in all three cell lines; binary combinations were cytotoxic to the MDBK cell line in the order [OTA/FB_1] > [AFB_1/FB_1] > [AFB_1/OTA]; tertiary combinations of AFB_1, FB_1, and OTA at the EU regulatory limits were not found to exhibit measurable cytotoxicity; in the concentrations above the legal limits, cytotoxicity was observed with up to 26% reduction in cell viability, and synergistic effects were evident with mitochondrial integrity	[206]
Sp2/0 mouse myeloma cells, 2C10 and 2E6 mouse hybridoma cells	AFB_1	25 µg, 100 ng	To describe the construction and expression of anti-AFB_1 scFvs 2E6 (scFv 2E6V_L-linker-2E6V_H and scFv 2E6V_H-linker-2E6V_L) in both V_H/V_L orientations	The yield of inclusion body of scFvs2E6 in either V_H/V_L orientation was similar; however, only the scFv in V_H-linker-V_L orientation showed anti-AFB_1 bioactivity after refolding	[207]

(Continued)

TABLE 31.3 (*Continued*)

Studies Using Human and Animal Cell Lines for the Examination of *Aspergillus* Mycotoxins Published Since 2010 Based on PubMed

Applied Cell Line	MT	Applied Concentration	Aim of the Study	Major Findings	References
SAOS-2 human osteosarcoma cells	AFB_1	5 and 50 ng/mL	To describe an unexpected toxicity of AFB_1 toward the vitamin D receptors	The expression of vitamin D receptor in osteosarcoma cell line SAOS-2 is significantly downmodulated by exposure to AFB_1	[208]
CYP19-overexpressing MCF-7aro human breast cancer cells, JEG-3 human placental cells, MCF-7 human breast adenocarcinoma cells, T98G human Caucasian glioblastoma cells	AFB_1, zeranol, ZEA	0.1% v/v	To evaluate the effect of MTs (AFB_1, zeranol, and ZEA) on aromatase using 4 cell lines	ZEA suppressed exon I.3 and II-specific expression of aromatase and competitively inhibited the enzyme activity; taken together, ZEA was a potential aromatase inhibitor among the three MTs	[209]
B-13 and B-13/H rat pancreatic progenitor cells	AFB_1	100, 200, and 400 nM	To establish the utility of B-13 cells in toxicity and genotoxicity screening, CYP induction, susceptibility to toxins, and transporter gene expression	B-13 cell has generated hepatocyte-like cells with functional drug metabolism and transporter activities that could alone—or in a humanized form—be used to screen for hepatotoxic and genotoxic end points *in vitro*	[210]
A549 human lung carcinoma cells	AFG_1	0.5, 2, 4, and 10 mg/L	To explore whether AFG_1 activates the ROS/MAPK/apoptosis pathway to cause cell damage in human AT-II cells like the cell line A549	AFG_1 induced the oxidative DNA damage and triggered apoptosis through ROS-mediated JNK and p38 signaling pathways in A549 cells, which resulted in AFG_1-induced AT-II cell damage	[211]

(*Continued*)

TABLE 31.3 (*Continued*)

Studies Using Human and Animal Cell Lines for the Examination of *Aspergillus* Mycotoxins Published Since 2010 Based on PubMed

Applied Cell Line	MT	Applied Concentration	Aim of the Study	Major Findings	References
INS-1 rat insulinoma cells	PAT	0, 1, 10, 100, 1,000, and 10,000 ng/mL	To analyze the effects of the microbial secondary metabolites pyrrolnitrin, phenazine, and PAT on INS-1 rat pancreatic β-cells	Of the three compounds tested, PAT was the most toxic; after both 24 and 72 h exposure, concentrations ≥10 ng/mL caused significant cell death with complete loss of viability at 10,000 ng/mL; PAT at 10 and 100 ng/mL caused a significant reduction of insulin secretion per live cell with 20 mM glucose; treatment of INS-1 cells with PAT had no effect on insulin mRNA levels; PAT showed no effect on the transcription of the insulin gene at any concentration tested	[212]
NCTC clone L929 murine subcutaneous connective tissue cells, and M12. C3.F6 murine B cell lymphoma cells	AFB_1	100, 500, and 1000 ng/plate	To detect the antimutagenic and antiproliferative compounds in octopus (*Paraoctopus limaculatus*)	The isolated fractions obtained from octopus contain compounds with chemoprotective properties that reduce the mutagenicity of AFB_1 and proliferation of cancer cell lines	[213]

AAI, aristolochic acid I; AF, aflatoxin; AFB_1, aflatoxin B_1; AFB_2, aflatoxin B_2; AFG_1, aflatoxin G_1; AFM_1, aflatoxin M_1; Ahr, aryl hydrocarbon receptor; AT-II alveolar type II; ATF-3, AMP-dependent transcription factor 3; BEN, benomyl; CFRD, cystic fibrosis-related diabetes; CTN, citrinin; CPA, cyclophosphamide; CYP, cytochrome P450; DNA, desoxyribonucleic acid; DON, deoxynivalenol; EAE, *Eclipta alba* hydroalcoholic extract; EGR-1, early growth response 1; EU, European Commission; FB_1, fumonisin B_1; HBV, hepatitis B virus; IdMOC, integrated discrete multiple organ co-culture system; IFN, interferon; IGF-IR, insulin-like growth factor receptor; IL, interleukin; IMA, imazalil; IRS2, insulin receptor substrate 2; JNK, Jun N-terminal kinase; MDR, multidrug resistance; MT, mycotoxin; NMR/MS, nuclear magnetic resonance/mass spectrometry; OTA, ochratoxin A; OTα, ochratoxin α; PAT, patulin; PXR, pregnane X receptor; RNA, ribonucleic acid; ROS, reactive oxygen species; TMX, tamoxifen; UGT, UDP-glucuronosyltransferase; VEGF, vascular endothelial growth factor; ZEA, zearalenone.

revealed that OTA causes significant deciliation in a model of the proximal tubule [196], that the hypoxia inducible factor 2α (but not 1α) plays a crucial role in prevention of diminishment of vascular endothelial growth factor production evoked by OTA in kidney proximal tubular epithelial cells [197], and that with its high antioxidant potential, quercetin protects Vero cells from OTA-induced oxidative stress and apoptosis [200]. The immunomodulatory or immunotoxic properties of *Aspergillus* mycotoxins were studied on J774A.1 murine macrophages [201] and human T-cell lines [202,203]. Studies on human colon carcinoma cell lines (Caco-2, HCT116, and SW620) revealed important data about the toxicity of AFB_1, OTA, and PAT [204–206]; for the latter, a new anticancer mechanism was proposed [205]. Further cell lines used since 2010 to study *Aspergillus* mycotoxins include Sp2/0 mouse myeloma cells and 2C10 and 2E6 mouse hybridoma cells [207]; SAOS-2 human osteosarcoma cells [208], MCF-7aro human breast cancer cells, JEG-3 human placental cells, MCF-7 human breast adenocarcinoma cells, and T98G human Caucasian glioblastoma cells [209]; B-13 and B-13/H rat pancreatic progenitor cells [210]; A549 human lung carcinoma cells [211]; INS-1 rat insulinoma cells [212]; as well as L929 murine subcutaneous connective tissue cells and M12.C3.F6 murine B cell lymphoma cells [213].

31.5 Conclusions

Since 2010, the most popular laboratory models to study various aspects of *Aspergillus* mycotoxins were animal models—primarily mice and rat lines—and human cell lines—mainly different liver and kidney cells. Studies performed with the application of these laboratory models provided large amount of important scientific information about the genotoxicity, organotoxicity, carcinogenicity, and metabolism of *Aspergillus* mycotoxins, as well as about the protective, antioxidant, and immunostimulatory effects of plant-derived metabolites, extracts, and oils against mycotoxicoses. The results from the studies discussed in this chapter clearly indicate that these laboratory models have the potential to further remain in the frontline of mycotoxin research.

Acknowledgments

This work was supported by the Indian National Science Academy and the Hungarian Academy of Sciences, as well as by projects NKFI-K115690 (National Research and Innovation Office, Hungary) and GINOP-2.3.3-15-2016-00006 (Széchenyi 2020 Programme). Csaba Vágvölgyi thanks the visiting professor program of King Saud University, Riyadh.

REFERENCES

1. Raper, K.B., Fennell, D.I., *The Genus Aspergillus*, Williams and Wilkins, Baltimore, MD, 1965.
2. Peterson, S.W., Phylogenetic analysis of *Aspergillus* species using DNA sequences from four loci, *Mycologia*, 100, 205, 2008.
3. Samson, R.A., Varga, J., What is a species in *Aspergillus*? *Med. Mycol.*, 47, S13, 2009.
4. Hope, W.W., Walsh, T.J., Denning, D.W., The invasive and saprophytic syndromes due to *Aspergillus* spp., *Med. Mycol.*, 43, S207, 2005.
5. Blout, W.P., Turkey "X" disease, *Turkeys*, 9, 52, 1961.
6. Van der Zijden, A.S.M., Blanche Koelensmid, W.A.A., Boldingh, J., Barrett, C.B., Ord, W.O., Philp, J., *Aspergillus flavus* and turkey X disease: Isolation in crystalline form of a toxin responsible for turkey X disease, *Nature*, 195, 1060, 1962.
7. Squire, R.A., Ranking animal carcinogens: A proposed regulatory approach, *Science*, 214, 877, 1981.
8. Varga, J., Frisvad, J.C., Samson, R.A., A reappraisal of fungi producing aflatoxins, *World Mycotoxin J.*, 2, 263, 2009.
9. Varga, J., Frisvad, J.C., Samson, R.A., Two new aflatoxin producing species, and an overview of *Aspergillus* section Flavi, *Stud. Mycol.*, 69, 57, 2011.

10. Goncalves, S.S. et al., *Aspergillus novoparasiticus*: A new clinical species of the section Flavi, *Med. Mycol.*, 50, 152, 2012.
11. Rank, C. et al., Distribution of sterigmatocystin in filamentous fungi, *Fungal Biol.*, 115, 406, 2011.
12. Olsen, M. et al., *Aspergillus nomius*, an important aflatoxin producer in Brazil nuts? *World Mycotoxin J.*, 1, 123, 2008.
13. Versilovskis, A., De Saeger, S., Sterigmatocystin: Occurrence in foodstuffs and analytical methods—An overview, *Mol. Nutr. Food Res.*, 54, 136, 2010.
14. Nielsen, K.F., Mycotoxin production by indoor molds, *Fungal Genet. Biol.*, 39, 103, 2003.
15. Chu, F.S., Mycotoxins: Food contamination, mechanisms, carcinogenic potential and preventive measures, *Mutat. Res.*, 259, 291, 1991.
16. Bennett, J.W., Klich, M., Mycotoxins, *Clin. Microbiol. Rev.*, 16, 497, 2003.
17. Nyikal, J. et al., Outbreak of aflatoxin poisoning—Eastern and Central Provinces, Kenya, January–July 2004, *Morbid. Mortal. Week. Rep.*, 53, 790, 2004.
18. Groopman, J.D., Johnson, D., Kensler, T.W., Aflatoxin and hepatitis B virus biomarkers: A paradigm for complex environmental exposures and cancer risk, *Cancer Biomarkers*, 1, 5, 2005.
19. Eaton, D.L., Groopman, J.D., *The Toxicology of Aflatoxins: Human Health, Veterinary, and Agricultural Significance*, Academic Press, San Diego, CA, 1994.
20. International Agency for Research on Cancer (IARC), *Some Traditional Herbal Medicine, Some Mycotoxins, Naphthalene and Styrene*, IARC Monographs on the Evaluation on the Carcinogenic Risks to Humans, Vol. 82, IARC Press, Lyon, 2002.
21. van Egmond, H.P., Schothorst, R.C., Jonker, M.A., Regulations relating to mycotoxins in food: Perspectives in a global and European context, *Anal. Bioanal. Chem.*, 389, 147, 2007.
22. van der Merwe, K.J., Steyn, P.S., Fourie, L., Scott, D.B., Theron, J.J., Ochratoxin A, a toxic metabolite produced by *Aspergillus ochraceus* Wilh, *Nature*, 205, 1112, 1965.
23. Moss, M.O., Mode of formation of ochratoxin A, *Food Addit. Contam.*, 13, S5, 1996.
24. Xiao, H., Marquardt, R.R., Frohlich, A.A., Ling, Y.Z., Synthesis and structural elucidation of analogs of ochratoxin A, *J. Agric. Food Chem.*, 43, 524, 1995.
25. Smith, J.E., Moss, M.O., *Mycotoxins: Formation, Analysis and Significance*, John Wiley & Sons, Chichester, 1985.
26. Krogh, P. et al. 1977. Balkan (endemic) nephropathy and foodborne ochratoxin A: Preliminary results of a survey of foodstuff, *Acta Pathol. Microbiol. Scand. Sect. B*, 85, 238, 1977.
27. Grollman, A.P. et al., Aristolochic acid and the etiology of endemic (Balkan) nephropathy, *Proc. Natl. Acad. Sci. USA*, 104, 12129, 2007.
28. O'Brien, E., Dietrich, D.R., Ochratoxin A: The continuing enigma, *Crit. Rev. Toxicol.*, 35, 33, 2005.
29. Davolos, D. et al., *Aspergillus affinis* sp. nov., a novel ochratoxin A-producing *Aspergillus* species (section Circumdati) isolated from decomposing leaves, *Int. J. Syst. Evol. Microbiol.*, 62, 1007, 2012.
30. Visagie, C.M. et al., Ochratoxin production and taxonomy of the yellow aspergilli (*Aspergillus* section Circumdati), *Stud. Mycol.*, 78, 1, 2014.
31. Varga, J. et al., Ochratoxin production by *Aspergillus* species, *Appl. Environ. Microbiol.*, 62, 4461, 1996.
32. Samson, R.A. et al., New ochratoxin or sclerotium producing species in *Aspergillus* section Nigri, *Stud. Mycol.*, 50, 45, 2004.
33. Perrone, G. et al., *Aspergillus niger* contains the cryptic phylogenetic species *A. awamori*, *Fungal Biol.*, 115, 1138, 2011.
34. Varga, J. et al., Recent advances in ochratoxin research I. Production, detection and occurrence of ochratoxins, *Cereal Res. Commun.*, 29, 85, 2001.
35. Noonim, P. et al., Fumonisin B_2 production by *Aspergillus niger* from Thai coffee beans, *Food Addit. Contam.*, 26, 94, 2009.
36. Bayman, P., Baker, J.L., Ochratoxins: A global perspective, *Mycopathologia*, 162, 215, 2006.
37. Varga, J., Kozakiewicz, Z., Ochratoxin A in grapes and grape-derived products, *Trends Food Sci. Technol.*, 17, 72, 2006.
38. Pel, H.J. et al., Genome sequencing and analysis of the versatile cell factory *Aspergillus niger* CBS 513.88, *Nat. Biotechnol.*, 25, 221, 2007.
39. Frisvad, J.C. et al., Fumonisin B_2 production by *Aspergillus niger*, *J. Agric. Food Chem.*, 55, 9727, 2007.
40. Varga, J. et al., Fumonisin contamination and fumonisin producing black aspergilli in dried vine fruits of different origin, *Int. J. Food Microbiol.*, 143, 143, 2010.

41. Varga, J. et al., Molecular diversity of agriculturally important *Aspergillus* species, *Eur. J. Plant Pathol.*, 110, 627, 2004.
42. Pitt, J. I., Hocking, A.D., *Fungi and Food Spoilage*, 2nd ed., Blackie Academic and Professional, London, UK, 1997.
43. Birkinshaw, J.H. et al., Patulin in the common cold. Collaborative research on a derivative of *Penicillium patulum* Bainier. II. Biochemistry and chemistry, *Lancet*, 245, 625, 1943.
44. Moake, M.M., Padilla-Zakour, O.I., Worobo, R.W., Comprehensive review of patulin control methods in foods, *Compr. Rev. Food Sci. Food Saf.*, 1, 8, 2005.
45. Chalmers, I., Clarke, M., The 1944 patulin trial: The first properly controlled multicentre trial conducted under the aegis of the British Medical Research Council, *Int. J. Epidemiol.*, 33, 253, 2004.
46. Walker, K., Wiesner, B.P., Patulin and clavicin, *Lancet*, 246, 294, 1944.
47. Steinman, R., Seigle-Murandi, F., Sage, L., Krivobok, S., Production of patulin by micromycetes, *Mycopathologia*, 105, 129, 1989.
48. Varga, J. et al., Taxonomic revision of *Aspergillus* section Clavati based on molecular, morphological and physiological data, *Stud. Mycol.*, 59, 89, 2007.
49. Joint FAO/WHO Expert Committee on Food additives (JECFA), *Position Paper on Patulin*, The Hague, The Netherlands, 9–13 March, 1998.
50. Scott, P.M., Patulin. In *Mycotoxins*, ed. I.F.H. Purchase, pp. 383–403. Elsevier, Amsterdam, The Netherlands, 1974.
51. Holzapfel, C.W., The isolation and structure of cyclopiazonic acid, a toxic metabolite of *Penicillium cyclopium* westling, *Tetrahedron*, 24, 2101, 1968.
52. Frisvad, J.C., Smedsgaard, J., Larsen, T.O., Samson, R.A., Mycotoxins, drugs and other extrolites produced by species in *Penicillium* subgenus *Penicillium*, *Stud. Mycol.*, 49, 201, 2004.
53. Larsen, T.O., Smedsgaard, J., Nielsen, K.F., Hansen, M.A., Samson, R.A., Frisvad, J.C., Production of mycotoxins by *Aspergillus lentulus* and other medically important and closely related species in section Fumigati, *Med. Mycol.*, 45, 225, 2007.
54. Betina, V., Mycotoxins—Chemical, biological and environmental aspects, *Bioact. Mol.*, 9, 27, 1989.
55. Samson, R.A. et al., New species in *Aspergillus* section Terrei, *Stud. Mycol.*, 69, 39, 2011.
56. Houbraken, J., Frisvad, J.C., Samson, R.A., Taxonomy of *Penicillium* section *Citrina*, *Stud. Mycol.*, 70 138, 2011.
57. Li, F.Q., Xu, G.R., Li, Y.W., Chen, Y., Ji, R., Natural occurrence of citrinin in *Monascus* products. *Wei Sheng Yan Jiu*, 34, 451, 2005.
58. Gardiner, D.M., Howlett, B.J., Bioinformatic and expression analysis of the putative gliotoxin biosynthetic gene cluster of *Aspergillus fumigatus*, *FEMS Microbiol. Lett.*, 248, 241, 2005.
59. Terabayashi, Y. et al., Identification and characterization of genes responsible for biosynthesis of kojic acid, an industrially important compound from *Aspergillus oryzae*, *Fungal Genet. Biol.*, 47, 953, 2010.
60. Klejnstrup, M.L. et al., Genetics of polyketide metabolism in *Aspergillus nidulans*, *Metabolites*, 2, 100, 2012.
61. Reiss, J., Insecticidal and larvicidal activities of the mycotoxins aflatoxin B_1 rubratoxin B, patulin and diacetoxyscirpenol towards *Drosophila melanogaster*, *Chem. Biol. Interact.*, 10, 339, 1975.
62. Kirk, H.D., Ewen, A.B., Emson, H.E., Blair, D.G., Effect of aflatoxin B_1 on development of *Drosophila melanogaster* (Diptera), *J. Invertebr. Pathol.*, 18, 313, 1971.
63. Lamb, M.J., Lilly, L.J., Induction of recessive lethals in *Drosophila melanogaster* by aflatoxin B_1, *Mutat. Res.*, 11, 430, 1971.
64. Chinnici, J.P., Melone, P.D., Genetic aspects of aflatoxin B_1 resistance in *Drosophila melanogaster*, *J. Hered.*, 76, 85, 1985.
65. Llewellyn, G.C., Chinnici, J.P., Variation in sensitivity to aflatoxin B_1 among several strains of *Drosophila melanogaster* (Diptera), *J. Invertebr. Pathol.*, 31, 37, 1978.
66. Chinnici, J.P., Booker, M.A., Llewellyn, G.C., Effect of aflatoxin B_1 on viability, growth, fertility, and crossing over in *Drosophila melanogaster* (Diptera), *J. Invertebr. Pathol.*, 27, 255, 1976.
67. Fahmy, M.J., Fahmy, O.G., Swenson, D.H., Aflatoxin B_1-2,3-dichloride as a model of the active metabolite of aflatoxin B_1 in mutagenesis and carcinogenesis, *Cancer Res.*, 38, 2608, 1978.
68. Chinnici, J.P., Erlanger, L., Charnock, M., Jones, M., Stein, J., Sensitivity differences displayed by *Drosophila melanogaster* larvae of different ages to the toxic effects of growth on media containing aflatoxin B_1, *Chem. Biol. Interact.*, 24, 373, 1979.

69. Chinnici, J.P., Genetic resistance to aflatoxin B_1 toxicity in *Drosophila melanogaster*: Chromosome substitution analysis, *J. Hered.*, 71, 275, 1980.
70. Fahmy, M.J., Fahmy, O.G., Intervening DNA insertions and the alteration of gene expression by carcinogens, *Cancer Res.*, 40, 3374, 1980.
71. Chinnici, J.P., Gunst, K., Llewellyn, G.C., Effects of mycotoxin pretreatment on aflatoxin B_1 post-treatment toxicity in *Drosophila melanogaster* (Diptera), *J. Invertebr. Pathol.*, 41, 321, 1983.
72. Foerster, R.E., Würgler, F.E., *In vitro* studies on the metabolism of aflatoxin B_1 and aldrin in testes of genetically different strains of *Drosophila melanogaster*, *Arch. Toxicol.*, 56, 12, 1984.
73. Belitsky, G.A., Khovanova, E.M., Budunova, I.V., Sharuptis, H.G., Mycotoxin induction of somatic mosaicism in *Drosophila* and DNA repair in mammalian liver cell cultures, *Cell Biol. Toxicol.*, 1, 133, 1985.
74. Melone, P.D., Chinnici, J.P., Selection for increased resistance to aflatoxin B_1 toxicity in *Drosophila melanogaster*, *J. Invertebr. Pathol.*, 48, 60, 1986.
75. Shibahara, T., Ogawa, H.I., Ryo, H., Fujikawa, K., DNA-damaging potency and genotoxicity of aflatoxin M_1 in somatic cells *in vivo* of *Drosophila melanogaster*, *Mutagenesis*, 10, 161, 1995.
76. Karekar, V., Joshi, S., Shinde, S.L., Antimutagenic profile of three antioxidants in the Ames assay and the *Drosophila* wing spot test, *Mutat. Res.*, 468, 183, 2000.
77. Sidorov, R.A., Ugnivenko, E.G., Khovanova, E.M., Belitsky, G.A., Induction of tumor clones in *D. melanogaster* wts/+ heterozygotes with chemical carcinogens, *Mutat. Res.*, 498, 181, 2001.
78. Sişman, T., The protective effect of hydrated sodium calcium aluminosilicate against the adverse effects of aflatoxin B_1 on *D. melanogaster*, *Toxicol. Ind. Health*, 22, 173, 2006.
79. Mutlu, A.G., Increase in mitochondrial DNA copy number in response to ochratoxin A and methanol-induced mitochondrial DNA damage in *Drosophila*, *Bull. Environ. Contam. Toxicol.*, 89, 1129, 2012.
80. Lee, S.E., Campbell, B.C., *In vitro* metabolism of aflatoxin B_1 by larvae of navel orangeworm, *Amyelois transitella* (Walker) (Insecta, Lepidoptera, Pyralidae) and codling moth, *Cydia pomonella* (L.) (Insecta, Lepidoptera, Tortricidae), *Arch. Insect. Biochem. Physiol.*, 45, 166, 2000.
81. Leung, M.C., Goldstone, J.V., Boyd, W.A., Freedman, J.H., Meyer, J.N., *Caenorhabditis elegans* generates biologically relevant levels of genotoxic metabolites from aflatoxin B_1 but not benzo[a]pyrene *in vivo*, *Toxicol. Sci.*, 118, 444, 2010.
82. González-Hunt, C.P. et al., Exposure to mitochondrial genotoxins and dopaminergic neurodegeneration in *Caenorhabditis elegans*, *PLoS One*, 9, e114459, 2014.
83. Shyamal, S. et al., Hepatoprotective effect of three herbal extracts on aflatoxin B_1-intoxicated rat liver, *Singapore Med. J.*, 51, 326, 2010.
84. Abdu, S., Ali, A., Ansari, S., Cytotoxic effect of ochratoxin A on the renal corpuscles of rat kidney: Could ochratoxin A cause kidney failure? *Histol. Histopathol.*, 26, 543, 2011.
85. Doktorova, T. et al., Comparison of genotoxicant-modified transcriptomic responses in conventional and epigenetically stabilized primary rat hepatocytes with *in vivo* rat liver data, *Arch. Toxicol.*, 86, 1703, 2012.
86. Jennings, P. et al., Transcriptomic alterations induced by ochratoxin A in rat and human renal proximal tubular *in vitro* models and comparison to a rat *in vivo* model, *Arch. Toxicol.*, 86, 571, 2012.
87. Tong, P.Z., Zhang, G.J., Zhang, X.H., Yan, X., Wang, J.L., Effects of sterigmatocystin on esophageal epithelium and experimental reflux esophagitis in rats, *Mol. Med. Rep.*, 8, 1043, 2013.
88. Akçam, M. et al., Caffeic acid phenethyl ester modulates aflatoxin B_1-induced hepatotoxicity in rats, *Cell Biochem. Funct.*, 31, 692, 2013.
89. Adeleye, A.O. et al., Phenolic extract of *Dialium guineense* pulp enhances reactive oxygen species detoxification in aflatoxin B_1 hepatocarcinogenesis, *J. Med. Food*, 17, 875, 2014.
90. Ou, C., Zheng, H.P., Su, J.J., Cao, J., Li, G.J., Li, L.Q., Effect of *Ginkgo biloba* extract on the expressions of Cox-2 and GST-Pi in rats with hepatocellular carcinoma risk, *Afr. Health Sci.*, 14, 37, 2014.
91. Supriya, C., Reddy, P.S., Prenatal exposure to aflatoxin B_1: Developmental, behavioral, and reproductive alterations in male rats, *Sci. Nat.*, 102, 1, 2015.
92. Suwannakul, N., Punvittayagul, C., Jarukamjorn, K., Wongpoomchai, R., Purple rice bran extract attenuates the aflatoxin B_1-induced initiation stage of hepatocarcinogenesis by alteration of xenobiotic metabolizing enzymes, *Asian Pac. J. Cancer Prev.*, 16, 3371, 2015.
93. Mantle, P., Kilic, M., Mor, F., Ozmen, O., Contribution of organ vasculature in rat renal analysis for ochratoxin A: Relevance to toxicology of nephrotoxins, *Toxins*, 7, 1005, 2015.
94. Mantle, P., Kulinskaya, E., Lifetime, low-dose ochratoxin A dietary study on renal carcinogenesis in male Fischer rats. *Food Addit. Contam. A. Chem. Anal. Control Expo. Risk Assess.*, 27, 1566, 2010.

95. Mantle, P.G., Nicholls, A.W., Shockcor, J.P., ^{1}H NMR spectroscopy-based metabolomic assessment of uremic toxicity, with toxicological outcomes, in male rats following an acute, mid-life insult from ochratoxin A, *Toxins*, 3, 504, 2011.
96. Corcuera, L.-A., Ibáñez-Vea, M., Vettorazzi, A., González-Peñas, E., Cerain, A.L.D., Validation of a UHPLC-FLD analytical method for the simultaneous quantification of aflatoxin B_1 and ochratoxin A in rat plasma, liver and kidney, *J. Chromatogr. B*, 879, 2733, 2011.
97. Koissi, N. et al., Lactone metabolite common to the carcinogens dioxane, diethylene glycol, and *N*-nitrosomorpholine: Aqueous chemistry and failure to mediate liver carcinogenesis in the F344 rat, *Chem. Res. Toxicol.*, 25, 1022, 2012.
98. Merrick, B.A. et al., Testing an aflatoxin B_1 gene signature in rat archival tissues, *Chem. Res. Toxicol.*, 25, 1132, 2012.
99. Lu, X. et al., Integrated analysis of transcriptomics and metabonomics profiles in aflatoxin B_1-induced hepatotoxicity in rat, *Food Chem. Toxicol.*, 55, 444, 2013.
100. Qi, X. et al., Ochratoxin A induces rat renal carcinogenicity with limited induction of oxidative stress responses, *Toxicol. Appl. Pharmacol.*, 280, 543, 2014.
101. Yang, W. et al., Genome-wide miRNA-profiling of aflatoxin B_1-induced hepatic injury using deep sequencing, *Toxicol. Lett.*, 226, 140, 2014.
102. Knipstein, B. et al., Dietary aflatoxin-induced stunting in a novel rat model: Evidence for toxin-induced liver injury and hepatic growth hormone resistance, *Pediatr. Res.*, 78(2), 120–127, 2015.
103. Hoffmann, D. et al., Evaluation of a urinary kidney biomarker panel in rat models of acute and subchronic nephrotoxicity, *Toxicology*, 277, 49, 2010.
104. Merrick, B.A. et al., RNA-seq profiling reveals novel hepatic gene expression pattern in aflatoxin B_1 treated rats, *PLoS One*, 8, e61768, 2013.
105. Johnson, N.M. et al., Complete protection against aflatoxin B_1-induced liver cancer with a triterpenoid: DNA adduct dosimetry, molecular signature, and genotoxicity threshold, *Cancer Prev. Res.*, 7, 658, 2014.
106. Yenilmez, A. et al., Antioxidant effects of melatonin and coenzyme Q_{10} on oxidative damage caused by single-dose ochratoxin A in rat kidney, *Chin. J. Physiol.*, 53, 310, 2010.
107. Kim, Y.-S. et al., Protective effect of Korean red ginseng against aflatoxin B(1)-induced hepatotoxicity in rat, *J. Ginseng Res.*, 35, 243, 2011.
108. Abdel-Aziem, S.H., Hassan, A.M., Abdel-Wahhab, M.A., Dietary supplementation with whey protein and ginseng extract counteracts oxidative stress and DNA damage in rats fed an aflatoxin-contaminated diet, *Mutat. Res. Genet. Toxicol. Environ. Mutagen.*, 723, 65, 2011.
109. Fiala, J.L.A. et al., Sulforaphane-mediated reduction of aflatoxin B_1-N7-guanine in rat liver DNA: Impacts of strain and sex, *Toxicol. Sci.*, 121, 57, 2011.
110. Shen, H. et al., Impairment of alveolar type-II cells involved in the toxicity of aflatoxin G_1 in rat lung, *Food Chem. Toxicol.*, 50, 3222, 2012.
111. Han, Z. et al. Combinatorial approach of LC–MS/MS and LC–TOF-MS for uncovering *in vivo* kinetics and biotransformation of ochratoxin A in rat, *J. Chromatogr. B*, 925, 46, 2013.
112. Palabiyik, S.S. et al., Protective effect of lycopene against ochratoxin A induced renal oxidative stress and apoptosis in rats, *Exp. Toxicol. Pathol.*, 65, 853, 2013.
113. Aydin, S. et al., The carotenoid lycopene protects rats against DNA damage induced by ochratoxin A, *Toxicon*, 73, 96, 2013.
114. Wei, W. et al., Evaluation of toxicological impacts on Sprague-Dawley rat by successively low dose of aflatoxin B_1, *J. Sci. Food Agric.*, 94, 3017, 2014.
115. Gong, B. et al., Transcriptomic profiling of rat liver samples in a comprehensive study design by RNA-Seq, *Sci. Data*, 1, 140021, 2014.
116. Hassan, A.M., Abdel-Aziem, S.H., El-Nekeety, A.A., Abdel-Wahhab, M.A., *Panax ginseng* extract modulates oxidative stress, DNA fragmentation and up-regulate gene expression in rats sub chronically treated with aflatoxin B_1 and fumonisin B_1, *Cytotechnology*, 67, 861–871, 2015.
117. Mantle, P.G., Dobrota, M., Gillett, C.E., Odell, E.W., Pinder, S.E., Oncological outcomes in rats given nephrocarcinogenic exposure to dietary ochratoxin A, followed by the tumour promoter sodium barbital for life: A pilot study, *Toxins*, 2, 552, 2010.
118. Singh, K., Maurya, B., Trigun, S., Activation of oxidative stress and inflammatory factors could account for histopathological progression of aflatoxin-B_1 induced hepatocarcinogenesis in rat, *Mol. Cell. Biochem.*, 401, 185, 2015.

119. Tozlovanu, M. et al., Glutathione conjugates of ochratoxin A as biomarkers of exposure, *Arh. Hig. Rada Toksikol.*, 63, 417, 2012.
120. Jennings-Gee, J.E., Tozlovanu, M., Manderville, R., Miller, M.S., Pfohl-Leszkowicz, A., Schwartz, G.G., Ochratoxin A: *In utero* exposure in mice induces adducts in testicular DNA, *Toxins*, 2, 1428, 2010.
121. Schütze, N., Lehmann, I., Bönisch, U., Simon, J.C., Polte, T., Exposure to mycotoxins increases the allergic immune response in a murine asthma model, *Am. J. Respir. Crit. Care Med.*, 181, 1188, 2010.
122. Li, C.-H. et al., Hepatic oval cell lines generate hepatocellular carcinoma following transfection with HBx gene and treatment with aflatoxin B_1 *in vivo*, *Cancer Lett.*, 311, 1, 2011.
123. Brahmi, D., Bouaziz, C., Ayed, Y., Ben Mansour, H., Zourgui, L., Bacha, H., Chemopreventive effect of cactus *Opuntia ficus indica* on oxidative stress and genotoxicity of aflatoxin B_1, *Nutr. Metab. (Lond.)*, 8, 73, 2011.
124. Eraslan, G., Kanbur, M., Aslan, Ö., Karabacak, M., The antioxidant effects of pumpkin seed oil on subacute aflatoxin poisoning in mice, *Environ. Toxicol.*, 28, 681, 2013.
125. Liu, Y. et al., Sterigmatocystin alters the number of $FoxP3^+$ regulatory T cells and plasmacytoid dendritic cells in BALB/c mice, *Food Chem. Toxicol.*, 50, 1920, 2012.
126. Zhang, Y., Yao, Z.G., Wang, J., Xing, L.X., Xia, Y., Zhang, X.H., Effects of sterigmatocystin on TNF-α, IL-6 and IL-12 expression in murine peripheral blood mononuclear cells and peritoneal macrophages *in vivo*, *Mol. Med. Rep.*, 5, 1318, 2012.
127. Larypoor, M., Bayat, M., Zuhair, M.H., Akhavan Sepahy, A., Amanlou, M., Evaluation of the number of CD4(+) CD25(+) FoxP3(+) treg cells in normal mice exposed to AFB_1 and treated with aged garlic extract, *Cell J.*, 15, 37, 2013.
128. Abbès, S., Salah-Abbès, J.B., Sharafi, H., Jebali, R., Noghabi, K.A., Oueslati, R., Ability of *Lactobacillus rhamnosus* GAF01 to remove AFM_1 *in vitro* and to counteract AFM_1 immunotoxicity *in vivo*. *J. Immunotoxicol.*, 10, 279, 2013.
129. Zhang, Z. et al., Cytochrome P450 2A13 mediates the neoplastic transformation of human bronchial epithelial cells at a low concentration of aflatoxin B_1, *Int. J. Cancer*, 134, 1539, 2014.
130. Ben Salah-Abbès, J., Abbès, S., Jebali, R., Haous, Z., Oueslati, R., Potential preventive role of lactic acid bacteria against aflatoxin M_1 immunotoxicity and genotoxicity in mice, *J. Immunotoxicol.*, 12, 107, 2015.
131. Sun, L.H., Lei, M.Y., Zhang, N.Y., Zhao, L., Krumm, C.S., Qi, D.S., Hepatotoxic effects of mycotoxin combinations in mice, *Food Chem. Toxicol.*, 74, 289, 2014.
132. Jebali, R., Abbès, S., Salah-Abbès, J.B., Younes, R.B., Haous, Z., Oueslati, R., Ability of *Lactobacillus plantarum* MON03 to mitigate aflatoxins (B_1 and M_1) immunotoxicities in mice, *J. Immunotoxicol.*, 12, 290, 2015.
133. Shen, H. et al., Enhanced phenotypic alterations of alveolar type II cells in response to aflatoxin G_1-induced lung inflammation, *J. Cell Physiol.*, 230, 1199, 2015.
134. Liu, C. et al., Oral administration of aflatoxin G_1 induces chronic alveolar inflammation associated with lung tumorigenesis, *Toxicol. Lett.*, 232, 547, 2015.
135. Cariddi, L.N. et al., Polyphenols as possible bioprotectors against cytotoxicity and DNA damage induced by ochratoxin A, *Environ. Toxicol. Pharmacol.*, 39, 1008, 2015.
136. Saxena, N., Ansari, K.M., Kumar, R., Chaudhari, B.P., Dwivedi, P.D., Das, M., Role of mitogen activated protein kinases in skin tumorigenicity of patulin, *Toxicol. Appl. Pharmacol.*, 257, 264, 2011.
137. Jha, A., Shah, K., Verma, R.J., Aflatoxin-induced biochemical changes in liver of mice and its mitigation by black tea extract, *Acta Pol. Pharm.*, 69, 851, 2012.
138. Jha, A., Krithika, R., Manjeet, D., Verma, R.J., Protective effect of black tea infusion on aflatoxin-induced hepatotoxicity in mice, *J. Clin. Exp. Hepatol.*, 3, 29, 2013.
139. Kumar, R. et al., Ochratoxin A-induced cell proliferation and tumor promotion in mouse skin by activating the expression of cyclin-D1 and cyclooxygenase-2 through nuclear factor-kappa B and activator protein-1, *Carcinogenesis*, 34, 647, 2013.
140. Chakraborty, D., Verma, R., Ameliorative effect of *Emblica officinalis* aqueous extract on ochratoxin-induced lipid peroxidation in the kidney and liver of mice, *Int. J. Occup. Med. Environ. Health*, 23, 63, 2010.
141. Naaz, F., Abdin, M.Z., Javed, S., Protective effect of esculin against prooxidant aflatoxin B_1-induced nephrotoxicity in mice, *Mycotoxin Res.*, 30, 25, 2014.
142. Gupta, R., Sharma, V., Ameliorative effects of *Tinospora cordifolia* root extract on histopathological and biochemical changes induced by aflatoxin-B_1 in mice kidney, *Toxicol. Int.*, 18, 94, 2011.

143. Choi, K.C. et al., Inhibitory effects of quercetin on aflatoxin B_1-induced hepatic damage in mice, *Food Chem. Toxicol.*, 48, 2747, 2010.
144. Choi, K.C. et al., Chemoprevention of a flavonoid fraction from *Rhus verniciflua* Stokes on aflatoxin B_1-induced hepatic damage in mice, *J. Appl. Toxicol.*, 31, 150, 2011.
145. Hou, Y.J. et al., Mycotoxin-containing diet causes oxidative stress in the mouse, *PLoS One*, 8, e60374, 2013.
146. Hsuuw, Y.D., Chan, W.H., Yu, J.S., Ochratoxin A inhibits mouse embryonic development by activating a mitochondrion-dependent apoptotic signaling pathway, *Int. J. Mol. Sci.*, 14, 935, 2013.
147. Hou, Y.J. et al., Oocyte quality in mice is affected by a mycotoxin-contaminated diet, *Environ. Mol. Mutagen.*, 55, 354, 2014.
148. Ferenczi, S. et al., A new ochratoxin A biodegradation strategy using *Cupriavidus basilensis* r16 strain, *PLoS One*, 9, e109817, 2014.
149. Song, E. et al., Hepatotoxicity and genotoxicity of patulin in mice, and its modulation by green tea polyphenols administration, *Food Chem. Toxicol.*, 71, 122, 2014.
150. Song, E. et al., Selenium supplementation shows protective effects against patulin-induced brain damage in mice via increases in GSH-related enzyme activity and expression, *Life Sci.*, 109, 37, 2014.
151. Miller, J.D., Sun, M., Gilyan, A., Roy, J., Rand, T.G., Inflammation-associated gene transcription and expression in mouse lungs induced by low molecular weight compounds from fungi from the built environment, *Chem. Biol. Interact.*, 183, 113, 2010.
152. Kim, J.E., Bunderson, B.R., Croasdell, A., Coulombe, R.A. Jr., Functional characterization of alpha-class glutathione S-transferases from the turkey (*Meleagris gallopavo*), *Toxicol. Sci.*, 124, 45, 2011.
153. de Melo et al., DNA damage in organs of mice treated acutely with patulin, a known mycotoxin, *Food Chem. Toxicol.*, 50, 3548, 2012.
154. Guindon-Kezis, K.A., Mulder, J.E., Massey, T.E., *In vivo* treatment with aflatoxin B_1 increases DNA oxidation, base excision repair activity and 8-oxoguanine DNA glycosylase 1 levels in mouse lung, *Toxicology*, 321, 21, 2014.
155. Al-Hazmi, M.A., Patulin in apple juice and its risk assessments on albino mice, *Toxicol. Ind. Health*, 30, 534, 2014.
156. Napoletano, M. et al., Ochratoxin A induces craniofacial malformation in mice acting on Dlx5 gene expression, *Front. Biosci.*, 2, 133, 2010.
157. Fox, J.G. et al., Gut microbes define liver cancer risk in mice exposed to chemical and viral transgenic hepatocarcinogens, *Gut*, 59, 88, 2010.
158. Liu, S.P. et al., Higher susceptibility to aflatoxin B_1-related hepatocellular carcinoma in glycine *N*-methyltransferase knockout mice, *Int. J. Cancer*, 128, 511, 2011.
159. Chawanthayatham, S. et al., Prenatal exposure of mice to the human liver carcinogen aflatoxin B_1 reveals a critical window of susceptibility to genetic change, *Int. J. Cancer*, 136, 1254, 2015.
160. Kensler, K.H. et al., Genetic or pharmacologic activation of Nrf2 signaling fails to protect against aflatoxin genotoxicity in hypersensitive GSTA3 knockout mice, *Toxicol. Sci.*, 139, 293, 2014.
161. Hoek-van den Hil, E.F., van Schothorst, E.M., van der Stelt, I., Hollman, P.C., Keijer, J., Rietjens, I.M., Quercetin tests negative for genotoxicity in transcriptome analyses of liver and small intestine of mice, *Food Chem. Toxicol.*, 81, 34, 2015.
162. Teoh, W.W. et al., Molecular characterization of hepatocarcinogenesis using mouse models, *Dis. Model Mech.*, 8, 743, 2015.
163. Wattanawaraporn, R. et al., A single neonatal exposure to aflatoxin B_1 induces prolonged genetic damage in two loci of mouse liver, *Toxicol. Sci.*, 128, 326, 2012.
164. Austin, K.J., Cockrum, R.R., Jons, A.M., Alexander, B.M., Cammack, K.M., Renin mRNA is upregulated in testes and testicular cells in response to treatment with aflatoxin B_1, *Theriogenology*, 77, 331, 2012.
165. Mulder, J.E., Bondy, G.S., Mehta, R., Massey, T.E., Up-regulation of nucleotide excision repair in mouse lung and liver following chronic exposure to aflatoxin B_1 and its dependence on p53 genotype, *Toxicol. Appl. Pharmacol.*, 275, 96, 2014.
166. Bondy, G.S. et al., Effects of chronic ochratoxin A exposure on p53 heterozygous and p53 homozygous mice, *Toxicol. Pathol.*, 43(5), 715–729, 2015.
167. Kuroda, K. et al., Role of p53 in the progression from ochratoxin A-induced DNA damage to gene mutations in the kidneys of mice, *Toxicol. Sci.*, 144, 65, 2015.

168. Mulder, J.E., Bondy, G.S., Mehta, R., Massey, T.E., The impact of chronic aflatoxin B_1 exposure and p53 genotype on base excision repair in mouse lung and liver, *Mutat. Res.*, 773, 63, 2015.
169. Guan, L.Z. et al., The aflatoxin-detoxifizyme specific expression in mouse parotid gland, *Transgenic Res.*, 24, 489, 2015.
170. Mulder JE, Turner PV, Massey TE. Effect of 8-oxoguanine glycosylase deficiency on aflatoxin B_1 tumourigenicity in mice, *Mutagenesis*, 30, 401, 2015.
171. Adedara, I.A., Owumi, S.E., Uwaifo, A.O., Farombi, E.O., Aflatoxin B_1 and ethanol co-exposure induces hepatic oxidative damage in mice, *Toxicol. Ind. Health*, 26, 717, 2010.
172. Zhu, C.C. et al., Effect of mycotoxin-containing diets on epigenetic modifications of mouse oocytes by fluorescence microscopy analysis, *Microsc. Microanal.*, 20, 1158, 2014.
173. Bastaki, S.A., Osman, N., Kochiyil, J., Shafiullah, M., Padmanabhan, R., Abdulrazzaq, Y.M., Toxicokinetics of aflatoxin in pregnant mice, *Int. J. Toxicol.*, 29, 425, 2010.
174. Ueta, E. et al., Gender-dependent differences in the incidence of ochratoxin A-induced neural tube defects in the *Pdn/Pdn* mouse, *Congenit. Anom. (Kyoto)*, 50, 29, 2010.
175. Koes, M.T., Forrester, L.J., Brown, H.D., Metabolism of aflatoxin B_1 by the guinea-pig, *Food Cosmet. Toxicol.*, 11, 463, 1973.
176. Peden, W.M., Richard, J.L., Thurston, J.R., Sacks, J.L., Effects of pre-treatment with aflatoxin on a second aflatoxin treatment in guinea pigs, *Mycopathologia*, 99, 107, 1987.
177. Thurston, J.R., Sacks, J.M., Richard, J.L., Peden, W.M., Driftmier, K., Complement, bacteriostatic, and enzymatic activities in sera from guinea pigs given aflatoxin and/or rubratoxin, *Am. J. Vet. Res.*, 50, 356, 1989.
178. Netke, S.P., Roomi, M.W., Tsao, C., Niedzwiecki, A., Ascorbic acid protects guinea pigs from acute aflatoxin toxicity, *Toxicol. Appl. Pharmacol.*, 143, 429, 1997.
179. Schiller, F., Lippold, U., Heinze, R., Hoffmann, A., Seffner, W., Liver fibrosis in guinea pigs experimentally induced by combined copper and aflatoxin application, *Exp. Toxicol. Pathol.*, 50, 519, 1998.
180. Stětinová, V., Grossmann, V., Květina, J., Changes in the gastrointestinal tract, cardiovascular function and some drug metabolizing processes in rats and guinea-pigs intoxicated with aflatoxin B_1, *Pol. J. Pharmacol.*, 50, 135, 1998.
181. Miranda, D.D. et al., Analysis of DNA damage induced by aflatoxin B_1 in Dunkin–Hartley guinea pigs, *Mycopathologia*, 163, 275, 2007.
182. Ungar, H., Sullman, S.F., Zuckerman, A.J., Acute and protracted changes in the liver of Syrian hamsters induced by a single dose of aflatoxin B_1. Observations on pathological effects of the solvent (dimethylformamide), *Br. J. Exp. Pathol.*, 57, 157, 1976.
183. Rajmon, R. et al., Combined effects of repeated low doses of aflatoxin B_1 and T-2 toxin on the Chinese hamster, *Vet. Med. Czech*, 46, 301, 2001.
184. Chu, C.C., Pan, K.L., Yao, H.T., Hsu, J.T., Development of a whole-cell screening system for evaluation of the human CYP1A2-mediated metabolism, *Biotechnol. Bioeng.*, 108, 2932, 2011.
185. Ratajewski, M., Walczak-Drzewiecka, A., Sałkowska, A., Dastych, J., Aflatoxin upregulate CYP3A4 mRNA expression in a process that involves the PXR transcription factor, *Toxicol. Lett.*, 205, 146, 2011.
186. Ma, Y., Kong, Q., Hua, H., Luo, T., Jiang, Y., Aflatoxin B_1 up-regulates insulin receptor substrate 2 and stimulates hepatoma cell migration, *PLoS One*, 7(10), e47961. doi: 10.1371/journal.pone.0047961, 2012.
187. Hanioka, N. et al., Effect of aflatoxin B_1 on UDP-glucuronosyltransferase mRNA expression in HepG2 cells, *Chemosphere*, 89, 526, 2012.
188. Tolosa, L., Donato, M.T., Pérez-Cataldo, G., Castell, J.V., Gómez-Lechón, M.J., Upgrading cytochrome P450 activity in HepG2 cells co-transfected with adenoviral vectors for drug hepatotoxicity assessment, *Toxicol. In Vitro*, 26, 1272, 2012.
189. Chaudhary, H., Jena, P.K., Seshadri, S., Evaluation of hydro-alcoholic extract of *Eclipta alba* for its multidrug resistance reversal potential: An *in vitro* study, *Nutr. Cancer*, 65, 775, 2013.
190. Lereau, M. et al., Interactions between hepatitis B virus and aflatoxin B_1: Effects on p53 induction in HepaRG cells, *J. Gen. Virol.*, 93, 640, 2012.
191. Jossé, R., Rogue, A., Lorge, E., Guillouzo, A., An adaptation of the human HepaRG cells to the *in vitro* micronucleus assay, *Mutagenesis*, 27, 295, 2012.
192. Li, A.P., Uzgare, A., LaForge, Y.S., Definition of metabolism-dependent xenobiotic toxicity with co-cultures of human hepatocytes and mouse 3T3 fibroblasts in the novel integrated discrete multiple organ co-culture (IdMOC) experimental system: Results with model toxicants aflatoxin B_1, cyclophosphamide and tamoxifen, *Chem. Biol. Interact.*, 199, 1, 2012.

193. Anninou, N., Chatzaki, E., Papachristou, F., Pitiakoudis, M., Simopoulos, C., Mycotoxins' activity at toxic and sub-toxic concentrations: Differential cytotoxic and genotoxic effects of single and combined administration of sterigmatocystin, ochratoxin A and citrinin on the hepatocellular cancer cell line Hep3B, *Int. J. Environ. Res. Public Health*, 11, 1855, 2014.
194. Mary, V.S., Valdehita, A., Navas, J.M., Rubinstein, H.R., Fernández-Cruz, M.L., Effects of aflatoxin B_1, fumonisin B_1 and their mixture on the aryl hydrocarbon receptor and cytochrome P4501A induction, *Food Chem. Toxicol.*, 75, 104, 2015.
195. Ellis, J.K. et al., Metabolic response to low-level toxicant exposure in a novel renal tubule epithelial cell system, *Mol. Biosyst.*, 7, 247, 2011.
196. Radford, R. et al., Carcinogens induce loss of the primary cilium in human renal proximal tubular epithelial cells independently of effects on the cell cycle, *Am. J. Physiol. Renal Physiol.*, 302, F905, 2012.
197. Stachurska, A., Kozakowska, M., Jozkowicz, A., Dulak, J., Loboda, A., Aristolochic acid I and ochratoxin A differentially regulate VEGF expression in porcine kidney epithelial cells—The involvement of SP-1 and HIFs transcription factors, *Toxicol. Lett.*, 204, 118, 2011.
198. Lei, M., Zhang, N., Qi, D., *In vitro* investigation of individual and combined cytotoxic effects of aflatoxin B_1 and other selected mycotoxins on the cell line porcine kidney 15, *Exp. Toxicol. Pathol.*, 65, 1149, 2013.
199. Bittner, A., Cramer, B., Harrer, H., Humpf, H.U., Structure elucidation and *in vitro* cytotoxicity of ochratoxin α amide, a new degradation product of ochratoxin A, *Mycotoxin Res.*, 31, 83, 2015.
200. Ramyaa, P., Padma, V.V., Ochratoxin-induced toxicity, oxidative stress and apoptosis ameliorated by quercetin–modulation by Nrf2, *Food Chem. Toxicol.*, 62, 205, 2013.
201. Bruneau, J.C., Stack, E., O'Kennedy, R., Loscher, C.E., Aflatoxins B_1, B_2 and G_1 modulate cytokine secretion and cell surface marker expression in J774A.1 murine macrophages, *Toxicol. In Vitro*, 26, 686, 2012.
202. Shao, J. et al., Toxicogenomics-based identification of mechanisms for direct immunotoxicity, *Toxicol. Sci.*, 135, 328, 2013.
203. Luongo, D., Russo, R., Balestrieri, A., Marzocco, S., Bergamo, P., Severino, L., *In vitro* study of AFB_1 and AFM_1 effects on human lymphoblastoid Jurkat T-cell model, *J. Immunotoxicol.*, 11, 353, 2014.
204. Ribonnet, L. et al., Potential of an *in vitro* toolbox combined with exposure data as a first step for the risk assessment of dietary chemical contaminants, *Food Addit. Contam. Part A. Chem. Anal. Control Expo. Risk Assess.*, 28, 1136, 2011.
205. Kwon, O. et al., Patulin induces colorectal cancer cells apoptosis through EGR-1 dependent ATF3 up-regulation, *Cell Signal.*, 24, 943, 2012.
206. Clarke, R., Connolly, L., Frizzell, C., Elliott, C.T., Cytotoxic assessment of the regulated, co-existing mycotoxins aflatoxin B_1, fumonisin B_1 and ochratoxin, in single, binary and tertiary mixtures, *Toxicon*, 90, 70, 2014.
207. Liu, A., Ye, Y., Chen, W., Wang, X., Chen, F., Expression of V(H)-linker-V(L) orientation-dependent single-chain Fv antibody fragment derived from hybridoma 2E6 against aflatoxin B_1 in *Escherichia coli*, *J. Ind. Microbiol. Biotechnol.*, 42, 255, 2015.
208. Costanzo, P., Santini, A., Fattore, L., Novellino, E., Ritieni, A., Toxicity of aflatoxin B_1 towards the vitamin D receptor (VDR), *Food Chem. Toxicol.*, 76, 77, 2015.
209. Wang, Y., Wong, T.Y., Chan, F.L., Chen, S., Leung, L.K., Assessing the effect of food mycotoxins on aromatase by using a cell-based system, *Toxicol. In Vitro*, 28, 640, 2014.
210. Probert, P.M et al., Utility of B-13 progenitor-derived hepatocytes in hepatotoxicity and genotoxicity studies, *Toxicol. Sci.*, 137, 350, 2014.
211. Shen, H. et al., Aflatoxin G_1-induced oxidative stress causes DNA damage and triggers apoptosis through MAPK signaling pathway in A549 cells, *Food Chem. Toxicol.*, 62, 661, 2013.
212. Nisr, R.B., Russell, M.A., Chrachri, A., Moody, A.J., Gilpin, M.L., Effects of the microbial secondary metabolites pyrrolnitrin, phenazine and patulin on INS-1 rat pancreatic β-cells, *FEMS Immunol. Med. Microbiol.*, 63, 217, 2011.
213. Moreno-Félix, C. et al., Bioactive lipidic extracts from octopus (*Paraoctopus limaculatus*): Antimutagenicity and antiproliferative studies, *Evid. Based Complement. Altern. Med.*, 2013, 273582, 2013.

32

Candida

María Jesús Andrade, Mar Rodríguez, Alicia Rodríguez, and Juan José Córdoba

CONTENTS

32.1 Introduction

The *Candida* genus comprises more than 200 recognized species[1,2] of which more than 40 are implicated in human infections.[3] *Candida* is a natural inhabitant of human skin, intestinal and genitourinary tracts, and mucosal surfaces. Being an opportunistic pathogen, *Candida* causes disease when the host defenses are impaired. With recent increase in immunosuppressed patients, *Candida* infections have been described with higher frequency.[4,5]

Nonetheless, the presence of *Candida* in foods has obvious positive effects, such as in fermented foods and beverages. This fungus also contributes to the synthesis of some healthy compounds such as vitamins and antioxidants.[6] On the contrary, its negative effects on food products relate to food spoilage, but not usually to food safety problems. Thus, *Candida* species are frequently included in daily food consumption of human beings, as disorders attributable to pathogenic *Candida* through foods have been described as improbable.[2,7] Consequently, its effect on human health has been scarcely evaluated in the literature and more attention has been paid to clinical *Candida* infections.

Candida albicans is the major species of the genus associated with human disease,[1] although it is not involved in foodborne infection. In addition, other *Candida* species that are frequently referred to as "non-*albicans*" *Candida* species have been isolated as being responsible for infection including *C. glabrata*, *C. tropicalis*, *C. parapsilosis*, and *C. krusei*.[2,5] In the recent years, the occurrence of infections caused by such "non-*albicans*" *Candida* species has increased. The abovementioned five species are the causative agents of more than 90% of *Candida* infections.[2,8] In spite of the fact that other *Candida* species like *C. kefyr*, *C. rugosa*, *C. guilliermondii*, and *C. famata* have been reported as causative agents of clinical infection, they have been rarely isolated from patients.[3,5,9]

Candida can cause a wide variety of infections in humans[1] that range from superficial infection of skin and mucosal membranes of intestinal and genital tracts,[2] including cutaneous and oropharyngeal candidiasis and vaginitis,[1] to systematic infections. The latter are of significant concern since they encompass life-threatening diseases such as candidemia, endophthalmitis, endocarditis, osteomyelitis, and disseminated and central nervous system infections.[1,8] Such severe invasive infections cause high morbidity and mortality rates.[8,10]

32.2 Foodborne *Candida* Infections

Candida species frequently associated with food have emerged as important pathogens in debilitated patients,[3] whose damaged intestinal mucosa may allow translocation of the yeast from the gastrointestinal tract to the blood system.[2,4,6] After colonizing the gastrointestinal tract, systematic dissemination causing a life-threatening infection could occur in vulnerable hosts. In spite of the fact that food contaminated with *Candida* is not the starting factor in these cases, vulnerable patients are advised to not consume foods with high yeast contents for preventing the contamination of catheters and the opportunistic infection.[6] It seems that *Candida*, which can be introduced into the hospital environment via food, has the ability to grow and form biofilms on the catheters and other medical invasive devices.[4,6,11]

Several *Candida* species are among the most common and widespread yeasts in food, but their distribution varies in relation to the type of food. *C. zeylanoides* has thus been reported as one of the main species during the ripening of Iberian dry-cured ham[12–14] and on fresh and processed poultry products.[15] This species has also been found in fermented sausages and other meat products,[16–18] cheeses,[19,20] and fruit juices.[21] Candidemia, endocarditis, arthritis, and other disorders have been reported in association with *C. zeylanoides*.[1,22]

Since the *Candida* genus is integrated by asexual yeast species, several other genera have *Candida* anamorphs. Thus, some *Candida* anamorphs are frequently found in some food such as *Debaryomyces hansenii* (anamorph *C. famata*) or *Kluyveromyces marxianus* (anamorph *C. kefyr*). It should be noted that both species are used as commercial starter cultures and are included in the QPS (Qualified Presumption of Safety) list developed by the European Food Safety Authority (EFSA).[23] *C. kefyr* appears to be an emerging pathogen.[3] Nevertheless, until now, there are no safety concerns about the use of *K. marxianus* because of the history of its apparent safe use and the rarity of its infections in human beings.[24] *C. famata*, found in food products such as cheese and meat products,[12,18,25] has been associated with occasional infections.[3,8,9,23–25]

C. krusei, one of the five most clinically relevant *Candida* species, has been found to be the predominant yeast in some African fermented products.[26,27]

Other *Candida* species related to infection have also been detected in food products. Specifically, *C. rugosa*, *C. glabrata*, and *C. tropicalis* have been found in soy sauce koji, accounting for about 40% of total yeasts, with *C. rugosa* being most predominant.[28] On the other hand, *C. tropicalis* isolated from sorghum beer has been proven to be adequate as starter culture for barley beer production.[29] *C. tropicalis* has also been found in orange fruit and juice.[21,30] The potential significant role of *C. krusei*, one of the dominant yeast species in the cocoa fermentation,[31,32] has been reported.[32] *C. krusei* isolated from fermented maize dough has been selected as a candidate to be a starter culture during fermentation of maize dough too.[33] Nevertheless, due to the fact that the main human pathogenic yeasts are included in the *Candida* genus and that several *Candida* species are considered as emergent pathogens, the genus cannot be considered suitable for QPS status.[24] Further investigations are needed to establish the role of foods in *Candida* infections.[11]

32.3 Investigation of Foodborne *Candida* Cases

A wide range of procedures to detect *Candida* can be found in the literature. Most of them have been specifically designed for clinical diagnosis but could be applied for analysis of *Candida* related to food safety (Figure 32.1).

Diagnosis of *Candida* from clinical and food samples causing infection is of critical importance since *Candida* species differ widely, both in their ability to cause infection and in their susceptibility to antifungal agents.[34] A rapid and accurate identification of *Candida* species is necessary to accelerate the beginning of the proper therapy, which would improve the patient outcomes.

In this section, some of the methods reported to detect and differentiate *Candida* of clinical and food origin will be reviewed.

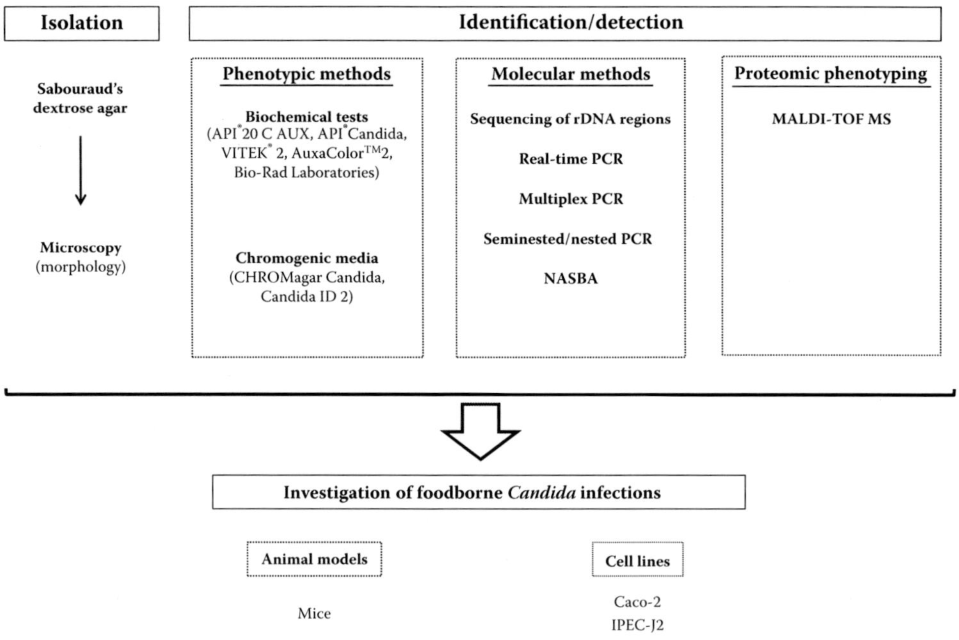

FIGURE 32.1 Procedures for analyzing foodborne *Candida* cases.

Diagnosis of *Candida* infections has traditionally relied on a combination of phenotypic features of the pathogenic yeast and the clinical signs of the infection.[2]

When analyzing the clinical presenting features, a combination of tests is necessary to obtain an adequate diagnosis.[2] However, the main problem is that clinical signs are often nonspecific for *Candida* disease.

Direct microscopic investigation of appropriate clinical samples and culture is included within the phenotypic techniques. Microscopic morphology of *Candida* shows important differences depending on environmental conditions and yeast species. Thus, *Candida* can produce round or elongated blastoconidia, pseudohyphae, true hyphae, or chlamydospores.[1] However, *C. glabrata* rarely produces pseudohyphae or hyphal structures.[2]

To study the other phenotypic features, the isolation of the causative agent is necessary. Sabouraud's dextrose agar (SDA) is the most commonly used primary isolation medium for *Candida*. However, it does not normally allow differentiation between species.[34] The colonies of *Candida* species are cream/white colored with smooth and wet looking. Nevertheless, *C. albicans* has the capability to reversibly change the surface appearance of its colony from smooth to rough as well as from creamy/white to opaque, a modification that corresponds to an alteration in cell shape from round (oval) to elongate.[2]

After isolating *Candida* as the causative agent of an infection, the identification at species level of the obtained isolates has to be performed. The assimilation and fermentation patterns can be studied for such purpose[35] by using the available commercial kits (API® 20 C AUX, API® Candida, VITEK® 2, bioMérieux; AuxaColor™ 2, Bio-Rad Laboratories). Such commercial techniques have been used to test for *Candida* of both food and clinical origin.

In addition, several chromogenic media for the direct and presumptive identification of *Candida* on the basis of the colony color have been developed. Thus, agars such as CHROMagar Candida (CHROMagar Company) and Candida ID 2 (bioMérieux) allow the differentiation of some of the most significant pathogenic *Candida* species.[36–38] The use of such media allows reducing the time of analysis in comparison with conventional phenotypic and biochemical methods. Despite the fact that the CHROMagar Candida was originally developed for identifying *Candida* isolates of clinical origin, it has been used for differentiation and presumptive identification of the most frequently isolated yeast species from different ingredients used for salads.[39]

Species identification of *Candida* based on the abovementioned phenotypic methods is not always sufficient in spite of using different tests[5] because of their possible ambiguous or inconclusive results. This fact together with the fact that they are often time consuming cause that infections are often not diagnosed or the diagnosis is obtained too late. Consequently, these methods are of limited clinical value, at least when they are used alone.

Nonculture diagnostic methods are available as alternative and supplementary to the conventional ones.

Thus, serological tests can be performed by detecting antigens and antibodies. Such tests are used in combination with conventional techniques for improving the diagnosis. The mannan and (1,3)-β-D-glucan detection is widely used. The mannan constituent of the *Candida* cell wall is a polysaccharide used as a target of many serological tests performed in serum or plasma specimens since it induces a strong antibody response.[40] Tests based on the detection of antibodies directed against the mannan antigen are recommended in combination with those based on *Candida* antigens for improving the sensitivity and the specificity of the diagnosis.[41] The commercial Platelia *Candida* antigen and antibody tests (Bio-Rad Laboratories) are the most commonly used serological methods for detecting *Candida* based on mannan.[41,42] However, since the antibodies could be not produced in immunocompromised patients, it is very complicated to diagnose *Candida* infections in them.[34,41] (1,3)-β-D-Glucan is also a cell wall component not only of *Candida* species but also of most of the other fungi. Several assays have been developed for quantifying such compounds in blood as a tool in the diagnosis of many pathogenic fungi including *Candida*.[40,43]

The detection of D-arabinitol, a metabolite produced by some of the *Candida* species with clinical importance, has also been proposed for diagnosis.[40]

The serological methods have limitations for the diagnosis of candidal infections due to their sensitivity and specificity. Besides, they cannot be used alone for reliable identification of clinical *Candida* isolates as previously described for traditional techniques.

To overcome the limitations of culture and nonculture methods, nucleic-acid-based procedures have been developed since they are powerful tool because of their rapidity and sensitivity. The literature contains a lot of data regarding molecular-based technology for clinical diagnosis of *Candida*. Lewis et al.[2,35] summarized the polymerase chain reaction (PCR) methods published for detecting *Candida* species from clinical samples, with most of them being based on rDNA regions as target. These authors considered that such methods could also be used for screening food products. However, the type of sample, enrichment processes, and DNA extraction should be optimized for the particular food and for removal of any interfering compounds within.[2]

Identification by sequencing of rRNA genes has been widely used for identifying *Candida* of both food and clinical origins.[28,44,45] However, such technique despite being a valid method for identifying *Candida* is time consuming, expensive, tedious, technically demanding, and inappropriate for routine applications.[46,47]

Among various molecular-based methods, real-time PCR (qPCR) has advantages of speed, sensitivity, and specificity.[48–50]

Multiplex PCR detecting different targets simultaneously and semi-nested/nested PCR incorporating two round of amplification reactions have also been developed.[51–54]

Other molecular-based methods for detecting *Candida* RNA include nucleic-acid-sequence-based amplification (NASBA),[55] etc.

Despite the vast number of suitable published molecular methods for detecting and/or identifying *Candida* species, they are still lacking standardization. Thus, it has been suggested that a combination of methods, including the molecular ones, may be performed to obtain a certain diagnosis.[5,43]

Novel technologies such as matrix-assisted laser desorption/ionization time-of-flight mass spectrometry (MALDI-TOF MS) have been reported as rapid and accurate methods for routine identification of *Candida* associated with human and food.[46,47,56,57] However, some of the authors have stated the necessity of improving the databases of the available systems for such technique.

The former methods can detect *Candida* involved in foodborne infection or in patients with clinical symptoms of disease. However, the study of candidal infections has also involved the subsequent investigation of the yeast in animals and other laboratory model systems.[58] They are necessary to provide new insights for clarifying and fully understanding the mechanisms leading to human *Candida* disease independently of whether the yeast is of clinical or food origin. Such powerful systems will be detailed in the next section.

32.4 Laboratory Models for Analyzing Foodborne *Candida* Infections

Many laboratory models have been developed for analyzing different aspects of clinical *Candida* infections, particularly those relating to *C. albicans*. Nevertheless, experimental models are not normally designed for investigating foodborne pathogen *Candida* species. As mentioned in Section 32.2, foodborne candidal infections could be found in humans with the altered integrity of the mucosal epithelium of the gastrointestinal tract or due to *Candida* biofilm formed in invasive devices. Since *Candida* species frequently associated with food are emerging as causative agents, the development of laboratory models for such species should be essential for understanding their pathogenicity. This section will thus be focused on studies using laboratory models related to the gastrointestinal tract (Table 32.1), the most relevant anatomic route of infection of foodborne *Candida*. Although the laboratory models reported until now have been developed for analyzing *Candida* of clinical origin, most of them could be adapted for being used with foodborne *Candida* (Figure 32.1).

A number of nonvertebrate and vertebrate animal models have been described as suitable to study *Candida* pathogenesis and virulence, immune response of the host, and efficacy of novel antifungal and

TABLE 32.1

Laboratory Models Used to Investigate Gastrointestinal *Candida* Infections

Model Type	*Candida* Species Detected	Main Measured Parameters	References
Mouse	*C. albicans* *C. glabrata*	Histopathology, yeast morphology and burden, inflammatory response	75
	C. albicans *C. tropicalis* *C. parapsilosis*	Yeast burden, histopathology	68
	C. albicans	Yeast burden, role of MAPK pathways	71
	C. albicans *C. tropicalis*	Yeast burden, colonization and invasion, mortality	70
	C. albicans	Yeast burden and morphology, immune response	74
	C. albicans	Yeast burden and morphology	72
Mouse, piglet	*C. albicans*	Yeast burden and morphology, gene expression	73
Drosophila larvae	*C. albicans*	Microscopy, systematic response	84
Caco-2 cells	*C. albicans* *C. tropicalis* *C. parapsilosis* *C. krusei*	LDH assay, genetic analysis, immune response	90
	C. tropicalis	Adhesion, cell damage, gene expression	89
	C. albicans	Yeast burden and morphology, adhesion, topography of cells, transcriptional regulation during adhesion	88
	C. albicans	Adhesion, cell damage, invasion, transmigration, yeast morphology, viability, and hydrolases	66
Caco-2 and IPEC-J2 cell lines	*C. albicans* *C. rugosa* *C. tropicalis* *C. krusei*	Yeast morphology, transepithelial electrical resistance	27

vaccination strategies. The advantages and limitations of both *in vivo* models have been summarized by Naglik et al.[58] The main advantage of nonvertebrate animal models consists of lower cost and fewer ethical concerns than vertebrate ones. Vertebrate animal model systems are more adequate tools for studying the host–*Candida* interactions than the nonvertebrate animal ones due to the fact that the used animals and their environment can be precisely controlled in a manner that is more relevant to humans. A good *in vivo* model for investigating *Candida* infections should be as reproducible as possible regarding the process of colonization and invasion at a specific portal of entry leading to infection as well as the major clinical symptoms detected in the human disease, relatively easy to set up, and closely match the specific immune defects or hormonal conditions associated with infection at that particular site.[59,60] The cost of the model should also be considered in the choice.[60] In addition, the host, the tested *Candida* strain, and the route of its administration are of critical importance to obtain reliable experimental conclusions when selecting the *in vivo* experimental model.[61]

Although many clinical animal models have been developed, rodents (mainly mice) are the most commonly used laboratory ones (Table 32.1) because of ethical reasons and their relative anatomic and immunological similarity to human.[60,62] Moreover, they are widely available, easy to handle, and relatively inexpensive.[63] However, since *C. albicans* is an opportunistic pathogen of humans but not a natural inhabitant of rodents, such models could give responses to infections in a different way than human beings, and so comparison between such models and human studies could be difficult.[58,64,65] Such difference should be considered when interpreting experimental data.[61]

In the case of modeling gastrointestinal *Candida* infections, mice are the animals selected by the majority of researchers,[65] and rats have been occasionally used.[58] However, most of them are mainly focused on studying the later dissemination of *Candida* because systematic dissemination normally originates from the gastrointestinal tract.[60,66]

In gastrointestinal models, it has to be taken into account that the colonization of the gastrointestinal tract by *Candida* seems to be apparently inhibited by the innate defenses of the host and mucosal immune response and also by competition of *Candida* with the native microbial population of the tract.[67] Thus, for establishing the gastrointestinal *Candida* infection the normal mucosal microbial population must be removed or the yeast must become implanted in the gastrointestinal tract before colonization by this microbial population.[67] Gastrointestinal models are thus commonly based on the use of immunosuppressed adult mice or neonatal mice.[68] For neonatal models, no pretreatment is necessary as for adult ones, which have to be treated for removing their natural gastrointestinal microbial population by using immunocompromising agents (such as by antibiotic administration) prior to *Candida* infection.[60,67] Germ-free or transgenic mice could also be used for this purpose.[58] The performed mice manipulations have to be considered when interpreting the obtained results.[58] Adult mice have been considered as more suitable since their handling and care are easier and less time consuming and their mortality due to the inoculation is minimal.[68] Adult mice can be inoculated either intragastrically or orally by using contaminated food or drinking water,[60] the second option being the most adequate for studying gastrointestinal infections related to foodborne *Candida* as the inoculum should be administered via the physiologically relevant portal of entry.[61] Since mice are coprophagous, the gastrointestinal mucosa of the animal fed with *Candida* contaminated food seems to be exposed to high levels of the yeast.[67] The majority of the clinical models based on gastrointestinal infection have carried out the injection of the yeast directly into the tract. After *Candida* inoculation, the maintenance of antibacterial antibiotic administration is necessary for achieving sustained gastrointestinal colonization[69] and allowing yeast dissemination to occur.[60]

The level of *Candida* inoculum used for gastrointestinal models ranges from about 10^7 to 10^8 cells suspended in a sterile saline solution, mainly phosphate buffered saline (PBS), when performing intragastric administration.[68,70–73] The inoculum level ranges from about 10^6 to 10^7 cells/mL via drink administration.[74,75]

The behavior of non-*albicans Candida* species, including *C. tropicalis*, *C. glabrata*, and *C. parapsilosis*, has also been analyzed by using gastrointestinal models, but scarcely. Several authors have compared the virulence of *C. albicans* and *C. tropicalis* in a gastrointestinal model in mice, with *C. tropicalis* appearing as more virulent than *C. albicans*.[60,70] When comparing the virulence of *C. tropicalis*, *C. glabrata*, and *C. albicans* in a gastrointestinal mouse model, the three species colonized the gastrointestinal tract, with *C. tropicalis* showing the least ability and the subsequent dissemination was only observed in

C. albicans and *C. tropicalis*.[68] The virulence of *C. glabrata* was also less than that shown by *C. albicans* by using this kind of models.[75] It seems that virulence mechanisms differ between *Candida* species, but more research using experimental animal models is necessary.

The most usually measured parameters in mice *Candida* gastrointestinal models during the colonization and the later dissemination include yeast burden at various sites, organ histopathology, and mortality of animals (Table 32.1). *Candida* burden has been determined in fecal and cecal contents and also in organs such as esophagus, stomach, intestines, cecum and colon, and internal organs (kidney, spleen, lungs, liver).[68,70,72,74,75] Visualization under microscope of stools and histologic examination of different organs for identifying yeast or filamentous forms have also been reported for most of gastrointestinal models. It seems that hyphae, but not yeast, are implied in the inflammatory and immune responses during colonization of the gastrointestinal tract,[74] having a significant role in the pathogenesis by invading epithelial cells and causing damage of tissues.[76] The ability to change morphology from yeast to hyphal form (dimorphism) in response to environmental changes is one of the best studied virulence attributes and characteristics of *C. albicans*.[77] However, the conversion of *C. albicans* to hyphae has been described by other authors as important but not essential for pathogenesis and tissue invasion.[78]

Regarding *Candida* immunity in gastrointestinal models, the host immune response during *Candida* colonization of the gastrointestinal tract has been scarcely studied. Vautier et al.[74] performed an immunoassay for analyzing cytokine levels after inoculating antibiotic-treated mice with *C. albicans* via their drinking water containing 10^7 yeast cells/mL together with antibiotics. Kidney, stomach, intestines, and caecum were used for such analysis. They reported that colonization of the gastrointestinal tract by *C. albicans* was not influenced by Th17 immunity.

The inflammatory response in gastrointestinal tissues was studied by Westwater et al.[75] in an assay based on phenotyping of GR1 granulocytes. *C. glabrata* and *C. albicans* were separately administrated to germ-free mice via drink. *C. glabrata* did not evoke a granulocytic inflammatory response in the host, in contrast to *C. albicans*.

Few studies have investigated the genes expressed in *C. albicans* during the gastrointestinal colonization in mice and also in piglets. White et al.[73] extracted RNA from different tissues of infected piglets and from cecum and ileum contents from mice, which was used for quantitative real-time reverse transcriptase PCR (qRT-PCR) and microarrays. Their aim was the study of the genes expressed by *C. albicans* during its growth in the gastrointestinal tract. In a separate work, Prieto et al.[71] used a gastrointestinal model in antibiotic-treated mouse to determine the role of the mitogen-activated protein kinase (MAPK) pathways of *C. albicans* in the colonization of the gastrointestinal tract. Such pathways represent one of the main mechanisms of adaptation to environmental changes in *C. albicans* and play an important role in virulence.[71]

In the above-described gastrointestinal models based on the host immune and inflammatory responses and genetic studies, *Candida* loads were also determined in different sites of the gastrointestinal tract and in stools.

Laboratory models focused on other vertebrates animals such as monkeys,[79] piglets,[80] rabbits,[81] and guinea pigs[82] have also been proposed.[60] In spite of the fact that the larger the animal model, the easier it is to sample repeatedly,[58] the use of those animals larger in size than mice implies higher cost and husbandry requirements, more intense monitoring during experimental infection, and fewer available molecular reagents.[61] In addition, they are not available as genetically defined strains.[61]

Due to the fact that research involving mammals is generally expensive, complex, and fraught with ethical concerns, alternative animal models for studying clinical *Candida* infection have been used.[83] Zebrafish (*Danio rerio*), a vertebrate minihost, and some nonvertebrate minihosts, including wax moth (*Galleria mellonella*) and silk worm (*Bombyx mori*) larvae, the fruit fly (*Drosophila melanogaster*), and a nematode (*Caenorhabditis elegans*), have been thus reported for studying *Candida* virulence.[60] A limitation of experimental models based on minihosts is that many of them lack an adaptive immune system.[60] A *Drosophila* larvae model has been reported as appropriate to assay the host–pathogen interaction due to an intestinal infection by *Candida* and the factors required for disseminated infection.[84] The similarity of mammals' and flies' immune response allows analyzing the genes involved.[84] However, the differences between fly and mammalian pathophysiology have to also be considered when using *Drosophila* intestinal models,[83] as previously mentioned for murine models. Since in the developed *D. melanogaster* model the larvae were fed with *Candida* contaminated food, such experimental model should be well suited for analyzing foodborne *Candida* infections. This characteristic is also attributable to the model described by Pukkila-Worley et al.,[85] who studied *C. albicans* pathogenesis using *C. elegans* as host, which was fed with artificially contaminated food, and described the utility of the assay for screening the genes involved in *Candida* morphogenesis and virulence. Since both the genome of *D. melanogaster* and *C. elegans* have been sequenced,[62] genetic analyses are available when performing experimental *Candida* models, which is not possible for wax moth and silk worm models because of lack of genome sequencing.[62] The use of invertebrate animal models has the disadvantage that they have to be kept at temperatures below the normal human body temperature, except for *Galleria*, which could affect the *Candida*–host interactions.[60]

On the other hand, *Candida*, including foodborne species, commonly grow as a biofilms in medical devices, thereby providing protection from the host immune system and antimicrobial therapies.[86] Scarce works based on experimental models for evaluating the formation of biofilms in invasive devices are available.[86]

Although animals offer a good model for analyzing *Candida* infections, they have some limitations,[87] such as they are not appropriate for wide laboratory testing and routine analysis since special facilities and special trained personnel are needed. Besides, ethical considerations should be taken into account when animals are used. The absence of *Candida* as native microorganisms in the gastrointestinal tract of rodent models should also be considered when searching for a good model for *Candida* analysis.

Several *in vitro* alternative methods based on the use of cell lines and epithelial organotypic models are available for such purpose, which are less expensive and time consuming to obtain results than animal models[87] (Table 32.1). They have been developed for *Candida* of clinical origin but not for foodborne *Candida*, as mentioned for animal models. However, such models could be adapted for such purpose. When selecting an *in vitro* model, its availability and easy handling for high-throughput testing as well as its good human predictive power must be checked.[87]

Most of the performed investigations based on the use of cell lines for host–*Candida* interaction studies have been focused on the ability of *Candida* (mainly *C. albicans*) to adhere to different epithelia. For studying foodborne *Candida*, the most interesting cell lines are those of intestinal origin. Specific studies must be carried out in gastrointestinal models since the environment of the different anatomic sites that could be colonized by *Candida* differs significantly, particularly with respect to pH, epithelium metabolism, or surface characteristics.[88] Due to the fact that the attachment of *Candida* to intestinal epithelium is the initial step in the infection, it is an essential parameter to be analyzed in a gastrointestinal model based on cell lines. The ability of *C. albicans* to adhere to and invade into intestinal cells for its translocation across the intestinal barrier and the subsequent dissemination through the body have been demonstrated by using intestinal cell lines.[66] To analyze the adhesion of *C. albicans* to an intestinal epithelium *in vitro*, the Caco-2 cell line has been the most commonly used. Caco-2 is a human epithelial colon adenocarcinoma cell line.[87] Intestinal cell models should be functionally similar to the *in vivo* situation,[87] and such cells have this characteristic since they resemble morphologically as well as biochemically primary enterocytes in many characteristics.[88]

Some examples of the use of Caco-2 cell line will be briefly described below.

Sohn et al.[88] infected monolayers of Caco-2 grown to confluency with *C. albicans*. Polystyrene plates were used for an adhesion assay. After applying *C. albicans* to the confluent monolayers, adherent *C. albicans* cells counting was performed. Scanning electron microscopy (SEM) was also carried out for visualizing the surface topographies of cells and the presence of hyphae. The Caco-2 cells had a brush border surface consisting of microvilli structures, which seemed to be tightly linked to *C. albicans* hyphae. The transcriptional regulation during adhesion of *C. albicans* to Caco-2 cells was analyzed by genome-wide *Candida* DNA microarray analysis using total RNA isolated from the adhesion assay in order to identify how *C. albicans* responds on the transcriptional level to the environmental stimuli during adhesion to intestinal cell. They found that the adherence of *C. albicans* to Caco-2 cells requires the expression of adhesive structures on the yeast surface with the ability of interacting very tightly with the host tissues.

Negri et al.[89] also evaluated the attachment of *C. tropicalis* to intestinal epithelial cells by using Caco-2 cells. Besides analyzing adhesion of *C. tropicalis* to a confluent layer of intestinal cells, the degree of the epithelial cell damage was assessed by measuring the lactate dehydrogenase (LDH) activity, and the inhibition of cellular activity was determined by tetrazolium salt reduction. *C. tropicalis* was able to adhere to and to damage Caco-2. However, differences were observed in this ability depending on the type of used cells. This fact was also strain dependent. Since the secretion of hydrolytic enzymes such as secreted aspartyl proteinases (SAPs) is considered an important factor for *C. tropicalis* virulence, the expression of four genes encoding SAPs (*SAPT1*, *SAPT2*, *SAPT3*, and *SAPT4*) present in *C. tropicalis* was also investigated by qPCR after RT-PCR, which was affected by *C. tropicalis* adhesion to the cells. The *SAP* expression was also strain and human cell line dependent. This fact highlights the necessity to check specific *Candida* strains in specific anatomic sites.

Dalle et al.[66] also investigated the potential of *C. albicans* to adhere to and invade into Caco-2 cells. They performed adherence assay by using epifluorescence after calcofluor white staining, SEM, and measurement of LDH release, among other techniques. These authors found that adhesion, invasion, and damage of *C. albicans* depended not only on yeast morphology and activity, but also on the epithelial cell type and the differentiation stage of the epithelial cells, indicating that epithelial cells differ in their susceptibility to *C. albicans*. Besides, they reported that hyphae seem to have a dominant role in penetration of monolayers of intestinal cells in relation to yeasts.

Intestinal Caco-2 cells have been recently used for analyzing the host immune response to *Candida* since intestinal epithelial cells not only represent a physiologic barrier for the yeasts, but also have a crucial role in the innate immune system as they, for example, produce cytokines and antimicrobial peptides.[90] The effect of *C. albicans* as well as *C. krusei*, *C. tropicalis*, and *C. parapsilosis* on the production of human β-defensin 2 (HBD-2), a key component of the innate immune system, by Caco-2 cell line has been studied by RT-q PCR.[90] To determine the HBD-2 secretion induction and HBD-2 peptide expression in Caco-2 cells infected with *Candida* spp., enzyme linked immunosorbent assay (ELISA) and Western blot analysis were carried out, respectively. *Candida* spp. stimulated HBD-2 expression in and release from Caco-2 cells, with *C. albicans* inducing the highest levels of HBD-2.

Cell lines have been interestingly used for investigating the probiotic potential of some yeasts from food including *Candida* species, but their pathogenic potential was also analyzed.[27] *C. rugosa*, *C. tropicalis*, and *C. krusei*, which are considered causative agents of infections, were studied. They were previously isolated from *Fura*, a spontaneously fermented pearl millet product consumed in West Africa. As control, the authors included a pathogenic yeast strain of *C. albicans*. These authors used two intestinal cell lines: Caco-2 and IPEC-J2. In spite of the fact that IPEC-J2 was isolated from the small intestine of neonatal pig,[87] it has been used for studying pathogen/beneficial interactions of microorganisms due to its similarity to the human intestinal epithelial cells.[27] In such study, after growing both cell lines in different media, the cells were grown at 37°C in a humidified atmosphere of 5% CO_2 and 95% air until a confluent monolayer was obtained. The culture media were changed every second day, and the cells were passaged every 5–7 days by trypsination with 1% trypsin–EDTA 10×. Cell passages used were P55–57 and P26–28 for Caco-2 and IPEC-J2, respectively. The transepithelial electrical resistance (TEER) assay was used for analyzing the potential pathogenic effect (and also the probiotic one) of used yeasts by means of evaluating the effect of yeast strains on cell polarity. To obtain a monolayer of polarized intestinal epithelial cells, both cell lines were cultured on polycarbonate membrane inserts. After reaching confluence, Caco-2 cells were cultured on the membranes for 21 days in order to obtain cell differentiation, while IPEC-J2 cells were cultured for 16 days in order to obtain a proper mucus layer. An inoculum of 10^6 cells/mL of yeasts was added and incubated at 37°C in a humidified atmosphere of 5% CO_2 and 95% air. The TEER of monolayers with Caco-2 or IPEC-J2 cell lines was then measured and the morphology of yeasts observed by microscopy. Moreover, *C. krusei*, *C. tropicalis*, *C. rugosa*, and *C. albicans* were able to survive under conditions simulating passage through the human gastrointestinal tract [37°C, pH 2.5, and 0.3% (w/v) bile salts]. The ability of potential pathogenic foodborne *Candida* to survive the conditions in the gastrointestinal tract should also be evaluated such as the survival and growth at 37°C and the presence of high bile salt concentration and low pH.

Epithelial organotypic models have been described as effective as an *in vitro* model for evaluating the invasion of *C. albicans* too. Such models allow studying the tissue degradation in *Candida* infection as well as conducting experiments to characterize pathogenesis-related genes. The mucosal epithelium is very important in host defense and immune surveillance, since it is the primary cell layer that first encounters potentially pathogenic microorganisms.[91] Reconstituted human epithelium that mimics human epithelium is commercially available. It has been used as model for oral and vaginal candidiasis, but not for foodborne *Candida* infections. After infecting the tissues, their level of colonization and invasion and morphological characteristics of *Candida* can be analyzed by microscopy.[92] Epithelial damage can be visualized by histological analysis and quantified by the extracellular activity of LDH.[91,92] Besides, the expression patterns of several *Candida* virulence genes, such as agglutinin-like sequence (*ALS*), hyphal wall protein (*HWP*), epithelial adhesion (*EPA*), phospholipase B (*PLB*), *PLD*, and *SAP* genes, can be analyzed by RT-PCR when using such *in vitro* models.[91–93] The lack of certain host immune responses is a limitation of such epithelium models, while proinflammatory cytokine responses by the tissue infected with *Candida* are associated with the yeast strain virulence.[91,93]

32.5 Conclusions

Several *Candida* species are commonly present in food and beverages. However, candidal infections are not normally associated with consumption of food containing *Candida*. In the last years, some *Candida* species found in food are emerging as causative agents of infections. Further researches are thus necessary for elucidating the role of foodborne *Candida* in human infections.

A variety of methods to detect and diagnose *Candida* infections have been described, most of which have been designed for clinical purpose but could also be applied for investigation of foodborne cases. Phenotypic methods have been traditionally used for detecting *Candida*; however, they are time consuming and can lead ambiguous or inconclusive results. Serological methods have limited sensitivity and specificity. More recently, molecular-based protocols have become available, although they are still in need of proper standardization. Consequently, a combination of methods for diagnosis of pathogenic *Candida* from both clinical and food origins should be performed for obtaining definitive results.

In order to analyze different aspects of clinical *Candida* infections, several laboratory models based on using vertebrate or nonvertebrate animals, cell lines, and commercial reconstituted human epithelium have been developed. Mice are the most commonly used animal models because of ethical reason and their relative anatomic and immunological similarity to human beings despite the fact that *Candida* is not a natural inhabitant of mice. The use of cell lines such as Caco-2 cells has less ethical concerns and other advantages compared to animal models. In addition, epithelial organotypic models used for analyzing oral and vaginal candidiasis could be adapted for *Candida* foodborne investigation.

As most currently available methods have focused on *Candida* of clinical origin and not on foodborne *Candida*, further investigations are required to develop adequate laboratory models for investigating foodborne *Candida* infections. These include refining the experimental gastrointestinal models designed for clinical *Candida*.

Acknowledgements

This work has been funded by the Spanish Comisión Interministerial de Ciencia y Tecnología with the projects AGL2013-45729P and Carnisenusa CSD2007-00016, Consolider Ingenio 2010 and GRU09162 of the Junta de Extremadura and FEDER.

REFERENCES

1. Eggimann, P., Garbino, J. and Pittet, D., Epidemiology of *Candida* species infections in critically ill non-immunosuppressed patients, *Lancet Infect. Dis.*, 3, 685, 2003.
2. Lewis, P. et al., *Candida*, in *Molecular Detection of Foodborne Pathogens*, p. 549, Liu, D. (Ed.), Taylor and Francis Group LLC, Boca Raton, 2010.
3. Johnson, E.M., Rare and emerging *Candida* species, *Curr. Fungal Infect. Rep.*, 3, 152, 2009.
4. Hobson, R., The global epidemiology of invasive *Candida* infections—is the tide turning? *J. Hosp. Infect.*, 55, 159, 2003.

5. Criseo, G., Scordino, F. and Romeo, O., Current methods for identifying clinically important cryptic *Candida* species, *J. Microbiol. Methods*, 111, 50, 2015.
6. Fleet, G.H. and Balia, R., The public health and probiotic significance of yeasts in foods and beverages, in *Yeasts in Food and Beverages*, p. 381, Querol, A., Fleet, G.H. (Eds.), Springer-Verlag, Berlin, Heidelberg, 2006.
7. Boekhout, T. and Robert, V. (Eds.), *Yeasts in Food. Beneficial and Detrimental Aspects*, Behr's Verlag, Hamburg, Germany, 2003.
8. Yapar, N., Epidemiology and risk factors for invasive candidiasis, *Ther. Clin. Risk Manag.*, 10, 95, 2014.
9. Brandt, M.E. and Lockhart, S.R., Recent taxonomic developments with *Candida* and other opportunistic yeasts, *Curr. Fungal Infect. Rep.*, 6, 170, 2012.
10. Quindós, G., Epidemiology of candidaemia and invasive candidiasis. A changing face, *Rev. Iberoam. Micol.*, 31, 42, 2014.
11. Fleet, G.H., Yeasts in foods and beverages: impact on product quality and safety, *Curr. Opin. Biotechnol.*, 18, 170, 2007.
12. Simoncini, N. et al., Dynamics and characterization of yeasts during ripening of typical Italian dry-cured ham, *Food Microbiol.*, 24, 577, 2007.
13. Núñez, F. et al., Yeast population during ripening of dry-cured Iberian ham, *Int. J. Food Microbiol.*, 29, 271, 1996.
14. Andrade, M.J. et al., Differentiation of yeasts growing on dry-cured Iberian ham by mitochondrial DNA restriction analysis, RAPD-PCR and their volatile compounds production, *Food Microbiol.*, 26, 578, 2009.
15. Ismail, S.A.S. et al., Presence and changes in populations of yeasts on raw and processed poultry products stored at refrigeration temperature, *Int. J. Food Microbiol.*, 62, 113, 2000.
16. Purriños, L. et al., Study of the counts, species and characteristics of the yeast population during the manufacture of dry-cured "lacón". Effect of salt level, *Food Microbiol.*, 34, 12, 2013.
17. Gardini, F. et al., A survey of yeasts in traditional sausages of southern Italy, *FEMS Yeast Res.*, 1, 161, 2001.
18. Encinas, J.P. et al., Yeast populations on Spanish fermented sausages, *Meat Sci.*, 54, 203, 2000.
19. Fadda, M.E. et al., Occurrence and characterization of yeasts isolated from artisanal Fiore Sardo cheese, *Int. J. Food Microbiol.*, 95, 51, 2004.
20. Prillinger, H. et al., Phenotypic and genotypic identification of yeasts from cheese, *Antonie Van Leeuwenhoek*, 75, 267, 1999.
21. Arias, C.R. et al., Yeast species associated with orange juice: evaluation of different identification methods, *Appl. Environ. Microbiol.*, 68, 1955, 2002.
22. Shokri, H., Genotypic variation and antifungal susceptibly of *Candida zeylanoides* clinical isolates, *J. Mycol. Med.*, 24, 179, 2014.
23. EFSA, Scientific Opinion on the maintenance of the list of QPS biological agents intentionally added to food and feed (2013 update), *EFSA J.*, 11, 3449, 2013.
24. EFSA, Introduction of a qualified presumption of safety (QPS) approach for assessment of selected microorganisms referred to EFSA, *EFSA J.*, 587, 1, 2007.
25. Beyda, N.D. et al., Treatment of *Candida famata* bloodstream infections: case series and review of the literature, *J. Antimicrob. Chemother.*, 68, 438, 2013.
26. Greppi, A. et al., Determination of yeast diversity in ogi, mawè, gowé and tchoukoutou by using culture-dependent and -independent methods, *Int. J. Food Microbiol.*, 165, 84, 2013.
27. Pedersen, L.L. et al., Biodiversity and probiotic potential of yeasts isolated from *Fura*, a West African spontaneously fermented cereal, *Int. J. Food Microbiol.*, 159, 144, 2012.
28. Yan, Y.Z. et al., Microbial composition during Chinese soy sauce koji-making based on culture dependent and independent methods, *Food Microbiol.*, 34, 189, 2013.
29. Alloue-Boraud, W.A.M. et al., Fermentation profile of *Saccharomyces cerevisiae* and *Candida tropicalis* as starter cultures on barley malt medium, *J. Food Sci. Technol.*, 1–7, 2014, doi: 10.1007/s13197-014-1526-0.
30. Las Heras-Vazquez, F.J. et al., Identification of yeast species from orange fruit and juice by RFLP and sequence analysis of the 5.8S rRNA gene and the two internal transcribed spacers, *FEMS Yeast Res.*, 3, 3, 2003.
31. Jespersen, L. et al., Occurrence and diversity of yeasts involved in fermentation of West African cocoa beans, *FEMS Yeast Res.*, 5, 441, 2005.

32. Nielsen, D.S. et al., Yeast populations associated with Ghanaian cocoa fermentations analysed using denaturing gradient gel electrophoresis (DGGE), *Yeast*, 22, 271, 2005.
33. Annan, N.T. et al., Volatile compounds produced by *Lactobacillus fermentum*, *Saccharomyces cerevisiae* and *Candida krusei* in single starter culture fermentations of Ghanaian maize dough, *J. Appl. Microbiol.*, 94, 462, 2003.
34. Raju, S.B. and Rajjappa, S., Isolation and identification of *Candida* from the oral cavity, ISRN Dent, article ID 487921, 2011.
35. Lewis, P., Perry, M.D. and Barnes, R.A., *Candida*, in *Molecular Detection of Human Fungal Pathogens*, p. 551, Liu, D. (Ed.), CRC Press, Boca Raton, FL 2011.
36. Willinger, B. et al., Performance of Candida ID, a new chromogenic medium for presumptive identification of *Candida* species, in comparison to CHROMagar Candida, *J. Clin. Microbiol.*, 39, 3793, 2001.
37. Odds, F.C. and Bernaerts, R., CHROMagar Candida, a new differential isolation medium for presumptive identification of clinically important *Candida* species, *J. Clin. Microbiol.*, 32, 1923, 1994.
38. Eraso, E. et al., Evaluation of the new chromogenic medium Candida ID 2 for isolation and identification of *Candida albicans* and other medically important *Candida* species, *J. Clin. Microbiol.*, 44, 3340, 2006.
39. Tornai-Lehoczki, J., Péter, G. and Dlauchy, D., CHROMagar Candida medium as a practical tool for the differentiation and presumptive identification of yeast species isolated from salads, *Int. J. Food Microbiol.*, 86, 189, 2003.
40. Malhotra, S. et al., Recent diagnostic techniques in mycology, *J. Med. Microbiol. Diagn.*, 3, 146, 2014.
41. Willinger, B. and Haase, G., State-of-the-art procedures and quality management in diagnostic medical mycology, *Curr. Fungal Infect. Rep.*, 7, 260, 2013.
42. Ellis, M. et al., Prospective evaluation of mannan and anti-mannan antibodies for diagnosis of invasive *Candida* infections in patients with neutropenic fever, *J. Med. Microbiol.*, 58, 606, 2009.
43. León, C., Ostrosky-Zeichner, L. and Schuster, M., What's new in the clinical and diagnostic management of invasive candidiasis in critically ill patients, *Intensive Care Med.*, 40, 808, 2014.
44. Miranda, L.N., et al., *Candida* colonisation as a source for candidaemia, *J. Hosp. Infect.*, 72, 9, 2009.
45. Andrade, M.J. et al., Efficiency of mitochondrial DNA restriction analysis and RAPD-PCR to characterize yeasts growing on dry-cured Iberian ham at the different geographic areas of ripening, *Meat Sci.*, 84, 377, 2010.
46. Jamal, W.Y. et al., Comparative evaluation of two matrix-assisted laser desorption/ionization time-of-flight mass spectrometry (MALDI-TOF MS) systems for the identification of clinically significant yeasts, *Int. J. Infect. Dis.*, 26, 167 2014.
47. Pulcrano, G. et al., Rapid and reliable MALDI-TOF mass spectrometry identification of *Candida* non-*albicans* isolates from bloodstream infections, *J. Microbiol. Methods*, 94, 262, 2013.
48. Fricke, S. et al., A real-time PCR assay for the differentiation of *Candida* species, *J. Appl. Microbiol.*, 109, 1150, 2010.
49. Arancia, S. et al., Rapid, simple, and low-cost identification of *Candida* species using high-resolution melting analysis, *Diagn. Microbiol. Infect. Dis.*, 69, 283, 2011.
50. Decat, E. et al., Rapid and accurate identification of isolates of *Candida* species by melting peak and melting curve analysis of the internally transcribed spacer region 2 fragment (ITS2-MCA), *Res. Microbiol.*, 164, 110, 2013.
51. Ahmad, S. et al., Seminested PCR for diagnosis of candidemia: comparison with culture, antigen detection, and biochemical methods for species identification, *J. Clin. Microbiol.*, 40, 2483, 2002.
52. Taira, C. L. et al., A multiplex nested PCR for the detection and identification of *Candida* species in blood samples of critically ill paediatric patients, *BMC Infect. Dis.*, 14, 406, 2014.
53. Romeo, O. et al., A multiplex PCR protocol for rapid identification of *Candida glabrata* and its phylogenetically related species *Candida nivariensis* and *Candida bracarensis*, *J. Microbiol. Methods*, 79, 117, 2009.
54. Carvalho, A. et al., Multiplex PCR identification of eight clinically relevant *Candida* species, *Med. Mycol.*, 45, 619, 2007.
55. Loeffler, J. et al., Development and evaluation of the nuclisens basic kit NASBA for the detection of RNA from *Candida* species frequently resistant to antifungal drugs, *Diagn. Microbiol. Infect. Dis.*, 45, 217, 2003.
56. Pavlovic, M. et al., MALDI-TOF MS based identification of food-borne yeast isolates, *J. Microbiol. Methods*, 106, 123, 2014.

57. Quiles-Melero, I. et al., Evaluation of matrix-assisted laser desorption/ionisation time-of-flight (MALDI-TOF) mass spectrometry for identification of *Candida parapsilosis*, *C. orthopsilosis* and *C. metapsilosis*, *Eur. J. Clin. Microbiol. Infect. Dis.*, 31, 67, 2012.
58. Naglik, J.R., Fidel, P.L. and Odds, F.C., Animal models of mucosal *Candida* infection, *FEMS Microbiol. Lett.*, 283, 129, 2008.
59. De Repentigny, L., Animal models in the analysis of *Candida* host–pathogen interactions, *Curr. Opin. Microbiol.*, 7, 324, 2004.
60. MacCallum, D.M., Hosting infection: experimental models to assay *Candida* virulence, *Int. J. Microbiol.*, 2012, Article ID 363764, 2012.
61. Hohl, T.M., Overview of vertebrate animal models of fungal infection, *J. Immunol. Methods*, 410, 100, 2014.
62. Chamilos, G. et al., Role of mini-host models in the study of medically important fungi, *Lancet Infect. Dis.*, 7, 42, 2007.
63. Samaranayake, Y.H. and Samaranayake, L.P., Experimental oral candidiasis in animal models, *Clin. Microbiol. Rev.*, 14, 398,2001.
64. Mestas, J. and Hughes, C.C.W., Of mice and not men: differences between mouse and human immunology, *J. Immunol.*, 172, 2731, 2004.
65. Naglik, J. R. et al., Quantitative expression of the *Candida albicans* secreted aspartyl proteinase gene family in human oral and vaginal candidiasis, *Microbiology*, 154, 3266, 2008.
66. Dalle, F. et al., Cellular interactions of *Candida albicans* with human oral epithelial cells and enterocytes, *Cell Microbiol.*, 12, 248, 2010.
67. Cole, G.T. et al., Gastrointestinal candidiasis: histopathology of *Candida*–host interactions in a murine model, *Mycol. Res.*, 97, 385, 1993.
68. Mellado, E. et al., Sustained gastrointestinal colonization and systemic dissemination by *Candida albicans*, *Candida tropicalis* and *Candida parapsilosis* in adult mice, *Diagn. Microbiol. Infect. Dis.*, 38, 21, 2000.
69. Koh, A.Y., Murine models of *Candida* gastrointestinal colonization and dissemination, *Eukaryot. Cell*, 12, 1416, 2013.
70. De Repentigny, L., Phaneuf, M. and Mathieu, L.G., Gastrointestinal colonization and systemic dissemination by *Candida albicans* and *Candida tropicalis* in intact and immunocompromised mice, *Infect. Immun.*, 60, 4907, 1992.
71. Prieto, D. et al., The HOG pathway is critical for the colonization of the mouse gastrointestinal tract by *Candida albicans*, *PLoS One*, 9, e87128, 2014.
72. Wiesner, S.M. et al., Gastrointestinal colonization by *Candida albicans* mutant strains in antibiotic-treated mice, *Clin. Diagn. Lab. Immunol.*, 8, 192, 2001.
73. White, S.J. et al., Self-regulation of *Candida albicans* population size during GI colonization, *PLoS Pathog.*, 3, e184, 2007.
74. Vautier, S. et al., *Candida albicans* colonization and dissemination from the murine gastrointestinal tract: the influence of morphology and Th17 immunity, *Cell Microbiol.*, 17, 445, 2015.
75. Westwater, C. et al., *Candida glabrata* and *Candida albicans*; dissimilar tissue tropism and infectivity in a gnotobiotic model of mucosal candidiasis, *FEMS Immunol. Med. Microbiol.*, 51, 134, 2007.
76. Yan, L., Yang, C. and Tang, J., Disruption of the intestinal mucosal barrier in *Candida albicans* infections, *Microbiol. Res.*, 168, 389, 2013.
77. Wächtler, B. et al., From attachment to damage: defined genes of *Candida albicans* mediate adhesion, invasion and damage during interaction with oral epithelial cells, *PLoS One*, 6, e17046, 2011.
78. Pacheco, M. et al., Attachment and entry of *Candida famata* in monocytes and epithelial cells, *Microsc. Res. Tech.*, 70, 975, 2007.
79. Steele, C., Ratterree, M. and Fidel, P.L., Differential susceptibility of two species of macaques to experimental vaginal candidiasis, *J. Infect. Dis.*, 180, 802, 1999.
80. Andrutis, K.A. et al., Intestinal lesions associated with disseminated candidiasis in an experimental animal model, *J. Clin. Microbiol.*, 38, 2317, 2000.
81. Filler, S.G. et al., Comparison of fluconazole and amphotericin B for treatment of disseminated candidiasis and endophthalmitis in rabbits, *Antimicrob. Agents Chemother.*, 35, 288, 1991.
82. Fransen, J. et al., Histopathology of experimental systemic candidosis in guinea-pigs, *Sabouraudia*, 22, 455, 1984.

83. Apidianakis, Y. and Rahme, L.G., *Drosophila melanogaster* as a model for human intestinal infection and pathology, *Dis. Model. Mech.*, 4, 21, 2011.
84. Glittenberg, M.T. et al., Pathogen and host factors are needed to provoke a systemic host response to gastrointestinal infection of *Drosophila* larvae by *Candida albicans*, *Dis. Model. Mech.*, 4, 515, 2011.
85. Pukkila-Worley, R. et al., *Candida albicans* hyphal formation and virulence assessed using a *Caenorhabditis elegans* infection model, *Eukaryot. Cell*, 8, 1750, 2009.
86. Nett, J.E., Marchillo, K. and Andes, D.R., Modeling of fungal biofilms using a rat central-vein catheter, *Methods Mol. Biol.*, 845, 547, 2012.
87. Cencič, A. and Langerholc, T., Functional cell models of the gut and their applications in food microbiology—a review, *Int. J. Food Microbiol.*, 141, S4, 2010.
88. Sohn, K. et al., An in vitro assay to study the transcriptional response during adherence of *Candida albicans* to different human epithelia, *FEMS Yeast Res.*, 6, 1085, 2006.
89. Negri, M. et al., An in vitro evaluation of *Candida tropicalis* infectivity using human cell monolayers, *J. Med. Microbiol.*, 60, 1270, 2011.
90. Gácser, A. et al., Induction of human defensins by intestinal Caco-2 cells after interactions with opportunistic *Candida* species, *Microbes Infect.*, 16, 80, 2014.
91. Schaller, M. et al., Models of oral and vaginal candidiasis based on in vitro reconstituted human epithelia, *Nat. Protoc.*, 1, 2767, 2006.
92. Alves, C.T. et al., *Candida albicans* promotes invasion and colonisation of *Candida glabrata* in a reconstituted human vaginal epithelium, *J. Infect.*, 69, 396, 2014.
93. Malic, S. et al., Characterisation of *Candida albicans* infection of an in vitro oral epithelial model using confocal laser scanning microscopy, *Oral Microbiol. Immunol.*, 22, 188, 2007.

33

Enterocytozoon bieneusi

Hirotake Mori and Aongart Mahittikorn

CONTENTS

Microsporidia are a large group of obligate intracellular parasites that infect almost all types of animals, including humans, domesticated animals, wildlife, fish, and even insects. To date, over 1400 microsporidian species have been described, with new species being discovered every year.[1,2] Of these, 14 species in 8 genera have been recognized as human pathogens.[3] Recently, members of the phylum Microsporidia have been shown to be related to fungi based on phylogenetic analysis of genes coding for tubulin and mitochondrial-like HSP70, amongst others.[4,5]

As the most common cause of intestinal microsporidiosis,[6] *Enterocytozoon bieneusi* was the only known member of the genus *Enterocytozoon* until the recent identification of *E. hepatopenaei* from shrimp.[7,8] Since the mid-1980s, *E. bieneusi* has become an increasingly important opportunistic pathogen, especially in HIV/AIDS patients.[9,10] While infective to both immunocompetent and immunocompromised hosts, *E. bieneusi* is associated with more persistent and severe diarrhea and wasting in immunocompromised individuals.[11] Given that *E. bieneusi* is unresponsive to therapies that are effective against many microsporidia,[12] there is an urgent need to utilize tissue culture and small animal models to better understand the infection, transmission, and pathogenesis of this parasite and to develop more efficient treatment strategies against *E. bieneusi* disease.

33.1 Structure and Life Cycle

Microsporidia possess eukaryotic (a true nucleus, an endomembrane system, and a cytoskeleton), prokaryotic (genome size and transcriptional apparatus), and fungal (chitin and trehalose) characteristics.[1] As shown in Figure 33.1, the life cycle of *E. bieneusi* includes several stages. In the early proliferative stage, the nucleus

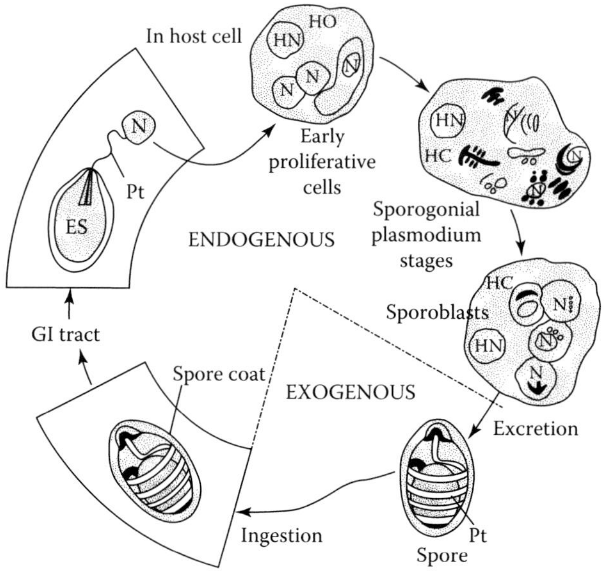

FIGURE 33.1 Diagram of *Enterocytozoon bieneusi* life cycle. HN, host cell nucleus; HC, host cell cytoplasm; N, nucleus of developing *E. bieneusi*; Pt, polar tube. (Reprinted from Santin, M. and Fayer, R., *Res. Vet. Sci.*, 90, 363–371, 2011. With permission from Elsevier.)

begins to divide, resulting in multiple elongated nuclei, and electron-dense inclusions appear in the cell. In the sporogonial plasmodium stage, polar tube elements are formed in the multinucleate plasmodium. Later, polar tube complexes and nuclei segregate into sporoblasts, which develop further to become spores that are shed with the host's feces. The spores of *E. bieneusi* are oval, 1–2 μm in size, and resistant to most environmental stresses. They have a cell wall consisting of three layers (plasma membrane, endospore, and exospore) and a long coiled tubular extrusion apparatus called the polar tube. The polar tube plays a critical role in invading and injecting infective material (sporoplasm) into the host cell.[6,13] In mature spores, the polar tubule has five to seven coils in two rows, as determined by transmission electron microscopy.

33.2 Biochemistry and Physiology

Microsporidia can survive outside of their hosts for many years in spore form.[14] Under appropriate conditions, the polar tube is expelled from the thin anterior part of the spore and the sporoplasm is transferred into a host cell.[15] The mechanisms of how *E. bieneusi* invade host enterocytes remain unclear, with the lack of an efficient *in vitro E. bieneusi* culture method hampering the development of a host–parasite interaction model. The rapid infection process of the polar tube has a high energy requirement; therefore, the process was assumed to be ATP independent. However, a recent genome survey of *E. bieneusi* revealed that the organism has a highly unusual and reduced biochemistry. Most of the glycolysis and trehalose metabolism, along with the pentose phosphate pathways, are missing, indicating that metabolism is largely dependent on the host. However, genes involved in other biochemical pathways, such as replication, translation, phosphate signaling, and amino acid metabolism, are intact. Therefore, the germination process may not be ATP independent.[16]

33.3 Immunology

Prior to the AIDS era, microsporidia were rarely reported in humans. Cell-mediated host immune responses seem to be key in defense against *E. bieneusi*. With the widespread outbreaks of AIDS, a significant number of microsporidiosis cases have been reported in HIV patients, especially in persons with ≤100 CD4+ T cells per μL of blood.[6,13,17,18] Highly activated antiretroviral therapy (HAART) has dramatically reduced opportunistic infections in HIV/AIDS patients, including those caused by microsporidia.[19] However, intestinal microsporidiosis is still a major problem in some parts of the world where access to HAART is limited. Additionally, *E. bieneusi* is being increasingly reported in non-HIV immunosuppressed patients, including the elderly, patients with diabetes, those with malignant disease undergoing chemotherapy, and those undergoing organ transplant.[20–22] The lack of an animal model and *in vitro* culture methods limit the work that can be carried out on *E. bieneusi*. As such, the detailed immunological characteristics and host–organism relationship remain unclear.

33.4 Host Range

E. bieneusi was originally only reported in humans, both immunocompromised and immunocompetent patients. However, many vertebrate animal hosts have now been shown to harbor *E. bieneusi*.[23] There remain a number of questions regarding the epidemiology of *E. bieneusi*, as the major host reservoirs and routes of transmission are not completely understood. Fortunately, recent development of molecular tools has allowed us to identify a wide range of animal hosts.[23,24] Host specificity and zoonotic potential of an organism can be evaluated by genotyping and phylogenetic analysis targeting the internal transcribed spacer of the ribosomal RNA gene.[23] Zoonotic and animal-specific *E. bieneusi* genotypes have been identified from domesticated animals, companion animals, and wildlife, including primates, marmosets, dogs, cats, pigs, cattle, horses, llamas, kudus, foxes, raccoons, otters, guinea pigs, beavers, rabbits, muskrats, falcons, and some other birds.[23] Cats and pigs are common hosts of zoonotic genotypes in Asian countries.[25–28] Phylogenetic analysis of *E. bieneusi* divides the ITS genotypes into eight groups.[1,29] Group 1 is the largest, containing all human-specific, zoonotic, and animal-specific genotypes. Groups 2–8 are mostly only found in specific hosts and wastewater.[1]

33.5 Transmission

Transmission routes of *E. bieneusi* are not fully elucidated. In humans, vertical transmission has not been reported, and the fecal–oral route is considered the main route of transmission. A report of *E. bieneusi* in the respiratory tract also suggests the possibility of airborne transmission.[30,31] Additionally, the spore enables the organism to survive in the environment for long periods of time, contributing to both water- and foodborne transmission. Risk factors associated with human microsporidiosis are homosexual practices, intravenous drug use, and water contact.[32,33] As summarized in Table 33.1, a foodborne outbreak in Sweden was attributed to contaminated cucumbers.[34] *E. bieneusi* spores also have been identified on other fresh food products and in milk.[35,36] Waterborne transmission is frequently reported, with *E. bieneusi* being isolated from surface water, waste water, irrigation water, swimming pools, and river water.[25,37–49] In 1999, there was a confirmed waterborne microsporidium outbreak associated with a water distribution system.[50] Numerous genotypes of *E. bieneusi* have also been identified from animals.[23] Cama and

TABLE 33.1

Enterocytozoon bieneusi Detected in Food

Produce Type	Country	Detection Method	Genotype	References
Raspberries, bean sprouts, lettuce	Poland	Microscopy (chromotrope 2R, calcofluor white), FISH	Unknown	35
Cucumbers in salad and sandwich	Sweden	PCR	C	34
Milk	Korea	PCR	D, I, J, CEbD, type IV	36

PCR, polymerase chain reaction; FISH, fluorescence *in situ* hybridization.

colleagues reported the transmission of *E. bieneusi* between a child and pet guinea pigs,[51] whereby an unusual genotype was identified in both the guinea pigs and a 2-year-old child in the same household.

33.6 Clinical Symptoms and Pathology

E. bieneusi causes intestinal microsporidiosis. Clinical symptoms include watery and nonbloody diarrhea, nausea, abdominal pain, and fever. In immunocompetent patients, the diarrhea is usually self-limiting, with symptoms lasting approximately 1 month or less.[52] However, the diarrhea can be persistent and severe in immunocompromised patients.[53] Respiratory infection, cholecystitis, and cholangitis are reported in rare cases,[53,54] and *E. bieneusi* has been linked to malnutrition in children.[55] Dissemination is usually not seen with *E. bieneusi* infection, distinguishing it from infection caused by *Encephalitozoon* species.

In humans, *E. bieneusi* is usually found in the small intestine, especially in the distal duodenum and proximal jejunum. The organism can also be found in the ileum, but is rarely present in the colon. The infection induces villous blunting and crypt hyperplasia, but does not induce ulceration. It also reduces mucosal surface area, leading to malabsorption.[56] *E. bieneusi* spores and proliferating forms are found on the apical surface of cells, but not on the basal surface.[57,58]

33.7 Techniques for Detection and Characterization

Several techniques for detection of *E. bieneusi* are now available. Stool, duodenal drainage, and biopsy samples are used for diagnosis. The various detection techniques are described below.

33.7.1 Light Microscopy

Microsporidia spores are difficult to identify in stool specimens by light microscopy because they are very small (1–2 μm) and can look similar to bacteria. Chromotrope-based techniques, initially described by Weber and colleagues,[59] and then modified and further improved by other groups,[60–63] are commonly applied. Using a 5-min rapid-hot Gram-chromotrope staining method, the microsporidial spores stain dark violet against a pale-green background, and a characteristic purple belt-like strip or bar in the middle or at the end of the body is enhanced. Microsporidian spores can also be visualized by ultraviolet microscopy using chemofluorescent stains such as Calcofluor White 2MR, Fungi-Fluor, or Uvitex 2B, which bind chitin in the endospore.[64] However, these types of stains are nonspecific and readily interact with some bacteria, small fungi, and artifactual materials, resulting in false-positive results.[1] Giemsa stains microsporidia a light blue color, but the organisms are difficult to differentiate from other stool elements, meaning this stain may only be of use in intestinal biopsies.[17,65] Histological examination, such as duodenal biopsy, can be used in some clinical situations. Various staining techniques for histological examination have been studied including fluorescent Uvitex 2B stain, Gram-derived stain, silver staining, Giemsa stain, and chromotrope 2R stain.[66–69]

33.7.2 Electron Microscopy

Transmission electron microscopy (TEM) remains the gold standard for genus and species identification of microsporidia. Structural features of the spores, intracellular proliferative forms, method of division, and host–parasite interface are required for diagnosis and differentiation of the species. All developmental stages are observed in infected tissue samples, but not in feces. Therefore, fecal samples alone are not sufficient for definitive classification of microsporidia.[70] The major disadvantages of TEM are low sensitivity, high cost, and laborious sample preparation. TEM is also not suitable for routine diagnosis.[71] The classification of microsporidia by TEM and detailed characteristics of *E. bieneusi* are described elsewhere.[6,72]

33.7.3 Immunofluorescence

Fluorescein-tagged antibodies are useful for detecting pathogens in clinical biopsy samples from humans and animals.[73,74] Immunofluorescence techniques, especially the use of polyclonal antibodies, allow visualization of all *E. bieneusi* development stages, including spores, intracellular developmental stages, and extruded polar tubes, whereas histochemical methods only stain the walls of sporoblasts or spores.[75,76] However, polyclonal immunofluorescence staining frequently cross-reacts with yeast and bacteria in fecal specimens, and the sensitivity of the technique is poor compared with chromotrope or chemofluorescent stains.[77] Monoclonal antibodies against *E. bieneusi* spores do not appear to cross react with *Encephalitozoon* species, but are not available for commercial distribution.[78,79]

33.7.4 *In Vitro* Culture and Animal Models

Although *E. bieneusi* infects a wide range of mammals and birds, it does not naturally infect rodents. Only short-term, self-limiting infections can be induced in severely immunosuppressed mice and rats.[80] *E. bieneusi* can cause intestinal microsporidiosis, especially hepatobiliary infections, in simian immunodeficiency virus-infected macaques, and experimental infections have been established.[81–83] *E. bieneusi* cannot be isolated *in vitro* in continuous tissue culture, with only short-term propagation reported to date.[84–86] In these experiments, adenovirus coinfection reportedly disturbed the short-term cultures.[86] Interestingly, both *E. bieneusi* and *Nucleospora salmonis*, a salmon pathogen previously assigned to the genus *Enterocytozoon*, were successfully cultured for a short time using rainbow trout kidney cells.[85]

33.7.5 Nucleic-Acid-Based Detection

Nucleic-acid-based detection methods are more sensitive and specific than microscopy, and therefore have been widely applied.[24,87] The most commonly used method is polymerase chain reaction (PCR). In PCR, the target pathogen DNA is bound by a specific set of primers, and the original few copies of DNA are amplified across several orders of magnitude, generating millions of copies of a particular DNA sequence. Molecular diagnostic tests for microsporidia are not routinely available in clinical diagnostic laboratories despite being widely used in research settings. PCR diagnosis of *E. bieneusi* was first reported by Zhu and colleagues.[88] PCR methods can also be used for more in-depth analyses, such as genotypic identification at the subspecies level. The primers generally used for the diagnosis of *E. bieneusi* target the small and large subunits and internal transcribed spacer (ITS) region of the rRNA gene. The ITS region in particular has been used in many studies for detecting and genotyping *E. bieneusi* because of the high degree of sequence diversity in this region. Although the ITS sequence remains the gold standard for the analysis of *E. bieneusi*, additional gene markers are being sought and, more recently, a multilocus sequence typing assay targeting three microsatellite and one minisatellite markers has been developed.[23,89]

Real-time PCR detects accumulating amplicons in real time via either nonspecific fluorochrome or specific fluorescently-labeled probes. Real-time PCR has the advantage of being quantitative over a broad dynamic range, but is relatively expensive. A few real-time PCR analyses have been reported for detection of *E. bieneusi*.[90,91]

Several studies have used fluorescent *in situ* hybridization (FISH)-based techniques to detect *E. bieneusi*. FISH is a cytogenetic technique using a fluorescently-labeled probe that binds to complementary nucleic acid (DNA or RNA) in the specimen.[92] It was used successfully to detect *E. bieneusi* from clinical samples, with probes targeting the small subunit or ITS regions of the rRNA.[93,94] FISH has the advantage of providing general morphological information, as the procedure is performed *in situ*. The major disadvantages of FISH are the laborious method and lower sensitivity than PCR.

A recently developed oligonucleotide microarray method can simultaneously detect *E. bieneusi* and *Encephalitozoon* species from clinical samples.[95] An array of target-complementary DNA fragments are spotted on a glass slide, and the nucleic acid from the sample is hybridized to the chip after being fluorescently labeled. This technique is somewhat quantitative; the abundance of the DNA in the sample is correlated with the intensity of fluorescence. Additionally, this method is high throughput.

TABLE 33.2
Enterocytozoon bieneusi Detected in Water

Water Source	Country	Detection Method	Genotype	References
River	France	PCR	Unknown	41
Surface water, ground water	USA	PCR	Unknown	37
River	Korea	Uvitex 2B, PCR	Unknown	42
River	Ireland	FISH, PCR	Unknown	43
Lake	France	PCR	Unknown	44
Recreational water	USA	FISH, PCR	Unknown	45
Surface inland and coastal waters	Ireland	FISH	Unknown	104
Wetlands	Ireland	FISH, PCR	K	46
Waste water	Ireland	FISH	Unknown	47
Waste water	China	PCR	Type IV, EbpA, EbpC, EbpD, Peru6, Peru8, Peru10, C, D, BEB6, PtEb IV, PigEBITS7, PigEBITS8, WL4, WL12, WL14, WW1–WW9	40
Waste water	Tunisia	PCR	D, IV, etc.	39
Storm water	USA	PCR	WL4, WL6, SW1 to SW3	48
Drinking water, waste water	Spain	Chromotrope stain, PCR	C, D, and D-like	49
River water	China	PCR	EbpA, EbpB, EbpC, D, CS-8, PtEb IX, Peru8, Peru11, PigEBITS4, G, O, RWSH1 to RWSH6	38

PCR, polymerase chain reaction; FISH, fluorescence *in situ* hybridization.

33.7.6 Detection from Environmental Samples

Microscopy or molecular techniques are generally used for detecting *E. bieneusi* from water samples after filtration and purification steps. There are no standard methods to purify and enrich microsporidia from water samples, although a few methods have been reported, including continuous separation channel centrifugation and continuous flow centrifugation.[96] The small size of the organism decreases the efficacy during the filtering of large volumes of water, as the smaller filter size increases the chance of clogging (Table 33.2).

33.8 Treatment

Albendazole, a benzimidazole drug used to treat of a variety of parasitic infections, is highly active against *Encephalitozoo*n species, but has shown only limited efficacy against *E. bieneusi.* When albendazole is used in patients infected with *E. bieneusi*, diarrhea may improve in some patients, but the excretion of the organism continues and diarrhea exacerbates rapidly after discontinuation of the drug.[97–99] Albendazole works by binding to β-tubulin, and variations in the amino acid sequence of *E. bieneusi* β-tubulin may be related to clinical resistance.[100] Fumagillin, originally isolated from *Aspergillus fumigatus,* is an antimicrobial agent that is effective against *E. bieneusi* and *Entamoeba histolytica*. Fumagillin has successfully treated *E. bieneusi* infection in AIDS and transplant patients.[101,102] The main toxicity is thrombocytopenia, which is reversible after cessation of treatment. However, more severe side effects, such as aseptic meningoencephalitis, have also been reported.[103]

33.9 Conclusion and Perspectives

E. bieneusi is the most common intestinal pathogen causing human microsporidiosis. Diarrhea is a major symptom of *E. bieneusi* infection in humans and can be severe and persistent in immunocompromised patients. *E. bieneusi* is characterized by the presence of a polar tube, which plays a central role in pathogenicity by injecting infective material into the host cell. The spore, which is the dormant form of the cell, is resistant to most environmental stresses and can survive outside of the host for many years. The diagnosis of *E. bieneusi* is difficult by conventional morphological methods because of its small size; therefore, molecular techniques have been widely used. However, the epidemiology and mode of transmission are not completely understood. Fecal–oral infection is thought to be the main route of transmission, and zoonotic transmission may be responsible in certain cases. An effective treatment for *E. bieneusi* has not yet been established. The lack of an animal model and *in vitro* culture methods limits the investigation of immunology, pathophysiology, and development of treatments for *E. bieneusi*. Establishing an animal model and *in vitro* culture methods will help us gain a better understanding of the transmission dynamics and help develop a comprehensive epidemiological picture for *E. bieneusi* infection.

REFERENCES

1. Thellier, M. & Breton, J. *Enterocytozoon bieneusi* in human and animals, focus on laboratory identification and molecular epidemiology. *Parasite* 15, 349–58 (2008).
2. Szumowski, S.C. & Troemel, E.R. Microsporidia-host interactions. *Curr Opin Microbiol* 26, 10–16 (2015).
3. Didier, E.S. Microsporidiosis: an emerging and opportunistic infection in humans and animals. *Acta Trop* 94, 61–76 (2005).
4. Vivares, C.P., Gouy, M., Thomarat, F. & Metenier, G. Functional and evolutionary analysis of a eukaryotic parasitic genome. *Curr Opin Microbiol* 5, 499–505 (2002).
5. Keeling, P.J. & Fast, N.M. Microsporidia: biology and evolution of highly reduced intracellular parasites. *Annu Rev Microbiol* 56, 93–116 (2002).
6. Desportes, I. et al. Occurrence of a new microsporidan: *Enterocytozoon bieneusi* n. g., n. sp., in the enterocytes of a human patient with AIDS. *J Protozool* 32, 250–4 (1985).
7. Tourtip, S. et al. *Enterocytozoon hepatopenaei* sp. nov. (Microsporida: Enterocytozoonidae), a parasite of the black tiger shrimp *Penaeus monodon* (Decapoda: Penaeidae): fine structure and phylogenetic relationships. *J Invertebr Pathol* 102, 21–9 (2009).
8. Tangprasittipap, A. et al. The microsporidian *Enterocytozoon hepatopenaei* is not the cause of white feces syndrome in whiteleg shrimp *Penaeus (Litopenaeus) vannamei. BMC Vet Res* 9, 139 (2013).
9. Cali, A. General microsporidian features and recent findings on AIDS isolates. *J Protozool* 38, 625–30 (1991).
10. Weiss, L.M. Microsporidia: emerging pathogenic protists. *Acta Trop* 78, 89–102 (2001).
11. Tumwine, J.K. et al. *Enterocytozoon bieneusi* among children with diarrhea attending Mulago Hospital in Uganda. *Am J Trop Med Hyg* 67, 299–303 (2002).
12. Didier, E.S. & Weiss, L.M. Microsporidiosis: current status. *Curr Opin Infect Dis* 19, 485–92 (2006).
13. Orenstein, J.M. Microsporidiosis in the acquired immunodeficiency syndrome. *J Parasitol* 77, 843–64 (1991).
14. Vavra, J. & Lukes, J. Microsporidia and 'the art of living together'. *Adv Parasitol* 82, 253–319 (2013).
15. Weidner, E. The microsporidian spore invasion tube. III. Tube extrusion and assembly. *J Cell Biol* 93, 976–9 (1982).
16. Keeling, P.J. et al. The reduced genome of the parasitic microsporidian *Enterocytozoon bieneusi* lacks genes for core carbon metabolism. *Genome Biol Evol* 2, 304–9 (2010).
17. van Gool, T. et al. Diagnosis of *Enterocytozoon bieneusi* microsporidiosis in AIDS patients by recovery of spores from faeces. *Lancet* 336, 697–8 (1990).
18. van Gool, T. et al. High prevalence of *Enterocytozoon bieneusi* infections among HIV-positive individuals with persistent diarrhoea in Harare, Zimbabwe. *Trans R Soc Trop Med Hyg* 89, 478–80 (1995).

19. Stark, D. et al. Limited genetic diversity among genotypes of *Enterocytozoon bieneusi* strains isolated from HIV-infected patients from Sydney, Australia. *J Med Microbiol* 58, 355–7 (2009).
20. Lono, A.R., Kumar, S. & Chye, T.T. Incidence of microsporidia in cancer patients. *J Gastrointest Cancer* 39, 124–9 (2008).
21. Antonios, S.N., Tolba, O.A., Othman, A.A. & Saad, M.A. A preliminary study on the prevalence of parasitic infections in immunocompromised children. *J Egypt Soc Parasitol* 40, 617–30 (2010).
22. Sak, B. et al. Unapparent microsporidial infection among immunocompetent humans in the Czech Republic. *J Clin Microbiol* 49, 1064–70 (2011).
23. Santin, M. & Fayer, R. Microsporidiosis: *Enterocytozoon bieneusi* in domesticated and wild animals. *Res Vet Sci* 90, 363–71 (2011).
24. Fedorko, D.P., Nelson, N.A. & Cartwright, C.P. Identification of microsporidia in stool specimens by using PCR and restriction endonucleases. *J Clin Microbiol* 33, 1739–41 (1995).
25. Li, W. et al. High diversity of human-pathogenic *Enterocytozoon bieneusi* genotypes in swine in northeast China. *Parasitol Res* 113, 1147–53 (2014).
26. Li, W. et al. Genotypes of *Enterocytozoon bieneusi* in livestock in China: high prevalence and zoonotic potential. *PLoS One* 9, e97623 (2014).
27. Mori, H. et al. Presence of zoonotic *Enterocytozoon bieneusi* in cats in a temple in central Thailand. *Vet Parasitol* 197, 696–701 (2013).
28. Karim, M.R. et al. Genetic diversity in *Enterocytozoon bieneusi* isolates from dogs and cats in China: host specificity and public health implications. *J Clin Microbiol* 52, 3297–302 (2014).
29. Karim, M.R. et al. Genetic polymorphism and zoonotic potential of *Enterocytozoon bieneusi* from nonhuman primates in China. *Appl Environ Microbiol* 80, 1893–8 (2014).
30. Didier, E.S. et al. Epidemiology of microsporidiosis: sources and modes of transmission. *Vet Parasitol* 126, 145–66 (2004).
31. Weber, R. et al. Pulmonary and intestinal microsporidiosis in a patient with the acquired immunodeficiency syndrome. *Am Rev Respir Dis* 146, 1603–5 (1992).
32. Hutin, Y.J. et al. Risk factors for intestinal microsporidiosis in patients with human immunodeficiency virus infection: a case-control study. *J Infect Dis* 178, 904–7 (1998).
33. Dascomb, K., Frazer, T., Clark, R.A., Kissinger, P. & Didier, E. Microsporidiosis and HIV. *J Acquir Immune Defic Syndr* 24, 290–2 (2000).
34. Decraene, V., Lebbad, M., Botero-Kleiven, S., Gustavsson, A.M. & Lofdahl, M. First reported foodborne outbreak associated with microsporidia, Sweden, October 2009. *Epidemiol Infect* 140, 519–27 (2012).
35. Jedrzejewski, S., Graczyk, T.K., Slodkowicz-Kowalska, A., Tamang, L. & Majewska, A.C. Quantitative assessment of contamination of fresh food produce of various retail types by human-virulent microsporidian spores. *Appl Environ Microbiol* 73, 4071–3 (2007).
36. Lee, J.H. Molecular detection of *Enterocytozoon bieneusi* and identification of a potentially human-pathogenic genotype in milk. *Appl Environ Microbiol* 74, 1664–6 (2008).
37. Dowd, S.E., Gerba, C.P. & Pepper, I.L. Confirmation of the human-pathogenic microsporidia *Enterocytozoon bieneusi*, *Encephalitozoon intestinalis*, and *Vittaforma corneae* in water. *Appl Environ Microbiol* 64, 3332–5 (1998).
38. Hu, Y., Feng, Y., Huang, C. & Xiao, L. Occurrence, source, and human infection potential of *Cryptosporidium* and *Enterocytozoon bieneusi* in drinking source water in Shanghai, China, during a pig carcass disposal incident. *Environ Sci Technol* 48, 14219–27 (2014).
39. Ben Ayed, L. et al. Survey and genetic characterization of wastewater in Tunisia for *Cryptosporidium* spp., *Giardia duodenalis*, *Enterocytozoon bieneusi*, *Cyclospora cayetanensis* and *Eimeria* spp. *J Water Health* 10, 431–44 (2012).
40. Li, N. et al. Molecular surveillance of *Cryptosporidium* spp., *Giardia duodenalis*, and *Enterocytozoon bieneusi* by genotyping and subtyping parasites in wastewater. *PLoS Negl Trop Dis* 6, e1809 (2012).
41. Sparfel, J.M. et al. Detection of microsporidia and identification of *Enterocytozoon bieneusi* in surface water by filtration followed by specific PCR. *J Eukaryot Microbiol* 44, 78S (1997).
42. Fournier, S. et al. Detection of microsporidia in surface water: a one-year follow-up study. *FEMS Immunol Med Microbiol* 29, 95–100 (2000).
43. Graczyk, T.K. et al. Human waterborne parasites in zebra mussels (*Dreissena polymorpha*) from the Shannon River drainage area, Ireland. *Parasitol Res* 93, 385–91 (2004).

44. Coupe, S. et al. Detection of *Cryptosporidium*, *Giardia* and *Enterocytozoon bieneusi* in surface water, including recreational areas: a one-year prospective study. *FEMS Immunol Med Microbiol* 47, 351–9 (2006).
45. Graczyk, T.K., Sunderland, D., Tamang, L., Lucy, F.E. & Breysse, P.N. Bather density and levels of *Cryptosporidium*, *Giardia*, and pathogenic microsporidian spores in recreational bathing water. *Parasitol Res* 101, 1729–31 (2007).
46. Graczyk, T.K. et al. Human zoonotic enteropathogens in a constructed free-surface flow wetland. *Parasitol Res* 105, 423–8 (2009).
47. Cheng, H.W., Lucy, F.E., Graczyk, T.K., Broaders, M.A. & Mastitsky, S.E. Municipal wastewater treatment plants as removal systems and environmental sources of human-virulent microsporidian spores. *Parasitol Res* 109, 595–603 (2011).
48. Guo, Y. et al. Host specificity and source of *Enterocytozoon bieneusi* genotypes in a drinking source watershed. *Appl Environ Microbiol* 80, 218–25 (2014).
49. Galvan, A.L. et al. Molecular characterization of human-pathogenic microsporidia and *Cyclospora cayetanensis* isolated from various water sources in Spain: a year-long longitudinal study. *Appl Environ Microbiol* 79, 449–59 (2013).
50. Cotte, L. et al. Waterborne outbreak of intestinal microsporidiosis in persons with and without human immunodeficiency virus infection. *J Infect Dis* 180, 2003–8 (1999).
51. Cama, V.A. et al. Transmission of *Enterocytozoon bieneusi* between a child and guinea pigs. *J Clin Microbiol* 45, 2708–10 (2007).
52. Sandfort, J. et al. *Enterocytozoon bieneusi* infection in an immunocompetent patient who had acute diarrhea and who was not infected with the human immunodeficiency virus. *Clin Infect Dis* 19, 514–6 (1994).
53. Kotler, D.P. & Orenstein, J.M. Clinical syndromes associated with microsporidiosis. *Adv Parasitol* 40, 321–49 (1998).
54. Remadi, S., Dumais, J., Wafa, K. & MacGee, W. Pulmonary microsporidiosis in a patient with the acquired immunodeficiency syndrome. A case report. *Acta Cytol* 39, 1112–6 (1995).
55. Mor, S.M., Tumwine, J.K., Naumova, E.N., Ndeezi, G. & Tzipori, S. Microsporidiosis and malnutrition in children with persistent diarrhea, Uganda. *Emerg Infect Dis* 15, 49–52 (2009).
56. Orenstein, J.M., Tenner, M. & Kotler, D.P. Localization of infection by the microsporidian *Enterocytozoon bieneusi* in the gastrointestinal tract of AIDS patients with diarrhea. *AIDS* 6, 195–7 (1992).
57. Schwartz, D.A. et al. Enteric Opportunistic Infections Working Group. The presence of *Enterocytozoon bieneusi* spores in the lamina propria of small bowel biopsies with no evidence of disseminated microsporidiosis. *Arch Pathol Lab Med* 119, 424–8 (1995).
58. Schwartz, D.A., Sobottka, I., Leitch, G.J., Cali, A. & Visvesvara, G.S. Pathology of microsporidiosis: emerging parasitic infections in patients with acquired immunodeficiency syndrome. *Arch Pathol Lab Med* 120, 173–88 (1996).
59. Weber, R. et al. The Enteric Opportunistic Infections Working Group. Improved light-microscopical detection of microsporidia spores in stool and duodenal aspirates. *N Engl J Med* 326, 161–6 (1992).
60. Kokoskin, E. et al. Modified technique for efficient detection of microsporidia. *J Clin Microbiol* 32, 1074–5 (1994).
61. Ryan, N.J. et al. A new trichrome-blue stain for detection of microsporidial species in urine, stool, and nasopharyngeal specimens. *J Clin Microbiol* 31, 3264–9 (1993).
62. Ignatius, R. et al. A new acid-fast trichrome stain for simultaneous detection of *Cryptosporidium parvum* and microsporidial species in stool specimens. *J Clin Microbiol* 35, 446–9 (1997).
63. Moura, H. et al. A new and improved "quick-hot Gram-chromotrope" technique that differentially stains microsporidian spores in clinical samples, including paraffin-embedded tissue sections. *Arch Pathol Lab Med* 121, 888–93 (1997).
64. van Gool, T. et al. Diagnosis of intestinal and disseminated microsporidial infections in patients with HIV by a new rapid fluorescence technique. *J Clin Pathol* 46, 694–9 (1993).
65. Rijpstra, A.C., Canning, E.U., Van Ketel, R.J., Eeftinck Schattenkerk, J.K. & Laarman, J.J. Use of light microscopy to diagnose small-intestinal microsporidiosis in patients with AIDS. *J Infect Dis* 157, 827–31 (1988).
66. Conteas, C.N. et al. Fluorescence techniques for diagnosing intestinal microsporidiosis in stool, enteric fluid, and biopsy specimens from acquired immunodeficiency syndrome patients with chronic diarrhea. *Arch Pathol Lab Med* 120, 847–53 (1996).

67. Field, A.S., Marriott, D.J. & Hing, M.C. The Warthin-Starry stain in the diagnosis of small intestinal microsporidiosis in HIV-infected patients. *Folia Parasitol (Praha)* 40, 261–6 (1993).
68. Franzen, C. et al. Tissue diagnosis of intestinal microsporidiosis using a fluorescent stain with Uvitex 2B. *J Clin Pathol* 48, 1009–10 (1995).
69. Kotler, D.P., Giang, T.T., Garro, M.L. & Orenstein, J.M. Light microscopic diagnosis of microsporidiosis in patients with AIDS. *Am J Gastroenterol* 89, 540–4 (1994).
70. Didier, E.S. et al. Identification and characterization of three *Encephalitozoon cuniculi* strains. *Parasitology* 111 (Pt 4), 411–21 (1995).
71. Field, A.S. Light microscopic and electron microscopic diagnosis of gastrointestinal opportunistic infections in HIV-positive patients. *Pathology* 34, 21–35 (2002).
72. Cali, A. & Owen, R.L. Intracellular development of *Enterocytozoon*, a unique microsporidian found in the intestine of AIDS patients. *J Protozool* 37, 145–55 (1990).
73. Weber, R., Bryan, R.T., Schwartz, D.A. & Owen, R.L. Human microsporidial infections. *Clin Microbiol Rev* 7, 426–61 (1994).
74. Weiss, L.M. & Vossbrinck, C.R. Microsporidiosis: molecular and diagnostic aspects. *Adv Parasitol* 40, 351–95 (1998).
75. Sheoran, A.S. et al. Purification of *Enterocytozoon bieneusi* from stools and production of specific antibodies. *J Clin Microbiol* 43, 387–92 (2005).
76. Sheoran, A.S. et al. Monoclonal antibodies against *Enterocytozoon bieneusi* of human origin. *Clin Diagn Lab Immunol* 12, 1109–13 (2005).
77. Garcia, L.S., Shimizu, R.Y. & Bruckner, D.A. Detection of microsporidial spores in fecal specimens from patients diagnosed with cryptosporidiosis. *J Clin Microbiol* 32, 1739–41 (1994).
78. Accoceberry, I. et al. Production of monoclonal antibodies directed against the microsporidium *Enterocytozoon bieneusi*. *J Clin Microbiol* 37, 4107–12 (1999).
79. Zhang, Q. et al. Production and characterization of monoclonal antibodies against *Enterocytozoon bieneusi* purified from rhesus macaques. *Infect Immun* 73, 5166–72 (2005).
80. Feng, X. et al. Serial propagation of the microsporidian *Enterocytozoon bieneusi* of human origin in immunocompromised rodents. *Infect Immun* 74, 4424–9 (2006).
81. Tzipori, S. et al. Transmission and establishment of a persistent infection of *Enterocytozoon bieneusi*, derived from a human with AIDS, in simian immunodeficiency virus-infected rhesus monkeys. *J Infect Dis* 175, 1016–20 (1997).
82. Sestak, K. et al. Quantitative evaluation of *Enterocytozoon bieneusi* infection in simian immunodeficiency virus-infected rhesus monkeys. *J Med Primatol* 32, 74–81 (2003).
83. Mansfield, K.G. et al. Localization of persistent *Enterocytozoon bieneusi* infection in normal rhesus macaques (*Macaca mulatta*) to the hepatobiliary tree. *J Clin Microbiol* 36, 2336–8 (1998).
84. Visvesvara, G. et al. Short-term in vitro culture and molecular analysis of the microsporidian, *Enterocytozoon bieneusi*. *J Eukaryot Microbiol* 42, 506–10 (1995).
85. Desportes-Livage, I. et al. Comparative development of two microsporidian species: *Enterocytozoon bieneusi* and *Enterocytozoon salmonis*, reported in AIDS patients and salmonid fish, respectively. *J Eukaryot Microbiol* 43, 49–60 (1996).
86. Visvesvara, G.S. et al. Adenovirus masquerading as microsporidia. *J Parasitol* 82, 316–9 (1996).
87. Franzen, C. & Muller, A. Molecular techniques for detection, species differentiation, and phylogenetic analysis of microsporidia. *Clin Microbiol Rev* 12, 243–85 (1999).
88. Zhu, X. et al. Small subunit rRNA sequence of *Enterocytozoon bieneusi* and its potential diagnostic role with use of the polymerase chain reaction. *J Infect Dis* 168, 1570–5 (1993).
89. Karim, M.R. et al. Multilocus sequence typing of *Enterocytozoon bieneusi* in nonhuman primates in China. *Vet Parasitol* 200, 13–23 (2014).
90. Wumba, R. et al. *Enterocytozoon bieneusi* identification using real-time polymerase chain reaction and restriction fragment length polymorphism in HIV-infected humans from Kinshasa Province of the Democratic Republic of Congo. *J Parasitol Res* 2012, 278028 (2012).
91. Verweij, J.J., Ten Hove, R., Brienen, E.A. & van Lieshout, L. Multiplex detection of *Enterocytozoon bieneusi* and *Encephalitozoon* spp. in fecal samples using real-time PCR. *Diagn Microbiol Infect Dis* 57, 163–7 (2007).
92. Langer-Safer, P.R., Levine, M. & Ward, D.C. Immunological method for mapping genes on *Drosophila* polytene chromosomes. *Proc Natl Acad Sci USA* 79, 4381–5 (1982).

93. Carville, A. et al. Development and application of genetic probes for detection of *Enterocytozoon bieneusi* in formalin-fixed stools and in intestinal biopsy specimens from infected patients. *Clin Diagn Lab Immunol* 4, 405–8 (1997).
94. Velasquez, J.N. et al. In situ hybridization: a molecular approach for the diagnosis of the microsporidian parasite *Enterocytozoon bieneusi*. *Hum Pathol* 30, 54–8 (1999).
95. Wang, Z., Orlandi, P.A. & Stenger, D.A. Simultaneous detection of four human pathogenic microsporidian species from clinical samples by oligonucleotide microarray. *J Clin Microbiol* 43, 4121–8 (2005).
96. Borchardt, M.A. & Spencer, S.K. Concentration of *Cryptosporidium*, microsporidia and other waterborne pathogens by continuous separation channel centrifugation. *J Appl Microbiol* 92, 649–56 (2002).
97. Blanshard, C., Ellis, D.S., Tovey, D.G., Dowell, S. & Gazzard, B.G. Treatment of intestinal microsporidiosis with albendazole in patients with AIDS. *AIDS* 6, 311–3 (1992).
98. Dieterich, D.T., Lew, E.A., Kotler, D.P., Poles, M.A. & Orenstein, J.M. Treatment with albendazole for intestinal disease due to *Enterocytozoon bieneusi* in patients with AIDS. *J Infect Dis* 169, 178–83 (1994).
99. Leder, K., Ryan, N., Spelman, D. & Crowe, S.M. Microsporidial disease in HIV-infected patients: a report of 42 patients and review of the literature. *Scand J Infect Dis* 30, 331–8 (1998).
100. Akiyoshi, D.E. et al. Analysis of the β-tubulin genes from *Enterocytozoon bieneusi* isolates from a human and rhesus macaque. *J Eukaryot Microbiol* 54, 38–41 (2007).
101. Champion, L. et al. Fumagillin for treatment of intestinal microsporidiosis in renal transplant recipients. *Am J Transplant* 10, 1925–30 (2010).
102. Molina, J.M. et al. Fumagillin treatment of intestinal microsporidiosis. *N Engl J Med* 346, 1963–9 (2002).
103. Audemard, A. et al. Fumagillin-induced aseptic meningoencephalitis in a kidney transplant recipient with microsporidiosis. *Transpl Infect Dis* 14, E147–9 (2012).
104. Lucy, F.E., Graczyk, T.K., Tamang, L., Miraflor, A. & Minchin, D. Biomonitoring of surface and coastal water for *Cryptosporidium*, *Giardia*, and human-virulent microsporidia using molluscan shellfish. *Parasitol Res* 103, 1369–75 (2008).

34

Fusarium

Palanisamy Manikandan, Coimbatore Subramanian Shobana, Mónika Homa, Sándor Kocsubé, János Varga†, Muthusamy Chandrasekaran, Naiyf S. Alharbi, Venkatapathy Narendran, Csaba Vágvölgyi, and László Kredics

CONTENTS

34.1 Introduction

The members of the genus *Fusarium* are hyaline filamentous fungi and are found largely as saprophytic organisms in soil. Fusaria cause a range of infections collectively known as fusariosis and have been documented as etiological agents in localized tissue infections, keratitis, endophthalmitis, septic arthritis, cystitis, peritonitis, brain abscesses, and breast abscess. The mycotoxins of these fungi are involved in the infectious processes and may serve as potential virulence factors. *Fusarium* is also one of the major fungal genera associated with maize and other cereals throughout the world. Several species are the most prolific producers of mycotoxins and are frequently associated with mycotoxicoses in both humans and animals. This chapter intends to provide an overview about the rodents (rats and mice) and mammalian cell lines that were most recently used as laboratory models to study *Fusarium* mycotoxicoses.

34.1.1 The Genus *Fusarium*

The members of the genus *Fusarium* were first described by Link [1] in 1809 as "*Fusisporium*." The name derives from the Latin word *fusus* (spindle) and refers to the typical macroconidial shape of *Fusarium* species. According to the latest molecular studies, this cosmopolitan genus comprises at least 20 species complexes and nine monotypic lineages including soil saprophytes and plant endophytes along with agriculturally and medically important fungi [2].

Fusaria are among the most important plant pathogens in the world. Just the *Fusarium solani* species complex (FSSC) alone affects more than 100 plant genera [3]. *Fusarium* spp. are responsible for a variety

† In memoriam Prof. János Varga

of devastating diseases, e.g., head blight of wheat and barley, wilt of bananas [4], sudden death syndrome of soybean [5], crown and root rot of tomato [6], and vascular wilt of tomato [7].

Fusarium spp. are also associated with a broad spectrum of human infections, including superficial (i.e., keratitis and onychomycosis), invasive localized, and disseminated infections, allergic diseases (i.e., sinusitis), and mycotoxicosis [8]. Out of the 20 species complexes of the genus, seven comprise clinically relevant Fusaria: the *F. chlamydosporum* species complex (FCSC), the *Fusarium dimerum* species complex (FDSC), the *F. fujikuroi* species complex (FFSC), the *F. incarnatum-equiseti* species complex (FIESC), the *F. oxysporum* species complex (FOSC), the *F. sambucinum* species complex (FSAMSC, including *Fusarium sporotrichioides*), and the FSSC. Among them, the FSSC—which comprises at least 60 haplotypes—is the most common group, accounting for 50%–60% of all fusarioses worldwide [9,10].

Certain *Fusarium* species are also able to produce mycotoxins, which may contribute to plant and human pathogenesis [11]. Besides the direct negative impact of fungal infections on crop yield, the mycotoxin contamination of animal feeds also causes significant financial loss and damage to agriculture worldwide. The major *Fusarium* mycotoxins frequently occurring in cereal grains and animal feeds are fumonisins, trichothecenes, and zearalenone.

34.1.2 Mycotoxins of *Fusarium* Species and Mycotoxicoses

34.1.2.1 Zearalenone

Zearalenone (ZEA, F-2 toxin) is a resorcyclic acid lactone mycotoxin. Its name was created as a combination of "*zea*" (for the main producer's name, *G. zeae*), "*-ral*" (for the initials of *r*esorcyclic *a*cid *l*actone), "*-ene*" (for the presence of a double bond at C1′-2′), and "*-one*" (for the ketone group at C6′) [12].

ZEA is produced by various *Fusarium* species commonly occurring on cereal grains, e.g., *F. graminearum* (teleomorph: *Gibberella zeae*), *F. culmorum*, *F. equiseti*, *F. cerealis* (synonym: *F. crookwellense*), *F. avenaceum*, *F. tricinctum*, and *F. oxysporum* [13,14]. ZEA is a frequent field contaminant of maize, barley, oats, rice, rye, soybeans, and wheat; however, poor storage conditions may contribute to its postharvest production as well [15,16].

It is assumed that ZEA has a role in sexual reproduction of the producing fungi [17], and it also affects the growth, development, and photosynthetic apparatus of the contaminated plant [18]. The nonsteroidal, estrogen-like structure also allows the toxin and its metabolites (α-zearalenol and β-zearalenol) to compete with mammalian endogenous estrogens for specific binding sites on their receptors. This estrogenic activity of ZEA causes hyperestrogenism and fertility disorders in farm animals after consumption of a high dose of affected crops (e.g., >0.25 mg/kg ZEA in maize) [14,19]. Pigs—particularly the prepubertal females—are the most sensitive to the higher concentrations of ZEA, whereas ruminants and poultry are reported as less sensitive [14]. Piglets can be affected as well, as ZEA can be excreted into sows' milk [19]. ZEA also represents a potential risk to human health. It is assumed that ZEA exposure is associated with precocious pubertal development in girls [15,20], and it was also mentioned as a possible causative agent of cervical cancer [21]. Besides its endocrine-disrupting effects on animals and humans, ZEA has been reported to be hepatotoxic, hematotoxic, immunotoxic, and genotoxic [12].

34.1.2.2 Fumonisins

Fumonisins have been discovered in 1988 in South Africa from cultures of the *F. moniliforme* strain MRC 826 (=*F. verticillioides*, teleomorph: *Gibberella moniliformis*) [22]. The chemical structure of this cancer-promoting compound was elucidated in the same by Bezuidenhout et al. [23]. The most important fumonisin-producing Fusaria are the members of section *Liseola,* including the widespread maize pathogens *F. verticillioides* and *F. proliferatum*. Other sections containing fumonisin-producing strains are *Dlaminia, Elegans,* and *Arthrosporiella*. In section *Elegans, F. oxysporum* produces only C-series fumonisins [24].

Fumonisins are polyketide-derived mycotoxins and can be divided into four groups (A, B, C, and P) based on their chemical structure with several isomers and stereomers identified during the past decade [25–28]. The most prevalent and naturally occurring fumonisins belong to type B analogues (FB). In Fusaria, usually FB_1 is predominant, accounting for 70%–80% of the total fumonisin content [29]. The A and B series of fumonisins consist of a 20 carbon atom long backbone, while the C-type fumonisins are 19 carbon atoms long. Fumonisins are sphingosine-analog compounds, which disrupt the biosynthesis of sphingolipids by the inhibition of the ceramide synthase enzyme [30]. The altered sphingolipid metabolism could lead to neural tube defects through the disrupted folate uptake [31]. Ingestion of fumonisin-contaminated feeds is associated with several fatal diseases in domestic animals, for example, equine leukoencephalomalacia (ELEM) and porcine pulmonary edema (PPE). Fumonisins can cause nephrotoxicity, hepatotoxicity, and hepatocarcinogenicity in laboratory animals [32]. Fumonisins are the possible causative agents of esophageal cancer in several countries like China and Transkei in South Africa [33,34]; therefore, fumonisin B_1 is considered as possibly carcinogenic to humans (Group 2B) by the International Agency for Research on Cancer (IARC) [35].

34.1.2.3 Trichothecenes

Fusaria and other mold fungi such as *Mycothecium, Trichoderma, Trichothecium, Stachybotrys, Verticimonosporium,* and *Cephalosporium* produce trichothecenes, which comprise a family of closely related low-molecular-weight, nonvolatile, and relatively water-insoluble compounds known as sesquiterpenoids. They are classified into four types—A, B, C, and D, among which Fusaria produce type A (like T-2 and HT-2 toxin) and type B [like deoxynivalenol (DON) and nivalenol (NIV)] trichothecenes [36]. *Fusarium* species are probably the most cited and among the most prolific trichothecene-producing fungi [37]. Trichothecenes contaminate many field crops across the world, and it is necessary to screen large number of food samples for the presence of these toxic fungal metabolites [38].

Trichothecenes are powerful inhibitors of eukaryotic protein synthesis and are phytotoxic, insecticidal, and toxic to animals, and some are among the most toxic non-nitrogenous compounds known to man. Several are commonly found in cereal grains, and the potential health risk from contaminated animal feed and human food is a major factor in stimulating research into this group of compounds [39–42]. Common manifestations of trichothecene toxicity are depression of immune responses and nausea, sometimes vomiting [43].

Wang et al. [44] reported human toxicosis in China caused by moldy rice contaminated with T-2 toxin of *F. heterosporum* and *F. graminearum*. The chief symptoms were nausea, dizziness, vomiting, chills, abdominal distension, abdominal pain, thoracic stuffiness, and diarrhea. T-2 mycotoxicosis—or "moldy corn disease"—in pigs is characterized by multiple hemorrhages on the serosa of the liver, stomach, and esophagus (at necropsy). T-2 also has an important impact on reproductive performance in pigs [45]. The *in vivo* toxicity of T-2 and H-2 is explicitly described by EFSA Panel on CONTAM [46]. Wu et al. [47] stated that in human cell lines, HT-2 and neosolaniol (NEO) are the major metabolites of T-2 toxin. The T-2 degradation products are less cytotoxic compared to T-2 toxin [48].

DON—or vomitoxin—is one of the most common mycotoxins produced by *F. graminearum* and *F. culmorum* found in grains. When ingested in high doses by agricultural animals, it causes nausea, vomiting, and diarrhea; at lower doses, pigs and other farm animals exhibit weight loss and food refusal [49]. Human or animal intoxications with either DON or NIV are much less likely to be fatal than those with T-2 [50].

34.2 Laboratory Models for Foodborne *Fusarium* Mycotoxicoses

34.2.1 Rodents

Rodents (rats and mice) are the most popular laboratory animal models for the examination of *Fusarium* mycotoxins; therefore, a broad search strategy was applied to retrieve full-text papers and—in cases when full-text access was not available—abstracts from the PubMed database. The search was performed in article titles and abstracts using the following keyword combinations in order to find specific literature: zearalenone OR fumonisin OR T-2 toxin OR deoxynivalenol AND rat OR mouse OR mice. The search was limited to publications from the year 2010 onwards in order to focus on recent literature. Articles and abstracts were screened to include studies with relevant information (exact definition of the mycotoxin studied and the animal model, details about the way of administration, and the dosage).

Table 34.1 gives an overview of the analyses of *Fusarium* mycotoxins utilizing rat models [51–74]. Out of 24 publications analyzed (1 DON, 10 T-2, 7 ZEA, and 6 FB_1), it was found that Sprague Dawley rats were mostly used, followed by Wistar rats. There is only one study included for DON [58] in 8-week-old axenic male Sprague Dawley rats, where the concentration of *E. coli* bacteria decreased in the gut at day 27 after subchronic exposure to DON.

Among the 10 studies analyzed for T-2 toxin, 7 studies utilized Wistar rats to investigate MRI in the tibial bone abnormalities [66,69], chondrocyte necrosis of Kashin Bede disease (KBD) [67], cardiac and autonomic nervous effect [68], reproductive toxicity [70], KBD [69,71,72], and apoptotic induction mechanism [70]. The remaining three studies with Sprague Dawley rats analyzed the effect of T-2 on lipid peroxidation in brain [53,63] and KBD [57].

Out of seven studies analyzing ZEA in rat models, the most popular model animals were Sprague Dawley rats in the cases the analyses were related to the effect of ZEA on the reproductive system [51,52,56,64] and modification of mRNA levels involved in ZEA detoxification [54]. Belli et al. [65] investigated the involvement of ZEA in breast endocrine disorders using Wistar female rats.

Of six studies analyzing the effects of FB_1, four used Sprague Dawley rats for the exploration of serum enzyme activities and DNA lesions [61], gene expression [62], general toxicity after nixtamalization of whole kernel corn [59], and gastric ulcer [60]. Riedel et al. [73] reported the effect of FB_1 in cancer promotion involving lipid changes utilizing Fischer rats. Pellanda et al. [74] investigated the global histone modification due to FB_1 administration in Wistar female and male rats.

Table 34.2 summarizes a total of 51 studies involving mice as animal models to study the possible effects of *Fusarium* mycotoxins viz., DON (22 studies), T-2 (9 studies), ZEA (16 studies), and FB_1 (10) [75–126]. Of 22 mice studies involving DON, B6C3F1 followed by BAL13/3 and C57136/6 mice were the most popular.

B6C3F1 mice were used for the analysis of the effect of DON on immunomodulation [98,99], insulin-like growth factor [93], and diet-related issues [94,95,96]. C57BL/6 mice were used to analyze the effect of DON on inflammation [104], diet issues [102], fetal skeletal development [103], and mouse thymus [100]. With BALB/c mice, the effects of DON were studied on hepatotoxicity [87], immunomodulation [76,81], and hemostability [77]. With the ICR mouse model, DON was analyzed for oocyte development [90,91] and oxidative stress [89]. With Swiss albino mice, Mishra et al. [115] found that topical application of DON increased cell proliferation, DNA synthesis, and inflammation, while Kouadio et al. [112] revealed that the oral administration of DON created disorders in systemic targets. Choi et al. [122] reported that DON in drinking water can disrupt the immune response of porcine parvovirus-vaccinated C3H mice. Behavioral changes due to inflammatory cytokines after DON intoxication in PGES wild-type and knockout mice were observed by Girardet et al. [125].

Chaudhary et al. [108,116] and Agarwal et al. [109] studied the oxidative stress due to T-2, while Agarwal et al. [110] reported the CC-2 formulation's effectiveness against topical exposure to T-2 toxin in Swiss albino mice. Using the BALB/c mouse model, Ahmadi et al. [88] showed that selenium protects the alteration of B lymphocytes after T-2 toxin exposure, while Maragos et al. [82] developed Mab2–13 antibody detection for T-2 and T-2glc. The harmful effect of T-2 toxin in early embryo development was explored by Somoskői et al. [123] in BDF1 mice. Yang et al. [118] demonstrated the decrease in the testosterone biosynthesis due to T-2 toxin in Kunming mice.

BALB/c mice were employed to investigate the hepatotoxicity [79,87] and immunotoxicity [75,78,80] of ZEA. Zhu et al. [90,92] and Hou et al. [91] explained the reduction in oocyte development, while Hou et al. [89] studied the oxidative stress induced by ZEA in ICR mice. Boeira et al. [111,113,114] clearly indicated the impaired testicular functions due to ZEA toxicity among male Swiss albino mice. ZEA was

TABLE 34.1

Studies Using Rat Models for the Examination of *Fusarium* Mycotoxins Published Since 2010 Based on PubMed

Rat Model	MT	Administration/Dose	Aim of the Study	Major Findings	References
10-w-old Sprague Dawley	ZEA	Intraperitoneal, 5 mg/kg	To investigate the effects of ZEA on spermatogenesis and possible mechanisms involved in germ cell injury	ZEA induces apoptosis in germ cells of male rats; this toxicity of ZEA is partially mediated through modulation of Fas and Fas-L systems, though ERα may not play a significant role	[51]
8-w-old male Sprague Dawley	ZEA	Intraperitoneal, 300 mg/kg	To investigate the effects of KRG extract on testicular toxicity induced by ZEA	Impaired spermatogenesis resulting from ZEA treatment was prevented by KRG through Fas-Fas L modulation	[52]
1-month-old male Sprague Dawley	T-2	Intragastric, 100 and 200 ng/g b.w./d	To compare antioxidant capacity and lipid peroxidation using a novel model	Increasing TBARS and decreasing antioxidants in serum and cartilage by T-2 toxin treatment with a selenium-deficient nutritional status may alter oxidative stress in joint tissues and contribute to the pathological process of cartilage damage in KBD	[53]
Male Sprague Dawley	ZEA	Intraperitoneal, 25 mg/kg b.w.	To determine the levels of expression of rat proteins that are involved in the ZEA detoxification pathway upon acute ZEA treatment	The initial modifications in mRNA levels indicate a close association with microsomal enzyme activity of the CYP2B, CYP2C, and CYP3A protein families	[54]
6-w-old female Sprague Dawley	ZEA	Intragastric, 3 mg/kg	To investigate the effect of ZEA supplementation on rat metabolism	ZEA exposure can cause oxidative stress and change common systemic metabolic processes, including cell membrane metabolism, protein biosynthesis, glycolysis, and gut microbiota metabolism	[55]
Pregnant Sprague Dawley	ZEA	Subcutaneous, 0, 1, 50, or 100 and 0, 0.75, 1.25, or 2.5 μg/d	To assess the impact of ZEA in adult rats exposed neonatally	ZEA neonatal exposure could affect the exposure of testis to ABC transporter substrates and negatively influence spermatogenesis and male fertility	[56]
1-month-old male Sprague Dawley	T-2	Intragastric, 100 and 200 ng/g b.w./d	To observe pathogenic lesions of joint cartilages in rats fed with T-2 toxin under a selenium-deficiency nutrition status in order to determine possible etiological factors causing KBD	Rat can be used as a suitable animal model for studying etiological factors contributing to the pathogenesis (chondronecrosis) observed in human KBD	[57]
8-w-old axenic male Sprague Dawley	DON	Gavage feeding, 100 μg/kg b.w.	To evaluate the impact of a subchronic NOAEL dose exposure of DON on the composition of human gut microbiota	A significant increase of 0.5 $\log_{10}$ was observed for the *Bacteroides/Prevotella* group during the first 3 w of administration, concentration levels for *Escherichia coli* decreased at d 27	[58]

(Continued)

TABLE 34.1 (*Continued*)

Studies Using Rat Models for the Examination of *Fusarium* Mycotoxins Published Since 2010 Based on PubMed

Rat Model	MT	Administration/Dose	Aim of the Study	Major Findings	References
3-w-old male Sprague Dawley	FB_1	FB_1 contaminated corn, 30 mg/kg	To assess how nixtamalization of whole kernel corn affects fumonisin toxicity, male rats were fed diets containing low, mid or high levels of uncooked (LU, MU, HU) or alkaline-cooked (LC, MC, HC) FB_1-contaminated corn for 3 week	Nixtamalization is an effective cooking method for reducing the potential toxicity of FB_1-contaminated corn	[59]
Male Sprague Dawley	FB_1	Gastric subserosa, 0.036–0.09 g/kg b.w.	To investigate whether accumulated ceramide could serve as the effector molecule of ulcer formation in a rat model of acetic-acid-induced gastric ulcer	The ceramide pathway, in particular the metabolites of ceramide, significantly contribute to acetic-acid-induced gastric damage, possibly via enhancing apoptosis	[60]
4- to 5-w-old adult male Sprague Dawley	FB_1	FB_1-contaminated corn, 50, 100, and 200 mg/kg diet	To assess changes in serum biochemical profile and DNA fragmentation of growing male rats fed a diet with FB_1-contaminated corn and the role of a *Lactobacillus delbrueckii* spp. *lactis* and *Pediococcus acidilactici* supplementation in counteracting the FB_1 effects in intoxicated rats	In rats consuming diets containing FB_1, there is a time- and dose-dependent increase in serum enzyme activities and DNA lesions	[61]
3-month-old female Sprague Dawley	FB_1	Oral feed—corn oil 100 μg/kg b.w.	To evaluate the protective role of PGE against the synergistic effect of subchronic administration of AFB_1 and FB_1 on DNA and gene expression in rat	AFB_1 and FB_1 have synergistic genotoxic effects; PGE induced protective effects against their oxidative stress and genotoxicity through its antioxidant properties	[62]
Early weaning male Sprague Dawley	T-2	Intragastric, 0.1 mg/kg/d 0.2 mg/kg/d	To explore the effects of T-2 toxin and it's synergy with low selenium on lipid peroxidation in brain	The single factor of T-2 toxin can cause lipid peroxidation in brain, lower the activity of GSH-Px, and increase the level of MDA	[63]
Female weanling Sprague Dawley	ZEA	Oral, 0 or 6 mg/kg	To evaluate the efficacy of an ADC in reducing the toxic effects of ZEA in the diet of rats and piglets	A long-term consumption of ZEA-contaminated diets stimulated growth of the reproductive tract in rats and piglets and the presence of ZEA residue in bile in piglets	[64]

(*Continued*)

TABLE 34.1 (***Continued***)

Studies Using Rat Models for the Examination of *Fusarium* Mycotoxins Published Since 2010 Based on PubMed

Rat Model	MT	Administration/Dose	Aim of the Study	Major Findings	References
Pregnant female Wistar	ZEA	Subcutaneous injection, 0.2 μg/kg to 5 mg/kg	(1) Morphometric analysis to evaluate the development of the mammary glands; (2) immunohistochemical study of mammary tissue slides to evaluate differentiation and the level of cell replication and apoptosis in the tissue; and (3) to test the presence of possible lesions in the mammary tissue	ZEA could contribute to the induction of breast endocrine disorders	[65]
4-w-old male and female weanling Wistar	T-2	Intragastric, 0.04 mg/kg/d	To investigate magnetic MRI in the tibial epiphyseal growth plate development	The MRI image of the rat epiphyseal plate is altered in the KBD model rats, and epiphyseal plate MRI appearance has been reproduced by using T-2 toxin and KBD-affected feed of an epidemic-affected district	[66]
Male and female Wistar	T-2	Intragastric, 0.1 mg/kg/d	To determine whether giving rats selenium- and iodine-deficient food low in protein and made with barley from an area where KBD is endemic, in combination with exposing them to T-2 toxin, would produce the characteristic chondrocyte necrosis of KBD, to establish an experimental KBD animal model	This animal model of KBD can be approached by feeding rats a low-nutrition diet (low levels of selenium, iodine, and protein) and exposing them to T-2 toxin. The pathological and radiographic changes observed were very similar to those in patients with KBD	[67]
10-w-old male Wistar	T-2	Subcutaneous, 0.1 and 0.5 mg/kg	To clarify and reevaluate the cardiac and autonomic nervous effects of T-2 toxin	T-2 toxin produced significant cardiac dysfunctions involving disturbance of the conduction pathway influenced by the autonomic nervous activity and also possible direct effects on cardiac myocytes	[68]
4-w-old-male and female weanling Wistar	T-2	0.04 mg/kg/d	To characterize the features of radiographic abnormalities of the tibial bone in rats that have been fed T-2 toxin and KBD epidemic district food	A low-nutrition diet may be involved in the etiology of KBD, and determination of this should be studied in the future	[69]
Wistar	T-2	0, 1, 10, and 100 nM	To investigate the reproductive toxicity and cytotoxicity of T-2 toxin and to explore its potential apoptotic induction mechanism	A possible underlying molecular mechanism for T-2 toxin is that it induces the apoptosis signaling pathway in rat granulosa cells by regulation of ROS-mediated mitochondrial pathway	[70]

(Continued)

TABLE 34.1 (*Continued*)

Studies Using Rat Models for the Examination of *Fusarium* Mycotoxins Published Since 2010 Based on PubMed

Rat Model	MT	Administration/Dose	Aim of the Study	Major Findings	References
Male and female weanling Wistar	T-2	Intragastric, 0.04 mg/kg/d	To investigate the effect of KBD-affected feed and T-2 toxin on bone development	KBD-affected feed rats had less weight gain than T-2 toxin intervention rats, which means there were other etiological factors in KBD-affected feed	[71]
Male and female Wistar rats	T-2	Intragastric, 1 mg/kg/d	To observe early lesions of rat epiphyseal plates and metaphysis caused by T-2 toxin and T-2 toxin combined with a low-nutrition diet to determine possible pathogenic factors of KBD	T-2 toxin combined with a low-nutrition diet could lead to more serious chondrocyte necrosis in the epiphyseal plate and disturb metaphyseal trabecular bone formation	[72]
Male Fischer	FB_1	250 mg/kg diet	To characterize the involvement of lipid changes during cancer promotion resulting in the development of preneoplastic lesions, altered lipid phenotype, and to compare FA profiles of two different cancer-promotion regimens	A typical lipid phenotype was observed, including increased membrane PE and cholesterol content, increased levels of C16:0 and monounsaturated fatty acids in PE and PC, as well as a decrease in C18:0 and long-chain polyunsaturated fatty acids in the PC fraction	[73]
3-month-old sexually mature virgin female and male Wistar	FB_1	Gavage feeding, 4 μg/kg b.w.	To investigate the synergistic impact of prenatal methyl donor deficiency and low dosage of FB_1 administration on the pattern of global histone modifications	Low doses of FB_1 interact with MDD, thus contributing to the disruption of the epigenetic landscape	[74]

ABC, ATP-binding cassette; ADC, activated diatomaceous clay; b.w., body weight; CYP, cytochrome P450; d, day; DON, deoxynivalenol; ER, estrogen receptor; FA, fatty acid; FB_1, fumonisin B_1; GSH, glutathione; KBD, Kashin–Beck disease; KRG, Korean red ginseng; MDA, malondialdehyde; MDD, methyl-deficient diet; MRI, magnetic resonance imaging; MT, mycotoxin; NOAEL, no observable adverse effect level; PC, phosphatidylcholine; PGE, *Panax ginseng* extract; PE, phosphatidyl ethanolamine; ROS, reactive oxygen species; TBARS, thiobarbituric acid reactive substances; w, week; ZEA, zearalenone.

TABLE 34.2

Studies Using Mouse Models for the Examination of *Fusarium* Mycotoxins Published Since 2010 Based on PubMed

Mouse Model	MT	Administration/Dose	Aim of the Study	Major Findings	References
6-w-old female BALB/c	ZEA	Oral, 40 mg/kg b.w.	To investigate the effects of radish extract on the b.w. gain, the lymphoid relative weight organs, hematological parameters, and the expression of antibody level and cytokines production LPS stimulation on intoxication	Radish extract was effective for the protection of high dose ZEA-immunotoxication in mice	[75]
6-w-old female BALB/c	DON	Intraperitoneal inoculation, 100–250 μg	To determine the optimal concentrations of antigen/antibody for DON detection	The detection limit was 0.01–100 μg/mL, and average recovery of DON from contaminated grain was 82%–93%	[76]
BALB/c	DON	10 μg/kg	To elucidate the reason for blood parameters and hemostatic effect after oral administration of DON	Significant decrease of hematocrit value and rise of blood clotting time and bleeding time; DON is a potential hematotoxin	[77]
5-w-old female BALB/c	ZEA	Oral, 40 mg/kg b.w.	To determine the abilities of the living *Lactobacillus plantarum* MON03 cells, TM clay, and their composite to accumulate ZEA from a liquid medium and elucidate the preventive effect of their composite in ZEA-contaminated mice	Both *L. plantarum* and TM are safe by themselves and their composite succeeded to exert a potential prevention by counteracting ZEA-immunotoxicity	[78]
BALB/c	ZEA	Oral, 50,100, 200 μg/kg b.w.	To establish whether ZEA produced hepatotoxicity via oral administration	ZEA is a potential hepatotoxin when administered by the oral route	[79]
8-w-old female BALB/c	ZEA	Oral, 40 mg/kg b.w.	To isolate a new ZEA-binding microorganism for use in biological detoxification and to examine its ability to remove ZEA in liquid medium and its potential for prevention of ZEA-induced immunomodulation in mice	ZEA induced toxicity in immunologic and hematologic parameters as indicated by the changes in lymphocyte cell numbers; *Lactobacillus paracasei* BEJ01 treatment prevents weight loss and reduces the immunotoxic effects caused by ZEA	[80]
Female BALB/c	DON	Oral, 0, 0.5, or 2 mg/kg	To investigate the differential immunomodulatory effects of DON	DON exposure differentially modulated IL-1, IL-10, and TNF-α production; DON can cause various immunomodulatory effects	[81]
Female BALB/c	T-2	Injection, 100 μg T2G-KLH	To develop an antibody capable of interacting with detection of T-2-Glc	Mab 2–13 will be useful for the simultaneous detection of T-2 toxin and T-2-Glc	[82]
7- to 8-w-old male and female BALB/c	FB_1	0.1 mg/kg 10 mg/kg	To evaluate the genotoxic potentials of FB_1 using a simple micronuclei test in a rodent model	FB_1 is nongenotoxic in nature, while the reduced ratio of PCE/NCE suggests the cytotoxic nature of FB_1	[83]

(Continued)

TABLE 34.2 (*Continued*)

Studies Using Mouse Models for the Examination of *Fusarium* Mycotoxins Published Since 2010 Based on PubMed

Mouse Model	MT	Administration/Dose	Aim of the Study	Major Findings	References
Female BALB/c	FB_1	Subcutaneous injection FB_1-BSA (0.2 mL, 100 mg)	To obtain a monoclonal antibody against FB_1 with high specificity and affinity	Anti-FB_1 mcAb excreted by 4G5 can be used to detect FB_1 in corn and related samples	[84]
10-w-old female BALB/c	FB_1	Intraperitoneal and gavage, 1.5 mg/kg FB_1 4.5 mg/kg FB_1	To investigate the effect of silymarin on experimental liver toxication induced by FB_1	Silymarin ameliorated toxic liver damage	[85]
6-w-old male BALB/c	T-2	Intraperitoneal, 1, 2, 3, 4, or 5 mg/kg	To check the effect of a sublethal dose of T-2 toxin on T lymphocyte subpopulation levels and the potential protective effects from treatment with selenium or vitamin E	Selenium could exert an important effect against the immunotoxic effects of T-2 toxin against T lymphocytes	[86]
4-w-old female BALB/c	DON, ZEA	5.0 mg/kg b.w.	To assess the individual and combined toxic effects of AFB_1, ZEA, and DON within the liver	The combination of AFB_1 + DON displayed a synergistic hepatotoxic effect, while AFB_1 + ZEA displayed an antagonistic hepatotoxic effect	[87]
BALB/c	T-2	NA	To investigate the toxic effect of T-2 toxin on the percentage of peripheral blood B lymphocytes and the potential protective role of selenium and vitamin E	Selenium plays a pivotal role on altered B lymphocyte subset induced by T-2 toxin compared with vitamin E	[88]
4-w-old female ICR	DON, ZEA	Maize feed DON (3.1 mg/kg) and ZEA (729 mg/kg)	To investigate the regulation of multiple mycotoxins on oxidative stress	MT-contaminated diet could result in liver damage, elevated GPx activity in the serum and liver tissues, and increased MDA level in the serum, indicative of oxidative stress	[89]
4-w-old female ICR	DON, ZEA	Low dose mycotoxin-containing diet (mass percentage: 15%) contaminated with DON (581 μg/kg) or ZEA (285 μg/kg), high dose mycotoxin-containing diet (mass percentage: 30%) contaminated with DON (1.163 μg/kg) or ZEA (569 μg/kg)	To identify any epigenetic effects of a mycotoxin-containing diet, including altered DNA methylation, H3K9 methylation, H3K27 methylation, and H4K20 methylation, on reduced oocyte developmental competence	DON affects chromatin compaction and the cell cycle progression of oocytes by reduced H4K20me2 and increased H4K20me3 levels; H4K20 methylations play important roles on the cell cycle, mitosis, and embryonal development; ZEA affects the level of H4K20 methylation, mitotic chromatin compaction, and the cell cycle progression of mouse eggs, which further affect the egg developmental competence	[90]

(*Continued*)

TABLE 34.2 (*Continued*)

Studies Using Mouse Models for the Examination of *Fusarium* Mycotoxins Published Since 2010 Based on PubMed

Mouse Model	MT	Administration/Dose	Aim of the Study	Major Findings	References
4-w-old female ICR	DON, ZEA	Contaminated maize feed 3.875 mg/kg of DON or 1,897 μg/kg of ZEA	To check the effect of DON on oocyte quality	DON and ZEA affect oocyte polarity during meiosis; abnormal mitochondrial distributions in the oocytes	[91]
Female ICR	ZEA	Germinal vesicle-intact oocytes harvested from ovaries of mice: low dose group—10 μM ZEA High dose group—50 μM ZEA	To study the effect of ZEA on mouse egg developmental competence	ZEA can affect chromatin compaction and the cell cycle progression of oocytes by reduced H4K20me2 and increased H4K20me3 levels	[92]
3- to 4-w-old female B6C3F1	DON	Oral, 0.1–12.5 mg/kg b.w.	To test the hypothesis that impairment of the GH axis precedes DON-induced growth retardation in the mouse	Oral DON exposure perturbs GH axis by suppressing two clinically relevant growth-related proteins, IGFALS and IGF1	[93]
11- to 12-w-old female B6C3F1	DON	Contaminated pellet feed, 0, 2, 5, or 10 ppm	To determine whether this mouse strain is similarly affected by DON during the process of obesity induction or when in the obese state	DON consumption lowered weight gain and produced weight loss in diet-induced obese mice at higher thresholds than that observed previously in normal B6C3F1 mice	[94]
10- to 11-w-old female, B6C3F1 mice	DON	Oral, 0, 0.1, 0.5, 1.0, or 2.5 mg/kg	To check whether the acute administration of DON to the mouse causes a rapid, measurable, and reproducible reduction in food intake	Mice had partial resistance to feed refusal when exposed to DON subsequently after 2 d but not after 7 d, suggesting that this modest tolerance was reversible	[95]
11-w-old female adult B6C3F1	DON	Oral, 10 mg/kg	To relate DON-induced b.w. loss in HF-induced obese mice to food intake, fat mass, lean mass, and obesity-related hormones	DON induced rapid decreases in b.w. and fat mass, which stabilized to those of the LF control within 11 d; DON-mediated effects on b.w., fat mass, food intake, and hormonal levels were consistent with a state of chronic energy restriction	[96]
3-w-old female B6C3F1	DON	0, 1, 2.5, 5, and 10 ppm	To explore the feasibility of using plasma IGFALS as a biomarker of effect for DON To demonstrate that in mice fed with 15 ppm DON diet there are significantly less plasma IGFALS than in mice fed identical amounts of control diet	Plasma IGFALS was significantly depressed; it might be a useful biomarker for DON's adverse effects on growth	[97]

(*Continued*)

TABLE 34.2 (*Continued*)

Studies Using Mouse Models for the Examination of *Fusarium* Mycotoxins Published Since 2010 Based on PubMed

Mouse Model	MT	Administration/Dose	Aim of the Study	Major Findings	References
8- to 10-w-old male B6C3F1	DON	Oral 5 mg/kg b.w.	To identify early protein phosphorylation changes in the immune system of DON-exposed mice and relate these to the toxin's downstream immunomodulatory effects	DON impacted phosphorylation of proteins within diverse immune cell populations, including monocytes, macrophages, T cells, B cells, dendritic cells, and mast cells	[98]
10- to 12-w-old female B6C3F1 (C57BL/6 × C3HeN)	DON	Oral, 2.5 mg/kg b.w. DON, D3G, 3-ADON, 15-ADON, FX, NIV, EN139528 or EN139544 in 100 μL PBS	To compare the effects of DON and its congeners on splenic TNF-α, interleukin and chemokine mRNA expression in the mouse	Naturally occurring and synthetic DON congeners to elicit aberrant mRNA upregulation of cytokines associated with acute and chronic trichothecene toxicity	[99]
7-w-old male C57BL/6	DON	5, 10, or 25 mg/kg b.w.	To understand the mechanism of action of DON in the thymus	DON downregulated genes involved in proliferation, mitochondria, protein synthesis, and ribosomal proteins Early precursor thymocytes, particularly at the double-positive CD4+ CD8+ stage, are more vulnerable to DON than very early or late precursor thymocytes	[100]
10-w-old C57BL/6	DON	Intraperitoneal injection 4, 6.5, or 10 mg/kg	To investigate the toxic effects of DON on fetal skeletal development	Various skeletal defects in fetuses, including misaligned or fused sternebrae and vertebrae, divided or fused ribs and polydactyly, hemivertebrae, short toe, and tail anomalies were observed	[101]
10- to 11-w-old C57BL/6	DON	10 ppm DON 20 ppm DON	To check the reversibility of DON-induced body weight loss and anorexia, to investigate the role of PKR	DON's effects on food consumption and body weight are not permanent; furthermore, PKR is not an essential signaling molecule for DON's anorectic and weight effects	[102]
C57BL/6J	ZEA	0, 0.8, 4, and 20 ppm	To investigate the potential cumulative effects of multiple pregnancy and multigenerational exposure to dietary ZEA on female puberty and reproduction	Exposure to a 20 ppm ZEA diet promoted female pubertal onset without obvious cumulative effect and diminished female fertility over generations	[103]
Adult male C57BL/6	DON	Oral, 1–25 mg/kg b.w.	To evaluate the impact of subchronic intoxication with DON, given at doses below the NOAEL, on inflammatory status	Subchronic administration of low DON doses produced a low-grade inflammation	[104]

(*Continued*)

TABLE 34.2 (*Continued*)

Studies Using Mouse Models for the Examination of *Fusarium* Mycotoxins Published Since 2010 Based on PubMed

Mouse Model	MT	Administration/Dose	Aim of the Study	Major Findings	References
Male and female CD1	ZEA	Oral, 1.3, 3.9, and 6.6 mg/kg	To analyze the effects on the transcriptome in testes exposed to mono-(2-ethylhexyl) phthalate, ZEA, lindane, bisphenol-A, or 17β-estradiol	Mono-(2-ethylhexyl)-phthalate and ZEA both produced specific alterations of gene signatures irrespective of the concentration of the toxicant or the developmental period	[105]
8-w-old CD1	FB_1	0, 25, 50, or 100 mg/mL of NSP	To examine toxicity on the development of embryos and its capacity to prevent teratogenesis-induced by FB_1	NSP had an unexpected high adsorption capacity *in vitro*; NSP is a feasible and effective agent for supplementary use in reducing the toxicity of FB_1 to animals	[106]
CD1 outbred male	ZEA	Oral, lower dose (0.15 μg/L, higher dose (150 μg/L)	To assess the effect of treatment with a low dose of ZEA on the male gonadal pathology, sperm quality, and expression of selected genes	Animals exposed to low ZEA concentration were affected considerably more than animals exposed to thigh ZEA concentration; a low concentration of ZEA is able to negatively influence the sperm parameters and testicular gene expression	[107]
Female Swiss albino	T-2	Dermal and subcutaneous, 5.94 mg/kg 1.54 mg/kg	To evaluate the acute toxicity of dermal and subcutaneous exposure of T-2 toxin on brain oxidative stress	Percutaneously and subcutaneously applied T-2 toxin can cause brain oxidative damage possibly after crossing blood–brain barrier by altering its permeability	[108]
Male Swiss albino	T-2	Percutaneous, 5.94 mg/kg b.w. 2.97, 5.94 and 11.88 mg/kg b.w.	To investigate the biochemical and histological alterations behind inflammation and cutaneous injury caused by T-2 toxin	Skin inflammation and cutaneous injury are mediated through oxidative stress, activation of myeloperoxidase, MMP activity, increase in inflammatory cytokines, activation of p38 MAPK, and apoptosis of epidermal cells, leading to degenerative skin histological changes	[109]
Male Swiss albino	T-2	Topical, 11.8 and 23.76 mg/kg	To evaluate the protective efficacy of CC-2 formulation against lethal topical doses; to check the effect of dose of T-2 toxin and time of CC-2 application on lethality; to evaluate the recovery profile of surviving animals	CC-2 formulation may be an effective decontaminant against topical exposure to T-2 toxin if treated within 5–15 min of T-2 toxin exposure	[110]
90-d-old male Swiss albino	ZEA	Gavage, 40 mg/kg	To investigate the effect of an acute dose of ZEA on reproductive and hematological parameters, as well as on markers of oxidative stress in liver, kidney, and testes	ZEA has acute toxic effects mainly in reproductive system of adult male mice, and its effect is probably related to a reduced activity of GST and increase in SOD activity in testes	[111]

(*Continued*)

TABLE 34.2 (*Continued*)

Studies Using Mouse Models for the Examination of *Fusarium* Mycotoxins Published Since 2010 Based on PubMed

Mouse Model	MT	Administration/Dose	Aim of the Study	Major Findings	References
7- to 8-w-old male and female Swiss	DON, FB_1	Oral 45 μg/kg	To determine whether association of DON and FB_1 cause an additive or synergistic toxic effect on their systemic targets	NOAEL of both DON and FB_1 should be lower than 45 μg/kg b.w./d and 110 μg/kg b.w./d; the oral repetitive administration of low dose of DON revealed disorders in lipid metabolism, renal filtration disturbance, and renal cell DNA methylation, rhabdomyolysis, and blood lymphocyte cell deaths	[112]
9-d-old male Swiss albino	ZEA	ZEA (40 mg/kg—8% of LD_{50})	To check the effects of lycopene on reproductive, hematological, histopathological, and oxidative stress parameters following acute exposure to ZEA	Lycopene prevented the testicular histopathological damage caused by ZEA, as well as its harmful effects on sperm count and motility	[113]
Male Swiss albino	ZEA	Oral, single dose (40 mg/kg—8% of LD50)	To study ZEA toxicity effects and the mechanism of lycopene's protective effects following ZEA exposure	ZEA impaired testicular functions (spermatozoa count and motility) and testosterone levels; lycopene can be considered a potential therapeutic nutrient in protection against male reproductive toxicity induced by ZEA	[114]
6- to 7-w-old female Swiss albino	DON	Topical application in concentrations of 84, 168, 336, and 672 nmol in 0.2 mL acetone	To understand the cellular events leading to DON-mediated *in vivo* dermal toxicity and to investigate whether DON causes induction of cell proliferation via activation of MAPK pathway, which may further lead to tumor formation	Topical application increased cell proliferation, DNA synthesis, and inflammation, which are mediated through PI3K/AKT and MAPK signaling pathways involving transcription factor NFκB and AP-1, further leading to transcriptional activation of downstream target proteins c-fos, c-jun, cyclin D1, iNOS, and cox-2	[115]
Swiss albino female	T-2	Percutaneous, subcutaneous, 5.94 mg/kg 1.57 mg/kg	To evaluate the effect of time course and route of T-2 toxin exposure on hepatic oxidative damage	Results showed oxidative stress as major underlying mechanism for T-2 toxin toxicity *in vivo*	[116]
25-d-old male Kunming	ZEA	Oral, 10 μg	To explore the effects of *Gynostemma pentaphyllum* on ZEA-induced apoptosis in germ cells	*Gynostemma pentaphyllum* protects against toxicity caused by ZEA through antioxidation and antiapoptosis via the regulation of Bax and Bcl-2 expression	[117]

(*Continued*)

TABLE 34.2 (*Continued*)

Studies Using Mouse Models for the Examination of *Fusarium* Mycotoxins Published Since 2010 Based on PubMed

Mouse Model	MT	Administration/Dose	Aim of the Study	Major Findings	References
60- to 90-d-old Kunming	T-2	10 ng/mL	To investigate the effects of T-2 toxin on testosterone biosynthesis in Leydig cells	T-2 toxin can directly decrease the testosterone biosynthesis in the primary Leydig cells derived from the testis	[118]
SWV and LM/Bc	FB_1	Intraperitoneal, 20 mg/kg	To determine the relationship between failure of neural tube closure and accumulation of sphingoid bases and their phosphorylated metabolites in maternal and embryonic tissue after administration of FB_1	Elevated sphingoid base-1-P after FB_1 or FTY720 suggest a potential role for these bioactive lipid ligands and activation of S1P receptor signaling pathways in the failure of neural tube closure after FB_1 or FTY720; Sa1P may represent a biomarker for FB_1-NTD risk assessment	[119]
15- to 18-w-old female acclimated LM/Bc	FB_1	25 mg/kg 1 mg/kg	To investigate whether the consumption of a diet deficient in folate and with reduced vitamin B_{12} would exacerbate NTD induction by FB_1	Folate deficiency does not exacerbate NTD induction by FB_1 in LM/Bc mice; interactions between folate, other nutritional factors, and FB_1 in this mouse model for NTDs are complex and require further investigations	[120]
Male LM/Bc	FB_1	Oral gavage, 5, 10, 15, 25, and 50 mg/kg b.w.	To develop and validate in an animal model, and ultimately in humans, a method to estimate the volume of blood collected as blood spots on absorbent paper so as to allow quantification of the molar concentration of sphingoid base 1-phosphates in blood	In both mouse and human, the A270, total protein, and blood volume were closely correlated and the volume of blood spotted was accurately estimated using only the A270 of the extracts; in mouse blood spots, as in tissues and embryos, the FB_1-induced changes in sphingolipids were correlated with urinary FB_1	[121]
6-w-old male C3H	DON	Contaminated drinking water, 2 mg/mL	To evaluate immune function of mice exposed to DON; to study the effects of exposure on the immune response after vaccination with inactivated porcine parvovirus vaccine	Exposure to DON at 2.0 mg/L via drinking water can disrupt the immune response in vaccinated mice by modulating cytokines and chemokines	[122]
6-w-old BDF1	T-2	0.5, 0.75, and 1 ng/mL	To investigate embryotoxicity of T-2 via the *in vitro* development of preimplantation mouse embryos and its effect on the nuclear chromatin status	T-2 mycotoxin has a harmful effect on early embryonal development, which results in decreased blastocyst proportion, delayed blastulation, and increased rate of chromatin damage	[123]

(*Continued*)

TABLE 34.2 (*Continued*)

Studies Using Mouse Models for the Examination of *Fusarium* Mycotoxins Published Since 2010 Based on PubMed

Mouse Model	MT	Administration/Dose	Aim of the Study	Major Findings	References
Female	FB_1	150 mg/kg	To examine the apoptotic and proliferative activity of gastric mucosa following administration of FB_1	Oral administration of FB_1 caused atrophy in gastric mucosa both via increasing of apoptosis and suppressing the mitotic activity of these cells	[124]
mPGES-1 wild-type and knock out adult male	DON	Oral, 6.25, 12.5, and 25 mg/kg	To demonstrate that DON-induced sickness-like behavior is independent of PGE2 action	Behavioral changes observed after DON intoxication differ from classical sickness behavior evoked by inflammatory cytokines	[125]
5- to 7-w-old p53+/– and p53+/+	FB_1	5, 50, and 150 mg/kg	To compare the responses to chronic dietary FB_1 exposure in p53+/– and p53+/+ mice to check whether FB_1 is a nongenotoxic carcinogen; to assess the role of p53 in protecting cells and tissues from FB_1-induced changes, including altered sphingolipid metabolism	Based on similar responses in p53+/– and p53+/+ mice, p53 and related pathways play a secondary role in responses to FB_1 toxicity and carcinogenesis	[126]

AFB_1, aflatoxin B_1; b.w., body weight; CC-2, N,N′-dichloro-bis(2,4,6-trichlorophenyl) urea; d, day; DNA, deoxyribonucleic acid; DON, deoxynivalenol; FB_1, fumonisin B_1; GH, growth hormone; GPx, glutathione peroxidase; GST, glutathione-S-transferase; HF, high-fat; IGF, insulin-like growth factor; IGFALS, insulin-like growth factor binding protein, acid labile subunit; LF, low-fat; LPS, lipopolysaccharide; Mab, monoclonal antibody; MAPK, mitogen-activated protein kinase; MDA, malondialdehyde; MT, mycotoxin; NCE, normochromatic erythrocytes; NOAEL, no observable adverse effect level; NSP, nanosilicate platelets; NTD, neural tube defect; TNF, tumor necrosis factor; PCE, polychromatic erythrocytes; PBS, phosphate-buffered saline; PGE2, prostaglandins E2; PKR, double-stranded RNA-activated protein kinase; SOD, superoxide dismutase; TM, Tunisian montmorillonite; w, week; ZEA, zearalenone.

confirmed to diminish female fertility in C57BL/6J mice [103]. Using CD1 males, Zatecka et al. [107] illustrated that ZEA has considerable effect on male fertility, and López-Casas et al. [105] proved the alterations of gene signatures due to ZEA. Yuan et al. [117] assessed the ZEA toxicity in male germ cells of Kunming mice.

FB_1-induced changes in sphingoids were analyzed by Riley et al. [121] and Gelineau-van Waes et al. [119] among LM/Bc mice. Voss et al. [120] established that folate deficiency does not exacerbate NTD induction by FB_1 in LM/Bc mice. Hepatotoxicity [85] and cytotoxicity [83] of FB_1 were confirmed using BALB/c mice. Ling et al. [84] utilized anti-FB_1 monoclonal antibody from BALB/c mice to detect FB_1 in corn and related samples. Liao et al. [106] evidenced that nanosilicate platelets reduced the toxicity of FB_1 in CD1 mice. Kouadio et al. [112] proved that no-observed-adverse effect level (NOAEL) of FB_1 should be lower than 110 µg/kg/b.w./day in Swiss albino mice. FB_1-induced carcinogenesis was reported to be similar in both p53 heterozygous cancer-prone model and p53 homozygous mice [126]. Alizadeh et al. [124] reported that atrophy in gastric mucosa was due to FB_1 in female mice.

34.2.2 Human and Animal Cell Lines

The literature review was restricted to papers published after 2010 and included in the PubMed database, with keyword combinations deoxynivalenol OR T-2 toxin OR zearalenone OR fumonisin AND cell line OR cell culture. Similar to the case of rat and mouse models, the information on exact definition of the mycotoxin and the cell line applied were collected from studies that used human and animal cell lines.

Table 34.3 furnishes the cumulative information of studies published since 2010 that have applied human or animal cell lines as laboratory models for the exploration of *Fusarium* mycotoxins [127–170]. Among the 44 relevant studies, the most frequently applied laboratory models were intestinal epithelial cell lines, primarily the IPEC-J2 nontransformed porcine intestinal epithelial cell line (9 studies), which was used to examine different aspects of *Fusarium* mycotoxins, including the effect of DON on functional characteristics of epithelial cells and intestinal barrier integrity [128,131,136], the effects of phytic acid as a possible inhibitor of cellular damage induced by DON [130], the gene-response profiles of epithelial cell layers in response to DON [129] and the cytotoxicity and combined effects of 4 *Fusarium* mycotoxins, DON, NIV, ZEA, and FB_1 [133]. Further intestinal cell lines used as laboratory models to study *Fusarium* mycotoxins included HT-29 human colon carcinoma cells for studying T-2 toxin [147,148] and FB_1 [155]; Caco-2 human colorectal adenocarcinoma cells for the examination of different properties of T-2 toxin [138], DON [139,140], and FB_1 [141]; as well as HCT116 and HCT116-Bax-KO human colon carcinoma cells [157], AGS human gastric epithelial cells, and SW742 human colon adenocarcinoma cells [156]; IEC-6 (CRL-1592) rat small intestine epithelium cells [159]; and IPEC-1 cells derived from the small intestine of a newborn unsuckled piglet [127,137].

Among the kidney cell lines, HepG2 human hepatocellular carcinoma cells were used to evaluate the role of mitogen-activated protein kinase phosphatases in prevention of DON-induced apoptosis [144], to investigate the effect of FB_1 on cytochrome P450 1B1 modulation [146], and to investigate the effect of DON on the cytosolic redox state and antioxidative system [145]. HL-7702 normal human liver cells were used to investigate the effect of FB_1 on the cell cycle and the expression of cell-cycle-related genes [169]. H4IIE rat hepatoma cells were used to evaluate the effects of FB_1 and AFB_1 alone or in combination on the activation and expression of cytochrome P450 1A and its transcription factor Ahr [168], while SHK-1 salmon head kidney cells were used to investigate the possible metabolization of ZEA in fish cell lines [170]. Vero African green monkey kidney cells were used to examine the ROS production induced by FB_1 on renal cells [167], and PK-15 porcine kidney cells were used to evaluate the nephrotoxicity of individual mycotoxins and combinations of AFB_1, ZEA, DON, and FB_1 to livestock [165], HEK293 human embryonic kidney cells were used to investigate the mechanisms of ZEA nephrotoxicity and DNA damage [166]. Further cell lines involved in *Fusarium* mycotoxin research during the past 5 years included MTEC1 murine thymic epithelial cells [142,143], Jurkat human T-lymphocyte cells [151,152], HL60 human promyelocytic leukemia cells [149,150], U937 human monocyte cells [154], CTLL-2 mouse blood lymphoblast cells [153], EL4 murine T-lymphocyte cells [160], STC-1 mouse intestinal neuroendocrine tumor cells [158], GT1-7 murine hypothalamic cells [161], NPTr newborn pig trachea

TABLE 34.3

Studies Using Human and Animal Cell Lines for the Examination of *Fusarium* Mycotoxins Published Since 2010 Based on PubMed

Applied Cell Line	MT	Administration/Dose	Aim of the Study	Major Findings	References
IPEC-1 cells derived from small intestine of a newborn unsuckled piglet	DON	30 μmol/L	To investigate the involvement of MAPK in the DON-induced loss of barrier function	DON-induced activation of the p44/42 ERK signaling pathway inhibits the expression of claudin-4 protein, which leads to impaired intestinal barrier function	[127]
IPEC-J2 non-transformed porcine intestinal columnar epithelial cells	DON	200 and 2000 ng/mL	To elucidate the impact of the direction of DON exposure on epithelial cell behavior and intestinal barrier integrity	Severity of impact of the DON on the intestinal epithelial barrier is dependent on route of application	[128]
IPEC-J2	DON	200 ng/mL 2000 ng/mL	To detect potential expression differences and to identify candidate genes for further investigations on cellular reaction mechanisms	Apical and basolateral challenge of epithelial cell layers trigger different gene-response profiles paralleled with a higher susceptibility towards basolateral challenge	[129]
IPEC-J2	DON		To evaluate the effects of IP6 as a possible inhibitor of cellular damage induced by toxic substances such as MTs	Phytic acid decreased the negative effects of DON on the membrane integrity	[130]
IPEC-J2	DON	10 μM	To assess short-term effects of DON on functional characteristics of the intestinal epithelial cells	A significant decrease in ATP levels was seen at 48 h in a dose-dependent manner and demonstrated that DON has a distinct cytotoxic effect on IPEC-J2 cells	[131]
IPEC-J2	DON	1 mg/mL	To determine the penetration of FOS in the presence and absence of DON	Nontoxic concentration of DON on IPEC-J2 cells after short-term exposure interferes with the pharmacokinetics of the antibiotic	[132]
IPEC-J2	DON NIV ZEA FB_1	2.5–40 μM	To investigate the cytotoxicity of four common *Fusarium* MTs	At noncytotoxic concentrations of individual MTs, all mixtures were cytotoxic with DON-NIV, DON-ZEA, DON-NIV-FB_1, DON-ZEA-FB_1, NIV-ZEA-FB_1, with the mixture of all four causing the greatest loss of cell viability	[133]
IPEC-J2	DON NIV ZEA FB_1	10 or 40 μM	To elucidate the modulatory capacity of individual *Fusarium* toxins and their mixtures on the mRNA expression of proinflammatory cytokines	Data suggest that individual *Fusarium* toxins or their mixtures could cause or exacerbate intestinal inflammation	[134]
IPEC-J2	DON NIV ZEA FB_1		To elucidate the individual and combined effects of four common *Fusarium* toxins	*Fusarium* toxins, individually and in mixtures, activate distinct antimicrobial defense mechanisms possessing the potential to alter the intestinal microbiota through diminished antimicrobial effects	[135]

(*Continued*)

TABLE 34.3 (*Continued*)

Studies Using Human and Animal Cell Lines for the Examination of *Fusarium* Mycotoxins Published Since 2010 Based on PubMed

Applied Cell Line	MT	Administration/Dose	Aim of the Study	Major Findings	References
IPEC-J2	DON	2 μg/mL	To examine whether *Bacillus subtilis* could improve barrier integrity and protection against DON-induced barrier disruption	*B. subtilis* may have potential to enhance epithelial barrier function and to prevent the cells from DON-induced barrier dysfunction	[136]
IPEC-1 cells derived from small intestine of a newborn unsuckled piglet	T-2	0, 0.3, 1, 3, 10, and 30 nM	To investigate how ENN modulates T-2 toxicity in the situation of co-contamination, by analysing the acute toxicity of T-2 on the digestive target	Downmodulation of the gastrointestinal toxicity of T-2 toxin by the emerging ENN B1 below toxic concentrations; confirmation of the relevance of the determinist approach for the analysis of toxin interactions, giving concordant results with the probabilistic approach in two different pig intestinal models	[137]
Caco-2 human colorectal adenocarcinoma cells	T-2	1 ng/mL	To elucidate the possible underlying molecular mechanism of conditional T-2-provoked IL-8 induction, a possible involvement of the AHR and the MAPK pathway	p38 MAPK is responsible for mediating the inflammatory properties of T-2	[138]
Caco-2	DON	3.2, 16, 80, 400, and 2000 ng/mL	To investigate dose-dependent AhR agonistic and antagonistic activities in luciferase reporter cells and to assess the dose-dependent agonistic and antagonistic activities on hER	Induction of CYP1A1 activity and inhibition of CYP3A4 activity occurred in Caco-2 cells at a realistic intestinal concentration of IMA, and some bacterial stress genes were induced in a range of realistic concentrations following exposure to DON	[139]
Caco-2	DON	0.1, 1, 10 1 g/mL	To evaluate the intestinal absorption and toxicity of the major derivatives of DON: 3ADON and 15ADON	15ADON caused the highest permeability and highest IL-8 secretion among DON, 3ADON, and 15ADON	[140]
Caco-2, MDBK bovine kidney cells and raw 264.7 murine macrophages	FB_1	0.0038 ng/mL 3.8 mg/mL 0.005 ng/mle10 mg/mL	To evaluate the effects of low levels of each MTs in combination at their EU regulatory limits	Only in the case of MDBK cells was a synergistic effect measured, and this was when all three toxins were present at levels in food above the legal limits (OTA 3 mg/mL; FB_1 8 mg/mL and AFB_1 1.28 mg/mL)	[141]

(*Continued*)

TABLE 34.3 (*Continued*)

Studies Using Human and Animal Cell Lines for the Examination of *Fusarium* Mycotoxins Published Since 2010 Based on PubMed

Applied Cell Line	MT	Administration/Dose	Aim of the Study	Major Findings	References
MTEC1 murine thymic epithelial cells	DON	100, 500, 1000, and 2000 ng/mL	To investigate the mRNA expression differences after treatment with DON in MTEC1 cells	Results provided molecular insights into the gene expression differences of DON-induced toxic effects and suggest that p53 signaling pathway may play an important role in the inhibition of MTEC1 cell proliferation	[142]
MTEC1	DON	500, 1000, and 2000 ng/mL	To investigate if DON could induce apoptosis and to elucidate the possible mechanism of the action	DON causes activation of p53, increases levels of ROS, and causes the induction of mitochondrial dysfunction, which may contribute to DON-induced apoptosis	[143]
HepG2 human hepatocellular carcinoma cells	DON	1 μM 10 μM	To evaluate the impact of MKPs, particularly DUSP1	DUSP1 is a novel target gene of DON, which is essential for the prevention of DON-induced apoptosis	[144]
HepG2, THP-1 human monocyte-like cells	DON	—	To investigate the effect of DON on the cytosolic redox state and antioxidative system	DON induced accumulation of Trx-1 in HepG2 cells, which plays one of the key roles in protection against cytotoxicity caused by DON; the mechanism may be mediated by the antioxidant properties of Trx-1	[145]
HepG2	FB_1	0–1000 μM	To investigate the effect of FB_1 on miR-27b suppression and its effect on CYP1B1 modulation	CYP1B1 is posttranscriptionally regulated by miR-27b after HepG2 exposure to FB_1; FB_1-induced modulation of miR-27b in hepatic cells may be an additional mode of hepatic neoplastic transformation	[146]
HT-29 human colon carcinoma cells and RPTEC human renal proximal tubular epithelial cells	T-2	1 nM to 200 μM	To analyze the metabolism of T-2 toxin	Both cell types metabolized T-2 toxin to a variety of compounds; cell cycle analysis in RPTEC proved the apoptotic effect of T-2 toxin and its metabolites HT-2 toxin and neosolaniol in micromolar concentrations	[147]
HT-29 human colon carcinoma cells and NHA normal human astrocytes	T-2	1 nM to 200 μM	To investigate the cytotoxic properties of T-2 toxin on cells derived from brain tissue	Results emphasize the neurotoxic potential of T-2 toxin in human astrocytes at low concentrations after short incubation times	[148]

(*Continued*)

TABLE 34.3 (*Continued*)

Studies Using Human and Animal Cell Lines for the Examination of *Fusarium* Mycotoxins Published Since 2010 Based on PubMed

Applied Cell Line	MT	Administration/Dose	Aim of the Study	Major Findings	References
HL60 human acute promyelocytic leukemia cells	DON NIV	—	To elucidate the mechanisms underlying the toxicities of the trichothecene mycotoxins deoxynivalenol and nivalenol, their effects on the secretion of antihematopoietic chemokines, macrophage inflammatory protein-1 α (MIP-1α), and MIP-1β	Deoxynivalenol induces the secretion of chemokines, whereas nivalenol has the opposite effect, clearly indicating that the toxicity mechanisms of deoxynivalenol and nivalenol differ	[149]
HL60	T-2	0, 4, 8, 16, and 32 μg/mL	To study the effects of T-2 toxin produced by *Fusarium* fungi on proliferation and apoptosis	T-2 toxin could inhibit proliferation and induce apoptosis *in vitro* in a dose-dependent manner	[150]
Jurkat human lymphoblastoid T-cells	DON	0.5 mM	Immunocytological and biochemical analyses of the effects of DON in Jurkat cells	Immune cells are more sensitive to DON than other cell types due to the induction of a T-cell activation response by increased intracellular calcium levels	[151]
Jurkat	DON	0.5 mM	To understand the mechanism of action of DON in immune cells; to understand why immune cells are more sensitive to DON than most other cell types	Immune cells are more sensitive to DON than other cell types due to the induction of a T-cell activation response by increased intracellular calcium levels	[152]
CTLL-2 murine cytotoxic T lymphocyte cells	DON	1 or 2 mM	To assess the usefulness of the mouse CTLL-2 cell line for immunotoxicity	Based on the results for TBTO and DON, the CTLL-2 cell line does not yield an added value for immunotoxicity compared to the human Jurkat T cell line	[153]
U937 human monocyte cultures	DON	500–1000 ng/mL	To elucidate linkages that exist between the ribosome and PKR, Hck, and p38 following stimulation with DON in human and mouse mononuclear phagocytes	PKR and Hck were critical for DON-induced ribosomal recruitment of p38, its subsequent phosphorylation, and p38-driven proinflammatory cytokine expression	[154]
HT-29 human colorectal adenocarcinoma cells	FB_1	1.1–69 μM	To investigate by dose- and time-dependent experiments the modifications induced by FB_1 at concentrations ranging from 0.25 to 69 μM	Lipid peroxidation followed by modifications to membrane microviscosity and inflammatory response was the main and most sensitive effect of FB_1	[155]
AGS gastric epithelial cells and SW742 human colon adenocarcinoma cells	FB_1	4.5–72 mg/L	To evaluate FB_1 effects on the production of inflammatory cytokines	FB_1 increases inflammatory cytokines production	[156]
HCT116 and HCT116-Bax-KO human colon adenocarcinoma cells	DON	0.2 mg/mL	To analyze the molecular mechanisms of DON-induced toxicity and explore the contribution of mitochondria to cell death	Mitochondria-related caspase-dependent apoptotic pathway is involved in this *in vitro* model of DON induced-cytotoxicity	[157]

(*Continued*)

TABLE 34.3 (*Continued*)

Studies Using Human and Animal Cell Lines for the Examination of *Fusarium* Mycotoxins Published Since 2010 Based on PubMed

Applied Cell Line	MT	Administration/Dose	Aim of the Study	Major Findings	References
STC-1 murine invasive small intestinal neuroendocrine carcinoma cells	DON	0–5 mM	To test the hypothesis that DON induces hormone exocytosis in EEC by GPCR-mediated Ca^{2+} signaling	DON evokes CCK and GLP-1 secretion in the STC-1 EEC model by activating CaSR- and TRPA1-mediated Ca^{2+} signaling pathways	[158]
IEC-6 (CRL-1592) rat small intestine epithelial cells	DON	0.5–80 mM	To test the effects of variable concentrations of NIV and/or DON on the nontumorigenic intestinal epithelial cells, IEC-6	Both nivalenol and deoxynivalenol significantly affected IEC-6 viability through a proapoptotic process	[159]
EL4 thymoma cells	DON	0.5 μM	To investigate whether proteomic analysis of thymoma cells treated with DON as compared to nontreated control cells would reveal differential protein expression	DON stimulated the expression levels of several proteins in thymoma cells	[160]
GT1-7 murine hypothalamic neuronal cells	T-2	0.01–1000 ng/mL	To find out at how low-dose T-2 toxin influences GnRH secretion, especially with the presence of kisspeptins	Low-dose T-2 toxin stimulates GnRH secretion and alter the expression of associated proteins *in vitro*, and the activation of cell response to T-2 toxin is increased after pretreatment with kisspeptins	[161]
MLTC-1 mouse Leydig cells	ZEA	0–200 μg/mL	To demonstrate the involvement of ER stress in ZEA-induced cell death	The activation of an ER stress pathway plays a key role in ZEA-induced apoptosis	[162]
MCF-7aro CYP19-overexpressing cells, T98G human brain glioblastoma cells and MCF-7 human breast adenocarcinoma cells	ZEA	0.1% v/v	To investigate two estrogenic MTs for their interference in aromatase by using a cell-based system	ZEA was a potential aromatase inhibitor among the three MTs tested, and this 4-cell line approach could be employed in principle to screen for compounds inhibiting or inducing estrogen synthesis	[163]
NPTr newborn pig trachea epithelial cells	DON	0, 70, 140, 280, 560, and 1200 ng/mL	To evaluate the impact of DON on *in vitro* and *in vivo* PCV2 pathogenesis	DON had no significant effect on clinical manifestation of PCVAD in PCV2b-infected animals	[164]
PK-15 porcine kidney cells	ZEA DON FB_1 AFB_1	10 mM	To evaluate the nephrotoxicity of individual MTs and combinations of AFB_1, ZEA, DON, and FB_1 to livestock	The combined effects of MTs acted in a concentration-dependent manner	[165]

(Continued)

TABLE 34.3 (*Continued*)

Studies Using Human and Animal Cell Lines for the Examination of *Fusarium* Mycotoxins Published Since 2010 Based on PubMed

Applied Cell Line	MT	Administration/Dose	Aim of the Study	Major Findings	References
HEK293 human embryonic kidney cells	ZEA	0–20 μM	To investigate the mechanisms of ZEA nephrotoxicity and DNA damage	Oxidative stress does not play a key role in DNA strand breaks induced by ZEA; lysosomal injury precedes DNA strand breaks; the lysosome may be a primary target for ZEA	[166]
Vero African green monkey kidney cells	FB_1	1, 5, and 10 μM	To study the ROS production using a procedure based on an antioxidant sensitive fluorescein probe H2-DCFDA and to detect intracellular ROS as early-stage marker for toxin-induced oxidative stress	Significant increase of ROS products in Vero cells at 10 IM dose; ROS production by FB_1 on renal cells is a mechanism of fumonisin-mediated toxicity	[167]
H4IIE rat hepatoma cells	FB_1 AFB_1	10 and 20 μM	To evaluate the effects of MTs, alone or combined, on activation and expression of CYP1A and its transcription factor AHR	AHR pathway activation as a toxicity mechanism of AFB_1 and FB_1; FB_1 may increase AFB_1 bioactivation	[168]
HL-7702 normal human liver cells	FB_1	0.0, 0.1, 1.0, 10.0, and 100 μmol/L	To investigate the effect of FB1 on the cell cycle and the expression of cell-cycle-related genes P21 and cyclin E	The underlying mechanism of action is associated with alterations in the expression levels of cyclin E and P21 induced by FB_1	[169]
RTL-W1 rainbow trout liver cells, RTgill-W1 gill cells and SHK-1 salmon head kidney cells	ZEA	300 ng/mL	To investigate the possible metabolization of ZEA in fish cell lines suggesting that mainly glucuronidation takes place	Results confirm a lysosomal pathway as a main target of ZEA in fish cells	[170]

ADON, acetyldeoxynivalenol; AFB_1, aflatoxin B_1; AHR, aryl hydrocarbon receptor; CaSR, GPCR Ca(2+)-sensing receptor; CCK, cholecystokinin; CYP, cytochrome P; DNA, deoxyribonucleic acid; DON, deoxynivalenol; DUSP1, dual specific phosphatase 1; EEC, enteroendocrine cell; ER, endoplasmic reticulum; ERK, extracellular signal-regulated kinase; GnRH, gonadotropin-releasing hormone; GLP-1, glucagon-like peptide-17–36 amide; GPCR, G protein-coupled receptor; ENN, enniatin; FOS, fosfomycin; FB_1, fumonisin B_1; H2-DCFDA, dichloro-dihydro-fluorescein diacetate; Hck, hematopoietic cell kinase; hER, the human oestrogenic receptor; IMA, imazalil; IP6, phytic acid; MAPK, mitogen activated protein kinases; MKP, mitogen activated protein kinase phosphatase; MT, mycotoxin; NIV, nivalenol; OTA, ochratoxin A; PCV2, porcine circovirus type 2; PCVAD, porcine circovirus-associated disease; PKR, RNA-activated protein kinase; RNA, ribonucleic acid; ROS, reactive oxygen species; TBTO, tributyltin oxide; TRPA1, transient receptor potential ankyrin-1 channel; Trx-1, thioredoxin 1; ZEA, zearalenone.

epithelial cells [164], MLTC-1 murine testis Leydig cells [162], as well as CYP19-overexpressing MCF-7aro human breast cancer cells and T98G human glioblastoma cells [163].

34.3 Conclusions

Based on the recent literature search (since 2010), mainly rats, mice, and cell lines—particularly IPEC-J2 nontransformed porcine intestinal epithelial cells—were extensively involved in *Fusarium* mycotoxin research. These laboratory models revealed substantial details of scientific and technical information related to the mycotoxins DON, T-2, ZEA, and FB_1. These available data and the particulars discussed in this review chapter obviously warrant that laboratory animal models and cell lines have the potential for the detection of fusarial mycotoxins in food and feed, as well as in the research leading to the reduction and control of these mycotoxins in the products for consumption by humans and animals, which may otherwise lead to deleterious effects.

Acknowledgments

This work was supported by the Indian National Science Academy and the Hungarian Academy of Sciences, by project GINOP-2.3.3-15-2016-00006 (Széchenyi 2020 Programme), and by King Saud University, Deanship of Scientific Research, College of Sciences Research Center.

REFERENCES

1. Link, H.F., Observationes in ordines plantarum naturales, Dissertatio I, *Mag. Ges. Naturf. Freunde Berlin*, 3, 3, 1809.
2. O'Donnell, K. et al., Phylogenetic analyses of *RPB1* and *RPB2* support a middle Cretaceous origin for a clade comprising all agriculturally and medically important fusaria, *Fungal Genet. Biol.*, 52, 20, 2013.
3. Coleman, J.J. et al., The genome of *Nectria haematococca*: contribution of supernumerary chromosomes to gene expansion, *PLoS Genet.*, 5, e1000618, 2009.
4. Summerell, B.A., Laurence, M.H., Liew, E.C.Y., Leslie, J.F., Biogeography and phylogeography of *Fusarium*: a review, *Fungal Divers.*, 44, 3, 2010.
5. Westphal, A., Abney, T.S., Xing, L.J., Shaner, G.E., Sudden death syndrome of soybean, *Plant Health Instr.*, doi: 10.1094/PHI-I-2008-0102-01, 2008.
6. Huang, C.H., Roberts, P.D., Datnoff, L.E., Silicon suppresses *Fusarium* crown and root rot of tomato, *J. Phytopathol.*, 159, 546, 2011.
7. Shanmugam, V., Kanoujia, N., Biological management of vascular wilt of tomato caused by *Fusarium oxysporum f.sp. lycospersici* by plant growth-promoting rhizobacterial mixture, *Biol. Control*, 57, 85, 2011.
8. Garnica, M., Nucci, M., Epidemiology of fusariosis, *Curr. Fungal Infect. Rep.*, 7, 301, 2013.
9. O'Donnell, K. et al., Internet-accessible DNA sequence database for identifying fusaria from human and animal infections, *J. Clin. Microbiol.*, 48, 3708, 2010.
10. van Diepeningen, A.D., Al-Hatmi, A.M.S., Brankovics, B., de Hoog, G.S., Taxonomy and clinical spectra of *Fusarium* species: where do we stand in 2014? *Curr. Clin. Microbiol. Rep.*, 1, 10, 2014.
11. Ma, L.J. et al., *Fusarium* pathogenomics, *Annu. Rev. Microbiol.*, 67, 399, 2013.
12. Zinedine, A., Soriano, J.M., Molto, J.C., Manes, J., Review on the toxicity, occurrence, metabolism, detoxification, regulations and intake of zearalenone: an oestrogenic mycotoxin, *Food Chem. Toxicol.*, 45, 1, 2007.
13. Maragos, C.M., Zearalenone occurrence and human exposure, *World Mycotoxin J.*, 3, 369, 2010.
14. Döll, S., Dänicke, S., The *Fusarium* toxins deoxynivalenol (DON) and zearalenone (ZON) in animal feeding, *Prev. Vet. Med.*, 102, 132, 2011.
15. Massart, F., Saggese, G., Oestrogenic mycotoxin exposures and precocious pubertal development, *Int. J. Androl.*, 33, 369, 2010.

16. EFSA Panel on Contaminants in the Food Chain (CONTAM), Scientific opinion on the risks for public health related to the presence of zearalenone in food, *EFSA J.*, 9, 2197, 2011.
17. Wolf, J.C., Mirocha, C.J., Control of sexual reproduction in *Gibberella zeae (Fusarium roseum* "Graminearum"), *Appl. Environ. Microbiol.*, 33, 546, 1977.
18. Kościelniak, J., Biesaga-Kościelniak, J., Janeczko, A., Filek, W., Kalaji, H.M., Can the *Gibberella zeae* toxin zearalenone affect the photosynthetic productivity and increase yield formation in spring wheat and soybean plants? *Photosynthetica*, 47, 586, 2009.
19. Wu, F., Measuring the economic impacts of *Fusarium* toxins in animal feeds, *Anim. Feed Sci. Technol.*, 137, 363, 2007.
20. Asci, A., Durmaz, E., Erkekoglu, P., Pasli, D., Bircan, I., Kocer-Gumusel, B., Urinary zearalenone levels in girls with premature thelarche and idiopathic central precocious puberty, *Minerva Pediatr.*, 66, 571, 2014.
21. Shephard, G., *Fusarium* mycotoxins and human health, *Plant Breed. Seed Sci.*, 64, 113, 2011.
22. Gelderblom, W.C. et al., Fumonisins—novel mycotoxins with cancer-promoting activity produced by *Fusarium moniliforme*, *Appl. Environ. Microbiol.*, 54, 1806, 1988.
23. Bezuidenhout, S.C. et al., Structure elucidation of the fumonisins, mycotoxins from *Fusarium moniliforme*, *J. Chem. Soc. Chem. Commun.*, 11, 743, 1988.
24. Seo, J.A., Kim, J.C., Lee, Y.W., Isolation and characterization of two new type C fumonisins produced by *Fusarium oxysporum*, *J. Nat. Prod.*, 59, 1003, 1996.
25. Rheeder, J.P., Marasas, W.F.O., Vismer, H.F., Production of fumonisin analogs by *Fusarium* species, *Appl. Environ. Microbiol.*, 68, 2101, 2002.
26. Bartók, T., Szécsi, Á., Szekeres, A., Mesterházy, Á., Bartók, M., Detection of new fumonisin mycotoxins and fumonisin-like compounds by reversed-phase high-performance liquid chromatography/electrospray ionization ion trap mass spectrometry, *Rapid Commun. Mass Spectrom.*, 20, 2447, 2006.
27. Bartók, T. et al., Detection and characterization of twenty-eight isomers of fumonisin B_1 (FB_1) mycotoxin in a solid rice culture infected with *Fusarium verticillioides* by reversed-phase high-performance liquid chromatography/electrospray ionization time-of-flight and ion trap mass spectrometry, *Rapid Commun. Mass Spectrom.*, 24, 35, 2010.
28. Bartók, T. et al., Detection of previously unknown fumonisin P analogue mycotoxins in a *Fusarium verticillioides* culture by high-performance liquid chromatography-electrospray ionization time-of-flight and ion trap mass spectrometry, *J. Chromatogr. Sci.*, 52, 508, 2014.
29. Marasas, W.F., Fumonisins: history, world-wide occurrence and impact, *Adv. Exp. Med. Biol.*, 392, 1, 1996.
30. Wang, E. et al., Inhibition of sphingolipid biosynthesis by fumonisins. Implications for diseases associated with *Fusarium moniliforme*, *J. Biol. Chem.*, 266, 14486, 1991.
31. Marasas, W.F. et al., Fumonisins disrupt sphingolipid metabolism, folate transport, and neural tube development in embryo culture and *in vivo*: a potential risk factor for human neural tube defects among populations consuming fumonisin-contaminated maize, *J. Nutr.*, 134, 711, 2004.
32. Stockmann-Juvala, H., Savolainen, K., A review of the toxic effects and mechanisms of action of fumonisin B_1, *Hum. Exp. Toxicol.*, 27, 799, 2008.
33. Rheeder, J.P. et al., *Fusarium moniliforme* and fumonisins in corn in relation to human esophageal cancer in Transkei, *Phytopathology*, 82, 353, 1992.
34. Chu, F.S., Li, G.Y., Simultaneous occurrence of fumonisin B_1 and other mycotoxins in moldy corn collected from the People's Republic of China in regions with high incidences of esophageal cancer, *Appl. Environ. Microbiol.*, 60, 847, 1994.
35. International Agency for Research on Cancer (IARC), *IARC Monographs on the Evaluation of Carcinogenic Risks to Humans.* Vol. 82. *Some Traditional Herbal Medicine, Some Mycotoxins, Naphthalene and Styrene*, IARC Press, Lyon, France, 2002.
36. Ueno, Y., Sawano, M., Ishii, K., Production of trichothecene mycotoxins by *Fusarium* species in shake culture, *Appl. Microbiol.*, 30, 4, 1975.
37. Rocha, O., Ansari, K., Doohan, F.M., Effects of trichothecene mycotoxins on eukaryotic cells: a review, *Food Addit. Contam.*, 22, 369, 2005.
38. Lancova, K., Bowens, P., Stroka, J., Gmuender, H., Ellinger, T., Naegeli, H., Transcriptomic-based bioassays for the detection of type A trichothecenes, *World Mycotoxin J.*, 2, 247, 2009.

39. Bennet, J.W., Klich, M., Mycotoxins, *Clin. Microbiol. Rev.*, 16, 497, 2003.
40. Grovey, J.F., The trichothecenes and their biosynthesis, *Prog. Chem. Org. Nat. Prod.*, 88, 63, 2007.
41. Bin-Umer, M.A., McLaughlin, J.E., Basu, D., McCormick, S., Tumer, N.E., Trichothecene mycotoxins inhibit mitochondrial translation—implication for the mechanism of toxicity, *Toxins*, 3, 1484, 2011.
42. Escrivá, L., Font, G., Manyes, L., *In vivo* toxicity studies of *Fusarium* mycotoxins in the last decade: a review, *Food Chem. Toxicol.*, 78, 185, 2015.
43. Peraica, M., Radić, B., Lucić, A., Pavlović, M., Toxic effects of mycotoxins in humans, *Bull. World Health Organ.*, 77, 754, 1999.
44. Wang, Z.G., Feng, J.N., Tong, Z., Human toxicosis caused by moldy rice contaminated with *Fusarium* and T-2 toxin, *Biomed. Environ. Sci.*, 6, 65, 1993.
45. Kanora, A., Maes, D., The role of mycotoxins in pig reproduction: a review, *Ved. Med-Czech.*, 54, 565, 2009.
46. EFSA Panel on Contaminants in the Food Chain (CONTAM), Scientific opinion on the risks for animal and public health related to the presence of T-2 and HT-2 toxin in food and feed, *EFSA J.*, 9, 2481, 2011.
47. Wu, Q., Dohnal, V., Kuca, K., Yuan, Z., Trichothecenes: structure-toxic activity relationships, *Curr. Drug Metab.*, 14, 641, 2013.
48. Beyer, M., Ferse, I., Mulac, D., Würthwein, E.U., Humpf, H.U., Structural elucidation of T-2 toxin thermal degradation products and investigations toward their occurrence in retail food, *J. Agric. Food Chem.*, 57, 1867, 2009.
49. Rotter, B.A., Prelusky, D.B., Pestka, J.J., Toxicology of deoxynivalenol (vomitoxin), *J. Toxicol. Environ. Health*, 48, 1, 1996.
50. Summerell, B.A., Leslie, J.F., Fifty years of *Fusarium*: how could nine species have ever been enough? *Fungal Divers.*, 50, 135, 2011.
51. Jee, Y., Noh, E.M., Cho, E.S., Son, H.Y., Involvement of the Fas and Fas ligand in testicular germ cell apoptosis by zearalenone in rat, *J. Vet. Sci.*, 11, 115, 2010.
52. Cho, E.S., Ryu, S.Y., Jung, J.Y., Park, B.K., Son, H.Y., Effects of red ginseng extract on zearalenone induced spermatogenesis impairment in rat, *J. Ginseng Res.*, 35, 294, 2011.
53. Chen, J.H. et al., Oxidant damage in Kashin-Beck disease and a rat Kashin-Beck disease model by employing T-2 toxin treatment under selenium deficient conditions, *J. Orthop. Res.*, 30, 1229, 2012.
54. Duca, R.C., Mabondzo, A., Bravin, F., Delaforge, M., *In vivo* effects of zearalenone on the expression of proteins involved in the detoxification of rat xenobiotics, *Environ. Toxicol.*, 27, 98, 2012.
55. Liu, G. et al., Biological system responses to zearalenone mycotoxin exposure by integrated metabolomic studies, *J. Agric. Food Chem.*, 61, 11212, 2013.
56. Koraïchi, F. et al., Neonatal exposure to zearalenone induces long term modulation of ABC transporter expression in testis, *Toxicology*, 310, 29, 2013.
57. Guan, F. et al., Histopathology of chondronecrosis development in knee articular cartilage in a rat model of Kashin-Beck disease using T-2 toxin and selenium deficiency conditions, *Rheumatol. Int.*, 33, 157, 2013.
58. Saint-Cyr, M.J., Perrin-Guyomard, A., Houée, P., Rolland, J.G., Laurentie, M., Evaluation of an oral subchronic exposure of deoxynivalenol on the composition of human gut microbiota in a model of human microbiota-associated rats, *PLoS One*, 8, e80578, 2013.
59. Voss, K.A., Riley, R.T., Moore, N.D., Burns, T.D., Alkaline cooking (nixtamalisation) and the reduction in the *in vivo* toxicity of fumonisin-contaminated corn in a rat feeding bioassay, *Food Addit. Contam. Part A Chem. Anal. Control Expo. Risk Assess.*, 30, 1415, 2013.
60. Nakashita, M. et al., Attenuation of acetic acid-induced gastric ulcer formation in rats by glucosylceramide synthase inhibitors, *Dig. Dis. Sci.*, 58, 354, 2013.
61. Khalil, A.A., Abou-Gabal, A.E., Abdellatef, A.A., Khalid, A.E., Protective role of probiotic lactic acid bacteria against dietary fumonisin B_1-induced toxicity and DNA-fragmentation in Sprague Dawley rats, *Prep. Biochem. Biotechnol.*, 45, 530, 2015.
62. Hassan, A.M., Abdel-Aziem, S.H., El-Nekeety, A.A., Abdel-Wahhab, M.A., *Panax ginseng* extract modulates oxidative stress, DNA fragmentation and up-regulate gene expression in rats sub chronically treated with aflatoxin B_1 and fumonisin B_1, *Cytotechnology*, 67, 861, 2015.
63. Zhang, R. et al., Combined effects of T-2 toxin and selenium deficiency on lipid peroxidation in rat's brain, *Wei Sheng Yan Jiu*, 43, 6, 2014.
64. Denli, M., Blandon, J.C., Guynot, M.E., Salado, S., Pérez, J.F., Efficacy of activated diatomaceous clay in reducing the toxicity of zearalenone in rats and piglets, *J. Anim. Sci.*, 93, 637, 2015.

65. Belli, P. et al., Fetal and neonatal exposure to the mycotoxin zearalenone induces phenotypic alterations in adult rat mammary gland, *Food Chem. Toxicol.*, 48, 2818, 2010.
66. Yan, D., Song, Y., Shen, B., Kang, P., Pei, F., Magnetic resonance imaging in the tibial epiphyseal growth plate development of Wistar rat, *J. Orthop. Surg. Res.*, 9, 39, 2014.
67. Kang, P., Yao, Y., Yang, J., Shen, B., Zhou, Z., Pei, F., An animal model of Kashin–Beck disease induced by a low-nutrition diet and exposure to T-2 toxin, *Osteoarthritis Cartilage*, 21, 1108, 2013.
68. Ngampongsa, S., Ito, K., Kuwahara, M., Ando, K., Tsubone, H., Reevaluation of arrhythmias and alterations of the autonomic nervous activity induced by T-2 toxin through telemetric measurements in unrestrained rats, *Toxicol. Mech. Methods*, 22, 662, 2012.
69. Yan, D. et al., Radiographic findings of Wistar rats fed with T-2 toxin and Kashin–Beck disease-affected diet, *Int. J. Rheum. Dis.*, 14, 92, 2011.
70. Wu, J., Jing, L., Yuan, H., Peng, S.Q., T-2 toxin induces apoptosis in ovarian granulosa cells of rats through reactive oxygen species-mediated mitochondrial pathway, *Toxicol. Lett.*, 202, 168, 2011.
71. Yan, D. et al., The effect of Kashin–Beck disease-affected feed and T-2 toxin on the bone development of Wistar rats, *Int. J. Rheum. Dis.*, 13, 266, 2010.
72. Yao, Y.F. et al., Study on the effect of T-2 toxin combined with low nutrition diet on rat epiphyseal plate growth and development, *Int. Orthop.*, 34, 1351, 2010.
73. Riedel, S., Abel, S., Swanevelder, S., Gelderblom, W.C., Induction of an altered lipid phenotype by two cancer promoting treatments in rat liver, *Food Chem. Toxicol.*, 78, 96, 2015.
74. Pellanda, H. et al., Fumonisin FB_1 treatment acts synergistically with methyl donor deficiency during rat pregnancy to produce alterations of H3- and H4-histone methylation patterns in fetuses, *Mol. Nutr. Food Res.*, 56, 976, 2012.
75. Salah-Abbès, J.B., Abbès, S., Abdel-Wahhab, M., Oueslati, R., Immunotoxicity of zearalenone in Balb/c mice in a high subchronic dosing study counteracted by *Raphanus sativus* extract, *Immunopharmacol. Immunotoxicol.*, 32, 628, 2010.
76. Ji, F., Li, H., Xu, J., Shi, J., Enzyme-linked immunosorbent-assay for deoxynivalenol (DON), *Toxins (Basel)*, 3, 968, 2011.
77. Chatopadhyay, P., Gupta, V., Gogoi, H.K., Singh, L., Hematotoxicity of deoxynivalenol in BALB/c mice, *J. Pharmacol. Pharmacother.*, 2, 115, 2011.
78. Abbès, S., Ben Salah-Abbès, J., Sharafi, H., Noghabi, K.A., Oueslati, R., Interaction of *Lactobacillus plantarum* MON03 with Tunisian montmorillonite clay and ability of the composite to immobilize zearalenone *in vitro* and counteract immunotoxicity *in vivo*, *Immunopharmacol. Immunotoxicol.*, 34, 944, 2012.
79. Chatopadhyay, P., Pandey, A., Chaurasia, A.K., Upadhyay, A., Karmakar, S., Singh, L., Hepatic hyperplasia and damages induces by zearalenone *Fusarium* mycotoxins in BALB/c mice, *Arq. Gastroenterol.*, 49, 77, 2012.
80. Abbès, S., Ben Salah-Abbès, J., Sharafi, H., Oueslati, R., Noghabi, K.A., *Lactobacillus paracasei* BEJ01 prevents immunotoxic effects during chronic zearalenone exposure in Balb/c mice, *Immunopharmacol. Immunotoxicol.*, 35, 341, 2013.
81. Islam, M.R., Roh, Y.S., Kim, J., Lim, C.W., Kim, B., Differential immune modulation by deoxynivalenol (vomitoxin) in mice, *Toxicol. Lett.*, 221, 152, 2013.
82. Maragos, C.M., Kurtzman, C., Busman, M., Price, N., McCormick, S., Development and evaluation of monoclonal antibodies for the glucoside of T-2 toxin (t2-glc), *Toxins (Basel)*, 5, 1299, 2013.
83. Karuna, R., Rao, B.S., Lack of micronuclei induction by fumonisin B_1 mycotoxin in BALB/c mice, *Mycotoxin Res.*, 29, 9, 2013.
84. Ling, S. et al., Preparation and identification of monoclonal antibody against fumonisin B_1 and development of detection by Ic-ELISA, *Toxicon*, 80, 64, 2014.
85. Sozmen, M., Devrim, A.K., Tunca, R., Bayezit, M., Dag, S., Essiz, D., Protective effects of silymarin on fumonisin B_1-induced hepatotoxicity in mice, *J. Vet. Sci.*, 15, 51, 2014.
86. Salimian, J., Arefpour, M.A., Riazipour, M., Poursasan, N., Immunomodulatory effects of selenium and vitamin E on alterations in T lymphocyte subsets induced by T-2 toxin, *Immunopharmacol. Immunotoxicol.*, 36, 275, 2014.
87. Sun, L.H., Lei, M.Y., Zhang, N.Y., Zhao, L., Krumm, C.S., Qi, D.S., Hepatotoxic effects of mycotoxin combinations in mice, *Food Chem. Toxicol.*, 74, 289, 2014.
88. Ahmadi, A., Poursasan, N., Amani, J., Salimian, J., Adverse effect of T-2 toxin and the protective role of selenium and vitamin E on peripheral blood B lymphocytes, *Iran. J. Immunol.*, 12, 64, 2015.

89. Hou, Y.J. et al., Mycotoxin-containing diet causes oxidative stress in the mouse, *PLoS One*, 8, e60374, 2013.
90. Zhu, C.C. et al., Effect of mycotoxin-containing diets on epigenetic modifications of mouse oocytes by fluorescence microscopy analysis, *Microsc. Microanal.*, 20, 1158, 2014.
91. Hou, Y.J. et al., Oocyte quality in mice is affected by a mycotoxin-contaminated diet, *Environ. Mol. Mutagen.*, 55, 354, 2014.
92. Zhu, C.C., Hou, Y.J., Han, J., Cui, X.S., Kim, N.H., Sun, S.C., Zearalenone exposure affects epigenetic modifications of mouse eggs, *Mutagenesis*, 29, 489, 2014.
93. Amuzie, C.J., Pestka, J.J., Suppression of insulin-like growth factor acid-labile subunit expression—a novel mechanism for deoxynivalenol-induced growth retardation, *Toxicol. Sci.*, 113, 412, 2010.
94. Amuzie, C.J., Flannery, B.M., Ulrich, A.M., Pestka, J.J., Effects of deoxynivalenol consumption on body weight and adiposity in the diet-induced obese mouse, *J. Toxicol. Environ. Health A*, 74, 658, 2011.
95. Flannery, B.M., Wu, W., Pestka, J.J., Characterization of deoxynivalenol-induced anorexia using mouse bioassay, *Food Chem. Toxicol.*, 49, 1863, 2011.
96. Kobayashi-Hattori, K., Amuzie, C.J., Flannery, B.M., Pestka, J.J., Body composition and hormonal effects following exposure to mycotoxin deoxynivalenol in the high-fat diet-induced obese mouse, *Mol. Nutr. Food Res.*, 55, 1070, 2011.
97. Flannery, B.M., He, K., Pestka, J.J., Deoxynivalenol-induced weight loss in the diet-induced obese mouse is reversible and PKR-independent, *Toxicol. Lett.*, 221, 9, 2013.
98. Pan, X., Whitten, D.A., Wu, M., Chan, C., Wilkerson, C.G., Pestka, J.J., Early phosphoproteomic changes in the mouse spleen during deoxynivalenol-induced ribotoxic stress, *Toxicol. Sci.*, 135, 129, 2013.
99. Wu, W. et al., Effects of oral exposure to naturally-occurring and synthetic deoxynivalenol congeners on proinflammatory cytokine and chemokine mRNA expression in the mouse, *Toxicol. Appl. Pharmacol.*, 278, 107, 2014.
100. van Kol, S.W., Hendriksen, P.J., van Loveren, H., Peijnenburg, A., The effects of deoxynivalenol on gene expression in the murine thymus, *Toxicol. Appl. Pharmacol.*, 250, 299, 2011.
101. Zhao, Y. et al., Evaluation of fetal skeletal malformations in deoxynivalenol-treated mice using microarray analysis, *Arch. Environ. Contam. Toxicol.*, 63, 445, 2012.
102. Flannery, B.M., Amuzie, C.J., Pestka, J.J., Evaluation of insulin-like growth factor acid-labile subunit as a potential biomarker of effect for deoxynivalenol-induced proinflammatory cytokine expression, *Toxicology*, 304, 192, 2013.
103. Zhao, F. et al., Postweaning exposure to dietary zearalenone, a mycotoxin, promotes premature onset of puberty and disrupts early pregnancy events in female mice, *Toxicol. Sci.*, 132, 431, 2013.
104. Tardivel, C. et al., The food born mycotoxin deoxynivalenol induces low-grade inflammation in mice in the absence of observed-adverse effects, *Toxicol. Lett.*, 232, 601, 2015.
105. López-Casas, P.P., Mizrak, S.C., López-Fernández, L.A., Paz, M., de Rooij, D.G., del Mazo, J., The effects of different endocrine disruptors defining compound-specific alterations of gene expression profiles in the developing testis, *Reprod. Toxicol.*, 33, 106, 2012.
106. Liao, Y.J. et al., Inhibition of fumonisin B_1 cytotoxicity by nanosilicate platelets during mouse embryo development, *PLoS One*, 9, e112290, 2014.
107. Zatecka, E. et al., Effect of zearalenone on reproductive parameters and expression of selected testicular genes in mice, *Reprod. Toxicol.*, 45, 20, 2014.
108. Chaudhary, M., Rao, P.V., Brain oxidative stress after dermal and subcutaneous exposure of T-2 toxin in mice, *Food Chem. Toxicol.*, 48, 3436, 2010.
109. Agrawal, M., Yadav, P., Lomash, V., Bhaskar, A.S., Lakshmana Rao, P.V., T-2 toxin induced skin inflammation and cutaneous injury in mice, *Toxicology*, 302, 255, 2012.
110. Agrawal, M., Pardasani, D., Lakshmana Rao, P.V., Evaluation of protective efficacy of CC-2 formulation against topical lethal dose of T-2 toxin in mice, *Food Chem. Toxicol.*, 50, 1098, 2012.
111. Boeira, S.P. et al., Possible role for glutathione-S-transferase in the oligozoospermia elicited by acute zearalenone administration in Swiss albino mice, *Toxicon*, 60, 358, 2012.
112. Kouadio, J.H., Moukha, S., Brou, K., Gnakri, D., Lipid metabolism disorders, lymphocytes cells death, and renal toxicity induced by very low levels of deoxynivalenol and fumonisin B_1 alone or in combination following 7 days oral administration to mice, *Toxicol. Int.*, 20, 218, 2013.
113. Boeira, S.P. et al., Lycopene treatment prevents hematological, reproductive and histopathological damage induced by acute zearalenone administration in male Swiss mice, *Exp. Toxicol. Pathol.*, 66, 179, 2014.

114. Boeira, S.P. et al., Lycopene protects against acute zearalenone-induced oxidative, endocrine, inflammatory and reproductive damages in male mice, *Chem. Biol. Interact.*, 230, 50, 2015.
115. Mishra, S., Tripathi, A., Chaudhari, B.P., Dwivedi, P.D., Pandey, H.P., Das, M., Deoxynivalenol induced mouse skin cell proliferation and inflammation via MAPK pathway, *Toxicol. Appl. Pharmacol.*, 279, 186, 2014.
116. Chaudhary, M., Bhaskar, A.S., Rao, P.V., Differential effects of route of T-2 toxin exposure on hepatic oxidative damage in mice, *Environ. Toxicol.*, 30, 64, 2015.
117. Yuan, H. et al., *Gynostemma pentaphyllum* protects mouse male germ cells against apoptosis caused by zearalenone via Bax and Bcl-2 regulation, *Toxicol. Mech. Methods*, 20, 153, 2010.
118. Yang, J., Zhang, Y., Jing, A., Ma, K., Gong, Q., Qin, C., Effects of T-2 toxin on testosterone biosynthesis in mouse Leydig cells, *Toxicol. Ind. Health*, 30, 873, 2014.
119. Gelineau-van Waes, J. et al., Increased sphingoid base-1-phosphates and failure of neural tube closure after exposure to fumonisin or FTY720, *Birth Defects Res. A Clin. Mol. Teratol.*, 94, 790, 2012.
120. Voss, K.A., Riley, R.T., Gelineau-van Waes, J., Fumonisin B_1 induced neural tube defects were not increased in LM/Bc mice fed folate-deficient diet, *Mol. Nutr. Food Res.*, 58, 1190, 2014.
121. Riley, R.T. et al., A blood spot method for detecting fumonisin-induced changes in putative sphingolipid biomarkers in LM/Bc mice and humans, *Food Addit. Contam. Part A Chem. Anal. Control Expo. Risk Assess.*, 10, 1, 2015.
122. Choi, B.K. et al., Effects of oral deoxynivalenol exposure on immune-related parameters in lymphoid organs and serum of mice vaccinated with porcine parvovirus vaccine, *Mycotoxin Res.*, 29, 185, 2013.
123. Somoskői, B., Kovács, M., Cseh, S., Effects of T-2 mycotoxin on *in vitro* development and chromatin status of mouse embryos being in preimplantation stages, *Toxicol. Ind. Health*, 32, 1260, 2016.
124. Alizadeh, A.M. et al., Apoptotic and proliferative activity of mouse gastric mucosa following oral administration of fumonisin B_1, *Iran. J. Basic Med. Sci.*, 18, 8, 2015.
125. Girardet, C. et al., Central inflammation and sickness-like behavior induced by the food contaminant deoxynivalenol: a PGE2-independent mechanism, *Toxicol. Sci.*, 124, 179, 2011.
126. Bondy, G. et al., Effects of long term exposure to the mycotoxin fumonisin B_1 in p53 heterozygous and p53 homozygous transgenic mice, *Food Chem. Toxicol.*, 50, 3604, 2012.
127. Pinton, P., Braicu, C., Nougayrede, J.P., Laffitte, J., Taranu, I., Oswald, I.P., Deoxynivalenol impairs porcine intestinal barrier function and decreases the protein expression of claudin-4 through a mitogen-activated protein kinase-dependent mechanism, *J. Nutr.*, 140, 1956, 2010.
128. Diesing, A.K. et al., Vulnerability of polarised intestinal porcine epithelial cells to mycotoxin deoxynivalenol depends on the route of application, *PLoS One*, 6, e17472, 2011.
129. Diesing, A.K. et al., Gene regulation of intestinal porcine epithelial cells IPEC-J2 is dependent on the site of deoxynivalenol toxicological action, *PLoS One*, 7, e34136, 2012.
130. Pacheco, G.D., Silva, C.A., Pinton, P., Oswald, I.P., Bracarense, A.P., Phytic acid protects porcine intestinal epithelial cells from deoxynivalenol (DON) cytotoxicity, *Exp. Toxicol. Pathol.*, 64, 345, 2012.
131. Awad, W.A., Aschenbach, J.R., Zentek, J., Cytotoxicity and metabolic stress induced by deoxynivalenol in the porcine intestinal IPEC-J2 cell line, *J. Anim. Physiol. Anim. Nutr. (Berl)*, 96, 709, 2012.
132. Martínez, G., Pérez, D.S., Soraci, A.L., Tapia, M.O., Penetration of fosfomycin into IPEC-J2 cells in the presence or absence of deoxynivalenol, *PLoS One*, 8, e75068, 2013.
133. Wan, L.Y., Turner, P.C., El-Nezami, H., Individual and combined cytotoxic effects of *Fusarium* toxins (deoxynivalenol, nivalenol, zearalenone and fumonisins B_1) on swine jejunal epithelial cells, *Food Chem. Toxicol.*, 57, 276, 2013.
134. Wan, L.Y., Woo, C.S., Turner, P.C., Wan, J.M., El-Nezami, H., Individual and combined effects of *Fusarium* toxins on the mRNA expression of pro-inflammatory cytokines in swine jejunal epithelial cells, *Toxicol. Lett.*, 220, 238, 2013.
135. Wan, L.Y., Woo, C.S., Allen, K.J., Turner, P.C., El-Nezami, H., Modulation of porcine β-defensins 1 and 2 upon individual and combined *Fusarium* toxin exposure in a swine jejunal epithelial cell line, *Appl. Environ. Microbiol.*, 79, 2225, 2013.
136. Gu, M.J., Song, S.K., Park, S.M., Lee, I.K., Yun, C.H., *Bacillus subtilis* protects porcine intestinal barrier from deoxynivalenol via improved zonula occludens-1 expression, *Asian-Australas. J. Anim. Sci.*, 27, 580, 2014.
137. Kolf-Clauw, M. et al., The emerging mycotoxin, enniatin B_1, down-modulates the gastrointestinal toxicity of T-2 toxin *in vitro* on intestinal epithelial cells and *ex vivo* on intestinal explants, *Arch. Toxicol.*, 87, 2233, 2013.

138. Kruber, P., Trump, S., Behrens, J., Lehmann, I., T-2 toxin is a cytochrome P450 1A1 inducer and leads to MAPK/p38- but not aryl hydrocarbon receptor-dependent interleukin-8 secretion in the human intestinal epithelial cell line Caco-2, *Toxicology*, 284, 34, 2011.
139. Ribonnet, L. et al., Potential of an *in vitro* toolbox combined with exposure data as a first step for the risk assessment of dietary chemical contaminants, *Food Addit. Contam. Part A Chem. Anal. Control Expo. Risk Assess.*, 28, 1136, 2011.
140. Kadota, T., Furusawa, H., Hirano, S., Tajima, O., Kamata, Y., Sugita-Konishi, Y., Comparative study of deoxynivalenol, 3-acetyldeoxynivalenol, and 15-acetyldeoxynivalenol on intestinal transport and IL-8 secretion in the human cell line Caco-2, *Toxicol. In Vitro*, 27, 1888, 2013.
141. Clarke, R., Connolly, L., Frizzell, C., Elliott, C.T., Cytotoxic assessment of the regulated, co-existing mycotoxins aflatoxin B_1, fumonisin B_1 and ochratoxin, in single, binary and tertiary mixtures, *Toxicon*, 90, 70, 2014.
142. Li, D. et al., Gene expression profiling analysis of deoxynivalenol-induced inhibition of mouse thymic epithelial cell proliferation, *Environ. Toxicol. Pharmacol.*, 36, 557, 2013.
143. Li, D. et al., Deoxynivalenol induces apoptosis in mouse thymic epithelial cells through mitochondria-mediated pathway, *Environ. Toxicol. Pharmacol.*, 38, 163, 2014.
144. Casteel, M., Nielsen, C., Kothlow, S., Dietrich, R., Märtlbauer, E., Impact of DUSP1 on the apoptotic potential of deoxynivalenol in the epithelial cell line HepG2, *Toxicol. Lett.* 199, 43, 2010.
145. Sugiyama, K., Kinoshita, M., Kamata, Y., Minai, Y., Tani, F., Sugita-Konishi, Y., Thioredoxin-1 contributes to protection against DON-induced oxidative damage in HepG2 cells, *Mycotoxin Res.*, 28, 163, 2012.
146. Chuturgoon, A.A., Phulukdaree, A., Moodley, D., Fumonisin B_1 modulates expression of human cytochrome P450 1b1 in human hepatoma (Hepg2) cells by repressing Mir-27b, *Toxicol. Lett.*, 227, 50, 2014.
147. Weidner, M., Welsch, T., Hübner, F., Schwerdt, G., Gekle, M., Humpf, H.U., Identification and apoptotic potential of T-2 toxin metabolites in human cells, *J. Agric. Food Chem.*, 60, 5676, 2012.
148. Weidner, M., Lenczyk, M., Schwerdt, G., Gekle, M., Humpf, H.U., Neurotoxic potential and cellular uptake of T-2 toxin in human astrocytes in primary culture, *Chem. Res. Toxicol.*, 26, 347, 2013.
149. Nagashima, H., Nakagawa, H., Kushiro, M., Opposite effects of two trichothecene mycotoxins, deoxynivalenol and nivalenol, on the levels of macrophage inflammatory protein (MIP)-1α and MIP-1β in HL60 cells, *Environ. Toxicol. Pharmacol.*, 34, 1014, 2012.
150. Hou, Y., Yuan, Z., Guo, W., Wang, H., Effects of T-2 toxin of *Fusarium* on proliferation and apoptosis of human acute promyelocytic leukemia cell line HL60, *Wei Sheng Yan Jiu*, 42, 381, 2013.
151. Katika, M.R., Hendriksen, P.J., Shao, J., van Loveren, H., Peijnenburg, A., Transcriptome analysis of the human T lymphocyte cell line Jurkat and human peripheral blood mononuclear cells exposed to deoxynivalenol (DON): new mechanistic insights, *Toxicol. Appl. Pharmacol.*, 264, 51, 2012.
152. Katika, M.R., Hendriksen, P.J., van Loveren, H., Peijnenburg, A.C.M., Characterization of the modes of action of deoxynivalenol (DON) in the human Jurkat T-cell line, *J. Immunotoxicol.*, 12, 206, 2015.
153. Schmeits, P.C., Volger, O.L., Zandvliet, E.T., van Loveren, H., Peijnenburg, A.A., Hendriksen, P.J., Assessment of the usefulness of the murine cytotoxic T cell line CTLL-2 for immunotoxicity screening by transcriptomics, *Toxicol. Lett.*, 217, 1, 2013.
154. Bae, H., Gray, J.S., Li, M., Vines, L., Kim, J., Pestka, J.J., Hematopoietic cell kinase associates with the 40S ribosomal subunit and mediates the ribotoxic stress response to deoxynivalenol in mononuclear phagocytes, *Toxicol. Sci.*, 115, 444, 2010.
155. Minervini, F. et al., Toxic mechanisms induced by fumonisin B_1 mycotoxin on human intestinal cell line, *Arch. Environ. Contam. Toxicol.*, 67, 115, 2014.
156. Mahmoodi, M. et al., Impact of fumonisin B_1 on the production of inflammatory cytokines by gastric and colon cell lines, *Iran. J. Allergy Asthma Immunol.*, 11, 165, 2012.
157. Bensassi, F., Gallerne, C., Sharaf El Dein, O., Lemaire, C., Hajlaoui, M.R., Bacha, H., Involvement of mitochondria-mediated apoptosis in deoxynivalenol cytotoxicity, *Food Chem. Toxicol.*, 50, 1680, 2012.
158. Zhou, H.R., Pestka, J.J., Deoxynivalenol (vomitoxin)-induced cholecystokinin and glucagon-like peptide-1 release in the STC-1 enteroendocrine cell model is mediated by calcium-sensing receptor and transient receptor potential ankyrin-1 channel, *Toxicol. Sci.*, 145, 407, 2015.
159. Bianco, G., Fontanella, B., Severino, L., Quaroni, A., Autore, G., Marzocco, S., Nivalenol and deoxynivalenol affect rat intestinal epithelial cells: a concentration related study, *PLoS One*, 7, e52051, 2012.

160. Osman, A.M., Pennings, J.L., Blokland, M., Peijnenburg, A., van Loveren, H., Protein expression profiling of mouse thymoma cells upon exposure to the trichothecene deoxynivalenol (DON): implications for its mechanism of action, *J. Immunotoxicol.*, 7, 147, 2010.
161. Liu, Y. et al., Effects of low dose T-2 toxin on secretion of gonadotropin-releasing hormone in the immortalized hypothalamic GT1-7 cell line, *Toxicon*, 100, 67, 2015.
162. Lin, P. et al., Mycotoxin zearalenone induces apoptosis in mouse Leydig cells via an endoplasmic reticulum stress-dependent signalling pathway, *Reprod. Toxicol.*, 52, 71, 2015.
163. Wang, Y., Wong, T.Y., Chan, F.L., Chen, S., Leung, L.K., Assessing the effect of food mycotoxins on aromatase by using a cell-based system, *Toxicol. In Vitro*, 28, 640, 2014.
164. Savard, C. et al., Effect of deoxynivalenol (DON) mycotoxin on *in vivo* and *in vitro* porcine circovirus type 2 infections, *Vet. Microbiol.*, 176, 257, 2015.
165. Lei, M., Zhang, N., Qi, D., *In vitro* investigation of individual and combined cytotoxic effects of aflatoxin B_1 and other selected mycotoxins on the cell line porcine kidney 15, *Exp. Toxicol. Pathol.*, 65, 1149, 2013.
166. Gao, F. et al., Genotoxic effects induced by zearalenone in a human embryonic kidney cell line, *Mutat. Res.*, 755, 6, 2013.
167. Meca, G., Ruiz, M.J., Fernandez-Franzón, M., Ritieni, A., Manes, J., Isolation, purification, LC-MS/MS characterization and reactive oxygen species induced by fumonisin B_1 in Vero cells, *Food Chem. Toxicol.*, 48, 2891, 2010.
168. Mary, V.S., Valdehita, A., Navas, J.M., Rubinstein, H.R., Fernández-Cruz, M.L., Effects of aflatoxin B_1, fumonisin B_1 and their mixture on the aryl hydrocarbon receptor and cytochrome P450 1A induction, *Food Chem. Toxicol.*, 75, 104, 2015.
169. Wang, S.K., Liu, S., Yang, L.G., Shi, R.F., Sun, G.J., Effect of fumonisin B_1 on the cell cycle of normal human liver cells, *Mol. Med. Rep.*, 7, 1970, 2013.
170. Pietsch, C., Noser, J., Wettstein, F.E., Burkhardt-Holm, P., Unraveling the mechanisms involved in zearalenone-mediated toxicity in permanent fish cell cultures, *Toxicon*, 88, 44, 2014.

35

Penicillium *and* Talaromyces

Elena Bermúdez, Félix Núñez, Josué Delgado, and Miguel A. Asensio

CONTENTS

35.1 Introduction

The genus *Penicillium* is a group of anamorphic fungi in the division Ascomycota, with over 250 recognized species, many of them ubiquitous in warm and moderate climates. The genus *Talaromyces* was described as a sexual state of *Penicillium* that produces soft-walled ascomata. Recently, the genus *Penicillium* has been redefined into *Penicillium sensu stricto* and *Talaromyces*, with all species of the former *Penicillium* subgenus *Biverticillium* being transferred to *Talaromyces*.[1,2] Species like *Talaromyces marneffei* lack a defined sexual cycle but possesses all genes believed to be required for mating.[3]

Penicillium species play important roles in the environment, agriculture, and industry. Some species of genus *Penicillium* are of economic importance to the food industry because they contribute to food ripening, while others are postharvest pathogens or cause spoilage. For example, *Penicillium camemberti* and *Penicillium roqueforti* are used for cheese manufacture; *Penicillium nalgiovense* and *Penicillium chrysogenum* contribute to ripening of dry-cured meat products. On the other hand, *Penicillium expansum* is the causal agent of blue mold postharvest rots of apples and is also able to produce patulin and other mycotoxins, as discussed later. *Penicillium digitatum* and *Penicillium italicum* are responsible for postharvest citrus decay. Heat-resistant ascospores produced by various *Talaromyces* spp. cause spoilage of pasteurized juices and other fruit-based products.[4]

Penicillia are generally soil saprophytes or plant pathogens with limited invasive ability for animal tissue, except for *Talaromyces marneffei* (formerly *Penicillium marneffei*[2]). However, production of toxic secondary metabolites is common in *Penicillium* spp. In addition, *Penicillium* is an environmental allergen of variable significance among asthmatic and atopic individuals.

35.2 Pathogens

35.2.1 General Description

The genera *Penicillium* and *Talaromyces* comprise animal pathogens. *Talaromyces radicus* and *Talaromyces helicus* have been described as rare causative agents of infectious diseases in dogs.[5,6] *Penicillium corylophilum* infects red snapper (*Lutjanus campechanus*), but infection seems to be confined to the swim bladder.[7]

Penicillium infections in humans are still called penicilliosis, even those due to *Talaromyces marneffei*. Members of the *P. chrysogenum–Penicillium rubens* complex can be responsible for rare opportunistic keratitis[8] and *Talaromyces piceum* has been rarely reported in chronic granulomatous disease.[9] However, the most outstanding infection by *Penicillium* or *Talaromyces* spp. is due to *T. marneffei*.

T. marneffei is responsible for one of the commonest opportunistic infections in humans with HIV-infected patients, causing fatal systemic mycosis in Southeast Asian countries. However, infection by *T. marneffei* has also been described as a result of other immunodeficiencies, immunosuppressive therapy, and genetic diseases leading to lymphocytes exhibiting defective IFN-γ production.[10]

T. marneffei was first identified as the causative agent of mycosis of the reticuloendothelial system.[11] *T. marneffei* can cause a focal infection that may lead to a progressive disseminated penicilliosis, mainly in immunocompromised patients. The focal infection can be cutaneous or bronchopulmonary, and it may spread to cause localized lymphadenopathy. However, *T. marneffei* can also be responsible for asymptomatic fungemia in HIV-infected patients.[12]

The mode of transmission is not well understood but is probably via ingestion or inhalation of the fungus.[13] By analogy with other endemic fungal pathogens, infection is presumably initiated by the inhalation of airborne conidia that are subsequently phagocytized by pulmonary alveolar macrophages. After an inflammatory response in macrophages, *T. marneffei* can lead to granulomatous or suppurative reactions in persons with normal immunity, whereas immunocompromised patients develop necrotizing reaction.[13] It is thought that the fungus is disseminated throughout the body to result in systemic infection. In some cases, the disease did not appear until immunosuppression allowed for the growth and dissemination throughout the host. Thus, unique skin lesions like abscesses and ulcers are more commonly noted in non-HIV-infected patients than in HIV-infected patients.[14] *T. marneffei* disseminates hematogenously causing generalized lymphadenopathy, even leading to osteolysis, which indicates severe systemic disturbance.[15]

Adaptation to the new environmental conditions during infection of human hosts is key to fungal survival.[16] *T. marneffei* is the only dimorphic species of *Penicillium* and *Talaromyces* that is capable of growing in a filamentous multinucleate hyphal form at 25°C or as a uninucleate pathogenic yeast form at 37°C.[17] When grown in air at 25°C, the saprophytic hyphae produce asexual spores (conidia), the infectious agents. Upon invasion of a human host, most environmental factors undergo immediate and drastic changes, including temperature, pH, nutrient sources, carbon dioxide, and oxygen concentrations. Temperature adaptation of *T. marneffei* within the host manifests itself as a morphological change from hyphal mold to budding yeast.

It seems that *T. marneffei* possesses the ability to evade a normally functioning immune system until *in vivo* conditions, due to immunosuppression, are favorable to dissemination throughout the host.[18] Upon switching to 37°C, *T. marneffei* undergoes a process termed arthroconidiation. Conidia germinate with the onset of extensive hyphal branching. Fragmentation occurs to generate single, uninucleated yeast cells that divide by fission. Switching from yeast growth at 37°C to hyphal growth at 25°C requires the polarized growth of yeast cells, septation, and branching, to form the characteristic multinucleate hyphal mycelium. Genes responsible for growth, morphogenesis, and development of *T. marneffei* have been characterized.[19] Conidial germination and temperature adaptation require both HHK and heterotrimeric G-protein-Ras signaling, as well as the establishment of actin-mediated polarized growth through a Rho GTPase. In addition, the p21-activated kinase pathway is involved in sensing the environment inside a host cell by *T. marneffei*, being central to eliciting the appropriate morphogenetic response to the host environment.[20] Systems for detecting and responding to changes in carbon sources also play a major role in adaptation to the host niche and are essential factors for persistence in a mammalian host. Many of the genes whose expression is upregulated during the mold-to-yeast transition are related to those genes involved in energy metabolism. The temperature-dependent regulation of isocitrate lyase, a key component of the glyoxylate cycle, activates the glyoxylate cycle for utilization of poor carbon sources.[16]

Other potential virulence factors include yeast phase-specific or upregulated genes that encode enzymes known to combat oxidative host defense responses. Potential virulence factors related to host-cell attachment have also been identified.[18]

Proteins secreted at different developmental stages and noncoding RNAs are suggested to play diverse roles in mycelium-to-yeast transition. RNA structural transition in response to temperature changes may be related to the control of thermal dimorphism.[21]

35.2.2 Diagnosis

The diagnosis of penicilliosis has been reviewed recently.[22] Patients suffering from penicilliosis may have skin, palatal, and pharyngeal lesions such as subcutaneous abscesses, molluscum contagiosum-like lesions and shallow papule-like ulcers. Respiratory signs include productive cough, dyspnea, and hemoptysis. The diagnosis is based on the identification of the organism on microscopy with the confirmation by the culture method through examination of cytology or biopsy specimens from skin, bone marrow, or lymph nodes.[23] In fungemia, yeast cells may be seen inside monocytes in peripheral blood smear. Detection of nonbudding yeast cells with characteristic central transverse septum should be confirmed by isolation of *T. marneffei* on Sabouraud's dextrose agar at 25°C. A diffusible dark red pigment that is produced when grown in the mycelial phase at 37°C is one of the first indications that an isolate may be *T. marneffei*.[24]

Diagnosis may also be made by the detection of antibodies produced against various fungal proteins, or by the detection of fungal proteins in sera or in urine.[25–27] A specific exoantigen test using the immunodiffusion technique was devised to identify *T. marneffei* cultures.[28]

Also, specific tests have been developed for detecting circulating *T. marneffei* antigens in serum and urine specimens by latex agglutination, enzyme immunoassay, and dot blot enzyme-linked immunosorbent assay (ELISA) with polyclonal antibodies; and a monoclonal antibody-based sandwich ELISA.[27]

Several serologic methods have also been developed for detecting specific antibodies against *T. marneffei* antigens in clinical specimens. These include an indirect fluorescent-antibody test for detecting IgG antibodies produced against germinating conidia and yeast forms, immunoblot assays for detecting at least two protein antigens produced during the growth phase of the yeast, Western blotting for three cytoplasmic antigens from yeast-form *T. marneffei*, and an ELISA-based antibody test developed with a recombinant *T. marneffei* mannoprotein.[27]

Specific oligonucleotide primers from the nuclear ribosomal DNA internal transcribed spacer region have been designed for the specific and selective amplification of *T. marneffei* DNA in a PCR-identification system from clinical material.[29] Similarly, oligonucleotide probes based on the 18S rRNA gene of *T. marneffei* have been used to identify *T. marneffei* DNA from clinical samples in PCR-hybridization reactions.[27]

35.2.3 Molecular Epidemiology

T. marneffei isolates have been classified using various methods that randomly sample for genetic variation, including restriction fragment length polymorphisms (RFLP), randomly amplified polymorphic

DNA (RAPD), and pulsed-field gel electrophoresis (PFGE) using restriction enzymes.[27] A number of types have been identified, but no correlation between restriction patterns and geographic region was observed. On the other hand, multilocus microsatellite typing allowed discrimination between isolates occurring within the "eastern" clade, from mainland China, Hong Kong, Indonesia, and Vietnam, or the "western" clade, from Thailand and India.[30]

T. marneffei life cycle has been related to the high prevalence of infection in the organs of the bamboo rats, including hoary bamboo rat (*Rhizomys pruinosus*), Chinese bamboo rat (*Rhizomys sinensis*), large bamboo rat (*Rhizomys sumatrensis*), and bay bamboo rat (*Cannomys badidus*), as well as the soil around their burrows.[14] Bamboo rats are the only known nonhuman host of *T. marneffei*, even though these animals showed no signs of illness.[25]

Bamboo rat and human coinfection can occur from a common environmental source, rather than the patients being infected from the rats.[27] However, *T. marneffei* has never been isolated from soil, water, vegetation, or air, other than from bamboo rats and their burrows. Multilocus genotypes show that *T. marneffei* isolates from humans are similar to those infecting rats and are in some cases identical, suggesting that bamboo rats can be a vector for human infections by acting as amplifiers of infectious dispersal stages.[31]

The asexual nature of *T. marneffei* would have led to the evolution of niche-adapted genotypes, which could explain the geographically restricted endemicity of this fungus.[18]

35.2.4 Treatment

Unlike some other emerging fungal pathogens, this organism remains sensitive to many antifungals.[32] *T. marneffei* is sensitive to amphotericin B, itraconazole, and other types of triazoles. However, long-term secondary prophylaxis with itraconazole has been recommended for HIV-infected patients to prevent reoccurrence.[33]

However, itraconazole interacts with antiretrovirals, which is an important issue in HIV-positive patients.[34]

35.3 Mycotoxins

35.3.1 General Description

The production of secondary metabolites is a characteristic of *Penicillium* species. Some of the extrolites produced by *Penicillium* spp. are toxic (Table 35.1).

Penicillium growth on foods and feeds poses a hazard to consumers due to these toxic metabolites. Mycotoxins produce toxic effects in different organisms, including mammals, birds, amphibians, arthropods, crustaceans, unicellular organisms, microorganisms, and plants. Taking advantage of this fact, different bioassays have been developed to study mycotoxin toxicity in foods and feeds.[41]

The sensitivity of bioassays is generally lower than that of chromatographic methods. However, bioassays do not require standards for the detection of mycotoxins. This is a major point, given that only a small part of the over 300 compounds documented as mycotoxins are commercially available. Therefore, analyses based on the biological action of mycotoxins allow testing various foods for the presence of both known and unknown *Penicillium* mycotoxins. On the other hand, biological methods are generally less reproducible, less sensitive, and require more time than physicochemical methods.[42]

The organisms used in biological methods for mycotoxins detection include microorganisms and invertebrate and vertebrate animals. The use of mammals for toxicity research is rather limited due to ethical objections as well as high costs for food, habitat, and animal welfare.

Organs and cell cultures have also been used. Cell lines and tissue cultures are the best alternatives to experimental animals in testing for toxic or pathogenic microorganisms.[43] Cell cultures are essential when seeking information on the effect on DNA or DNA-binding molecules, cell organelles, or cell membranes. But such studies are usually not necessary to detect toxicity in molds from foods. For this, simpler and easier tests are used, such as those with microorganisms or invertebrates.

TABLE 35.1

Mycotoxins and Potentially Toxic Extrolites Produced by *Penicillium* spp. and *Talaromyces* spp.

***Penicillium* spp. and *Talaromyces* spp.**	**Metabolites**
P. aethiopicum	Griseofulvin, viridicatumtoxin
P. albocoremium	Cyclopenol, cyclopenin, roquefortine C
P. allii	Cyclopenol, cyclopenin, roquefortine C
P. atramentosus	Roquefortine C, rugulovasine
P. aurantiogriseum	Chaetoglobosin A, penicillic acid, roquefortine C, terrestric acid, verrucosidin, viomellein, xanthomegnin
P. bialowiezense	Mycophenolic acid
P. brevicompactum	Brevianamide A, mycophenolic acid, patulin
P. camemberti	Cyclopiazonic acid
P. canescens	Griseofulvin, penitrem A
P. carneum	Mycophenolic acid, patulin, penicillic acid, penitrem A, roquefortine C
P. caseifulvum	Cyclopenin, rugulovasine
P. chrysogenum	Cyclopiazonic acid, ochratoxin A, patulin, penicillic acid, PR-toxin, roquefortine C, secalonic acid
P. clavigerum	Cyclopiazonic acid, patulin, penitrem A, xanthomegnin
P. citreonigrum	Citreoviridin
P. citrinum	Citreoviridin, citrinin
P. citreoviride	Citreoviridin
P. commune	Cyclopenin, cyclopiazonic acid, ochratoxin A, rugulovasine
P. concentricum	Patulin, roquefortine C
P. confertum	Asteltoxin
P. coprobium	Patulin, roquefortine C
P. coprophilum	Griseofulvin
P. corymbiferum	Patulin, roquefortine C
P. crustosum	Cyclopenol, cyclopenin, cyclopiazonic acid, penitrem A
P. cyaneo-fulvum	Patulin
P. cyclopium	Cyclopenol, cyclopiazonic acid, ochratoxin A, patulin, penicillic acid, roquefortine C, rugulosin, verrucofortine, viomellein, vioxanthin, xanthomegnin
P. dipodomyicola	Cyclopiazonic acid, griseofulvin, patulin
P. discolor	Cyclopenol, chaetoglobosins

(Continued)

TABLE 35.1 (*Continued*)

Mycotoxins and Potentially Toxic Extrolites Produced by *Penicillium* spp. and *Talaromyces* spp.

***Penicillium* spp. and *Talaromyces* spp.**	**Metabolites**
P. expansum	Chaetoglobosins, citrinin, cyclopiazonic acid, patulin, penitrem A, roquefortine C
P. flavigenum	Penicillin, penitrem A, roquefortine C, secalonic acids
P. formosanum	Patulin
P. freii	Cyclopenol, penicillic acid, xanthomegnin
T. funiculosus[a]	Patulin
P. glandicola	Patulin, penitrem A, roquefortine C
P. gladioli	Patulin
P. granulatum	Patulin
P. griseofulvum	Cyclopiazonic acid, griseofulvin, patulin, roquefortine C
P. hirsutum	Cyclopenol, cyclopenin, roquefortine C, terrestric acid
P. hordei	Roquefortine C, terrestric acid
T. islandicus[a]	Luteoskirin
P. janczewskii	Griseofulvin, penitrem A
P. marinum	Chaetoglobosins, patulin, roquefortine C
P. melanoconidium	Penicillic acid, penitrem A, verrucosidin, xanthomegnin
P. mononematosum	Fumitremorgins, isochromantoxin, verrucologen, viriditoxin
P. nalgiovense	Cyclopiazonic acid
P. neoechinulatum	Cyclopenol, penicillic acid
P. nordicum	Ochratoxin A, viridic acid
P. oxalicum	Secalonic acids
P. palitans	Cyclopenin, cyclopiazonic acid, fumigaclavine
P. paneum	Patulin, roquefortine C
P. paxilli	Verruculogen
P. persicinum	Griseofulvin, roquefortine C
P. polonicum	Cyclopenol, penicillic acid, verrucofortine. Verrucosidin
P. puberulum	Cyclopiazonic acid
T. purpurogenus[a]	Ochratoxin A, rubratoxin B

(*Continued*)

TABLE 35.1 (*Continued*)

Mycotoxins and Potentially Toxic Extrolites Produced by *Penicillium* spp. and *Talaromyces* spp.

***Penicillium* spp. and *Talaromyces* spp.**	**Metabolites**
P. radicicola	Citrinin, cyclopenol, cyclopenin, roquefortine C, terrestric acid
P. raistrickii	Griseofulvin
P. roqueforti	Mycophenolic acid, patulin, penicillic acid, PR-toxin, roquefortine C
P. rubrum	Rubratoxin, xanthoviridicatin
T. radicus[a]	Rugulosin, skyrin
T. rugulosus[a]	Ochratoxin A, rugulosin
P. simplicissimum	Penicillic acid, verruculogen
P. solitum	Cyclopenol
P. sclerotigenum	Griseofulvin, patulin, roquefortine C
P. tricolor	Asteltoxin, terrestric acid, verrucofortine, xanthomegnin
P. tularense	Janthitrem, paspalinine, paxilline
P. tulipae	Cyclopenol, cyclopenin, penitrem A, roquefortine C, terrestric acid
P. venetum	Cyclopenol, cyclopenin, roquefortine C, terrestric acid
P. verrucosum	Citrinin, ochratoxin A
P. viridicatum	Brevianamide A, citrinin, cyclopiazonic acid, ochratoxin A, penicillic acid, viomellein, viopurpurin, vioxanthin, viridamine, viridic acid, viridicatin, viridicatum toxin, xanthomegnin, xanthoviridicatin
P. vulpinum	Patulin, roquefortine C
T. wortmannii[a]	Rugulosin, skyrin

Source: Adapted from Yilmaz, N. et al., *Stud. Mycol.*, 78, 175, 2014; Moss, M.O., The environmental factors controlling mycotoxin formation, in *Mycotoxins and Animal Foods*, p. 37, Smith J.F. and Henderson R.S. (Eds.), CRC Press, Boca Raton, 1991; Mantle, P.G., Miscellaneous toxigenic fungi, in *Mycotoxins and Animal Foods*, p. 141, Smith J.E. and Henderson R.S. (Eds.), CRC Press, Boca Raton, 1991; Bullerman, L.B., Fusaria and toxigenic molds and other Aspergilli and Penicillia, in *Food Microbiology: Fundamentals and Frontiers*, p. 419, Doyle M.P., Beuchat L.R. and Montville T.J. (Eds.), ASM Press, Washington, 1997; Pitt, J.I., Toxigenic Penicillium species, in *Food Microbiology: Fundamentals and Frontiers*, p. 406, Doyle M.P., Beuchat L.R. and Montville T.J. (Eds.), ASM Press, Washington, 1997; Frisvad, J.C. and Samson, R.A., *Stud. Mycol.*, 49, 1, 2004; Barkai-Golan, R., Penicillium mycotoxins, in *Mycotoxins in Fruits and Vegetables*, p. 153, Barkai-Golan R. and Paster N. (Eds.), Academic Press, San Diego, 2008.

[a] Formerly *Penicillium*.[2]

35.3.2 Invertebrate Organisms in Toxicity Bioassays

Invertebrate organisms have been frequently used for toxicity tests because of their ease of handling. They do not need special care, are very prolific, and have a short generation time. These characteristics improve the statistical interpretation of the results and enhance the statistical significance without increased costs.[43]

Tests based on the use of aquatic animals, including crustaceans and protozoa, are among the most extensively used in screening bioassays to detect mycotoxins (Table 35.2). The brine shrimp test using *Artemia salina*, an aquatic crustacean, is perhaps one of the most extensively used screening bioassay to detect mycotoxins. Harwig and Scott[44] observed that most of the *Penicillium* mycotoxins showed low toxicity to *Artemia salina*, with mortality rates lower than 50%, whereas rubratoxin B reached over 90% mortality. *Moina macrocopa*, another fresh water crustacean used for bioassays, was more sensitive than *Artemia salina* to ochratoxin A and rubratoxin B.[45] Other crustaceans have been also used, such as *Daphnia magna* to detect toxicity of patulin[46] and *Cyclops fuscus* that is highly sensitive to the lethal effects of rubratoxin B and patulin at low doses.[47]

Tetrahymena pyriformis is a unicellular ciliate protozoan widely used as a test organism for *Penicillium* toxins. This organism has been used in toxicological bioassays because its metabolism is quite similar to that of higher animals, as was shown in studies on the toxicity of patulin in rats and *T. pyriformis*.[48,49] Several researchers have used this protozoan to detect the toxic effect of *Penicillium* mycotoxins (Table 35.2). However, *T. pyriformis* did not detect ochratoxin A at low concentrations.[54] Dive et al.[53] reported that some mycotoxins were responsible for growth inhibition of the ciliated protozoan *Colpidium campylum* (Table 35.2).

TABLE 35.2

Invertebrate Organisms Used in Bioassay Tests to Detect *Penicillium* Mycotoxins

Organism	Organism Type	Mycotoxins/Mold Detected	Test Type	References
Artemia salina	Aquatic crustacean	Patulin, citrinin, griseofulvin, penicillic acid, rubratoxin B	Disc screening method	44
Moina macrocopa	Aquatic crustacean	Ochratoxin A, rubratoxin B	Mortality	45
Daphnia magna	Aquatic crustacean	Patulin	Bioluminescent test	46
Cyclops fuscus	Aquatic crustacean	Rubratoxin B, patulin	Mortality	47
Tetrahymena pyriformis	Protozoan	Patulin Penicillic acid, gliotoxin Rubratoxin B Patulin	Mortality	48–52
Colpidium campylum	Protozoan	Patulin, ochratoxin B, roquefortin	Mortality	53
Anagasta kuehniella	Insects	*Penicillium* spp. and *Penicillium* mycotoxins	Mortality	41
Attagenus megatoma			Growth inhibition	
Corcyra cephalonica			Reduced larval growth	
Drosophila melanogaster				
Glycyphagus domesticus			Decrease in larva and adult size	
Heliothis zea				
Lasioderma serricorne			Fertility	
Lucilia sericata				
Spodoptera frugiperda				
Spodoptera littoralis				
Tenebrio molitor				
Tribolium confusum				

Other invertebrate animals used in toxicological bioassays are *insects*. Panigrahi[41] reviewed the tests using insects to detect the toxicity of mycotoxins produced by *Penicillium* spp. These tests are based on the effect of feeds intentionally contaminated with mycotoxins on growth or mortality of adults or larvae (Table 35.2).

35.3.3 Microorganisms in Toxicity Bioassays

The use of microorganisms to detect the toxicity of different species of *Penicillium* involves different tests (Table 35.3). Working with microorganisms is simple and usually faster and less expensive than using cell cultures or animals. Microorganisms have been used for quite some time to detect mycotoxins[42,55] and toxic effects from *Penicillium* spp.[56]

Tests based on *Bacillus thuringiensis* growth inhibition were optimized to detect low concentrations of *Penicillium* mycotoxins.[57,58] Watson and Lindsay[42] reviewed the use of *Bacillus* spp. to detect the toxic effect of certain mycotoxins, including some produced by *Penicillium* spp. Skaug et al.[59] used *Bacillus subtilis* to detect toxicity in conidial extracts from *Penicillium verrucosum* containing ochratoxin A.

TABLE 35.3

Microorganisms Used in Bioassay Tests to Detect *Penicillium* Mycotoxins

Organism	Organism Type	Mycotoxins Detected	Test Type	References
Azospirillum brasilense	Motile bacteria	Patulin, rubratoxin, penicillic acid, citrinin	Swarming inhibition assay	57
Bacillus thuringiensis	Bacteria	Kojic acid	Cup plate assay	57,58
		Citrinin, kojic acid, luteoskyrin, ochratoxin A, patulin, penicillic acid	Cup plate assay	
Bacillus subtilis	Bacteria	Ochratoxin A	Growth inhibition	59
		Patulin, rubratoxin, penicillic acid, citrinin		
Bacillus cereus	Bacteria	Ochratoxin A		42
Bacillus megaterium	Bacteria	Patulin, ochratoxin A		
Escherichia coli	Bacteria	Patulin, penicillic acid, ochratoxin A, rubratoxin B, kojic acid, citrinin, PR toxin	Genotoxicity in the SOS spot test	60,61
		Ochratoxin A		
Photobacterium phosphoreum	Bacteria	Rubratoxin B, penicillic acid, citrinin, ochratoxin A, PR-toxin, patulin	Bioluminescence assay	62
Proteus mirabilis	Motile bacteria	Patulin, rubratoxin, penicillic acid, citrinin	Swarming inhibition assay	57
Pseudomonas syringae	Bacteria	Patulin	Growth inhibition	52
Saccharomyces cerevisiae	Yeast	Rubratoxin B, penicillic acid, citrinin, ochratoxin A, PR-toxin, patulin	Mutagenicity test	62
Salmonella typhimurium	Bacteria	Ochratoxin A	Ames test (mutagenicity)	61,63,64,65,66
		Patulin		
		Ochratoxin A, B, citrinin		
		Citrinin, patulin, cyclopiazonic acid, luteoskyrin, griseofulvin, mycophenolic acid, ochratoxin A, penicillic acid, secalonic acid D		
Several microorganisms		Kojic acid	Growth inhibition	67

Another bioassay with bacteria is based on the motility inhibition of *Proteus mirabilis* and *Azospirillum brasilense* caused by some *Penicillium* toxins.[57] *Escherichia coli* is widely used in bioassays for toxicity and for genotoxic or mutagenic activity detection. Auffray and Boutibones[60] evaluated the genotoxic activity of some mycotoxins produced by strains of *Penicillium* using *Escherichia coli* in the SOS spot test. When comparing the efficiency to detect toxigenic activity between SOS spot test and other tests, including mutagenicity to *Salmonella typhimurium* (Ames test), *Bacillus subtilis* (Rec assay), and *in vivo* carcinogenicity, similar results were obtained for ochratoxin A and rubratoxin B, but different results were observed for patulin, penicillic acid, kojic acid, citrinin, and PR toxin. Tests with microorganisms have also been used to evidence the toxicity of kojic acid,[67] but mutagenicity could only be evidenced at high concentrations of the mycotoxin.

The Ames test is a classic assay used to assess the mutagenic potential of molds or its metabolites that produce genetic damage leading to gene mutations.[68] The test uses a number of histidine-auxotrophic *Salmonella typhimurium* strains carrying different point mutations in genes of the histidine operon. These mutations can revert to wild type when a mutagen is present. Many strains of *Penicillium* or their mycotoxins led to negative results in Ames test, even when rat liver S9 mix was used as an external metabolizing enzyme system or when a preincubation of toxins in cultured rat hepatocytes was performed. This enzyme treatment is necessary for some compounds to be mutagenic because they need a metabolic liver transformation that bacterial systems do not possess. Ochratoxin A showed no evidence of mutagenic activity by SOS spot and Ames tests,[61] in contrast to the results obtained with more sensitive bioassays, such as cell cultures or experimental animals, as discussed later.

Penicillium toxins were tested using the bioassay based on the bacterial bioluminescence of *Photobacterium phosphoreum*,[62] which is a reliable short-term method for assessing the toxicity of mycotoxins. The order of toxicity determined by bacterial bioluminescence parallels that reported for mammalian cell cultures.

Growth of *Pseudomonas syringae* in presence of patulin was evaluated measuring optical density of culture broth. The sensitivity of this microorganism to the mycotoxin was even higher than that obtained with rats.[52]

35.3.4 Vertebrate Animals in Toxicity Bioassays

Different vertebrate animals can be used to test the toxicity of mycotoxins from *Penicillium* spp. (Table 35.4). Tests with vertebrate animals to assay mycotoxins toxicity require careful consideration of various factors.[41] The different animal species have different sensitivity to mycotoxins. Age is another important factor in bioassays, with younger animals being usually more susceptible. Males are generally less resistant to mycotoxins than females. The route of administration also affects the toxicity observed. Lower mycotoxin doses are needed when the route of administration is intraperitoneal instead of oral. The nutritional and health status of animals may influence toxicity symptoms, with high-protein diets having a protective effect against mycotoxins.

A classic animal toxicity test is the chicken embryos bioassay, which is very sensitive to numerous mycotoxins from *Penicillium* strains.[67,72] After the administration of samples into the air sac of eggs, toxicological effects are detected by observing death and teratogenic abnormalities.

Rats and mice have also been used to evaluate ochratoxin A carcinogenicity.[61,69,70] The sensitivity to ochratoxin A carcinogenicity is higher in rodents (higher in rats than in mice, and in males than in females). Effects of kojic acid in mammals have been reviewed by Burdock et al.[67] Both rodents and dogs showed convulsions and died in few hours.

The carcinogenic effect of ochratoxin A has been assessed in studies on chicks. Degenerative changes in liver and kidney were detected after exposure of chicks to this mycotoxin.[71]

Zebrafish (*Danio rerio*) succeeded as a vertebrate animal model because of its high developmental similarity to mammals. The larvae and embryo of this fish have been used in toxicity bioassays for *Penicillium* mycotoxins. Larvae proved to be sensitive to ochratoxin A and patulin, but not to penicillic acid.[73] Zebrafish embryos revealed profound nephrotoxicity in histological structure and biological function after citrinin and patulin exposure, as well as a severe impact on heart development affecting heart morphogenesis after citrinin exposure.[74,75]

TABLE 35.4

Vertebrate Animals Used in Several Bioassay Tests to Detect *Penicillium* Mycotoxins

Organism	Organism Type	Mycotoxins	Test Type	References
Dog	Mammals	Kojic acid	Rate of death	67
Rats and mice	Mammals	Ochratoxin A	Kidney and liver lesions	61,69,70
		Kojic acid	Rate of death	67
Chicks	Birds	Ochratoxin A	Kidney and liver lesions	71
Chicken embryos	Birds	Mycotoxins from *Penicillium* spp.	Rate of death	67,72
		Kojic acid	Teratogenic abnormalities	
Zebrafish (*Danio rerio*)	Fish	Ochratoxin A, patulin, penicillic acid	Toxicity in larvae	73–75
		Citrinin, patulin	Nephrotoxicity in embryos	
		Patulin	Cardiotoxicity in embryos	

35.3.5 Cell Cultures in Toxicity Bioassays

Several mammalian cell lines have been employed to evaluate the ability of some *Penicillium* mycotoxins to induce cytotoxicity and DNA damage (Table 35.5). The neutral red (NR) and the tetrazolium (MTT) assays are two *in vitro* cytotoxicity tests that use colorimetric measurements for the quantification of viable cells after incubation with mycotoxins.[76] The NR assay is based on the capture and accumulation of the neutral red dye in the lysosomes of uninjured living cells. The MTT assay is based on the reduction of the soluble yellow MTT tetrazolium salt to a blue insoluble MTT formazan product by mitochondrial succinic dehydrogenase of uninjured living cells. These tests have been developed with different types of cell lines, such as fibroblasts, liver cells, keratinocytes, neural cells, corneal epithelium, and tumoral cells.

The sister chromatid exchange method (SCE) is based on the differential staining of sister chromatids during replication and permits the direct visualization of genetic exchanges between sister chromatids.[86] The generation of SCE depends on homologous recombination that increases on DNA exposure to damaging agents like mycotoxins.

The *in vitro* micronucleus MN assay is a robust quantitative assay of chromosome damage and discriminates between once-divided cells, which are accumulated and recognized by their binucleated appearance, and mononucleated cells, which did not divide during the *in vitro* culturing period.[87]

The comet assay (single-cell gel electrophoresis) is a simple method for measuring DNA strand breaks in eukaryotic cells.[88]

35.3.6 Culture Media for Mycotoxin Production

Growth of toxigenic *Penicillium* spp. in culture media can result in the production of various toxic metabolites. Different substrates and culture conditions have been tested to produce mycotoxins in higher amounts than in foods. Dombrink-Kurtzman and Blackburn[89] evaluated six different culture media, both with and without manganese supplementation, to determine maximum patulin production by eleven *Penicillium* species. For most strains, potato dextrose broth supplemented with manganese was optimal for maximum patulin production.

Kokkonen et al.[90] evaluated the effect of substrate on mycotoxin production of selected *Penicillium* strains using yeast extract sucrose (YES) agar, as well as cheese and bread analogues media. The composition of the substrates was adequate to support the growth of *Penicillium* but not mycotoxin production in all cases. Cheese analog was not favorable for mycotoxin production by *P. verrucosum*, but favored roquerfortin C by *P. crustosum*. However, *P. nordicum* synthesized ochratoxin A on all three media.

TABLE 35.5

Cell Lines Used in Bioassay Tests to Study *Penicillium* Mycotoxins

Cell lines	Cell Type	Mycotoxins Detected	Test Type	References
CHO-K1-BH4	Chinese hamster ovary cells	Ochratoxin A	Cytotoxicity (*in vitro* micronucleus MN and Comet assays)	77
TK6	Human lymphoblastoid cells		DNA damage	
CHO-K1	Chinese hamster ovary cells	Patulin	Cytotoxicity (NR and MTT assays)	78
Vero	Green monkey kidney cells	Ochratoxin A, citrinin	Cytotoxicity (MTT assay) and DNA fragmentation	79
A-549	Human lung cancer cells	Sterigmatocystin, verruculogen, roquefortine C, penitrem A, mycophenolic acid	Cytotoxicity (NR assay)	80
Hep-G2	Human hepatocellular carcinoma cells			
L-929	Murine fibroblasts			
Neuro-2a	Murine neuroblastoma cells			
Hep G2	Human hepatocellular carcinoma cells	Ochratoxin A	Cytotoxicity (MTT assay)	81,82
		Patulin	Comet assay	
LLC-PK1	Porcine renal cells	Ochratoxin A and B, patulin, citrinin	Cytotoxicity (MTT assay)	83
CHOK1	Chinese hamster ovary cells	Patulin, citrinin	Sister chromatid exchange test	84
HEK293	Human embryonic kidney cells		Cytotoxicity (MTT assay)	
V79 cells	Chinese hamster lung fibroblasts cells	Ochratoxin A, citrinin	Cytotoxicity (NR assay and *In vitro* micronucleus MN assays)	85

Rao et al.[91] evaluated the effect of 19 synthetic and flour media of several *Penicillium* species in the ochratoxin A production. A rice-flour-based medium supported maximum amount of OTA production. OTA production was stimulated by addition of Zn^{2+} and Mg^{2+} to the medium. However, when Fe^{3+} was added, OTA production was suppressed.

35.4 Allergens

35.4.1 *Penicillium* Allergies

Penicillium is among the four major genera of allergenic fungi, and its species are commonly isolated from indoor environments.[92] Some *Penicillium* species pose a respiratory health hazard in susceptible populations, and the increasing exacerbation of current asthma symptoms has been associated with increased levels of *Penicillium* and other fungal genera.[93] According to allergen database of the World Health Organization and International Union of Immunological Societies[94] five *Penicillium* species produce a total of 17 allergens (Table 35.6), including proteases, ribosomal proteins, membrane proteins, enolases, and heat shock proteins. Moreover, some fungal volatile metabolites could act as irritants in susceptible individuals.[95]

Penicillia are common in a wide variety of food, mainly those of low and intermediate water activity, such as dry-cured ham, dry sausages, and ripened cheese. Also, airborne spores can be found in all environments where these foods are manufactured, handled, or stored. Some operations, such as brushing and coating, spread a high amount of *Penicillium* spores into the air, which may cause occupational respiratory diseases or urticaria to food industry workers, including those at pork butcheries,[96–99] and cheese factories.[100] The most frequent problems associated with fungal allergy are respiratory disorders,

TABLE 35.6

Allergens from *Penicillium* Species According to WHO/IUIS Allergen Standardization Committee

Species	Allergen	Biochemical Name
Penicillium brevicompactum	Pen b 13	Alkaline serine protease
	Pen b 26	Acidic ribosomal protein P1
Penicillium chrysogenum	Pen ch 13	Alkaline serine protease
	Pen ch 18	Vacuolar serine protease
	Pen ch 20	*N*-acetyl-glucosaminidase
	Pen ch 31	Calreticulin
	Pen ch 33	16 kDa allergen
	Pen ch 35	Transaldolase
Penicillium citrinum	Pen c 3	Peroxisomal membrane protein
	Pen c 13	Alkaline serine protease
	Pen c 19	Heat shock protein Hsp70
	Pen c 22	Enolase
	Pen c 24	Elongation factor 1 β
	Pen c 30	Catalase
	Pen c 32	Pectate lyase
Penicillium crustosum	Pen cr 26	60S acidic ribosomal phosphoprotein P1
Penicillium oxalicum	Pen o 18	Vacuolar serine protease

Source: WHO/IUIS, Allergen nomenclature. http://www.allergen.org/index.php

TABLE 35.7

Penicillium Species Causing Food Allergies

Species	Food	References
Penicillium camemberti	Cheese	102
Penicillium chrysogenum	Dry-sausage	103
Penicillium italicum	Dry-sausage and cheese	104
Penicillium spp.	Dry-sausage	101,105

including rhinitis and asthma, atopic dermatitis, allergic bronchopulmonary mycosis, allergic fungal sinusitis, and hypersensitivity pneumonitis.[92]

Several *Penicillium* species have also been considered as causes of food allergy, and this occurs on intake of dry-ripened foods coated with molds, such as dry-fermented sausages and cheese (Table 35.7). These food allergies can cause facial angioedema, allergic rhinitis, urticaria, oropharyngeal pruritus, or anaphylactic shock immediately upon ingestion.[101–105]

In addition, mold-ripened foods can be colonized by *Penicillium* species such as *P. chrysogenum*, *P. nalgiovense*, *P. dipodomis*, *P. griseofulvum*, and *P. flavigenum* that are well-known penicillin producers. Small amounts of penicillin in foods have been proposed as a potential causative factor in chronic urticarial reactions,[106] and it can also provoke allergic reaction in penicillin-sensitive individuals.[107,108]

Sensitivity to mold allergens in individuals can be studied using skin prick and intradermal testing or *in vitro* testing for specific IgE antibodies with fungal extracts.[92] However, to assess the potential allergenic effect of molds, some laboratorial models have been used.

35.4.2 Animal Models to Study Allergies Caused by *Penicillium* spp.

Animal models, mainly mice, have been employed to study allergenic potential and other different aspects related with *Penicillium* allergies (Table 35.8). The use of these models allows studying the effect of fungal inhalation in airways and its correlation with lung inflammation or allergic airway responses, such as rhinitis and asthma. Data on allergenicity of fungi and the organic effects can be obtained by

TABLE 35.8

Animal Models Used to Study Allergenicity of *Penicillium* Species

Animal Model	Mold Species	Exposure	Main Observed Responses	References
BALB/c mice	*P. chrysogenum*	Involuntary aspiration of extract four times over a 4-week period.	Dose-dependent increases in BALF total cell and eosinophil, serum and BALF IgE, BALF IL-5, and lungs lesions Single exposure produced edema and cellular damage but not immune responses	110
		Intratracheal aspiration four times during 4 weeks of exposure to *P. chrysogenum* or house dust mite extracts	Increase in BALF macrophage, neutrophil, eosinophil, total cell counts, and antigen-specific IgE Allergic response more robust than *P. chrysogenum*	111
	P. citrinum	Intratracheal inoculation of Pen c 13 protease	Junctional structure alterations and cytoskeletal rearrangements in lungs Increased total IgE and Pen c 13-specific IgE and IgG1	112
	P. chrysogenum and other molds	Intratracheal aspiration four times during 4 weeks of exposure to of mold or house dust mite extracts.	Differences in the threshold dose for allergy induction among mold species	113
BALB/c and C57 Black/6 mice	*P. chrysogenum*	Intranasal instillation of 10^2 viable conidia weekly for 11 weeks	No significant lung inflammation or increased serum immunoglobulins	114
C57 Black/6 mice	*P. chrysogenum*	Intraperitoneal injections of conidia protease extract for 6 weeks followed by intranasal challenge with protease extract, viable, or nonviable conidia for 2 weeks	Increases serum IgE and IgG1, eosinophilia, and mucus hyperproduction	114
		Intraperitoneal injections of Pen ch for 4 weeks and Pen ch intranasal inoculation the final 2 weeks	Increases of serum IgE and IgG1, increase of BALF eosinophils and neutrophils Perivascular inflammation by eosinophils and neutrophils and increased mucus production in lungs	115
		Intranasal instillation of viable and nonviable conidia	Dose-dependent inflammatory reactions to instillation of viable conidia No reaction to instillation of nonviable conidia	116
		Intranasal instillation of 10^4 viable or nonviable conidia once a week for 6 weeks	Viable conidia increased serum-specific IgE and IgG1, peripheral eosinophils, airway eosinophilia, and IL-4 and IL-5 in lung lavages Nonviable conidia increase serum IgG_{2a} and BALF levels of IFN-γ	117
Swiss Webster Carworth Farms white mice	*P. brevicompactum* *P. chrysogenum*	Intratracheal instillation of a single dose of brevianamide A, mycophenolic acid, or roquefortine C	Increase of macrophage, neutrophil, MIP-2, TNF, IL-6, and LDH concentrations in BALF with dose-dependent-like responses	118
Guinea pigs	*P. chrysogenum*	Aerosol inhalation of a glycoprotein from mold culture during 12 weeks	Production of specific serum IgM, IgG, and IgE and leukotriene CD4 cells Interstitial infiltrates of macrophages and leukotriene cells, cellular bronchiolitis, and single nonnecrotizing granulomas in lungs	119

BALF, bronchoalveolar lavage fluid.

provoking the responses in animals. The main responses evaluated include eosinophil, specific IgE and IgG antibodies, interferon, and interleukins (ILs) levels; bronchoalveolar and lung lavage fluid production; and histopathological lesions. Animal assays have also aided in the characterization of the cell receptors involved in the immune recognition of fungi.[109]

In addition to mold species, animal strains, dose, and exposure method are important factors affecting allergic responses. Animal strains have different ability to respond to fungal allergens depending to their genetic susceptibility. It has been reported that BALB/c mice exhibited stronger inflammatory response to *Stachybotrys chartarum* than other congenic mouse strains, such as C3H/HeJ and C56BL/6.[120] In contrast, intranasal instillation of viable *P. chrysogenum* conidia increased serum IgE levels in C57 Black/6 mice but not in BALB/c mice.[117]

Different exposure and sensitization techniques can be used, including inhalation by aerosol of fungal spores or extracts in suspension, involuntary aspiration, intranasal instillation, and intratracheal instillation. The advantages and disadvantages of each type of method with regard to modelling fungal exposure and allergic sensitization in human have been reviewed by Templeton et al.[109] Involuntary aspiration and intranasal or intratracheal instillation are reproducible and allow for a homogenous suspension of fungi to be administered, but they are not representative of natural exposure. With aerosol, animals are exposed by normal breathing, but lung exposure concentration is unknown, and its reproducibility is lower.[109]

Animal studies have demonstrated the ability of *Penicillium* spp. to provoke allergic and asthma response. Most of these studies involve acute or chronic exposure of animals to conidia, spore extracts, hyphae, subcellular fragments, or fungal proteins from *P. chrysogenum*.

Female C57 Black/6 mouse model was established to examine the *in vivo* effect in allergic processes of different levels of *P. chrysogenum* viable and nonviable conidia by nasal instillation.[114,116,117] The acute instillation of viable *P. chrysogenum* conidia induced inflammatory reactions in a dose-dependent manner, while the instillation of nonviable conidia did not.[116] Low levels of *P. chrysogenum* conidia (10^2 spores) neither provoked lung inflammation nor increased serum immunoglobulins.[114] About 18% of the viable conidia instilled intranasally were deposited and retained in the lungs and could produce substances that induce allergic reactions.[116] Viable conidia could remain in the lungs long enough to produce an unknown protein responsible for the chronic Th2-mediated airway inflammation by increasing total and specific IgE and IgG1, bronchoalveolar lavage fluid levels of IL-4 and IL-5, and eosinophilia.[117] In contrast, repeated exposures to nonviable conidia of *P. chrysogenum* induced type Th1 helper responses in mice, with increases in total serum IgG_{2a} and levels of interferon (IFN)-γ in bronchoalveolar lavage fluid.[117]

The allergic and inflammatory response of BALB/c mice to mycelium and spores showed differences in the threshold dose for allergy induction among species.[113] These authors classified molds according to their allergenic potential: G1 molds induced low-to-moderate responses requiring higher doses than house dust mite; G2 molds, including *P. chrysogenum*, required lower doses to induce a similar response.[113] In this sense, the allergic and inflammatory response of BALB/c mice to intratracheal aspiration of mycelium and spores of *P. chrysogenum* was more robust than that produced by house dust mite.[111] Therefore, molds must be evaluated individually for allergic or asthmatic potential.

Some experiments have been performed to study the response of animals to different extract or compounds from *P. chrysogenum*. A dose-dependent allergic asthma-like response was observed after involuntary aspiration of soluble components of *P. chrysogenum* by BALB/c mice.[110] There are some differences in the effect between single and multiple exposures to high doses. While single exposure resulted in edema and cellular damage but not immune responses, multiple exposures showed increased allergen-triggered immediate respiratory responses.[110]

C57 Black/6 mice have been used to study the effect of allergen protease extract Pen ch released by viable *P. chrysogenum* conidia.[114,115] Mice previously sensitized with Pen ch or viable *P. chrysogenum* conidia by intraperitoneal injections and exposed to intranasal challenge showed increased allergic airway inflammation.[114] In these mice, significant increases in serum IgE and IgG1, eosinophilia, and mucus hyperproduction in bronchoalveolar lavage and lung tissue was recorded.[114] Furthermore, intraperitoneal sensitization and posterior instillation of Pen ch produce a strong allergic inflammatory response characterized by increasing serum IgE and IgG1, eosinophils and neutrophils counts in bronchoalveolar lavage, as well as mucous production and perivascular inflammation by eosinophils and neutrophils in lungs.[115]

Guinea pigs have been used for experimental induction of hypersensitivity pneumonitis by using aerosol inhalation for 12 weeks of a glycoprotein from *P. chrysogenum* (formerly *P. notatum*).[119] In exposed animals, specific serum IgM, IgG, and IgE antibodies and sensitized leukotriene CD4 cells were detected. Moreover, interstitial infiltrates of macrophages and leukotriene cells, cellular bronchiolitis, and single nonnecrotizing granulomas were observed in lungs. This response is a typical delayed-type reaction due to chronic contact with the heterologous glycoprotein of *Penicillium*.[119]

Swiss Webster Carworth Farms white (CFW) male mice were used to elucidate the role of bioactive constituents of spores from *P. brevicompactum* and *P. chrysogenum* that mediate allergenic responses.[118] Inflammatory and cytotoxic responses to intratracheal instillation of brevianamide A and mycophenolic acid from *P. brevicompactum*, and roquefortine C from *P. chrysogenum* were investigated. High doses of these three metabolites induced a dose-dependent-like inflammatory response expressed as increased macrophage, neutrophil, MIP-2, TNF, and IL-6 concentrations in the bronchoalveolar lavage fluid in exposed mice. The analysis of this fluid revealed some compound-specific toxic responses: brevianamide A and mycophenolic acid provoked vascular leakage according to albumin concentration, and brevianamide A induced cytotoxicity, increasing lactate dehydrogenase (LDH) concentration.[118]

BALB/c mice have also been used to study the effect of intratracheal inoculation of Pen c 13 protease from *P. citrinum*. An increase in airway hyperresponsiveness, inflammatory cell infiltration, mucus overproduction, and collagen deposition in the lung, as well as serum levels of total IgE and Pen c 13-specific IgE and IgG1 were observed.[112] Moreover, the exposure to Pen c 13 provoked changes in lung proteome, with increase of proteins involved in leukocyte extravasation signalling, oxidative stress response, and actin cytoskeleton organization, and also decrease of vinculin, a junctional protein between cells.[112] Therefore, Pen c 13 exposure causes structure alterations and actin cytoskeletal rearrangements, resulting in increased permeability and airway structural changes.[112]

35.4.3 Cell Cultures to Study Allergies Caused by *Penicillium*

Several cell cultures have been used to study the effect of allergens from *Penicillium* spp. The effects of the Pen ch 13 allergen from *P. chrysogenum* on respiratory epithelial cells have been demonstrated on A549 cells, a human alveolar type II epithelium-like cell line; 16HBE14o-, an immortalized bronchial epithelial cell line; and HBEpC, primary cells derived from normal human bronchi.[121] Pen ch 13 induced in all three cell cultures the secretion of mediators linked to local immune responses and inflammatory process such as prostaglandin-E2 (PGE_2), IL-8, and transforming growth factor (TGF)-β1. Additionally, Pen ch 13 degraded the protein occludin of cultured 16HBE14o-, contributing to disruption of the tight junctions, damaging the barrier formed by the airway epithelium.[121]

Spleen cell culture obtained from BALB/c mice as well as NCI-H441cell line derived from human lung epithelial adenocarcinoma have been used to study the effect of Pen c 13 from *P. citrinum*. The *in vitro* stimulation of splenocytes with Pen c 13 resulted in an immunologic response by increasing production of Th2 cytokines IL-4, IL-5, and IL-13.[112]

NCI-H441 cell line was used to mimic the effect of Pen c 13 on pulmonary epithelial barrier integrity. This antigen produced disruption of tight junctions integrity and increased permeability, which can be associated with the loss of pulmonary epithelium integrity observed in Pen c 13-sensitized BALB/c mice.[112]

35.5 Conclusions

Given that penicillia only show a limited invasive ability for animal tissue, they are mainly soil saprophytes or plant pathogens, being responsible for mycotoxin production and postharvest rotting of fruits and vegetables. However, *Penicillium* and *Talaromyces* spp. have been described as rare causative agents of infectious diseases in humans, dogs, and fish, as well as environmental allergens among asthmatic and atopic individuals. *T. marneffei*, the only dimorphic species of this group, is capable of growing as a uninucleate pathogenic yeast at 37°C upon invasion of a human host, leading to localized reactions in persons with normal immunity, while causing fatal systemic mycosis in immunocompromised patients. Diagnosis of penicilliosis is based on the confirmation by culture methods through specimen examination, detection of antibodies produced against various fungal proteins, the detection of fungal proteins in sera or in urine, or by selective amplification of *T. marneffei* DNA. Bamboo rats are the only known nonhuman host of *T. marneffei*, but these animals show no signs of illness. Therefore, laboratory models to study the pathogenicity of *T. marneffei* are extremely limited.

In contrast to pathogenic penicillia, mycotoxins produce toxic effects in various organisms, allowing the use of different bioassays to study the toxicity of such secondary metabolites. Vertebrate animals have been used to test the toxicity of *Penicillium* mycotoxins. Tests with microorganisms and invertebrates are preferred to the use of vertebrates due to ethical objections and high maintenance costs. Organs and cell cultures have also been used, particularly when obtaining information on the effect on DNA, cell organelles, or cell membranes. In addition, different culture media have been tested to study the ability and efficiency of *Penicillium* spp. to produce mycotoxins, including food analogues media.

Some *Penicillium* species produce allergens that pose a respiratory health hazard to susceptible individuals. Airborne spores from dry-ripened foods may cause food allergies or occupational respiratory diseases to food industry workers. Sensitivity to penicillia allergens can be studied using common tests with fungal extracts, but rodents and cell cultures are required to assess organic effects, such as eosinophil, specific IgE and IgG antibodies, interferon, and ILs levels; lung lavage fluids production; and histopathological lesions. The genetic susceptibility of the different animal strains to allergens makes it necessary to select the strain exhibiting the adequate response to the mold, according to the exposure and sensitization techniques used. Different types of respiratory epithelial cell cultures have been used to study the effect of allergens form *Penicillium* spp. on secretion of mediators linked to local immune responses. However, the use of cell cultures is otherwise rather limited.

REFERENCES

1. Houbraken, J. and Samson, R.A., Phylogeny of *Penicillium* and the segregation of *Trichocomaceae* into three families, *Stud. Mycol.*, 70, 1, 2011.
2. Samson, R.A. et al., Phylogeny and nomenclature of the genus *Talaromyces* and taxa accommodated in *Penicillium* subgenus *Biverticillium*, *Stud. Mycol.*, 70, 159, 2011.
3. Dyer, P.S. and O'Gorman, C.M., A fungal sexual revolution: *Aspergillus* and *Penicillium* show the way, *Curr. Opin. Microbiol.*, 14, 649, 2011.
4. Yilmaz, N. et al., Polyphasic taxonomy of the genus *Talaromyces*, *Stud. Mycol.*, 78, 175, 2014.
5. Vos, J.P. et al., Disseminated *Penicillium radicum* infection in a dog, clinically resembling multicentric malignant lymphoma, *Vlaams Diergeneeskundig Tijdschrift*, 78, 183, 2009.
6. Tomlinson, J.K. et al., Case report: Granulomatous lymphadenitis caused by *Talaromyces helicus* in a labrador retriever, *Vet. Clin. Pathol.*, 40, 553, 2011.
7. Blaylock, R.B., Overstreet, R.M. and Klich, M.A., Mycoses in red snapper (*Lutjanus campechanus*) caused by two deuteromycete fungi (*Penicillium corylophilum* and *Cladosporium sphaerospermum*), *Hydrobiologia*, 460, 221, 2001.
8. Ahearn, D.G. and Stulting, R.D., Fungi associated with drug recalls and rare disease outbreaks, *J. Ind. Microbiol. Biotechnol.*, 41, 1591, 2014.
9. Santos, P.E. et al., *Penicillium piceum* infection: Diagnosis and successful treatment in chronic granulomatous disease, *Med. Mycol.*, 44, 749, 2006.
10. Ho, M.H.K. et al., *Penicillium marneffei* infection and impaired IFN-g immunity in humans with autosomal-dominant gain-of-phosphorylation STAT1 mutations, *J. Allergy Clin. Immunol.*, 133, 894, 2014.
11. Segretain, G., *Penicillium marneffei* agent of mycosis of reticuloendothelial system, *Mycopathologia*, 11, 327, 1959.
12. Wang, T.K.F., Yuen, K.Y. and Wong, S.S.Y., Asymptomatic *Penicillium marneffei* fungemia in an HIV-infected patient, *Int. J. Inf. Dis.*, 11, 280, 2007.
13. Walsh, T.J. et al., Infections due to emerging and uncommon medically important fungal pathogens, *Eur. Soc. Clin. Microbiol. Infect. Dis., CMI*, 10 (Suppl. 1), 48, 2004.
14. Plonga, R. et al., *Penicilliosis marneffei* infection in non-HIV-infected patients, *J. Infect. Dis. Antimicrob. Agents*, 28, 71, 2011.

15. Qiu, Y. et al., Retrospective analysis of 14 cases of disseminated *Penicillium marneffei* infection with osteolytic lesions, *BMC Infect. Dis.*, 15, 47, 2015.
16. Cooney, N.M. and Klein, B.S., Fungal adaptation to the mammalian host: It is a new world, after all, *Curr. Opin. Microbiol.*, 11, 511, 2008.
17. Andrianopoulos, A., Control of morphogenesis in the human fungal pathogen *Penicillium marneffei*, *Int. J. Med. Microbiol.*, 292, 331, 2002.
18. Cooper, C.R. and Vanittanakom, N., Insights into the pathogenicity of *Penicillium marneffei*, *Future Microbiol.*, 3, 43, 2008.
19. Boyce, K.J. and Andrianopoulos, A., Morphogenetic circuitry regulating growth and development in the dimorphic pathogen *Penicillium marneffei*, *Eukaryot. Cell*, 12, 154, 2013.
20. Boyce, K.J. and Andrianopoulos, A., A p21-activated kinase is required for conidial germination in *Penicillium marneffei*, *PLoS Pathog.*, 3, e162, 2007.
21. Yang, E. et al., Signature gene expression reveals novel clues to the molecular mechanisms of dimorphic transition in *Penicillium marneffei*, *PLoS Genet.*, 10, 1, 2014.
22. Deepa, A.G. et al., Uncommon opportunistic fungal infections of oral cavity: A review, *J. Oral Maxillofac. Pathol.*, 18, 235, 2014.
23. Supparatpinyo, K. et al., Disseminated *Penicillium marneffei* infection in Southeast Asia, *Lancet*, 344, 110, 1994.
24. Lupi, O., Tyring, S.K. and McGinnis, M.R., Tropical dermatology: Fungal tropical diseases, *J. Am. Acad. Dermatol.*, 53, 931, 2005.
25. Sirisanthana, T. and Supparatpinyo, K., Epidemiology and management of penicilliosis in human immunodeficiency virus-infected patients, *Int. J. Infect. Dis.*, 3, 48, 1998.
26. Hamilton, A., *Penicillium marneffei*: Penicilliosis and the red peril in the east, *Mycologist*, 17, 84, 2003.
27. Vanittanakom, N., *Penicillium marneffei* infection and recent advances in the epidemiology and molecular biology aspects, *Clin. Microbiol. Rev.*, 19, 95, 2006.
28. Sekhon, A.S., Li, J.S.K. and Garg, A.K., *Penicilliosis marneffei*: Serological and exoantigen studies, *Mycopathologia*, 77, 51, 1982.
29. LoBuglio, K.F. and Taylor, J.W., Phylogeny and PCR identification of the human pathogenic fungus *Penicillium marneffei*, *J. Clin. Microbiol.*, 33, 85, 1995.
30. Fisher, M.C. et al., Multilocus microsatellite typing system for *Penicillium marneffei* reveals spatially structured populations, *J. Clin. Microbiol.*, 42, 5065, 2004.
31. Cao, C. et al., Common reservoirs for *Penicillium marneffei* infection in humans and rodents, China, *Emerg. Infect. Dis.*, 17, 209, 2011.
32. Steinbach, W.J. and Perfect, J.R., Newer antifungal therapy for emerging fungal pathogens, *Int. J. Infect. Dis.*, 7, 5, 2003.
33. Sirisanthana, T. et al., Amphotericin B and itraconazole for treatment of disseminated *Penicillium marneffei* infection in human immunodeficiency virus-infected patients, *Clin. Infect. Dis.*, 26, 1107, 1998.
34. Wong, S.Y.N. and Wong, K.F., *Penicillium marneffei* infection in AIDS, *Pathol. Res. Int.*, 2011, 1, 2011.
35. Moss, M.O., The environmental factors controlling mycotoxin formation, in *Mycotoxins and Animal Foods*, p. 37, Smith J.F. and Henderson R.S. (Eds.), CRC Press, Boca Raton, 1991.
36. Mantle, P.G., Miscellaneous toxigenic fungi, in *Mycotoxins and Animal Foods*, p. 141, Smith J.E. and Henderson R.S. (Eds.), CRC Press, Boca Raton, 1991.
37. Bullerman, L.B., Fusaria and toxigenic molds and other Aspergilli and Penicillia, in *Food Microbiology: Fundamentals and Frontiers*, p. 419, Doyle M.P., Beuchat L.R. and Montville T.J. (Eds.), ASM Press, Washington, 1997.
38. Pitt, J.I., Toxigenic *Penicillium* species, in *Food Microbiology: Fundamentals and Frontiers*, pp. 406, Doyle M.P., Beuchat L.R. and Montville T.J. (Eds.), ASM Press, Washington, 1997.
39. Frisvad, J.C. and Samson, R.A., Polyphasic taxonomy of *Penicillium* subgenus *Penicillium*: A guide to identification of food and air-borne terverticillate Penicillia and their mycotoxins, *Stud. Mycol.*, 49, 1, 2004.
40. Barkai-Golan, R., *Penicillium* mycotoxins, in *Mycotoxins in Fruits and Vegetables*, p. 153, Barkai-Golan R. and Paster N. (Eds.), Academic Press, San Diego, 2008.
41. Panigrahi, S., Bioassay of mycotoxins using terrestrial and aquatic, animal and plant species, *Food Chem. Toxicol.*, 31, 767, 1993.
42. Watson, D.H. and Lindsay, D.G., A critical review of biological methods for the detection of fungal toxins in foods and foodstuffs, *J. Sci. Food Agric.*, 33, 59, 1982.

43. Lagadic, L. and Caquet, T., Invertebrates in testing of environmental chemicals: Are they alternatives? *Environ. Health Perspect.*, 106 (Suppl 2), 593, 1998.
44. Harwig, J. and Scott, P.M., Brine shrimp (*Artemia salina* L.) larvae as a screening system for fungal toxins, *Appl. Microbiol.*, 21, 1011, 1971.
45. Tanaka, K., Goto, T. and Manabe, M., Biological test using *Moina macrocopa* Strauss (Part 1). Rearing conditions of *Moina macrocopa* Strauss and influence of mycotoxins on this animal plankton, *Rep. Nat. Food Res. Inst.*, 48, 36, 1986.
46. Starodub, N.F. et al., Express control of toxicity and content of patulin by optical biosensors, in *Nanotechnology 2010: Bio Sensors, Instruments, Medical, Environment and Energy*, p. 137, Nano Science and Technology Institute, Danville, 2010.
47. Reiss, J., Toxic effects of mycotoxins, aflatoxin B1, rubratoxin B, patulin, and diacetoxyscirpenol on the crustacean *Cyclops fuscus*, *J. Assoc. Off. Anal. Chem.*, 55, 895, 1972.
48. Escoula, L., More, J. and Baradat, C., The toxins of *Byssochlamys nivea* Westling. 1. Acute toxicity of patulin in adult rats and mice, *Ann. Rech. Vet.*, 8, 41, 1977.
49. McKinley, E.R. and Carlton, W.W., Patulin mycotoxicosis in Swiss ICR mice, *Food Cosmet Toxicol.*, 18, 181, 1980.
50. Gräbsch, C. et al., Cytotoxicity assessment of gliotoxin and penicillic acid in *Tetrahymena pyriformis*, *Environ. Toxicol.*, 21, 111, 2006.
51. Hayes, A.W., Antiprotozoal activity of rubratoxin B, *Antimicrob. Agents Chemother.*, 4, 80, 1973.
52. Skrobek, A. et al., Evaluation of different biological test systems to assess the toxicity of metabolites from fungal biocontrol agents, *Toxicol. Lett.*, 161, 43, 2006.
53. Dive, D., Moreau, S. and Cacan, M., Use of a ciliate protozoan for fungal toxins studies, *Bull. Environ. Contam. Toxicol.*, 19, 489, 1978.
54. Bijl, J.P. et al., Potentials of a synchronized culture of *Tetrahymena pyriformis* for toxicity studies of mycotoxins, *J. Assoc. Off. Anal. Chem.*, 71, 282, 1988.
55. Rao, V.K. and Girisham, S., Antimicrobial and DNA damaging activity of ochratoxin A extracted from *Penicillium* species, *Int. J. Pharm. Bio Sci.*, 5, 335, 2014.
56. Gharaei-Fathabad, E., Tajick-Ghanbary, M.A. and Shahrokhi, N., Antimicrobial properties of *Penicillium* species isolated from agricultural soils of northern Iran, *Res. J. Toxins*, 6, 1, 2014.
57. Lenz, P. and Süssmuth, R., A highly sensitive bacterial assay for toxins based on swarming inhibition, and comparison with the cup plate assay based on growth inhibition, *Toxicology*, 45, 185, 1987.
58. Lenz, P., Süssmuth, R. and Seibel, E., Development of sensitive bacterial tests, exemplified by two mycotoxins, *Toxicology*, 40, 199, 1986.
59. Skaug, M.A., Eduard, W. and Størmer, F.C., Ochratoxin A in airborne dust and fungal conidia, *Mycopathologia*, 151, 93, 2000.
60. Auffray, Y. and Boutibones, P., Evaluation of the genotoxic activity of some mycotoxins using *Escherichia coli* in the SOS spot test, *Mutat. Res.*, 171, 79, 1986.
61. Haighton, L.A. et al., A reassessment of risk associated with dietary intake of ochratoxin A based on a lifetime exposure model, *Crit. Rev. Toxicol.*, 42, 147, 2012.
62. Yates, I.E. and Porter, J.K., Bacterial bioluminescence as a bioassay for mycotoxins, *Appl. Environ. Microbiol.*, 44, 1072, 1982.
63. Föllmann, W. and Lucas, S., Effects of the mycotoxin ochratoxin A in a bacterial and a mammalian in vitro mutagenicity test system, *Arch. Toxicol.*, 77, 298, 2003.
64. Puel, O., Galtier, P. and Oswald, I.P., Biosynthesis and toxicological effects of patulin, *Toxins*, 2, 613, 2010.
65. Wurgler, F.E., Friederich, U. and Schlatter, J., Lack of mutagenicity of ochratoxin A and B, citrinin, patulin and cnestine in *Salmonella typhimurium* TA102, *Mutat. Res.*, 261, 209, 1991.
66. Wehner, F.C. et al., Mutagenicity to *Salmonella typhimurium* of some *Aspergillus* and *Penicillium* mycotoxins, *Mutat. Res.*, 58, 193, 1978.
67. Burdock, G.A., Soni, M.G. and Carabin, I.G., Evaluation of health aspects of kojic acid in food. *Regul. Toxicol. Pharmacol.*, 33, 80, 2001.
68. Mortelmans, K. and Zeiger, E., The Ames *Salmonella*/microsome mutagenicity assay, *Mutat. Res.*, 455, 29, 2000.
69. Huff, J., Carcinogenicity of ochratoxin A in experimental animals, in *Mycotoxins, Endemic Nephropathy and Urinary Tract Tumours*, p. 229, Castegnaro, M., Plestina, R., Dirheimer, G., Chernozemsky, I.N. and Bartsch, H. (Eds.), International Agency for Research on Cancer, Lion, 1991.

70. Vettorazzi, A., González-Peñas, E. and López de Cerain, A., Ochratoxin A kinetics: A review of analytical methods and studies in rat model, *Food Chem. Toxicol.*, 72, 273, 2014.
71. Stoev, S.D., Studies on carcinogenic and toxic effects of ochratoxin A in chicks, *Toxins*, 2, 649, 2010.
72. Vesely, D., Vesela, D. and Jelinek, R., Use of chick embryo in screening for toxin-producing fungi, *Mycopathologia*, 88, 135, 1984.
73. Abedi, Z.H. and Scott P.M., Detection of toxicity of aflatoxins, sterigmatocystin and other fungal toxins by lethal action on zebra fish larvae, *J. Assoc. Off. Anal. Chem.*, 52, 963, 1969.
74. Wu, T.-S. et al., Evaluation of nephrotoxic effects of mycotoxins, citrinin and patulin, on zebrafish (*Danio rerio*) embryos, *Food Chem. Toxicol.*, 50, 4398, 2012.
75. Wu, T.-S. et al., Cardiotoxicity of mycotoxin citrinin and involvement of microRNA-138 in zebrafish embryos, *Toxicol. Sci.*, 136, 402, 2013.
76. Borenfreund, E., Babich, H. and Martin-Alguacil, N., Comparisons of two in vitro cytotoxicity assays-the neutral red (NR) and tetrazolium (MTT) tests, *Toxicol. In Vitro*, 2, 1, 1988.
77. Ali, R. et al., Comparative analysis of micronuclei and DNA damage induced by Ochratoxin A in two mammalian cell lines, *Mutat. Res.*, 723, 58, 2011.
78. Ferrer, E. et al., Reactive oxygen species induced by beauvericin, patulin and zearalenone in CHO-K1 cells, *Toxicol. In Vitro*, 23, 1504, 2009.
79. Bouslimi, A. et al., Individual and combined effects of ochratoxin A and citrinin on viability and DNA fragmentation in cultured Vero cells and on chromosome aberrations in mice bone marrow cells, *Toxicology*, 251, 1, 2008.
80. Bünger, J. et al., Cytotoxicity of occupationally and environmentally relevant mycotoxins, *Toxicology*, 202, 199, 2004.
81. Corcuera, L.A. et al., Ochratoxin A reduces aflatoxin B1 induced DNA damage detected by the comet assay in Hep G2 cells, *Food Chem. Toxicol.*, 49, 2883, 2011.
82. Zhou, S. et al., Patulin-induced oxidative DNA damage and p53 modulation in HepG2 cells, *Toxicon*, 55, 390, 2010.
83. Heussner, A.H., Dietrich, D.R. and Brien, E.O., In vitro investigation of individual and combined cytotoxic effects of ochratoxin A and other selected mycotoxins on renal cells, *Toxicol. In Vitro*, 20, 332, 2006.
84. Liu, B. et al., Evaluation of genotoxic risk and oxidative DNA damage in mammalian cells exposed to mycotoxins, patulin and citrinin, *Toxicol. Appl. Pharmacol.*, 191, 255, 2003.
85. Föllmann, W., Behm, C., and Degen, G.H., Toxicity of the mycotoxin citrinin and its metabolite dihydrocitrinone and of mixtures of citrinin and ochratoxin A in vitro, *Arch. Toxicol.*, 88, 1097, 2014.
86. Simpson, L.J. and Sale, J.E., Sister chromatid exchange assay, *Subcell. Biochem.*, 40, 399, 2006.
87. Kirsch-Volders, M. et al., In vitro genotoxicity testing using the micronucleus assay in cell lines, human lymphocytes and 3D human skin models, *Mutagenesis*, 26, 177, 2011.
88. Collins, A.R., The comet assay for DNA damage and repair. Principles, applications, and limitations, *Mol. Biotechnol.*, 26, 249, 2004.
89. Dombrink-Kurtzman, M.A. and Blackburn, J.A., Evaluation of several culture media for production of patulin by *Penicillium* species, *Int. J. Food Microbiol.*, 98, 241, 2005.
90. Kokkonen, M., Jestoi, M. and Rizzo, A., The effect of substrate on mycotoxin production of selected *Penicillium* strains, *Int. J. Food Microbiol.*, 99, 207, 2005.
91. Rao, V.K. et al., Culture media and factors influencing ochratoxin A production by two species of *Penicillium* isolated from poultry feeds, *Natl. Acad. Sci. Lett.*, 36, 101, 2013.
92. Portnoy, J.M. and Jara, D., Mold allergy revisited, *Ann. Allergy Asthma Immunol.*, 114, 83, 2015.
93. Sharpe, R.A. et al., Indoor fungal diversity and asthma: A meta-analysis and systematic review of risk factors, *J. Allergy Clin. Immunol.*, 135, 110, 2015.
94. WHO/IUIS, Allergen nomenclature. http://www.allergen.org/index.php (Accessed May 27, 2015), 2015.
95. Gunschera, J. et al., Formation and emission of chloroanisoles as indoor pollutants, *Environ. Sci. Pollut. Res. Int.*, 11, 147, 2004.
96. Feo, F. et al., *Penicillium nalgiovense* as an occupational and contact allergen, *J. Allergy Clin. Immunol.*, 112, 213, 2003.
97. Merget, R. et al., Occupational immunoglobulin E-mediated asthma due to *Penicillium camemberti* in a dry-sausage packer, *Respiration*, 76, 109, 2008.
98. Rouzaud, P. et al., Symptoms and serum precipitins in workers exposed to dry sausage mould: Consequences of exposure to sausage mould, *Int. Arch. Occup. Environ. Health*, 74, 371, 2001.

99. Wantke, F. et al., Contact dermatitis caused by salami skin, *Contact Dermat.*, 64, 111, 2011.
100. Campbell, J.A. et al., Cheese worker's hypersensitivity pneumonitis, *Am. Rev. Respir. Dis.*, 127, 495, 1983.
101. Bobolea, I. et al., Allergy to dry fermented sausage, *J. Invest. Allergol. Clin. Immunol.*, 19, 324, 2009.
102. Boisnault, A. et al., Allergie alimentaire à une moisissure présente sur une croûte de fromage: À propos d'une observation pédiatrique, *Rev. Fr. d'Allergol. d'Immunol. Clin.*, 45, 597, 2005.
103. González-de-Olano, D. et al., Different patterns of sensitization in allergy to dry fermented sausage, *J. Invest. Allergol. Clin. Immunol.*, 22, 152, 2012.
104. Guillet, M.H. et al., Urticaire et choc anaphylactique par allergie alimentaire à *Penicillium italicum*, *Rev. Fr. d'Allergol. d'Immunol. Clin.*, 43, 520, 2003.
105. Morisset, M. et al., Food allergy to moulds: Two cases observed after dry fermented sausage ingestion, *Allergy Eur. J. Allergy Clin. Immunol.*, 58, 1203, 2003.
106. Ormerod, A.D., Reid, T.M.S. and Main, R.A., Penicillin in milk—its importance in urticaria, *Clin. Exp. Allergy*, 17, 229, 1987.
107. Padinjakara, R.N.K., Ashawesh, K. and Patel, V., Allergic reaction to blue cheese: Serendipity or actual causation? *N. Z. Med. J.*, 121, 102, 2008.
108. Raison-Peyron, N. et al., Anaphylaxis to beef in penicillin-allergic patient, *Allergy*, 56, 796, 2001.
109. Templeton, S.P. et al., Murine models of airway fungal exposure and allergic sensitization, *Med. Mycol.*, 48, 217, 2010.
110. Chung, Y.J. et al., Dose-dependent allergic responses to an extract of *Penicillium chrysogenum* in BALB/c mice, *Toxicology*, 216, 73, 2005.
111. Ward, M.D.W. et al., A comparison of the allergic responses induced by *Penicillium chrysogenum* and house dust mite extracts in a mouse model, *Indoor Air*, 20, 380, 2010.
112. Chen, J.C. et al., The protease allergen Pen c 13 induces allergic airway inflammation and changes in epithelial barrier integrity and function in a murine model, *J. Biol. Chem.*, 286, 26667, 2011.
113. Ward, M.D.W. et al., Assessing the allergenic potential of molds found in water-damaged homes in a mouse model, *Inhal. Toxicol.*, 26, 474, 2014.
114. Schwab, C.J. et al., Characterization of exposure to low levels of viable *Penicillium chrysogenum* conidia and allergic sensitization induced by a protease allergen extract from viable *P. chrysogenum* conidia in mice, *Int. Arch. Allergy Immunol.*, 130, 200, 2003.
115. Schwab, C.J. et al., Allergic inflammation induced by a *Penicillium chrysogenum* conidia-associated allergen extract in a murine model, *Allergy Eur. J. Allergy Clin. Immunol.*, 59, 758, 2004.
116. Cooley, J. et al., Cellular and humoral responses in an animal model inhaling *Penicillium chrysogenum* conidia, in *Bioaerosols, Fungi and Mycotoxins : Health Effects, Assessment, Prevention and Control*, p. 403, Johanning, E. (Ed.), Boyd Printing Company Inc., Albany, NY, 1999.
117. Cooley, J.D. et al., An animal model for allergic penicilliosis induced by the intranasal instillation of viable *Penicillium chrysogenum* conidia, *Thorax*, 55, 489, 2000.
118. Rand, T.G. et al., Inflammatory and cytotoxic responses in mouse lungs exposed to purified toxins from building isolated *Penicillium brevicompactum* Dierckx and *P. chrysogenum* Thom, *Toxicol. Sci.*, 87, 213, 2005.
119. Alonso, A. et al., Hypersensitivity pneumonitis induced by *Penicillium notatum* antigens in guinea pigs, *J. Invest. Allergol. Clin. Immunol.*, 8, 52, 1998.
120. Lichtenstein, J.H.R. et al., Strain differences influence murine pulmonary responses to *Stachybotrys chartarum*, *Am. J. Respir. Cell Mol. Biol.*, 35, 415, 2006.
121. Tai, H.Y. et al., Pen ch 13 allergen induces secretion of mediators and degradation of occludin protein of human lung epithelial cells, *Allergy Eur. J. Allergy Clin. Immunol.*, 61, 382, 2006.

Section V

Foodborne Infections due to Protozoa

36

Acanthamoeba

Dongyou Liu

CONTENTS

36.1 Introduction

Acanthamoeba is a unicellular organism that was first detected by Castellani from *Cryptococcus pararoseus* cultures. As a free-living bacterivore, *Acanthamoeba* is present in a diversity of environments, including soil, dust, water, air, plants, animals, and humans. However, *Acanthamoeba* has the ability to take advantage of temporary weakness (trauma or suppressed immune function) in human host, causing keratitis, encephalitis, and skin infections, often with severe consequences. In view of the limited treatment options for *Acanthamoeba* infections at present, there is a need to improve our understanding of the pathogenicity of this devastating disease, and to devise novel, highly effective intervention strategies. Utilization of laboratory models to *Acanthamoeba* infections will be undoubtedly beneficial in this regard.

36.1.1 Classification, Morphology, and Genomics

36.1.1.1 Classification

The genus *Acanthamoeba* is classified in the family Acanthamoebidae, order Centramoebida, subclass Longamoebia, class Discosea, kingdom Amoebozoa, domain Eukaryota, kingdom Animalia. Beside *Acanthamoeba*, the family Acanthamoebidae contains two other genera (*Comandonia* and *Protacanthamoeba*). Another former genus (*Balamuthia*) of the family Acanthamoebidae, order Centramoebida, has been moved to a newly created family Balamuthiidae, order Centramoebida.

Currently, the genus *Acanthamoeba* consists of at least 25 named species [i.e., *A. astronyxis* (T7), *A byersi* (T18), *A. castellanii* (T4), *A. comandoni* (T9), *A. culbertsoni* (T10), *A. divionensis* (T4), *A griffini* (T3), *A. hatchetti* (T11), *A. healyi* (T12), *A. jacobsi* (T15), *A. lenticulata* (T3), *A. lugdunensis* (T4), *A. mauritaniensis* (T4), *A. palestinensis* (T2), *A. paradivionensis* (T4), *A. pearcei* (T3), *A. polyphaga* (T4), *A. pustulosa* (T2), *A. quina* (T11), *A. rhysodes* (T4), *A. royreba* (T4), *A. stevensoni* (T11), *A. terricola* (T4), *A. triangularis* (T4), and *A. tubiashi* (T8)] in addition to many unassigned species/isolates [1–3].

On the basis of their morphological features, members of the genus *Acanthamoeba* are placed into three groups (designated Groups I, II, and III). *Acanthamoeba* Group I consists of five species [e.g., *A. astronyxis* (T7), *A. tubiashi* (T8), *A. comondani* (T9)], which produce large trophozoites of 25–35 μm and cysts (with distinctly stellate endocysts and smooth spherical ectocysts) of >18 μm in size, with cysts that either have four or less arms or more than six arms. This group is mostly environmental and not convincingly associated with infections in humans or animals. *Acanthamoeba* Group II comprises 10 species such as *A. castellanii* (T4), *A. griffinii* (T3), *A. hatchetti* (T11), *A. polyphaga*, and *A. stevensoni* (T11), which produce cysts (with polygonal to stellate endocysts and irregular or wrinkled ectocysts) of <18 μm in size with rounded arms. This group is responsible for the majority of reported human infections [e.g., amebic keratitis (AK), granulomatous amebic encephalitis (GAE), cutaneous acanthamoebiasis, and sinusitis]. *Acanthamoeba* Group III comprises species *A. culbertsoni* (T10), *A. healyi* (T12), *A. palestinensis* (T1), *A. pustulosa* (T2), and *A. lenticulata* (T5), which possess cysts of <15 μm in size (smaller than those of Group II) with three to five points. *A. culbertsoni* from Group III is a recognized pathogen causing both keratitis and encephalitis [2].

Sequencing analysis of the nuclear small-subunit (18S) rRNA gene indicates that members of the genus *Acanthamoeba* may be distinguished into 18 genotypes (T-type 1–18, or T1–18), with different genotypes exhibiting 5% or more sequence divergence (Table 36.1) [3,4]. However, some genotypes may form clusters (i.e., T4/T3/T11, T2/T6, T10/T12/T14, T7/T8/T9/T17, and T13/T16) in this analysis, and careful examination of the sequence data is essential for correct identification. In addition, as *A. divionensis*, *A. mauritaniensis*, and *A. paradivionensis* display 18S rRNA gene sequence differences of 0.2%–1.1% to *A. rhysodes* and with each other, they may be regarded as synonyms of *A. rhysodes*. Furthermore, recent 18S rRNA gene sequence analysis suggests that some T16 isolates may form distinct, novel genotypes (T19 and T20) within the genus *Acanthamoeba* [3,5]. Although T4 isolates have accounted for a majority of clinical cases of AK and GAE, isolates belonging to other genotypes (e.g., T1, T2, T3, T5, T6, T10, T11, T12, T15, and T18) have also been implicated in human infections (Table 36.1) [6–11].

36.1.1.2 Morphology

Acanthamoeba trophozoites (usually <30 μm, rarely >65 μm in size) are amoeboid in shape (or wide and tongue-shaped), with irregular margins and short and fine pseudopodia (or acanthopodia; Greek "acanth" means "spikes"). Continuously forming and reabsorbing, acanthopodia (formed by bundles of actin microfilaments extending as rigid cores) protrude from every area of the body's surface, contributing to *Acanthamoeba* locomotion. However, in *A. astronyxis* and *A. comandoni*, the acanthopodia may be quite long.

Under adverse environmental conditions (e.g., food deprivation, desiccation, and changes in temperature and pH), *Acanthamoeba* forms a double-walled wrinkled cyst (of 13–20 μm in size). The outer wall is made up of proteins and polysaccharides, and the inner wall contains cellulose, with the two walls being separated by a space forming the so-called ectocyst and endocyst. Cysts are resistant to

TABLE 36.1

Correlation of *Acanthamoeba* T-Types with Clinical Diseases

T-Type	Species	Clinical Disease
T1	*Acanthamoeba* sp.	Encephalitis
T2	*A. palestinensis*, *A. pustulosa*	Keratitis, encephalitis, sinusitis
T3	*A. griffini*, *A. pearcei*	Keratitis
T4	*A. castellanii*, *A. divionensis*, *A. lugdunensis*, *A. mauritaniensis*, *A. polyphaga*, *A.quina*, *A. rhysodes*	Keratitis, encephalitis
T5	*A. lenticulata*	Keratitis, encephalitis
T6	*Acanthamoeba* sp.	Keratitis
T7	*A. astronyxis*	Unknown
T8	*A. tubiashi*	Unknown
T9	*A. comandoni*	Unknown
T10	*A. culbertsoni*, *Acanthamoeba* sp.	Keratitis, encephalitis
T11	*A. hatchetti*, *A. stevensoni*	Keratitis
T12	*A. healyi*	Encephalitis
T13	*Acanthamoeba* sp.	Unknown
T14	*Acanthamoeba* sp.	Unknown
T15	*A. jacobsi*	Keratitis
T16	*Acanthamoeba* sp.	Unknown
T17	*Acanthamoeba* sp.	Unknown
T18	*A. byersi*	Encephalitis, skin infection

Sources: Crary, M.J. Genetic variability and its relationship to *Acanthamoeba* pathogenesis. Thesis, The Ohio State University, 2012; Sente, C., et al., *Parasite Vectors*, 9(1), 127, 2016.

biocides, chlorination, and antibiotics and survive low temperatures (0°C–2°C). There is evidence that *Acanthamoeba* cysts remain viable for more than 24 years after storage in water at 4°C and sustain desiccation for more than 20 years. However, *Acanthamoeba* cysts are sensitive to treatment with Freon or methylene oxide or autoclaving. When condition improves [e.g., inoculation on nonnutrient agar (NNA) containing bacteria], excystment takes place with trophozoites emerging from the cyst. As *Acanthamoeba* cyst formation (encystment) and excystation may be induced using nonnutrient media, *Acanthamoeba* offers a valuable tool to study cellular differentiation [12].

Electron microscopy examination reveals the presence of a Golgi complex, smooth and rough endoplasmic reticula, free ribosomes, digestive vacuoles, mitochondria, and microtubules in *Acanthamoeba* trophozoites. Apart from the existence of a trilaminar plasma membrane surrounding the cytoplasmic contents, characteristic spiny surface projections (i.e., acanthopodia) are observable. Within the cytoplasm are one or more prominent contractile vacuoles with osmotic function, and a nucleus (with a large central nucleolus) that is approximately one-sixth the size of trophozoite. Other notable types of vacuoles in the cytoplasm include lysosomes, digestive vacuoles, and glycogen-containing vacuoles [2].

36.1.1.3 Genomics

From the sequence data of 14 *Acanthamoeba* species available in GenBank, it is clear that *Acanthamoeba* genomes vary from 42.02 to 120.42 Mb in size, with GC contents ranging from 42.7% to 59.3% (Table 36.2). Interestingly, *A. castellanii*, a solitary free-living amoebozoan of T4 genotype, possesses a genome of 42.02 Mb including 15,455 compact intron-rich genes. Besides encoding various tyrosine kinase signaling proteins, *A. castellanii* genome also generates a diverse repertoire of predicted pattern recognition receptors, many of which show predicted orthologous functions in the innate immune systems of higher organisms. The size of *A. castellanii* mitochondrial DNA genome is 41,591 bp [13].

TABLE 36.2

Genomic Comparison of *Acanthamoeba* Species

Species	T-Type	Genome Size (Mb)	GC Content (%)
Acanthamoeba astronyxis	T7	83.43	42.9
Acanthamoeba castellanii	T4	42.02–45	57.8–58.4
Acanthamoeba culbertsoni	T10	55.54	50.4
Acanthamoeba divionensis	T4	84.77	42.7
Acanthamoeba healyi	T12	75.32	58.7
Acanthamoeba lenticulata	T3	66.03	56.9
Acanthamoeba lugdunensis	T4	99.42	59.1
Acanthamoeba mauritaniensis	T4	106.84	58.7
Acanthamoeba palestinensis	T2	103.48	59.1
Acanthamoeba pearcei	T3	115.61	59
Acanthamoeba polyphaga	T4	120.42	59.3
Acanthamoeba quina	T11	83.59	59.2
Acanthamoeba rhysodes	T4	75.82	57.9
Acanthamoeba royreba	T4	79.54	51.3

36.1.2 Life Cycle and Epidemiology

36.1.2.1 Life Cycle

Acanthamoeba undergoes two stages of development in its life cycle: motile trophozoite and dormant cyst. Trophozoites (Greek "tropho" means "to nourish") actively feed on bacteria (e.g., *Escherichia coli* K-12, *Klebsiella aerogenes*), algae, yeast, and other protozoa in the environment through pseudopod formation, phagocytosis, or food cup formation, and are infective to humans. In addition, trophozoites are capable of living axenically on nutrients in liquid through pinocytosis. The locomotion of trophozoites involves the formation of hyaline pseudopodia (acanthopodia), and trophozoites divide mitotically (through binary fission) under optimal conditions (food supply, neutral pH, ~30°C, and 50–80 mOsmol). *Acanthamoeba* forms dormant cyst with a resistant double-layer wall under harsh conditions, facilitating its long-term survival under extreme temperatures and pH, desiccation, and chemical exposure [2].

In amebic or acanthamoebic keratitis (AK), which is a more common form of *Acanthamoeba* infections, *Acanthamoeba* enters into the cornea following a corneal trauma. The abrased corneal epithelium secretes high concentrations of mannose-glycoprotein, to which *Acanthamoeba* binds with high affinity. After penetration into the corneal epithelium, *Acanthamoeba* invades the underlying stroma (a collagenous matrix), in which *Acanthamoeba* produces a collagenase to dissolve the collagenous matrix [2].

In GAE, *Acanthamoeba* enters into the body via an open wound (in the skin or the upper respiratory tract) or via inhalation of airborne cysts, and spreads hematogenously into the central nervous system (CNS). Inside the brain, *Acanthamoeba* generates proteases, which in association with host immune systems induce proinflammatory responses and cause massive brain swelling and neuronal damage, leading to death in ~95% of the infected individuals within few days [2].

36.1.2.2 Epidemiology

As a free-living organism, *Acanthamoeba* is ubiquitously distributed in nature and has been found in the environment (e.g., soil, dust, air, natural and treated water, seawater, swimming pools, sewage, sediments, air-conditioning units, domestic tap water, drinking water treatment plants, and bottled water), health care facilities (e.g., dental treatment units, hospitals and dialysis units, eyewash stations, contact lenses, and lens cases), vegetation, animals (e.g., fish, amphibia, reptiles, and mammals), and humans (skin, nasal cavities, throat, intestines, brain, lungs, and cornea) [14–16].

In addition, *Acanthamoeba* may harbor various microbial endosymbionts (e.g., *Candidatus Caedibacter acanthamoebae*, *Candidatus Odyssella thessalonicensis*, *Candidatus Paracaedibacter acanthamoebae*, *Candidatus Paracaedibacter symbiosus*, *Comamonas acidovorans*, *Legionella pneumophila*, *Pseudomonas aeruginosa*, mimivirus, megavirus, and pandoravirus). Further, a large number of other bacterial species (e.g., *Aeromonas, Bacillus cereus*, *Bartonella, Burkholderia, Campylobacter jejuni*, *Chlamydia pnuemoniae*, *Coxiella burnetii*, *Cytophaga*, *E. coli* O157:H7, *Flavobacterium, Francisella tularensis*, *Helicobacter pylori*, *Listeria*, *Mycobacterium*, *Pasteurella multocida*, *Prevotella intermedia*, *Porphyromonas gingivalis*, *Rickettsia*, *Salmonella* Typhimurium, *Shigella*, *Simkania negevensis*, *Staphylococcus aureus*, *Vibrio*, and *Waddlia chondrophila)* have been shown to survive and multiply within *Acanthamoeba*. This highlights the potential role of *Acanthamoeba* in serving as bacterial reservoirs for human infections [2,12].

36.1.3 Clinical Features and Pathogenesis

Despite their reputation as free-living organisms, *Acanthamoeba* spp. are opportunistic pathogens with the ability to cause AK, GAE, cutaneous acanthamoebiasis, and sinusitis in humans.

36.1.3.1 Amoebic (Acanthamoebic) Keratitis

As a most common form of *Acanthamoeba* infections, AK is a debilitating, painful, and vision-impairing disease of cornea that usually occurs in immunocompetent individuals with corneal abrasions (as a result of contact lens wearing or poor lens hygiene). Characterized by a ring-like opaque infiltrate underlying an epithelial ulcer, with clinical symptoms ranging from redness, tearing, photophobia, lid edema, corneal ulcers, to blindness, AK may be caused by several *Acanthamoeba* species, including *A. castellanii*, *A. polyphaga*, *A. hatchetti*, *A. culbertsoni*, *A. rhysodes*, *A. griffini*, *A. quina*, and *A. lugdunensis*. Histopathologically, trophozoites are initially present in the corneal epithelium and subsequently invade the underlying stroma, inducing a ring-like stromal infiltrate (composed of neutrophils and other inflammatory cells) that leads to extensive damage (e.g., conjunctival hyperemia, corneal inflammation, episcleritis, and scleritis). Further spread of trophozoites into corneal nerves or the retina may result in neuritis and necrosis, or chorioretinitis [17,18].

36.1.3.2 Granulomatous Amoebic Encephalitis

As a less common but more severe form of *Acanthamoeba* infection, GAE is usually associated with immunodeficiency (e.g., HIV), diabetes, malignancies, malnutrition, systemic lupus erythematosus, renal failure, cirrhosis, tuberculosis, skin ulcers, Hodgkin's disease, and other predisposing factors (alcoholism, drug abuse, steroid treatment, cancer chemotherapy, radiotherapy, and organ transplantation). Characterized by a chronic protracted slowly progressive CNS infection, GAE shows a range of subacute symptoms, from headaches, confusion, nausea, vomiting, fever, lethargy, stiff neck, focal neurological deficits (such as cranial nerve palsies and coma), seizures, to rapid death (within 1 week to several months). GAE may also involve the lungs [19–21].

36.1.3.3 Cutaneous Acanthamoebiasis and Sinusitis

Cutaneous acanthamoebiasis and sinusitis may occur in AIDS patients with or without CNS involvement. Characterized by the formation of hard erythematous nodules or skin ulcers, cutaneous acanthamoebiasis causes high mortalities in infected patients (73% without CNS involvement, and 100% with CNS involvement).

Postmortem biopsy from patients with GAE reveals severe edema, hemorrhagic necrosis, fibrin thrombi, and inflammation, with multifocal lesions present in the midbrain, brain stem, corpus callosum, and cerebellum. Chronic inflammatory exudate (composed mainly of polymorphonuclear leucocytes and mononuclear cells) is evident over the cortex. Numerous trophozoites (and microscopic cysts) may be found in neurological tissues and occasionally other organs (e.g., the liver, kidneys, trachea, adrenals, and lungs). Histological findings in cutaneous lesions include foci of necrosis surrounded by inflammatory cells, vasculitis, trophozoite, and cyst forms, the latter of which may be mistaken for yeast forms of *Blastomyces dermatitidis*, sporangia of *Rhinosporidium seeberi*, *Cryptococcus neoformans*, or *Prototheca wickerhamii* [2].

A central mechanism of *Acanthamoeba* pathogenesis relates to its ability to attach to the host cell, a process mediated chiefly by a 130-kDa mannose-binding protein (MBP) expressed on the surface of *Acanthamoeba*. Other adhesins produced by *Acanthamoeba* include a 28.2-kDa laminin-binding protein, a 55-kDa laminin-binding protein, and a >207-kDa adhesin. Following initial binding, *Acanthamoeba* engages in other activities (e.g., phagocytosis and toxin production) that lead to host cell death [22,23]. Other factors that influence *Acanthamoeba* pathogenesis include cytolytic enzyme production (e.g., serine and cysteine metalloproteinases, phospholipases, glycosidases, elastases, sphingomyelinase, collagenase, fibrinolytic enzyme, and ecto-ATPases, many of which are secreted only by clinical isolates), temperature tolerance, and immune evasion capability, in addition to various host determinants (interleukin-β, interleukin-α, interferon-γ, tumor necrosis factor-α, host cell apoptosis) [24,25].

36.1.4 Diagnosis

Traditionally, identification of *Acanthamoeba* is dependent on morphological characterization through microscopic examination of clinical specimens [direct wet mounts of cerebrospinal fluid (CSF) or bronchoalveolar lavage (BAL) fluid cytospin preparations, and stained smears of CSF sediment, brain, cutaneous lesion scrape/biopsy, or corneal scrape/biopsy]. The sizes and shape of trophozoites and cysts as well as distinct nuclear structure [characterized by a prominent nucleolus, contractile vacuole, and cytoplasmic vacuoles, which may be visualized more readily using trichrome or hematoxylin and eosin (H&E) stains on fixed preparations after cytocentrifugation] help distinguish *Acanthamoeba* from host macrophages and other immune cells. In addition, *Acanthamoeba* cyst wall turns red after periodic acid-Schiff staining, and its cyst becomes black after Gomori methenamine silver staining. Calcofluor white (a chemofluorescent dye with an affinity for the polysaccharide polymers) stains amebic cyst walls bright apple green, which, with Evans blue counterstaining the background, is useful for identification of *Acanthamoeba* cysts in brain or corneal tissue. Acridine orange staining of corneal scrapings or CSF represents another simple and reliable method for rapid histological diagnosis of AK or GAE. Furthermore, transmission electron microscopy of infected tissues, immunofluorescent or immunoperoxidase cytochemical staining of cryostat sections, or infected tissues embedded in paraffin offer other approaches for identification of *Acanthamoeba* [26].

To enhance its detection and identification, *Acanthamoeba* is grown on NNA, 1.5% containing a lawn of *E. coli* or *E. aerogenes*, axenically in PYG medium (2% proteose peptone, 0.2% yeast extract, and 0.1 M glucose), or in Oxoid medium (Cline medium, containing serum and hemin) at 28°C–35°C for 10 days or more (to allow sufficient time for excystment). Alternatively, *Acanthamoeba* may be cultured on mammalian cell monolayers [e.g., African green monkey kidney (Vero), human embryonic lung (HEL), human embryonic kidney (HEK), HeLa, B103 rat neuroblastoma, and L929 fibroblasts]. *Acanthamoeba* ingestion of bacteria or cells produces clear plaques after a week. Microscopic examination of *Acanthamoeba* culture isolates provides valuable confirmation [2].

With molecular techniques moving from laboratory bench to clinical setting, a variety of nucleic acid amplification procedures have been developed for *Acanthamoeba* identification. These methods exploit variations of complete or partial nuclear 18S rRNA gene, complete mitochondrial 16S rRNA, complete mitochondrial genome, and randomly amplified polymorphic DNA patterns. In particular, analysis of nuclear 18S ribosomal RNA (18S rRNA) gene sequence has made fast, reliable, and repeatable identification of *Acanthamoeba* feasible. By setting a cutoff of <5% sequence divergences, the nuclear 18S ribosomal RNA-based method has enabled separation of *Acanthamoeba* isolates into 18 genotypes (T-types 1–18 or T1–18). Typically, primers CRN5 (5′-TGGTTGATCCTGCCAGTAG-3′) and SSU2-TRUN

(5′-TGATCCCTCCGCAGGTTCAC-3′), or primers JDP1 (5′-GGCCCAGATCGTTTACCGTGAA-3′) and JDP2 (5′-TCTCACAAGCTGCTAGGGGAGTCA-3′), are used to amplify nearly complete *Rns* sequence, or a partial *Rns* region (ASA.S1) that contains a diagnostic fragment with respect to genotype, respectively. The amplification products are then sequenced by using primers JDP1/JDP2 and internal primers 892 (5′-CCAAGAATTTCACCTCTGAC-3′) and 892C (5′-GTCAGAGGTGAAATTCTTGG-3′). The resulting sequences are aligned with other *Acanthamoeba* sequences in *Rns* database [6].

36.1.5 Treatment and Prevention

Due to the resilient nature of *Acanthamoeba* cysts, current treatment of AK involves topical applications of a cationic antiseptic agent (e.g., polyhexamethylene biguanide 0.02% or chlorhexidine 0.02%) with or without a diamidine (e.g., propamidine 0.1% or hexamidine 0.1%) for several weeks. A combination of 0.1% propamidine isethionate (Brolene) topically with 0.15% dibromopropamidine is also effective if treatment is initiated early in infection. For pain relief, topical cyclopegic solutions and oral nonsteroidal medications may be administered. Penetrating keratoplasty may help restore visual acuity but is unnecessary if AK patients are treated within 6 weeks of presentation [27–31].

For disseminated *Acanthamoeba* infection such as GAE, which evolves rapidly, it is important to initiate treatment early. Trimethoprim–sulfamethoxazole therapy together with ketoconazole and rifampin has proven effective for treating two immunocompetent pediatric patients with CNS involvement. A 4-week course of IV pentamidine isethionate, topical chlorhexidine gluconate, and 2% ketoconazole cream has also been successfully applied for treatment of disseminated *Acanthamoeba* infection in a renal transplant patient. For CNS infection, 5-fluorocytosine rather than pentamidine is recommended because the latter demonstrates nephrotoxicity and fails to cross the blood–brain barrier. In addition, 40 mg of 5-fluorocytosine per kg for 2 weeks appears to be useful for treating cutaneous acanthamoebiasis (e.g., AIDS patient with cutaneous and sinus lesions). A combination of lipid formulation of amphotericin B and voriconazole may also be applied for granulomatous dermatitis secondary to *Acanthamoeba* infection [32].

For effective prevention of AK, decontamination of contact lens and lens cases with 3% solution of hydrogen peroxide is valuable. Other precautions include (1) wearing and replacing contact lenses according to the schedule, (2) removing contact lenses before any activity involving contact with water (showering, using a hot tub, or swimming), (3) washing hands with soap and water and drying before handling contact lenses, (4) cleaning contact lenses according to instructions, (5) rubbing and rinsing storage cases with sterile contact lens solution (not tap water), and (6) replacing storage cases at least once every 3 months.

36.2 Laboratory Models

36.2.1 Animal Models

36.2.1.1 Mice

Mice represent a model of choice for CNS infection, as mice infected via intranasal inoculation of *Acanthamoeba* develop symptoms of head tilt, circling, twirling, limb paresis, and convulsive seizures, along with amebic rhinitis and pneumonitis. In the capillaries of the lungs and brains, trophozoites and cysts are present. In addition, mice may also be used for modeling AK, with neutrophil being the predominant cell type during early stage of AK pathogenesis [33].

36.2.1.2 Rats

Wistar rats are a preferred model for AK because of a low death rate and larger corneas for inoculation. Intrastromal injection of *Acanthamoeba* or coinjection with bacteria (e.g., *Corynebacterium*) provides an efficient way to initiate AK [34–36].

36.2.1.3 Chinese Hamsters

Application of "contact lenses" containing *Acanthamoeba* to the abraded corneal surfaces of Chinese hamsters for at least 5–7 days leads to acute and self-limiting corneal infection, with clinical features of neutrophil infiltration, epithelial ulceration, edema, corneal opacity, and neovascularization [37–40].

36.2.1.4 Rabbits

Microinjection of $1 \times 10^4/100\,\mu L$ *Acanthamoeba healyi* trophozoites between the corneal epithelium and Bowman's layer, anterior to the corneal stroma of New Zealand white rabbits, enabled successful establishment of AK, which displayed an efficient immune response with less severe pathology and was strikingly similar to AK in humans [41,42].

36.2.1.5 Pigs

Due to the anatomical similarities of the pig eye to the human eye, pigs offer a useful model for ocular infections such as AK. However, spontaneous resolution of AK in pigs occurs in 8–10 weeks, in contrast to the prolonged infection in humans [43].

36.2.2 *In Vitro* Models

In vitro osmotolerance (0.5 M mannitol) and temperature tolerance (30°C) assays offer a valuable way to assess *Acanthamoeba* pathogenicity [44]. Human epithelial cells (e.g., corneal epithelial cells), stromal keratocytes, and stromal cell homogenates may be used for *in vitro* assessment of AK, with clearing of cell monolayers (CPE) being an indication of *Acanthamoeba* pathogenicity. *In vitro* experiment with human brain microvascular endothelial cells (HBMEC), which constitute the blood–brain barrier, indicates that *Acanthamoeba* abolishes the HBMEC transendothelial electrical resistance by degrading occludin and zonula occludens-1 tight junction proteins, leading to increased permeability.

36.3 Conclusion

The genus *Acanthamoeba* consists of more than 25 species of free-living protozoa, which are characterized by the generation of pseudopodia (acanthopodia) for locomotion during trophozoite stage and the formation of a double-walled wrinkled cyst under adverse environmental conditions. While being seemingly harmless creatures in the environment, certain *Acanthamoeba* species are surprisingly dangerous, with the ability to take advantage of the temporary weakness in human host for its own gain, and cause painful, sight-threatening AK in contact lens wearers and deadly GAE and skin infections in immunocompromised individuals. In addition, *Acanthamoeba* has the potential to serve as reservoirs for a number of human bacterial pathogen due to the fact that many bacteria have been shown to survive and multiply within the parasite. Following the development of molecular test targeting nuclear 18S rRNA gene, rapid and reliable identification of *Acanthamoeba* is possible [45]. However, considering the current lack of options for treating and controlling *Acanthamoeba* infections, especially GAE, further research using laboratory models to uncover the molecular details of *Acanthamoeba* pathogenesis is critical.

REFERENCES

1. Qvarnstrom Y, Nerad TA, Visvesvara GS. Characterization of a new pathogenic *Acanthamoeba* species, *A. byersi* n. sp., isolated from a human with fatal amoebic encephalitis. *J Eukaryot Microbiol.* 2013;60(6):626–33.
2. Marciano-Cabral F, Cabral G. *Acanthamoeba* spp. as agents of disease in humans. *Clin Microbiol Rev.* 2003;16:273–307.

3. Crary MJ. Genetic variability and its relationship to *Acanthamoeba* pathogenesis. PhD Thesis, The Ohio State University, 2012. Available at https://etd.ohiolink.edu/rws_etd/document/get/osu1343831120/inline.
4. Corsaro D, Venditti D. Phylogenetic evidence for a new genotype of *Acanthamoeba* (Amoebozoa Acanthamoebida). *Parasitol Res.* 2010;107(1):233.
5. Fuerst PA, Booton GC, Crary M. Phylogenetic analysis and the evolution of the 18S rRNA gene typing system of *Acanthamoeba. J Eukaryot Microbiol.* 2015;62(1):69–84.
6. Booton GC, Visvesvara GS, Byers TJ, Kelly DJ, Fuerst PA. Identification and distribution of *Acanthamoeba* species genotypes associated with nonkeratitis infections. *J Clin Microbiol.* 2005;43(4):1689–93.
7. Booton GC, Joslin CE, Shoff M, Tu EY, Kelly DJ, Fuerst PA. Genotypic identification of *Acanthamoeba* sp. isolates associated with an outbreak of *Acanthamoeba* keratitis. *Cornea.* 2009;28(6):673–6.
8. Kong HH. Molecular phylogeny of *Acanthamoeba. Korean J Parasitol.* 2009;47:S21–8.
9. Nuprasert W, Putaporntip C, Pariyakanok L, Jongwutiwes S. Identification of a novel T17 genotype of *Acanthamoeba* from environ-mental isolates and T10 genotype causing keratitis in Thailand. *J Clin Microbiol.* 2010;48(12):4636–40.
10. Maciver SK, Asif M, Simmen MW, Lorenzo-Morales J. A systematic analysis of *Acanthamoeba* genotype frequency correlated with source and pathogenicity: T4 is confirmed as a pathogen-rich genotype. *Eur J Protistol.* 2013;49(2):217.
11. Mirjalali H, Niyyati M, Abedkhojasteh H, Babaei Z, Sharifdini M, Rezaeian M. Pathogenic assays of *Acanthamoeba* belonging to the T4 genotype. *Iran J Parasitol.* 2013;8(4):530–5.
12. Siddiqui R, Khan NA. Biology and pathogenesis of *Acanthamoeba. Parasite Vectors.* 2012;5:6.
13. Clarke M, et al. Genome of *Acanthamoeba castellanii* highlights extensive lateral gene transfer and early evolution of tyrosine kinase signaling. *Genome Biol.* 2013;14(2):R11.
14. Lasjerdi Z, et al. Potentially pathogenic free-living amoebae isolated from hospital wards with immunodeficient patients in Tehran Iran. *Parasitol Res.* 2011;109(3):575.
15. Rahdar M, et al. Isolation and genotyping of *Acanthamoeba* strains from environmental sources in Ahvaz city, Khuzestan province, southern Iran. *Iran J Parasitol.* 2012;7(4):22–6.
16. Sente C, et al. Occurrence and genetic characterisation of *Acanthamoeba* spp. from environmental and domestic water sources in Queen Elizabeth Protected Area, Uganda. *Parasite Vectors.* 2016;9(1):127.
17. Polat ZA, Ozcelik S, Vural A, Yildiz E, Cetin A. Clinical and histologic evaluations of experimental *Acanthamoeba* keratitis. *Parasitol Res.* 2007;101(6):1621–5.
18. Knickelbein JE, Kovarik J, Dhaliwal DK, Chu CT. *Acanthamoeba* keratitis: a clinicopathologic case report and review of the literature. *Hum Pathol.* 2013;44(5):918–22.
19. McKellar MS, et al. Fatal granulomatous *Acanthamoeba* encephalitis mimicking a stroke, diagnosed by correlation of results of sequential magnetic resonance imaging, biopsy, in vitro culture, immunofluorescence analysis, and molecular analysis. *J Clin Microbiol.* 2006;44(11):4265–9
20. Khan NA. *Acanthamoeba* invasion of the central nervous system. *Int J Parasitol.* 2007;37(2):131–8.
21. Mortazavi PN, Goldsworthy G, Kirk R, Khan NA. *Acanthamoeba* produces disseminated infection in locusts and traverses the locust blood-brain barrier to invade the central nervous system. *BMC Microbiol.* 2010;10:186.
22. Clarke DW, Niederkorn JY. The pathophysiology of *Acanthamoeba* keratitis. *Trends Parasitol.* 2006;22(4):175–80.
23. Panjwani N. Pathogenesis of *Acanthamoeba* keratitis. *Ocul Surf.* 2010;8(2):70–9.
24. Khan NA. Pathogenicity, morphology, and differentiation of *Acanthamoeba. Curr Microbiol.* 2001;43(6):391–5.
25. Khan NA. Pathogenesis of *Acanthamoeba* infections. *Microb Pathog.* 2003;34(6):277–85.
26. Walochnik J, Obwaller A, Aspock H. Correlations between morphological, molecular biological, and physiological characteristics in clinical and nonclinical isolates of *Acanthamoeba* spp. *Appl Environ Microbiol.* 2000;66(10):4408.
27. Gooi P, Lee-Wing M, Brownstein S, El-Defrawy S, Jackson WB, Mintsioulis G. *Acanthamoeba* keratitis: persistent organisms without inflammation after 1 year of topical chlorhexidine. *Cornea.* 2008;27:246–8.
28. Roberts CW, Henriquez FL. Drug target identification, validation, characterization and exploitation for treatment of *Acanthamoeba* (species) infections. *Ext Parasite.* 2010; 126(1):91–6.
29. Kashiwabuchi RT, et al. Assessing efficacy of combined riboflavin and UV-A light (365 nm) treatment of *Acanthamoeba* trophozoites. *Invest Ophthalmol Vis Sci.* 2011;52(13):9333–8.

30. Berra M, et al. Treatment of *Acanthamoeba* keratitis by corneal cross-linking. *Cornea*. 2013;32(2):174–8.
31. Polat ZA, Walochnik J, Obwaller A, Vural A, Dursun A, Arici MK. Miltefosine and polyhexamethylene biguanide: a new drug combination for the treatment of *Acanthamoeba* keratitis. *Clin Exp Ophthalmol*. 2014;42(2):151–8.
32. Walia R, Montoya JG, Visvesvera GS, Booton GC, Doyle RL. A case of successful treatment of cutaneous *Acanthamoeba* infection in a lung transplant recipient. *Transpl Infect Dis*. 2007;9(1):51–4.
33. Massilamany C, et al. SJL mice infected with *Acanthamoeba castellanii* develop central nervous system autoimmunity through the generation of cross-reactive T cells for myelin antigens. *PLoS One*. 2014;9(5):e98506.
34. Ren M, Wu X. Evaluation of three different methods to establish animal models of *Acanthamoeba* keratitis. *Yonsei Med J*. 2010;51(1):121–7.
35. Ren MY, Wu XY. Toll-like receptor 4 signalling pathway activation in a rat model of *Acanthamoeba* keratitis. *Parasite Immunol*. 2011;33(1):25–33.
36. Veríssimo Cde M, Maschio VJ, Correa AP, Brandelli A, Rott MB. Infection in a rat model reactivates attenuated virulence after long-term axenic culture of *Acanthamoeba* spp. *Mem Inst Oswaldo Cruz*. 2013;108(7):832–5.
37. Van Klink F, Leher H, Jager MJ, Alizadeh H, Taylor W, Niederkorn JY. Systemic immune response to *Acanthamoeba* keratitis in the Chinese hamster. *Ocul Immunol Inflamm*. 1997;5:235–44.
38. Hurt M, Apte S, Leher H, Howard K, Niederkorn J, Alizadeh H. Exacerbation of *Acanthamoeba* keratitis in animals treated with anti-macrophage inflammatory protein 2 or antineutrophil antibodies. *Infect Immun*. 2001;69:2988–95.
39. Polat ZA, Obwaller A, Vural A, Walochnik J. Efficacy of miltefosine for topical treatment of *Acanthamoeba* keratitis in Syrian hamsters. *Parasitol Res*. 2012;110(2):515–20.
40. Suryawanshi A, Cao Z, Sampson JF, Panjwani N. IL-17A-mediated protection against *Acanthamoeba* keratitis. *J Immunol*. 2015;194(2):650–63.
41. Said NA, Shoeir AT, Panjwani N, Garate M, Cao Z. Local and systemic humoral immune response during acute and chronic *Acanthamoeba* keratitis in rabbits. *Curr Eye Res*. 2004;29:429–39.
42. Feng X, Zheng W, Wang Y, Zhao D, Jiang X, Lv S. A Rabbit model of *Acanthamoeba* keratitis that better reflects the natural human infection. *Anat Rec (Hoboken)*. 2015;298(8):1509–17.
43. Alizadeh H, et al. Successful immunization against *Acanthamoeba* keratitis in a pig model. *Cornea*. 1995;14:180–6.
44. De Jonckheere JF. Growth characteristics, cytopathic effect in cell culture, and virulence in mice of 36 type strains belonging to 19 different *Acanthamoeba* spp. *Appl Environ Microbiol*. 1980;39(4):681–5.
45. Niyyati M, et al. Genotyping of *Acanthamoeba* isolates from clinical and environmental specimens in Iran. *Exp Parasitol*. 2009;121(3):242–5.

37

Cryptosporidium

Dongyou Liu

CONTENTS

37.1 Introduction

The genus *Cryptosporidium* comprises nearly 30 known species of apicomplexan protozoans that infect birds, reptiles, fish, and mammals, including humans. Of these, *C. parvum* and *C. hominis* along with a few other species are responsible for producing watery diarrhea (intestinal cryptosporidiosis) with or without a persistent cough (respiratory cryptosporidiosis) in both immunocompetent and immunodeficient humans.

Although *Cryptosporidium* was first identified from the stomach of mice in 1907, its role in human infection was only established in 1976. Since then, cryptosporidiosis has been recognized as a major cause of chronic diarrhea in patients with AIDS, as a cause of zoonotic and waterborne outbreaks of diarrhea, and as a cause of diarrhea and unexplained cough in immunocompetent children.

37.1.1 Classification, Morphology, and Genome Structure

37.1.1.1 Classification

Taxonomically, the genus *Cryptosporidium* belongs to the family Cryptosporidiidae, suborder Eimeriorina, order Eucoccidiorida, subclass Coccidiasina, class Conoidasida, phylum Apicomplexa, domain Eukaryota, kingdom Animalia.

The suborder Eimeriorina includes 12 established families (Aggregatidae, Atoxoplasmatidae, Barrouxiidae, Calyptosporidae, Caryotrophidae, Cryptosporidiidae, Eimeriidae, Elleipsisomatidae, Lankesterellidae, Sarcocystidae, Selenococcidiidae, and Spirocystidae).

Being the only member of the family Cryptosporidiidae, the genus *Cryptosporidium* currently consists of about 30 recognized species, i.e., *Cryptosporidium andersoni*, *Cryptosporidium baileyi*,

Cryptosporidium bovis, *Cryptosporidium canis*, *Cryptosporidium cuniculus*, *Cryptosporidium erinacei*, *Cryptosporidium fayeri*, *Cryptosporidium felis*, *Cryptosporidium fragile*, *Cryptosporidium galli*, *Cryptosporidium hominis*, *Cryptosporidium huwi*, *Cryptosporidium macropodum*, *Cryptosporidium meleagridis*, *Cryptosporidium molnari*, *Cryptosporidium muris*, *Cryptosporidium parvum*, *Cryptosporidium proliferans*, *Cryptosporidium rubeyi*, *Cryptosporidium ryanae*, *Cryptosporidium saurophilum*, *Cryptosporidium serpentis*, *Cryptosporidium suis*, *Cryptosporidium tyzzeri*, *Cryptosporidium ubiquitum*, *Cryptosporidium varanii*, *Cryptosporidium viatorum*, *Cryptosporidium wrairi*, and *Cryptosporidium xiaoi* [1–4].

Due the fact that members of the genus *Sarcocystis* in the family Sarcocystidae produce oocysts with thin walls that often rupture to release sporocysts, each of which possesses four sporozoites similar to *Cryptosporidium* oocysts, a number of *Cryptosporidium* species were erroneously assigned to the genus *Sarcocystis* in the early days. However, subsequent ultrastructural examinations revealed the presence of a unique attachment organelle as a defining feature for the genus *Cryptosporidium* (indeed, the name *Cryptosporidium* refers to the absence of sporocysts within the oocyst) and the family Cryptosporidiidae. In addition, biological and molecular studies indicated that *Cryptosporidium* species are more closely related to gregarine parasites of the subclass Gregarinasina, class Conoidasida rather than to Coccidians (e.g., *Isospora*, *Cyclospora*, and *Sarcocystis*), since *Cryptosporidium* species undergo gregarine-like gamont stages and have the ability to complete their life cycles in the absence of host cells [1,2,5,6].

Cryptosporidium species are capable of infecting many vertebrates, ranging from mammals (including man) and birds to reptiles and fish [7]. Within the genus *Cryptosporidium*, species of medical and veterinary importance consist of *C. andersoni* (cattle), *C. baileyi* (chicken and turkey), *C. bovis* (cattle), *C. canis* (dogs), *C. felis* (cats), *C. galli* (birds), *C. hominis* (humans), *C. meleagridis* (birds and humans), *C. molnari* (fish), *C. muris* (rodents, other mammals, and humans), *C. parvum* (cattle, sheep, goats, and humans), and *C. suis* (pigs and humans). It is noteworthy that *C. parvum* and *C. hominis* are responsible for a majority (90%) of human clinical cases of cryptosporidiosis, while *C. canis*, *C. felis*, *C. suis*, *C. meleagridis*, *C. muris*, and *C. andersoni* as well as chipmunk genotype I (W17) and skunk genotype may be occasionally involved in human illnesses (Table 37.1) [1,2,8].

Based on their organ specificity (localization of endogenous development in the host), *Cryptosporidium* species and genotypes may be distinguished into two major groups: the larger intestinal group (which has affinity for the intestine, as well as the lungs and bursa of Fabricius), and the smaller gastric group (which has affinity for the glands of the glandular stomach). Within the gastric group, two species (*C. muris* and *C. andersoni*) are specific for mammals. Interestingly, *C. muris* strain TS03 from a naturally infected East African mole rat (*Tachyoryctes splendens*) in Kenya was recently renamed as *Cryptosporidium proliferans* n. sp. Oocysts of *C. proliferans* [of 7.7 × 5.3 μm (range 6.8–8.8 × 4.8–6.2 μm) in size, with a length/width ratio of 1.48] are morphologically distinct from *C. parvum*, *C. muris* HZ206, *C. andersoni*, *C. suis*, and *C. xiaoi*. Phylogenetic analyses on the basis of small subunit rRNA, *Cryptosporidium* oocyst wall protein, thrombospondin-related adhesive protein of *Cryptosporidium*-1, heat shock protein 70, actin, heat shock protein 90 (MS2), MS1, MS3, and M16 gene sequences further highlighted the genetic differences of *C. proliferans* from *C. muris* and other *Cryptosporidium* species [9,10].

37.1.1.2 Morphology

Cryptosporidium oocysts are round to oval in shape and measure about 4–6 μm in diameter (Table 37.1). Under a phase-contrast microscope, *Cryptosporidium* oocysts are highly refractile in wet smear and contain prominent dark granules (black dot) or small vacuoles, in contrast to yeast, which are not refractile and contain no granules. Further, *Cryptosporidium* oocysts do not stain with Lugol's iodine solution, but yeast do. Excystment of *Cryptosporidium* oocysts release four sporozoites each.

37.1.1.3 Genome Structure

The genome sequences of *C. parvum* IOWA, *C. hominis* TU502, and *C. muris* were released on CryptoDB in 2003 and 2011, respectively [11–13]. Composed of eight chromosomes each, these genomes are of 9.1–92 Mb in size, with 26%–30% GC content and 94%–97% nucleotide identity. Altogether, about 4000

TABLE 37.1

Morphological and Biological Characteristics of Human-Infecting *Cryptosporidium* Species

	Former Designation	Oocyst Dimension (μm)	Oocyst Length/ Width Ratio	Main Susceptible Hosts	Predilection Site
Cryptosporidium parvum senso stricto	Bovine genotype, genotype 2	5.0 × 4.5 (4.5–5.4 × 4.2–5.0) or 5.2 × 4.6 (4.8–5.6 × 4.2–4.8)	1.16 or 1.15 (1.04–1.22)	Cattle (especially preweaned calves), sheep, goats, deer, and humans	Small intestine
Cryptosporidium hominis	Human genotype, genotype 1, genotype H	5.2 × 4.9 (4–6 × 4–5)	1.1 (1.0–1.1)	Humans, dugong, and lamb	Small intestine
Cryptosporidium canis	Dog genotype	4.95 × 4.71 (3.68–5.88 × 3.68–5.88)	1.05 (1.04–1.06)	Dogs, coyotes, foxes, and humans	Small intestine
Cryptosporidium felis		5 × 4.5 (6.0–5.0 × 5.0–4.5)	1.19	Cats	Small intestine
Cryptosporidium suis		4.6 × 4.2 (4.9–4.4 × 4.0–4.3)	1.1	Pigs and humans	Small and large intestine
Cryptosporidium meleagridis		5.2 × 4.6 (4.5–6.0 × 4.2–5.3)	1.13 (1.00–1.33)	Chicken and turkey	Small intestine
Cryptosporidium muris		7.4 × 5.6 (7–8 × 5–6.5) or 8.4 × 6.3 (7.5–10 × 5.5–7)	1.3 (1.1–1.5) or 1.4	Rodents, other mammals, and humans	Stomach
Cryptosporidium andersoni	*C. muris*-like	7.4 × 5.5 (6.0–8.1 × 5.0–6.5)	1.35	Cattle, bactrian camels, sheep, rodents, and humans	Abomasum

protein-encoding genes are present. *Cryptosporidium* genomes appear to be unusual among eukaryotes in having a degenerate “mitosome” (located in the posterior end of sporozoites) instead of a mitochondrion, with accompanied loss of many mitochondrial proteins, including those required for the TCA cycle, oxidative phosphorylation, and fatty acid oxidation as well as de novo biosynthesis of amino acids, nucleotides, and sugars. These gene losses render *Cryptosporidium* species heavily reliant on scavenging nutrients from the host rather than de novo biosynthesis [12–15]. In addition, sequencing analyses of *C. hominis* IbA10G2 (the most virulent subtype responsible for all outbreaks in Europe and Australia) and IaA28R4 (a dominant outbreak subtype in the United States) uncovered major differences in the 5′ and 3′ ends of chromosome 6 and the putative virulence determinant gp60 region, suggesting that genetic recombination plays a potential role in the emergence of hypertransmissible *C. hominis* subtypes [16,17].

37.1.2 Life Cycle and Epidemiology

As an obligate intracellular monoxenous parasite with the ability to carry out its entire biological cycle in a single host, *Cryptosporidium* produces oocysts (each containing four naked sporozoites) for transmission to humans via various routes (man-to-man, animal-to-man, contaminated water, food, or air). Inside the stomach, sporozoites exit through a suture along one side of the oocyst wall, pass through the microvillus brush border, enter into epithelial cells of the jejunum and duodenum, and complete their biological cycle with both asexual and sexual phases [1,2].

During asexual reproduction, sporozoites multiply several times; and in subsequent sexual reproduction, sporozoites form oocysts (via trophozoites and merozoites) at the luminal apex of the enterocytes. The newly formed oocysts gather in an extracytoplasmatic parasitophorous vacuole whose membrane is made from the enteric epithelium. It is notable that while about 80% of oocysts have a thicker, strong wall and are shed in the feces and become immediately infective, the remainder 20% of oocysts have a thin, incomplete wall, rupture in the intestine, and release four infectious sporozoites, resulting in a new cycle of infection (endogenous autoinfection). The autoinfection allows *Cryptosporidium* to sporulate and persist within the same host indefinitely, accounting for the severe chronic forms of cryptosporidiosis in immunocompromised persons in the absence of exogenous reinfection. In respiratory cryptosporidiosis, as seen in immune-deficient patients, the infective oocysts spread via nasal secretions into the environment [1,2].

Cryptosporidium oocysts are resistant to environmental adversaries and are highly infectious; as few as 10 *C. hominis* oocysts are able to produce disease in healthy adults. Although fecal–oral transmission represents the main route of infection, transmission via coughing and fomites is also possible [1,2].

Most (90%) of the human infections are caused by *C. hominis* and *C. parvum*, with *C. hominis* being found almost exclusively in humans, while *C. parvum* is readily transmitted between humans and animals as well as between humans. Other *Cryptosporidium* species (*C. meleagridis*, *C. muris*, *C. felis*, *C. canis*, *C. suis*, and *C. andersoni*) may also be involved in human infections [1,2].

As a most common cause of severe diarrhea in humans, cryptosporidiosis poses a serious public health risk worldwide. In industrialized nations, cryptosporidiosis is often associated with recreational water use, animals on petting farms, and day care centers. In developing countries (e.g., sub-Saharan Africa, Asia, and Latin America), cryptosporidiosis may result from contaminated food or drinking water, or contaminated recreational water (as in swimming pools) and occurs mostly in children younger than 5 years (with peak occurrence in children younger than 2 years) [1,2].

The population groups with an elevated risk of being exposed to *Cryptosporidium* include: (1) people who swim regularly in pools with insufficient sanitation (as some *Cryptosporidium* strains are chlorine-resistant); (2) child-care workers; (3) parents of infected children; (4) people caring for others with cryptosporidiosis; (5) backpackers, hikers, and campers who drink unfiltered, untreated water; (6) people, including swimmers, who swallow water from contaminated sources; (7) people handling infected cattle; and (8) people exposed to human feces [1,2].

37.1.3 Clinical Features and Pathogenesis

Cryptosporidium spp. (notably *C. parvum* and *C. hominis*) are apicomplexan parasites that commonly infect the microvillus border of the gastrointestinal epithelium of humans as well as other vertebrate hosts, typically producing moderate-to-severe watery diarrhea, abdominal pain, vomiting, nausea, fever, anorexia, dehydration, and weight loss. While immunocompetent individuals may experience transient gastroenteritis lasting up to 2 weeks and recover without treatment, young children (especially those <2 years of age) and individuals with immune deficiencies (e.g., HIV/AIDS), renal failure, inflammatory bowel disease, or undergoing immunosuppressive therapy may develop intractable diarrhea, growth stunting, and wasting, with fatal consequence [1,2].

Patients with respiratory cryptosporidiosis show inflammation of the nasal mucosa, sinuses, larynx, and trachea, accompanied by nasal discharge and voice change (upper respiratory cryptosporidiosis), or productive cough, dyspnea, fever, and hypoxemia (lower respiratory cryptosporidiosis). However, diarrhea is absent and fecal examination for oocysts is negative [18].

Occasionally, *Cryptosporidium* parasites may settle in the hepatobiliary system, pancreas, and urinary bladder, particularly in patients with dysfunctional immune systems.

In intestinal cryptosporidiosis, *Cryptosporidium* spp. are predominately located in the epithelial cells of the jejunum and ileum as well as other parts of the gastrointestinal tract (including biliary organ) in immunocompromised patients. The parasite stays within a parasitophorous vacuole that protrudes out of the host cytoplasm into the intestinal lumen. This provides a unique intracellular but extracytoplasmic niche for the parasite, and segregates it from direct interaction with other cell types. Host responses to *Cryptosporidium* spp. and disease outcomes are therefore dependent on the functionality of mucosal epithelial cells, including expression of pathogen pattern recognition receptors (PRRs) [e.g., the Toll-like receptors (TLRs), which recognize microbes on the cell surface and in endosomes; and nucleotide binding and oligomerization domain-like receptors (NLRs), which sense microbial molecules in the cytosol], production of antimicrobial peptides (e.g., β-defensins and LL-37), release of inflammatory cytokines/chemokines [e.g., interleukin-8 (IL-8)], production of epithelial-cell-derived exosomes, and feedback regulation of immune homoeostasis [19,20]. In immunocompetent individuals, mucosal epithelial cells are capable of initiating adequate immune responses that contribute to the elimination of invading *Cryptosporidium* parasites within a relatively short period [21]. In immunocompromised individuals and young children, immune responses of mucosal epithelial cells are insufficient to overcome the parasites, which affect host enteric nervous system (ENS) (e.g., efferent and afferent neurons, as well as interneurons) and disturb water and electrolyte balance in the gastrointestinal tract, leading to malabsorption and hypersecretion [22].

37.1.4 Diagnosis

Traditional method for diagnosing human intestinal cryptosporidiosis is based on microscopic detection of *Cryptosporidium* oocysts in stool samples with acid-fast, auramine, or indirect immunofluorescence stains. Use of fluorescence microscopy (e.g., light-emitting diode light sources) together with fluorescent stains (e.g., auramine-rhodamine, which makes oocysts appear yellow-orange) provides a more sensitive detection than the modified acid-fast stain [e.g., modified Ziehl–Neelsen (hot) or Kinyoun (cold), which demonstrates the acid resistance of oocysts]. Further, application of formalin-ether or formalin-ethyl acetate sedimentation concentration procedures, or Sheather's flotation-concentration method (using a sucrose gradient) improves the sensitivity of microscopic detection. Nonetheless, *Cryptosporidium* oocysts (of 4–6 μm in diameter) must be differentiated from other partially acid-fast organisms (e.g., *Cyclospora cayetanensis*) and other similar looking yeast or fungal spores [23,24].

Similarly, human respiratory cryptosporidiosis may be confirmed by microscopic detection of *Cryptosporidium* organisms in respiratory secretions [e.g., sputum, tracheal aspiration, bronchoalveolar lavage (BAL)], bronchial and lung biopsy, or autopsy specimens. Examination of respiratory secretions using acid-fast, auramine or indirect immunofluorescence stains may show thick-walled *Cryptosporidium* oocysts and invasive forms of the parasite (sporozoites and merozoites). Sporozoites and merozoites may be also detected from BAL fluid specimens using Giemsa stain. Hematoxylin–eosin staining of tissue specimens may uncover sporozoites, merozoites, and oocysts lining the mucosal epithelium at the luminal surface of the trachea, bronchi, and bronchioles, as well as within bronchial mucous glands. Bronchial and lung biopsy specimens may yield intracellular and extracellular cryptosporidia, including sporozoites, merozoites, or oocysts on the bronchoepithelial surface [25].

Serological assays are important for detection of specific antibodies in both symptomatic and asymptomatic infections. Although antibodies to Cp23 correlate with distant infection, those to Cp17 (also called gp15) suggest recent infection, and those to P2 are associated with repeated infection. In addition, several serological assays [e.g., direct immunofluorescence assay (DFA), enzyme immune assays (e.g., EIA, ELISA), and immunochromatographic assays] allow detection of oocyst-specific antigens [26].

In recent years, nucleic amplification techniques [e.g., polymerase chain reaction (PCR), PCR-restriction fragment length polymorphism (RFLP) analysis] are increasingly used for sensitive detection and subtyping of *Cryptosporidium* species [27–30]. This involves disruption of oocysts by bead-beating, freeze-thaw, boiling, or chemical lysis for nucleic acid extraction and use of oligonucleotide primers from selective gene targets [e.g., 18S rRNA, actin, oocyst wall protein (COWP), thrombospondin-related adhesive protein 1 (TRAP-C1), tubulin, HSP70, *gp60*, and gp900 (also known as polythreonine protein or poly-T) in addition to *noncoding* internal transcribed spacer 1 and microsatellites] for amplification, and detection of amplified products on various platforms. Application of molecular procedures has enabled the differentiation of *C. andersoni* from *C. muris*, the discrimination of *C. canis* from *C. parvum*, and the establishment of *C. hominis* as a species distinct from *C. parvum* [1,2].

37.1.5 Treatment and Prevention

Being the only drug approved by the FDA for the treatment of intestinal cryptosporidiosis in immunocompetent hosts, nitazoxanide (NTZ, marketed under the name Alinia, which is a thiazolide drug with broad antiparasitic activities) shows efficacy of 56%–96%. Supplemental zinc may improve symptoms, particularly in recurrent or persistent infections or in patients at risk for zinc deficiency. Nonetheless, the relatively high cost of nitazoxanide limits its clinical adoption. Further, nitazoxanide is ineffective against respiratory cryptosporidiosis and has proven unsuccessful in eradicating the parasite from the guts of chronically infected children with HIV [31].

In individuals coinfected with HIV, antiretroviral therapy (ART) is useful for controlling chronic diarrhea and wasting due to cryptosporidiosis. Essentially, ART helps restore immune function of human patients, leading to parasitological clearance after treatment for several months. Protease inhibitors, including indinavir, have also been shown to directly interfere with *Cryptosporidium* development. Alternatively, paromomycin may be considered for AIDS patients with respiratory cryptosporidiosis [31].

Prevention of human cryptosporidiosis should aim to interrupt fecal–oral transmission of *Cryptosporidium* oocysts through provision of clean water and sanitation, maintenance of scrupulous hygiene in communal settings such as day care centers, implementation of food safety practice, and development of effective vaccines [32].

Use of filtration technologies (e.g., slow sand filters, diatomaceous earth filter, membranes, bag- and cartridge-filter products) is useful for the removal of *Cryptosporidium* in water supply. Ultrafiltration and UV irradiation are also efficient. However, high-rate filtration and chlorine disinfection are largely ineffective in the reduction of *Cryptosporidium* oocysts from water supply. For wastewater treatment, stabilization ponds and constructed wetlands are valuable for the elimination of *Cryptosporidium* oocysts [33].

37.2 Laboratory Models

Given the ability of *Cryptosporidium* to perpetuate in human hosts, use of animal models and *in vitro* cultivation system is vital for characterization of *Cryptosporidium* life cycle, elucidation of host immune mechanisms, evaluation of potential drugs, and development of vaccines [34–36].

37.2.1 Animal Models

A number of laboratory animals are susceptible to *Cryptosporidium* infection, although numerous parameters (e.g., isolate employed, dosage, oocyst age, oocyst storage conditions, chemical pretreatments, and host genetics) may have a bearing on the outcome.

Out of the 19 different strains of adult mice examined, *C. parvum* has been shown to produce the highest levels of infection in the beige mouse (C57BL/6J-bgJ). In addition, similar acute patterns of *C. parvum* infection in C57BL/6 wild-type and T- and B cell-deficient $Rag2^{-/-}$ newborn mice were also observed. Using dexamethasone-treated or untreated adult severe combined immunodeficiency (SCID) mice, it was shown that *C. parvum* strains (of both animal and human origins) are capable of inducing intraepithelial neoplasia and invasive adenocarcinoma in the stomach, ileocecal region, and intrahepatic biliary tree [37]. Further, use of the interleukin-12 (IL-12) knockout mouse model, which mimics acute human cryptosporidiosis, facilitated the identification of a compound (P131) with significant antiparasitic activity [36]. Use of various animal models has helped to confirm that *C. hominis* at high doses may also infect calves, lambs, and piglets.

C. muris oocysts obtained from the gastric glands of wild rat stomach have been successfully transmitted to uninfected rats, mice, guinea pigs, rabbits, dogs, and cats. Indeed, *C. muris* seems to be infective to hamsters, squirrels, Siberian chipmunks, wood mice (*Apodemus sylvaticus*), bank voles (*Clethrionomys glareolus*), *Dolichotis patagonum*, rock hyrax, bactrian camels, mountain goats, cynomolgus monkeys, and humans. Interestingly, *C. muris* demonstrates notable differences in prepatent and patent periods in two laboratory rodents—BALB/c mice and the southern multimammate rat (*Mastomys coucha*). Specifically, *C. muris* infection progressed more rapidly in BALB/c mice (with a prepatent period of 7.5–10 days) than *M. coucha* (with prepatent period of 18–21 days) [38].

In contrast to *C. muris*, *C. andersoni* (previously *C. muris*-like) is not infective to outbred, inbred, neonatal, or immunocompetent mice (including common field mice, BALB/c mice, and SCID mice) as well as common voles, bank voles, desert gerbils, guinea pigs, rats, rabbits, or goats. However, *C. andersoni* has the capacity to infect Mongolian gerbils.

37.2.2 *In Vitro* Models

C. parvum and *C. hominis* are capable of replicating in murine macrophages, and *C. parvum* has been shown to grow in the human ileocecal epithelial cell line (HCT-8) [39]. *C. hominis* was found to proliferate in cell-free culture [40]. A novel method for prolonged *in vitro* cultivation of primary human intestinal epithelial cells (PECs) from small intestinal crypts enabled the detailed study of *Cryptosporidium* infection [41]. In addition, a reproducible and quantitative *C. parvum* infection model based on a noncarcinomatous, human small intestinal epithelial cell type, named FHs 74 Int, was developed [42]. An optimal protocol for transfection of *C. parvum* sporozoites in tissue culture was also described.

Using the hollow fiber technology that mimics the gut by delivering nutrients and oxygen from the basal layer upward while allowing separate redox and nutrient control of the lumen for parasite development, *Cryptosporidium* oocyst production was maintained for more than 6 months, yielding approximately 1×10^8 oocysts/mL/day [43].

A chicken embryo model supports the development of all the endogenous life cycle stages of *C. baileyi*, and provides a new and effective *in vitro* cultivation system for studies on antigens, virulence, infectivity, metabolites, and sensitivity of drugs against the parasite [44]. In a separate report, chick embryo tracheal organ (TOC) was used to cultivate oocysts or sporozoites of *C. baileyi* [45].

37.3 Conclusion

Of nearly 30 recognized species within the genus *Cryptosporidium*, *C. parvum* and *C. hominis* are important waterborne and foodborne pathogens of humans. These parasites cause villus atrophy, crypt hyperplasia, infiltration of the lamina propria, chloride hypersecretion, glucose malabsorption, and reduced barrier function, leading to diarrhea, wasting, stunt growth, and explained cough. Considering its high infectivity (as few as 10 oocysts are sufficient to establish infection in immunocompetent persons) and strong resistance to chlorine disinfection, large-scale outbreaks of human cryptosporidiosis attributable to contaminated drinking or recreational water have been reported. In order to develop improved control measures, it is essential to uncover the molecular mechanisms of *Cryptosporidium* infection. Application of laboratory models will be fundamental to achieving this goal.

REFERENCES

1. Xiao L, Fayer R, Ryan U, Upton SJ. *Cryptosporidium* taxonomy: recent advances and implications for public health. *Clin Microbiol Rev.* 2004;17:72–97.
2. Xiao L, Feng Y. Zoonotic cryptosporidiosis. *FEMS Immunol Med Microbiol.* 2008;52(3):309–23.
3. Li X, et al. *Cryptosporidium rubeyi* n. sp. (Apicomplexa: Cryptosporidiidae) in multiple *Spermophilus* ground squirrel species. *Int J Parasitol Parasites Wildl.* 2015;4(3):343–50.
4. Ryan U, et al. *Cryptosporidium huwi* n. sp. (Apicomplexa: Eimeriidae) from the guppy (*Poecilia reticulata*). *Exp Parasitol.* 2015;150:31–5.
5. Ryan U, Fayer R, Xiao L. *Cryptosporidium* species in humans and animals: current understanding and research needs. *Parasitology.* 2014;141(13):1667–85.
6. Ryan U, Hijjawi N. New developments in *Cryptosporidium* research. *Int J Parasitol.* 2015;45(6):367–73.
7. Nakamura AA, Meireles MV. *Cryptosporidium* infections in birds—a review. *Rev Bras Parasitol Vet.* 2015;24(3):253–67.
8. Jiang Y, et al. *Cryptosporidium andersoni* as a novel predominant *Cryptosporidium* species in outpatients with diarrhea in Jiangsu Province, China. *BMC Infect Dis.* 2014;14:555.
9. Vinayak S, et al. Genetic modification of the diarrhoeal pathogen *Cryptosporidium parvum. Nature.* 2015;523(7561):477–80.
10. Kváč M, et al. *Cryptosporidium proliferans* n. sp. (Apicomplexa: Cryptosporidiidae): molecular and biological evidence of cryptic species within gastric *Cryptosporidium* of mammals. *PLoS One.* 2016;11(1):e0147090.
11. Puiu D, Enomoto S, Buck GA, Abrahamsen MS, Kissinger JC. CryptoDB: the *Cryptosporidium* genome resource. *Nucleic Acids Res.* 2004;32:D329–31.

12. Widmer G, Sullivan S. Genomics and population biology of *Cryptosporidium* species. *Parasite Immunol.* 2012;34(2–3):61–71.
13. Widmer G, Lee Y, Hunt P, Martinelli A, Tolkoff M, Bodi K. Comparative genome analysis of two *Cryptosporidium parvum* isolates with different host range. *Infect Genet Evol.* 2012;12(6):1213–21.
14. Abrahamsen MS, et al. Complete genome sequence of the apicomplexan *Cryptosporidium parvum. Science.* 2004;304:441–5.
15. Xu P, et al. The genome of *Cryptosporidium hominis. Nature.* 2004;431:1107–12.
16. Guo Y, et al. Comparative genomic analysis reveals occurrence of genetic recombination in virulent *Cryptosporidium hominis* subtypes and telomeric gene duplications in *Cryptosporidium parvum. BMC Genomics.* 2015;16:320.
17. Hadfield SJ, et al. Generation of whole genome sequences of new *Cryptosporidium hominis* and *Cryptosporidium parvum* isolates directly from stool samples. *BMC Genomics.* 2015;16:650.
18. Sponseller JK, Griffiths JK, Tzipori S. The evolution of respiratory cryptosporidiosis: evidence for transmission by inhalation. *Clin Microbiol Rev.* 2014;27(3):575–86.
19. Kothavade RJ. Challenges in understanding the immunopathogenesis of *Cryptosporidium* infections in humans. *Eur J Clin Microbiol Infect Dis.* 2011;30(12):1461–72.
20. Zhou R, Feng Y, Chen XM. Non-coding RNAs in epithelial immunity to *Cryptosporidium* infection. *Parasitology.* 2014;141(10):1233–43.
21. Petry F, Jakobi V, Tessema TS. Host immune response to *Cryptosporidium parvum* infection. *Exp Parasitol.* 2010;126(3):304–9.
22. Halliez MC, Buret AG. Gastrointestinal parasites and the neural control of gut functions. *Front Cell Neurosci.* 2015;9:452.
23. Chalmers RM, Katzer F. Looking for *Cryptosporidium*: the application of advances in detection and diagnosis. *Trends Parasitol.* 2013;29(5):237–51.
24. Laude A, et al. Is real-time PCR-based diagnosis similar in performance to routine parasitological examination for the identification of *Giardia intestinalis, Cryptosporidium parvum/Cryptosporidium hominis* and *Entamoeba histolytica* from stool samples? Evaluation of a new commercial multiplex PCR assay and literature review. *Clin Microbiol Infect.* 2016;22(2):190.e1–8.
25. Checkley W, et al. A review of the global burden, novel diagnostics, therapeutics, and vaccine targets for *Cryptosporidium. Lancet Infect Dis.* 2015;15(1):85–94.
26. Kothavade RJ. Potential molecular tools for assessing the public health risk associated with waterborne *Cryptosporidium* oocysts. *J Med Microbiol.* 2012;61(8):1039–51.
27. Bankier AT, et al. Integrated mapping, chromosomal sequencing and sequence analysis of *Cryptosporidium parvum. Genome Res.* 2003;13:1787–99.
28. Skotarczak B. Progress in the molecular methods for the detection and genetic characterization of *Cryptosporidium* in water samples. *Ann Agric Environ Med.* 2010;17(1):1–8.
29. Robinson G, Chalmers RM. Assessment of polymorphic genetic markers for multi-locus typing of *Cryptosporidium parvum* and *Cryptosporidium hominis. Exp Parasitol.* 2012 132(2):200–15.
30. Lendner M, Daugschies A. *Cryptosporidium* infections: molecular advances. *Parasitology.* 2014;141(11):1511–32.
31. Gargala G. Drug treatment and novel drug target against *Cryptosporidium. Parasite.* 2008;15(3):275–81.
32. Mead JR. Prospects for immunotherapy and vaccines against *Cryptosporidium. Hum Vaccines Immunother.* 2014;10(6):1505–13.
33. Nasser AM. Removal of *Cryptosporidium* by wastewater treatment processes: a review. *J Water Health.* 2016;14(1):1–13.
34. Qin HY, Wu JC, Tong XD, Sung JJ, Xu HX, Bian ZX. Systematic review of animal models of post-infectious/post-inflammatory irritable bowel syndrome. *J Gastroenterol.* 2011;46(2):164–74.
35. Costa LB, et al. Novel in vitro and in vivo models and potential new therapeutics to break the vicious cycle of *Cryptosporidium* infection and malnutrition. *J Infect Dis.* 2012;205(9):1464–71.
36. Gorla SK, et al. Validation of IMP dehydrogenase inhibitors in a mouse model of cryptosporidiosis. *Antimicrob Agents Chemother.* 2014;58(3):1603–14.
37. Benamrouz S, et al. *Cryptosporidium parvum*-induced ileo-caecal adenocarcinoma and Wnt signaling in a mouse model. *Dis Model Mech.* 2014;7(6):693–700.
38. Melicherová J, Ilgová J, Kváč M, Sak B, Koudela B, Valigurová A. Life cycle of *Cryptosporidium muris* in two rodents with different responses to parasitization. *Parasitology.* 2014;141(2):287–303.

39. Borowski H, Thompson RC, Armstrong T, Clode PL. Morphological characterization of *Cryptosporidium parvum* life-cycle stages in an in vitro model system. *Parasitology.* 2010;137(1):13–26.
40. Hijjawi N, Estcourt A, Yang R, Monis P, Ryan U. Complete development and multiplication of *Cryptosporidium hominis* in cell-free culture. *Vet Parasitol.* 2010;169(1–2):29–36.
41. Castellanos-Gonzalez A, Cabada MM, Nichols J, Gomez G, White AC Jr. Human primary intestinal epithelial cells as an improved in vitro model for *Cryptosporidium parvum* infection. *Infect Immun.* 2013;81(6):1996–2001.
42. Varughese EA, Bennett-Stamper CL, Wymer LJ, Yadav JS. A new in vitro model using small intestinal epithelial cells to enhance infection of *Cryptosporidium parvum. J Microbiol Methods.* 2014;106:47–54.
43. Morada M, et al. Continuous culture of *Cryptosporidium parvum* using hollow fiber technology. *Int J Parasitol.* 2016;46(1):21–9.
44. Huang L, et al. An in vitro model of infection of chicken embryos by *Cryptosporidium baileyi. Exp Parasitol.* 2014;147:41–7.
45. Zhang S, et al. Chick embryo tracheal organ: a new and effective in vitro culture model for *Cryptosporidium baileyi. Vet Parasitol.* 2012;188(3–4):376–81.

38

Cystoisospora belli

Chaturong Putaporntip and Somchai Jongwutiwes

CONTENTS

38.1 Introduction

Cystoisospora belli, formerly known as *Isospora belli*, is an important enteric coccidian protozoa that is responsible for diarrheal illness in humans. Belonging to phylum Apicomplexa, coccidians parasitize diverse species of hosts ranging from higher invertebrates to vertebrates. First observed as globular and ovoidal structures in the bile of a cow by Leeuwenhoek in 1674 [1], coccidian protozoa of the apicomplexan parasites are known to cause diseases in humans. These include the genera *Plasmodium* and *Babesia* infecting mainly erythrocytes, *Toxoplasma* and *Sarcocystis* inhabiting various tissues, and enteric coccidia in the genera *Cystoisospora*, *Cyclospora*, and *Cryptosporidium*, mainly involving intestinal epithelial cells. All enteric coccidian protozoa produce gastrointestinal symptoms indistinguishable from one another; therefore, laboratory investigation is crucial to make definite diagnosis that is mandatory for a proper management. Although a number of species in the genus *Cystoisospora* are known to infect mammalian hosts, *C. belli* is the only known pathogenic species to humans, causing cystoisosporiasis. The description of characteristic oocysts of *C. belli* in stools was initially observed among military personnel stationed in Egypt, the Middle East, and eastern Mediterranean countries during 1914 and 1921 [2]. Nowadays, it is well recognized that cystoisosporiasis has a cosmopolitan distribution. Although oocysts of *Isospora natalensis* were reportedly found in human fecal samples, several studies have failed to reaffirm the presence of these oocysts as a causative agent of diarrheal illness in humans since its first report in 1953, raising the possibility that it may not be a valid pathogenic species in humans [3]. Most of the enteric coccidian parasites of humans are considered to be opportunistic pathogens except the genus *Sarcocystis*. Thus, the prevalence of these coccidian infections seems to increase as the number of immunocompromised patients increases, especially after AIDS became pandemic. In general, severity of illness caused by *C. belli* seems to be more aggressive in patients with compromised immunity than those with normal immune status. However, chronic emaciating infections have been reported in immunocompetent patients, suggesting that host immunity *per se* may not determine the severity of cystoisosporiasis [4,5]. Meanwhile, difference in intrinsic virulence of this organism has not yet been explored.

38.1.1 Classification and Phylogeny

Members of the phylum Apicomplexa are intracellular organisms that share common structural features by possessing apical complex at certain stages of their life cycles. Apical complex is subcellular organelles consisting of polar ring, rhoptries, micronemes, conoid, dense granules, and subpellicular microtubules that can be identified under a transmission electron microscope. Of these, micronemes, rhoptries, and dense granules are secretory organelles involved in ligand–receptor recognition and host cell invasion. Rhoptries are club-shaped or teardrop membrane-bound organelles with a connecting duct to the anterior end of the cell. Micronemes are elliptical-shaped vesicles located in close proximity to polar ring. Dense granules are secretory vesicles found clustering around the anterior portion of the organism. Apical organelles exist only at the invasive stages of apicomplexan protozoa and become gradually disintegrated after host cell invasion. Parasitic protozoa in the phylum Apicomplexa inhabit tissues, blood, or lymphoid cells of the host, and most of them, including genus *Cystoisospora*, belong to the class Sporozoea. Previous taxonomic classification based on biology and characters of protozoa placed the formerly known mammalian *Isospora* species, currently classified as genus *Cystoisospora*, in subclass Coccidia, suborder Eimeriina, and family Eimeriidae whose members include genera *Eimeria* and *Cyclospora*. These coccidian parasites require single hosts to complete both sexual and asexual reproductions, termed monoxenous life cycle. Although the mature oocysts of mammalian *Cystoisospora* contain two sporocysts (diplosporocystic) and each with four sporozoites (tetrasporozoic) akin to some members of protozoa in the family Eimeriidae, the absence of a polar Stieda body in each sporocyst and recent phylogenetic inference from the small subunit ribosomal RNA sequences of several members of apicomplexan protozoa have repositioned members of mammal host species of *Cystoisospora* within tissue cyst-forming coccidia that belong to the family Sarcocystidae [6]. The members in this family include genera *Toxoplasma*, *Neospora*, *Hammondia*, and *Sarcocystis* as a monophyletic group. On the other hand, avian host species *Isospora robini* and *Isospora gryphoni* are clustered within another monophyletic group with genera *Caryospora*, *Eimeria*, and *Cyclospora*, all of which are non-tissue-cyst-forming coccidia in the family Eimeriidae (Figure 38.1) [7–9]. The presence of tissue cyst in the genus *Cystoisospora* favors phylogenetic rather than phonetic classification [10–12]. The genus *Cystoisospora* was created by Frenkel in 1977 based on the characteristics of oocysts and the presence of tissue cysts in paratenic hosts that are distinct from those of the genus *Isospora* [13]. Meanwhile, *Isospora hominis* described over seven decades ago was misclassified, which *de facto* belongs to the genus *Sarcocystis* based on the life cycle and structures of the developmental stages. Other species of *Cystoisospora* are parasitic in wild and domestic animals, for example, *C. arctopitheci* in nonhuman primates, *C. canis* and *C. ohioensis* in dogs, *C. rivolta* and *C. felis* in cats, and *C. suis* in pigs [14]. More than 300 species of *Cystoisospora* have been proposed and found to infect a wide range of animals that includes amphibians, reptiles, birds, and mammals. However, some of these species may require further validation.

38.1.2 Biology, Morphology, and Life Cycle

Developmental stages of *Cystoisospora* comprise asexual multiplication, sexual reproduction, and sporogony. The life cycle of *C. belli* completes within the human host without known animal reservoirs. Humans get infected by ingestion of food or water contaminated with oocysts (Figure 38.2). The infective

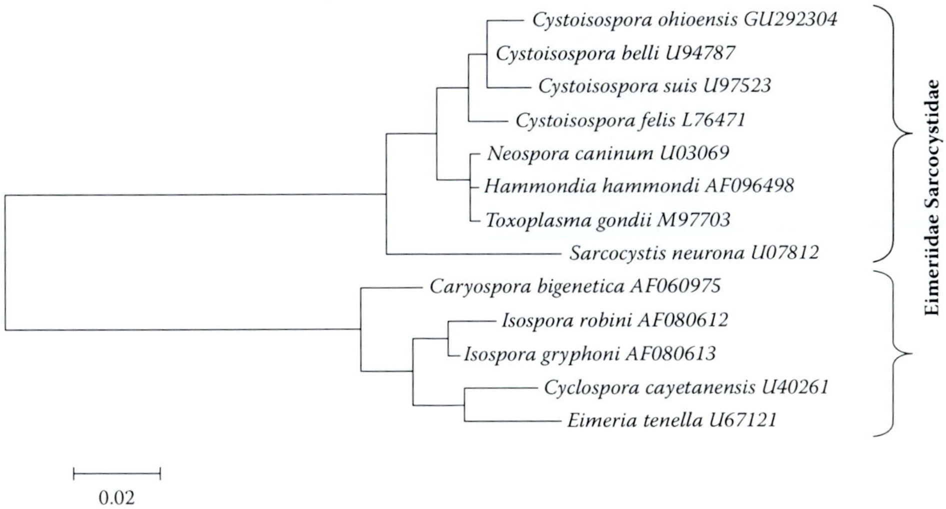

FIGURE 38.1 Phylogeny of *Cystoisospora* and other apicomplexan protozoa inferred by using the Maximum Likelihood method based on the small subunit ribosomal RNA sequences. Tree was constructed by using the MEGA6 program [15]. Scale indicates the number of base substitutions per site. GenBank accession numbers are listed after species.

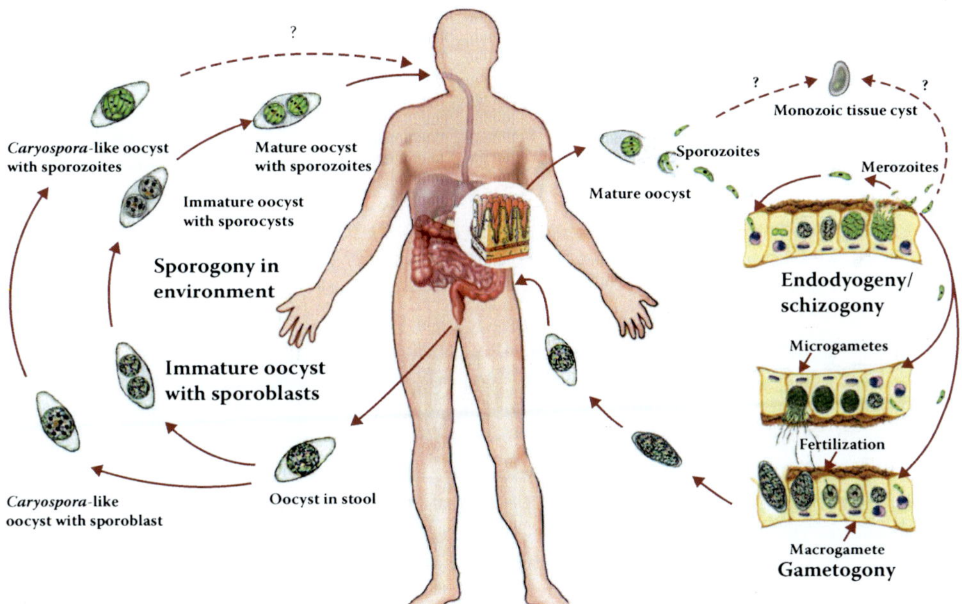

FIGURE 38.2 Life cycle of *Cystoisospora belli* depicting asexual multiplication, sexual development, and sporogony. (Illustration by Urassaya Pattanawong.)

stage of *C. belli* is characterized by an oval or rugby-shaped oocyst with a wide range of size variation in terms of the maximum length and width, both within and between isolates. The average length measured from more than 700 oocysts is 28.3 μm (range = 17–37). A remarkable narrowing with a neck-like appearance may be observed at one end of some oocysts, whereas most oocysts are nearly symmetric at both ends. The maximum width of oocysts averages 13.5 μm (range = 8–21). The mean shape index or the ratio of length to width of oocysts is 2.1 (range = 1.3–3.3) [9]. The oocyst wall is bilayer, thin, smooth, and transparent for all stages of development. The inner wall is membranous, and the outer wall is rigid and relatively impermeable to external fluids [16]. When freshly passed in feces, the oocyst remains immature, containing a spherical or slightly elongated, oval-shaped sporoblast that develops within another rigid cyst wall called sporocyst. Occasionally, some freshly passed oocysts in stool may contain two sporocysts, each of which possesses one sporoblast. The sporocyst of *C. belli* lacks a proteinaceous plug, known as a Stieda body, a structure existing in the genus *Isospora*. The dimension of sporocyst is 12–14 μm × 7–9 μm.

Sporogony, the process of producing infective sporozoites of *Cystoisospora*, usually occurs outside the host (Figure 38.2). The maturation period of oocysts usually completes within 24 h to 10 days [9]. The mature oocyst possesses two sporocysts, each with four crescentic or banana-shaped sporozoites clustering together with a clump of granular mass or residual bodies surrounded by sporocyst wall (Figure 38.3). Detailed investigation of *ex vivo* development has shown that the percentage of oocyst maturation varies between isolates. *In vitro* studies reveal that only 27%–30% of oocysts excreted from stools of patients become fully sporulated at ambient temperature. However, sporulation process requires optimum conditions that could be variable among studies. Importantly, despite the frequently identified characteristic mature oocysts containing two sporocysts, about 5% of the oocysts that undergo the maturation process possess eight sporozoites in a single sporocyst akin to oocysts of *Caryospora*, a member of family Eimeriidae [9,17]. The formation of *Caryospora*-like oocysts requires 5–14 days after excretion in feces [9]. Both types of mature oocysts contain a number of granular residual bodies clumping together. Environmental factors such as moisture, temperature, and oxygen composition in the atmosphere have been suggested to influence the maturation of *Cystoisospora* oocysts. It is noteworthy that incubation of *C. rivolta* oocysts at 50°C for 5 min enhances development of *Caryospora*-like oocysts, and they are infective to mice and cats [18]. It is therefore likely that *Caryospora*-like oocysts of *C. belli* could also be infective to humans.

Sporozoites are released from the oocysts and become free in the lumen of small intestine, especially around distal duodenum and jejunum. Thereafter, sporozoites invade small intestinal epithelial cells where asexual multiplication ensues. All endogenous stages of *Cystoisospora* reside within parasitophorous vacuoles in the cytoplasm of host cells. Sporozoites of mammalian *Cystoisospora* species, for example, *C. suis*, *C. rivolta*, and *C. felis*, initially undergo asexual multiplication by endodyogeny, a cellular division process in which two daughter cells are generated from a mother cell so that these daughter cells are surrounded by the mother cell membrane [19–21]. Repeated endodyogeny probably occurs and generates at least three structurally distinct types of merozoites in *C. suis*, whereas schizogonic development with repeated nuclear division within a single cytoplasmic mass possibly occurs in epithelium of upper small intestine [22,23]. However, the number of generations of merozoites remains unknown. Eventually, some merozoites develop into sexual stages consisting of female and male gametocytes. Although detailed knowledge on asexual development of *C. belli* remains to be elucidated, it could resemble these porcine and feline species of *Cystoisospora*. *In vitro* studies have shown that endodyogenic development of *C. belli* occurs after inoculation of sporozoites into cell cultures resulting in the formation of merozoites, while histopathological studies demonstrated multinucleate schizonts in infected enterocytes [11–13,24–27]. Fully developed merozoites disintegrate host cells, invade other epithelial cells, and undergo further asexual multiplication cycles (Figure 38.4).

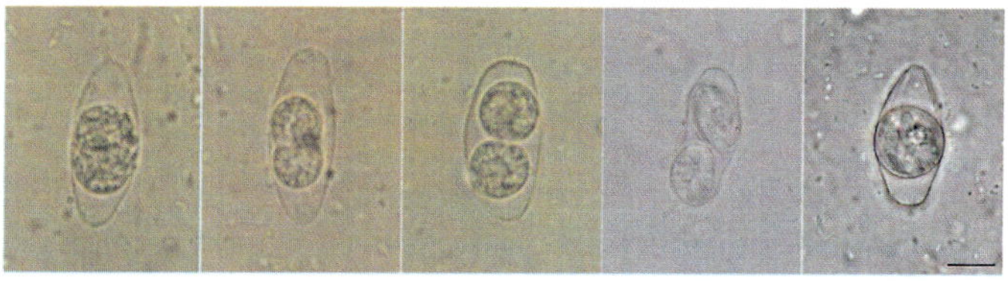

(A) (B) (C) (D) (E)

FIGURE 38.3 Oocysts of *Cystoisospora belli*. (A) Oocyst from freshly passed stool containing one sporocyst. (B) and (C) Developing oocysts with two sporoblasts. (D) Mature oocyst with two sporocysts, each with four sporozoites. (E) *Caryospora*-like oocyst with eight sporozoites in one sporocyst. Scale = 10 μm.

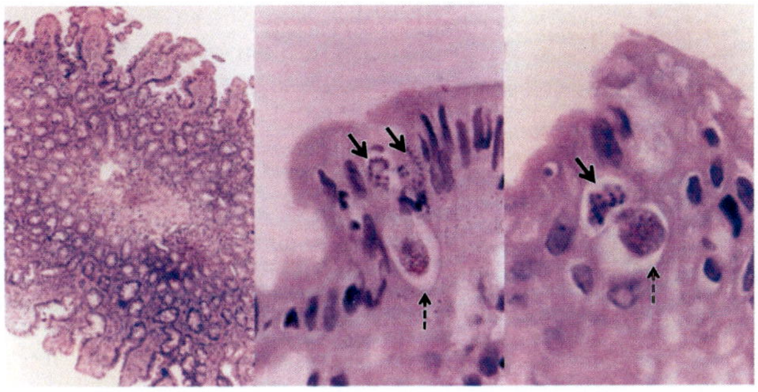

FIGURE 38.4 Upper intestinal biopsy showing (A) shortening of villi, (B) and (C) endogenous stages of *Isospora belli* in epithelium cells. Single and broken arrows indicate merozoites/schizonts and oocysts, respectively.

Extraintestinal monozoic tissue cysts of *C. belli* have been found in sections of lymph nodes, spleen, and liver at autopsy of AIDS patients, suggesting that these stages may represent hypnozoites or dormant forms that could be responsible for relapse of symptoms [11,12,28–30]. Ultrastructural studies reveal that the extraintestinal monozoic tissue cyst is surrounded by a parasitophorous vacuole membrane and characterized by oval or ellipsoidal body measuring about 11 μm × 8 μm. A conspicuous cyst wall is observed with a thickness varying from 0.7 to 4 μm, enclosing a crescent-shaped zoite. Occasionally, the surface of monozoic tissue cyst possesses grooves or small projections. The zoite inside monozoic tissue cyst resembles the sporozoite stage and contains a single rather than multiple nuclei. The anterior part of these zoites possesses polysaccharide-like and lipid-like granules, and apical organelles consisting of rhoptries, micronemes, dense bodies, and a conoid. Each zoite is surrounded by a parasitophorous vacuole and granular materials forming the cyst wall. Monozoic tissue cysts enlarge over time without cellular division. The monozoic tissue cysts of *C. canis* in mouse, a paratenic host, are infective to dogs similar to ingestion of oocysts [31]. Although the origin of this developmental process remains elusive, abundant numbers of tissue cysts found at autopsy favor that they probably originated from merozoites rather than sporozoites from the oocyst stage [28–30]. It is noteworthy that extraintestinal stages and monozoic tissue cysts have also been found in *C. felis* after oral inoculation of the oocysts into kittens [10,32,33].

Sexual reproduction or gametogony usually occurs after initiation of the asexual cycle for about 1 week. Female gamont develops intracellularly without nuclear division and becomes macrogamete, whereas male gamont undergoes multiple nuclear and cytoplasmic divisions, resulting in the production of a number of microgametes. Microgametes disrupt host cells and enter into other epithelial cells where macrogametes are located. After fertilization, zygote is formed within the epithelial cell and sequentially develops into an oocyst. Intraepithelial oocysts containing granular bodies are excreted in feces upon rupture of host cell. The oocysts are resistant to the hostile environment, including common disinfectants, and remain viable in cool and moist condition for months [14].

38.1.3 Epidemiology

Transmission of *C. belli* is solely anthroponotic because humans are the only known natural hosts. Before 1935, the prevalence of human cystoisosporiasis seemed to be have been low as only 200 cases were reported [34]. However, in certain circumstances when living condition is overcrowded or sanitation is below the standard level, cases of *C. belli* infections could be encountered. The majority of infected cases were identified mainly among military or related personnel during 1914–1921 [2]. Until 1961, after awareness of the infection along with the competency of microscopic detection, more than 800 cases were detected in various localities in South America [34]. Before the pandemic of HIV infections, sporadic or endemic cystoisosporiasis has been reported from several countries both in tropical and temperate zones. The infection rates of *C. belli* among immunocompetent individuals are usually

<2%, while a remarkable higher rate was observed among those with underlying immunodeficiency. The prevalence of cystoisosporiasis in AIDS patients in Zaire, Haiti, Venezuela, Brazil, and Thailand is 19%, 15%, 14%, 9.9%, and 7.6%, respectively [9,35–38]. Nevertheless, endemicity, public sanitation, preexisting anticoccidial drug treatment, and other factors may also affect such prevalence. An 8-year surveillance of intestinal infections among AIDS patients in Los Angeles County revealed the presence of *C. belli* in 1% of pathogen-positive stool samples [39]. Interestingly, most of the *C. belli*-positive samples were from foreign-born patients, especially those from Central America. Travel-related infections may be a contributing factor.

38.1.4 Pathology and Clinical Aspects

Infection with *C. belli* is usually confined to mucosal epithelial cells of upper small intestine. Endogenous stages vary in size and structure depending on their growth and development. Both asexual multiplication and sporogonic developmental cycles may persist if untreated or probably due to the failure of gut immune response to control the progression of infection. No obvious pathological change is observed in the early phase of infection except for some leucocyte infiltration in mucosa and lamina propria, most of which are plasma cells, lymphocytes, and eosinophils [11,12]. In chronic cystoisosporiasis, pathological features include shortening or blunting of villi or even atrophy, hypertrophied crypts, and lamina propria infiltrated with eosinophils, lymphocytes, and neutrophils [4,40,41]. In some cases, histopathology of cystoisosporiasis may masquerade eosinophilic enteritis [5]. However, administration of corticosteroid cannot alleviate the symptoms or even worsen its clinical course. Postmortem examinations reveal that some cystoisosporiasis cases who had underlying AIDS exhibit extraintestinal involvement of lymphoid tissues, causing enlargement of mesenteric lymph nodes, liver, and spleen [11,12,28,29].

A spectrum of clinical manifestations of cystoisosporiasis has been noted. Although both immunocompetent and immunocompromised patients are susceptible to *C. belli* infections, higher prevalence of cystoisosporiasis afflicts the latter, especially in AIDS patients. Incubation period takes about 1 week after ingestion of infective oocysts [42,43]. Asymptomatic infection with spontaneous clearance of oocysts is rarely observed. Symptomatic infections with *C. belli* in immunocompetent hosts present with watery diarrhea, abdominal discomfort, colicky abdominal pain, low-grade fever, malaise, anorexia, nausea, vomiting, headache, and dehydration. Experimental infections in human volunteers show that symptoms precede patency of oocysts in stool for a few days. The duration of symptoms continues for about 1 day, whereas oocysts are discharged in stool for more than 1 month [42]. Diarrheal episode in some patients can be severe with up to 20 stools per day [44]. Diarrheal symptoms usually become chronic if left undiagnosed. Chronic infection may turn to malnutrition, steatorrhea, and cachexia. The excretion of oocysts may last for months or years in chronic cystoisosporiasis. A few reports have shown chronic diarrheal symptoms lasting for over a decade in immunocompetent patients [4,5,41].

Cystoisosporiasis in immunosuppressed patients, such as those who receive immunosuppressive drugs, prolonged corticosteroid treatment, and more commonly in AIDS patients, is generally more severe than those with normal immune status in terms of duration of symptoms and volume of stools. The diarrheal symptoms caused by *C. belli* seem to be a secretory process that may lead to massive fluid loss, resulting in dehydration and electrolyte imbalance. Fever is frequently observed in dehydrated cases. Coinfections with other enteric pathogens can be detected, especially in AIDS patients [9,36,37,45].

Recurrent symptomatic cystoisosporiasis occurs in both immunocompetent and immunodeficient individuals. In studies of cystoisosporiasis in Haitian AIDS patients, relapse occurs in almost half of the cases [36,45]. Extraintestinal cystoisosporiasis identified in lymphoid tissues at postmortem in AIDS patients has suggested that these stages could reinvade mucosal epithelial cells and are responsible for clinical relapse [11,28,29]. Although relapse is believed to be caused by reactivation of these dormant forms or monozoic tissue cysts, its pathogenesis remains to be elucidated. An unusual case of endometrial cystoisosporiasis has been reported in an otherwise immunologically uncompromised host presenting with granulomatous endometritis. Numerous intracytoplasmic cysts resembling *C. belli* oocysts, but not unicellular zoite, were found in the endometrial layer of the uterus [46].

TABLE 38.1

Comparative Characteristics of *Cystoisospora belli*, *Cyclospora cayetanensis*, *Cryptosporidium* spp., and *Sarcocystis hominis*

Feature	*C. belli*	*C. cayetanensis*	*Cryptosporidium* spp.	*S. hominis*
First report in humans	1914	1977–1978	1974	1925
Location in enterocytes of small intestine	Intracytoplasmic in apical supranuclear region	Intracytoplasmic in apical supranuclear region	Intracellular, extracytoplasmic at luminal surface	Lamina propria
Zoonotic potential	No	Unknown	Yes	Yes
Complete life cycle within humans	Yes	Yes	Yes	No
Average size of oocyst	13.5 × 28.3 μm	8–10 μm	4.5 to >6 μm	13.5 × 19.5 μm (sporocyst: 9.3 × 14.7 μm)
Acid-fast staining of oocysts	Acid-fast	Variably acid-fact	Acid-fast	Not acid-fast (due to thick sporocyst wall)
Number of sporocysts in mature oocyst	2 (1 for *Caryospora*-like oocyst)	2	No sporocyst	2
Number of sporozoites per sporulated oocyst	8	4	4	8
Sporulation outside host	Yes	Yes	No	No
Autofluorescence of oocyst wall	Yes	Yes	No	Unknown. Oocyst usually disrupted when passed in stool
Number of oocysts in stools	Low	Low to moderate	Low to very high	Low
Infectivity of oocysts in freshly passed stool	No	No	Yes	Yes
Resistance of oocysts to chlorine	Yes	Yes	Yes	Unknown

Source: Herwaldt, B.L., *Clin. Infect. Dis.*, 31, 1040, 2000.

Laboratory test frequently shows eosinophilia in the majority of infected cases, but a remarkable alteration of other hematological profiles is not observed. Patients with normal immune status tend to have a higher level of eosinophils than those with suppressed immune system [9]. Charcot–Leyden crystals are frequently observed in stools of cystoisosporiasis patients, indicating active proliferation and degradation of eosinophils in gut tissues. Because endogenous stages of *C. belli* neither cause apparent ulceration of intestinal epithelial cells nor induce intense inflammatory response, a remarkable number of erythrocyte or leucocyte is not found in stool. Identification of characteristic oocysts in stool allows definite diagnosis in routine laboratory practice. The efficiency of detection depends on various factors such as the microscopist's experience, shedding pattern of oocysts that may be fluctuating during the course of infection, as well as the choice of diagnostic techniques. Some differential features of coccidian protozoa that shed their oocysts in host stool samples are listed in Table 38.1 [47].

38.1.5 Laboratory Diagnosis

38.1.5.1 Stool Examination

Definite diagnosis of cystoisosporiasis relies on identification of characteristic oocysts in stool or intestinal content. Collection of stool samples containing or suspicious of having oocysts of *C. belli* essentially follows standard guidelines for parasitological diagnosis. Although immediate examination of stool specimen is not a crucial issue as those for amoebic trophozoites, it is recommended to examine the stool samples within 24 h after passage. If examination of stool sample is not possible within 1 day, stool should be kept at less than 10°C, but not frozen, in order to minimize bacterial overgrowth and decay of fecal materials [48]. Although coccidian oocysts can be preserved in 5% or 10% formalin solution for microscopy-based diagnostic purpose, preservation of coccidian oocysts in ethanol also gives good results for subsequent acid-fast stain. Structures of *C. belli* and *Cryptosporidium* oocysts kept in 80% ethanol at ambient temperature for more than 2 years are also well preserved for acid-fast stains [9,49].

Microscopic examination of stool samples should be done by both direct smear and concentration methods. For fresh stool samples containing mucus, direct smear sampling from the mucus portion may yield better discovery than the watery part of stool as coccidian oocysts tend to stick to mucoid substance. Meanwhile, *C. belli* oocysts usually do not excrete in large amount; therefore, they are often diluted in voluminous diarrhea precluding efficient diagnosis by direct smear method. Oocysts of *C. belli* can be concentrated without morphological alteration by formalin-ethylacetate (or ether) sedimentation procedure to increase diagnostic yield [50]. Despite superiority of formalin-ethylacetate sedimentation to simple smear method, simple sedimentation by centrifugation of previously sieved stool samples may yield higher diagnostic probability than conventional formalin-ethylacetate sedimentation, because some oocysts are lost to the fatty layer in the latter method [51]. It is noteworthy that oocysts of *C. belli* can be overlooked in unstained stool samples because of their transparency when adjustment of microscope does not make good contrast. Alternatively, various acid-fast staining methods such as modified Kinyoun acid-fast and modified Ziehl–Neelsen stains that intentionally deployed for detection of *Cryptosporidium* spp. and *Cyclospora* oocysts are also useful for diagnosing *C. belli* oocysts. Specimens can be versatile ranging from fresh stool, sediment from formalin-ethylacetate sedimentation, ethanol-preserved stool sample, duodenal fluid, and other body fluids [52].

Oocysts of *C. belli*, *Cyclospora cayetanensis*, and *Cryptosporidium* spp. become autofluorescent bright green under violet excitation (405 nm), bluish violet autofluorescent under ultraviolet excitation (365 nm), and green autofluorescent under blue-violet light excitation (436 nm) [53]. Faint reddish fluorescence is also observed under green light (546 nm). Characteristic size, shape, and structure of *C. belli* oocysts under autofluorescence microscopy differ from those of *C. cayetanensis* and *Cryptosporidium* spp., and thus, the method is useful for diagnosis. Autofluorescence is proven to be more sensitive than bright-field microscopy in detecting *Cystoisospora* oocysts [54,55].

38.1.5.2 Examination of Duodenal Content

Duodenal content can be obtained from various procedures such as duodenal capsule technique or duodenal aspirate during endoscopic examination of upper gastrointestinal mucosa [56]. Like *Giardia intestinalis* and *Strongyloides stercoralis*, aspiration of duodenal content for direct examination of *C. belli* can be an adjunctive source other than stool samples that should be submitted to parasitology laboratory without preservative. Examination using wet smear method should be done soon to visualize motility of parasites such as falling leaf pattern of *G. intestinalis* trophozoites and rapid sine-curve movement of *S. stercoralis* larvae. If the sample cannot be examined within a few hours after collection, keeping at <10°C is recommended. Preservation of the content with 5%–10% formalin solution is adequate for parasitological detection but not for molecular analysis. Oocysts of *C. belli* can be seen by direct smear of duodenal content or the acid-fast staining procedure. Mucus in the intestinal tract attracts a number of oocysts so that the sampling of content containing mucus material may help in increasing the recovery rate. In case of voluminous content, examination of sediment after centrifugation at $500 \times g$ for 10 min should be done. If the volume of the content is small and precludes several preparative diagnostic methods, some authors suggest processing the sample for staining procedure or molecular diagnosis, instead of preparing for direct wet smear examination [56].

38.1.5.3 Histopathological Diagnosis

Biopsy material from upper small intestine during endoscopic examination provides a useful diagnostic sample for *C. belli* as well as other pathogens that parasitize these regions of the gastrointestinal tract such as *G. intestinalis*, *Cryptosporidium* spp., and *C. cayetanensis*. Biopsy material should never be allowed to dry but should be fixed with 10% formalin solution, embedded in paraffin or equivalent materials, and stained with hematoxylin-eosin. Biopsy materials are also good sample sources for molecular detection provided that they should be kept frozen at less than –20°C without preservatives. The endogenous stages of *C. belli* consisting of asexual and sexual development can be seen in the cytoplasmic mass of enterocytes or at the lamina propria of upper small intestine, although involvement of biliary ducts and large intestine has been reported [11,12,28]. Duodenal mucosal architecture of patients with chronic cystoisosporiasis usually presents with blunting of villi and hypertrophied crypts. Inflammatory responses in upper intestinal tissues of both cystoisosporiasis and cyclosporiasis are similar, showing mixed infiltration of polymorphonuclear cells, plasma cells, lymphocytes, and eosinophils, although the eosinophilic infiltration seems to be more prominent in cystoisosporiasis [40,57]. The presence of characteristic immature oocysts of *C. belli* in the intestinal epithelium cells assists in definite diagnosis. Meanwhile, all endogenous stages of *Cryptosporidium* spp. are located at the intracellular rim at the luminal surface of mucosal epithelium, and the diameters of all stages do not exceed 6 μm, making them easily differentiable from *C. belli* [58,59]. Extraintestinal stages of *C. belli*, existing predominantly in mesenteric lymph nodes, liver, and spleen of some AIDS patients, are mostly crescent-shaped, centrally located, and contain only one zoite in the tissue cyst. The presence of longitudinal grooves or projections of the zoite surface may represent simultaneous invasion of two zoites in the same host cell [60]. Neither multinucleate nor sexual stages exist in the tissue cyst of cystoisosporiasis [30]. Although stool examination often gives more sensitive results than histopathology, the sampling areas of biopsy are limited and may not include the affected regions. However, a large study involving 118 adult AIDS patients who suffered from chronic diarrhea identified *C. belli* from duodenal biopsy samples in two patients who had negative results in stool examinations [28].

38.1.5.4 Molecular Diagnosis

Amplification of a specific region in the small subunit ribosomal RNA gene (18S rDNA) of *C. belli* by nested polymerase chain reaction (PCR) has been successfully applied to duodenal biopsy and bile and stool samples without cross-reactivity to other enteric protozoa [8,61]. PCR targeting the 5.8S ribosomal RNA gene and the internal transcribed spacer II (ITS2) region is proven to be more sensitive than microscopy in diagnosing *C. belli* [62]. Likewise, quantitative PCR assay with melting curve analysis using primers derived from the 18S rDNA can differentiate human enteric coccidian protozoa [63]. Diagnosis of *C. belli* infection by PCR involves the isolation of *C. belli* DNA from clinical specimens that can be stool, duodenal fluid, tissue sample, or other potential sources of infections. The source of *C. belli* DNA is usually from oocyst stage in stool, while tissue sample contains any stages of endogenous development. Freshly collected stool specimen, frozen stool sample, or stool preserved in 80% ethanol can be used for extraction of *C. belli* DNA. Ethanol in preserved stool samples should be removed before disruption of oocysts. Duodenal fluid or tissue samples can be kept at –20°C, and ethanol should not be used for preservation of these samples. Because the oocyst wall of coccidian protozoa is solid and resistant to diverse environmental conditions, it is recommended that the oocyst wall should be ruptured so that internal contents, either sporoblasts or sporozoites, are exposed to DNA extraction solution. Examples of simple procedures to disrupt oocyst wall of *C. belli* are mechanical disruption and freeze-thaw method.

38.1.6 Treatment and Prevention

The standard therapy of cystoisosporiasis is a combination of trimethoprim (160 mg) and sulfamethoxazole (800 mg) given orally 4 times a day for 10 days [36,45]. Usually, the patients respond well to treatment with symptomatic and parasitological clearance within a few days after initiation of therapy. Pyrimethamine (50–75 mg/day) is an alternative drug for those who are intolerant or allergic to anti-sulfa agents. The pharmacological effects of these compounds involve inhibition of parasite folate biosynthesis pathway. Sulfamethoxazole is absorbed well after oral administration and excreted relatively slowly with a serum half-life of 11 h [64]. Like other sulfonamides, the drug acts by competitive inhibition of dihydropteroate synthase, resulting in interruption of 7,8 hydropteroate synthesis from para-amino benzoic acid and pteridine in the folate biosynthesis pathway [6,64–66]. The drug is usually marketed in fixed-dose combinations with trimethoprim, which is a potent competitive inhibitor of dihydrofolate reductase that converts dihydrofolate to tetrahydrofolate in the folate pathway; therefore, these combinations exert synergistic antimicrobial effect [67]. Although resistance to these combinations has been observed in several pathogenic bacteria, no clear evidence is reported for treatment of cystoisosporiasis. However, a high rate of relapse has been observed in cystoisosporiasis, which requires maintenance therapy with trimethoprim–sulfamethoxazole given twice daily for 3 weeks [36,45]. Untoward effects of these compounds include gastrointestinal disturbance, folate deficiency in those who have low folate reserve in tissues, and various forms of dermatological adverse drug reactions. In some obstinate cases, long-term administration with alternative drugs such as pyrimethamine, a dihydrofolate reductase inhibitor, can alleviate or confer long-term remission of symptoms [5].

There is no known animal reservoir of *C. belli*, and transmission is likely to be foodborne or waterborne; therefore, taking food safety precautions and drinking properly treated water should prevent acquiring *C. belli* infection as well as other enteric pathogens. Good personal hygiene and appropriate sanitary disposal of infected stools can interrupt transmission of this enteric coccidian.

38.2 Laboratory Models

38.2.1 Animal Models

A number of animals have been tested for the susceptibility of infection with orally administered oocysts of *C. belli*. These include kittens, puppies, guinea pigs, mice, rats, rabbits, and rhesus macaques [14,68,69]. However, no consistent evidence of infection has been demonstrated in these studies.

38.2.2 *In Vitro* Growth in Mammalian Cells

38.2.2.1 Preparation of Infective Oocysts

The fundamental step for *in vitro* growth study of *C. belli* in mammalian cells is to obtain viable sporozoites that are suitable for inoculation into cell culture system. The lack of known reservoir hosts of *C. belli* and no available standard frozen strain of this protozoa make it necessary to isolate viable oocysts from infected individuals. After obtaining fresh stool samples from infected human subjects prior to treatment, add equal or more volume of 2.5% potassium dichromate solution to the samples, and stir until complete mixing occurs. Stools containing a lot of fecal debris or substances should be filtered through cheesecloth before adding dichromate solution to stool filtrate. Potassium dichromate confers antimicrobial activity but does not affect viability of coccidian oocysts while it also provides additional oxygen source during the process of oocyst sporulation. During storage of stool samples, it is recommended not to fill the tube or collection container all the way to the top, but leave a layer of air between the top of the feces-dichromate mixture and the cap of collection tube or container to allow the oocysts expose to some atmospheric oxygen. Sporulation is better achieved by spreading fecal-dichromate mixture into a Petri dish. Complete sporulation of *C. belli* oocysts usually requires 2 days to 1 week by keeping samples at room temperature (20°C–25°C) [70]. After complete sporulation, potassium dichromate is removed, and sporulated oocysts are purified. The following method for purification of viable *C. belli* oocysts has been described by Oliveira-Silva et al. [24].

1. Removal of 2.5% potassium dichromate from sporulated oocyst suspension is done by adding more than one volume of phosphate buffer saline solution (PBS, pH 7.2) containing 2% Tween 20 and is centrifuged at $1500 \times g$ for 10 min at 4°C.
2. After discarding supernatant, PBS-Tween 20 solution and diethylether (2:1 v/v) are added to the pellet for removal of lipid portion of the sample, and centrifugation is repeated under the same condition. The supernatant is discarded.

3. Further purification of oocysts is done by discontinuous gradient flotation. To obtain better purification, it requires prior knowledge of the approximate specific gravity of the oocysts. However, the sucrose gradient of 1.05 and 1.15 g/mL reportedly offers acceptable results. For gradient preparation, sucrose solutions are allowed to run down steadily inside the 15 mL tube. More dense sucrose solution must be added to the tube before layering less dense sucrose solution. The aqueous suspension containing oocysts is then layered on top of the gradient solution and centrifuged at 1500 × *g* for 20 min. The oocysts will form a thin layer at the interface between the two sucrose gradient layers that also separates the oocysts from both heavier and lighter contaminants.
4. The layer containing oocysts is carefully pipetted to a new tube, and PBS is added for washing by centrifugation at 1500 × *g* for 10 min. The washing step is repeated twice.
5. After the purification step, the oocysts are subject to further cleaning by adding 1% sodium hypochlorite solution and allowed to stand for 10 min at 4°C.
6. Subsequent washing of the pellets is done thrice by adding normal saline solution and centrifuged at 1500 × *g* for 10 min. Optionally, a small aliquot of the pellet containing purified oocysts can be sampled for enumeration of oocysts using Neubauer chamber.
7. The purified oocysts are diluted with PBS (pH 7.2) containing 1.5% sodium taurocholate and 0.5% trypsin. Excystation is done by incubation at 37°C for 30 min.
8. To neutralize trypsin, Eagle's minimum essential medium (MEM) supplemented with 10% fetal bovine serum is added to the excysted sporozoites and is centrifuged at 1500 × *g* for 10 min.
9. The number of sporozoites is estimated using the Neubauer chamber.
10. The viability of sporozoites can be assessed by observation of their movement. Alternatively, trypan blue exclusion test is more feasible to determine the viability of sporozoites by adding 1 volume of 0.4% solution of trypan blue in PBS (pH 7.2) to 10 volumes of cell suspension. After loading the sample to Neubauer chamber, it is immediately examined under light microscope. The number of blue staining cells should be counted, and total number of cells is examined to estimate percentage of viable sporozoites. For each inoculation, the number of sporozoites should be about 10^4 sporozoites per well or 10^6 sporozoites per 12.5 cm^2 culture flask.

Alternatively, the following similar protocol has been described by Siripanth et al. for purification of infective oocysts of *C. belli* [25].

1. Fecal samples containing oocysts of *C. belli* that have been sporulated in 2.5% potassium dichromate solution are filtered through a few layers of gauze or cheesecloth. Subsequently, oocysts are purified from the remaining fecal materials by sugar flotation. Preparation of sugar solution is done by dissolving 454 g of table sugar in 355 mL water using gentle or indirect heat. To keep the viability of oocysts, the originally recommended preservatives such as formaldehyde and phenol are not added in the sugar solution. The specific gravity of sugar solution should be about 1.27. After adding filtered stool sample to a 15 mL centrifuge tube, the tube is filled with sugar solution until the level reaches below the top of the tube. A lid or cap is attached if the tube has screw cap, and it is then centrifuged at 280 × *g* for 5 min. A centrifuge that has swinging cup should be used. Slowly sugar solution is added by carefully inserting tip of pipette below the surface to create a slightly inverted meniscus. Oocysts at the surface of the solution can be subsequently harvested [71].
2. After obtaining the oocysts from the flotation procedure, the sample is washed thrice by adding distilled water and is centrifuged at 1000 × *g* for 10 min.
3. 2% sodium hypochlorite solution is added to the sediment containing oocysts and is left on ice for 10 min.
4. The sample with PBS is washed by centrifugation at 1000 × *g* for 10 min. This step is repeated twice.

5. The sediment is resuspended in Eagle's MEM medium containing 80 µg/mL gentamicin and kept at 4°C until use.
6. A Teflon-coated tissue grinder is used to break oocyst wall for releasing sporocysts.
7. The sporocysts are resuspended in sterile 3% sodium taurocholate solution containing 1% trypsin in MEM medium. It is then incubated at 37°C in a CO_2 incubator for 2 h.
8. Sporozoites are harvested by centrifugation at 1000 × *g* for 10 min and washed twice with PBS.
9. The number of oocysts is estimated using Neubauer chamber.
10. Sporozoite viability is checked by using trypan blue exclusion test.

38.2.2.2 Inoculation of Sporozoites into Cell Lines

Mammalian cells reportedly supported *in vitro* multiplication of *C. belli* sporozoites are human ileocecal adenocarcinoma (Hct-8), human larynx carcinoma (Hep-2), human lung carcinoma (A-549), African monkey kidney (VERO cells), bovine epithelial kidney (BEK), and rhesus monkey kidney cells (MK2). These cell lines are maintained per standard cell culture protocol using appropriate culture media. Oliveira-Silva, Resende, and colleagues report that VERO cells can support the growth of *C. belli* better than Hct-8 and A-549 cells in which the parasites invade these cell lines between 4 and 12 h post inoculation and undergo asexual multiplication after 24 h as demonstrated by the presence of paired merozoites inside the cytoplasm of these cells [24,27]. Meanwhile, Siripanth et al. observed complete schizogonic development in Hct-8 cells after 5 days of sporozoite inoculation into the cell culture. After the second passage of the culture, sporogonic development and formation of microgametocytes reportedly occurred within 5 days [25]. Furthermore, it is also noted that Hep-2 cells could support the growth of *C. belli* as the parasites developed to early schizogony, whereas monozoic cysts could be observed in both BEK and VERO cells. However, complete developmental cycle with the formation of oocysts could not be achieved [25].

38.2.2.2.1 Preparation of Cell Culture

Usually, standard cell lines are available from cell banks as frozen stabilates kept in liquid nitrogen. Initiation of cell culture essentially follows the standard procedure as follows:

1. The frozen cells are thawed by placing the vial immediately in a water bath preset at 37°C.
2. Before opening, the vial is cleaned thoroughly with 70% ethanol.
3. The content is transferred from the vial to a 15 mL tube containing appropriate culture medium.
4. It is centrifuged at 200 × *g* for 5 min and supernatant is discarded.
5. The remaining pellets containing cells are transferred to a new culture flask preadded with appropriate cell culture medium. It is mixed by gently pipetting.
6. It is incubated at 37°C. The culture flask is examined using inverted microscope.

38.2.2.3 Inoculation of Sporozoites into Human and Murine Macrophages

Resende et al. succeeded in growing *C. belli* sporozoites using human and murine macrophages as host cells [26]. Sporozoites reportedly invaded host cells after 15–17 h post inoculation and multiplied by endodyogeny within 24 h. Procedures for preparation of human and murine macrophages have been described by Resende et al. as follows [26].

38.2.2.3.1 Preparation of Human Macrophages

1. Fresh venous blood sample treated with an anticoagulant (e.g., heparin, citrate) is prepared.
2. One volume of Ficoll-Hypaque gradient (v/v) solution is added to a sterile tube.
3. One volume of the blood sample is carefully and slowly layered over the density gradient medium and centrifuged at 400 × *g* for 20 min at room temperature in a swinging-bucket rotor without brake.

4. After centrifugation, four layers appear consisting of plasma at the top, followed by a thin layer of peripheral blood mononuclear cells (PBMCs), Ficoll-Hypaque layer, and erythrocytes mixed with granulocytes at the bottom. PBMCs are collected either by gently removing the upper plasma layer until the PBMCs can be harvested or by inserting the pipette tip directly through the upper plasma layer to collect cells at the interface. The PBMCs are kept in a new tube.
5. The harvested cells are washed with RPMI-1640 supplemented with 20% fetal calf serum and centrifuged at 200 × *g* for 20 min.
6. After discarding the supernatant, the cell pellet is resuspended with RPMI-1640 supplemented with 20% fetal calf serum and kept in culture flask for 20 days or until complete differentiation to macrophages with a density of 3×10^6 cells per 12.5 cm^2 culture flask.

38.2.2.3.2 Preparation of Murine Macrophages

1. 10 mL of 0.9% NaCl solution containing heparin sodium 10 units/mL is aseptically injected into peritoneal cavity of an albino mouse. The mouse abdomen is massaged for 30 s.
2. Fluid from mouse peritoneal cavity is aspirated using sterile syringe, and the content is transferred to a sterile 15 mL tube. It is then centrifuged at 720 × *g* for 10 min.
3. A small volume of pellet is used for counting the macrophages using a Neubauer chamber, and viability of cells can be simultaneously evaluated by trypan blue exclusion test. Unstained cells indicate viability.
4. RPMI-1640 supplemented with 20% fetal calf serum is added to resuspend macrophages. It is kept in a 12.5 cm^2 culture flask. A density of 3×10^6 cells/flask or 3×10^5 cells/well is appropriate for sporozoite inoculation.

About 10^5–10^6 sporozoites are inoculated to the cell culture flask when host cells reach appropriate density. The culture flask is incubated at 37°C in a CO_2 incubator. Growth and development of sporozoites should be monitored every day under inverted microscope until no further development can be observed. Infected cells can be examined by histopathological and ultrastructural studies.

38.3 Conclusion and Future Perspectives

Taxonomic repositioning of the formerly known *Isospora belli* to *Cystoisospora belli* has been based on biological and phylogenetical characters of this diplosporocystic, tetrasporozoic oocyst-producing coccidian protozoa recognized as a human pathogen almost a century ago. The genus *Isospora*, whose sporocyst contains Stieda body, undergoes monoxenous life cycle development without any demonstrable monozoic tissue cysts; therefore, no intermediate host exists. The avian parasites in the genus *Isospora*, for example, *I. robini* and *I. gryphoni*, are phylogenetically closely related with the genera *Eimeria* and *Cyclospora* (Figure 38.1). On the other hand, the genus *Cystoisospora* includes the formerly known mammalian *Isospora* species and *I. belli*, whose sporocyst lacks Stieda body and contains monozoic tissue cysts. A number of optional intermediate or paratenic hosts harboring monozoic tissue cysts have been reported in canine and feline *Cystoisospora* [60]. Monozoic tissue cysts that are the dormant forms in these intermediate hosts can serve as an infective stage for transmission. The efficacy of monozoic tissue cysts of *C. felis*, *C. rivolta*, and *C. canis* from murine paratenic hosts to produce intestinal infections in their respective feline and canine definitive hosts was comparable to that produced by infective oocysts [31,72]. Despite the presence of monozoic tissue cysts in cystoisosporiasis in AIDS patients, it remains unknown whether these tissue cysts generally occur during infections in immunocompetent persons. Meanwhile, optional intermediate or paratenic hosts have not been clearly demonstrated in *C. belli*, although attempts to infect a calf and chickens with human-derived oocysts failed to generate monozoic tissue cysts in these animals [29]. The presence of monozoic tissue cysts observed in BEK and VERO cells after inoculation of *C. belli* sporozoites may not directly indicate that bovine and simian hosts are paratenic hosts if host cell entry by *Cystoisospora* resembles an active invasion process of *Toxoplasma gondii*, rendering the parasites to infect a variety of host cells. Undoubtedly, identification of natural cryptic optional intermediate or paratenic hosts for *C. belli* has significant biological and medical implications.

Developmental cycle of *C. belli* consists of both asexual and sexual reproductions in a single host. Freshly excreted oocysts from infected human stools are not readily infective but require some time in environment to reach maturation, producing eight infective, crescentic-shaped sporozoites per oocyst. The oocysts are environmentally resistant and can retain viability for months at low temperature. Invasion of upper small intestinal enterocytes by sporozoites initiates asexual developmental cycle where endodyogeny and possibly schizogony ensue. However, extraintestinal unicellular zoites or monozoic tissue cysts, found mostly in lymphoid tissues of some AIDS patients, are suggestive of dormant stage that could be responsible for relapse in almost half of symptomatic isosporiasis cases. The high rate of relapse in *C. belli* infection suggests the presence of resistant cryptic stage that is not responsive to treatment. It remains to be explored whether monozoic tissue cysts of *C. belli* directly contribute to relapses and what reactivates their development. Although the majority of multiple relapses in cystoisosporiasis occurred in humans with compromised immunity, some immunocompetent individuals reportedly suffered from recurrent relapsing symptoms. It seems that the interplay between host immunity and parasite virulent factors determines clinical course of *C. belli* infection. However, the lack of natural reservoir host for *C. belli* has hindered experimental investigation to elucidate this issue. Therefore, modern molecular biology techniques will be important alternative strategies to unravel several unknown aspects of pathogenesis of human cystoisosporiasis. However, establishment of complete developmental cycle of this coccidian parasite in cell culture system is mandatory to shed light on this neglected but important human enteric coccidian pathogen.

REFERENCES

1. Manwell, R.D. Coccidia. In Manwell, R.D. (ed.), *Introduction to Protozoology*, 1st ed. St. Martin's Press, New York, pp. 515–538, 1961.
2. Wenyon, C.M. Coccidiosis of cats and dogs and the status of the *Isospora* of man. *Ann. Trop. Med. Parasitol.* 17, 231, 1923.
3. Elsdon-Dew, R. *Isospora natalensis* (sp. Nov.) in man. *J. Trop. Med. Hyg.* 56, 149, 1953.
4. Ravenel, J.M., Suggs, J.L., and Legerton, C.W. Human coccidiosis. Recurrent diarrhea of 26 years duration due to *Isospora belli*: A case report. *J. S. Carolina Med. Assoc.* 72, 217, 1976.
5. Jongwutiwes, S., Sampatanukul, P., and Putaporntip, C. Recurrent isosporiasis over a decade in an immunocompetent host successfully treated with pyrimethamine. *Scand. J. Infect. Dis.* 34, 859, 2002.
6. Barta, J.R. et al. The genus *Atoxoplasma* (Garnham 1950) as a junior objective synonym of the genus *Isospora* (Schneider 1881) species infecting birds and resurrection of *Cystoisospora* (Frenkel 1977) as the correct genus for *Isospora* species infecting mammals. *J. Parasitol.* 91, 726, 2005.
7. Franzen, C. et al. Taxonomic position of the intestinal protozoan parasite *Isospora belli* as based on ribosomal RNA sequences. *Parasitol. Res.* 86, 669, 2000.
8. Morrison, D.A. et al. The current status of the small subunit rRNA phylogeny of the coccidia (Sporozoa). *Int. J. Parasitol.* 34, 501, 2004.
9. Jongwutiwes, S. et al. Morphologic and molecular characterization of *Isospora belli* oocysts from patients in Thailand. *Am. J. Trop. Med. Hyg.* 77, 107, 2007.
10. Dubey, J.P., and Frenkel, J.K. Extra-intestinal stages of *Isospora felis* and *I. rivolta* (Protozoa: Eimeriidae) in cats. *J. Protozool.* 19, 89, 1972.
11. Restrepo, C., Macher, A.M., and Radany, E.H. Disseminated extraintestinal isosporiasis in a patient with acquired immune deficiency syndrome. *Am. J. Clin. Pathol.* 87, 536, 1987.
12. Michiels, J.F. et al. Intestinal and extraintestinal *Isospora belli* infection in an AIDS patient. A second case report. *Pathol. Res. Pract.* 190, 1089, 1994.
13. Frenkel, J.K. *Besnoitia wallacei* of cats and rodents: With a reclassification of other cyst-forming isosporoid coccidia. *J. Parasitol.* 63, 611, 1977.
14. Lindsay, D.S., Dubey, J.P., and Blagburn, B.L. Biology of *Isospora* spp. from humans, nonhuman primates, and domestic animals. *Clin. Microbiol. Rev.* 10, 19, 1997.
15. Tamura, K. et al. MEGA6: Molecular evolutionary genetics analysis version 6.0. *Mol. Biol. Evol.* 30, 2725, 2013.

16. Beaver, P.C., Jung, R.C., and Cupp, E.W. Coccidia, Microsporidia and *Pneumocystis*. In Beaver, P.C., Jung, R.C., and Cupp, E.W. (eds.), *Clinical Parasitology*, 9th ed. Lea and Febiger, Philadelphia, PA, pp. 149–173, 1984.
17. Zaman, V. Observations on human *Isospora*. *Trans. R. Soc. Trop. Med. Hyg.* 62, 556, 1968.
18. Matsui, T. et al. Infectivity and sporogony of *Caryospora*-type oocyst of *Isospora rivolta* obtained by heating. *Parasitol. Res.* 79, 599, 1993.
19. Dubey, J.P. Life cycle of *Isospora rivolta* (Grassi, 1879) in cats and mice. *J. Protozool.* 26, 433, 1979.
20. Ferguson, D.J.P. et al. Ultrastructural observations on multiplication of *Cystoisospora* (*Isospora*) *felis* by endodyogeny. *Z. Parasitenkd.* 63, 289, 1980.
21. Matuschka, F.R. Ultrastructural evidence of endodyogeny in *Isospora suis* from pigs. *Z. Parasitenkd.* 67, 27, 1982.
22. Shah, H.L. The life cycle of *Isospora felis* Wenyon, 1923, a coccidium of the cat. *J. Protozool.* 18, 3, 1971.
23. Fayer, R., and Thompson, D.E. *Isospora felis*: Development in cultured cells with some cytological observations. *J. Parasitol.* 60, 160, 1974.
24. Oliveira-Silva, M.B. et al. *Cystoisospora belli*: In vitro multiplication in mammalian cells. *Exp. Parasitol.* 114, 189, 2006.
25. Siripanth, C. et al. Development of *Isospora belli* in Hct-8, Hep-2, human fibroblast, BEK and Vero culture cells. *Southeast Asian J. Trop. Med. Public. Health.* 35, 796, 2004.
26. Resende, D.V. et al. Experimental infection of murine and human macrophages with *Cystoisospora belli*. *Acta. Trop.* 111, 177, 2009.
27. Resende, D.V. et al. Ultrastructural aspects of *Cystoisospora belli* (syn. *Isospora belli*) in continuous cell lines. *Microsc. Res. Tech.* 77, 472, 2014.
28. Velásquez, J.N. et al. Isosporosis and unizoite tissue cysts in patients with acquired immunodeficiency syndrome. *Hum. Pathol.* 32, 500, 2001.
29. Frenkel, J.K. et al. *Isospora belli* infection: Observation of unicellular cysts in mesenteric lymphoid tissues of a Brazilian patient with AIDS and animal inoculation. *J. Eukaryot. Microbiol.* 50(Suppl), 682, 2003.
30. Lindsay, D.S. et al. Examination of extraintestinal tissue cysts of *Isospora belli*. *J. Parasitol.* 83, 620, 1997.
31. Dubey, J.P. Experimental *Isospora canis* and *Isospora felis* infection in mice, cats, and dogs. *J. Protozool.* 22, 416, 1975.
32. Houk, A.E., and Lindsay, D.S. *Cystoisospora canis* (Apicomplexa: Sarcocystidae): Development of monozoic tissue cysts in human cells, demonstration of egress of zoites from tissue cysts, and demonstration of repeat monozoic tissue cyst formation by zoites. *Vet. Parasitol.* 197, 455, 2013.
33. Mitchell, S.M., Zajac, A.M., and Lindsay, D.S. Development and ultrastructure of *Cystoisospora canis* Nemeséri, 1959 (syn, *Isospora canis*) monozoic cysts in two noncanine cell lines. *J. Parasitol.* 95, 793, 2009.
34. Faust, E.C. et al. Human isosporosis in the western hemisphere. *Am. J. Trop. Med. Hyg.* 10, 343, 1961.
35. Henry, M.C. et al. Parasitological observations of chronic diarrhoea in suspected AIDS adult patients in Kinshasa (Zaire). *Trans. R. Soc. Trop. Med. Hyg.* 80, 309, 1986.
36. DeHovitz, J.A. et al. Clinical manifestations and therapy of *Isospora belli* infection in patients with the acquired immunodeficiency syndrome. *N. Engl. J. Med.* 315, 87, 1986.
37. Certad, G. et al. Isosporiasis in Venezuelan adults infected with human immunodeficiency virus: Clinical characterization. *Am. J. Trop. Med. Hyg.* 69, 217, 2003.
38. Lainson, R., and da Silva, B.A. Intestinal parasites of some diarrhoeic HIV-seropositive individuals in North Brazil, with particular reference to *Isospora belli* Wenyon, 1923 and *Dientamoeba fragilis* Jepps & Dobell, 1918. *Mem. Inst. Oswaldo. Cruz.* 94, 611, 1999.
39. Sorvillo, F.J. et al. Epidemiology of isosporiasis among persons with acquired immunodeficiency syndrome in Los Angeles County. *Am. J. Trop. Med. Hyg.* 53, 656, 1995.
40. Brandborg, L.L., Goldberg, S.B., and Breidenbach, W.C. Human coccidiosis—A possible cause of malabsorption. *N. Engl. J. Med.* 283, 1306, 1970.
41. Trier, J.S. et al. Chronic intestinal coccidiosis in man: Intestinal morphology and response to treatment. *Gastroenterology* 66, 923, 1974.
42. Matsubayashi, H., and Nozawa, T. Experimental infection of *Isospora hominis* in man. *Am. J. Trop. Med.* 28, 633, 1948.

43. McCracken, A.W. Natural and laboratory-acquired infection by *Isospora belli*. *South. Med. J.* 65, 800, 1972.
44. Syrkis, I. et al. A case of severe human coccidiosis in Israel. *Israel. J. Med. Sci.* 1, 373, 1975.
45. Pape, J.W., Verdier, R.I., and Johnson, W.D. Jr. Treatment and prophylaxis of *Isospora belli* infection in patients with the acquired immunodeficiency syndrome. *N. Engl. J. Med.* 320, 1044, 1989.
46. de Otazu, R.D. et al. Endometrial coccidiosis. *J. Clin. Pathol.* 57, 1104, 2004.
47. Herwaldt, B.L. *Cyclospora cayetanensis*: A review, focusing on the outbreaks of cyclosporiasis in the 1990s. *Clin. Infect. Dis.* 31, 1040, 2000.
48. Garcia, L.S. Collection, preservation, and shipment of fecal specimens. In Garcia, L.S. (ed.), *Diagnostic Medical Parasitology*, 4th ed. ASM Press, Washington, DC, pp. 723–740, 2001.
49. Jongwutiwes, S. et al. Simple method for long-term copro-preservation of *Cryptosporidium* oocysts for morphometric and molecular analysis. *Trop. Med. Int. Health.* 7, 257, 2002.
50. Garcia, L.S. Macroscopic and microscopic examination of fecal specimens. In Garcia, L.S. (ed.), *Diagnostic Medical Parasitology*, 4th ed. ASM Press, Washington, DC, pp. 741–785, 2001.
51. Pacheco, F.T. et al. Differences in the detection of *Cryptosporidium* and *Isospora* (*Cystoisospora*) oocysts according to the fecal concentration or staining method used in a clinical laboratory. *J. Parasitol.* 99, 1002, 2013.
52. Ma, P., and Soave, R. Three-step stool examination for cryptosporidiosis in 10 homosexual men with protracted watery diarrhea. *J. Infect. Dis.* 147, 824, 1983.
53. Varea, M. et al. Fuchsin fluorescence and autofluorescence in *Cryptosporidium*, *Isospora* and *Cyclospora* oocysts. *Int. J. Parasitol.* 28, 1881, 1998.
54. Daugschies, A. et al. Autofluorescence microscopy for the detection of nematode eggs and protozoa, in particular *Isospora suis*, in swine feces. *Parasitol. Res.* 87, 409, 2001.
55. Bialek, R. et al. Comparison of autofluorescence and iodine staining for detection of *Isospora belli* in feces. *Am. J. Trop. Med. Hyg.* 67, 304, 2002.
56. Garcia, L.S. Sputum, aspirates, and biopsy material. In Garcia, L.S. (ed.), *Diagnostic Medical Parasitology*, 4th ed. ASM Press, Washington, DC, pp. 809–828, 2001.
57. Sun, T. et al. Light and electron microscopic identification of *Cyclospora* species in the small intestine. Evidence of the presence of asexual life cycle in human host. *Am. J. Clin. Pathol.* 105, 216, 1996.
58. Nime, F.A. et al. Acute enterocolitis in a human being infected with the protozoan *Cryptosporidium*. *Gastroenterology* 70, 592, 1976.
59. Meisel, J.L. et al. Overwhelming watery diarrhea associated with a *Cryptosporidium* in an immunosuppressed patient. *Gastroenterology* 70, 1156, 1976.
60. Lindsay, D.S. et al. Developmental biology of *Cystoisospora* (Apicomplexa: Sarcocystidae) monozoic tissue cysts. *J. Parasitol.* 100, 392, 2014.
61. Müller, A. et al. Detection of *Isospora belli* by polymerase chain reaction using primers based on small-subunit ribosomal RNA sequences. *Eur. J. Clin. Microbiol. Infect. Dis.* 19, 631, 2000.
62. Taniuchi, M. et al. Multiplex polymerase chain reaction method to detect *Cyclospora*, *Cystoisospora*, and microsporidia in stool samples. *Diagn. Microbiol. Infect. Dis.* 71, 386, 2011.
63. Lalonde, L.F., Reyes, J., and Gajadhar, A.A. Application of a qPCR assay with melting curve analysis for detection and differentiation of protozoan oocysts in human fecal samples from Dominican Republic. *Am. J. Trop. Med. Hyg.* 89, 892, 2013.
64. Petri, W.A. Sulfonamides, trimethoprim-sulfamethoxazole, quinolones, and agents for urinary tract infections. In Brunton, L.L., Chabner, B.A., and Knollmann, B.C. (eds.), *Goodman & Gilman's: The Pharmacological Basis of Therapeutics*, 12th ed. McGraw-Hill, New York, NY, pp. 1462–1476, 2011.
65. Gangjee, A., Kurup, S., and Namjoshi, O. Dihydrofolate reductase as a target for chemotherapy in parasites. *Curr. Pharm. Des.* 13, 609, 2007.
66. Capasso, C.L., and Supuran, C.T. Sulfa and trimethoprim-like drugs—Antimetabolites acting as carbonic anhydrase, dihydropteroate synthase and dihydrofolate reductase inhibitors. *J. Enzyme Inhib. Med. Chem.* 29, 379, 2014.
67. Bushby, S.R., and Hitchings, G.H. Trimethoprim, a sulphonamide potentiator. *Br. J. Pharmacol. Chemother.* 33, 72, 1968.
68. Foner, A. An attempt to infect animals with *Isospora belli*. *Trans. R. Soc. Trop. Med. Hyg.* 33, 357, 1939.

69. Jeffery, G.M. Human coccidiosis in South Carolina. *J. Parasitol.* 42, 491, 1956.
70. Duszynski, D.W., and Conder, G.A. External factors and self-regulating mechanisms which may influence the sporulation of oocysts of the rat coccidium, *Eimeria nieschulzi*. *Int. J. Parasitol.* 7, 83, 1977.
71. Ryley, J.F. et al. Methods in coccidiosis research: Separation of oocysts from faeces. *Parasitology* 73, 311, 1976.
72. Dubey, J.P., and Streitel, R.H. *Isospora felis* and *I. rivolta* infections in cats induced by mouse tissue or oocysts. *Br. Vet. J.* 132, 649, 1976.

39

Entamoeba histolytica

Mineko Shibayama, Nidia León-Sicairos, Jesús Serrano-Luna, and Mireya de la Garza

CONTENTS

39.1 Introduction

39.1.1 Human Amoebiasis

Amoebiasis is a parasitic infection that is caused by an extracellular enteric protozoan named *Entamoeba histolytica*. This parasite is able to coexist with the human intestine through a tolerogenic/hyporesponsive immune reaction, and the intestinal barrier is not broken. Through an unknown mechanism, the host switches perception of the parasites from commensal to invaders and mounts an acute inflammatory response that leads to colonic lesions of diverse magnitude. The symptoms of intestinal amoebiasis (IA) are abdominal pain, dysentery, and ulcerative colitis with mucous and blood, which can, in some cases, evolve into appendicitis and amoeboma, which is an acute manifestation of the disease. Rarely, the parasites can spread through portal circulation to the liver and produce the main extraintestinal infection, amoebic liver abscess (ALA). *E. histolytica* can also reach the lungs and the brain, primarily in immunocompromised patients. Both IA and ALA can be fatal if left untreated. Amoebiasis is the third leading cause of death due to parasites after malaria and schistosomiasis. This disease presents a high index of morbidity and mortality, mainly in developing countries with poor hygiene conditions, where it can be endemic. It can also affect travelers who visit those countries. According to the World Health Organization (WHO) [1], 500 million people are infected with amoebae worldwide; 10% of them have virulent *E. histolytica*, which leads to 40,000–100,000 deaths annually. Epidemiological studies demonstrated the presence of antibodies (Abs) against *E. histolytica* in the populations of Mexico, Brazil, Venezuela, Bangladesh, South Africa, and Egypt, showing the high prevalence of this parasite [2]. The closely related nonpathogenic species *E. dispar* and *E. moshkovskii* are morphologically identical to *E. histolytica* in standard stool microscopy analysis and are the cause of asymptomatic amoebiasis [3–8]. Studies of colon microbiota have been valuable in understanding the interplay among amoeba, the host, and disease status [9–11]. Other host factors also influence the pathogenicity of amoeba, such as diet, sex, and age [12].

39.1.2 Life Cycle of *E. histolytica*

The *E. histolytica* life cycle does not include a vector and consists of two main stages—trophozoite and cyst. The cycle begins with the ingestion of mature quadrinucleate cysts (resistant forms, infective phase), which tolerate the acidic pH of the stomach and excyst in the terminal ileum, where they produce eight uninucleate trophozoites per cyst. Trophozoites, which are also called amoebae (motile forms, invasive phase), travel to and colonize the large intestine. In response to unknown stimuli, the amoebae undergo morphological and biochemical changes that lead to the formation of cysts, which are eliminated in the feces and transmitted to other hosts via the ingestion of contaminated water or foods, thus reinitiating the cycle [3,4,13].

39.1.3 Isolation and Culture Conditions

The excystation of *E. histolytica* cysts was first carried out in a medium with bacteria, but further amoebae could be cultured axenically [14]. Currently, amoebae are cultured *in vitro* in BI-S-33 (TYI-S-33) medium, an axenic medium designed by Diamond et al. [15], which contains peptone, glucose, ferric citrate, and ascorbic acid and cysteine as reducing compounds; the medium is also supplemented with bovine serum and vitamins. The most frequently used strain in research is HM-1:IMSS, which was isolated in 1971 by De la Torre et al. from feces of a patient suffering from IA [16]. Because this strain has been grown in culture for years, it is less virulent, and in the case of *in vivo* studies, amoebae are periodically passed through a susceptible animal, mainly hamster (see Section 39.2.2).

39.1.4 Virulence Factors

E. histolytica is an active protist parasite that can adhere to the human epithelial mucosa and extracellular matrix proteins, phagocytose erythrocytes and cell debris, secrete toxins and proteolytic enzymes, cause apoptosis in target cells, and scavenge iron and other nutrients from the host [17–21]. Amoebiasis is a multifactorial disease because host pathogen recognition receptors (PRRs), immune cells, amoeba virulence factors, and microenvironment all contribute to the damage of the intestinal mucosa [4,22]. The contact between amoebae and target cells appears to be the first step in cell lysis and phagocytosis. Intestinal flask-shaped ulcers, which are a hallmark of amoebic colitis, are characterized by severe damage to enteric cells and migration to the lamina propria and blood vessels; an inflammatory reaction is also present [20,23,24].

A wide variety of molecules have been described as *E. histolytica* virulence factors that act in concert and favor amoebae invasion. For example, disruption of the colonic mucosa through amoebic enzymes leads to direct contact between the parasites and the epithelium mainly through the amoebic Gal/GalNAc lectin (Gal-lectin), which activates the inflammasome. The IL-1β released during invasion produces a severe inflammatory response [25]. Amoebapore polypeptides can lyse bacterial and host cell membranes and are involved in liver abscess and cell apoptosis [26–29]. Several membrane and secreted cysteine proteases (CPs) that degrade host proteins have been described in *E. histolytica*; the most studied are CP1 and CP5 [30–33]. Currently, more than 50 genes are known to encode CPs, and their expression depends on conditions such as the amoeba strain and culture conditions [34,35]. Lipophosphopeptidoglycan is a surface molecule that activates NKT cells in a CD1d-dependent manner, and this interaction limits ALA development [25,36,37].

39.2 Animal Models of Amoebiasis

Presently, there is no animal model that can reproduce the whole life cycle of *E. histolytica* as seen in humans. Therefore, the studies of experimental amoebiasis have been performed either with intestinal or hepatic models as separate systems. However, the use of different animal strains that are susceptible and resistant to amoebiasis has led to a better understanding of the host–parasite relationship regarding amoeba-secreted products and pathogenicity mechanisms, the role of neutrophils and other host cells and molecules affecting invasiveness, sequential stages of tissue damage, and immune response. Animal models have also enabled the testing of new therapeutic targets to treat amoebiasis. In the case of IA, several strains of susceptible mice have been employed; hamsters and gerbils have been mainly utilized to reproduce ALA [38–41]. In this chapter, we briefly describe the main models used to study intestinal and hepatic amoebiasis, including the lesions caused by amoebae and treatments for the disease.

39.2.1 Intestinal Amoebiasis

The first *in vivo* model of IA was developed by Lesh in 1875 [42]. He produced intestinal lesions in a dog inoculated with feces from a patient suffering from dysentery. Since then, many studies have been performed in different animals (dogs, cats, rabbits, hamsters, and monkeys); however, the use of these models is limited at present due to several factors, including a lack of uniformity, difficulty in handling, and an insufficient number of animals [41]. IA in animals has yielded inconsistent and poor results, and in many cases, this difficulty has been simply attributed to their natural immune resistance [41,43]. Moreover, although experimental animals are inoculated with several million trophozoites of a known specific virulent strain, this inoculum then mixes with poorly characterized intestinal flora consisting of additional protozoa, bacteria, and other microorganisms. This mixture of intestinal organisms, and our lack of knowledge of the physicochemical characteristics of the intestinal milieu, are important factors that have not been considered in the past and are likely the primary causes of the heterogeneity of results [41].

Differences in susceptibility have been reported by different groups, even when using similar laboratory animals under apparently identical conditions. Despite these limitations, interesting information regarding the early stages of IA, pathology, and host immune response has been obtained [41,44–46]. At the molecular level, various research groups have studied the genes or gene products of *E. histolytica* related to those of humans through the evaluation of severe combined immunodeficient (SCID) and SCID-modified mice, microarray analysis, and examination of several aspects of the mechanisms of the immune response. In the forthcoming sections, we analyze some of the models used to establish IA [43]. Table 39.1 presents the foremost models used in IA.

39.2.1.1 Rats

Wistar strain rats (mainly weanling), or colon loops of adult Sprague Dawley rats, have been used to model IA. However, due to the great variability of the macroscopic appearance related to the damage induced by *E. histolytica* trophozoites, this model is no longer used [47,48].

TABLE 39.1

The Foremost Models Used to Study Intestinal Amoebiasis

Animal Model	Method	Advantages/Main Findings	Limitations	Reference
Dogs	Inoculation with feces from a patient suffering from dysentery	Intestinal lesions in some animals	Results were not reproducible, the animals resolved the infection	[42]
Newborn guinea pigs, hamsters	Intracecal inoculation axenic cultures	Lesions in cecum	Results were not reproduced in a larger number of animals	[49]
C3H/HeCr, BALB/c, NZB/BIN, B10.A, DBA/2, and C57BL/6 inbred mice	Intracecal inoculation of axenic cultures	Susceptibility of mice to infection depended on the genetic charge of the strains	Only early states of ameobiasis can be studied	[56]
Conventionally raised animals (hamsters or Guinea pigs)	Inoculation of axenic or monoxenic cultures on a closed loop of the cecum	Early stages of amoeba invasion inside the mucosal and submucosal layers with a substantial inflammatory reaction	Only early states of amoebiasis can be studied	[50]
Adult Sprague Dawley rats	Inoculation of axenic cultures into *in vivo* colon loops	The mucus blanket provided a significant barrier to trophozoite access to intestinal epithelium target	Only early states of amoebiasis can be studied	[47]
Washed-closed cecal loop model	An artificial cecal loop inoculated with axenic or monoxenic cultures	It helped reported the possible role of leukocytes in the development of the amoebic ulcer	Only early states of amoebiasis can be studied	[24,50]
Balb/c mice	Intracecal inoculation of axenic cultures in the cecal loop of animals	Trophozoites did not survive in the animals, but they colonize longer in animals with the cecal loop	Some mice were resistant to disease	[54]
Rabbit colon preparations	Full-thickness rabbit colon preparations mounted in Ussing-type chambers, incubated with monoxenic cultures	Lesions in colon were demonstrated by increased decay rates for potential difference, short-circuit current and transmural resistance	Only early events of amoebiasis pathogenesis can be studied	[44]

(*Continued*)

TABLE 39.1 (*Continued*)

The Foremost Models Used to Study Intestinal Amoebiasis

Animal Model	Method	Advantages/Main Findings	Limitations	Reference
Mongolian gerbils	Intracecal inoculation of gerbils with monoxenic cultures	Microulcerative mucosal lesions appeared 24–72 h postinoculation, inflammatory infiltrates and edema of the lamina propria	Normal cecal mucosa at 96 h postinfection	[51]
C3H/HeJ mice	The cecum was directly inoculated with monoxenic cultures	Amoebae release toxic factors that contributed to the inflammatory disease	Only intestinal amoeabiasis developed	[125]
SCID mouse-human intestinal xenograft model	Human intestinal xenografts were infected with monoxenic cultures	Extensive tissue damage, an early inflammatory response composed primarily of neutrophils. IL-1 and IL-8 were produced	Only early intestinal amoebiasis can be studied	[59]
Porcine colonic fragments	Trophozoites cocultured with porcine colonic fragments	Severe acute ulcerative jejunitis, with large hemorrhagic lesions. Typical large-sized hepatic abscesses	Lesions in two of the four animals	[52]
Human colon model *ex vivo*		Parasite penetration into the mucus and mucosa, cell lysis, and an inflammatory response	Human colonic explants are far less available	[63]

39.2.1.2 Guinea Pigs and Hamsters

Only amoebae cultured in axenic conditions have been able to produce ulcers in intestines of guinea pigs. Diamond induced lesions in newborn guinea pigs by inoculating intracecally; however, the results could not be reproduced consistently in a larger number of animals [49]. Anaya-Velazquez et al. described typical intestinal ulcers in guinea pigs and hamsters by producing an artificial cecal loop inoculated with axenic or monoxenic *E. histolytica* trophozoites [50]. With this model, the authors showed the early stages of amoeba invasion of the mucosal and submucosal layers with a substantial inflammatory reaction. Electron microscopy analysis of the ulcers showed the possible role of leukocytes in the development of IA [41]. This model is particularly useful for the study of early intestinal lesions produced by virulent amoebae, and may also be applied to studies on the immunology of invasive IA [50].

39.2.1.3 Rabbits

To develop a model capable of reproducing the events that occur during the initial interaction of *E. histolytica* trophozoites with the mucosa of the large intestine, Navarro-García used full-thickness rabbit colon preparations ($0.28\,cm^2$) mounted in Ussing-type chambers. The untreated samples had electrophysiological properties (potential difference, short-circuit current, and electrical resistance) that were similar in magnitude and duration to those reported for stripped colonic mucosa. However, in samples exposed to amoeba lysates for up to 80 min, dose-dependent lesions in the colon were observed and consisted of (1) increased decay rates for potential difference, short-circuit current, and transmural resistance and (2) mucosal lesions involving vacuolization at the bases and shortening of epithelial cells, loss of intercellular junctions, destruction of microvilli, and necrosis of interglandular epithelial zones. The authors discussed the specificity and speed of the electrophysiological effects and their correlation with the microscopic lesions and suggested that this model will help increase our understanding of the initial pathogenic events of IA [44].

39.2.1.4 Gerbils

By using monoxenic cultures of *E. histolytica* trophozoites inoculated in Mongolian gerbils, Shibayama observed that an increase in mucus production occurred during the first hours of the interaction. Furthermore, microulcerative mucosal lesions appeared 24–72 h postinoculation. Inflammatory infiltrate and edema of the lamina propria were associated with necrotic foci. Unfortunately, at 96 h, the cecal mucosa was normal, amoebae were no longer detected, and the gerbils healed spontaneously [51]. However, overall, it was concluded that gerbils are useful as experimental models for studying the early stages of invasive IA.

39.2.1.5 Pigs

More recently, assays with trophozoites of virulent *E. histolytica* cocultured with porcine colonic fragments have shown that outbred pigs can be used to reproduce some lesions associated with human IA. A detailed analysis showed that loops inoculated with virulent amoebae showed severe acute ulcerative jejunitis 14 days postinoculation, with large hemorrhagic areas associated with the presence of amoebae in the depth of the mucosa in two of the four animals [52]. Furthermore, typical large hepatic abscesses were observed in the liver of one animal at 7 days postinfection after inoculation by the portal vein or directly into the liver parenchyma, which showed that the pig model could be useful in simultaneously studying IA and ALA. Moreover, human colonic explants have been identified as valuable assets in the study of host–parasite interactions. However, because human colonic explants are far less available, porcine colonic explants have been investigated as an alternative to human tissues. Porcine colonic explants cultured with virulent *E. histolytica* (HM1:IMSS) or an avirulent strain (Rahman) showed that explants cultured with virulent trophozoites react similar to their human counterparts, including tissue invasion by amoebae and the triggering of a typical innate immune response against the parasite. In contrast, explants cultured with avirulent amoebae were healthy. The authors suggest that this study opens the way for the use of porcine colonic explants in the study of the complex interactions between the parasite and the host [53].

39.2.1.6 Mice Models

In the past, mice were not commonly used as a model for amoebiasis, primarily since these rodents were always considered naturally resistant to *E. histolytica* infection [54]. Depending on the strain, mice presented different susceptibilities to developing IA; thus, the authors concluded that a genetic factor could be involved in mouse susceptibility to intestinal infection. It was also postulated that other factors such as a cholesterol-rich diet, association with bacteria, and the presence of other protozoa could influence the results found in the different mice strains [43,54]. A neutropenic BALB/c mouse model was developed, in which the production of a granulomatous inflammatory reaction in the intestinal wall of mice inoculated with axenically cultured amoebae was reported for the first time [55]. Assays in other strains corroborated that the establishment of IA depends on the mouse strain; for example, C57BL/6 mice are highly resistant, whereas C3H/HeJ mice are relatively susceptible. The amoebic colitis in these mice was limited to the cecum, and the morphology of the inflammatory infiltrate was similar to that observed in humans. This model of resistant versus susceptible mice could provide useful insight into the human variability of parasite clearance when compared with invasive disease [56]. In the next section, we analyze the mice strains that are susceptible to intestinal amoebiasis.

39.2.1.7 C3H/HeJ Mice

In 2002, it was shown that C3H/HeJ mice, which have a mutation at the lipopolysaccharide response locus, were 60% infected with amoeba after intracecal inoculation, whereas C57BL/6 or BALB/c mice were resistant, including mice that were deficient for IL-12, IFN-γ, or inducible NO synthase. Infection was a chronic cecum inflammation that pathologically mirrored the human disease. This model revealed important immune factors that influence susceptibility to infection and, for the first time, established the pathological contribution of the host immune response in amoebiasis [57]. The early steps in IA, such as parasite adhesion to the mucosa, can be investigated in this model, and chronic infection can be obtained after mechanical injury of the cecal epithelium.

39.2.1.8 C57BL/6 IL-10-Deficient Mouse

Use of this model showed that nonhematopoietic cells mediate the natural resistance of mice to IA, but this resistance depends on hematopoietic IL-10 activity [58].

39.2.1.9 SCID-HU-INT Mouse

Infection of the SCID mouse-human intestinal xenograft (SCID-HU-INT) model of disease with amoebae resulted in extensive tissue damage, which was associated with an early inflammatory response composed primarily of neutrophils [59]. In this model, it was evidenced that human intestinal epithelial cells can produce inflammatory cytokines in response to infection *in vivo*, and it was established as a system for studying the interactions between *E. histolytica* and the human intestine [59,60]. Furthermore, research on an *E. histolytica* substrain in which the expression of several CPs was downregulated by an antisense transcript showed that these enzymes play a major role in the advancement of IA in the SCID-HU-INT mouse [61]. One of them, EhCP-A5, which is not expressed in *E. dispar*, has been shown to degrade the cysteine-rich domains of the MUC2 mucin, which are the major structural component of the colonic mucus gel in the human digestive tract [61].

39.2.1.10 Human Colonic Explants Model

Recently, human colonic explants have been used to study host–parasite interactions in IA [62]. This *ex vivo* model allows identifying the first steps of invasion and the comparison of the reaction to different strains in the same colon sample. It is possible to examine both sides of the host–parasite interaction by determining the kinetics of parasite penetration into the mucus and mucosa, structural changes in the mucosa, cell lysis and the inflammatory response to the virulent wild-type *E. histolytica* strain, and to compare them with those observed after infection with the nonpathogenic parasite *E. dispar* [63]. Based on recent results in work with this model, the Gal-lectin and amoebapores are not required for the invasion of human colon explants, thus suggesting that CP-A5, which is involved in extracellular matrix degradation, is not required for crossing the mucus but rather contributes directly or indirectly to penetrating the lamina propria and inducing inflammation [64].

39.2.2 Amoebic Liver Abscess

The most common extraintestinal amoebiasis is ALA. Macroscopically, liver lesions are formed by necrotic tissue, which has a yellowish color and creamy consistency, and are surrounded by congestive parenchyma and a discrete fibrosis. When the damage is extensive, the necrotic material can be substituted by liquid necrosis, thus creating a cavitary aspect. Microscopically, trophozoites are localized in the border of the necrotic areas and close to the apparently normal hepatic tissue. Occasionally, the amoebae are mixed with the necrotic parenchyma and contain erythrocytes and ingested cell debris. The inflammatory foci in the liver tissue vary but show an important chronic inflammatory reaction in the portal areas.

Reports of early liver lesions caused by *E. histolytica* in humans are very scarce; when ALA is well established, there is extensive necrotic tissue. The information on early liver lesions is based on experimental studies using susceptible animals such as hamsters and gerbils that were inoculated by intraportal or intrahepatic routes with virulent amoebae. The first laboratory animal used successfully to produce ALA was the hamster [65,66]. This was done by inoculating axenic trophozoites via the portal vein and producing large amoebic abscesses. From these pioneering studies, data have been obtained regarding pathogenesis, immunopathology, cell and molecular biology, and genetics, among other topics involved in the host–parasite interaction.

An excellent work in 1984 [23] described the sequential histopathological study of ALA formation from very early stages in hamsters inoculated by the intraportal route with *E. histolytica* trophozoites. The authors demonstrated the role of the host inflammatory reaction in the liver tissue damage for the first time. They suggested that the amoebae do not produce ALA by direct lysis of the hepatocytes but rather the inflammatory cell lysis contributes to liver damage. This finding broke the paradigm that amoebic liver lesions produced by *E. histolytica* do not produce inflammation in human ALA. These findings were later confirmed by transmission electron microscopy [39]. Figure 39.1 shows macroscopic and histopathological analysis of ALA in hamsters. Table 39.2 lists the susceptibility of different rodents to IA and ALA.

Another set of rodents highly susceptible to hepatic amoebiasis are the gerbils [67]. Hepatic lesions are similar to those reported in hamsters, although amoebae show less virulent behavior in gerbils than they do in hamsters. In gerbils, parasites were in direct contact with the hepatocytes and were associated with inflammatory cells and the liver parenchyma and showed severe damage in both conditions. The production of ALA and the possibility of producing intestinal lesions have led some researchers to consider this rodent model of disease as being more similar to human hepatic pathology [41]. Studies of host immune responses and the effect of several vaccine candidates from purified proteins from *E. histolytica* (recombinant) administered orally or parenterally have been reported in gerbils [68].

E. histolytica infects only humans and nonhuman primates; studies in laboratory mice models suggest the existence of natural resistance to amoebiasis. However, the use of different mice strains has increased the understanding of the cellular and molecular basis of the mechanisms involved in the host–parasite relationship. Genetically modified animals, such as the SCID mice reported by Cieslak et al. [69], developed ALA when the animals were challenged intrahepatically with virulent *E. histolytica* trophozoites. When the immunocompetent congenic CB-17 (used as control) was used, only one of seven mice developed an abscess.

To study immune responses, the role of macrophages in the resistance to invasive amoebiasis has been examined in several species. Mice treated with silica are susceptible to developing ALA, probably due to impairment in macrophage function. However, the absence of macrophages or their activation does not reduce the natural resistance of mice to the development of ALA [70].

Data on the role of lymphocytes are contradictory. CBA/nu/nu athymic or Balb/c mice treated with an anti-lymphocyte Ab showed that these cells are not important in the resistance to *E. histolytica* infection. However, C3H/mg mice treated with the same Ab developed hepatic lesions, which suggests that

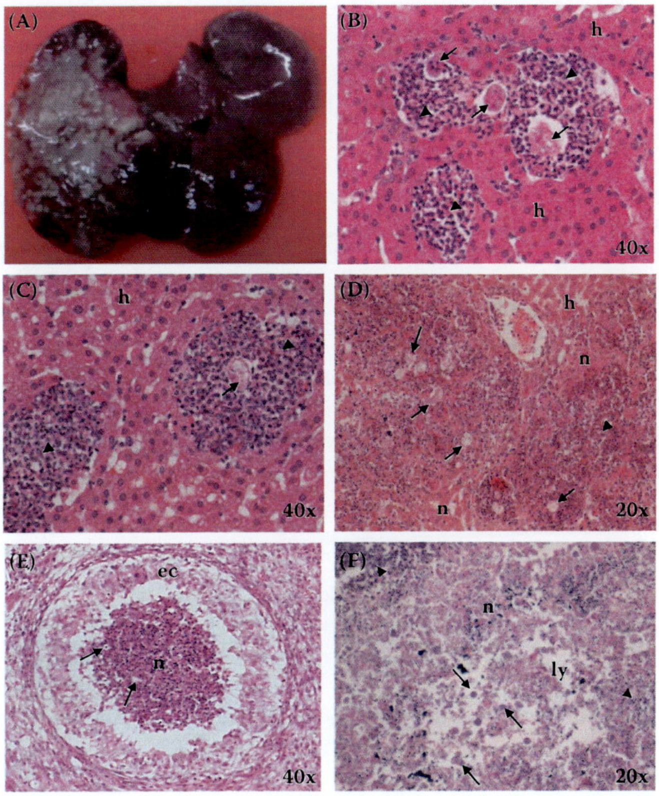

FIGURE 39.1 (a) Amoebic liver abscess. Macroscopic aspect. Whitish lesions covering the entire right lobule (visceral face); 7 days postintrahepatic inoculation with 1 × 106 trophozoites of *Entamoeba histolytica*. (b–f) Histopathological analysis. (b) Three hours postinfection. Trophozoites (arrows), inflammatory cells (arrow-heads), and hepatocytes (h). 40×. (c) Six hours postinoculation. The inflammatory foci increase in number; amoeba (arrow), inflammatory reaction (arrow-heads) and hepatocytes (h) 40×. (d) Two days postintrahepatic inoculation. The necrotic areas are evident (n), the amoebae increase in number (arrows), inflammatory cells (arrow-head). Some hepatocytes appear apparently normal (h). 20×. (e) Granulomatous reaction. Five days postinfection. Necrotic center (n), epithelioid cells (ec), and trophozoites (arrows). 40×. (f) Seven days postinoculation. The necrotic areas increased in the liver. Some amoebae are seen (arrows) intermixed with the lytic necrosis (ly), inflammation (arrow-heads). 20×.

TABLE 39.2

Amoebiasis: Susceptibility in Rodents

	Intestinal	Liver
Hamster	–	+
Gerbil	+/–	+
Guinea pig	+/–	–
Mouse	+/–	–
Rat	+/–	–

lymphocytes are important in the resistance to hepatic amoebiasis [69,71]. More recently, studies using Balb/c mice treated with a monoclonal Ab (RB6–8C5) against neutrophils were challenged by intrahepatic administration of amoebae. The animals were sacrificed at different times postinoculation, and livers were analyzed histologically. The results showed that the neutropenic animals developed more severe ALA than did the controls. The authors concluded that neutrophils are important to natural resistance in amoebiasis, in contrast to the results found in hamster and gerbil models that develop ALA [72].

Information about the participation of cytokines, complement, and Abs is still incomplete. Cytokines such as IFN-γ, TNF-α, and MCSF-1 increase the amoebicidal effect of macrophages (peritoneal and Kupffer cells) [73]. Proteinases and amoebapores of *E. histolytica* are also involved in ALA production. To actively invade, amoebae must degrade different components of the extracellular matrix, which occurs via the action of proteinases secreted by the parasite. These enzymes degrade collagen, elastin, fibrinogen, fibronectin, and laminin [21,30]. Another factor that is known to contribute to tissue invasion is the inflammatory reaction produced by the presence of amoebae in the liver. The expression of TNF-α, IFN-γ, IL-1β, IL-8, and IL-10 was studied throughout ALA evolution in hamster. The authors showed that neutrophils and macrophages that infiltrate the liver parenchyma in the acute and chronic stages of ALA are also responsible for the tissue damage. IL-10 does not regulate the local production of proinflammatory cytokines. The results showed that the exacerbated inflammatory milieu contributes to damage and probably supports the survival of *E. histolytica* [74].

It is important to mention that the nonpathogenic *E. dispar* strain SAW760 cultured in axenic conditions showed early inflammatory reaction around the amoebae. However, hepatic lesions did not progress to ALA formation. In other recent studies using different strains of *E. dispar* in xenic or monoxenic culture conditions, the production of amoebic liver lesions has been reported [75,76].

39.2.3 Animal Models for Amoebiasis Treatment

As in the case of every drug for human use, it is desirable that antiamoebic compounds act exclusively against the parasite and do not affect human cells or metabolism, although this is not always possible. The first study of amoebiasis in rats used to evaluate new antiamoebic compounds was reported by Jones in 1946 [77]. Since then, numerous investigations have been performed in different animal species to test the efficacy of antiamoebics because results in animals are more validly extrapolated to the effects in humans than those performed in cultures *in vitro*.

Several drugs have been tested for both IA and ALA. Metronidazole (Mtz) has been used for 50 years for trichomoniasis treatment, and its effect in ALA treatment in hamsters was evaluated [78]. Derivatives of Mtz obtained through minor changes in the molecule have also been assayed in animal models and, in some cases, are more effective than Mtz. They also extend the parasiticidal effect of Mtz to other protozoa [79–84]. Treatment of amoebiasis should include luminal and/or extraluminal agents (some of them are known as cysticides), depending on the site of infection [13,85,86]. Symptomatic amoebiasis requires a two-drug regimen: a tissue-amoebicidal agent, such as Mtz, and a luminal-acting agent, such as iodoquinol, diloxanide furoate, or paromomycin [2,87,88]. The treatment for ALA is generally Mtz; this azole and several luminal drugs have also been used against IA. Mtz has toxic adverse effects because it produces nausea, vomiting, and abdominal pain; thus, patients often discontinue the treatment. Mtz induces DNA breaks and chromosomal aberrations in mouse peripheral lymphocytes, and it could be carcinogenic [85,86]. In addition, strains resistant to Mtz have been found in *E. histolytica* cultures [89]. Thus, the need for an effective and safe antiamoebic compound is increasing, and several new drugs have been tested in animal models against the parasite. Plant crude extracts that have been utilized for amoebiasis in traditional medicine have been examined against *E. histolytica* infections; in some cases, the compounds responsible for the antiamoebic effect have been purified, and their mode of action has been elucidated [90–93]. Generally, a number of compounds are assessed in *in vitro* cultures, and afterward the compound that presented the best effect is evaluated in *in vivo* models. In both cases, Mtz is used as positive control [94,95]. For example, the steroidal alkaloid chonemorphine was identified as an antiamoebic during the course of a screening program for novel antiparasitic agents from plant sources. Interestingly, this alkaloid led to a 100% cure of ALA in hamsters and cleared 90% of the intestinal infection in weanling Wistar rats [96].

A comparative study of experimental cecal amoebiasis to evaluate amoebicides in the mouse, hamster, and rat was developed in India. The authors found different responses for each animal and proposed the mouse model as the most useful for the primary screening of antiamoebic compounds [97,98]. In another study, the cytotoxic effect of amide derivatives of trifluoromethionine (TFM) against *E. histolytica* was evaluated. Amoebae but not mammals possess L-methionine γ-lyase, an enzyme that hydrolyzes TFM and its derivatives, which makes it a good target for amoebicidal activity [99]. Interestingly, a high-throughput screening for compounds effective against amoeba identified auranofin, an FDA-approved drug used therapeutically against rheumatoid arthritis. Auranofin is 10 times more potent than Mtz, which makes it a promising therapy for amoebiasis [100]. The innate immune response by protein lactoferrin, an iron-chelating molecule that is responsible for avoiding pathogens to acquire iron in mucosae and infection sites, has been tested in ALA and IA [101,102]. Remarkably, bovine lactoferrin was able to cure ALA and synergize with Mtz in biological action. This observation is particularly important given that a reduced dose of Mtz can be used if combined with lactoferrin, with the same effect on the parasite, thus diminishing the side effects and toxicity of the drug. In addition, other properties of lactoferrin such as its anti-inflammatory activity can help resolve the amoebic abscesses [101].

39.2.4 Animal Models for Amoebic Vaccine Research

The development of a vaccine against *E. histolytica* is imperative and represents a great opportunity to prevent and even eradicate amoebiasis. In general, vaccines are cost-effective and safe and provide high protection rates in many infectious diseases. However, to date, no amoebic vaccine has been tested in clinical trials. Recently, several vaccines have been tested successfully in animal models; researchers hope to obtain good results [103,104].

An important point in the human colonic-defense mechanism is the production of mucosal immunoglobulins (Igs), which have important roles in maintaining intestinal integrity [105]. Secretory IgA (sIgA) is one of the most abundant Abs produced by plasma cells within the lamina propria, preventing pathogens from adhering to the intestinal mucosal barrier. There is evidence that mucosal anti-Gal-lectin responses are critical for amoebic colonization and invasion [106]. IgA titers correlate with protection against amoebiasis, and other studies suggest that the presence of IgG has a detrimental effect [107]. Analysis of asymptomatic carriers of *E. histolytica* showed that these persons had high levels of IFN-γ, which reflects a Th1 response, but that patients with invasive amoebiasis had higher levels of IL-4, which resembled Th2-type responses [108]. The uncertainty regarding the role of adaptive immunity in providing protection against *E. histolytica* IA and ALA is important to resolve to develop a vaccine against *E. histolytica*.

Clinical studies in Bangladesh suggest that children with amoebic colitis develop mucosal IgA Abs to the surface *E. histolytica* Gal-lectin [107]. However, immunity was short-lived, and 20% of the children had a new case of amoebiasis [109]. In contrast, a group in South Africa with a previous history of ALA and with mucosal Abs to amoebic lectin was less predisposed to *E. histolytica* or *E. dispar* infection [110]. The data from a Bangladesh study suggest that mucosal immunity against Gal-lectin can reduce the risk of amoebic reinfection. However, in Vietnam, it was reported that individuals with a prior history of ALA present the same or higher risk of a second case of ALA [111]. This is important, because patients with ALA produce high quantities of *E. histolytica*-specific serum Abs and develop T cell proliferative responses that recognize amoebic antigens [112]. In addition, there is evidence that indicates that reinfection rates may be lower in persons with mucosal Abs against Gal-lectin.

The major requirements of an effective vaccine include immunological memory, which depends on the production of a strong immune response, identification of a protective antigen, and an appropriate delivery route. A basic concept for vaccine development is to use a live attenuated strain of a virus or bacterium to stimulate immunity to prevent subsequent reinfection with virulent strains. This approach requires that the vaccine strain be immunogenic and stably attenuated, with little risk of reversion. Although there is no animal model that completely mimics human IA and ALA, various *in vivo* models have been developed and contribute to advance the development of a potential vaccine. In guinea pigs, previous infection with a noninvasive strain of *E. histolytica* resulted in some protection against subsequent intracecal challenge with a virulent strain of the parasite [113]. It is also possible to genetically modify *E. histolytica* to create attenuated strains or to use naturally occurring *Entamoeba* strains that show reduced virulence compared with the wild-type, such as *E. histolytica* Rahman or *E. dispar* [103]. One approach to attenuation is to target *E. histolytica* virulence factors. Mirelman et al. were able to silence the expression of the *E. histolytica* amoebapore A gene and showed that intraperitoneal challenge in hamster with these amoebae induced IgG antiamoebic Abs [114].

Other possible vaccine candidates are antigenic proteins, such as Gal-lectin adhesin, which is the best characterized *E. histolytica* protein. This protein is a 260-kDa heterodimer localized on the amoeba surface [115]. Gal-lectin is an attractive candidate for a vaccine because of its immunogenicity and its importance in the development of the disease. The first experiment to use purified Gal-lectin showed 86% protection against ALA in the gerbil model [116]. Another potential protein target is the serine-rich *E. histolytica* protein (SREHP). Previous studies of this protein have reported its success as an antigen to protect gerbils against ALA [68]. Intradermal immunization with SREHP as a maltose-binding protein induced 100% protection against ALA in gerbils [68]. The *E. histolytica* 29-kDa molecule (Eh29) is also considered an important antigen for an amoebiasis vaccine. Eh29 is an alkyl hydroperoxide reductase that is involved in the detoxification of reactive oxygen species secreted by microflora or immune cells [117]. One study of this protein in the hamster model showed 54% protection against ALA [118]. Another study using Eh-29 conjugated with cholera toxin B subunit (CTB) conferred protection against intracecal amoebiasis in C3H/heJ mice (IA), which was associated with anti-Eh29 IgA Abs in the intestine and anti-Eh29 IgG-specific Abs in the serum [119].

The development of DNA vaccines is recent and involves the induction of DNA of known sequences of some antigens of interest into a bacterial plasmid. DNA vaccines have been shown to exert strong humoral and cell-mediated responses and were successful in conferring protection against parasites. One example of this method is the generation of a codon-optimized DNA vaccine encoding a portion of the Gal-lectin of *E. histolytica*. When Balb/c mice were vaccinated intradermally with the DNA plasmid, the vaccine stimulated a Th1-type Gal-lectin-specific cellular immune response as well as the development of serum Abs that recognized a recombinant portion of the heavy subunit and inhibited the adherence of trophozoites to target cells *in vitro* [120]. Another type of DNA vaccine is the multivalent vaccine. One example of this is the EhCPADH complex, which is formed by two surface molecules: CP112 (EhCP112) and an adhesin (EhADH112). Compared with immunization with each plasmid alone (EhCP112) or EhADH112), the coimmunization of hamsters with the two plasmids induced a significantly greater level of anti-*E. histolytica* IgG. Interestingly, protection against liver abscesses was detected only in animals that received the plasmid mixture, and no protection was observed in hamsters independently inoculated with plasmid pcDNA-Ehcp112 or pcDNA-Ehadh112 [121].

There are other amoebic targets such as the heparin sulfate-binding protein (HSBP) and the 30-kDa collagen-binding protein (CBP30) recombinant fused to portions of the *Trypanosoma cruzi* heat shock protein of 70kDa, which have been assayed as vaccine antigens in guinea pigs [122] and in hamsters [123], respectively. Beyond rodent models, only one study has explored the utility of vaccines in a nonhuman primate model of amoebiasis [124]. In this model, Gal-lectin was administered with CTB as an adjuvant in baboons by colonoscopy into the lumen of small bowel and cecum. This vaccine resulted in a moderate level of protection against *E. histolytica* reinfection. However, the nonhuman primate study represents a significant advance toward an anti-*E. histolytica* vaccine. It is evident that more studies are needed in this field. The identification of *E. histolytica* immunogenic proteins and the correct combination of doses, boosts, and adjuvants is a priority to obtain long-term immunological memory in experimental animal models to advance beyond the preclinical stage in humans.

39.3 Conclusions and Future Directions

Amoebiasis is the third leading cause of death by parasitic diseases; this problem mainly affects developing countries. In recent years, knowledge of the aspects involved in amoebiasis, the virulence factors of *E. histolytica*, and the human–amoeba interaction have advanced rapidly. However, despite the abundant data obtained using *in vitro* and *in vivo* experimental procedures, there are still many unknown features related to the mechanisms involved in the invasion of this parasite to the intestine and liver. Drug therapies that only affect the parasite and a vaccine that protects people living in zones where amoebiasis is endemic are also necessary. Improvements in sanitation and hygiene practices are particularly important to eliminate amoebiasis.

Acknowledgments

This work was supported by grants No. 1674231 (Serrano-Luna), 179251 (de la Garza), and 237523 (Shibayama-Salas) from Conacyt, México.

REFERENCES

1. W.H.O., Amebiasis. *Wkly. Epidemiol. Rec.*, 72: 97–9, 1997.
2. Chacin-Bonilla, L., Current pharmacotherapy of amebiasis, advances in new drugs, and design of a vaccine. *Invest. Clin.*, 53: 301–14, 2012.
3. Ximenez, C., Epidemiology of amebiasis in Mexico: a molecular approach. *Arch. Med. Res.*, 37: 263–5, 2006.
4. Espinosa-Cantellano, M. and Martinez-Palomo, A., Pathogenesis of intestinal amebiasis: from molecules to disease. *Clin. Microbiol. Rev.*, 13: 318–31, 2000.
5. Stanley, S.L., Jr., Amoebiasis. *Lancet*, 361: 1025–34, 2003.
6. Haque, R., et al., Amebiasis. *N. Engl. J. Med.*, 348: 1565–73, 2003.
7. Ali, I.K., Clark, C.G., and Petri, W.A., Jr., Molecular epidemiology of amebiasis. *Infect. Genet. Evol.*, 8: 698–707, 2008.
8. Mortimer, L. and Chadee, K., The immunopathogenesis of *Entamoeba histolytica. Exp. Parasitol.*, 126: 366–80, 2010.
9. Phillips, B.P., et al., Studies on the ameba-bacteria relationship in amebiasis; comparative results of the intracecal inoculation of germfree, monocontaminated, and conventional guinea pigs with *Entamoeba histolytica. Am. J. Trop. Med. Hyg.*, 4: 675–92, 1955.
10. Mirelman, D., Ameba-bacterium relationship in amebiasis. *Microbiol. Rev.*, 51: 272–84, 1987.
11. Galvan-Moroyoqui, J.M., et al., The interplay between *Entamoeba* and enteropathogenic bacteria modulates epithelial cell damage. *PLoS Negl. Trop. Dis.*, 2: e266, 2008.
12. Soh, C.T., Factors influencing the pathogenicity of *Entamoeba histolytica. Yonsei. Med. J.*, 29: 1–10, 1988.
13. Martinez-Palomo, A., The pathogenesis of amoebiasis. *Parasitol. Today*, 3: 111–8, 1987.
14. Diamond, L.S., Axenic cultivation of *Entamoeba hitolytica. Science*, 134: 336–7, 1961.
15. Diamond, L.S., Harlow, D.R., and Cunnick, C.C., A new medium for the axenic cultivation of *Entamoeba histolytica* and other *Entamoeba. Trans. R. Soc. Trop. Med. Hyg.*, 72: 431–2, 1978.
16. De la Torre, M., et al., Cultivos axénicos de *Entamoeba histolytica. Arch. Invest. Med.*, 2: 165–172, 1971.
17. Mann, B.J., Mirelman, D., and Petri, W.A., Jr., The D-galactose-inhibitable lectin of *Entamoeba histolytica. Carbohydr. Res.*, 213: 331–8, 1991.
18. Lopez-Soto, F., et al., Use and endocytosis of iron-containing proteins by *Entamoeba histolytica* trophozoites. *Infect. Genet. Evol.*, 9: 1038–50, 2009.
19. Christy, N.C. and Petri, W.A., Jr., Mechanisms of adherence, cytotoxicity and phagocytosis modulate the pathogenesis of *Entamoeba histolytica. Future Microbiol.*, 6: 1501–19, 2011.
20. Ralston, K.S. and Petri, W.A., Jr., Tissue destruction and invasion by *Entamoeba histolytica. Trends Parasitol.*, 27: 254–63, 2011.
21. Pina-Vazquez, C., et al., Host-parasite interaction: parasite-derived and -induced proteases that degrade human extracellular matrix. *J. Parasitol. Res.*, 2012: 748206, 2012.
22. Faust, D.M. and Guillen, N., Virulence and virulence factors in *Entamoeba histolytica*, the agent of human amoebiasis. *Microbes Infect.*, 14: 1428–41, 2012.
23. Tsutsumi, V., et al., Cellular bases of experimental amebic liver abscess formation. *Am. J. Pathol.*, 117: 81–91, 1984.
24. Tsutsumi, V., Anaya-Velazquez, F., and Martinez-Palomo, A., Experimental intestinal amebiasis: invasion and extension of the amebic lesion. *Arch. Invest. Med. (Mex)*, 21 Suppl 1: 47–52, 1990.
25. Mortimer, L., et al., Gal-lectin-dependent contact activates the inflammasome by invasive *Entamoeba histolytica. Mucosal Immunol.*, 7: 829–41, 2014.
26. Leippe, M., Amoebapores. *Parasitol. Today*, 13: 178–83, 1997.
27. Bracha, R., Nuchamowitz, Y., and Mirelman, D., Transcriptional silencing of an amoebapore gene in *Entamoeba histolytica*: molecular analysis and effect on pathogenicity. *Eukaryot. Cell*, 2: 295–305, 2003.

28. Leippe, M., et al., Ancient weapons: the three-dimensional structure of amoebapore A. *Trends Parasitol.*, 21: 5–7, 2005.
29. Andra, J., Herbst, R., and Leippe, M., Amoebapores, archaic effector peptides of protozoan origin, are discharged into phagosomes and kill bacteria by permeabilizing their membranes. *Dev. Comp. Immunol.*, 27: 291–304, 2003.
30. Que, X. and Reed, S.L., Cysteine proteinases and the pathogenesis of amebiasis. *Clin. Microbiol. Rev.*, 13: 196–206, 2000.
31. Stanley, S.L., Pathophysiology of amoebiasis. *Trends Parasitol.*, 17: 280–5, 2001.
32. Bruchhaus, I., et al., The intestinal protozoan parasite *Entamoeba histolytica* contains 20 cysteine protease genes, of which only a small subset is expressed during *in vitro* cultivation. *Eukaryot. Cell*, 2: 501–9, 2003.
33. Serrano-Luna, J., et al., Proteases from *Entamoeba* spp. and pathogenic free-living amoebae as virulence factors. *J. Trop. Med.*, 2013: 890603, 2013.
34. Loftus, B., et al., The genome of the protist parasite *Entamoeba histolytica*. *Nature*, 433: 865–8, 2005.
35. Loftus, B.J. and Hall, N., *Entamoeba*: still more to be learned from the genome. *Trends Parasitol.*, 21: 453, 2005.
36. Wong-Baeza, I., et al., The role of lipopeptidophosphoglycan in the immune response to *Entamoeba histolytica*. *J. Biomed. Biotechnol.*, 2010: 254521, 2010.
37. Padilla-Vaca, F. and Anaya-Velazquez, F., Insights into *Entamoeba histolytica* virulence modulation. *Infect. Disord. Drug Targets*, 10: 242–50, 2010.
38. Perez-Tamayo, R., Martinez-Villegas, J.E., and Perez-Montfort, R., Effect of cellular immunity on the interaction between the peritoneal cell and the ameba *in vitro*. *Arch. Invest. Med. (Mex)*, 17 Suppl 1: 259–67, 1986.
39. Tsutsumi, V. and Martinez-Palomo, A., Inflammatory reaction in experimental hepatic amebiasis. An ultrastructural study. *Am. J. Pathol.*, 130: 112–9, 1988.
40. Chadee, K. and Meerovitch, E., *Entamoeba histolytica*: diffuse liver inflammation in gerbils (*Meriones unguiculatus*) with experimentally induced amebic liver abscess. *J. Protozool.*, 36: 154–8, 1989.
41. Tsutsumi, V. and Shibayama, M., Experimental amebiasis: a selected review of some *in vivo* models. *Arch. Med. Res.*, 37: 210–20, 2006.
42. Lesh, F.A., Massive development of amebas in the large intestine. Fedor Aleksandrovich Lesh (Losch). *Am. J. Trop. Med. Hyg.*, 24: 383–92, 1975.
43. Ivory, C., Kammanadiminti, S., and Chadee, K., Innate resistance to *Entamoeba histolytica* in murine models. *Trends Parasitol.*, 23: 46–8, 2007.
44. Navarro-Garcia, F., et al., Model of intestinal amebiasis: structural and functional lesions to the rabbit colon mucosa by *Entamoeba histolytica* lysates. *Arch. Med. Res.*, 23: 197–201, 1992.
45. Ghosh, P.K., Castellanos-Barba, C., and Ortiz-Ortiz, L., Intestinal amebiasis: cyclic suppression of the immune response. *Parasitol. Res.*, 81: 475–80, 1995.
46. Ghosh, P.K., et al., Experimental amebiasis: immunohistochemical study of immune cell populations. *J. Eukaryot. Microbiol.*, 47: 395–9, 2000.
47. Leitch, G.J., et al., *Entamoeba histolytica* trophozoites in the lumen and mucus blanket of rat colons studied *in vivo*. *Infect. Immun.*, 47: 68–73, 1985.
48. Neal, R.A. and Harris, W.G., Proceedings: attempts to infect inbred strains of rats and mice with *Entamoeba histolytica*. *Trans. R. Soc. Trop. Med. Hyg.*, 69: 429–30, 1975.
49. Diamond, L.S., Tanimoto Weki, M., and Martinez-Palomo, A., Production of cecal lesions in newborn guinea pigs with axenically cultivated *Entamoeba histolytica*. *Arch. Invest. Med. (Mex)*, 9 Suppl 1: 223–8, 1978.
50. Anaya-Velazquez, F., et al., Intestinal invasive amebiasis: an experimental model in rodents using axenic or monoxenic strains of *Entamoeba histolytica*. *Am. J. Trop. Med. Hyg.*, 34: 723–30, 1985.
51. Shibayama-Salas, M., Tsutsumi, V., and Martinez-Palomo, A., Early invasive intestinal amebiasis in Mongolian gerbils. *Arch. Med. Res.*, 23: 187–90, 1992.
52. Girard-Misguich, F., et al., Towards the establishment of a porcine model to study human amebiasis. *PLoS One*, 6: e28795, 2011.
53. Girard-Misguich, F., et al., Porcine colon explants in the study of innate immune response to *Entamoeba histolytica*. *Vet. Immunol. Immunopathol.*, 145: 611–7, 2012.
54. Anaya-Velazquez, F. and Underdown, B.J., Early expression in mice of genetic resistance to intestinal amebiasis. *Arch. Invest. Med. (Mex)*, 21 Suppl 1: 53–6, 1990.

55. Rivero-Nava, L., et al., *Entamoeba histolytica*: acute granulomatous intestinal lesions in normal and neutrophil-depleted mice. *Exp. Parasitol.*, 101: 183–92, 2002.
56. Ghadirian, E. and Kongshavn, P.A., Genetic control of susceptibility of mice to infection with *E. histolytica. Parasite Immunol.*, 6: 349–60, 1984.
57. Houpt, E.R., et al., The mouse model of amebic colitis reveals mouse strain susceptibility to infection and exacerbation of disease by CD4+ T cells. *J. Immunol.*, 169: 4496–503, 2002.
58. Hamano, S., et al., Resistance of C57BL/6 mice to amoebiasis is mediated by nonhemopoietic cells but requires hemopoietic IL-10 production. *J. Immunol.*, 177: 1208–13, 2006.
59. Seydel, K.B., et al., Human intestinal epithelial cells produce proinflammatory cytokines in response to infection in a SCID mouse-human intestinal xenograft model of amebiasis. *Infect. Immun.*, 65: 1631–9, 1997.
60. Seydel, K.B., et al., Epithelial cell-initiated inflammation plays a crucial role in early tissue damage in amebic infection of human intestine. *Gastroenterology*, 115: 1446–53, 1998.
61. Zhang, Z., et al., *Entamoeba histolytica* cysteine proteinases with interleukin-1 β converting enzyme (ICE) activity cause intestinal inflammation and tissue damage in amoebiasis. *Mol. Microbiol.*, 37: 542–8, 2000.
62. Bansal, D., et al., An *ex-vivo* human intestinal model to study *Entamoeba histolytica* pathogenesis. *PLoS Negl. Trop. Dis.*, 3: e551, 2009.
63. Thibeaux, R., et al., Identification of the virulence landscape essential for *Entamoeba histolytica* invasion of the human colon. *PLoS Pathog.*, 9: e1003824, 2013.
64. Thibeaux, R., et al., The parasite *Entamoeba histolytica* exploits the activities of human matrix metalloproteinases to invade colonic tissue. *Nat. Commun.*, 5: 5142, 2014.
65. Reinertson, J.W. and Thompson, P.E., Experimental amebic hepatitis in hamsters. *Proc. Soc. Exp. Biol. Med.*, 76: 518–21, 1951.
66. Tanimoto, M., Sepúlveda, B., and Vázquez, J.A., Lesiones producidas en el hígado de hámsteres por inoculación de *E. histolytica* cultivada en medio axénico. *Arch. Invest. Med. (Mex)*, 2: 275–84, 1971.
67. Chadee, K. and Meerovitch, E., The pathogenesis of experimentally induced amebic liver abscess in the gerbil (*Meriones unguiculatus*). *Am. J. Pathol.*, 117: 71–80, 1984.
68. Zhang, T. and Stanley, S.L., Jr., Protection of gerbils from amebic liver abscess by immunization with a recombinant protein derived from the 170-kilodalton surface adhesin of *Entamoeba histolytica. Infect. Immun.*, 62: 2605–8, 1994.
69. Cieslak, P.R., Virgin, H.W.T., and Stanley, S.L., Jr., A severe combined immunodeficient (SCID) mouse model for infection with *Entamoeba histolytica. J. Exp. Med.*, 176: 1605–9, 1992.
70. Stern, J.J., Graybill, J.R., and Drutz, D.J., Murine amebiasis: the role of the macrophage in host defense. *Am. J. Trop. Med. Hyg.*, 33: 372–80, 1984.
71. Wijesundera Mde, S., Hepatic amoebiasis in immunodepressed mice. *Trans. R. Soc. Trop. Med. Hyg.*, 74: 216–20, 1980.
72. Velazquez, C., et al., Role of neutrophils in innate resistance to *Entamoeba histolytica* liver infection in mice. *Parasite Immunol.*, 20: 255–62, 1998.
73. Denis, M. and Chadee, K., Cytokine activation of murine macrophages for *in vitro* killing of *Entamoeba histolytica* trophozoites. *Infect. Immun.*, 57: 1750–6, 1989.
74. Pacheco-Yepez, J., et al., Expression of cytokines and their regulation during amoebic liver abscess development. *Parasite Immunol.*, 33: 56–64, 2011.
75. Shibayama, M., et al., A Brazilian species of *Entamoeba dispar* (ADO) produces amoebic liver abscess in hamsters. *Ann. Hepatol.*, 6: 117–8, 2007.
76. Dolabella, S.S., et al., Amoebic liver abscess production by *Entamoeba dispar. Ann. Hepatol.*, 11: 107–17, 2012.
77. Jones, W.R., The experimental infection of rats with *Entamoeba histolytica*; with a method for evaluating the anti-amoebic properties of new compounds. *Ann. Trop. Med. Parasitol.*, 40: 130–40, 1946.
78. Hernandez-Lopez, H.R. and Escobedo-Salinas, A., Effects of metronidazole on amebic hepatic abscess in hamsters. *Arch. Invest. Med. (Mex)*, 1: Suppl:125–8, 1970.
79. Kradolfer, F. and Jarumilinta, R., Ciba 32,644-Ba, a new systemically active amoebicide. *Ann. Trop. Med. Parasitol.*, 59: 210–8, 1965.
80. Cuckler, A.C., Malanga, C.M., and Conroy, J., Therapeutic efficacy of new nitroimidazoles for experimental trichomoniasis, amebiasis, and trypanosomiasis. *Am. J. Trop. Med. Hyg.*, 19: 916–25, 1970.

81. Ray, D.K., et al., Comparative studies on the amoebicidal activity of known 5-nitroimidazole derivatives and CG 10213-Go in golden hamsters, *Mesocricetus auratus*, infected in the liver or caecum or both with trophozoites of *Entamoeba histolytica*. *Ann. Trop. Med. Parasitol.*, 77: 287–91, 1983.
82. Pargal, A., et al., Pharmacokinetics and amoebicidal activity of (±)-(*E*)-3-(4-methylsulphinylstyryl)-1,2,4-oxadiazole (BTI 2286^{E}) and its sulphone metabolite (BTI 2571^{E}) in the golden hamster, *Mesocricetus auratus*. *J. Antimicrob. Chemother.*, 32: 109–15, 1993.
83. Pargal, A., et al., Comparative pharmacokinetics and amoebicidal activity of metronidazole and satranidazole in the golden hamster, *Mesocricetus auratus*. *J. Antimicrob. Chemother.*, 32: 483–9, 1993.
84. Kradolfer, F., Jarumilinta, R., and Sackmann, W., The amoebicidal, trichomonicidal, and antibacterial effects of niridazole in laboratory animals. *Ann. N. Y. Acad. Sci.*, 160: 740–8, 1969.
85. Roe, F.J., Toxicologic evaluation of metronidazole with particular reference to carcinogenic, mutagenic, and teratogenic potential. *Surgery*, 93: 158–64, 1983.
86. Dobias, L., et al., Genotoxicity and carcinogenicity of metronidazole. *Mutat. Res.*, 317: 177–94, 1994.
87. Kitchen, L.W., Case studies in international travelers. *Am. Fam. Physician*, 60: 471–4, 1999.
88. Showler, A.J. and Boggild, A.K., *Entamoeba histolytica*. *Can. Med. Assoc. J.*, 185: 1064, 2013.
89. Upcroft, P. and Upcroft, J.A., Drug targets and mechanisms of resistance in the anaerobic protozoa. *Clin. Microbiol. Rev.*, 14: 150–64, 2001.
90. Sharma, G.L. and Bhutani, K.K., Plant based antiamoebic drugs; Part II. Amoebicidal activity of parthenin isolated from *Parthenium hysterophorus*. *Planta Med.*, 54: 120–2, 1988.
91. Sohni, Y.R., Kaimal, P., and Bhatt, R.M., The antiamoebic effect of a crude drug formulation of herbal extracts against *Entamoeba histolytica in vitro* and *in vivo*. *J. Ethnopharmacol.*, 45: 43–52, 1995.
92. Sawangjaroen, N., Sawangjaroen, K., and Poonpanang, P., Effects of *Piper longum* fruit, *Piper sarmentosum* root and *Quercus infectoria* nut gall on caecal amoebiasis in mice. *J. Ethnopharmacol.*, 91: 357–60, 2004.
93. Avila-Blanco, M.E., et al., Amoebicidal activity of essential oil of *Dysphania ambrosioides* (L.) Mosyakin & Clemants in an amoebic liver abscess hamster model. *Evid. Based Complement. Alternat. Med.*, 2014: 930208, 2014.
94. Ghosh, S., et al., Effects of bisphosphonates on the growth of *Entamoeba histolytica* and *Plasmodium* species *in vitro* and *in vivo*. *J. Med. Chem.*, 47: 175–87, 2004.
95. Ordaz-Pichardo, C., et al., Antiamoebic and toxicity studies of a carbamic acid derivative and its therapeutic effect in a hamster model of hepatic amoebiasis. *Antimicrob. Agents Chemother.*, 49: 1160–8, 2005.
96. Chatterjee, D.K., Iyer, N., and Ganguli, B.N., Antiamoebic activity of chonemorphine, a steroidal alkaloid, in experimental models. *Parasitol. Res.*, 74: 30–3, 1987.
97. Bhopale, K.K., et al., A comparative study of experimental caecal amoebiasis and the evaluation of amoebicides. *Ann. Trop. Med. Parasitol.*, 89: 253–9, 1995.
98. Bhopale, K.K., et al., Additive effect of diloxanide furoate and metronidazole (Entamizole) in experimental mouse caecal amoebiasis. *Indian. J. Exp. Biol.*, 33: 73–4, 1995.
99. Sato, D., et al., Cytotoxic effect of amide derivatives of trifluoromethionine against the enteric protozoan parasite *Entamoeba histolytica*. *Int. J. Antimicrob. Agents*, 35: 56–61, 2010.
100. Debnath, A., et al., A high-throughput drug screen for *Entamoeba histolytica* identifies a new lead and target. *Nat. Med.*, 18: 956–60, 2012.
101. Ordaz-Pichardo, C., et al., Effect of bovine lactoferrin in a therapeutic hamster model of hepatic amoebiasis. *Biochem. Cell. Biol.*, 90: 425–34, 2012.
102. Leon-Sicairos, N., et al., Oral lactoferrin treatment resolves amoebic intracecal infection in C3H/HeJ mice. *Biochem. Cell. Biol.*, 90: 435–41, 2012.
103. Stanley, S.L., Jr., Vaccines for amoebiasis: barriers and opportunities. *Parasitology*, 133 Suppl 2: S81–6, 2006.
104. Quach, J., St-Pierre, J., and Chadee, K., The future for vaccine development against *Entamoeba histolytica*. *Hum. Vaccines Immunother.*, 10: 1514–21, 2014.
105. Lamm, M.E., Interaction of antigens and antibodies at mucosal surfaces. *Annu. Rev. Microbiol.*, 51: 311–40, 1997.
106. Abd-Alla, M.D., et al., Mucosal immunity to asymptomatic *Entamoeba histolytica* and *Entamoeba dispar* infection is associated with a peak intestinal anti-lectin immunoglobulin A antibody response. *Infect. Immun.*, 74: 3897–903, 2006.

107. Haque, R., et al., Innate and acquired resistance to amebiasis in Bangladeshi children. *J. Infect. Dis.*, 186: 547–52, 2002.
108. Sanchez-Guillen Mdel, C., et al., Differentiation of *Entamoeba histolytica/Entamoeba dispar* by PCR and their correlation with humoral and cellular immunity in individuals with clinical variants of amoebiasis. *Am. J. Trop. Med. Hyg.*, 66: 731–7, 2002.
109. Haque, R., et al., Amebiasis and mucosal IgA antibody against the *Entamoeba histolytica* adherence lectin in Bangladeshi children. *J. Infect. Dis.*, 183: 1787–93, 2001.
110. Ravdin, J.I., et al., Intestinal antilectin immunoglobulin A antibody response and immunity to *Entamoeba dispar* infection following cure of amebic liver abscess. *Infect. Immun.*, 71: 6899–905, 2003.
111. Blessmann, J., et al., Epidemiology of amebiasis in a region of high incidence of amebic liver abscess in central Vietnam. *Am. J. Trop. Med. Hyg.*, 66: 578–83, 2002.
112. Jackson, T.F., Anderson, C.B., and Simjee, A.E., Serological differentiation between past and present infection in hepatic amoebiasis. *Trans. R. Soc. Trop. Med. Hyg.*, 78: 342–5, 1984.
113. Jain, P., Sawhney, S., and Vinayak, V.K., Experimental amoebic infection in guinea-pigs immunized with low grade amoebic infection. *Trans. R. Soc. Trop. Med. Hyg.*, 74: 347–50, 1980.
114. Bujanover, S., et al., A virulence attenuated amoebapore-less mutant of *Entamoeba histolytica* and its interaction with host cells. *Int. J. Parasitol.*, 33: 1655–63, 2003.
115. Petri, W.A., Jr., et al., Subunit structure of the galactose and *N*-acetyl-D-galactosamine-inhibitable adherence lectin of *Entamoeba histolytica*. *J. Biol. Chem.*, 264: 3007–12, 1989.
116. Petri, W.A., Jr. and Ravdin, J.I., Protection of gerbils from amebic liver abscess by immunization with the galactose-specific adherence lectin of *Entamoeba histolytica*. *Infect. Immun.*, 59: 97–101, 1991.
117. Bruchhaus, I., Richter, S., and Tannich, E., Removal of hydrogen peroxide by the 29 kDa protein of *Entamoeba histolytica*. *Biochem. J.*, 326 (Pt 3): 785–9, 1997.
118. Soong, C.J., et al., Protection of gerbils from amebic liver abscess by immunization with recombinant *Entamoeba histolytica* 29-kilodalton antigen. *Infect. Immun.*, 63: 472–7, 1995.
119. Carrero, J.C., et al., Protection against murine intestinal amoebiasis induced by oral immunization with the 29 kDa antigen of *Entamoeba histolytica* and cholera toxin. *Exp. Parasitol.*, 126: 359–65, 2010.
120. Gaucher, D. and Chadee, K., Construction and immunogenicity of a codon-optimized *Entamoeba histolytica* Gal-lectin-based DNA vaccine. *Vaccine*, 20: 3244–53, 2002.
121. Madriz, X., et al., Expression in fibroblasts and in live animals of *Entamoeba histolytica* polypeptides EhCP112 and EhADH112. *Microbiology*, 150: 1251–60, 2004.
122. Kaur, U., et al., Immunogenicity and protective efficacy of heparan sulphate binding proteins of *Entamoeba histolytica* in a guinea pig model of intestinal amoebiasis. *Exp. Parasitol.*, 135: 486–96, 2013.
123. Gonzalez-Vazquez, M.C., et al., Obtaining of three recombinant antigens of *Entamoeba histolytica* and evaluation of their immunogenic ability without adjuvant in a hamster model of immunoprotection. *Acta. Trop.*, 122: 169–76, 2012.
124. Abd Alla, M.D., et al., Efficacy of a Gal-lectin subunit vaccine against experimental *Entamoeba histolytica* infection and colitis in baboons (*Papio* sp.). *Vaccine*, 30: 3068–75, 2012.
125. Ghosh, P.K., Mancilla, R., and Ortiz-Ortiz, L., Intestinal amebiasis: histopathologic features in experimentally infected mice. *Arch. Med. Res.*, 25: 297–302, 1994.

40

Giardia lamblia

Steven M. Singer, Jenny G. Maloney, and Camila H. Coelho

CONTENTS

40.1 Introduction

The protozoan parasite *Giardia lamblia* (Protista, Diplomonadida, Hexamitidae, syn. *G. intestinalis*, *G. lamblia*) is the etiologic agent of giardiasis, a common diarrheal disease worldwide. *G. lamblia* can infect humans and many other mammals, with prevalence rates in humans ranging from 2% to 7% in developed countries to 20%–30% in developing countries and 100% prevalence reported in some populations.[1,2] Human outbreaks of *G. lamblia* are most often associated with fecal contamination of drinking water. However, several foodborne outbreaks have been documented.[3] Until recently, giardiasis was not the subject of significant scientific investigation, but after its inclusion in the Neglected Disease Initiative by WHO in 2004 and its reemergence in developed countries, the landscape has changed significantly.[4–6] In a bibliometric review covering all studies on giardiasis published between 1971 and 2010, Escobedo et al.[7] reported that especially in the last decade, there has been an increased number of publications and clinical research studies involving giardiasis. Comparative studies, clinical trials, and pharmacotherapy assessment were the main identified research areas. This chapter will discuss several aspects of the biology of *Giardia* and the pathology associated with infection while highlighting the laboratory models currently used to study this parasite.

40.2 History of *G. lamblia*

Historically, *G. lamblia* was so common that it was once considered a commensal organism of the human intestine. *G. lamblia* was first discovered in 1681 by Antony van Leeuwenhoek during an examination of his own stool.[5] Almost 200 years later in 1859, the protozoan was given the name *Cercomonas intestinalis* by Lambl and then later renamed *Giardia lamblia* by Stiles in 1915 in honor of Professor A. Giard of Paris and Dr F. Lambl of Prague, who both described the parasite.[8] The first association between *G. lamblia* and the disease was found in 1954, when it was demonstrated that the ingestion of *G. lamblia* cysts by humans could cause diarrheal disease even after ingestion of as few as 10 cysts.[9] It was not until 1981, 300 years after its first description, that *G. lamblia* was added to the WHO's list of parasitic pathogens.[8] Soon after its addition to the WHO's list of parasitic pathogens, Koch's postulates were fulfilled, confirming *G. lamblia* as a pathogen of humans.[10]

40.3 Life Cycle of *G. lamblia* and Diagnosis

The life cycle of *G. lamblia* includes an infectious cyst form and a noninfectious, rapidly multiplying trophozoite form. The cyst form is extremely hardy in the environment and can be transmitted to humans by contaminated food, water, hands, or fomites via the fecal–oral route.[11] Once ingested, exposure to stomach acid and digestive enzymes triggers excystation, and the noninfective trophozoite form takes residence on the mucosal surface of the small intestine where it multiplies rapidly. As the trophozoites are passed through the intestines, they transform back into the infectious and environmentally hardy cyst form and are shed with the feces (Figure 40.1). In the environment, cysts can survive in harsh conditions and may persist for several months in cold water.[11] Cysts can survive water temperatures of <10°C for 2–3 months, and some cysts can survive a single freeze–thaw cycle.[11] As water temperature increases, survival rates decrease; at 21°C, cysts can survive in water for about 1 month. At a water temperature of 54°C, cysts tend to die within 1 min, whereas cysts tend to die immediately in boiling water.[11] Infected individuals can shed up to 10 billion cysts a day, and ingestion of as few as 10 cysts can yield infection.[9] Because of the low infectious dose of *G. lamblia* as well as the environmentally hardy nature of the infectious cyst, *G. lamblia* is included on the National Institute of Allergy and Infectious Diseases' list of category B pathogens.

Diagnosis of infection in both humans and animals is predominately based on microscopic identification of *Giardia* cysts in fecal samples, although in recent years, ELISA- and PCR-based techniques have been developed.[12,13] The course of disease following *G. lamblia* infection can be highly variable and ranges in severity from asymptomatic infections to infections that result in severe diarrhea and painful abdominal cramping.[1,2] These clinical outcomes can have a disproportionate effect on children. Infection in children has been associated with stunting and impaired development of cognitive function in populations where giardiasis is prevalent.[14,15] Infections are treatable, but current treatment for *G. lamblia* involves administration of toxic drugs to which parasite resistance is reported.[16]

40.4 Morphology of *G. lamblia*

The two forms of the parasite, the trophozoite and the cyst, are highly specialized for survival within the host and within the environment, respectively. Therefore, by studying the morphology of *G. lamblia,* we can better understand both infection and transmission of the parasite. Most of what we know about *G. lamblia* morphology is drawn from studies utilizing electron microscopy to visualize the surface, cytoskeletal, and intracellular components of the parasite. More gross features of the parasite can easily be visualized using light microscopy. These studies rely on *in vitro* culture techniques, which allow for the production of both trophozoites and cysts outside of a host. The trophozoites of *G. lamblia* can be grown in an axenic medium originally created for cultivation of *Entamoeba histolytica*.[17] Axenization of new strains can be somewhat problematic. Protocols for excystation of cysts *in vitro* (Sections 40.6 and 40.7) can be used to initiate cultures, although a more common procedure is to collect trophozoites from the small intestines of infected animals.[18] Addition of antibiotics to culture media then helps propagation of newly axenized strains.[19] Cyst production is more complicated, but several protocols exist (Table 40.1).

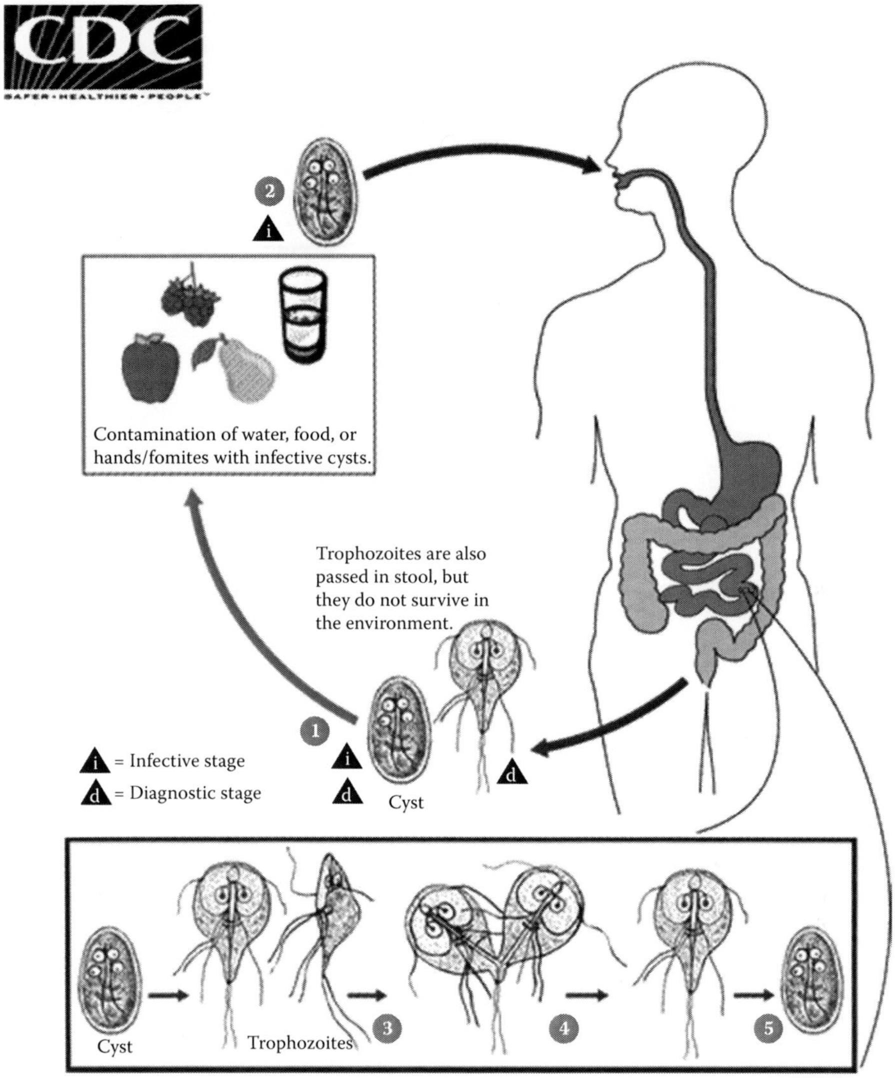

FIGURE 40.1 Life cycle of *Giardia*. Transition from noninfectious, rapidly multiplying trophozoites to infectious, environmentally hardy cysts. (Courtesy of US CDC.)

TABLE 40.1

Methods Used for Encystation of *Giardia lamblia In Vitro*

Pre-encystation Medium	Encystation Medium	Timing	References
TYI-S33 without bile	TYI-S33 + 10 mM lactic acid, pH 7.8 and 10 mg/mL porcine bile	72 h pre-encystation 48 h encystation	52,54
None	TYI-S33 + 12.5 mg/mL bovine bile, no lactic acid	72 h encystation	55
None	TYI-S33 with delipidated serum	72 h encystation	57
None	TYI-S33, pH 7.8 + 10 mg/mL bovine bile	96 h encystation	165

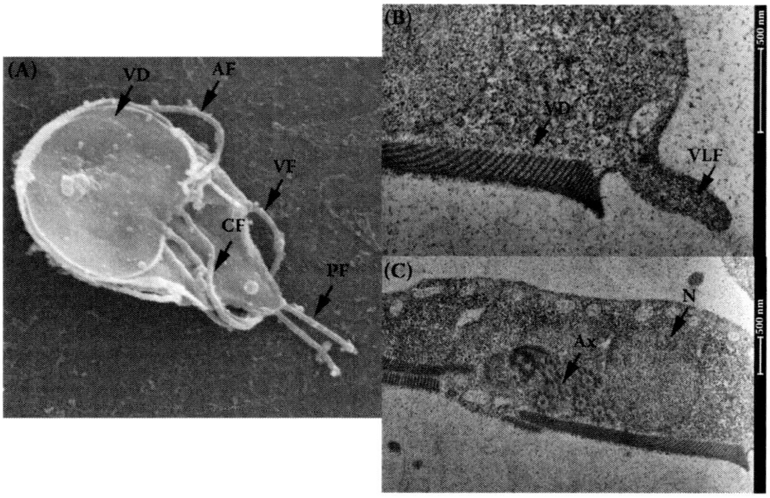

FIGURE 40.2 Electron micrographs of *Giardia lamblia* trophozoites. (A) Scanning electron micrograph of the ventral surface showing the ventral disc (VD), anterior (AF), ventral (VF), caudal (CF), and posterior (PF) flagellar pairs. (B and C) Transmission electron micrographs showing the ventrolateral flange (VLF), ventral disc (VD), flagellar axonemes (Ax), and nuclei (N).

Trophozoites of *G. lamblia* have a distinctive pear shape and exhibit bilateral symmetry. Each cell is 12–15 µm in length and 5–9 µm in width with a flattened ventral surface, a rounded dorsal surface, and four pairs of flagella: anterior, posterior, caudal, and ventral (Figure 40.2). A major component of the *G. lamblia* cytoskeleton is the ventral disc. It is a concave structure that covers the ventral surface of the trophozoite and is thought to play a major role in attachment of the parasite to the intestinal mucosa. Another prominent feature of the trophozoite cytoskeleton is the median body, which serves as the microtubule-organizing center of the cell. Differences in appearance of the median body contribute to distinction among the species of *Giardia*. Other morphological features of the trophozoite include the lateral crest and ventrolateral flange, which are structures with unknown function, although believed to be participating in adhesion.[20–22]

Like other eukaryotic cells, *Giardia* also contain numerous cytoplasmic organelles. In the trophozoites, mitosomes, peripheral vesicles, and ribosome granules can be visualized.[23] Interestingly, while a Golgi complex is easily visualized during the process of encystation, vegetatively growing trophozoites contain only vesicular structures suggestive of a Golgi complex.[24–26] Unlike most other eukaryotic cells, trophozoites have two nuclei that are localized anteriorly and are symmetric across the longitudinal axis of the cell. During encystation, the nuclei replicate to produce mature cysts with four nuclei. Following excystation, each cyst gives rise to two trophozoites.

The transformation of trophozoites of *G. lamblia* into cysts occurs in response to physiological stimuli, leading to a process characterized by decreased cell metabolic rate, internalization of flagella, cytoplasmic condensation, peripheral vesicle formation, and production of a cystic membrane.[22,27] Encystation is fundamental for survival, transmission, and pathogenesis of *G. lamblia*. It is considered an important virulence factor because it allows *G. lamblia* to survive outside the host and to be transmitted.[23,28]

Cysts are ellipsoid or oval and have a size of 7–12 µm. They have a refractive cyst wall, which is 0.3–0.5 µm thick and is composed of an outer layer and an inner filamentous membranous layer. Median bodies and structural elements of flagella, the axonemes, and four nuclei in mature and infectious cysts are found in the cytoplasm.[29,30] The cyst wall is composed mainly of *N*-acetylgalactosamine (GalNAc) homopolymers.[31] The precursor for the GalNAc homopolymer is UDP-GalNAc, an amino sugar phosphate, which is produced from glucose. Enzymes required for the synthesis of UDP-GalNAc include glucosamine-6-phosphate isomerase, phosphoacetyl-glucosamine mutase, UDP-GlcNAc 49-epimerase, and glucosamine-6-phosphate *N*-acetylase.[32] During encystation, transcript levels and enzyme activity for some of these enzymes increase significantly.[33] The external portion of the cyst wall is covered by a network of filaments, 7–20 nm in diameter, composed primarily of GalNAc homopolymer, along with cyst wall proteins. Considering that this is the stable life cycle stage, the cysts have a metabolic rate of only 10%–20% relative to trophozoites.[34] Some cysts are resistant to chlorination and radiation. Boiling is the most efficient way to eliminate cysts; freezing requires several days to reduce cyst viability.[35–37]

40.5 Epidemiology

Within the genus *Giardia*, there are currently six recognized species:[38] *G. agilis* (a parasite of amphibians), *G. ardeae* and *G. psittaci* (both of which are found in birds), *G. microti* (found in voles and muskrats), *G. muris* (found in rodents), and *G. lamblia* (which infects a wide range of mammalian species including humans). The aforementioned taxonomic divisions are based on morphological differences that have been historically used to define the six *Giardia* species.[38] More recently, *G. lamblia*, the only species known to infect humans, has been divided into assemblages A–H based on molecular evidence.[5] Assemblages A and B are known to infect a wide range of mammals including humans, but assemblages C–G appear to have some degree of host specificity.[39] Assemblages C and D are most commonly found in dogs, assemblage E in hoofed animals, assemblage F in cats, and assemblage G in rats.[40] The most recent addition, assemblage H, was identified in marine animals.[39] The classification of these assemblages is supported by studies using both enzyme electrophoretic data and phylogenetic analyses of molecular sequence data.[5]

There are striking genetic differences between the two assemblages, assemblage A and assemblage B, associated with human infections. Whole genome sequencing of two strains of *G. lamblia* isolated from humans, WB (assemblage A strain) and GS (assemblage B strain), revealed that these two assemblages share only 77% nucleotide identity and 78% amino acid identity in protein-coding regions.[41] Because these two assemblages are so genetically different, it is tempting to speculate that they should contribute to different geographical distributions or clinical outcomes in humans. However, data to support these geographical and clinical differences are often conflicting. Reports from Asia, Europe, Africa, and the Americas all demonstrate that each assemblage is found in pockets throughout these areas.[42] Symptomatic differences attributable to assemblage follow a similar trend. While assemblage A is found to account for more symptomatic infection in some studies, assemblage B is found to have a greater association with symptoms in others, with some studies reporting no association between patient symptoms and assemblage causing the infection.[42] In a laboratory infection model using adult mice, it was demonstrated that while the GS (assemblage B) strain caused measurable disaccharidase deficiency, infections with the WB (assemblage A) strain did not.[43] Similarly, another assemblage B strain, H3, caused chronic infections that contributed to enhanced weight loss in mice on a protein-deficient diet, while the assemblage A strain, WB, did not.[44] The use of antibiotics to allow the WB strain to infect mice in the former paper[43] and the use of H3 cysts rather than WB trophozoites to initiate infections in the latter paper[44] require caution in interpreting these results. Nevertheless, they suggest that mouse models may be useful for identifying strain- and assemblage-specific differences in virulence. Better cohort studies in humans could also help to distinguish the potential differences in pathogenicity of these two assemblages.

Because *G. lamblia* infects such a diverse range of mammals, it is conceivable that zoonotic transmission could occur. However, the zoonotic risk of *G. lamblia* infection is unclear. Some studies have found an increased risk for human infection in households that harbor domestic pets.[45,46] Molecular evidence has linked human infections to assemblages found only in cattle, cats, and dogs, indicating that cross-species transmission can occur.[47–50] A large-scale European study found that only 1% of human *G. lamblia* infection could be attributed to assemblages C, D, E, and F, indicating that the zoonotic potential of *G. lamblia* is low for these assemblages.[51] Interestingly, this same study demonstrated that while assemblage B infections are mostly restricted to humans, assemblage A appears quite frequently in companion animals, wildlife, and livestock. So, while humans may serve as the main source of assemblage B, assemblage A could have a greater zoonotic potential.

40.6 Encystation

Encystation is the process of differentiation of the trophozoite into the resistant cyst, which occurs due to external stimuli. During encystation, the main transformations that occur are internalization of flagella creating an oval shape, formation of encystation-specific vesicles (ESVs), the appearance of a refractory cyst wall, and replication of the nuclei. The cell also has diminished adherence capacity owing to fragmentation of the ventral disc.[22] Much of the work on *G. lamblia* encystation has been done by *in vitro* manipulation of axenically cultured trophozoites. By altering the culture conditions of trophozoites, it is possible to mimic some of the signals a cell might encounter in the host small intestine that are necessary to drive encystation (Table 40.1). *In vitro* studies have shown that encystation begins when important modifications of the environment occur such as shifts in pH, concentrations of bile salts, availability of lipids, and production of lactic acid by intestinal bacteria.[52] In a suckling mouse model of *G. lamblia* infection, cysts were found in the mid to lower portion of the jejunum, while later the majority of cysts were in the large intestine and cecum, suggesting that bile concentration may be an important component of the encystation process.[53] It is thought that bile stimulates encystation by sequestering cholesterol. Common methods used to achieve encystation are based on parasite growth in conventional culture media without bile, and after 2–3 days, the parasites are switched to a culture media containing a high concentration of porcine bile with lactic acid and a pH of 7.8.[52,54,55] Using these methods, trophozoites can be induced to transform to viable cysts. However, the ability of *in vitro*-derived cysts to excyst, either in animals or in culture, is often poor.

G. lamblia has little capacity for synthesizing lipid molecules *de novo* and instead depends on environmental sources. It has been hypothesized that most of the lipid uptake by the cells is then used for energy production and biosynthesis of organelles.[56] Because lipid metabolism is central to the process of encystation, these pathways are of interest as a potential drug target.[52,57] Recently, attention has been focused on glycosylceramide pathways in encystation.[58] Conversely, specific lipid products from the host have also been shown to kill trophozoites.[59,60]

The role of ESVs in the encystation process and their relationship with the normal functions of the Golgi complex have been the subject of significant interest. Stefanic et al.[61] used two-dimensional electrophoresis to analyze Golgi-like vesicles isolated from encysting *G. lamblia*. These vesicles are highly dynamic spaces that act as hosts of the cyst wall materials and increase only during the induction of encystation. The authors have shown that the vesicles do not have sorting functions characteristic of mature Golgi, but do retain protein quality control functions. They also suggested that the ESVs can be involved in anterograde and retrograde trafficking with the endoplasmic reticulum, similar to the function of the Golgi.[61]

40.7 Excystation

In vivo, excystation starts following ingestion by the host when the cysts are exposed to the acidic pH of the stomach. Mobile parasites emerge from within the cyst wall in the host small intestine, but at this stage, the ventral disc is not fully reorganized. Soon after emergence, the cell divides rapidly generating two trophozoites capable of adhering to intestinal cells by the ventral disc, which is now fully reorganized.[62] Parasite viability requires that cysts produce trophozoites that can emerge from the cyst wall, perform cytokinesis, and adhere to the intestinal epithelium of a newly infected host. Several protocols for excystation have been published (e.g.,[54,63–65]), but the success of the excystation process may vary depending on whether the source of the cysts is feces or *in vitro* generation.

40.8 Adhesion

The ability of the parasite to adhere to the intestinal epithelium is an essential factor in its ability to reproduce and cause disease. *In vivo*, *Giardia* trophozoites can adhere to the surface of the small intestine and *in vitro* to surfaces like glass and polymers.[66] Understanding the mechanisms that mediate parasite adhesion can explain how *Giardia* cells can attach to many surfaces as well as the intestinal wall of many host animals.[67] Several approaches have been developed to evaluate *Giardia* cell adhesion. These include adhesion of trophozoites to different glass surfaces, scanning electron microscopy to analyze attachment, and a microarray-based approach.[67–69]

Many studies have been published in an attempt to elucidate the adhesion mechanisms of *Giardia* cells. The structure responsible for the attachment is the ventral disc, a spiral array of microtubules associated with microribbons and found just below the plasma membrane in the anterior and ventral portion of the trophozoite.[70] Several auxiliary proteins are found associated with the ventral disc including β-giardin, γ-giardin, δ-giardin, and SALP-1. The periphery of the disc has contractile proteins including actin, myosin, and tropomyosin, which may serve as the biochemical basis for contraction of the disc, a fundamental process in parasite adhesion. Hagen et al. used a proteomic approach to detect more than 100 protein candidates associated with the ventral disc as well as the first proteins localized to the lateral flange.[71] Because adhesion is central to parasite viability and these structures are unique to the parasite, strategies targeting these molecules could be developed as novel therapeutics.

40.9 Pathology

Studies of the pathology associated with giardiasis often use animal models that aim to recapitulate human infection conditions. Several animal models for studying *G. lamblia* exist including mouse, gerbil, hamster, rat, and recently, nematodes (Table 40.2). Infection with *G. muris* is a common model of human giardiasis because the host specificity of this species in mice and rats can be useful for simplifying infection protocols. Both human assemblages of *G. lamblia* are commonly used in experimental giardiasis, which can be useful for understanding assemblage-level differences in the human host. *In vitro* models are also used to study *G. lamblia* pathology by exposing human intestinal cell lines to live parasites or their byproducts. There is a bias in the literature for mouse infections performed with

TABLE 40.2

Common Models Used for Studies of Giardiasis

Model	Advantages	Disadvantages
Axenic culture of trophozoites	• Rapid • Inexpensive	• Limited applicability *in vivo*
Coculture of *Giardia* with epithelial cell lines	• More rapid than *in vivo* • Can test hypotheses without using animals • Easy to analyze interactions	• Limited applicability *in vivo*
Adult mouse or rat infection with *G. muris*	• Robust infections • Natural infection cycle	• Limited applicability to human infection • *G. muris* cannot be cultured axenically—cyst stocks may have fecal contamination
Neonatal mouse infection with *G. lamblia*	• Can use human relevant strains • Can study more complex host–parasite interactions	• Limited immunity in neonatal mice
Adult mouse infection with *G. lamblia*	• Abundant reagents for analysis of immunity • Many molecular tools available	• Not all findings translate to human infections • Infections have little overt pathology[a]
Gerbil infection	• Highly susceptible to infection and development of symptoms	• Fewer reagents and mutant strains available
Hamster infection	• Highly susceptible to infection and development of symptoms	• Fewer reagents and mutant strains available
C. elegans	• Fewer ethical considerations compared to vertebrates	• Limited applicability to human infection

[a] This is actually similar to most human infections, however.

assemblage B strains (particularly GS) and for *in vitro* experiments performed with assemblage A strains (e.g., Portland-1 and WB).

There are several pathological mechanisms thought to contribute to the diarrheal disease caused by *G. lamblia* infection. These include epithelial barrier breakdown, defects in the epithelial brush border, and increased intestinal motility.[1,2] Changes in epithelial barrier function have been linked to both tight junction proteins and epithelial apoptosis in epithelial cell lines and in human infections.[72–74] Parasite strain could be important in determining the severity of this pathology. When the human epithelial cell line HCT-8 was exposed to different assemblages of *G. lamblia*, the assemblage was seen to influence both epithelial barrier integrity and apoptosis.[73] Similarly, Chin et al. found that some strains of *G. lamblia* were superior to others in inducing apoptosis in the SCBN cell line (reported at the time to be a human small intestinal cell, but later genotyped as being of canine origin).[75] Hardin et al. first showed increased macromolecular uptake in *Giardia*-infected gerbils.[76] Zhou et al. showed that transepithelial electrical resistance of mucosa isolated from infected mice was reduced compared to mucosa from non-infected animals, and Chen et al. showed epithelial barrier breakdown and translocation of commensal bacteria in infected mice.[77,78] Reductions in transepithelial electrical resistance were first shown *in vitro* using cell lines exposed to *Giardia*, and these changes were further associated with alterations in tight junction proteins.[79] Parasite factors responsible for changes to the intestinal epithelium have not been identified at the molecular level, although new structures related to adhesion and potential toxins have been suggested.[68,80,81]

Giardiasis is also associated with reduced disaccharidase activity in humans, and this has been associated with shortening of the intestinal microvilli in animal models. This enzyme deficiency can contribute to diarrhea by creating a gradient that drives water out of the tissue and into the intestinal lumen. In a mouse infection model using the murine-specific species *G. muris*, infection and CD8+ T cell responses were shown to contribute to microvillous shortening and a deficiency in disaccharidases.[82,83] This same deficiency in small intestinal disaccharidases was confirmed in mice using human strains of *G. lamblia*.[43] In the latter study, parasite assemblage was again shown to be an important factor contributing to intestinal pathology, as disaccharidase deficiency was observed in assemblage-B-infected mice but not assemblage-A-infected mice.[43]

G. lamblia infection has been linked with other changes in bowel function in humans and animal models. Humans infected with *Giardia* are prone to postinfectious irritable bowel syndrome.[1] Gerbils infected with *G. lamblia* had increased intestinal transit and hypercontractility of intestinal smooth muscle.[84] Studies using *G. muris* and *G. lamblia* infections in adult mice showed that increased intestinal transit required adaptive immune responses as well as the neuronal isoform of nitric oxide synthase.[85,86] These studies also showed that increased transit helped facilitate elimination of the parasite. Follow-up studies using the adult mouse model of *G. lamblia* infection found that altered motility is due at least in part to increased contractility of intestinal smooth muscle and that the gut hormone cholecystokinin (CCK) contributes to this phenotype.[87] CCK normally acts to induce contraction of the gall bladder and release of bile into the digestive tract in response to ingestion of fats. *Giardia* requires bile for growth and, thus, appears to have evolved a mechanism to induce the host to provide it with an essential nutrient.

Infection with *G. lamblia* is also associated with long-term health consequences. These diseases can manifest well after infection clearance and have a range of severity. *G. lamblia* has been linked to several metabolic outcomes including lowered cognitive function, lower weight and height, failure to thrive, and nutritional deficiencies.[1] Furthermore, acute *G. lamblia* disease severity is associated with nutritional deficiency in animals as both mice and gerbils on nutrient-limited diets presented with worsened intestinal pathology as compared to non-nutrient-limited controls.[44,88] These studies indicate that the relationship between *G. lamblia* infection and disease could be complex and bidirectional. The long-term health consequences of *G. lamblia* infection represent an intriguing area of investigation as these associations can be strong but mechanisms are lacking. For example, one study of humans with giardiasis found that one-third of infected individuals presented with extraintestinal manifestations of infections including symptoms of the eye, skin, joints, and urinary tract.[89] *G. lamblia* has also been linked with postinfection development of allergies, muscular complications, chronic fatigue syndrome, and irritable bowel syndrome,[1] but good animal models of these long-term sequelae have not been developed.

40.10 Immunology

G. lamblia infection leads to pathological alterations of the host's intestinal epithelia, yet intestinal inflammation is not often seen during infection.[90] This could indicate that the host immune response is protective against inflammatory effects. Furthermore, several studies demonstrate the host immune response having a role in driving pathology in the intestine during infection.[43,83] Thus, it is important to understand which elements of the host immune response contribute to pathology versus those which aid in clearance and protection during *G. lamblia* infection.

To successfully colonize the small intestine, *G. lamblia* must contend with the hosts' natural barriers to infection. Both intestinal mucus and epithelial cell turnover could serve as obstacles for trophozoite attachment, and *in vitro* studies have shown that mucin can inhibit the ability of trophozoites to attach to a substrate.[91] Recent work has also demonstrated that the host microbiome can play an important role in successful colonization. Infections of mice from different suppliers with *G. lamblia* showed that intestinal microbiota conferred protection against infection and that this protective effect could be transmitted through cohousing of mice.[92] This same study demonstrated that antibiotic treatment could ablate the protective effect of the host microbiome, demonstrating that intestinal bacteria composition plays a role in infection susceptibility.[92] Whether the influence of the host microbiome on *G. lamblia* infection is direct or a result of bacteria-driven immune stimulation remains unclear. *Lactobacillus johnsonii* culture supernatants have been shown *in vitro* to have a cytostatic effect on *G. lamblia*.[93] This same bacterium was shown to inhibit *G. lamblia* infection of gerbils.[94] In a different study, mouse infection with *Enterococcus faecium* led to immunological stimulations that decreased *G. lamblia* colonization success in infected mice, demonstrating that bacteria could alter host immunity to prevent establishment of *G. lamblia* infection.[95]

The innate immune system serves as an early line of defense against pathogens as it does not require the specificity of adaptive immunity to exert its protective effects. The importance of innate immunity in the control of *G. lamblia* infection is not well characterized. However, several studies have shown that innate immune effectors do have a role in *G. lamblia* infection. Antimicrobial peptides like defensins and lactoferrin have been shown to kill *G. lamblia in vitro*.[96,97] Another potent antimicrobial, nitric oxide (NO), has also been shown to have inhibitory effects on *G. lamblia* growth and survival in an *in vitro* model using human intestinal epithelia cells.[98] Studies in *G. lamblia*-infected mice have shown that while NO produced by inducible NO synthase (NOS2) may not be effective in controlling infection alone, it may have a redundant role with MMP-7, a protease required for cleavage of defensins.[85,99] These studies demonstrated that while a loss of either NOS2 or MMP-7 did not affect infection clearance, mice lacking both MMP-7 and NOS2 had a defect in parasite clearance.

Interestingly, it has also been shown that *G. lamblia* can inhibit innate cellular immune responses. *G. lamblia* extracts are able to inhibit the production of proinflammatory cytokines by murine bone-marrow-derived dendritic cells stimulated with LPS and other innate immune stimuli.[100] At the same time, production of the anti-inflammatory cytokine IL-10 was increased. The importance of dendritic cells in stimulating anti-*Giardia* immunity *in vivo* was shown using cell transfer systems.[101] *Giardia* infection can also modulate macrophage responses in the intestinal lamina propria. We recently showed that adult mice infected with *G. lamblia* have an increase in the proportion of macrophages in the small intestine and that these cells express both NOS2 and arginase (ARG1).[102] In addition, these macrophages express more IL-10 and less TNF following infection, consistent with an anti-inflammatory phenotype (JM and SMS, unpublished observation). *Giardia* has also been reported to actively block neutrophil recruitment by degrading the chemokine IL-8, which was induced in epithelial cells following exposure to the parasite *in vitro*.[103] Combined analysis using *in vitro* and animal models continue to elucidate interactions between the host and parasite.

Another important role of the innate immune response is to begin to mobilize the adaptive immune response. Innate immune recognition of a pathogen drives the release of chemical signals that recruit and activate other immune cells. *In vitro* studies have shown that several intestinal chemokines are secreted in response to *G. lamblia* infection. CCL2, CCL20, and CXCL1-3 are all secreted when Caco-2 cells, a colon cell line, are stimulated with *G. lamblia*.[104] The *in vivo* importance of these chemokines remains to be explored.

Communication between the parasite and host is bidirectional, and parasites alter their behavior in response to intestinal epithelial cells. Ringqvist et al. cocultured *G. lamblia* trophozoites with the human adenocarcinoma line, $CaCo_2$, and used proteomics to identify three parasite proteins (arginine deiminase, ornithine carbamoyl transferase, and enolase) in culture medium.[105] All three proteins were previously known as antigens recognized by giardiasis patient sera. The authors showed that *G. lamblia* contact with epithelial cells triggers release of metabolic enzymes, which would facilitate effective parasite colonization of the human small intestine.[105]

Adaptive immunity has been shown to be essential for control of *Giardia* infection, with both antibodies and T cells having important roles. Studies using a mouse model of *G. muris* infection have demonstrated that B cells and specifically IgA produced by these cells are essential for parasite control and elimination.[106–109] B cells appear to be less important for resolving *G. lamblia* infection. In a study using B-cell- and T-cell-deficient mice, it was demonstrated that while T cells were required for elimination of *G. lamblia* infection in mice, B cells were not.[110] However, in a study on the role of IL-6 in *G. lamblia* infection, IL-6-deficient mice had a defect in parasite clearance until 60 days after infection when parasite clearance correlated with the production of antibodies that were reactive to a diverse population of parasites.[111] These studies indicate that while host antibody may not be required for *G. lamblia* clearance, it could play an important role in parasite population control throughout the course of infection. These results are consistent with studies on the role of antibody in human infection where patients with antibody deficiency are at only a slightly increased risk for *G. lamblia* infection.[112] Unlike B cells, T cells seem to be required for the clearance of *G. lamblia* infection in mice. In fact, CD4 T cells have been shown to be essential for parasite clearance as mice with specific deficiency in CD4 T cells are unable to control infections.[43,110,113]

The role of several T cell cytokines has been analyzed in animal models. IFN-γ has been shown to be important for parasite clearance during mouse infection with either *G. muris* or *G. lamblia*.[110,114,115] IFN-γ production and T cell proliferation were also observed in *ex vivo* stimulations of human lymphocytes with *G. lamblia*, supporting a role for this cytokine in human infections.[116] TNF-α has also been shown to be important in early control of *G. lamblia* infection as mice with defects in TNF-α production demonstrated increased parasite burden during infection.[77] Recent studies on the inflammatory cytokine IL-17 have demonstrated that this cytokine is important for clearance of infection in mouse models of both *G. muris* and *G. lamblia*.[117,118] The role of IL-6 in *Giardia* infection has been analyzed in IL-6-deficient mice. *G. lamblia* infections of IL-6-deficient mice have demonstrated that this cytokine is required for infection control and parasite clearance.[111,119] As mentioned earlier, these mice did not have a defect in antibody production, but instead, defects in dendritic cell function may be responsible for the observed phenotype.[101]

Intestinal mast cells also have important functions in control of *Giardia* infection. Mouse infection models using *G. muris* and *G. lamblia* have demonstrated that mast cell deficiencies lead to deficiencies in infection clearance.[120,121] Mast cell hyperplasia has also been observed in Mongolian gerbils infected with *G. muris*.[76] Furthermore, *G. lamblia*-infected mast cell-deficient mice produce less IL-6 mRNA and IgA than their wild-type counterparts.[121] Mast-cells-also appear to contribute to intestinal smooth muscle contractions during *G. lamblia* infection in mice and could further support a protective role for these cells in infection.[87]

One important goal of immunological studies of *G. lamblia* is the development of a human vaccine. Currently, no human *Giardia* vaccine exists; however, a veterinary vaccine, GiardiaVax, has been used in a variety of domestic mammals. Studies in humans have found that previous *Giardia* infection could convey some level of protection.[122,123] This protection indicates that immunological memory to *Giardia* is possible in humans and supports the development of a human vaccine. Recently, several studies in mice have been done with the goal of better understanding how to develop a human vaccine. A reinfection model in mice with *G. lamblia* was able to recapitulate the protection observed in humans, indicating that mouse models could serve as a useful tool for human vaccine development.[124] In a mouse model of *G. lamblia* infection, a vaccine against parasite cyst wall protein was able to reduce cyst shedding.[125–127] In another mouse vaccine study, vaccination against a *G. lamblia* protein, α1-giardin, was able to confer protection against infection.[128] This protein is highly conserved between both human assemblages of *G. lamblia*, indicating it could provide coverage against multiple strains of the parasite.[128] The ability of *Giardia* to undergo antigenic variation of the variant-specific surface proteins (VSPs) (discussed in Section 40.11) may be a major impediment to the design of a successful vaccine. Rivero et al. used parasites engineered to express multiple VSPs simultaneously as a vaccine in gerbils and achieved significant levels of protection.[129] Combinations of approaches may be necessary to generate an effective vaccine for humans and/or animals.

40.11 Antigenic Variation in *G. lamblia*

Many pathogenic bacteria and protozoans develop strategies to evade the host immune response and maintain chronic infections. One of them, present in microorganisms like *Neisseria*, *Candida*, and *Trypanosoma*, is the frequent interchange of antigens on the surface of the cell. This phenomenon is known as antigenic variation. There are three important criteria defining this phenomenon: Only one antigen can be expressed at the surface at a time; there must be a family of homologous genes for encoding the variant proteins that will be expressed; and there must be a mechanism for altering the pattern of the expression of this antigen in the cell.[130]

Antigenic variation in *Giardia* was first observed *in vitro* as an alteration in the pattern of expression of highly abundant surface molecules (later termed VSPs) during the *in vitro* growth of the parasite.[131,132] This process was then verified during infections in humans and animals.[133–135] According to Nash et al., switching among VSPs occurs every 6–16 generations, depending on the strain.[136] The expression of VSPs in trophozoite cells is controlled through post-transcriptional gene silencing, involving a large number of miRNAs.[137,138]

Similar to the role of antigenic variation in other pathogens, reports indicate that by altering VSPs on the cell surface during infection, the parasite can avoid the immune response and prolong infection.[135,139] Additional roles for VSPs have also been postulated. More than 200 VSP genes have been described,[23] and proteomics studies showed that the VSP expression profile can differ considerably by strain and genotype.[140] This variation may allow for repeated infections with giardia, irrespective of antigenic switching that occurs during a single infection. Earlier, Nash et al. showed that VSPs differed in sensitivity to host proteases and might prefer a growth advantage depending on their environment.[141] This was validated *in vivo* when parasite clones expressing different VSPs were shown to exhibit growth preferences in immunodeficient mice and gerbils.[142] Astiazarán-Garcia suggested that VSPs, based on their ability to bind zinc, could inhibit the function of zinc- or metal-requiring intestinal enzymes, or compete with the host for zinc and contribute to zinc malnutrition.[143] Recently, proteomic comparison of virulent and avirulent strains of *Giardia* suggested that differences in the VSP repertoire might contribute to the different levels of virulence observed in animals.[144]

40.12 *Giardia* and Arginine

The relationship between *Giardia* and arginine within the host is of interest because the parasite consumes arginine as an energy source, which could have important implications for immune function and parasite survival during infection. *Giardia* consumes arginine through a three-step enzymatic pathway that involves arginine deiminase (ADI), ornithine carbamoyltransferase (OCT), and carbamate kinase (CK) with ATP generated as an end product.[145] This pathway also serves as a means by which *Giardia* could reduce arginine availability in the host and limit its conversion to nitric oxide (NO). In fact, arginine is the sole amino acid substrate for NO production by the enzyme nitric oxide synthase (NOS), making it a limiting factor for NO production.[146]

The production of NO from arginine is an important element of the innate immune response as NO is toxic to many pathogens. *In vitro* infection models using human intestinal epithelial cells and activated mouse macrophages have demonstrated that NO has the ability to inhibit trophozoite growth and is potentially lethal to the parasite and may block cyst formation.[98,147] Human intestinal epithelial cells have increased NOS expression following coculture with *G. lamblia in vitro*, and increased NOS expression is also observed in small intestinal tissue from *G. lamblia*-infected mice.[85,102,148] The importance of host

NO production *in vivo* is unclear. Recent studies indicate that host-produced NO may be redundant with α-defensins as only a loss of the expression of enzymes required for the production of both influenced infection outcomes in mice.[85,99] Furthermore, it is likely that *G. lamblia* influences arginine availability within the host indirectly. Stimulation of a mouse macrophage cell line with parasite lysate leads to increased expression of arginase 1 (ARG1), a host-produced arginine-consuming enzyme.[102] This same study found that small intestinal tissue from infected mice also showed increased ARG1 expression compared to uninfected controls.[102]

The consumption of arginine through parasite ADI is advantageous for the *G. lamblia* in potentially two ways. First, it serves as a means of energy production through conversion of arginine to ATP, and second, it provides a means by which *Giardia* might evade the host immune response by limiting NO production. Limiting arginine availability in the gut could also affect the host immune system in other ways. For example, it is possible that T cell function is altered as arginine is an important modulator of T cell proliferation. Arginine depletion could result in the limitation of T cell proliferation and may also cause an impairment of CD3ζ chain turnover of the T cell receptor after antigen stimulation, inhibiting signaling through the T cell receptor complex.[149] Both of these scenarios would result in limiting the T cell response during *Giardia* infection, providing an additional benefit for the parasite through arginine depletion. Thus, giardial ADI or host ARG1 may directly interfere with the host's ability to generate a strong, protective cellular immune response against the parasite while limiting the innate immune response by inhibiting NO production.

40.13 Drug Resistance in *Giardia*

Current giardiasis treatments cause a variety of side effects, treatment failures are common, and a considerable incidence of drug resistance occurs. Many drugs have been tested for a therapeutic approach against *G. lamblia*. Albendazole, quinacrine, metronidazole, nitazoxanide, and isoflavones are important drugs used in some studies involving drug resistance in giardiasis.[19,150–161] Uzlikova and Nohynkova showed by flow cytometry analysis that sublethal drug concentrations of metronidazole affect the replication phase of the cell cycle of *Giardia*.[162] Cells incubated with lethal drug concentrations lose the ability to adhere to a surface after few hours of incubation. A recent study of drug resistance in *Giardia*[19] identified the minimum lethal concentrations of 28 drugs, and compared the results with metronidazole, the drug most used to treat giardiasis. Fumagillin, carbadox, and tioxidazole were able to eliminate metronidazole-resistant *G. lamblia* isolates. In the same studies, the authors evaluated the dose-dependent efficacy of fumagillin in a mouse model, and showed that the effective dose of fumagillin was much lower than that for metronidazole.

Nitroheterocyclic drugs like metronidazole are the first line of therapy for giardiasis. These compounds are redox-active and are believed to cause damage to protein and DNA after being activated by oxidoreductase enzymes in metabolically active cells. In a recent study, Ansell et al. reviewed the molecular phenotype of nitroheterocyclic-resistant *G. lamblia* isolated from patients. The authors believe that resistance mechanisms are related to enzymes commonly associated with drug resistance, like pyruvate ferredoxin, oxidoreductases, and nitroreductases. These authors also highlighted new approaches using systems biology and advanced bioinformatics to evaluate mechanisms of nitroheterocyclic resistance in *Giardia*.[163]

Using proteomic and transcriptomic analysis, researchers have identified and characterized differentially expressed genes in albendazole-resistant clones of *G. lamblia*. The expression of proteins and their corresponding mRNA-resistant clones were analyzed at different concentrations, and these were compared with albendazole-sensitive clones using two-dimensional electrophoresis and mass spectrometry. The authors identified eight differentially expressed proteins in albendazole-resistant clones that are involved in the following systems: cytoskeleton, antioxidant metabolism, and energy metabolism. In addition, it was suggested that resistance to albendazole in *G. lamblia* is involved in response to gene regulation and may have an important role in maintaining cellular stability and oxidative stress.[164]

40.14 Conclusion

The relationship between *G. lamblia* and its host is complex and multifactorial with host immunity, intestinal microbial composition, nutritional status, and parasite strain all having potential roles in host susceptibility and infection severity. As a disease of the developing world, many of these aspects of infection remain understudied and poorly understood. Studies that aim to illuminate these relationships will provide information that could lead to new strategies for combating parasitic disease. These types of investigations could yield not only better treatments but also better control of a disease with a global distribution that infects millions of humans and other animals.

REFERENCES

1. Halliez, M.C. & Buret, A.G. Extra-intestinal and long term consequences of *Giardia duodenalis* infections. *World J Gastroenterol* 19, 8974–85 (2013).
2. Solaymani-Mohammadi, S. & Singer, S.M. *Giardia duodenalis*: the double-edged sword of immune responses in giardiasis. *Exp Parasitol* 126, 292–7 (2010).
3. Dawson, D. Foodborne protozoan parasites. *Int J Food Microbiol* 103, 207–27 (2005).
4. Savioli, L., Smith, H. & Thompson, A. *Giardia* and *Cryptosporidium* join the "Neglected Diseases Initiative." *Trends Parasitol* 22, 203–8 (2006).
5. Monis, P.T., Caccio, S.M. & Thompson, R.C. Variation in *Giardia*: towards a taxonomic revision of the genus. *Trends Parasitol* 25, 93–100 (2009).
6. Thompson, R.C. & Monis, P.T. Variation in *Giardia*: implications for taxonomy and epidemiology. *Adv Parasitol* 58, 69–137 (2004).
7. Escobedo, A.A. et al. A bibliometric study of international scientific productivity in giardiasis covering the period 1971–2010. *J Infect Dev Ctries* 9, 76–86 (2015).
8. Lipoldová, M. Giardia and Vilém Dušan Lambl. *PLoS Negl Trop Dis* 8, e2686 (2014).
9. Rendtorff, R.C. & Holt, C.J. The experimental transmission of human intestinal protozoan parasites. IV. Attempts to transmit *Endamoeba coli* and *Giardia lamblia* cysts by water. *Am J Hyg* 60, 327–38 (1954).
10. Nash, T.E., Herrington, D.A., Losonsky, G.A. & Levine, M.M. Experimental human infections with *Giardia lamblia*. *J Infect Dis* 156, 974–84 (1987).
11. Robertson, L.J. & Lim, Y.A.L. Waterborne and environmentally-borne giardiasis. in *Giardia: A Model Organism* (eds. Lujan, H.D. & Svard, S.) 29–69 (Springer-Verlag, Vienna, 2011).
12. Guy, R.A., Payment, P., Krull, U.J. & Horgen, P.A. Real-time PCR for quantification of *Giardia* and *Cryptosporidium* in environmental water samples and sewage. *Appl Environ Microbiol* 69, 5178–85 (2003).
13. Upjohn, M. et al. Prevalence, molecular typing and risk factor analysis for *Giardia duodenalis* infections in dogs in a central London rescue shelter. *Vet Parasitol* 172, 341–6 (2010).
14. Berkman, D.S., Lescano, A.G., Gilman, R.H., Lopez, S.L. & Black, M.M. Effects of stunting, diarrhoeal disease, and parasitic infection during infancy on cognition in late childhood: a follow-up study. *Lancet* 359, 564–71 (2002).
15. Nematian, J., Gholamrezanezhad, A. & Nematian, E. Giardiasis and other intestinal parasitic infections in relation to anthropometric indicators of malnutrition: a large, population-based survey of schoolchildren in Tehran. *Ann Trop Med Parasitol* 102, 209–14 (2008).
16. Gardner, T.B. & Hill, D.R. Treatment of giardiasis. *Clin Microbiol Rev* 14, 114–28 (2001).
17. Keister, D.B. Axenic culture of *Giardia lamblia* in TYI-S-33 medium supplemented with bile. *Trans R Soc Trop Med Hyg* 77, 487–8 (1983).
18. Meloni, B.P. & Thompson, R.C. Comparative studies on the axenic in vitro cultivation of *Giardia* of human and canine origin: evidence for intraspecific variation. *Trans R Soc Trop Med Hyg* 81, 637–40 (1987).
19. Kulakova, L. et al. Discovery of novel antigiardiasis drug candidates. *Antimicrob Agents Chemother* 58, 7303–11 (2014).
20. Benchimol, M. The nuclei of *Giardia lamblia*—new ultrastructural observations. *Arch Microbiol* 183, 160–8 (2005).

21. Piva, B. & Benchimol, M. The median body of *Giardia lamblia*: an ultrastructural study. *Biol Cell* 96, 735–46 (2004).
22. Midlej, V. & Benchimol, M. *Giardia lamblia* behavior during encystment: how morphological changes in shape occur. *Parasitol Int* 58, 72–80 (2009).
23. Adam, R.D. Biology of *Giardia lamblia*. *Clin Microbiol Rev* 14, 447–75 (2001).
24. Gillin, F.D., Reiner, D.S. & McCaffery, J.M. Cell biology of the primitive eukaryote *Giardia lamblia*. *Annu Rev Microbiol* 50, 679–705 (1996).
25. Lanfredi-Rangel, A., Kattenbach, W.M., Diniz, J.A., Jr. & de Souza, W. Trophozoites of *Giardia lamblia* may have a Golgi-like structure. *FEMS Microbiol Lett* 181, 245–51 (1999).
26. Soltys, B.J., Falah, M. & Gupta, R.S. Identification of endoplasmic reticulum in the primitive eukaryote *Giardia lamblia* using cryoelectron microscopy and antibody to Bip. *J Cell Sci* 109 (Pt 7), 1909–17 (1996).
27. Sulemana, A., Paget, T.A. & Jarroll, E.L. Commitment to cyst formation in *Giardia*. *Microbiology* 160, 330–9 (2014).
28. Svard, S.G., Hagblom, P. & Palm, J.E. *Giardia lamblia*—a model organism for eukaryotic cell differentiation. *FEMS Microbiol Lett* 218, 3–7 (2003).
29. Adam, R.D. The biology of *Giardia* spp. *Microbiol Rev* 55, 706–32 (1991).
30. Carpenter, M.L., Assaf, Z.J., Gourguechon, S. & Cande, W.Z. Nuclear inheritance and genetic exchange without meiosis in the binucleate parasite *Giardia intestinalis*. *J Cell Sci* 125, 2523–32 (2012).
31. Jarroll, E.L., Manning, P., Lindmark, D.G., Coggins, J.R. & Erlandsen, S.L. *Giardia* cyst wall-specific carbohydrate: evidence for the presence of galactosamine. *Mol Biochem Parasitol* 32, 121–31 (1989).
32. Macechko, P.T., Steimle, P.A., Lindmark, D.G., Erlandsen, S.L. & Jarroll, E.L. Galactosamine-synthesizing enzymes are induced when *Giardia* encyst. *Mol Biochem Parasitol* 56, 301–9 (1992).
33. Lopez, A.B., Sener, K., Jarroll, E.L. & van Keulen, H. Transcription regulation is demonstrated for five key enzymes in *Giardia intestinalis* cyst wall polysaccharide biosynthesis. *Mol Biochem Parasitol* 128, 51–7 (2003).
34. Paget, T.A., Jarroll, E.L., Manning, P., Lindmark, D.G. & Lloyd, D. Respiration in the cysts and trophozoites of *Giardia muris*. *J Gen Microbiol* 135, 145–54 (1989).
35. Kasprzak, W. & Majewska, A.C. Infectivity of *Giardia* sp. cysts in relation to eosin exclusion and excystation in vitro. *Tropenmed Parasitol* 34, 70–2 (1983).
36. Ortega, Y.R. & Adam, R.D. *Giardia*: overview and update. *Clin Infect Dis* 25, 545–9; quiz 550 (1997).
37. Wolfe, M.S. Giardiasis. *Clin Microbiol Rev* 5, 93–100 (1992).
38. Ballweber, L.R., Xiao, L., Bowman, D.D., Kahn, G. & Cama, V.A. Giardiasis in dogs and cats: update on epidemiology and public health significance. *Trends Parasitol* 26, 180–9 (2010).
39. Lasek-Nesselquist, E., Welch, D.M. & Sogin, M.L. The identification of a new *Giardia duodenalis* assemblage in marine vertebrates and a preliminary analysis of *G. duodenalis* population biology in marine systems. *Int J Parasitol* 40, 1063–74 (2010).
40. Caccio, S.M. & Ryan, U. Molecular epidemiology of giardiasis. *Mol Biochem Parasitol* 160, 75–80 (2008).
41. Franzen, O. et al. Draft genome sequencing of *Giardia intestinalis* assemblage B isolate GS: is human giardiasis caused by two different species? *PLoS Pathogens* 5, e1000560 (2009).
42. Laishram, S., Kang, G. & Ajjampur, S.S. Giardiasis: a review on assemblage distribution and epidemiology in India. *Indian J Gastroenterol* 31, 3–12 (2012).
43. Solaymani-Mohammadi, S. & Singer, S.M. Host immunity and pathogen strain contribute to intestinal disaccharidase impairment following gut infection. *J Immunol* 187, 3769–75 (2011).
44. Bartelt, L.A. et al. Persistent *G. lamblia* impairs growth in a murine malnutrition model. *J Clin Invest* 123, 2672–84 (2013).
45. Traub, R.J. et al. Epidemiological and molecular evidence supports the zoonotic transmission of *Giardia* among humans and dogs living in the same community. *Parasitology* 128, 253–62 (2004).
46. Sackey, M.E., Weigel, M.M. & Armijos, R.X. Predictors and nutritional consequences of intestinal parasitic infections in rural Ecuadorian children. *J Trop Pediatr* 49, 17–23 (2003).
47. Souza, S.L. et al. Molecular identification of *Giardia duodenalis* isolates from humans, dogs, cats and cattle from the state of Sao Paulo, Brazil, by sequence analysis of fragments of glutamate dehydrogenase (*gdh*) coding gene. *Vet Parasitol* 149, 258–64 (2007).

48. Thompson, R.C. The zoonotic significance and molecular epidemiology of *Giardia* and giardiasis. *Vet Parasitol* 126, 15–35 (2004).
49. Lalle, M., Jimenez-Cardosa, E., Caccio, S.M. & Pozio, E. Genotyping of *Giardia duodenalis* from humans and dogs from Mexico using a β-giardin nested polymerase chain reaction assay. *J Parasitol* 91, 203–5 (2005).
50. Johnson, D. et al. Serum biochemistry, serology, and parasitology of boreal caribou (*Rangifer tarandus* caribou) in the Northwest Territories, Canada. *J Wildl Dis* 46, 1096–107 (2010).
51. Sprong, H., Caccio, S.M., van der Giessen, J.W., & ZOOPNET network & partners. Identification of zoonotic genotypes of *Giardia duodenalis*. *PLoS Negl Trop Dis* 3, e558 (2009).
52. Gillin, F.D., Boucher, S.E., Rossi, S.S. & Reiner, D.S. *Giardia lamblia*: the roles of bile, lactic acid, and pH in the completion of the life cycle in vitro. *Exp Parasitol* 69, 164–74 (1989).
53. Gillin, F.D. et al. Encystation and expression of cyst antigens by *Giardia lamblia* in vitro. *Science* 235, 1040–3 (1987).
54. Boucher, S.E. & Gillin, F.D. Excystation of in vitro-derived *Giardia lamblia* cysts. *Infect Immun* 58, 3516–22 (1990).
55. Sun, C.H., McCaffery, J.M., Reiner, D.S. & Gillin, F.D. Mining the *Giardia lamblia* genome for new cyst wall proteins. *J Biol Chem* 278, 21701–8 (2003).
56. Yichoy, M. et al. Lipid metabolism in *Giardia*: a post-genomic perspective. *Parasitology* 138, 267–78 (2011).
57. Lujan, H.D., Mowatt, M.R., Byrd, L.G. & Nash, T.E. Cholesterol starvation induces differentiation of the intestinal parasite *Giardia lamblia*. *Proc Natl Acad Sci USA* 93, 7628–33 (1996).
58. Mendez, T.L. et al. Glucosylceramide transferase activity is critical for encystation and viable cyst production by an intestinal protozoan, *Giardia lamblia*. *J Biol Chem* 288, 16747–60 (2013).
59. Reiner, D.S., Wang, C.S. & Gillin, F.D. Human milk kills *Giardia lamblia* by generating toxic lipolytic products. *J Infect Dis* 154, 825–32 (1986).
60. Das, S. et al. Killing of *Giardia lamblia* trophozoites by human intestinal fluid in vitro. *J Infect Dis* 157, 1257–60 (1988).
61. Stefanic, S., Palm, D., Svard, S.G. & Hehl, A.B. Organelle proteomics reveals cargo maturation mechanisms associated with Golgi-like encystation vesicles in the early-diverged protozoan *Giardia lamblia*. *J Biol Chem* 281, 7595–604 (2006).
62. Bernander, R., Palm, J.E. & Svard, S.G. Genome ploidy in different stages of the *Giardia lamblia* life cycle. *Cell Microbiol* 3, 55–62 (2001).
63. Bingham, A.K. & Meyer, E.A. *Giardia* excystation can be induced in vitro in acidic solutions. *Nature* 277, 301–2 (1979).
64. Rice, E.W. & Schaefer, F.W., III. Improved in vitro excystation procedure for *Giardia lamblia* cysts. *J Clin Microbiol* 14, 709–10 (1981).
65. Hautus, M.A., Kortbeek, L.M., Vetter, J.C. & Laarman, J.J. In vitro excystation and subsequent axenic growth of *Giardia lamblia*. *Trans R Soc Trop Med Hyg* 82, 858–61 (1988).
66. Holberton, D.V. Mechanism of attachment of *Giardia* to the wall of the small intestine. *Trans R Soc Trop Med Hyg* 67, 29–30 (1973).
67. Hansen, W.R., Tulyathan, O., Dawson, S.C., Cande, W.Z. & Fletcher, D.A. *Giardia lamblia* attachment force is insensitive to surface treatments. *Eukaryot Cell* 5, 781–3 (2006).
68. Maia-Brigagao, C. & de Souza, W. Using environmental scanning electron microscopy (ESEM) as a quantitative method to analyse the attachment of *Giardia duodenalis* to epithelial cells. *Micron* 43, 494–6 (2012).
69. Pickering, H., Wu, M., Bradley, M. & Bridle, H. Analysis of *Giardia lamblia* interactions with polymer surfaces using a microarray approach. *Environ Sci Technol* 46, 2179–86 (2012).
70. Holberton, D.V. Arrangement of subunits in microribbons from *Giardia*. *J Cell Sci* 47, 167–85 (1981).
71. Hagen, K.D. et al. Novel structural components of the ventral disc and lateral crest in *Giardia intestinalis*. *PLoS Negl Trop Dis* 5, e1442 (2011).
72. Buret, A.G., Mitchell, K., Muench, D.G. & Scott, K.G. *Giardia lamblia* disrupts tight junctional ZO-1 and increases permeability in non-transformed human small intestinal epithelial monolayers: effects of epidermal growth factor. *Parasitology* 125, 11–9 (2002).
73. Koh, W.H. et al. *Giardia duodenalis* assemblage-specific induction of apoptosis and tight junction disruption in human intestinal epithelial cells: effects of mixed infections. *J Parasitol* 99, 353–8 (2013).

74. Troeger, H. et al. Effect of chronic *Giardia lamblia* infection on epithelial transport and barrier function in human duodenum. *Gut* 56, 328–35 (2007).
75. Chin, A.C. et al. Strain-dependent induction of enterocyte apoptosis by *Giardia lamblia* disrupts epithelial barrier function in a caspase-3-dependent manner. *Infect Immun* 70, 3673–80 (2002).
76. Hardin, J.A., Buret, A.G., Olson, M.E., Kimm, M.H. & Gall, D.G. Mast cell hyperplasia and increased macromolecular uptake in an animal model of giardiasis. *J Parasitol* 83, 908–12 (1997).
77. Zhou, P., Li, E., Shea-Donohue, T. & Singer, S.M. Tumour necrosis factor α contributes to protection against *Giardia lamblia* infection in mice. *Parasite Immunol* 29, 367–74 (2007).
78. Chen, T.L. et al. Persistent gut barrier damage and commensal bacterial influx following eradication of *Giardia* infection in mice. *Gut Pathog* 5, 26 (2013).
79. Teoh, D.A., Kamieniecki, D., Pang, G. & Buret, A.G. *Giardia lamblia* rearranges F-actin and α-actinin in human colonic and duodenal monolayers and reduces transepithelial electrical resistance. *J Parasitol* 86, 800–6. (2000).
80. Rodriguez-Fuentes, G.B. et al. *Giardia duodenalis*: analysis of secreted proteases upon trophozoite-epithelial cell interaction in vitro. *Mem Inst Oswaldo Cruz* 101, 693–6 (2006).
81. Shant, J., Bhattacharyya, S., Ghosh, S., Ganguly, N.K. & Majumdar, S. A potentially important excretory-secretory product of *Giardia lamblia*. *Exp Parasitol* 102, 178–86 (2002).
82. Scott, K.G., Logan, M.R., Klammer, G.M., Teoh, D.A. & Buret, A.G. Jejunal brush border microvillous alterations in giardia muris-infected mice: role of T lymphocytes and interleukin-6. *Infect Immun* 68, 3412–8 (2000).
83. Scott, K.G., Yu, L.C. & Buret, A.G. Role of $CD8^+$ and $CD4^+$ T lymphocytes in jejunal mucosal injury during murine giardiasis. *Infect Immun* 72, 3536–42 (2004).
84. Deselliers, L.P., Tan, D.T., Scott, R.B. & Olson, M.E. Effects of *Giardia lamblia* infection on gastrointestinal transit and contractility in Mongolian gerbils. *Dig Dis Sci* 42, 2411–9 (1997).
85. Li, E., Zhou, P. & Singer, S.M. Neuronal nitric oxide synthase is necessary for elimination of *Giardia lamblia* infections in mice. *J Immunol* 176, 516–21 (2006).
86. Andersen, Y.S., Gillin, F.D. & Eckmann, L. Adaptive immunity-dependent intestinal hypermotility contributes to host defense against *Giardia* spp. *Infect Immun* 74, 2473–6 (2006).
87. Li, E., Zhao, A., Shea-Donohue, T. & Singer, S.M. Mast cell mediated changes in smooth muscle contractility during mouse giardiasis. *Infect Immun* 75, 4514–8 (2007).
88. Ventura, L.L. et al. Impact of protein malnutrition on histological parameters of experimentally infected animals with *Giardia lamblia*. *Exp Parasitol* 133, 391–5 (2013).
89. Cantey, P.T. et al. Study of nonoutbreak giardiasis: novel findings and implications for research. *Am J Med* 124, 1175 e1–8 (2011).
90. Oberhuber, G., Kastner, N. & Stolte, M. Giardiasis: a histologic analysis of 567 cases. *Scand J Gastroenterol* 32, 48–51 (1997).
91. Roskens, H. & Erlandsen, S.L. Inhibition of in vitro attachment of *Giardia* trophozoites by mucin. *J Parasitol* 88, 869–73 (2002).
92. Singer, S.M. & Nash, T.E. The role of normal flora in *Giardia lamblia* infections in mice. *J Infect Dis* 181, 1510–2 (2000).
93. Perez, P.F. et al. Inhibition of *Giardia intestinalis* by extracellular factors from Lactobacilli: an in vitro study. *Appl Environ Microbiol* 67, 5037–42 (2001).
94. Humen, M.A. et al. *Lactobacillus johnsonii* La1 antagonizes *Giardia intestinalis* in vivo. *Infect Immun* 73, 1265–9 (2005).
95. Benyacoub, J. et al. *Enterococcus faecium* SF68 enhances the immune response to *Giardia intestinalis* in mice. *J Nutr* 135, 1171–6 (2005).
96. Aley, S.B., Zimmerman, M., Hetsko, M., Selsted, M.E. & Gillin, F.D. Killing of *Giardia lamblia* by cryptdins and cationic neutrophil peptides. *Infect Immun* 62, 5397–403 (1994).
97. Eckmann, L. Mucosal defences against *Giardia*. *Parasite Immunol* 25, 259–70 (2003).
98. Eckmann, L. et al. Nitric oxide production by human intestinal epithelial cells and competition for arginine as potential determinants of host defense against the lumen-dwelling pathogen *Giardia lamblia*. *J Immunol* 164, 1478–87 (2000).
99. Tako, E.A., Hassimi, M.F., Li, E. & Singer, S.M. Transcriptomic analysis of the host response to *Giardia duodenalis* infection reveals redundant mechanisms for parasite control. *MBio* 4 (6), e00660–13 (2013).

100. Kamda, J.D. & Singer, S.M. Phosphoinositide 3-kinase-dependent inhibition of dendritic cell interleukin-12 production by *Giardia lamblia*. *Infect Immun* 77, 685–93 (2009).
101. Kamda, J.D., Nash, T.E. & Singer, S.M. *Giardia duodenalis*: dendritic cell defects in IL-6 deficient mice contribute to susceptibility to intestinal infection. *Exp Parasitol* 130, 288–91 (2012).
102. Maloney, J., Keselman, A., Li, E. & Singer, S.M. Macrophages expressing arginase 1 and nitric oxide synthase 2 accumulate in the small intestine during *Giardia lamblia* infection. *Microbes Infect* 17, 462–7 (2015).
103. Cotton, J.A. et al. *Giardia duodenalis* cathepsin B proteases degrade intestinal epithelial interleukin-8 and attenuate interleukin-8-induced neutrophil chemotaxis. *Infect Immun* 82, 2772–87 (2014).
104. Roxstrom-Lindquist, K., Ringqvist, E., Palm, D. & Svard, S. *Giardia lamblia*-induced changes in gene expression in differentiated Caco-2 human intestinal epithelial cells. *Infect Immun* 73, 8204–8 (2005).
105. Ringqvist, E. et al. Release of metabolic enzymes by *Giardia* in response to interaction with intestinal epithelial cells. *Mol Biochem Parasitol* 159, 85–91 (2008).
106. Snider, D.P., Gordon, J., McDermott, M.R. & Underdown, B.J. Chronic *Giardia muris* infection in anti-IgM-treated mice. I. Analysis of immunoglobulin and parasite-specific antibody in normal and immunoglobulin-deficient animals. *J Immunol* 134, 4153–62 (1985).
107. Skea, D.L. & Underdown, B.J. Acquired resistance to *Giardia muris* in X-linked immunodeficient mice. *Infect Immun* 59, 1733–8 (1991).
108. Langford, T.D. et al. Central importance of immunoglobulin A in host defense against *Giardia* spp. *Infect Immun* 70, 11–8 (2002).
109. Davids, B.J. et al. Polymeric immunoglobulin receptor in intestinal immune defense against the lumen-dwelling protozoan parasite *Giardia*. *J Immunol* 177, 6281–90 (2006).
110. Singer, S.M. & Nash, T.E. T-cell-dependent control of acute *Giardia lamblia* infections in mice. *Infect Immun* 68, 170–5 (2000).
111. Zhou, P. et al. Role of interleukin-6 in the control of acute and chronic *Giardia lamblia* infections in mice. *Infect Immun* 71, 1566–8 (2003).
112. Lai Ping So, A. & Mayer, L. Gastrointestinal manifestations of primary immunodeficiency disorders. *Semin Gastrointest Dis* 8, 22–32 (1997).
113. Heyworth, M.F., Carlson, J.R. & Ermak, T.H. Clearance of *Giardia muris* infection requires helper/inducer T lymphocytes. *J Exp Med* 165, 1743–8 (1987).
114. Venkatesan, P., Finch, R.G. & Wakelin, D. A comparison of mucosal inflammatory responses to *Giardia muris* in resistant B10 and susceptible BALB/c mice. *Parasite Immunol* 19, 137–43 (1997).
115. Jimenez, J.C. et al. Antibody and cytokine responses to *Giardia* excretory/secretory proteins in *Giardia intestinalis*-infected BALB/c mice. *Parasitol Res* 113, 2709–18 (2014).
116. Ebert, E.C. *Giardia* induces proliferation and interferon γ production by intestinal lymphocytes. *Gut* 44, 342–6 (1999).
117. Dreesen, L. et al. *Giardia muris* infection in mice is associated with a protective interleukin 17A response and induction of peroxisome proliferator-activated receptor alpha. *Infect Immun* 82, 3333–40 (2014).
118. Dann, S.M. et al. IL-17A promotes protective IgA responses and expression of other potential effectors against the lumen-dwelling enteric parasite *Giardia*. *Exp Parasitol* 156, 68–78 (2015).
119. Bienz, M., Dai, W.J., Welle, M., Gottstein, B. & Muller, N. Interleukin-6-deficient mice are highly susceptible to *Giardia lamblia* infection but exhibit normal intestinal immunoglobulin A responses against the parasite. *Infect Immun* 71, 1569–73 (2003).
120. Erlich, J.H., Anders, R.F., Roberts-Thomson, I.C., Schrader, J.W. & Mitchell, G.F. An examination of differences in serum antibody specificities and hypersensitivity reactions as contributing factors to chronic infection with the intestinal protozoan parasite, *Giardia muris*, in mice. *Aust J Exp Biol Med Sci* 61, 599–615 (1983).
121. Li, E., Zhou, P., Petrin, Z. & Singer, S.M. Mast cell-dependent control of *Giardia lamblia* infections in mice. *Infect Immun* 72, 6642–9 (2004).
122. Istre, G.R., Dunlop, T.S., Gaspard, G.B. & Hopkins, R.S. Waterborne giardiasis at a mountain resort: evidence for acquired immunity. *Am J Public Health* 74, 602–4 (1984).
123. Isaac-Renton, J.L., Lewis, L.F., Ong, C.S. & Nulsen, M.F. A second community outbreak of waterborne giardiasis in Canada and serological investigation of patients. *Trans R Soc Trop Med Hyg* 88, 395–9 (1994).

124. Li, E., Liu, M. & Singer, S.M. Resistance to reinfection in mice as a vaccine model for giardiasis. *Hum Vaccines Immunother* 10, 1536–43 (2014).
125. Abdul-Wahid, A. & Faubert, G. Mucosal delivery of a transmission-blocking DNA vaccine encoding *Giardia lamblia* CWP2 by *Salmonella typhimurium* bactofection vehicle. *Vaccine* 25, 8372–83 (2007).
126. Abdul-Wahid, A. & Faubert, G. Characterization of the local immune response to cyst antigens during the acute and elimination phases of primary murine giardiasis. *Int J Parasitol* 38, 691–703 (2008).
127. Lee, P., Abdul-Wahid, A. & Faubert, G.M. Comparison of the local immune response against *Giardia lamblia* cyst wall protein 2 induced by recombinant *Lactococcus lactis* and *Streptococcus gordonii*. *Microbes Infect* 11, 20–8 (2009).
128. Jenikova, G. et al. α1-giardin based live heterologous vaccine protects against *Giardia lamblia* infection in a murine model. *Vaccine* 29, 9529–37 (2011).
129. Rivero, F.D. et al. Disruption of antigenic variation is crucial for effective parasite vaccine. *Nat Med* 16, 551–7 (2010).
130. Prucca, C.G., Rivero, F.D. & Lujan, H.D. Regulation of antigenic variation in *Giardia lamblia*. *Annu Rev Microbiol* 65, 611–30 (2011).
131. Adam, R.D. et al. Antigenic variation of a cysteine-rich protein in *Giardia lamblia*. *J Exp Med* 167, 109–18 (1988).
132. Aggarwal, A., Merritt, J.W., Jr. & Nash, T.E. Cysteine-rich variant surface proteins of *Giardia lamblia*. *Mol Biochem Parasitol* 32, 39–47 (1989).
133. Aggarwal, A. & Nash, T.E. Antigenic variation of *Giardia lamblia* in vivo. *Infect Immun* 56, 1420–3 (1988).
134. Nash, T.E., Herrington, D.A., Levine, M.M., Conrad, J.T. & Merritt, J.W., Jr. Antigenic variation of *Giardia lamblia* in experimental human infections. *J Immunol* 144, 4362–9 (1990).
135. Nash, T.E. Antigenic variation in *Giardia lamblia* and the host's immune response. *Philos Trans R Soc Lond B Biol Sci* 352, 1369–75 (1997).
136. Nash, T.E., Banks, S.M., Alling, D.W., Merritt, J.W., Jr. & Conrad, J.T. Frequency of variant antigens in *Giardia lamblia*. *Exp Parasitol* 71, 415–21 (1990).
137. Prucca, C.G. et al. Antigenic variation in *Giardia lamblia* is regulated by RNA interference. *Nature* 456, 750–4 (2008).
138. Saraiya, A.A., Li, W., Wu, J., Chang, C.H. & Wang, C.C. The microRNAs in an ancient protist repress the variant-specific surface protein expression by targeting the entire coding sequence. *PLoS Pathog* 10, e1003791 (2014).
139. Prucca, C.G. & Lujan, H.D. Antigenic variation in *Giardia lamblia*. *Cell Microbiol* 11, 1706–15 (2009).
140. Emery, S.J., Lacey, E. & Haynes, P.A. Quantitative proteomic analysis of *Giardia duodenalis* assemblage A: a baseline for host, assemblage, and isolate variation. *Proteomics* 15(13), 2281–5 (2015).
141. Nash, T.E., Merritt, J.W., Jr. & Conrad, J.T. Isolate and epitope variability in susceptibility of *Giardia lamblia* to intestinal proteases. *Infect Immun* 59, 1334–40 (1991).
142. Singer, S.M., Elmendorf, H.G., Conrad, J.T. & Nash, T.E. Biological selection of variant-specific surface proteins in *Giardia lamblia*. *J Infect Dis* 183, 119–24 (2001).
143. Astiazaran-Garcia, H., Inigo-Figueroa, G., Quihui-Cota, L. & Anduro-Corona, I. Crosstalk between zinc status and *Giardia* infection: a new approach. *Nutrients* 7, 4438–52 (2015).
144. Emery, S.J., van Sluyter, S. & Haynes, P.A. Proteomic analysis in *Giardia duodenalis* yields insights into strain virulence and antigenic variation. *Proteomics* 14, 2523–34 (2014).
145. Schofield, P.J., Costello, M., Edwards, M.R. & O'Sullivan, W.J. The arginine dihydrolase pathway is present in *Giardia intestinalis*. *Int J Parasitol* 20, 697–9 (1990).
146. Popovic, P.J., Zeh, H.J., III & Ochoa, J.B. Arginine and immunity. *J Nutr* 137, 1681S–86S (2007).
147. Fernandes, P.D. & Assreuy, J. Role of nitric oxide and superoxide in *Giardia lamblia* killing. *Braz J Med Biol Res* 30, 93–9 (1997).
148. Stadelmann, B., Merino, M.C., Persson, L. & Svard, S.G. Arginine consumption by the intestinal parasite *Giardia intestinalis* reduces proliferation of intestinal epithelial cells. *PLoS One* 7, e45325 (2012).
149. Choi, B.S. et al. Differential impact of L-arginine deprivation on the activation and effector functions of T cells and macrophages. *J Leukoc Biol* 85, 268–77 (2009).
150. Lindquist, H.D. Induction of albendazole resistance in *Giardia lamblia*. *Microb Drug Resist* 2, 433–4 (1996).

151. Upcroft, J., Mitchell, R., Chen, N. & Upcroft, P. Albendazole resistance in *Giardia* is correlated with cytoskeletal changes but not with a mutation at amino acid 200 in β-tubulin. *Microb Drug Resist* 2, 303–8 (1996).
152. Upcroft, J.A., Campbell, R.W. & Upcroft, P. Quinacrine-resistant *Giardia duodenalis*. *Parasitology* 112 (Pt 3), 309–13 (1996).
153. Arguello-Garcia, R., Cruz-Soto, M., Romero-Montoya, L. & Ortega-Pierres, G. Variability and variation in drug susceptibility among *Giardia duodenalis* isolates and clones exposed to 5-nitroimidazoles and benzimidazoles in vitro. *J Antimicrob Chemother* 54, 711–21 (2004).
154. Arguello-Garcia, R., Cruz-Soto, M., Romero-Montoya, L. & Ortega-Pierres, G. In vitro resistance to 5-nitroimidazoles and benzimidazoles in *Giardia duodenalis*: variability and variation in gene expression. *Infect Genet Evol* 9, 1057–64 (2009).
155. Boreham, P.F., Phillips, R.E. & Shepherd, R.W. Altered uptake of metronidazole in vitro by stocks of *Giardia intestinalis* with different drug sensitivities. *Trans R Soc Trop Med Hyg* 82, 104–6 (1988).
156. Upcroft, J.A. & Upcroft, P. Drug resistance and *Giardia*. *Parasitol Today* 9, 187–90 (1993).
157. Upcroft, J.A. & Upcroft, P. Drug susceptibility testing of anaerobic protozoa. *Antimicrob Agents Chemother* 45, 1810–4 (2001).
158. Leitsch, D., Schlosser, S., Burgess, A. & Duchene, M. Nitroimidazole drugs vary in their mode of action in the human parasite *Giardia lamblia*. *Int J Parasitol Drugs Drug Resist* 2, 166–70 (2012).
159. Muller, J., Sterk, M., Hemphill, A. & Muller, N. Characterization of *Giardia lamblia* WB C6 clones resistant to nitazoxanide and to metronidazole. *J Antimicrob Chemother* 60, 280–7 (2007).
160. Sterk, M., Muller, J., Hemphill, A. & Muller, N. Characterization of a *Giardia lamblia* WB C6 clone resistant to the isoflavone formononetin. *Microbiology* 153, 4150–8 (2007).
161. Escobedo, A.A., Hanevik, K., Almirall, P., Cimerman, S. & Alfonso, M. Management of chronic *Giardia* infection. *Expert Rev Anti Infect Ther* 12, 1143–57 (2014).
162. Uzlikova, M. & Nohynkova, E. The effect of metronidazole on the cell cycle and DNA in metronidazole-susceptible and -resistant *Giardia* cell lines. *Mol Biochem Parasitol* 198, 75–81 (2014).
163. Ansell, B.R. et al. Drug resistance in *Giardia duodenalis*. *Biotechnol Adv* 33(6), 888–901 (2015).
164. Paz-Maldonado, M.T., Arguello-Garcia, R., Cruz-Soto, M., Mendoza-Hernandez, G. & Ortega-Pierres, G. Proteomic and transcriptional analyses of genes differentially expressed in *Giardia duodenalis* clones resistant to albendazole. *Infect Genet Evol* 15, 10–7 (2013).
165. Kane, A.V., Ward, H.D., Keusch, G.T. & Pereira, M.E. In vitro encystation of *Giardia lamblia*: large-scale production of in vitro cysts and strain and clone differences in encystation efficiency. *J Parasitol* 77, 974–81 (1991).

41

Toxoplasma: *Animal and* In Vitro *Models on Toxoplasmosis*

Renato Augusto DaMatta, Andrea Cristina Vetö Arnholdt, and Farlen José Bebber Miranda

CONTENTS

41.1 Introduction

41.1.1 The Disease

Toxoplasmosis is a disease that infects nearly all warm-blooded vertebrates.[1,2] It is estimated that about a third of the world's human population is serologically positive for this infection.[2–4] Transmission of the parasite has been historically implicated by the ingestion of undercooked or raw meat containing the parasite in tissue cysts.[5] However, in the last two decades, accumulated evidence point also to the parasite infection by oocysts ingestion through contaminated water, soil, or food—mainly vegetables and fruits.[5,6] Food consumption is culturally based; thus, the venue of infection may vary greatly in different human populations, explaining the variance in different geographical locations.[3] Toxoplasmosis outbreaks through contaminated water in human populations have been reported in detail[7–10] showing how relevant environmental contamination can be.

Most often, toxoplasmosis is asymptomatic in immunocompetent individuals.[2–4,11] However, depending on the *Toxoplasma gondii* strain,[2] the genetic constitution of the host, parasite load, and the mode of transmission,[12,13] infection can cause severe clinical manifestations including ocular toxoplasmosis,[14] cerebral toxoplasmosis,[15] and congenital toxoplasmosis.[1] When acquired by healthy individuals, toxoplasmosis most often causes lymphadenopathy, usually of the head and neck, but pneumonia, retinochoroiditis, myocarditis, neurological disorders, and severe ocular infection have also been reported in severe cases especially in French Guiana[16] and in Brazil.[17] Congenital toxoplasmosis can cause mental retardation, encephalitis, neonatal mortality, and abortion of the fetuses.[2,17,18] In immunocompromised patients, the reactivation of the parasite into fast replicating forms in the central nervous system (CNS) causes cerebral toxoplasmosis.[5] In addition, *T. gondii* infection has been implicated in schizophrenia and other neurological disorders.[15,19] Domestic animals also develop toxoplasmosis upon infection, causing relevant economic losses, and wild animals are also infected by *T. gondii*.[1,20] Due to its severe manifestations, worldwide distribution, and diverse host infection, toxoplasmosis is a relevant public health problem.[21,22]

41.1.2 Biology of the Parasite

The etiological agent of toxoplasmosis is *Toxoplasma gondii*, an obligate intracellular protozoan parasite of the Apicomplexa phylum.[13,23] This parasite presents three infection forms. Tachyzoites (Figure 41.1) replicate fast, rupturing host cells, causing lytic cycle during the acute phase of the disease.[13,24] Bradyzoites are slow-multiplying forms found within tissue cysts (Figure 41.2) mainly located in the CNS and skeletal muscles.[13] Sporozoites are found in sporulated oocysts shed by the definitive host.[13] In healthy intermediate and definitive hosts, tachyzoites convert to bradyzoites, and the infection becomes latent. *In vitro* treatment of infected cells with stress factors, such as proinflammatory cytokines, NO, serum from infected laboratory animals, drugs, low nutrient, alkaline or acid pH, and extreme temperature cause conversion of tachyzoites into bradyzoites.[13,25–27] It was first thought that the immune host response was responsible for inducing this conversion.[13,25,26,28] However, spontaneous conversion of tachyzoites into bradyzoites in cultures of cells from the CNS[29,30] and in skeletal muscle cells[31] suggests that this phenomenon may be orchestrated by the host cell type.[31,32] This highlights the possible tissue tropism found in infected hosts.[13,33]

The parasite life cycle is complex and involves different host species (Figure 41.3). The definitive hosts are members of the Felidae family.[34,35] After ingestion of the parasite, the infection of the Felidae intestinal epithelium cells (enterocytes) results first in schizogony followed by the sexual phase of the cycle.[35,36] Infection may occur by feline oral ingestion of bradyzoites (in tissue cysts) or sporozoites (in sporulated oocysts).[35,36] These forms infect enterocytes, and some are converted into tachyzoites in the lamina propria cells,[35] but different generations of merozoites,[37] with five distinct forms (named from A to E), are found in these cells before the appearance of gametes.[35] Fertilization generates a diploid zygote that becomes the unsporulated oocyst shed in the feline's feces.[35] The released oocyst sporulates

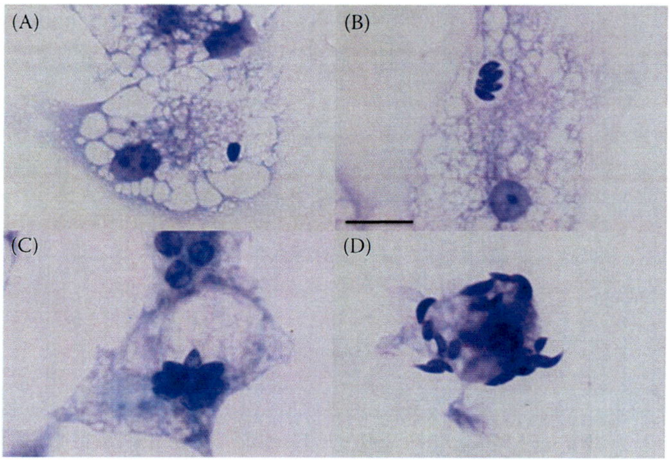

FIGURE 41.1 Tachyzoites infecting ostrich macrophages. Crescent shape of the parasite can be noted after infection (2 h—A), replication (24 h—B; 48 h—C) and egress (48 h—D). During egress, host cells are usually destroyed (D). Bar: 20 μm.

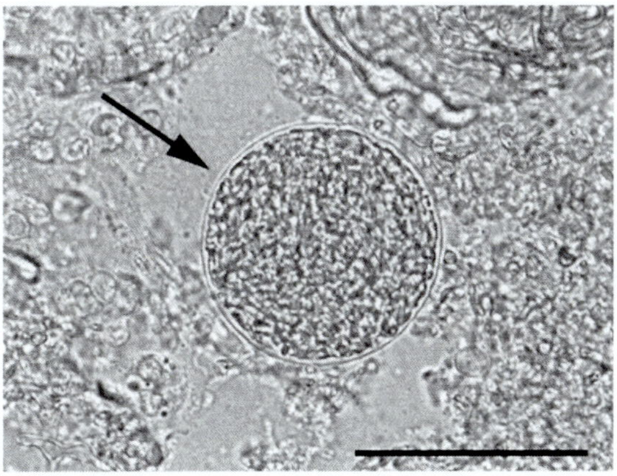

FIGURE 41.2 *Toxoplasma gondii* cyst (arrow) found in the brain of mice inoculated with heart tissue of an infected pig. Bar: 50 μm.

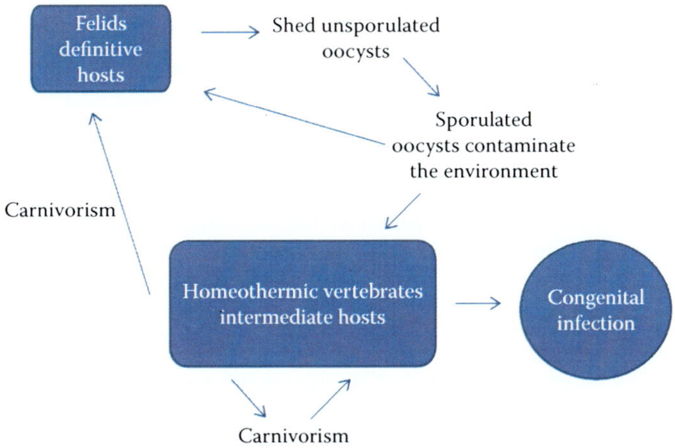

FIGURE 41.3 Schematic representation of the *Toxoplasma gondii* life cycle. In the enterocytes of the members of the Felidae family, the parasite differentiates in gametocytes that generate a zygote that is shed in the feces as unsporulated oocysts. After 2–5 days in the environment, oocysts sporulate to form two esporocysts, each with four sporozoites that contaminate the environment. Hosts get infected by the ingestion of oocysts. Oocysts are digested liberating sporozoites that first infect enterocytes differentiating in tachyzoites that later infect lamina propria cells and multiply rapidly. These forms disseminate through the organism by the circulatory system. Tachyzoites convert into bradyzoites forming tissue cysts. Tachyzoites can migrate to the placenta in pregnant hosts causing congenital infection. Carnivorism among the hosts disseminates the parasite through the ingestion of bradyzoites in tissue cysts. As with oocysts, tissue cysts are digested liberating bradyzoites that are resistant to this process. These forms infect host enterocytes and differentiate in tachyzoites similar to sporozoites.

by meiosis and mitosis in the environment and disseminates. Infection of intermediate hosts occurs by ingestion of oocysts through the consumption of contaminated water or food, or by the ingestion of tissue cysts by carnivorism. In these hosts, the infective forms, sporozoites or bradyzoites, pass through the enterocytes and convert into tachyzoites[38] that infect and multiply in cells of the lamina propria. At this stage, tachyzoites disseminate through the organism and then convert into bradyzoites, forming tissue cysts. During their life cycle, tachyzoites may pass from the blood to the placenta and infect the fetus causing congenital toxoplasmosis (Figure 41.3).

Sexual reproduction of the parasite in Felidae provides genetic variability that results in different recombinant strains.[24,39,40] Isolated strains from North America and Europe have an unusual population structure with basically three clonal lineages (types I, II, and III).[41] However, recombinant strains are found in different parts of the world and have diverse virulence.[20,39,42] Studies on isolates from Brazil, Colombia, and French Guiana have shown that most *T. gondii* strains are recombinant, suggesting greater sexual recombination of this parasite in South America,[16,39,43] thus explaining the wider spectra of clinical manifestations.[16,17,43] This indicates that the parasite population structure varies in different locations around the world, being composed by distinct genotypes with important implications to infection, clinical manifestations, and development of distinct diagnosis and treatment strategies.

41.2 Laboratory Models to Study *Toxoplasma gondii*

The first description of *T. gondii* was actually made independently by two groups using two different laboratory experimental animals: rabbits in Brazil,[44] and the gundis (*Ctenodactylus gundi*), a rodent, in Tunisia.[45,46] Since then, many different laboratory models have been used to study different aspects of this parasite. Some of them will be described here.

41.2.1 *In Vitro* Models

In vitro studies are crucial to understand the biology of cells, including intracellular parasitism. All fields studying cells have adopted cell culture techniques[47] resulting in most of what we know, *T. gondii* biology included. The first laboratory procedure to obtain tachyzoites of *T. gondii* for experimental infection was to pass parasites into mice by the inoculation of macerated tissue, usually of the brain.[48] Later, parasites were maintained by peritoneal wash of infected mice and used for further experimentation.[49,50] At that moment, some key aspects of the biology of the parasite were studied: its intracellular nature,[48] drug treatment,[49–51] and transmission.[52–54] Today, tachyzoites are mostly maintained by culture passage in cell monolayers. However, the use of animals is still necessary to obtain bradyzoites, merozoites, or sporozoites, since *in vitro* system does not generate these forms yet.

The first reproducible culture system for *T. gondii* was obtained by Guimarães and Meyer.[55–57] Tachyzoites can be maintained in cell culture allowing relevant studies on the biology of the parasite.[24,46] The infective forms of these parasites are haploid, facilitating transient and stable transfections.[58] These characteristics transformed this parasite in an Apicomplexan and intracellular parasitism model.[24] Furthermore, host cells can also be genetically changed,[59] allowing a precise questioning about the interplay between parasite and host cells. Thus, numerous *in vitro* laboratory models are used nowadays and can be created in the future to come.

Aided by these models, relevant information was obtained on the host cell–*T. gondii* interaction (motility, invasion, replication, and egress), infective form conversion (tachyzoites to bradyzoites and vice versa), gene expression, diagnostic methods, screening of new therapeutic compounds, etc. An active field of study is how *T. gondii* escapes the cell's autonomous immunity and also the immune system response as a whole. One of the first evasion mechanisms described for *T. gondii* involves the structure in which this parasite resides in the host cell: the parasitophorous vacuole (PV). Using transmission electron microscopy, it was found that live parasites reside in the PV that does not fuse with lysosomes.[60] However, parasites that are dead or opsonized with antibodies do fuse, and the parasites are destroyed by lysosomal digestion.[61,62] Further work has been performed in this cell compartment revealing interesting parasite adaptations.[63,64] These early findings described the PV harboring *T. gondii* as "invisible" to the endolysosomal system, allowing its persistence and multiplication in the host cell. This demonstrates a cellular escape mechanism, clearly showing that the evasion concept could explain *T. gondii*'s success as a parasite.

Another classical evasion mechanism of *T. gondii* is its capacity to resist killing by reactive oxygen species (ROS) produced by host cells, especially macrophages. By the end of the 1970s, it was clear that ROS could be produced not only by neutrophils, but also by macrophages and monocytes with clear negative correlation to bacterial growth. It was found that in cell-free systems, *T. gondii* was 100-fold more resistant to H_2O_2

than *Trypanosoma cruzi*.[65] This ROS resistance was correlated with high catalase activity of *T. gondii*,[65] which was later confirmed by the activity of superoxide dismutase showing the robust antioxidant arsenal of this parasite.[66,67] As a result, production of ROS is not detected in the *T. gondii* macrophage contact sites, explaining parasite survival, persistence, and multiplication in host cells that are professional ROS producers.[68,69] All this work was done with mammalian macrophages, but the similar result was obtained when chicken macrophages were examined,[70] suggesting that the same mechanism operates in birds. Although this parasite can evade from host cell ROS production *in vitro*, it was later demonstrated that these radicals are implicated in the *in vivo* control of *T. gondii*[71] suggesting that this system still needs further work.

More recently, it has been demonstrated that IRGs (immunity-related GTPases) are involved in the growth control of *T. gondii* in mice.[72–75] This is significant because most of the data from different aspects of *T. gondii* biology, including IRGs, point to rodents as the most relevant intermediate host for the parasite, in a clear example of coevolution of host cell resistance mechanisms and *T. gondii* virulence factors.[73,75] Interferon-gamma (IFN-γ) induces IRGs in rodent host cells of myeloid and nonmyeloid origin. If an IFN-γ-activated cell is infected by *T. gondii*, the GTPases hierarchically and cooperatively accumulate in the PV membrane, changing their form until disruption, allowing *T. gondii* to end up in the cytosol where it is killed.[73,76] However, some *T. gondii* strains can escape the IRG system.[40,75] Virulence of *T. gondii* strains to mice is directly related to its repertory of rhoptry proteins kinases and pseudokinases against the IRGs microbicidal system.[40] Rhoptry is a secretory organelle of *T. gondii* that releases its protein contents into the host cell during invasion. However, there is high polymorphism of these proteins between strains.[77,78] Virulent strains of *T. gondii*, such as type I, have rhoptry proteins that efficiently phosphorylate IRGs components, impairing their association to the PV, resulting in parasite survival and virulence capacity.[40,79] Less-virulent strains have fewer or no effective rhoptry proteins against IRGs, being destroyed by this system. In virulent strains, these proteins form complexes that inactivate IRGs.[80] These complexes also involve proteins from another secretory organelle of the parasite, the dense granules.[81] The composition of these secretory protein complexes and how they deactivate IRGs is currently an important area of *T. gondii* research.

In the last decade, it has also been shown that *T. gondii* hijacks cells from the immune system, allowing the parasite to cross host barriers such as gut epithelial barrier, blood–brain, and blood–retina barriers in a mechanism known as Trojan horse.[82–84] Some of the molecular events underlying this mechanism have been described by our group. *In vitro* infection of unstimulated murine macrophages leads to downregulation of integrin alphaL, alpha4, and alpha5 immediately after infection, allowing cells to de-attach from extracellular matrix components.[82] On the other hand, *T. gondii* infection of human or murine macrophages impairs upregulation of CD86, CD80, CD40, and CD1a, while maintaining their migratory ability.[85] *T. gondii* seems to regulate the migratory machinery similar to what metastatic cells do as demonstrated by increased expression of host metalloproteinases MT1-MMP and ADAM-10 upon infection of murine macrophages, in parallel with increased secretion of active MMP-9.[84] Active MMP-9 is secreted in association with its cell membrane acceptor CD44, and with uPAR and TIMP-1.[86] In fact, secretion of this multiprotein complex is partially dependent on the extracellular plasminogen urokinase pathway (uPA/uPAR)[86] and ERK.[84] *In vivo* expression of MMP-9 is also increased in the ilea and lungs of orally infected C57Bl/6 mice (Pimentel, P. M., et al., unpublished data). Adoptive transfer experiments of MMP knockdown cells infected with *T. gondii* have been developed in order to access its contribution to *in vivo* dissemination of host cells carrying the parasite.

Nitric oxide (NO) produced by inducible NO synthase (iNOS or NOS2) expressed in classically activated macrophages is another microbicidal molecule of vertebrates[87,88] involved in *T. gondii* control. Infection of iNOS knockout mice or mice treated with iNOS inhibitors (aminoguanidine) have shown that NO is important to control *T. gondii* growth, especially during the chronic phase of the disease.[89–91] However, NO may also induce host death in the acute phase, being one of the responsible events of the classical histopathology found in the ileum.[91,92] Macrophages in culture classically activated by IFN-γ and lipopolysaccharide express iNOS and produce high amounts of NO that control *T. gondii* growth.[90,93,94] After gut infection, *T. gondii* disseminates throughout the host, and immune response initiated locally becomes systemic and, depending on the strain, controls parasite growth leading to the latent infection. During this phase, tachyzoite will come across activated macrophages producing high amounts of NO. How do these parasites cope with these encounters? We have been working on this question by studying *T. gondii* infection of activated macrophages of mice and chickens *in vitro*. It was shown that tachyzoites of *T. gondii* inhibit NO production after infection of activated macrophages.[70,95–99] NO production inhibition is caused by iNOS degradation.[70,96,97] Interestingly enough, this degradation occurs right at the beginning (2 h) of the infection[70,97,99] and is caused by the proteasome pathway.[99] It was determined that an autocrine secretion of TGF-β1 caused by the parasite infection was responsible for iNOS degradation.[97] Similar to amastigotes of *Leishmania amazonensis*,[100] *T. gondii* tachyzoites expose phosphatidylserine (PS) without cell death, which, when blocked by annexin-V, abolishes the inhibition of NO production caused by the infection.[97] This indicates that tachyzoites of *T. gondii* may, indeed, use what is known as "apoptotic mimicry," a concept established for *L. amazonensis* that basically states that the parasites that expose PS mimic the anti-inflammatory response induced by apoptotic cells.[101] We then separated the PS-positive and -negative tachyzoite subpopulations using annexin-V conjugated with magnetic beads and found that only the PS-positive subpopulation was able to inhibit NO production in infected macrophages.[98] In addition, mice infected with positive or negative PS subpopulations died faster than the ones infected with both populations. Mice infected with the positive PS subpopulation died of high parasite burden, while the ones infected with the negative PS subpopulation died of uncontrolled inflammatory response.[98] These results indicate that both subpopulations are necessary for a balanced response of the host, allowing survival and parasite transmission. Recently, we have found that four strains of *T. gondii* with distinct virulence exposed PS and inhibited iNOS expression immediately after the infection of activated macrophages (Damasceno-Sá, J. C. et al., unpublished data). Thus, it seems that apoptotic mimicry is a general escape mechanism to NO microbicidal system that is present in mammals and birds.

Tachyzoite conversion to bradyzoite and formation of tissue cysts may be a naturally occurring phenomenon (not necessarily induced by the immune response) in the course of *T. gondii*'s life cycle and occurs mainly in the CNS and in skeletal and cardiac muscles, defining the latent stage of *T. gondii* infection.[13,31,32,102] Thus, there is great interest in the understanding of this conversion and the need of a well-established model in order to test drugs, because most treatments are not effective against the bradyzoite stage. An interesting *in vitro* model involves the infection of primary cultures of embryonic-derived skeletal muscle cells by *T. gondii*.[103] The most impressive result concerns the spontaneous capacity of some *T. gondii* strains to convert tachyzoites into bradyzoites, forming *in vitro* tissue cyst-like structures in these muscular cells.[31,32] The infection of these cells with bradyzoites, instead of tachyzoites, accelerates the formation of cyst-like structures.[102] Skeletal muscle cells become microbicidal against *T. gondii* after activation with IFN-γ, controlling parasite growth by NO production and IRGs, indicating that these cells may also help to control *T. gondii* growth in mice.[104] This model has also been used to show the induction of lipid bodies by *T. gondii* infection in these cells and its association to the PV.[105] The PV association to lipid bodies was also seen in macrophages infected with *T. gondii* in a lipid body model developed by our group.[106] Embryonic-derived skeletal muscle cells should be further explored as a model of *in vitro* infection to study other aspects of the biology of this parasite.

In vitro models lead to important findings on *T. gondii*. The lytic cycle and different aspects of the parasite biology are well studied using tachyzoites to infect distinct cell types in culture. Tachyzoite is the most easily obtained *T. gondii* form, which explains why this is, by far, the most used model. Infection of culture cells with bradyzoites has been used to study the formation of tissue cyst-like structures *in vitro* (see paragraph above). Bradyzoites are obtained from mice but may be replaced by an efficient *in vitro* system as earlier stated. More recently, spontaneous cystogenesis was demonstrated in an established cell line of renal feline kidney cells infected with bradyzoites, suggesting that cell culture system may eventually replace infected animals for studies with cysts and bradyzoites.[107] The enteroepithelial part of the life cycle of *T. gondii* in Felidae is less well known because there is no established cell culture model, with most findings deriving from cat experiments.[35] However, primary cultures of cat intestinal cells have been infected with bradyzoites resulting in syncytial parasite structures suggesting the *in vitro* reproduction of schizogony.[108] These results indicate that it is a matter of time for the different parts of *T. gondii* cell cycle to be obtained *in vitro*.

41.2.2 Animal Models

The use of animal models to study toxoplasmosis is extremely important. *T. gondii* infects a plethora of warm-blooded animals; thus, most animal infections are unique and important to understand the parasite cell cycle and the coevolution of host mechanisms that control the parasite and of virulence strategies for the evasion of these mechanisms. Owing to ethical reasons, the use of animals is always controversial and must follow standard and strict protocols. In addition, some host vertebrates are difficult and expensive to maintain, including wild, and large or medium-sized domestic animals. Many different hosts have been used to study infections by *T. gondii*, but the mouse is by far the most common (Table 41.1). A PubMed search using common host names (chosen by the authors) and the terms "toxoplasma" and later "experimental" or "infection" or "epidemiology" (Table 41.1) showed the highest number of articles for the mouse followed by the rat, cat, pig, chicken, *Calomys*, and fish. The term "infection" had the highest number of articles, and since infection encompasses "epidemiology," this term was also included in the survey to better understand the numbers of published articles. This reveals how important the cat and the pig are in the study of *T. gondii* epidemiology (Table 41.1). Here, we will refer to some relevant findings of the models mentioned in Table 41.1.

41.2.2.1 Detection of the Parasite in Experimental Infected Animals

Initially, *T. gondii* infection induces a humoral response that is characterized by the production of IgM, and later IgG, which gives protection to subsequent infections.[109] The most used serological tests in experimental and epidemiological studies are the ELISA (enzyme-linked immunosorbent assay), DT (dye test), IHA (indirect hemagglutination antibody), IFA (indirect fluorescent antibody), LAT (latex agglutination test), and MAT (modified agglutination test). Histopathological diagnosis is highly sensitive when cysts are observed in tissues stained with H&E. However, due to the sparse distribution of cysts in organs, particularly in medium-sized animals, a greater number of slides need to be analyzed.

Inoculation of mice with pepsin-digested tissue from possible infected animals (bioassay) has good specificity, although the sensitivity may be diminished by the amount of the processed material and parasitemia. Inoculation of leukocytes and plasma in the peritoneum of mice is a relatively simple method, confirming the infection by the detection of tachyzoites in the peritoneal cavity, brain cysts, or serology of mice.[110] Another interesting method to detect *T. gondii* is to seed culture cells with the blood leukocyte and detect the parasite by direct visualization due to growth.[111]

Polymerase chain reaction (PCR) is a method used as a diagnostic tool for the presence of parasites in various organs. In addition, nested PCR is preferable due to greater specificity. However, real-time PCR can quantify parasite DNA in tissue samples.[111,112]

The pig as a medium-sized animal model has great potential for use in experimental *T. gondii* infections filling the gap left by the murine model that is not able to respond adequately due to physiological limitations, such as the IRGs microbicidal system that is inexistent in humans. In our opinion, this study model can answer important issues of great interest and refinement.

TABLE 41.1

Number of Published Articles in a PubMed Search (September 2015) Crossing the Animal Models of the First Column with "Toxoplasma" and One of the Terms on the Second, Third, or Fourth Column

Animal Host	Experimental	Infection	Epidemiology
Mouse[a]	520	2358	442
Rats	81	263	76
Cat	70	746	437
Pigs	53	315	270
Chicken	14	90	90
Calomys	8	14	1
Fish	3	28	34

[a] The use of plural or singular for the animal host denomination was based on the higher number of articles obtained on the "experimental" column.

41.2.2.2 Rodent

Rodents are abundantly present in different parts of the world.[113] Species of this order are usually infected with *T. gondii*, although depending on the location, the prevalence varies and is probably related to the lack of seropositivity in congenital infected animals of this order.[114,115] These animals are probably the most common intermediate hosts of *T. gondii*,[73,75] with small- or medium-sized Felidae being the furthermost natural definitive hosts.[116,117] In addition, the mouse is the most used laboratory animal, and numerous articles describe experimental infections with *T. gondii*. As a result, most of what is known about *T. gondii* biology comes from laboratory models involving rodents, especially mice.

41.2.2.2.1 Mice

Infection of mice has been crucial for the understanding of many aspects of *T. gondii*. One of the greatest advantages of mice as animal models for parasitic infection is the availability of a vast number of mice strains with different genetic backgrounds, deficiencies, and the capacity to genetically manipulate these animals.[118] This has allowed the discovery of important aspects of *T. gondii* biology. Some relevant findings concerning characteristics of the *T. gondii* infection will be described in the following paragraphs.

IFN-γ is crucial to control *T. gondii* infection.[119] The first evidence that this cytokine was important came from *in vitro* studies showing that the presence of this cytokine increases the microbicidal potential of different cell types after infection.[26,90,93,94,104] Studies using mice treated with the IFN-γ[120] and antibodies that neutralize this cytokine[121] proved that IFN-γ is essential to control *T. gondii* growth *in vivo*. The importance of IFN-γ in controlling *T. gondii* infection was also shown in knockout mice that died rapidly after infection.[122]

The interleukin (IL)-12/IFN-γ axis was first shown in mice with severe combined immunodeficiency (SCID) with the demonstration that IL-12 was mainly produced by dendritic cells (DCs) upon IL-1β stimulation, resulting in the induction of IFN-γ production by NK cells in this model.[123,124] The importance of IFN-γ and IL-12 and their signaling pathway was demonstrated by the infection of mice lacking the transcription factor interferon consensus sequence binding protein (ICSBP) with an avirulent strain of *T. gondii* that lead mice to rapid death after 14 days of infection.[125] Also, the regulatory role of IL-10 was demonstrated in knockout mice that succumbed to infection after 2 weeks due to elevated circulating IFN-γ and IL-12 and intense liver pathology.[126] The importance of TLRs in immune response to *T. gondii* was revealed by the use of knockout mice to MyD88, a crucial adaptor molecule for TLR signaling transduction.[127,128]

Another interesting use of the mous model for *T. gondii* infection came from studies showing that oral infection of these animals causes death by a typical pathology of the ileum that involves tumor necrosis factor-α (TNF-α), IFN-γ, and NO production.[92] This is relevant because oral infection is the natural route of *T. gondii* infection. Depending on the strain of the parasite, the genetic background of the mice, and the parasite burden, the ilea pathology varies. When the C57BL/6 mice receive 100 cysts of the ME-49 strain of *T. gondii* by the oral route, a potent Th1 response is induced leading to the destruction of the ileum by intense necrosis. The typical ileum villus morphology is lost due to the high inflammatory response (Figure 41.4). Eventually, the ileum ruptures, and the mice die by sepsis. The infection of C57BL/6 mice with the ME-49 strain of *T. gondii* has been proposed as a model for intestinal bowel disease.[129] In addition, this pathology is not exclusive to mice, since it has also been described in different hosts.[130]

Although TNF-α and IFN-γ are involved in the ileum pathology, another crucial player is the gut flora.[131] Oral infection causes an increase in Gram-negative bacteria, antibiotic treatment reduces IFN-γ levels and NO production in the small intestine, and gnotobiotic mice do not develop the ileum pathology.[131] Thus, the presence of LPS from the gut flora and the increase in IFN-γ, and possibly of TNF-α, caused by the immune response against the parasite, induces a Th1 response that ends up destroying the ileum. In this model, IL-22 induced MMP-2 expression in an IL-17-independent manner, and infection of MMP-2 KO mice showed decreased immunopathology scores of the gut.[132] However, oral infection of MIF KO mice led to reduced pathogenesis in the ileum, with decreased amounts of IL-12, IFN-γ, and TNF-α and also MMP-9, whereas no modification in MMP-2 expression was observed.[133] Our group has demonstrated that oral infection of C57BL/6 with the ME-49 *T. gondii* strain induced high amounts of MMP-9 but no MMP-2 in the ilea and lungs after 7 days of infection (Pimentel, P. M., et al., unpublished data).

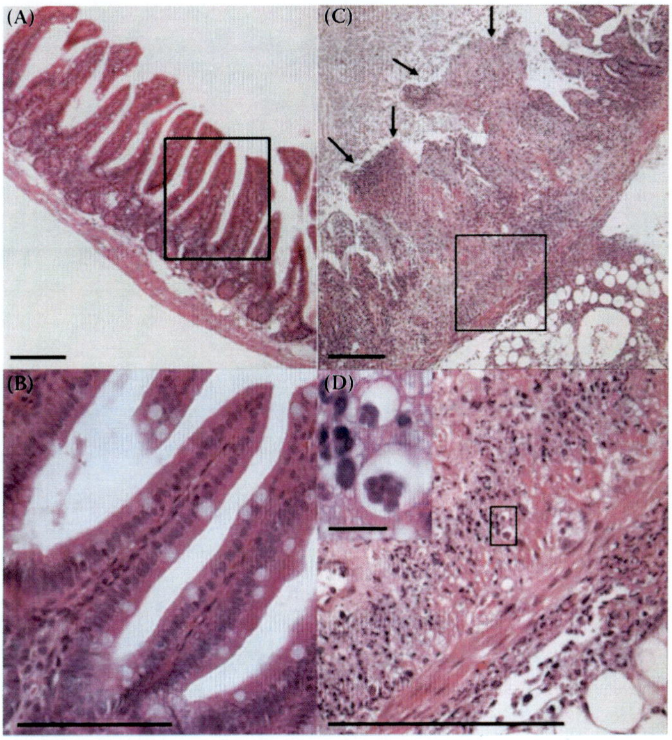

FIGURE 41.4 General aspects of the ilea of C57BL/6 mice infected orally with tissue cysts of *Toxoplasma gondii* of the ME-49 strain. (A) Noninfected ileum with apparent normal villi. (B) Higher magnification of the box image "A." (C) Ileum from infected mice with necrotic areas disrupting the villi architecture (arrows). (D) Higher magnification of the box of image "C"; parasites can be seen in the inset. Bars = 200 μm and 10 μm in inset.

Mice models have recently been used to further explore the role of the microbiota and IL-17 in the gut upon *T. gondii* infection. Some models show that IL-17 is involved in the immunopathogenesis at the gut mucosal barrier.[134,135] However, it has not been clearly identified whether IL-17 is produced by Th17 or ROR-γ innate lymphoid cells or both. Elimination of Paneth cells by IFN-γ produced by Th1 cells induces decreased amounts of antimicrobial peptides, leading to expansion of bacteria of the Enterobacteriaceae family, amplifying immunopathology at the gut after infection.[136] On the other hand, mice models have also identified mechanisms of containment of spreading of microflora during infection. Translocation into the luminal space of inflammatory macrophages and neutrophils is observed early in *T. gondii* infection, where Fprl (*N*-formyl peptide receptor)-expressing neutrophils promote encapsulation of γ-proteobacteria,[137] leading to bacterial death probably by NETosis.[138]

Finally, it would be interesting to see in the coming years the exploration of diverse mice models combined with infection with transgenic parasites designed to verify the cross-talk between *T. gondii* virulence factors and host immune response.

41.2.2.2.2 Rats

Infection of rats with *T. gondii* has been performed since the 1950s. Most studies on rats are performed with two species, *Rattus norvegicus* and *R. rattus*, with the latter being the most used owing to the different strains available.[139] Rats may have an important epidemiological role for toxoplasmosis, since they can be a source of infection to different hosts, especially pigs and possibly the domestic cats.[114,140] Epidemiological studies involving rats are complex, since *T. gondii* has been isolated by bioassay (in mice) from tissue of serologically negative rodents[140] including from experimental congenital infections.[114] In addition, various cases have been reported where inoculation with different strains of *T. gondii* resulted in different serology findings, the antibodies titers varied greatly with the type of test used, and negative serology was not uncommon.[140] Thus, epidemiological studies of rats for *T. gondii* based on serology must be analyzed with great care.

Most *T. gondii* strains have been isolated by bioassay with mice using pepsin-digested tissue from distinct hosts. This standard procedure to isolate *T. gondii* strains directly classifies the isolates as virulent or avirulent to mice. Naturally, these strains have been used to experimentally infect other hosts, such as rats. Rats have a natural resistance to *T. gondii* that has been known for a long time.[56,140–142] Strains that are virulent to mice may not necessarily cause the disease in other host species. Immunocompetent rats infected with tachyzoites or tissue cysts of different strains of *T. gondii* do not develop clinical manifestations being resistant to the parasite,[140] but infection with oocysts of the VAG strain induced clinical manifestations and, depending on the dosage, killed the rats.[143] Because immunocompetent humans are normally resistant to *T. gondii* infection, the rat *T. gondii* infection model is considered by different authors to be a closer animal model for human toxoplasmosis. The rat *T. gondii* infection model has been used as a congenital transmission model showing that the parasite first colonizes the placenta and later the fetus.[140] It was found that the transmission to the fetus may be low,[140] but higher transmission has also been reported,[144,145] although the lactation route is not relevant.[144] Differences in the transmission to the fetus in the distinct congenital rat *T. gondii* models may be related to the strain of rat and to the strain of parasite used.

One of the most interesting outcomes with the rat *T. gondii* infection model is the heterogeneous resistance of the different strains of rats to this parasite. After experimental infection, no parasite cysts are found in the brain of the LEW strain in contrast to the Fisher strain.[146] The lack of parasite cysts in the brain of the LEW rats was further confirmed, and no detectable humoral response was found when compared to the *T. gondii* susceptible Fisher and Brown Norway strains where a clear IgG response was observed.[147] With the aid of bone marrow chimeras between these rats strains, the resistance of the LEW rat was linked to cells derived from the myeloid lineage.[146] Using congenic lines of the LEW and Brown Norway rats, the resistance of the LEW strain was determined to be on chromosome 10 in a locus named *Toxo1*.[148] In addition, *T. gondii* grew in infected fibroblasts of all congenic rat strains of this study but not in macrophages from rats with the LEW *Toxo1* locus, indicating that these cells are the main immune component related to the resistance of this rat strain,[148] and that possible genetic polymorphisms of the genes at this locus exist in the rat strains. Although the locus responsible for LEW resistance was found, the gene or genes were not characterized at that time, and the killing mechanism was not determined, but it was not NO-dependent.[148]

By using classical and molecular genetic tools with the rat strains that are susceptible or resistant for *T. gondii* and distinct studies on infected macrophages, two research groups have recently suggested the identity of a gene in the *Toxo1* locus related to the capacity of the resistant rat macrophage to kill the parasite.[149,150] These groups found that the NOD-like receptor 1 (Nlrp1), which is part of the inflammasome, is essential for the resistant rat against the parasite. After bone marrow macrophage infection, this intracellular receptor senses *T. gondii* and activates the inflammasome, host cells die, the parasite is killed, and the tissue becomes inflamed. Inflammasomes are high-molecular-weight cytosolic multimeric protein complexes, which, upon assemblage, activate caspase-1 leading to a lytic cell death program (pyroptosis) and also to the secretion of IL-1β and IL-18, caused by the cleavage of their precursors, inducing a tissue proinflammatory response that controls infections.[151] It was found that Nlrp1 sequence varies among rat strains and that a variant from the resistant rat strains was responsible for macrophage pyroptosis upon infection and parasite killing; on the other hand, two other variants from *T. gondii* susceptible rat strains allowed macrophage survival and parasite growth.[150] Activation of caspase-1 of infected macrophages depends on the *Nlrp1* variants, with the ones from resistant rats releasing higher levels of IL-18 and IL-1β. Knockdown and over expression of the *Nlrp1* variant from the resistant rat in macrophages establish the importance of this receptor in pyroptosis initiated by the infection and subsequent killing of *T. gondii*.[150] Genetic and infection studies using the different susceptible or resistant *T. gondii* rat strains and their congenics also predicted the *Nlrp1* gene to confer the rat macrophages resistance to *T. gondii*.[149] *In vitro* infection caused parasite and peritoneal macrophage death only of resistant rat strains. The death of the infected macrophages of the resistant rats was neither apoptotic nor autophagic. Higher production of ROS and cleavage of caspase-1 were shown on infected macrophages of the resistant rat strain, clearly correlating *T. gondii* death with host cells death by inflammasome activation.[149] The use of pharmacological inhibitors of caspase-1 cleavage abrogated parasite killing in the resistance rat macrophages, but the growth was not totally restored, indicating that the control of *T. gondii* in resistance rat macrophages involves more mechanisms.[149] The identification of the *Nlrp1* gene in the Toxo1 locus of the rat and its possible relation to resistance rat to *T. gondii* infection is an example of how different host models are important to better understand parasite biology.

Inflammasome is also important in *T. gondii* control in mice.[152,153] In addition, human *Nlrp1* has been implicated in human congenital toxoplasmosis[154,155] and monocyte capacity to control *T. gondii* growth.[153] In addition, GRA15 from a specific *T. gondii* strain was responsible for triggering IL-1β release from infected human macrophages.[156] Thus, inflammasome activation by sensing different strains of the parasites in distinct hosts resulting in broad killing potential seems to be a widespread strategy to control *T. gondii* growth that varies within host species. The killing of *T. gondii* by the inflammasome is rapid and may explain the low serology results seen in rodent epidemiological surveys. If this microbicidal mechanism is also operating in humans and varies depending on gene polymorphism, there may be low estimations of *T. gondii* infection in the human population as well.

41.2.2.2.3 Calomys callosus

The genus *Calomys* (Waterhouse, 1937) (Rodentia, Sigmodontinae) comprises small rodents that are endemic in South America, originally being observed from northern Argentina to Peru, including Brazil, Bolivia, and Paraguay. Currently, 10 species are described by considering morphology and karyotype.[157] *Calomys callosus* mainly inhabits the central region of Brazil and has the capacity to live in different biomes, giving it a great adaptive capacity.[158]

Such a rodent has been successfully domesticated and presents advantages as an animal model being well adapted to the animal-housing environment.[159] They are very prolific throughout the year and are easy to handle. Common infections of rats, mice, and guinea pigs (*Cavia aperea*) do not affect this rodent, which is a robust animal.[159]

High genetic diversity has been observed in isolates of *T. gondii* from South America.[17] Recombinant strains isolated in Brazil were more pathogenic to mice than clonal isolates from North America and Europe.[160] Because *C. callosus* has been found in environments with more pathogenic recombinant strains of *T. gondii*, this model becomes quite interesting for the study of the parasite–host relationship.[161]

C. callosus is highly susceptible to infection with the RH strain of *T. gondii*,[162] which proliferates in large numbers in the newly implanted trophoblasts in pregnant females,[163] and especially in the labyrinthine zone of the placenta.[164] The same applies if the females are in the acute phase of the infection with the ME-49 strain. However, females with chronic infection with the ME-49 strain are, at least temporarily, protected against RH infection with no transplacental infection.[165] In contrast, the same chronic infection does not protect against transplacental migration when reinfection is performed with Brazilian recombinant strains.[166] Thus, this is a good model to study neonatal infection.[167]

Few studies have been done with blood leukocytes in this model, thus demonstrating great potential for further research. Intraperitoneal infection with the RH strain of *T. gondii* induces migration of neutrophils, monocytes, and mast cells after 24 h. In addition, basophils with distinct morphology were identified after 48 h in the ileum submucosa. This cellular process coincided with the presence of dead extracellular parasites. In lymphoid tissues, *T. gondii* was seen in mast cells, with or without the PV.[168] In intraconjunctival infection with the RH parasites, mast cell influx was detected within 5 and 24 h, with correlation to the inflammatory process.[169]

Pregnant *Calomys* were orally infected with the ME-49 strain resulting in ocular injury and the presence of cysts in 40% of the fetuses. Ocular lesions were found in 25% of the females and 75% of males, while 25% of each group had binocular lesions.[170] These results indicate that the *Calomys* model is promising for studying ocular and congenital toxoplasmosis.

41.2.2.3 Pigs

Knowledge obtained with laboratory animals (especially mice) needs to be validated in medium-sized animal models before clinical human trials.[171] Because pigs and humans have similar anatomy, genetics, and physiology, this animal is an interesting human infectious disease model[172] and has been used as a biomedical model for preclinical experimentation.[173] However, few functional experiments have been carried out in these animals, including *T. gondii* infection. Thus, there is great potential concerning this experimental model.

Besides being an interesting animal model for *T. gondii* infection, pigs are important in the epidemiology of this parasite, since the ingestion of tissue cysts from undercooked pig meat is a relevant source for *T. gondii* human infection.[174] Most pigs are infected by ingesting oocysts,[175] but due to their omnivorous behavior, the ingestion of birds and rodents with *T. gondii* cysts is important for the cycle.[1,175] Pig infection has a direct correlation with the environment and animal management.[176] Low management, access of other animals to where food and water is offered to pigs, and low control of rodent access are directly related to the prevalence of *T. gondii* infection in pig-breeding farms, especially the noncertified ones.[177]

In pigs, cysts can be found in organs such as the brain, lungs, liver, and spleen, with the possible occurrence of retinopathy.[178] In addition, these animals do not usually develop severe pathological clinical manifestations when infected with *T. gondii* if the immune system is not compromised, similar to what occurs in humans with competent immune system.[110]

Although there is potential use of the pig model for *T. gondii* infection involving the reproductive system, only a few groups have been working on this issue. The fact that these animals develop congenital infection strengthens the advantages of their experimental use. In naturally infected pregnant pigs, there may be damage to the placenta,[179] encephalitis, liver necrosis, and pneumonia in fetuses.[180] Similar changes may occur in human fetuses as a result of parasitemia and placental migration, particularly if the mother has contact with the parasite for the first time during pregnancy.[181] In addition, infective tachyzoites of *T. gondii* can be found in porcine semen for up to 49 days postinfection, demonstrating a potential for sexual transmission in this species.[182] Surprisingly, the hypothesis of congenital toxoplasmosis as a consequence of infection through human semen has also been recently proposed.[183]

An aspect evaluated via experimental infection with *T. gondii* in swine relates to its presence in blood. Biological evidence in infected mice demonstrates the viability of this parasite in the pig blood, which can be observed in neutrophils and monocytes, as well as outside these cells.[110] In fact, the parasitemia is persistent in pigs, and an accurate observation can detect tachyzoites for up to 63 days postinfection in pigs that show no clinical symptoms.[110]

Some studies have demonstrated important clinical and hematological alterations in pigs infected with *T. gondii*. However, such studies typically use highly infective parasite inoculum or very young animals,[184] factors that possibly enhance the pathology.[110] In our study, pigs with developed immune system were infected with parasite dosages close to a natural infection with the ME-49 strain, and typical clinical manifestations of acute toxoplasmosis were observed, such as general apathy, low fever, and, rarely, lymphadenopathy. Diarrhea was not observed, although the possibility exists and may depend on the strain and the infecting dose.[184] When orally infected, an increase in the number of band neutrophil followed by a tendency to monocytopenia was observed. Under these experimental infection conditions, no anemia or thrombocytopenia were observed, although liver injury was found as seen by increased serum alanine aminotransferase readings.[110] In addition, a persistent parasitemia without exacerbation of pathology was found, which is similar to human toxoplasmosis infections caused by some Brazilian strains.[185] This strengthens pig as an animal model to study human toxoplasmosis in immunocompetent patients.

41.2.2.4 Cat

Felids were only discovered as definitive hosts of *T. gondii* in the last half of the 1960s and in the beginning of the 1970s. Excellent reviews report how the life cycle of *T. gondii* was discovered.[35,186] With the advent of the serological tests, it was realized that *T. gondii* infection was disseminated around the globe in different hosts, including humans. This led to the assumption that transmission through carnivorism alone could not explain the broad prevalence in the world and how herbivores were being infected; in addition, transmission by invertebrate vectors was unsuccessful.[35,186] Ferguson[186] reported that Hutchison was intrigued by these questions and hypothesized that *T. gondii* could be fecal–orally transmitted by domestic animals. By using basic parasitology techniques, cats were fed with brain from infected mice; the parasite material was purified by flotation from cat feces, stored in water, and then fed to mice that later became ill with toxoplasmosis.[187] This work opened the field to others; the *T. gondii* oocyst was discovered, the sexual cycle of *T. gondii* in cats revealed, and the whole life cycle of this parasite solved.[34,188–191]

The discovery of the *T. gondii* life cycle explained the broad dissemination of this parasite with relevant importance to the epidemiology of the disease. Oocyst shedding contaminated the environment, ended up infecting herbivores, and was also responsible for human toxoplasma outbreaks by water contamination.[192] This led to proper management in farms and reduced the contamination of these animals.[35] In addition, more cats release faster and higher number of oocysts if fed with the bradyzoite form when compared to the tachyzoites or oocysts.[193] This indicates that the natural life cycle of the parasite is indeed by the transmission to the definitive host by carnivorism of small intermediate hosts (like mice) containing tissue cysts.

Naturally, the cat became an animal model to better understand the sexual cycle of *T. gondii*, for oocysts obtainment for further infection of other hosts, and to answer how the parasite behaves in the definitive host. Experimental manipulation of the cat with corticosteroids indicated that the immune response is important to stop oocysts shedding and to control parasite growth in the organs of these animals.[194] Young cats (no older than 12 months) shed more oocysts and also present more parasites in their tissue.[195]

Simultaneous infection of cats with two distinct strains of *T. gondii* has been done showing that classical genetic crosses do occur in this parasite.[196] These crosses have been used to create linkage maps of this parasite,[197] identifying 11 chromosomes.[198] Later, this technique was used to study virulent factors of the parasites, and a few have been identified.[40] Infected cats have also been used to genetically characterize the merozoites forms that multiply in cat enterocytes before gametogenesis. A complete distinct set of genes was identified in these forms when compared to tachyzoites.[37] Thus, the cat is fundamental for the development of *T. gondii* biology.

41.2.2.5 Chicken

Chickens are less used in experimental trials compared to the murine model. However, in some specific cases, chickens are an extremely suitable model for experimental studies.[199] The ease in handling, low maintenance cost, and the prolific characteristics of chickens favor their use in experimentations. In addition, chickens are the world's largest population of domestic animals with great economic importance and, therefore, there is interest in studying them.

Chickens may be 100% infected with *T. gondii* when raised in backyards, whereas those raised in free-range organic rearing can present prevalence as high as 50%.[200] Because of their feeding habits, chickens are characterized for being indicators of soil contamination by *T. gondii* oocysts. Moreover, these animals may be as important in the biological cycle as mice, due to their greater longevity.[200] In chickens, tissue cysts are more common in muscles than in the brain,[1] strengthening the importance of this animal as a source of human infection.[200]

Chickens do not normally develop a severe clinical disease in natural infections[200] and may not develop the disease when inoculated with the virulent *T. gondii* strains for mice.[201] However, diarrhea, emaciation, blindness, and sudden death have been reported in chickens naturally infected with *T. gondii*, with isolation of the parasite in the heart, liver, brain, and lungs. Interestingly enough, the parasite isolates were avirulent for mice.[202,203]

Prior to the discovery of *T. gondii* life cycle and the importance of oral transmission, studies emphasized the parenteral infection route, which was not accompanied by clinical manifestations.[202–204] After this period, oral infections have shown variable results with chickens depending on the strain used. Chickens inoculated with oocysts of the ME-49 strain (Type II) did not develop any clinical symptoms, whereas chickens infected with the GT1 strain (Type I) became ill, resulting in 20% (one of five animals) death.[205] In addition, oral infection with oocysts of the E strain caused mild fever only in highly infective doses.[206] Although the relative acute phase of the *T. gondii* infection in chickens does not result in clinical manifestations, the chronicity of the disease may cause physiological changes and possible production losses.[207] Chicken infection with the M-7741 strain caused diarrhea and inflammatory cell infiltration in the intestinal wall, with a strong reduction of the density of myenteric neurons. The chronic phase was correlated with neuronal death and atrophy of the intestinal wall, causing gastrointestinal disorders.[207]

The relatively mild acute phase followed by a chronic phase with potential epidemiological importance indicates that the experimental infection of chickens with *T. gondii* is indeed an interesting model, particularly with respect to immunological studies. Therefore, interaction with immune cells from primary cultures, or with the established cell lines such as the HD11 of the chicken macrophages,[70,208] may generate interesting results.

41.2.2.6 Fish

T. gondii is known to infect nucleated cells of warm-blooded animals, with mice being the most common and important intermediate host.[73,75] However, some studies have indicated that fish and shellfish may be involved in the natural infection of different hosts by *T. gondii*. In China, parasitic DNA was identified in a small number of crayfish *Procambarus clarkii*, fish *Hypophthalmichthys molitrix*, and shrimp *Macrobrachium nipponense*.[209] The authors indicated that the consumption of raw meat of these animals may be considered a health threat related to *T. gondii* infection.[209] In Brazil, it was shown that oysters (*Crassostrea rhizophorae*) may retain oocysts of *T. gondii* as these animals filter water, and the parasite DNA was found in a low percentage of individuals.[210] There is a need for more studies on the participation of these animals as carriers of *T. gondii*.[210] However, the importance of these cold-blooded animals as an effective intermediate host has not been clearly demonstrated, since the presence of DNA in the sample does not mean viable parasites. It is necessary to differentiate the participation of a potential host or parasite carrier from a mechanical vector.

Grizzly bears (*Ursus arctos*) in Alaska showed a *T. gondii* serology prevalence up to 37% depending on the area studied.[211] Although this animal feeds on fish, no established relationship with these cold-blooded animals was established, because toxoplasmosis is also a waterborne disease.[6] Thus, one cannot exclude the infection of these animals through contaminated water.

Experimental studies have demonstrated the inability of *T. gondii* to persist in tissues of goldfish maintained at 37°C.[212] However, in appropriate conditions, especially temperature, this can be used as a new experimental infection model for *T. gondii* in the zebrafish (*Danio rerio*).[213] Proliferation of *T. gondii* of the ME-49 and VEG strains was observed in the cardiac myocytes, endothelium, lumen of peripheral vessels, liver, spleen, brain, pancreas, ovaries, skeletal musculature, and the eye of this animal infected after acclimation at 37°C. Two weeks after infection, cystic structures were observed in the brain, suggesting that this model may also be fit for chronic studies. The authors report that the only limitation to the study of toxoplasmosis in this fish was temperature. This is a promising model because this fish can be kept at temperatures of 37°C, used in high-throughput screenings for new compound, and be genetically modified.[213]

41.3 Concluding Remarks

Toxoplasmosis is a disease with worldwide distribution affecting about 2×10^9 people. First considered a foodborne disease, it is now also recognized as a waterborne disease. It has been suggested that cultural customs dictate the way of transmission, especially in regard to the consumption of raw or undercooked meat. This disease can cause different clinical manifestations, with serious implications to immunocompromised patients. The disease also affects a large number of wild and domestic warm-blooded animals causing great economic losses. The etiological agent of this disease is the obligate intracellular protozoan *Toxoplasma gondii*, which infects a vast number of hosts and has an elaborate life cycle that helps to explain its broad dissemination. With the establishment of *T. gondii* cell culture, the biological characteristics of the parasite have been delineated through the infection of primary cells, cell lines, and transfected cells with distinct strains and mutant parasites. Because of its broad host range, most animal models are suited to investigate this disease.

With a complex life cycle involving various hosts and a diversity of genotypes made up of multiple strains, *T. gondii* is a peculiar parasite. Thus, it is imperative to use different models to better understand the many distinct steps of infection leading to acute and chronic stages of this disease. *In vitro* models are in constant development and yield interesting outcomes, but the obtained results must be confirmed via *in vivo* models. To date, no animal model that mimics the human infection by *T. gondii* exists; however, recent advances on secreted virulence factors by distinct *T. gondii* strains and specific antimicrobial mechanisms of host cells from different host animals that are strain-dependent have increased our knowledge and shown how adapted this parasite is. Further research is clearly required to fully elucidate the clinical variations of toxoplasmosis and to unravel if specific microbicidal mechanisms operate in humans and what parasite genes code for virulence factors that help its evasion of host immune functions. In addition, its genetic diversity also demands that diagnosis and treatment of toxoplasmosis need to be geographically specific. There is no doubt that the models described here and others to come will contribute to improved understanding of this important parasite and the disease it causes.

Acknowledgments

The authors would like to thank Andrèa Carvalho César for proofing the manuscript. This study was supported by the following Brazilian agencies: Conselho Nacional de Desenvolvimento Científico e Tecnológico (CNPq), Fundação Carlos Chagas Filho de Amparo à Pesquisa do Estado do Rio de Janeiro (FAPERJ), and Fundação de Coordenação de Pessoal de Nível Superior (CAPES).

REFERENCES

1. Dubey, J. P., and Beattie, C. P., *Toxoplasmosis of Animals and Man*, Boca Raton, FL: CRC Press, pp. 1–220, 1988.
2. Robert-Gangneux, F., and Dardé, M, L., Epidemiology of and diagnostic strategies for toxoplasmosis, *Clin. Microbiol. Rev.*, 25, 264, 2012.
3. Tenter, A. M., Heckeroth, A. R., and Weiss, L. M., *Toxoplasma gondii*: from animals to humans, *Int. J. Parasitol.*, 30, 1217, 2000.
4. Dubey, J. P., The history of *Toxoplasma gondii*—the first 100 years, *J. Eukaryot. Microbiol.*, 55, 467, 2008.
5. Weiss, L. M., and Dubey, J. P., Toxoplasmosis: a history of clinical observations, *Int. J. Parasitol.*, 39, 895, 2009.
6. Jones, J. L., and Dubey, J. P., Waterborne toxoplasmosis: recent developments, *Exp. Parasitol.*, 124, 10, 2010.
7. Bowie, W. R., et al., Outbreak of toxoplasmosis associated with municipal drinking water. The BC Toxoplasma Investigation Team, *Lancet*, 350, 173, 1997.
8. Bahia-Oliveira, L. M., et al., Highly endemic, waterborne toxoplasmosis in north Rio de Janeiro state, Brazil, *Emerg. Infect. Dis.*, 9, 55, 2003.
9. de Moura, L., et al., Waterborne toxoplasmosis, Brazil, from field to gene, *Emerg. Infect. Dis.*, 12, 326, 2006.
10. Vaudaux, J. D., et al., Identification of an atypical strain of *Toxoplasma gondii* as the cause of a waterborne outbreak of toxoplasmosis in Santa Isabel do Ivai, Brazil, *J. Infect. Dis.*, 202, 1226, 2010.
11. Diana, J., et al., Migration and maturation of human dendritic cells infected with *Toxoplasma gondii* depend on parasite strain type, *FEMS Immunol. Med. Microbiol.*, 42, 321, 2004.
12. Johnson, A. M., Strain-dependent, route of challenge-dependent, murine susceptibility to toxoplasmosis, *Z. Parasitenkd.*, 70, 303, 1984.
13. Dubey, J. P., Lindsay, D. S., and Speer, C. A., Structures of *Toxoplasma gondii* tachyzoites, bradyzoites, and sporozoites and biology and development of tissue cysts, *Clin. Microbiol. Rev.*, 11, 267, 1998.
14. Vasconcelos-Santos, D. V., Ocular manifestations of systemic disease: toxoplasmosis, *Curr. Opin. Ophthalmol.*, 23, 543, 2012.
15. Halonen, S. K., and Weiss, L. M., Toxoplasmosis, *Handb. Clin. Neurol.*, 114, 125, 2013.
16. Demar, M., et al., Acute toxoplasmoses in immunocompetent patients hospitalized in an intensive care unit in French Guiana, *Clin. Microbiol. Infect.*, 18, E221, 2012.
17. Dubey, J. P., et al., Toxoplasmosis in humans and animals in Brazil: high prevalence, high burden of disease, and epidemiology, *Parasitology*, 139, 1375, 2012.
18. Waleska, T. C., et al., Toxoplasmosis and mental retardation. Report of a case control study, *Mem. Inst. Oswaldo Cruz*, 88, 253, 1993.
19. Sutterland, A. L., et al., Beyond the association. *Toxoplasma gondii* in schizophrenia, bipolar disorder, and addiction: systematic review and meta-analysis, *Acta Psychiatr. Scand.*, 132, 161, 2015.
20. Sibley, L. D., et al., Genetic diversity of *Toxoplasma gondii* in animals and humans, *Philos. Trans. R. Soc. Lond. B. Biol. Sci.*, 364, 2749, 2009.
21. Frenkel, J. K., *Toxoplasma* in and around us, *Bioscience*, 23, 343, 1973.
22. Torrey, E. F., and Yolken, R. H., *Toxoplasma* oocysts as a public health problem, *Trends Parasitol.*, 29, 380, 2013.

23. Levine, N. D., *The Protozoan Phylum Apicomplexa*, Boca Raton, FL: CRC Press, 1988.
24. Black, M. W., and Boothroyd, J. C., Lytic cycle of *Toxoplasma gondii*, *Microbiol. Mol. Biol. Rev.*, 64, 607, 2000.
25. Shimada, K., et al., Cyst formation by *Toxoplasma gondii* (RH strain) *in vitro*. The role of immunologic mechanisms, *Arch. Ophthalmol.*, 92, 496, 1974.
26. Jones, T. C., Bienz, K. A., and Erb, P., *In vitro* cultivation of *Toxoplasma gondii* cysts in astrocytes in the presence of γ-interferon, *Infect. Immun.*, 51, 147, 1986.
27. Bohne, W., Holpert, M., and Gross, U., Stage differentiation of the protozoan parasite *Toxoplasma gondii*, *Immunobiology*, 201, 248, 1999.
28. Skariah, S., Mcintyre, M. K., and Mordue, D. G., *Toxoplasma gondii*: determinants of tachyzoite to bradyzoite conversion, *Parasitol. Res.*, 107, 253, 2010.
29. Fischer, H. G., et al., Host cells of *Toxoplasma gondii* encystation in infected primary culture from mouse brain, *Parasitol. Res.*, 83, 637, 1997.
30. Lüder, C. G., et al., *Toxoplasma gondii* in primary rat CNS cells: differential contribution of neurons, astrocytes, and microglial cells for the intracerebral development and stage differentiation, *Exp. Parasitol.*, 93, 23, 1999.
31. Ferreira da Silva Mda, F., et al., Stress-related and spontaneous stage differentiation of *Toxoplasma gondii*, *Mol. Biosyst.*, 4, 824, 2008.
32. Ferreira da Silva Mda, F., et al., Spontaneous stage differentiation of mouse-virulent *Toxoplasma gondii* RH parasites in skeletal muscle cells: an ultrastructural evaluation, *Mem. Inst. Oswaldo Cruz*, 104, 196, 2009.
33. Dubey, J. P., Tissue cyst tropism in *Toxoplasma gondii*: a comparison of tissue cyst formation in organs of cats, and rodents fed oocysts, *Parasitology*, 115, 15, 1997.
34. Frenkel, J. K., Dubey, J. P., and Miller, N. L., *Toxoplasma gondii* in cats: fecal stages identified as coccidian oocysts, *Science*, 167, 893, 1970.
35. Dubey, J. P., History of the discovery of the life cycle of *Toxoplasma gondii*, *Int. J. Parasitol.*, 39, 877, 2009.
36. Fayer, R., Toxoplasmosis update and public health implications, *Can. Vet. J.*, 22, 344, 1981.
37. Hehl, A. B., et al., Asexual expansion of *Toxoplasma gondii* merozoites is distinct from tachyzoites and entails expression of non-overlapping gene families to attach, invade, and replicate within feline enterocytes, *BMC Genomics*, 16, 66, 2015.
38. Speer, C. A., and Dubey, J. P., Ultrastructure of early stages of infections in mice fed *Toxoplasma gondii* oocysts, *Parasitology*, 116, 35, 1998.
39. Dubey, J. P., et al., Genetic diversity of *Toxoplasma gondii* isolates from chickens from Brazil, *Vet. Parasitol.*, 157, 299, 2008.
40. Hunter, C. A., and Sibley, L. D., Modulation of innate immunity by *Toxoplasma gondii* virulence effectors, *Nat. Rev. Microbiol.*, 10, 766, 2012.
41. Howe, D. K., and Sibley, L. D., *Toxoplasma gondii* comprises three clonal lineages: correlation of parasite genotype with human diseases, *J. Infect. Dis.*, 172, 1561, 2005.
42. Shwab, E. K., et al., Geographical patterns of *Toxoplasma gondii* genetic diversity revealed by multilocus PCR-RFLP genotyping, *Parasitology*, 141, 453, 2014.
43. Cañón-Franco, W. A., et al., An overview of seventy years of research (1944–2014) on toxoplasmosis in Colombia, South America, *Parasite Vectors*, 7, 427, 2014.
44. Splendore, A., Um nuovo protozoa parassita dei conigli incontrato nelle lesioni anatomiche duna malattia che ricorda in molti ponti il kala-azar delluomo, *Rev. Soc. Sci. São Paulo*, 3, 109, 1908.
45. Nicolle, C., and Manceaux, L., Sur une infection a corps de Leishman (ou organisms voisins) du gondi, *C. R. Acad. Sci.*, 147, 763, 1908.
46. Ferguson, D. J., *Toxoplasma gondii*: 1908–2008, homage to Nicolle, Manceaux and Splendore, *Mem. Inst. Oswaldo Cruz*, 104, 133, 2009.
47. Carrel, A., and Montrose, T. B., Cultivation of tissues *in vitro* and its technique, *J. Exp. Med.*, 13, 387, 1911.
48. Sabin, A. B., and Olitsky, P. K., *Toxoplasma* and obligate intracellular parasitism, *Science*, 85, 336, 1937.
49. Beverley, J. K., and Fry, B. A., Sulphadimidine, pyrimethamine and dapsone in the treatment of toxoplasmosis in mice, *Br. J. Pharmacol. Chemother.*, 12, 189, 1957.
50. Maloney, E. D., and Kaufman, H. E., Multiplication and therapy of *Toxoplasma gondii* in tissue culture, *J. Bacteriol.*, 88, 319, 1964.
51. Summers, W. A., Antagonism of sulfonamide inhibition by para-aminobenzoic acid and folic acid in *Toxoplasma* infected mice, *Proc. Soc. Exp. Biol. Med.*, 66, 509, 1947.

52. Wolf, A., Cowen, D., and Paige, B., Human toxoplasmosis: occurrence in infants as an encephalomyelitis verification by transmission to animals, *Science*, 89, 226, 1939.
53. Wolf, A., Cowen, D., and Paige, B. H., Toxoplasmic encephalomyelitis: IV. Experimental transmission of the infection to animals from a human infant, *J. Exp. Med.*, 71, 187, 1940.
54. Nakayama, I., and Matsubayashi, H., Experimental transmission of *Toxoplasma gondii* in mice, *Keio J. Med.*, 10, 163, 1961.
55. Guimarães, F. N., and Meyer, H. Cultivo de "Toxoplasma" Nicolle & Manceaux, 1909, em cultura de tecidos, *Ver. Bras. Biol.*, 2, 123, 1942.
56. Jacobs, L., Propagation, morphology, and biology of *Toxoplasma*, *Ann. N. Y. Acad. Sci.*, 64, 154, 1956.
57. de Souza, W., DaMatta, R. A., and Attias, M., Brazilian contribution for a better knowledge on the biology of *Toxoplasma gondii*, *Mem. Inst. Oswaldo Cruz*, 104, 149, 2009.
58. Jiménez-Ruiz, E., et al., Advantages and disadvantages of conditional systems for characterization of essential genes in *Toxoplasma gondii*, *Parasitology*, 141, 1390, 2014.
59. Hartl, D. L., and Jones, E. W., *Genetics: Principles and Analysis*, Boston, MA: Jones and Bartlett Publishers, 1998.
60. Jones, T. C., and Hirsch, J. G., The interaction between *Toxoplasma gondii* and mammalian cells: II. The absence of lysosomal fusion with phagocytic vacuoles containing living parasites, *J. Exp. Med.*, 136, 1173, 1972.
61. Jones, T. C., Yeh, S., and Hirsch, J. G., The interaction between *Toxoplasma gondii* and mammalian cells: I. Mechanism of entry and intracellular fate of the parasite, *J. Exp. Med.*, 136, 1157, 1972.
62. Sibley, L. D., Weidner, E., and Krahenbuhl, J. L., Phagosome acidification blocked by intracellular *Toxoplasma gondii*, *Nature*, 315, 416, 1985.
63. Sinai, A. P., Biogenesis of and activities at the *Toxoplasma gondii* parasitophorous vacuole membrane, *Subcell. Biochem.*, 47, 155, 2008.
64. Boyle, J. P., and Radke, J. R., A history of studies that examine the interactions of *Toxoplasma* with its host cell: emphasis on *in vitro* models, *Int. J. Parasitol.*, 39, 903, 2009.
65. Murray, H. W., and Cohn, Z. A., Macrophage oxygen-dependent antimicrobial activity. I. Susceptibility of *Toxoplasma gondii* to oxygen intermediates, *J. Exp. Med.*, 150, 938, 1979.
66. Sibley, L. D., Lawson, R., and Weidner, E., Superoxide dismutase and catalase in *Toxoplasma gondii*, *Mol. Biochem. Parasitol.*, 19, 83, 1986.
67. Ding, M., et al., The antioxidant systems in *Toxoplasma gondii* and the role of cytosolic catalase in defence against oxidative injury, *Mol. Microbiol.*, 51, 47, 2004.
68. Wilson, C. B., Tsai, V., and Remington, J. S., Failure to trigger the oxidative burst by normal macrophages. Possible mechanism for survival of intracellular pathogens, *J. Exp. Med.*, 151, 328, 1980.
69. Murray, H. W., and Cohn, Z. A., Macrophage oxygen-dependent antimicrobial activity. III. Enhanced oxidative metabolism as an expression of macrophage activation, *J. Exp. Med.*, 152, 1596, 1980.
70. Guillermo, L. V., and DaMatta, R. A., Nitric oxide inhibition after *Toxoplasma gondii* infection of chicken macrophage cell lines, *Poult. Sci.*, 83, 776, 2004.
71. Shrestha, S. P., et al., Proliferation of *Toxoplasma gondii* in inflammatory macrophages *in vivo* is associated with diminished oxygen radical production in the host cell, *Int. J. Parasitol.*, 36, 433, 2006.
72. Collazo, C. M., et al., Inactivation of LRG-47 and IRG-47 reveals a family of interferon γ-inducible genes with essential, pathogen-specific roles in resistance to infection, *J. Exp. Med.*, 194, 181, 2001.
73. Howard, J. C., Hunn, J. P., and Steinfeldt, T., The IRG protein-based resistance mechanism in mice and its relation to virulence in *Toxoplasma gondii*, *Curr. Opin. Microbiol.*, 14, 414, 2011.
74. Liesenfeld, O., et al., The IFN-γ-inducible GTPase, Irga6, protects mice against *Toxoplasma gondii* but not against *Plasmodium berghei* and some other intracellular pathogens, *PLoS One*, 6, e20568, 2011.
75. Gazzinelli, R. T., et al., Innate resistance against *Toxoplasma gondii*: an evolutionary tale of mice, cats, and men, *Cell Host Microbes*, 15, 132, 2014.
76. Fentress SJ, and Sibley, L. D., The secreted kinase ROP18 defends *Toxoplasma's* border, *Bioessays*, 33, 693, 2011.
77. Khan, A., et al., Selection at a single locus leads to widespread expansion of *Toxoplasma gondii* lineages that are virulent in mice, *PLoS Genet.*, 5, e1000404, 2009.
78. Reese, M. L., et al., Polymorphic family of injected pseudokinases is paramount in *Toxoplasma* virulence, *Proc. Natl. Acad. Sci. USA*, 108, 9625, 2011.

79. Steinfeldt, T., et al., Phosphorylation of mouse immunity-related GTPase (IRG) resistance proteins is an evasion strategy for virulent *Toxoplasma gondii*, *PLoS Biol.*, 8, e1000576, 2010.
80. Etheridge, R. D., et al., The *Toxoplasma* pseudokinase ROP5 forms complexes with ROP18 and ROP17 kinases that synergize to control acute virulence in mice, *Cell Host Microbes*, 15, 537, 2014.
81. Alaganan, A., et al., *Toxoplasma* GRA7 effector increases turnover of immunity-related GTPases and contributes to acute virulence in the mouse, *Proc. Natl. Acad. Sci. USA*, 111, 1126, 2014.
82. Da Gama, L. M., et al., Reduction in adhesiveness to extracellular matrix components, modulation of adhesion molecules and *in vivo* migration of murine macrophages infected with *Toxoplasma gondii*, *Microbes Infect.*, 6, 1287, 2004.
83. Lambert, H., et al., The *Toxoplasma gondii*-shuttling function of dendritic cells is linked to the parasite genotype, *Infect. Immun.*, 77, 1679, 2009.
84. Seipel, D., et al., *Toxoplasma gondii* infection positively modulates the macrophages migratory molecular complex by increasing matrix metalloproteinases, CD44 and alpha v beta 3 integrin, *Vet. Parasitol.*, 169, 312, 2010.
85. Seipel, D., et al., Monocytes/macrophages infected with *Toxoplasma gondii* do not increase co-stimulatory molecules while maintaining their migratory ability, *APMIS*, 117, 672, 2009.
86. Schuindt, S. H., et al., Secretion of multi-protein migratory complex induced by *Toxoplasma gondii* infection in macrophages involves the uPA/uPAR activation system, *Vet. Parasitol.*, 186, 207, 2012.
87. MacMicking, J., Xie, Q. W., and Nathan, C., Nitric oxide and macrophage function, *Annu. Rev. Immunol.*, 15, 323, 1997.
88. Bogdan, C., Nitric oxide synthase in innate and adaptive immunity: an update, *Trends Immunol.*, 36, 161, 2015.
89. Hayashi, S., et al., Contribution of nitric oxide to the host parasite equilibrium in toxoplasmosis, *J. Immunol.*, 156, 1476, 1996.
90. Khan, I. A., et al., A dichotomous role for nitric oxide during acute *Toxoplasma gondii* infection in mice, *Proc. Natl. Acad. Sci. USA*, 94, 13955, 1997.
91. Scharton-Kersten, T. M., et al., Inducible nitric oxide is essential for host control of persistent but not acute infection with the intracellular pathogen *Toxoplasma gondii*, *J. Exp. Med.*, 185, 1261, 1997.
92. Liesenfeld, O., et al., TNF-α, nitric oxide and IFN-γ are all critical for development of necrosis in the small intestine and early mortality in genetically susceptible mice infected perorally with *Toxoplasma gondii*, *Parasite Immunol.*, 21, 365, 1999.
93. Adams, L. B., et al., Microbiostatic effect of murine-activated macrophages for *Toxoplasma gondii*. Role for synthesis of inorganic nitrogen oxides from L-arginine, *J. Immunol.*, 144, 2725, 1990.
94. Nibbering, P. H., et al., Nitrite production by activated murine macrophages correlates with their toxoplasmastatic activity, Ia antigen expression, and production of H_2O_2, *Immunobiology*, 184, 93, 1991.
95. DaMatta, R. A., et al., Nitric oxide is not involved in the killing of *Trypanosoma cruzi* by chicken macrophages, *Parasitol. Res.*, 86, 239, 2000.
96. Seabra, S. H., de Souza, W., and DaMatta, R. A., *Toxoplasma gondii* partially inhibits nitric oxide production of activated murine macrophages, *Exp. Parasitol.*, 100, 62, 2002.
97. Seabra, S. H., de Souza, W., and DaMatta, R. A., *Toxoplasma gondii* exposes phosphatidylserine inducing a TGF-beta1 autocrine effect orchestrating macrophage evasion, *Biochem. Biophys. Res. Commun.*, 324, 744, 2004.
98. Santos, T. A., et al., Phosphatidylserine exposure by *Toxoplasma gondii* is fundamental to balance the immune response granting survival of the parasite and of the host, *PLoS One*, 6, e27867, 2011.
99. Padrão, J. C., et al., *Toxoplasma gondii* infection of activated J774-A1 macrophages causes inducible nitric oxide synthase degradation by the proteasome pathway, *Parasitol. Int.*, 63, 659, 2014.
100. de Freitas Balanco, J. M., et al., Apoptotic mimicry by an obligate intracellular parasite downregulates macrophage microbicidal activity, *Curr. Biol.*, 11, 1870, 2001.
101. Wanderley, J. L., and Barcinski, M. A., Apoptosis and apoptotic mimicry: the Leishmania connection, *Cell Mol. Life Sci.*, 67, 1653, 2010.
102. Guimarães, E. V., de Carvalho, L., and Barbosa, H. S., Primary culture of skeletal muscle cells as a model for studies of *Toxoplasma gondii* cystogenesis, *J. Parasitol.*, 94, 72, 2008.
103. Andrade, E. F., et al., Do microtubules around the *Toxoplasma gondii*-containing parasitophorous vacuole in skeletal muscle cells form a barrier for the phagolysosomal fusion? *J. Submicrosc. Cytol. Pathol.*, 33, 337, 2001.

104. Takács, A. C., Swierzy, I. J., and Lüder, C. G., Interferon-γ restricts *Toxoplasma gondii* development in murine skeletal muscle cells via nitric oxide production and immunity-related GTPases, *PLoS One*, 7, e45440, 2012.
105. Gomes, A. F., et al., *Toxoplasma gondii*-skeletal muscle cells interaction increases lipid droplet biogenesis and positively modulates the production of IL-12, IFN-g and PGE_2, *Parasite Vectors*, 7, 47, 2014.
106. Mota, L. A., et al., Culture of mouse peritoneal macrophages with mouse serum induces lipid bodies that associate with the parasitophorous vacuole and decrease their microbicidal capacity against *Toxoplasma gondii*, *Mem. Inst. Oswaldo Cruz*, 109, 767, 2014.
107. de Muno, R. M., et al., Spontaneous cystogenesis of *Toxoplasma gondii* in feline epithelial cells *in vitro*, *Folia Parasitol. (Praha)*, 61, 113, 2014.
108. Moura, Mde. A., Amendoeira, M. R., and Barbosa, H. S., Primary culture of intestinal epithelial cells as a potential model for *Toxoplasma gondii* enteric cycle studies, *Mem. Inst. Oswaldo Cruz*, 104, 862, 2009.
109. Lind, P., et al., The time course of the specific antibody response by various ELISAs in pigs experimentally infected with *Toxoplasma gondii*, *Vet. Parasitol.*, 71, 1, 1997.
110. Miranda, F. J., et al., Experimental infection with the *Toxoplasma gondii* ME-49 strain in the Brazilian BR-1 mini pig is a suitable animal model for human toxoplasmosis, *Mem. Inst. Oswaldo Cruz*, 110, 95, 2015.
111. Dias, R. R., et al., *Toxoplasma gondii* oral infection induces intestinal inflammation and retinochoroiditis in mice genetically selected for immune oral tolerance resistance, *PLoS One*, 9(12), e113374, 2014.
112. Lin, M. H., et al., Real-time PCR for quantitative detection of *Toxoplasma gondii*, *J. Clin. Microbiol.*, 38, 4121, 2000.
113. Meerburg, B. G., Singleton, G. R., and Kijlstra, A., Rodent-borne diseases and their risks for public health, *Crit. Rev. Microbiol.*, 35, 221, 2009.
114. Dabritz, H. A., et al., Risk factors for *Toxoplasma gondii* infection in wild rodents from central coastal California and a review of *T. gondii* prevalence in rodents, *J. Parasitol.*, 94, 675, 2008.
115. Hide, G., et al., Evidence for high levels of vertical transmission in *Toxoplasma gondii*, *Parasitology*, 136, 1877, 2009.
116. Lilue, J., et al., Reciprocal virulence and resistance polymorphism in the relationship between *Toxoplasma gondii* and the house mouse, *Elife*, 2, e01298, 2013.
117. Lélu, M., et al., When should a trophically and vertically transmitted parasite manipulate its intermediate host? The case of *Toxoplasma gondii*, *Proc. Biol. Sci.*, 280, 20131143, 2013.
118. Yap, G. S., and Sher, A., The use of germ line-mutated mice in understanding host-pathogen interactions, *Cell Microbiol.*, 4, 627, 2002.
119. Subauste, C. S., and Remington, J. S., Role of γ interferon in *Toxoplasma gondii* infection, *Eur. J. Clin. Microbiol. Infect. Dis.*, 10, 58, 1991.
120. McCabe, R. E., Luft, B. J., and Remington, J. S., Effect of murine interferon-γ on murine toxoplasmosis, *J. Infect. Dis.*, 150, 961, 1984.
121. Suzuki, Y., et al., Interferon-γ: the major mediator of resistance against *Toxoplasma gondii*, *Science*, 240, 516, 1988.
122. Scharton-Kersten, T. M., et al., In the absence of endogenous IFN-γ, mice develop unimpaired IL-12 responses to *Toxoplasma gondii* while failing to control acute infection, *J. Immunol.*, 157, 4045, 1996.
123. Hunter, C. A., et al., Studies on the role of interleukin-12 in acute murine toxoplasmosis, *Immunology*, 84, 16, 1995.
124. Hunter, C. A., Chizzonite, R., and Remington, J. S., IL-1 beta is required for IL-12 to induce production of IFN-γ by NK cells. A role for IL-1 beta in the T cell-independent mechanism of resistance against intracellular pathogens, *J. Immunol.*, 155, 4347, 1995.
125. Scharton-Kersten, T., et al., Interferon consensus sequence binding protein-deficient mice display impaired resistance to intracellular infection due to a primary defect in interleukin 12 p40 induction, *J. Exp. Med.*, 186, 1523, 1997.
126. Gazzinelli, R. T., et al., In the absence of endogenous IL-10, mice acutely infected with *Toxoplasma gondii* succumb to a lethal immune response dependent on CD4+ T cells and accompanied by overproduction of IL-12, IFN-γ and TNF-α, *J. Immunol.*, 157, 798, 1996.
127. Scanga, C. A., et al., Cutting edge: MyD88 is required for resistance to *Toxoplasma gondii* infection and regulates parasite-induced IL-12 production by dendritic cells, *J. Immunol.*, 168, 5997, 2002.

128. Yarovinsky, F., and Sher, A., Toll-like receptor recognition of *Toxoplasma gondii*, *Int. J. Parasitol.*, 36, 255, 2006.
129. Liesenfeld, O., Oral infection of C57BL/6 mice with *Toxoplasma gondii*: a new model of inflammatory bowel disease? *J. Infect. Dis.*, 185 Suppl 1, S96, 2002.
130. Schreiner, M., and Liesenfeld, O., Small intestinal inflammation following oral infection with *Toxoplasma gondii* does not occur exclusively in C57BL/6 mice: review of 70 reports from the literature, *Mem. Inst. Oswaldo Cruz*, 104, 221, 2009.
131. Heimesaat, M. M., et al., Gram-negative bacteria aggravate murine small intestinal Th1-type immunopathology following oral infection with *Toxoplasma gondii*, *J. Immunol.*, 177, 8785, 2006.
132. Muñoz, M., et al., Interleukin (IL)-23 mediates *Toxoplasma gondii*-induced immunopathology in the gut via matrixmetalloproteinase-2 and IL-22 but independent of IL-17, *J. Exp. Med.*, 206, 3047, 2009.
133. Cavalcanti, M. G., et al., MIF participates in *Toxoplasma gondii*-induced pathology following oral infection, *PLoS One*, 6(9), e25259, 2011.
134. Guiton, R., et al., Interleukin 17 receptor signaling is deleterious during *Toxoplasma gondii* infection in susceptible BL6 mice, *J. Infect. Dis.*, 202, 427, 2010.
135. Passos, S. T., et al., IL-6 promotes NK cell production of IL-17 during toxoplasmosis, *J. Immunol.*, 184, 1776, 2010.
136. Raetz, M., et al., Parasite-induced TH1 cells and intestinal dysbiosis cooperate in IFN-γ-dependent elimination of Paneth cells, *Nat. Immunol.*, 14, 136, 2013.
137. Molloy, M. J., et al., Intraluminal containment of commensal outgrowth in the gut during infection-induced dysbiosis, *Cell Host Microbes*, 14, 318, 2013.
138. Abi Abdallah, D. S., and Denkers, E. Y., Neutrophils cast extracellular traps in response to protozoan parasites, *Front. Immunol.*, 3, 382, 2012.
139. Mouse Nomenclature Home Page, http://www.informatics.jax.org/mgihome/nomen/strains.shtml (accessed September 2015).
140. Dubey, J. P., and Frenkel, J. K., Toxoplasmosis of rats: a review, with considerations of their value as an animal model and their possible role in epidemiology, *Vet. Parasitol.*, 77, 1, 1998.
141. Chinchilla, M., Alfaro, M., and Guerrero, O. M., Natural adaptation of the white rat to *Toxoplasma gondii*, *Rev. Biol. Trop.*, 29, 273, 1981.
142. Chinchilla, M., Guerrero, O. M., and Solano, E., Lack of multiplication of *Toxoplasma* in macrophages of rats *in vitro*, *J. Parasitol.*, 68, 952, 1982.
143. Dubey, J. P., Pathogenicity and infectivity of *Toxoplasma gondii* oocysts for rats, *J. Parasitol.*, 82, 951, 1996.
144. Freyre, A., et al., *Toxoplasma gondii*: an improved rat model of congenital infection, *Exp. Parasitol.*, 120, 142, 2008.
145. Zenner, L., et al., Rat model of congenital toxoplasmosis: rate of transmission of three *Toxoplasma gondii* strains to fetuses and protective effect of a chronic infection, *Infect. Immun.*, 61, 360, 1993.
146. Kempf, M. C., et al., Different manifestations of *Toxoplasma gondii* infection in F344 and LEW rats, *Med. Microbiol. Immunol.*, 187, 137, 1999.
147. Sergent, V., et al., Innate refractoriness of the Lewis rat to toxoplasmosis is a dominant trait that is intrinsic to bone marrow-derived cells, *Infect. Immun.*, 73, 6990, 2005.
148. Cavaillès, P., et al., The rat Toxo1 locus directs toxoplasmosis outcome and controls parasite proliferation and spreading by macrophage-dependent mechanisms, *Proc. Natl. Acad. Sci. USA*, 103, 744, 2006.
149. Cavaillès, P., et al., A highly conserved Toxo1 haplotype directs resistance to toxoplasmosis and its associated caspase-1 dependent killing of parasite and host macrophage, *PLoS Pathog.*, 10(4), e1004005, 2014.
150. Cirelli, K. M., et al., Inflammasome sensor NLRP1 controls rat macrophage susceptibility to *Toxoplasma gondii*, *PLoS Pathog.*, 10(3), e1003927, 2014.
151. Lamkanfi, M., and Dixit, V. M., Mechanisms and functions of inflammasomes, *Cell*, 157, 1013, 2014.
152. Ewald, S. E., Chavarria-Smith, J., Boothroyd, J. C., NLRP1 is an inflammasome sensor for *Toxoplasma gondii*, *Infect. Immun.*, 82, 460, 2014.
153. Gorfu, G., et al., Dual role for inflammasome sensors NLRP1 and NLRP3 in murine resistance to *Toxoplasma gondii*, *mBio*, 5(1), pii, e01117-13, 2014.
154. Witola, W. H., et al., ALOX12 in human toxoplasmosis, *Infect. Immun.*, 82, 2670, 2014.
155. Witola, W. H., et al., NALP1 influences susceptibility to human congenital toxoplasmosis, proinflammatory cytokine response, and fate of *Toxoplasma gondii*-infected monocytic cells, *Infect. Immun.*, 79, 756, 2011.

156. Gov, L., et al., Human innate immunity to *Toxoplasma gondii* is mediated by host caspase-1 and ASC and parasite GRA15, *mBio*, 4(4) pii, e00255-13, 2013.
157. Bonvicino, C. R., Lima, J. F. S., and Almeida, F. C., A new species of *Calomys* Waterhouse (Rodentia, Sigmodontinae) from the cerrado of central Brazil, *Rev. Bras. Zool.*, 20, 301, 2003.
158. Hershkovitz, P., The evolution of mammals of the neotropical region. A zoogeographic and ecological review, *Q. Ver. Biol.*, 44, 1, 1969.
159. Mello, A. D., *Calomys callosus* Rengger, 1830 (Rodentia-Cricetidadae): Sua caracterização, distribuição, biologia, criação e manejo de uma cepa em laboratório, *Mem. Inst. Oswaldo Cruz*, 79, 37, 1984.
160. Dubey, J. P., et al., Characterization of *Toxoplasma gondii* isolates in free-range chickens from Amazon, Brazil, *J. Parasitol.*, 92, 36, 2006.
161. Franco, P. S., et al., Experimental infection of *Calomys callosus* with atypical strains of *Toxoplasma gondii* shows gender differences in severity of infection, *Parasitol. Res.*, 113, 2655, 2014.
162. Favoreto-Junior, S., et al., Experimental Infection of *Calomys callosus* (Rodentia, Cricetidae) by *Toxoplasma gondii*, *Mem. Inst. Oswaldo Cruz*, 93, 103, 1998.
163. Ferro, E. A. V., et al., *Callomys callosus* (Rodentia: Cricetidae) trophoblast cells as host cells to *Toxoplasma gondii* in early pregnancy, *Parasitol. Res.*, 85, 647, 1999.
164. Ferro, E. A. V., et al., Effect of *Toxoplasma gondii* infection kinetics on trophoblast cell population in *Calomys callosus*, a model of congenital toxoplasmosis, *Infect. Immun.*, 70, 7089, 2002.
165. Franco, P. S., et al., Evaluation of vertical transmission of *Toxoplasma gondii* in *Calomys callosus* model after reinfection with heterologous and virulent strain, *Placenta*, 32, 116, 2011.
166. Franco, P. S., et al., *Calomys callosus* chronically infected by *Toxoplasma gondii* clonal type II strain and reinfected by Brazilian strains is not able to prevent vertical transmission, *Front. Microbiol.*, 6, 181, 2015.
167. Barbosa, B. F., et al., Susceptibility to vertical transmission of *Toxoplasma gondii* is temporally dependent on the preconceptional infection in *Calomys callosus*, *Placenta*, 28, 624, 2007.
168. Ferreira, G. L. S, et al., *Toxoplasma gondii* and mast cell interactions *in vivo* and *in vitro*: experimental infection approaches in *Calomys callosus* (Rodentia, Cricetidae), *Microbes Infect.*, 6, 172, 2004.
169. Gil, C. D., et al., Mast cells in the eyes of *Calomys callosus* (Rodentia: Cricetidae) infected by *Toxoplasma gondii*, *Parasitol. Res.*, 88, 557, 2002.
170. Pereira, M. F., et al., Acquired and congenital ocular toxoplasmosis experimentally induced in *Calomys callosus* (Rodentia, Cricetidae), *Mem. Inst. Oswaldo Cruz*, 1, 103, 1999.
171. Pearce, A. I., et al., Animal models for implant biomaterial research in bone: a review, *Eur. Cell. Mater.*, 13, 1, 2007.
172. Meurens, F., et al., The pig: a model for human infectious diseases, *Trends Microbiol.*, 20, 50, 2012.
173. Vodicka, P., et al., The miniature pig as an animal model in biomedical research, *Ann. N. Y. Acad. Sci.*, 1049, 161, 2005.
174. Dubey, J. P., Toxoplasmosis, *J. Am. Vet. Med. Assoc.*, 189, 166, 1986.
175. Penkert, R. A., Possible spread of toxoplasmosis by feed contaminated for cats, *J. Am. Vet. Med. Assoc.*, 162, 924, 1973.
176. Kijlstra, A., and Jongert, E., Toxoplasma-safe meat: close to reality? *Trends Parasitol.*, 25, 18, 2009.
177. Piassa, F. R., et al., Prevalence and risk factors for *Toxoplasma gondii* infection in certified and non-certified pig breeding farms in the Toledo microregion, PR, Brazil, *Rev. Bras. Parasitol. Vet.*, 19, 152, 2010.
178. Giraldi, N., et al., Estudo da toxoplasmose congênita natural em granjas de suínos em Londrina, PR, *Arq. Bras. Med. Vet. Zootec.*, 48, 83, 1996.
179. Chang, G. N., et al., Serological survey of swine toxoplasmosis in Taiwan, *Southeast Asian J. Trop. Med. Public Health.*, 22 Suppl, 111, 1991.
180. Haritani, M., et al., Demonstration of *Toxoplasma gondii* antigen in stillborn piglets using immunoperoxidase technique, *Nihon Juigaku Zasshi*, 50, 954, 1988.
181. Lopes, F. M. R., et al., *Toxoplasma gondii* infection in pregnancy, *Braz. J. Infect. Dis.*, 11, 496, 2007.
182. Moura, A. B., et al., Occurrence of anti-*Toxoplasma gondii* antibodies in swine and ovine slaughtered at municipality of Guarapuava in the State of Paraná, Brazil, *Rev. Bras. Parasitol. Vet.*, 16, 54, 2007.
183. Flegr, J., Klapilová, K., and Kaňková, S., Toxoplasmosis can be a sexually transmitted infection with serious clinical consequences. Not all routes of infection are created equal, *Med. Hypotheses*, 83, 286, 2014.
184. Wingstrand, A., et al., Clinical observations, pathology, bioassay in mice and serological response at slaughter in pigs experimentally infected with *Toxoplasma gondii*, *Vet. Parasitol.*, 72, 129, 1997.

185. Silveira, C., et al., *Toxoplasma gondii* in the peripheral blood of patients with acute and chronic toxoplasmosis, *Br. J. Ophthalmol.*, 95, 396, 2011.
186. Ferguson, D. J., Identification of faecal transmission of *Toxoplasma gondii*: small science, large characters, *Int. J. Parasitol.*, 39, 871, 2009.
187. Hutchison, W. M., Experimental transmission of *Toxoplasma gondii*, *Nature*, 206, 961, 1965.
188. Hutchison, W. M., et al., Coccidian-like nature of *Toxoplasma gondii*, *Br. Med. J.*, 1, 142, 1970.
189. Dubey, J. P., Miller, N. L., and Frenkel, J. K., The *Toxoplasma gondii* oocyst from cat feces, *J. Exp. Med.*, 132, 636, 1970.
190. Overdulve, J. P., The identity of *Toxoplasma* Nicolle and Manceaux, 1909 with *Isospora* Schneider, 1881. *I. Proc. K. Ned. Akad. Wet. C.*, 73, 129, 1970.
191. Sheffield, H. G, and Melton, M. L., *Toxoplasma gondii*: the oocyst, sporozoite, and infection of cultured cells, *Science*, 167, 892, 1970.
192. Dubey, J. P., Advances in the life cycle of *Toxoplasma gondii*, *Int. J. Parasitol.*, 28, 1019, 1998.
193. Dubey, J. P., and Frenkel, J. K., Feline toxoplasmosis from acutely infected mice and the development of *Toxoplasma* cysts, *J. Protozool.*, 23, 537, 1976.
194. Dubey, J. P., and Frenkel, J. K., Immunity to feline toxoplasmosis: modification by administration of corticosteroids, *Vet. Pathol.*, 11, 350, 1974.
195. Dubey, J. P., Hoover, E. A., and Walls, K. W., Effect of age and sex on the acquisition of immunity to toxoplasmosis in cats, *J. Protozool.*, 24, 184, 1977.
196. Pfefferkorn, E. R., and Kasper, L. H., *Toxoplasma gondii*: genetic crosses reveal phenotypic suppression of hydroxyurea resistance by fluorodeoxyuridine resistance, *Exp. Parasitol.*, 55, 207, 1983.
197. Sibley, L. D., et al., Generation of a restriction fragment length polymorphism linkage map for *Toxoplasma gondii*, *Genetics*, 132, 1003, 1992.
198. Sibley, L. D., and Boothroyd, J. C., Construction of a molecular karyotype for *Toxoplasma gondii*, *Mol. Biochem. Parasitol.*, 51, 291, 1992.
199. de Macchi, B. M., et al., Chickens treated with a nitric oxide inhibitor became more resistant to *Plasmodium gallinaceum* infection due to reduced anemia, thrombocytopenia and inflammation, *Vet. Res.*, 44, 8, 2013.
200. Dubey, J. P., *Toxoplasma gondii* infections in chickens (*Gallus domesticus*): prevalence, clinical disease, diagnosis and public health significance, *Zoonoses Public Health*, 57, 60, 2010.
201. Dubey, J. P., et al., Biological and genetic characterization of *Toxoplasma gondii* isolates from chickens (*Gallus domesticus*) from São Paulo, Brazil: unexpected findings, *Int. J. Parasitol.*, 32, 99, 2002.
202. Erichsen, S., and Harboe, A., Toxoplasmosis in chickens. I. An epidemic outbreak of toxoplasmosis in a chicken flock in South-Eastern Norway, *Acta Pathol. Microbiol. Scand.*, 33, 381, 1953.
203. Erichsen, S., and Harboe, A., Toxoplasmosis in chickens. II. So-called gliomas observed in chickens infected with toxoplasms, *Acta Pathol. Microbiol. Scand.*, 33, 381, 1953.
204. Jones, F. E., et al., Experimental toxoplasmosis in chickens, *J. Parasitol.*, 45, 31, 1959.
205. Dubey, J. P., et al., Serologic and parasitologic responses of domestic chickens after oral inoculation with *Toxoplasma gondii* oocysts, *Am. J. Vet. Res.*, 54, 1668, 1993.
206. Kaneto, C. N., et al., Experimental toxoplasmosis in broiler chicks, *Vet. Parasitol.*, 69, 203, 1997.
207. Bonapaz, R. S., et al., Effects of infection with *Toxoplasma gondii* oocysts on the intestinal wall and the myenteric plexus of chicken (*Gallus gallus*), *Pesq. Vet. Bras.*, 30, 787, 2010.
208. Ong, Y.C., Boyle, J.P., and Boothroyd, J.C., Strain-dependent host transcriptional responses to *Toxoplasma* infection are largely conserved in mammalian and avian hosts, *PLoS One*, 6(10), e26369, 2011.
209. Zhang, M., et al., Detection of *Toxoplasma gondii* in shellfish and fish in parts of China, *Vet. Parasitol.*, 200, 85, 2014.
210. Esmerini, P. O., Gennari, S. M, and Pena, H. F., Analysis of marine bivalve shellfish from the fish market in Santos City, São Paulo State, Brazil, for *Toxoplasma gondii*, *Vet. Parasitol.*, 170, 8, 2010.
211. Zarnke, R. L., et al., Serologic survey for *Toxoplasma gondii* in grizzly bears from Alaska, *J. Wildl. Dis.*, 33, 267, 1997.
212. Omata, Y., et al., *Toxoplasma gondii* does not persist in goldfish (*Carassius auratus*), *J. Parasitol.*, 91, 1496, 2005.
213. Sanders, J. L., et al., The zebrafish, *Danio rerio*, as a model for *Toxoplasma gondii*: an initial description of infection in fish, *J. Fish Dis.*, 38, 675, 2015.

Section VI

Foodborne Infections due to Helminths

42

Anisakis

Mauricio Afonso Vericimo, Gerlinde Teixeira, Israel Figueiredo Jr., Janaina Ribeiro, Maria Augusta Moulin Fantezia, and Sergio Carmona São Clemente

CONTENTS

42.1 Introduction

Parasites from the marine environment have historically been overlooked as a risk for human disease and are thus not in the main stream of basic or clinical investigation, although they can infect humans, thereby causing anthropozoonosis, and are therefore a public health risk. Within the marine worms with clinical importance are those pertaining to the Anisakidae family (*Anisakis*, *Pseudoterranova*, and *Contracaecum*) and Raphidascarididae family (*Hysterothylacium*) causing anisakidosis.[1] Anisakids are nematodes whose definitive hosts are marine mammals; intermediate hosts are crustaceans (L2), fish, and cephalopods (L3), and they have a worldwide distribution. Humans become accidental hosts after ingestion of raw or undercooked infected seafood.[2]

There is an estimate of 20,000 human cases of anisakidosis, with an annual registration of 2000 new cases. The highest incidence with approximately 90% of all reported cases occurs in Japan,[3] probably owing to the routine habit of eating raw fish in dishes like sushi and sashimi.[4] Other countries that habitually consume raw or undercooked seafood also record an expressive number of cases of the disease. This is the case of European countries, mainly in the coastal areas of Germany, Netherlands, and Scandinavian countries that consume salted, pickled, and smoked herring, or Spain, where typical appetizers are ceviche (fresh seafood marinated in lemon juice) and *Boquerón's en vinegar* (pickled anchovies).[1] In the Americas, there has been an increase in the number of reported anisakidosis cases, probably because of the popularization of oriental cuisine and the consumption of dishes like Lomi-lomi salmon and ceviche.[1,4,5] The improvement of diagnostic methods is probably another explanation for the increase in the report of new cases.

Since the first descriptions of human cases, the number of researchers who investigate actual and potential human marine infections has increased, and several animal models have been developed in order to understand the host–parasite relationship associated with the sensitization of individuals who accidentally ingest anisakid larvae. In this chapter, we will contextualize several *in vitro* and *in vivo* experimental models that are employed to reproduce and understand the natural history of human disease and explore the molecular and biological aspects of these parasites.

42.2 Taxonomy, Life Cycle, and World Distribution of *Anisakis* Species

The taxonomic classification of anisakids[6] consists of:

- ⇨ Kingdom: Animalia
  - ⇨ Phylum: Nematoda
    - ⇨ Class: Rabdititia (=Secernentea)
      - ⇨ Subclass: Rabditia (=Phasmidea)
        - ⇨ Order: Ascarida
          - ⇨ Superfamily: Ascaridoidea
            - ⇨ Family: Anisakidae
              - ⇨ Genus: *Anisakis*
Pseudoterranova
Contracaecum
            - ⇨ Family: Raphidascarididae
              - ⇨ Genus: *Hysterothylacium*

Within the Ascaridoidea superfamily, the Anisakidae family is considered the largest and includes species that can parasitize fish, reptiles, mammals, and fish-eating birds. The representatives of this family are dependent on the aquatic environment for the development of their biological cycle and usually involve invertebrates and fish as intermediate or paratenic hosts.

Among the Anisakidae family, the species of the genus *Anisakis* have low specificity for the definitive host, but, in general, live in the stomach of cetaceans such as whales, dolphins, and porpoises. Those of the *Pseudoterranova* genus are more specific having the pinnipeds (seals, walruses, and sea lions) as definitive hosts. The species belonging to the *Contracaecum* genus have as definitive hosts fish-eating birds and pinnipeds, and unlike the other genera, *Contracaecum* larvae can parasitize both marine and freshwater fish. Finally, the definite hosts of the species belonging to the *Hysterothylacium* genus of the Raphidascarididae family are pinnipeds, fish, and shellfish.[7–10] Because the biological cycles of the four genera are similar, and the aim here is the experimental approach to study these worms, we will only depict the *Anisakis* life cycle.

Adult worms release their eggs in the gut of the definite host. Through the feces, the eggs gain access to the seawater, where they embryonate and form the first larval stage (L1) and progress to the second stage (L2). The L2 are eaten by small crustaceans such as krill (first intermediate host), where they progress to the third-stage larvae (L3), the infective stage for the definitive host. Second intermediate hosts (fish or shellfish) ingest the crustaceans, which in turn are eaten by bigger fish, transferring L3 through the food chain and resulting in their accumulation in the larger fish until eaten by sea mammals, which are their definite hosts.[11,12] Once L3 have been eaten by their definite hosts, they progress to the fourth larval stage (L4) and finally become adults[13,14] (Figure 42.1).

When fish are captured, soon after their death, L3 migrate to the viscera, peritoneal cavity, and muscles. The degree of migration depends on environmental conditions, the parasite, and fish species. When humans consume raw or undercooked infected fish or shellfish, they may become accidental hosts. As the parasites are not adapted to humans, they do not reach sexual maturity although they may cause mild irritation to anaphylactic shock.[5,15,16]

At least 200 fish and 25 cephalopods species have been described as being infected with anisakid larvae.[17–19] Within all anisakids, species pertaining to the *Anisakis* genus are considered the most

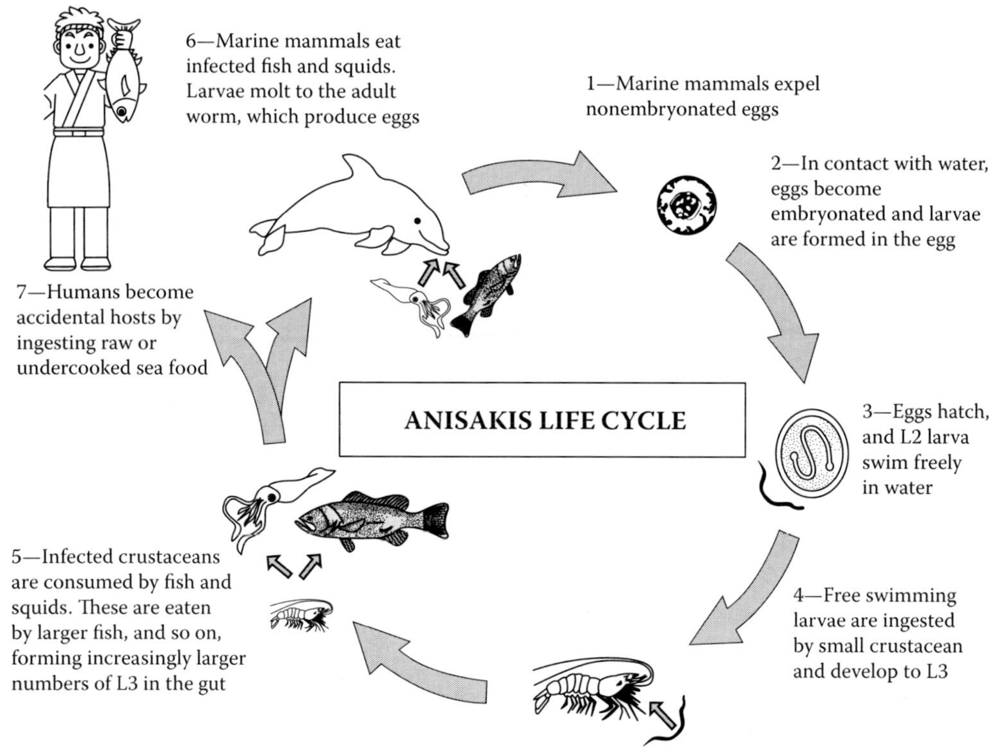

FIGURE 42.1 Life cycle of anisakids.

pathogenic and cause the largest number of human occurrences.[20,21] Although morphologically very similar, the genus has nine species that have been identified by molecular technologies and have distinct definitive host distribution worldwide.[22–24] As depicted in Figure 42.2, larvae are classified by their morphology and genetic characteristics in "clades (I and II)." Clade I contains the *Anisakis simplex* complex, which includes *A. simplex* (*strict sense*), *A. pegreffi*, and *A. simplex* (complex); the other sister species in this clade are *A. typica*, *A. ziphidarum*, and *A. nascettii*. The definitive host of the Clade I species are mainly distributed in the Atlantic and Pacific oceans. *A. simplex* (ss) and *A. pegreffi* are also found in the Mediterranean, Arctic, and Antarctic Seas.[2,12] *Anisakis* species pertaining to Clade II are classified as *A. physeteris* complex, which includes *A. physeteris*, *A. brevispiculata*, and *A. paggiae* species. Although considered to have a cosmopolitan distribution, they are mainly found in the Atlantic Ocean.[2,25]

The *Pseudoterranova decipiens* complex consists of species that include the *P. decipiens* (*sensu stricto*) or *P. decipiens B*, *P. krabbei*, *P. bulbous*, *P. azarasi*, and *P. cattani*. They are considered cosmopolitan and are very abundant in the Atlantic Ocean, occurring from the Arctic to Antarctica.[26]

The *Contracaecum* genus has species that are able to parasitize both freshwater and marine organisms. From this genus, the species that most frequently cause anisakidosis pertain to the *C. osculatum* complex, which is a set of five members including *C. osculatum* types A, B, C, D, and E, where *C. osculatum sensu stricto* corresponds to type C.[27] The most frequent geographic distribution of these species is the Alaskan and Japanese waters, the Baltic Sea, and Antarctic and Atlantic Ocean.[7,28]

Although the *Hysterothylacium* genus has a worldwide distribution, it is described as a rarely occurring causative agent of anisakidosis. Apparently, the first human case of *H. aduncum* was registered in 1996 in Japan.[29] Molecular identification of *Anisakis* and *Hysterothylacium* larvae from marine fish of the East China Sea and the Pacific coast of central Japan showed that approximately 10% of the larvae pertained to the *Hysterothylacium* genus (*H. amoyense*—5.0%, *H. aduncum*—1.6%, *H. fabri*—3.4%, and *H.* spp.—2.9%), while the majority of the remaining Anisakidae nematodes belong to the *Anisakis* genus.[30] This result correlates well with the clinical finding.

Using classical techniques (morphological taxonomy), human anisakidosis is most frequently described as being caused by *Anisakis* and *Pseudoterranova* genus.[29,31–33] Among the *Anisakis* species, *A. simplex* (ss) is reported as responsible for the highest number of human cases. However, after the introduction of molecular biology in taxonomy, *A. pegreffi* has been more frequently described as the agent responsible for anisakiosis in some countries such as Italy.[18,34,35]

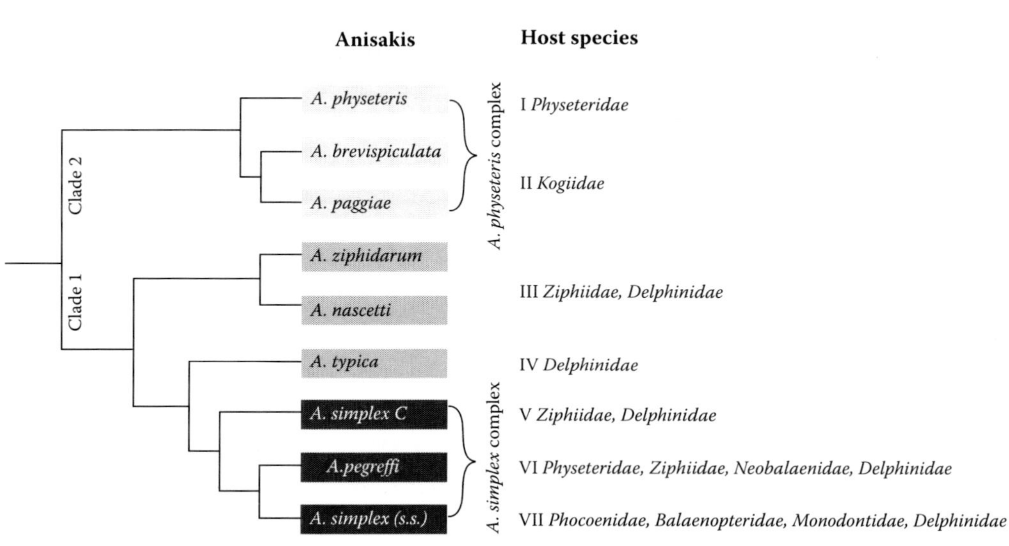

FIGURE 42.2 Cladistic distribution of anisakid larvae.

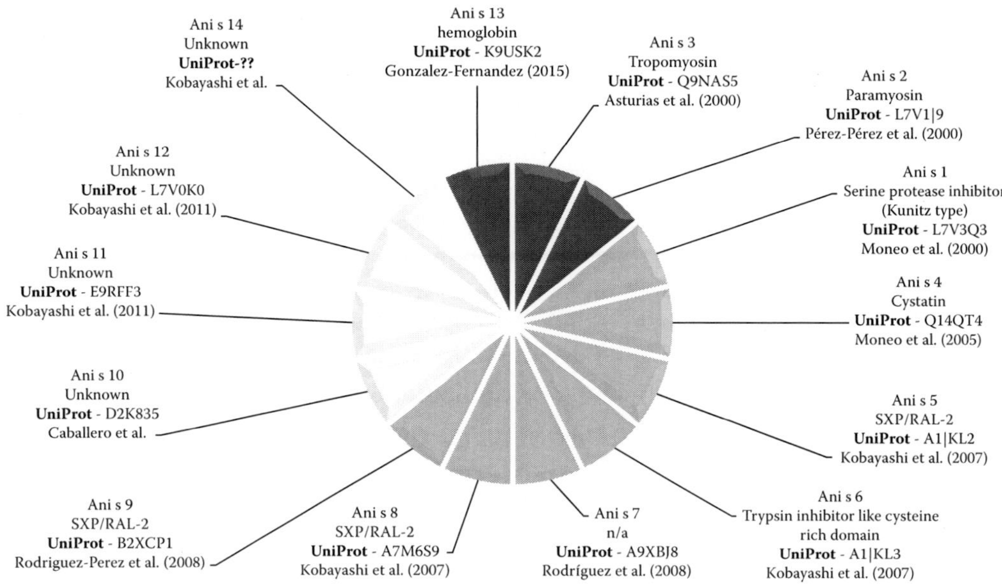

FIGURE 42.3 Updated *Anisakis* allergen compiled mainly from data extracted from the Allergome database in combination with published literature. The colors of the wedges indicate the origin of the antigens: Dark gray—somatic antigens; medium gray—excretory-secretory antigens; light gray—unknown origin. (http://www.allergen.org/treeview.php.)

42.3 Allergen Nomenclature

The abbreviation of the name of the gender (first three letters) and of the species (first letter) followed by a number indicating the chronology of the allergen purification was adopted as the systematic nomenclature of allergens, implemented by the Nomenclature Sub-Committee of the World Health Organization (WHO) and International Union of Immunological Societies. So, the *A. simplex* allergens are called "Ani s #" (e.g., Ani S1).[36] *A. simplex* (ss) has 14 allergens characterized by origin and molecular aspects. The immunoreactivity pattern for these allergens has been studied both with human sera and with experimental animals. A synthesis of the structural classification of *Anisakis* allergens is presented in Figure 42.3 based mainly on the allergen database AllFam,[37] which can be accessed on the web at http://www.meduniwien.ac.at/allergens/allfam. The data was complemented from other published literature. For example, data from the Conserved Domain Database (CDD) and from the domain of unknown function (PF; DUF, Pfam) were used.

42.4 Pathogenesis, Immunological Response, and Clinical Signs and Symptoms

In humans, the ingestion of the live anisakid larvae causes distinct clinical forms of illness: gastric, intestinal and/or ectopic anisakidosis, and/or allergic reactions, which may vary from mild to severe reactions. Although not a very common finding, gastroallergic anisakiosis (GAA) is a well-established clinical entity, characterized by acute IgE-mediated urticaria, angioedema, or anaphylaxis shortly after an *A. simplex* acute infection. The immunologic response that accompanies this parasite presents a significant polyclonal stimulation of different immunoglobulin isotypes comprising a mixed Th1- and Th2-mediated reaction.[38] The ingestion of dead anisakid larvae or proteins derived from the larvae may also trigger mild to severe allergic reactions. Therefore, anisakid extracts should be included in the standard sets of allergens used to investigate undefined allergies and anaphylactic reactions.[39]

The insertion of the cephalic portion of the larva in the mucosal wall and the secretion of proteases that permit its fixation result in a local inflammatory reaction that leads to the clinical symptoms such as epigastric pain, nausea, diarrhea, vomiting, and fever.[40] Furthermore, when larvae penetrate the submucosa, it may sensitize the host with its excretory-secretory (ES) products by stimulating the development of a predominantly Th2 immune response, which favors the production of IgE antibodies, responsible for allergies.[41] Persistence of larvae in the tissue can result in direct damage and, in turn, the development of a eosinophilic granuloma, characterized by an inflammatory infiltrate of eosinophils and neutrophils associated with a diffuse interstitial edema and proliferation of connective tissue around the body of the larva.[42–46]

There is evidence in the literature that the continued exposure to *Anisakis* antigens by fish factory workers, anglers, and their families can sensitize them through inhalation or direct contact.[47,48] Farmers are another group of workers who can become sensitized to *Anisakis* antigens when in direct contact with the corresponding allergens, e.g., fish meal.[47] Gastrointestinal conditions, asthma, conjunctivitis, and occupational contact dermatitis have been frequently described in *Anisakis*-sensitized patients.[47,49–53] Signs and symptoms can range from discreet allergic symptoms, urticaria up to angioedema, and fatal anaphylactic reactions with or without gastrointestinal symptoms.[54,55]

The human immune response to *Anisakis* sp. antigens is highly heterogeneous, varying both in quantity and in quality between individuals.[56] Studies in patients showed that infection with *Anisakis* larvae induces a strong immune response with the production of specific antibodies reaching maximum titers within the first month of infection.[57] Infections with low numbers of larvae and continuous exposure frequently result in the production of high levels of IgE, whereas the exposure to high numbers of larvae frequently results in the production of IgG.[58,59] The analysis of the cytokine profile obtained from the peripheral blood and intestinal biopsy samples of newly infected patients reinforces the concept that the Th2 response plays an important role in the immunopathogenesis of anisakiosis.[38] Further detail pertaining to the immune response shall be presented during the experimental section.

42.5 Laboratorial Diagnosis

Initially, specific IgG was used to diagnose anisakiosis; however, as the IgG titers persist elevated for a relatively long period, it is not a good parameter to differentiate current from previous *A. simplex* infections. Another observation is that anisakid allergy is frequently associated with high levels of specific IgG4.[60–62]

A good diagnostic tool can be the use of the proportion of specific IgE and IgG4 titers. This strategy has been used to evaluate allergic diseases caused by a variety of other nematodes even if the nematode is not observed by a gastroscopy.[60,63,64] Thus, the serological diagnosis of a gastroallergic anisakiosis can be a good alternative.[57,60,63,65] Chronic urticaria (CU) associated with anisakiosis is another clinical setting in which IgG4 can be used for diagnosis and follow-up. Unlike patients who continue their exposure to the fish, those that are subjected to a fish-free diet experience a significant reduction in CU symptoms accompanied by significant reduction of IgG4 levels.[65] The comparison of the levels of IgE, IgG, and IgG4 to *A. simplex* in CU and GAA patients showed that the latter presented significantly higher levels of all tested immunoglobulins.[66]

42.6 Experimental Models

42.6.1 General Considerations

Even if the conditions that are used in animal experimentation do not exactly match those that occur in the natural history of disease, this is a widely used method for acquiring knowledge of various diseases in human and veterinary medicine. Through *in vivo* experimentation, it is possible to answer specific questions about the pathophysiology of diseases generating information that can then be extrapolated to the clinical setting, permitting a better understanding of the disease, leading to better prevention and better treatment.

Since the discovery of the first human anisakiosis cases in the 1960s, many animal species have been used as a model for this disease. The first studies used rabbits and guinea pigs to understand the migration trajectory of the larvae to the tissues and granuloma formation. However, to study the allergic reactions induced by *Anisakis* larvae, most researchers prefer to use rats and mice. We chose to present the animal models by species and route of infection/sensitization.

42.6.2 Guinea Pigs

42.6.2.1 Intradermic Route

In order to evaluate the *in vivo* chemotactic effect of *A. simplex* larvae extract, Tanaka and Torisu[67] used guinea pigs as experimental animals. These researchers found that a few hours after intradermal injection of crude larvae extract (CE), a dose-dependent accumulation of eosinophils occurred at the site of injection. To confirm this effect, these authors carried out *in vitro* chemotaxis assays using Boyden chambers.[68] Using the same concentration of the extract with which eosinophil chemotaxis was observed, no chemotactic activity was found for neutrophils, supporting the idea that the CE plays an important role in the development of eosinophilia in anisakiosis.

42.6.2.2 Intraperitoneal Route

Early in the 1980s, in the attempt to determine the etiologic mechanism of the allergic reactions associated with anisakiosis, guinea pigs were sensitized by implanting live *Anisakis* sp. larvae in the peritoneal cavity.[69] The Schultz–Dale[70,71] reaction was used to determine the presence of type I reactivity. In short, intestinal fragments of intraperitoneal-sensitized guinea pigs with live *Anisakis* larvae responded intensely when stimulated with *Anisakis* larvae hemoglobin and with less intensity when stimulated with CE from other anisakids (*Contracaecum* and *Pseudoterranova*), whereas no response was observed when *Toxocara canis* or *Ascaris suum* extracts were used. These results confirm the IgE-mediated etiology of the allergic reactions associated with anisakiosis.

42.6.2.3 Intragastric Route

To determine the migratory pattern and viability of live *Anisakis* larvae, these were delivered to the gastric cavity. Larvae gained different organs and tissues passing the stomach wall through an active migration mechanism without a preestablished migratory pattern. Live larvae without any morphological changes were recovered up to the fifth day after administration. These were able to reinfect another guinea pig maintaining the same migration capability. However, as of the sixth day post infection, all larvae disappeared leaving no hint of its presence in any part of the body.

Guinea pigs experimentally infected with *A. simplex* have also been used to test drugs.[72] For example, oral treatment with ivermectin or albendazole was tested and found to present high *in vivo* efficacy against the larvae present in different organs of the guinea pigs.[73]

42.6.3 Pigs

42.6.3.1 Oral Route

Anisakis larvae infection in pigs was studied by feeding the animals with fish offal contaminated with L3. In these studies, researchers observed that the severity of injury was proportional to the number of larvae ingested. Histological alterations due to larvae interaction with the mucosa included primary mechanical damage accompanied with bleeding, ulceration of the mucosa and submucosa, and intense cellular infiltration with connective tissue proliferation around the larva.[74] The histological changes of the stomach mucosa from experimentally infected pigs with *Anisakis* sp. and *Pseudoterranova* sp. larvae involved intense inflammatory reaction around the larva with the presence of numerous eosinophilic cells.[75] That is, feeding pigs L3 of *C. osculatum* results in the same histopathological findings that corresponded to findings of infections caused by other *Anisakis* sp. pathogens.[76]

42.6.4 Rabbit

The histological aspects of intestinal sections of experimentally infected rabbits resemble those of accidentally infected humans, suggesting a similarity of the pathogenesis. Thus, rabbits were successfully introduced as experimental anisakiosis models soon after the publication of the first human anisakiosis cases.[77]

42.6.4.1 Intragastric Route

In the early 1970s, the experimental determination of the pathogenesis of anisakiosis was performed by administrating live larvae to the stomach of rabbits and semiquantitatively grading the inflammatory reaction of the surrounding tissue where larvae penetrated.[78] Three days after the oral administration of 40 *Anisakis* larvae, only a very small number entered the stomach wall, many of which were still alive, and the degree of the inflammatory reactions of the gastric mucosa surrounding the distinct larvae varied between mild, moderate, and severe in an individual animal and between individuals.

Necrosis and massive amounts of granulocytes, including eosinophils, were the main findings on day 3 after infection. On day 5, the larval viability declined, and an infiltrate of plasma cells and immunoblasts was observed along with the granulocytes in the center of the reaction, while fibroblasts were already present in the periphery. After 7 days, the fibroblast infiltrate became more intense; by 10 days, granulation tissue was observed; and by a month, the necrotic tissue was substituted by new connective tissue surrounded predominantly by mononuclear cells with moderate amounts of eosinophils.

The serological reactivity in association with the histopathological pattern was also studied in rabbits infected with 30 live *A. simplex* larvae through the oral rout. Although most larvae were recovered in the stomach, some migrated from the gastrointestinal tract and reached extragastric tissues, resulting in the formation of abscess that contained dead larvae. By 30 days, the reactions progressed to granulomatous abscesses followed by calcification of the larvae.[79]

From the serological point of view, IgG peaked by 30 days, coinciding with the granuloma resolution and calcification of the larva followed by an abrupt decline. Another study that infected rabbits with 10 larvae showed a peak of IgM on the 11th day, whereas IgG peaked approximately a month later.[80]

Intragastric sensitization of rabbits with *Anisakis* larvae was also employed to assess the recognition pattern of somatic and secreted antigens of infective *Anisakis* larvae comparing possible relationships with antigens from other nematodes of Ascaroidea family using radioimmunoprecipitation techniques.[81] Such as in serum derived from *Anisakis* infected patients, infected rabbits preferentially respond to somatic antigens, and the recognition sequence occurs to different components of secreted antigens. The differences in the recognition of secreted/excreted antigens and somatic components may be due to the duration of sensitization and the degree of penetration by nematodes in the tissues. Kennedy et al.[81] also demonstrated that a 14-kDa component derived from *A. simplex* cross-reacts with a homologues component derived from *Ascaris suum, Ascaris lumbricoides, and Toxocara canis*, species from the Ascaroidea family.

42.6.4.2 Subcutaneous Route

A chemotactic factor selectively attractive for eosinophils found in the extract from *Anisakis* larva was termed eosinophil chemotactic factor of parasites (ECF-P).[67] To determine whether the eosinophilic phlegmonous inflammation typically observed in human anisakiosis could be experimentally reproduced, normal and subcutaneously immunized rabbits received intraserosal injection of ECF-P into the ileum of rabbits. All rabbits developed a significant eosinophilic inflammation at the injection site in a dose-dependent manner. Although immunized rabbits presented high anti-ECF-P antibody titers while normal animals had no detectable antibody, there was no significant histological difference between the lesions observed in either group of rabbits. These results support the argument that, especially in the early phase of primary infection with anisakiosis, ECF-P may contribute to the development of eosinophilic phlegmonous inflammation without any immunological intervention.[67,82]

42.6.4.3 *Intramuscular Route*

One of the experimental protocols involves the intramuscular route to investigate if larval antigens of *A. simplex* present molecular similarity to interleukin IL-4. The resulting rabbit anti-mouse IL-4 antibodies were tested against *A. simplex* ES and CE antigens in ELISA. The anti-IL-4 antibodies showed a strong cross-reactivity, which was confirmed by western blot analysis. A complementary assay, the absorption of the anti-IL-4 sera with *A. simplex* antigen, demonstrated a 70%–80% inhibition of antigen binding when retested in ELISA. These results support the hypothesis that *A. simplex* proteins share several epitopes with IL-4 or conversely that *A. simplex* larval ES and somatic products present IL-4-like molecules. This finding implies that the parasite may control and modulate the mucosal Th1-Th2 dichotomy for its own benefit in an attempt to avoid its expelling.[83]

Currently, experimental *Anisakis* research has not used rabbits as a model to study allergic reactions caused by this nematode. However, intramuscular inoculation with *Anisakis* antigens has been employed when the aim is to characterize allergens and to produce laboratory reagents.[84–88]

42.6.5 Rats

Rats have been used extensively to investigate the immune response to *Anisakis* larvae. Although the oral route is the natural form of infection, in the experimental scenario, investigators have shown a limited usefulness of *per os* administration due to the difficulty in accurately determining the parasite load, since many larvae are expelled through the anus, hampering the establishment of the relationship between parasite load and immune response.[89] Although the surgical implant may appear to be an inadequate route of infection, the argument used to validate this technique and to expect that the antibody production profile would be the same regardless of the route is that orally administered larvae pass from the intestinal lumen into the peritoneal cavity after infection.[90,91] Another observation that supports this hypothesis is that extragastrointestinal anisakiosis has also been observed in humans who are infected.[92]

42.6.5.1 *Intraperitoneal Larval Implant*

Immunization of rats by intraperitoneal L3 larvae implantation was used to determine the immune response to SE and CE. In contrast to oral infection in rabbits, intraperitoneal implantation of live larvae in rats induced a strong response to both SE and CE antigens. After 63 days of implantation, no larva was found alive; thus, the hypothesis is that the immune response was due to the release of somatic antigens in the peritoneal cavity.[81]

ES-specific IgM and IgG titers of rats inoculated with increasing *A. simplex* L3 load (1, 5, or 20 larvae) show a positive correlation after the primary inoculum but not to the secondary inoculum. IgM and IgG titers of animals inoculated with 20 larvae did not further increase. However, after the second inoculation, those animals that received one or five larvae presented antibody titers comparable to the 20 L3 inoculation.

The primary inoculation induced low ES-specific IgE antibody titers in all groups, and in the secondary inoculation, a negative correlation was obtained. In other words, rats receiving one larva developed higher IgE titers than rats receiving larger inoculums. IgE titers of single larvae-inoculated rats peaked at 3–5 days after secondary inoculation and disappeared by day 14, which is consistent with the duration of infection. Thus, monitoring ES-specific IgE may be a useful diagnostic tool for human intestinal anisakiosis, because in the natural scenario, infections typically course with low larvae loads.[93]

42.6.5.2 *Intragastric Infection*

Let us return to the intragastric/intraperitoneal duel. Authors argue that despite the importance of the live larvae intraperitoneal inoculum studies, the human natural history of gastroallergic anisakiosis is given orally; so experiments using this pathway are important.[59] Rats were infected with L3 *Anisakis* by the oral route twice with an interval of 9 weeks to investigate the kinetics of isotype-specific antibody expression, and it was found that IgM's peak with similar titers after primary and reinfection presents the same antigenic recognition. After reinfection, as expected, IgG1 and IgG2a levels were higher and showed accelerated kinetics; however, IgG2b level was substantially lower. The biological allergy state peaked earlier (1 week) than the immunochemical allergy state (2 weeks). Since no meaningful correlation between specific IgE avidity and biological allergy state was found and elevated IgM levels at reinfection occurred, the hypothesis is that the allergic response induced by oral L3 infection might not be related to specific IgE avidity.[94]

A procedure developed recently to deliver live larvae directly to the stomach of mice by an esophageal catheterization[95] was adapted to perform live *Anisakis* spp. infection in rats.[96] The aim of this study was to understand the histopathological effects of acute (single) and chronic (multiple reinfections—24, 48, 72, and 96 h intervals) *Anisakis* infection. Live larvae were found anchored to the mucosa at different locations (whose milieu varied from a very acidic to basic pH gradient), passing through the stomach wall and in organs out of the gastrointestinal tract. The histopathology showed an acute inflammatory reaction, with eosinophil predominance and a mild fibrotic reaction. Even though not all larvae were recovered as previously placed as an obstacle to the oral route, this protocol can be considered a good experimental model because the histopathological alterations are similar to those described in human anisakiosis.[97]

Although there are reported cases of allergic reactions due to the ingestion of cooked and frozen seafood, there is also evidence that only live larvae trigger the allergic reactions. Consequently, the debate on the risk of *Anisakis*-associated hypersensitivity by ingestion of properly cooked and frozen fish remains. To elucidate this fact, an experimental model was designed to study the antibody production kinetics after oral inoculation with live or dead *Anisakis* L3. The results show that animals produce specific IgM, IgG, and IgE to ES antigens after primary and secondary inoculation with live L3 but not after dead L3 (frozen, heated, cut, or homogenized). These results suggest that the ingestion of cooked or frozen seafood containing *Anisakis* L3 is safe even for allergic individuals.[98]

42.6.5.3 In Vivo *L3–L4 Transformation Model in Rats*

To study the morphological transformations of L3 to L4 *Anisakis* type I, *P. decipiens*, *Contracaecum* type B and *Hysterothylacium* L3, recovered after experimental infection in rats, and *Anisakis* type I L4 derived from humans were examined with the aid of scanning electron microscopy to examine the anterior and posterior extremities and the cuticular structures of the larvae. Rats were sacrificed at different times after oral administration, and a careful search in the digestive tract, abdominal cavity, muscles, and viscera was performed. Molting from L3 to L4 was observed as of the third day onward in rats that received *Anisakis* type I and *P. decipiens*. *Anisakis* larvae penetrated the stomach and the intestinal wall, and a single larva of *Pseudoterranova* penetrated to muscularis mucosa of the stomach. No *Contracaecum* larvae were recovered. Electron microscopy revealed that L4 of *Anisakis* type I from rat and man were similar, while the L4 of *Anisakis* type I and *P. decipiens* showed ultrastructural differences, which might be of clinical value for the identification of fragments recovered during endoscopy in man.[99]

42.6.6 Mice

Because of the accumulated data in the last decades, especially concerning IgE synthesis of the antibody associated with allergic reactions, mice are considered better animal models than other species to investigate allergic reactions.[100–102] In food-associated allergies, it is still unclear what conditions make certain foods strong IgE inductors. There are reports of more than 170 foods causing food allergies, but only 8 (peanut, tree nuts, milk, egg, wheat, soy, fish, and shellfish) account for 90% of all food-allergic reactions.[103] It is known that the major reaction to food proteins when ingested in physiological conditions usually is a phenomenon called oral tolerance, while the parenteral administration of the same food proteins in experimental models induces sensitization.[104–110] This intriguing dichotomy has interested immunologists.

It is also known that in natural helminthic infections, IgE and eosinophilia are major hallmarks of the immunological response as the consequence of a Th2 lymphocyte profile activation by helminths. Among other interleukins, Th2 cells secrete IL-4 and IL-5; the former promotes immunoglobulin class switching to IgE, and the latter stimulates eosinophil development and activation. Furthermore, in IgE experimental models, animals are immunized with the antigenic preparations mixed with adjuvants such as aluminum hydroxide[102] or pertussis toxin[111,112] and commonly administered by a parenteral route. In experimental models where the aim is to develop oral sensitization and food allergy, antigens are associated to cholera toxin.[113] Complete or incomplete Freund's adjuvant is another commonly used adjuvant, which is considered a good IgG inducer.[114]

The humoral and cellular immune responses to live *A. simplex* larvae observed in mice models share similarities with those observed in human disease. However, due to the difficulty in introducing live larva into the gastric cavity of mice, the majority of the experiments have been conducted by immunizing the animals with CE or with ES antigens of cultivated larvae, thus indicating the relevance of the investigation of the immune mechanisms that control the allergic responses to live and dead *Anisakis* spp. larvae.[115–118]

42.6.6.1 Intraperitoneal Larva Implant

A. simplex L3 were surgically implanted into the abdominal cavity of mice to investigate histopathological alterations.[119] Necropsy performed at 7, 14, or 21 days post infection evidenced that larvae were mostly found embedded in the gut mesentery and only rarely invaded the viscera. On day 7, adjacent to viable parasites, an intense neutrophil aggregation characterizing an acute inflammatory reaction was observed; by day 14, this reaction evolved to a mature eosinophilic granuloma with large numbers of fibroblasts and associated collagen. Granulocytes and occasionally multinucleate giant cells were observed at the still viable host–parasite interface. By day 21, the L3 were dead, invaded by inflammatory cells, and the lesions displayed the predominance of connective tissue. Multinucleate giant cells and eosinophils adjacent to parasite remnants or scattered within the walls of the granulomata were frequent. Hematological findings, regardless of the number of implanted worms, showed that on days 7 and 14 mice presented neutrophilia of varying magnitude accompanied with an eosinopenia that began to return to normal values by day 21. Both hematological and histological findings are consistent with those seen in human anisakiosis.

42.6.6.2 Intraperitoneal Immunization

To help understand some of the unknown immune interactions between helminth infection and allergy, mice were intraperitoneally sensitized to develop a hypersensitivity reaction with *A. simplex* proteins, followed by an intravenous or oral *A. simplex* challenge. The sensitized mice presented as of the third week specific IgE, IgG1, and IgG2a to numerous *A. simplex* allergens, some of which were similar to those found in human serum. When challenged with intravenous *A. simplex* antigens but not after an oral antigen challenge, anaphylaxis and plasma histamine release were observed. The cellular and molecular profile showed that *A. simplex* stimulated splenocytes to release IL-10, IFN-γ, IL-4, IL-13, and IL-5 with a mixed Th1/Th2 pattern.[120] This seems to be a good model to investigate the peculiar allergic reactions to parasitic proteins.

42.6.6.3 Intragastric Infection

Live *Anisakis* L3 were orally inoculated in C57BL/10 and BALB/c mice to investigate isotype-specific immune responses to ES and CE products. The C57BL mouse strains typically produce a Th1-Type cytokine profile, while BALB/c mice produce a Th2-Type cytokine profile. Both ES and CE antigens stimulated similar antibody patterns; however, CE stimulated the production of higher antibody levels.

BALB/c mice produced a faster IgM response than C57BL/10 mice, while the latter produced higher IgG1 and IgG2b antibodies with practically undetectable IgG2a levels.[121] Further anisakiosis studies using BALB/c mice showed that after multiple immunizations using Freund's adjuvant, mice showed a single maximum peak of IL-4 between weeks 8 and 14, whereas animals inoculated with a single larva *per os* showed two IL-4 peaks—the first with moderate levels between days 6 and 12 p.i. and the second maintained from week 3 to 9.[122] After Perteguer and Cuellar showed the consequences of natural sensitization,[121,122] the authors of this chapter proposed a simplified method to introduce live larvae with an intragastric tube. This technique results in similar data as those published in the literature.[95,118]

42.6.6.4 Epicutaneous Immunization

As cited before, contact dermatitis is one of the consequences of antigen exposure to *Anisakis* proteins in seafood-processing workers. Therefore, to understand the basic mechanisms in the development of allergic sensitization through the skin, repeated epicutaneous exposure of *Anisakis* proteins in wild-type (WT), IL-4-, IL-4Rα-, IL-13-, and IL-4/IL-13-deficient mice was evaluated by following the systemic signs and symptoms. Epicutaneous sensitization with *Anisakis* larval antigens induced the WT localized inflammation, epidermal hyperplasia, production of TH2 cytokines, antigen-specific IgE and IgG1, and anaphylactic shock after intravenous challenge. IL-13-deficient mice failed to develop epidermal hyperplasia and inflammation, and in IL-4-, IL-4/IL-13-, and IL-4Rα-deficient mice, anaphylaxis was reduced. These results suggest that interleukin-13 plays a central role in contact dermatitis development, whereas IL-4 drives the Th2 profile and resultant anaphylactic reactions.[48]

42.6.6.5 Subcutaneous Immunization

The subcutaneous route is a technique frequently utilized in immunological studies. The footpad is a frequently used location, since the draining lymph nodes are easily removed making it possible to study the local immunological response. Taking the advantages of this strategy, the cellular immune response to *A. simplex* L3 antigens was compared in mice that were infected either after being pre-sensitized, with homologous CE antigen or sham sensitized. As the pre-sensitization simulates a primary immune response the authors addressed the difference between a primary infection and a simulation of a reinfection. This footpad sensitization protocol induced an increase in the size and weight of the popliteal lymph nodes (PLN). A high proportion of systemic $CD4^+$ and $TCR\alpha\beta^+$ T cells in both groups was observed. A reduction in B cells accompanied by a decrease of $CD8\alpha^+$ T cells was observed in pre-sensitized and infected mice, while those only exposed to infection present the greatest increase in $CD8\alpha^+$ and $TCR\alpha\beta^-$ T cells.

42.6.6.6 Intranasal Immunization

To examine the immunological mechanisms underlying the development of allergic airway inflammation, WT and interleukin-4 receptor alpha (IL-4Rα)-deficient mice were sensitized to *Anisakis* antigens through different routes.[123] Live or heat-killed *Anisakis* larvae were administered intraperitoneally, while *Anisakis* extract was administered by the intranasal route. Subsequently, all animals were challenged intranasally with an *Anisakis* extract. Allergen-specific antibodies developed only in intraperitoneally sensitized mice; however, both routes of sensitization induced IL-4Rα-dependent allergic airway responses in WT mice in an IL-4/IL-13-dependent pathway. Unexpectedly, infection with live *Anisakis* larvae induced airway hyperresponsiveness that was abrogated when IFN-γ was neutralized *in vivo*. Thus, infection leads to IL-4/IL-13-independent, IFN-γ-dependent airway hyperresponsiveness. Together, these results demonstrate that both infection with larvae and inhalational exposure to *Anisakis* proteins are potent routes of allergic sensitization, explaining food- and work-related allergies in humans, which can involve either IL-4/IL-13 or IFN-γ. Importantly for diagnosis, detectable *Anisakis*-specific antibodies may not accompany allergic airway inflammation.

In vitro studies demonstrated that a 24-kDa protein (22U homologous; As22U) derived from *A. simplex* larva elicits several Th2-related chemokine gene expression, meaning that it may be one of the important allergens for the clinical setting. In order to examine their hypothesis, six intranasal applications of ovalbumin (OVA) or recombinant As22U (rAs22U) and OVA were performed. When compared to the group that only received OVA, the animals challenged with rAs22U associated to OVA presented severe airway inflammation, immune cell recruitment in special, eosinophils, increased levels of IL-4, IL-5, and IL-13 in the Bronchoalveolar lavage fluid (BALF), significantly increased airway hyperresponsiveness, and significantly higher anti-OVA-specific IgE and IgG1. After receiving rAs22U, the GRO-α (CXCL1) gene expression increased immediately while eotaxin (CCL11) and TARC (CCL17) gene expressions increased significantly at 6h. Thus, rAs22U may be responsible for a Th2/Th17-mediated airway allergic inflammation.[124] Using the same experimental protocol, two other *Anisakis* antigens (Ani s 1 Ani s 9) were tested, eliciting similar results expressing Th2 (IL-4, IL-5, IL-13, e IL-25) and Th17 (IL-6 e IL-17) cytokines because of the intranasal exposure.[125]

42.6.6.7 Nematode Molecules as Immunoregulators

In the last decades, a variety of immunoregulatory molecules have been isolated from a number of nematodes. The identified biological activities include actions equivalent to cytokines, protease inhibitors, macrophage migration inhibitory factor-like protein (MIF), proteins as poison expressed sequence tags (ESTs), and allergen.[126–132] The *A. simplex* macrophage MIF protein obtained from third-stage larvae of *A. simplex* was cloned (rAs-MIF) and tested in a murine OVA/Alum-induced asthma model.[129] The rAs-MIF treatment coupled with OVA/alum induced a complete inhibition of eosinophilia, reduced lung goblet cell hyperplasia, profoundly improved lung hyperactivity, and reduced the quantity of Th2-related cytokines (IL-4, IL-5, and IL-13) in the BALF and allergen-specific IgG2a in sera. Conversely, the BALF of the rAs-MIF-treated group contained significantly higher of IL-10 and TGF-β than controls. In addition, rAs-MIF recruited regulatory T cells ($CD4^+CD25^+Foxp3^+$) to the spleen and lungs.

These authors evaluated the function of rAs-MIF on a dextran sodium sulphate (DSS)-induced intestinal inflammation. Mice treated with rAs-MIF recovered weight loss and the disease activity index (DAI) value. The cytokine profile evaluation showed that rAs-MIF-treated mice presented higher levels of splenic and mesenteric lymph nodes (MLN) TGF-β and IL-10 with lower levels of IFN-γ, IL-6, and IL-13. In addition, Treg were greatly increased in the MLNs of the rAs-MIF-treated mice. *In vitro* experiments showed that rAs-MIF stimulated IL-10 production via toll-like receptor 2.[133]

Further studies on rAs-MIF also showed that TLR2 gene expression was significantly increased following rAs-MIF treatment. To further understand the relation between TLR2 and the amelioration mechanisms of rAs-MIF, the OVA/Alum allergic airway inflammation protocol was induced with or without rAs-MIF associated or not to anti-TLR2-specific antibody and comparing WT and TLR2 knockout mice. As a result, the amelioration effects of rAs-MIF in allergic airway inflammation model, as previously described, were diminished to fewer than two of the TLR2 blocking models. The expression of TLR2 on the surface of lung epithelial cell was significantly elevated by rAs-MIF or Pam3CSK (TLR2-specific agonist) treatment.[134] Pretreatment with α-mTLR2 Ab or Pam3CSK inhibited the elevation of IL-10 gene expression by rAs-MIF, suggesting that the anti-inflammatory effects of rAs-MIF might be closely related to TLR2.

42.6.7 Fish

Many marine fish are infected with third-stage larvae of *A. simplex* (*sensu lato*). To ensure food safety, it is important to determine whether these larvae are present in the flesh of commercial fish species. However, there is little information regarding the tissue specificity of anisakid species. Thus, the rationale for the use of fish as an experimental model to study Anisakidae nematode is to understand the infective capacity in commercially relevant fish species, the parasite mechanisms of aggression, and the host's immunological response.

42.6.7.1 Oral Infection

Rainbow trout (*Oncorhynchus mykiss*) and olive flounder (*Paralichthys olivaceus*) received L3 larvae of two sibling species of *A. simplex per os* and were accompanied for 5 weeks. In the rainbow trout, *A. simplex s.s.*

predominantly migrated into the body muscle while a small number of freely moving *A. pegreffi* larvae were recovered within the body cavity. In the olive flounder, *A. simplex s.s.* larvae were found both in the body cavity and the muscle, while *A. pegreffi* larvae were found only in the body cavity encapsulated in lumps.[135]

In another set of *in vivo* investigations, *A. simplex* was used to experimentally infect rainbow trout (*Oncorhynchus mykiss*), Baltic salmon (*Salmo salar*), and brown trout (*Salmo trutta*). Of the three species, Baltic salmon was the most susceptible, presenting the highest number of successfully established nematodes, whereas brown and rainbow trout had a higher natural resistance. The preferred *A. simplex* larvae microhabitat in the brown trout was the stomach, pyloric caeca, and intestine, while the majority of larvae found in rainbow trout were located at the pyloric caeca. In the Baltic salmon, the most susceptible fish species, nematodes were dispersed in and on the spleen, head, kidney, liver, swim bladder, and musculature. $CD8^+$ cells were present while IgM^+-bearing cells were absent in the inflammatory tissue around the nematodes of all three fish species. MHCII-bearing cells were present in the encapsulated *A. simplex* in rainbow and the brown trout, but not in Baltic salmon.[136]

Yet, another set of recent experiments show that closely related salmonids differ in their susceptibility toward different anisakid larvae and agree that parasites select different microhabitats in the hosts.[137] Orally infected rainbow trout (*Oncorhynchus mykiss*), brown trout (*Salmo trutta*), and Atlantic salmon (*Salmo salar*) with larval stages of *H. aduncum*, *C. osculatum*, or *A. simplex* were studied to determine parasite survival and location up to14 days post infection (dpi). Although the most prevalent and numerous nematode in brown trout at 2 dpi was *H. aduncum*, a large proportion of the worms were already recovered dead with no tissue penetration. This fish species exhibited the highest natural resistance to *A. simplex*. Rainbow trout exhibited the highest susceptibility to *C. osculatum* larvae at 2, 7, and 14 dpi with eventual pyloric cecum penetration. *A. simplex* larvae established a more successful infection in salmon compared to rainbow trout, although at 2 and 7 dpi this fish showed the highest intensity and abundance of larvae, but not after 14 days. Although the pyloric ceca was the preferred microhabitat for *Anisakis* in both rainbow trout and salmon, larval penetration into muscle and liver was found.

42.6.7.2 Intraperitoneal

Since hydrolytic enzymes play an important role in the nematode host tissue penetration, determination of what enzymes are present within the ES proteins seems important. Lipase, esterase/lipase, valine and cysteine arylamidases, naphthol-AS-BI-phosphohydrolase, and α-galactosidase activities were found. To further elucidate the influence of intraperitoneally injected ES, substances on the immune system of fish-specific gene expression in spleen and liver of the rainbow trout (*Oncorhynchus mykiss*) were measured. The results demonstrate a generalized downregulation of immune-related gene expression, suggesting a suppressive immunomodulatory role for ES proteins. From the ecological point of view, this makes biological sense. One can argue that when worm enzymes directly target the host's immune molecules, a decreased immune response with an increased worm survival is the consequence.[138]

42.6.8 *In Vitro* Cultivation

The *in vitro* cultivation of nematodes has been for long a goal of the field of parasitology. These techniques permit the understanding of parasite behavior, physiology, and metabolism as well as the molecular nature of the ES products and their relationship with the host. This, in turn, permits more adequate vaccine production designs, vaccine efficacy testing, antigen production for serological reagents, detection of drug-resistance, screening of potential therapeutic agents, and conducting epidemiological studies. However, the complexity of the parasite's life cycle involving different host species for their developmental stages frequently makes their cultivation a difficult task. Each parasite requires different cultivation conditions with specific nutrients, temperature, and incubation conditions.

A search in biological databases indicate that the first papers regarding parasite cultivation, in general, were published in the 1910s[139] and the first *Anisakis* cultivation papers in the 1970s.[140] An important systematization of the developed technique was performed by Silverman,[141] which has been updated in a diversity of technical books.[142,143]

For many clinically important parasites, *in vitro* cultivation is an important diagnosis tool. An array of commercial systems, which have been developed, such as the Harada-Mori culture technique for larval-stage nematodes, permit rapid diagnosis. In comparison, although *in vitro* cultivation techniques are used more often than *in vivo* techniques, the *in vivo* techniques are sometimes used for diagnosing parasitic infections such as trypanosomiasis and toxoplasmosis. Parasite cultivation continues to be a challenging diagnostic option. Thus, an overview of intricacies of parasitic culture and an update on popular methods used for cultivating parasites are presented.

42.6.8.1 Culture Media

The first description of Anisakidae nematode cultivation occurred in the early 1960s. *P. decipiens* larvae removed from the flesh of fresh fish were immediately transferred to 199 culture media enriched with glucose, beef embryo extract, beef liver extract, and antibiotics. In this study, the authors obtained larvae that reached morphological changes consistent with adult worms.[144] Subsequently, with adjustments of the initial conditions, *A. marina* developed successfully to adult worms. The first larval developmental changes were observed within 4 days after the release of cuticles in the medium. The development of gonadal tissue characterizing the preadult stage occurred between 26 and 98 days. The complete maturation characterized by the worm wall thickening and gonadal maturation can be distinguished *in vitro*. The first free larvae were observed after 4–8 days at a temperature of 13°C–18°C and after 20–27 days at a temperature of 5°C–7°C. The larvae are very active, and their mobility has no fixed direction. In seawater, they can live for 3–4 weeks at temperatures of 13°C–18°C and for 6–7 weeks at 5°C–7°C; temperatures above 20°C led to increased mortality, and a temperature of 34°C was absolutely inadequate, indicating that the first intermediate host must be cold-blooded.[145]

42.6.8.2 L3–L4 Transformation Model

Improvements of *in vitro Anisakis* L3 culture conditions were introduced in 1976; these allowed to explore the formation of cuticles and ecdysis.[140] Different culture media (199, Krebs-Ringer), carbon dioxide concentration, temperature, and storage conditions were tested. Among the tested conditions, culture media 199 gave the best results, with the highest number of molts and viability. The carbon dioxide concentration of 5% in low concentrations is more efficient in the first 40 h of cultivation. Using fluorescent tracers, it was determined that larvae do not feed (salt and glucose) until their digestive tract is complete, in other words, when they enter the fourth stage of development (L4).

To simulate the natural conditions of the fish's body, where the larvae remain for long periods in anabiose, and to determine the temporal resistance, *Anisakis* L3 were collected from herring and kept in saline solution culture (0.65% NaCl) at about 5°C. The mortality of *Anisakis* in culture presented three phases: Phase 1 (months 1–2)—low mortality; Phase 2 (months 3–5)—significant increase in mortality rate; and Phase 3 (months 6–8)—only the strongest survive larvae. Thus, the larvae kept in saline solution survived for about 35 weeks.[146]

CO_2 fixation is an important metabolic process for many organisms. In anisakid nematodes, CO_2 has been shown to be required for its development, at least *in vitro*. Comparing culture conditions, molting to L4 was reduced to one-third, after a 30-day culture in air, which corresponds to one-third of the survival of L3 cultivated in air + 5% CO_2. Thus, at suitable temperatures, a high pCO_2 is vital for the optimum development of L3 to adult (M3). Regarding the activity of the CO_2-fixing enzymes, phosphoenolpyruvate carboxykinase (PEPck) activity (305 nmol/min·mg protein) was much higher than that of PEPC (6.8 nmol/min·mg protein). The activity of these enzymes in the worms cultivated in air + 5% CO_2 was highest during M3 and, in general, was higher than that of those cultivated in air only, especially during molting from L3 to L4. The presence of CO_2 stimulates the molting from L3 to L4 and prolongs the survival at least *in vitro*.[147–153]

A. simplex L3 larvae tend to prefer fish tissues with high lipid content.[154] *In vitro* tests were carried out to study the behavior of *A. simplex* L3 in response to different concentrations of cod liver oil lipids. Larvae were placed into culture dishes containing agar separated into three segments, containing

0.2%–7% of cod liver oil. The results demonstrate that although L3 move randomly, they do not stop randomly. The tendency to move out of a certain area was inversely correlated with lipid concentration. A second observation indicates that the intentional migration range of larvae is short. In conclusion, L3 prefer high fat content and seek it over short distances. These *in vitro* data agree with previous observations that *A. simplex* L3 randomly tend to migrate out of the fish gut into the flesh.[155]

42.6.9 Larvicidal Models

With the growing number of human anisakiosis cases, an alternative was the search for active larvicidal compounds. *In vitro* and *in vivo* assays were undertaken to evaluate herbs used to season fish based on epidemiological observations that prevalence of anisakidosis in the Chinese regions where raw fish is often seasoned with ginger (rhizome of *Zingiber officinale*) and/or "perilla mint," "Chinese basil," or "wild basil" (common names for *Perilla frutescens* [Lamiaceae]) is smaller. [156,157]

In vitro studies showed that [6]-Shogaol and [6]-gingerol components derived from ginger rhizome induced an important reduction in larvae mobility and altered both their cuticle and digestive tract.[158] Further studies revealed that [10]-gingerol, [10]-shogoal, and other compounds derived from ginger also has a very effective larvicidal effect.[159]

In vivo protocols where rats were infected by delivering larvae directly to the stomach through the use of a gavage were used to evaluate the action of essential oils on *Anisakis* L3. Simultaneously or 2 h after infection, each rat received one of five monoterpenes. To determine the localization and viability of the larvae and to determine gastrointestinal histopathological changes, rats were sacrificed at various time points. The order of *in vivo* larvicidal activity was peril aldehyde > citral > citronellol > cuminaldehyde > carvacrol. When peril aldehyde, citral, and citronellol were given together with the nematodes, no hemorrhages were observed, leading to the conclusion that these monoterpenes somehow inhibit the fixation and/or penetration capacity of the larvae. The time gap of 2 h between the infestation and the administration of any of the tested compounds is sufficient for the larvae to develop their pathogenicity in the rats.[160]

Three sesquiterpenic derivatives (nerolidol, farnesol, and elemolto) were studied to determine their *in vivo* larvicidal activity. The order of *in vivo* larvicidal activity was nerolidol > farnesol > elemolto; the first two caused the death of all nematodes, which showed cuticle changes and intestinal wall rupture. Only 20% of infected rats treated with nerolidol or farnesol showed gastric wall lesions in comparison to 86.6% of control animals, suggesting that nerolidol and farnesol are good candidates for further research as biocidal agents against L(3) larvae of *Anisakis* type I.[161]

The histological parameters to evaluate the effect of potentially larvicidal compounds were the analysis of the cuticle and intestinal wall structure. Fixed formalin *A. simplex* L3 was assessed by optical microscopy study of transverse thin sections (0.5–1 μm) stained with hematoxylin eosin, Masson's trichromic dyes, or toluidine blue.

Knowing that essential oils can irritate the mucosa, gut inflammatory reaction was studied after oral administration of the tested compounds.[161] A marker of neutrophilic infiltration is the titration of myeloperoxidase activity (MPO), determined by solubilization of myeloperoxidase with hexadecyltrimethylammonium bromide and measured with a dianisidine-H_2O_2 assay.[162]

42.7 Conclusion

A range of laboratory models are available to investigate foodborne infectious diseases, including those due to *Anisakis* nematodes. As presented in the short epidemiological and taxonomical review of the anisakid family along with the review of laboratory models used to study anisakiosis, there are still many open questions regarding the life cycle, host–pathogen interaction, pathogenesis, and immune response of the anisakid family larvae. These questions should be addressed because these nematodes are more frequently contaminating foods due to the diffusion of oriental and Spanish cuisine, making this an emerging anthropozoonosis of clinical importance.

Acknowledgments

We thank Maira Platais for help in the translation process.

REFERENCES

1. Audicana, M.T. & Kennedy, M.W. *Anisakis simplex*: from obscure infectious worm to inducer of immune hypersensitivity. *Clin Microbiol Rev* 21, 360–79 (2008).
2. Baird, F.J., Gasser, R.B., Jabbar, A. & Lopata, A.L. Foodborne anisakiasis and allergy. *Mol Cell Probes* 28, 167–74 (2014).
3. Pravettoni, V., Primavesi, L. & Piantanida, M. *Anisakis simplex*: current knowledge. *Eur Ann Allergy Clin Immunol* 44, 150–6 (2012).
4. Chai, J.Y., Darwin Murrell, K. & Lymbery, A.J. Fish-borne parasitic zoonoses: status and issues. *Int J Parasitol* 35, 1233–54 (2005).
5. Hochberg, N.S. & Hamer, D.H. Anisakidosis: perils of the deep. *Clin Infect Dis* 51, 806–12 (2010).
6. Gibbons, L. *Keys to the Nematode Parasite of Vertebrates*. Supplementary Volume. CAB International, Wallingford, UK (2010).
7. Koie, M. & Fagerholm, H.P. The life cycle of *Contracaecum osculatum* (Rudolphi, 1802) sensu stricto (Nematoda, Ascaridoidea, Anisakidae) in view of experimental infections. *Parasitol Res* 81, 481–9 (1995).
8. Navone, G.T., Etchegoin, J.A. & Cremonte, F. *Contracaecum multipapillatum* (Nematoda: Anisakidae) from *Egretta alba* (Aves: Ardeidae) and comments on other species of this genus in Argentina. *J Parasitol* 86, 807–10 (2000).
9. Navone, G.T., Sardella, N.H. & Timi, J. T. Larvae and adults of *Hysterothylacium aduncum* (Rudolphi, 1802) (Nematoda: Anisakidae) in fishes and crustaceans in the South West Atlantic. *Parasite* 5, 127–36 (1998).
10. Palm, H.W. Ecology of *Pseudoterranova decipiens* (Krabbe, 1878) (Nematoda: Anisakidae) from Antarctic waters. *Parasitol Res* 85, 638–46 (1999).
11. Cipriani, P. et al. Genetic identification and distribution of the parasitic larvae of *Anisakis pegreffii* and *Anisakis simplex* (s. s.) in European hake *Merluccius merluccius* from the Tyrrhenian Sea and Spanish Atlantic coast: implications for food safety. *Int J Food Microbiol* 198, 1–8 (2015).
12. Klimpel, S. & Palm, H.W. Anisakid nematode (Ascaridoidea) life cycles and distribution: increasing zoonotic potential in the time of climate change? in *Progress in Parasitology* (ed. Melhorn, H.) 201–22 (Springer, Berlin and Heidelberg, Germany, 2011).
13. Kuhn, T., Hailer, F., Palm, H.W. & Klimpel, S. Global assessment of molecularly identified *Anisakis* Dujardin, 1845 (Nematoda: Anisakidae) in their teleost intermediate hosts. *Folia Parasitol (Praha)* 60, 123–34 (2013).
14. Mattiucci, S. & Nascetti, G. Genetic diversity and infection levels of anisakid nematodes parasitic in fish and marine mammals from Boreal and Austral hemispheres. *Vet Parasitol* 148, 43–57 (2007).
15. Ishikura, H. & Kikuchi, Y. Infection by larvae of *Anisakis* in humans. *Nihon Ishikai Zasshi* 57, 1649–55 (1967).
16. Ishikura, H. et al. Transition of occurrence of anisakiasis and its paratenic host fishes in Japan, with pathogenesis of anisakiasis. *Hokkaido Igaku Zasshi* 63, 376–91 (1988).
17. Abollo, E., Gestal, C. & Pascual, S. *Anisakis* infestation in marine fish and cephalopods from Galician waters: an updated perspective. *Parasitol Res* 87, 492–9 (2001).
18. D'Amico, P. et al. Evolution of the *Anisakis* risk management in the European and Italian context. *Food Res Int* 64, 348–362 (2014).
19. Klimpel, S., Palm, H.W., Ruckert, S. & Piatkowski, U. The life cycle of *Anisakis simplex* in the Norwegian Deep (northern North Sea). *Parasitol Res* 94, 1–9 (2004).
20. Andersen, K. *Hysterothylacium aduncum* (Rudolphi, 1862) infection in cod from the Oslofjord: seasonal occurrence of third- and fourth-stage larvae as well as adult worms. *Parasitol Res* 79, 67–72 (1993).
21. Yorimitsu, N. et al. Colonic intussusception caused by anisakiasis: a case report and review of the literature. *Intern Med* 52, 223–6 (2013).
22. Klimpel, S., Kellermanns, E. & Palm, H.W. The role of pelagic swarm fish (Myctophidae: Teleostei) in the oceanic life cycle of *Anisakis* sibling species at the Mid-Atlantic Ridge, Central Atlantic. *Parasitol Res* 104, 43–53 (2008).

23. Mattiucci, S. et al. First molecular identification of the zoonotic parasite *Anisakis pegreffii* (Nematoda: Anisakidae) in a paraffin-embedded granuloma taken from a case of human intestinal anisakiasis in Italy. *BMC Infect Dis* 11, 82 (2011).
24. Valentini, A. et al. Genetic relationships among *Anisakis* species (Nematoda: Anisakidae) inferred from mitochondrial *cox2* sequences, and comparison with allozyme data. *J Parasitol* 92, 156–66 (2006).
25. Mattiucci, S. et al. Metazoan parasite infection in the swordfish, *Xiphias gladius*, from the Mediterranean Sea and comparison with Atlantic populations: implications for its stock characterization. *Parasite* 21, 35 (2014).
26. McClelland, G. The trouble with sealworms (*Pseudoterranova decipiens* species complex, Nematoda): a review. *Parasitology* 124 Suppl, S183–203 (2002).
27. Mehrdana, F. et al. Occurrence of zoonotic nematodes *Pseudoterranova decipiens*, *Contracaecum osculatum* and *Anisakis simplex* in cod (*Gadus morhua*) from the Baltic Sea. *Vet Parasitol* 205, 581–7 (2014).
28. Koie, M. Experimental infections of copepods and sticklebacks *Gasterosteus aculeatus* with small ensheathed and large third-stage larvae of *Anisakis simplex* (Nematoda, Ascaridoidea, Anisakidae). *Parasitol Res* 87, 32–6 (2001).
29. Yagi, K. et al. Femal worm *Hysterothylacium aduncum* excreted from human: a case report. *Jpn J Parasitol* 45, 12–23 (1996).
30. Kong, Q. et al. Molecular identification of *Anisakis* and *Hysterothylacium* larvae in marine fishes from the East China Sea and the Pacific coast of central Japan. *Int J Food Microbiol* 199, 1–7 (2015).
31. Deardorff, T.L. & Overstreet, R.M. *Contracaecum multipapillatum* (=*C. robustum*) from fishes and birds in the northern Gulf of Mexico. *J Parasitol* 66, 853–6 (1980).
32. Ishikura, H. et al. Anisakidae and anisakidosis. *Prog Clin Parasitol* 3, 43–102 (1993).
33. Sakanari, J.A., Staunton, C.E., Eakin, A.E., Craik, C.S. & McKerrow, J.H. Serine proteases from nematode and protozoan parasites: isolation of sequence homologs using generic molecular probes. *Proc Natl Acad Sci USA* 86, 4863–7 (1989).
34. Fumarola, L. et al. *Anisakis pegreffi* etiological agent of gastric infections in two Italian women. *Foodborne Pathog Dis* 6, 1157–9 (2009).
35. Lim, H. et al. Molecular diagnosis of cause of anisakiasis in humans, South Korea. *Emerg Infect Dis* 21, 342–4 (2015).
36. Chapman, M.D., Pomés, A., Breiteneder, H. & Ferreira, F. Nomenclature and structural biology of allergens. *J Allergy Clin Immunol* 119, 414–420 (2007).
37. Radauer, C., Bublin, M., Wagner, S., Mari, A. & Breiteneder, H. Allergens are distributed into few protein families and possess a restricted number of biochemical functions. *J Allergy Clin Immunol* 121, 847–852.e7 (2008).
38. del Pozo, V. et al. Immunopathogenesis of human gastrointestinal infection by *Anisakis simplex*. *J Allergy Clin Immunol* 104, 637–43 (1999).
39. Chung, Y.B. & Lee, J. Clinical characteristics of gastroallergic anisakiasis and diagnostic implications of immunologic tests. *Allergy Asthma Immunol Res* 6, 228–33 (2014).
40. Shrestha, S. et al. Intestinal anisakiasis treated successfully with conservative therapy: importance of clinical diagnosis. *World J Gastroenterol* 20, 598–602 (2014).
41. Gonzalez-Munoz, M., Rodriguez-Mahillo, A.I. & Moneo, I. Different Th1/Th2 responses to *Anisakis simplex* are related to distinct clinical manifestations in sensitized patients. *Parasite Immunol* 32, 67–73 (2010).
42. Hiramatsu, K. et al. A case of acute gastric anisakiasis presenting with malignant tumor-like features: a large gastric vanishing tumor accompanied by local lymph node swelling. *Digest Dis Sci* 49, 965–9 (2004).
43. Bouree, P., Paugam, A. & Petithory, J.C. Anisakidosis: report of 25 cases and review of the literature. *Comp Immunol Microbiol Infect Dis* 18, 75–84 (1995).
44. Caramello, P. et al. Intestinal localization of anisakiasis manifested as acute abdomen. *Clin Microbiol Infect* 9, 734–7 (2003).
45. Choi, S.C., Lee, S.Y., Song, H.O., Ryu, J.S. & Ahn, M.H. Parasitic infections based on 320 clinical samples submitted to Hanyang University, Korea (2004–2011). *Korean J Parasitol* 52, 215–20 (2014).
46. Takeuchi, K., Hanai, H., Iida, T., Suzuki, S. & Isobe, S. A bleeding gastric ulcer on a vanishing tumor caused by anisakiasis. *Gastrointest Endosc* 52, 549–51 (2000).
47. Anibarro, P.C. et al. Protein contact dermatitis caused by *Anisakis simplex*. *Contact Dermatitis* 37, 247 (1997).

48. Nieuwenhuizen, N., Herbert, D.R., Brombacher, F. & Lopata, A.L. Differential requirements for interleukin (IL)-4 and IL-13 in protein contact dermatitis induced by *Anisakis*. *Allergy* 64, 1309–18 (2009).
49. Anibarro, B. & Seoane, F.J. Occupational conjunctivitis caused by sensitization to *Anisakis simplex*. *J Allergy Clin Immunol* 102, 331–2 (1998).
50. Armentia, A. et al. Occupational asthma by *Anisakis simplex*. *J Allergy Clin Immunol* 102, 831–4 (1998).
51. Armentia, A. et al. *Anisakis simplex* allergy after eating chicken meat. *J Investig Allergol Clin Immunol* 16, 258–63 (2006).
52. Estrada Rodriguez, J.L. & Gozalo Reques, F. Sensitization to *Anisakis simplex*: an unusual presentation. *Allergol Immunopathol (Madr)* 25, 95–7 (1997).
53. Scala, E. et al. Occupational generalised urticaria and allergic airborne asthma due to *Anisakis simplex*. *Eur J Dermatol* 11, 249–50 (2001).
54. Alonso, A., Daschner, A. & Moreno-Ancillo, A. Anaphylaxis with *Anisakis simplex* in the gastric mucosa. *N Engl J Med* 337, 350–1 (1997).
55. Moreno-Ancillo, A. et al. Allergic reactions to *Anisakis simplex* parasitizing seafood. *Ann Allergy Asthma Immunol* 79, 246–50 (1997).
56. Mattiucci, S. et al. Anisakiasis and gastroallergic reactions associated with *Anisakis pegreffii* infection, Italy. *Emerg Infect Dis* 19, 496–9 (2013).
57. Daschner, A., Cuellar, C., Sanchez-Pastor, S., Pascual, C.Y. & Martin-Esteban, M. Gastro-allergic anisakiasis as a consequence of simultaneous primary and secondary immune response. *Parasite Immunol* 24, 243–51 (2002).
58. Daschner, A., Alonso-Gomez, A., Cabanas, R., Suarez-de-Parga, J.M. & Lopez-Serrano, M.C. Gastroallergic anisakiasis: borderline between food allergy and parasitic disease-clinical and allergologic evaluation of 20 patients with confirmed acute parasitism by *Anisakis simplex*. *J Allergy Clin Immunol* 105, 176–81 (2000).
59. Alonso-Gomez, A. et al. *Anisakis simplex* only provokes allergic symptoms when the worm parasitises the gastrointestinal tract. *Parasitol Res* 93, 378–84 (2004).
60. Daschner, A. et al. Specific IgG4: possible role in the pathogenesis and a new marker in the diagnosis of *Anisakis*-associated allergic disease. *Scand J Immunol* 79, 120–6 (2014).
61. del Pozo, M.D. et al. Laboratory determinations in *Anisakis simplex* allergy. *J Allergy Clin Immunol* 97, 977–84 (1996).
62. Desowitz, R.S., Raybourne, R.B., Ishikura, H. & Kliks, M.M. The radioallergosorbent test (RAST) for the serological diagnosis of human anisakiasis. *Trans R Soc Trop Med Hyg* 79, 256–9 (1985).
63. Gonzalez-Fernandez, J. et al. Haemoglobin, a new major allergen of *Anisakis simplex*. *Int J Parasitol* 45, 399–407 (2015).
64. Lorenzo, S. et al. Human immunoglobulin isotype profiles produced in response to antigens recognized by monoclonal antibodies specific to *Anisakis simplex*. *Clin Exp Allergy* 29, 1095–101 (1999).
65. Daschner, A., Vega de la Osada, F. & Pascual, C.Y. Allergy and parasites reevaluated: wide-scale induction of chronic urticaria by the ubiquitous fish-nematode *Anisakis simplex* in an endemic region. *Allergol Immunopathol (Madr)* 33, 31–7 (2005).
66. Daschner, A., De Frutos, C., Valls, A. & Vega, F. *Anisakis simplex* sensitization-associated urticaria: short-lived immediate type or prolonged acute urticaria. *Arch Dermatol Res* 302, 625–9 (2010).
67. Tanaka, J. & Torisu, M. *Anisakis* and eosinophil. I. Detection of a soluble factor selectively chemotactic for eosinophils in the extract from *Anisakis* larvae. *J Immunol* 120, 745–9 (1978).
68. Boyden, S. The chemotactic effect of mixtures of antibody and antigen on polymorphonuclear leucocytes. *J Exp Med* 115, 453–66 (1962).
69. Asaishi, K. et al. Studies on the etiologic mechanism of anisakiasis. 1. Immunological reactions of digestive tract induced by *Anisakis* larva. *Gastroenterol Jpn* 15, 120–7 (1980).
70. Dale, H.H. The anaphylactic reaction of plain muscle in the guinea-pig. *J Pharmacol Exp Ther* 4, 167–223 (1913).
71. Schultz, W. Physiological studies in anaphylaxis. II. Reaction of smooth muscle from guinea-pigs rendered tolerant to large doses of serum. *J Pharmacol Exp Ther* 2, 221–9 (1910).
72. Myers, B.J. The migration of Anisakis-type larvae in experimental animals. *Can J Zool* 41, 147–8 (1963).
73. Dziekonska-Rynko, J., Rokicki, J. & Jablonowski, Z. Effects of ivermectin and albendazole against *Anisakis simplex in vitro* and in guinea pigs. *J Parasitol* 88, 395–8 (2002).

74. Smith, J.W. & Wootten, R. *Anisakis* and anisakiasis. in *Advances in Parasitology*, Vol. 16 (eds. Lumsden, W.H.R., Muller, R. & Baker, J.R.) 93–163 (Academic Press, London, 1978).
75. Bier, J.W., Jackson, G.J., Earl, F.L. & Knollenberg, W.G. Gross and microscopic pathology with larval *Anisakis* sp. and *Phocanema* sp. Nematodes from fish. *Trans Am Microsc Soc* 95, 264–265 (1976).
76. Strom, S.B. et al. Third-stage nematode larvae of *Contracaecum osculatum* from Baltic cod (*Gadus morhua*) elicit eosinophilic granulomatous reactions when penetrating the stomach mucosa of pigs. *Parasitol Res* 114, 1217–20 (2015).
77. Kuipers, F.C. Eosinophilic phlegmonous inflammation of the alimentary canal caused by a parasite from the herring. *Pathol Microbiol (Basel)* 27, 925–30 (1964).
78. Ruitenberg, E.J., Berkvens, J.M. & Duyzings, M.J. Experimental *Anisakis marina* infections in rabbits. *J Comp Pathol* 81, 157–63 (1971).
79. Hong, S.T. & Lee, S.H. Histopathological and serological observations on experimental anisakiasis of rabbits. *Kisaengchunghak Chapchi* 25, 168–180 (1987).
80. Yang, H.J., Cho, Y.J. & Paik, Y.H. Changes of IgM and IgG antibody levels in experimental rabbit anisakiasis as observed by ELISA and SDS-PAGE/immunoblot. *Kisaengchunghak Chapchi* 29, 389–96 (1991).
81. Kennedy, M.W. et al. The secreted and somatic antigens of the third stage larva of *Anisakis simplex*, and antigenic relationship with *Ascaris suum*, *Ascaris lumbricoides*, and *Toxocara canis*. *Mol Biochem Parasitol* 31, 35–46 (1988).
82. Iwasaki, K. & Torisu, M. *Anisakis* and eosinophil. II. Eosinophilic phlegmon experimentally induced in normal rabbits by parasite-derived eosinophil chemotactic factor (ECF-P). *Clin Immunol Immunopathol* 23, 593–605 (1982).
83. Cuellar, C., Perteguer, M.J. & Rodero, M. Presence of IL-4-like molecules in larval excretory-secretory products and crude extracts from *Anisakis simplex*. *Scand J Immunol* 53, 483–8 (2001).
84. Kobayashi, Y., Ishizaki, S., Nagashima, Y. & Shiomi, K. Ani s 1, the major allergen of *Anisakis simplex*: purification by affinity chromatography and functional expression in *Escherichia coli*. *Parasitol Int* 57, 314–9 (2008).
85. Rodero, M., Jimenez, A., Chivato, T., Laguna, R. & Cuellar, C. Purification of *Anisakis simplex* antigen by affinity chromatography. *Parasitol Res* 87, 736–40 (2001).
86. Rodero, M., Jimenez, A. & Cuellar, C. Evaluation by ELISA of *Anisakis simplex* larval antigen purified by affinity chromatography. *Mem Inst Oswaldo Cruz* 97, 247–52 (2002).
87. Perez-Perez, J. et al. Molecular cloning of paramyosin, a new allergen of *Anisakis simplex*. *Int Arch Allergy Immunol* 123, 120–9 (2000).
88. Iglesias, R., Leiro, J., Santamarina, M.T., Sanmartin, M.L. & Ubeira, F.M. Monoclonal antibodies against diagnostic *Anisakis simplex* antigens. *Parasitol Res* 83, 755–61 (1997).
89. Ishikura, H. Clinical features of intestinal anisakiasis. in *Intestinal Anisakiasis in Japan*, Vol. III (eds. Ishikura, H. et al.) 100 (Springer-Verlag, Tokyo, 1990).
90. Asami, K. & Inoshita, Y. Experimental anisakiasis in guinea pigs: factors influencing infection of larvae in the host. *Jpn J Parasitol* 16, 415–22 (1967).
91. Oshima, T. *Anisakis* and anisakiasis in Japan and adjacent area. *Progr Med Parasitol Japan* 4, 305–93 (1972).
92. Yoshimura, H. Clinical patho-parasitology of extra-gastrointestinal anisakiasis. in *Intestinal Anisakiasis in Japan* (eds. Ishikura, H. & Kikuchi, K.) 145–54 (Springer-Verlag, Tokyo, 1990).
93. Amano, T., Nakazawa, M., Sugiyama, H., Secor, W.E. & Oshima, T. Specific antibody patterns of Wistar rats inoculated with third stage larvae of *Anisakis simplex*. *J Parasitol* 81, 536–42 (1995).
94. Cho, T.H. et al. The time course of biological and immunochemical allergy states induced by *Anisakis simplex* larvae in rats. *Clin Exp Immunol* 143, 203–8 (2005).
95. Figueiredo, I., Jr. et al. A technique for the intra-gastric administration of live larvae of *Anisakis simplex* in mice. *Exp Parasitol* 130, 285–7 (2012).
96. Zuloaga, J. et al. A rat model of intragastric infection with *Anisakis* spp. live larvae: histopathological study. *Parasitol Res* 112, 2409–11 (2013).
97. Arizono, N., Yamada, M., Tegoshi, T. & Yoshikawa, M. *Anisakis simplex* sensu stricto and *Anisakis pegreffii*: biological characteristics and pathogenetic potential in human anisakiasis. *Foodborne Pathog Dis* 9, 517–21 (2012).

98. Abe, N. & Teramoto, I. Oral inoculation of live or dead third-stage larvae of *Anisakis simplex* in rats suggests that only live larvae induce production of antibody specific to *A. simplex*. *Acta Parasitol* 59, 184–8 (2014).
99. Weerasooriya, M.V., Fujino, T., Ishii, Y. & Kagei, N. The value of external morphology in the identification of larval anisakid nematodes: a scanning electron microscope study. *Z Parasitenkd* 72, 765–78 (1986).
100. Bowman, C.C. & Selgrade, M.K. Utility of rodent models for evaluating protein allergenicity. *Regulat Toxicol Pharmacol* 54, S58–61 (2009).
101. Katz, D.H. Regulation of IgE antibody production by serum molecules. III. Induction of suppressive activity by allogeneic lymphoid cell interactions and suppression of IgE synthesis by the allogeneic effect. *J Exp Med* 149, 539–44 (1979).
102. Katz, D.H. Recent studies on the regulation of IgE antibody synthesis in experimental animals and man. *Immunology* 41, 1–24 (1980).
103. Boyce, J.A. et al. Guidelines for the diagnosis and management of food allergy in the United States: summary of the NIAID-sponsored expert panel report. *J Allergy Clin Immunol* 126, 1105–18 (2010).
104. Pabst, O. Trafficking of regulatory T cells in the intestinal immune system. *Int Immunol* 25, 139–43 (2013).
105. Pabst, O. & Mowat, A.M. Oral tolerance to food protein. *Mucosal Immunol* 5, 232–9 (2012).
106. Schippers, A. et al. β7 integrin controls immunogenic and tolerogenic mucosal B cell responses. *Clin Immunol* 144, 87–97 (2012).
107. Campos, S.M. et al. Maternal immunomodulation of the offspring's immunological system. *Immunobiology* 219, 813–21 (2014).
108. Paschoal, P.O. et al. Food allergy/hypersensitivity: antigenicity or timing? *Immunobiology* 214, 269–78 (2009).
109. Teixeira, G. et al. Diet selection in immunologically manipulated mice. *Immunobiology* 213, 1–12 (2008).
110. Cardoso, C.R. et al. Modulation of mucosal immunity in a murine model of food-induced intestinal inflammation. *Clin Exp Allergy* 38, 338–49 (2008).
111. Suko, M., Ogita, T., Okudaira, H. & Horiuchi, Y. Preferential enhancement of IgE antibody formation by *Bordetella pertussis*. *Int Arch Allergy Appl Immunol* 54, 329–37 (1977).
112. Tada, T., Okumura, K., Ochiai, T. & Iwasa, S. Effect of lymphocytosis-promoting factor of *Bordetella pertussis* on the immune response. II. Adjuvant effect for the production of reaginic antibody in the rat. *Int Arch Allergy Appl Immunol* 43, 207–16 (1972).
113. Li, X.M. et al. A murine model of peanut anaphylaxis: T- and B-cell responses to a major peanut allergen mimic human responses. *J Allergy Clin Immunol* 106, 150–8 (2000).
114. Ishizaka, K. & Ishizaka, T. Mechanisms of reaginic hypersensitivity and IgE antibody response. *Immunol Rev* 41, 109–148 (1978).
115. Iglesias, R., Leiro, J., Ubeira, F.M., Santamarina, M.T. & Sanmartin, M.L. *Anisakis simplex*: antigen recognition and antibody production in experimentally infected mice. *Parasite Immunol* 15, 243–50 (1993).
116. Iglesias, R., Leiro, J., Ubeira, F.M., Santamarina, M.T. & Sanmartin, M.L. *Anisakis simplex*: stage-specific antigens recognized by mice. *J Helminthol* 69, 319–24 (1995).
117. Perteguer, M.J., Rodero, M., Flores, J.M., Dorea, R.C. & Cuellar, C. Cellular immune responses in mice immunized with *Anisakis simplex* larval antigens. *Parasitol Res* 87, 396–404 (2001).
118. Vericimo, M.A., Figueiredo, I., Teixeira, G.A. & Clemente, S.C. Experimental anisakid infections in mice. *J Helminthol*, 1–5 (2014).
119. Jones, R.E., Deardorff, T.L. & Kayes, S.G. *Anisakis simplex*: histopathological changes in experimentally infected CBA/J mice. *Exp Parasitol* 70, 305–13 (1990).
120. Baeza, M.L. et al. *Anisakis simplex* allergy: a murine model of anaphylaxis induced by parasitic proteins displays a mixed Th1/Th2 pattern. *Clin Exp Immunol* 142, 433–40 (2005).
121. Perteguer, M.J. & Cuellar, C. Isotype-specific immune responses in murine experimental anisakiasis. *Zentralbl Veterinarmed B* 45, 603–10 (1998).
122. Perteguer, M.J. & Cuellar, C. Interleukin-4 production in BALB/c mice immunized with *Anisakis simplex*. *Mem Inst Oswaldo Cruz* 96, 979–82 (2001).
123. Kirstein, F. et al. *Anisakis pegreffii*-induced airway hyperresponsiveness is mediated by γ-interferon in the absence of interleukin-4 receptor α responsiveness. *Infect Immun* 78, 4077–86 (2010).

124. Park, H.K. et al. A 24 kDa excretory-secretory protein of *Anisakis simplex* larvae could elicit allergic airway inflammation in mice. *Korean J Parasitol* 49, 373–80 (2011).
125. Cho, M.K. et al. Allergenicity of two *Anisakis simplex* allergens evaluated *in vivo* using an experimental mouse model. *Exp Parasitol* 146, 71–7 (2014).
126. Falcone, F.H., Zang, X., MacDonald, A.S., Maizels, R.M. & Allen, J.E. A *Brugia malayi* homolog of macrophage migration inhibitory factor reveals an important link between macrophages and eosinophil recruitment during nematode infection. *J Immunol* 167, 5348–54 (2001).
127. Gregory, W.F., Blaxter, M.L. & Maizels, R.M. Differentially expressed, abundant trans-spliced cDNAs from larval *Brugia malayi*. *Mol Biochem Parasitol* 87, 85–95 (1997).
128. Murray, J., Gregory, W.F., Gomez-Escobar, N., Atmadja, A.K. & Maizels, R.M. Expression and immune recognition of *Brugia malayi* VAL-1, a homologue of vespid venom allergens and *Ancylostoma* secreted proteins. *Mol Biochem Parasitol* 118, 89–96 (2001).
129. Park, S.K. et al. Macrophage migration inhibitory factor homologs of *Anisakis simplex* suppress Th2 response in allergic airway inflammation model via $CD4^+CD25^+Foxp3^+$ T cell recruitment. *J Immunol* 182, 6907–14 (2009).
130. Pastrana, D.V. et al. Filarial nematode parasites secrete a homologue of the human cytokine macrophage migration inhibitory factor. *Infect Immun* 66, 5955–63 (1998).
131. Yenbutr, P. & Scott, A.L. Molecular cloning of a serine proteinase inhibitor from *Brugia malayi*. *Infect Immun* 63, 1745–53 (1995).
132. Yu, H.S. et al. *Anisakis simplex*: analysis of expressed sequence tags (ESTs) of third-stage larva. *Exp Parasitol* 117, 51–6 (2007).
133. Cho, M.K., Lee, C.H. & Yu, H.S. Amelioration of intestinal colitis by macrophage migration inhibitory factor isolated from intestinal parasites through toll-like receptor 2. *Parasite Immunol* 33, 265–75 (2011).
134. Cho, M.K. et al. TLR2-dependent amelioration of allergic airway inflammation by parasitic nematode type II MIF in mice. *Parasite Immunol* 37, 180–91 (2015).
135. Quiazon, K.M., Yoshinaga, T. & Ogawa, K. Experimental challenge of *Anisakis simplex* sensu stricto and *Anisakis pegreffii* (Nematoda: Anisakidae) in rainbow trout and olive flounder. *Parasitol Int* 60, 126–31 (2011).
136. Bahlool, Q.Z., Skovgaard, A., Kania, P., Haarder, S. & Buchmann, K. Microhabitat preference of *Anisakis simplex* in three salmonid species: immunological implications. *Vet Parasitol* 190, 489–95 (2012).
137. Haarder, S., Kania, P.W. & Buchmann, K. Comparative infectivity of three larval nematode species in three different salmonids. *Parasitol Res* 112, 2997–3004 (2013).
138. Bahlool, Q.Z., Skovgaard, A., Kania, P.W. & Buchmann, K. Effects of excretory/secretory products from *Anisakis simplex* (Nematoda) on immune gene expression in rainbow trout (*Oncorhynchus mykiss*). *Fish Shellfish Immunol* 35, 734–9 (2013).
139. Bass, C.C. & Johns, F.M. The cultivation of malarial plasmodia (*Plasmodium vivax* and *Plasmodium falciparum*) *in vitro*. *J Exp Med* 16, 567–79 (1912).
140. Sommerville, R.I. & Davey, K.G. Stimuli for cuticle formation and ecdysis *in vitro* of the infective larva of *Anisakis* sp. (Nematoda: Ascaridoidea). *Int J Parasitol* 6, 433–9 (1976).
141. Silverman, P.H. *In vitro* cultivation procedures for parasitic helminths. in *Advances Parasitology* (ed. Dawes, B.) 159–222 (Academic Press, London, 1965).
142. Smyth, J.D. *In Vitro Cultivation of Parasitic Helminths*, (CRC Press, Boca Raton, 1990).
143. Yasuraoka, K. & Hata, H. *In vitro* cultivation of parasitic helminths. in *Progress of Medical Parasitology Japan*, Vol. 7 (eds. Osturu, M., Kamegai, S. & Hayashi, S.) 211–26 (Meguro Parasitological Museum, Tokyo, 2003).
144. Townsley, P., Wight, H., Scott, M. & Hughes, M. The in-vitro maturation of the parasitic nematode, *Terranova decipiens*, from cod muscle. *J Fish Board Can* 20, 743–7 (1963).
145. Van Banning, P. Some notes on a successful rearing of the herring-worm, *Anisakis marina* L. (Nematoda: Heterocheilidae). *J Conseil* 34, 84–8 (1971).
146. Grabda, J. Studies on survival and development in vitro of *Anisakis simplex* stage 3 larvae in time [Worm parasite of herring and other marine fish, Nematoda]. *Acta Ichthyol Piscatoria* 12, 1 (1982).
147. Davila, C., Malagon, D., Valero, A., Benitez, R. & Adroher, F.J. *Anisakis simplex*: CO(2)-fixing enzymes and development throughout the *in vitro* cultivation from third larval stage to adult. *Exp Parasitol* 114, 10–5 (2006).

148. Adroher, F.J., Malagon, D., Valero, A. & Benitez, R. *In vitro* development of the fish parasite *Hysterothylacium aduncum* from the third larval stage recovered from a host to the third larval stage hatched from the egg. *Dis Aquat Organ* 58, 41–5 (2004).
149. Iglesias, L., Valero, A. & Adroher, F.J. Some factors which influence the *in vitro* maintenance of *Anisakis simplex* (Nematoda). *Folia Parasitol (Praha)* 44, 297–301 (1997).
150. Iglesias, L., Valero, A., Benitez, R. & Adroher, F.J. *In vitro* cultivation of *Anisakis simplex*: pepsin increases survival and moulting from fourth larval to adult stage. *Parasitology* 123, 285–91 (2001).
151. Iglesias, L., Valero, A., Galvez, L., Benitez, R. & Adroher, F.J. *In vitro* cultivation of *Hysterothylacium aduncum* (Nematoda: Anisakidae) from 3rd-stage larvae to egg-laying adults. *Parasitology* 125, 467–75 (2002).
152. Malagon, D., Benitez, R., Valero, A. & Adroher, F.J. CO_2-fixing enzymes and phosphoenolpyruvate metabolism in the fish parasite *Hysterothylacium aduncum* (Ascaridoidea, Anisakidae). *Dis Aquat Organ* 85, 217–23 (2009).
153. Mladineo, I., Simat, V., Miletic, J., Beck, R. & Poljak, V. Molecular identification and population dynamic of *Anisakis pegreffii* (Nematoda: Anisakidae Dujardin, 1845) isolated from the European anchovy (*Engraulis encrasicolus* L.) in the Adriatic Sea. *Int J Food Microbiol* 157, 224–9 (2012).
154. Stromnes, E. An *in vitro* study of lipid preference in whaleworm (*Anisakis simplex*, Nematoda, Ascaridoidea, Anisakidae) third-stage larvae. *Parasitol Res* 113, 1113–8 (2014).
155. Young, P. The relationship between the presence of larval anisakine nematodes in cod and marine mammals in British home waters. *J Appl Ecol*, 459–85 (1972).
156. Goto, C., Kasuya, S., Koga, K., Ohtomo, H. & Kagei, N. Lethal efficacy of extract from *Zingiber officinale* (traditional Chinese medicine) or [6]-shogaol and [6]-gingerol in *Anisakis* larvae *in vitro. Parasitol Res* 76, 653–6 (1990).
157. Kasuya, S. et al. Lethal efficacy of leaf extract from *Perilla frutescens* (traditional Chinese medicine) or perillaldehyde on *Anisakis* larvae *in vitro. Jpn J Parasitol* 39, 220–5 (1990).
158. Goto, C., Kasuya, S., Koga, K., Ohtomo, H. & Kagei, N. Lethal efficacy of extract from *Zingiber officinale* (traditional Chinese medicine) or [6]-shogaol and [6]-gingerol in *Anisakis* larvae *in vitro. Parasitol Res* 76, 653–6 (1990).
159. Lin, R.-J., Chen, C.-Y., Chung, L.-Y. & Yen, C.-M. Larvicidal activities of ginger (*Zingiber officinale*) against *Angiostrongylus cantonensis. Acta Trop* 115, 69–76 (2010).
160. Hierro, I., Valero, A. & Navarro, M.C. *In vivo* larvicidal activity of monoterpenic derivatives from aromatic plants against L3 larvae of *Anisakis simplex* s.l. *Phytomedicine* 13, 527–31 (2006).
161. Navarro-Moll, M.C., Romero, M.C., Montilla, M.P. & Valero, A. *In vitro* and *in vivo* activity of three sesquiterpenes against L(3) larvae of *Anisakis* type I. *Exp Parasitol* 127, 405–8 (2011).
162. Krawisz, J.E., Sharon, P. & Stenson, W.F. Quantitative assay for acute intestinal inflammation based on myeloperoxidase activity. Assessment of inflammation in rat and hamster models. *Gastroenterology* 87, 1344–50 (1984).

43

Clonorchis sinensis

Bayissa Chala Legissa and Sung-Tae Hong

CONTENTS

43.1 Introduction

43.1.1 History

Clonorchis sinensis was first discovered by James McConnell at the autopsy of a Chinese male in 1874. Following a detailed description of the fluke, Spencer Cobbold suggested that the species be named *Distoma sinense* in 1875. Afterward, its nomenclature was changed to *Opisthorchis sinensis* by Blanchard in 1895 and finally renamed as *Clonorchis sinensis* by Arthur Looss in 1907. Commonly called the "Chinese or oriental liver fluke," *C. sinensis* belongs to the family Opisthorchiidae. Similar to those in the families Dicrocoeliidae and Fasciolidae, members of the family Opisthorchiidae infect the biliary tree of mammals including humans. Currently, *C. sinensis* is recognized as a major parasite of the human intrahepatic bile duct causing clonorchiasis [1].

43.1.2 Life Cycle and Morphology

The adult flukes of *C. sinensis* are leaf-shaped and measure 15–20 mm long and 3–5 mm wide (Figures 43.1 and 43.2). The human and mammalian reservoir hosts (dogs, pigs, cats, and rats) acquire the infection by ingestion of raw or undercooked fish containing infectious metacercariae. Inside these mammals, the metacercariae are liberated from the fish flesh by pepsin digestion in the stomach and the

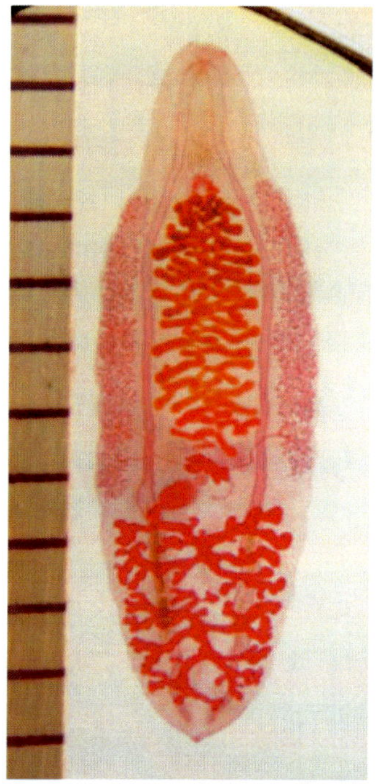

FIGURE 43.1 Adult worm of *C. sinensis*, acetocarmine stained.

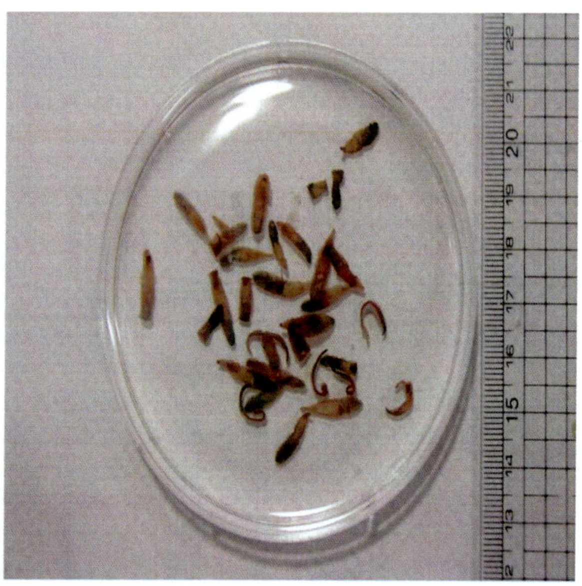

FIGURE 43.2 Gross view of recovered adult worms of *Clonorchis sinensis* from a man after chemotherapy in Korea. The uterus of worms that are red in color is full of eggs.

metacercariae excyst in the duodenum after trypsin digestion [2]. The excysted juvenile flukes migrate into the common bile duct through the ampulla of Vater to reach the intrahepatic biliary duct. One experimental study using rabbits observed that the juvenile flukes moved to the intrahepatic bile duct as early as 10–20 min after they excysted and entered into the duodenum [3]. The flukes develop into adults within 4 weeks and begin to produce eggs, which mix in feces and pass through the intestine [1].

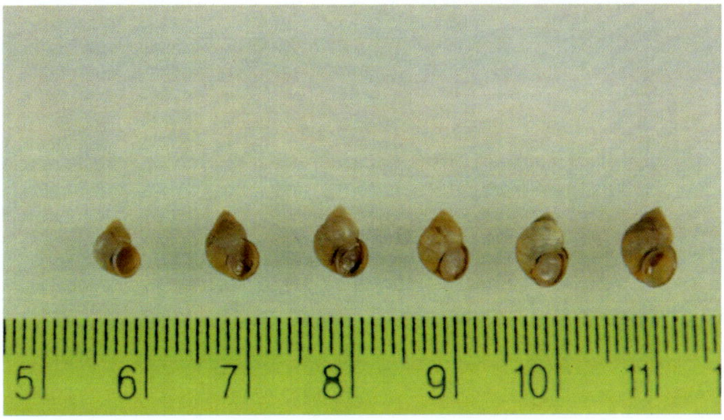

FIGURE 43.3 Gross view of the snail host, *Parafossarulus manchouricus*.

After their discharge by mammalian hosts, the eggs are ingested by freshwater snail, which is the intermediate hosts, in the rivers or lakes. The first intermediate snail hosts are *Parafossarulus* or *Bithynia* species (Figure 43.3). In the snail host, the miracidium hatches out in the rectum and penetrates the wall to reach the perirectal tissue. In the perirectal tissue, the miracidium undergoes metamorphoses into a sporocyst, which in turn gives rise to rediae and then cercariae. One miracidium may amplify to 1000–1500 cercariae, which emerge into the water [4]. The emerging cercariae freely swim and penetrate the skin, gills, fins, or muscles of freshwater fish within 24–48 h. In the fish, the cercariae encyst to become metacercariae. Numerous species of small-sized freshwater fish such as *Pseudorasbora parva*, *Pungtungia herzi*, *Pseudogobio esocinus*, *Acheilognathus intermedia*, *Odontobutis interrupta*, and others serve as the second intermediate host of *C. sinensis*, harboring the metacercariae [5,6]. A study investigating freshwater fish in their infection status with metacercariae of *C. sinensis* in Korea reported that the number of metacercariae per fish was found to be 48 (1–1142) in *P. parva* and 60 (1–412) in *P. herzi* (Figure 43.4), implying that the two species are able to be the index fish for the estimation of *C. sinensis* transmission in a certain locality [5]. When humans or mammals consume infected raw freshwater fish or poorly cooked ones (including smoked or pickled fish), the metacercariae infect the host and develop into adults.

43.1.3 Pathogenesis and Pathology

Excysted *C. sinensis* metacercariae migrate into the end of the bile ducts of their host and use the bile ducts as their nest [1]. Pathogenesis of liver-fluke-mediated tissue damage may be directly via mechanical and chemical irritation and also by immune-mediated response during their reforming process. Mechanical injury caused by oral suckers of the parasite during feeding activities and motility stimulates the biliary epithelial cells. Simultaneously, chemical and immunological irritation by the excretory–secretory products induces cell proliferation and inflammation in addition to the mechanical stimulation [7].

The infected bile duct is widened as a sac-like nest over 5–7 mm in diameter and becomes tortuous (Figure 43.5). The endothelium of the dilated ducts undergoes several changes. The extent of ductal changes is proportional to the number of worms present, with some infections running into the thousands; more than 6000 adult worms have been recovered from a single patient [8]. In an experimental study using Sprague Dawley rats infected by *C. sinensis*, the bile duct endothelial cells proliferated more than 20-fold at the base of the mucosal layer [9]. The bile duct mucosal layer revealed severe hyperplasia. At the base of the hyperplastic tissue, many mucin-secreting cells appear as a metaplasia. Some dysplastic cells are also noticed among the hyperplastic cells. Eosinophilic inflammation is commonly found in the mucosa and along the bile ducts. Fibroblasts proliferate, and collagen fibers are deposited surrounding the duct [10]. The pathological changes were observed 1 year after chemotherapy in experimentally infected rabbits [11].

FIGURE 43.4 Second intermediate host. (A) *Pseudorasbora parva*. (B) *Pungtungia herzi*.

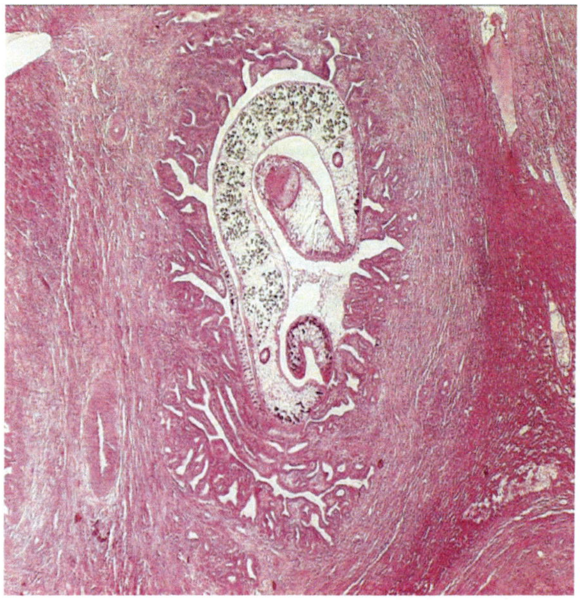

FIGURE 43.5 Histopathology of the *Clonorchis sinensis*-infected intrahepatic bile duct in an experimentally infected cat. HE stained, original magnification, ×40. The duct is enlarged with adenomatous hyperplasia and periductal fibrosis.

43.1.4 Clinical Manifestations

Clonorchiasis is rather well adapted to human hosts, and most of the light infected cases show no clinical findings. When present, clinical features of clonorchiasis are often nonspecific and related to inflammation and intermittent obstruction of the biliary ducts. Adult worms cause obstruction and blockage of the bile flow, which may induce epigastric discomfort or pain, swelling, and jaundice. In severe cases with heavy worm burdens, abdominal pain, nausea, and diarrhea can occur. However, in long-standing and untreated infections, chronic inflammation of the biliary system can lead to a neoplastic change as cholangiocarcinoma (CCA) [1].

43.1.5 Epidemiology

Globally, clonorchiasis is a common foodborne zoonosis distributed mainly in East Asia including China, Korea, Vietnam, the Philippines, and the far eastern part of Russia [1,12]. It may also occur in other regions where there are immigrants from endemic areas [13]. It is estimated that more than 200×10^6 people are at risk of infection, $15–20 \times 10^6$ people are infected, and $1.5–2 \times 10^6$ show symptoms or complications [1].

Several studies in Korea reported that the infection status of *C. sinensis* has shown only a little decrease despite wide use of praziquantel, an effective anthelmintic, for 30 years [14,15]. Cho et al. [14] reported that the prevalence of *C. sinensis* was 11.1% among 24,075 inhabitants in the riverside areas of southern Korea. Recently, several small endemic foci were reported in southern rural communities where eating raw fish is common [15].

The Korea Centers for Disease Control and Prevention (KCDC) performed the eighth nationwide survey on prevalence of intestinal parasitic infections in Korea and reported that the overall egg-positive rate of *C. sinensis* was 1.9% in 2012 compared to 4.6% in 1971, 1.8% in 1976, 2.6% in 1981, 2.7% in 1986, 2.2% in 1992, 1.4% in 1997, and 2.4% in 2004 [16]. The egg-positive rates, which are regarded as average prevalence of clonorchiasis in the general population in the surveyed year, have been decreasing slowly. Hong and Fang [1] explained that the slow decrease is due mainly to frequent treatment failure or reinfection after treatment. Still, about 1×10^6 residents who live along the river beds are infected in southern endemic areas in Korea [1,16].

In a study conducted during 2000–2004 in Korea, *C. sinensis* prevalence (2.1% in Chuncheon, 7.8% in Chungju, and 31.3% in Haman) was significantly correlated with CCA incidence rates (0.3, 1.8, and 5.5 per 100,000 persons, respectively) [17]. Furthermore, earlier case–control study in Busan, Korea, including 41 patients with CCA and 203 cases of hepatocellular carcinoma (HCC) revealed that CCA was significantly correlated with *C. sinensis* eggs in stool, past history of liver flukes, hepatitis history, and a history of heavy consumption of alcoholic beverages [18]. Shin et al. [19] described that the odds ratio of CCA due to *C. sinensis* infection was 4.7, and about 10% of CCA cases in Korea were due to this organism. The overall odds ratio of CCA is 4.47 for clonorchiasis, and about 5000 CCA cases are estimated annually in the world [20].

In Guangdong, China, where fish farming is common, the prevalence of human clonorchiasis is about 14%, and it may reach 80% in rural endemic areas [8]. A recent analysis of data from three parasitic disease surveys conducted in Hengxian County, China, in the last 22 years showed substantial increases in the patterns of clonorchiasis prevalence and infection intensity and decrease in trends of prevalence of soil-transmitting helminths [21]. Another recent study in Guangdong, China, reported that incidence rates of clonorchiasis showed a direct increasing trend by years with temperature change [22].

The distribution of clonorchiasis is determined not only by the distribution of snail intermediate hosts but also by the fish-consuming custom of residents in the endemic community. Fishing and eating of raw fish is an old popular tradition that has remained habitual and constant for thousands of years in endemic areas. The eating habit has resulted in persistent transmission of the liver fluke to humans [23,24]. There are several well-known risk factors that are associated with the transmission of clonorchiasis such as poor educational level of local residents, lack of sanitation, habit of eating raw or undercooked freshwater fish, development of freshwater aquaculture, and lack of systematic control activities in many endemic areas. In particular, freshwater aquaculture has rapidly expanded, with a resulting increase in fish contamination that has resulted from a lack of quarantine measures for fish products [25].

Regarding accumulated infection, infection rate and burden as measured by egg-positive rate and number of eggs per gram of feces (EPGs) have been recorded as increasing with age in men [23,26]. The reason for its increase among adult men reflects behavioral patterns of fishing as an occupation or recreation. Since the lifespan of *C. sinensis* is estimated to be around 30 years, the infection occurs repeatedly through the life of humans [23]. That finding suggests that host immunity is not so efficient to prevent reinfection or superinfection in humans.

43.1.6 Diagnosis

Diagnosis of clonorchiasis is essential in the control programs as well as for treatment. Early diagnosis and treatment is important to prevent serious complications of clonorchiasis in humans. Detection of eggs or adult worms in feces, bile, or duodenal fluids is a definite diagnosis. However, since detection of worms from a human body needs invasive procedures, the gold standard diagnosis of *C. sinensis* infection is microscopic identification of eggs in feces. The Kato-Katz (KK) method is reliable for diagnosis of clonorchiasis by egg detection [27], but the formalin-ether concentration technique is more sensitive than the KK method in cases of mild infection [28]. Furthermore, Choi et al. [28] reported that is good correlation between the EPGs determined by KK method and by direct smear. Since the KK method is far from straightforward for differential diagnosis between eggs of *C. sinensis* and heterophyid flukes, well-trained technicians are required for mixed infections [6]. Morphological identification of these eggs by fecal examination is very tricky and requires much more experience (Figure 43.6).

Methods other than fecal examination include serology tests, detection of DNA, and image diagnosis. There have been many serological methods used for diagnosis of *C. sinensis* infection, but ELISA is more commonly used [29]. The ELISA diagnosis using excretory-secretory antigen (ESA) and crude antigen (CA) of *C. sinensis* showed sensitivities of 92.5% and 88.2% and specificities of 93.1% and 87.8%, respectively [30]. The data implied that ESA was a better serodiagnostic antigen than CA for ELISA. ELISA detects specific IgG antibodies in serum, and the antibodies may last over 2 years after treatment, which may make it less specific.

Clonorchiasis has been diagnosed using ultrasound scanning, computerized tomography (CT), and magnetic resonance imaging (MRI) [31,32]. Choi and Hong [31] reported that radiological images employing CT or MRI of the liver were correlated with worm burdens (EPG counts) frequency and also severity. The images of the liver and bile ducts can be used as a good practical alternative diagnostic method of clonorchiasis. Although most of the positive images are acceptable for diagnosis, their sensitivity and specificity are lower than those of the fecal examination. This is mainly because those images recognize the tissue changes caused by the worms but not the worms itself. The images of cholangiocarcinoma associated with clonorchiasis show both the tumor with obstruction images

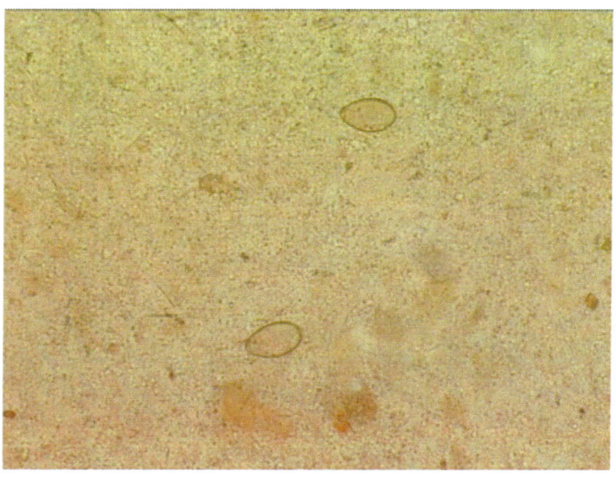

FIGURE 43.6 Eggs of *C. sinensis* by the Kato-Katz smear. Original magnification, ×400.

and diffuse dilatation of the peripheral intrahepatic bile ducts [31]. The remaining ductal dilatation with thick wall after chemotherapy further reduces the image specificity [11]. In endemic areas, it is recommended to make diagnosis of clonorchiasis by any feasible method and to treat positives more actively.

43.1.7 Chemotherapy and Control Measures

Praziquantel exhibits satisfactory efficacy and represents the first-line drug of choice for clonorchiasis [4]. The recommended treatment regimen by WHO is 25 mg/kg three times daily for two consecutive days [13]. In heavily endemic areas of China, using mass chemotherapy with praziquantel, it was reported that four selective annual treatments for egg-positive subjects reduced the egg-positive rates from 54.9% in 2001 to 15.0% in 2004 or from 73.2% in 2001 to 12.3% in 2004 [26]. In general, thorough cooking of fish, proper diagnosis, health education, and praziquantel treatment are major requirements for its control. Unfortunately, control of snail intermediate hosts by molluscicides was not considered feasible because of widespread distribution of snails, low prevalence of fluke infection among snails even in endemic communities, hazards to surrounding ecology, and economic considerations [26].

As clonorchiasis is one of the important foodborne neglected tropical diseases (NTDs) in Asia, it is included as a foodborne zoonosis in control programs of NTDs by the World Health Organization [33]. Currently, fish farming and neighboring environmental ponds that are routinely contaminated by untreated sewage have resulted in the establishment of infection in fish populations at large. This phenomenon along with the involvement of animal reservoir hosts will make control of the liver fluke infection even more challenging.

43.2 Laboratory Models for *C. sinensis* Research

Much of our knowledge and understanding on pathogenic mechanism of foodborne pathogens have been obtained by animal experiments or *in vitro* cultures. Animals are used as basic research models of helminths to study their life cycle in the laboratory as well as to study pathology, immunology, diagnosis, or treatment. *In vitro* cultivation of helminths or cell culture models, especially the secondary cell line, is a valuable tool that may help us in determining its virulence potential. The studies on *C. sinensis* have developed several appropriate animal models for specific aspects of the parasite, such as host susceptibility for infection and reinfection, host immune response, diagnosis, chemotherapy, pathogenesis, and carcinogenesis.

43.2.1 Rationale on Models

Animal research has provided valuable information about many physiological processes that are relevant to humans and human pathogens. Rodents (e.g., mice, rats) and rabbits are the most commonly used laboratory animal models for biomedical studies. Of course, several other mammals have been used. The animal model for *C. sinensis* research may depend on host–parasite relationship, feasibility of animal quarters, research purpose, and cost.

43.2.2 Laboratory Animals for Host Susceptibility to *C. sinensis*

Most mammals have been known to be susceptible to *C. sinensis* infection. Sohn et al. [34] investigated worm recovery rates of six laboratory mammals after infection and reinfection with *C. sinensis* metacercariae. The recovery rates of the mammals were from 17.6% in mice to 68.0% in hamsters (Table 43.1). The recovery rates suggest that rats and hamsters are highly susceptible; guinea pigs, rabbits, and dogs are moderately susceptible but mice are not. Rats, mice, guinea pigs, and rabbits were recognized to be resistant to reinfection after chemotherapy [34]. Another study reported that rats were found susceptible to *C. sinensis* infection with an 83.3% recovery rate [35].

TABLE 43.1
Recovery Rates of *C. sinensis* in Mammals by Experimental Infection

Mammals	No. of Heads	Recovery Rates (%)
		Mean ± SD
Rats	10	63.9 ± 20.4
Mice	11	17.6 ± 13.0
Hamsters	9	68.0 ± 15.0
Guinea pigs	10	34.7 ± 13.7
Rabbits	7	35.0 ± 7.6
Dogs	3	41.6 ± 14.5

Source: Sohn, W.M., et al., *Korean J. Parasitol.*, 44, 163–166, 2006.

Mice are the most commonly used laboratory animal model in biomedical research, but they are not susceptible to *C. sinensis* infection [34,36]. Many strains of mice have been developed for diverse research purposes, and the susceptibility was different in the strains [36]. ICR, Balb/c, C57BL/6, DDY, CBA/N, and C3H/HeN mouse strains were not susceptible to *C. sinensis* infection and showed very low recovery rates below 10%. Only the FVB strain showed 17.6% recovery rate, which is little higher than other strains [34]. Since the FVB strain is relatively susceptible compared to other strains, this strain has been used for *in vivo* studies of *C. sinensis* as a mouse model [37,38]. The FVB mice show cystic dilatation of the infected bile duct, which is a unique pathological change [38].

On the other hand, there have been reports that rats are resistant (to some degree) to reinfection by *C. sinensis*. The significant increase in the levels of bile IgA antibodies and serum IgE antibodies in resistant reinfected rats might play a role in the development of resistance to reinfection by *C. sinensis* [39].

Rabbits, dogs, and cats are susceptible hosts for *C. sinensis* infection. Rabbits are commonly used to maintain adult worms in the laboratory. Dogs, cats, and pigs are not used commonly for experimental infection of *C. sinensis*, but their natural infection has been reported to be common in China [26,40]. They are good reservoir hosts of *C. sinensis* in nature.

43.2.3 Animal Models for Immune Response and Reinfection

C. sinensis stimulates the host cells and tissues in the bile duct by both mechanical means and releasing excretory-secretory proteins (ESPs) [41]. The ESPs contain mostly enzymes, but various other soluble immunogenic components have been identified and have been found to be good serodiagnostic antigens by ELISA [30]. A recent study using mice reported that T2 ribonuclease in trematode ESPs has been identified as a potent regulator of dendritic cells (DCs) [42]. A T2 ribonuclease in ESPs of *C. sinensis*, named CsRNASET2, could modulate maturation of DCs and might play an important role in *C. sinensis*-associated immune regulation in mice. In another recent mouse model study, it has been reported that CA from *C. sinensis* plays a role in the anti-inflammatory function of dendritic DCs by inducing IL-10 and TGF-β through activation of extracellular signal-regulated kinase, which is a mitogen-activated protein kinase [43].

Previous studies demonstrated that *C. sinensis*-derived CAs suppress development of allergic responses. Jeong et al. [44] reported that *C. sinensis* venom allergen-like (CsVAL) peptide treatment inhibited activation of protein kinase C-α and extracellular signal-regulated kinase 1/2, which were involved in degranulation of IgE-sensitized mast cells. Furthermore, immunization with CsVAL suppressed development of skin inflammation, and this was demonstrated by assessing ear thickness and cutaneous infiltration by eosinophils and mast cells in oxazolone-induced contact hypersensitivity *in vivo* mouse model.

In an attempt to evaluate immunological responses and to develop an appropriate model, C57BL/6 and CBA/N mice were experimentally infected with *C. sinensis* [45]. The result of the study showed high proliferation of splenocytes, increased IL-17, and more severe gross and histopathological changes in CBA/N mice as compared to C57BL/6 mice. Based on these results, it has been suggested that CBA/N could be an appropriate mouse model for studying early immune responses in liver fluke infection [45].

A few experimental studies on reinfection or superinfection of *C. sinensis* used mostly rats because they are susceptible to primary infection but resistant to reinfection [34,39,46]. Meanwhile, hamsters, dogs, and humans are susceptible to primary and also reinfection or superinfection, and the infection is accumulated in those hosts. The resistance of the rats was mainly modulated by the local immune response in the infected tissue [39].

The findings of laboratory studies suggest that *C. sinensis* infection in mammals may suppress host immunity, which makes the host susceptible. The host immunity provokes the host resistant to reinfection or superinfection in rats but not in humans. Humans are repeatedly reinfected.

43.2.4 Animal Models for Metabolism Study

Another remarkable issue in host–parasitic interaction is the energy-acquiring mechanism of the parasite from its host. In this perspective, glycolytic enzymes are crucial molecules for trematode survival and have been targeted for drug development. It has been reported that hexokinase of *C. sinensis* (CsHK), the first key regulatory enzyme of the glycolytic pathway, was significantly inhibited compared to that of the corresponding negative control by praziquantel and anti-rCsHK serum [47]. In addition, the study pinpointed that there were differences in both mRNA and protein levels of CsHK among the life stages of the adult worm, metacercaria, excysted metacercaria, and eggs of *C. sinensis*, suggesting different energy requirements during different development stages. Another recent report showed that subcutaneous immunization with recombinant *C. sinensis* hexokinase (rCsHK) decreased worm burden and EPGs in challenged Sprague Dawley rats, implying that CsHK is a potential vaccine candidate and is a promising drug target for clonorchiasis [48]. Those studies used rats because they are susceptible and easy to handle.

43.2.5 Animal Models for Chemotherapy

One study evaluated cure of *C. sinensis* infection and recovery of the bile duct pathology after praziquantel medication with a dose of 50 mg/kg × 2/day × 2 days in rabbits [11]. The rabbit is an optimal model of *C. sinensis* infection because the liver of rabbits is big enough and its pathology is similar to that of humans.

On the other hand, there have been several reports on *in vivo* study to evaluate other chemotherapeutic agents than praziquantel to treat clonorchiasis. Those studies investigated cure rates, reduction of worm burdens, and morphological damage of *C. sinensis* adults using rats [49,50]. The new agents were artemether, artesunate, OZ78, and tribendimidine, and they were screened in various doses in rats or rabbits. Of them, tribendimidine was found promising and was used to a human trial. One clinical trial in China reported a cure rate of 44% with a 400 mg single dose of tribendimidine and of 58% after therapy for 3 days [51].

The chemotherapy study using experimental animals should evaluate cure rates or worm reduction rates after medication. Rats or rabbits are commonly used for this purpose because they are susceptible and because *C. sinensis* can grow well in the host. It should be noted that doses of antihelmintics vary between rodents and humans.

43.2.6 Animal Models for Histopathology and Carcinogenesis

As adult worms commonly parasitize the intrahepatic ducts, they provoke pathological changes in the biliary tracts, which may lead to further complications. The induced pathological changes of the infected bile duct start to present different clinical symptoms, although severity of the pathology depends upon intensity and chronicity of the infection. Both experimental findings and epidemiological evidence have implied that long-term infections by liver flukes lead to chronic pathological changes, including cholangitis, cholecystitis, cholelithiasis, cholangiectasis, adenomatous hyperplasia, hepatomegaly, hepatic fibrosis, and CCA [52–54]. In extreme cases, liver enlargement, thickening of the bile ducts, fibrosis, and some destruction of liver parenchyma are evident. However, unlike *Fasciola, Clonorchis* does not invade liver tissues and therefore, does not cause extensive liver necrosis [8].

Several well-documented epidemiological, histopathological, and experimental studies of *C. sinensis* have provided convincing evidence of a relationship between this trematode infection and the tendency for malignant transformation of the biliary epithelium in humans and experimentally infected animals. The International Agency for Research on Cancer (IARC) of the World Health Organization recategorized *C. sinensis* as a Group 1 biological carcinogen with *Opisthorchis viverrini*, as it produces human CCA [55,56].

The neoplastic response that is associated with clonorchiasis is CCA, which is a highly malignant epithelial cancer of the biliary tract [57]. In humans, primary sclerosing cholangitis, hepatolithiasis, fibropolycystic diseases of the biliary tract, Caroli's disease, and liver fluke infection have been considered as predisposing conditions for development of CCA [18,57,58].

Concerning history of experimental animals for induction of CCA, Thamavit et al. [59] first reported the possible development of CCA in the hamster model by administering *Opisthorchis viverrini* and administration of *N*-nitrosodimethylamine (NDMA; exogenously) in drinking water. Since then, similar experiments were also performed following *C. sinensis* infection in combination with 2-acetylaminofluorene or *N*-nitroso compounds (NDMA or *N*-nitrosodihydroxydi-*n*-propylamine) in hamsters [60–62] and rats [63]. These studies supported the role of *C. sinensis* as a promotor instead of an initiator. Experimental infection with *C. sinensis* alone in animals has never induced bile duct tumors [62]. During the past two decades, the primary experimental model for CCA induced by *C. sinensis* [60–62] as well as *O. viverrini* is the Syrian golden hamster [64,65]. The study using rat model on the possible modifying potential of *C. sinensis* infection and NDMA suggested that *C. sinensis* infection might facilitate the proliferation of NDMA-induced preneoplastic lesions of the liver [63].

One experimental animal study using animals other than hamsters or rats reported that intrahepatic cholangiocarcinomas (ICCs) could originate from fully differentiated hepatocytes. Using a mouse model of hepatocyte fate tracing, they found that activated NOTCH and AKT signaling cooperated to convert normal hepatocytes into neoplastic biliary cells that acted as precursors of rapidly progressing, lethal ICCs [66].

43.2.7 *In Vitro* Models

One study suggested that it was possible to establish and maintain the complete life cycle of *C. sinensis* in the laboratory [67]. It is evident that *in vitro* maintenance or culturing of *C. sinensis* in an active state (for the whole life cycle) is helpful for further characterization of its physiology and important regulatory molecules. In this regard, it was reported that bile acids and conjugated bile salts favored the survival of newly excysted juvenile *C. sinensis* (CsNEJ) under *in vitro* model. In a similar fashion, the study also indicated that NCTC 109 medium was found to be optimal for the *in vitro* maintenance of CsNEJs and 1× Locke's solution to be suitable for analyzing the biological effects of bioactive compounds [68]. *C. sinensis* worms were maintained *in vitro* using 0.85% normal saline (NaCl) or 1× PBS, mostly to collect ESP [69]. Normal saline and PBS lack essential nutrients, and the worms in those solutions may pass all reserved ESP and eggs within a few days but not produce new ESP [70]. It has been reported that the RPMI-1640 medium maintained living *C. sinensis* adults better and longer (up to 114 days) *in vitro* than other media. On the other hand, 1× Locke's solution best supported the worms alive among inorganic solutions for 57 days, confirming Locke's solution and RPMI-1640 media could maintain *C. sinensis* adult worms much longer than other inorganic solutions or nutrient media [70]. These studies helped define the best conditions for *in vitro* maintenance of the organism for long durations in different inorganic solutions and nutrient media under laboratory conditions. *In vitro* cultivation in any nutrient media could not foster adult worms from larvae.

It has already been reported that the ESPs of *C. sinensis* play important roles in host–parasite interactions. Recently, it has been revealed that the protein compositions of ESPs at different secretory times are variable. Moreover, it was observed, using the MTT assay, that the ESPs were found to inhibit proliferation of hepatic stellate cells (LX-2) [41].

For the biliary hyperplasia, several *in vitro* experiments provided strong evidence that *C. sinensis* possessed mitogenic factors in its ESPs and exhibited different levels of cell proliferations. For instance, *C. sinensis* ESPs were reported to exhibit antiapoptotic effect in human CCA cells [71]. Likewise, exposure of human CCA cells (HuCCT1) to *C. sinensis* ESPs demonstrated increased free radical generation through activation of NADPH oxidase, inducible nitric oxide synthase (iNOS), and xanthine oxidase, subsequently leading to nuclear factor-kappa B (NF-κB)-mediated inflammatory processes [72].

There has been a well-established body of knowledge about these highly reactive free radicals that can damage biologically relevant molecules of cells. Consequently, oxidative stress arising as a result of an imbalance between free radical production and antioxidant defenses is associated with damage to a wide range of molecular species including lipids, proteins, and nucleic acids [73,74]. The damage to cells caused by free radicals, especially the damage to DNA, may trigger the development of neoplasms and also other health-related disorders.

43.3 Conclusion

C. sinensis, a foodborne trematode infecting intrahepatic bile duct of humans, is considered as one of the target NTDs by WHO. Since its host specificity is low, most of the laboratory mammals except for mice are susceptible, and rats or rabbits are commonly used for common experimental infection. For experimental CCA studies, hamster is the only successful animal model. It is recommended to develop other animal models such as murine model for experimental studies on immunology and oncology of *C. sinensis*. *In vitro* cell cultivation is a good option for molecular studies of ESPs and generates molecular insights into its carcinogenesis.

REFERENCES

1. Hong ST, Fang Y. *Clonorchis sinensis* and clonorchiasis, an update. *Parasitol Int*. 2012; 61: 17–24.
2. Li S, Chung YB, Chung BS, Choi MH, Yu JR, Hong ST. The involvement of the cysteine proteases of *Clonorchis sinensis* metacercariae in excystment. *Parasitol Res*. 2004; 93: 36–40.
3. Kim TI, Yoo WG, Kwak BK, Seok JW, Hong SJ. Tracing of the bile-chemotactic migration of juvenile *Clonorchis sinensis* in rabbits by PET-CT. *PLoS Negl Trop Dis*. 2011; 5: e1414.
4. Rim HJ. Current pathobiology and chemotherapy of clonorchiasis. *Korean J Parasitol*. 1986; 24(Suppl): 1–141.
5. Kim EM, et al. Infection status of freshwater fish with metacercariae of *Clonorchis sinensis* in Korea. *Korean J Parasitol*. 2008; 46: 247–51.
6. Chai JY. Fish-borne parasitic diseases. *Hanyang Med Rev*. 2010; 30: 223–31.
7. Kim YJ, Choi MH, Hong ST, Bae YM. Proliferative effects of excretory/secretory products from *Clonorchis sinensis* on the human epithelial cell line HEK293 via regulation of the transcription factor E2F1. *Parasitol Res*. 2008; 102: 411–7.
8. Bogitsh BJ, Carter CE, Oeltmann TN. *Human Parasitology*. 4th ed. Amsterdam: Elsevier Inc; 2013.
9. Hong ST, Kho WG, Kim WH, Chai JY, Lee SH. Turnover of biliary epithelial cells in *Clonorchis sinensis* infected rats. *Korean J Parasitol*. 1993; 31: 83–9.
10. Rim HJ. Clonorchiasis: an update. *J Helminthol*. 2005; 79: 269–81.
11. Lee SH, Hong ST, Kim CS, Sohn WM, Chai JY, Lee YS. Histopathological changes of the liver after praziquantel treatment in *Clonorchis sinensis* infected rabbits. *Korean J Parasitol*. 1987; 25: 110–22.
12. Fürst T, Keiser J, Utzinger J. Global burden of human food-borne trematodiasis: a systematic review and meta-analysis. *J Lancet Infect Dis*. 2012; 12: 210–21.
13. WHO. Control of foodborne trematode infections. Report of a WHO study group. *WHO Tech Rep Ser*. 1995; 849: 1–157.
14. Cho SH, et al. Prevalence of clonorchiasis in southern endemic areas of Korea in 2006. *Korean J Parasitol*. 2008; 46: 133–7.
15. Park DS, Na SJ, Cho SH, June KJ, Cho YC, Lee YH. Prevalence and risk factors of clonorchiasis among residents of riverside areas in Muju-gun, Jeollabuk-do, Korea. *Korean J Parasitol*. 2014; 52: 391–7.

16. Korea Centers for Disease Control and Prevention. The 8th National surveys on the prevalence of intestinal parasitic infections. Seoul, Korea; 2013. pp. 35–68. http://www.cdc.go.kr/CDC/contents/CdcKrContentLink.jsp?fid=31&cid=24152&ctype=6.
17. Lim MK, et al. *Clonorchis sinensis* infection and increasing risk of cholangiocarcinoma in the Republic of Korea. *Am J Trop Med Hyg.* 2006; 75: 93–6.
18. Shin HR, et al. Hepatitis B and C virus, *Clonorchis sinensis* for the risk of liver cancer: a case-control study in Pusan, Korea. *Int J Epidemiol.* 1996; 25: 933–40.
19. Shin HR, et al. Descriptive epidemiology of cholangiocarcinoma and clonorchiasis in Korea. *J Korean Med Sci.* 2010; 25: 1011–6.
20. Qian MB, Chen YD, Liang S, Yang GJ, Zhou XN. The global epidemiology of clonorchiasis and its relation with cholangiocarcinoma. *Infect Dis Poverty.* 2012; 1: 4.
21. Qian MB, et al. Prevalence and intensity of foodborne clonorchiasis, Hengxian County, China, 1989–2011. *Emerg Infect Dis.* 2014; 20: 1872–5.
22. Li T, Yang Z, Wang M. Correlation between clonorchiasis incidences and climatic factors in Guangzhou, China. *Parasite Vectors.* 2014; 7: 29.
23. Hong ST. *Clonorchis sinensis.* In: Miliotis MD, Bier JW, editors. *International Handbook of Foodborne Pathogens.* New York, Basel: Marcel Dekker, Inc.; 2003. pp. 581–92.
24. Seo M, et al. Paleoparasitological report on the stool from a medieval child mummy in Yangju. *Korean J Parasitol.* 2007; 93: 589–92.
25. Fang Y, et al. Investigation and analysis on epidemic status of clonorchiasis in Guangdong province. *J Pathog Biol.* 2007; 2: 241–3.
26. Choi MH, et al. Effect of control strategies on prevalence, incidence and re-infection of clonorchiasis in endemic areas of China. *PLoS Negl Trop Dis.* 2010; 4: e601.
27. Hong ST, Choi MH, Kim CH, Chung BS, Ji Z. The Kato-Katz method is reliable for diagnosis of *Clonorchis sinensis* infection. *Diagn Microbiol Inf Dis.* 2003; 47: 345–9.
28. Choi MH, Ge T, Yuan S, Hong ST. Correlation of egg counts of *Clonorchis sinensis* by three methods of fecal examination. *Korean J Parasitol.* 2005; 43: 115–7.
29. Lee MK, Hong SJ, Kim HR. Seroprevalence of tissue invading parasitic infections diagnosed by ELISA in Korea. *J Korean Med Sci.* 2010; 25: 1272–6.
30. Choi MH, Park IC, Li S, Hong ST. Excretory-secretory antigen is better than crude antigen for the serodiagnosis of clonorchiasis by ELISA. *Korean J Parasitol.* 2003; 41: 35–9.
31. Choi D, Hong ST. Imaging diagnosis of clonorchiasis. *Korean J Parasitol.* 2007; 45: 77–85.
32. Lim JU, Joo KR, Shin HP, Cha JM, Lee JI, Lim SJ. Obstructive jaundice caused by clonorchiasis-associated duodenal papillitis: a case report. *J Korean Med Sci.* 2011; 26: 135–7.
33. World Health Organization (WHO). First WHO report on neglected tropical diseases 2010: working to overcome the global impact of neglected tropical diseases. http://www.who.int/neglected_diseases/2010report/en/.
34. Sohn WM, Choi MH, Hong ST. Susceptibility of experimental animals to reinfection with *Clonorchis sinensis. Korean J Parasitol.* 2006; 44: 163–6.
35. Wang X, et al. Experimental model in rats for study on transmission dynamics and evaluation of *Clonorchis sinensis* infection immunologically, morphologically, and pathologically. *Parasitol Res.* 2009; 106: 15–21.
36. Uddin MH, Li S, Bae YM, Choi MH, Hong ST. Strain variation in the susceptibility and immune response to *Clonorchis sinensis* infection in mice. *Parasitol Int.* 2012; 61: 118–23.
37. Yoon BI, et al. Infectivity and pathological changes in murine clonorchiasis: comparison in immuno-competent and immunodeficient mice. *J Vet Med Sci.* 2001; 63: 421–5.
38. Kim EM, Bae YM, Choi MH, Hong ST. Cyst formation, increased anti-inflammatory cytokines and expression of chemokines support for *Clonorchis sinensis* infection in FVB mice. *Parasitol Int.* 2012; 61: 124–9.
39. Zhang H, Chung BS, Li S, Choi MH, Hong ST. Changing patterns of serum and bile antibodies in re-infected rats with *Clonorchis sinensis. Korean J Parasitol.* 2008; 46: 17–22.
40. Lin RQ, et al. Prevalence of *Clonorchis sinensis* infection in dogs and cats in subtropical southern China. *Parasite Vectors.* 2011; 4: 180.

41. Zheng M, Hu K, Liu W, Li H, Chen J, Yu X. Proteomic analysis of different period excretory secretory products from *Clonorchis sinensis* adult worms: molecular characterization, immunolocalization, and serological reactivity of two excretory secretory antigens-methionine aminopeptidase 2 and acid phosphatase. *Parasitol Res.* 2013; 112: 1287–97.
42. Xu Y, et al. Molecular characterization and immune modulation properties of *Clonorchis sinensis*-derived RNASET2. *Parasite Vectors.* 2013; 6: 360.
43. Jin Y, Wi HJ, Choi MH, Hong ST, Bae YM. Regulation of anti-inflammatory cytokines IL-10 and TGF-β in mouse dendritic cells through treatment with *Clonorchis sinensis* crude antigen. *Exp Mol Med.* 2014; 46: e74.
44. Jeong YI, et al. Identification of anti-allergic effect of *Clonorchis sinensis*-derived protein venom allergen-like proteins (CsVAL). *Biochem Biophys Res Commun.* 2014; 445: 549–55.
45. Jin Y, Wi HJ, Choi MH, Hong ST, Bae YM. Differential regulation of pathogenesis and immune reactions was performed in C57BL/6 and CBA/N mice infected with *Clonorchis sinensis. J Immunol.* 2012; 188: 164.22.
46. Chung BS, et al. Development of resistance to reinfection by *Clonorchis sinensis* in rats. *Korean J Parasitol.* 2004; 42: 19–26.
47. Chen T, et al. Sequence analysis and molecular characterization of *Clonorchis sinensis* hexokinase, an unusual trimeric 50-kDa glucose-6-phosphate-sensitive allosteric enzyme. *PLoS One.* 2014; 9: e107940.
48. Chen T, et al. Advanced enzymology, expression profile and immune response of *Clonorchis sinensis* hexokinase show its application potential for prevention and control of clonorchiasis. *PLoS Negl Trop Dis.* 2015; 9: e0003641.
49. Keiser J, et al. Effect of artesunate and artemether against *Clonorchis sinensis* and *Opisthorchis viverrini* in rodent models. *Int J Antimicrob Agents.* 2006; 28: 370–3.
50. Keiser J, Xiao SH, Smith TA. Utzinger J. Combination chemotherapy against *Clonorchis sinensis*: experiments with artemether, artesunate, OZ78, praziquantel, and tribendimidine in a rat model. *Antimicrob Agents Chemother.* 2009; 53: 93770–6.
51. Qian MB, et al. Efficacy and safety of tribendimidine against *Clonorchis sinensis. Clin Infect Dis.* 2013; 56: e76–e82.
52. Choi D, et al. Cholangiocarcinoma and *Clonorchis sinensis* infection: a case-control study in Korea. *J Hepatol.* 2006; 44: 1066–73.
53. Sripa B, et al. Liver fluke induces cholangiocarcinoma. *PLoS Med.* 2007; 4: e201.
54. Fried B, Reddy A, Mayer D. Helminths in human carcinogenesis. *Cancer Lett.* 2011; 305: 239–49.
55. International Agency for Research on Cancer. *A Review of Human Carcinogens Part B: Biological Agents, Vol. 100 B.* IARC Monographs on the Evaluation of Carcinogenic Risks to Humans. Lyon, France: IARC; 2009.
56. Grosse Y, et al. A review of human carcinogens-part A: pharmaceuticals. *Lancet Oncol.* 2009; 10: 13–4.
57. Sirica AE. Cholangiocarcinoma: molecular targeting strategies for chemoprevention and therapy. *Hepatology.* 2005; 41: 5–15.
58. Khan SA, Thomas HC, Davidson BR, Taylor-Robison SD. Cholangiocarcinoma. *Lancet.* 2005; 366: 1303–14.
59. Thamavit W, Bhamarapravati N, Sahaphong S, Vajrasthira S, Angsubhakorn S. Effects of dimethylnitrosamine on induction of cholangiocarcinoma in *Opisthorchis viverrini*-infected Syrian golden hamsters. *Cancer Res.* 1978; 38: 4634–9.
60. Iida H. Experimental study of the effects *of Clonorchis sinensis* infection on induction of cholangiocarcinoma in Syrian golden hamsters administered 0.03% *N*-2-fluorenylacetamide (FAA). *Jpn J Parasitol.* 1985; 34: 7–16.
61. Lee JH, Rim HJ, Bak UB. Effect of *Clonorchis sinensis* infection and dimethylnitrosamine administration on the induction of cholangiocarcinoma in Syrian golden hamsters. *Korean J Parasitol.* 1993; 31: 21–30.
62. Lee JH, Yang HM, Bak UB, Rim HJ. Promoting role of *Clonorchis sinensis* infection on induction of cholangiocarcinoma during two-step carcinogenesis. *Korean J Parasitol.* 1994; 32: 13–8.
63. Jang JJ, Cho KJ, Myong NH, Chai JY. Enhancement of dimethylnitrosamine-induced glutathione S-transferase P-positive hepatic foci by *Clonorchis sinensis* infestation in F344 rats. *Cancer Lett.* 1990; 52: 133–8.

64. Loilome W, et al. Altered gene expression in *Opisthorchis viverrini*-associated cholangiocarcinoma in hamster model. *Mol Carcinog.* 2006; 45: 279–87.
65. Boonmars T, Boonjaraspinyo S, Kaewsamut B. Animal models for *Opisthorchis viverrini* infection. *Parasitol Res.* 2009; 104: 701–3.
66. Fan B, et al. Cholangiocarcinomas can originate from hepatocytes in mice. *J Clin Invest.* 2012; 122: 2911–5.
67. Liang C, et al. Experimental establishment of life cycle of *Clonorchis sinensis. Int J Automation Comput.* 2009; 27: 148–50.
68. Li S, Kim TI, Yoo WG, Cho PY, Kim TS, Hong SJ. Bile components and amino acids affect survival of the newly excysted juvenile *Clonorchis sinensis* in maintaining media. *Parasitol Res.* 2008; 103: 1019–24.
69. Kang JM, et al. A family of cathepsin F cysteine proteases of *Clonorchis sinensis* is the major secreted proteins that are expressed in the intestine of the parasite. *Mol Biochem Parasitol.* 2010; 170: 7–16.
70. Uddin MH, Li S, Bae YM, Choi MH, Hong ST. *In vitro* maintenance of *Clonorchis sinensis* adult worms. *Korean J Parasitol.* 2012; 50: 309–15.
71. Kim YJ, Choi MH, Hong ST, Bae YM. Resistance of cholangiocarcinoma cells to parthenolide-induced apoptosis by the excretory-secretory products of *Clonorchis sinensis. Parasitol Res.* 2009; 104: 1011–6.
72. Nam JH, et al. Free radicals enzymatically triggered by *Clonorchis sinensis* excretory–secretory products cause NF-κB-mediated inflammation in human cholangiocarcinoma cells. *Int J Parasitol.* 2012; 42: 103–13.
73. McCord JM. The evolution of free radicals and oxidative stress. *Am J Med.* 2000; 108: 652–9.
74. Young IS, Woodside JV. Antioxidants in health and disease. *J Clin Pathol.* 2001; 54: 176–86.

44

Fasciola *and Fasciolosis*

Antonio Muro and Jose Rojas-Caraballo

CONTENTS

44.1 Introduction

Classified in the phylum Platyhelminthes, subphylum Neodermata, class Trematoda, subclass Digenea, order Echinostomida, and family Fasciolidae, the genus *Fasciola* consists of two liver fluke species that cause fasciolosis in human and animal hosts, i.e., *Fasciola hepatica* and *Fasciola gigantica*. Jean DeBrie was the first person who referred fasciolosis to "putrefaction from liver of sheep" in 1379. However, no association between fasciolosis and the causal agent, the trematode *F. hepatica*, was mentioned at that time. On the contrary, there was a belief that the disease could be attributed to toxic substances from plants that were consumed by animals.[1] The discovery of *F. hepatica* and its life cycle in the intermediate host (*Galba truncatula*) was the result of many observations made by several researchers worldwide. The first observation was by Sir Anthony Fitzherbert (1523) and the Italian physicist Fanensi Gabucin (1549), who described the presence of worms similar to pumpkin seeds in the blood vessels of sheep and goats. At this time, it was thought that the onset of the disease was spontaneous. In 1688, the Italian physicist Francesco Redi demonstrated that the disease was not a spontaneous phenomenon, but was caused by oviposition of adult worms. In 1698, Govert Bidloo, a professor of anatomy, documented the presence of worms in the bile ducts of sheep, deer, and calves and also described for the first time the presence of worms in the liver of human beings.

During the second half of the 18th century, the intermediate phases in the life cycle of this parasite were described. These observations were made by Johann Swammerdam (1758) while he was dissecting a snail (*Paludina vivipara*), and he noted living organisms that evidently did not belong to the snail. Otto Müller (1798) found microscopic living organisms similar to tadpoles in backwaters, and he called them cercariae, but the life cycle of these worms remained a mystery, and nobody thought that more than one host will be required for completing its life cycle. In 1803, Johan Zeder noted egg hatching from different trematode species and also the release of a ciliated embryo in water. In 1807, Christian Nitzsch described for the first time the encystment of cercariae. He noted that after a period of time, cercariae stick on the surface of aquatic plants, they lose the tail, and then they were covered with a gelatinous substance. Guido Wagener (1857) noted miracidia penetration into snail and subsequent development of rediae. The first person who suggested that larval stages of liver worm were developed in the *G. truncatula* snail was the German helminthologist David Weinlad (1875) when he noted cercariae inside the snail. He also described for the first time the encystment of the cercariae on aquatic plants and the subsequent digestion by sheep, and he argued that these cercariae were the juvenile stages of liver worm. However, how the encysted cercariae reached the liver and bile ducts remained unknown. In 1914, the Russian helminthologist Dimitry Sinitsin demonstrated, using rabbits as animal models, that juvenile stages of worms, once released in the small intestine, penetrate the intestinal wall and migrate to the liver by the peritoneal cavity. By then, it was clearly evident that *F. hepatica* involves the snail as an intermediate host in its life cycle, resulting in the so-called liver-putrefaction disease, now known as fasciolosis.[1–5]

During the past decades, the epidemiology of human fasciolosis has markedly changed, and the number of reported cases has increased since 1990, and currently, it is considered as a reemerging disease worldwide. The geographic distribution of the two parasitic species causing fasciolosis is quite different. *F. hepatica* is cosmopolitan, while *F. gigantica* is mainly distributed in Africa and Asia, although its presence has also been reported in some regions of Europe and Russia.[6] Although the European origin of *F. hepatica* is well accepted, the incidence of human fasciolosis has been reported in 51 countries worldwide. It is estimated that more than 1 million people are currently infected, and there are more than 90 million people living in high-risk areas of acquiring the disease. The distribution of *F. hepatica* around the world has been mainly caused by cattle exportation from Europe to other continents and also because *F. hepatica* has been able to easily adapt into other mammalian hosts.

The ability of *F. hepatica* to successfully adapt into other mammalian hosts has been well described in the literature. For example, *F. hepatica* has been well adapted to camelids in Africa and the South of America, in some marsupials in Australia, and also in black rats, otters, and pigs,[7–9] which increases fasciolosis transmission rates. *F. hepatica* has also been able to adapt to different intermediate hosts. The main species responsible for *F. hepatica* transmission is the snail *G. truncatula*, with an important role in the distribution of fasciolosis around the world, as these snails inhabit ponds or waters that are formed, for example, in the rainy season, and its presence has also been detected in all continents. The snail responsible for the transmission of *F. gigantica* (*Radix*) lives mainly in standing and deep waters rich in aquatic vegetation, and its presence is limited to certain geographic areas.

The incidence of human fasciolosis could be underestimated, as not all cases are described in international reports, and they remain as internal reports, thesis, or publications in local journals, without international diffusion. Also, fasciolosis is not a mandatory notifiable disease in some countries,[10] and in many instances, they are not diagnosed, as they remain asymptomatic.

Andean (Bolivia, Perú, Chile, and Ecuador), Caribbean (Cuba), north African (Egypt), western European (Portugal, France, and Spain), and Caspian Sea (Iran) countries have the highest prevalence of fasciolosis in the world. In Bolivia, the north plateau is the area most affected with human fasciolosis,[11] where the prevalence is higher than 72%, and in some individuals even about 5000 eggs have been found in per gram of feces.[12,13] High prevalence has also been noted in the Puno plateau, Cajamarca, and Mantaro Valley (Perú).[14,15]

44.2 Life Cycle

An infected mammalian host releases *F. hepatica* eggs to the environment through the feces. They are ovoid, are yellow-brown in color, and measure about 130–145 μm long and 70–90 μm wide. Development of eggs and hatching are closely related to physical and chemical factors such as relative humidity, temperature, oxygen concentration, and pH. It has been demonstrated that the optimal temperature for egg development is 23°C–26°C; under these conditions, the eggs are embryonated in 2–3 weeks at pH 7.0. The precise mechanisms involved in egg hatching still remain unknown. Miracidia (hatched from eggs), which measure about 130 μm long and have great mobility due to their ciliated tegument, swim quickly to find the intermediate host (snails). And this happens within 24 h; otherwise, they die. Several factors are involved in the localization and penetration of miracidia into the snail, the most important being the presence of stimulatory molecules in the snail mucose, such as glucose, amino acids, and lipids. Miracidia penetrate into snail, lose cilia, and transform into sporocyst. Sporocysts are transformed into rediae, which in turn transform into cercariae. It has been estimated that one miracidia can develop up to 600 cercariae, through parthenogenetic proliferation. The cercariae have a propeller tail for swimming actively, which is subsequently lost before its encystment and posterior transformation into metacercariae, which attach to aquatic plants. A gelatinous substance covers the metacercariae to protect from environmental conditions until they are ingested by the mammalian host. The infection occurs after digesting metacercariae attached on the surface of aquatic plants. Approximately 1 h after the ingestion of metacercariae, the excystment occurs in the small intestine. After that, excysted metacercariae perforate the intestinal wall, migrate to the peritoneal cavity, perforate the Glisson's capsule, and penetrate into the hepatic parenchyma, reaching the inside of the liver, causing hemorrhage and fibrosis. In this stage, the parasite attains maximum growth. Finally, the parasites reach the bile ducts (7 weeks after the infection), and they remain here until full maturity. Occasionally, juvenile parasites can also be found in rare anatomical locations such as lungs, pancreas, lymph nodes, and thymus. Eight weeks after the infection, eggs can be found in the bile and later in the feces, thus completing the entire life cycle. Adult worms of *F. hepatica* can remain alive in the liver of the mammalian host even for several years and can also produce approximately 20,000 eggs daily.[2–5,16] The main phases of the life cycle of *F. hepatica* are shown in Figure 44.1.

44.3 Genome, Transcriptome, and Proteome

The availability of the genome sequence of different pathogens, including some helminths, during the last few years, has allowed a better and deeper understanding concerning relationships between host and pathogen as well as the identification of target molecules essential for host invasion, molecules involved in drug resistance generation, and also molecules with potential as vaccine candidates to prevent the diseases.[17–19]

Recently, a draft of the *F. hepatica* genome has been published.[20] This genome has a length of approximately 1.3 Gb, being larger than other trematodes. *F. hepatica* genome is three times the size of *Schistosoma* (363–397 Mb) or two times that of *Clonorchis sinensis* (547 Mb) and *Opisthorchis viverrini* (634 Mb). There is no evidence of genome duplication or repeat expansion to explain the large genome. It is possible that much of the noncoding gene is involved in regulation.

Through the investigation of the transcriptome of *F. hepatica*, more than 44,597 open reading frame (ORF) sequences have been identified using a 454-Roche-based methodology, starting with RNA isolated from adult worms.[21,22] A subsequent annotation and sequence analysis using the Basic Local Alignment Search Tool (BLAST) in search for both orthologous and paralogous proteins with the related trematodes

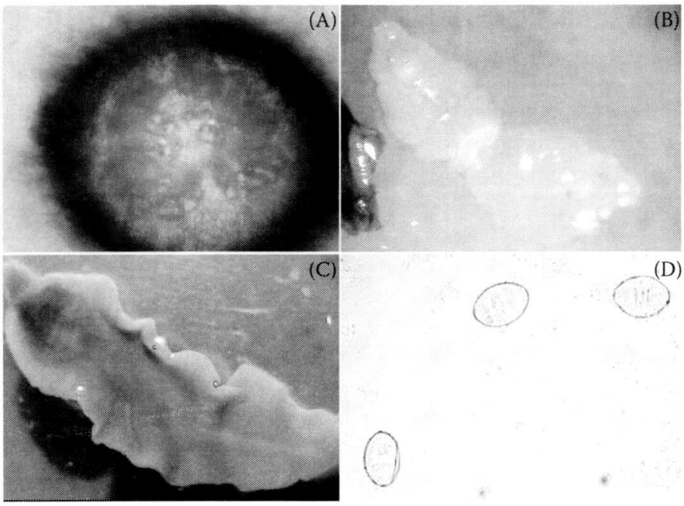

FIGURE 44.1 The biology of *F. hepatica*. (A) Metacercariae. (B) Adolescariae. (C) Adult. (D) Eggs.

Schistosoma mansoni (http://schistodb.net/) and *S. japonicum* (http://schistodb.net/) was performed together with a functional profile using the Gene Ontology (GO) and the Kyoto Encyclopedia of Genes and Genomes (KEGG). A great similarity between *F. hepatica* and *S. mansoni* proteins was identified, which was not only at the amino acid sequence level but also at the structural and functional level.

Nevertheless, data concerning the genome and transcriptome from *F. hepatica* still remain scarce, and several studies have been performed to describe *F. hepatica*'s proteome in more detail using different larval stages of its life cycle. Two-dimensional gel electrophoresis (2D-E) and mass spectrometry analysis (MS) have been the two techniques used more widely for the characterization of *F. hepatica* proteome. A proteomic analysis from the tegument of *F. hepatica* demonstrated high similarity with the *S. mansoni* proteome, resulting in the identification of 229 proteins, classified as follows: cytoskeleton (18), defense (12), metabolism (12), unknown (11), calcium-binding proteins (12), chaperones (7), proteases and inhibitors (7), mitochondrion (3), secretion (2), and nuclear (1).[23] A proteomic study from the surface proteins using an ultra-high performance liquid chromatography (UPLC)-MS-based technique allowed the identification of 64 proteins, which are the fatty-acid-binding proteins (FABPs) found in high abundance, and among them the FABP1 was the most abundant. Glutathione-*S*-transferase (GST) was also identified as a high-abundance protein from the surface of the tegument and has been considered as an important vaccine candidate against *F. hepatica*. Juvenile flukes have also been subjected to proteomic characterization, allowing the identification of a group of proteolytic proteins (cathepsins) (44%) together with proteins associated with detoxification, energy, and metabolism (16.9%), such as enolase, thioredoxin, and malate dehydrogenase.[24] Cathepsins have also been identified as the most abundant proteins in the products of excretion-secretion (ES) from *F. hepatica*, previously obtained by *in vitro* cell culture. This group of proteins has a molecular weight close to 30 kDa and an isoelectric point between 5.0 and 6.0. Other proteins such as enolase, GST, thioredoxin, and FABPs were also identified in the ES antigen.[25]

A special interest concerning cross-immunoprotective antigens between *F. hepatica* and *S. mansoni* has been raised during the last few years. Using adult worms, a proteomics approach allows the identification of 28 cross-reactive proteins present in both parasites, 10 out of them being highly expressed in *S. mansoni*, 9 out of them with high expression in *F. hepatica*, and the others with a similar expression pattern. Cross-reactive proteins were classified as FABPS, cathepsins, thioredoxin, and GST.[26]

44.4 Host–Pathogen Interaction

After ingestion of the metacercariae, the newly excysted juveniles perforate the intestinal wall and migrate through the hepatic parenchyma to reach the bile ducts. The induced damage to the mammalian host during its migration is caused mainly by factors such as mechanical abrasion by its spicules or the suction caused by their suckers. The hepatic parenchyma suffers a huge destruction with several internal injuries and bleeding during parasite migration, which are caused not only by the juvenile flukes but also by the inflammatory and immunological responses that are generated by the mammalian host. Peritonitis can also be produced as a result of secondary bacterial infections.

Adult worms are located in the bile ducts, but they can also be found in the cystic duct, gall bladder, Vater blister, and also in the choledochus. Bile ducts are highly dilated with hyperplasia and obstruction of bile flow.[27,28]

44.4.1 Immunity

Although one of the biggest challenges in the study and understanding of the immunology of infections caused by parasites is the vaccine development, the extreme importance of function and regulation of different types of T-lymphocytes and cytokines they produce help us to gain a better understanding concerning the immunological reactions between the host and the parasite and figure out the mechanisms evolved by the parasites not only to successfully invade and establish the cells and tissues from the mammalian host but also to evade the immune response from the host. Helminths are organisms that can measure from few centimeters to even meters. This is the reason why they do not invade cells but tissues, and the induced immune response by the mammalian host is also different from the common immune response generated by other pathogens. Immunomodulation from the host immune response becomes the most important factor by which helminths successfully survive for long periods of time in the host. During helminthic infections, the cells of the innate immune system are activated, resulting in the generation of different cellular phenotypes that promote the differentiation of Th2 cells with secretion of IL-4, IL-5, and IL-13, which stimulate the production of IgG1, IgE, and IgA. At the same time, the differentiation of Th1 cells that produce high levels of IFN-γ and are supposed to be highly protective is suppressed.[29–31]

The chronic infection occurs when the helminths are well established in the mammalian host. At this time, the Th2 phenotype still remains through the secretion of IL-4, but the T-regulator (Treg) phenotype is generated with secretion of IL-10 together with other cell types that secrete TGF-β with the expression of the cytotoxic T-lymphocyte antigen (CTLA).[32] Treg cells also play an important role in the suppression of Th1 cells and also regulate Th2 response that prevent tissue damage in the mammalian host caused by migration and feeding of parasite.[33] It has also been proposed that in helminthic infections, the mammalian host suppresses Th1 responses with destructive potential to the parasite and generates a Th2 response instead, which prevents tissue damage.[29,34] However, not all the infections caused by helminths induce the same pattern in the mammalian host.[35] Thus, infection caused by *Schistosoma* spp. induces a Th1 response at early stages (acute). Only when the helminth has established in the host and egg liberation by adult worms occurs, Th1 response switches to Th2 as a consequence of immunomodulation by the antigens from the egg parasites.[36] A similar behavior has also been observed in the infection caused by *B. malayi* using a murine model. Microfilariae induce a Th1 response in the acute phase, and once the adult worm has been established in the host, a Th2 phenotype differentiation occurs.[37]

Once *F. hepatica* enters into the intestinal wall, a complex network of cellular and molecular interactions between the parasite and different cell types from the host, such as dendritic cells, macrophages, and mastocytes, occurs. The parasite secretes a great number of molecules, which induce both Th2 and Treg nonprotective immune response in the host, and favors adaptation and survival of the parasite inside the host. It has been estimated that one adult worm from *F. hepatica* can remain alive in the host for 2 years and for even 20 years in sheep.[1,32]

Using a murine model of fasciolosis, it has been reported that the first attack from the immune system of the host to the parasite occurs during the next 4 h after the infection when the metacercariae have excysted in the duodenum. It consists of the generation of immunoglobulins that attempt to perform an effector mechanism, but the parasite releases cysteine proteases that are able to break these immunoglobulins, thereby preventing its effector action.[38–40] It has also been demonstrated that peritoneal macrophages induce the expression of Th2 phenotype cells together with a decrease in the Th1 phenotype at 24 h after infection. Simultaneously in this early stage of infection, the juvenile parasites also release immunomodulatory molecules, which have a direct effect on the function of the innate immune system such as dendritic cells and macrophages allocated in the intestinal wall and peritoneal cavity.[41,42]

Once the infection has been well established, there is a secretion of Th2-type cytokines such as IL-4, IL-5, and IL-13 by the splenocytes; in later stages of infection (3 weeks), there is also generation of Treg cytokines IL-10 and TGF-β by macrophages and dendritic cells. It has also been demonstrated that in the peritoneum, most of the T-CD4+ cells secrete only IL-10 without IL-4 or IFN-γ, thus inducing a suppression of both Th1 and Th2 responses, which are ineffective against the parasite.[43] Chronic fasciolosis in cattle has a similar immune response pattern as that of the murine model of fasciolosis, with predominance of Th2 and suppression of Th1 phenotype. Serological studies have demonstrated high levels of IgG1 and little or no IgG2a,[44–46] and the susceptibility to *F. hepatica* has been correlated with an increase in both the IgG1/IgG2a and the IL-4/IFN-γ ratio.[47] Owing to such immunoregulation/immunosuppression, the host infected with *F. hepatica* becomes more susceptible to acquire other infections, for example, in those infections where a Th1 response is necessary to prevent acquiring a disease. More specifically, it has been observed that mice infected with *F. hepatica* and coinfected with *Bordetella pertussis* could not develop a strong Th1 immune response which in turn becomes protective.[48–50] However, not all coinfections with *F. hepatica* present the same behavior. For example, *Toxoplasma gondii* induces a high Th1 response, with high levels of IFN-γ even in the presence of *F. hepatica*.[51]

44.4.2 Evasion Mechanisms

F. hepatica has developed several mechanisms to evade the immune response raised by the host. Anatomic location of adult worms in the bile ducts can be considered as an evasion mechanism as it represents, immunologically speaking, an inaccessible site. Furthermore, *F. hepatica* releases eggs and antigens together with the bile fluid, which also protect them from the immune response of the host. The tegument of *F. hepatica* also plays an important role in immune evasion, which is composed by the glycocalyx, which changes its chemical composition during the phases of migration and maturation of the parasite.

In the adolescariae stage, the tegumental change occurs every 3 h, thus avoiding the contact and attack of ligated antibody effector cells such as eosinophils and neutrophils. Juvenile flukes are highly resistant to the action of the complement of the host, and it is supposed to be due to the presence of sialic acid in surface glycoproteins, which activates the alternative pathway of complement. *F. hepatica* has mechanisms to prevent the action of nitric oxide, free oxygen radicals, superoxide dismutase (SOD), glutathione peroxidase, and GST. The parasite also secretes molecules that suppress or immunomodulate the immune response from the host.[27,28]

44.5 Clinical Manifestations

44.5.1 Acute Phase

The acute or invasive phase from fasciolosis is caused by the migration of the worms through the peritoneum and the hepatic parenchyma, causing mechanical damage and allergic and toxic reactions within the first 2–4 months after infection. The more frequent clinical manifestations include fever, abdominal pain, flatulence, nausea, diarrhea, constipation, urticaria, and respiratory symptoms such as cough, dyspnea, and chest pain. A physical exploration of patients with fasciolosis reveals hepatomegaly, ascites, and jaundice, while analytical data reveal leukocytosis, eosinophilia, anemia, and high activity from hepatic enzymes.[27,28]

44.5.2 Chronic Phase

The chronic phase from fasciolosis (also called obstructive phase) could occur within the next months or years after the infection. Adult worms migrate and establish within the bile ducts, causing inflammation and hyperplasia of epithelium and also thickening and dilatation of bile ducts and gallbladder, leading to cholangitis, cholecystitis, and gallbladder obstruction. Analytical data reveal leukocytosis, eosinophilia, middle anemia, high activity from hepatic enzymes, hypoalbuminemia, and hypergammaglobulinemia.[27,28]

44.6 Diagnosis

The diagnosis of human fasciolosis is difficult, as there are no specific symptoms and also because there is no egg-parasite detection in the early phases of infection. Two methods are widely used for the diagnosis of *F. hepatica*: parasitological- and immunological-based. There are also other methods based on the detection of parasite DNA, such as polymerase chain reaction (PCR), which are not commonly used because of technical difficulties.

44.6.1 Parasitological Techniques

The detection of *F. hepatica* eggs in feces is performed through sedimentation techniques, and it represents the most common methodology used for the diagnosis of fasciolosis.[27] The eggs released in the feces have an ovoid morphology, normally with a green-yellow coloration (caused by the direct contact with bile), and it is 60 μm in width and 130–150 μm in length. Parasitological techniques have several advantages, as they are not time consuming and are economical. However, these methods cannot be used in the acute phase of the disease, as eggs are released at 3–4 months after infection. In areas with a high prevalence of animal fasciolosis, a false-positive detection could occur, which is caused by the ingestion of bovine-liver-containing eggs from *F. hepatica*.[52]

44.6.2 Immunological Techniques

The antibodies produced against the antigens of *F. hepatica* have high sensitivity, and can be diagnosed even in the acute phase of the disease. ELISA assay is used, with antigens from *F. hepatica*, mainly excretory-secretory antigens,[53,54] cathepsin L1,[55,56] proteins from the tegument,[57] saposin-like proteins,[58] leucine aminopeptidase,[59] and other recombinant antigens.[60] Although there are several antigens for such detection, cathepsins remain the main source of antigens for the detection of circulating antibodies.[61] Monoclonal antibodies have also been used for the detection of antigens in sera samples and also for the detection of coproantigens. Some examples of monoclonal antibodies currently used are ES78 and MM3.[62,63] A new lateral flow test was constructed with a recombinant cathepsin L1, using protein A and MM3 monoclonal antibodies as detector reagents.[64]

44.6.3 Molecular Techniques

Although molecular techniques for the detection of parasitic DNA, such as PCR, have high specificity and sensibility, because of technical difficulties in areas with high incidence, PCR is not commonly used for the diagnosis of *F. hepatica* infection. Also, this is the only technique that identifies the causative organism both in the intermediate host and in the mammalian one. The molecular techniques more widely used for the detection of fasciolosis are (1) conventional PCR,[65,66] (2) real-time PCR,[65,66] (3) LAMP (loop-mediated isothermal amplification),[67] and (4) multiplex PCR.[68,69]

44.7 Treatment

To date, the number of chemotherapeutic agents for the successful treatment of fasciolosis is very scarce, as most of them are effective only against adult worms and ineffective or partially effective against immature worms. Triclabendazole (TBZ) is currently the drug of choice for the treatment of fasciolosis, as it is highly active against both mature and immature worms. Furthermore, it is a safe compound when administered and is well tolerated.[70] The recommended dosage is 10 mg/kg/day for 2 consecutive days, obtaining a curative rate between 92.2% and 93.9%. Adverse effects caused by TBZ are normally minimal, the more frequent being abdominal pain and sweating. In some cases, it can also induce nausea, vomiting, chills, cough, fever, and itching. In animals, the dose regimen may vary depending on the infected species; thus, bovines receive a single dose of 12 mg/kg, whereas sheep and goats receive a single dose of 10 mg/kg. For animals that are used for human consumption, such treatment should not be applied during the 28 days prior to sacrifice. On the other hand, milk from treated animals should not be consumed for the next 4 days after treatment.[71]

Structurally, TBZ belongs to the benzimidazole's family, and its mechanism of action is associated with the β-tubulin function, attaching to and hampering biological process associated with microtubules.[72] The resistance to TBZ was reported for the first time in Australia in the mid-1990s.[73] Since then, resistance to TBZ has also been reported in European countries such as Ireland, Scotland, Wales, Spain, and Netherlands.[74–76] The alternative treatment for the treatment of fasciolosis caused by TBZ-resistant strains is based on the administration of artemisinin-derivative compounds.

44.8 Experimental Models in Fasciolosis

F. hepatica naturally affects grazing ruminants, including sheep, cows, lambs, and goats, and also affects human beings. Currently, fasciolosis is considered as an emerging pathogen because of the high number of human infections reported. Because of the difficulty in investigating the immunology, pathology, and cellular and molecular mechanisms associated with fasciolosis in grazing ruminants, several experimental models that mimic fasciolosis have been used for decades in the laboratory. The main species currently used to study fasciolosis include mice, rats, and rabbits, although several studies have also been performed in sheep. The use of mice, rats, and rabbits in scientific experimentation has been well documented.

44.8.1 Fasciolosis in Mice

Mice have been used for a long time to study the immunology of the infection and are also used as an experimental model to assess vaccine efficacy. Furthermore, the use of natural infection models to study fasciolosis has been hampered because of technical issues such as cost, space, manipulation, and the availability of specific reagents for immunological studies. The University of Salamanca's Research Center for Tropical Diseases (CIETUS) has been working for the last decade on the development of an appropriate laboratory model to study the infection caused by *F. hepatica*, including both BALB/c and CD1 mice. Thus, a hyperinfection model using lethal doses of *F. hepatica* metacercariae to study both the immunopathology induced by the infection and the immunoprotective efficacy of several vaccine candidates was described.[77] The current approaches to study fasciolosis in mice are discussed in the following paragraphs.

44.8.1.1 Experimental Infection

Mice are orally infected with five to seven *F. hepatica* metacercariae, which are suspended in water or phosphate buffered solution (PBS). This suspension is then administered to each mouse using an intragastric gavage technique.[78,79] Nonlethal dosage in mice is frequently used to study the dynamics of the infection as mice remain alive during the whole study.[80] Other authors have used a large number of metacercariae to infect mice, including 10 and 15 metacercariae. In both cases, a human endpoint is established at 3 weeks after experimental infection.[49]

44.8.1.2 Physicopathological Changes

It is well known that helminth-associated infections induce a strong Th2 immune response in the mammalian host.[35] Thus, typically the success of the infection is investigated by measuring the level of antibodies raised against the excretory-secretory antigens. It is worth noting that the hyperinfection in mice with five to seven metacercariae induces death of animals between days 28 and 34.[77,79]

44.8.1.2.1 Measuring Antibody Levels

After experimental infection, mice are weekly bled to investigate the dynamics of antibody production. Blood is recovered on heparin-containing microtubes, centrifuged, and the plasma is carefully removed from the surface of the microtube. An ELISA technique is then used to investigate the levels of IgG, IgG1, and IgG2a, according to previously established protocols.[79] Normally, the levels of both IgG and IgG1 increase as the infection occurs, and little or no IgG2a is detected.[49,77]

44.8.1.2.2 Hepatic Transaminases

Alterations in hepatic transaminase levels are often used to assess the course of infection, and such measures can be done either in the serum or in the liver of infected mice. Typically, aspartate transaminase (AST), alanine transaminase (ALT), and γ-glutamyl transpeptidase are measured.

44.8.1.2.3 Eosinophilia Alteration during Fasciolosis

The infection caused by *F. hepatica* also induces changes in the number of eosinophils in peripheral blood. It has been demonstrated that eosinophilia occurs immediately after infection, reaching its maximum level at 7 days postinfection, where it begins to decrease.[81]

44.8.1.2.4 Cytokines

To assess cellular immune responses induced by *F. hepatica*, the most representative Th1- and Th2-Treg and Th17-associated cytokines are measured. For this purpose, infected mice are humanely euthanized and necropsied for spleen recovery. Later, splenocytes are recovered, *in vitro* cultured, and stimulated to quantify the levels of the aforementioned cytokines. The results clearly demonstrate that *F. hepatica* induces a Th2-polarized immune response with high levels of IL-4 and IL-5 and little or no IFN-γ.[49]

44.8.1.2.5 Hepatic Damage

Measuring the hepatic damage induced after infection also assesses the success of the infection by *F. hepatica*. At the time of necropsy, the liver is isolated and macroscopically evaluated considering the size, consistency, color, and dilatation of the bile ducts and repletion of the vessels, and scored as intense (+++), mild (++), moderate, (+) or without lesions (–).[77,79] There is no consensus in the quantitation of hepatic damage, and most of the times authors use their own scale and parameters. It is worth noting that not all of the mouse strains are equally susceptible/resistant to the infection with *F. hepatica*.[49] An example of liver lesions in healthy and infected mice is shown in Figure 44.2.

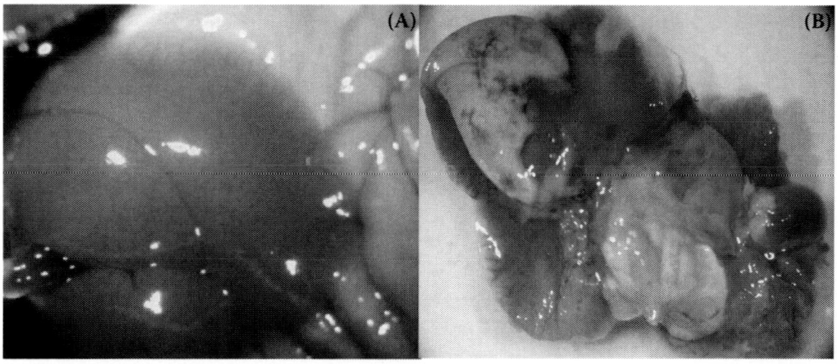

FIGURE 44.2 The hepatic damage induced by the experimental infection of *F. hepatica* in mouse. (A) Liver of one uninfected healthy mouse. (B) Liver of one mouse infected and necropsied at 21 days after infection. In this stage, bile duct enlargement is observed together with scars on the hepatic surface and irregular yellowish white area on the liver and parenchyma.

44.8.1.2.6 Worm Burden

Although determining worm burden is very useful to evaluate the success of infection with *F. hepatica*, the results of this approach are slightly inappropriate when using mouse as an experimental model, as they could not harbor too many adult worms in the liver, thereby leading to inconclusive statistical inference. Furthermore, when using lethal doses, mice die before worms reach the liver and bile ducts.

44.8.1.3 Microarray Analysis

Recently, we have investigated the gene expression profile during *F. hepatica* infection at 7 and 21 days after experimental infection using a microarray-based methodology. Differential expression of genes occurs as the infection progresses. For this purpose, we have isolated the RNA from the liver of infected mice at 7 and 21 days postinfection and investigated the gene expression profile compared to untreated mice. Using functional profiling, we identified several upregulated key genes involved in the induction of hepatic injury.[78]

44.8.2 Fasciolosis in Rats

Rats have also been widely used to study not only the immunology of the infection but also the protective efficacy of several vaccine candidates. Rats are more resistant to the infection, and typically they do not die, thus allowing a more detailed study during advanced stages of the infection. Moreover, previous reports have demonstrated the important role of black rats, *Rattus rattus*, in the epidemiology of fasciolosis, mainly in the Mediterranean island of Corsica. It has been demonstrated that these rats become naturally infected with the trematode *F. hepatica*. Although human infections of fasciolosis have also been reported in Corsica, the prevalence of the disease remains low.[82] Thus, the investigation of the immune response and the assessment of vaccines in rats have attracted interest, and several studies have been developed since then.

44.8.2.1 Experimental Infection

Rats are orally infected with the metacercariae of *F. hepatica* using the intragastric gavage technique. However, there is no consensus regarding the number of metacercariae that are needed for a successful infection, and this decision depends almost exclusively on each researcher's criteria and experience.[83–85]

44.8.2.2 Physicopathological Changes

44.8.2.2.1 Antibody Response

Using rats as laboratory experimental model allows a more compressive study of the dynamics of antibody production during infection with *F. hepatica* for a long period of time. Typically, IgG, IgG1, IgG2a, IgG2b, and IgG2c are measured during infection in rats by the ELISA technique using sera or plasma samples. IgG antibodies are detected from 1 week after experimental infection, and an increase occurs during the following 5 weeks. From 7 to 10 weeks after infection, IgG remains unchanged.[86] IgG1 and IgG2a are also detected from the first week after the infection, and this increases during the next 10 weeks, with similar patterns. High levels of IgM are induced by 2 weeks after infection, which slowly decreases as the infection occurs. However, high levels still remain by 10 weeks after infection. IgE has a biphasic behavior, with peaks being detected at 5 and 9 weeks after infection.[87]

44.8.2.2.2 Cytokines

Most of the experiments aimed at the study of the production of cytokines after the experimental infection with *F. hepatica* are performed in the spleen cells at different times of infection. An increase in the percentage of T-CD4 cells after the experimental infection is observed while the level of T-CD8 cells remains unchanged. It has also been demonstrated that the amount of cells producing IFN-γ remains unchanged during the first week after the infection, but they suffer a decrease at 2 weeks after the infection,[84] thus demonstrating a downregulation of the Th1 phenotype as the infection moves along. This behavior has also been reported in mice infections.[49] An increase in the level of both IL-4 and IL-10 is observed at 1 and 2 weeks after the infection.[83,84]

44.8.2.2.3 Reactive Oxygen Species

It is well known that the infection of rats with *F. hepatica* is characterized by the formation of reactive oxygen species (antioxidants). Thus, the study of both enzymatic and nonenzymatic antioxidants in the serum of infected rats is frequently performed to study the time-course of the infection of fasciolosis.[88]

44.8.2.2.4 Worm Burden

Using rats as an experimental model in the study of fasciolosis allows a more accurate investigation of worm burden than using mice. From third week after infection until the seventh week, flukes are harbored in the hepatic parenchyma. Later, they continue migrating and reach the bile ducts at week 11, where they remain for a long period of time.[89]

44.8.2.2.5 Hepatic Damage

As in mice, the hepatic damage induced by experimental infection is assessed considering the size, consistency, color, dilatation of the bile ducts, and repletion of the vessels, and scored as intense (+++), mild (++), moderate (+), or without lesions (–).[77] However, the pathology in the liver has previously been described from the first week up to 12 weeks after infection.[90]

44.8.3 Fasciolosis in Rabbits

The use of rabbits as an experimental model to study the immunology of the infection caused by *F. hepatica* has been very limited mainly because immunological assays for this species remain unavailable. However, rabbits have been widely used for the production of monoclonal antibodies (as large quantities of sera could be obtained) and also for the evaluation of the protective efficacy of vaccine candidates. Currently, there is only one report concerning the cytokine profile in rabbits immunized with a saposin-like (SAP2) recombinant protein by means of quantification of mRNA by real-time PCR.[91]

As in rats, rabbits could harbor a large number of adult worms in the liver, making them suitable to statistically assess worm burden.

44.8.3.1 Experimental Infection

To perform the experimental infection, rabbits are orally infected with the metacercariae of *F. hepatica* suspended in either distilled water or PBS. As in the rat model, the number of metacercariae required for the infection is variable and depends on the researcher's criteria and experience. Because a large number of metacercariae are commonly used to infect rabbits, the administration protocol could also be different. In some cases, the metacercariae of *F. hepatica* are contained within a gelatin capsule, thus avoiding the count of the metacercariae. The number of metacercariae required to infect rabbits is usually in the range of 25 to 50.[91–95]

44.8.3.2 Physicopathological Changes

44.8.3.2.1 Antibody Response

After experimental infection, rabbits are weekly bled to investigate the dynamics of antibodies raised against the parasite. ELISA technique is used to measure antibodies. From the third week after experimental infection, IgG is detectable, reaching a maximum peak at sixth week after infection. From week 6 onward, antibody levels remain unchanged until 10th week.[91]

44.8.3.2.2 Worm Burden

Rabbits are humanely slaughtered between 12 and 20 weeks after experimental infection, and the flukes are isolated from bile ducts. Both immature and mature flukes are isolated from the liver. For this purpose, livers are cut, soaked in water or PBS at 37°C, squeezed, and passed through 300-μm mesh sieve, according to previously described methodologies.[91,92]

44.8.3.2.3 Hepatic Damage

At necropsy, independent, experienced pathologists assess hepatic damage. They macroscopically examine the livers and assign a score according to the intensity of the damage. There is no consensus regarding the scale used in the quantitation of hepatic damage, but researchers have their own scales, which are applied independently. Typically, hepatic damage is classified as mild, moderate, and intense.[96]

44.9 Experimental Assessment of Vaccine Candidates

During the last decade, we performed several studies, in which we assessed the immunoprotective efficacy of different vaccine candidates. The FABPs and cathepsins from *F. hepatica* have been considered many times as promissory vaccine candidates.[97] The immunoprotective efficacy from the so-called Fh12, a *F. hepatica* FABP-derived protein weighing 12 kDa and obtained in its native form, was assessed in mice and sheep using the ADAD vaccination system. Results in mice showed a survival rate of 40% compared to the control group, whereas immunized sheep displayed lower fluke recovery (24.5%), a significant reduction in the number of eggs in the bile (58.1%) and feces (40.3%) compared to control groups.[98] Later, we evaluated the efficacy of the so-called Fh15, another FABP from *F. hepatica* having a molecular weight of 15 kDa, in mice and sheep as experimental models. Here, we demonstrated a survival rate of approximately 50% in mice, with low titers of IgG1. In sheep, a significant reduction in worm burden was achieved (43%) with less hepatic damage than in unimmunized control sheep.[99]

By means of a bioinformatics approach, we have identified two T-cell-containing epitopes derived from the amino acid sequence from the recombinant Fh15 protein (termed IKMVSSLKTKIT and VKAVTTLLKA) and produced as recombinant-GST-linked proteins. Its protective efficacy was then assessed in mice and rabbits, demonstrating that those peptides induced survival rates of 48.2% and 59.1% in mice, respectively. In rabbits, the immunization procedure induced a reduction in worm burden (46%), showing its potential as vaccine candidates.[100]

Other studies with the Fh12 protein in mice and sheep were performed by introducing a lipidic aminoalcohol as immunomodulator into the ADAD vaccination system. The mice showed survival rates ranging from 40% to 50%, while vaccination in sheep induced lower fluke recovery (42%), lower adult worm counts (57%), lower fecal egg count (38%), and less hepatic damage.[101] In another study with the recombinant Fh15 protein, we have demonstrated that the protocol of immunization with the ADAD vaccination system improves the survival rates in a murine model of mice. The immunization procedure induced high levels of IgG2a and IFN-γ, survival rates of about 50%, and less hepatic damage.[16]

More recently, we have used a bioinformatics approach to select vaccine candidates against *F. hepatica*. From the *F. hepatica* available protein sequences in databases, we selected those with the following criteria: (1) proteins having a signal peptide sequence according to the SignalP4.0 server and (2) proteins with no transmembrane domain according to the TMHMM2.0 server. The criteria highly fit to select those proteins that seem to be secreted are considered as important vaccine candidates. Once selected, the proteins are grouped into common families, and its amino acid sequences are aligned using the BLAST. Then, we identified peptides containing B and T cell epitopes within the conserved or semiconserved amino acid sequenced from proteins. The immune response induced in mice by the immunization with the peptides containing B and T cell epitopes is well characterized by means of the production of antibodies (IgG, IgG1, IgG2a, IgE, IgM), cytokines (IFN-γ, TNF-α, IL-1α, IL-4, IL-5, IL-6, IL-10, IL-13), and T cell populations (CD4, CD8, CD27, CD62L, CD197). Those peptides inducing a high immune response in mice are then selected to assess their immunoprotective efficacy in mice. We have demonstrated that peptides containing B and T cell epitopes are highly immunogenic, as high levels of IgG, IgG1, and IgG2 are reached, with high levels of IFN-γ, IL-4, IL-17, and IL-10 together with increased CD62L T cell populations. Those peptides also resulted in the induction of immunoprotection, as high levels of survival rates were achieved.[79] The success of the vaccination trials conducted by our group in different experimental models of fasciolosis is summarized in Table 44.1.

TABLE 44.1

Vaccination Trials Carried Out by Our Group to Assess the Immunoprotective Efficacy of Vaccine Candidates against Experimental Infection with *F. hepatica*

Experimental	Immunization			Infection	Protection Assessment			
Model	**Antigen**	**Vaccine**	**Dose (μg)**	**mc**	**Survival Rate (%)**	**Worm Burden Reduction (%)**	**Hepatic Damage Reduction (%)**	**References**
Mice	B, T epitopes	Synthetic	10	5	33–67	ND	22–33	79
	Fh15	Recombinant	20	5	11–50	NS	14–44	16
	Fh12	Native	20	5	40–50	ND	ND	101
	Fh15	Recombinant peptides	20	5	12–43	ND	ND	100
	Fh15	Recombinant	20	5	40–63	ND	NS	99
	Fh12	Native	20	5	10–42	NS	ND	98
Rabbit	Fh15	Recombinant peptides	50	20	ND	48–59	NS	100
	Fh15	Peptides	50	20	ND	5–46	NS	100
Sheep	Fh12	Native	150	100	ND	42–57	25–42	101
	Fh15	Recombinant	150	100	ND	24–43	NS	99
	Fh12	Native	150	100	ND	24–34	NS	98

mc, metacercariae; ND, not detected; NS, not significant.

44.10 Conclusion and Future Perspective

Fasciolosis remains a neglected tropical disease, with an increasing reemergence in many parts of the world. It has been estimated that 17×10^6 people are currently infected, with most of the cases being reported in Southeast Asia, the Middle East, Africa, and some countries in the South of America, and more than 170×10^6 people are in risk of acquiring the disease. Moreover, fasciolosis caused by *F. hepatica* also represents a serious veterinary problem causing economic losses of more than 3×10^9 dollars per year.

Despite the availability of an effective treatment for *F. hepatica* infection, issues related to the emergence of resistant strains and the toxicity and side effects induced by the current treatment have decreased its efficacy. Although the immunology and the immune response induced in the infected mammalian host have been well described, little is known about the precise molecular mechanisms during *F. hepatica* infection and establishment in the host. Recently, a draft of the genome of *F. hepatica* has been published, which represents a valuable source of information to explore issues related to coding and regulatory genes and its implication in the success of the infection. It is also important to apply new technologies such as microarray and RNA-seq-based methodologies in experimental models of fasciolosis to improve the measurement and quantitation of both immunological and pathological key markers and to gain a better understanding of mechanisms involved during the infection, migration, and establishment of the parasite in the mammalian host.

Currently, there is no effective vaccine for the prevention of fasciolosis, and for this reason, use of experimental models to investigate naturally infected hosts is indispensable for vaccine development. Existing data indicate that experimental models are valuable not only to study the immunology of the infection, but also to assess the immunoprotective efficacy of several vaccine candidates against *F. hepatica*. Compared to the natural models of infection, experimental models have several obvious advantages, especially in the areas of material availability and technical simplicity.

REFERENCES

1. Andrews, S.J. The life cycle of *Fasciola hepatica*. in *Fasciolosis*. Dalton, J.P., ed. 1–29 (CAB International, Wallingford, Oxon, 1999).
2. Galaktionov, K.V. & Dobrovolskij, A.A. *The Biology and Evolution of Trematodes. An Essay on the Biology, Morphology, Life Cycles, Transmissions, and Evolution of Digenetic Trematodes*. 215–346 (Kluwer Academic Publishers, Dordrecht, the Netherlands, 2003).
3. Gunn, A. & Pitt, S.J. Helminth parasites. in *Parasitology: An Integrated Approach* (Wiley-Blackwell, Chichester, 2012).
4. Kendall, S.B. Relationships between the species of *Fasciola* and their molluscan hosts. *Adv Parasitol* 8, 251–8 (1970).
5. Olsen, O.W. *Animal Parasites: Their Life Cycles and Ecology*, 199–267 (General Publishing Company, Toronto, Canada, 1974).
6. Mas-Coma, S. & Bargues, M.D. Human liver flukes: a review. *Res Rev Parasitol* 57, 145–218 (1997).
7. Menard, A. et al. Myocastor coypus as a reservoir host of *Fasciola hepatica* in France. *Vet Res* 32, 499–508 (2001).
8. Valero, M.A., Marcos, M.D., Fons, R. & Mas-Coma, S. *Fasciola hepatica* development in the experimentally infected black rat *Rattus rattus*. *Parasitol Res* 84, 188–94 (1998).
9. Valero, M.A., Panova, M., Comes, A.M., Fons, R. & Mas-Coma, S. Patterns in size and shedding of *Fasciola hepatica* eggs by naturally and experimentally infected murid rodents. *J Parasitol* 88, 308–13 (2002).
10. Mas-Coma, S. Epidemiology of fascioliasis in human endemic areas. *J Helminthol* 79, 207–16 (2005).
11. Mas-Coma, S. et al. Human fascioliasis in Bolivia: a general analysis and critical review of existing data. *Res Rev Parasitol* 55, 73–79 (1995).
12. Esteban, J.G. et al. A population-based coprological study of human fascioliasis in a hyperendemic area of the Bolivian Altiplano. *Trop Med Int Health* 2, 695–9 (1997).
13. Hillyer, G.V. et al. Use of the Falcon assay screening test—enzyme-linked immunosorbent assay (FAST-ELISA) and the enzyme-linked immunoelectrotransfer blot (EITB) to determine the prevalence of human fascioliasis in the Bolivian Altiplano. *Am J Trop Med Hyg* 46, 603–9 (1992).

14. Esteban, J.G. et al. High fascioliasis infection in children linked to a man-made irrigation zone in Peru. *Trop Med Int Health* 7, 339–48 (2002).
15. Marcos-Raymundo, L.A. et al. Hiperendemicidad de fasciolosis humana en el valle del mantaro, perú: Factores de riesgo de la infección por *Fasciola hepatica. Rev Gastroenterol Perú* 24, 158–64 (2004).
16. Lopez-Aban, J. et al. Adaptive immune stimulation is required to obtain high protection with fatty acid binding protein vaccine candidate against *Fasciola hepatica* in Balb/C mice. *J Parasitol* 98, 527–35 (2012).
17. Berriman, M. et al. The genome of the blood fluke *Schistosoma mansoni. Nature* 460, 352–8 (2009).
18. Tsai, I.J. et al. The genomes of four tapeworm species reveal adaptations to parasitism. *Nature* 496, 57–63 (2013).
19. Zheng, H. et al. The genome of the hydatid tapeworm *Echinococcus granulosus. Nat Genet* 45, 1168–75 (2013).
20. Cwiklinski, K. et al. The *Fasciola hepatica* genome: gene duplication and polymorphism reveals adaptation to the host environment and the capacity for rapid evolution. *Genome Biol* 16, 71 (2015).
21. Young, N.D., Hall, R.S., Jex, A.R., Cantacessi, C. & Gasser, R.B. Elucidating the transcriptome of *Fasciola hepatica*—a key to fundamental and biotechnological discoveries for a neglected parasite. *Biotechnol Adv* 28, 222–31 (2010).
22. Margulies, M. et al. Genome sequencing in microfabricated high-density picolitre reactors. *Nature* 437, 376–80 (2005).
23. Wilson, R.A. et al. Exploring the *Fasciola hepatica* tegument proteome. *Int J Parasitol* 41, 1347–59 (2011).
24. Hernandez-Gonzalez, A., Valero, M.L., del Pino, M.S., Oleaga, A. & Siles-Lucas, M. Proteomic analysis of in vitro newly excysted juveniles from *Fasciola hepatica. Mol Biochem Parasitol* 172, 121–8 (2010).
25. Morphew, R.M., Wright, H.A., LaCourse, E.J., Woods, D.J. & Brophy, P.M. Comparative proteomics of excretory-secretory proteins released by the liver fluke *Fasciola hepatica* in sheep host bile and during in vitro culture ex host. *Mol Cell Proteomics* 6, 963–72 (2007).
26. Boukli, N.M., Delgado, B., Ricaurte, M. & Espino, A.M. *Fasciola hepatica* and *Schistosoma mansoni*: identification of common proteins by comparative proteomic analysis. *J Parasitol* 97, 852–61 (2011).
27. López-Abán, J., Pardo, L.J., Pérez-Arellano, J.L. & Muro, A. Infecciones difícilmente transmisibles en el inmigrante V: Otras trematodosis. in *Manual de Enfermedades Importadas.* Muro Álvarez, A. & Pérez Arellano, J.L., eds (Elsevier, Barcelona, España, 2012).
28. Muro, A., Pérez del Villar, L., Velasco, V. & Pérez-Arellano, J.L. Infecciones por trematodos. *Medicine—Programa de Formación Médica Continuada Acreditado* 10, 3717–3728 (2010).
29. Allen, J.E. & Maizels, R.M. Diversity and dialogue in immunity to helminths. *Nat Rev Immunol* 11, 375–88 (2011).
30. McSorley, H.J., Hewitson, J.P. & Maizels, R.M. Immunomodulation by helminth parasites: defining mechanisms and mediators. *Int J Parasitol* 43, 301–10 (2013).
31. van Riet, E., Hartgers, F.C. & Yazdanbakhsh, M. Chronic helminth infections induce immunomodulation: consequences and mechanisms. *Immunobiology* 212, 475–90 (2007).
32. Dalton, J.P., Robinson, M.W., Mulcahy, G., O'Neill, S.M. & Donnelly, S. Immunomodulatory molecules of *Fasciola hepatica*: candidates for both vaccine and immunotherapeutic development. *Vet Parasitol* 195, 272–85 (2013).
33. Hill, J.A., Benoist, C. & Mathis, D. Treg cells: guardians for life. *Nat Immunol* 8, 124–5 (2007).
34. Allen, J.E. & Wynn, T.A. Evolution of Th2 immunity: a rapid repair response to tissue destructive pathogens. *PLoS Pathog* 7, e1002003 (2011).
35. Maizels, R.M. & Yazdanbakhsh, M. Immune regulation by helminth parasites: cellular and molecular mechanisms. *Nat Rev Immunol* 3, 733–44 (2003).
36. Pearce, E.J. & MacDonald, A.S. The immunobiology of schistosomiasis. *Nat Rev Immunol* 2, 499–511 (2002).
37. Lawrence, R.A., Allen, J.E., Osborne, J. & Maizels, R.M. Adult and microfilarial stages of the filarial parasite *Brugia malayi* stimulate contrasting cytokine and Ig isotype responses in BALB/c mice. *J Immunol* 153, 1216–24 (1994).
38. Berasain, P., Carmona, C., Frangione, B., Dalton, J.P. & Goni, F. *Fasciola hepatica*: parasite-secreted proteinases degrade all human IgG subclasses: determination of the specific cleavage sites and identification of the immunoglobulin fragments produced. *Exp Parasitol* 94, 99–110 (2000).
39. Berasain, P. et al. Proteinases secreted by *Fasciola hepatica* degrade extracellular matrix and basement membrane components. *J Parasitol* 83, 1–5 (1997).

40. Carmona, C., Dowd, A.J., Smith, A.M. & Dalton, J.P. Cathepsin L proteinase secreted by *Fasciola hepatica* in vitro prevents antibody-mediated eosinophil attachment to newly excysted juveniles. *Mol Biochem Parasitol* 62, 9–17 (1993).
41. Donnelly, S., O'Neill, S.M., Sekiya, M., Mulcahy, G. & Dalton, J.P. Thioredoxin peroxidase secreted by *Fasciola hepatica* induces the alternative activation of macrophages. *Infect Immun* 73, 166–73 (2005).
42. Donnelly, S. et al. Helminth 2-Cys peroxiredoxin drives Th2 responses through a mechanism involving alternatively activated macrophages. *FASEB J* 22, 4022–32 (2008).
43. Walsh, K.P., Brady, M.T., Finlay, C.M., Boon, L. & Mills, K.H. Infection with a helminth parasite attenuates autoimmunity through TGF-β-mediated suppression of Th17 and Th1 responses. *J Immunol* 183, 1577–86 (2009).
44. Hoyle, D.V. & Taylor, D.W. The immune response of regional lymph nodes during the early stages of *Fasciola hepatica* infection in cattle. *Parasite Immunol* 25, 221–9 (2003).
45. Mulcahy, G. et al. Immune responses of cattle to experimental anti-*Fasciola hepatica* vaccines. *Res Vet Sci* 67, 27–33 (1999).
46. Mulcahy, G. et al. Correlation of specific antibody titre and avidity with protection in cattle immunized against *Fasciola hepatica*. *Vaccine* 16, 932–9 (1998).
47. Pleasance, J., Wiedosari, E., Raadsma, H.W., Meeusen, E. & Piedrafita, D. Resistance to liver fluke infection in the natural sheep host is correlated with a type-1 cytokine response. *Parasite Immunol* 33, 495–505 (2011).
48. Brady, M.T., O'Neill, S.M., Dalton, J.P. & Mills, K.H. *Fasciola hepatica* suppresses a protective Th1 response against *Bordetella pertussis*. *Infect Immun* 67, 5372–8 (1999).
49. O'Neill, S.M. et al. *Fasciola hepatica* infection downregulates Th1 responses in mice. *Parasite Immunol* 22, 147–55 (2000).
50. Vukman, K.V. et al. The effects of *Fasciola hepatica* tegumental antigens on mast cell function. *Int J Parasitol* 43, 531–9 (2013).
51. Miller, C.M. et al. Immunological interactions between 2 common pathogens, Th1-inducing protozoan *Toxoplasma gondii* and the Th2-inducing helminth *Fasciola hepatica*. *PLoS One* 4, e5692 (2009).
52. Taira, N., Yoshifuji, H. & Boray, J.C. Zoonotic potential of infection with *Fasciola* spp. by consumption of freshly prepared raw liver containing immature flukes. *Int J Parasitol* 27, 775–9 (1997).
53. El-Aziz, M.M., Ghazy, A.A. & Effat, M.M. Immunodiagnosis of bovine fasciolosis using *Fasciola hepatica* excretory-secretory antigens ELISA. *J Egypt Soc Parasitol* 31, 327–34 (2001).
54. Espino, A.M., Dumenigo, B.E., Fernandez, R. & Finlay, C.M. Immunodiagnosis of human fascioliasis by enzyme-linked immunosorbent assay using excretory-secretory products. *Am J Trop Med Hyg* 37, 605–8 (1987).
55. O'Neill, S.M., Parkinson, M., Strauss, W., Angles, R. & Dalton, J.P. Immunodiagnosis of *Fasciola hepatica* infection (fascioliasis) in a human population in the Bolivian Altiplano using purified cathepsin L cysteine proteinase. *Am J Trop Med Hyg* 58, 417–23 (1998).
56. Rokni, M.B., Massoud, J., O'Neill, S.M., Parkinson, M. & Dalton, J.P. Diagnosis of human fasciolosis in the Gilan province of Northern Iran: application of cathepsin L-ELISA. *Diagn Microbiol Infect Dis* 44, 175–9 (2002).
57. Morales, A. & Espino, A.M. Evaluation and characterization of *Fasciola hepatica* tegument protein extract for serodiagnosis of human fascioliasis. *Clin Vaccine Immunol* 19, 1870–8 (2012).
58. Figueroa-Santiago, O., Delgado, B. & Espino, A.M. *Fasciola hepatica* saposin-like protein-2-based ELISA for the serodiagnosis of chronic human fascioliasis. *Diagn Microbiol Infect Dis* 70, 355–61 (2011).
59. Marcilla, A. et al. Leucine aminopeptidase is an immunodominant antigen of *Fasciola hepatica* excretory and secretory products in human infections. *Clin Vaccine Immunol* 15, 95–100 (2008).
60. Caban-Hernandez, K., Gaudier, J.F., Ruiz-Jimenez, C. & Espino, A.M. Development of two antibody detection enzyme-linked immunosorbent assays for serodiagnosis of human chronic fascioliasis. *J Clin Microbiol* 52, 766–72 (2014).
61. Cordova, M. et al. *Fasciola hepatica* cysteine proteinases: immunodominant antigens in human fascioliasis. *Am J Trop Med Hyg* 57, 660–6 (1997).
62. Espino, A.M. & Finlay, C.M. Sandwich enzyme-linked immunosorbent assay for detection of excretory secretory antigens in humans with fascioliasis. *J Clin Microbiol* 32, 190–3 (1994).
63. Ubeira, F.M. et al. MM3-ELISA detection of *Fasciola hepatica* coproantigens in preserved human stool samples. *Am J Trop Med Hyg* 81, 156–62 (2009).

64. Martinez-Sernandez, V. et al. Development and evaluation of a new lateral flow immunoassay for serodiagnosis of human fasciolosis. *PLoS Negl Trop Dis* 5, e1376 (2011).
65. Ai, L. et al. Specific PCR-based assays for the identification of *Fasciola* species: their development, evaluation and potential usefulness in prevalence surveys. *Ann Trop Med Parasitol* 104, 65–72 (2010).
66. Rokni, M.B. et al. Identification and differentiation of *Fasciola hepatica* and *Fasciola gigantica* using a simple PCR-restriction enzyme method. *Exp Parasitol* 124, 209–13 (2010).
67. Ai, L. et al. Rapid identification and differentiation of *Fasciola hepatica* and *Fasciola gigantica* by a loop-mediated isothermal amplification (LAMP) assay. *Vet Parasitol* 174, 228–33 (2010).
68. Magalhaes, K.G. et al. Isolation and detection of *Fasciola hepatica* DNA in *Lymnaea viatrix* from formalin-fixed and paraffin-embedded tissues through multiplex-PCR. *Vet Parasitol* 152, 333–8 (2008).
69. Magalhaes, K.G., Passos, L.K. & Carvalho Odos, S. Detection of *Lymnaea columella* infection by *Fasciola hepatica* through multiplex-PCR. *Mem Inst Oswaldo Cruz* 99, 421–4 (2004).
70. Boray, J.C., Jackson, R. & Strong, M.B. Chemoprophylaxis of fascioliasis with triclabendazole. *N Z Vet J* 33, 182–5 (1985).
71. Ibarra-Velarde, F., Vera-Montenegro, Y., Najera-Fuentes, R. & Sanchez-Albarran, A. Efficacy of combined chemotherapy against gastrointestinal nematodes and *Fasciola hepatica* in cattle. *Vet Parasitol* 99, 199–204 (2001).
72. Brennan, G.P. et al. Understanding triclabendazole resistance. *Exp Mol Pathol* 82, 104–9 (2007).
73. Overend, D.J. & Bowen, F.L. Resistance of *Fasciola hepatica* to triclabendazole. *Aust Vet J* 72, 275–6 (1995).
74. Brockwell, Y.M. et al. Confirmation of *Fasciola hepatica* resistant to triclabendazole in naturally infected Australian beef and dairy cattle. *Int J Parasitol Drugs Drug Resist* 4, 48–54 (2014).
75. Fairweather, I. Triclabendazole: new skills to unravel an old(ish) enigma. *J Helminthol* 79, 227–34 (2005).
76. Martinez-Valladares, M., Cordero-Perez, C. & Rojo-Vazquez, F.A. Efficacy of an anthelmintic combination in sheep infected with *Fasciola hepatica* resistant to albendazole and clorsulon. *Exp Parasitol* 136, 59–62 (2014).
77. López-Abán, J., Nogal-Ruíz, J.J., Muñoz-Pera, B.C., Martínez-Fernández, A.R. & Muro, A. Evaluation of a fasciolosis hyper-infection model in mice for vaccination trials. *Rev Ibérica Parasitol* 65, 71–78 (2005).
78. Rojas-Caraballo, J. et al. Gene expression profile in the liver of BALB/c mice infected with *Fasciola hepatica*. *PLoS One* 10, e0134910 (2015).
79. Rojas-Caraballo, J. et al. In vitro and in vivo studies for assessing the immune response and protection-inducing ability conferred by *Fasciola hepatica*-derived synthetic peptides containing B- and T-cell epitopes. *PLoS One* 9, e105323 (2014).
80. Bundesen, P.G. & Janssens, P.A. Biochemical tracing of parasitic infections. I. *Fasciola hepatica* L. in mice—a qualitative study. *Int J Parasitol* 1, 7–14 (1971).
81. Milbourne, E.A. & Howell, M.J. Eosinophil responses to *Fasciola hepatica* in rodents. *Int J Parasitol* 20, 705–8 (1990).
82. Mas-Coma, S. et al. Small mammals as natural definitive hosts of the liver fluke, *Fasciola hepatica* Linnaeus, 1758 (Trematoda: Fasciolidae): a review and two new records of epidemiologic interest on the island of Corsica. *Riv Parassitol* 5, 73–78 (1987).
83. Cervi, L., Cejas, H. & Masih, D.T. Cytokines involved in the immunosuppressor period in experimental fasciolosis in rats. *Int J Parasitol* 31, 1467–73 (2001).
84. Tliba, O. et al. Early hepatic immune response in rats infected with *Fasciola hepatica*. *Vet Res* 33, 261–70 (2002).
85. Tliba, O., Sibille, P., Boulard, C. & Chauvin, A. Local hepatic immune response in rats during primary infection with *Fasciola hepatica*. *Parasite* 7, 9–18 (2000).
86. Poitou, I., Baeza, E. & Boulard, C. Humoral and cellular immune responses in rats during a primary infestation with *Fasciola hepatica*. *Vet Parasitol* 45, 59–71 (1992).
87. Poitou, I., Baeza, E. & Boulard, C. Kinetic responses of parasite-specific antibody isotypes, blood leucocyte pattern and lymphocyte subsets in rats during primary infestation with *Fasciola hepatica*. *Vet Parasitol* 49, 179–90 (1993).
88. Kolodziejczyk, L., Siemieniuk, E. & Skrzydlewska, E. *Fasciola hepatica*: effects on the antioxidative properties and lipid peroxidation of rat serum. *Exp Parasitol* 113, 43–8 (2006).
89. Keegan, P.S. & Trudgett, A. *Fasciola hepatica* in the rat: immune responses associated with the development of resistance to infection. *Parasite Immunol* 14, 657–69 (1992).

90. Thorpe, E. The pathology of experimental fascioliasis in the albino rat. *J Comp Path* 75, 39–44 (1965).
91. Espino, A.M. & Rivera, F. Quantitation of cytokine mRNA by real-time RT-PCR during a vaccination trial in a rabbit model of fascioliasis. *Vet Parasitol* 169, 82–92 (2010).
92. Acosta, D. et al. *Fasciola hepatica* leucine aminopeptidase, a promising candidate for vaccination against ruminant fasciolosis. *Mol Biochem Parasitol* 158, 52–64 (2008).
93. Espino, A.M. & Hillyer, G.V. A novel *Fasciola hepatica* saposinlike recombinant protein with immunoprophylactic potential. *J Parasitol* 90, 876–9 (2004).
94. Maggioli, G. et al. A recombinant thioredoxin-glutathione reductase from *Fasciola hepatica* induces a protective response in rabbits. *Exp Parasitol* 129, 323–30 (2011).
95. Muro, A., Ramajo, V., Lopez, J., Simon, F. & Hillyer, G.V. *Fasciola hepatica*: vaccination of rabbits with native and recombinant antigens related to fatty acid binding proteins. *Vet Parasitol* 69, 219–29 (1997).
96. Casanueva, R. et al. Immunoprophylaxis against *Fasciola hepatica* in rabbits using a recombinant Fh15 fatty acid-binding protein. *J Parasitol* 87, 697–700 (2001).
97. Hillyer, G.V. *Fasciola* antigens as vaccines against fascioliasis and schistosomiasis. *J Helminthol* 79, 241–7 (2005).
98. Martinez-Fernandez, A.R. et al. Vaccination of mice and sheep with Fh12 FABP from *Fasciola hepatica* using the new adjuvant/immunomodulator system ADAD. *Vet Parasitol* 126, 287–98 (2004).
99. Lopez-Aban, J. et al. Progress in the development of *Fasciola hepatica* vaccine using recombinant fatty acid binding protein with the adjuvant adaptation system ADAD. *Vet Parasitol* 145, 287–96 (2007).
100. Muro, A. et al. Identification of *Fasciola hepatica* recombinant 15-kDa fatty acid-binding protein T-cell epitopes that protect against experimental fascioliasis in rabbits and mice. *J Parasitol* 93, 817–23 (2007).
101. Lopez-Aban, J. et al. The addition of a new immunomodulator with the adjuvant adaptation ADAD system using fatty acid binding proteins increases the protection against *Fasciola hepatica*. *Vet Parasitol* 153, 176–81 (2008).

45

Haplorchis

Dongyou Liu

CONTENTS

45.1 Introduction

The genus *Haplorchis* contains several small intestinal flukes that require mollusks (first intermediate host), fish (second intermediate host), humans, other mammals, or birds (definitive hosts) for completing life cycles. In comparison to the liver flukes *Opisthorchis viverrini* and *Clonorchis sinensis*, which are well known and extensively investigated (as shown by 1320 and 1166 entries in recent PubMed searches using the terms *Opisthorchis* and *Clonorchis*, respectively), *Haplorchis* species are relatively understudied (as shown by 149 entries in PubMed search using the term *Haplorchis*). Nevertheless, this does not make *Haplorchis* a less significant pathogen. In fact, intestinal flukes *Haplorchis taichui* and *Metagonimus yokogawai* (see Chapter 46) of the family Heterophyidae are listed by WHO as foodborne trematodes of medical importance and public health concern in Asia, along with liver flukes *O. viverrini* (see Chapter 47) and *C. sinensis* (see Chapter 43) of the family Opisthorchiidae. Given the medical prominence of *H. taichui*, this chapter will focus on the genus *Haplorchis*, despite the presence of several other human-infecting genera (e.g., *Heterophyes*, *Centrocestus*, *Pygidiopsis*, *Stellantchasmus*, and *Procerovum*) in the family Heterophyidae.

45.1.1 Classification and Morphology

45.1.1.1 Classification

The family Heterophyidae encompasses a large group of intestinal flukes that are classified in the superfamily Opisthorchioidea, suborder Opisthorchiata, order Opisthorchiida, subclass Digenea, class Trematoda, phylum Platyhelminthes, kingdom Animalia.

Of the three families (Opisthorchiidae, Heterophyidae, and Cryptogonimidae) within the superfamily Opisthorchioidea, the families Opisthorchiidae and Heterophyidae include many human pathogenic flukes such as *O. viverrini*, *O. felineus*, *C. sinensis*, *Haplorchis* spp., and *Metagonimus* spp. Recent molecular phylogenetic analyses indicated that the families Opisthorchiidae and Heterophyidae share a paraphyletic relationship, with the former nested within the latter [1].

The family Heterophyidae is composed of at least 30 genera, including *Adleriella*, *Apophalloides*, *Apophallus*, *Ascocotyle*, *Centrocestus*, *Cercarioides*, *Cryptocotyle*, *Euhaplorchis*, *Euryhelmis*, *Galactosomum*, *Haplorchis*, *Heterophyes*, *Heterophyopsis*, *Leighia*, *Metagonimoides*, *Metagonimus*, *Neostictodora*, *Phagicola*, *Phocitrema*, *Phocitremoides*, *Pholeter*, *Pricetrema*, *Procerovum*, *Pseudascocotyle*, *Pseudogalactosoma*, *Pygidiopsis*, *Pygidiopsoides*, *Scaphanocephalus*, *Stellantchasmas*, and *Stictodora* [2,3].

To date, more than 20 species have been identified in the genus *Haplorchis*, with 5 implicated in human infections: *Haplorchis pleurolophocerca*, *H. pumilio* (synonym: *Monorchotrema taihokui*), *H. taichui* (synonyms: *M. taichui*, *M. microrchia*, *H. microrchis*), *H. vanissimus*, and *H. yokogawai* (synonym: *M. yokogawai*) [2].

45.1.1.2 Morphology

Similar to other members of the family Heterophyidae (often referred to as heterophyids), *Haplorchis* species are small flukes (of 0.3–1 mm in length and 0.14–0.2 mm in width), with tegument covered by spines. In addition, they possess a small oral sucker (which is armed with spines) and a ventrogenital sucker complex (also with spines). Other organs present in *Haplorchis* species include pharynx, simple intestinal system, ceca, one testis (compared to two testes seen in some other heterophyids), and vitellaria (both located in posterior part of the body); however, cirrus and bursa are absent.

Within the *Haplorchis* genus, *H. pleurolophocerca* adult worms (of 0.32–0.42 × 0.14–0.17 mm in size) are characterized by having only one testis and a ventrogenital sucker complex armed with spines. Freshwater snails *Melanoides tuberculata* and *Cleopatra bulimoides* are the first intermediate host (harboring cercariae); freshwater fish *Gambusia affinis* is the second intermediate host (harboring metacercariae); and cats are the natural definitive hosts. *H. pleurolophocerca* infection in humans has been reported in Egypt.

H. pumilio adult worms (of 0.45–0.89 × 0.2–0.4 in size) are characterized by the presence of only one testis and a ventrogenital sucker complex armed with 27–39 (average 32) gonotyl and chitinous spines. The freshwater snail *Melania reiniana* var. *hitachiens* appears to act as the first intermediate host; freshwater fish belonging to the families Cyprinidae, Siluridae, and Cobitidae are the second intermediate host; and dogs and cats serve as the natural definitive hosts. This parasite is present in the Philippines, Thailand, Laos, Vietnam, South China, Taiwan, Malaysia, India, Sri Lanka, Iraq and Egypt; and *H. pumilio* infection in humans was first documented in Thailand in 1983 [4].

H. taichui adult worms (of 0.47–0.64 × 0.18–0.22 mm in size) are characterized by the presence of only one testis and 14–20 (av. 15) large chitinous, fan-shaped spines (hooklets) on the ventrogenital sucker complex. Under scanning electron microscope (SEM), the whole body surface of newly excysted juvenile *H. taichui* shows numerous transverse rows of scale-like spines and two types of sensory papillae (type I, ciliated knob-like swellings and type II, round swellings of the tegument) [5]. The first intermediate hosts are freshwater snails *Melania obliquegranosa*, *Melania juncea*, or *M. tuberculata* [6]; the second intermediate hosts are freshwater fish, including *Cyclocheilichthys repasson*, *Cyprinus auratus*, *Cyprinus carpio*, *Gambusia affinis*, *Hampala dispar*, *Labiobarbus leptocheila*, *Puntius binotatus*, *Puntius brevis*, *Puntius gonionotus*, *Puntius leicanthus*, *Puntius orphoides*, *Puntius palata*, *Pseudorasbora parva*, *Rhodeus ocellatus*, and *Zacco platypus* in addition to *Raiamas guttatus*, *Mystacoleucus marginatus*, and *Henichoryhnchus siamensis*; and the natural definitive hosts are dogs, cats, and birds. This parasite is found in Taiwan, the Philippines, Bangladesh, India, Sri Lanka, Malaysia, Thailand, Laos, Vietnam, South China, Iraq, Palestine and Egypt. The first human case of *H. taichui* infection was reported in the Philippines [4,7].

H. vanissimus adult worms (of 0.38–0.51 × 0.25–0.51 mm in size) utilize freshwater fish as the second intermediate host, and pelicans and wild mammals as the definitive hosts. Human infection with *H. vanissimus* was described in the Philippines [4].

H. yokogawai adult worms (of 0.47–0.64 × 0.18–0.22 mm in size) are characterized by having only one testis and numerous (uncountable) minute chitinous spines on the ventrogenital sucker complex. The first intermediate host is freshwater snail (*M. tuberculata* or *Stenomelania newcombi*); the second intermediate host is freshwater fish, belonging to *Cyclocheilichthys armatus*, *Hampala dispar*, *Labiobarbus leptocheila*, *Misgurnus* sp., *Mugil* spp., *Onychostoma elongatum*, *Ophiocephalus striatus*, and *Puntius* spp.; and the natural definitive hosts are dogs, cats, cattle, and other mammals. The fluke is known to occur in the Philippines, South China, Malaysia, Indonesia, Thailand, Laos, Vietnam, India, Australia, and Egypt; and human infection with *H. yokogawai* was first noted in the Philippines [4].

The complete mitochondrial genome of *H. taichui* measures 15,130 bp in length and contains 12 protein-coding genes, 2 ribosomal RNAs (rRNAs, a small and a large subunit), and 22 transfer RNAs (tRNAs). Similar to other trematodes, the *atp8* gene is absent, although a single long noncoding region is present between *trnE* and *trnG* in *H. taichui* mitochondrial genome. On the whole, *H. taichui* appears to be more closely related to Opisthorchiidae than other trematode groups [8].

45.1.2 Life Cycle and Epidemiology

Haplorchis eggs (each containing a miracidium) produced by adult worms residing in the intestine of the definitive hosts (e.g., humans, cats, dogs, foxes, wolves, and pelicans) are discharged into fresh or brackish water, and subsequently taken up by the first intermediate host (snails). In the gut of snails, miracidia (contained within eggs) evolve via three stages (sporocysts, rediae, and cercariae) to become cercariae, which upon release by snails penetrate the epithelium of the second intermediate host (fish). Encysting in the muscles (and sometimes in the head) [9], cercariae develop into metacercariae. Finally, infected fish (harboring metacercariae) are consumed by humans and other fish-eating mammals and birds, in which metacercariae burrow between the villi of the small intestine and mature into adult worms [10].

Due to the abundance of intermediate hosts (snails and fish) and the customs of eating raw, incompletely cooked, smoked, or pickled fish, heterophyds such as *Haplorchis* spp. are prevalent in Asia (including Korea, Japan, China, Thailand, Vietnam, Laos, the Philippines, Indonesia, and India), Mediterranean countries, Egypt, Ukraine, Hawaii, and Brazil [9,11–14]. In the past 20 years, human cases of heterophyd infections have been also reported in Canada and USA. It is estimated that about 50 million people are infected with heterophyds worldwide, and children appear to be more susceptible to the infections than adults.

Using Kato-Katz fecal smear technique, Sohn et al. [15] confirmed the high prevalence of the intestinal fluke *H. taichui* among humans and fish in Luang Prabang Province, Laos. The worm load for *H. taichui* averaged 7691 per infected person, and 138 (67%) of 207 freshwater fish (17 species) purchased in a market in Luang Prabang District harbored *H. taichui* metacercariae (average 520 metacercariae per fish).

45.1.3 Clinical Features

With an incubation period of between 1 and 15 days, *Haplorchis* infections in humans are generally asymptomatic or mild and transient in immunocompetent individuals. However, in heavily infected or immunocompromised individuals, *Haplorchis* infestations may cause weakness, discomfort, abdominal pain, diarrhea, loss of appetite, nausea, and vomiting [16].

In case that *Haplorchis* eggs released in the intestine by adult worms gain entry into the blood and lymph vascular systems via mucosa, and migrate to the valves and myocardium, the heart, the brain, or spinal cord, they may induce cardiac failure, neurological disorders, and occasional fatalities.

Microscopic examination of the small intestine from patients with *H. taichui* infection may reveal mucosal ulceration, mucosal and submucosal hemorrhages, fusion and shortening of villi, chronic inflammation, and fibrosis of the submucosa [17].

45.1.4 Diagnosis

Diagnosis of *Haplorchis* infections has traditionally relied on microscopic observation of adult worms or eggs in feces. This is sometimes problematic as adult worms may not always be present and because *Haplorchis* eggs are indistinguishable from those of *O. viverrini* and *C. sinensis* [18].

The development of molecular methods based on nucleic acid amplification has offered new opportunities to improve the identification of *Haplorchis* adult worms, metacercariae, and eggs. By targeting ribosomal RNA gene and mitochondrial cytochrome c oxidase subunit I (mCOI) gene, polymerase chain reaction (PCR) and its derivatives [e.g., PCR-restriction fragment length polymorphism (RFLP), multiplex PCR, nested PCR, real-time PCR, and PCR pyrosequencing] have made rapid, sensitive, and precise determination and phylogenetic analysis of heterophyid trematodes possible [19–23].

Thaenkham et al. [24] showed that using primers (COI-OV-Hap F&R primers) from mitochondrial cytochrome c oxidase subunit I (COI) gene, two common fish-borne trematodes in Thailand, *O. viverrini* and *H. taichui*, can be rapidly and precisely differentiated. In addition, Thaenkham et al. [25] described a PCR-RFLP based on 28S ribosomal RNA gene and restriction enzyme *Mbo*II digestion for identification of metacercaria of *C. sinensis*, *O. viverrini*, *H. taichui*, *H. pumilio*, and *H. yogokawai*.

Similarly, Sato et al. [19] designed primers from ITS 1 and ITS 2 regions in ribosomal DNA that enabled amplification of 800, 820, 1250, and 930 bp products (ITS1) and 380, 390, 380, and 530 bp products (ITS2) from the eggs of *O. viverrini*, *C. sinensis*, *H. pumilio*, and *H. taichui*, respectively, allowing their effective discrimination.

Furthermore, Wongsawad et al. [26,27] developed species-specific primers for *H. taichui* (Hapt_F 5′-GGCCAACGCAATCGTCATCC-3′ and Hapt_R1 5′-CTCTCGACCTCCTCTAGAAT-3′, which yield a 170 bp PCR product) and *O. viverrini* (OpV-1F: 5′-AATCGGGCTGCATATTGACCGAT-3′ and OpV-1R: 5′-CGGTGTTGCTTATTTTGCAGACAA-3′, which generate a 319 bp PCR product). Use of these primers facilitated the specific confirmation of both parasites in fish and snail intermediate hosts.

45.1.5 Treatment and Prevention

Haplorchis infection in humans can be treated with praziquantel (75 mg/kg body weight) in three doses of 25 mg/kg for 1 day [28]. Alternatively, a single dose of praziquantel (40 mg/kg) and pyrantel pamoate (10 mg/kg) followed by purge with magnesium sulfate will eliminate adult worms.

Prevention of human *Haplorchis* infections requires conducting campaigns to educate people about the dangers of eating raw fish, the proper way to prepare fish (e.g., skinning fish and removal of gills before cooking), and to avoid eating raw, pickled, and smoked fish from endemic areas.

45.2 Laboratory Models

45.2.1 Animal Models

45.2.1.1 Rodents

Mice (*Mus musculus*) are a suitable experimental definitive host for adult *H. taichui*. Sukontason et al. [29] noted that mice infected orally with 200 active *H. taichui* metacercaria developed mature adult worms in a few days, with egg production as early as 3 days postinfection, and increasing daily to about 50–60 eggs. Kay et al. [30] also found mice highly susceptible to *H. pumilio* infection, with peak adult worm recovery 7 days after oral inoculation with metacercariae.

Rats may be used as definitive host for *H. taichui* infection. Saenphet et al. [31] observed that male rats orally fed with 300 *H. taichui* metacercariae produced adult worms as early as 3 days postinfection.

Syrian golden hamster (*Mesocricetus auratus*) is also capable of supporting growth of *Haplorchis* parasites, producing adult worms in the intestine 14 days after being fed metacercariae via a stomach tube.

45.2.1.2 Dogs

Nissen et al. [32] showed that dogs (3–6 months old) infected with 500 *H. pumilio* metacercariae yielded adult flukes without clinical symptoms. The infection lasted for at least 2 months, and up to 2 eggs/g feces were excreted. The predilection site of the flukes appeared to be the lower part of jejunum (93% of total worm burden).

45.2.1.3 Foxes

Nissen et al. [33] reported that foxes orally infected with 2000 *H. pumilio* metacercariae displayed anorexia at 12 days postinfection (which last for approximately a week), produced between 116 and 2070 adult flukes per animal mostly in the lower part of the jejunum, and excreted *H. pumilio* eggs (initially about 16–20 with a maximum of 1443 ± 1176 eggs/g of feces).

45.2.1.4 Chicks

Kumchoo et al. [34] demonstrated the suitability of chicks (*Gallus gallus domesticus*) as definitive host for infection with *H. taichui*, since chick infected orally with 200 *H. taichui* metacercariae generated mature adult worms as early as days 3 postinfection. Kay et al. [30] also confirmed that chicks are useful host for *H. pumilio* metacercariae infection, although mice appear to be significantly more susceptible to infection than chicks.

45.2.2 *In Vitro* Models

Newly excysted metacercariae of *H. taichui* survived for 12–14 days in a candle jar at 37°C using RPMI 1640 with blood agar, with ovary and testes observed after 3 days of incubation [35].

45.3 Conclusion

The genus *Haplorchis* consists of more than 20 intestinal trematode species whose complex life cycles involve two intermediate hosts (snail and fish) and a definitive host (mammals and birds). Of the five *Haplorchis* species (*H. pleurolophocerca*, *H. pumilio*, *H. taichui*, *H. vagabundi*, and *H. yokogawai*) implicated in human infections, *H. taichui* is particularly common and poses a major public health threat in Asia, parts of Africa, and the Americas. In order to decipher the pathogenic mechanisms and immunobiology of haplorchiasis, use of laboratory models is crucial. Fortunately, a number of laboratory animals (e.g., mice, rats, hamsters, dogs, foxes, and chicks) have been shown to support the growth and maturation of *Haplorchis* adult worms from metacercaria recovered from fish. Further experimentation will undoubtedly contribute to an improved understanding of human haplorchiasis and to the development of novel intervention strategies against *Haplorchis* parasites.

REFERENCES

1. Thaenkham U, Blair D, Nawa Y, Waikagul J. Families Opisthorchiidae and Heterophyidae: are they distinct? *Parasitol Int*. 2012;61(1):90–3.
2. Pearson JC, OwYang CK. New species of *Haplorchis* from Southeast Asia, together with keys to the *Haplorchis*-group of heterophyid trematodes of the region. *Southeast Asian J Trop Med Public Health*. 1982;13:35–60.
3. ITIS report. Available at http://www.itis.gov/servlet/SingleRpt/SingleRpt?search_topic=TSN&search_value=57114. Accessed on February 12, 2016.

4. Chai JY, Shin EH, Lee SH, Rim HJ. Foodborne intestinal flukes in Southeast Asia. *Korean J Parasitol.* 2009;47(Suppl):S69–102.
5. Sukontason KL, et al. Surface ultrastructure of excysted metacercariae of *Haplorchis taichui* (Trematoda: Heterophyidae). *Southeast Asian J Trop Med Public Health.* 2000;31(4):747–54.
6. Chontananarth T, Wongsawad C. Prevalence of *Haplorchis taichui* in field-collected snails: a molecular approach. *Korean J Parasitol.* 2010;48(4):343–6.
7. Sommerville C. The life-history of *Haplorchis pumilio* (Looss, 1896) from cultured tilapias. *J Fish Dis.* 1982;5:233–41.
8. Lee D, et al. Complete mitochondrial genome of *Haplorchis taichui* and comparative analysis with other trematodes. *Korean J Parasitol.* 2013;51(6):719–26.
9. Kumchoo K, Wongsawad C, Chai JY, Vanittanakom P, Rojanapaibul A. High prevalence of *Haplorchis taichui* metacercariae in cyprinoid fish from Chiang Mai Province, Thailand. *Southeast Asian J Trop Med Public Health.* 2005;36(2):451–5.
10. Skov J, Kania PW, Dalsgaard A, Jorgensen TR, Buchmann K. Life cycle stages of heterophyid trematodes in Vietnamese freshwater fishes traced by molecular and morphometric methods. *Vet Parasitol.* 2009;160:66–75.
11. Nithikathkul C, Wongsawad C. Prevalence of *Haplorchis taichui* and *Haplorchoides* sp. metacercariae in freshwater fish from water reservoirs, Chiang Mai, Thailand. *Korean J Parasitol.* 2008;46(2):109–12.
12. Chai JY, et al. Prevalence of the intestinal flukes *Haplorchis taichui* and *H. yokogawai* in a mountainous area of Phongsaly Province, Lao PDR. *Korean J Parasitol.* 2010;48(4):339–42.
13. Wijit A, Morakote N, Klinchid J. High prevalence of haplorchiasis in Nan and Lampang provinces, Thailand, proven by adult worm recovery from suspected opisthorchiasis cases. *Korean J Parasitol.* 2013;51(6):767–9.
14. Krailas D, Veeravechsukij N, Chuanprasit C, Boonmekam D, Namchote S. Prevalence of fish-borne trematodes of the family Heterophyidae at Pasak Cholasid Reservoir, Thailand. *Acta Trop.* 2016;156:79–86.
15. Sohn WM, et al. Prevalence of *Haplorchis taichui* among humans and fish in Luang Prabang Province, Lao PDR. *Acta Trop.* 2014;136:74–80.
16. Watthanakulpanich D, et al. *Haplorchis taichui* as a possible etiologic agent of irritable bowel syndrome-like symptoms. *Korean J Parasitol.* 2010;48:225–9.
17. Sukontason K, Unpunyo P, Sukontason KL, Piangjai S. Evidence of *Haplorchis taichui* infection as pathogenic parasite: three case reports. *Scand J Infect Dis.* 2005;37:388–90.
18. Sato M, Sanguankiat S, Pubampen S, Kusolsuk T, Maipanich W, Waikagul J. Egg laying capacity of *Haplorchis taichui* (Digenea: Heterophyidae) in humans. *Korean J Parasitol.* 2009;47:315–8.
19. Sato M, Thaenkham U, Dekumyoy P, Waikagul J. Discrimination of *O. viverrini*, *C. sinensis*, *H. pumilio* and *H. taichui* using nuclear DNA-based PCR targeting ribosomal DNA ITS regions. *Acta Trop.* 2009;109:81–3.
20. Sato M, et al. Copro-DNA diagnosis of *Opisthorchis viverrini* and *Haplorchis taichui* infection in an endemic area of Lao PDR. *Southeast Asian J Trop Med Public Health.* 2010;41:28–35.
21. Dung DT, Hop NT, Thaenkham U, Waikagul J. Genetic differences among Vietnamese *Haplorchis taichui* populations using the COI genetic marker. *J Helminthol.* 2013;87(1):66–70.
22. Chontananarth T, Wongsawad C, Chomdej S, Krailas D, Chai JY. Molecular phylogeny of trematodes in Family Heterophyidae based on mitochondrial cytochrome c oxidase subunit I (mCOI). *Asian Pac J Trop Med.* 2014;7(6):446–50.
23. Tantrawatpan C, et al. Development of a PCR assay and pyrosequencing for identification of important human fish-borne trematodes and its potential use for detection in fecal specimens. *Parasites Vectors.* 2014;7:88.
24. Thaenkham U, Visetsuk K, Dung Do T, Waikagul J. Discrimination of *Opisthorchis viverrini* from *Haplorchis taichui* using COI sequence marker. *Acta Trop.* 2007;103:26–32.
25. Thaenkham U, et al. Rapid and simple identification of human pathogenic heterophyid intestinal fluke metacercariae by PCR-RFLP. *Parasitol Int.* 2011;60:503–6.
26. Wongsawad C, Wongsawad P, Chai JY, Anuntalabhochai S. Development of a HAT-RAPD marker for the detection of minute intestinal fluke infection. *Exp Parasitol.* 2009;123(2):158–61.

27. Wongsawad C, Wongsawad P. *Opisthorchis viverrini* and *Haplorchis taichui*: development of a multiplex PCR assay for their detection and differentiation using specific primers derived from HAT-RAPD. *Exp Parasitol.* 2012;132:237–42.
28. Belizario VY Jr., de Leon WU, Bersabe MJ, Purnomo, Baird JK, Bangs MJ. A focus of human infection by *Haplorchis taichui* (Trematoda: Heterophyidae) in the southern Philippines. *J Parasitol.* 2004;90(5):1165–9.
29. Sukontason K, Sukontason KL, Boonsriwong N, Chaithong U, Piangjai S, Choochote W. Development of *Haplorchis taichui* (Trematoda: Heterophyidae) in *Mus musculus* mice. *Southeast Asian J Trop Med Public Health.* 2001;32(Suppl 2):43–7.
30. Kay H, et al. Optimization of an experimental model for the recovery of adult *Haplorchis pumilio* (Heterophyidae: Digenea). *J Parasitol.* 2009;95:629–33.
31. Saenphet S, Wongsawad C, Saenphet K, Rojanapaibul A, Vanittanakom P, Chai JY. *Haplorchis taichui*: worm recovery rate and immune responses in infected rats (*Rattus norvegicus*). *Exp Parasitol.* 2008;120:175–9.
32. Nissen S, Nguyen LA, Dalsgaard A, Thamsborg SM, Johansen MV. Experimental infection with the small intestinal trematode, *Haplorchis pumilio*, in young dogs. *Vet Parasitol.* 2013;191(1–2):138–42.
33. Nissen S, Thamsborg SM, Kania PW, Leifsson PS, Dalsgaard A, Johansen MV. Population dynamics and host reactions in young foxes following experimental infection with the minute intestinal fluke, *Haplorchis pumilio. Parasites Vectors.* 2013;6:4.
34. Kumchoo K, Wongsawad C, Chai JY, Vanittanakom P, Rojanapaibul A. Recovery and growth of *Haplorchis taichui* (Trematoda: Heterophyidae) in chicks. *Southeast Asian J Trop Med Public Health.* 2003;34(4):718–22.
35. Chaithong U, Sukontason K, Boonsriwong N, Sukontason KL, Piangjai S. In vitro development of *Haplorchis taichui* (Trematoda: Heterophyidae). *Southeast Asian J Trop Med Public Health.* 2001;32(Suppl 2):31–5.

46

Metagonimus

Jong-Yil Chai

CONTENTS

46.1 Introduction

Twelve genera (29 species) of the family Heterophyidae are so far known to infect humans around the world: *Apophallus*, *Ascocotyle*, *Centrocestus*, *Cryptocotyle*, *Haplorchis*, *Heterophyes*, *Heterophyopsis*, *Metagonimus*, *Procerovum*, *Pygidiopsis*, *Stellantchasmus*, and *Stictodora*.[1] Among them, *Haplorchis*, *Heterophyes*, and *Metagonimus* are the three most important genera in public health significance. The genus *Metagonimus* is distinct morphologically from *Haplorchis* and *Heterophyes* and also differs in its life cycle and geographical distribution.[1,2]

The genus *Metagonimus* was established by F. Katsurada in 1912 with *M. yokogawai* as the type species.[3,4] Later, six more species of *Metagonimus* have been described in the literature: *M. takahashii* Suzuki, 1930[5]; *M. minutus* Katsuta, 1932[6]; *M. katsuradai* Izumi, 1935[7]; *M. otsurui* Saito and Shimizu, 1968[8]; *M. miyatai* Saito et al., 1997[9]; and *M. hakubaensis* Shimazu, 1999.[10] From public health points of view, *M. yokogawai*, *M. takahashii*, and *M. miyatai* are the three major species frequently causing human infections.[2,11] *M. minutus* was also listed as a human-infecting species[12]; however, no literature background is available to verify it. With regard to *M. katsuradai*, an experimental human infection was reported[7]; however, no natural human infections have been documented yet. *M. otsurui* was originally described from experimental hamsters infected with metacercariae from freshwater fishes,[8] and later, natural infection of the Japanese water shrew was discovered[13] but never from humans. *M. hakubaensis* was first found from experimental rats that were fed the metacercariae in lampreys in Japan,[10] but human infection is yet unknown.

Human *Metagonimus* spp. infections have been found exclusively in the Far East, including the Republic of Korea (=Korea), China, Japan, and the Far Eastern Russia.[1,2,11] However, the life cycle of *M. yokogawai* was also detected in Taiwan and eastern Europe. The potential for human infections with other species of *Metagonimus* remains to be elucidated.

The pathogenicity of *Metagonimus* spp. to humans and animals has been reported to be generally mild.[1] The habitat of the adult flukes in immunocompetent mice was confined to the mucosal layer, mainly in the intervillous space and crypt of the small intestine, causing mucosal inflammations, but did not extend beyond the submucosa and underneath.[14] However, in immunosuppressed mice, the adult flukes could invade into a deeper layer of the submucosa where mesenteric vessels are available.[15] This suggests a potential deep invasion of *Metagonimus* flukes to elicit significant pathogenesis and pathology in the human host. To verify this suggestion, experiments with proper laboratory animal models are strongly needed. In this chapter, laboratory models so far used to study the life cycle, host–parasite relationships, pathogenesis, immunity, and other related aspects of *Metagonimus* infection are briefly reviewed.

46.1.1 Classification and Morphology

The characteristic morphological features of *Metagonimus* spp. include a minute body (0.5–1.5 mm in length), a small laterally deviated ventral sucker with no ventrogenital apparatus and no genital sucker, a medially located ovary, and two testes located almost side by side or a little oblique near the posterior extremity.[2,3,16] A brief history of the discovery, taxonomic debates if any, and morphological characteristics of each *Metagonimus* species are given as follows.

46.1.1.1 Metagonimus yokogawai *Katsurada, 1912*

(syn. *Loxotrema ovatum* Kobayashi, 1912; *Metagonimus ovatus* Yokogawa, 1913; *Loossia romanica* Ciurea, 1915; *Loossia parva* Ciurea, 1915; *Loossia dobrogensis* Ciurea, 1915)[16]

This species was originally reported from an experimental dog fed with metacercariae from the sweetfish (*Plecoglossus altivelis*) in 1912 from Taiwan and named as *Heterophyes yokogawai*.[3,30] However, the adult worm morphology was significantly different from *Heterophyes*, and subsequently it was renamed as *M. yokogawai* in the same year.[3,4] It is the most highly prevalent of all *Metagonimus* spp. and has a most wide geographical distribution, including the Republic of Korea, Japan, China, Taiwan, Russia, Ukraine, India, Romania, Serbia, Bulgaria, Israel, Egypt, the Balkan States, and Spain.[2,3,12,17,18–22] Some old literature on *M. yokogawai* were actually referring to *M. takahashii* or *M. miyatai*, and caution is required when reviewing *M. yokogawai sensu stricto*.[3]

The characteristic morphologies of *M. yokogawai* include the presence of two testes, which are closely adjacent to each other near the posterior end of the body.[2,16] However, in *M. miyatai* and *M. takahashii*, the two testes are more or less separated from each other (see below).[2,3,23] Another differential character of *M. yokogawai* is the distribution of vitelline follicles; they extend in lateral fields from the level of the ovary down to the posterior end of the posterior testis, but not beyond the posterior testis.[16] By contrast, in *M. takahashii*, vitelline follicles are abundant from the level of the ovary down to the posterior testis and usually extend beyond the posterior testis level.[9,23] In *M. miyatai*, vitelline follicles distribute from the level of the ovary down to the anterior level of the posterior testis; there are no vitellaria distribution beyond the posterior testis.[9,23] In addition, *M. yokogawai* has the uterine tubule, which does not overlap or cross over the middle portion of the anterior testis, whereas *M. takahashii* and *M. miyatai* have the uterine tubules that overlap the whole anterior testis.[9,23]

The adult flukes of *M. yokogawai* are slightly smaller ($0.80–1.32 \times 0.42–0.54$ mm) than those of *M. takahashii* ($0.86–1.19 \times 0.44–0.57$ mm) and *M. miyatai* ($1.00–1.30 \times 0.46–0.63$ mm).[3,9] The eggs of *M. yokogawai* are also smaller ($0.026–0.030 \times 0.015–0.018$ mm) than those of *M. takahashii* ($0.030–0.036 \times 0.017–0.020$ mm) and *M. miyatai* ($0.029–0.032 \times 0.017–0.020$ mm), although there are some overlaps in the range.[3,23] The adult flukes of *M. minutus* differ from those of *M. yokogawai*, *M. takahashii*, and *M. miyatai* by having a smaller body (0.46×0.28 mm) and smaller egg size (0.023×0.013 mm).[6,23] The adult flukes of *M. katsuradai*, *M. otsurui*, and *M. hakubaensis* differ from those of *M. yokogawai*, *M. takahashii*, *M. miyatai*, and *M. minutus* in that the formers have a larger ventral sucker than the oral sucker.[7,10,23]

46.1.1.2 Metagonimus takahashii *Suzuki, 1930*

(syn. *Metagonimus yokogawai* var. *takahashii* Asada, 1934; *Metagonimus yokogawai* Koga type Koga, 1938)[16]

This species was first reported by S. Takahashi in 1929 from the small intestine of mice and dogs fed with metacercariae encysted in various freshwater fish species in Japan.[24] It was at that time called a larger egg-type *M. yokogawai* because the worm was morphologically very similar to *M. yokogawai*, except that its eggs were larger than those of *M. yokogawai*.[24] However, in 1930, *M. takahashii* was reported as a new species, accepting the larger egg size as a specific character.[5] However, its validity was questioned because the difference from *M. yokogawai* was minor, and thus, the name was compromised as *M. yokogawai* var. *takahashii*.[25] Later, however, the taxonomic validity of *M. takahashii* was strongly supported by differential morphological characters of larval and adult stages and also by the different host specificities of the two species at experimental infection with the cercariae.[26,27]

M. takahashii morphologically differs from *M. yokogawai* and *M. miyatai* in the position of the two testes, the distribution of vitelline follicles, and the size of their eggs.[23] The so-called Koga type of *Metagonimus* encysting in the dace *Tribolodon hakonensis*[28] is regarded a synonym of *M. takahashii*.[3,16,29] The adults of *M. takahashii* differ from those of *M. katsuradai*, *M. otsurui*, and *M. hakubaensis* in that the latter three species have a smaller ventral sucker than the oral sucker.[10,23] *M. minutus* has a smaller body and egg size compared to *M. takahashii*.[10,23]

46.1.1.3 Metagonimus miyatai *Saito, Chai, Kim, Lee, and Rim, 1997*

This species was first described by F. Katsurada[9,30] (only by a figure drawing) together with *M. yokogawai* in 1912 in Taiwan and then by I. Miyata in 1941 in Japan.[31] However, its specific status was not acknowledged, and no special name was given until 1984 when S. Saito began to call it as the "*Metagonimus* Miyata type."[9] It was in 1997 when the specific level of this fluke became acknowledged and described as a new species, *M. miyatai*.[9] The new species description was based on adult flukes collected from dogs and hamsters experimentally fed with the metacercariae from the sweetfish, dace, common fat-minnow *Morocco steindachneri*, pale chub *Zacco platypus*, and dark chub *Zacco temmincki* in Korea and Japan.[3,9] Human infections have been reported from Korea and Japan.[9,23]

The specific morphological characters of *M. miyatai* include the two markedly separated testes from each other, the posterior one being just touching the terminal end of the worm, its vitelline follicles never distributing beyond the posterior testis, and the egg size, which is intermediate between those of *M. yokogawai* and *M. takahashi*.[2,9,16] *M. miyatai* is also genetically distinct from *M. yokogawai* and *M. takahashii*.[32–34] *M. miyatai* differs from *M. minutus* in its larger body and egg size.[6,23] It also differs from *M. katsuradai*, *M. otsurui*, and *M. hakubaensis* in that the latter three species have a smaller ventral sucker compared with the oral sucker, whereas in *M. miyatai*, the ventral sucker is larger than its oral sucker.[10,23]

46.1.1.4 Metagonimus minutus *Katsuta, 1932*

This species was originally described in Taiwan in 1932 based on adult flukes from cats and mice experimentally fed with the metacercariae in the brackish water mullet.[6,16] It has body and egg sizes smaller than those of *M. yokogawai*, *M. takahashii*, and *M. miyatai*.[2,3] The body size of *M. minutus* is similar to that of *M. katsuradai*, but its egg size is smaller than that of *M. katsuradai*.[3,6,7] The relative size of the oral and ventral suckers is also a characteristic feature; the oral sucker is bigger than the ventral sucker in *M. katsuradai*, and the oral sucker is smaller than the ventral sucker in *M. minutus*.[3,6,7] *M. minutus* is listed as a human-infecting species; however, no relevant background literature is available.[12,16]

46.1.1.5 Metagonimus katsuradai *Izumi, 1935*

This species was first described in Japan based on adult flukes from experimental rats, mice, rabbits, dogs, and cats fed with the metacercariae from freshwater fish, including *Pseudorasbora parva*, *Zacco platypus*, and *Tanakia lanceolata*.[7,16] The possibility of human infection was experimentally proved (by the author himself and family).[7] *M. katsuradai* differs morphologically from *M. yokogawai*, *M. takahashii*, *M. miyatai*, and *M. minutus* in having a smaller ventral sucker than the oral sucker.[2,7,16] It also differs from *M. otsurui* in the position of the seminal receptacle—on the left side of the ovary in *M. katsuradai* and on the right side of the ovary in *M. otsurui*.[8] *M. katsuradai* is distinct from *M. hakubaensis* in that the former has long ceca that enter the posttesticular region, whereas the latter has short ceca ending at the mid-level of the posterior testis.[10]

46.1.1.6 Metagonimus otsurui *Saito and Shimizu, 1968*

This species was described in Japan in 1968 based on adult flukes from golden hamsters experimentally fed with the metacercariae in freshwater fishes, including *Tridentiger obscurus*, *Chaenogobius castanea*, and *C. urotaenia* in a lake in Ibaragi Prefecture.[8] Later, this species was reported again from another fish species, *Tridentiger brevispinis*, in Aomori Prefecture, Japan.[35] *M. otsurui* has a smaller ventral sucker compared to its oral sucker, which is similar to *M. katsuradai* and *M. hakubaensis* but differs from *M. yokogawai*, *M. takahashii*, and *M. minutus*, which have a smaller oral sucker than their ventral sucker.[8] *M. otsurui* differs from *M. katsuradai* in that the seminal vesicle is at the left side of the ovary in the latter, whereas it is at the right side of the ovary in the former.[8] *M. otsurui* has long ceca, whereas *M. hakubaensis* has short ceca.[8,10]

46.1.1.7 Metagonimus hakubaensis *Shimazu, 1999*

This species was described in Japan in 1999 based on adult flukes from experimental rats fed with the metacercariae in a lamprey, *Lethenteron reissneri*, a fish species collected in Hakuba Village, Nagano

Prefecture.[10] *M. hakubaensis* differs from *M. yokogawai*, *M. takahashii*, *M. miyatai*, and *M. minutus* in that the former has a larger oral sucker than its ventral sucker, whereas the later four species have a larger ventral sucker than their oral sucker.[8,10] The ceca of *M. hakubaensis* are short, but the ceca of *M. otsurui* and *M. katsuradai* are long.[8]

46.1.2 Biology and Life Cycle

Three types of hosts are required to complete the life cycle of *Metagonimus* spp. The first one is the snail intermediate host, in particular, freshwater or brackish water snails of the genus *Semisulcospira* or other related genera.[2] The second host is fish, including a wide variety of freshwater and brackish water species.[1,2] The definitive hosts include humans and animals, including mammals and birds.[2]

Semisulcospira coreana or *Semisulcospira libertina* has been reported to be the molluscan intermediate host of *M. yokogawai*,[36] and *S. libertina*, *S. coreana*, or *Koreanomelania nodifila* have been recognized as the snail hosts for *M. takahashii*.[26,36] The snail host for *M. miyatai* includes *S. libertina*, *S. dolorosa*, or *S. globus*.[37,38] *M. katsuradai* takes *S. libertina*[39] and *Juga tegulata*[40] as the molluscan host in Japan and Russia, respectively. The snail host for *M. hakubaensis* is *S. dolorosa*,[10] and that for *M. otsurui* is *S. libertina* and *S. reiniana*.[13] However, the molluscan host of *M. minutus* has not been reported. *Metagonimus* spp. produce cercariae of the ophthalmo-pleuro-lophocercous type, which are characterized by a pair of eyespots and a long slender tail without bifurcation but covered with membranous fin-like structure.[16] When they are shed in water, they swim freely and infect freshwater fish, i.e., the second intermediate host.[4]

The major fish host for *M. yokogawai* is the sweetfish, *P. altivelis*, in Korea and Japan.[1–4] The dace *Tribolodon hokonensis* or *T. taczanowskii* and perch *Lateolabrax japonicus* have also been reported to be infected with *M. yokogawai* metacercariae.[2,4,16,41] In comparison, the fish hosts for *M. takahashii* were reported to be the crussian carp *Carassius carassius*, carp *Cyprinus carpio*, dace *T. taczanowskii*, and perch *L. japonicus*.[1,2,4] The metacercariae of *M. miyatai* were detected in *Zacco platypus*, *Z. temminckii*, *P. altivelis*, *T. hakonensis*, *T. taczanowskii*, *Opsariichthys bidens*, *Morocco steindachneri*, and *Phoxinus lagowskii steindachneri*.[9,37] The fish host for *M. minutus* is the brackish water fish, *M. cephalus*,[7] and that of *M. katsuradai* are *Tanakia lanceolata* (formerly *Acheilognathus lanceolata*), *T. limbata*, *T. rhombea*, *T. moriokae*, *Pseudorasbora parva*, *Z. platypus*, and *Gnathopogon* sp.[3,4,7] With regard to *M. otsurui*, the second host includes the freshwater fish, including *Tridentiger obscurus*, *T. brevispinis*, *Chaenogobius castanea*, *Chaenogobius urotaenia*, and *Rhinogobius flumineus*.[8,13,35] The fish host for *M. hakubaensis* is a lamprey, *Lethenteron reissneri*.[10] The cercariae of *Metagonimus* spp. attach to the skin or scale of the fish host and penetrate into the dermis and then the muscles leaving their tail outside the body of the fish.[4] The metacercariae usually encyst in the fish muscle but rarely under the scale or in the fin.[4,16,35] The metacercariae can persist in the fish muscle for at least 2.5 years and can be alive throughout the life span of the fish host.[3,4,16]

The definitive hosts of *Metagonimus* spp. are fish-eating birds and mammals. With regard to *M. yokogawai*, dogs,[42] rats,[43] cats,[44] foxes,[45] and kites (bird)[45] have been known to be natural definitive hosts. However, the significance of each animal host as the source of human infections has not been properly verified.[2,4,16] Mice, rats, cats, dogs, gerbils, hamsters, and ducks are experimental definitive hosts for *M. yokogawai*.[2,46,47] The usefulness of these animal hosts as laboratory models for *Metagonimus* spp. infection will be discussed later in this chapter. As for *M. takahashii*, pelicans, kites, and other species of birds,[48] and mice, rats, dogs, cats, and other mammals[24,44,49–51] were reported to be the natural and experimental definitive hosts. Natural definitive hosts for *M. miyatai* include the dog, red fox, raccoon dog, and black-eared kite.[4,9] Mice, rats, hamsters, and dogs are experimental definitive hosts for *M. miyatai*.[9,29,37,38,52,53] For *M. minutus*, vertebrate animals (detailed animal species is unknown) were reported to be natural definitive hosts,[54] and rats and mice were used as experimental definitive hosts.[6] With regard to *M. katsuradai*, dogs are the only reported natural definitive hosts,[3] whereas its experimental definitive hosts include mice, white mice, rats, rabbits, puppies, kittens,[7] and golden hamsters.[54] As for *M. otsurui*, the natural definitive host is Japanese water shrews,[13] and the experimental hosts include golden hamsters and rats.[13,35] *M. hakubaensis* was experimentally infected to the hamster, rat, mouse, dog, cat, chicken, and quail,[10,55] and Japanese water shrews were proved to be a natural definitive host.[56]

46.1.3 Epidemiology and Distribution

The principal mode of human *M. yokogawai* infection is ingestion of raw or improperly cooked flesh of the fish host, notably the sweetfish (*P. altivelis*) and the dace (*T. taczanowskii*), in Far Eastern countries.[16] Human infections are widely scattered along riverside areas, where local people traditionally eat these raw fish.[4,16]

In the Republic of Korea, almost all large and small rivers and streams in eastern and southern coastal areas are endemic foci of *M. yokogawai*.[1,2,17] The Sumjin, Tamjin and Boseong Rivers, Geoje Island, and Osip Stream (Gangwon-do) are hyperendemic areas with 20%–70% egg-positive rates of the riparian residents.[2,17,57,58] This high prevalence is persisting in many of these areas, although in some areas the intensity of infection is slightly reduced.[59] The national prevalence of heterophyid eggs (mostly *M. yokogawai*) in 2012 was 0.26%; thus, the number of infected people in the Republic of Korea is about 130,000.[60]

With regard to the infection status of sweetfish, a survey undertaken in 2011 revealed that 88 (60.7%) of 145 sweetfish from 10 streams of eastern and southern coastal areas were found infected, with the average number of metacercariae per fish of 61.[61] Four (9.8%) of 41 stray cats, a natural definitive host, purchased from a market in Seoul had *M. yokogawai* infection,[44] whereas 78 (17.8%) of 438 cats from a market in Busan were found infected with *M. yokogawai* (under the name *Metagonimus* sp.),[62] and 61 (30.0%) of 188 stray cats caught near the five large rivers in Korea were positive for *Metagonimus* spp. worms.[63] Among 170 wild rats captured from various localities in Korea, 4 (2.4%) were infected with *M. yokogawai*.[43]

Interestingly, *M. yokogawai* eggs were detected from Korean mummies of the 17th century, Joseon Dynasty.[64,65] It indicates that the life cycle of *M. yokogawai* was actively maintained at about 400 years ago in Korea.[64] The area where the mummy was found[64] is now a well-known high-endemic area of *M. yokogawai* infection.[4,17]

In Japan, a lot of epidemiological studies have been performed since 1912 when *M. yokogawai* was first described.[3] It was thought that the infection was distributed nationwide, except in Hokkaido.[3] Later, however, its presence in Hokkaido was also confirmed.[45] Until the 1960s, the reported prevalence in Japanese people had been 0.5%–35.1% depending on the locality.[3] However, in some areas, for example, along the Takatsu River, Shimane Prefecture, the prevalence in residents was high up to 71.9% in 1965,[66] and in Akita Prefecture, two areas surveyed in 1972 revealed 8.5% and 45.9% egg prevalence among the residents, respectively.[67] In addition, an epidemiological survey was performed around the Hamamatsu Lake, Shizuoka Prefecture, during 1982–1988, in which a 13.2% egg-positive rate among 4524 lakeside people examined was found infected.[68] Thereafter, it seems that the prevalence has been decreasing steadily in Japan.[4]

In 1966–1967, 11 prefectures, including Gifu, Kyoto, Hyogo, Okayama, Shimane, Yamaguti, Ehime, Kochi, Miyazaki, Kumamoto, and Kagoshima, were reported as high-endemic areas of metagonimiasis with heavy metacercarial burdens in the fish host (>1000 metacercariae per sweetfish).[69] Recently, in 2006, the prevalence of metacercariae ranging from 0% to 100% was observed in sweetfish caught from 18 small rivers in Shizuoka Prefecture, with variable numbers of metacercariae detected per fish.[70]

In Taiwan, where *M. yokogawai* was originally described as a new species from the sweetfish *P. altivelis*,[3] surprisingly few investigations have been conducted on adult fluke infections in humans and animals.[4] Studies on larval infections are also scarce.[46,47] Freshwater fish species, including *Z. platypus*, *Varicorhinus barbatulus*, *Distoechodon turmirostris*, *Hemibarbus labeo*, *Sinibrama macrops*, *Opsariichthys pachycephalus*, and *Acrosscheilus formosanus*, were added as the fish intermediate hosts in Taiwan.[46,47]

In mainland China, little information is available on human *M. yokogawai* infection, although it was briefly mentioned that the infection exists in Guangdong, Anhui, Hubei, and Zhejiang Province.[12] However, it is of interest to note that eight human infections were diagnosed by genetic analysis of fecal eggs in Guangxi Province in 2012.[71] In Heilongjiang Province, 6.2% of 178 farm dogs were found to harbor the adult flukes (2–7850 specimens per dog).[72] In addition, in Yangshuo County of Guangxi Province, 18 of 31 fish species examined, including *Z. platypus* and *Hemibarbus maculatus*, were infected with *M. yokogawai* metacercariae.[73] Meanwhile, in Yunnan Province, the cercariae of *M. yokogawai* were detected from *S. libertina* snails.[4,74]

In Russia, human *M. yokogawai* infection is highly endemic in the Amur and Ussuri valleys of the Khabarovsk Territory.[12] The prevalence among the ethnic minority was reported between 20% and 70%.[12] In the north of Sakhalin Island, the infection rate was 1.5% among Russians and 10% among ethnic minorities, and sporadic cases were also reported in the Amur district and the Primorye Territory.[12,16] The population at risk is estimated to be about 859,000, which is 14.7% of the total population in these territories.[12]

In eastern Europe, its existence in fish hosts and wild animals has been reported in Ukraine,[21] Serbia,[18,22] Bulgaria,[19,75] and Czech Republic,[20] although no human infections have been confirmed so far.

Regarding *M. takahashii* infection, in the Republic of Korea, the adult flukes were first recovered from rabbits experimentally fed with the metacercariae in 1960.[76] Thereafter, the presence of human infections was first demonstrated from riparian people along the Hongcheon River, Gangwon-do, in 1988.[77] Subsequently, in 1993, an endemic focus was discovered from Umsong-gun, Chunchungnam-do, along the upper reaches of the Namhan River.[23] The riparian people along this river were mixed-infected with *M. miyatai*, with an egg-positive rate of 9.7% for both species.[23] A year later, in 1994, *M. takahashii* adult flukes (8–402 in number) were recovered in six residents living in Kochang-gun, Gyeongsangnam-do.[78] The major sources of human infections were the crucian carp *C. carassius*.[26] The perch, *L. japonicus*, was also identified as the fish host in Gyeongsangnam-do.[79]

In Japan, several articles were published on *M. takahashii* after this species was first recorded in 1930.[3,4] However, before the studies of S. Saito in the 1970s–1990s,[13,35,36] its precise epidemiological characteristics, including the prevalence and geographical distribution of human infections, have not been clearly defined. However, it is worthwhile to refer to Takahashi,[24] who originally discovered *M. takahashii*; in 1929 in Okayama City, 43 (0.64%) of 6680 residents were infected with *M. takahashii*, whereas 54 (0.81%) were infected with *M. yokogawai*.[3] Later, in 1957 in Fuchu City, Hiroshima Prefecture, 11 (4.8%) of 231 residents examined were infected with *M. takahashii*, whereas 81 (35.1%) were infected with *M. yokogawai*.[3] Quite recently, in 2003, *Semisulcospira* snails collected around the Lake Biwa were found infected with the cercariae of *M. takahashii*, together with those of *M. yokogawai*, *M. hakubaensis*, *M. otsurui*, and *M. katsuradai*.[80]

In the Republic of Korea, before *M. miyatai* was designated as a distinct species, its human infections were noticed by detecting eggs in the feces, which were slightly larger in size than those of *M. yokogawai*.[52] Later, adult flukes (under the name *Metagonimus* Miyata type) were recovered from 32 riverside people living along the Namhan River in Umsong-gun (those people were concurrently infected with *M. takahashii*) and Yongwol-gun (infected only with *M. miyatai*).[23] In Yongwol-gun, the egg-positive rate of *M. miyatai* among 77 riparian residents was 48.1%.[23] Adult flukes were also recovered from residents along the Hantan River, Chorwon-gun[81] and Talchon River, Chungwon-gun.[82]

In Japan, epidemiological studies, particularly on human infections, are scarce. Saito et al.[9] enlisted dogs, foxes, raccoon dogs, and black-eared kites for animal definitive hosts in Shimane, Kochi, and Yamagata Prefectures. In Hiroi River basin, Nagano Prefecture, cercariae and rediae were found in *Semisulcospira* snails, and metacercarial infection was detected in *P. lagowski steindachneri* fish.[37] In Shizuoka Prefecture, many small rivers were found to have *M. miyatai* metacercariae in fish.[70]

The geographical distribution of *M. katsuradai* is confined to Japan.[3,7] In Hyogo Prefecture, the larval infection rate of *Semisulcospira* snails was 2.0% for *M. yokogawai*, 8.6% for *M. takahashii*, and 1.9% for *M. katsuradai* in 1939.[3] In Kobe City, dogs were reported as a natural definitive host.[3] Around the Lake Biwa, cercariae of *M. katsuradai* (having six oral spines) were discovered in 2003.[80] Natural human infections have never been documented.[4]

46.1.4 Pathogenesis and Clinical Features

In laboratory animal models using mice, rats, cats, and dogs, adult *M. yokogawai* flukes inhabited the middle part of the small intestine; they invaded the crypt of Lieberkühn by day 2–3 postinfection and localized between villi in later stages.[2,14,15,83,84] The most prominent histopathological findings were villous atrophy and crypt hyperplasia, accompanied by inflammatory reactions in the stroma.[14] The infected intestinal mucosa showed blunting and fusion of the villi, edema of the villus tip, congestion, goblet cell hyperplasia, mastocytosis, and inflammatory cell infiltrations in the stroma, and markedly decreased villus/crypt height ratio.[2,14,15,84] Almost the same intestinal histopathology seems to occur in human *M. yokogawai* infection.[2,16,85]

In immunocompetent hosts, the location of *M. yokogawai* worms is confined to the intestinal mucosa, not beyond the submucosa or muscularis mucosa.[14] However, immunosuppression could allow a deeper invasion of worms into the submucosa or underneath.[15] Immunosuppression can also enhance survival of the worms and prolong their life spans.[15,86] Poor absorption of intestinal secretions from the secretory crypt cells seems to induce watery diarrhea in humans or animal hosts.[2,16] In *M. miyatai* infection in mice, the intestinal histopathology was similar to that in *M. yokogawai* infection, although the degree of mucosal damage was less severe than in *M. yokogawai* infection.[87]

A big question regarding the pathogenicity of *M. yokogawai* is whether this fluke can cause extraintestinal infections in humans and animal hosts.[16] It was reported in 1940 in the Philippines that several species of heterophyid flukes, including *Stellantchasmus falcatus*, *Haplorchis* spp., and *Procerovum* spp., caused erratic parasitism in the heart, brain, and spinal cord in humans, which was often fatal.[2,17,88] Such erratic parasitism may have occurred in immunocompromised patients.[2,16] In *M. yokogawai*-infected human patients, no direct evidence of extraintestinal infections has been reported.[16] However, it is worth noting that intracerebral hemorrhage and diabetes mellitus occurred in a metagonimiasis patient, and the intracerebral hemorrhage may be an acute complication, whereas diabetes mellitus may be a chronic sign.[89]

The clinical symptoms in human metagonimiasis, unless complicated by other diseases and there are no signs of immunosuppression, include mild to moderate degrees of abdominal pain, diarrhea, lethargy, anorexia, and weight loss.[2,16] However, the severity of symptoms may vary depending on various host-side factors.[4,16] One factor is the intensity of infection—the number of worms infected in each patient; heavier infection results in more severe illness.[2,16,17] The second factor is the immune status of the host.[16,17] In immunocompromised patients, *M. yokogawai* infection can possibly cause severe clinical manifestations, including erratic parasitism in vital organs.[4,16]A severe clinical case of *M. yokogawai* infection complicated with multiple intracerebral hemorrhages and diabetes mellitus may have been immunosuppressed.[89] A third factor is the history of previous infections that might confer some degree of acquired immunity in individual patients.[16] New visitors to an endemic area are likely to suffer from a severe illness from a primary infection.[2,16] On the other hand, individuals living in an endemic area for a long time, who inevitably are infected repeatedly, may be protected from severe pathology and symptoms caused by *M. yokogawai* worms.[4,16]

46.1.5 Diagnosis

Human *Metagonimus* infection can be diagnosed by detecting eggs in fecal samples.[16] The direct smear, cellophane thick smear, and Kato-Katz thick smear techniques can be applied in field conditions, whereas in the laboratory equipped with centrifuges, concentration techniques, for example, formalin-ether sedimentation technique and brine (or zinc sulfate) floatation techniques, can be performed.[4,16]

However, morphological differentiation of eggs in fecal samples may be problematic in areas having coprevalence of *Clonorchis sinensis* and/or heterophyid fluke (*Metagonimus* spp., *Heterophyes* spp., *Pygidiopsis summa*, and *Haplorchis* spp.) infections.[4] Eggs of *M. yokogawai* can be differed from eggs of other heterophyids by their length of 26.9–31.6 μm, elliptical shape with length/width ratio of 1.5–2.1, clean shell surface (without muskmelon patterns), less prominent operculum, lack of shoulder rims, and dark yellow or brown color.[90,91] Eggs of *M. takahashii* and *M. miyatai* are very similar to those of *M. yokogawai*, except their sizes being larger in the former and smaller in *M. yokogawai*.[23,90,91] In patients infected with less than 100 worms, the probability of detecting eggs in one Kato-Katz smear is almost zero,[4,16] assuming that the daily number of eggs produced per *M. yokogawai* worm in the human host is 14–64.[92] In such cases, serological tests, including ELISA, are helpful.[16]

Molecular techniques can be applied to detect *Metagonimus* spp. infections in the feces[71] or food materials including fish.[93]

46.1.6 Treatment and Control

Praziquantel in a single oral dose of 10–20 mg/kg is the drug of choice for treating human metagonimiasis.[16] Praziquantel is safe without particular adverse reactions at this dose even in children and pregnant women.[94] Control measures for metagonimiasis can include control of the snail host, control of the fish host, and mass chemotherapy of residents in endemic areas.[4] However, snail control and fish control are not feasible.[4] Mass chemotherapy can temporarily reduce the prevalence and infection intensity (individual worm load), but reinfection continues to occur in the area unless the consumption of raw or improperly cooked fish is stopped.[4] Therefore, prevention is the best way. Health education not to eat raw fish will help prevention of the disease.[4] The infectivity of *M. yokogawai* metacercariae in fish can be controlled by gamma-irradiation at 200 Gy.[95] However, this method is not feasible in the field due to various reasons such as the necessity of an irradiator, high cost, and low preference of irradiated fish by the consumers.[95]

46.2 Laboratory Models

46.2.1 General Considerations

Metagonimus spp. are zoonotic, and various animals are known to serve as natural definitive hosts. In endemic areas of metagonimiasis due to *M. yokogawai* or *M. miyatai*, for example, humans are the principal host and the most important epidemiological element. The infected humans may suffer from mild to moderate degrees of abdominal discomfort, diarrhea, indigestion, and weight loss.[1,2,4,17] However, the mechanisms of diarrhea and other clinical symptoms as well as host immune responses against *Metagonimus* spp. infection are poorly understood. Therefore, for proper understanding of human metagonimiasis, laboratory models, in particular, experimental metagonimiasis in laboratory animals (Table 46.1), or *in vitro* culture models, are indispensable. For this purpose, mammalian animals like mice, rats, dogs, and hamsters have been most commonly used, and rarely cats, rabbits, chickens, and ducks were used. The study subjects using laboratory animals have been the morphology, taxonomy, growth and sexual development of worms, their longevity and fecundity, comparative susceptibility of different animal species, pathogenesis and pathology, host–parasite interactions, host immune responses, chemotherapy, and control.

TABLE 46.1

Species of *Metagonimus* Reported in the Literature and Laboratory Animals Used

Species	Human Infection	Source of Infection	Laboratory Animals Models
Metagonimus yokogawai	Natural	Sweetfish, dace, perch	Dog, cat, hamster, rat, mouse, rabbit, guinea pig, gerbil, mink, raccoon dog, fox, duck, chicken (unsuccessful)
Metagonimus takahashii	Natural	Crussian carp, common carp, dace, perch	Dog, cat, hamster, rat, mouse, rabbit
Metagonimus miyatai	Natural	Sweetfish, dace, minnow, pale chub, dark chub	Dog, hamster, rat, mouse, duck (unsuccessful)
Metagonimus minutus	Uncertain[a]	Mullet	Cat, mouse
Metagonimus katsuradai	Experimental	Carp, pale chub, *Acheilognathus lanceolata*	Dog, cat, hamster, rat, mouse, rabbit, duck
Metagonimus otsurui	None	Freshwater fish	Dog, hamster, chicken (unsuccessful)
Metagonimus hakubaensis	None	Freshwater fish	Dog, hamster, rat

[a] Enlisted among the species infecting humans,[12] but without literature background.

46.2.2 Animal Models Used for Studies on Morphology and Taxonomy

46.2.2.1 Dogs

When *M. yokogawai* was first described as a new species in 1912, dogs were used as the principal laboratory host.[30,96,97] Thereafter, a lot of researchers used dogs for experimental *M. yokogawai* infection.[24,41,98–101] Muto[98] studied the morphology of *M. yokogawai* using dogs as the laboratory host, and Saito[101] compared the morphological and biological characteristics of *M. yokogawai*, *M. yokogawai* Miyata type (now *M. miyatai*) and Koga type (now considered synonym of *M. takahashii*), and *M. takahashii*. Kagei and Kihata[99] studied the morphology as well as development and egg-laying capacity of *M. yokogawai* in various animals including dogs. Ahn[41] morphologically confirmed *M. yokogawai* adults recovered from dogs fed with the flesh of the perch from Gurye-gun (Seomjin-gang River), Korea. Ahn et al.[100] studied the egg-laying capacity of *M. yokogawai*.

For studies on *M. miyatai*, dogs were also popularly used.[9,50,53] Ahn[50] used a dog to obtain adult flukes of *M. miyatai* (as *Metagonimus* Miyata type) after feeding the metacercariae from *Z. platypus* in Korea. Saito et al.[9] obtained adult flukes of *M. miyatai* from dogs experimentally fed with freshwater fish including *Z. temmincki* and *Morocco steindachneri* in Japan and established *M. miyatai* as a new species. To study *M. takahashii*, Takahashi[24] used dogs to compare the morphology of adult flukes and susceptibility of different laboratory animals. Surface ultrastructure of *M. takahashii* was also studied using worms recovered from experimental dogs.[102] Dogs were further used to obtain adult flukes and establish *M. katsuradai*,[7] *M. otsurui*,[8] and *M. hakubaensis*[55] as new species.

The use of dogs as a laboratory definitive host for *Metagonimus* spp. includes several merits.[24,41,98–101] One is the high recovery rate of worms with a long-time establishment in the small intestine of dogs. They are highly susceptible to *Metagonimus* infection with larger worm size and longer-time survival compared to mice and rats. Another merit is full growth and maturation of worms as comparable as those that develop and mature in the human host. A third is the relatively large body size of dogs as an alternative host that can mimic the human body. A fourth may include easier laboratory management compared to cats.

46.2.2.2 Cats

Cats have been used rarely as a laboratory host for *Metagonimus* spp.[7,97] Kobayashi[97] infected cats and dogs with the metacercariae of *M. yokogawai* isolated from sweetfish muscle to obtain the adult flukes. Kim et al.[79] infected a cat with metacercariae from perches and obtained *M. takahashii* adult flukes. Katsuta[6] also infected cats with metacercariae of *M. minutus* and obtained adult flukes. Izumi[7] used cats and several other mammalian animals to recover the adult flukes of *M. katsuradai*. Cats were also used for pathology[103,104] and immunology studies[105,106] in *Metagonimus* spp. infection (see 46.2.4 and 46.2.5).

46.2.2.3 Hamsters

Hamsters were occasionally used as an experimental definitive host for *Metagonimus* spp. Yokogawa and Sano[107] infected hamsters, mice, rats (cotton rats and Wistar rats), and guinea pigs with the metacercariae of *M. yokogawai* to obtain the adult flukes and to observe the development of worms in different hosts. Kagei and Kihata[99] also studied the morphology, recovery, and the life span of *M. yokogawai* in hamsters. Miyamoto and Kutsumi[108] infected hamsters and dogs to recover *M. yokogawai* adult flukes. A similar study was performed recently by Li et al.[46] in Taiwan. Hamsters were also used to differentiate morphological characters of *M. yokogawai* from those of *M. miyatai*[9,29,37,70,109] and *M. takahashii*.[26,29,109] Moreover, unique morphological features of *M. katuradai*[54], *M. otsurui*,[8,13] and *M. hakubaensis*[55] were also studied using hamsters.

The merits of using hamsters for an experimental definitive host include their high susceptibility to *Metagonimus* spp. infection with high worm recovery and longer worm survival.[46,107] Suitability as a laboratory animal and easy handling, compared with dogs and cats, are also among the merits.

46.2.2.4 Mice and Rats

Takahashi[24] used mice and rats to study the morphology and life cycle of *M. yokogawai* in comparison to *M. takahashii*. Koga,[28] Gushima,[110] Komiya et al.,[111] and Yokogawa and Sano[107] also used mice and rats to infect *M. yokogawai*, although the susceptibility of these animals was less satisfactory compared with animals like dogs, cats, and hamsters. Chun[112] infected mice, rabbits, and guinea pigs to obtain the adult flukes of *M. yokogawai*, and Hong and Seo[113] observed worm development in mice. Kagei and Kihata[99] compared the developmental status of *M. yokogawai* in mice, hamsters, and dogs.

For studies on *M. miyatai* and/or *M. takahashii*, mice or rats were used by Kim,[52] Kim et al.,[38] Ahn,[50] Ahn and Ryang,[53,77] and Lee et al.[114] In addition, Chai et al.[29] used rats and hamsters to obtain the adult flukes of *M. yokogawai*, *M. miyatai* (as *Metagonimus* Miyata type), and *M. takahashii* (as *Metagonimus* Koga type). Mice and cats were used to study the morphology of *M. minutus*,[6] and mice, rats, rabbits, cats, dogs, and/or ducks were used to infect the metacercariae of *M. katsuradai*.[7,39] For the new species description of *M. hakubaensis*,[10] rats were used as the experimental host.

Rats were also used to study the surface ultrastructure of the adult flukes of *M. yokogawai*,[115] *M. miyatai*,[116] and *M. takahashii*.[117] The surface ultrastructure was similar to each other, but a few unique findings were found, for example, the presence of type III sensory papillae in *M. yokogawai* but not in two other species.[115–117] Molecular and genetic differentiation and taxonomic studies were performed using adult flukes of *M. yokogawai*, *M. miyatai*, and *M. takahashii* recovered from experimental rats.[32,34,118] These three species were genetically unique and considered to be varied from each other.

The use of mice and rats as an experimental definitive host of *Metagonimus* spp. was in consideration of fairly good morphology of the recovered adult flukes and easy handling in the laboratory.

46.2.2.5 Rabbits, Guinea Pigs, and Other Mammals

Rabbits were not uncommonly used to obtain the adult flukes of *Metagonimus* spp., *M. yokogawai*,[24,112,119] *M. takahashii*,[119] and *M. katsuradai*.[7] Moreover, guinea pigs[107,112] and gerbils[46] were also used in studies of *M. yokogawai* for the same purpose. Other mammalian animals used for studies on *M. yokogawai* included American minks, raccoon dogs, and red foxes.[45]

46.2.2.6 Ducks and Chickens

Studies on the morphology of *Metagonimus* spp. using avian hosts have been scarce. As for *M. yokogawai*, a recent study in Taiwan[47] reported that domestic ducks can host this fluke within 14 days after infection but were considered not susceptible to this fluke infection. Similarly, *M. miyatai* metacercariae were experimentally fed to ducks, but no adult flukes could be recovered.[53] However, Kurokawa[39] succeeded in obtaining two adult flukes at 7 days after experimental infection of a duck with *M. katsuradai* metacercariae. On the other hand, Saito and Shimizu[8] failed to obtain adult flukes of *M. otsurui* from two chickens experimentally fed with the metacercariae.

46.2.3 Animal Models Used for Studies on Worm Development and Host Susceptibility

46.2.3.1 Dogs

Dogs were most popularly used as a laboratory definitive host to complete the development and sexual maturity of *M. yokogawai*[24,41,98–100,108] and other *Metagonimus* spp.[7,8,24,28,38,39,50,53] Yokogawa[96] reported that *M. yokogawai* attained its full development in dogs by day 5 (120 h) after infection with the metacercariae. Muto[98] found eggs of *M. yokogawai* discharged in the feces of dogs experimentally infected with the metacercariae from day 9 after infection. Kagei and Kihata[99] reported that the development and egg-laying capacity of *M. yokogawai* were better in dogs compared with mice. They observed eggs in the uterus of worms recovered from dogs as early as day 3 after infection and eggs in the feces of dogs from day 6 after infection. Ahn et al.[100] reported that the daily egg output (EPDPW; eggs per day per worm) of *M. yokogawai* was 35–45 eggs at day 7 postinfection in the dog host. Miyamoto and Kutsumi[108] confirmed that the dace *Tribolodon* spp. is a second intermediate host of *M. yokogawai* by recovery of adult flukes from experimentally infected dogs from day 8 to 129 postinfection. Kang et al.[84] infected dogs with the metacercariae of *M. yokogawai* and obtained the worm recovery rate of about 20% even at week 6 postinfection; a repeated infection led to displacement of worms from the duodenum and jejunum down to the ileum.

With regard to *M. miyatai*, dogs were used for studies on the worm development and/or fecundity.[9,38,50,53] Ahn and Ryang[53] reported that the worm development of *M. miyatai* and susceptibility was better in dogs than in rats; the worm maturity was the best at day 35 after infection in dogs. In *M. takahashii*, few studies have been performed using dogs as the definitive host. Takahashi[24] observed better development of *M. takahashii* worms in dogs compared to those developed in mice. Similarly, Koga[28] studied the satisfactory worm development of *M. takahashii* (under the name, large egg-type *Metagonimus*) in dogs and cats. Dogs were further used to study the worm development and host susceptibility of *M. katsuradai*,[7,39] *M. otsurui*,[8] and *M. hakubaensis*.[55] Kudo et al.[55] reported that dogs and hamsters were more susceptible to *M. hakubaensis* infection than cats, rats, mice, chickens, and quails.

Regardless of the species of *Metagonimus*, dogs were found to be a highly susceptible host to experimental *Metagonimus* infection. The worms grow quickly (within 3–6 days) in the small intestine of dogs compared with other laboratory hosts including mice and rats, and survive remarkably a longer time. Their fecundity in dogs was fairly good.

46.2.3.2 Cats

Kobayashi[97] first used cats to infect the metacercariae of *M. yokogawai* isolated from the sweetfish. Thereafter, only a few researchers[73,103–106] used cats to study *M. yokogawai*. These studies were mainly to recover the adult flukes or to observe the intestinal pathology and host immune responses[103–106] rather than to observe the life cycle, development, and sexual maturity of worms. Sohn et al.[73] infected cats and dogs to recover adult *M. yokogawai* after an experimental infection with the metacercariae in freshwater fish from China.

Studies using cats were also scarce for other *Metagonimus* spp. As for *M. takahashii*, Kim et al.[79] infected a cat with the metacercariae in the perch *Lateolabrax japonicus* from Korea and obtained adult flukes 21 days later. Similarly, Katsuta[6] obtained *M. minutus* adult flukes from cats experimentally infected with the metacercariae in Taiwan, and Izumi[7] recovered adult flukes of *M. katsuradai* from experimental cats.

46.2.3.3 Hamsters

Hamsters were popularly used to study the development and longevity of *Metagonimus* spp. worms. Yokogawa and Sano[107] reported that hamsters were a highly suitable animal host for *M. yokogawai* infection compared with mice, rats, and guinea pigs. However, Kagei and Kihata[99] reported that although some *M. yokogawai* worms survived up to a year in dogs, no worms were retained in the intestine of hamsters after 5 months. Miyamoto and Kutsumi[108] reported similar findings; in dogs, a considerable number of worms survived at day 129 postinfection, but in hamsters, only a few worms survived at day 29 postinfection. However, Li et al.[46] obtained a high worm recovery rate (75.3%) of *M. yokogawai* from hamsters at day 14 postinfection. Compared with hamsters or dogs, mice were not a suitable host for *M. yokogawai* infection.[99,107] Therefore, it can be said that the suitability of laboratory animals to infection with *M. yokogawai* is the highest for dogs, followed by hamsters, rats, mice, and other rodent animals. Hamsters were also used for studies on the development and host susceptibility of *M. miyatai*[37] and *M. otsurui*.[13]

46.2.3.4 Mice and Rats

Mice and rats have been popularly used as laboratory animal models for *Metagonimus* spp. infection, although the susceptibility of these animals is lower compared with bigger animals like dogs. Komiya et al.[111] reported that *M. yokogawai* developed well in the intestine of mice by day 6 after infection and survived until day 26. Yokogawa and Sano[107] compared the recovery and growth of *M. yokogawai* in mice, Wistar rats, cotton rats, hamsters, and guinea pigs. They reported that the worms obtained full maturation by week 1 after infection in all of these animals except guinea pigs, which showed no worms even at 1 week. Hamsters and cotton rats retained more worms until week 6 after infection, whereas mice and Wistar rats retained smaller number of worms.[107] Hong and Seo[113] observed the full growth and development of *M. yokogawai* in mice by day 7 postinfection, although the rate of growth varied individually. Kagei and Kihata[99] reported that *M. yokogawai* developed well in mice, hamsters, and dogs; however, the worm survived shorter than 1 month in mice, whereas it survived until 4 months in hamsters and up to a year in dogs. On the other hand, Li et al.[46] obtained a fairly high worm recovery rate (70.0%) of *M. yokogawai* from ICR mice at day 14 postinfection, comparable with that (75.3%) obtained in hamsters. Chai et al.[86] infected *M. yokogawai* to various strains of mice (CBH, A, C57BL, DBA, and KK) and observed the susceptibility of each mouse strain judged by the worm development and worm recovery. KK mice (diabetic mice) showed the highest worm recovery at day 7 after infection.[86]

Mice and rats were also used for studies on the development and maturation of *M. miyatai* and/or *M. takahashii*.[29,38,50,51–53,77] Kim et al.[38] observed the development and site distribution of *M. miyatai* worms (under the name *Metagonimus* Miyata type) in the intestine of mice. Ahn and Ryang[53] reported that *M. miyatai* (under the name *Metagonimus* Miyata type) continuously increased the body size in rats until day 30 postinfection when the number of uterine eggs also peaked. Some retarded worms were recovered until day 55 postinfection in rats.[53]

Guk et al.[51] studied on the comparative growth and development of *M. yokogawai*, *M. miyatai*, and *M. takahashii* in different strains of mice (BALB/c, ddY, C57BL/6J, C3H/HeN, and A/J). They found that ddY mice were the most suitable host for *M. yokogawai* with the highest worm recovery, that BALB/c and ddY were equally suitable for *M. miyatai* infection, and that BALB/c, ddY, and C3H/HeN mice were equally suitable for *M. takahashii* infection.[51] Generally, mice were fairly susceptible to *M. yokogawai* infection with high worm recovery, good worm dimension, and high worm fecundity but less susceptible to *M. miyatai* and *M. takahashii* infection.[51]

Mice were also used to study the growth and development of *M. minutus*; the sexual maturation was completed by day 7 postinfection.[6] By comparison, the development of *M. katsuradai* was completed earlier in mice, by day 5 postinfection; moreover, eggs were seen in the feces of mice from day 4 (90 h) postinfection.[7] Fully mature specimens of *M. hakubaensis* were recovered in experimentally infected rats at days 7–22 postinfection.[10]

46.2.3.5 Rabbits, Guinea Pigs, and Other Mammals

Rabbits were seldom used to observe the growth and development of *Metagonimus* spp. However, sexually mature worms of *M. yokogawai* were obtained from rabbits fed with the flesh of the sweetfish after days 7–12 postinfection,[112] although the susceptibility of rabbits to *Metagonimus* spp. infection is estimated to be fairly low. Successful development of *M. takahashii*[76] and *M. katsuradai*[7] was also observed in rabbits.

Guinea pigs[107,112] and gerbils[46] were used to infect *Metagonimus* spp., but the results were generally not satisfactory. Some eggs of *M. yokogawai* were detected in the feces of guinea pigs from day 10 to 35 postinfection; however, no adult flukes were recovered at day 45 postinfection.[112] Gerbils were infected with the metacercariae of *M. yokogawai*, and a 6.0% worm recovery rate was obtained at day 14 postinfection.[46]

46.2.3.6 Ducks and Chickens

Ducks were used to observe the infectivity and development of *M. yokogawai*.[47] About 20% of worms attained their sexual development by day 7 after infection; however, the worms were quickly expelled thereafter.[47] However, in ducks experimentally infected with *M. miyatai*, a small number of worms were recovered at day 2 postinfection, and no worms were recovered at day 10 postinfection.[53] Kurokawa[39] succeeded in recovery of two adult flukes at day 7 after infection of a duck with *M. katsuradai* metacercariae. Chicks were used for infection with *M. yokogawai*[73] and *M. otsurui*,[8] but no worms were recovered.

46.2.4 Animal Models Used for Studies on Pathogenesis and Pathology

46.2.4.1 Dogs, Cats, Rats, and Mice

The pathogenesis and pathology of metagonimiasis have been studied mainly in dogs,[28,84] cats,[28,103,104] rats,[14,83,120,121] and mice,[15] exclusively in *M. yokogawai* infection. Koga[28] was the first who described the pathogenesis and intestinal pathology of dogs and cats experimentally infected with *M. yokogawai* metacercariae. He[28] reported transient diarrhea in light infections and severe catarrh sometimes with bloody mucous diarrhea in heavy infections in these animals. However, worms were confined to the mucosa of the small intestine and never invaded into the deeper layers such as the submucosa and underneath.[28] Chai[14] observed the clinical course and intestinal pathology of experimental rats infected with the metacercariae; diarrhea occurred from day 6 after infection and continued until sacrificed at week 4 after infection. The gross intestinal pathology included marked dilatation of the intestinal loop, foamy watery content in the lumen, and sometimes mucosal hyperemia.[14] Such watery content in metagonimiasis is considered to be due to poor absorption of intestinal secretion from crypt cells.[103] Microscopically, the worms were located in the intervillous space or on the top of the villi, and the worm location was confined to the mucosal layer; the pathological changes could be categorized largely into two kinds: villous atrophy and crypt hyperplasia.[14] Villous atrophy included shortening of villi, blunting and edema at the tip of the villi, villous adhesion, and inflammation with edema and congestion of the villous stroma.[14]

Mast cell hyperplasia is another prominent feature in the intestine of *M. yokogawai*-infected rats; it peaked at week 3 postinfection.[120] It was further observed in experimental rats that the number of intraepithelial lymphocytes within the epithelial layer of the small intestinal villi increased rapidly at day 5 postinfection but decreased to a lower level during days 10–24 postinfection and then normalized.[83] Moreover, the position of the intraepithelial lymphocytes changed significantly at day 5 postinfection toward the apical side of the villi; from this time, active worm expulsion seemed to occur.[83] When rats were coinfected with *M. yokogawai* and another kind of intestinal fluke *Neodiplostomum seoulense* (under the name *Fibricola seoulensis*), the intestinal lesions became severer than in rats infected with only one fluke species.[121] In cats,[103] the intestinal pathology due to *M. yokogawai* infection was similar to that reported in rats,[14] and some of the pathological changes continued until week 10 postinfection. As an early phase mechanism of inducing intestinal lesions observed in dogs, uncompensated enterocytes deficit due to marked destruction of the villi and upper parts of the crypt or villous–crypt junction was proposed.[104] In dogs, similar intestinal pathology was observed until week 6 after infection.[84] There is one paper reporting the intestinal pathology of metagonimiasis in a human patient; *M. yokogawai* worms were found free in the jejunal lumen or impacted in the intervillous spaces, and massive lymphoplasmacytic and eosinophilic infiltration in the stroma, erosion of nearby enterocytes, goblet cell depletion, and occasional villous edema were observed.[85]

The location of worms was strictly confined to the mucosal layer (intervillous space, crypt, or on the top of villi) in laboratory animals including dogs, cats, and rats.[14,28,84,103] However, in immunosuppressed mice, the location of worms became deeper into the submucosa and muscular layer of the intestinal wall.[15] This strongly suggests that the invasiveness of *Metagonimus* worms may be related to the host immunity, protecting them from deeper invasion.

46.2.5 Animal Models Used for Studies on Immunity and Host–Parasite Interaction

46.2.5.1 Dogs, Cats, Rats, Mice, and Hamsters

Immunological characteristics including acquired resistance, antigenicity of worms, and host–parasite relationships have been studied using dogs,[84,122,123] cats,[105,106] rats,[83,110,120,124] mice,[86,87,110,125–127] and hamsters[70] mostly in *M. yokogawai* and rarely in *M. miyatai* infection. Gushima[110] was the first to study immune responses of rats and mice after experimental infection with *M. yokogawai* in 1939.

He performed repeated infections of rats (2–3 times) and mice (2–6 times) with *M. yokogawai* at 4- to 30-day intervals and reported the following results: Acquired resistance was obtained in the reinfected animals, and the resistance became stronger in accordance with the frequency of reinfection and the lapse of the days after reinfection. Worm development was retarded in reinfected animals, particularly in genital organs, and animals possessing immature worms got reinfected more easily, whereas those possessing fully developed worms got reinfected with a markedly lower incidence rate.[110] He also notified a tendency of changing parasitic locations from the upper part of the small intestine to the lower part of the intestine.[110] He further notified that the resistance could be acquired by an injection of emulsified worms after refrigeration, drying, and powdering.[110] The shifting of the parasitic locations from the upper part of the small intestine to the lower part of the intestine was also observed in 1983 in dogs reinfected with *M. yokogawai*.[84]

Chai et al.[86] reported that the susceptibility of mice to *M. yokogawai* infection was highly dependent on the immune status of the host. In immunocompetent ICR mice experimentally infected with *M. yokogawai*, most of the worms were expelled spontaneously before day 7 after infection.[86] However, ICR mice immunosuppressed with prednisolone injection allowed a longer survival (until day 21 postinfection) of a greater number of worms than in immunocompetent control mice.[86] Eosinophils were suggested to have an important role to induce such protective immunity against *M. yokogawai* infection.[86,126] Other effector mechanisms conferring protective immune responses of the host may include intestinal intraepithelial lymphocytes (IEL), lamina propria lymphocytes (LPL), mucosal mast cells (MMC), and goblet cells.[1,2,4,11] The kinetic of intestinal IEL was studied by Chai et al.[83]; the IEL number was markedly increased at day 5 postinfection, which decreased thereafter until day 24 postinfection and then normalized. The day 5 postinfection was well corresponded with the time of active worm expulsion from the host intestine.[86] Moreover, the location of the IEL was moved to the apical portion of the villi from the original basal or intermediate portion around the day 5 after infection.[83] Mucosal mastocytosis was observed throughout the infection period, from week 1 to 4, in rats,[120] and, from week 1 to 3, in mice,[125] experimentally infected with *M. yokogawai*.[120] Thus, MMC seemed also a significant factor for inducing immunopathological effects and damages on the mucosal regions and finally, expulsion of worms. Increased mucosal permeability was also suggested to be a factor related to the expulsion of worms from mice.[125]

Antigenicity of body portions of *M. yokogawai* was studied using adult worms recovered from experimental cats.[105,106] Tegumental cells and tegumental syncytium as well as the intestinal content of the worms were suggested to be the major antigens of *M. yokogawai*.[105,106] However, a purified 100-kDa antigen reacting to the tegumental and subtegumental layers of *M. yokogawai* appeared to be not a species-specific antigen but a common antigen crossly recognizing even the liver fluke (*Clonorchis sinensis*), lung fluke (*Paragonimus westermani*), and another intestinal fluke species (*Gymnophalloides seoi*).[124] The antigenicity of *M. yokogawai* was generally stronger than that of *M. miyatai* as assessed by the lower levels of the villus/crypt height ratio and the lowered expression patterns of the proliferating cell nuclear antigen in the former.[87] In addition, the parasitic location of *M. yokogawai* was generally at the upper level of the small intestine in experimental hamsters, whereas that of *M. miyatai* was generally at the middle level of the small intestine.[70]

An interesting observation of the worm posture within the ecological niche of *M. yokogawai* in the dog's small intestine was that the worms, during the early stage of infection (around day 3 postinfection), invade the slit of the Lieberkhün's crypts by making their anterolateral bodies as one or more reversible tube-like structures.[122] Thereafter, until week 10 of observation, worms with such protruded anterior bodies decreased continuously in number when most of the worms were floating on the top of the villi.[122] Another point of interest in host–parasite interactions includes the mechanisms of diarrhea in metagonimiasis. It has been suggested that diarrhea is caused by excessive water content in the small intestine of *M. yokogawai*-infected hosts.[14,103,127] The excessive water content is considered to be due to poor absorption of intestinal secretions resulting from mucosal inflammation in the affected small intestine.[103] The poor absorption of intestinal secretions is likely resulting from the reduced absorptive surface due to the decreased villous height; however, this may be a reversible phenomenon.[123] It was also reported that the activities of brush border membrane-bound enzymes decreased in the small intestine of *M. yokogawai*-infected mice and participated in the generation of diarrhea and malabsorption.[127]

46.2.6 Animal Models Used for Studies on Chemotherapy and Control

46.2.6.1 Dogs and Rats

In vivo experimental studies have never been performed to study on the effects of anthelmintic drugs against *Metagonimus* spp. infection. Instead, there were a few *in vitro* studies to evaluate the effects of anthelmintics such as praziquantel,[128] bithionol, and menichlopholan.[129] *M. yokogawai* adult flukes were harvested from experimentally infected rats fed with the metacercariae and exposed to 0 (control), 1, 10, and 100 μg praziquantel/mL *in vitro*.[128] The tegumental surface of the drug-treated *M. yokogawai* showed vacuolization and bleb formation followed by worm contraction, which led to death of the worms.[128] Similarly, *M. takahashii* adult flukes recovered from experimental dogs and rats lost their motility and tegumental damage after *in vitro* exposure to bithionol and menichlopholan.[129] For the purpose of controlling metagonimiasis, Chai et al.[95] tried irradiation of sweetfish infected with *M. yokogawai* metacercariae by 200 Gy, which was highly effective to control infectivity as well as development of the metacercariae in rats.

46.2.7 *In Vitro* Culture and Others

46.2.7.1 In Vitro *Culture*

Yasuraoka and Kojima[130] tried *in vitro* cultivation of *M. yokogawai* metacercariae using a culture medium containing chick embryo extract, human serum, and NCTC 109 medium. The trial appeared to be successful as fully mature flukes possessing 70–90 eggs in their uteri could be obtained by this *in vitro* cultivation.[130] The maximum worm size was reached at 16 days after beginning of the culture, and the maximum worm survival was 35 days by *in vitro* culture.[130]

46.2.7.2 Experimental Infection of Fish

Saito[27] experimentally infected sweetfish (*P. altivelis*) and goldfish (*Carassius auratus*) with the cercariae of both *M. yokogawai* and *M. takahashii* and found that the sweetfish serves as the second intermediate host of only *M. yokogawai* and the goldfish serves as that of only *M. takahashii*. This host (fish) specificity of *M. yokogawai* and *M. takahashii* has been approved by other studies.[76,101]

46.3 Conclusion and Future Perspectives

Metagonimus spp. are important foodborne parasites and are mainly distributed in the Far Eastern countries, including Korea, China, Taiwan, Japan, and Russia. The major etiological agents of human metagonimiasis are *M. yokogawai*, *M. takahashii*, and *M. miyatai*. However, four other species, namely *M. minutus*, *M. katsuradai*, *M. otsurui*, and *M. hakubaensis*, are potentially infective to humans. Laboratory models for studies on *Metagonimus* spp. infection include mammalian animals (e.g., dogs, cats, hamsters, rats, and mice) and avian animals (e.g., ducks). The topics of research using these laboratory models encompass morphology and taxonomy, worm development and host susceptibility, pathogenesis and pathology, immunity and host–parasite interaction, and chemotherapy and control, among others. *In vitro* cultures using culture medium have also been employed for studies on development and maturation of worms.

The adult flukes parasitize the small intestine of humans and also animal hosts. They can cause mucosal pathology that consists of villous atrophy, crypt hyperplasia, and mucosal inflammations, but the symptoms in human metagonimiasis are generally mild unless heavily infected. However, in immunocompromised patients, complications such as erratic parasitism in the heart, brain, and spinal cord, or intracerebral hemorrhage, may occur, as reported in other heterophyid fluke infections. The possibility of erratic parasitism in metagonimiasis should be ruled out. For this reason, new laboratory models are urgently needed, in particular, larger animal models including monkeys and pigs that may allow passage of eggs and/or worms into the mesenteric venule, resulting in erratic parasitism.

REFERENCES

1. Chai, J.Y., Intestinal flukes, in *Food-Borne Parasitic Zoonoses: Fish and Plant-Borne Parasites*, Murrell, K.D. and Fried, B., Eds., Springer, New York, 2007.
2. Chai, J.Y., Shin, E.H., Lee, S.H., and Rim, H.J., Foodborne intestinal flukes in Southeast Asia, *Korean J. Parasitol.*, 47, S103, 2009.
3. Ito, J., *Metagonimus* and other human heterophyid trematodes, *Progr. Med. Parasitol. Jpn.*, 1, 314, 1964.
4. Chai, J.Y., *Metagonimus*, in *Biology of Foodborne Parasites*, Xiao, L., Ryan, U. and Feng, Y., Eds., CRC Press, Taylor & Francis Group, Boca Raton, FL, 427–443, 2007.
5. Suzuki, S., Yokogawa's *Metagonimus*. List of publications on special animals in Okayama prefecture, *Okayama Prefectural Rep.*, 1, 146, 1930.
6. Katsuta, I., Studies on the metacercariae of Formosan brackish water fishes (2) On a new species, *Metagonimus minutus* n. sp. parasitic in *Mugil cephalus*, *Taiwan Iggakai Zasshi*, 31, 26, 1932 (in Japanese).
7. Izumi, M., Studies on a new species of *Metagonimus* and its life cycle, *Tokyo Iji Shinshi*, 2929, 1224, 1935 (in Japanese).
8. Saito, S. and Shimizu, T., A new trematode, *Metagonimus otsurui* sp. nov. from the fresh-water fishes (Trematoda: Heterophyidae), *Jpn. J. Parasitol.*, 17, 167, 1968.
9. Saito, S., et al., *Metagonimus miyatai* sp. nov. (Digenea: Heterophyidae), a new intestinal trematode transmitted by freshwater fishes in Japan and Korea, *Korean J. Parasitol.*, 35, 223, 1997.
10. Shimazu, T. *Metagonimus hakubaensis* sp. n. (Digenea, heterophyidae) from Nagano, Japan: morphology and life cycle, *Bull. Nat. Sci. Mus., Tokyo Ser. A*, 25, 87, 1999.
11. Chai, J.Y., Murrell, K.D., and Lymbery, A.J., Fish-borne parasitic zoonoses: status and issues, *Int. J. Parasitol.*, 35, 1233, 2005.
12. Yu, S.H. and Mott, K.E., Epidemiology and morbidity of food-borne intestinal trematode infections, *Trop. Dis. Bull.*, 91, R125, 1994.
13. Shimazu, T. and Urabe, M. Morphology and life cycle of *Metagonimus otsurui* (Digenea: Heterophyidae) from Nara, Japan, *Bull. Nat. Sci. Mus., Tokyo Ser. A*, 28, 21, 2002.
14. Chai, J.Y., Study on *Metagonimus yokogawai* (Katsurada, 1912) in Korea V. Intestinal pathology in experimentally infected albino rats, *Seoul J. Med.*, 20, 104, 1979.
15. Chai, J.Y., Kim, J., and Lee, S.H., Invasion of *Metagonimus yokogawai* into the submucosal layer of the small intestine of immunosuppressed mice, *Korean J. Parasitol.*, 33, 313, 1995.
16. Yu, J.R. and Chai, J.Y., *Metagonimus*. In *Molecular Detection of Human Parasitic Pathogens*, Liu, D. (Ed.), CRC Press, Boca Raton, FL, 2013.
17. Chai, J.Y. and Lee, S.H., Food-borne intestinal trematode infections in the Republic of Korea, *Parasitol. Int.*, 51, 129, 2002.
18. Cakiç, P., et al., *Metagonimus yokogawai*, a new parasitic trematoda species in ichthyoparasitofauna of the Serbia, *Acta Vet. (Beograd)*, 57, 537, 2007.
19. Nachev, N. and Sures, B., The endohelminth fauna of barbel (*Barbus barbus*) correlates with water quality of the Danube River in Bugaria, *Parasitology*, 136, 545, 2009.
20. Francová, K., et al., Parasite fauna of native and non-native populations of *Neogobius melanostomus* (Pallas, 1814) (Gobiidae) in the longitudinal profile of the Danube River, *J. Appl. Ichthyol.*, 27, 879, 2011.
21. Davydov, O.N., Lysenko, V.N., and Kurovskaya, L.Y., Species diversity of carp, *Cyprinus carpio* (Cypriniformes, Cyprinidae), parasites in some cultivation regions, *Vestn. Zool.*, 45, e9, 2011.
22. Djikanovic, V., et al. Parasitofauna of freshwater fishes in the Serbian open waters: a checklist of parasites of freshwater fishes in Serbian open waters, *Rev. Fish Biol. Fish.* 22, 297, doi:10.1007/s11160-011-9226-6, 2012.
23. Chai, J.Y., et al., An epidemiological study of metagonimiasis along the upper reaches of the Namhan River, *Korean J. Parasitol.*, 31, 99, 1993.
24. Takahashi, S., On the life history of *Metagonimus yokogawai*, a new species of *Metagonimus*, and *Exorchis major*, *Okayama Igakkai Zasshi*, 41, 2687, 1929 (in Japanese).
25. Asada, S., On *Metagonimus yokogawai* and its related species, *Rinsho Igaku (Clin. Med.)*, 22, 43, 1934 (in Japanese).
26. Saito, S., On the differences between *Metagonimus yokogawai and Metagonimus takahashii* I. The morphological comparisons, *Jpn. J. Parasitol.*, 21, 449, 1972 (in Japanese).

27. Saito, S., On the differences between *Metagonimus yokogawai and Metagonimus takahashii* II. The experimental infections to the second intermediate hosts, *Jpn. J. Parasitol.*, 22, 39, 1973 (in Japanese).
28. Koga, G., Studies on the genus *Metagonimus*, *Igaku Kenkyu*, 12, 3471, 1938 (in Japanese).
29. Chai, J.Y., et al., Three morphological types of the genus *Metagonimus* encysted in the dace, *Tribolodon taczanowskii*, caught from the Sumjin River, *Korean J. Parasitol.*, 29, 217, 1991.
30. Katsurada, F., On the genus *Heterophyes* in Japan, *Okayama Igakkai Zasshi* (no. 268), 373, 1912 (in Japanese).
31. Miyata, I., Comments on the classification of trematodes of the genus *Metagonimus*, *Dobutsugaku Zasshi*, 56, 16, 1944 (in Japanese).
32. Yang, H.J., et al., Molecular differentiation of three species of *Metagonimus* by simple sequence repeat anchored polymerase chain reaction (SSR-PCR) amplification, *J. Parasitol.*, 86, 1170, 2000.
33. Yu, J.R., et al., PCR-RFLP pattern of three kinds of *Metagonimus* in Korea, *Korean J. Parasitol.*, 35, 271, 1997.
34. Lee, S.U., et al., A cytogenetic study on human intestinal trematodes of the genus *Metagonimus* (Digenea: Heterophyidae) in Korea, *Korean J. Parasitol.*, 37, 237, 1999.
35. Oyamada, T., et al., *Metagonimus otsurui* metacercarial infection in a gobiid fish (*Tridentiger brevispinis*) collected from Lake Ogawara in Aomori Prefecture, Japan, *Jpn. J. Parasitol.*, 45, 275, 1996.
36. Cho, S.Y., Kang, S.Y., and Lee, J.B., Metagonimiasis in Korea, *Arzneimittelforschung*, 34(9B), 1211, 1984.
37. Shimazu, T., Life cycle and morphology of *Metagonimus miyatai* (Digenea: Heterophyidae) from Nagano, Japan, *Parasitol. Int.*, 51, 271, 2002.
38. Kim, C.H., et al., Studies on the *Metagonimus* fluke in the Daecheong reservoir and the upper stream of Geum River, Korea, *Korean J. Parasitol.*, 25, 69, 1987 (in Korean).
39. Kurokawa, T., Studies on the genus *Metagonimus*, especially on the life history of *M. katsuradai* Izumi, 1935 and determination of its first intermediate host, *Tokyo Iji Shinshi*, 3161, 2877, 1939 (in Japanese).
40. Besprozvannykh, V.V., et al., Occurrence of *Metagonimus katsuradai* Izumi, 1935 (Trematoda: Heterophyidae) in the southern Primor'e region. In *Gel'minty I vyzyvaemye imi zabollevaniya*, Yu, L. (Ed.), Mamaev, Vladivostock, Russia, 1987.
41. Ahn, Y.K., *Lateolaborax japonicus*, a role of the second intermediate host of *Metagonimus yokogawai*, *New Med. J.*, 26, 135, 1983 (in Korean).
42. Cho, S.Y., Kang, S.Y., and Ryang, Y.S., Helminthes infections in the small intestine of stray dogs in Ejungbu City, Kyunggi Do, Korea, *Korean J. Parasitol.*, 19, 55, 1981.
43. Seo, B.S., et al., Studies on parasitic helminths of Korea 5. Survey on intestinal trematodes of house rats, *Korean J. Parasitol.*, 19, 131, 1981.
44. Huh, S., Sohn, W.M., and Chai, J.Y., Intestinal parasites of cats purchased in Seoul, *Korean J. Parasitol.*, 31, 371, 1993.
45. Miyamoto, K., Studies on zoonoses in Hokkaido 7. Survey of natural definitive hosts of *Metagonimus yokogawai*, *Jpn. J. Parasitol.*, 34, 371, 1985 (in Japanese).
46. Li, M.H., *Metagonimus yokogawai*: metacercariae survey in fishes and its development to adult worms in various rodents, *Parasitol. Res.*, 112, 1647, 2013.
47. Li, M.H., et al., Infectivity and development of *Metagonimus yokogawai* in experimentally infected domestic ducks, *Vet. Parasitol.*, 168, 45, 2010.
48. Yamaguti, S., Digenetic trematodes of vertebrates, *Systema Helminthum*, Volume 1, Interscience Publishers Inc., New York, 1958.
49. Kamiya, H., et al., Helminths of stray dogs in Sapporo. II, *Jpn. J. Parasitol.*, 24, 41, 1975 (in Japanese).
50. Ahn, Y.K., Intestinal flukes of genus *Metagonimus* and their second intermediate hosts in Kangwon-do, *Korean J. Parasitol.*, 31, 331, 1993 (in Korean).
51. Guk, S.M., et al., Susceptibility of inbred mouse strains to infection with three species of *Metagonimus* prevalent in the Republic of Korea, *J. Parasitol.*, 91, 12, 2005.
52. Kim, C.H., Study on the *Metagonimus* sp. in Gum River basin, Chungchung-nam Do, Korea, *Korean J. Parasitol.*, 18, 215, 1980 (in Korean).
53. Ahn, Y.K. and Ryang, Y.S., Growth, egg-laying and life span of Miyata type of genus *Metagonimus* (Trematoda: Heterophyidae) in the final hosts, *J. Wonju Med. Coll.*, 8, 1, 1995 (in Korean).
54. Shimazu, T., Morphology of metacercariae and adults of *Metagonimus katsuradai* Izumi (Digenea, Heterophyidae) from Shiga, Japan, *Bull. Nat. Sci. Mus. Tokyo Ser. A*, 29, 47, 2003.

55. Kudo, N., et al., Experimental final hosts of *Metagonimus hakubaensis* (Trematoda: Heterophyidae) and their susceptibility to the fluke, *J. Vet. Med. Sci.*, 76, 1651, 2014.
56. Kudo, N., et al., Discovery of natural infection by *Metagonimus hakubaensis* Shimizu, 1999 (Trematoda: Heterophyidae) in Japanese water shrews (*Chimarrogalus platycephala*) *J. Vet. Med. Sci.*, 76, 1531, 2014.
57. Chai, J.Y., et al., High endemicity of *Metagonimus yokogawai* infection along residents of Samcheok-shi, Kangwon-do, *Korean J. Parasitol.*, 38, 33, 2000.
58. Chai, J.Y., et al., Heterophyid trematodes recovered from people residing along the Boseong River, South Korea, *Acta Trop.*, 148, 142, 2015.
59. Lee, J.J., et al., Decrease of *Metagonimus yokogawai* endemicity along the Tamjin River basin, *Korean J. Parasitol.*, 46, 289, 2008.
60. Center for Diseases Prevention and Control, Ministry of Health and Welfare (Korea), Statistics of intestinal helminths nationwide (the 8th Report), Osong, Korea, pp. 1–210, 2013.
61. Cho, S.H., et al., Prevalence of *Metagonimus* metacercariae in sweetfish, *Plecoglossus altivelis*, from eastern and southern coastal areas of Korea, *Korean J. Parasitol.*, 49, 161, 2011.
62. Sohn, W.M. and Chai, J.Y., Infection status with helminthes in feral cats purchased from a market in Busan, Republic of Korea, *Korean J. Parasitol.*, 43, 93, 2005.
63. Shin, S.S., et al., Zoonotic intestinal trematodes in stray cats (*Felis catus*) from riverside areas of the Republic of Korea, *Korean J. Parasitol.*, 53, 209, 2015.
64. Seo, M., et al., *Gymnophalloides seoi* eggs from the stool of a 17th century female mummy found in hadong, Republic of Korea, *J. Parasitol.*, 94, 467, 2008.
65. Shin, D.H., et al., Scanning electron microscope study of ancient parasite eggs recovered from Korean mummies of the Joseon Dynastry, *J. Parasitol.*, 95, 137, 2009.
66. Kagei, N., Epidemiological studies of metagonimiasis in Japan. I. Epidemological survey of metagonimiasis among the Takatsu River, Shimane Prefecture, *Bull. Inst. Public Health, Japan*, 14, 213, 1965 (in Japanese).
67. Yoshimura, S., et al., Epidemological studies of metagonmiasis in Chokai village, Akita Prefecture, *Jpn. J. Parasitol.*, 21, 400, 1972 (in Japanese).
68. Ito, J., et al., On the prevalence of *Metagonimus* sp. among the inhabitants at Hamamatsu basin in Shizuoka Prefecture, Japan, *Jpn. J. Parasitol.*, 40, 274, 1991.
69. Kagei, N. and Oshima, T., Nationwide epidemiological surveys on the metacercariae of *Metagonimus yokogawai* in "ayu", *Plecoglossus altivelis*, in Japan, *Jpn. J. Parasitol.*, 17, 461, 1968 (in Japanese).
70. Kino, H. et al., Geographical distribution of *Metagonimus yokogawai* and *M. miyatai* in Shizuoka Prefecture, Japan, and their site preferences in the sweetfish, *Plecoglossus altivelis*, and hamsters, *Parasitol. Int.*, 55, 201, 2006.
71. Jeon, H.K., et al., Human infections with liver and minute intestinal flukes in Guangxi, Chia: analysis by DNA sequencing, ultrasonography, and immunoaffinity chromatography, *Korean J. Parasitol.*, 50, 391, 2012.
72. Wang, C.R., et al., Prevalence of helminthes in adult dogs in Heilongjiang Province, the People's Republic of China, *Parasitol. Res.*, 99, 627, 2006.
73. Sohn, W.M., et al., Fishborne trematode metacercariae in freshwater fish from Guangxi Zhuang Autonomous Region, China, *Korean J. Parasitol.*, 47, 249, 2009.
74. Chen, Y., et al., Investigation of snails transmitting parasitic diseases in Yunnan Province, *J. Pathog. Biol. (China)*, 4, 211, 2009 (in Chinese).
75. Ondračkova, M., et al., Shoreline distribution and parasitic infection of black-striped pipefish *Syngnathus abaster* Risso, 1827 in the low River Danube, *J. Appl. Ichthyol.*, 28, 590, 2012.
76. Chun, S.K., A study on the metacercaria of *Metagonimus takahashii* and *Exorchis oviformis* from *Carassius carassius*, *Bull. Pusan Fish Coll.*, 3, 31, 1960 (in Korean).
77. Ahn, Y.K. and Ryang, Y.S., Epidemiological studies on *Metagonimus* infection along the Hongcheon river, Kangwon Province, *Korean J. Parasitol.*, 26, 207, 1988 (in Korean).
78. Son, W.Y., Intestinal trematode infections in the villagers in Koje-myon, Kochang-gun, Kyongsangnam-do, Korea, *Korean J. Parasitol.*, 32, 149, 1994.
79. Kim, D.G., et al., Heterophyid metacercarial infections in brackish water fishes from Jinju-man (Bay), Kyongsangnam-do, Korea, *Korean J. Parasitol.*, 44, 7, 2006.
80. Urabe, M., Trematode fauna of prosobranch snails of the genus *Semisulcospira* in Lake Biwa and the connected drainage system, *Parasitol. Int.*, 52, 21, 2003.

81. Park, M.S., et al., Intestinal parasite infections in the inhabitants along the Hantan River, Chorwon-gun, *Korean J. Parasitol.*, 31, 375, 1993.
82. Yu, J.R., Kwon, S.O., and Lee, S.H., Clonorchiasis and metagonimiasis in the inhabitants along Talchongang (River), Chungwon-gun, *Korean J. Parasitol.*, 32, 267, 1994.
83. Chai, J.Y., et al., Chronological observation on intestinal histopathology and intraepithelial lymphocytes in the intestine of rats infected with *Metagonimus yokogawai*, *Korean J. Parasitol.*, 32, 215, 1994 (in Korean).
84. Kang, S.Y., et al., A study on intestinal lesions of experimentally reinfected dogs with *Metagonimus yokogawai*. *Korean J. Parasitol.*, 21, 58, 1983.
85. Chi, J.G., et al., Intestinal histopathology in human metagonimiasis with ultrastructural observations of parasites, *J. Korean Med. Sci.*, 3, 171, 1988.
86. Chai, J.Y., Seo, B.S., and Lee, S.H., Study on *Metagonimus yokogawai* (Katsurada, 1912) in Korea VII. Susceptibility of various strains of mice to *Metagonimus* infection and effect of prednisolone, *Korean J. Parasitol.*, 22, 153, 1984.
87. Yu, J.R., Myong, N., and Chai, J.Y., Expression patterns of proliferating cell nuclear antigen in the small intestine of mice infected with *Metagonimus yokogawai* and *Metagonimus* Miyata type, *Korean J. Parasitol.*, 35, 239, 1997.
88. Africa, C.M., de Leon, W., and Garcia, E.Y., Visceral complications in intestinal heterophyidiasis of man. *Acta Med. Philipp. Monogr. Ser.*, 1, 1–132, 1940.
89. Yamada, S.M., et al., A case of metagonimiasis complicated with multiple intracerebral hemorrhages and diabetes mellitus, *J. Nippon Med. Sch.*, 75, 32–35, 2008.
90. Lee, J.J., et al. Comparative morphology of minute intestinal fluke eggs that can occur in human stools in the Republic of Korea, *Korean J. Parasitol.*, 50, 207–213, 2012.
91. Lee, S.H., et al., Comparative morphology of eggs of heterophyids and *Clonorchis sinensis* causing human infections in Korea, *Korean J. Parasitol.*, 22, 171, 1984.
92. Ahn, Y.K., Epidemiological studies on *Metagonimus yokogawai* infection in Samcheok-gun, Kangwon-do, Korea, *Korean J. Parasitol.*, 22, 161, 1984 (in Korean).
93. Pyo, K.H., et al., Species identification of medically important trematodes in aquatic food samples using PCR-RFLP targeting 18S rRNA, *Foodborne Pathog. Dis.*, 10, 1, 2013.
94. Chai, J.Y., Praziquantel in treatment of trematode and cestode infections, *Infect. Chemother.*, 45, 1, 2013.
95. Chai, J.Y., et al., Effects of gamma-irradiation on the survival and development of *Metagonimus yokogawai* metacercariae in rats, *Korean J. Parasitol.*, 33, 297, 1995.
96. Yokogawa, S., On a new trematode which takes *Plecoglossus altivelis* as second intermediate host-*Metagonimus*, *Tokyo Igakkai Zashii*, 27(9), 685, 1913 (in Japanese).
97. Kobayashi, H., On a new genus of trematode (a preliminary report), *Dobutsugaku Zasshi*, 24(289), 1, 1912 (in Japanese).
98. Muto, S., Studies on human trematodes caused by their intermediate host, freshwater fish in Biwa-Lake, especially on *Metagonimus yokogawai* from Cyprinidae, *Nippon Shokakibyo Gakkai Zasshi*, 16(2), 135, 1917 (in Japanese).
99. Kagei, N. and Kihata, M., On the development of *Metagonimus yokogawai* (Katsurada, 1912) (Heterophyidae, Trematoda) in laboratory animals, *Rep Natl Inst Public Hyg (Japan)*, 19(1), 48, 1970 (in Japanese).
100. Ahn, Y.K., Soh, C.T., and Lee, S.K., Egg laying capacity of *Metagonimus yokogawai*, *Yonsei Rep. Trop. Med.*, 12, 11, 1981.
101. Saito, S., On the synonymy of the trematodes of the genus *Metagonimus*, Report of Parasite Classification and Morphology Seminar (Japan), 2, 1, 1984 (in Japanese).
102. Inatomi, S., et al., The ultrastructure of helminth 2. The body wall of *Metagonimus takahashii* Suzuki, 1930, *Jpn. J. Parasitol.*, 17, 455, 1968 (in Japanese).
103. Lee, J.B., et al., Study on pathology of metagonimiasis in experimentally infected cat intestine, *Korean J. Parasitol.*, 19, 109, 1981.
104. Rho, I.H., et al., Observation on the pathogenesis of villous changes in early phase of experimental metagonimiasis, *Chung-Ang J. Med.*, 9, 67, 1984 (in Korean).
105. Ahn, H., Rim, H.J., and Kim, S.J., Antigenic localities in the tissues of *Metagonimus yokogawai* observed by immunogold labeling method, *Korean J. Parasitol.*, 29, 245, 1991 (in Korean).

106. Rim, H.J., Kim, S.J., and Yang, M.G., Antigenic localities in the tissues of *Metagonimus yokogawai* in the period of growth, *Korean J. Parasitol.*, 30, 309, 1992 (in Korean).
107. Yokogawa, M. and Sano, M., Studies on the intestinal flukes IV. On the development of the worm in the experimentally infected animals with metacercariae of *Metagonimus yokogawai*, *Jpn. J. Parasitol.*, 17, 540, 1968 (in Japanese).
108. Miyamoto, K. and Kutsumi, H., Studies on zoonoses in Hokkaido, Japan 2. On the second intermediate hosts of *Metagonimus yokogawai* in Asahikawa City, Kamikawa District, *Jpn. J. Parasitol.*, 27, 445, 1978 (in Japanese).
109. Rim, H.J., Kim, K.H., and Joo, K.H., Classification of host specificity of *Metagonimus* spp. from Korean freshwater fish, *Korean J. Parasitol.*, 34, 7, 1996.
110. Gushima, M., Studies on the immunity of *Metagonimus yokogawai*, *Igaku Kenkyu*, 13, 1713, 1939 (in Japanese).
111. Komiya, Y., Ito, J., and Yamamoto, S., Studies on *Metagonimus yokogawai* infected in a brackish water fish *Salangichthys microdon* from the lake Kasumigaura, *Jpn. J. Parasitol.*, 7, 7, 1958 (in Japanese).
112. Chun, S.K., A study on *Metagonimus yokogawai* from *Plecoglossus altivelis* in the Miryang River, *Bull. Pusan Fish Coll.*, 3, 24, 1960 (in Korean).
113. Hong, N. T. and Seo, B.S., Study on *Metagonimus yokogawai* (Katsurada, 1912) in Korea, *Korean J. Parasitol.*, 7, 129, 1969.
114. Lee, S.U., et al., Sequence comparisons of 28S ribosomal DNA and mitochondrial cytochrome c oxidase subunit I of *Metagonimus yokogawai*, *M. takahashii* and *M. miyatai*, *Korean J. Parasitol.*, 42, 129, 2004.
115. Lee, S.H., et al., Study on *Metagonimus yokogawai* (Katsurada, 1912) in Korea VII. Electron microscopic observation on the tegumental structure, *Korean J. Parasitol.*, 22, 1, 1984 (in Korean).
116. Chai, J.Y., et al., Surface ultrastructure of *Metagonimus miyatai* metacercariae and adults, *Korean J. Parasitol.*, 36, 217, 1998.
117. Chai, J.Y., et al., Surface ultrastructure of *Metagonimus takahashii* metacercariae and adults, *Korean J. Parasitol.*, 38, 9, 2000.
118. Yu, J.R., Chung, J.S., and Chai, J.Y., Different RAPD patterns between *Metagonimus yokogawai* and *Metagonimus* Miyata type, *Korean J. Parasitol.*, 35, 295, 1997.
119. Kogame, Y., On *Metagonimus yokogawai ovatus* (Kobayashi, 1913) found from *Tribolodon hakonensis* in Daiseiji River, Ishikawa Prefecture, Japan, *Tokyo Iji Shinshi*, (3127), 793, 1939 (in Japanese).
120. Chai, J.Y., et al., Mucosal mast cell responses to experimental *Metagonimus yokogawai* infection in rats, *Korean J. Parasitol.*, 31, 129, 1993.
121. Hong, S.J., et al., Worm recovery and small intestinal lesions of albino rats coinfected with *Fibricola seoulensis* and *Metagonimus yokogawai*, *Korean J. Parasitol.*, 31, 109, 1993.
122. Jang, Y.K., et al., In situ posture of anterior body of *Metagonimus yokogawai* in experimentally infected dog, *Korean J. Parasitol.*, 23, 203, 1985.
123. Cho, S.Y., et al., A preliminary observation on water content of small intestine in *Metagonimus yokogawai* infected dog, *Korean J. Parasitol.*, 23, 175, 1985.
124. Han, E.T., et al., *Metagonimus yokogawai*: a 100-kDa somatic antigen commonly reacting with other trematodes, *Korean J. Parasitol.*, 52, 201, 2014.
125. Ohnishi, Y. and Taufen, M., Increases in the intestinal mucosa of mice infected with *Metagonimus yokogawai*, *Jpn. J. Vet. Sci.*, 46, 885, 1984.
126. Ohnishi, Y., Eosinophil responses in the mice infected with *Metagonimus yokogawai*, *Jpn. J. Parasitol.*, 36, 271, 1987.
127. Hong, S.T., et al., Activities of brush border membrane bound enzymes of the small intestine in *Metagonimus yokogawai* infection in mice, *Korean J. Parasitol.*, 29, 1, 1991.
128. Mehlhorn, H., et al., Ultrastructural investigations on the effects of praziquantel on human trematodes from Asia: *Clonorchis sinensis*, *Metagonimus yokogawai*, *Opisthorchis viverrini*, *Paragonimus westermani* and *Schistosoma japonicum*, *Arzneimit-Forsch/Drug Res.*, 33, 91, 1983.
129. Hamajima, F., et al., Studies on the *in vitro* effects of bithionol and menichlopholan on flukes of *Clonorchis sinensis*, *Metagonimus takahashii* and *Paragonimus miyazakii*, *Int. J. Parasitol.*, 9, 241, 1979.
130. Yasuraoka, K. and Kojima, K., In vitro cultivation of the heterophyid trematode, *Metagonimus yokogawai*, from the metacercaria to adult, *Jpn. J. Parasitol.*, 23, 199, 1970.

47

Opisthorchis viverrini

Thidarut Boonmars

CONTENTS

47.1 Introduction

Opisthorchis viverrini is a trematode (commonly called liver fluke) that is classified in the family Opisthorchiidae, order Opisthorchiata, subclass Digenea, class Trematoda, phylum Platyhelminthes, kingdom Animalia (Figure 47.1). With a high prevalence in Southeast Asia, particularly Thailand, Lao PDR, Cambodia, and Vietnam,[1] this parasite lives in the hepatobiliary tract of humans, causing hepatobiliary diseases such as cholangitis, lithiasis, gallstone, hepatitis, and also cholangiocarcinoma. Humans normally acquire infection with this liver fluke by ingestion of uncooked cyprinoid fish that contains or is contaminated by the infective stage (namely "metacercaria") of *O. viverrini*. After ingestion, the cyst wall of metacercaria is digested by gastric juice and duodenum juice, and the excysted juvenile migrates to the biliary tract and grows to adult in the hepatobiliary tract where it dwells (Figure 47.2). This infection (often called "opisthorchiasis" or "opisthorchiosis") induces many changes in hepatobiliary tract, such as dilatation of the bile ducts with hyperplasia, desquamation, proliferation of the bile duct epithelial cells, glandular formation, and fibrous connective tissue infiltration of the walls, and other complication of hepatobiliary diseases (e.g., cholangitis, cystitis, cholecystitis including cholangiocarcinoma).[2–4]

Besides *O. viverrini*, which mainly occurs in Thailand, Lao PDR, Vietnam, and Cambodia,[1] *Clonorchis sinensis* and *Opisthorchis felineus* within the family Opisthorchiidae also cause liver fluke infections in humans, dogs, and cats. While *C. sinensis* infection appears in China, Korea, Japan, and Vietnam,[5] *O. felineus* infection is present in Russia and European countries.[6]

47.2 Morphological Characteristics of *O. viverrini*

The adult worm of *O. viverrini* is transparent and flat; its anterior part is small and slender; and its size is about 5.4–10.2 mm × 0.8–1.9 mm. Oral sucker and ventral sucker are located at the anterior part of the body. Two testes (deep lobes) are tandemly located at the posterior part. The excretory bladder is S-shaped and located between two testes. Vitelline glands are located at both sides of the trunk with

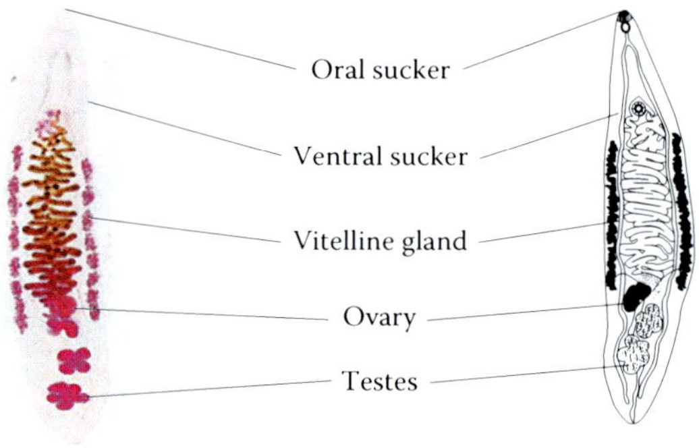

FIGURE 47.1 Adult worm of *O. viverrini*.

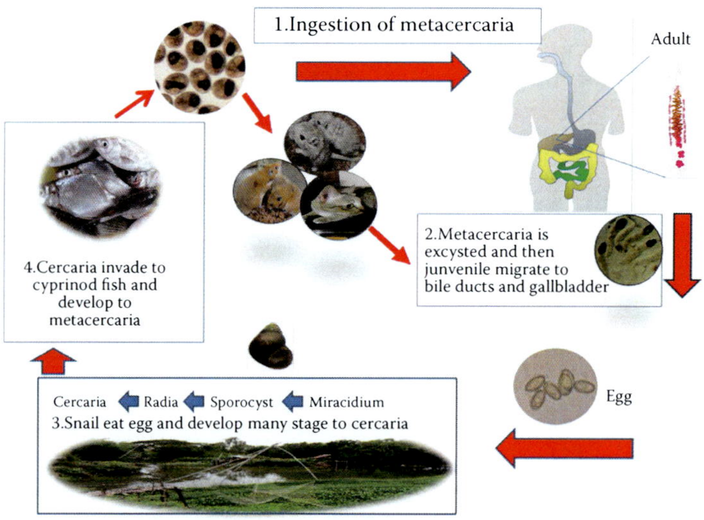

FIGURE 47.2 The life cycle of *O. viverrini*.

columns (Figure 47.1). *In vivo* study showed that *O. viverrini* in different hosts show varied shape and size.[7] Rat and mouse are resistant to *O. viverrini,* whereas hamster and gerbil are susceptible to this parasite. In general, worm recovery from hamster is higher than that from the gerbil, and the recovered worm from gerbil is larger than that from hamster. The worm size correlates to the size of the reproductive organs (testes, uterus, and vitelline gland), eggs per worm, and worms per gram of feces as shown in Figure 47.3 and Table 47.1.[8]

47.3 Animal Models for *O. viverrini*

A study of liver fluke infection in animals showed that many animals, including dogs, rabbits, hamsters, gerbils, and mice, display varied susceptibility to *O. viverrini* infection and pathological changes.[7–9] One hundred percent worm recovery was found in hamster and gerbil, and the highest number of worm recovery was observed from the hamster. Severe pathology was found in gerbil, as evidenced by cirrhosis and abdominal edema.[9] Therefore, hamsters are the appropriate experimental animals for studying *O. viverrini* and cholangiocarcinoma.

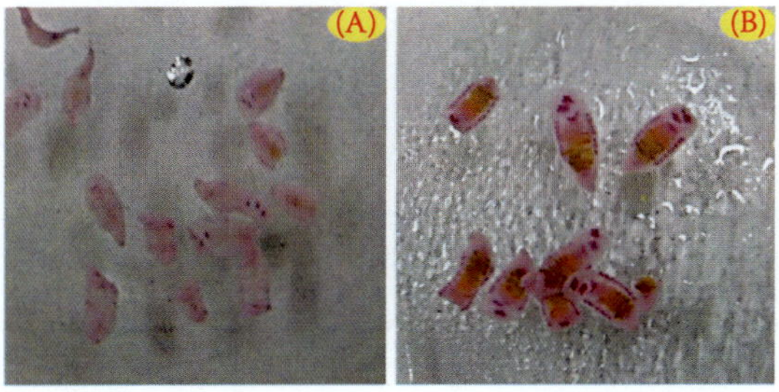

FIGURE 47.3 The representative adult of *O. viverrini* from hamster (A) and gerbil (B) at 1.5 month.

TABLE 47.1
The *O. viverrini* Infection in Animal Models

Animal Model	Number of Metacercariae	Size of Adult Worm (Length × Height) in mm	% Infection	References
Hamster	50	1.43 ± 0.08 × 4.4 ± 0.3	28 ± 9.8	7
Gerbil	50	1.81 ± 0.06 × 5 ± 0.12	60.5 ± 14.44	
Mouse	50	—	0	8
Rat	50	—	0	

47.3.1 Establishment of *O. viverrini* Infection in Animal Models

Normally, *O. viverrini* metacercaria from infected freshwater cyprinoid fish is needed for animal infection. Typically, fish is blended into a small size and digested with the pepsin-HCl solution at 37°C for 1 h. The digested solution is filtered through a sieve and then washed with normal saline multiple times until the sediment is easy to observe under stereomicroscope, and then 50 *O. viverrini* metcercariae are gastric-intubated (Figure 47.4).

47.3.2 Gross Pathology of Hamster and Gerbil Liver

Based on the uninfected control, the color of the infected liver appears reddish-brown, similar to uninfected control in the early stages of infection. In a chronic infection, the liver color is pale, and the infected bile duct is much larger than the uninfected one. Moreover, bile duct size and thickness increase with the time of infection. Normally, the color of the bile ducts outside is colorless, but the infected bile duct and gallbladder are creamy. As mentioned earlier, host differences have an impact on parasite development and pathology.[8,10,11] Liver flukes recovered from infected gerbils are larger than those recovered from infected hamster, correlating with the parasite reproductive development and the liver pathology. Large parasites cause blockage of the bile duct, leading to duct enlargement. The inflammation at the liver tissue quickly leads to cirrhosis (Figure 47.5). The bile flow of the bile duct is interrupted and easy to break, resulting in egg granulomas in the liver tissue (Figure 47.6).

47.3.3 Pathology of Liver Fluke Infection in Hamster and Gerbil

Pathology within the liver is correlated to the gross pathology. The liver of uninfected hamster does not contain inflammatory cell aggregation surrounding the hepatic bile ducts, which are different from those of the infected hamster. The peak of inflammatory cell aggregation surrounding the hepatic bile ducts is observed during 14–30 days postinfection and then gradually decreases at 2, 3, and 6 months postinfection. During 14–30 days postinfection, neutrophils and eosinophils in the liver tissue are increased. During chronic infection, monocytes are increased. Moreover, the fibrosis surrounding the hepatic

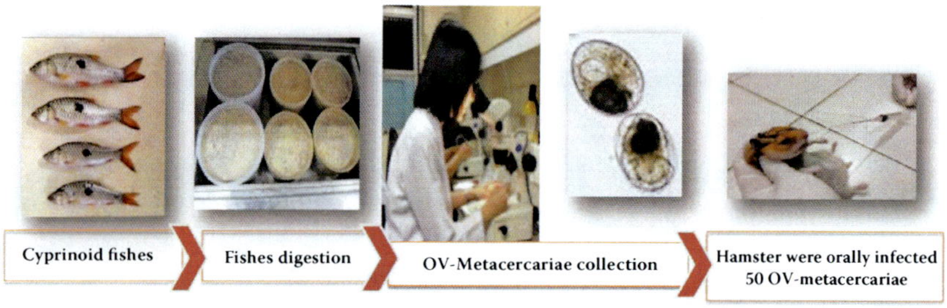

FIGURE 47.4 Metacercariae preparation and animal infection. OV, *O. viverrini*.

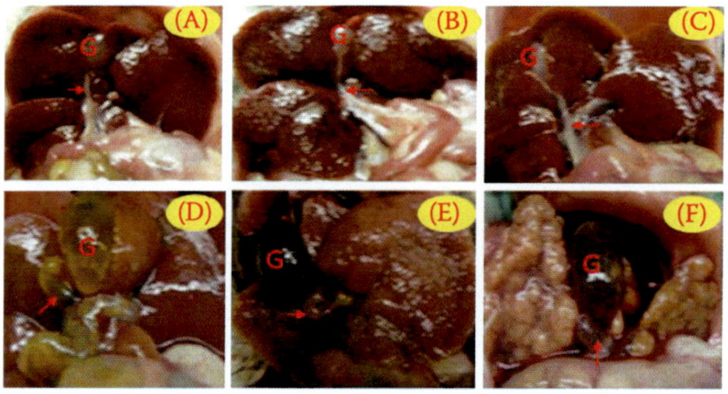

FIGURE 47.5 Representative gross liver pathology of the infected hamster (A–C) and gerbil (D–F) at 30 days (A,D), 60 days (B,E), and 90 days (C,F). Gall bladder (G); common bile duct (arrow). Note: Liver obstruction and cirrhosis could be observed at all time points.

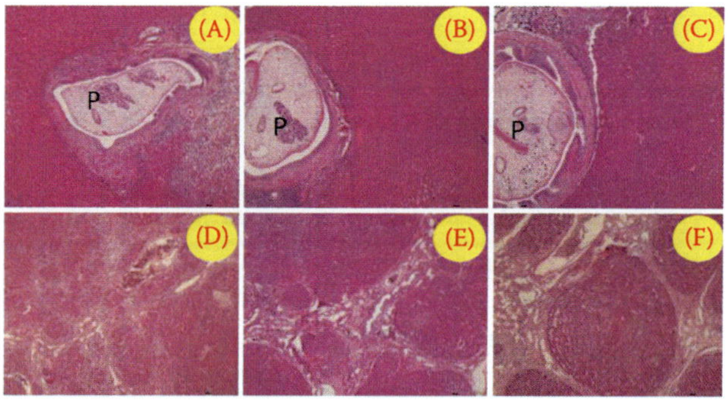

FIGURE 47.6 Representative liver pathology of the infected hamster (A–C) and gerbil (D–F) at 30 days (A,D), 60 days (B,E), and 90 days (C,F). Parasite (P). Note: Fibrosis could be observed in infected gerbil at all time points.

bile ducts appears and thickens in line with the duration of the infection, which corresponds to gross pathological bile ducts (Figure 47.7).

47.3.4 The Level of White Blood Cells in the Liver Fluke Infection in Hamsters

The levels of white blood cells in infected hamster in the first period (<2 months postinfection) are significantly increased in comparison with that in the uninfected hamster (Table 47.2). Among various types of white blood cells, lymphocytes are higher in all phases of the infection. Neutrophils significantly increase at 6 months postinfection. Eosinophils peak at the first period of ~1 month and decrease at

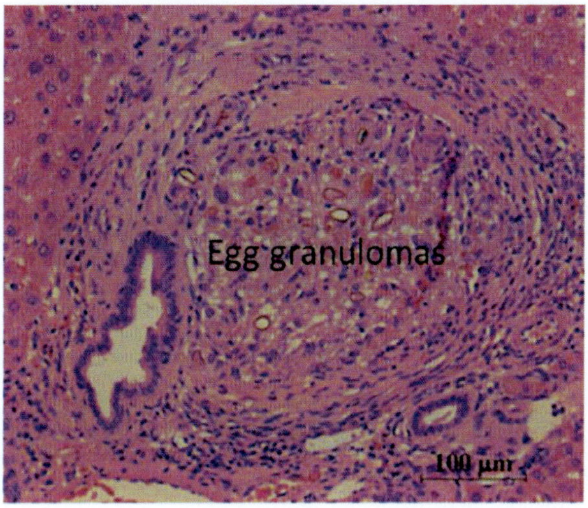

FIGURE 47.7 Egg granulomas.

TABLE 47.2

Blood Cell Counts in Uninfected/Infected Hamster

		Estimated White Blood Cell Count (Months Postinfection)			
		1	2	3	6
	Experimental Group	Mean ± SEM	Mean ± SEM	Mean ± SEM	Mean ± SEM
Total white blood cell count	Normal	2040.0 ± 160.0	2012.5 ± 173.7	1800 ± 115.5	1266.7 ± 176.4
	OV	2600.0 ± 739.4	2960.0 ± 865.8	2850.0 ± 895.8	2250.0 ± 639.7
Neutrophil	Normal	32.0 ± 3.3	26.7 ± 2.1	25.7 ± 5.4	23.9 ± 2.0
	OV	42.2 ± 4.7	20.9 ± 1.8	31.5 ± 6.7	33.5 ± 2.8*
Monocyte	Normal	2.8 ± 1.0	4.5 ± 1.0	3.8 ± 1.9	0.0 ± 0.0
	OV	17.3 ± 1.6*	21.0 ± 3.5*	14.0 ± 2.4*	0.5 ± 0.5
Eosinophil	Normal	0.0 ± 0.0	0.0 ± 0.0	0.0 ± 0.0	0.0 ± 0.0
	OV	3.8 ± 1.6*	1.2 ± 0.5*	1.0 ± 0.6	0.0 ± 0.0
Lymphocyte	Normal	65.2 ± 4.0	68.8 ± 1.6	72.0 ± 4.0	76.1 ± 2.0
	OV	35.0 ± 5.0*	56.9 ± 4.4*	54.0 ± 5.1*	66.0 ± 2.9*

*$p = 0.05$. OV, *O. viverrini*.

2 months postinfection. Monocytes are significantly higher during the first 1–3 months and significantly decrease at 6 months postinfection.

47.3.5 Opisthorchiasis-Associated Cholangiocarcinoma

Cholangiocarcinoma (CCA) is defined as a carcinoma arising from bile duct epithelium within the liver, from the large hilar duct and the extrahepatic duct. CCA may arise anywhere in the biliary tree, but the most frequent sites are the ampulla of Vater, lower end of the common bile duct, hepatic duct, and the junction of hepatic duct to the common bile duct. CCA is usually small, extending for 1–2 cm along the duct, thickening the of the duct.[12] CCA is a relatively rare cancer in the western world, with the incidence of 1–2 cases per 100,000 people[13] but rising worldwide over the past several decades. Recent epidemiological studies revealed that the incidence of CCA has increased in the United States, the United Kingdom, and Australia.[14,15] The prevalence of CCA is higher in Thailand, Loa PDR, and Cambodia where opisthorchiasis is endemic.[3,16] CCA is a crucial public health problem in the Northeastern part of Thailand. The incidence of CCA is approximately >50% of all liver cancer in Thailand and ~90% of all liver cancer in the northeastern part of Thailand.[17] This incidence rate is the highest in the world.[18] It is well known that both liver flukes *O. viverrini* and *C. sinensis* have been classified as class I carcinogens by the International Agency for Research on Cancer owing to their association with CCA and hepatocarcinoma in humans.[3,19,20]

The risk factors for CCA in northeastern part of Thailand was a clear association with past or present infection with *O. viverrini*, as indicated by raised serum antibodies, and at least two-thirds of cases can be attributed to this cause, and males may be at a higher risk than females.[21,22] CCA patients who have been infected with a liver fluke develop this cancer, and patients who had been infected and treated for opisthorchiasis have increased risk of developing CCA than the noninfected population.[23] Several experiments and clinical reports revealed the correlation between *O. viverrini* and CCA development, including clinical studies[3,19,20,21,24] and experimental models.[25–28]

The mechanism of *O. viverrini*-associated CCA relates to chronic inflammation resulting in a combination of mechanical damage, parasite secretions, and immunopathology.[28] The primary pathological change in epithelial desquamation may cause mechanical irritation by the liver fluke and/or produce metabolic products from the parasite. The mechanical damage is caused by oral sucker and ventral sucker of fluke during activities of feeding and migrating in the bile duct. The fluke eggs are trapped in the lumen of the bile duct, leading to ulcerates, granulomatous inflammation, and cholelithiasis.[29] The excretory-secretory (ES) products released by *O. viverrini* cause cell proliferation.[30,31] After infection for 2 weeks, the bile duct epithelium hyperplasia and periductal inflammatory cells, predominantly eosinophils and lymphocytes, are observed. Furthermore, immunopathological processes lead to the long-standing hepatobiliary damage. Host immune response can induce the inflammatory cell infiltrations surrounding the hepatic bile ducts for killing the parasite by reactive oxygen species (ROS) and reactive nitrogen species (RNS).[32] ROS not only destroy parasite, but also destroy host tissue and DNA leading to DNA fragmentation, which is well known during apoptosis. Moreover, RNS, which are produced during inflammation, may play a role in initiation and subsequent modulation stage of CCA development through DNA mutation leading to cancer development.[33,34] In *O. viverrini*-associated CCA, Sripa and Kaewkes found *O. viverrini* antigens located in epithelium bile ducts in hamster model, and Thuwajit et al.[30,31] reported that the ES products of *O. viverrini* increased NIH3T3 cell proliferations, which may enhance bile duct proliferation and account for nitrative and oxidative DNA damage in hamsters infected with *O. viverrini*.[34] The development of cholangiocarcinogenesis involves many changes of cancer-relevant genes. Several studies on the molecular mechanisms of *O. viverrini*-associated CCA have highlighted the potential influences of KRAS and TP53 mutations in a large number of biological processes, including hamster CCA,[35] interleukin 6 (IL-6),[36] transforming growth factor-β (TGF-β), IL-8, tumor necrosis factor-α (TNF-α) and platelet-derived growth factor (PDGF),[37] proteomic profiling of CCA,[38] Rb1 and related genes in CCA,[39] comparative protein expression profiles (MUC5AC, Akt2, CK8, annexin II, and VEGF A) of hilar and peripheral hepatic CCA,[40] MMP-7 and MMP-9 in CCA,[41] p38delta/MAPK13 as a diagnostic marker for CCA,[42] and c-ski and related genes in CCA.[43] However, infection with *O. viverrini* alone may not produce CCA, which requires a combination of factors such as *O. viverrini* infection, nitrosamine, host immune response, telomerase, alcohol, and hepatobiliary disease.

47.4 Conclusion

Opisthorchis viverrini is a liver fluke living in the hepatobiliary tract that causes opisthorchiasis in human and animal reservoir host. The highest prevalence of this disease was found in Southeast Asia, especially in Thailand, Cambodia, and Lao PDR. Chronic infection with *O. viverrini* enhances the hepatobiliary diseases including CCA, which is a rare cancer but a highly fatal disease in the world. However, the highest prevalence of CCA is in northeastern Thailand, which correlates to the *O. viverrini* infection and other combination factors. To date, one effective anthelmintic drug is praziquantel, but the CCA remains high. Therefore, use of animal models for the investigation of *O. viverrini* pathogenesis is important. Syrian hamster is the best animal model for the study of *O. viverrini* infection and CCA.

REFERENCES

1. WHO. Control of food borne trematode infections. *WHO Tech Rep Ser* 849, 1–157 (1995).
2. Srivatanakul, P. et al. *Opisthorchis viverrini* infestation and endogenous nitrosamines as risk factors for cholangiocarcinoma in Thailand. *Int J Cancer* 48, 821–5 (1991).
3. IARC. Infection with liver flukes (*Opisthorchis viverrini*, *Opisthorchis felineus* and *Clonorchis sinensis*). *IARC Monogr Eval Carcinog Risks Hum* 61, 121–75 (1994).
4. Sriamporn, S., Parkin, D.M., Pisani, P., Suwanrungruang, K. & Pengsaa, P. Behavioural risk factors for cervical cancer from a prospective study in Khon Kaen, Northeast Thailand. *Cancer Detect Prev* 28, 334–9 (2004).
5. Rim, H.J. et al. Fishborne trematode metacercariae detected in freshwater fish from Vientiane Municipality and Savannakhet Province, Lao PDR. *Korean J Parasitol* 46, 253–60 (2008).
6. Pozio, E., Armignacco, O., Ferri, F. & Gomez Morales, M.A. *Opisthorchis felineus*, an emerging infection in Italy and its implication for the European Union. *Acta Trop* 126, 54–62 (2013).
7. Boonmars, T., Boonjaraspinyo, S. & Kaewsamut, B. Animal models for *Opisthorchis viverrini* infection. *Parasitol Res* 104, 701–3 (2009).
8. Wonkchalee, O. et al. Comparative studies on animal models for *Opisthorchis viverrini* infection: host interaction through susceptibility and pathology. *Parasitol Res* 110, 1213–23 (2012).
9. Wonkchalee, O. et al. *Opisthorchis viverrini* infection causes liver and biliary cirrhosis in gerbils. *Parasitol Res* 109, 545–51 (2011).
10. Juasook, A. et al. Immunosuppressive prednisolone enhances early cholangiocarcinoma in Syrian hamsters with liver fluke infection and administration of *N*-nitrosodimethylamine. *Pathol Oncol Res* 19, 55–62 (2013).
11. Sriraj, P. et al. Does a combination of opisthorchiasis and ethyl alcohol consumption enhance early cholangiofibrosis, the risk of cholangiocarcinoma? *Parasitol Res* 112, 2971–81 (2013).
12. Gatto, M. & Alvaro, D. Cholangiocarcinoma: risk factors and clinical presentation. *Eur Rev Med Pharmacol Sci* 14, 363–7 (2010).
13. Landis, S.H., Murray, T., Bolden, S. & Wingo, P.A. Cancer statistics, 1998. *CA Cancer J Clin* 48, 6–29 (1998).
14. Patel, T. Increasing incidence and mortality of primary intrahepatic cholangiocarcinoma in the United States. *Hepatology* 33, 1353–7 (2001).
15. Gores, G.J. A spotlight on cholangiocarcinoma. *Gastroenterology* 125, 1536–8 (2003).
16. Sithithaworn, P. & Haswell-Elkins, M. Epidemiology of *Opisthorchis viverrini*. *Acta Trop* 88, 187–94 (2003).
17. Srivatanakul, P., Sriplung, H. & Deerasamee, S. Epidemiology of liver cancer: an overview. *Asian Pac J Cancer Prev* 5, 118–25 (2004).
18. Vatanasapt, V. et al. Cancer incidence in Thailand, 1988–1991. *Cancer Epidemiol Biomarkers Prev* 4, 475–83 (1995).
19. de Martel, C., Plummer, M. & Franceschi, S. Cholangiocarcinoma: descriptive epidemiology and risk factors. *Gastroenterol Clin Biol* 34, 173–80 (2010).
20. Fried, B., Reddy, A. & Mayer, D. Helminths in human carcinogenesis. *Cancer Lett* 305, 239–49 (2011).
21. Parkin, D.M. et al. Liver cancer in Thailand. I. A case-control study of cholangiocarcinoma. *Int J Cancer* 48, 323–8 (1991).
22. Srivatanakul, P. et al. Endogenous nitrosamines and liver fluke as risk factors for cholangiocarcinoma in Thailand. *IARC Sci Publ* 105, 88–95 (1991).
23. Haswell-Elkins, M.R., Satarug, S. & Elkins, D.B. *Opisthorchis viverrini* infection in northeast Thailand and its relationship to cholangiocarcinoma. *J Gastroenterol Hepatol* 7, 538–48 (1992).
24. Kurathong, S. et al. *Opisthorchis viverrini* infection and cholangiocarcinoma. A prospective, case-controlled study. *Gastroenterology* 89, 151–6 (1985).

25. Thamavit, W., Kongkanuntn, R., Tiwawech, D. & Moore, M.A. Level of *Opisthorchis* infestation and carcinogen dose-dependence of cholangiocarcinoma induction in Syrian golden hamsters. *Virchows Arch B Cell Pathol Incl Mol Pathol* 54, 52–8 (1987).
26. Thamavit, W. et al. Promotion of cholangiocarcinogenesis in the hamster liver by bile duct ligation after dimethylnitrosamine initiation. *Carcinogenesis* 14, 2415–7 (1993).
27. Thamavit, W., Pairojkul, C., Tiwawech, D., Shirai, T. & Ito, N. Strong promoting effect of *Opisthorchis viverrini* infection on dimethylnitrosamine-initiated hamster liver. *Cancer Lett* 78, 121–5 (1994).
28. Sripa, B. et al. Liver fluke induces cholangiocarcinoma. *PLoS Med* 4, e201 (2007).
29. Sripa, B. Pathobiology of opisthorchiasis: an update. *Acta Trop* 88, 209–20 (2003).
30. Thuwajit, C. et al. Increased cell proliferation of mouse fibroblast NIH-3T3 in vitro induced by excretory/secretory product(s) from *Opisthorchis viverrini*. *Parasitology* 129, 455–64 (2004).
31. Thuwajit, C. et al. Gene expression profiling defined pathways correlated with fibroblast cell proliferation induced by *Opisthorchis viverrini* excretory/secretory product. *World J Gastroenterol* 12, 3585–92 (2006).
32. Coussens, L.M. & Werb, Z. Inflammation and cancer. *Nature* 420, 860–7 (2002).
33. Ohshima, H. et al. Increased nitrosamine and nitrate biosynthesis mediated by nitric oxide synthase induced in hamsters infected with liver fluke (*Opisthorchis viverrini*). *Carcinogenesis* 15, 271–5 (1994).
34. Pinlaor, S. et al. Mechanism of NO-mediated oxidative and nitrative DNA damage in hamsters infected with *Opisthorchis viverrini*: a model of inflammation-mediated carcinogenesis. *Nitric Oxide* 11, 175–83 (2004).
35. Tangkawattana, S., Kaewkes, S., Pairojkul, C., Tangkawattana, P. & Sripa, B. Mutations of KRAS and TP53 in a minor proportion of *Opisthorchis viverrini*-associated cholangiocarcinomas in a hamster model. *Asian Pac J Cancer Prev* 9, 101–6 (2008).
36. Sripa, B. et al. Advanced periductal fibrosis from infection with the carcinogenic human liver fluke *Opisthorchis viverrini* correlates with elevated levels of interleukin-6. *Hepatology* 50, 1273–81 (2009).
37. Berthiaume, E.P. & Wands, J. The molecular pathogenesis of cholangiocarcinoma. *Semin Liver Dis* 24, 127–37 (2004).
38. Scarlett, C.J. et al. Proteomic classification of pancreatic adenocarcinoma tissue using protein chip technology. *Gastroenterology* 130, 1670–8 (2006).
39. Boonmars, T. et al. Alterations of gene expression of RB pathway in *Opisthorchis viverrini* infection-induced cholangiocarcinoma. *Parasitol Res* 105, 1273–81 (2009).
40. Guedj, N. et al. Comparative protein expression profiles of hilar and peripheral hepatic cholangiocarcinomas. *J Hepatol* 51, 93–101 (2009).
41. Leelawat, K., Sakchinabut, S., Narong, S. & Wannaprasert, J. Detection of serum MMP-7 and MMP-9 in cholangiocarcinoma patients: evaluation of diagnostic accuracy. *BMC Gastroenterol* 9, 30 (2009).
42. Tan, F.L. et al. p38delta/MAPK13 as a diagnostic marker for cholangiocarcinoma and its involvement in cell motility and invasion. *Int J Cancer* 126, 2353–61 (2010).
43. Boonmars, T. et al. Involvement of *c-Ski* oncoprotein in carcinogenesis of cholangiocacinoma induced by *Opisthorchis viverrini* and *N*-nitrosodimethylamine. *Pathol Oncol Res* 17, 219–27 (2011).

48

Paragonimus

Dongyou Liu

CONTENTS

48.1 Introduction

The genus *Paragonimus* covers a large group of flatworms that are associated with parenchymal and/or pleural lung infections (thus collectively known as lung flukes). Although Naterer made initial observation of lung flukes in 1828, it was Coenraad Kerbert who first described *Paragonimus westermani* (originally named *Distoma westermani*) infection in a Bengal tiger suffering from pneumonia at the Amsterdam zoological gardens in 1878. In the subsequent year, B. S. Ringer identified *P. westermani* in the lungs of a human patient who died of an aortic aneurysm, thus confirming the first case of human paragonimiasis. Following the establishment of the genus *Paragonimus* ("para"—Greek word meaning "on the side of" and "gonimos"—Greek word meaning "gonads or genitalia") by Max Braun in 1899, in which *P. westermani* was the type species, much work has been done toward the identification of many other species in this genus, including the independent, near-simultaneous discovery of the only North American *Paragonimus* species (*P. kellicotti*) by Henry Ward and D. S. Kellicott in 1894 [1,2].

48.1.1 Classification and Morphology

48.1.1.1 Classification

Taxonomically, the genus *Paragonimus* belongs to the family Paragonimidae, suborder Troglotremata, order Plagiorchiida, class Trematodea, phylum Platyhelminthes (i.e., flatworms), kingdom Animalia.

Currently, the genus *Paragonimus* comprises more than 50 species, including *Paragonimus africanus*, *Paragonimus bangkokensis*, *Paragonimus caliensis*, *Paragonimus compactus*, *Paragonimus ecuadoriensis*, *Paragonimus harinasutai*, *Paragonimus heterotremus*, *Paragonimus hokuoensis*, *Paragonimus hueitugensis*, *Paragonimus iloktsuenensis*, *Paragonimus kellicotti*, *Paragonimus macrorchis*, *Paragonimus mexicanus*, *Paragonimus miyazakii*, *Paragonimus ohirai*, *Paragonimus paishuihoensis*, *Paragonimus peruvianus*, *Paragonimus proliferus*, *Paragonimus pseudoheterotremus*, *Paragonimus pulmonalis*, *Paragonimus sadoensis*, *Paragonimus siamensis*, *Paragonimus skrjabini*, *Paragonimus uterobilateralis*, *Paragonimus vietnamensis*, and *Paragonimus westermani*, in addition to a number of unassigned species [3–6]. Of these, predominant human-infecting *Paragonimus* species (and their geographic distribution) are *P. westermani* (Asia, India, Philippines, and New Guinea), *P. miyazakii* (Japan), *P. skrjabini* (China, Southeast Asia), *P. heterotremus* (Thailand, China, Southeast Asia), *P. hueitungensis* (China), *P. uterobilateralis* (West Africa), *P. africanus* (West Africa), *P. kellicotti* (North America), and *P. mexicanus* (Central America, South America) [1,2,7–16].

48.1.1.2 Morphology

Paragonimus adult worm has an oval- shaped body (of 7.5–12 mm in length, 4–6 mm in width, and 3.5–5 mm in thickness), composed of smooth muscle that is protected by a thick tegument, which is in turn covered with variably scattered spines. Both the oral terminal sucker (0.19 mm in diameter) and ventral sucker (or acetabulum, 0.12 mm in diameter) are round and muscular. The lobed ovary is located behind the acetabulum and anterior to the two branched testes. Between the ovary and the acetabulum are the seminal receptacle, the tightly coiled uterus, and the thick-walled terminal part (metraterm). The digestive system comprises a truncated pharynx and esophagus that bifurcates early into paired ceca [17].

Paragonimus eggs (80–120 × 45–70 μm) are ovoid, thick-shelled (measuring 2–4 μm in thickness), golden brown, and operculate (showing a shoulder- or ridge-like structure). The eggs are produced by self-fertilization (hermaphrodites), and more often by cross-fertilization in lung or pleura-encysted pairs. Following release into bronchioles, the eggs are coughed up along with sputum, and then either expectorated or swallowed for later discharge in feces.

Morphological features useful for the identification of *Paragonimus* adult worm include size and shape, the patterns of lobation of the ovary and testes, and the appearance of the tegumental spines. For example, the *P. kellicotti* group has tegumental spines arranged in groups, whereas the *P. africanus* group shows individual spines. Microscopic observation of the thick shell-walled, unembryonated, operculated eggs within sputum or feces is a clear indication of paragonimiasis, since other helminthic parasites that produce operculate eggs (e.g., trematodes *Clonorchis sinensis*, *Opisthorchis* species, *Fasciola hepatica*, and *Fasciolopsis buski*, as well as cestode *Diphyllobothrium latum*) do not cause pulmonary infections. In addition, the morphological features of *Paragonimus* adult, cercariae, metacercariae, and metecercarial cyst may also be exploited for differentiation from other trematodes [17].

48.1.2 Life Cycle and Epidemiology

Paragonimus species exhibit a circuitous life cycle that involves a mammal as definitive host (in which the adult resides and undergoes sexual reproduction) and aquatic snails and crustaceans as specific and sequential intermediate hosts.

The life cycle begins when fertilized, operculate eggs produced by sexually competent adult worms are discharged in sputum or feces from mammalian definitive host, enter fresh or brackish water, and hatch (embryonate) into ciliated miracidia. The miracidia then invade the soft parts of freshwater snails (the first intermediate host), and undergo two asexual reproductive cycles (including the formation of sac-like sporocysts in the hemocoel of the snail and two generations of rediae in the lymphatic system of the snail) to become infective cercariae (with anterior stylets and short tails). The infective cercariae released by snails into freshwater (or infected snails eaten by freshwater crustaceans) penetrate the gills and other soft tissue sites on the exoskeletons of crustaceans (the second intermediate host) and develop into metacercariae [18].

Following consumption of raw, undercooked, or alcohol-pickled freshwater crustaceans (crabs or crayfish or undercooked infected tissues from a paratenic/nonpermissive mammalian host such as rats, pigs, and possibly water birds), the permissive mammalian definitive host acquires the metacercariae, which move from the intestine into the abdomen and the lungs [19]. In the lungs, the metacercariae mature into adult worm, encyst (being spherical to ovoid in shape and up to 2 cm in diameter, each containing two or more adult flukes or a diploid or triploid variant as in the exceptional case of *P. westermani*), and cross-fertilize each other. With the eventual rupture of the cyst in the lungs, the eggs are released and coughed up or swallowed and excreted in the feces for the next cycle of infection.

The first molluscan intermediate hosts of *Paragonimus* species include 54 species of freshwater snails from the families Pleuroceridae and Thiaridae of the superfamilies Cerithioidea and Rissooidea; the second crustacean intermediate hosts consist of freshwater crabs or crayfish (belonging to 53 species of 21 genera); and the definitive hosts are omnivorous and carnivorous crustacean-eating mammals such as dogs, cats, tigers, and humans [20]. In the case of *P. kellicotti*, besides domestic dogs and cats, wild animals [including skunks (*Melphitis mephitis*), red foxes (*Vulpes vulpes*), coyotes (*Canis latrans*), mink (*Mustela vison*), and bobcats (*Felis rufus*)] also function as competent definitive hosts.

Reflecting the natural distribution of permissive definitive and intermediate hosts that support infection and the customs of consuming raw or undercooked seafood or wild boar or using raw seafood preparation as folk medicine, *Paragonimus* species have an endemic presence in East and South Asia (e.g., China, Korea, Japan, Laos, Thailand, Vietnam, Malaysia, and the Philippines), sub-Saharan Africa (e.g., Cameroon), and in the Americas (from Peru to Canada). It is of interest to note that *P. westermani* and other Asian species have been traditionally referred to as Asian lung flukes, and *P. kellicotti* and Latin American species have been called American lung flukes [11,21]. Worldwide, about 293 million people are at risk of *Paragonimus* infection, and nearly 20 million people (of both sexes and all age) are being infected [2].

48.1.3 Clinical Features

In comparison with other taxonomically related flukes in the class Trematoda (intestinal fluks, liver flukes, and blood flukes or schistosomes), lung flukes (formerly oriental lung fluke, pulmonary distomatosis, and benign endemic hemoptysis) of the genus *Paragonimus* demonstrate a unique tropism in their definitive hosts for encysting within the parenchyma or on the pleural surfaces of the lungs, and occasionally in ectopic sites such as the brain (cerebral paragonimiasis), mesentery and visceral organs, abdominal wall, muscles, and subcutaneous tissues (trematode larva migrans). Thus, human paragonimiasis may be categorized as acute paragonimiasis (early-stage disease), chronic pleuropulmonary paragonimiasis (late-stage disease), and ectopic paragonimiasis (with the parasite in a location other than the lungs).

48.1.3.1 Acute Paragonimiasis (Early-Stage Disease)

Following the ingestion of the infective metacercariae and subsequent migration to the pleural space, the acute paragonimiasis may be asymptomatic or subclinical in about 20% of patients, while presenting with fever, abdominal and chest pain, diarrhea, fatigue, and urticaria in other patients (often with heavy worm burdens) with the possibility of eosinophilia (as high as $500–80{,}000 \times 10^9$ eosinophils/mm^3 compared to the normal reference range of up to 0.45×10^9/mm^3).

The differential diagnosis of acute paragonimiasis with fever, abdominal pain, and diarrhea may include viral and bacterial causes of acute gastroenteritis (e.g., caliciviruses, *Salmonella*, *Shigella*, *Vibrio*, and *Campylobacter* species, as well as *Giardia* parasite). The occurrence of eosinophilia as seen in acute paragonimiasis may also be attributed to other migrating parasites (e.g., *Ascaris*) [22].

48.1.3.2 Chronic Pleuropulmonary Paragonimiasis (Late-Stage Disease)

The chronic pleuropulmonary paragonimiasis reflects host responses to *Paragonimus* worms that migrate to their final destination in the pulmonary parenchyma, encyst, and produce fertilized eggs. Depending on the precise location and number of the parasitic cysts and associated sequelae, cough, blood-tinged sputum (with hemorrhage into the airspace resulting in hemoptysis, and into the pleural space leading to hemothorax), dyspnea, pleuritic chest pain, fever, crepitation, weakness, rhonchi, hoarseness, and breathlessness may be evident (implying bronchiectasis, interstitial pneumonitis, or bronchopneumonia).

The radiological findings based on computed tomography (CT) or magnetic resonance imaging (MRI) scans in patients with pleuropulmonary paragonimiasis include consolidation, pleural effusions (due to worm penetration into the pleural cavity and the lung), cystic lesions, linear streaks (due to worm penetration and migration of the lung parenchyma, resulting in burrow tracts of 2–4 × 2–7 mm), nodules (due to cyst formation), pleural thickening, ring shadow, calcified lesions, and adenopathy [23,24].

For differential diagnosis, chronic pleuropulmonary paragonimiasis showing cough and hemoptysis in conjunction with cavitary changes (cyst production) may be confused with tuberculosis. In this case, use of acid-fast stains for respiratory specimens and *Paragonimus* serological testing are valuable. Other causes of lung disease that may be misdiagnosed with pleuropulmonary paragonimiasis include bacterial pneumonia, lung abscess, and echinococcosis. In addition, the presence of eosinophilia and an elevated IgE level observable in pleuropulmonary paragonimiasis may also be found in other parasitic infections (e.g., strongyloidiasis, ascariasis, toxocariasis, and ancylostomiasis), fungal infections of the lungs (e.g., coccidioidomycosis and bronchopulmonary aspergillosis), and noninfectious diseases (e.g., Churg–Strauss syndrome, which is an autoimmune vasculitis predominantly involving blood vessels of the lungs).

48.1.3.3 Ectopic Paragonimiasis

Besides the pleurae and lungs, *Paragonimus* parasites may migrate to other locations (e.g., the brain, skin, breast, adrenal gland, heart, mediastinum, and genital organs), especially during heavy infections. More common in children than in adults, cerebral paragonimiasis often produces two manifestations: an expansive, space-occupying lesion in the brain (particularly the cerebral cortex) and a meningitis or meningoencephalitis due to the migration of the worm, with common clinical signs ranging from headache, vomiting, seizures, personality changes, decline of cognitive function, coma, to death (through herniation caused by increased intracranial pressure). The next common site for ectopic lesions of paragonimiasis is the skin, resulting in multifocal skin lesions (e.g., subcutaneous nodules).

The pathogenic mechanisms of paragonimiasis largely reflect the host's immune responses to *Paragonimus* worms and eggs that enter and migrate from the intestines to the lungs, causing edema, effusion, and subsequently fibrosis that entraps and restricts the lungs, contributing to the associated signs and symptoms (e.g., shortness of breath). While nodules most likely contain immature cysts, the mature cysts harboring the adult worms may gradually become fibrotic and die. The cysts with dead worms degenerate, form scars that hold residual eggs entrapped in the fibrous tissue, and become calcified.

48.1.4 Diagnosis

Given its largely nonspecific symptoms, clinical diagnosis of paragonimiasis based on clinical observations, radiological findings, dietary history, and travel history is insufficient and inconclusive [23,24]. Therefore, several conventional laboratory procedures have been routinely used to confirm the presence of *Paragonimus* eggs spp. in sputum (through alkaline decontamination and centrifuge sedimentation) and/or feces (through formalin–ether concentration). Considering that microscopic examination of respiratory and/or stool specimens has limited sensitivity (30%–40% for a single sputum specimen and 54%–89% for multiple sputum specimens in comparison to 11%–15% for a single stool specimen and about 25% for three stool specimens) and relatively low specificity, serological tests (e.g., ELISA for IgG in serum and/or pleural effusion fluid) are helpful when clinical suspicion for paragonimiasis is high but *Paragonimus* eggs are not detected [14,25]. In addition, molecular methods based on nucleic acid amplification have been applied in recent years to further enhance the sensitivity, specificity, and speed of laboratory diagnosis of paragonimiasis [26–30].

Among various nucleic acid amplification techniques that have been developed to date, polymerase chain reaction (PCR) has been a standout due to its unprecedented sensitivity, specificity, speed, and robustness. In a relatively short period since its inception, PCR (and its variants such as conventional PCR, random amplified polymorphic DNA-PCR, multiplex PCR, and real-time PCR) has proven to be highly valuable for improved detection and/or species discrimination of *Paragonimus* parasites directly from clinical samples (e.g., lung biopsy specimens, bronchoalveolar lavage fluid, sputum, pleural fluid, or feces) [31]. The commonly used genetic targets for *Paragonimus* diagnosis include the internal transcribed spacer (ITS) regions of ribosomal genetic complexes, particularly ITS2, and the mitochondrial cytochrome *c* oxidase gene (CO1) [32–36].

Sugiyama et al. [37] established PCR-restriction fragment length polymorphism (RFLP) and specific PCR with primers from the second internal transcribed spacer (ITS2) region of ribosomal DNA to discriminate the metacercariae of the lung flukes, *P. westermani* and *P. miyazaki*, both of which are found in the same freshwater crab species and are morphologically similar. In addition, Sugiyama et al. [38] successfully applied primers from the second internal transcribed spacer (ITS2) region of the nuclear ribosomal DNA in PCR-RFLP and multiplex PCR for precise differentiation between individual metacercariae of *P. heterotremus* and *P. westermani* that occur in Thailand. Using a similar approach, Sugiyama et al. [39] developed multiplex PCR and PCR-RFLP (with ScrFI) that enabled differentiation of metacercaria from five *Paragoniumus* species occurring in Thailand, including *P. heterotremus* (ca. 310 and 520 bp), *P. westermani* (ca. 140 and 520 bp), *P. siamensis* (520 bp), *P. bangkokensis* (520 bp), and *P. harinasutai* (520 bp). Digestion of the 520 bp products with restriction enzyme ScrFI yielded three bands (ca. 60, 210, and 250 bp) for *P. harinasutai*, two bands (ca. 250 and 270 bp) for *P. bangkokensis*, and an uncut band (520 bp) for *P. siamensis*. In a more recent study, Tantrawatpan et al. [40] described a real-time fluorescence resonance energy transfer PCR (real-time FRET PCR) with melting curve analysis for specific and sensitive detection of *P. heterotremus* eggs in the feces of infected cats, with the detection limit of 10(–3) ng of *P. heterotremus* genomic DNA or 10 eggs of *P. heterotremus* per gram of cat feces.

48.1.5 Treatment and Prevention

Being able to separate or disrupt the tegument of *Paragonimus* species, praziquantel (75 mg/kg/day given in three doses for 2 days) represents the therapeutic agent of choice for pulmonary paragonimiasis, with a 71%–75% cure rate after 1 day, and 86%–100% cure rate after 2 days, and a complete (100%) cure rate after 3 days. The side effects of praziquantel may include dizziness, headache, and gastrointestinal disturbance. Other medications such as bithionol (at 30–50 mg/kg on alternate days for 10–15 doses), niclofan, and triclabendazole (at 10 mg/kg twice a day) also provide high cure rates for paragonimiasis [41,42]. However, surgery may be necessary for patients with complicated pleural disease or cerebral paragonimiasis. Prevention of human paragonimiasis should be centered on thorough cooking of an infected crustacean (crab or crayfish) to kill all stages of the parasite, and avoiding eating raw, even if pickled, crustaceans.

48.2 Laboratory Models

48.2.1 Animal Models

A number of laboratory animals have been attempted as experimental models for *Paragonimus* species, including rodents (rats, mice, Syrian hamsters, and Mongolian gerbils), dogs, and cats, with varied success [43]. While dogs and cats, especially puppies and kittens, are susceptible to most *Paragonimus* species prevalent in Asia, rats and mice serve well as paratenic hosts when infected naturally or experimentally with some *Paragonimus* species.

48.2.1.1 *Rodents*

Rats have been found to be susceptible to *P. westermani* and *P. ohirai* infection, with metacercaria developing to adult worms that are capable of producing eggs [44,45].

Narain et al. [46] demonstrated that outbred Wistar rats infected with *P. heterotremus* metacercariae produced adult worms in the lungs and pleural cavity, with some immature flukes being present in the skeletal muscles, highlighting the utility of outbred Wistar rats as a model for pulmonary *P. heterotremus* infection.

Weina et al. [47] showed that Syrian hamsters infected with 3–16 metacercariae of *P. kellicotti* displayed acute pleuritis, reactive mesothelial hyperplasia, subpleural accumulations of reactive and mature plasma cells, neovascularization, fibrohistiocytic thickening with and without giant cells, raised fibroconnective tissue lesions, and granulomatous inflammation with hemorrhage within 35 days post infection (dpi). In addition, perivascular plasmacytic (lymphocytic) infiltrate, multifocal bronchopneumonia, and parenchymal necrotizing suppurative granulomatous inflammation, hemorrhagic pneumonia, and diffuse sprinkling of eosinophils, neutrophils, and intraalveolar macrophages were also observed. These results confirmed the suitability of Syrian hamsters as a small animal mortality model for *P. kellicotti*.

Fischer et al. [48] reported that Mongolian gerbils (*Meriones unguiculatus*) infected intraperitoneally or by oral gavage with three to eight metacercariae of *P. kellicotti*, developed signs of apathy, weight loss, dehydration as early as 14 dpi, and yielded mature, gravid lung flukes as early as 39 dpi, with 69% of infected gerbils succumbing to the infection by 49 dpi. Necropsies revealed pulmonary hemorrhage with necrosis, and flukes as long as 8 mm were recovered from intrathoracic tissues. Thus, gerbils appear to be a suitable small animal for producing *P. kellicotti* parasite material and for studying parasite migration, immunobiology, and pathogenesis. Sanpool et al. [49] also described that Mongolian gerbils injected intraperitoneally with *P. macrorchis* metacercariae formed adult worms in the lungs 45 dpi.

48.2.1.2 *Cats*

Weina et al. [50] noted that cats infected orally with 2–30 *P. kellicotti* metacercariae underwent changes in the lungs of intense eosinophilic pneumonia, granulomatous pneumonitis, squamous epithelial-lined cyst formation of bronchogenic origin, and partial resolution of the host response. Fan et al. [51] showed that cats fed with *P. pulmonalis* metacercariae developed mature adult worms in the lungs and the pleural cavity. Interestingly, while eggs from worms encased in a cyst were passed in the feces, those from worms in the pleural cavity were only found on the lung surface and pleural cavity.

48.2.1.3 *Dogs*

Xue et al. [41,42] observed that dogs infected with 100 *P. westermani* metacercariae developed adult worms in the lungs. In early study, Yokogama et al. [44] demonstrated dogs' susceptibility to pulmonary *P. ohirai* infection after animals fed with metacercariae developed adult worms in lungs. Tsubokawa et al. [45] also noted that dogs infected with *P. westermani* metacercaria yielded egg-producing adult worms. Furthermore, Singh et al. [43] found that only puppies appeared suitable for pulmonary *P. heterotremus* infection whereas rats, mice, guinea pigs, and rabbits were largely refractory to oral infection with metacercariae.

48.2.2 *In Vitro* Cultures

Yokogawa et al. [52] reported that excysted metacercariae of *P. westermani* maintained at 37°C in equal parts of cat serum and Tyrode's solution for 3 weeks, then supplemented with chick embryo extract and cat blood cells, remained alive after 98 days. The average length of metacercaria increased from 0.6 to 2.0–2.5 mm at rest (and up to 4 mm when stretching) along with the development of the male reproductive system (especially the testes).

48.3 Conclusion

The genus *Paragonimus* comprises more than 50 trematode species whose adult worms inhabit the lungs of various mammalian hosts (so called lung flukes), and whose eggs and larval stages are found in snail and crustacean hosts. Of the nine human-infecting *Paragonimus* species, five occur mainly in Asia (*P. westermani*, *P. miyazaki*, *P. skrjabini*, *P. heterotremus*, and *P. hueitungensis*), two in Africa (*P. uterobilateralis* and *P. africanus*), and two in the Americas (*P. kellicotti* and *P. mexicanus*). Humans usually acquire *Paragonimus* infection through consumption of raw or undercooked crustaceans or meat from a paratenic host, with serious consequence. Clearly, continued application of rapid, sensitive, and specific laboratory diagnostic procedures and prompt implementation of appropriate treatment regimens are critical for limiting the harmful effects of human paragonimiasis. However, to keep an upper hand over *Paragonimus* parasites, it is essential to use appropriate laboratory models that will help elucidate the molecular mechanisms of host–parasite interactions and pathogenesis, and contribute to the development of innovative control strategies against paragonimiasis.

REFERENCES

1. Procop GW. North American paragonimiasis (caused by *Paragonimus kellicotti*) in the context of global paragonimiasis. *Clin Microbiol Rev.* 2009;22(3):415–46.
2. Blair D. Paragonimiasis. *Adv Exp Med Biol.* 2014;766:115–52.
3. Blair D, et al. A molecular perspective on the genera *Paragonimus* Braun, *Euparagonimus* Chen and *Pagumogonimus* Chen. *J Helminthol.* 1999;73:295–9.
4. Waikagul J. A new species of *Paragonimus* (Trematoda: Troglotrematidae) from a cat infected with metacercariae from mountain crabs *Larnaudia larnaudii*. *J Parasitol.* 2007;93:1496–500.
5. Iwagami M, Rajapakse RP, Paranagama W, Okada T, Kano S, Agatsuma T. Ancient divergence of *Paragonimus westermani* in Sri Lanka. *Parasitol Res.* 2008;102:845–52.
6. Uniprot. www.uniprot.org/taxonomy (accessed June 30, 2016).
7. Iwagami M, et al. A molecular phylogeographic study based on DNA sequences from individual metacercariae of *Paragonimus mexicanus* from Guatemala and Ecuador. *J Helminthol.* 2003;77(1):33–8.
8. Aka NA, Adoubryn K, Rondelaud D, Dreyfuss G. Human paragonimiasis in Africa. *Ann Afr Med.* 2008;7(4):153–62.
9. Liu Q, Wei F, Liu W, Yang S, Zhang X. Paragonimiasis: an important food-borne zoonosis in China. *Trends Parasitol.* 2008;24(7):318–23.
10. Singh TS, Sugiyama H, Rangsiruji A. *Paragonimus* & paragonimiasis in India. *Indian J Med Res.* 2012;136(2):192–204.
11. Diaz JH. Paragonimiasis acquired in the United States: native and nonnative species. *Clin Microbiol Rev.* 2013;26(3):493–504.
12. Doanh PN, Horii Y, Nawa Y. *Paragonimus* and paragonimiasis in Vietnam: an update. *Korean J Parasitol.* 2013;51(6):621–7.
13. Calvopiña M, Romero D, Castañeda B, Hashiguchi Y, Sugiyama H. Current status of *Paragonimus* and paragonimiasis in Ecuador. *Mem Inst Oswaldo Cruz.* 2014;109(7):849–55.
14. Fischer PU, Weil GJ. North American paragonimiasis: epidemiology and diagnostic strategies. *Expert Rev Anti Infect Ther.* 2015;13(6):779–86.
15. Nagayasu E, Yoshida A, Hombu A, Horii Y, Maruyama H. Paragonimiasis in Japan: a twelve-year retrospective case review (2001–2012). *Intern Med.* 2015;54(2):179–86.
16. Yatera K, et al. A rare case of paragonimiasis miyazaki with lung involvement diagnosed 7 years after infection: a case report and literature review. *Parasitol Int.* 2015;64(5):274–80.
17. Devi KR, et al. Morphological and molecular characterization of *Paragonimus westermani* in northeastern India. *Acta Trop.* 2010;116:31–8.
18. Chai JY. Paragonimiasis. *Handb Clin Neurol.* 2013;114:283–96.
19. Lane MA, Barsanti MC, Santos CA, Yeung M, Lubner SJ, Weil GJ. Human paragonimiasis in North America following ingestion of raw crayfish. *Clin Infect Dis.* 2009;49(6):e55–61.
20. Vargas-Arzola J, et al. Detection of *Paragonimus mexicanus* (Trematoda) metacercariae in crabs from Oaxaca, Mexico. *Acta Trop.* 2014;137:95–8.

21. Rekha Devi K, et al. Presence of three distinct genotypes within the *Paragonimus westermani* complex in northeastern India. *Parasitology* 2013;140:76–86.
22. Boland JM, et al. Pleuropulmonary infection by *Paragonimus westermani* in the United States: a rare cause of eosinophilic pneumonia after ingestion of live crabs. *Am J Surg Pathol.* 2011;35(5):707–13.
23. Shim SS, Kim Y, Lee JK, Lee JH, Song DE. Pleuropulmonary and abdominal paragonimiasis: CT and ultrasound findings. *Br J Radiol.* 2012;85(1012):403–10.
24. Henry TS, Lane MA, Weil GJ, Bailey TC, Bhalla S. Chest CT features of North American paragonimiasis. *AJR Am J Roentgenol.* 2012;198(5):1076–83. Erratum in: AJR Am J Roentgenol. 2013;201(5):1165.
25. McNulty SN, Fischer PU, Townsend RR, Curtis KC, Weil GJ, Mitreva M. Systems biology studies of adult *Paragonimus* lung flukes facilitate the identification of immunodominant parasite antigens. *PLoS Negl Trop Dis.* 2014;8(10):e3242.
26. Le TH, Van De N, Blair D, McManus DP, Kino H, Agatsuma T. *Paragonimus heterotremus* Chen and Hsia (1964), in Vietnam: a molecular identification and relationships of isolates from different hosts and geographical origins. *Acta Trop.* 2006;98:25–33.
27. Tandon V, Prasad PK, Chatterjee A, Bhutia PT. Surface fine topography and PCR-based determination of metacercaria of *Paragonimus* sp. from edible crabs in Arunachal Pradesh, Northeast India. *Parasitol Res.* 2007;102(1):21–8.
28. Chen MX, et al. Sensitive and rapid detection of *Paragonimus westermani* infection in humans and animals by loop-mediated isothermal amplification (LAMP). *Parasitol Res.* 2011;108(5):1193–8.
29. Intapan PM, et al. Molecular identification of a case of *Paragonimus pseudoheterotremus* infection in Thailand. *Am J Trop Med Hyg.* 2012;87(4):706–9.
30. Sanpool O, et al. Molecular variation in the *Paragonimus heterotremus* complex in Thailand and Myanmar. *Korean J Parasitol.* 2013;51(6):677–81.
31. Iwagami I, et al. Molecular phylogeographic studies on *Paragonimus westermani* in Asia. *J Helminthol.* 2000;74:315–22.
32. Park GM, Im KI, Yong TS. Phylogenetic relationship of ribosomal ITS2 and mitochondrial CO1 among diploid and triploid *Paragonimus westermani* isolates. *Korean J Parasitol.* 2003;41:47–55.
33. Doanh PN, Shinohara A, Horii Y, Habe S, Nawa Y. Discovery of *Paragonimus westermani* in Vietnam and its molecular phylogenetic status in *P. westermani* complex. *Parasitol Res.* 2009;104:1149–55.
34. Doanh PN, Dung DT, Thach DT, Horii Y, Shinohara A, Nawa Y. Human paragonimiasis in Viet Nam: epidemiological survey and identification of the responsible species by DNA sequencing of eggs in patients' sputum. *Parasitol Int.* 2011;60:534–7.
35. Prasad PK, Goswami LM, Tandon V, Chatterjee A. PCR-based molecular characterization and insilico analysis of food-borne trematode parasites *Paragonimus westermani*, *Fasciolopsis buski* and *Fasciola gigantica* from Northeast India using ITS2 rDNA. *Bioinformation* 2011;6(2):64–8.
36. Biswal DK, Chatterjee A, Bhattacharya A, Tandon V. The mitochondrial genome of *Paragonimus westermani* (Kerbert, 1878), the Indian isolate of the lung fluke representative of the family Paragonimidae (Trematoda). *Peer J.* 2014;2:e484.
37. Sugiyama H, Morishima Y, Kameoka Y, Kawanaka M. Polymerase chain reaction (PCR)-based molecular discrimination between *Paragonimus westermani* and *P. miyazaki* at the metacercarial stage. *Mol Cell Probes* 2002;16(3):231–6.
38. Sugiyama H, et al. Molecular discrimination between individual metacercariae of *Paragonimus heterotremus* and *P. westermani* occurring in Thailand. *Southeast Asian J Trop Med Public Health* 2005;36(Suppl 4):102–6.
39. Sugiyama H, Morishima Y, Rangsiruji A, Binchai S, Ketudat P, Kawanaka M. Application of multiplex pcr for species discrimination using individual metacercariae of *Paragonimus* occurring in Thailand. *Southeast Asian J Trop Med Public Health* 2006;37(Suppl 3):48–52.
40. Tantrawatpan C, et al. Application of a real-time fluorescence resonance energy transfer polymerase chain reaction assay with melting curve analysis for the detection of *Paragonimus heterotremus* eggs in the feces of experimentally infected cats. *J Vet Diagn Invest.* 2013;25(5):620–6.
41. Xue J, Utzinger J, Zhang YN, Tanner M, Keiser J, Xiao SH. Artemether and tribendimidine lack activity in experimental treatment of *Paragonimus westermani* in the dog. *Parasitol Res.* 2008;102(3):537–40.
42. Xiao SH, Xue J, Li-li X, Zhang YN, Qiang HQ. Effectiveness of mefloquine against *Clonorchis sinensis* in rats and *Paragonimus westermani* in dogs. *Parasitol Res.* 2010;107(6):1391–7.

43. Singh TS, Sugiyama H, Devi KR, Singh LD, Binchai S, Rangsiruji A. Experimental infection with *Paragonimus heterotremus* metacercariae in laboratory animals in Manipur, India. *Southeast Asian J Trop Med Public Health* 2011;42(1):34–8.
44. Yokogama M, et al. Chemotherapy of paragonimiasis with bithionol. I. Experimental chemotherapy on the animals infected with *Paragonimus westermani* or *P. ohirai*. *Jpn J Parasitol.* 1961;10(2):302–16.
45. Tsubokawa D, et al. Collection methods of trematode eggs using experimental animal models. *Parasitol Int.* 2016;65:584–87.
46. Narain K, Rekha Devi K, Mahanta J. A rodent model for pulmonary paragonimiasis. *Parasitol Res.* 2003;91(6):517–9.
47. Weina PJ, Burns WC. Mortality in Syrian hamsters infected with *Paragonimus kellicotti*. *J Parasitol.* 1992;78(2):378–80.
48. Fischer PU, Curtis KC, Marcos LA, Weil GJ. Molecular characterization of the North American lung fluke *Paragonimus kellicotti* in Missouri and its development in Mongolian gerbils. *Am J Trop Med Hyg.* 2011;84(6):1005–11.
49. Sanpool O, et al. Morphological and molecular identification of a lung fluke, *Paragonimus macrorchis* (Trematoda, Paragonimidae), found in central Lao PDR and its molecular phylogenetic status in the genus *Paragonimus*. *Parasitol Int.* 2015;64(6):513–8.
50. Weina PJ, England DM. The American lung fluke, *Paragonimus kellicotti*, in a cat model. *J Parasitol.* 1990;76(4):568–72.
51. Fan PC, Lu H, Lin LH. Egg production capacity of *Paragonimus pulmonalis* in cats. *J Parasitol.* 1998;84(6):1282–5.
52. Yokogama M, Oshima T, Kihata M. Studies to maintain excysted metacercariae of *Paragonimus westermani* in vitro. *J Parasitol.* 1955;41(6):28.

49

Taenia

Dongyou Liu

CONTENTS

49.1 Introduction

The genus *Taenia* encompasses a large group of parasitic tapeworms that undergo three distinct stages of development (i.e., adult worm, eggs, and larva/metacestode) during their life cycles and require two mammalian hosts for transmission. The ribbon-like adult worm (so called tapeworm) resides in the intestines of carnivores or omnivores, causing intestinal taeniasis, whereas the cyst-like larva (metacestode) occurs in the internal organs of herbivores or omnivores, causing cysticercosis or neurocysticercosis (representing different forms of taeniasis).

Of the 45 described *Taenia* species, three [i.e., *Taenia solium* (pork tapeworm), *Taenia saginata* (beef tapeworm), and *Taenia asiatica* (Asian tapeworm)] are known to infect humans. Indeed, as the only definitive host for these three species, human infection with the adult worms leads to intestinal parasitism, which is largely asymptomatic. However, the larva (metacestode) of *T. solium* causes cysticercosis (cysticercus cellulosae) in pigs and humans, that of *T. saginata* causes cysticercosis (cysticercus bovis) in cattle, and that of *T. asiatica* causes cysticercosis (cysticercus viscerotropica) in pigs (Table 49.1).

Although tapeworms were mentioned in ancient Egyptian works dated back to 2000 BC, as well as in the *History of Animals* written by Greek author Aristotle (384–322 BC), DNA analyses point out the likelihood that ancestors of modern humans in Africa had acquired parasite through consumption of infected antelope and bovid meats more than 10,000 years ago and subsequently passed it to other animals such as domestic pigs. A more recent description of cysticercosis was made by Johannes Udalric Rumler in 1555; however, the connection between tapeworm and cysticercosis was not established until the middle of the 19th century.

49.1.1 Classification and Morphology

49.1.1.1 Classification

The genus *Taenia* belongs to the family Taeniidae, order Cyclophyllidea (terrestrial cycles, scolex with suckers), subclass Eucestoda (segmented, hermaphroditic), class Cestoda (tapeworms), phylum Platyhelminthes (flatworms), kingdom Animalia.

Members of the family Taeniidae are characterized by their terrestrial life cycle, which includes three developmental stages (eggs, larva, and adult) and involves a carnivorous/omnivorous definitive host for adult worm, and a herbivorous/omnivorous intermediate host for larva (metacestode). Of the two well-recognized genera (*Echinococcus* and *Taenia*) within the family Taeniidae, the genus *Echinococcus* is monophyletic, with a notable similarity in morphology, developmental processes, and genetic makeup, and is separated into nine species: *E. granulosus* sensu stricto (including the former genotypic variants G1–3), *E. felidis* (the former "lion strain"), *E. equinus* (the "horse strain," genotype G4), *E. ortleppi* (the "cattle strain," genotype G5), *E. canadensis*, *E. multilocularis*, *E. oligarthra*, *E. vogeli*, and *E. shiquicus* [1]. On the other hand, the genus *Taenia* is a highly diverse, paraphyletic group with 45 species identified to date, including *T. arctos*, *T. asiatica* (synonym *T. saginata asiatica*), *T. brachyacantha*, *T. crassiceps*, *T. dinniki*, *T. gonyamai*, *T. hydatigena*, *T. jaipurensis*, *T. kotlani*, *T. krabbei*, *T. krepkogorski*, *T. laticollis*, *T. madoquae*, *T. martis*, *T. multiceps*, *T. mustelae*, *T. omissa*, *T. ovis*, *T. parva*, *T. pisiformis*, *T. regis*, *T. retracta*, *T. rileyi*, *T. saginata* (synonym *Taeniarhynchus saginata*), *T. saigoni*, *T. serialis*, *T. simbae*, *T. solium*, *T. taeniaeformis*, *T. twitaelli*, and *T. ursina* [2,3]. Among these, *T. solium*, *T. saginata*, and *T. asiatica* are infective to humans.

Recent examination of nuclear and mitochondrial genes enables further clarification of the genetic relationships among the representative members of the *Taenia* genus. Indeed, the nuclear phylogenetic trees of 18S ribosomal DNA and concatenated exon regions of protein-coding genes (phosphoenolpyruvate carboxykinase and DNA polymerase delta) revealed that both *T. mustelae* and a clade formed by *T. parva*, *T. krepkogorski*, and *T. taeniaeformis* are only distantly related to other *Taenia* species. Similar topologies were also observed in mitochondrial genomic analyses using 12 complete protein-coding genes. In addition, a sister relationship between *T. mustelae* and *Echinococcus* spp. was noted. These findings led to the proposals to resurrect the genus *Hydatigera* for *T. parva*, *T. krepkogorski*, and *T. taeniaeformis* and to create a new genus, *Versteria*, for *T. mustelae* as well as *T. brachyacantha* [4].

49.1.1.2 Morphology

Taenia adult worms (Taenia is derived from Greek ταίνια, tainia meaning ribbon, bandage, or stripe) are flat, ribbon-like in appearance, and white or yellowish-white in color. and consist of a knob-like head called scolex, a short neck, and a ribbon-like body (also known as strobila). The scolex (hold-fast organ) often possesses suckers (acetabula), rostellum, and spiny hooks. It is noteworthy that the scolex of *T. solium* is spheroidal, with a rostellum (of 1 mm in diameter), which is armed with two rows of 22–32 hooks; that of *T. saginata* is cuboidal (of 2 mm in width) without rostellum or hooks; and that of *T. asiatica* is spheroidal, with a cuspidal rostellum (of 0.8 mm in width), which has no hooks (Table 49.1). The strobila (up to 22 m long depending on the species) is composed of a chain of segments called proglottids, each of which contains a set of male and female reproductive organs (thus hermaphroditic). The reproductive organs are made up of tubular unbranched uterus (filled with eggs), ovary, genital pore, testes, and vitelline gland, with testes and ovary opening into a common genital pore located on the side. Being an acoelomate animal, the adult worm has no body cavity or digestive system, and relies entirely on its penetrable tegument to absorb nutrients. As body growth starts from the neck region, immature proglottids are found near the neck, mature proglottids in the middle, and gravid (oldest) proglottids at the posterior end. On average, *T. solium* adult worm has about 1000 proglottids, each producing 50,000 eggs; *T. saginata* adult worm has about 1000–2000 proglottids, each producing 100,000 eggs; *T. asiatica* adult worm has 700–900 proglottids, each producing 80,000 eggs [5].

Taenia eggs are spherical, of about 40–48 μm in diameter, and surrounded by a thick striated wall, which encases an embryo (oncosphere). After gravid proglottids detach from the strobila or discharge

TABLE 49.1

Morphological and Biological Characteristics of Human-Infecting *Taenia* Species

	Taenia solium **(Pork Tapeworm)**	***Taenia saginata*** **(Beef Tapeworm)**	***Taenia asiatica*** **(Asian Tapeworm)**
Adult worm	Adult worm of 2–8 m in length resides in human small intestine; its scolex has four suckers that surround the rostellum, with 22–32 hooks arranged in two rows; its strobila contains 1000 proglottids; its gravid proglottid has no posterior protuberances	Adult worm of 4–12 m in length (as long as 25 m) resides in human small intestine; its scolex has four suckers, and an apical pit instead of a rostellum and hooks; its strobila contains 1000–2000 proglottids; its gravid proglottid has no posterior protuberances	Adult worm of about 3.5 m in length resides in human small intestine; its scolex has four suckers surrounding the rostellum, without hooks; its strobila contains 700–900 proglottids; its gravid proglottid has posterior protuberances
Metacestode (*Cysticercus*)	*Cysticercus cellulosae* (5–8 × 3–6 mm) occurs mostly in pig muscle, as fluid-filled cyst (0.5–1.5 cm in diameter) containing single scolex with hooks; its external surface lacks wart-like formations	*Cysticercus bovis* (6–10 × 4–6 mm) occurs inside muscle, liver, and lungs of cattle, as fluid-filled cyst containing single scolex without hooks; its external surface lacks wart-like formations (or not prominent if present)	*Cysticercus viscerotropica* (2 × 2 mm) occurs in liver and lungs in pigs, wild boars, and occasionally cattle, as fluid-filled cyst containing single scolex with two rows of rudimentary hooks; its external surface has wart-like formations
Genome	122–131 Mb in size, with 11,902–12,481 coding genes	169 Mb in size, with 13,161 coding genes	168 Mb in size, with 13,323 coding genes
Clinical features	Humans infected with adult worms show minimal symptoms; accidental ingestion of embryonated eggs or proglottids (via autoinfection or contaminated food) by humans leads to cysticercosis, which shows three morphologically distinct forms [the ordinary "cellulose" cysticercus, with a fluid-filled cyst and an invaginated scolex; the intermediate form with a scolex; the "racemose" form (20 cm in length and 60 mL of fluid) with no evident scolex]	Humans infected with adult worms are usually asymptomatic, although heavy infection may lead to weight loss, dizziness, abdominal pain, diarrhea, headaches, nausea, constipation, chronic indigestion, and loss of appetite; cysticercosis is not observed in humans	Humans infected with adult worms are largely asymptomatic
Distribution	Cosmopolitan, and highly prevalent in Mexico, Latin America, West Africa, Russia, India, Pakistan, China, and Southeast Asia	Prevalent in Africa, Europe, Southeast Asia, South Asia, and Latin America	Prevalent in Taiwan, South Korea, Indonesia, the Philippines, Thailand, China, Vietnam, Japan, and Nepal

Sources: Parija, S.C. and Ponnambath, D.K., *Trop. Parasitol.*, 3, 120–124, 2013; Wang, S. et al., Nature Commun., 7, 12845, 2016.

along with feces, eggs are released. Being highly resistant to desiccation and sewage treatment, these eggs can survive on pastures for weeks. Upon ingestion by intermediate host, the eggs hatch and develop into infective cysticerci in selective organs (e.g., muscle, liver, and lungs).

Taenia larvae (metacestodes) such as those of *T. solium*, *T. saginata*, and *T. asiatica* form small, pearly-white, fluid-filled cysts (cysticerci) (hence the common name bladder worms) within which a single invaginated protoscolex (infective stage) is located [5]. The larvae of other *Taenia* spp. may show varied appearance, with some forming a strobilocercus (containing noninvaginated protoscolex) and others forming a large coenurus (containing several invaginated protoscoleces).

49.1.2 Life Cycle and Epidemiology

49.1.2.1 Life Cycle

Members of the family *Taeniidae* are unique among the subclass Eucestoda in requiring two obligate mammalian hosts for their life cycles: a carnivorous/omnivorous definitive host for the adult stage, and a herbivorous/omnivorous intermediate host for the larval stage (metacestode).

Typically, the life cycle begins when vegetation, feed, or water contaminated with taeniid eggs (or gravid proglottids) are consumed by a herbivorous/omnivorous intermediate host. Once inside, the embryonated eggs hatch into motile oncospheres within the duodenum and penetrate the intestinal wall, enter the bloodstream, and differentiate into metacestodes within 70 days (either cysticercoid, cysticercus, or a hydatid cyst) in selective internal organs (liver, lungs, brains, and muscles, etc.). When raw or undercooked pork and beef containing the fluid-filled cysticerci are eaten by a carnivorous/omnivorous final host, the cyst (bladder) is digested away, the inverted scolex evaginates under stimuli from the digestive enzymes of the host, the scolex then embeds itself into the intestinal wall, and the neck buds off segments to form the strobila. New eggs may appear in the feces of the definitive host within 6–9 weeks, starting the next cycle of infection [6].

49.1.2.2 Epidemiology

With humans as the definitive host and pigs as the intermediate host, *T. solium* is present in many parts of the world where humans live in close proximity with pigs and eat undercooked pork [7]. In addition, humans may function as aberrant intermediate host after accidental ingestion of embryonated eggs, either through autoinfection or consumption of contaminated food, leading to neurocysticercosis, which demonstrates a predilection for brain tissue and other soft muscle tissues [8]. Geographically, *T. solium* infection is particularly common in Mexico, Latin America, West Africa, India, Pakistan, Southeast Asia, China, Russia, and Slavic countries of Europe. Previous reports indicate that the seroprevalence of *T. solium* in areas of Guatemala, Bolivia, and Peru may be as high as 20% in humans and 37% in pigs.

In the case of *T. saginata*, humans and cattle act as the definitive and intermediate hosts, respectively. This parasite is found in places where beef is eaten, and is relatively common in Africa, the Middle East, the Philippines, Eastern Europe, and Latin America.

T. asiatica utilizes humans as the definitive host, and pigs, wild boar, and possibly cattle as the intermediate host. While the adult worm is found in the small intestine of humans, pea-sized, fluid-filled cyst (cysticercus) is present in liver, serosa, and lungs of pigs and in the liver of cattle. Geographically, the parasite is essentially restricted to Taiwan, Korea, Indonesia, Nepal, Thailand, and China, in addition to Japan, the Philippines, and Vietnam [9,10].

49.1.3 Clinical Features

Infection with *T. solium* adult worm in humans generally involves one to two worms and is often asymptomatic, although heavy infection may lead to mild intermittent diarrhea or constipation, anemia, indigestion, inappetite, and emaciation, together with urticaria, anal pruritus, and eosinophilia. However, accidental consumption of *T. solium* eggs from contaminated vegetables/water or ingestion of *T. solium* eggs or proglottids, which rupture within the host intestines, may result in cysticercosis or neurocysticercosis [11]. In the case of cysticercosis, solid lumps of 1–2 cm in size may develop under the skin near the trunk and extremities, become painful and swollen, and then resolve. In the case of neurocysticercosis, cysts of 5–20 mm in diameter may form in the parenchyma of the brain. In more severe cases (racemose neurocysticercosis), lesions as large as 6 cm in diameter, lobulated, and may occur in subarachnoid space and fissures. Clinical presentations may include severe headaches, dizziness, epilepsy, seizures, dementia, hypertension, lesions in the brain, blindness, tumor-like growths, hydrocephalus, paraplegy, meningitis, convulsions, and even death.

Similarly, human infection with *T. saginata* adult worm is asymptomatic, but heavy infection causes dizziness, abdominal pain, diarrhea, headaches, nausea, constipation, chronic indigestion, inappetite, and weight loss. Other rare clinical signs include ileus, pancreatitis, cholecystitis, and cholangitis.

Further, human infection with *T. asiatica* adult worm is usually asymptomatic. Nevertheless, in severe cases, damage and bleeding in the stomach and intestine may be observed.

49.1.4 Diagnosis

Identification of *Taenia* species (*T. saginata*, *T. asiatica*, and *T. solium*) in the definitive host has traditionally relied on microscopic detection of eggs or proglottids in feces. However, this approach lacks desired sensitivity and specificity given that the eggs from different *Taenia* species are morphologically indistinguishable [12].

Use of computed tomography (CT) or magnetic resonance imaging (MRI) enables detection of cysticerci in the brain, as in the case of human neurocysticercosis. In addition, application of, X-rays facilitates detection of calcified cysticerci in the subcutaneous and muscle tissues.

Serological methods (ELISA and IET) based on adult worm excretion-secretion antigens (ES Ag) (ES33 and ES38) of *T. solium* have proven useful for confirmation of *T. solium* taeniasis, with sensitivity and specificity of >97% and 91%, respectively [13]. Furthermore, serological detection of coproantigens in fecal specimens allows genus-specific discrimination. Enzyme-linked immunoelectrotransfer blot (EITB) detects an immunoblot band of 21.5 kDa from *T. asiatica*, permitting effective differentiation of asiatica from other taeniid infections.

Utilization of molecular methods targeting various gene regions of *Taenia* species has greatly enhanced the diagnosis of taeniasis and cysticercosis [14–25]. Polymerase chain reaction (PCR) amplification of ribosomal 5.8S and HDP2 gene sequences enables differentiation among *T. solium*, *T. saginata*, and *T. asiatica* [26,27]. Sequencing analysis of the variable regions (V1–V5) within nuclear 18S rRNA gene reveals pertinent phylogenetic details within the genus *Taenia*. The mitochondrial gene sequence (mtDNA) provides a valuable molecular marker for uncovering evolutionary relationships among distantly related taxa, and also for investigating the phylobiogeography of closely related species. A multiplex PCR based on the primers Ta4978F, Ts5058F, Tso7421F, and Rev7915 from the mitochondrial gene sequence has streamlined differential diagnosis, molecular characterization, and epidemiological surveys of *Taenia* species [28]. Moreover, use of primers (Ta7216F, Ts7313F, Tso7466F, and Rev7915) from valine transfer RNA and NADH dehydrogenase subunit two genes in multiplex PCR generates species-specific products of 706, 629, and 474 bp for *T. asiatica*, *T. saginata*, and *T. solium* in a sensitive, specific, and speedy manner [29].

49.1.5 Treatment and Prevention

49.1.5.1 Treatment

Praziquantel (at 5–10 mg/kg orally once for adults and children) is frequently used to treat active taeniasis. It should be noted that praziquantel is cysticidal and may cause inflammation around dying cysts in patients with cysticercosis, leading to seizures. Niclosamide (at 2 g orally once for adults, 1.5 g orally once for children weighing >34 kg, 1 g orally once for children weighing 11–34 kg, and 500 mg orally once for children under 2 years of age) represents an alternative medication for treating *Taenia* infection. Stools from treated patients should be collected for 3 days and examined for tapeworm proglottids for species verification. Furthermore, stools from patients should be reexamined for *Taenia* eggs 1 and 3 months after treatment to ensure the complete clearance of infection. Another useful medication is albendazole, which is an effective and safe drug for treating neurocysticercosis. Due to lower cost, fewer drug interactions, and higher efficiency for both adult worm and larvae, albendazole and mebendazole are generally preferred over praziquantel [30]. In cases of intraventricular, racemose, or spinal neurocysticercosis, surgical intervention (e.g., direct excision of ventricular cysts, shunting procedures, and removal of cysts via endoscopy) may be needed.

49.1.5.2 Prevention

The key for preventing *Taenia* infections in humans is to break their transmission cycle through stringent meat inspection, condemnation of infected carcasses for human consumption, proper cooking (80°C) or freezing (–10°C for 9 days) of meat, sanitary disposal of feces, prohibiting the use of sewage for fertilizing pastures, washing salad vegetables, strict personal hygiene, and improved access to clean water.

To prevent *Taenia* infections in pigs and cattle, it is important to avoid the exposure of pigs and cattle to environments contaminated with human feces; to eliminate the use of untreated sewage effluent to irrigate land that is later used by pigs and cattle for forage and food crops; and to vaccinate pigs with genetically engineered 45W-4B antigens or S3PVAC consisting of three protective peptides (KETc12, KETc1, and GK1) against *T. solium* infection [31,32].

49.2 Laboratory Models

As humans act as the only definitive host for *T. solium*, *T. saginata*, and *T. asiatica*, use of laboratory models is not only necessary but also important in helping elucidate the pathogenesis, host–parasite relationship, and other related issues concerning taeniasis and cysticercosis [33,34]. From published data, it is clear that pigs, dogs, cats, and rabbits, even in immunosuppressed status, are relatively poor hosts for experimental investigation of taeniid infections, On the other hand, various rodent hosts (rats, mice, hamsters, gerbils, and chinchillas) support the maintenance, growth, and maturation of taeniid tapeworms, contributing to our improved understanding of taeniasis and cysticercosis [35,36].

In an early study, Wang et al. [37] showed that immunosuppressed Syrian hamsters are susceptible to *T. solium* infection, with adult worm developing in the intestine after oral ingestion of cysticerci contained in severe combined immunodeficiency (SCID) mouse muscles. Separately, Avila et al. [38] demonstrated that hamsters, gerbils, and chinchillas are adequate experimental models for investigating the growth and immune responses of *T. solium* adult worm infection. Furthermore, chinchillas and, to a lesser extent, hamsters appear to be more permissive than gerbils, allowing longer-term survival and development of *T. solium* parasites. In a more recent report, Verastegui et al. [39] described a useful neurocysticercosis model based on Holtzman rats. It was noted that rats injected extraparenchymally as well as intraparenchymally with activated *T. solium* oncospheres developed cysticerci in the brain tissue after 4 months; that the cysticerci were present in the parenchymal, ventricle, or submeningeal brain tissue; and that epilepsy was observed in 9% of rats with neurocysticercosis. The age of rats appeared to be crucial for the success of infection, although the sites of injection and *T. solium* oncosphere dosages were largely irrelevant.

Several reports showed that nonobese diabetic SCID (NOD/Shi-scid) mice of both sexes are susceptible to infection with *in vitro* hatched oncospheres of human-infecting *Taenia* species (*T. solium*, *T. saginata*, and *T. asiatica*), producing cysticerci comparable to those developed in their known intermediate host animals [40–44]. On the other hand, only female SCID mice of BALB/c, C57BL, or C.B-17 inbred strains allow the development of cysticerci from oncospheres of human taeniid species [44].

Additionally, normal C3H/HeN female mice were showed to initiate Th1 cells immune response against *T. asiatica* during the early stages of oncosphere infection [45]. Similarly, hamsters are also capable of eliciting cytokine expression at the anchor site of *T. solium* [46–48].

Chowdhury et al. [49] successfully induced neurocysticercosis in rhesus monkey with 12,000 *T. solium* eggs, resulting in development of hyperexcitability, epileptic seizures, muscular tremors, digital cramps at 10 days postinfection, and paralysis of limbs and death days post infection. Necropsy of the infected monkey revealed numerous cysticerci in the brain, and histopathological examination uncovered liquefactive necrosis and formation of irregular cystic cavities lined by atrophied parenchymal septa with remnants of neuropil of the cerebrum.

49.3 Conclusion

Members of the genus *Taenia* are parasitic tapeworms that involve two mammalian hosts (a carnivorous/omnivorous definitive host and a herbivorous/omnivorous intermediate host) during their life cycles.

Taenia infection usually begins when *Taenia* eggs and proglottids produced by adult worm that reside in the small intestine of definitive hosts are discharged in stools, and subsequently ingested by intermediate host. Inside intermediate host, *Taenia* eggs hatch to release oncospheres, which then penetrate through intestinal wall, migrate via blood circulation to various internal organs and muscles, and mature into cyst-like larvae (metacestodes). Once taken up by definitive host, *Taenia* larvae (metacestodes) release protoscoleces, which utilize their hooks to anchor onto the intestinal wall and grow into adult worms that produce eggs for the next cycle of infection.

Of 45 *Taenia* species identified to date, three (*T. solium*, *T. saginata*, and *T. asiatica*) cause taeniasis in humans and cysticercosis in pigs and cattle as well as humans. Given that the adult worms of these human-infecting *Taenia* species only occur in humans, and the larvae are present in pigs and cattle, detailed examination of taeniid pathogenesis and host–parasite relationship have been hampered by ethical and cost concerns. For this reason, use of laboratory models is critically important in helping to improve our understanding of taeniasis and cysticercosis, and leading to innovative control measures against these parasites. Fortunately, of various types of laboratory animals available, rodents (e.g., rats, mice, hamsters, gerbils, and chinchillas) have been shown to support the maintenance, growth, and maturation of human taeniid tapeworms (in both taeniasis and cysticercosis/neurocysticercosis forms) and to elicit competent immune responses against these parasites. It is envisaged that in our ongoing effects to elucidate the molecular basis of host–parasite interactions and to develop novel, tailor-made control strategies against *Taenia* parasites, rodents will continue to play an indispensable role [35,50].

REFERENCES

1. Romig T, Ebi D, Wassermann M. Taxonomy and molecular epidemiology of *Echinococcus granulosus* sensu lato. *Vet Parasitol.* 2015;213(3–4):76–84.
2. Hoberg EP. Phylogeny of *Taenia*: species definitions and origins of human parasites. *Parasitol Int.* 2006;55(Suppl):S23–30.
3. Haukisalmi V, Lavikainen A, Laaksonen S, Meri S. *Taenia arctos* n. sp. (Cestoda: Cyclophyllidea: Taeniidae) from its definitive (brown bear *Ursus arctos* Linnaeus) and intermediate (moose/elk *Alces* spp.) hosts. *Syst Parasitol.* 2011;80(3):217–30.
4. Nakao M, et al. Molecular phylogeny of the genus *Taenia* (Cestoda: Taeniidae): proposals for the resurrection of *Hydatigera* Lamarck, 1816 and the creation of a new genus *Versteria*. *Int J Parasitol.* 2013;43(6):427–37.
5. Eom KS, Rim HJ. Morphologic descriptions of *Taenia asiatica* sp. n. *Korean J Parasitol.* 1993;31:1–6.
6. Parija SC, Ponnambath DK. Laboratory diagnosis of *Taenia asiatica* in humans and animals. *Trop Parasitol.* 2013;3(2):120–4.
7. Singh AK, Prasad KN, Prasad A, Tripathi M, Gupta RK, Husain N. Immune responses to viable and degenerative metacestodes of *Taenia solium* in naturally infected swine. *Int J Parasitol.* 2013;43(14):1101–7.
8. Bobes RJ, et al. Evolution, molecular epidemiology and perspectives on the research of taeniid parasites with special emphasis on *Taenia solium*. *Infect Genet Evol.* 2014;23:150–60.
9. Eom KS, Jeon HK, Rim HJ. Geographical distribution of *Taenia asiatica* and related species. *Korean J Parasitol.* 2009;47:S115–24.
10. Ooi HK, Ho CM, Chung WC. Historical overview of *Taenia asiatica* in Taiwan. *Korean J Parasitol.* 2013;51(1):31–6.
11. Garcia HH, Gonzalez AE, Gilman RH. Cysticercosis of the central nervous system: how should it be managed? *Curr Opin Infect Dis.* 2011;24(5):423–7.
12. Rodrignez-Hidalgo R, Geysen D, Benitez-Ortiz W, Geerts S, Brandt J. Comparison of conventional techniques to differentiate between *Taenia solium* and *Taenia saginata* and an improved polymerase chain reaction-restriction fragment length polymorphism assay using a mitochondrial 12S rDNA fragment. *J Parasitol.* 2002;88:1007–11.
13. Ochoa-Sánchez A, Jiménez L, Landa A. The hamster model for identification of specific antigens of *Taenia solium* tapeworms. *J Biomed Biotechnol.* 2011;2011:504959.
14. Mayata H, et al. Differentiating *Taenia solium* and *Taenia saginata* infections by simple hematoxylin-eosin staining and PCR-restriction enzyme analysis. *J Clin Microbiol.* 2000;38:133–7.

15. Jeon HK, Kim KH, Eom KS. Complete sequence of the mitochondrial genome of *Taenia saginata*: comparison with *T. solium* and *T. asiatica*. *Parasitol Int*. 2007;56:243–6.
16. Jeon HK, Kim KH, Eom KS. Molecular identification of *Taenia* specimens after long-term preservation in formalin. *Parasitol Int*. 2011;60:203–5.
17. Gonzales LM, et al. Differential diagnosis of *Taenia saginata* and *Taenia saginata asiatica* taeniasis through PCR. *Diagn Microbiol Infect Dis*. 2004;49:183–8.
18. Yamasaki H, et al. DNA differential diagnosis of taeniasis and cysticercosis by multiplex PCR. *J Clin Microbiol*. 2004;42:548–53.
19. Yamasaki H, Nakaya K, Nakao M, Sako Y, Ito A. Significance of molecular diagnosis using histopathological specimens in cestode zoonoses. *Trop Med Health*. 2007;35:307–21.
20. McManus DP. Molecular discrimination of taeniid cestodes. *Parasitol Int*. 2006;55(Suppl):31–7.
21. Lavikainen A, Laaksonen S, Beckmen K, Oksanen A, Isomursu M, Meri S. Molecular identification of *Taenia* spp. in wolves (*Canis lupus*), brown bears (*Ursus arctos*) and cervids from North Europe and Alaska. *Parasitol Int*. 2011;60(3):289–95.
22. Tsai IJ, et al. The genomes of four tapeworm species reveal adaptations to parasitism. *Nature*. 2013;496(7443):57–63.
23. Zhang G, et al. Utility of DNA barcoding in distinguishing species of the family *Taeniidae*. *J Parasitol*. 2014;100(4):542–6.
24. Pajuelo MJ, et al. Identification and characterization of microsatellite markers derived from the whole genome analysis of *Taenia solium*. *PLoS Negl Trop Dis*. 2015;9(12):e0004316.
25. Roelfsema JH, Nozari N, Pinelli E, Kortbeek LM. Novel PCRs for differential diagnosis of cestodes. *Exp Parasitol*. 2016;161:20–6.
26. González LM, et al. Characterization of the *Taenia* spp HDP2 sequence and development of a novel PCR-based assay for discrimination of *Taenia saginata* from *Taenia asiatica*. *Parasites Vectors*. 2010;3:51.
27. Yan H, et al. The nuclear 18S ribosomal RNA gene as a source of phylogenetic information in the genus *Taenia*. *Parasitol Res*. 2013;112(3):1343–7.
28. Jeon HK, Eom KS. Molecular approaches to *Taenia asiatica*. *Korean J Parasitol*. 2013;51(1):1–8.
29. Jeon HK, et al. Differential diagnosis of *Taenia asiatica* using multiplex PCR. *Exp Parasitol*. 2009;121:151–6.
30. Palomares-Alonso F, Palencia Hernández G, Rojas-Tomé IS, Jung-Cook H, Pinzón-Estrada E. Murine cysticercosis model: influence of the infection time and the time of treatment on the cysticidal efficacy of albendazole and praziquantel. *Exp Parasitol*. 2015;149:1–6.
31. Lightowlers MW. Fact or hypothesis: *Taenia crassiceps* as a model for *Taenia solium*, and the S3Pvac vaccine. *Parasite Immunol*. 2010;32(11–12):701–9.
32. Salazar AM, et al. Genotoxicity induced by *Taenia solium* and its reduction by immunization with calreticulin in a hamster model of taeniosis. *Environ Mol Mutagen*. 2013;54(5):347–53.
33. Siles-Lucas M, Hemphill A. Cestode parasites: application of in vivo and in vitro models for studies on the host-parasite relationship. *Adv Parasitol*. 2002;51:133–230.
34. Nash TE, et al. Neurocysticercosis: a natural human model of epileptogenesis. *Epilepsia*. 2015;56(2):177–83.
35. Flisser A, et al. *Taenia solium:* current understanding of laboratory animal models of taeniosis. *Parasitology*. 2010;137(3):347–57.
36. Hamamoto Filho PT, et al. Development of an experimental model of neurocysticercosis-induced hydrocephalus. Pilot study. *Acta Cir Bras*. 2015;30(12):819–23.
37. Wang IC, Guo JX, Ma YX, Chung WC, Lu SC, Fan PC. Sexual development of *Taenia solium* in hamsters from rodent-derived cysticerci. *J Helminthol*. 1999;73(4):347–50.
38. Avila G, et al. Laboratory animal models for human *Taenia solium*. *Parasitol Int*. 2006;55 (Suppl):S99–103.
39. Verastegui MR, et al, Novel rat model for neurocysticercosis using *Taenia solium*. *Am J Pathol*. 2015;185(8):2259–68.
40. Ito A, et al. Human *Taenia* eggs develop into cysticerci in SCID mice. *Parasitology*. 1997;114 (1):85–8.
41. Ito A, Ito M. Human *Taenia* in severe combined immunodeficiency (SCID) mice. *Parasitol Today*. 1999;15(2):64–7.

42. Ito A, Nakaya K, Sako Y, Nakao M, Ito M. NOD-SCID mouse as an experimental animal model for cysticercosis. *Southeast Asian J Trop Med Public Health.* 2001;32(Suppl 2):85–9.
43. Chang SL, et al. Development of *Taenia saginata asiatica* metacestodes in SCID mice and its infectivity in human and alternative definitive hosts. *Parasitol Res.* 2005;96(2):95–101.
44. Nakaya K, et al. Usefulness of severe combined immunodeficiency (SCID) and inbred mice for studies of cysticercosis and echinococcosis. *Parasitol Int.* 2006;55(Suppl):S91–7.
45. Peng SY, et al. Infection of normal C3H/HeN mice with *Taenia saginata asiatica* oncospheres. *Res Vet Sci.* 2009;86(2):261–6.
46. Cruz-Rivera M, et al. Cytokine expression at the anchor site in experimental *Taenia solium* infection in hamsters. *Vet Parasitol.* 2014;200(3–4):299–302.
47. Garcia HH, Rodriguez S, Friedland JS; Cysticercosis Working Group in Peru. Immunology of *Taenia solium* taeniasis and human cysticercosis. *Parasite Immunol.* 2014;36(8):388–96.
48. Peón AN, Ledesma-Soto Y, Terrazas LI. Regulation of immunity by taeniids: lessons from animal models and in vitro studies. *Parasite Immunol.* 2016;38(3):124–35.
49. Chowdhury N, Saleque A, Sood NK, Singla LD. Induced neurocysticercosis in rhesus monkeys (*Macaca mulatta*) produces clinical signs and lesions similar to natural disease in man. *Sci World J.* 2014;2014:248049.
50. Ito A, et al. Recent advances in basic and applied science for the control of taeniasis/cysticercosis in Asia. *Southeast Asian J Trop Med Public Health.* 2002;33(Suppl 3):79–82.

50

Trichinella

Ljiljana Sofronic-Milosavljevic, Natasa Ilic, and Alisa Gruden-Movsesijan

CONTENTS

50.1 *Trichinella* and Trichinellosis

Trichinellosis is a foodborne disease caused by the consumption of raw or undercooked meat of different animal origins, e.g., pork, horse, or game, infected with helminth of the *Trichinella* genus [1,2]. The genus consists of two clades that inhabit mainly striated muscles of the host. Parasites from the first clade encapsulate, and those from the second do not [3]. Among the so-far identified 12 taxa of the genus, the species and genotypes of the first clade parasitize only mammals and include *T. spiralis*, *T. britovi* with its closely related genotype T8, *T. nativa* and related T6, *T. murrelli* and related T9, *T. nelsoni*, and *T. patagoniensis*. The second clade comprises three species: *T. pseudospiralis*, which infects mammals and birds, and *T. papuae* and *T. zimbabwensis*, which parasitize mammals and reptiles, respectively [3].

Although *Trichinella* has cosmopolitan distribution and infects more than 2500 people annually [4], its infection results in minimal health burden and low mortality rate in human population [5]. Of all *Trichinella* taxa, *T. spiralis* accounts for most human infections [2].

Trichinella spp. are unique among various helminthes in that all three life cycle stages of the parasite-infective muscle larvae, adult, and newborn larvae develop in one host. Infection is acquired by consumption of infected raw or undercooked meat. Under the influence of gastric juice, larvae are released from infected meat in the stomach, molt, and develop into the adult stage inside the enterocytes of small intestine. After mating, newborn larvae move into the circulation and spread throughout the tissues and organs, and only those that penetrate striated muscles mature into muscle larvae [6]. This parasite has ability to transform the infected muscle cell into a new type of cell in the host body, the so-called nurse cell (NC) [7,8]. From this immunologically privileged place, parasite achieves long-lasting communication with the host through muscle larvae excretory-secretory products (ES).

Unlike what happens in humans, *T. spiralis* can reach a high worm burden in animals without inducing clinical symptoms [2]. Although infectivity for laboratory animals differs among various *Trichinella* taxa, infection provides valuable *in vivo* models for basic research.

This chapter reviews the most significant observations gained in biological, pathological, and immunological studies performed on rodent models (inbred mice and rats, or genetically manipulated), which have the benefit of being easy to handle and relatively cost-effective. It also covers data from some larger animal models using domestic and wild swine, horses, and other extremely costly animals that host *Trichinella* species (including even marine mammals and reptiles).

50.2 Mouse

Mice are the most frequently used mammalian model system due to their close genetic (over 95% of the mouse genome is identical to human genome) and physiological similarities to humans. Creation of various inbred strains by genetic manipulation, as well as knock-out and transgenic animals, enabled studies of complex phenomenon like susceptibility to the infection and immune responses elicited.

Trichinella is not a picky nematode, as it invades and completes its life cycle in the variety of host organisms, including mice. Among *Trichinella* species, the most extensively studied is *T. spiralis*. Immune response to primary infection with this intestinal and tissue-dwelling parasite starts with recognition of the antigens and proceeds to the inflammation of the intestinal mucosa, leading to the reduction of female fecundity, cytopathological damage, and expulsion of adults from intestine during primary infection, whereas in challenge infection, it induces accelerated expulsion of larvae and adults from the intestine and reduction of growth and fecundity [9,10]. Differences in the resistance to *T. spiralis* infection between various strains have a genetic background. Some strains possess alleles that enable rapid (strong) response, which means that they expel worms more quickly, reduce female worm fecundity, and as a consequence have lower muscle larvae burden than slow (weak) responders, which possess another set of alleles [9,11]. Although Wakelin [9] used time of expulsion as a parameter for determining the response phenotypes of inbred strains of mice infected with *T. spiralis*, the number of recovered larvae from the muscles should also be considered for distinguishing between susceptible and resistant strains, since it takes into account all variables that affect the response to infection.

50.2.1 Contribution of MHC-Linked and Background Genes to the Control of the Response

Response to primary *T. spiralis* infection is governed by more than one gene, located within and outside the mouse major histocompatibility complex *H-2*, judged by the results of crossing experiments involving various types of responders [12,13]. *H-2* genes code for proteins that are directly or indirectly involved in antigen presentation to T cells, thus controlling lymphocyte responsiveness to infection. First experiments were conducted using panel of B10-background congenic mice, baring various alleles at *H-2* loci [12]. Mouse strain B10.BR carrying $H\text{-}2^k$ haplotype was shown to be associated with susceptibility, whereas mouse strain bearing *H-2q* (B10.Q) and *H-2s* (B10.S) were both resistant to *T. spiralis* infection. Investigations conducted on *H-2* recombinant mice revealed that at least two genes, mapping within this locus, were involved in the resistance to *T. spiralis*, and they were designated as *Ts-1* and *Ts-2* genes [12]. The rapid expulsion responses after challenged infection are also controlled by *H-2* and non-*H-2* genes [14].

Another set of experiments explored the role of non-*H-2* genes in the rejection process, and for that reason, mice with identical *H-2* genes but different backgrounds were used [13,15]. Wassom et al. [15] using susceptible $H\text{-}2^k$ strains, and Wakelin and Donachie [13] using resistant $H\text{-}2^q$ haplotype, came to similar conclusion that the time of expulsion is controlled by non-*H-2* genes.

For discovering genes involved in the resistance to *T. spiralis* infection, crosses among strains with large phenotypic differences (strong responder—NFS, intermediate responder—C3H, and weak responder—B10.BR) and segregation analyses of crosses, intercrosses, and backcrosses among phenotypic generations were analyzed [16]. The results pointed out that there are at least two dominant non-*H-2* genes that control rejection of adult worms. The presence of both of them generates strong responding phenotype, while each gene on its own is responsible for intermediate phenotype.

50.2.2 Response of the Host during Intestinal and Muscle Phase of the Infection

Response of the host implies immunological and nonimmunological mechanisms at the level of intestine and muscles, engaged in the process of defense from *T. spiralis* invasion. Using genetically defined strains of mice or mice with targeted deletions in cytokine or cytokine receptor genes, it is possible to identify particular components and mechanisms of the host response important for the outcome of *T. spiralis* infection. Effective immunity in intestinal phase of *T. spiralis* infection is crucial for the establishment of parasitism, since it controls elimination of adult worms from the gut, fecundity of female parasites, and therefore the number of newborn larvae that migrate toward muscle cells. Parasite provokes T-cell-mediated immune response at intestinal level, which is accelerated upon reinfection [17]. Investigation encompassing four strains of inbred mice that exhibit a range of responsiveness to *T. spiralis* infection—NIH (high), CBA (intermediate), C57BL/10 (intermediate to low), and B10.BR (low)—revealed that the resistance is correlated with the intensity of intestinal inflammatory responses. NIH and CBA mice were the most resistant to *T. spiralis* and have the most intense intestinal inflammatory responses [18]. Intestinal inflammation is absent in congenitally athymic mice infected with *T. spiralis* [19], indicating an important role of T cells, especially CD4+ T cells [20] in host-protective immunity. *T. spiralis* induce inflammatory responses by distorting the gut epithelial cell layer. This is a signal that generates release of Th1 cytokines such as IFN-γ within 2 days [21], followed by predominant Th2 response that occurs between days 2 and 8 in all the aforementioned strains [18].

Th1 type of response leads to susceptibility to worm infection. Mice deficient in cytokine IFN-γ expel *T. spiralis* more rapidly than wild-type mice [22]. Th1 type responses are under control of proinflammatory cytokine IL-12, produced by dendritic cells and macrophages. Transgenic mice overexpressing IL-12 show delayed worm expulsion, accompanied with increased IFN-γ and decreased IL-4 and IL-13 production [23].

Resistance to *T. spiralis* infection is mediated by Th2 type of response [24], primarily IL-4, IL-9, and IL-13 cytokines [17]. The presence of type-2 cytokines causes eosinophilia, intestinal mastocytosis, and elevated IgE production. Primary infection of C57BL/6 IL-4 knock-out (KO) mice with *T. spiralis* pointed out the importance of IL-4 in the expulsion of parasites [25], and treatment with anti-IL-4 receptor monoclonal antibody caused prolonged adult infections and higher muscle larvae burdens [26]. IL-4 deficiency also leads to impaired Th2 antibody response, both IgE and IgG1. However, BALB/c IL-4 KO mice showed no difference in the time of expulsion or in enteropathy from wild-type mice. Moreover, expulsion was delayed in wild-type C57BL/6 mice compared to BALB/c mice with no significant differences in Th2 responses, which indicates that the function of cytokines depends on genetic background [27]. Studies with IL-4-receptor-α deficient mice proved that expression of this receptor on bone-marrow- and non-bone-marrow-derived cells has a crucial role in worm expulsion and thereby in host protection [28]. Another important cytokine in Th2 response—IL-13—which shares the same IL-4 receptor α chain (IL-4Rα), stimulates B cell proliferation and antibody class switching to IgE, eosinophilia, and mastocytosis [29,30]. Binding of either of these cytokines to this receptor results in the phosphorylation and activation of signal transducer and activator of transcription factor 6 (Stat6), which leads to eosinophilia, mastocytosis, and expulsion of *T. spiralis* [22,31]. Degranulation of mast cells that occur in response to *T. spiralis* antigens increases the permeability of intestinal epithelial cells, thus promoting parasite expulsion [32]. Signaling via Stat6 also induces intestinal smooth muscle cell hypercontractility, another mechanism necessary for efficient worm expulsion [33]. In *T. spiralis* infection, intestinal muscle hypercontractility is under control of Th2 responses [34]. Hypercontractility does not occur in athymic mice and in animals with major histocompatibility class II or CD4+ cell deficiencies, implicating T cell control [35,36]. A shift to Th1 response via an overexpression of IL-12 significantly inhibits *T. spiralis*-induced muscle hypercontractility and delays worm expulsion [23].

Th2 cytokines IL-3 and IL-5 are also produced during *T. spiralis* infection [24]; however, treatment with anti-IL-3 and IL-5 antibodies, or using IL-3 KO mice, revealed that these cytokines are not significant for host protection. Since IL-5 controls eosinophilia provoked by *T. spiralis* [37], it could be concluded that eosinophils are not essential for the expulsion of the parasites and thus do not affect host protection, although they exert larval cytotoxicity in *T. spiralis* infection [38]. However, experiments with IL-5-deficient mice revealed that IL-5 and eosinophils are involved in rapid expulsion during challenge (secondary) *T. spiralis* infections [39,40]. Another role for IL-5 and eosinophils was suggested in a model of *T. spiralis* infection in mouse strains that constitutively overexpress IL-5. It appears that high levels of IL-5 and eosinophilia may be advantageous for the parasite, since the number of muscle larvae is much higher in transgenic mice than in wild-type [41]. This was confirmed by the work of Huang et al. [42], who pointed out immunoregulatory role of eosinophils in the early muscle phase of the infection. They have found that eosinophils produce IL-10 necessary for the expansion of IL-10-producing DCs and CD4+ T cells, which then inhibit inducible nitric oxide synthase (iNOS) expression and the production of NO, thus protecting intracellular larvae.

Using IL-9 transgenic mice that constitutively overexpress this cytokine, the role of IL-9 in resistance to *T. spiralis* was revealed for the first time [43]. Enhanced level of IL-9 *in vivo* caused high levels of parasite-specific IgG1, marked intestinal mastocytosis, and accelerated worm expulsion. IL-9 might mediate effective expulsion of *T. spiralis* by enhancing the release of mouse mast cell protease-1 (mMCPT-1). The involvement of IL-9 in mucosal immunity, i.e., recruitment and/or survival of mast cells, was suggested [44]. These data imply that IL-9 is a specific effector molecule involved in host defense against *T. spiralis* infection.

Mast cells appear to be involved in the expulsion in nematode infections [45]. Mice in which mast cell depletion was induced either by genetic manipulation [46] or by application of specific antibodies [47] exhibit a significant delay in expulsion of *T. spiralis*. Mast cells contribute to intestinal inflammation through the production of proinflammatory cytokines, including tumor necrosis factor, proteinases, and other inflammatory mediators such as histamine, leukotrienes, and prostaglandins [48]. It appears that mastocytosis in the infected mucosa may represent an immunopathological rather than a protective response. Mice deficient in TNF receptor, infected with *T. spiralis*, exhibit minimal pathology in the gut and the absence of intestinal mastocytosis, and yet they are able to expel adult worms. TNF may mediate pathological effects in gastrointestinal helminth infections via the production of iNOS [49]. iNOS-deficient mice infected with *T. spiralis* show reduced pathogenic but not protective response to *T. spiralis* [50].

When newborn larvae reach striated muscle tissue, they invade muscle cells and rearrange them to form a completely new entity, the NC [7]. NC is also composed of satellite cells that undergo cell division and join the invaded muscle cell [8,51,52]. Tissue damage provoked initially by parasite penetration accounts also to the presence of inflammatory cells, which produce high levels of reactive oxygen species and other free radicals, after activation [53]. These actions result in an inflammatory response, which causes transient myositis.

Inflammatory response to various encapsulated and nonencapsulated *Trichinella* species differs [54]. *T. pseudospiralis*, a nonencapsulated species, induce a lower inflammatory reaction around the parasites than encapsulated species (*T. spiralis* and *T. britovi*). There is also difference in the level of inflammation between encapsulated species; that is, *T. spiralis* is accompanied by the higher inflammatory response, compared to that induced by *T. britovi* [55].

The immunological response to *T. spiralis* muscle invasion in mice is mixed, although mainly Th2 type, characterized by the production of IL-5, IL-10, IL-13, and IFN-γ [56] and by the presence of parasite-specific IgG1 and IgE during the chronic infection [57]. Eighty percent of IgG1 specific for larval antigens recognize a single shared epitope, highly immunogenic sugar, tyvelose [58]. Although chronic, lifelong infection in mice is characterized by persistent B cell response, local inflammatory response to the *T. spiralis*–NC complex is limited, suggesting the existence of suppressive parasite and/or host factors.

The cellular infiltrate surrounding the NC is composed of macrophages, able to invade the cytoplasm of the NC [57]. Among other cells (CD4+ T cells, CD8+ T cells, Treg, B cells), eosinophils enter sites of infection immediately after tissue invasion by *T. spiralis* larvae, and by producing IL-10, control the activation of proinflammatory macrophages and neutrophils that otherwise kill parasite larvae by releasing NO [42]. Inflammatory infiltrates fail to form around NCs in T-lymphocyte-deficient mice, identifying these cells as the coordinators of the cellular response to muscle infection [59]. In IL-10 KO mice, the extent of the inflammatory infiltrate around the NC was markedly increased compared to control mice during the acute phase of the infection, although the cellular composition remained the same [57,60]. The other source of IL-10 that participates in reduction of inflammation is CD4+ CD25-T effector cells [56]. Another cytokine with immunosuppressive effect on T cells is TGF-β, produced by regulatory T cells (Treg) [61], which acts together with IL-10 to control local inflammation. When IL-10 KO mice were treated with anti-TGF-β, they develop a strong inflammation around the NC–parasite complex with a worm burden significantly reduced compared to control mice [56]. Control of inflammation in the chronic phase of the infection is IL-10-independent and could be ascribed to strong Th2 response [57].

50.2.3 Applied Studies on Mouse Model for Trichinellosis

Mouse models of *Trichinella* infection represent useful tool either for investigation of efficacy of anti-helminthic drugs [62–64] or for study of mechanisms involved in immunomodulation of inflammatory diseases or tumors by parasite (reviewed in Ref. [65]).

To control trichinellosis, transmission from animal hosts (usually pigs) to humans should be prevented, and this could be achieved by successful prophylactic vaccination. However, this goal has not yet been reached. Most of the studies concerning development of vaccines against *Trichinella* infection were conducted on a mouse model using at first whole extracts or ES products of the parasite. These vaccines were more efficient (50%–80% of protection) than those composed of isolated parasite antigens, as seen by the reduction in larval or adult worm burden [66,67]. Implementation of more refined vaccines using recombinant or synthetic proteins or peptides, with different immunogenic properties, or third generation of DNA vaccines made of plasmids containing sequences coding for foreign proteins, resulted in variable protective effect (reviewed in Ref. [68]).

50.3 Rat

While carnivores are considered as particular and permissive species for *Trichinella* infections, rats are considered as selective hosts in which only a restricted number of species other than *T. spiralis* reproduce through either natural [69,70] or experimental infection [71]. As experimental model, rats are used for the investigation of *Trichinella* life cycle, host–parasite relationship, and immune response provoked by this parasite as well as for investigation of parasite's potential to manipulate the immune response of the host and to suppress some severe diseases.

Since they are characterized as selective host species for *Trichinella* infection, rats have been used as a model for investigation of *Trichinella* species-specific defense. These experiments were performed by comparing *T. spiralis* and *T. nativa* infections in Wistar rats [71,72]. In *T. spiralis*-infected rats, muscle larvae burden and *in vitro* newborn larvae production were many fold higher compared to *T. nativa*-infected rats, which indicates that the infectivity of *T. nativa* is lower compared to *T. spiralis*. No differences between two *Trichinella* species were observed in intestine histology, serological response, and weight dynamics of infected rats. This helped understanding that defense against different *Trichinella* species not only is solely enteral, but also happens in the parenteral phase of the parasite's life cycle.

Investigations of the influence of host genetics on the susceptibility to primary *T. spiralis* infection on rat model system gave valuable contribution to understanding of parasitological and parasite-induced immunological differences between strains. Rat strain variation in response to *T. spiralis* infection examined on five inbred strains (AO, LOU, PVG, WKA/H, and LEWIS) was evaluated through different rate of worm elimination in each strain [73]. LOU and LEWIS rats were the most efficient in expelling *T. spiralis*. In these two rat strains, the lowest level of serum IgE was found, while the intestinal IgE response was the most pronounced. Variations in rat response to *T. spiralis* were amended by investigation of interstrain differences with respect to muscle larvae burden between AO and DA rats [74], and it was shown that DA rats express higher susceptibility in comparison with AO rats.

Immune responses of rats to *Trichinella* infection during the intestinal phase involve both Th1 and Th2 cell subsets [75]. During the muscle stage of infection, with the emphasis on rat strain differences, *T. spiralis* provoked the increased level of anti-inflammatory cytokines (IL-4 and IL-10), indicating a shift toward Th2 and regulatory response, without the restriction of Th1 type of response in DA rats, while the response in AO rats remained in favor of Th1 type [74]. *In vivo* application of dendritic cells (key cells in the initiation and regulation of immune response) primed with *T. spiralis* ES products creates an immune status that resembles the one observed during *Trichinella* infection [76].

Research on effector mechanisms involved in protection during the intestinal phase of the *T. spiralis* infection indicated that the rapid worm expulsion is IgE-mediated and T-cell-mediated [10]. Extensive investigation of the relationship between intestinal immune responses and worm expulsion, which occur upon the infection with *T. spiralis*, performed on several rat strains (BN strain—mast cell and IgE high responders; F344—low responders on the level of mast cells and IgE activation; DA strain—intermediate responders comparing to BN and F344; Ws/Ws strain—mast cell deficient rats), revealed that mast cells in addition to IgE could contribute to the worm expulsion following *T. spiralis* infection in rats [77].

Studies of host–parasite relationships in *Trichinella*-infected animals with respect to different mediators involved in infected muscle cell transformation also used rat model system. Potential differences in signals derived or provoked by different *Trichinella* species (encapsulated and nonencapsulated) were studied regarding the production of NO, a known cell-to-cell mediator. It was found that host macrophages differ in NO production between encapsulated and nonencapsulated *Trichinella* species [78]. Macrophages stimulated with *T. pseudospiralis* muscle larvae antigens produced increased levels of NO only at the highest antigen concentrations, while *T. spiralis* antigens induced increased production of NO even when applied in much lower concentrations compared to doses of nonencapsulated *Trichinella* antigens.

Investigation of the dynamics of antibody response was performed on Wistar rats using different doses of *T. spiralis* muscle larvae. It was demonstrated that rats survive the high infection doses (up to 16,000 muscle larvae) and that low doses (10 muscle larvae) can be infective and can cause detectable levels of specific antibodies [79]. Time of seroconversion and antibody titers showed to be dose-dependent in rats.

Significant contribution to the body of evidence considering biochemical parameters of the host and their changes during the infection with *Trichinella* was given by the study of lipid status, levels of enzyme with antiatherogenic properties (paraoxonase-1), and oxidative stress in rats infected with this parasite [80]. It was demonstrated that during the infection, serum activity of cardioprotective enzyme paraoxanase-1 was significantly reduced accompanied with the increased levels of triglycerides and LDL cholesterol and decreased levels of HDL.

Very important information regarding the potential of *Trichinella* infection or its products to manipulate the host immune response and to suppress some severe diseases like experimental autoimmune encephalomyelitis (EAE), as experimental model of multiple sclerosis, were obtained in experiments performed on DA rat model system. It was demonstrated that chronic infection with this parasite as well as the application of *Trichinella* ES products, or dendritic cells stimulated with these products, has a strong impact on the course of EAE, i.e., on the reduction of the disease severity by shifting the immune response toward Th2 and regulatory type [81,82].

50.4 Swine

Although human *Trichinella* infections can be a consequence of the consumption of infected meat originating from different animal species including feral animals, *T. spiralis*-infected pork meat still remains the main source of the infection [2,83].

Owing to the model of experimentally infected pigs, predilection sites for *T. spiralis* and other *Trichinella* species were discovered in these animals [84,85]. It was established that all *Trichinella* spp. accumulate preferentially in diaphragm, tongue, and masseter of infected pigs. Investigation of infectivity of the *Trichinella* genotypes to pigs showed that *T. spiralis* is highly infective and *T. britovi* and *T. nelsoni* are moderately infective but as persistent as *T. spiralis*, while *T. nativa*, *T. murrelli*, *T. pseudospiralis*, and *Trichinella* (T6) are low infective and nonpersistent [86].

Several methods such as trichinelloscopy, artificial digestion, immunofluorescence assay (IFA), and ELISA were compared for establishing reliable test for detection of experimental porcine trichinellosis. It was concluded that protection of public health from *Trichinella* in pork and prevention of the infection spreading can be ensured only by application of parasitological methods and, in particular, artificial digestion [87,88]. This was due to the finding that digestion detects infection as early as the larvae enter and encapsulate in muscle tissue i.e., 17–21 days post infection (p.i.), while serological methods cannot detect infection before seroconversion occurs, i.e., mean times were 5–6 weeks p.i. [88]. Although serology cannot be used as a procedure to certify the safety of pork meat, it could be a useful method for surveillance and epidemiological studies of swine *Trichinella* infection [89–91].

Trichinella-specific antibodies in swine could be detected by 1.5 years p.i. [92], and specific antibodies raised by the infection with any of 12 different taxa of the genus *Trichinella* could be reliably detected by one common antigen since all taxa share similar antigenic pattern [2,86,90,93]. The investigations performed on experimentally infected pigs provided significant impact on the development of serological tests, and defining their performances and the identification of highly immunogenic antigens enabled the improvement of diagnostic tests. Diagnostic tools for antibody detection in swine evolved from application of the whole larvae in IFA, via total larvae extract and up to ES or its components that belong to TSL-1 in ELISA and western blot [87,94,95]. The specificity of anti-*Trichinella* antibodies relies on recognition of antigens from TSL-1 group, which consists of protein and unique sugar component (tyvelose) [96]. Experimentally infected pigs are currently used for development of international biological standards or reference materials in an aim to validate in-house tests or commercial kits and to improve the interlaboratory comparability for the serological detection of anti-*Trichinella* IgG in pigs [97].

Trichinella larvae recovered from experimentally infected pigs were used during development and evaluation of specificity and sensitivity of polymerase chain reaction methods for the identification of species or genotypes of *Trichinella* single muscle larvae [98,99].

Besides the use of the aforementioned outbred swine, few studies that used SLA-inbred miniature swine proved to be an important experimental model for mapping of genetic resistance to *Trichinella* infection. The influence of genes within the MHC (SLA) versus background (non-MHC) genes on the initiation, maturation, and effector arm of host antiparasite defense mechanisms was investigated. Swine selectively bred so that they are homozygous at their *SLA* complex for three *SLA* haplotypes, *aa*, *cc*, or *dd*, were used to determine whether those genes regulate swine immune response to *T. spiralis* [100]. The obtained results indicated that swine with $SLA^{c/c}$ haplotype have lower worm burden correlating with the earlier development of humoral antibody response. This and work of other authors implied that this "resistance" to infection relies not only on *SLA* locus (expression of one allele from SLA^{a} haplotype) but also on other non-*SLA*-encoded genes [101].

Some extension of this work performed on outbred swine indicated that besides the role of SLA class II molecules, SLA class I phenotype also influences mechanisms that mediate the presentation of *Trichinella* antigens and consequently the host immune response. Swine that expressed particular SLA I phenotype (2–12–3 positive) found to be more resistant to infection (with lower larvae burden), compared to negative animals [102].

50.4.1 Applied Studies on Swine Model for *Trichinella* Infection

Experimental infection in swine was used for evaluation of some other aspects of *Trichinella* infection model. One of them is the effect of concomitant infection of *T. spiralis* with two other helminthes *Ascaris suum* and *Metastrongylus apri*. Data obtained in this study indicated that *T. spiralis* had a synergistic interaction with *M. apri* (higher establishment of larvae in the lungs), and an antagonistic interaction in concurrent infection with *A. suum* (reduced numbers of migrating larvae and juvenile worms) [103]. Another model combined *T. spiralis* and *T. gondii* and was used for evaluation and validation of bead-based assay for simultaneous detection of anti-*T. spiralis* and anti-*T. gondii*-specific antibodies in swine serum samples [104]. Experimentally infected swine were also the source of sera used for identifying several immunodominant *T. spiralis* adult antigens, which may present promising candidates for development of an early diagnostic tool [105].

Among the trials that aimed to develop vaccine against establishment of *Trichinella* infection in swine, Marti et al. [106] showed that vaccine based on newborn larvae (NBL) antigen was considerably more protective than the one based on muscle larvae stichosomal fractions or ES proteins [107,108]. Experimentally infected swine are currently in use for determination of specific epitopes on NBL [109], adult and muscle larvae antigens [110], targeted by host antibodies, which could potentially be used for new approach to vaccine development as well as for early diagnosis of *Trichinella* infection.

50.5 Wild Boar

Wild boar infections with both domestic and sylvatic *Trichinella* spp. occur quite frequently in Europe, North America, Asia, and Africa, and infected meat of this animal species presents an important risk factor for human *Trichinella* infection [111,112]. First experimental *Trichinella* infection was performed to determine infectivity, muscle larvae distribution, and ability of investigated species to generate immune response in wild boars [113]. This study revealed that *T. spiralis* is highly infective for wild boars; *T. britovi*, *T. nelsoni*, and *T. pseudospiralis* are moderately infective, while *T. nativa*, *T. murrelli*, and *Trichinella* T6 are not well adapted for this host species. Predilection sites, independently of *Trichinella* genotype and infective dose, are diaphragm and tongue [113,114]. Antibody response increased rapidly from week 3 to 5 p.i. for all investigated *Trichinella* genotypes except for *T. pseudospiralis* where it increased gradually. For encapsulated species of *Trichinella*, high antibody level remained until the end of examined period (10 weeks), whereas for nonencapsulated species, antibody levels declined during the postinfection period. Data concerning wild boars correlate with those observed in domestic pig model system [86].

A long-lasting experiment was designed using wild boars to investigate the freezing tolerance of *Trichinella*-encapsulated muscle larvae *T. spiralis* and *T. britovi* [115]. It was demonstrated that muscle larvae of both *Trichinella* species could be inactivated after 1 week on $-21 \pm 2°C$. In contrast to this finding, Gari-Toussaint et al. [116] reported six cases of human trichinellosis originating from the consumption of *T. britovi*-infected wild boar meat that had been frozen for 1 week at $-35°C$. It has been suggested that wild animals' muscles contain substances that protect *Trichinella* muscle larvae from freezing [117]. Hence, although the study on experimentally infected wild boars provided some new data on *Trichinella* freezing tolerance, freezing method is not recommended as a control measure to inactivate this parasite in game meat [89].

50.6 Horse

Infected horse meat has been recognized as a source of human trichinellosis since 1975 when the outbreaks were reported for the first time in Italy and France [118,119]. Because horse has been considered as an atypical host for *Trichinella* spp., many aspects of the biology and epidemiology of *Trichinella* infection in the horse were poorly investigated and understood during the previous century. Still, being recognized as a serious food safety risk for consumers of fresh horse meat, the infection was started to be investigated in experimental horse model from the 1970s. Those very early studies were aimed to detect muscle larvae and specific antibodies presence during the course of the infection, but latter investigations included the route of infection, tissue distribution of muscle larvae, parasite life span, and assessment of reliability of parasitological and serological diagnostic methods for the detection of infected animals.

The first report on distribution of *Trichinella* muscle larvae (L1) in individual muscles was published by Pampiglione et al. [120] referring to the examination performed on 6-month-old horses infected with 10,000 L1 in which the highest muscle larvae burden was detected in tongue, masseter, and diaphragm. In this millennium, all three aforementioned muscles were confirmed to be predilection sites for *Trichinella* muscle larvae [114]. The study included 30 horses experimentally infected with three *Trichinella* genotypes: *T. spiralis*, *T. britovi*, and *T. pseudospiralis*. Limited study on three ponies, [121] as well as studies of other authors [92,122,123], corroborated that the tongue is the main predilection site. Results obtained in experimentally infected horses were in accordance with those observed in naturally infected animals [124], and they supported the contention that food safety inspection of the horse meat should target the tongue as a tissue most promising to yield positive results in horses infected with *Trichinella*.

Several studies have been undertaken to describe the infection in this atypical host [123,125–127]. Horses mostly do not develop the disease during *Trichinella* infection, except the rare transient muscular disturbance observed between 2 and 6 weeks p.i. in case of animals infected with a high dose (50,000) of L1 larvae. Moreover, hematological parameters were not significantly changed, except for the level of eosinophils, which was slightly increased (up to 15%) independently of infective larvae dose. Levels of enzymes that indicate the muscle damage (lactate-dehydrogenase, aldolase, and creatine phosphokinase) were increased from 2 to 17 weeks p.i. [125].

Experimental infection of horses with *Trichinella* spp. was used for the examination of kinetics of antibody production and dependence of antibody response on different *Trichinella* spp. that cause the infection in horses. Detection of specific, anti-*Trichinella* antibodies depend on *Trichinella* species (*T. spiralis* induced higher level of anti-*Trichinella* IgG than *T. murrelli* due to its antigenicity), the infective dose applied for horse infection, as well as on the antigen used for antibody detection [124]. A number of studies demonstrated that in high-dose infection (more than 20,000 L1), anti-*Trichinella*-specific IgG appears between 2 and 5 weeks p.i. and disappears between 16 and 40 weeks p.i. [120,125,128], judging by indirect IFA and ELISA tests. Low doses of infective larvae (1100 L1) also induce antibody production detectable by ELISA for up to 18 weeks p.i. [129]. The short period of time (in average up to 26 weeks p.i.) during which anti-*Trichinella* antibodies in infected horses could be detected by ELISA makes this kind of test unreliable for detection of *Trichinella* infection in horses [119,125,130]. On the other hand, other serological methods were proven to be effective in specific antibody detection for a longer period (IFA of paraffin sections of L1 and western blot analysis of ES and tyvelose-BSA, at least 8 months) [129]. It was supposed that for diagnosis of horse trichinellosis, additional tests including antigen detection should be performed. One such test was a dot-blot test for the detection of *Trichinella* ES muscle larvae antigens in circulation, which allowed the detection between weeks 4 and 32 p.i. in the sera of all experimentally infected horses [131].

The most extensive and long-term experimental study was performed on 35 horses infected with 1000, 5000, or 10,000 *T. spiralis* muscle larvae in which the course of infection was followed for 1 year [92]. It was shown that the persistence of infective muscle larvae is at least 1 year, but serological response could be detected only by 26 weeks p.i. (ELISA with ES). This study offered the conclusion that parasite recovery methods are the only suitable detection assays for both meat inspection and epidemiological studies of *Trichinella* infection in horses.

The question considering the enigma of early disappearance of detectable antibody response in horses versus the existence of live muscle larvae raised by this research could partly find the answer in data showing the presence of seven IgG subclasses in horse and five other isotypes [132,133] and the presumption that anti-horse IgG used in previous studies may be inappropriate for detecting subclass-specific anti-*Trichinella* reactivity in the horse [92].

Since horses are herbivores and nevertheless, natural infections with *Trichinella* occur from time to time, an intriguing question considering the source of infection emerged. Eating behavior of these animals studied by intentional feeding of horse with meat products [134] revealed that 31% of horses readily consumed it and that this feeding practice could be a source of *Trichinella* infection.

50.7 Other Animal Species as Models for *Trichinella* Infection

In order to investigate the susceptibility to *Trichinella* infection, define predilection sites as well as freeze tolerance for muscle larvae and determine serological response parameters in wildlife. Some animal species that are, conditionally speaking, unusual for experimental purposes were used as animal models in such surveys.

Studies comparing the muscle distribution of different *Trichinella* genotypes were performed, among other species, on foxes. In these animals, where all *Trichinella* species were established in high numbers, the encapsulating species were found primarily in the tongue, extremities, and diaphragm, whereas the nonencapsulating species were found primarily in the diaphragm [114]. Foxes were one of the experimental models used for studies of *Trichinella* larvae freeze tolerance. Since *T. nativa* is the most frequent *Trichinella* species in arctic wildlife and well adapted to the cold climate, it was useful for testing the tolerance on repeated freezing and thawing. Results obtained in experimentally infected foxes showed that *T. nativa* isolates originating from carnivores from higher northern latitudes expressed highest tolerance to freezing and that temperature fluctuation around freezing point did not influence larval infectivity [135]. Foxes were also experimentally infected with *T. nativa* for the purpose of serological survey, which showed that western blot is a more reliable method compared to ELISA for antibody detection in this animal species [136]. Potential of *Trichinella* to cross the placental barrier was investigated on experimentally infected foxes, and it was found that, unlike some other animal species (ferrets, guinea pigs, and mice), vertical transmission, measured as recovery of muscle larvae in the offspring, does not happen in foxes [137].

Since marine mammals may acquire *Trichinella* infection in nature by scavenging even small amounts of infected tissue left by hunters or predators, the important data considering infection in this group of animals were obtained on seals experimentally infected with *T. nativa* in order to determine the susceptibility of seals to this parasite infection and freeze tolerance of larvae in the seal meat [138]. Seals showed high susceptibility to *Trichinella* infection and measurable antibody response unrelated to the infection dose. Experiments investigating freeze tolerance of muscle larvae in the meat indicated that there was correlation between resistance to freezing and the age of host–parasite tissue complex; that is, freeze tolerance increases with time post infection. The results obtained in this study were complemented with research in which the meat of seals experimentally infected with *T. nativa* was used for the preparation of traditional northern foods in an aim to assess the infectivity of larvae present in those foods [139]. Data indicated significant food safety risk associated with the food containing infected seal meat.

One of the most unusual experimental models for trichinellosis is caiman. Notification of *Trichinella zimbabwensis* (nonencapsulated species) infection in farmed crocodiles in Zimbabwe due to the feeding practice [140] raised a question of susceptibility of reptiles to infection with other *Trichinella* spp. The experimental infection of caimans with different *Trichinella* isolates showed that these animals can be infected with nonencapsulated *T. pseudospiralis*, which is, from evolutionary point of view, likely to infect reptiles since it can be infective for both birds and mammals, while the infection with encapsulated species was impossible [141].

50.8 Conclusion

Animal models for *Trichinella* and trichinellosis are a tool that allows detailed investigation of host–parasite relationships. The host's susceptibility to different *Trichinella* genotypes indicates the impact of genetic background of the host and the parasite on the parasitism establishment. The vast range of experimental animal models reveals insight into parasite muscle distribution and predilection sites, which form the basis of meat sample properties for accurate parasitological detection and diagnosis. Studies using experimental *Trichinella* infection yield valuable information on the course of innate and adaptive immune response. Knowledge about dynamic of specific antibody synthesis provides a guide for designing and implementing specific and sensitive serological tests in surveillance, monitoring, and control programs. The rodent models highlight the importance of antigen-presenting cells to drive T-helper type 2 and T-regulatory responses in *Trichinella* infection that are responsible for parasite long-lasting survival and creation of tolerogenic environment in the host body. Animal models have been, and will continue to be, vital in helping elucidate the mechanisms that underline the host–parasite coevolution, which is, among others, responsible for the training of host immune systems for accurate discrimination between foreign and self-antigens.

REFERENCES

1. Gottstein B, Pozio E, Nöckler K. Epidemiology, diagnosis, treatment, and control of trichinellosis. *Clin Microbiol Rev.* 2009; 22:127–45.
2. Pozio E, Zarlenga DS. New pieces of the *Trichinella* puzzle. *Int J Parasitol.* 2013; 43:983–97.
3. Pozio E, Hoberg E, La Rosa G, Zarlenga DS. Molecular taxonomy, phylogeny and biogeography of nematodes belonging to the *Trichinella* genus. *Infect Genet Evol.* 2009; 9:606–16.
4. Murrell KD, Pozio E. Worldwide occurrence and impact of human trichinellosis, 1986–2009. *Emerg Infect Dis.* 2011; 17:2194–202.
5. Devleesschauwer B, et al. The low global burden of trichinellosis: evidence and implications. *Int J Parasitol.* 2015; 45:95–9.
6. Bruschi F, Gómez-Morales MA. The translational immunology of trichinellosis: from rodents to humans. In: Jirillo E, Magrone T, Miragliotta G. (eds) *Immune Response to Parasitic Infections: Immunity to Helminths and Novel Therapeutic Approaches.* Bantham E-Books, Beijing, 2014; 2:125–61.
7. Despommier DD. How does *Trichinella spiralis* make itself at home. *Parasitol Today.* 1998; 14:318–23.
8. Wu Z, Sofronic-Milosavljevic L, Nagano I, Takahashi Y. *Trichinella spiralis*: nurse cell formation with emphasis on analogy to muscle cell repair. *Parasites Vectors.* 2008; 1:27–41.
9. Wakelin D. Helminth infections. In: Wakelin D, Blackwell JM. (eds) *Genetics of Resistance to Bacterial and Parasitic Infection.* Taylor & Francis, London, 1988; 153–224.
10. Bell, RG. The generation and expression of immunity to *Trichinella spiralis* in laboratory rodents. In: Baker JR. (ed) *Advance in Parasitology.* Academic press, London, 1998; 149–217.
11. Bell RG, Adams LS, Ogden RW. *Trichinella spiralis*: genetics of worm expulsion in inbred and F1 mice infected with different worm doses. *Exp Parasitol.* 1984; 58:345–55.
12. Wassom DL, David CS, Gleich GJ. Genes within the major histocompatibility complex influence susceptibility to *Trichinella spiralis* in the mouse. *Immunogenetics.* 1979; 9:491–6.
13. Wakelin D, Donachie AM. Genetic control of immunity to parasites: adoptive transfer of immunity between inbred strains of mice characterized by rapid and slow immune expulsion of *Trichinella spiralis. Parasite Immunol.* 1980; 2:249–60.
14. Wassom DL, et al. Genetic control of immunity to *Trichinella spiralis* infection of mice. Hypothesis to explain the role of H-2 genes in primary and challenge infections. *Immunology.* 1984; 51:625–31.
15. Wassom DL, Brooks BO, Cypess RH. *Trichinella spiralis*: role of Non-*H-2* genes in resistance to primary infection in mice. *Exp Parasitol.* 1983; 55:153–8.
16. Bell RG. Genetic analysis of expulsion of adult *Trichinella spiralis* in NFS, C3H/He, and 810 1 BR mice. *Exp Parasitol.* 1988; 66:57–65.
17. Artis D, Grencis RK. T helper cell cytokine responses during intestinal nematode infection: induction, regulation and effector function. In: Kennedy MW, Harnett W. (eds) *Parasitic Nematodes. Molecular Biology, Biochemistry and Immunology.* CABI Publishing, Wallingford, UK, 2001; 331–71.
18. Dehlawi MS, Goyal PK. Responses of inbred mouse strains to infection with intestinal nematodes. *J Helminthol.* 2003; 77:119–24.
19. Ruitenberg EJ, Elgersma A. Absence of intestinal mast cell response in congenitally athymic mice during *Trichinella spiralis* infection. *Nature.* 1976; 264(5583):258–60.
20. Grencis RK, Riedlinger J, Wakelin D. L3T4-positive T lymphoblasts are responsible for transfer of immunity to *Trichinella spiralis* in mice. *Immunology.* 1985; 56:213–18.
21. Ishikawa N, et al. Early cytokine responses during intestinal parasitic infections. *Immunology.* 1998; 93:257–63.
22. Urban JF, Jr., et al. Stat6 signaling promotes protective immunity against *Trichinella spiralis* through a mast cell- and T cell-dependent mechanism. *J Immunol.* 2000; 164:2046–52.
23. Khan WI, et al. IL-12 gene transfer alters gut physiology and host immunity in nematode-infected mice. *Am J Physiol Gastrointest Liver Physiol.* 2001; 81:G102–10.
24. Grencis RK, Hültner L, Else KJ. Host protective immunity to *Trichinella spiralis* in mice: activation of Th cell subsets and lymphokine secretion in mice expressing different response phenotypes. *Immunology.* 1991; 74:329–32.
25. Lawrence CE, et al. IL-4 regulated enteropathy in an intestinal nematode infection. *Eur J Immunol.* 1998; 28:2672–84.

26. Finkelman FD, et al. Cytokine regulation of host defense against parasitic gastrointestinal helminths: lessons from studies with rodent models. *Ann Rev Immunol.* 1997; 15:505–33.
27. Scales HE, Ierna MX, Lawrence CE. The role of IL-4, IL-13 and IL-4Rα in the development of protective and pathological responses to *Trichinella spiralis. Parasite Immunol.* 2007; 29:81–91.
28. Urban JF, Jr, et al. Cutting edge: IL-4 receptor expression by non-bone marrow derived cells is required to expel gastrointestinal nematode parasites. *J Immunol.* 2001; 167:6078–81.
29. Barner M, Mohrs M, Brombacher F, Kopf M. Differences between IL-4R α-deficient and IL-4-deficient mice reveal a role for IL-13 in the regulation of Th2 responses. *Curr Biol.* 1998; 8:669–72.
30. Emson CL, et al. Interleukin (IL)-4-independent induction of immunoglobulin (Ig) E, and perturbation of T cell development in transgenic mice expressing IL-13. *J Exp Med.* 1998; 188:399–404.
31. Khan WI, et al. Stat6 dependent goblet cell hyperplasia during intestinal nematode infection. *Parasite Immunol.* 2001; 23:39–42.
32. McDermott JR, et al. Mast cells disrupt epithelial barrier function during enteric nematode infection. *Proc Natl Acad Sci USA.* 2003; 100:7761–6.
33. Khan WI, et al. Critical role for signal transducer and activator of transcription factor 6 in mediating intestinal muscle hypercontractility and worm expulsion in *Trichinella spiralis*-infected mice. *Infect Immun.* 2001; 69:838–44.
34. Khan WI, et al. Disruption of CD40-CD40 ligand pathway inhibits the development of intestinal muscle hypercontractility and protective immunity in nematode infection. *Am J Physiol Gastrointest Liver Physiol.* 2005; 288:G15–22.
35. Vallance BA, Croitoru K, Collins SM. T lymphocyte-dependent and -independent intestinal smooth muscle dysfunction in the *T. spiralis*-infected mouse. *Am J Physiol.* 1998; 275:G1157–65.
36. Vallance BA, Galeazzi F, Collins SM, Snider DP. CD4 T cells and major histocompatibility complex class II expression influence worm expulsion and increased intestinal muscle contraction during *Trichinella spiralis* infection. *Infect Immun.* 1999; 67:6090–97.
37. Herndon FJ, Kayes SG. Depletion of eosinophils by anti-IL-5 monoclonal antibody treatment of mice infected with *Trichinella spiralis* does not alter parasite burden or immunologic resistance to reinfection. *J Immunol.* 1992; 149:3642–47.
38. Gurish MF, et al. CCR3 is required for tissue eosinophilia and larval cytotoxicity after infection with *Trichinella spiralis. J Immunol.* 2002; 168:5730–6.
39. Vallance BA, Matthaei KI, Sanovic S, Young IG, Collins SM. Interleukin-5 deficient mice exhibit impaired host defence against challenge *Trichinella spiralis* infections. *Parasite Immunol.* 2000; 22:487–92.
40. Huang L, et al. Eosinophils mediate protective immunity against secondary nematode infection. *J Immunol.* 2015; 194:283–90.
41. Dent LA, et al. Immune responses of IL-5 transgenic mice to parasites and aeroallergins. *Mem Inst Oswaldo Cruz.* 1997; 92:45–54.
42. Huang L, et al. Eosinophil-derived IL-10 supports chronic nematode infection. *J Immunol.* 2014; 193(8):4178–87.
43. Faulkner HF, et al. Interleukin-9 is involved in host-protective immunity to intestinal nematode infection. *Eur J Immunol.* 1997; 27:2536–40.
44. Veldhoen M, et al. Transforming growth factor-β "reprograms" the differentiation of T helper 2 cells and promotes an interleukin 9-producing subset. *Nat Immunol.* 2008; 9:1341–6.
45. Brown JK, Donaldson DS, Wright SH, Miller HR. Mucosal mast cells and nematode infection: strain specific differences in mast cell precursor frequency revisited. *J Helminthol.* 2003; 77:155–61.
46. Ha TY, Reed ND, Crowle PK. Delayed expulsion of adult *Trichinella spiralis* by mast cell-deficient W/Wv mice. *Infect Immun.* 1983; 41:445–7.
47. Donaldson LE, et al. A critical role for stem cell factor and c-kit in host protective immunity to an intestinal helminth. *Int Immunol.* 1996; 8:559–67.
48. Garside P, Kennedy MW, Wakelin D, Lawrence CE. Immunopathology of intestinal helminth infection. *Parasite Immunol.* 2000; 22:605–12.
49. James SL. Role of nitric oxide in parasitic infections. *Microbiol Rev.* 1995; 59:533–47.
50. Lawrence CE, et al. Nitric oxide mediates intestinal pathology but not immune expulsion during *Trichinella spiralis* infection in mice. *J Immunol.* 2000; 164:4229–34.

51. Matsuo A, Wu Z, Nagano I, Takahashi Y. Five types of nuclei present in the capsule of *Trichinella spiralis*. *Parasitology*. 2000; 121:203–10.
52. Wu Z, et al. Different response of satellite cells in the kinetics of myogenic regulatory factors and ultrastructural pathology after *Trichinella spiralis* and *T. pseudospiralis* infection. *Parasitology*. 2001; 123:85–94.
53. Chiumiento L, Bruschi F. Enzymatic antioxidant systems in helminth parasites. *Parasitol Res*. 2009; 105:593–603.
54. Boonmars T, Wu Z, Nagano I, Takahashi Y. *Trichinella pseudospiralis* infection is characterized by more continuous and diffuse myopathy than *T. spiralis* infection. *Parasitol Res*. 2005; 97:13–20.
55. Bruschi F, Marucci G, Pozio E, Masetti M. Evaluation of inflammatory responses against muscle larvae of different *Trichinella* species by an image analysis system. *Vet Parasitol*. 2009; 159:258–62.
56. Beiting BP, et al. Coordinated control of immunity to muscle stage *Trichinella spiralis* by IL-10, regulatory T cells, and TGF-β. *J Immunol*. 2007; 178:1039–47.
57. Beiting BP, et al. Interleukin-10 limits local and body cavity inflammation during infection with muscle-stage *Trichinella spiralis*. *Infect Immun*. 2004; 72:3129–37.
58. Reason AJ, et al. Novel tyvelose-containing tri- and tetra-antennary *N*-glycans in the immunodominant antigens of the intracellular parasite *Trichinella spiralis*. *Glycobiology*. 1994; 4:593–603.
59. Walls RS, Carter RL, Leuchars E, Davies AJ. The immunopathology of trichiniasis in T-cell deficient mice. *Clin Exp Immunol*. 1973; 13:231–42.
60. Fabre MV, Beiting DP, Bliss SK, Appleton JA. Immunity to *Trichinella spiralis* muscle infection. *Vet Parasitol*. 2009; 159:245–48.
61. Nakamura K, Kitani A, Strober W. Cell contact–dependent immunosuppression by $CD4^+ CD25^+$ regulatory T cells is mediated by cell surface–bound transforming growth factor β. *J Exp Med*. 2001; 194:629–44.
62. García A, et al. Novel albendazole formulations given during the intestinal phase of *Trichinella spiralis* infection reduce effectively parasitic muscle burden in mice. *Parasitol Int*. 2013; 62:568–70.
63. Basyoni MM, El-Sabaa AA. Therapeutic potential of myrrh and ivermectin against experimental *Trichinella spiralis* infection in mice. *Korean J Parasitol*. 2013; 51:297–304.
64. Codina AV, et al. Efficacy of albendazole:β-cyclodextrin citrate in the parenteral stage of *Trichinella spiralis* infection. *Int J Biol Macromol*. 2015; 77:203–6.
65. Sofronic-Milosavljevic L, Ilic N, Pinelli E, Gruden-Movsesijan A. Secretory products of *Trichinella spiralis* muscle larvae and immunomodulation: implication for autoimmune diseases, allergies, and malignancies. *J Immunol Res*. 2015; 2015:523875.
66. Dea-Ayuela MA, Rama-Iniguez S, Torrado-Santiago S, Bolas-Fernandez F. Microcapsules formulated in the enteric coating copolymer Eudragit L100 as delivery systems for oral vaccination against infections by gastrointestinal nematode parasites. *J Drug Target*. 2006; 14:567–75.
67. Ortega-Pierres G, Muniz E, Coral-Vazquez R, Parkhouse RM. Protection against *Trichinella spiralis* induced by purified stage-specific surface antigens of infective larvae. *Parasitol Res*. 1989; 75:563–7.
68. Ortega-Pierres G, et al. Induction of protection in murine experimental models against *Trichinella spiralis*: an up-to-date review. *J Helminthol*. 2015; 89(5):526–39.
69. Oivanen L, et al. Associations between *Trichinella* species and host species in Finland. *J Parasitol*. 2002; 88:84–8.
70. Dick TA, Pozio E. *Trichinella* spp. and trichinellosis. In: Samuel WM, Pybus MJ, Kocan AA. (eds) *Parasitic Diseases of Wild Mammals*, 2nd edn. Iowa State University Press, Ames, IA, 2008; 380–96.
71. Pozio E, La Rosa G, Rossi P, Murrell KD. Biological characterization of *Trichinella* isolates from various host species and geographical regions. *J Parasitol*. 1992; 78:647–53.
72. Airas N, et al. The different infectivity of *Trichinella spiralis* and *Trichinella nativa* in rat does not solely localize to enteral or parenteral phase. *Parasitol Res*. 2012; 111:2281–8.
73. Negrao-Correa D, Adams LS, Bell RG. Variability of the intestinal immunoglobulin E response of rats to infection with *Trichinella spiralis*, *Heligmosomoides polygyrus* or *Nippostrongylus brasiliensis*. *Parasite Immunol*. 1999; 21:287–97.
74. Vasilev S, Gruden-Movsesijan A, Ilic N, Sofronic-Milosavljevic L. Strain difference in susceptibility to *Trichinella spiralis* infection between dark agouti and albino oxford rats. *Vet Parasitol*. 2009; 159:229–31.

75. Stewart GL, Na H, Smart L, Seelig Jr. LL. The temporal relationship among anti-parasite immune elements expressed during the early phase of infection of the rat with *Trichinella spiralis*. *Parasitol Res*. 1999; 85:672–7.
76. Gruden-Movsesijan A, et al. The impact of *Trichinella spiralis* excretory-secretory products on dendritic cells. *Comp Immunol Microbiol Infect Dis*. 2011; 34:429–39.
77. Suzuki T, et al. The effectors responsible for gastrointestinal nematode parasites, *Trichinella spiralis*, expulsion in rats. *Parasitol Res*. 2008; 103:1289–95.
78. Andrade AM, et al. *Trichinella*: differing effects of antigens from encapsulated and non-encapsulated species on in vitro nitric oxide production. *Vet Parasitol*. 2007; 143:86–90.
79. Franssen FJ, et al. Antibody response against *Trichinella spiralis* in experimentally infected rats is dose dependent. *Vet Res*. 2011; 42:113–22.
80. Mido S, et al. *Trichinella spiralis*: infection changes serum paraoxonase-1 levels, lipid profile, and oxidative status in rats. *Exp Parasitol*. 2012; 131:190–4.
81. Gruden-Movsesijan A, et al. *Trichinella spiralis*: modulation of experimental autoimmune encephalomyelitis in DA rats. *Exp Parasitol*. 2008; 118:641–7.
82. Radovic I, et al. Immunomodulatory effects of *Trichinella spiralis*-derived excretory—secretory antigens. *Immunol Res*. 2015; 61:312–25.
83. Murrell KD. Strategies for the control of human trichinosis transmitted by pork. *Food Technol*. 1985; 39:65–8.
84. Zimmermann WJ. Reproductive potential and muscle distribution of *Trichinella spiralis* in swine. *J Am Vet Med Assoc*. 1970; 156:770–4.
85. Nöckler K, et al. Experimental studies in pigs on *Trichinella* detection in different diagnostic matrices. *Vet Parasitol*. 2005; 132:85–90.
86. Kapel CM, Gamble HR. Infectivity, persistence, and antibody response to domestic and sylvatic *Trichinella* spp. in experimentally infected pigs. *Int J Parasitol*. 2000; 30:215–21.
87. Van Knapen F, et al. Comparison of four methods for early detection of experimental *Trichinella spiralis* infection in pigs. *Vet Parasitol*. 1981; 9:117–23.
88. Gamble HR. Sensitivity of artificial digestion and enzyme immunoassay methods of inspection for trichinae in pigs. *J Food Prot*. 1998; 61:339–343.
89. Gamble HR, et al. International Commission on *Trichinellosis*: recommendations on methods for the control of *Trichinella* in domestic and wild animals intended for human consumption. *Vet Parasitol*. 2000; 93:393–408.
90. Gamble HR, et al. 2004. International Commission on Trichinellosis: recommendations on the use of serological tests for the detection of *Trichinella* infection in animals and man. *Parasite*. 2004; 11:3–13.
91. European Commission. Commission Regulation (EC) No. 2075/2005 of the European Parliament and of the Council of 5 December 2005 Laying down specific rules on official controls for *Trichinella* in meat. *Off J Eur Comm L*. 2005; 338:60–82.
92. Hill DE, et al. Larval viability and serological response in horses with long-term *Trichinella spiralis* infection. *Vet Parasitol*. 2007; 146:107–16.
93. Ortega-Pierres G, et al. Workshop on a detailed characterization of *Trichinella spiralis* antigens: a platform for future studies on antigens and antibodies to this parasite. *Parasite Immunol*. 1996; 18:273–84.
94. Smith HJ. Evaluation of the ELISA for serological diagnosis of trichinosis in Canadian swine. *Can J Vet Res*. 1987; 51:194–7.
95. Ilic N, Gruden-Movsesijan A, Zivojinovic M, Sofronic-Milosavljevic L. Characteristic band pattern in western blots for specific detection of anti-*Trichinella spiralis* antibodies in different host species. *Acta Vet (Beograd)*. 2014; 64:33–43.
96. Yepez-Mullia L, et al. Contributions to the study of *Trichinella spiralis* TSL-1 antigens in host immunity. *Parasite Immunol*. 2007; 29:661–70.
97. Gómez-Morales MA, Ludovisi A, Amati M, Pozio E. Candidates for reference swine serum with anti-*Trichinella* antibodies. *Vet Parasitol*. 2015; 208:218–24.
98. Pozio E, Kapel CMO, Gamble HR. Specificity and sensitivity of amplified polymorphic DNA analyses for the identification of single muscle larvae of *Trichinella* after experimental infection of pigs. *Parasitol Res*. 1999; 85:504–6.
99. Guenther S, et al. Detection of *Trichinella spiralis*, *T. britovi* and *T. pseudospiralis* in muscle tissue with real-time PCR. *J Microbiol Methods*. 2008; 75:287–92.

100. Lunney JK, Murrell KD. Immunogenetic analysis of *Trichinella spiralis* infection in swine. *Vet Parasitol.* 1988; 29:179–93.
101. Madden KB, et al. *Trichinella spiralis*: genetic basis and kinetics of the anti-encysted muscle larval response in miniature swine. *Exp Parasitol.* 1993; 77:23–35.
102. Bugarski D, Sofronic-Milosavljevic L, Popovic L, Cuperlovic K. Changes in peripheral blood mononuclear cell subsets during *Trichinella spiralis* infection in pigs. In: Cambell V, Pozio E, Bruschi F. (eds) *Proceedings of 8th ICT on Trichinellosis.* Istituto Superiore di Sanita Press, Rome, 1993; 413–6.
103. Frontera E, et al. Concurrent infection with *Trichinella spiralis* and other helminths in pigs. *Vet Parasitol.* 2007; 146:50–7.
104. Bokken G, Bergwerff A, van Knapen F. A novel bead-based assay to detect specific antibody responses against *Toxoplasma gondii* and *Trichinella spiralis* simultaneously in sera of experimentally infected swine. *BMC Vet Res.* 2012; 8:36.
105. Liu P, et al. Screening of early antigen genes of adult-stage *Trichinella spiralis* using pig serum from different stages of early infection. *Vet Parasitol.* 2013; 194:222–5.
106. Marti HP, Murrell KD, Gamble HR. *Trichinella spiralis*: immunization of pigs with newborn larval antigens. *Exp Parasitol.* 1987; 63:68–73.
107. Murrell KD, Despommier DD. Immunization of swine against *Trichinella spiralis. Vet Parasitol.* 1984; 15:263–70.
108. Gamble HR, Murrell KD, Marti HP. Immunization of pigs against *Trichinella spiralis* using excretory-secretory antigens. *Am J Vet Res.* 1986; 47:2396–9.
109. Yang Y, et al. Identification and characterization of immunodominant linear epitopes on the antigenic region of a serine protease in newborn *Trichinella* larvae. *J Helminthol.* 2015; 90(2):232–7. doi:10.1017/S0022149X15000267.
110. Bien J, Cabaj W, Moskwa B. Proteomic analysis of potential immunoreactive proteins from muscle larvae and adult worms of *Trichinella spiralis* in experimentally infected pigs. *Folia Parasitol (Praha).* 2015; 62. pii:2015.022.
111. Jongwutiwes S, et al. First outbreak of human trichinellosis caused by *Trichinella pseudospiralis. Clin Infect Dis.* 1998; 26:111–5.
112. Pozio E, Kapel CMO. *Trichinella nativa* in sylvatic wild boar. *J Helminthol.* 1999; 73:87–9.
113. Kapel CM. Sylvatic and domestic *Trichinella* spp. in wild boars; infectivity, muscle larvae distribution, and antibody response. *J Parasitol.* 2001; 87:309–14.
114. Kapel CM, Webster P, Gamble HR. Muscle distribution of sylvatic and domestic *Trichinella* larvae in production animals and wildlife. *Vet Parasitol.* 2005; 132:101–5.
115. Lacour SA, et al. Freezing tolerance of *Trichinella* muscle larvae in experimentally infected wild boars. *Vet Parasitol.* 2013; 194:175–8.
116. Gari-Toussaint M, et al. Human trichinellosis due to *Trichinella britovi* in southern France after consumption of frozen wild boar meat. *Euro Surveill.* 2005; 10:117–18.
117. Dick TA. Infectivity of isolates of *Trichinella* and the ability of an arctic isolate to survive freezing temperatures in the raccoon, *Procyon lotor*, under experimental conditions. *J Wildl Dis.* 1983; 19:333–6.
118. Mantovani A, Filippini I, Bergomi S. Indagini su un'epidemia di trichinellosi umana verificatasi in Italia. *Parassitologia (Rome).* 1980; 22:107–32.
119. Boireau P, et al. *Trichinella* in horses: a low frequency infection with high human risk. *Vet Parasitol.* 2000; 93:309–20.
120. Pampiglione S, et al. Infezione sperimentale del cavallo con larve di trichina. *Parassitologia (Rome).* 1978; 20:183–93.
121. Smith HJ, Snowdon KE. Detection of *Trichinella spiralis* native antibodies in porcine sera by ELISA using *T. spiralis* spiralis excretory-secretory antigen. *Can J Vet Res.* 1987; 51:413–4.
122. Gamble HR, Gajadhar AA, Solomon MB. Methods for the detection of trichinellosis in horses. *J Food Prot.* 1996; 59:420–5.
123. Pozio E, et al. Predilection sites of *Trichinella spiralis* larvae in naturally infected horses. *J Helminthol.* 1999; 73:233–7.
124. Pozio E, et al. Evaluation of ELISA and western blot analysis using three antigens to detect anti-*Trichinella* IgG in horses. *Vet Parasitol.* 2002; 108:163–78.
125. Soule C, et al. Experimental trichinellosis in horses: biological and parasitological evaluation. *Vet Parasitol.* 1989; 31:19–36.

126. Soule C, et al. Biological and parasitic variations in horses infested and reinfected by *Trichinella spiralis*. *Vet Res*. 1993; 24:21–31.
127. Pozio E, et al. Distribution of *Trichinella spiralis* larvae in muscles from a naturally infected horse. *Vet Parasitol*. 1998; 74:19–27.
128. Van Knapen F, Franchimont JH, Hendriks WML, Eysker M. Experimental *Trichinella spiralis* infection in two horses. In: Geerts S, Kumar V, Brandt J. (eds) *Helminth Zoonoses*. Martinus Nijhoff Publisher, Dordrecht, 1987; 192–201.
129. Sofronic-Milosavljevic L, et al. Anti-*Trichinella* antibodies detected in chronically infected horses by IFA and western blot, but not by ELISA. *Vet Parasitol*. 2005; 132:107–11.
130. Yepez-Mullia L, et al. Detection of *Trichinella* infection in slaughter horses by ELISA and western blot analysis. *Vet Parasitol*. 1999; 81:57–68.
131. Ilic N, Petrovic M, Djordjevic M, Sofronic-Milosavljevic L. A novel dot blot test for *Trichinella spiralis* antigen detection. *Acta Veterinaria (Beograd)*. 2004; 54:301–10.
132. Sugiura T, Kondo T, Imagawa H, Kamada M. Production of monoclonal antibodies to six isotypes of horse immunoglobulin. *Vet Immunol Immunopathol*. 1998; 62:145–51.
133. Keggan A, Freer H, Rollins A, Wagner B. Production of seven monoclonal equine immunoglobulins isotyped by multiplex analysis. *Vet Immunol Immunopathol*. 2013; 153:187–93.
134. Murrell KD, et al. Epidemiology of *Trichinella* infection in the horse: the risk from animal product feeding practices. *Vet Parasitol*. 2004; 123:223–33.
135. Davidson RK, Handeland K, Kapel CM. High tolerance to repeated cycles of freezing and thawing in different *Trichinella* nativa isolates. *Parasitol Res*. 2008; 103:1005–10.
136. Davidson RK, Opertveit I, Moller L, Kapel CM. Serological detection of anti-*Trichinella* antibodies in wild foxes and experimentally infected farmed foxes in Norway. *Vet Parasitol*. 2009; 163:93–100.
137. Webster P, Kapel CM. Studies on vertical transmission of *Trichinella* spp. in experimentally infected ferrets (*Mustela putorius furo*), foxes (*Vulpes vulpes*), pigs, guinea pigs and mice. *Vet Parasitol*. 2005; 130:255–62.
138. Kapel CM, et al. Experimental *Trichinella* infection in seals. *Int J Parasitol*. 2003; 33:1463–70.
139. Forbes LB, Measurel L, Gajadhar A, Kapel CM. Infectivity of *Trichinella nativa* in traditional northern (country) foods prepared with meat from experimentally infected seals. *J Food Prot*. 2003; 66:1857–63.
140. Pozio E, et al. *Trichinella zimbabwensis* n.sp. (Nematoda), a new non-encapsulated species from crocodiles (*Crocodylus niloticus*) in Zimbabwe also infecting mammals. *Int J Parasitol*. 2002; 32:1787–99.
141. Kapel CM, et al. *Trichinella spiralis*, *Trichinella britovi*, and *Trichinella nativa*: infectivity, larval distribution in muscle, and antibody response after experimental infection of pigs. *Parasitol Res*. 1998; 84:264–71.

Index

B

C

F

G

H

I

K

L

M

N

T

U

V